W9-CMD-974

ANNUAL REVIEW OF MICROBIOLOGY

ANNUAL REVIEW OF MICROBIOLOGY

VOLUME 47, 1993

L. NICHOLAS ORNSTON, *Editor*

Yale University

ALBERT BALOWS, *Associate Editor*

Centers for Disease Control, Atlanta

E. PETER GREENBERG, *Associate Editor*

University of Iowa, Iowa City

ANNUAL REVIEWS INC. 4139 EL CAMINO WAY P.O. BOX 10139 PALO ALTO, CALIFORNIA 94303-0897

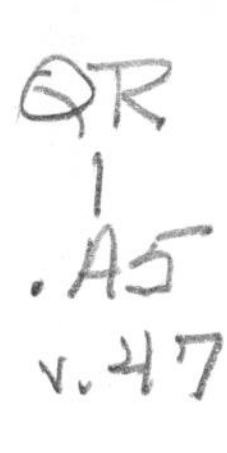

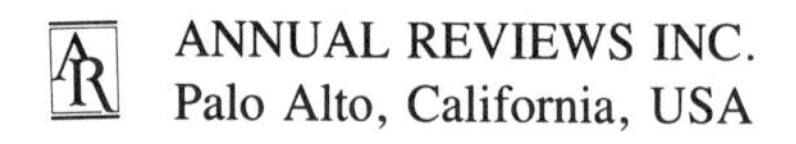

ANNUAL REVIEWS INC.
Palo Alto, California, USA

International Standard Serial Number: 0066–4227
International Standard Book Number: 0–8243–1147-7
Library of Congress Catalog Card Number: 49-432

♾ The paper used in this publication meets the minimum requirements of American National Standard for Information Sciences—Permanence of Paper for Printed Library Materials, ANSI Z39.48-1984.

Typesetting by Kachina Typesetting Inc., Tempe, Arizona; John Olson, President; Marty Mullins, Typesetting Coordinator; and by the Annual Reviews Inc. Editorial Staff

PRINTED AND BOUND IN THE UNITED STATES OF AMERICA

PREFACE

The constancy of biology is achieved by constant change, processes of renewal and adaptation that are represented at every level of biological organization. We should not be surprised that the smallest organisms exhibit the most extensive changes. A microbial cell must be a specialist because its limited genetic information will allow no more. Specialists cannot survive alone in a changing world; thus microorganisms exist in the wild as components of complex interacting populations, as we microbiologists are increasingly aware. Relatively few members of these populations have been grown in the laboratory, and it is by no means certain that the properties of such domesticated strains are representative of microbial populations as a whole. The biological give and take within natural populations has produced astonishing microbial diversity, which is only beginning to be appreciated.

Within the realm of microbial diversity lies genetic potential that is largely untapped. Microorganisms compensate in number for what they lack in size, and a consequence of their abundance is a reservoir of variation that is drawn upon by natural selection. Genetic variants improbable in the laboratory are almost certain to be represented in the huge populations achieved by members of the microbial world. Genetic exchange allows successful adaptations to be shared by different cell lines as exemplified by the extensive spread of antibiotic resistance. Presumably the spread of other forms of genetic variation has helped microorganisms adapt to chemical changes that have been visited upon the environment in the past century. Given the possibilities for change, perhaps the most remarkable characteristic of any single microorganism is its sharply defined biological identity, a manifestation of the specialization that is both its history and its destiny.

Again unity within diversity is the theme of *Annual Review of Microbiology*, and again it is a pleasure to thank the authors of the *Review* for the insights they have shared. Authors and topics were selected by the editorial board, which welcomed its new members Mary Lidstrom and Graham Walker as deliberations for this volume began. From inception to publication, the *Review* was nurtured by the many talents of its production editor, Amanda Suver. Rosemarie Hansen continued the daunting task of minimizing chaos in the editor's office.

L. NICHOLAS ORNSTON
EDITOR

Annual Review of Microbiology
Volume 47, 1993

CONTENTS

INDEXES

OTHER REVIEWS OF INTEREST TO MICROBIOLOGISTS

From the *Annual Review of Biochemistry,* Volume 62 (1993)

From Bacterial Nutrition to Enzyme Structure: A Personal Odyssey, E. E. Snell
Nucleocytoplasmic Transport in the Yeast Saccharomyces cerevisiae, M. A. Osborne and P. A. Silver
Introns as Mobile Genetic Elements, A. M. Lambowitz and M. Belfort

From the *Annual Review of Genetics,* Volume 27 (1993)

The Pheromone Response Pathway in Saccharomyces cerevisiae, J. Kurjan
The Molecular Genetics of Nitrate Assimilation in Fungi and Plants, N. M. Crawford and H. N. Arst, Jr
DNA Repair and Proteins of Saccharomyces cerevisiae, S. Prakash, P. Sung, and L. Prakash
Genetics of Poliovirus, E. Wimmer, C. U. T. Hellen, and X. Cao

From the *Annual Review of Immunology,* Volume 11 (1993)

Defensins: Antimicrobial and Cytotoxic Peptides of Mammalian Cells, R. I. Lehrer, A. K. Lichtenstein, and T. Ganz
Immunity to Intracellular Bacteria, S. H. E. Kaufmann

From the *Annual Review of Medicine,* Volume 44 (1993)

Antibiotic Therapy for Chronic Infection of Pseudomonas *in the Lung,* N. Høiby
Role of Bacterial Cytotoxins in Hemolytic Uremic Syndrome and Thrombotic Thrombocytopenic Purpura, S. Ashkenazi
Pathogensis and Pathophysiology of Bacterial Meningitis, A. R. Tunkel and W. M. Scheld
Prophylactic Antibiotics in Surgery, K. A. Ludwig, M. A. Carlson, and R. E. Condon

From the *Annual Review of Plant Physiology and Plant Molecular Biology,* Volume 44 (1993)

Molecular Genetics of Cyanobacterial Development, W. J. Buikema and R. Haselkorn

ANNUAL REVIEWS INC. is a nonprofit scientific publisher established to promote the advancement of the sciences. Beginning in 1932 with the *Annual Review of Biochemistry*, the Company has pursued as its principal function the publication of high quality, reasonably priced *Annual Review* volumes. The volumes are organized by Editors and Editorial Committees who invite qualified authors to contribute critical articles reviewing significant developments within each major discipline. The Editor-in-Chief invites those interested in serving as future Editorial Committee members to communicate directly with him. Annual Reviews Inc. is administered by a Board of Directors, whose members serve without compensation.

ANNUAL REVIEWS OF
Anthropology
Astronomy and Astrophysics
Biochemistry
Biophysics and Biomolecular Structure
Cell Biology
Computer Science
Earth and Planetary Sciences
Ecology and Systematics
Energy and the Environment
Entomology
Fluid Mechanics
Genetics
Immunology
Materials Science
Medicine
Microbiology
Neuroscience
Nuclear and Particle Science
Nutrition
Pharmacology and Toxicology
Physical Chemistry
Physiology
Phytopathology
Plant Physiology and Plant Molecular Biology
Psychology
Public Health
Sociology

SPECIAL PUBLICATIONS

Excitement and Fascination of Science, Vols. 1, 2, and 3

Intelligence and Affectivity, by Jean Piaget

For the convenience of readers, a detachable order form/envelope is bound into the back of this volume.

Norbert Pfennig

Annu. Rev. Microbiol. 1993. 47:1–29

REFLECTIONS OF A MICROBIOLOGIST, OR HOW TO LEARN FROM THE MICROBES[1]

Norbert Pfennig
Primelweg 12, D-88662 Überlingen, Germany

KEY WORDS: purple sulfur bacteria, green sulfur bacteria, sulfur-reducing bacteria, sulfate-reducing bacteria, syntrophic cocultures

---

CONTENTS

[1]In memory of my parents.

0066-4227/93/1001-0001$02.00

ABSTRACT

This autobiographical chapter summarizes the author's work with a defined mineral medium for fastidious sulfide-oxidizing phototrophic purple and green sulfur bacteria that were known already from Winogradsky's and Lauterborn's descriptions. The pure cultures, isolated from natural mud deposits, revealed interesting new cytological and biochemical features. In the wake of these studies, new anaerobic bacteria with unusual metabolic capacities were isolated and characterized. Ecologically most significant is the dehydrogenation of acetate to carbon dioxide. Electron acceptors are sulfur for the sulfur reducers and sulfate for the new sulfate reducers obtained by Widdel. Thauer and Fuchs showed that a modified TCA-cycle and the new acetyl-CoA:carbon monoxide dehydrogenase pathway operates in the oxidation of acetate. Many aliphatic and aromatic compounds were shown to be completely degradable by marine sulfate reducers. The biogeochemical transformations of the anoxic sulfur cycle are now understood in terms of the capacities of the phototrophic and chemotrophic bacterial species involved in the cycle.

* * *

> Thus it seems to me that the real problem is whether the enthusiasm of the best students can be stimulated so that they become aware of the importance of non-molecular biology, not as something of interest only to dilettantes and collectors, but as an indispensable source of inspiration, opening up new and broader areas for investigation.
>
> C. B. van Niel, 1966 (119c)

When I received the invitation to contribute a prefatory chapter to the *Annual Review of Microbiology,* I hesitated to accept it. Were there any good reasons to write about my activities as a microbiologist? It occurred to me that a review of this kind gives me the chance to tell of my never-ending appreciation for the strive for scientific clarity. And in particular, I thought it timely to acknowledge how ecology-oriented thinking—so strongly exemplified by S. N. Winogradsky (134)—remained stimulating for my and my coworkers' research. My studies have made me aware of how deeply nature is entrusted to us. Don't we all somehow feel the dignity and preciousness of things, especially the living? Unless we ever try to uncover the truth inherent in the being of things, we should not think that we will come close to our world. Imagination and concepts that we inherit or form may not always be adequate to comprehend what we perceive. Therefore, I have found it worthwhile to look quietly at what is open to us; it often tells us more than we anticipate. This is why I feel that I learned not only from the people in my life but also from the microbes.

PROTECTION AND LASTING INFLUENCES AT HOME, DURING WARTIME, AND AS A PRISONER OF WAR

My parents were practicing art lovers. From his early years, my father liked to paint landscapes in oil. Because professionally he was a master furrier in a traditional family firm in Kassel (central Germany), Sundays were the time for his hobby. My mother had started to learn pharmacy during World War I, but also became an active sculptress. I was born July 8, 1925, as the second boy. In retrospect two things stick most firmly in my mind: first the walls of all the rooms in our apartment were covered almost completely with pleasant oil paintings, not only my father's works but also paintings from various artist friends. Second, the whole family regularly spent Sundays somewhere in the surroundings of Kassel with my father doing his painting. On a few outstanding Sundays we went with our parents to exhibitions of paintings or to the art gallery. Everyone was well dressed, and there was an impressively quiet and respectful atmosphere in the rooms that I found very pleasant. On the way home we talked about what we saw, and our various impressions were considered seriously. My parents usually spent their evenings reading literature about paintings and painters.

At the age of ten I inherited an old but quite good microsope from an uncle. I learned to use it by playing with it and trying out all the functions. The entirely new world of shapes, colors, and motile creatures in samples of pond water in particular awakened my curiosity. What I saw appeared more simple, and yet the admirably artistic and ornate forms were inexplicably different from what I saw in the usual world. One of my schoolmates knew how to concentrate the floating plankton with a net made from ladies' stockings. Before long, with the help of books, I learned to know the planktonic algae in the ponds of our surrounding countryside. However, the various kinds of rotifers in the different ponds attracted my attention most. It excited me to find some kind of correlation between the type of pond and the rotifer species present at a given time of the year.

My parents supported my interests in every respect and thus protected me from many of the stupid activities of the Nazi organizations. They completely disagreed with the ideas and activities of this regime and kept company with artist friends who thought and lived as they did. The microscopic studies kept me busy in my free time until 1941. Then I suddenly lost the microscope when our house was destroyed during a bombing attack.

In 1944, I finished high school as a boarder far outside Kassel. Thereafter I had to enter military service and was trained as a radio operator, signaling by Morse code. We stayed in various places in southern France. My comrades and I felt the injustice and desolation of war and really did not want to be involved in it. In February 1945, we got the chance to surrender and became prisoners of war in France. After the capitulation of Germany in May 1945, groups of prisoners were allowed to work for French patrons. I had to cooperate with a 63-year-old forest worker in a charcoal-burner company. We got along very well and soon became good friends, particularly because my master loved to teach me French during work. After a while I could talk like a native forest worker. Apart from that, I devoted myself to all kinds of practical work and repairs. I found satisfaction in how properly I performed what I did and didn't ask what or for whom it was done.

With the cold, rainy, and snowy season, the reconciling beauty of the leaf colors disappeared and the defoliated forest became dreary. One day I became aware of the impressively clear silhouettes of the stems and branches of trees at some distance, particularly if they stood out against the sky. To follow the gestures of the different silhouettes was consoling and stimulating during this uneasy time of the year. I marvelled at the perseverence of the trees' gradual and steady development through favorable and hard times. I intimately followed the various gestures and saw in them strength (oak), uprightness (ash, poplar), openness (beach, linden tree), cheerfulness (willow), or grace (birch). In this way, in addition to the people around me, the trees became my particular friends when working in the forest, and this remains so today.

In summer 1946, my time as a prisoner of war ended and I traveled home to my parents in Kassel. I gratefully accepted their offer to support me while I studied at a university. Of course, my interest was biology. In particular, I wanted to learn all about the little creatures in ponds and lakes that I had begun to study during my high-school years.

UNIVERSITY STUDY IN GÖTTINGEN

The nearest university to Kassel was only 40 miles away in Göttingen, which turned out to be the only German university offering microbiology at that time. I was extremely lucky because I was able to begin my studies in the winter term 1946/1947. In this very difficult post-war time, it was a great help to me that my dear school friend Ernst-August Müller was likewise accepted in Göttingen. He studied mathematics and later specialized in hydro- and aerodynamics. Although studying in different disciplines, we got together at least once a week and discussed our ideas on basic questions of science and humanity. All our experiences during the past totalitarian Nazi regime and war greatly stimulated us to cultivate the interpersonal human relations to which we all owe life and humanity. We became familiar with various schools of thought, but only R. Steiner's *Philosophy of Freedom* clearly expressed and extended our experience of thinking and life. After the thorough study of this book, in which the foundation for an ethical individualism is laid, we cooperated with the like-minded persons of the Göttingen branch of the Anthroposophical Society and stood up for a free, private school system. My appreciation and search for the union of science and art has been greatly promoted ever since by meeting with persons who also strive for an ever better understanding of our world.

At the university in Göttingen, the course of studies in microbiology supplemented a complete study of biology, including physics, chemistry, zoology, and botany. To my great pleasure, the number of students choosing microbiology was small. This subject was less known and lay somewhat off the mainstream of the well-established life sciences.

The Institut für Mikrobiologie was a small two-story building nicely situated at the edge of the city center, surrounded by agricultural experimental fields. In the building, the atmosphere was quiet, friendly, and inviting; to me it was something like Sleeping Beauty's castle. The head of the institute was Prof. August Rippel-Baldes, who did not create the impression of a self-centered chairman; rather he behaved like a father or grandfather to the people around him.

In winter 1946, I took part in the basic practical course in microbiology held by Dr. R. Meyer. He was a real expert in what he was teaching and, therefore, I got the impression that I belonged in this subject as well as in this institute. Of the other subjects, plant ecology with its instructive excursions guided by Prof. Firbas attracted my attention. I participated in a one-week botanical excursion and collected many soil samples from well-characterized plant associations. When the samples were later studied for the presence of streptomycetes by a selective plate-count method, we obtained clear correlations, e.g. streptomycetes were qualitatively and quantitatively most abundant in dry and warm limestone soils of the Rhine valley and practically absent in acidic conifer-forest soils on the primary rocks of the Black Forest.

The microbiology lecture courses gave me the impression that I had to extend my knowledge in organic chemistry/biochemistry for graduate work in microbiology. In summer 1949, I therefore went to the chairman of the organic chemistry laboratories, Prof. Hans Brockmann, and asked him to allow me to participate in an organic chemistry course. His reception was very friendly, and he explained to me that I would best aquire the necessary experience by joining his laboratory and learning from his coworkers rather than by taking a course. He told me that his field of work was the structure of new pigments and antibiotics of streptomycetes and that he would like to have a microbiologist among his coworkers. He finished by saying: "Think on it and let me know your decision at the beginning of the winter term."

The longer I thought about it the more attractive seemed the chance to become familiar with the analytical methods of the organic chemists. In

October 1949 I started my experimental thesis in Brockmann's laboratory with the full blessing of Rippel-Baldes.

It was my task to isolate chemically pure actinomycins C out of a naturally produced mixture of three very similar compounds. Since microbiological approaches did not yield the desired result, I used countercurrent distribution to separate the actinomycin C components and to crystallize them for further analysis with paper chromatography. This approach was fully successful (9, 60), and I finished my thesis in summer 1952. During my stay, I enjoyed doing color photomicrographs of new crystalline compounds obtained by the chemists. When I left their labs, I realized how much experience I had gained and I gratefully remembered Brockmann for his wise, liberal, and friendly guidance.

POSTDOCTORAL WORK

Göttingen 1952–1958

In fall 1952, I accepted Rippel-Baldes' offer to join the microbiology institute as a research assistant. I had to teach two three-hour practical courses, of which the first dealt with various enrichment and pure-culture techniques for aerobic and anaerobic bacteria and molds. This course gave me the chance to extend my experience and knowledge of microbial diversity and ecology. The other course was designed to introduce the students to various biochemical activities of microbes and to provide test methods used in industry and waste-water control.

Following a suggestion of Rippel-Baldes, I started a research project on symbiotic nitrogen fixation. We wanted to know what nitrogen compounds from the fixed nitrogen of the pea plant's symbiotic rhizobia would be used by the plant. Using the highly sensitive paper chromatography, I was able to obtain quite conclusive results (61).

Apart from these studies, my research interest remained with the streptomycetes. I had learned that in nature they are most abundant in dry limestone grassland soils. I also knew how pure cultures of *Streptomyces* strains grow in small and large batches. Now I wanted to learn how these mycelium-forming bacteria actually develop in their native habitat, the soil. Stimulated by Winogradsky's views on soil microbiology (134), I applied Cholodny's method (13) using buried glass slides. It was quite surprising how much could be seen of the mycelium development and spore formation of streptomycetes in soil with this method (62). Also, I got a first glimpse of the role of protozoa as grazers of bacteria in natural environments.

ETH Zürich 1955–1956

A one-year fellowship from the Swiss Federal Institute of Technology and the Deutscher Akademischer Austauschdienst gave me the opportunity for research and advanced training at the Eidgenössische Technische Hochschule (ETH) in Zürich, Switzerland. I gratefully accepted work in Prof. T. O. Wiken's laboratory on the fermentative activity of a wine yeast under various defined experimental conditions. During this work I fully recognized how significant the age of a culture, the physiological state of the cells, and the test conditions were for various metabolic activities of a microbe population (131, 132). We thought varying the experimental conditions was quite important to learn about the microbe's actual activities from its responses. Only if the physiological state of the cells is considered, can biochemical explanations be applied properly.

Wiken and his coworkers maintained a stimulating and friendly atmosphere in the department, and I appreciated the good international relationships at the ETH. One of the friends I made during this time was Heikki Suomalainen from Helsinki, Finland. When we discovered that we both loved the fine arts, we happily visited art exhibitions and exchanged impressions to learn from each other. Some years later in Helsinki, Heikki sympathetically introduced me to the world of Finnish painters.

To my great pleasure, practial courses in limnology were offered at the ETH by Prof. Otto Jaag, who at that time was Director of the Eidgenössische Anstalt für Wasserversorgung, Abwasserreinigung und Gewässerschutz (EAWAG). My old love for the world of plankton revived, and I extended my knowledge in this field as well as in the latest methods for waste-water treatment. Jaag stood up as one of the early warners in Switzerland against the increasing pollution of lakes and rivers by municipal, industrial, and agricultural waste waters. His dedication to finding solutions for these environmental problems stimulated me to include this subject in my teaching programs ever since. By 1957, I became a university lecturer in Göttingen.

Microbes and Open-Flow Systems (Göttingen)

The physiological studies in Wiken's lab made me aware of the fact that microbes cannot properly be studied without considering their growth conditions. As a closed system, batch culture particularly appeared to represent an artificial, effortless laboratory system in which microbes were finally doomed to perish. Therefore in 1956/1957, I studied with keen interest the impressively clear and detailed publication by Herbert et al (37) on the principles of homocontinuous chemostat culture. The multiplying bacterial population as an open living system could be grown for long periods under various defined conditions in this open-culture system. Growth rate, population density, and growth-limiting substrate are chosen at will, and the effects of these factors on the quality of the cell substance are assessed. In comparison to batch culture, entirely different competition experiments became feasible in which the slow-growing microbes with the highest substrate affinity outcompete the fast-growing ones. My interest in the literature on microbial growth and open-flow systems was primarily theoretical and served to improve my concepts on metabolic regulation and cellular biosynthesis for my lecture courses. I tried to think dynamically about living organisms and to extend this thinking into microbial ecology. Toward the latter I was encouraged by Holger W. Jannasch who worked in the Göttingen microbiology institute in the late 1950s. I owe him for elucidating discussions on Winogradsky's concepts of soil microflora (134) and on microbial life in aquatic habitats. Holger started using chemostats for experimental work on questions that arose in microbial ecology very early, whereas I did so only much later (50). Anyhow, we considered the basic questions around the chemostat so attractive for microbiology that we compiled results and concepts available at that time in a review (73).

BECOMING FAMILIAR WITH THE ANAEROBIC PURPLE AND GREEN SULFUR BACTERIA

Imitation of the Natural Habitat (Göttingen)

After Rippel-Baldes' retirement in 1957, Dr. H. G. Schlegel became chairman of the Institut für Mikrobiologie in fall 1958. As a student of J. Buder, Halle, he had studied phototaxis and irritability by light in the large purple bacterium *Chromatium okenii*. This species, like *Thiospirillum jenense* and most other species and genera described by Winogradsky in 1888 (133), could not be grown in laboratory media. An exception was the small *Chromatium vinosum*, currently the best-studied pure culture (14). Researchers in Halle harvested their cells from swarming mass populations that annually developed in the garden ponds of the small castle Ostrau near Halle/Saale, East Germany. Winogradsky had carefully studied the purple sulfur bacteria in various samples from natural habitats and mud columns maintained in the laboratory (133). In 1955 van Niel (119b), the best connoisseur of the phototrophic sulfur bacteria, expressed the situation as follows (p. 205):

> At present we are completely ignorant of the factors that permit the occasionally observed mass development of the red sulfur spirilla. A more complete knowledge of their ecology should go far towards solving the problem of culturing these interesting organisms;...

In spring 1959, Schlegel obtained a one-liter bottle of mainly *C. okenii* and accompanying *T. jenense* in pond water from Ostrau and brought it through the iron curtain of the Soviet bloc to Göttingen. The valuable sample was stored in the refrigerator and used as an inoculum source for a large variety of growth experiments. Most of the synthetic media for purple sulfur bacteria that had been published since van Niel's studies in 1932 (119a) were tried; however, neither growth nor even survival of the cells could be achieved. Frustratingly, within 24 hours the cells of the inoculum became nonmotile and settled at the bottom of the culture bottles.

At this point I got interested in the problem. I was convinced that in principle it should be possible to grow these anaerobic bacteria in defined media. But instead of directly applying the bacteria to a given media with mineral nutrient concentrations far above those of freshwater habitats, I intended to follow a more gentle course. First, the Ostrau purple bacteria should be transferred to conditions that closely resembled their oxic-anoxic pondwater habitat. We collected mud and water from several ponds in the surroundings of Göttingen. With this material I prepared about 30 small mud-water columns representing variations of the Winogradsky column (85, 133). These raw cultures were inoculated from the rest of the Ostrau cell suspension and incubated near the window at room temperature. I hoped that at least one of the columns might provide conditions under which our favorite bacteria from Ostrau would stay alive or even grow.

It was quite a new and attractive pursuit for me to follow the gradual changes and successions of organisms in the columns by microscopic inspections of Pasteur-pipette samples. This sharpened my understanding of the differences between oxic and anoxic areas in the mud columns and comparable situations in natural habitats.

Several weeks after I started the columns, the Ostrau purple sulfur bacteria began to grow in a special type of mud column that contained equal volumes of undigested sewage sludge, garden soil, and calcium sulfate. This surprisingly positive response of our purple friends was gratefully accepted and stimulated further experiments. The advantage of the successful raw culture over those prepared with pond mud and calcium sulfate were: (*a*) no other phototrophic organisms came up in these columns except those being introduced on purpose; (*b*) sulfide formation started gently and reliably and lasted for months; and (*c*) batches of these columns were prepared in advance, stored in the dark, and could be used as raw media for phototrophic sulfur bacteria if required. Thus, it was possible to propagate the enriched populations of *C. okenii*, *Chromatium warmingii*, and *T. jenense* from the Ostrau inoculum in larger batches.

For a fine arts–lover like me, handling the beautifully colored purple bacterial cultures induced an ever-present desire to care for the cultures that has remained with me ever since. *T. jenense* exhibits a pale beige-yellow to orange, *C. okenii* a royal purple, and *C. warmingii* a pale to deep violet.

During growth these colors changed from being pale and chalky to clear and intense, depending on the presence or absence of the highly refractile globules of stored sulfur inside the cells (78, 85).

The most reliable criterion for the well being of the cells was their motility. The distribution and swarming of the purple red cells in the whole culture vessel and the absence of a cell sediment made this action apparent to the naked eye (64). With increasing population density, well growing and swimming purple sulfur bacteria caused slow vertical ribbon-like currents in their cultures, so-called bioconvection patterns (64, 79). In columns with high population density, purple stripes moved slowly downward, while columns with low population density contained upward-swimming cells.

Later I found that vertical movements of the cells could also be observed under the microscope. I realized that the conventional way of inspecting cell suspensions on horizontally positioned slides forced the large *Chromatium* cells to swim in all horizontal directions at random. If I oriented the microscope horizontally with the slide in a vertical position, most cells swam up and down, always with their polar flagellum up. For a cell with a long flagellum, the centers of geometry and gravity do not coincide. Therefore, gravity provides a torque that tends to keep the cell axis vertical, particularly if the cell swims upward. In effect the result is similar to negative geotaxis of eukaryotic organisms.

Synthetic Culture Medium and Pure Cultures

The raw cultures of the large Ostrau phototrophs served as inoculum sources for further growth studies in which the clear supernatant sludge water was used as culture medium in screw cap bottles. With these cultures, optimum conditions were established to successfully prepare pure cultures. In addition, I was eager to develop a defined synthetic medium in order to get rid of the sludge water as well as to resolve the minimum growth requirements and further physiological characteristics.

The swarming behavior of the cells and their growth rate remained the best criteria for judging good culture conditions. Well correlated with this was the microscopic picture of the cells: their regular or irregular shape and size, their speed of swimming, and further details.

It became quite obvious that the continuous illumination and bright incandescent light that were previously used for phototrophic sulfur bacteria (33, 119a) were unsuitable for *C. okenii* and *T. jenense*. The cells looked sick under these conditions; they were irregularly shaped and filled with very large sulfur globules and polysaccharide granules. They moved slowly if at all and settled to the bottom of the bottles. If incubated in dim light and diurnal cycles of 16 h light and 8 h dark at room temperature, cultures gradually recovered.

For the development of the defined synthetic medium, I used calcium sulfate as a mineral base, because this salt was present to saturation in the supernatent water of the successful sludge columns. The nutrient minerals, particularly sulfide, phosphate, ammonium, and carbonate were added in low concentrations and increased stepwise as far as tolerated. The same was done with the trace elements. When it became apparent that growth could still be improved by changes in the concentrations of the trace elements, I developed a new formula experimentally (74). If inoculated from raw cultures in sludge water, *C. okenii, C. warmingii,* and *T. jenense* grew well at first; in subcultures, however, growth gradually declined, and swollen, polysaccharide-stuffed, slow-moving cells appeared. To me this indicated lack of a growth factor that was present in the sludge water. The most typical vitamin of organically rich anoxic sludge is vitamin B_{12} (5). When I added cyanocobalamin (20 μg per liter) to my cultures, the negative symptoms were relieved and all three Ostrau purple sulfur bacterial species grew well continuously in the otherwise purely mineral culture medium (63–65, 74). Growth factors characteristic of purple nonsulfur bacteria (41) were not required. Because the large members of the family *Chromatiaceae* were inhibited by sulfide concentrations above 2–3 mM, the initial amount of sulfide had to be limited and replenished each time after use (64, 97). Thus, I was able to achieve high population densities under photoautotrophic conditions in a gentle way.

When I used the new culture medium to isolate more *C. okenii* strains from other habitats, I found that phototrophic green sulfur bacteria of the *Chlorobium* type developed extremely well if incidentally introduced with the inoculum. In fact they turned out to be a weed in my *Chromatium* enrichments and outgrew the purples within short time. In order to selectively enrich bacteriochlorophyl a–containing purple bacteria, we successfully applied two types of infrared radiation filters: (*a*) a filter absorbing all radiation below 700 nm and therefore excluding the chlorophyll a–containing algae and (*b*) a filter absorbing all radiation below 800 nm that excluded the green sulfur bacteria and algae (e.g. cyanobacteria) (66, 85, 103). With these filters selective enrichments became readily possible.

A full year of exciting work, with ups and downs in the interplay between our purple friends from Ostrau and myself, had passed. I learned a lot from them, and only by taking them very seriously did I confidently arrive at my initial aim of establishing a defined mineral medium for reliable cultivation. Now, the way was open for a second phase of studies in which several others of Winogradsky's genera of purple sulfur bacteria could be isolated in pure culture and characterized physiologically and biochemically.

On Learning at Hopkins Marine Station, Pacific Grove

A research grant from the Deutsche Forschungsgemeinschaft gave me the opportunity to spend a year (1962/1963) in van Niel's laboratory at Hopkins Marine Station of Stanford University, California. Van Niel was a frank science teacher from the bottom of his heart who at the same time knew of the strong links between science, with its strengths and limitations, the humanities, and the arts. Therefore, the exchange of ideas was very sincere, delightful, and stimulating. I soon felt at home at the lab because all we van Niel students—John Bennett, Ercole Canale-Parola, Ed Leadbetter, Lynn Miller, and Roger Whittenbury—were such happily working individuals, each one of us available to listen to the others' little problems or pleasures of daily lab work.

I continued my studies on the described but hitherto uncultured purple sulfur bacteria. When I complained about the green sulfur bacteria being a nuisance in my enrichment cultures, van Niel replied: "Oh no, by all means keep them! We had a hard time to get and grow them." So I isolated pure cultures of *Chlorobium* species that came up in deep agar cultures directly inoculated with black mud from marine and freshwater habitats. The fact that most strains required vitamin B_{12} for growth (74) partly explained why isolation had not been easy in the past (45). The new pure cultures should soon become a valuable stock for comparative fine structure studies.

During my stay in Pacific Grove, van Niel held the last of his famous summer courses and I was allowed to participate as a guest. It was an exciting experience, and we all got the impression that this is the way in which science should be taught. In the lecture courses van Niel represented with great devotion the conceptual development of microbiology and microbial biochemistry since the days of Antonie van Leeuwenhoek in the seventeenth century. He revived the questions raised and the answers worked out by individual scientists at their time and presented them in the context of the prevailing knowledge of that time. This way we could fully appreciate the individual achievement. Growth of knowledge depended on precise perceptions, diligent experiments, and careful conclusions of the interested researchers. Van Niel liked to show that "conclusion jumpers" often failed to reach the correct answer. It became quite clear that questions could be long-lasting while answers often reflected present-day knowledge and had to be expanded upon later.

In his lectures, van Niel followed his view that the main task of the academic teacher is to present, together with the established knowledge, the open questions, unsolved problems, and neglected fields of work. Thus, instead of overwhelming his audience with established facts as piled up in textbooks, van Niel roused the curiosity of the students and offered them abundant opportunities for their mental and experimental activities.

Nowadays some university teachers consider it inevitable to become travelers for science and to leave students and labs to themselves. Did they ever notice how encouraging the sheer presence or interest of an experienced person is for a student? I was pleased to read that Ralph Wolfe had also noticed this positive effect (135). And it may be good to recall that van Niel was present for his collaborators like the liberal father of a family (119d).

Fine Structures of Phototrophs (Berkeley)

During my stay at Hopkins Marine Station, Drs. Germaine Cohen-Bazire and Roger Y. Stanier from the University of California at Berkeley came to visit van Niel and the lab. When they saw my well-growing *Chlorobium* cultures, they expressed their interest in a cooperation for fine-structure studies and invited me to Berkeley. Cohen-Bazire had just published impressively good electron micrographs of the development of intracytoplasmic photosynthetic membrane systems of purple nonsulfur bacteria (15). However, little was known about the photosynthetic structures of the green sulfur bacteria. Although no internal membrane systems had been detected in *Chlorobium* (120), Stanier & Cohen-Bazire believed that the molecular components of photosynthesis would probably be membrane bound, for functional reasons. In the course of several weeks of exciting electron microscopic work in Cohen-Bazire's lab, we discovered in all of my six *Chlorobium* strains oblong vesicular bodies that we called "*Chlorobium* vesicles." They were located adjacent to the inner side of the cytoplasmic membrane (17). The vesicles were also visible in negatively stained preparations of the main pigmented fraction isolated from *Chlorobium* extracts, thus precluding the possibility that they were artifacts of the thin-section preparation technique, as argued by other investigators. The vesicles were later named chlorosomes (101), and it was shown that they contain all of the light-harvesting bacteriochlorophyll c or d, whereas reaction centers and electron-transport components are mainly housed in the adjacent cytoplasmic membrane (87, 101).

During subsequent years, I gladly and gratefully accepted Germaine Cohen-Bazire's cooperation to elucidate the fine structure of my new isolates of phototrophic bacteria. It became quite clear that the structure of the photosynthetic apparatus represented a taxonomically significant feature. In the course of these studies, we also obtained the first thin-section pictures of the gas vesicles in the gas-vacuolated purple and green sulfur bacteria of the genera *Amoebobacter* (*Rhodothece*), *Lamprocystis, Thiodictyon,* and *Pelodictyon* (16, 72).

Characteristics of Phototrophic Sulfur Bacteria, Carotenoids, Taxonomy (Göttingen)

During the 1960s, it became increasingly clear that most of the species of Winogradsky's 13 genera (7, 66, 133) of the purple sulfur bacteria could be isolated in pure culture using the new synthetic culture medium (65). The same was true for Lauterborn's green sulfur bacteria (7). Therefore, my task apparently was to establish pure cultures of all these species and to confirm in pure culture that they could be upheld taxonomically and were not just naturally occurring growth forms of a few stable types. Winogradsky's characterizations of the genera were based on the size, shape, and mode of development of cell aggregates from natural habitats (7, 133). In order to identify my new pure-culture isolates with Winogradsky's descriptions, the conditions under which cell aggregates are formed or disintegrate had to be determined experimentally. The results of these studies showed that sulfide concentration and light intensity were the primary regulating factors; pH and incubation temperature had to be considered as well (12, 66, 102). Thus I was able to identify most pure-culture isolates as Winogradsky's species and genera. Also apparent, however, was that some species gradually lost their typical aggregate formation after one or a few years maintenance in pure culture, e.g. *Thiodictyon elegans* and *Thiopedia rosea* (23, 102). Therefore, morphological descriptions were primarily based on the characteristics of the individual cells (12, 102). From an ecological point of view, however, the possible changes between life in fixed aggregates or as free single cells must be considered as advantageous reactions to the environmental factors described.

The biochemical characterization of the purple and green sulfur bacteria was an enormous task for years and could not be done without cooperation. Manley Mandel was so kind to determine the DNA base ratios in all samples of cell material that I mailed to him in Texas (51, 52). Hans Hermann Thiele tested the utilization of inorganic and organic substrates by our purples (111), and Eike Siefert studied various aspects of dinitrogen fixation (95, 96). The phototrophic sulfur bacteria cultures exhibited a whole range of colors, so that the presence of various different carotenoids could be expected. Dr. Synnøve Liaaen-Jensen from Trondheim, Norway had just successfully elucidated the structure and possible biosynthesis of several carotenoids of the purple nonsulfur bacteria (33, 46). We were therefore extremely glad that she was willing to study the structure of new carotenoids from our phototrophs. Synnøve introduced Karin Schmidt from our institute—as well as myself—to the analytical methods, and she cooperated by doing the chemistry, identification, and naming (48). The carotenoids of the spirilloxanthin and rhodopinal series, which also occur in purple nonsulfur bacteria, were identified in many species of purple sulfur bacteria. Unique to several species of the latter group were the carotenoids of the okenone series (75, 86, 88), which give the cells and cultures an unrivaled bright purple color. Typical of the green sulfur bacteria were the carotenoids of the chlorobactene and isorenieratene series (47, 49, 86, 89). With the exception of one species (24, 25), all purple sulfur bacteria contained bacteriochlorophyll a with phytol as the esterifying alcohol. The same was true for the minor component bacteriochlorophyll a of the reaction centers of the green sulfur bacteria (34). During these studies, Axel Gloe discovered the new bacteriochlorophyll e in all brown-colored species of the green sulfur bacteria (35). The structure of this new bacteriochlorophyll e was elucidated in cooperation with Prof. H. Brockmann Jr. (8).

It was a great pleasure and an indispensable support for the necessary taxonomic work when Dr. Hans Trüper joined my group in 1969. Hans had worked on the carbon and sulfur metabolism of *C. okenii* (112, 116) and subsequently remained true to the phototrophic sulfur bacteria (113, 114). Our cooperation was strongly stimulated when Drs. B. Buchanan and N. E. Gibbons invited us to revise all groups of phototrophic bacteria with anoxygenic photosynthesis for the eighth edition of *Bergey's Manual* that appeared in 1974 (12). The purple and green sulfur bacteria had to become *Chromatiaceae* and *Chlorobiaceae,* respectively (76, 115), and type or neotype strains for 41 species had to be designated (77). I discussed subsequent advances and taxonomic problems in my comparative systematic survey of 1977 (68). That the *Chromatiaceae,* originally established as a systematic group by Molisch in 1907, indeed form a phylogenetically coherent group was later shown by 16S rRNA oligonucleotide cataloging (29). The closest relatives (that are sufficiently separate) of the *Chromatiaceae* are the species of the genus *Ectothiorhodospira,* which advanced to a family (99). Both families and several groups of well-known gram-negative chemotrophic bacteria, e.g. the enteric-vibrio group, are now part of the γ-subclass of the *Proteobacteria*. This class comprises the purple nonsulfur bacteria and their nonphototrophic relatives in its α- and β-subclass (100, 102).

The anoxygenic phototrophic purple bacteria exemplify the problems with which bacterial taxonomy has been confronted during the 15 years since Woese and his coworkers (30) established molecular phylogeny as a new basis for taxonomy. The phenotypic character of phototrophic metabolism, so important in traditional taxonomy, lost weight when investigators realized that, according to those criteria, chemotrophic bacteria of other physiological groups would be much more closely related to phototrophs than various groups of phototrophs are related among themselves (93). Certainly, bacterial taxonomy could not be reformed as fast as intended by Woese because identification of bacteria is an indispensable task. As a member of Bergey's Manual Trust for 11 years, I was well aware of the problems. We felt a strong responsibility to keep *Bergey's Manual of Systematic Bacteriology* a standard reference book that is as useful as possible for working bacteriologists. At the same time, molecular phylogeny had to be recognized as much as possible. In his personal review (54), R. G. E. Murray vividly described the existing problems for the "taxonomy in transition."

COOPERATION WITH FRIENDS OF ANAEROBES IN URBANA, ILLINOIS

In 1967 Dr. L. Leon Campbell, Department of Microbiology, University of Illinois at Urbana, invited me to teach Ralph Wolfe's microbial diversity course in the winter term while Wolfe was on a sabbatical. I did not hesitate to accept the offer because I found it challenging to teach in English and to hold a detailed course of nine hours per week. Also, I was glad to meet Campbell, who was working on sulfate reducers, and Drs. Mike Wolin, Marvin Bryant, and possibly also Ralph Wolfe, all three of whom were involved in successful work on methanogenic bacteria (10, 135). I was glad that the course went as well as I had hoped. The enrichments for purple nonsulfur bacteria even yielded a new species: *Rhodopseudomonas acidophila,* which turned out to be widespread in acidic freshwater habitats such as peat bogs (67).

Mike Wolin and Marv Bryant invited me over to their lab in Dairy Science after I finished the course. Marv Bryant successfully cultivated his strict anaerobes with the ingenious Hungate technique and was very happy to introduce me to it. I had no problems in applying it; my cultures grew splendidly, and we spent a wonderful time together. It was a good start to more cooperation in the future.

Ralph Wolfe joined my lab in Göttingen for five months in 1975. He got interested in the sulfur-reducing bacteria on which we were working at that time and isolated and characterized the facultative sulfur reducer "Spirillum" 5175 (136). This metabolically most versatile bacterium (91) was recently recognized as the new genus and species *Sulfurospirillum deleyianum* (92). Ralph joined us again in spring 1984 in Konstanz where he continued his studies on the marvellous magnetotactic bacteria. At this time, plans took shape for my participation as a coordinator in the 1986 Microbiology Course at the Marine Biological Laboratory, Woods Hole, Mass. with Ralph Wolfe as Director. When I saw and smelled Sippewissett salt marsh near Woods Hole, I was convinced that it is a real paradise for the phototrophic sulfur bacteria; indeed I had no problems in finding and introducing them successfully at the summer course.

Marv Bryant joined my group for six months in winter 1976/1977 in Göttingen. With him we became fully enthusiastic about the degradation powers of anaerobes cooperating by interspecies hydrogen transfer. Marv obtained his first methanogenic associations capable of degrading C4 to C8 fatty acids to acetate and methane; the defined coculture was later characterized by his coworker McInerney (53). Once more the limits of pure-culture techniques and the importance of microbial interactions became obvious.

EXPANSION OF RESEARCH IN GÖTTINGEN, 1970–1979

Dr. F. Göbel's Quantostat

Instead of following a call to a chair for Microbiology at the University in Braunschweig, I accepted a position as Head of a Division in the Institut für Mikrobiologie of the Gesellschaft für Strahlen- und Umweltforschung in Göttingen in 1970. In the large new University building my group obtained ample room and gradually acquired the equipment necessary for our research.

A project with the phototrophic bacteria that I had in mind since the days when I worked in van Niel's lab now became feasable: determination of the number of quanta required per gram of biomass formed under steady-state growth conditions in the chemostat using monochromatic radiation as the growth-limiting factor. This project would have been impossible without the scientific and practical capacities of Florian Göbel, an experienced chemical process engineer, who joined my group and designed all the necessary instruments and supervised their manufacture. Most significant for the quantitative treatment of light-dependent growth, Göbel introduced the new and indispensable parameter: the mean irradience ψm. Göbel summarized his results with the quantostat in a chapter of the *Photosynthetic Bacteria* (36). Autotrophic growth of *Rhodopseudomonas acidophila* on H_2 plus CO_2 required 0.69 mol quanta (λ = 860 nm) per gram of dry cell mass formed. With acetate, the quantum requirement decreased about 50%. Göbel's results also showed that anoxygenic photosynthesis with H_2 is by no means more efficient than oxygenic photosynthesis with water as the electron donor. *Anabaena cylindrica* strain 7120 required 0.62 mol quanta (λ = 682 nm) per gram of dry cell mass if ammonium was the nitrogen source. With dinitrogen, the quantum requirement increased 1.6-fold. Corresponding results were obtained with *R. acidophila*.

How I Became Familiar with Sulfur-Reducing Eubacteria

Colleagues often sent me a phototrophic bacterium for identification or tests for purity. In 1972 I received a green sulfur bacterium culture that grew very well in sulfide-reduced medium on ethanol in the light. Using various media, I prepared pure cultures of all bacterial types present in the culture. The isolated *Chlorobium* did not use ethanol, but a *Desulfovibrio* strain from the culture grew well on it. If I mixed the two pure cultures, the coculture grew well on ethanol, as the *Chlorobium* grew on the sulfide and acetate formed by *Desulfovibrio*. I was satisfied and the problem was solved.

In 1974 I received another green sulfur bacterium culture from Japan, also growing well on ethanol. This time I isolated *Prosthecochloris aestuarii* as the phototroph and a *Streptococcus,* but no sulfate reducer came up on ethanol. Thin slender rods also visible in the original culture did not grow. The situation was puzzling and presented a real challenge for me, all the more so because the original mixture grew well on acetate, but the *P. aestuarii* pure culture used neither acetate nor ethanol. I decided to play with subcultures of the original mixture, hoping that they might give me a hint. In addition to acetate and ethanol for robust growing controls, I tested 10 simple organic substrates for utilization in the light and dark. None of them yielded a better growing culture than the control on sulfide alone. However, the light cultures showed a strong initial turbidity as is known for pure cultures of green sulfur bacteria and is caused by globules of extracellular elemental sulfur transiently formed from sulfide in the light. For unknown reasons, the robustly growing controls on ethanol and acetate did not show this turbidity. Why didn't they form sulfur globules as well? Another quite puzzling result of the test was that the unknown slender rods had multiplied significantly in the light and dark cultures with 10 mM malate, but *P. aestuarii* was not promoted in the light and had only grown as much as it did in the sulfide control. Did this mean that the thin slender rod was not the mysterious growth promoter of the green mixed culture? I decided to add wet sterile elemental sulfur (S^0) to the ethanol and acetate dark control cultures. Of course no growth of the mixed culture could be expected in the dark; however, after a few days of incubation I noticed a terribly strong smell of hydrogen sulfide in both cultures! Repeated subcultures in mineral media with sulfur flower and either ethanol or acetate in the dark continued to form sulfide strongly and showed growth of the thin slender rods. In controls without sulfur flower neither sulfide formation nor growth occurred. In collaboration with Dr. Hanno Biebl, the experiments were subsequently repeated with pure cultures of the well-growing thin slender rods that we named *Desulfuromonas acetoxidans* (71). We were able to show that the oxidation of acetate or ethanol to CO_2 is stoichiometrically linked to the reduction of elemental sulfur to sulfide. Sulfur compounds more oxidized than S^0 cannot be reduced by *Desulfuromonas* species. Determinations of the dry cell mass yields per mol of substrate oxidized convinced us beyond all doubt that the bacteria were able to conserve the necessary energy for growth by the reduction of sulfur (S^0) to sulfide (71). And finally we fulfilled Koch's postulates by establishing robust growing stable cocultures on acetate or ethanol by adding *D. acetoxidans* to pure cultures of green sulfur bacteria (6, 71). Now the reason was clear why the acetate and ethanol cultures of my initial substrate tests did not form the turbidity-causing sulfur globules: sulfur formed from sulfide and excreted by the green sulfur bacteria in the light became immediately reduced back to sulfide by the acetate-oxidizing *Desulfuromonas* spp. Both bacteria mutually recycle their metabolic products (sulfide, sulfur) to substrates. And we were able to show that only small amounts of sulfide are required to sustain growth of such robustly growing syntrophic cocultures (6).

By doing our experimental work very carefully, Biebl and I had learned tremendously from these delicate microbes, of which we isolated several types from anoxic sediments. It was, however, difficult to talk to colleagues about their existence. The process appeared disturbing to them for two reasons, given established biochemical knowledge: First, the free-energy change of the reduction of S^0 to sulfide is very small and was considered to be hardly sufficient to support energy conservation for growth, and second, the oxidation of acetate to two CO_2s was considered to be possible only with electron acceptors of a much more positive redox potential than S^0/H_2S, e.g. oxygen or nitrate. In this situation meeting Prof. Rolf Thauer with his profound knowledge of energy conservation in anaerobic bacteria was quite gratifying (110). Instead of precluding this kind of metabolism theoretically on the basis of known biochemical reactions, he was so open-minded as to consider the process as a challenge for biochemists. Eventually, we learned of hitherto unknown reactions successfully exploited by *Desulfuromonas* spp. That new reactions were involved was later shown by Thauer and coworkers (32, 109): The bacterium could dehydrogenate acetate via the citric-acid cycle by linking the critical S^0 reduction with succinate to a reversed electron transport (109).

F. Widdel Discovers Diversity Among Sulfate-Reducing Bacteria

Friedrich Widdel completed his masters thesis in spring 1976 at a time when I was interested in photosynthetic hydrogen production from organically rich waste waters. I was extremely glad that Fritz was ready to do his doctoral thesis in our group. His marvellously attentive and original style toward his work had convinced me of his extraordinary capacities for research. He accepted the project to search for phototrophic bacteria forming hydrogen from acetate, as this was the most important intermediate product of anaerobic decomposition. In his enrichment cultures on acetate as sole organic substrate, Fritz surprisingly detected not only purple nonsulfur bacteria, but also small *Chromatium* cells with internal sulfur globules, indicating the presence of sulfide. Indeed, the culture bottles had a strong hydrogen sulfide smell! Because all known pure cultures of sulfate reducers did not oxidize acetate and this process was considered to be energetically unlikely (123), I urged Fritz to grow subcultures of these enrichments in the dark; he reproducibly obtained sulfide formation on acetate. Large motile, rod-shaped bacteria with pointed ends were dominant in his sulfidogenic enrichments. As I knew that acetate oxidation by sulfate reduction would be a most important process ecologically (42), Fritz and I decided to stop the work on phototrophs and to continue his thesis on the new sulfate reducers. With the pure culture, Fritz convincingly determined the stoichiometry of the process, the dry cell mass yield on acetate, and classified the new sporing bacterium as *Desulfotomaculum acetoxidans* (129, 130). The publication appeared in 1977, only one year after the discovery of acetate oxidation by sulfur reduction.

Fritz Widdel continued to be exceptionally productive in enriching, isolating, and characterizing new types of sulfate reducers that mineralized reduced aliphatic and aromatic compounds, the complete oxidation of which had been considered impossible under anoxic conditions. His thesis of 1980 became an extraordinary book presenting not only five new genera and nine new species of sulfate reducers, but also a thorough evaluation of the history, ecology, physiology, and biochemical problems of dissimilatory sulfate-reducing bacteria (124). More recently, he wrote a general review on the group for A. J. B. Zehnder's book (127, 137). In their laboratories, Rolf Thauer and G. Fuchs showed that two different pathways of acetate dehydrogenation exist in the sulfate-reducing bacteria. While *Desulfobacter postgatei* uses a modified citric acid cycle (31, 109), *D. acetoxidans* and *Desulfobacterium* species have a novel reaction sequence: the acetyl-CoA:carbonmonoxide dehydrogenase pathway (70, 80, 98, 109).

MICROBIAL ECOLOGY AND LIMNOLOGY IN KONSTANZ, 1980–1990

In order to be free to continue our studies on anaerobic aquatic bacteria, I left Göttingen in 1979/1980 to join the Faculty of Biology at the University of Konstanz, where I accepted a chair for Limnology in 1981. I was ready to concentrate on teaching and to leave research mainly to coworkers and students. However, I brought my collection of purple and green sulfur bacteria along, used it for various projects, and took care of it until I retired in 1990. Fritz Widdel joined me at Konstanz and continued his studies on new metabolic capacities of new sulfate reducers (11, 125, 126, 128) until he joined Ralph Wolfe in Urbana, Illinois.

In lectures and practical courses, we tried to integrate existing knowledge of biochemistry, physiology, and microbiology into the representation of microbial processes proceeding in water and sediment of aquatic habitats (69). Consequently we attracted diploma and PhD students with a broad interest and willingness to work experimentally in microbial ecology. Each research project left sufficient room for individual approaches and eventual changes in case of unexpected results. And usually one could correct one's ideas or handling to care for the wellbeing of the microbes and their activity. It goes without saying that all the work was devoted to anaerobic microbes and processes.

Cooperation with Dr. B. Schink, and Dr. H. Cypionka and PhD Students

Fortunately, Dr. Bernhard Schink joined my group as a coworker in 1981. He wanted to shift his research from aerobic to anaerobic microbial processes. In particular he concentrated on fermentative degradations and transformations of simple organic substances whose fate had not been studied so far. Once again the enrichment-culture techniques with defined mineral media proved to be extremely useful, and Schink achieved surprising results with them. Within a few years he described 4 new genera and 14 new species of anaerobic fermentative bacteria with hitherto unknown degradation capacities and fermention pathways (81). Several fermentations became energetically feasible only in cocultures with hydrogen-consuming methanogenic bacteria. We considered it likely that Schink would someday detect a fermentative energy-conserving reaction whose free-energy change was thought too small to sustain growth. Indeed he discovered *Propionigenium modestum,* which can grow at the expense of the membrane-bound decarboxylation of succinate to propionate (38, 82).

Four of the 13 PhD students who took their degree in my department in Konstanz cooperated directly in B. Schink's research program. Andreas Tschech studied the degradation of hydroxybenzoates, 2-aminobenzoate, resorcylates, and resorcinol in newly isolated pure cultures or methanogenic cocultures (117–119). Marion Stieb obtained pure cultures and methanogenic cocultures degrading polyethylene glycol, fatty acids (4–11 carbon atoms), 3-hydroxybutyrate, and isovalerate (83, 104–106). The degradation of hydroquinone, gentisate, and benzoate by a defined coculture was demonstrated by Ulrich Szewzyk (108). And Jürgen Seitz determined the growth dependencies of selected species of hydrogen-forming fermentative bacteria and hydrogen-consuming methanogenic, acetogenic, or sulfidogenic bacteria on the hydrogen concentrations and terminal-electron acceptors in cocultures in the chemostat (94).

When Schink accepted a Professorship in Marburg, Dr. Endre Laczko from Switzerland joined us for five years. He supervised diploma students working on sediment protozoa and constructively cooperated in our practical courses.

Dr. Heribert Cypionka joined our group in summer 1983 and concentrated on the energy metabolism and sulfate transport in sulfate-reducing bacteria. He established low sulfide-controlled continuous culture (sulfidostat) and proved, contrary to current belief, that *Desulfotomaculum orientis* can grow autotrophically in a defined mineral medium with hydrogen and sulfate as the sole energy source (18, 21). He also built a special measuring cell to determine proton movements and sulfide and hydrogen concentrations by electrodes with on-line recording. Thus, Cypionka became the first who succeeded in characterizing sulfate transport and electron transport–driven proton translocation in sulfate reducers (19, 20). These studies were feasible only with use of intact cells freshly harvested from continuous culture. Robert Fitz did his doctoral thesis with Cypionka on electron transport–dependent energy conservation in *Desulfovibrio* (26–28).

Rainer Kreikenbohm, my first PhD student in Konstanz, was interested in constructing theoretical models for defined commensalistic cocultures in the chemostat. His most impressive example was the complete degradation of trimethoxybenzoate by coculture of *Acetobacterium woodii, Pelobacter acidigallici,* and *Desulfobacter postgatei* (44). Regine Szewzyk studied the competition for ethanol between various sulfate-reducing and fermenting bacteria. In substrate-sufficient batch cultures, the fermenting bacteria outcompeted the sulfate reducers, but in ethanol-limited chemostat cultures, the situation was reversed (107). Friedhelm Bak determined the in situ sulfate-reduction activities in littoral sediment profiles (Lake Constance) and the metabolic types and species of sulfate reducers involved (3, 4). And he discovered a novel type of energy metabolism, the disproportionation of thiosulfate or sulfite by *Desulfovibrio sulfodismutans* (1, 2). The capacity for microaerobic chemotrophic growth on thiosulfate in the dark was shown for certain species of purple sulfur bacteria by Charlotte Kämpf (43). Barbara Eichler followed the seasonal development and bloom formation of the phototrophic sulfur bacterium *Thiopedia rosea* in a holomictic lake (22). She also isolated new pure cultures of this delicate species and contributed to a proper characterization (23). Jörg Overmann combined in his thesis several competently performed subjects of which I mention only two: his illuminating studies on the reasons for the dense purple bacterial layer of Mahoney Lake (58) and his precise characterization of an extremely low light–adapted green sulfur bacterium from the chemocline (80 m depth) of the Black Sea (59). With the thesis of Stefan Wagener, we entered the relatively unknown area of eukaryotic bacterial antagonists: anaerobic ciliates. Stefan was kindly introduced to this world by a connoisseur, Dr. Claudius Stumm, Nijmegen (122). Stefan comprehended so well that he was soon on his own way, learning continuously from his companion, the ciliate *Trimyema compressum*. He established the first monoxenic culture of an anaerobic ciliate (121) and uncovered its surprising food specialization in certain bacterial species (90). My last PhD student, Stephan Holler, continued Wagener's innovative work on *Trimyema* by studying the role of an endosymbiotic bacterium in strain N as well as the ciliate's fermentation products (39, 40).

EPILOGUE

S. N. Winogradky's and L. Pasteur's studies more than a hundred years ago mark the beginning of microbial ecology. Both were aware of the fact that the real nature of a microbe cannot truly be understood if it is not considered in the context of the biogeochemical processes on earth. For a lecture on "the role of the microbes in the general cycle of life" held 1896 in St. Petersburg, Winogradsky chose as a motto a quotation of Pasteur: "...le rôle des infiniment petits m'apparaissait infiniment grand..." (...the role of the infinitely small appears to me infinitely large...).

Wherever their living conditions arise, microbes will be present and start their transforming activity. This is also true for the anaerobes of which I talked in my review. However, their world is mostly hidden to us. We have to descend to deeper water layers and sediments to enter the anoxic world of fermentative degradations where the substrates for the methanogenic, sulfur- and sulfate-reducing bacteria are formed. These microbes degrade these compounds to carbon dioxide, methane, and hydrogen sulfide and establish the iron sulfide–blackened, reduced end of the biosphere. Further transformations depend on the access of either light or oxygen. Even in dim light at some depth, the purple, red, brown, or green phototrophic sulfur bacteria bloom, e.g. in stratified lakes, all kinds of lagoons, and salt marshes (84). Because phototrophic sulfur bacteria oxidize hydrogen sulfide to sulfate anaerobically in the light, they close the anoxic sulfur cycle, initiated by the sulfate reducers, and form a colorful blooming sulfuretum. Compared to other ecosystems, the sulfuretum appears simple with respect to the interactions of its microbes. And the individual sulfureta in freshwater and saline environments are variations on a theme, each one proceeding by different species of the same metabolic types involved (84).

In retrospect it fills me with gratitude that the research of my coworkers and myself more and more centered on the microbes of the anoxic sulfur cycle. For two reasons this could not have been planned theoretically: first, access to the unexplored realm was gained only after certain microbes appeared and grew in our lab and so became our guidestars to unusual companions, and second, microbes with metabolic capacities were discovered that until then were considered energetically or biochemically unlikely.

Also involved in research on certain microbes around the sulfur cycle were several scientists whom I have not yet mentioned, but to whom I feel connected. By their interest and stimulating conversations, they encouraged our work: Helge Larsen, Hans Veldkamp, Hans van Gemerden, Riks Laanbroek, Jerry Ensign, John Ormerod, Moshe Shilo,[2] Yehuda Cohen, Fuad Hashwa, Bo Jørgensen, Rod Herbert, Robert Matheron, Pierre Caumette, John Breznak, Kazuhiro Tanaka, little Ma Haiyan, Ly Renhao, and Waltraud Dilling.

My wife Helga and our children, finally, had to look at my work from a different point of view. They could tell you that a father who cares for fastidious microbe cultures and tries to achieve something with them often spends more time at the lab than anticipated. Therefore, I am extremely grateful to my dear family for having accepted my timetable and shown favorable understanding for my clinging to research on microbes.

Literature Cited

1. Bak, F., Cypionka, H. 1987. A novel type of energy metabolism involving fermentation of inorganic sulphur compounds. *Nature* 326:891–92
2. Bak, F., Pfennig, N. 1987. Chemolithotrophic growth of *Desulfovibrio sulfodismutans* sp. nov. by disproportionation of inorganic sulfur compounds. *Arch. Microbiol.* 147:184–89
3. Bak, F., Pfennig, N. 1991. Microbial sulfate reduction in littoral sediment of Lake Constance. *FEMS Microbiol. Ecol.* 85:31–42
4. Bak, F., Pfennig, N. 1991. Sulfate-reducing bacteria in littoral sediment of Lake Constance. *FEMS Microbiol. Ecol.* 85:43–52
5. Bernhauer, K., Friedrich, W. 1954. Über die Vitamine der B_{12}-Gruppe. *Angew. Chem.* 66:776–80
6. Biebl, H., Pfennig, N. 1978. Growth yields of green sulfur bacteria in mixed cultures with sulfur and sulfate reducing bacteria. *Arch. Microbiol.* 117:9–16
7. Breed, R. S., Murray, E. G. D., Smith, N. R. 1957. In *Bergey's Manual of Determinative Bacteriology*, pp. 35–67. Baltimore: Williams and Wilkins. 1094 pp. 7th ed.
8. Brockmann, H. Jr. 1976. Bacteriochlorophyll e: structure and stereochemistry of a new type of chlorophyll from Chlorobiaceae. *Philos. Trans. R. Soc. London Ser. B* 273:277–85
9. Brockmann, H., Pfennig, N. 1953. Die Gewinnung reiner Actinomycine durch Gegenstromverteilung (V. Mitteil. über Actinomycin). *Z. Physiol. Chem.* 292:77–88
10. Bryant, M. P., Wolin, E. A., Wolin, M. J., Wolfe, R. S. 1967. *Methanobacillus omelianskii,* a symbiotic association of two species of bacteria. *Arch. Microbiol.* 59:20–31
11. Brysch, K., Schneider, C., Fuchs, G., Widdel, F. 1987. Lithoautotrophic growth of sulfate-reducing bacteria, and description of *Desulfobacterium autotrophicum* gen. nov., sp. nov. *Arch. Microbiol.* 148:264–74
12. Buchanan, R. E., Gibbons, N. E., eds. 1974. *Bergey's Manual of Determinative Bacteriology,* pp. 24–64. Baltimore: Williams and Wilkins. 1246 pp. 8th ed.
13. Cholodny, N. 1930. Über eine neue Methode zur Untersuchung der Bodenmikroflora. *Arch. Mikrobiol.* 1: 620–52
14. Clayton, R. K., Sistrom, W. R., eds. 1978. *The Photosynthetic Bacteria.* New York: Plenum. 946 pp.
15. Cohen-Bazire, G., Kunisawa, R. 1963. The fine structure of *Rhodospirillum rubrum*. *J. Cell. Biol.* 16:401–19
16. Cohen-Bazire, G., Kunisawa, R., Pfennig, N. 1969. Comparative study of the structure of gas vacuoles. *J. Bacteriol.* 100:1049–61
17. Cohen-Bazire, G., Pfennig, N., Kunisawa, R. 1964. The fine structure of green bacteria. *J. Cell. Biol.* 22:207–25
18. Cypionka, H. 1986. Sulfide-controlled continuous culture of sulfate-reducing bacteria. *J. Microbiol. Methods* 5:1–9

[2]Deceased.

19. Cypionka, H. 1987. Uptake of sulfate, sulfite and thiosulfate by proton-anion symport in *Desulfovibrio desulfuricans*. *Arch. Microbiol.* 148:144–49
20. Cypionka, H. 1989. Characterization of sulfate transport in *Desulfovibrio desulfuricans*. *Arch. Microbiol.* 152: 237–43
21. Cypionka, H., Pfennig, N. 1986. Growth yields of *Desulfotomaculum orientis* with hydrogen in chemostat culture. *Arch. Microbiol.* 143:396–99
22. Eichler, B., Pfennig, N. 1990. Seasonal development of anoxygenic phototrophic bacteria in a holomictic drumlin lake (Schleinsee, FRG). *Arch. Hydrobiol.* 119:369–92
23. Eichler, B., Pfennig, N. 1991. Isolation and characteristics of *Thiopedia rosea* (neotype). *Arch. Microbiol.* 155:210–16
24. Eimhjellen, K. E. 1970. *Thiocapsa pfennigii* sp. nov., a new species of phototrophic sulfur bacteria. *Arch. Mikrobiol.* 73:193–94
25. Eimhjellen, K. E., Steensland, H., Traetteberg, J. 1967. A *Thiococcus* sp. nov. gen., its pigments and internal membrane system. *Arch. Mikrobiol.* 59:82–92
26. Fitz, R. M., Cypionka, H. 1989. A study on electron transport-driven proton translocation in *Desulfovibrio desulfuricans*. *Arch. Microbiol.* 152: 369–76
27. Fitz, R. M., Cypionka, H. 1990. Formation of thiosulfate and trithionate during sulfite reduction by washed cells of *Desulfovibrio desulfuricans*. *Arch. Microbiol.* 154:400–6
28. Fitz, R. M., Cypionka, H. 1991. Generation of a proton gradient in *Desulfovibrio vulgaris*. *Arch. Microbiol.* 155:444–48
29. Fowler, V. J., Pfennig, N., Schubert, W., Stackebrandt, E. 1984. Towards a phylogeny of phototrophic purple sulfur bacteria—16S rRNA oligonucleotide cataloguing of 11 species of Chromatiaceae. *Arch. Microbiol.* 139: 382–87
30. Fox, G. E., Pechmann, K. R., Woese, C. R. 1977. Comparative cataloging of 16S ribosomal ribonucleic acid: molecular approach to procaryotic systematics. *Int. J. Syst. Bacteriol.* 27:44–57
31. Gebhardt, N. A., Linder, D., Thauer, R. K. 1983. Anaerobic acetate oxidation to CO_2 by *Desulfobacter postgatei*. 2. Evidence from ^{14}C-labelling studies for the operation of the citric acid cycle. *Arch. Microbiol.* 136:230–33
32. Gebhardt, N. A., Thauer, R. K., Linder, D., Kaulfers, P.-M., Pfennig, N. 1985. Mechanism of acetate oxidation to CO_2 with elemental sulfur in *Desulfuromonas acetoxidans*. *Arch. Microbiol.* 141:392–98
33. Gest, H., San Pietro, A., Vernon, L. P., eds. 1963. *Bacterial Photosynthesis*, pp. 19–34. Yellow Springs, OH: Antioch. 523 pp.
34. Gloe, A., Pfennig, N. 1974. Das Vorkommen von Phytol und Geranylgeraniol in den Bakteriochlorophyllen roter und grüner Schwefelbakterien. *Arch. Microbiol.* 96:93–101
35. Gloe, A., Pfennig, N., Brockmann, H. Jr., Trowitzsch, W. 1975. A new bacteriochlorophyll from brown-colored Chlorobiaceae. *Arch. Microbiol.* 102: 103–9
36. Göbel, F. 1978. Quantum efficiencies of growth. See Ref. 14, pp. 907–25
37. Herbert, D., Elsworth, R., Telling, R. C. 1956. The continuous culture of bacteria: a theoretical and experimental study. *J. Gen. Microbiol.* 14:601–22
38. Hilpert, W., Schink, B., Dimroth, P. 1984. Life by a new decarboxylation-dependent energy conservation mechanism with Na^+ as coupling ion. *EMBO J.* 3:1665–70
39. Holler, S., Pfennig, N. 1991. Fermentation products of the anaerobic ciliate *Trimyema compressum* in monoxenic cultures. *Arch. Microbiol.* 156:327–34
40. Holler, S., Pfennig, N. 1992. Two types of endocytobiotic bacteria of the anaerobic ciliate *Trimyema compressum* and their effect on the viability of the ciliate. *Endocytobiosis Cell Res.* 8:201–3
41. Hutner, S. H. 1950. Anaerobic and aerobic growth of purple bacteria (Athiorhodaceae) in chemically defined media. *J. Gen. Microbiol.* 4:286–93
42. Jørgensen, B. B., Fenchel, T. M. 1974. The sulfur cycle of a marine sediment model system. *Mar. Biol.* 24:189–201
43. Kämpf, Ch., Pfennig, N. 1986. Chemoautotrophic growth of *Thiocystis violacea, Chromatium gracile* and *C. vinosum* in the dark at various O_2-concentrations. *J. Basic Microbiol.* 26: 517–31
44. Kreikenbohm, R., Pfennig, N. 1985. Anaerobic degradation of 3,4,5-trimethoxybenzoate by a defined mixed culture of *Acetobacterium woodii, Pelobacter acidigallici,* and *Desulfobacter postgatei*. *FEMS Microbiol. Ecol.* 31:29–38
45. Larsen, H. 1952. On the culture and

general physiology of the green sulfur bacteria. *J. Bacteriol.* 64:187–96
46. Liaaen-Jensen, S. 1965. Biosynthesis and function of carotenoid pigments in microorganisms. *Annu. Rev. Microbiol.* 19:163–82
47. Liaaen-Jensen, S. 1965. Aryl-carotenes from Phaeobium. *Acta Chem. Scand.* 19:1025
48. Liaaen-Jensen, S. 1978. Chemistry of carotenoid pigments. See Ref. 14, pp. 233–47
49. Liaaen-Jensen, S., Hegge, E., Jackman, L. M. 1964. Bacterial carotenoids XVIII. The carotenoids of photosynthetic green bacteria. *Acta Chem. Scand.* 18:1703–18
50. Lippert, K.-D., Pfennig, N. 1969. Die Verwertung von molekularem Wasserstoff durch *Chlorobium thiosulfatophilum*. Wachstum und CO_2-Fixierung. *Arch. Mikrobiol.* 65: 29–47
51. Mandel, M., Bergendahl, J. C., Pfennig, N. 1965. Deoxyribonucleic acid base composition of isolates of the genus *Chlorobium*. *J. Bacteriol.* 89:917–18
52. Mandel, M., Leadbetter, E. R., Pfennig, N., Trüper, H. G. 1971. Deoxyribonucleic acid base compositions of phototrophic bacteria. *Int. J. Syst. Bacteriol.* 21:222–30
53. McInerney, M., Bryant, M. P., Pfennig, N. 1979. Anaerobic bacterium that degrades fatty acids in syntrophic association with methanogens. *Arch. Microbiol.* 122:129–35
54. Murray, R. G. E. 1988. A structured life. *Annu. Rev. Microbiol.* 42:1–34
55. Deleted in proof
56. Deleted in proof
57. Deleted in proof
58. Overmann, J., Beatty, J. T., Hall, K. J., Pfennig, N., Northcote, T. G. 1991. Characterization of a dense, purple sulfur bacterial layer in a meromictic salt lake. *Limnol. Oceanogr.* 36:846–59
59. Overmann, J., Cypionka, H., Pfennig, N. 1992. An extremely low-light-adapted phototrophic sulfur bacterium from the Black Sea. *Limnol. Oceanogr.* 37:150–55
60. Pfennig, N. 1953. Untersuchungen an Actinomycin-bildenden Strahlenpilzstämmen und deren Actinomycinen. *Arch. Mikrobiol.* 18:327–41
61. Pfennig, N. 1956. Freie und gebundene Aminosäuren in knöllchentragenden und knöllchenfreien Erbsenpflanzen verschiedener Altersstadien. *Arch. Mikrobiol.* 24:8–30
62. Pfennig, N. 1958. Beobachtungen des Wachstumsverhaltens von Streptomyceten auf Rossi-Cholodny-Aufwuchsplatten im Boden. *Arch. Mikrobiol.* 31:206–16
63. Pfennig, N. 1961. Eine vollsynthetische Nährlösung zur selektiven Anreicherung einiger Schwefelpurpurbakterien. *Naturwissenschaften* 48:136
64. Pfennig, N. 1962. Beobachtungen über das Schwärmen von *Chromatium okenii*. *Arch. Mikrobiol.* 42:90–95
65. Pfennig, N. 1965. Anreicherungskulturen für rote und grüne Schwefelbakterien. *Zentralbl. Bacteriol. Parasitenk. Abt. I* 1:179–89, 503–4 (Suppl.)
66. Pfennig, N. 1967. Photosynthetic bacteria. *Annu. Rev. Microbiol.* 21:285–324
67. Pfennig, N. 1969. *Rhodopseudomonas acidophila,* sp. n., a new species of the budding purple nonsulfur bacteria. *J. Bacteriol.* 99:597–602
68. Pfennig, N. 1977. Phototrophic green and purple bacteria: a comparative, systematic survey. *Annu. Rev. Microbiol.* 31:275–90
69. Pfennig, N. 1984. Microbial behaviour in natural environments. *Symp. Soc. Gen. Microbiol.* 36(Part 2):23–50
70. Pfennig, N. 1989. Metabolic diversity among the dissimilatory sulfate-reducing bacteria. *Antonie van Leeuwenhoek J. Microbiol. Serol.* 56:127–38
71. Pfennig, N., Biebl, H. 1976. *Desulfuromonas acetoxidans* gen. nov. and sp. nov., a new anaerobic, sulfur-reducing, acetate-oxidizing bacterium. *Arch. Microbiol.* 110:3–12
72. Pfennig, N., Cohen-Bazire, G. 1967. Some properties of the green bacterium *Pelodictyon clathratiforme*. *Arch. Mikrobiol.* 59:226–36
73. Pfennig, N., Jannasch, H. W. 1962. Biologische Grundfragen bei der homokontinuierlichen Kultur von Mikroorganismen. *Ergeb. Biol.* 25:93–135
74. Pfennig, N., Lippert, K. D. 1966. Über das Vitamin B_{12}-Bedürfnis phototropher Schwefelbakterien. *Arch. Mikrobiol.* 55:245–56
75. Pfennig, N., Markham, M. C., Liaaen-Jensen, S. 1968. Carotenoids of Thiorhodaceae. 8. Isolation and characterization of a *Thiothece, Lamprocystis* and *Thiodictyon* strain and their carotenoid pigments. *Arch. Mikrobiol.* 62:178–91
76. Pfennig, N., Trüper, H. G. 1971. Conservation of the family name Chromatiaceae Bavendamm 1924 with the type genus *Chromatium* Perty 1852.

Request for an opinion. *Int. J. Syst. Bacteriol.* 21:15–16

77. Pfennig, N., Trüper, H. G. 1971. Type and neotype strains of the species of phototrophic bacteria maintained in pure culture. *Int. J. Syst. Bacteriol.* 21:19–24
78. Pfennig, N., Trüper, H. G. 1989. Anoxygenic phototrophic bacteria. See Ref. 102, pp. 1635–1709
79. Platt, J. R. 1961. Bioconvection patterns in cultures of free-swimming organisms. *Science* 133:1766
80. Schauder, R., Eikmanns, B., Thauer, R. K., Widdel, F., Fuchs, G. 1986. Acetate oxidation to CO_2 in anaerobic bacteria via a novel pathway not involving reactions of the citric acid cycle. *Arch. Microbiol.* 145:162–72
81. Schink, B. 1988. Principles and limits of anaerobic degradation: environmental and technological aspects. See Ref. 137, pp.771–846
82. Schink, B., Pfennig, N. 1982. *Propionigenium modestum* gen. nov., sp. nov., a new strictly anaerobic, nonsporing bacterium growing on succinate. *Arch. Microbiol.* 133:209–16
83. Schink, B., Stieb, M. 1983. Fermentative degradation of polyethylene glycol by a new, strictly anaerobic, Gram-negative, non-sporeforming bacterium, *Pelobacter venetianus* sp. nov. *Appl. Environ. Microbiol.* 45:1905–13
84. Schlegel, H. G., Bowien, B., eds. 1989. *Autotrophic Bacteria,* pp. 97–116. Madison, WI: Science Tech. Publ. 528 pp.
85. Schlegel, H. G., Pfennig, N. 1961. Die Anreicherungskultur einiger Schwefelpurpurbakterien. *Arch. Mikrobiol.* 38:1–39
86. Schmidt, K. 1978. Biosynthesis of carotenoids. See Ref. 14, pp. 729–50
87. Schmidt, K. 1980. A comparative study on the composition of chlorosomes (Chlorobium vesicles) and cytoplasmic membranes from *Chloroflexus aurantiacus* strain Ok-70-fl and *Chlorobium limicola* f. *thiosulfatophilum* strain 6230. *Arch. Microbiol.* 124:21–31
88. Schmidt, K., Pfennig, N., Liaaen-Jensen, S. 1965. Carotenoids of Thiorhodaceae. IV. The carotenoid composition of 25 pure isolates. *Arch. Mikrobiol.* 52:132–46
89. Schmidt, K., Schiburr, R. 1970. Die Carotinoide der grünen Schwefelbakterien: Carotinoidzusammensetzung in 18 Stämmen. *Arch. Mikrobiol.* 74: 350–55
90. Schulz, S., Wagener, S., Pfennig, N. 1990. Utilization of various chemotrophic and phototrophic bacteria as food by the anaerobic ciliate *Trimyema compressum. Eur. J. Protistol.* 26:122–31
91. Schumacher, W., Kroneck, P. M. H. 1992. Anaerobic energy metabolism of the sulfur-reducing bacterium "Spirillum" 5175 during dissimilatory nitrate reduction to ammonia. *Arch. Microbiol.* 157:464–70
92. Schumacher, W., Kroneck, P. M. H., Pfennig, N. 1992. Comparative systematic study on "Spirillum" 5175, *Campylobacter* and *Wolinella* species. Description of "Spirillum" 5175 as *Sulfurospirillum deleyianum* gen. nov., spec. nov. *Arch. Microbiol.* 158:287–93
93. Seewaldt, E., Schleifer, K. H., Bock, E., Stackebrandt, E. 1982. The close phylogenetic relationship of *Nitrobacter* and *Rhodopseudomonas palustris. Arch. Microbiol.* 131:287–90
94. Seitz, H. J., Schink, B., Pfennig, N., Conrad, R. 1990. Energetics of syntrophic ethanol oxidation in defined chemostat cocultures. 1. Energy requirement for H_2 production and H_2 oxidation. *Arch. Microbiol.* 155:82–88
95. Siefert, E., Pfennig, N. 1978. Hydrogen metabolism and nitrogen fixation in wild type and nif^- mutants of *Rhodopseudomonas acidophila. Biochimie* 60:261–65
96. Siefert, E., Pfennig, N. 1980. Diazotrophic growth of *Rhodopseudomonas acidophila* and *Rhodopseudomonas capsulata* under microaerobic conditions in the dark. *Arch. Microbiol.* 125:73–77
97. Siefert, E., Pfennig, N. 1984. Convenient method to prepare neutral sulfide solution for cultivation of phototrophic sulfur bacteria. *Arch. Microbiol.* 139:100–1
98. Spormann, A. M., Thauer, R. K. 1988. Anaerobic acetate oxidation to CO_2 by *Desulfotomaculum acetoxidans*. Demonstration of enzymes required for the opertion of an oxidative acetyl-CoA/carbon monoxide dehydrogenase pathway. *Arch. Microbiol.* 150:374–80
99. Stackebrandt, E., Fowler, V. J., Schubert, W., Imhoff, J. F. 1984. Towards a phylogeny of phototrophic purple sulfur bacteria—the genus *Ectothiorhodospira. Arch. Microbiol.* 137:366–70
100. Stackebrandt, E., Murray, R. G. E., Trüper, H. G. 1988. Proteobacteria classis nov., a name for the phylogenetic taxon that includes the "purple

bacteria and their relatives". *Int. J. Syst. Bacteriol.* 38:321–25

101. Staehelin, L. A., Golecki, J. R., Fuller, R. C., Drews, G. 1978. Visualization of the supramolecular architecture of chlorosomes (*Chlorobium*-type vesicles) in freeze-fractured cells of *Chloroflexus aurantiacus*. *Arch. Microbiol.* 119:269–77
102. Staley, J. T., Bryant, M. P., Pfennig, N., Holt, J. G., eds. 1989. *Bergey's Manual of Systematic Bacteriology,* 3:1635–1709. Baltimore: Williams and Wilkins. 1601–2298 pp.
103. Starr, M. P., Stolp, H., Trüper, H. G., Balows, A., Schlegel, H. G., eds. 1981. *The Prokaryotes,* 1:279–89. Berlin/Heidelberg/New York: Springer. 1102 pp.
104. Stieb, M., Schink, B. 1984. A new 3-hydroxybutyrate fermenting anaerobe, *Ilyobacter polytropus,* gen. nov. sp. nov., possessing various fermentation pathways. *Arch. Microbiol.* 140: 139–46
105. Stieb, M., Schink, B. 1985. Anaerobic oxidation of fatty acids by *Clostridium bryantii* sp. nov., a spore-forming, obligately syntrophic bacterium. *Arch. Microbiol.* 140:387–90
106. Stieb, M., Schink, B. 1986. Anaerobic degradation of isovalerate by a defined methanogenic coculture. *Arch. Microbiol.* 144:291–95
107. Szewzyk, R., Pfennig, N. 1990. Competition for ethanol between sulfate-reducing and fermenting bacteria. *Arch. Microbiol.* 153:470–77
108. Szewzyk, U., Schink, B. 1989. Degradation of hydroquinone, gentisate, and benzoate by a fermenting bacterium in pure or defined mixed culture. *Arch. Microbiol.* 151:541–45
109. Thauer, R. K. 1988. Citric-acid cycle, 50 years on. Modifications and an alternative pathway in anaerobic bacteria. *Eur. J. Biochem.* 176:497–508
110. Thauer, R. K., Jungermann, K., Decker, K. 1977. Energy conservation in chemotrophic anaerobic bacteria. *Bacteriol. Rev.* 41:100–80
111. Thiele, H. H. 1968. Die Verwertung einfacher organischer Substrate durch Thiorhodaceae. *Arch. Mikrobiol.* 60: 124–38
112. Trüper, H. G. 1964. CO_2-Fixierung und Intermediärstoffwechsel bei *Chromatium okenii* Perty. *Arch. Mikrobiol.* 49:23–50
113. Trüper, H. G. 1968. *Ectothiorhodospira mobilis* Pelsh, a photosynthetic sulfur bacterium depositing sulfur outside the cells. *J. Bacteriol.* 95:1910–20
114. Trüper, H. G., Jannasch, H. W. 1968. *Chromatium buderi* nov. spec., eine neue Art der grossen Thiorhodaceae. *Arch. Mikrobiol.* 61:363–72
115. Trüper, H. G., Pfennig, N. 1971. Family of phototrophic green sulfur bacteria: Chlorobiaceae Copeland, the correct family name; rejection of *Chlorobacterium* Lauterborn; and the taxonomic situation of the consortium-forming species. *Int. J. Syst. Bacteriol.* 21:8–10
116. Trüper, H. G., Schlegel, H. G. 1964. Sulfur metabolism in Thiorhodaceae. I. Quantitative measurements on growing cells of *Chromatium okenii*. *Antonie van Leeuwenhoek J. Microbiol. Serol.* 30:225–38
117. Tschech, A., Schink, B. 1985. Fermentative degradation of resorcinol and resorcylic acids. *Arch. Microbiol.* 143: 52–59
118. Tschech, A., Schink, B. 1986. Fermentative degradation of monohydroxybenzoates by defined syntrophic cocultures. *Arch. Microbiol.* 145:396–402
119. Tschech, A., Schink, B. 1988. Methanogenic degradation of anthranilate (2-aminobenzoate). *Syst. Appl. Microbiol.* 11:9–12

119a. van Niel, C. B. 1932. On the morphology and physiology of the purple and green sulphur bacteria. *Arch. Mikrobiol.* 3:1–112

119b. van Niel, C. B. 1955. Natural selection in the microbial world. *J. Gen. Microbiol.* 13:201–17

119c. van Niel, C. B. 1966. Symposium on stability in dynamic microbial systems. IV. Microbiology and molecular biology. *Q. Rev. Biol.* 41:105–12

119d. van Niel, C. B. 1967. The education of a microbiologist; some reflections. *Annu. Rev. Microbiol.* 21:1–30

120. Vatter, A. E., Wolfe, R. S. 1958. The structure of photosynthetic bacteria. *J. Bacteriol.* 75:480–88
121. Wagener, S., Pfennig, N. 1987. Monoxenic culture of the anaerobic ciliate *Trimyema compressum* Lackey. *Arch. Microbiol.* 149:4–11
122. Wagener, S., Stumm, C. K., Vogels, G. D. 1986. Electromigration, a tool for studies on anaerobic ciliates. *FEMS Microbiol. Ecol.* 38:197–203
123. Wake, L. V., Christopher, R. K., Rickard, P. A. D., Andersen, J. E., Ralph, B. J. 1977. A thermodynamic assessment of possible substrates for

sulphate-reducing bacteria. *Aust. J. Biol. Sci.* 30:155–72
124. Widdel, F. 1980. *Anaerober Abbau von Fettsäuren und Benzoesäure durch neu isolierte Arten Sulfat-reduzierender Bakterien*. PhD thesis. Univ. Göttingen. 443 pp.
125. Widdel, F. 1983. Methods for enrichment and pure culture isolation of filamentous gliding sulfate-reducing bacteria. *Arch. Microbiol.* 134:282–85
126. Widdel, F. 1987. New types of acetate-oxidizing, sulfate-reducing *Desulfobacter* species, *D. hydrogenophilus* sp. nov., *D. latus* sp. nov., and *D. curvatus* sp. nov. *Arch. Microbiol.* 148:286–91
127. Widdel, F. 1988. Microbiology and ecology of sulfate- and sulfur-reducing bacteria. See Ref. 137, pp. 469–585
128. Widdel, F., Kohring, G. W., Mayer, F. 1983. Studies on dissimilatory sulfate-reducing bacteria that decompose fatty acids. III. Characterization of the filamentous gliding *Desulfonema limicola* gen. nov. sp. nov., and *Desulfonema magnum* sp. nov. *Arch. Microbiol.* 134:286–94
129. Widdel, F., Pfennig, N. 1977. A new anaerobic, sporing, acetate-oxidizing, sulfate-reducing bacterium, *Desulfotomaculum* (emend.) *acetoxidans*. *Arch. Microbiol.* 112:119–22
130. Widdel, F., Pfennig, N. 1981. Sporulation and further nutritional characteristics of *Desulfotomaculum acetoxidans*. *Arch. Microbiol.* 129:401–2
131. Wikén, T., Pfennig, N. 1957. Untersuchungen über die Physiologie der Weinhefen. VII. Mitteilung. *Antonie van Leeuwenhoek J. Microbiol. Serol.* 23:113–49
132. Wikén, T., Pfennig, N. 1959. Untersuchungen über die Physiologie der Weinhefen. VIII. Mitteilung. *Antonie van Leeuwenhoek J. Microbiol. Serol.* 25:193–229
133. Winogradsky, S. N. 1888. *Beiträge zur Morphologie und Physiologie der Schwefelbakterien,* Vol. 1. Leipzig: Felix. 120 pp.
134. Winogradsky, S. N. 1949. *Microbiologie du sol. Oeuvres Complètes.* Paris: Masson et Cie. 861 pp.
135. Wolfe, R. S. 1991. My kind of biology. *Annu. Rev. Microbiol.* 45:1–35
136. Wolfe, R. S., Pfennig, N. 1977. Reduction of sulfur by spirillum 5175 and syntrophism with *Chlorobium*. *Appl. Environ. Microbiol.* 33: 427–33
137. Zehnder, A. J. B., ed. 1988. *Biology of Anaerobic Microorganisms.* New York: Wiley. 872 pp.

Annu. Rev. Microbiol. 1993. 47:31–55

REGULATION OF SULFUR AND NITROGEN METABOLISM IN FILAMENTOUS FUNGI

George A. Marzluf

Department of Biochemistry, The Ohio State University, Columbus, Ohio 43210

KEY WORDS: nitrogen metabolism, sulfur metabolism, *Aspergillus nidulans, Neurospora crassa,* genetic regulation

---

CONTENTS

0066-4227/93/1001-0031$02.00

Abstract

In the filamentous fungi, *N. crassa* and *A. nidulans,* complex regulatory circuits control nitrogen metabolism and sulfur metabolism. The expression of entire sets of unlinked structural genes that encode metabolic enzymes is repressed when favored sulfur or nitrogen sources are available. These structural genes are coregulated by global positive-acting regulatory proteins and often are also controlled by metabolic inducers and pathway-specific regulatory proteins. The recent isolation of regulatory genes and representative structural genes of these circuits has provided significant new insight into the operation of both the nitrogen and the sulfur regulatory circuits, which involve sequence-specific DNA binding proteins, promoter control elements, metabolic inducers and repressors, and autogenous regulation.

OVERVIEW

In the filamentous fungi, complex regulatory circuits control entire areas of metabolism; for example, expression of enzymes involved in carbon, nitrogen, sulfur, or phosphorus metabolism, in each case, is subject to precise control by a defined set of regulatory genes and specific metabolic repressors. The sulfur and nitrogen regulatory circuits of filamentous fungi each mediate element-specific catabolite repression and operate to insure that favored sources of sulfur or nitrogen, respectively, are utilized. Conventional genetic and biochemical studies carried out in the past several decades have yielded substantial evidence that related catabolic enzymes are coregulated and have provided glimpses, through glasses fogged with uncertainty, of the responsible control mechanisms. In just the past several years, molecular approaches have illuminated these regulatory circuits and have yielded many answers, some anticipated, some completely unexpected. Various models and proposed mechanisms, of which a critical evaluation was previously virtually impossible, are now being directly examined at an amazing pace, finally vindicating some long-standing views, at last discarding others. This review summarizes certain of the earlier biochemical genetic studies to provide a backdrop, but primarily focuses upon the recent molecular studies that almost completely redefine this field and provide important new insights about the complex control circuits that regulate sulfur and nitrogen metabolism. Additional information can be gleaned from earlier reviews that emphasized the genetic and biochemical aspects of these regulatory circuits (2, 11, 41, 42).

THE SULFUR REGULATORY CIRCUIT

The sulfur-containing amino acid methionine initiates the synthesis of nearly all proteins in all organisms. Cysteine plays a critical role in the structure, stability, and function of many proteins; however, in many other less dramatic but important instances, sulfur is crucial for function of a biological molecule. Thus, considerable regulatory circuitry is devoted to insure that the cells can secure a steady supply of sulfur. Sulfur regulation has been studied primarily in *Neurospora crassa,* beginning with the pioneering work of Metzenberg & Parson (44). *N. crassa* preferentially utilizes certain favored sulfur sources, such as methionine. However, this fungus can use proteins or various secondary sulfur sources (e.g. choline-0-sulfate or aromatic sulfates), but utilization of these compounds requires de novo synthesis of various sulfur catabolic enzymes. Control occurs primarily at the level of entry of various potential sulfur sources into the assimilatory pathway, which itself does not appear to be subject to any marked degree of regulation. The sulfur regulatory circuit of *N. crassa* contains a set of unlinked but coregulated structural genes that specify these sulfur catabolic enzymes. Synthesis of this entire family of enzymes, which includes aryl sulfatase, choline sulfate permease and choline sulfatase, an extracellular alkaline protease, a methionine-specific permease, and two distinct sulfate-permease species, only occurs during conditions of sulfur limitation (24, 39, 44, 51). The expression of these sulfur catabolic

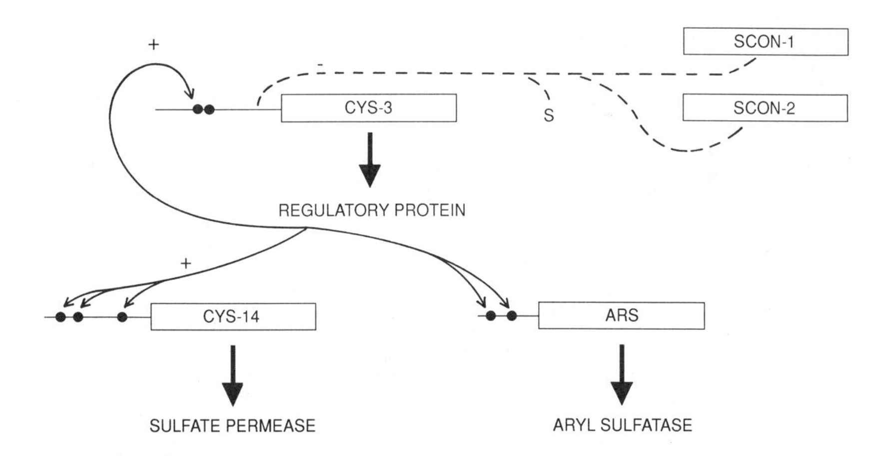

Figure 1 Molecular model for the sulfur regulatory circuit. The exact function of the two *scon* genes is unknown but they presumably code for products that negatively control the expression of the *cys-3* regulatory gene, which is expressed only when the cells become limited for sulfur. CYS3 is a *trans*-acting DNA-binding protein that via autogenous control increases its own expression and also turns on *cys-14, ars,* and other structural genes of the circuit. Three CYS3 binding sites in the 5′ promoter region of *cys-14* and a single duplex site upstream of *cys-3* have been identified.

enzymes is controlled by three regulatory genes, *scon-1, scon-2,* and *cys-3* (5, 43, 48). Mutant cells with *cys-3* cannot grow on any of the various secondary sulfur sources and display a pleiotropic loss of the entire family of relevant catabolic enzymes, suggesting that the *cys-3*$^+$ gene product acts in a positive way to turn on the expression of these enzymes (Figure 1). In contrast, both *scon-1* and *scon-2* appear to act in a negative fashion, and mutants of either show constitutive synthesis of the sulfur catabolic enzymes (5, 48).

Several genes of the sulfur circuit have recently been cloned, including two representative structural genes, *cys-14,* which encodes sulfate permease II and *ars,* which encodes aryl sulfatase (34, 47). In addition the positive-acting *cys-3* and the negative-acting *scon-2,* regulatory genes have been isolated and characterized (19, 23, 48, 50). The cellular content of *cys-14* mRNA and of *ars* mRNA is highly regulated by sulfur repression and by the *cys-3* and *scon-1* genes, i.e. these mRNAs are detected in wild-type only upon sulfur limitation, are missing entirely in *cys-3* mutants, and are expressed in a constitutive fashion in *scon-1* mutants (34, 47). These findings strongly suggest that both *cys-14* and *ars,* as well as each of the other structural genes of the sulfur circuit, are subject to transcriptional control. Nuclear run-on experiments have confirmed that *ars* expression is controlled at the level of initiation of transcription (47).

Sulfate Permeases

Two unlinked genes, *cys-13* and *cys-14,* which appear to specify distinct sulfate permease species, were identified through mutation over 20 years ago (39). Either single mutant strain grows readily on inorganic sulfate and retains one sulfate uptake system; the *cys-13–cys-14* double mutant lacks all sulfate transport activity and cannot utilize sulfate. Sulfate permease I, encoded by *cys-13,* is found predominately in conidia whereas sulfate permease II occurs more prominently in mycelia (39). The *cys-14*$^+$ gene has been cloned, and conceptual translation of its nucleotide sequence indicates that it encodes a protein of approximately 90 kDa (33). The predicted amino acid sequence of the CYS14 protein reveals 12 highly hydrophobic helical regions, which may represent membrane-spanning domains. The use of anti-CYS14 antibodies, raised to a segment of the CYS14 protein expressed in *Escherichia coli,* revealed that the CYS14 protein of the expected size is found in *N. crassa* and localized within the plasma membrane fraction (29). Moreover, the cellular level of this protein is highly regulated by de novo synthesis and turnover, in good agreement with the kinetics of appearance and disappearance of sulfate transport activity (29). These features thus strongly suggest that CYS14 functions as a membrane-bound sulfate transporter. The utilization of inorganic sulfate is highly regulated at the level of its entry into the cell.

Control is exerted at the level of transcription of the sulfate permease genes, but also in the dynamic turnover of both sulfate permease species, with a half-life of approximately two hours (40).

The cys-3 Regulatory Gene Encodes a Bzip Protein

The *cys-3* gene encodes a *trans*-acting regulatory protein that acts in a positive fashion to turn on the expression of *cys-14* and *ars* and presumably each of the other coregulated structural genes. Null mutants of *cys-3* fail to express any *cys-14* or *ars* mRNA, whereas temperature-sensitive *cys-3* mutants also completely lack these mRNAs at the conditional temperature (37°C) but express both *cys-14* and *ars* mRNA at 25°C, the permissive temperature (34, 47). The CYS3 protein is composed of 236 amino acids and is a member of the bzip class of DNA-binding proteins, which include *S. cerevisiae* GCN4 protein and the mammalian proteins FOS and JUN (19, 23, 31, 32). The CYS3 DNA-binding domain is bipartite and consists of a leucine zipper responsible for dimerization of two CYS3 monomers and an immediately adjacent upstream basic region that appears to make direct contact with DNA (19, 31). Cross-linking experiments have shown that the CYS3 protein exists as a dimer in the presence or absence of DNA (32). A truncated CYS3 protein, encompassing only the bzip region, is also capable of dimer formation and DNA binding. Moreover, when coexpressed in vivo or in vitro, hybrid dimers, consisting of one full-length monomer and one truncated CYS3 monomer, are formed and are active in DNA binding (31, 32).

DNA Binding by the CYS3 Protein

The full-length CYS3 protein has been expressed in *E. coli* and used to examine DNA binding with gel mobility shift experiments and with DNA-footprint analyses (19). CYS3 binds in vitro to three distinct sites upstream of *cys-14*, the sulfate permease structural gene, and to a single site in the 5′ promoter region of the *cys-3* gene itself (19). A smaller CYS3 protein containing only the bzip domain was found to bind to these same sites with the same specificity and affinity as the full-length protein, demonstrating that DNA recognition by CYS3 is determined solely by the bzip domain and does not involve other regions of the protein (31).

The three CYS3 binding sites upstream of *cys-14* and the single site upstream of *cys-3* show only limited sequence homology, except that each site contains repeats of a sequence, CAT. These repeated CAT sequences appear to provide a limited dyad symmetry, which may represent the central core of a CYS3 binding site. The proximal CYS3 binding site upstream of *cys-14* has two CAT sequences at its center, and nucleotide substitutions for either of these core elements resulted in approximately an 80% loss in recognition by the CYS3 protein (33). Because CYS3 is a wide-domain regulatory protein responsible for turning on an entire family of coregulated genes, presumably to different extents and with different kinetics, we should not be surprised that it can recognize different sequence elements. Two of the CYS3 binding sites detected in vitro are approximately twice the size of other sites and appear to represent duplex sites. The single binding site in the promoter of the *cys-3* gene, believed to be responsible for autogenous control, is such a duplex site and is bound by CYS3 with high affinity (19). Still unknown is the minimum requirement for a CYS3 binding site and the basis for different binding affinities.

Mutation Studies of CYS3

The polymerase chain reaction (PCR) was employed to clone several conventional *cys-3* mutant genes. The nucleotide sequence of a *cys-3* null mutant revealed that glutamine had replaced two basic amino acids within the DNA-binding domain of the CYS3 protein (23). This mutant protein, expressed in *E. coli,* was incapable of DNA binding (19). A *cys-3* mutant with a temperature-sensitive phenotype in vivo has a glutamine substitution of a single arginine residue in the bzip motif; this mutant CYS3 protein displays temperature-dependent DNA binding in vitro (M. N. Kanaan & G. A. Marzluf, in preparation).

Site-directed mutagenesis was used to alter several highly conserved amino acids in the bzip domain of CYS3 to determine whether they are essential for function in vivo or for CYS3 dimerization or DNA binding in vitro. Substitution of glutamine for basic amino acids within the positively charged region upstream of the zipper abolished *cys-3* function in vivo (31). Moreover, *E. coli* expressed CYS3 proteins corresponding to each of the basic region mutants formed dimers but were completely deficient in DNA binding (31). These results support the concept that the basic region constitutes the DNA contact surface and that substitution of even a single amino acid in this region can entirely eliminate CYS3 DNA-binding activity; in contrast, CYS3 dimerization depends upon the leucine zipper motif and does not require a functional basic region.

Immediately downstream of the basic region, the CYS3 protein contains a heptad repeat motif, the hallmark of a leucine zipper, although, interestingly, its second position is occupied by methionine. The effect of amino acid substitutions within this putative zipper structure was found to depend upon the position within the zipper and the nature of the substitutions, some replacements being compatible with function, others eliminating *cys-3* function (31). Single valine substitutions for residues at zipper positions 1 or 2 eliminated CYS3 function, whereas CYS3 proteins with valine replacements at positions 3 or 4 retained function. The simultaneous introduction of valine at any two positions within the zipper led to a nonfunctional CYS3 protein.

In contrast, CYS3 proteins with two or even three methionine residues in zipper positions possessed full *cys-3* activity. Perhaps surprisingly, a CYS3 protein that had a pure methionine zipper was functional in vivo and displayed significant, although reduced, dimer formation and DNA binding in vitro. Thus, although leucine appears particularly suited for the heptad positions in a so-called leucine zipper, certain other amino acids, e.g. methionine or isoleucine (31, 32, 36), can function in a zipper motif, which explains the natural occurrence of methionine in a zipper position of CYS3.

Autogenous Regulation of CYS3

The *cys-3* regulatory gene is itself subject to a high degree of regulation. Thus, the *cys-3* gene is usually turned off when the cells have a sufficient sulfur source and is expressed only when sulfur catabolite repression is lifted (23). Apparently the products of the sulfur-controller genes, *scon-1* and *scon-2*, negatively control *cys-3* and prevent its expression, probably at the level of transcription, when an adequate cellular supply of sulfur is available. When expression of CYS3 is achieved with a heterologous promoter in cells subject to repressing levels of sulfur, the target enzyme aryl sulfatase is expressed (49). This result implies that *cys-3* is controlled at the level of transcription. Cysteine may be the repressing sulfur metabolite, although this point has not been rigorously demonstrated.

Several lines of evidence indicate that CYS3 autogenously regulates its own expression in a positive fashion. Thus, when the cells become limited for sulfur, the *cys-3* control gene is turned on, and as the pool of CYS3 protein increases, it may bind at an element in the promoter region of its own structural gene to strongly enhance CYS3 expression (19, 23). Mobility-shift and DNA-footprinting experiments demonstrate that CYS3 binds in vitro with high affinity to this promoter element (19, 31). Furthermore, *cys-3* missense mutants that encode a nonfunctional CYS3 protein are greatly deficient in synthesis of both *cys-3* mRNA and CYS3 protein, fully consistent with the concept that the mutant protein cannot turn on its own expression (19).

CYS3 Is a Nuclear Protein

An important question to consider is whether the CYS3 protein, predicted from the single open reading frame of the *cys-3* gene, actually exists in vivo in *N. crassa*. To address this question, specific anti-CYS3 polyclonal antibodies were utilized in Western blot analyses. These experiments revealed that wild-type nuclear extracts, but not the cytoplasmic fraction, indeed possess a protein of the expected size that specifically reacts with the antibody (M. N. Kanaan & G. A. Marzluf, unpublished results). Furthermore, mobility-shift experiments conducted with the wild-type nuclear extract revealed a DNA-protein complex at the identical position found with the *E.*

coli–expressed CYS3 protein. Significantly, nuclear extracts of a *cys-3* missense mutant lack detectable CYS3 protein (M. N. Kanaan & G. A. Marzluf, unpublished results). These results demonstrate clearly that CYS3 is indeed a nuclear DNA-binding protein and provide additional evidence for the concept that *cys-3* expression is subject to positive autogenous regulation. The mutant CYS3 protein is presumably not expressed at a detectable level because it cannot enhance its own synthesis, although the possibility that such a mutant protein is unstable and degraded in vivo has not yet been rigorously excluded.

Molecular Dissection of the CYS3 Protein

The CYS3 protein has several potentially important regions in addition to the well-defined bzip DNA-binding domain; as noted in other regulatory proteins, CYS3 has distinct regions that are highly enriched in one or more amino acids. Near the amino terminus of CYS3, a segment rich in threonine and serine is followed by a proline-rich region, then an acidic domain, the bzip region, and finally, an alanine-rich segment. Any of these regions might be important for CYS3 function in transactivation or in recognition of sulfur catabolite signals.

A deleted form of the *cys-3* gene, which encodes a truncated protein consisting primarily of the bzip domain, binds DNA in vitro but does not function in vivo. Moreover, it also strongly inhibited the function of a cotransformed wild-type *cys-3* gene. The truncated CYS3 protein thus appears to compete with the full-length protein for DNA binding but lacks any transactivation function, which must reside in regions of CYS3 other than the DNA-binding domain. A deletion analysis was employed to identify such possible activation domains. A CYS3 protein lacking all residues C terminal to the bzip domain, including the alanine-rich region, retains partial function in vivo. All deletions N terminal to the bzip region completely eliminated CYS3 function, suggesting that structural integrity of the N terminus is important for proper function of the activation domains or for stability of the CYS3 protein. Alanine saturation mutagenesis, used to identify functionally important amino acids in this region of CYS3, demonstrated that the acidic region is essential but only as a structural or spacer region; its acidic nature is not important for CYS3 function. In contrast, the proline-rich region appears to represent an activation domain of CYS3. The substitution of alanine for different proline residues in this region completely abolished CYS3 function (M. N. Kanaan & G. A. Marzluf, unpublished results).

Operation of the Sulfur Circuit

Many intriguing questions concerning the sulfur regulatory circuit remain unanswered. These include the mechanism by which the presence of repressing cellular levels of sulfur is detected and conveyed as a molecular signal to control the expression of the *cys-3* gene. This signal transduction almost certainly involves both the *scon-1* and *scon-2* gene products (5, 48), but exactly how they act remains a mystery. The evidence now available suggests a regulatory mechanism in which the negative-acting *scon* genes repress *cys-3* expression when sulfur is readily available. Upon sulfur limitation, this negative control is lifted and *cys-3* transcription is turned on, giving rise to the CYS3 protein, which itself further increases *cys-3* expression, yielding an increasing pool of CYS3 protein that activates the expression of the various unlinked structural genes. Although it now seems certain that the CYS3 protein is a *trans*-acting DNA-binding protein, the detailed mechanism by which it activates structural gene expression remains unknown, e.g. whether it might interact directly with RNA polymerase or with accessory factors such as the TATA-box protein, TFIID. TFIID does bind to TATA-boxes in the upstream promoter region of both the *cys-14* and *cys-3* genes (M. N. Kanaan, B. Tyler & G. A. Marzluf, unpublished results).

Sulfur Regulation in Yeast

In *Saccharomyces cerevisiae,* expression of the entire set of genes that encode enzymes required for sulfate assimilation requires the lifting of sulfur catabolite repression and a functional *met-4*$^+$ gene; *met-4* mutants lack all of the enzymes required for assimilation of inorganic sulfate. The MET4 protein is a positive-acting leucine-zipper protein with significant homology to CYS3 that is, however, restricted to the DNA-binding domain. MET4 binds to a recognition element within the sequence 5′-TGGCAAATG-3′ that resembles the core sequences responsible for CYS3 binding (58).

A surprising finding is that expression of some, but not all, of the sulfur-related structural genes of *S. cerevisiae* also requires a second DNA-binding protein, known as centromere-binding factor 1 (CBF1). For example, the expression of MET16 is reduced to a near-undetectable level in a *cbf1* mutant, whereas that of MET3 is unaffected. CBF1 appears to be a member of a class of yeast multifunctional DNA-binding proteins that bind to elements found throughout the genome, such as centromeres, telomeres, and replication origins. CBF1 binds to a sequence element present in all yeast centromeres and is required for proper chromosome segregation. What then is the importance of CBF1 in expression of certain sulfur-related structural genes? Thomas et al (58) have suggested that CBF1 plays a cooperative role, increasing the accessibility of promoter regions to MET4, perhaps by preventing nucleosome formation. Interestingly, MET3, which is fully activated by MET4 in the absence of CBF1, contains a long poly$(AT)_{19}$ tract near its TATA box. Such a poly(AT) tract may exclude nucleosomes, thus making the MET3 promoter accessible to the positive-acting MET4 protein, without the need for CBF1.

NITROGEN REGULATION

Nitrogen is found in many of the simple compounds and nearly all of the complex macromolecules of all living cells and is a major component of proteins and of nucleic acids. Thus, it is not at all surprising that in filamentous fungi, a major investment is made in metabolic machinery, constituting a myriad of different nitrogen catabolic pathways plus a global, hierarchical regulatory system to govern their efficient use. This complex regulatory circuit functions to insure that a constant nitrogen supply is readily available for growth, even in the face of widely different or rapidly changing environments. This description focuses almost exclusively upon *Aspergillus nidulans* and *Neuropsora crassa* because nearly all of our understanding of nitrogen metabolic regulation in filamentous fungi has been obtained with these two organisms. However, we can anticipate that many other fungal species may possess similar nitrogen-control systems.

Certain primary nitrogen sources, e.g. ammonia, glutamine, or glutamate, are preferentially utilized by *N. crassa* and by *A. nidulans*. However, in the absence of these preferred nitrogenous compounds, many secondary nitrogen sources, e.g. nitrate, purines, proteins, peptides, amino acids, and amides, are readily utilized (2, 11, 41). The use of nearly all secondary nitrogen sources requires the de novo synthesis of permeases and catabolic enzymes, whose expression is highly regulated by the nitrogen regulatory circuit in both *A. nidulans* and *N. crassa*.

Positive-acting regulatory genes predominate in these nitrogen circuits and include a single global regulatory gene that mediates nitrogen catabolite derepression plus a series of minor or pathway-specific regulators. In most cases, turning on the expression of enzymes of a particular nitrogen catabolic pathway requires two distinct positive signals, one that is system wide and indicates nitrogen limitation and, second, a pathway-specific signal. This feature allows for the selective activation of a specific metabolic pathway from an impressive repertoire of players and special teams.

Nitrate Assimilation

Inorganic nitrate is an excellent nitrogen source for *A. nidulans* and *N. crassa* but it will not be utilized unless the cells are depleted for the favored compounds ammonia, glutamate, or glutamine (11, 52). Nitrate is converted to nitrite in a two-electron transfer reaction, catalyzed by the enzyme, nitrate reductase, which is a dimeric protein composed of two identical subunits (9). Nitrate reductase is a large multicenter redox enzyme and possesses three separate domains, an amino terminal domain that contains an unusual molybdopterin cofactor, a smaller central heme-containing domain, and a carboxy terminal domain that contains a flavin cofactor (FAD). Electrons derived from NADPH are transferred stepwise from the flavin to the heme domain, and then to the molybdopterin cofactor, and finally utilized to reduce nitrate to nitrite (9). Mutants in the structural genes that specify nitrate reductase or nitrite reductase lead to the inability to utilize nitrate. Moreover, mutants in the positive-acting major regulatory gene (*areA* in *A. nidulans, nit-2* in *N. crassa*), as well as mutants within a pathway-specific control gene, lack both nitrate reductase and nitrite reductase (11, 27). Mutants that lack nitrate reductase can be selected directly because, unlike wild-type, they are resistant to the toxic analogue chlorate. Thus, efficient two-way selection is possible. Kinghorn, Unkles, and their colleagues (62) have expolited this feature to develop transformation systems for various filamentous fungi.

In *A. nidulans,* the structural genes encoding nitrate reductase and nitrite reductase, *niaD* and *niiA,* respectively, are closely linked to each other but transcribed in opposite directions and are presumably coregulated by the common central region (30). In *N. crassa,* the structural genes for these two enzymes are unlinked but are also controlled in a parallel fashion. Northern blot experiments have demonstrated that the mRNA that encodes nitrate reductase in *A. nidulans* and *N. crassa* accumulates only under conditions of nitrogen limitation and nitrate induction, consistent with control at the level of transcription (17, 30, 46). Both the synthesis and turnover of nitrate reductase mRNA occurs very rapidly in *N. crassa;* a steady-state level of mRNA is present within 15 min of induction, whereas upon repression, the mRNA turns over with a half-life of approximately 5 min (46). The *nit-6* gene of *N. crassa,* which encodes nitrite reductase, was recently cloned and characterized (G. E. Exley & R. H. Garrett, in preparation). The expression of a 3.5-kb *nit-6* transcript requires both nitrogen derepression and nitrate induction and depends upon functional *nit-2* and *nit-4* gene products; thus *nit-6* is very clearly coregulated with the unlinked *nit-3* structural gene (G. E. Exley & R. H. Garrett, in preparation).

Global Nitrogen Regulatory Genes areA and nit-2

A single positive-acting regulatory gene, designated *areA* in *A. nidulans* and *nit-2* in *N. crassa,* mediates global nitrogen repression/derepression. The majority of *areA* and *nit-2* mutants display a loss of many or all nitrogen catabolic enzymes, as expected for loss of function of a positive control factor. However, in rare *areA* mutants, expression of certain nitrogen catabolic enzymes is insensitive to nitrogen catabolic repression, whereas other enzymes may be very poorly expressed or missing altogether (2, 6, 35). These features imply that the AREA protein differentially recognizes promoter elements of various structural genes that specify nitrogen catabolic enzymes. The molecular cloning of *nit-2* of *N. crassa* (16, 55) and of *areA* of *A. nidulans* (8) represents a major step forward in understanding nitrogen control in filamentous fungi.

The AREA and NIT2 Regulatory Proteins

NIT2 and AREA are global *trans*-acting regulatory proteins responsible for turning on many different coregulated structural genes of the nitrogen circuit in *N. crassa* and *A. nidulans,* respectively. One can anticipate that NIT2 and AREA have several functional regions, including a DNA-binding domain, one or more activation regions, and recognition sites that somehow mediate nitrogen repression.

AREA and NIT2 each contain a single Cys2/Cys2-type zinc-finger motif with a central loop of 17 amino acids, which together with an immediately adjacent downstream basic region constitute a DNA-binding domain (20–22, 35). AREA and NIT2 have 98% amino acid identity within their DNA binding domains, a stretch of 50 residues that encompasses the finger structure plus the adjacent basic region (Figure 2). The *N. crassa nit-2* gene can substitute for the *A. nidulans areA* gene (13), and the NIT2 protein binds in vitro to four different sites in the control region between the divergently transcribed structural genes that encode nitrate and nitrite reductase in *A. nidulans* (21, 31). Interestingly, the DNA-binding domain of NIT2 and AREA as well as GLN3 of *S. cerevisiae* (45) shows strong homology with two adjacent finger motifs in the mammalian and avian GATA-factors, the best characterized of which, GATA-1, is responsible for tissue-specific expression of genes of the erythropoietic cell lineage (Figure 2). One particularly interesting *A. nidulans* mutant, *areA-300,* elevates the expression of certain structural genes but reduces the expression of others; it possesses an in-frame tandem duplication of 417 bp that resulted in an AREA protein with two fingers. The characteristics of *areA-300* suggest that the two fingers act independently and alternately, rather than jointly, which led Caddick & Arst (7) to speculate that the presence of two zinc fingers might allow a *trans*-acting protein to diverge and gain control over new sets of structural genes.

A central region of the NIT2 protein, which includes the DNA-binding domain, has been expressed as a fusion protein in *E. coli*. This protein shows sequence-specific DNA binding to three different recognition elements upstream of the *nit-3* structural gene (21, 22); to two sites 5′ of the *lao* gene, which encodes L–amino acid oxidase (X. D. Xiao & G. A. Marzluf, in preparation); and to a single site in front of *alc,* which specifies allantoicase (38).

Site-directed mutagenesis was employed to obtain *nit-2* mutants that specify different NIT2 proteins with substitutions of highly conserved amino acids

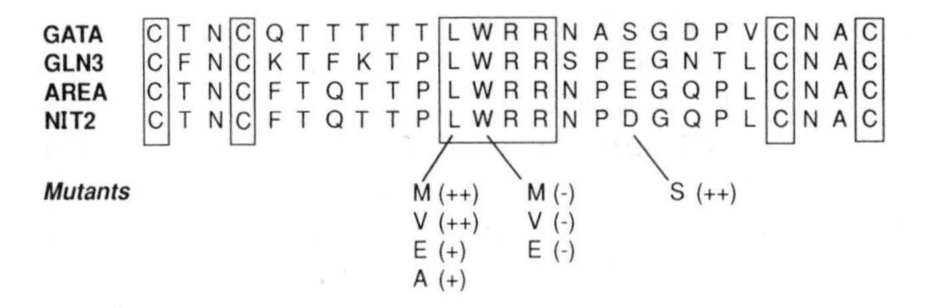

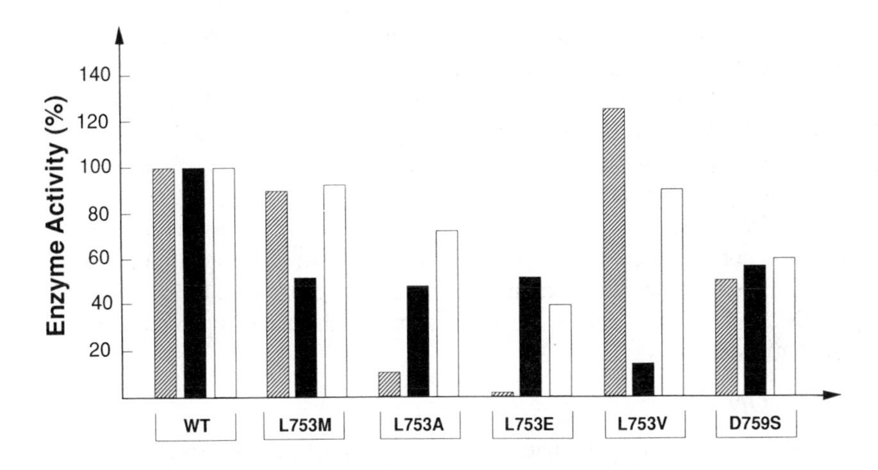

Figure 2 (*Upper panel*) The single Cys2/Cys2-type zinc-finger motif with a central loop of 17 amino acids of the *S. cerevisiae* GLN3, the *A. nidulans* AREA, and the *N. crassa* NIT2 proteins are shown aligned with the second finger of a mammalian GATA-binding protein. The conserved cysteine residues and a central block of four conserved amino acids are boxed. Mutant NIT2 proteins with specific amino acid substitutions in the finger structure obtained via site-directed mutagenesis are shown: (++), fully active; (+), active; (−), nonfunctional. (*Lower panel*) Activities of enzymes encoded by target genes as determined in transformants that have wild-type NIT2 protein (Leu at residue 753) compared with transformants containing NIT2 proteins with four different substitutions at amino acid 753, and one protein with serine for Asp769. Symbols: Cross hatched, nitrate reductase; solid, allantoicase; open, L–amino acid oxidase.

within the putative zinc finger or basic region (22). Each of these mutants proved to be nonfunctional in vivo (22) and was completely inactive in DNA binding in vitro. These results clearly demonstrate that the zinc finger and adjacent basic region of NIT2 represent its DNA-binding domain.

Certain mutant NIT2 proteins with single amino acid substitutions for highly conserved residues within the DNA binding domain are functional in vivo. The promoter regions of structural genes that specify nitrate reductase, allantoicase, and L-amino acid oxidase all have NIT2 binding sites that share common repeated GATA core sequences, but otherwise are different from one another (see below). An intriguing possibility was that certain mutant NIT2 proteins might display altered recognition of different DNA binding elements. The results of substitution individually of two highly conserved adjacent residues within the loop region of the zinc finger motif, Leu753 and Trp754, are of particular interest (X. D. Xiao & G. A. Marzluf, in preparation). All substitutions of Trp754 resulted in a nonfunctional NIT2 protein. In contrast, Leu753 could be substituted by three hydrophobic residues, alanine, valine, and methionine, which differ significantly in the bulkiness of their side chains, and also by the charged amino acid aspartate, and still retain function in vivo and in vitro. These mutant NIT2 proteins displayed dramatic differences in their ability to turn on three different coregulated genes, *nit-3, alc,* and *lao,* in vivo in *N. crassa* (Figure 2). These same NIT2 proteins show corresponding differences in their ability to bind in vitro to the promoter elements of these three structural genes. Thus, change of a single amino acid within the DNA binding domain of NIT2 is sufficient to alter the specificity of DNA element recognition. The results of sophisticated genetic studies with *A. nidulans* reinforce this interpretation, because certain *areA* mutants that alter the pattern of nitrogen source utilization affect amino acid residues within the loop region of the zinc finger motif of the AREA protein (6, 35).

NIT2 DNA-Recognition Elements

As global *trans*-acting regulatory proteins, AREA and NIT2 are responsible for turning on many different coregulated genes within the nitrogen control circuit; not surprisingly, individual structural genes may be expressed at considerably different levels or with different kinetics of activation. The binding sites for NIT2 in the promoter region of different genes differ markedly in nucleotide sequence but have in common the presence of two or more copies of a core element, GATA, similar to the binding sites for the mammalian GATA-binding proteins (21). These mammalian factors are tissue-specific proteins that possess two zinc fingers, each with remarkable homology to the single zinc fingers of NIT2 and of AREA (21, 35, 61). The minimum requirement for a specific NIT2-recognition element appears to be the presence of two (or more) GATA core sequences, which can face in the same or opposite directions (Figure 3). A single GATA element shows only very weak NIT2 binding. The spacing between two natural or artificial GATA elements that constitute a strong binding site can vary from 3 bp to at least 30 bp (T. Y. Chiang & G. A. Marzluf, in preparation). In fact, when the distance between two synthetic GATA elements was varied, with spacings of 5, 10, 15, 20, or 30 base pairs, only relatively modest differences in binding affinity (two- to fourfold) for NIT2 were detected. However, when the two GATA sequences were more than 40 bp from another, only very weak NIT2

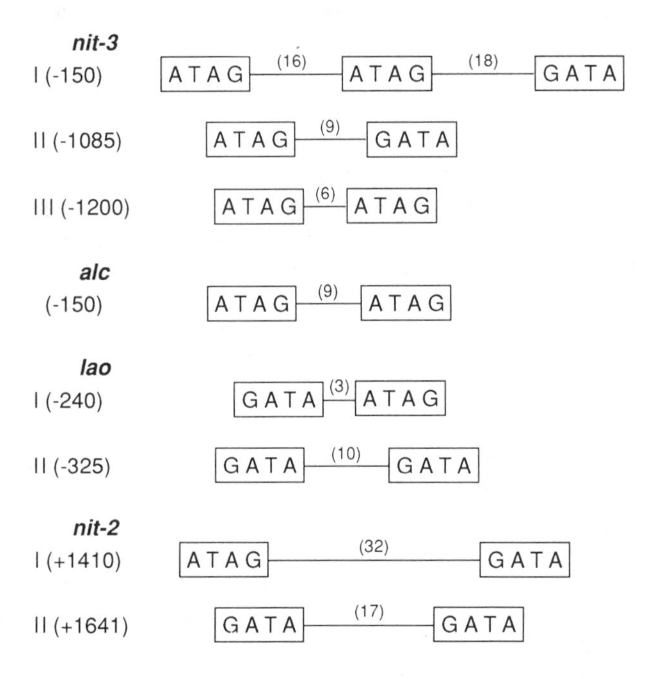

Figure 3 NIT2 binding sites. The location, orientation, and spacing of the GATA core elements of NIT2 binding sites in the upstream promoter region of three structural genes, *nit-3* (encodes nitrate reductase), *alc* (allantoicase), and *lao* (L-amino acid oxidase), identified with DNA footprinting are shown; a NIT2 binding site within the second intron and one in the 5′ region of the third exon of the *nit-2* gene itself are also shown. GATA core sequences shown in reverse orientation occur on the complementary strand; the spacing in base pairs between the core GATA elements is indicated. All NIT2 binding sites have two or three GATA core sequences.

binding in vitro was observed, similar to that detected with only a single GATA element. Substitution of bases 5′ or 3′ of one or both GATA elements only minorly affected relative binding affinity (T. Y. Chiang, T. G. Cooper & G. A. Marzluf, in preparation). These results indicate that the primary requirement for a NIT2 binding site is the presence of two adjacent GATA core elements, which argues for a type of cooperative binding, suggesting that NIT2 (and AREA) may function as dimeric proteins. Alternatively, the single-fingered NIT2 or AREA proteins may bind as monomers to a GATA element, but only weakly. However, if two monomers bind to nearby core sequences, they may interact and strongly stabilize their joint binding. It is still not clear what other factors may affect the affinity of individual NIT2 DNA-binding sites, nor how the differential recognition of various promoter elements occurs. Another area of importance is an understanding of how the location and spacing of multiple NIT2 binding sites, as found in the upstream region of the *nit-3* structural gene, contribute to a strongly regulated promoter.

Expression of AREA and NIT2

Many regulatory genes are themselves subject to regulation, including autogenous control. Although a detailed understanding is not yet available, apparently the expression of *areA* and of *nit-2* is controlled. The presence of different *areA* transcripts dependent upon nitrogen availability has been reported; strong expression of a single 2.9-kb *areA* transcript is observed during conditions of nitrogen derepression (6). In contrast, during nitrogen catabolite repression, two larger *areA* transcripts were observed and appear to arise from two constitutively expressed upstream promoters (6). In the case of *nit-2,* transcript heterogeneity also occurs (16). Is expression of *areA* or of *nit-2* controlled by still other regulatory genes? No additional authentic control genes have been identified in either *A. nidulans* or *N. crassa;* two possible candidates in *N. crassa, nmr-1* and *gln-1,* do not appear to control *nit-2* expression (16). Interestingly, two strong NIT2 protein–binding sites occur within the *nit-2* gene itself, one in the second intron, the other 200 base pairs farther downstream in a coding region (exon 3) of the gene (T. Y. Chiang & G. A. Marzluf, in preparation). This feature suggests that NIT2 may be subject to autogenous control and hints that alternative initiation sites may also exist for *nit-2* transcription or translation, as suggested for *areA* (6). An interesting, but completely speculative possibility, is that different forms of the AREA and NIT2 proteins might be produced from the various transcripts that arise during contrasting nitrogen conditions, some of which may serve as activators, whereas others could function as repressors. In this regard, Cunningham & Cooper (12) have shown that the yeast DAL80 protein contains a region with strong homology to the DNA-binding finger motif of GLN3, AREA, and NIT2; however, DAL80 acts as a negative regulator of multiple nitrogen catabolic genes. The expression of *areA* and of *nit-2* appears complex, but obviously, much research is needed to understand the regulated expression of these global regulatory proteins and how it affects the activation of the many structural genes of the nitrogen control circuit.

Homologies Between AREA and NIT2 Proteins

As described above, AREA and NIT2 have an almost identical amino acid sequence in the zinc finger and adjacent basic region that comprise their DNA binding domains. Do the AREA and NIT2 proteins have any other common features and homology? These two proteins may possess at least two other common functions, activation domains that turn on target gene expression and a control site that is sensitive to nitrogen catabolite repression. In addition to the nearly identical finger region, AREA and NIT2 share 13 other blocks of homology, which occur in the same relative position (in every case but one) along the length of the proteins between regions with no detectable homology. In total, AREA has 22% amino acid identity with NIT2 (160 identical residues out of a total of 719). NIT2 appeared to be larger at its amino terminus by approximately 300 residues; however, recently Langdon & Arst (see 6) identified an upstream *areA* exon that shows signficant amino acid homology with the first exon of NIT2, which implies that AREA and NIT2 are even more similar than originally thought.

Transcriptional activators seem to possess different types of activation domains; e.g. acidic regions as found in GAL4 and GCN4 are common (6), but other activation domains seem to be glutamine-rich or proline-rich, or enriched in alanine or in the hydroxy amino acids, serine and threonine (10, 37). The nature, structure, and function of activation domains is an area of intense investigation, and at present only very limited precise information is available. Can potential activation domains be identified in AREA and NIT2? These proteins each possess two comparable acidic regions; preliminary evidence suggests that these may have an activation function. Arst and his coworkers (35) have demonstrated that an AREA protein lacking all residues carboxy terminal to the DNA-binding domain retains activation function. Other truncated forms of AREA that retain function lack many of the amino terminal residues, such that the essential region or AREA for gene activation is only 223 amino acid residues (amino acids 343 to 566). This region primarily consists of the DNA-binding domain plus an amphiphatic acidic region. In the case of NIT2, a protein that lacks either of the two acidic domains retains function whereas a NIT2 protein lacking both acidic regions shows loss of function (Y. H. Fu, H. G. Pan & G. A. Marzluf, unpublished data). These results must be considered as preliminary but are consistent in suggesting that AREA and NIT2 possess acidic activation regions.

Deleted forms of the *areA* and of the *nit-2* genes that specify proteins lacking residues carboxy terminal to the DNA binding domain retain the gene activation function but appear to become insensitive to nitrogen catabolite repression. This suggests that a site within the carboxy terminus of these proteins is in some fashion sensitive to nitrogen repression, either by directly binding a metabolite such as glutamine (52) or by serving as a recognition region for some other protein that senses the nitrogen status of the cells, perhaps in a manner similar to the GAL4-GAL80 protein-protein interaction identified *S. cerevisiae*. Kudla et al (35) have suggested that a 16–amino acid segment of AREA, which has strong homology with a corresponding segment of NIT2, may represent the nitrogen-sensitive site. This region could constitute a glutamine-binding site, although as yet no direct evidence suggests that either AREA or NIT2 bind glutamine. In fact, a full-length NIT2 protein, expressed in *E. coli,* which is functional in DNA binding, failed to display any glutamine binding in vitro in equilibrium dialysis experiments (X. D. Xiao & G. A. Marzluf, unpublished data). Of course, the true metabolic repressor may be a derivative of glutamine rather than the amino acid itself. Additional experiments to delete or replace residues within this suspected control site of AREA and NIT2 will be instructive, and if this region is indeed implicated in nitrogen control, identifying the molecules that signal nitrogen repression to these global-control proteins will be very important. It is plausible that the nitrogen repressor metabolite, whether glutamine or a derivative, is recognized by a second negative-acting regulatory protein that, in turn, interacts with the global positive-acting proteins. However, the extensive genetic analysis by Arst & Scazzocchio (2) has not uncovered a second control locus, which certainly argues against such a possibility.

Is the N. crassa nmr Gene Regulatory?

Sorger & coworkers (53) and Garrett & colleagues (59) independently discovered that mutation of a gene designated as *nmr* results in the constitutive synthesis of nitrate reductase and other nitrogen metabolic enzymes in the presence of sufficient primary nitrogen sources to fully repress synthesis of these enzymes in nmr^+ strains. This phenotype of course suggests that *nmr* is a negative-acting regulatory gene; an NMR protein might sense the glutamine content of the cell and in turn modulate the activity of NIT2. However, a regulatory-type phenotype can also result from mutations that simply block the synthesis of a small-molecular-weight effector or that affect the uptake of metabolic repressors or inducers. A concern in this regard was that, despite the intensive genetic analysis of nitrogen regulation in *A. nidulans,* mutants analogous to *nmr* that result in constitutivity have not been recovered, despite selective systems that should have detected them.

The *nmr* gene has been isolated (54, 63) and appears to encode a protein of 488 amino acid residues; however, the putative NMR protein has no distinctive features such as obvious DNA binding domains nor is there any significant homology with proteins in the various data bases (63). The NMR protein, expressed in *E. coli,* displayed neither glutamine binding nor DNA-binding activities. *N. crassa nmr* might be equivalent to the *meaB* mutations of *A. nidulans*. Both nmr^- and $meaB^-$ strains appear generally derepressed in the presence of NH_4^+ and may affect membrane functions; i.e. they may mimic regulatory-type functions by preventing the uptake of nitrogen compounds that lead to catabolite repression in wild-type strains (6). Further characterization of the physiological and molecular properties of *meaB* and *nmr* is needed to determine whether they are indeed related and to critically evaluate their role, if any, in nitrogen control.

PATHWAY-SPECIFIC CONTROL GENES

One of the most intriguing aspects of AREA and NIT2 is that they not only have a global role in controlling the expression of many nitrogen catabolic genes but also must cooperate in some way with different minor positive-acting control genes that are required, in addition to AREA or NIT2, for the selective activation of sets of structural genes that specify enzymes of a particular catabolic pathway. Nitrate is required for induction of synthesis of enzymes of nitrate assimilation, amino acids for that of L-amino acid oxidase, proline for proline catabolic enzymes, and uric acid for the purine catabolic enzymes. Thus, in general, expression of the enzymes of a particular nitrogen catabolic pathway requires specific induction, which appears to be mediated by pathway-specific control genes, such as *nit-4* for nitrate induction in *N. crassa* or *ua-Y* for purine induction in *A. nidulans*.

Expression of *nit-3* and *nit-6*, which specify enzymes for nitrate assimilation in *N. crassa*, requires a functional NIT2 protein and also completely depends upon nitrate induction and the product of the pathway-specific regulatory gene, *nit-4*. The putative NIT4 protein is composed of 1090 amino acids and contains a single GAL4-like Cys6/Zn2 binuclear type of zinc finger that serves as the DNA-binding domain for a number of fungal *trans*-acting factors (64). The corresponding gene of *A. nidulans, nirA,* encodes a protein whose amino-terminal half, approximately 600 amino acids, has approximately 60% amino acid identity with the amino-terminal half of NIT4 (3, 4). In particular, the 50 amino acids that constitute the DNA-binding motifs of NIT4 and NIRA are 90% identical with one another. However, the carboxy-terminal portion of NIT4 and NIRA are completely different. NIT4 contains two segments that are extremely rich in glutamine and appear to function in transactivation (64, 65); in contrast, the carboxy terminus of NIRA lacks any such glutamine-rich regions but instead has a proline-rich region and is more acidic. Nevertheless, the *N. crassa nit-4* gene can substitute for the *A. nidulans nirA* gene when transformed into a mutant host strain (25).

The combination of NIT2 plus NIT4 in *N. crassa* (and AREA and NIRA in *A. nidulans*) appears to constitute a two-way switch. Investigators believed that NIT2 might control NIT4 expression (similarly, AREA controls that of NIRA), which in turn activated structural gene expression. However, this pattern of a sequential gene expression does not seem possible because *nit-4* is not controlled by *nit-2* nor is it subject to nitrogen repression (15). Rather it appears that NIT2 and NIT4 jointly strongly activate *nit-3* expression, whereas neither factor alone promotes any detectable *nit-3* expression (16). Although a direct NIT2-NIT4 protein-protein interaction is possible, NIT2 and NIT4 more likely bind at independent sites upstream of *nit-3* and together strongly enhance the expression of this highly regulated structural gene. It is very important to identify the number of NIT2 and NIT4 binding sites and their relative positions with respect to one another as a first step in understanding the mechanism by which these two proteins together evoke strong gene expression.

In this context an important observation is that AREA and NIT2 must also cooperate with still other positive-acting pathway-specific regulatory proteins in the case of genes that encode enzymes of other nitrogen catabolic pathways. Although each pathway-specific factor might be expected to be a *trans*-acting DNA-binding protein, this feature has yet to be established. It will be of considerable interest to determine the nature of the DNA binding and activation domains of each of these specific factors. Of equal importance is the need to compare the pattern of binding sites in the respective promoters for the various specific factors and to determine the mechanism by which the global control proteins work in cooperation with such a host of specific factors.

Autogenous Regulation by Nitrate Reductase?

The possibility that the enzyme nitrate reductase plays a regulatory role in controlling its own synthesis and that of nitrite reductase, first suggested in the pioneering studies of Cove (11), has intrigued others (2, 18, 60), but a resolution has remained elusive. In most, but not all, mutants of the nitrate reductase structural gene (*nit-3* in *N. crassa; niaD* in *A. nidulans*), expression of nitrate reductase mRNA and inactive protein (and also that of nitrite reductase) does not require nitrate induction. In contrast, expression of these same activities in the wild-type strains has an absolute requirement for nitrate induction. A long-standing speculative model is that the pathway specific NIT4 (or NIRA) protein cannot enter the nucleus because it is bound by nitrate reductase, present in low basal levels in uninduced cells. Thus the structural genes remain off. In the presence of nitrate, perhaps detected by nitrate reductase, the NIT4 (or NIRA) protein is freed and enters the nuclei, thus turning on structural gene expression. The availability of the cloned structural and regulatory genes implies that a direct test of this hypothesis should at long last be possible.

Purine Catabolism

In *A. nidulans* and in *N. crassa,* the synthesis of enzymes that allow various purines to be used as nitrogen sources is subject to a high degree of regulation. In *A. nidulans,* where this pathway has been studied extensively, at least eight unlinked structural genes that specify permeases and enzymes involved in purine uptake and degradation are controlled by a positive-acting pathway-specific regulatory gene, *uaY,* which mediates uric acid induction (56, 57). Most *uaY* mutants, including alleles with deletions or insertions, have a null phenotype and are completely lacking or severly reduced in all activities, whereas rare *uaY* mutants are constitutive for some or all of these same activities. This fact strongly supports the interpretation that this gene encodes an authentic positive-acting regulatory protein. The rare constitutive mutants may alter the UAY protein so that it mimics the wild-type protein when in an active conformation resulting from binding the inducer, uric acid; alternatively, the constitutive mutations may alter a recognition site for a negative-acting protein that otherwise interacts with UAY, similar to the situation for the GAL4-GAL80 interacting proteins in yeast. Expression of the genes that encode the purine catabolic enzymes also requires an active AREA protein and nitrogen derepression. Two structural genes of the purine catabolic pathway have been cloned—*uap,* which encodes a purine-specific permease, and *uaZ,* which specifies urate oxidase (14). Northern analyses have demonstrated that the accumulation of mRNA for these two genes is controlled by uric acid induction and nitrogen derepression and requires functional *uaY* and *areA* products; thus, the purine catabolic genes are apparently controlled at the level of transcription. A particularly interesting mutation, *uap-100,* is *cis*-dominant and results in a 2.5-fold up-promoter effect, eliminates the requirement for induction, and partially suppresses a specific mutation, *areA102,* in the wide-domain regulatory gene. The *uap-100* mutation results from a duplication of 139 bp in the promoter of the *uap* gene; the duplicated segment includes two GATA sequences, the target sites for NIT2 and AREA (14).

The *uaY* regulatory gene specifies a 4-kb transcript that is constitutively synthesized; *uaY* transcription does not require induction, nitrogen derepression, nor a functional *areA* product (57). Thus, as discussed above for *nirA* and for *nit-4,* these results exclude the possibility of sequential expression wherein AREA activates *uaY,* whose product in turn activates by itself the purine catabolic structural genes. Rather, binding of both the AREA and UAY proteins to the cognate promoters is probably necessary for expression of the structural genes. Instances of sequential action of regulatory genes have been demonstrated in fungi (23, 42).

Acetamidase Gene Control

The control of acetamidase gene expression in *A. nidulans* represents a particularly complex situation in which multiple *trans*-acting factors can act within the promoter region of this structural gene. Acetamide, which is hydrolyzed by acetamidase, can serve as the sole nitrogen source and also as the sole carbon source for *A. nidulans*. The regulation of expression of the acetamidase structural gene, *amdS,* is surprisingly complex. The elegant studies by Hynes & colleagues (1, 28) have revealed that in addition to global control by either nitrogen or carbon limitation, *amdS* expression can be induced in multiple ways, e.g. by either w-amino acids, acetyl CoA, or benzoate (1). Thus, the *amdS* promoter must be extremely modular and complex, being responsive to the positive-acting AREA or negative-acting CREA global regulatory proteins plus a variety of pathway-specific control factors. An understanding of the multiple protein-protein and DNA-protein interactions that control acetamidase gene expression will certainly provide new insight concerning the mechanisms that govern truly complex genetic regulation. Other gene sets that encode enzymes of additional pathways of nitrogen catabolism, e.g. proline metabolism (26) or protein utilization, are controlled in related and equally interesting ways, but cannot be discussed here.

POSTSCRIPT

Several decades of sophisticated genetic analyses and intensive biochemical studies provided a general picture of nitrogen control in the filamentous fungi. However, in most cases, important details were missing and direct tests of possible control mechanisms were virtually impossible. The application of molecular genetic approaches just in the past few years has led to a dramatic increase in our understanding of nitrogen regulation. We can now anticipate definitive answers to long-standing mysteries. Indeed, the future promises to be rewarding with important new information and insights and exciting surprises.

ACKNOWLEDGMENTS

Research in the author's laboratory is supported by grant GM-23367 from the National Institutes of Health. I thank all former and present members of our laboratory for their many important contributions, with special thanks to Tso-Yu Chiang, Ying-Hui Fu, Gabor Jarai, Moien Kanaan, James Ketter, Hakjoo Lee, Patricia Okamoto, Xiao-Dong Xiao, James Young, and Gwo-Fang Yuan.

Literature Cited

1. Andrianopoulos, A., Hynes, M. J. 1990. Sequence and functional analysis of the positively acting regulatory gene *amdR* from *Aspergillus nidulans*. *Mol. Cell. Biol.* 10:3194–3203
2. Arst, H. N., Scazzocchio, C. 1985. Formal genetics and molecular biology of the control of gene expression in *Aspergillus nidulans*. In *Gene Manipulations in Fungi*, ed. J. Bennett, L. Lasure, pp. 309–43. Orlando, FL: Academic
3. Burger, G., Strauss, J., Scazzocchio, C., Lang, B. 1991. *nirA*, the pathway-specific regulatory gene of nitrate assimilation in *Aspergillus nidulans*, encodes a putative GAL4-type zinc finger protein and contains introns in highly conserved regions. *Mol. Cell. Biol.* 11:5746–55
4. Burger, G., Tilburn, J., Scazzocchio, C. 1991. Molecular cloning and functional characterization of the pathway-specific regulatory gene *nirA*, which controls nitrate assimilation in *Aspergillus nidulans*. *Mol. Cell. Biol.* 11: 795–802
5. Burton, E. G., Metzenberg, R. L.

1972. Novel mutation causing derepression of several enzymes of sulfur metabolism in *Neurospora crassa*. *J. Bacteriol.* 109:140–50

6. Caddick, M. X. 1992. Characterization of a major *Aspergillus* regulatory gene, *areA*. In *Molecular Biology of Filamentous Fungi,* ed. U. Stahl, P. Tudzynski, pp. 141–52. Weinheim: VCH Press
7. Caddick, M. X., Arst, H. N. 1990. Nitrogen regulation in *Aspergillus:* are two fingers better than one? *Gene* 95: 123–27
8. Caddick, M. X., Arst, H. N., Taylor, L. H., Johnson, R. I., Brownlee, A. G. 1986. Cloning of the regulatory gene *areA* mediating nitrogen metabolite repression in *Aspergillus nidulans*. *EMBO J.* 5:1087–90
9. Campbell, W. H., Kinghorn, J. R. 1990. Functional domains of assimilatory nitrate reductases and nitrite reductases. *Trends Biochem. Sci.* 15: 315–19
10. Courey, A. J., Tjian, R. 1988. Analysis of Spl in vivo reveals multiple transcriptional domains, including a novel glutamine-rich activation motif. *Cell* 55:887–98
11. Cove, D. J. 1979. Genetic studies of nitrate assimilation in *Aspergillus nidulans*. *Biol. Rev.* 54:291–327
12. Cunningham, T. S., Cooper, T. C. 1991. Expression of the DAL80 gene, whose product is homologous to the GATA factors and is a negative regulator of multiple nitrogen catabolic genes in *Saccharomyces cerevisiae,* is sensitive to nitrogen catabolite repression. *Mol. Cell. Biol.* 11:6205–15
13. Davis, M. A., Hynes, M. J. 1987. Complementation of *areA*$^-$ regulatory gene mutations of *Aspergillus nidulans* by the heterologous regulatory gene *nit-2* of *Neurospora crassa*. *Proc. Natl. Acad. Sci. USA* 84:3753–57
14. Diallinas, G., Scazzocchio, C. 1989. A gene coding for the uric acid-xanthine permease of *Aspergillus nidulans:* inactivational cloning, characterization, and sequence of a *cis*-acting mutation. *Genetics* 122:341–50
15. Fu, Y. H., Knessi, J. Y., Marzluf, G. A. 1989. Isolation of *nit-4,* the minor nitrogen regulatory gene which mediates nitrate induction in *Neurospora crassa*. *J. Bacteriol.* 171:4067–70
16. Fu, Y. H., Marzluf, G. A. 1987. Characterization of *nit-2,* the major nitrogen regulatory gene of *Neurospora crassa*. *Mol. Cell. Biol.* 7:1691–96
17. Fu, Y. H., Marzluf, G. A. 1987. Molecular cloning and analysis of the regulation of *nit-3,* the structural gene for nitrate reductase in *Neurospora crassa*. *Proc. Natl. Acad. Sci. USA* 84:8243–47
18. Fu, Y. H., Marzluf, G. A. 1988. Metabolic control and autogenous regulation of *nit-3,* the nitrate reductase structural gene of *Neurospora crassa*. *J. Bacteriol.* 170:657–61
19. Fu, Y. H., Marzluf, G. A. 1990. *cys-3,* the positive-acting sulfur regulatory gene of *Neurospora crassa,* encodes a sequence-specific DNA-binding protein. *J. Biol. Chem.* 265:11942–47
20. Fu, Y. H., Marzluf, G. A. 1990. *nit-2,* the major nitrogen regulatory gene of *Neurospora crassa,* encodes a protein with a putative zinc finger DNA-binding domain. *Mol. Cell. Biol.* 10:1056–65
21. Fu, Y. H., Marzluf, G. A. 1990. *nit-2, the* major positive acting nitrogen regulatory gene of *Neurospora crassa,* encodes a sequence-specific DNA-binding protein. *Proc. Natl. Acad. Sci. USA* 87:5331–35
22. Fu, Y. H., Marzluf, G. A. 1990. Site-directed mutagenesis of the zinc finger DNA-binding domain of the nitrogen regulatory protein NIT2 of *Neurospora*. *Mol. Microbio.* 4:1847–52
23. Fu, Y. H., Paietta, J. V., Mannix, D. G., Marzluf, G. A. 1989. *cys-3,* the positive-acting sulfur regulatory gene of *Neurospora crassa,* encodes a protein with a putative leucine zipper DNA-binding element. *Mol. Cell. Biol.* 9:1695–99
24. Hanson, K. A., Marzluf, G. A. 1975. Control of the synthesis of a single enzyme by multiple regulatory circuits in *Neurospora crassa*. *Proc. Natl. Acad. Sci. USA* 72:1240–44
25. Hawker, K. L., Montague, P., Marzluf, G. A., Kinghorn, J. R. 1991. Heterologous expression and regulation of the *Neurospora crassa nit-4* pathway-specific regulatory gene for nitrate assimilation in *Aspergillus nidulans*. *Gene* 100:237–40
26. Hull, E. P., Green, P. M., Arst, H. N., Scazzocchio, C. 1989. Cloning and characterization of the L-proline catabolism gene cluster of *Aspergillus nidulans*. *Mol. Microbiol.* 3:553–60
27. Hurlburt, B. K., Garrett, R. H. 1988. Nitrate assimilation in *Neurospora crassa:* enzymatic and immunoblot analysis of wild-type and *nit* mutant protein products in nitrate-induced and glutamine-repressed cultures. *Mol. Gen. Genet.* 211:35–40
28. Hynes, M. J., Corrick, C. M., Kelly, J. M., Littlejohn, T. G. 1988. Iden-

tification of the sites of action for regulatory genes controlling the *amdS* gene of *Aspergillus nidulans*. *Mol. Cell. Biol.* 8:2589–96

29. Jarai, G., Marzluf, G. A. 1991. Sulfate transport in *Neurospora crassa:* regulation, turnover, and cellular localization of the CYS14 protein. *Biochemistry* 30:4768–73
30. Johnstone, I. L., McCabe, P. C., Greaves, P., Cole, G. E., Brow, M. A. et al. 1990. The isolation and characterization of the *crnA-niiA-niaD* gene cluster for nitrate assimilation in the filamentous fungus *Aspergillus nidulans*. *Gene* 90:181–92
31. Kanaan, M. N., Marzluf, G. A. 1991. Mutational analysis of the DNA-binding domain of the CYS3 regulatory protein of *Neurospora crassa*. *Mol. Cell. Biol.* 11:4356–62
32. Kanaan, M. N., Fu, Y. H., Marzluf, G. A. 1992. The DNA-binding domain of the Cys-3 regulatory protein of *Neurospora crassa* is bipartite. *Biochemistry* 31:3197–3203
33. Ketter, J. S., Jarai, G., Fu, Y. H., Marzluf, G. A. 1991. Nucleotide sequence, messenger RNA stability, and DNA recognition elements of *cys-14,* the structural gene for sulfate permease II in *Neurospora crassa*. *Biochemistry* 30:1780–87
34. Ketter, J. S., Marzluf, G. A. 1988. Molecular cloning and analysis of the regulation of *cys-14*$^{+}$, a structural gene of the sulfur regulatory circuit of *Neurospora crassa*. *Mol. Cell. Biol.* 8: 1504–8
35. Kudla, B., Caddick, M. X., Langdon, T., Martinez-Rossi, N. M., Bennett, C. F., et al. 1990. The regulatory gene *areA* mediating nitrogen metabolite repression in *Aspergillus nidulans*. Mutations affecting specificity of gene activation alter a loop residue of a putative zinc finger. *EMBO J.* 9:1355–64
36. Landschulz, W. H., Johnson, P. F., McKnight, S. L. 1989. The DNA binding domain of the rat liver nuclear protein C/EBP is bipartite. *Science* 243: 1681–88
37. Laurent, B. C., Treitel, M. A., Carlson, M. 1990. The SNF5 protein of *Saccharomyces cerevisiae* is a glutamine- and proline-rich transcriptional activator that affects expression of a broad spectrum of genes. *Mol. Cell. Biol.* 10:5616–25
38. Lee, H. J., Fu, Y. H., Marzluf, G. A. 1990. Nucleotide sequence and DNA recognition elements of *alc,* the structural gene which encodes allantoicase, a purine catabolic enzyme of *Neurospora crassa*. *Biochemistry* 29:8779–87
39. Marzluf, G. A. 1970. Genetic and biochemical studies of distinct sulfate permease species in different developmental stages of *Neurospora crassa*. *Arch. Biochem. Biophys.* 138:254–63
40. Marzluf, G. A. 1972. Control of the synthesis, activity, and turnover of enzymes of sulfur metabolim in *Neurospora crassa*. *Arch. Biochem. Biophys.* 150:714–24
41. Marzluf, G. A. 1981. Regulation of nitrogen metabolism and gene expression in fungi. *Microbiol. Rev.* 45:437–61
42. Marzluf, G. A., Fu, Y. H. 1989. Molecular analyses of the nitrogen and sulfur regulatory circuits of *Neurospora crassa*. In *Genetics and Molecular Biology of Industrial Organisms,* ed. C. Hershberger, S. Queener, G. Hegeman, pp. 279–87. Washington, DC: Am. Soc. Microbiol.
43. Marzluf, G. A., Metzenberg, R. L. 1968. Positive control by the *cys-3* locus in regulation of sulfur metabolism in *Neurospora*. *J. Mol. Biol.* 33:423–37
44. Metzenberg, R. L., Parson, J. W. 1966. Altered repression of some enzymes of sulfur utilization in a temperature-conditional lethal mutant of *Neurospora*. *Proc. Natl. Acad. Sci. USA* 55:629–35
45. Minehart, P. L., Magasanik, B. 1991. Sequence and expression of GLN3, a positive nitrogen regulatory gene of *Saccharomyces cerevisiae* encoding a protein with a putative zinc finger DNA-binding domain. *Mol. Cell. Biol.* 12:6216–68
46. Okamoto, P. M., Fu, Y. H., Marzluf, G. A. 1991. *Nit-3,* the structural gene of nitrate reductase in *Neurospora crassa:* nucleotide sequence and regulation of mRNA synthesis and turnover. *Mol. Gen. Genet.* 227:213–23
47. Paietta, J. V. 1989. Molecular cloning and regulatory analysis of the arylsulfatase structural gene of *Neurospora crassa*. *Mol. Cell. Biol.* 9:3630–37
48. Paietta, J. V. 1990. Molecular cloning and analysis of the *scon-2* negative regulatory gene of *Neurospora crassa*. *Mol. Cell. Biol.* 10:5207–14
49. Paietta, J. V. 1992. Production of the CYS3 regulator, a bZIP DNA-binding protein, is sufficient to induce sulfur gene expresson in *Neurospora crassa*. *Mol. Cell. Biol.* 12:1568–77
50. Paietta, J. V., Akins, R. A., Lambowitz, A. M., Marzluf, G. A.

1987. Molecular cloning and characterization of the *cys-3* regulatory gene of *Neurospora crassa*. *Mol. Cell. Biol.* 5:1554–59

51. Pall, M. L. 1971. Amino acid transport in *Neurospora crassa* II: properties and regulation of a methionine transport system. *Biochim. Biophys. Acta* 233: 201–14
52. Premakumar, R., Sorger, G. J., Gooden, D. 1979. Nitrogen metabolite repression of nitrate reductase in *Neurospora crassa*. *J. Bacteriol.* 137:1119–26
53. Premakumar, R., Sorger, G. J., Gooden, D. 1980. Physiological characterization of a *Neurospora crassa* mutant with impaired regulation of nitrate reductase. *J. Bacteriol.* 144:542–51
54. Sorger, G. J., Brown, D., Farzannejad, M., Guerra, A., Jonathan, M., et al. 1989. Isolation of a gene that downregulates nitrate assimilation and influences another regulatory gene in the same system. *Mol. Cell. Biol.* 9:4113–17
55. Stewart, V., Vollmer, S. J. 1986. Molecular cloning of *nit-2*, a regulatory gene required for nitrogen metabolite repression in *Neurospora crassa*. *Gene* 46:291–95
56. Suaarez, T., Oestreicher, N., Kelly, J., Ong, G., Sankarsingh, R., Scazzocchio, C. 1991. The *uaY* positive control gene of *Aspergillus nidulans:* fine structure, isolation of constitutive mutants and reversion patterns. *Mol. Gen Genet.* 230:359–68
57. Suaarez, T., Oestreicher, N., Penalva, M. A., Scazzocchio, C. 1991. Molecular cloning of the *uaY* regulatory gene of *Aspergillus nidulans* reveals a favoured region for DNA insertions. *Mol. Gen. Genet.* 230:369–75
58. Thomas, D., Jacquemin, I., Surdin-Kerjan, Y. H. 1992. MET4, a leucine zipper protein, and centromere-binding factor 1 are both required for transcriptional activation of sulfur metabolism in *Saccharomyces cerevisiae*. *Mol. Cell. Biol.* 12:1719–27
59. Tomsett, A. B., Dunn-Coleman, N. S., Garrett, R. H. 1981. The regulation of nitrate assimilation in *Neurospora crassa*. The isolation and genetic analysis of *nmr-1* mutants. *Mol. Gen. Genet.* 182:229–33
60. Tomsett, A. B., Garrett, R. H. 1981. Biochemical analysis of mutants defective in nitrate assimilation in *Neurospora crassa:* evidence for autogenous control by nitrate reductase. *Mol. Gen. Genet.* 184:183–90
61. Tsai, S. F., Martin, D. I., Zon, L. I., D'Andrea, A. D., Wong, G. G., Orkin, S. H. 1989. Cloning of cDNA for the major DNA-binding protein of the erythroid lineage through expression in mammalian cells. *Nature* 339:446–51
62. Whitehead, M. P., Unkles, S. E., Ramsden, M., Campbell, E. I., Gurr, S. J., et al. 1989. Transformation of a nitrate reductase deficient mutant of *Penicillium chrysogenum* with the corresponding *Aspergillus niger* and *Aspergillus nidulans niaD* genes. *Mol. Gen. Genet.* 216:408–11
63. Young, J. L., Jarai, G., Fu, Y. H., Marzluf, G. A. 1990. Nucleotide sequence and analysis of NMR, a negative-acting regulatory gene in the nitrogen circuit of *Neurospora crassa*. *Mol. Gen. Genet.* 222:120–28
64. Yuan, G. F., Fu, Y. H., Marzluf, G. A. 1991. *nit-4*, a pathway-specific regulatory gene of *Neurospora crassa*, encodes a protein with a putative binuclear zinc DNA-binding domain. *Mol. Cell. Biol.* 11:5735–45
65. Yuan, G. F., Marzluf, G. A. 1992. Molecular characterization of mutations of *nit-4*, the pathway-specific regulatory gene which controls nitrate assimilation in *Neurospora crassa*. *Mol. Microbiol.* 6:67–73

Annu. Rev. Microbiol. 1993. 47:57–87

AGROACTIVE COMPOUNDS OF MICROBIAL ORIGIN

Yoshitake Tanaka and Satoshi Ōmura

Research Center for Biological Function, The Kitasato Institute, Minato-ku, Tokyo 108, Japan

KEY WORDS: insecticide, herbicide, fungicide, microbial metabolite, screening, avermectin, milbemycin

---

CONTENTS

ABSTRACT

Microbial metabolites attract increasing attention as potential pesticides. They are expected to overcome the resistance and pollution that have accompanied the use of synthetic pesticides. Several microbial metabolites, such as avermectin, have proved useful as agroactive agents. In this review, we attempt to identify newer agroactive microbial metabolites with feasible activity or interesting action sites from those reported in recent years. In addition, microbial and chemical modifications of existing microbial agrochemicals are discussed to illustrate the usefulness of these technologies in potentiating agroactivity and stability. We discuss the possibility of future discovery of excellent microbial agrochemicals, and the importance of efforts to promote positive public perception and public acceptance of pesticide chemicals.

0066-4227/93/1001-0057$02.00

INTRODUCTION

Pesticides have been of great help to crop production. They protect crop plants from injuries and damage caused by harmful insects and mites, infectious fungi and bacteria, and invasive weeds. Without pesticides, crop production decreases by an estimated 20–40%. This includes decreases in amounts at harvest and additional losses occurring after the harvest and during transportation. The harmful organims are almost countless: insects and mites number at least 5000 species, fungi at least 8000, bacteria at least 70, and viruses at least 500. In addition, 1800 weed species may be harmful, 200 of which account for 95% of damage (45).

The world market for pesticides had increased to $17.9 billion US in 1986, increasing further to $20.4 billion in 1988 (end-user-value basis), according to the 1990 *Wood Mackenzie Report* (77a), a well-known annual report on agroindustry and the world market for agrochemicals and veterinary products. Of global sales in 1988, herbicides comprised 43.6%, insecticides 29.7%, fungicides 20.5%, and others 6.1%.

The development of pesticides has not been without problems. One of the most serious problems arising in association with pesticide use has been the adverse effects of residual chemicals on environmental ecosystems. Another problem is the emergence of pesticide-resistant insects and fungi. In view of the very strict requirements for approval from regulatory authorities, and the efforts of industry, the former problem should soon be less serious, if appropriate ways of use are followed. However, the general public does not recognize the fact that recently approved agrochemicals are far safer than early ones (28). In addition, even with those compounds approved recently, uncontrolled (repeated and heavy) pesiticide use currently practiced in some areas, and possible misuses, will likely pollute ground water. Therefore, public concern today is still focused on safety of pesticides or alternative measures of pest control.

Here lies a dilemma. Although public concern over safety runs high, current and future problems involving many uncontrolled crop-harming insects, mites, and fungi urgently require a solution. Besides, the expected increase in global population in the coming twenty-first century requires an increase in crop production by about 30% above the present level (25, 28).

Interest in microbial metabolites as pesticides arises from attempts to solve the above dilemma. Microbial cultures are a treasure box, as is often mentioned and borne out by pharmacoactive drugs and antibiotics. Microbial metabolites are expected to conquer the resistance and pollution likely caused by synthetic pesticides. The merits of microbial metabolites as pesticides would be: (*a*) They are versatile in structure and activity. An unexpected structure with new agroactivity showing no cross-resistance is very likely to be identified through effective screening systems, which have progressed greatly in the long history of antibiotic screening. And (*b*) they are biodegradable. They degrade usually within a month, or even a few days, when exposed to firm soils. Thus they are expected to stress and pollute ecosystems less.

Pesticides of microbial origin introduced into field applications 30 years ago include blasticidin S, polyoxin, kasugamycin, validamycin, and mildiomycin as fungicides, and tetranactin as a miticide. Recent examples include avermectin, milbemycin, and bialaphos. The excellent activity of these compounds suggests that other desirable pesticides will be discovered from microbial metabolites.

This review presents an overview of the discovery of agroactive compounds from microorganisms reported in the past 10 years, emphasizing the methods of screening and sites of action of these active compounds. We attempt to identify areas of agroactivity of new and growing interest in the hope that screening of microbial metabolites will be more successful if efforts are focused toward particular types of agroactivity in which synthetic chemicals were ineffective.

Use of antibiotics as crop protectants has been reviewed (59, 83, 85) as has the use of microbial products with herbicidal activity (4, 45). Reviews on bioactive microbial metabolites and the methods for their screening (100, 102, 103, 117) have also appeared. For progress in pesticide sciences, readers can refer to the books published on the occasion of the International Conference of Pesticide Chemistry, held every four years (see 32 for the latest issue, 1991). A book describing strategies and methods for screening of bioactive microbial metabolites appeared very recently (104). This is a good source of information, screening methods, and techniques for insecticides, herbicides, and fungicides, as well as other pharmacoactives and antibiotics.

DISCOVERY OF AGROACTIVE COMPOUNDS FROM MICROORGANISMS

Current Trends in the Search for Agroactive Microbial Metabolites

A survey of recent literature indicates a continual increase in the discovery of new agroactive compounds from microorganisms. Most of them were obtained by conventional activity-monitoring biological screening with *Streptomyces* spp. as the major microbial source. Chemical-screening techniques were adopted less frequently. Fundamentally, the increased discovery comes from an enhanced interest toward microbial metabolites as potential pesticides.

More importantly, it arises from keen needs for excellent pesticides. Technically, the increase is a consequence of the following:

1. Establishment and improvement of screening technologies.
   a. Exploitation of novel microbial groups as sources for production of active compounds. Basidiomycetes, blue-green algae, and myxobacteria were shown to be rich producers of agroactive metabolites.
   b. Development of unique fermentation techniques and their application to screening programs.
   c. Development of new methods for detection of agroactivity. Unconventional test organisms were introduced to find new compounds.
2. Application of genetic techniques for breeding mutants that produce an altered spectrum of metabolites. Microbial conversion and mutational biosynthetic techniques were applied.
3. Progress in chemistry and biochemistry of pesticide sciences and its application to the synthetic design and the construction of new screening methods.
4. Others. The chemical basis of plant-pathogen interaction was elucidated, and many compounds likely to be involved in the interaction were identified. In addition, interesting pharmacological activities were uncovered for several agroactive microbial metabolites. These findings stimulated the interest in microbial metabolites.

From the standpoint of agrochemical discovery, construction and improvement of screening methods (1a–c) are most important and are described in some detail below.

NOVEL MICROBIAL SOURCES FOR AGROCHEMICAL PRODUCTION *Streptomyces* spp. have been, and remain, the most fruitful source of microorganisms for all types of bioactive metabolites, including agroactives. In fact, about 60% of new insecticides and herbicides reported in the past five years are of *Streptomyces* origin. With *Streptomyces* spp., however, the frequency of rediscovery of known compounds has become fairly high. In an effort to reduce this replication, the use of organisms other than *Streptomyces* spp., which include rare actinomycetes, fungi, basidiomycetes, and other taxa, has increased steadily.

Rare actinomycetes We (113, 114) discovered that *Kitasatosporia,* a strain of the order *Actinomycetales* produces setamycin, an insecticidal 16-membered macrolide (123). This genus is characterized by its abundant aerial mycelia formation and by unique amino acid composition in the cell walls. The cell walls of aerial spores and submerged spores contain LL-diaminopimelic acid, the isomeric type of which is of taxonomic importance, whereas aerial and submerged mycelia contain the *meso*-isomer. Members of this genus produce phosalacine, an herbicidal tripeptide (108, 140); cystargin, an antifungal peptide (156); and a variety of antibiotics (103). Strains of *Kitasatosporia* and other species of rare actinomycetes and bacteria can be isolated from soil samples with selective isolation techniques (37, 40, 139).

Bacidiomycetes and filamentous fungi Anke and his group focus their attention on bacidiomycete cultures for identifying bioactive metabolites (2). Strains of basidiomycetes produce many new antifungal agents with agrochemical interest, such as oudemansin (3) produced by a strain of *Oudemansiella radicata,* aleurodiscal (77) produced by a strain of *Aleurodiscus mirabilis,* strobilurin (2) produced by a strain of *Strobilurus* sp., pilatin (43) produced by a strain of *Flagelloscypha pilatii,* and herbicidal agents such as pereniporin (68) produced by a strain of *Perenniporia madullaenpanis.* Xerulin and dihydroxerulin produced by a strain of *Xerula melanotricha* are unique in structure and activity. These compounds possess an α,β-unsaturated γ-lactone and conjugated poly-ene-yne moieties. Xerulin inhibits cholesterol biosynthesis in HeLa cells (72).

Blue-green algae, Myxobacteria, *and red algae* Blue-green algae, such as *Anabaena* spp., produce a variety of antifungal compounds (30, 86). Another blue-green alga, *Lyngbya majuscula,* produced malyngolide, a δ-lactone with a long alkyl chain (48). Myxobacterial strains were found to produce antifungal, antibacterial, and cytocidal compounds (127). Among them, phenoxan (74), ambruticin, and pyrrolnitrin exhibit agricultural antifungal activity. A methanol extract of the red alga *Laurencia nipponica* was used to isolate a family of insecticides, Z-leurentin, Z-isoleurentin, and deoxyprepaciferol (157).

The isolation, growth, and preservation of these new groups of microorganisms requires specific techniques different from those used for *Streptomyces* spp. None of their products so far identified has reached a marketable stage. Nevertheless, these results appear to warrant further screening.

UNIQUE FERMENTATION TECHNIQUES The uniqueness of the fermentation technique is another factor that often favors discovery of new compounds (144). A medium containing a high content of inorganic phosphate was employed for production of an antifungal compound, FR-900848 (159). This medium refutes a widely accepted notion that a low content of inorganic phosphate favors antibiotic production. FR-900848 is a nucleoside, and an unusual linear chain of five cyclopropane rings is attached to its sugar moiety. It showed potent in vitro activity against *Sclerotinia arachidis, Fusarium oxysporum,* and *Penicillium chrysogenum.* Our laboratory (79, 115, 116) developed phosphate ion– and ammonium ion–depressed fermentation using trapping agents such as the minerals allophane and zeolite, respectively. This fermentation technique led to the discovery of phthoxazolin (Figure 3, below) (119), jietacin (52, 112) (Figure 2, below), globopeptin (145), and others (101).

TARGETED AGROACTIVITY AND NEW DETECTION METHODS Synthetic chemicals are in use or are under development in many areas of crop protection. Microbial metabolites as pesticides are later developments. In view of this situation, and the possible advantages of microbial metabolites over synthetic chemicals, targeted agroactivities and screening methods are of strategic importance in their screening. Two approaches were taken in the choice of target activity and therefore in constructing and improving screening methods, as illustrated in Tables 1–3.

In one approach, efforts centered on compounds of possible use in existing areas of agroactivity where synthetic chemicals are available but must be improved. The new microbial metabolites discovered were expected to serve as new leads, because of novel structure and activity with no cross-resistance and low residual secondary effects. In this approach, improved or modified screening methods are critical to obtaining better compounds. New compounds thus discovered are jietacin (112), discovered by using the pine wood nematode; altemicidin (138) and dioxapyrrolomycin (20), detected using brine shrimp; and allosamidin (130), found as an enzyme inhibitor for chitinase. This category also includes a new herbicidal inhibitor of starch synthesis, 6241-B substance (67).

In the second approach, attention was focused toward compounds with possible uses in novel agrochemical areas, namely, compounds with activity toward pests for which previous synthetic chemicals were either not effective or insufficiently effective. The resultant compounds were expected to open a new area of pesticide application, as did avermectin. Examples of these include inhibition of cellulose biosynthesis (phthoxazolin) (119) and growth inhibition of *Phytophthora* (phthoramycin) (118) and *Puccinia* (rustmicin) (96).

Microbial products already put into practice are concentrated in a few areas, namely, antifungal agents for rice-disease control and miticides. The use of microbial products in these areas is well established, while products for other areas are under development. Consideration of the pest to be targeted and the site of action of the compounds may help select the most appropriate screening method for finding novel agroactive microbial metabolites.

Described below are recently discovered microbial metabolites with feasible activity, or with interesting action sites. In order to exemplify the current trends, selected methods of screening are also discussed.

Table 1 Insecticides of chemical and microbial origins

Harmful insects	Insecticides: Chemical[b]	Insecticides: Microbial
Hemiptera		
Brown planthopper, green rice leafhopper, aphids	**Organophosphorus,** buprofezin, **carbamates,** imidacloprid, pyrethroids	
Lepidoptera		
Rice stem borer, corn borer, potato tuberworm, tobacco cutworm, cabbage armyworm, diamondback moth, apple leaf miner	**Organophosphorus, carbamates,** pyrethroids, benzoylphenylureas	(Avermectin = EMA)
Diptera		
Rice leaf miner, wheat thigh chloropid fly, seed maggot	**Organophosphorus, carbamates,** pyrethroids	
Coleoptera		
Soybean beetle, cucurbit leaf beetle, 28-spotted lady beetle	**Organophosphorus, carbamates**	
Acarina		
Kanzawa spider mite, two-spotted spider mite, citrus red mite	Benzilates, organophosphorus, pyrethroids, hexythiazox	Avermectin, milbemycin, polynactins (allosamidin)

[a] Commercialized compounds are listed. Compounds of recent discovery with noticeable activity are included in parentheses.
[b] Compounds showing decreased efficacies due to resistance are in bold.

Insecticides

AVERMECTINS AND MILBEMYCINS Avermectin (Figure 1) represents a family of fused 16-membered macrolides produced by *Streptomyces avermitilis* (11, 12, 29). The potent nematicidal, insecticidal, and acaricidal activities were discovered by a joint research effort of Merck, Sharp and Dohme Research Laboratories and the Kitasato Institute (135). Soon after the introduction of an avermectin derivative, ivermectin, to field applications as a veterinary drug, the marked antiectoparasitic effect was demonstrated in sheep, cattle, horse, pig, dog, and so on. Ivermectin is currently also used as a human drug for control of onchocerciasis, or African river blindness, one of the most serious endemics caused by a microfilaria, *Onchocerca volvulus* (12, 13).

Avermectin is a neuroactive substance with a fast-acting effect. It inhibits

Table 2 Agricultural fungicides of chemical and microbial origins

Fungal diseases	Fungicides[a]	
	Chemical[b]	Microbial
Phycomycetes		
Late blight	Fosetyl, dithiocarbamates, **acylalanines**	
Downy mildew	Fosetyl, dithiocarbamates, **acylalanines,** Bordeaux mixtures	(Phthoramycin, valclavam)
Ascomycetes		
Powdery mildew	**Triazoles, benzimidazoles**	Mildiomycin
Bitter rot	**Benzimidazoles,** dithiocarbamates	
Scab	**Benzimidazoles, triazoles,** dithiocarbamates	
Brown rot	**Benzimidazoles, dicarboximides**	
Basidiomycetes		
Rust	**Triazoles,** anilides, dithiocarbamates	(Rustmicin)
Sheath blight	Anilides	Validamycin (Dapiramicin)
Fungi imperfecti		
Blast	Organophosphorus, probenazole, tricyclazole, isoprothiolane	Blasticidin S Kasugamycin
Anthracnose	**Benzimidazoles,** dithiocarbamates	
Gray mold	**Benzimidazoles, dicarboximides**	
Leaf mold	**Benzimidazoles,** dithiocarbamates	
Leaf spot	Dithiocarbamates, dicarboximides, Bordeaux mixtures	**Polyoxins**
Wilt diseases	Benzimidazoles	

[a,b] See footnotes to Table 1.

Table 3 Herbicides and plant growth regulators (PGR) of chemical and microbial origins and their action sites.

Action site	Herbicides or PGRs[a]	
	Chemical	Microbial
Photosynthesis and respiration	Triazines, acylanilides, pyrazolate, diphenylethers, dinitrophenols	(6241-B)
Biosynthesis or function of:		
Auxin	2,4-D	
Gibberellin	Triazoles	Gibberellin
Fatty acid	Aryloxyphenoxypropionates, cyclohexanediones	
Amino acid	Sulfonylureas, Imidazolinones, glyphosate, phosphinothricin	Bialaphos (Phosalacine, vulgamycin)
Cellulose	(Isoxaben)	(Phthoxazolin)

[a] See footnote a to Table 1.

signal transmission at the GABA (γ-aminobutyric acid) receptor level, although the precise mechanism of inhibition is not yet defined. In *Ascaris* sp., avermectin functioned as a GABA receptor agonist, stimulating GABA release from presynaptic inhibitory membranes. Kass et al (62) proposed that in the presence of avermectin, chloride ion channels remain open, allowing chloride and sodium ions to flow out. The resulting ion imbalance results in the blockade of signal transmission. Recently, avermectin-binding proteins were detected. By using photoaffinity labeling techniques with a biologically active, radiolabeled azidoavermectin derivative, Rohrer et al (129a) identified three (53, 47, and 8 kDa) specific avermectin-binding polypeptides in membranes of a nematode, *Caenorhabditis elegans,* and one (~47 kDa) in an insect, *Drosophila melanogaster*. Studies on these polypeptides, including cloning, are expected to lead to a better understanding on the mode of avermectin action.

In parallel to the veterinary and medical uses of an avermectin derivative, agricultural usefulness of avermectin itself was also examined. Avermectin showed extremely high efficacies at low rates of applications. It is used as an acaricide in agriculture and horticulture and as an antiarthropod agent for both crop and noncrop applications (76). It is active against mites, leafminers, thrips, armyworms, aphids, and psyllids. Mites (Tetranychidae and Eriophyidae) were highly sensitive with IC_{90} values of 0.02–0.24 ppm, a level some orders of magnitude lower than that for existing synthetic acaricides. It also kills eggs of a tetranychid spider mite. Avermectin's effects on Lepidoptera (hornworms and armyworms) vary somewhat: an LC_{50} of 0.02 ppm for tobacco hornworm and 6 ppm for southern armyworm. Lasota & Dybas (76) suggest that avermectin use in agriculture does not adversely affect beneficial arthropods and agricultural ecosystems [see also Campbell (12), chapters 11–14]. A controversial interpretation of these effects was reported recently (136).

Milbemycins are a structurally related family of compounds produced by *Streptomyces hygroscopicus* subsp. *aureolacrimosus* (142). They are 16-membered macrolides very similar to the aglycone of the avermectins but lacking the disaccharide moiety attached to the C-13 hydroxyl of the aglycone (Figure 1). Milbemycins are available commercially as an agricultural acaricide and as a veterinary antiparasite. An oxime derivative has appeared on the market as an antiparasite for dog (151).

Avermectin's uses have had a positive impact on the chemistry and biochemistry of natural products, including avermectin-milbemycin itself. The compound undoubtedly encouraged those who were searching for the utility of natural products, supporting the strategy of pest control by microbial metabolites. The *Journal of Antibiotics* published four papers in 1980 on microbial metabolites with insecticidal or herbicidal activity. This number increased to 7 in 1986, and 22 in 1992. Discoveries of antifungal compounds with agrochemical activity also increased.

To date, attempts to isolate new avermectin analogues directly from soil microorganisms have been unsuccessful. A later section describes products of microbial and chemical modifications. On the other hand, new milbemycins, totaling 27 components, have been discovered; Figure 1 depicts some of them. The first paper on milbemycins by Takiguchi et al (142) described ten members of the α series and three members of the β series. From a mutant of the original milbemycin-producing culture, Mishima et al (84) isolated five new members of the α series, (D, F, G, J, K) and two members of the β series (E, H). Four VM compounds corresponding to the milbemycin α series were discovered from *Streptomyces* sp. E 255 (47). VM-44857 (Figure 1) showed in vivo efficacy against *Trichostrongylus* spp. at 0.25 mg/kg. Naturally acquired roundworm infections by *Haemonchus, Trichostrongylus,* and *Chabertia* spp. were cleared by more than 99% (with a few cases of 96% and 80%) at 0.2 mg/kg. In comparative testing, ivermectin afforded 100% clearance at the same dose level. Another four compounds, LL-F28249 substances named nemadectins, were isolated from *Streptomyces*

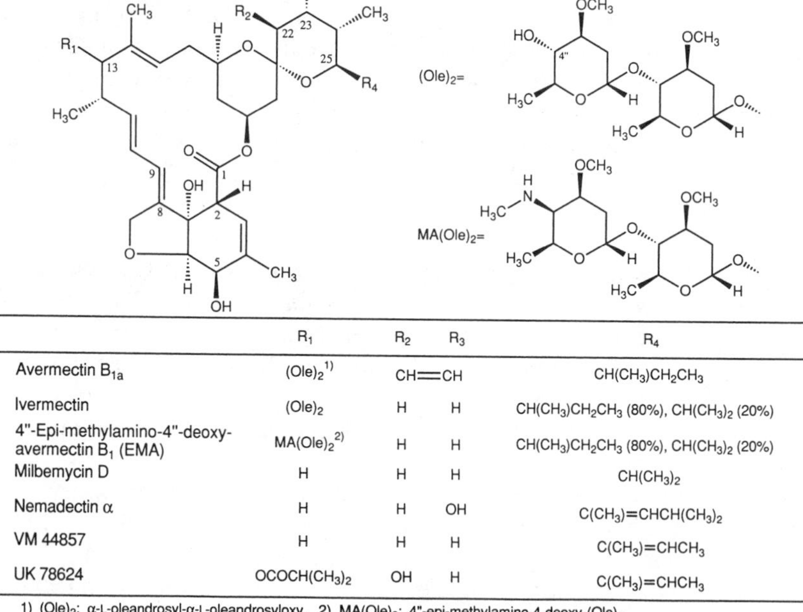

	R_1	R_2	R_3	R_4
Avermectin B_{1a}	$(Ole)_2$[1)]	CH═CH		$CH(CH_3)CH_2CH_3$
Ivermectin	$(Ole)_2$	H	H	$CH(CH_3)CH_2CH_3$ (80%), $CH(CH_3)_2$ (20%)
4"-Epi-methylamino-4"-deoxy-avermectin B_1 (EMA)	$MA(Ole)_2$[2)]	H	H	$CH(CH_3)CH_2CH_3$ (80%), $CH(CH_3)_2$ (20%)
Milbemycin D	H	H	H	$CH(CH_3)_2$
Nemadectin α	H	H	OH	$C(CH_3)=CHCH(CH_3)_2$
VM 44857	H	H	H	$C(CH_3)=CHCH_3$
UK 78624	$OCOCH(CH_3)_2$	OH	H	$C(CH_3)=CHCH_3$

1) $(Ole)_2$: α-L-oleandrosyl-α-L-oleandrosyloxy, 2) $MA(Ole)_2$: 4"-epi-methylamino-4-deoxy-$(Ole)_2$

Figure 1 Structures of avermectins and new milbemycin homolgues discovered recently.

cyaneogriseus (15, 150). They were compounds of the milbemycin α series. Nemadectin α at a single dose of 0.2 mg/kg [given orally (or)] was effective in vivo against *Trichostrongylus colubriformis* in gerbils.

One of the curious discoveries resulting from these screenings was the isolation C-13–substituted derivatives from natural sources, such as avermectin aglycone. Unfortunately, the above milbemycin components bore no hydroxyl at the C-13 position. Haxel et al (39) of the Pfizer group discovered a series of 12 C-13 β-acyloxylated molecules, known as UK substances, produced by *Streptomyces hygroscopicus* ATCC 53718. UK-78624 showed in vitro more than 95% killing of *C. elegans* at 10 ng/ml. Thus, a total of 40 milbemycin components were added to the arsenal of chemicals available for pest control.

The C-13 acyloxyl group of UK substances exhibited a β configuration, in contrast to the α configuration of the C-13 hydroxyl in avermectin, the oxygen of which originated biosynthetically from propionate (14). Haxel et al suggested that the C-13 oxygen of the former UK compounds was derived oxidatively from 13 unsubstituted intermediates in *S. hygroscopicus* ATCC 53718. They verified this hypothesis by the conversion of nemadectin γ, which is not a product of ATCC 53718, to 13β-isobutyloxylated nemadectin γ by *S. hygroscopicus* ATCC 53718 (39). No biosynthetic experiments with molecular oxygen were reported.

NEW INSECTICIDES DETECTED BY NEWLY CONSTRUCTED SCREENING METHODS

Otoguro et al (122, 152) devised a screening method for antinematode activity in which the pine wood nematode *Bursaphelenchus lignicolus* was used as the test organism, instead of the conventionally employed free-living nematode, *C. elegans* (128). Nematicidal activity was assayed by counting motile nematodes that could actively pass through a thin filter paper. These authors found jietacins (Figure 2) from a *Streptomyces* strain that was isolated from a soil sample collected at the Jie-tai temple in Beijing. The compound showed an in vitro activity comparable to that of avermectin. This assay system could detect avermectin B_1 (IC_{50} 0.5 μg/ml), as well as gramicidin S (0.1 μg/ml), aureothin (0.5 μg/ml), colistin (1.0 μg/ml), and staurosporine (3.5 μg/ml).

Jietacin is unique in structure in that the molecule possesses an α,β-unsaturated azoxy moiety. Synthetic derivatives of jietacin suggested that the vinyl-azoxy moiety was essential for activity, but the carbonyl moiety could be reduced to a hydroxyl, or to a methylene functionality (152).

Several insecticides were detected by using brine shrimp. Eggs of the brine shrimp *Artemia salina* are available commercially (e.g. at a pet shop). Eggs hatch out within 1–2 days in 3–5% aqueous NaCl solution, or in artificial sea water (which is better and is also commercially available), with or without forced aeration at room temperature (below 28°C). They are then ready for

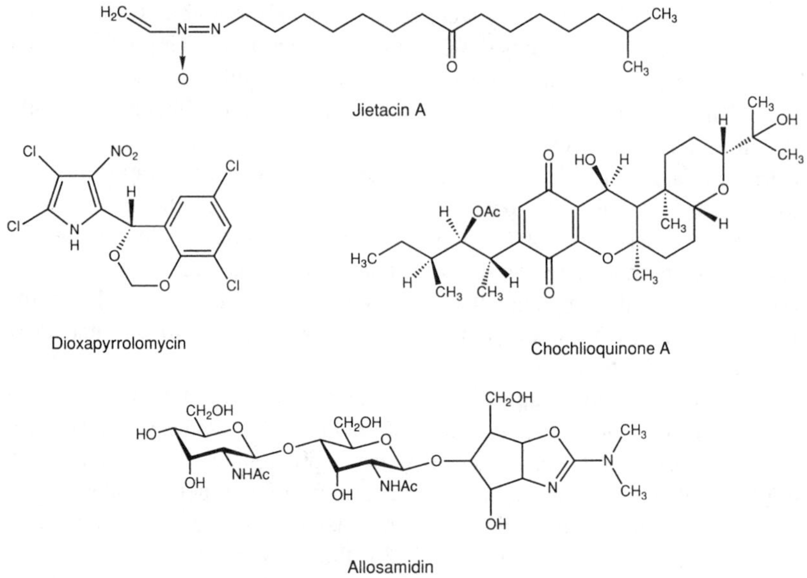

Jietacin A

Dioxapyrrolomycin

Chochlioquinone A

Allosamidin

Figure 2 Insecticidal microbial metabolites discovered recently.

use in an insecticide assay. Blizzard et al (9) took advantage of this simplicity to evaluate hundreds of semisynthetic avermectins. The result was roughly parallel to that from a conventional assay method, which uses two-spotted spider mites (*Tetranychus urticae*) and kidney bean leaves and takes 1–3 days. Also using brine shrimp, Conder et al (20) identified dioxapyrrolomycin as an insecticidal compound produced by a strain of *Streptomyces* sp. (95).

ALLOSAMIDIN AND OTHER INSECTICIDES Many other insect-active compounds have been identified from microorganisms. In many cases, the screening methods were conventional or were not reported in detail. Some of them are in an early stage of evaluation. These are chochlioquinone A (131), paraherquamide and recently described related compounds (8), okaramines (41), aculeximycin (88), W-719 substance (126), allosamidins (97), all insecticides, and the miticides AB-3217-A substance (61) and bagougeramin (139).

Chochlioquinone A (Figure 2), which is produced by *Helminthosporium sativum* and was originally discovered as a metabolite of the plant pathogen *Cochliobolus miyabeanus,* is a competitive inhibitor of [^{3}H]-ivermectin binding to membranes of *C. elegans* (131).

A synthetic chemical, buprofezin, was postulated to be a chitinase inhibitor, but allosamidin is the first such compound obtained from a microorganism. Chitin is a homopolymer of *N*-acetylglucosamine with a β-1,4 linkage. It is involved in the envelope of insects and fungi. Mammalian cells do not contain chitin; hence, chitin synthesis or degradation have attracted much attention as a promising target site for insecticidal and fungicidal agents (18, 58). Sakuda et al (97, 130) exploited allosamidins (Figure 2), which are produced by *Streptomyces* No. 1713. A chitinase preparation from the silkworm (*Bombyx mori*) was used in screening. Allosamidins are oligosaccharides composed of an *N*-acetylyglucosamine dimer (chitobiose) and a new bicyclic aminocyclitol named allosamizolin. Allosamidin and demethylallosamidin inhibited chitinases from *B. mori* equally (IC_{50} of 0.20 and 0.25 μg/ml, respectively). Allosamidin inhibited ecdysis of *B. mori* and *Leucania* sp. when injected at 2–4 μg/insect. No contact inhibition was observed. *Saccharomyces cerevisiae* cells exposed to allosamidin were heavily attached to each other.

Herbicides

Herbicides compose the largest portion (44%) of total pesticide sales. Currently applied herbicides consist mostly of synthetic chemicals. A few of them were designed with microbial and other natural products as a model. NK-049, for example, was synthesized based on the structure of anisomycin, an herbicidal metabolite from *Streptomyces* sp.

Bialaphos is the first microbial product put into practice as an herbicide. Keys to this great success were the potent and broad-spectrum activity, the high fermentation yield achieved during the development of this compound (141), and the basic studies on biochemistry and genetics of bialaphos biosynthesis and resistance (38, 73). It is an inhibitor of glutamine synthetase.

Discoveries of new herbicidal microbial metabolites increased recently, but are slower than discoveries of insecticides and fungicides (4, 45). Many phytotoxins and plant-growth regulators were identified that were produced by plant-pathogenic fungi as chemicals likely to be involved in plant-pathogen interactions. However, these toxins are not included in the above reckoning, because the relevant plants are beneficial ones like rice or ornamentals and thus the reported phytotoxins have less herbicidal value. Phytotoxins produced by weed-parasitic fungi have more potential as herbicides (146). An example is alteichin produced by *Alternaria eichorniae* (129).

NEW HERBICIDAL COMPOUNDS WITH PROMISING ACTIVITY Three new microbial metabolites showed promising activity. Hydantocidin (Figure 3) produced by *S. hygroscopicus* SANK 13584 is a unique sugar-spiro-hydantoin with potent herbicidal activity (82, 93). Chinese cabbage seeds in small test tubes were used for herbicidal assays. A spore suspension of *S. hygroscopicus* was used as inoculant for constant fermentation yields. It inhibited nonselectively

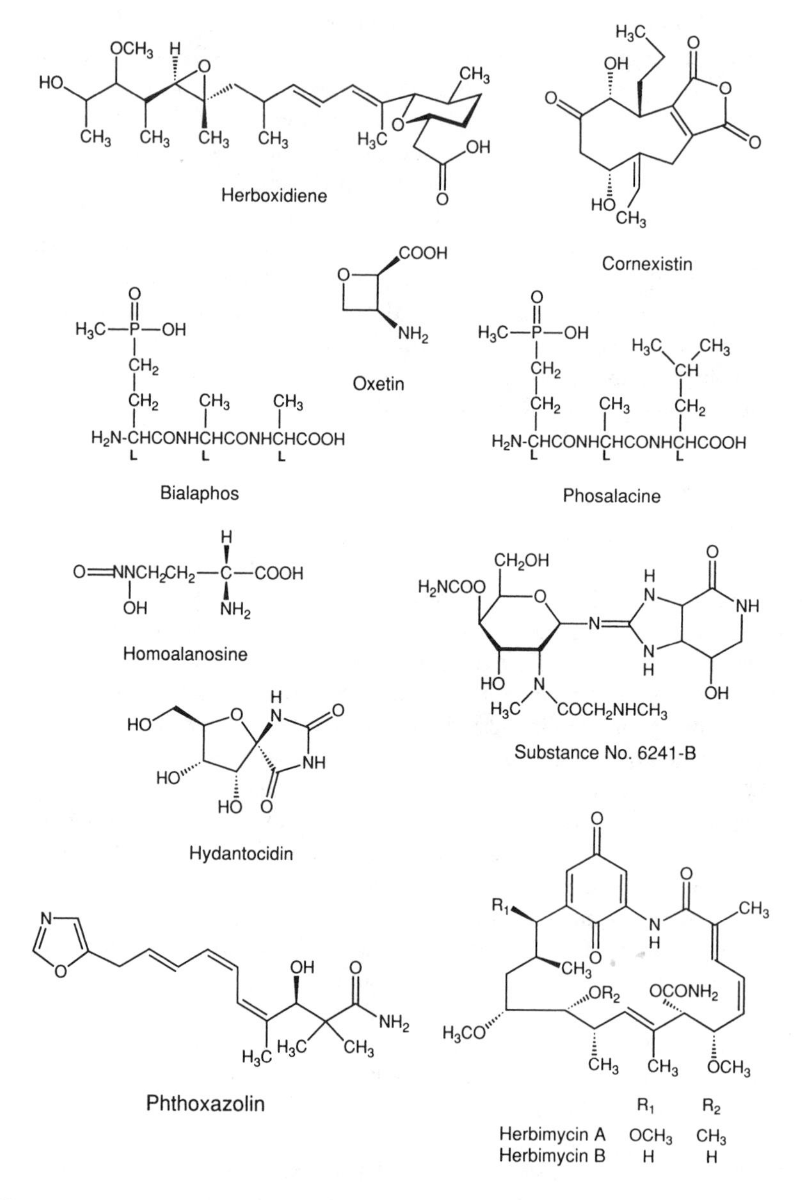

Figure 3 Recently discovered herbicidal microbial metabolites. Bialaphos is included as reference.

the growth of mono- and dicotyledonous annuals and perennials. Under pot-test conditions, the effect of hydantocidin at 500 ppm was comparable to glyphosate (at 500 ppm) and somewhat superior to bialaphos (at 500 ppm). Deoxy derivatives were synthesized but showed no activity (82).

Cornexistin is a lipophilic herbicidal metabolite with a nonadride structure (Figure 3) isolated from *Paecilomyces variotii* SANK 21086 (94). Pot tests demonstrated activity against common annual species of weeds on postemergence treatment at 100 and 500 ppm. Corn seedlings tolerated 500 ppm of cornexistin well. Thus, cornexistin is probably useful for postemergence weed control in corn fields. A weak herbicidal effect of the structurally related rubratoxin B was reported earlier (94).

Herboxidiene was discovered as an herbicidal metabolite of *Streptomyces chromofuscus* (81). Postemergence treatment demonstrated a selective effect at 69 g/ha. Many weeds (rape, wild buckwheat, morning glory) were well controlled, whereas rice, soybean, and wheat exhibited little or no sensitivity. Fermentation yields have been increased by about 20-fold, reaching 160 mg/liter. Screening method and mode of action were not reported for these compounds.

HERBICIDAL COMPOUNDS WITH INTERESTING MODES OF ACTION Herbicidal screening usually requires more effort and a longer time period for the assay than does antifungal screening. A conventional spray-assay in-pot test requires two weeks or so, which implies a smaller number of test samples per year, and perhaps less chance of discovering new compounds. These difficulties may account, at least in part, for the observed slow discovery of new herbicidal compounds from microorganisms.

A gap between in vitro effect and in vivo efficacy also hampers herbicidal screening. In addition, rediscovery confounds screening, especially when a random screening technique is employed. In fact, many known compounds have been observed more than once with herbicidal activity, such as several nucleosides including aristeromycin, coaristeromycin, 5-deoxy-toyocamycin, coformycin, Ara-A, toyocamycin, tubercidin, sangivamycin (4, 55), isoxazole-4-carboxylate, harman and norharman, arabenoic acid, α-methylene-β-alanine (56, 57, 71, 158), homoalanosine (34), and so on.

An alternative approach to finding new herbicides would be biorational mechanism–based in vitro screening, followed by in vivo pot tests, the latter of which is absolutely necessary. Microbiological assay methods or enzyme inhibition assays (149) are convenient when available. This approach enables herbicidal assays to be smaller in apparatus size and larger in number of test samples. Recent trials along this line led to the discovery of new herbicides of interesting modes of action. These are amino acid–synthesis inhibitors, cellulose-biosynthesis inhibitors, and starch-synthesis inhibitors.

Ōmura et al (64, 105, 108, 109) found the glutamine antagonists, phosalacine and oxetin, both with potent herbicidal activity (Figure 3). A bacterium, *Bacillus subtilis*, grown in a synthetic medium (Davis' medium) was employed for the initial detection of glutamine antagonism. Phosalacine is a tripeptide, phosphinothrycyl (PT)-alanylleucine, similar to bialaphos (PT-Ala-Ala). The original culture of bialaphos-producing *S. hygroscopicus* SF-1293 produced, besides bialaphos, PT-Ala, PT-Gly-Ala, and PT-Ala-Val (54). Another strain of *S. hygroscopicus* produced trialaphos (PT-Ala-Ala-Ala) (63). These compounds are probably also glutamine antagonists.

L-Homoalanosine is a synthetic analogue of L-alanosine, an ecdysis inhibitor of microbial origin (80). L-Homoalanosine was also found in a culture of a *Streptomyces* sp. and showed selective herbicidal activity at 400 g/ha. It is an aspartic acid and/or glutamic acid antagonist.

Vulgamycin, a lipophilic metabolite with no structural similarity to amino acids reported earlier, was found to act as an isoleucine antagonist. It showed selective protection of cotton, barley, and maize from weeds on postemergence treatment at 125 and 500 g/ha (6). In *E. coli*, vulgamycin inhibited acetolactate synthase (isozyme I). In a cell suspension of *Catharanthus roseus*, chlorotic symptoms caused by vulgamycin (10 μg/ml) reverted upon supplementation with isoleucine plus valine. L-Methionine-reversible chlorosis was also observed with a *Streptomyces* product, 2-amino-1-hydroxyl-cyclobutane-1-acetic acid (5). Herbicidal effects of aromatic amino acid analogues, 2,5-dihydrophenylalanine (54) and 1,2,4-triazolealanine (53) were reported.

Kishore & Shah (70) pointed out four pathways to amino acid groups as feasible target sites of herbicides: glutamate and glutamine, branched-chain amino acids, aromatic amino acids, and histidine. Apparently other pathways are also important, although the mechanism by which amino acids reverse chlorosis is not known.

A cellulose biosynthesis inhibitor, phthoxazolin (Figure 3), was discovered by our group as a potent herbicide (119). Cellulose resides in all the plant systems, but not in mammalian cells. Cellulose biosynthesis therefore is one of the promising target sites with excellent selective toxicity. Ōmura et al employed the fungus *Phytophthora parasitica* as a test organism. This fungus contains cellulose in the cell walls, but common fungi such as *Candida*, *Pyricularia* spp., etc, do not. The active microbial cultures showed anti-*Phytophthora* activity, but no growth inhibition against common fungi. Phthoxazolin showed potent herbicidal activity in pot tests and inhibited cell-free cellulose synthesis. This is the first microbial metabolite isolated as a cellulose-biosynthesis inhibitor. Recently, a synthetic chemical, isoxaben, was shown to have this activity in *Arabidopsis thaliana* (42).

Photolinked starch synthesis occurs only in plants and limited genera of bacteria (photosynthetic bacteria). This is another target site warranting excellent selective toxicity for herbicides. In fact, half of synthetic herbicides currently available in the marketplace are photosynthetic electron-transfer inhibitors. A new inhibitor of starch synthesis, 6241-B substance (Figure 3), of *Streptomyces* origin was exploited as the first example of this class of microbial herbicides (67). This substance showed at 500 ppm selective herbicidal activity against barnyard millet (*Panicum crus-galli*) with no toxicity to rice. The screening was based on the initial detection of de novo starch-synthesis inhibition in leaf segments of barnyard millet, followed by determination of O_2 evolution from cells of a green alga, *Scenedesmus obliquus*.

Fungicides

Fungal diseases of crop plants are extremely numerous. Because of narrow host-pathogen specificity, different antifungal pesticides are required for individual diseases. Hence, many kinds of fungicides are available (see Table 2). However, the effectiveness of antifungal agents changes (7) because of the emergence of resistance, and antifungal agents that control fungal diseases are still limited. A few serious fungal diseases remain uncontrolled today.

A great many antifungal metabolites have been isolated from microorganisms. Many of them were studied mainly as medicines; some were evaluated as agricultural fungicides. This section discusses those with promising crop-protection action in pot tests.

ANTIFUNGAL COMPOUNDS WITH FEASIBLE ACTIVITY A family of phosphate ester antifungal agents, phoslactomycins, was isolated from *Streptomyces nigrescens* (Figure 4) (35). Structurally related phosphazomycins (148) have different substituents on the terminal cyclohexane ring. Phoslactomycin E exhibited potent in vitro activity (MIC 0.3–3.0 μg/ml) against *Botrytis cinerea, Rhizoctonia solani,* and *Alternaria kikuchiana*. Phoslactomycin E controlled *Botrytis* infection at 10 ppm with no phytotoxicity in pot tests, whereas other derivatives were somewhat phytotoxic to wheat.

A 7-deaza purine nucleoside analogue, dapiramicin (98) (Figure 4), produced by a *Micromonospora* sp. showed weak in vitro antifungal activity against *Rhizoctonia, Pyricularia,* and *Botrytis* spp., which was monitored by observing mycelial elongation of *Rhizoctonia solani* on pieces of water agar under a microscope. However, dapiramicin potently suppressed rice-sheath blight caused by *R. solani*. The efficacy (95% protection at 50 ppm) was comparable to validamycin.

Our group (111, 120) discovered irumamycin (Figure 4), a 20-membered macrolide produced by a *Streptomyces flavus* subsp. *irumaensis*. It showed potent in vivo and in vitro activity against *Pyricularia* and *Botrytis,* and it

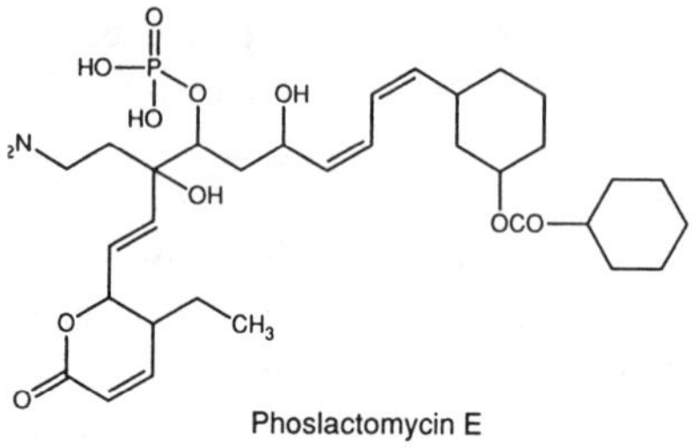

Phoslactomycin E

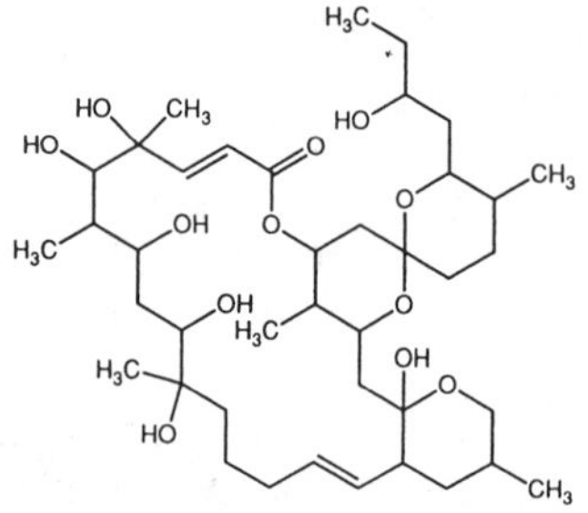

Phthoramycin

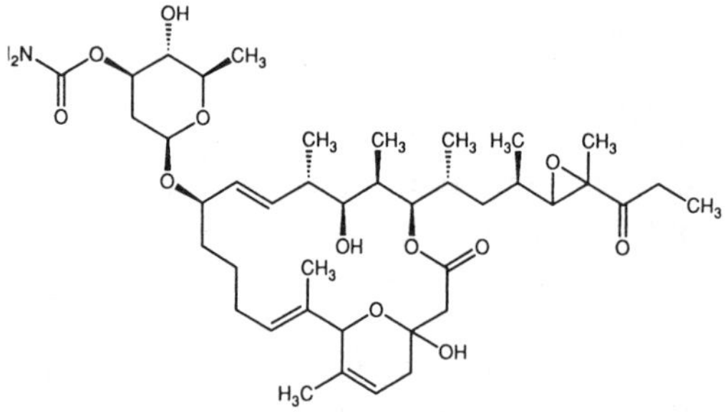

Irumamycin

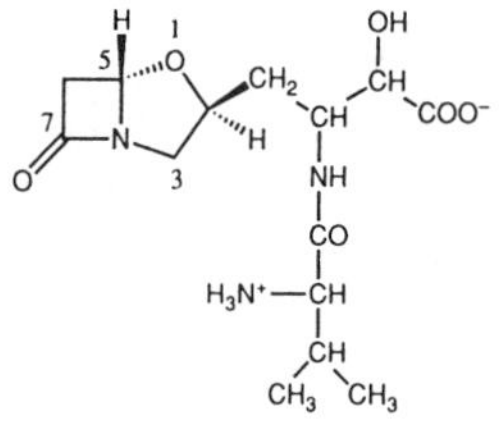

Valclavam

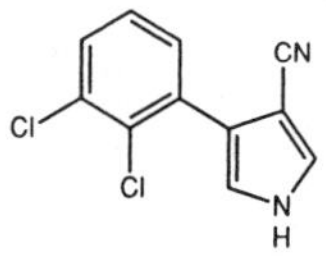

Fenpiconil

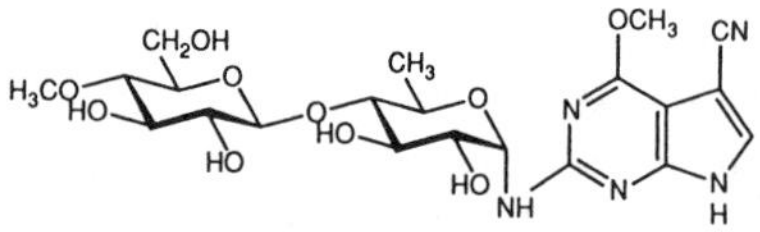

Dapiramicin

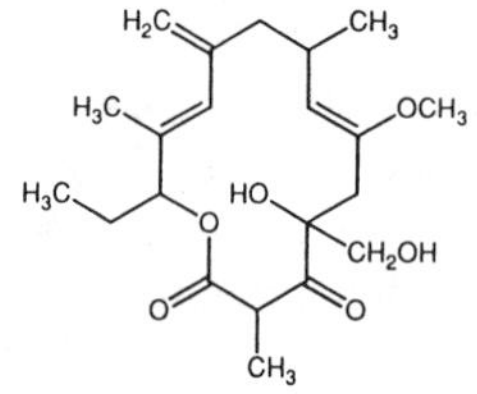

Rustmicin

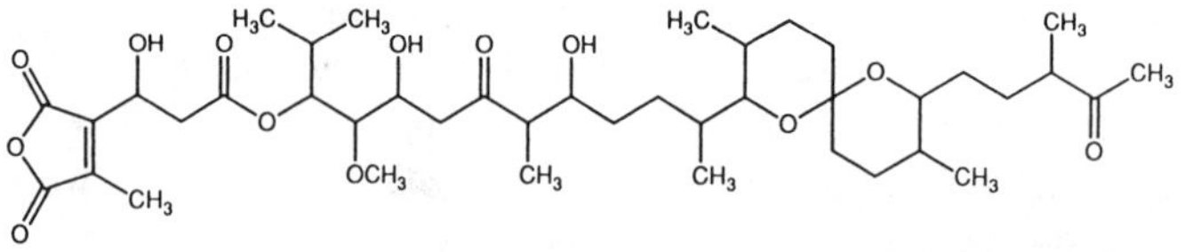

Tautomycin

Figure 4 Recently discovered fungicidal microbial metabolites and their synthetic derivatives.

tolerated sunlight. The application of irumamycin to *Botrytis* control is under study.

FUNGAL CELL-WALL INHIBITORS AND OTHER ANTIFUNGAL COMPOUNDS WITH NOVEL TYPE OF AGROACTIVITY Fungal cell walls provide one of the best target sites for selective toxicity of fungicides. Polyoxins represent this class of fungicides and have been used in agricultural fields (58). The emergence of polyoxin-resistant fungal strains have restricted the effect of polyoxins.

Nikkomycins are structurally related nucleosides with antifungal, insecticidal, and miticidal activities (27). Unfortunately, the development of nikkomycins as an agricultural agent has been discontinued because of lower efficacy than expected. In a search for new nikkomycin components, mutants of *Streptomyces tendae* were found to produce new nikkomycins: K_2, K_x, O_z, O_x, ϕ_z, ϕ_J, W_z, W_x, S_z, S_x, S_{oz}, S_{ox} (10, 132).

Peptide antifungal agents such as globopeptin (145), neopeptin (153), and cystargin (156) inhibit cell-wall glycan biosynthesis and induce swollen morphology of plant-pathogenic fungal cells. Asperfuran (124) was isolated as an inhibitor of chitin synthase from *Coprinus* sp.

Widely spread fungal diseases of crop plants that have not been controlled well include, among many others, potato late blight (caused by *Phytophthora infestans*), grape downy mildew (*Plasmopara viticola*), barley stem rust (*Puccinia graminis*), and apple scab (*Venturia inaequalis*). Bacterial infection by *Erwinia* spp. has also been poorly controlled. Microbial metabolites active against these pathogens are of agricultural interest and importance (19, 133). Omura (90, 118) found a 22-membered macrolide named phthoramycin, which was active against *Phytophthora* spp. A related glycoside, cytovaricin (69), and a stereoisomer, kaimonolide (44), were also reported. The latter compound was herbicidal. Phthoramycin is now under evaluation as a pesticide.

Many antifungal β-lactams, valclavam, clavamycin, Ro 22-5417, and hydroxyethylclavam (89, 125) have been discovered. Valclavam (Figure 4) showed potent in vitro activity against an oomycete, *Pythium,* and other phytopathogenic fungi. Antibacterial activity varied. Unfortunately, an immediate application of valclavam as fungicide could not be achieved because of its instability; either of two stereoisomers can occupy the C-5 position. Valclavam exhibited 5S configuration, whereas 5R configuration was evident in clavulanic acid, other antibacterial β-lactamase inhibitors, and penicillins. Hydroxyethylclavam inhibited RNA synthesis in *S. cerevisiae,* but acted as a methionine antagonist in *E. coli* and *B. subtilis*.

New 14-membered macrolides, rustmicin and neorustmicin, were isolated from *Micromonospora chalcea* (Figure 4) (96). Galbonolides isolated independently from *Streptomyces galbus* were identical to rustmicins (26).

Rustmicin potently inhibited the proliferation of the wheat stem rust fungus *Puccinia graminis* in a greenhouse. This is the first microbial metabolite that shows anti-*Puccinia* activity in vitro and in vivo. An in vitro assay method for this obligate parasite was developed and used in the screening.

MICROBIAL AND CHEMICAL MODIFICATIONS OF COMMERCIAL MICROBIAL AGROCHEMICALS

Investigators have used chemical and microbial methods to modify agroactive microbial metabolites to obtain derivatives with potentiated activity and increased photostability. Many semisynthetic avermectin derivatives have been synthesized and evaluated (12, 21, 22, 24, 87). Among them, 4″-*epi*-methylamino-4″-deoxy avermectin B_1 (EMA) (Figure 1) exhibited a dramatically altered activity spectrum (24). EMA was one tenth as active against spider mites as avermectin B_1, but 1500 times more active against armyworms.

The biosynthetic pathway to avermectins (12, 49–51), its regulation (99), and selective production (106) have been elucidated at the biochemical and genetic levels. Studies on aglycone biosynthesis showed that the C-25 substituents originated from isobutyrate or sec-valerate provided by valine or isoleucine metabolism, respectively. A mutant, *S. avermitilis* ATCC 53568, was isolated that was defective in the ability to synthesize precursors for aglycone synthesis (23, 36). This mutant restored avermectin production when the culture medium was supplemented with the precursor fatty acids. Dutton et al (23) applied a mutational biosynthesis technique using this mutant to prepare avermectin analogues. Feeding about 800 fatty acids to cultures of ATCC 53568 yielded 17 new avermectins possessing an altered substitutent at C-25 that included 4-tetrahydropyranyl, cyclohexyl, 2(or 3)-thienyl, 1-methylbut-1-enyl. All showed 100% killing of *C. elegans* at 0.1 μg/ml (23). By similar techniques, analogues with a sec-valeryl, or a sec-hexyl substituent at C-25, were obtained (16).

Chemical modification of milbemycin D led to the discovery of 5-keto-5-oxime derivatives with increased activity against a microfilaria, *Dirofilaria immitis* in dogs (151).

Microbial conversion of milbemycins allows hydroxylation at C-13. This conversion can be accomplished using various microorganisms, including *Streptomyces violascens* ATCC 31560, *Streptomyces cavourensis* SANK 67386, *Amycolata autotrophica* ATCC 35204, *Cunninghamella echinulata* ATCC 9244, and *Syncephalastrum racemosus* SANK 62672 (91, 92, 147). Thus milbemycins A_3, A_4, D, and nemadectin α were monohydroxylated at C-13, and in some cases dihydroxylated at C-13 and C-24, 28, 29, or 30. *Amycolatopsis mediterranei* IFO 13415 hydroxylated 22,23-dihydro-avermectin B_{1a} at C-30.

The photostability of avermectins is very poor, as it is for other macrolides with a diene-lactone ring system. Avermectin loses activity within one day under sunlight. Conversion of avermectin to an 8,9-epoxide increased the stability significantly without decreasing the miticidal activity (24). The same modification of 13-deoxyaglycone, a milbemycin, resulted in a loss of activity.

Pyrrolnitrin is an antifungal metabolite from a *Pseudomonas* species. It is also photolabile. Nyfeler (28) synthesized a stable derivative, fenpiconil (Figure 4), by replacing Cl and NO_2 with CN and Cl, respectively. Fenpiconil is now under development as a seed treatment.

By feeding 5-fluorocytosine to a culture of the blasticidin S–producing *Streptomyces griseochromogenes,* Kawashima et al (65) obtained 5-fluoroblasticidin S. Antifungal activity did not change. The fluorinated product is expected to show long-lasting activity in field applications because of resistance to enzymatic 4-deamination, a mechanism of blasticidin S inactivation caused by blasticidin S–resistant bacteria (60). Fluorovulgamycins were prepared in a similar manner (66).

UNCOVERED PHARMACOLOGICAL ACTIVITY OF AGROACTIVE MICROBIAL METABOLITES

Several compounds identified as insecticides, herbicides, or fungicides were later found to exhibit entirely different, but pharmacologically interesting, activities (102).

Herbimycins, an ansamycin group of compounds produced by a *Streptomyces* species, were discovered by our group (33, 34, 110, 134) as herbicidal, antitumor, and antiviral compounds. Recently herbimycin A (Figure 3) was demonstrated to be an inhibitor of oncogene function, inhibiting protein-tyrosine kinase. Herbimycin A was effective in restoring normal cell morphology from abnormal NRK cells that were previously transformed by tyrosine kinase oncogenes such as *src, ras, myc,* etc (154). Antitumor activity of herbimycin A and its semisynthetic derivatives are under evaluation (46, 155).

Tautomycin (Figure 4) was initially reported by Isono's group as an antifungal compound of agrochemical interest (17). Later studies revealed that it induced bleb formation in K562 cells. The bleb formation resulted from inhibition of protein phosphatase by tautomycin and other phosphatase inhibitors (75).

Sangivamycin is an herbicidal nucleoside with a 7-deaza adenosine structure. Osada et al (121) showed that sangivamycin inhibited protein kinase C in K562 cells. The inhibition was less potent than that of staurosporine (107), a microbial alkaloid with insecticidal activity (122) and the most potent inhibitor of protein kinase C (143). Tubercidin and toyocamycin are related 7-deaza adenosines, but exhibited weaker inhibition than sangivamycin.

Dioxapyrrolomycin (Figure 2) and related pyrrolomycins are substance P antagonists (78). Substance P and neurokinin A are neuropeptides of the tachykinin family. Substance P was proposed to be a neurotransmitter of pain or an internal mediator in inflamatory and immune responses.

How the above activities are related to the initially described agroactivity of the compounds is not known. Nevertheless, these examples demonstrate the diversity in activity and structure of microbial secondary metabolites, and at the same time, the importance of screening technology.

PERSPECTIVES

The avermectin story encourages optimism about further discovery of excellent pesticides from microorganisms. This review has described many microbial metabolites exhibiting a variety of activities and structures. The described compounds probably include candidates for the next success story, as supported by the considerations below.

The number of microorganisms, and the secondary metabolites they produce, is almost uncountable. About 180,000 species have been identified, but an estimated 5 to 10 times more species probably exist. Applied microbiology is currently dealing with only 10% or less of the total microbial population. To date, 16,500 individual microbial metabolites have been isolated and characterized. One percent of them, 130–150 compounds, have found practical uses. Ten or so are used in agriculture, several tens in veterinary fields, and all the rest in humans. In medicines, approximately 20% of sales come from microbial compounds, to which semisynthetic penicillins and cephalosporins contribute much. One wonders why microbial metabolites do not exceed this rate in the pesticide market. In any case, approximately one practicable pesticide is found at a rate of 10^{-2}, an efficiency two orders of magnitude higher than the corresponding rate, 10^{-4}, for synthetic agrochemicals (45). One should note that the 10 already commercialized agricultural compounds have been selected out, not from 16,500, but from a smaller number of agroactive microbial metabolites discovered in pesticidal screening. Obviously, a great many agroactive compounds produced by a huge number of microorganisms await discovery. An immediate question is how to identify them.

In the development of a biological screening program, one considers the target pest to be controlled first, then the mode of action of newly discovered compounds. Based on such considerations, combined with an estimate of need, impact, market size, etc, appropriate screens are constructed and optimized. Tables 1–3 suggest examples of the pests to be tackled by future screening programs. In choosing the expected sites of action exhibited by the compounds discovered, selective toxicity is the major concern. In this sense, targets of action of possible new inhibitors include chitin synthesis and metabolism; cellulose biosynthesis; amino acid biosynthesis, in paticular that of essential amino acids; sterol biosynthesis and metabolism; photosynthetic reactions; and so on. These metabolic pathways represent fundamental structural and physiological differences between harmful insects, mites, fungi, weeds, and humans. The advantage of cell envelopes as targets of pesticide action is well documented (18, 58). In addition, insects and mites are not capable of de novo synthesis of cholesterol. Insects therefore require phytosterols, from which they derive ecdysteroids for growth and reproduction (137). These insect-specific steps are rational targets for selective toxicity, but no microbial agrochemicals of this class have been reported. Alternate uses of two pesticides with different modes of action is one way to reduce the emergence of resistant pests. Therefore, microbial metabolites with varying actions are necessary. Compounds with multiple action sites may be of interest because a low frequency of resistant pests should result.

Biological pest control is based on the chemicals involved in plant-pest and natural enemy–pest interactions (1, 31, 146). Despite its potentially great advantages, biochemical pest control plays only a compensatory role for chemical means because it is too strict in host or pest specificity and the efficacy is insufficient to protect crops from simultaneous attacks by multiple types of harmful pests. This technology will find more utility in future plant factories, where gene-engineered plants are grown in clean houses.

Pesticide chemistry has progressed by unexpected discovery, concentrated research, basic science, and interdisciplinary cooperation. Further progress will come from these sources. Of particular importance is the cooperation of microbiologists, plant physiologists, biochemists, entomologists, and chemists.

Finally, both industry and academia must exert more effort to promote positive public perception and public acceptance of pesticide chemicals (28). One way can be a campaign about roles of pesticides in crop production, correct uses, production process, and regulation, and about the global necessity and ecological safety of modern pesticides of both microbial and synthetic origins. These efforts should lead to an increase in the variety of chemicals available for field applications, as well as increase general interest in microbial pesticides. The shortage of research scientists and research funds in pesticide sciences would be solved in this way. In total, these efforts should accelerate progress in pest control.

ACKNOWLEDGMENTS

We are indebted to our colleagues in the Kitasato Institute. In particular, Dr. Hiroshi Yoshida, Ageo Research Laboratory, Nippon Kayaku Co., and Dr. Kazuro Shiomi are gratefully acknoweldged for their help and discussion in the preparation of this manuscript.

Literature Cited

1. Adams, P. B. 1990. The potential of mycoparasites for biological control of plant diseases. *Annu. Rev. Phytopathol.* 28:59–72
2. Anke, T., Steglich, W. 1988. Neue Wirkstoffe aus Basidiomyceten. *Forum Mikrobiol.* 11:21–25
3. Anke, T., Werle, A. 1990. Antibiotics from basidiomycetes XXXIII. Oudemansin X, a new antifungal *E*-β-methoxyacrylate from *Oudemansiella radicata* (Relhan ex FR.) Sing. *J. Antibiot.* 43:1010–11
4. Ayer, S. W., Isaac, B. G., Krupa, D. M., Crosby, K. E., Letendre, L. J., et al. 1989. Herbicidal compounds from microorganisms. *Pestic. Sci.* 27: 221–23
5. Ayer, S. W., Isaac, B. G., Luchsinger, K., Makkar, N., Tran, M., et al. 1991. *cis*-2-Amino-1-hydroxycyclobutane-1-acetic acid, a herbicidal antimetabolite produced by *Streptomyces rochei* A13018. *J. Antibiot.* 44:1460–62
6. Babczinski, P., Dorgerloh, M., Lobberding, A., Santel, H.-J., Schmidt, R. R., et al. 1991. Herbicidal activity and mode of action of vulgamycin. *Pestic. Sci.* 33:439–46
7. Baldwin, B. C., Rathmell, W. G. 1988. Evolution of concepts for chemical control of plant disease. *Annu. Rev. Phytopathol.* 26:265–83
8. Blanchflower, S. E., Banks, R. M., Everett, J. R., Manger, B. R., Reading, C. 1991. New paraherquamide antibiotics with anthelmintic activity. *J. Antibiot.* 44:492–97
9. Blizzard, T. A., Ruby, C. L., Mrozik, H., Preiser, F. A., Fisher, M. H. 1989. Brine shrimp (*Artemia salina*) as a convenient bioassay for avermectin analogs. *J. Antibiot.* 42:1304–7
10. Bormann, C., Mattern, S., Schrempf, H., Fiedler, H.-P., Zahner, H. 1989, Isolation and *Streptomyces tendae* mutants with an altered nikkomycin spectrum. *J. Antibiot.* 42:913–18
11. Burg, R. W., Miller, B. M., Baker, E. E., Birnbaum, J., Ōmura, S., et al. 1979. Avermectins, new family of potent anthelmintic agents: producing organism and fermentation. *Antimicrob. Agents Chemother.* 15:361–67
12. Campbell, W. C., ed. 1989. *Ivermectin and Abamectin.* New York/Tokyo: Springer-Verlag. 393 pp.
13. Campbell, W. C. 1991. Ivermectin as an antiparasitic agent for use in humans. *Annu. Rev. Microbiol.* 45:445–74
14. Cane, D. E., Liang, T.-C., Kaplan, L., Nallin, M. K., Schulman, M. D., et al. 1983. Biosynthetic origin of the carbon skeleton and oxygen atoms of the avermectins. *J. Am. Chem. Soc.* 105:4110–12
15. Carter, G. T., Nietsche, J. A., Hertz, M. R., Williams, D. R., Siegel, M. M., et al. 1988. LL-F28249 antibiotic complex: a new family of antiparasitic macrocyclic lactones. Isolation, characterization and structures of LL-F28249 α, β, γ, λ. *J. Antibiot.* 41: 519–29
16. Chen, T. S., Inamine, E. S., Hensens, O. D., Zink, D., Ostlind, D. A. 1989. Directed biosynthesis of avermectins. *Arch. Biochem. Biophys.* 269:544–47
17. Cheng, X.-C., Kihara, T., Kusakabe, H., Magae, J., Kobayashi, Y., et al. 1987. A new antibiotic, tautomycin. *J. Antibiot.* 40:907–9
18. Cohen, E. 1987. Chitin biochemistry: synthesis and inhibition. *Annu. Rev. Entomol.* 32:71–93
19. Cohen, Y., Coffey, M. D. 1986. Systemic fungicides and the control of oomycetes. *Annu. Rev. Phytopathol.* 24:311–38
20. Conder, G. A., Zielinski, R. J., Johnson, S. S., Kuo, M.-S. T., Cox, D. L., et al. 1992. Anthelmintic activity of dioxapyrrolomycin. *J. Antibiot.* 45: 977–83
21. Danishefsky, S. J., Selnick, H. G.,

Armistead, D. M., Wincott, F. E. 1987. The total synthesis of avermectin A_{1a}. New protocols for the synthesis of novel 2-deoxypyranose systems and their axial glycosides. *J. Am. Chem. Soc.* 109:8119–20
22. Davies, H. G., Green, R. H. 1986. Avermectins and milbemycins. *Nat. Prod. Rep.* 3:87–121
23. Dutton, C. J., Gibson, S. P., Goudie, A. C., Holdom, K. S., Pacey, M. S., et al. 1991. Novel avermectins produced by mutational biosynthesis. *J. Antibiot.* 44:357–65
24. Dybas, R. A., Hilton, N. J., Babu, J. R., Preiser, F. A., Dolce, G. J. 1989. Novel second-generation avermectin insecticides and miticides for crop protection. In *Novel Microbial Products for Medicine and Agriculture,* ed. A. L. Demain, G. A. Somkuti, J. C. Hunter-Cevera, H. W. Rossmoore, pp. 203–12. Amsterdam: Elsevier. 266 pp.
25. Eue, L. 1986. World challenges in weed science. *Weed Sci.* 34:155–60
26. Fauth, U., Zahner, H., Muhlenfeld, A., Achenbach, H. 1986. Galbonolides A and B—Two non-glycosidic antifungal macrolides. *J. Antibiot.* 39:1760–64
27. Fiedler, H.-P. 1989. Nikkomycins and polyoxins. In *Natural Products Isolation. Separation Methods for Antimicrobials, Antivirals and Enzyme Inhibitors,* ed. G. H. Wagman, R. Cooper, pp. 153–89. Amsterdam: Elsevier
28. Finney, J. R. 1991. Where do we stand—where do we go? See Ref. 32, pp. 555–76
29. Fischer, M. H., Mrozik, H. 1984. The avermectin family of macrolide-like antibiotics. In *Macrolide Antibiotics. Chemistry, Biology, and Practice,* ed. S. Ōmura, pp. 553–606. Orlando/Tokyo: Academic Press. 653 pp.
30. Frankmolle, W. P., Larsen, L. K., Caplan, F. R., Patterson, G. M. L., Knubel, G., et al. 1992. Antifungal cyclic peptides from the terrestrial blue-green alga *Anabaena laxa.* I. Isolation and biological properties. *J. Antibiot.* 45:1451–57
31. Fravel, D. R. 1988. Role of antibiotics in the biocontrol of plant diseases. *Annu. Rev. Phytopathol.* 26:75–92
32. Frehse, H., ed. 1991. *Pesticide Chemistry. Advances in International Research, Development, and Legislation.* Weinheim/Cambridge: VCH Verlagsgesellschaft. 666 pp.
33. Furusaki, A., Matsumoto, T., Nakagawa, A., Ōmura, S. 1980. Herbimycin A: an ansamycin antibiotic; X-ray crystal structure. *J. Antibiot.* 33:781–82
34. Fushimi, S., Nishikawa, S., Mito, N., Ikemoto, M., Sasaki, M., et al. 1989. Studies on a new herbicidal antibiotic, homoalanosine. *J. Antibiot.* 42:1370–78
35. Fushimi, S., Nishikawa, S., Shimazu, A., Seto, H. 1989. Studies on new phosphate ester antifungal antibiotics phoslactomycins. I. Taxonomy, fermentation, purification and biological activities. *J. Antibiot.* 42:1019–25
36. Hafner, E. W., Holley, B. W., Holdom, K. S., Lee, S. E., Wax, R. G., et al. 1991. Branched-chain fatty acid requirement for avermectin production by a mutant of *Streptomyces avermitilis* lacking branched-chain 2-oxo acid dehydrogenase activity. *J. Antibiot.* 44:349–56
37. Hanka, L. J., Schaadt, R. D. 1988. Methods for isolation of *Streptoverticillia* from soils. *J. Antibiot.* 41: 576–78
38. Hara, O., Anzai, H., Imai, S., Kumada, Y., Murakami, T., et al. 1988. The bialaphos biosynthetic genes of *Streptomyces hygroscopicus:* cloning and analysis of the genes involved in the alanylation step. *J. Antibiot.* 41: 538–47
39. Haxell, M. A., Bishop, B. F., Bryce, P., Gration, K. A. F., Kara, H., et al. 1992. C-13β-Acyloxymilbemycins, a new family of macrolides. Discovery, structural determination and biological properties. *J. Antibiot.* 45:659–70
40. Hayakawa, M., Kajiura, T., Nonomura, H. 1991. New methods for the highly selective isolation of *Streptosporangium* and *Dactylosporangium* from soil. *J. Ferment. Bioeng.* 72:327–33.
41. Hayashi, H., Takiuchi, K., Murao, S., Arai, M. 1989. Structure and insecticidal activity of new indole alkaloids, okaramines A and B, from *Penicillium simplicissimum* AK-40. *Agric. Biol. Chem.* 53:461–69
42. Heim, D. R., Roberts, J. L., Pike, P. D., Larrinua, I. M. 1990. A second locus, *Ixr* B1 in *Arabidopsis thaliana,* that confers resistance to the herbicide isoxaben. *Plant Physiol.* 92:858–61
43. Heim, J., Anke, T., Mocek, U., Steffan, B., Steglich, W. 1988. Antibiotics from basidiomycetes. XXIX: Pilatin, a new antibiotically active marasmane derivative from cultures of

Flagelloscypha pilatii Agerer. *J. Antibiot.* 34:1752–57
44. Hirota, A., Okada, H., Kanza, T., Nakayama, M., Hirota, H., et al. 1989. Structure of kaimonolide A, a novel macrolide plant growth inhibitor from a *Streptomyces* strain. *Agric. Biol. Chem.* 53:2831–33
45. Hoagland R. E. 1990. Microbes and microbial products as herbicides. An overview. In *ACS Symposium Series 439, Microbes and Microbial Products as Herbicides,* ed. R. E. Hoagland, pp. 2–52. Washington, DC: Am. Chem. Soc.
46. Honma, Y., Kasukabe, T., Hozumi, M., Shibata, K., Ōmura, S. 1992. Effects of herbimycin A derivatives on growth and differentiation of K562 human leukemic cells. *Anticancer Res.* 12:189–92
47. Hood, J. D., Banks, R. M., Brewer, M. D., Fish, J. P., Manger, B. R., et al. 1989. A novel series of milbemycin antibiotics from *Streptomyces* strain E225. I. Discovery, fermentation and anthelmintic activity. *J. Antibiot.* 42:1593–98
48. Ichimoto, I., Machiya, K., Kirihata, M., Ueda, H. 1990. Stereoselective synthesis of marine antibiotic (–)-malyngolide and its stereoisomer. *Agric. Biol. Chem.* 54:657–62
49. Ikeda, H., Kotaki, H., Ōmura, S. 1987. Genetic studies of avermectin biosynthesis in *Streptomyces avermitilis. J. Bacteriol.* 169:5615–21
50. Ikeda, H., Kotaki, H., Tanaka, H., Ōmura, S. 1988. Involvement of glucose catabolism in avermectin production by *Streptomyces avermitilis. Antimicrob. Agents Chemother.* 32: 282–84
51. Ikeda, H., Ōmura, S. 1991. Strategic strain improvement of antibiotic producer. *Actinomycetologica* 5:86–99
52. Imamura, N., Kuga, H., Otoguro, K., Tanaka, H., Ōmura, S. 1989. Structures of jietacins: unique α,β-unsaturated azoxy antibiotics. *J. Antibiot.* 42:156–58
53. Imamura, N., Murata, M., Yao, T., Oiwa, R., Ōmura, S., et al. 1985. Occurrence of 1,2,4-triazole ring in acetinomycetes. *J. Antibiot.* 38:1110–11
54. Inouye, S., Sezaki, M. 1990. Antagonistic amino acids and carbohydrates from microbial sources. *Sci. Rep. Meiji Seika Kaisha.* 29:43–122 (in Japanese)
55. Isaac, B. G., Ayer, S. W., Letendre, L. J., Stonard, R. J. 1991. Herbicidal nucleosides from microbial sources. *J. Antibiot.* 44:729–32
56. Isaac, B. G., Ayer, S. W., Stonard, R. J. 1991. Arabenoic acid, a natural product herbicide of fungal origin. *J. Antibiot.* 44:793–94
57. Isaac, B. G., Ayer, S. W., Stonard, R. J. 1991. The isolation of α-methylene-β-alanine, a herbicidal microbial metabolite. *J. Antibiot.* 44: 795–96
58. Isono, K. 1988. Nucleoside antibiotics: structure, biological activity, and biosynthesis. *J. Antibiot.* 41:1711–39
59. Isono, K. 1990. Antibiotics as nonpollution agricultural pesticides. In *Comments in Agricultural and Food Chemistry,* 2:123–42. Science Publishers S. A.
60. Kamakura, T., Kobayashi, K., Tanaka, T., Yamaguchi, I., Endo, T. 1987. Cloning and expression of a new structural gene for blasticidin S deaminase, a nucleoside aminohydrolase. *Agric. Biol. Chem.* 51:3165–68
61. Kanbe, K., Takahashi, A., Tamamura, T., Sato, K., Naganawa, H., et al. 1992. Isolation and structures of two novel anti-mite substances, AB3217-B and C. *J. Antibiot.* 45:568–71
62. Kass, I. S., Wang, C. C., Walrond, J. P., Stretton, A. O. W. 1980. Avermectin B_{1a}, a paralyzing anthelmintic that affects interneurons and inhibitory motorneurons in *Ascaris suum. Proc. Natl. Acad. Sci. USA* 77:6211–15
63. Kato, H., Nagayama, K., Abe, H., Kobayashi, R., Ishihara, E. 1991. Isolation, structure and biological activity of trialaphos. *Agric. Biol. Chem.* 55: 1133–34
64. Kawahata, Y., Takatsuto, S., Ikekawa, N., Murata, M., Ōmura, S. 1986. Synthesis of a new amino acid-antibiotic, oxetin and its three stereoisomers. *Chem. Pharm. Bull.* 34:3102–10
65. Kawashima, A., Seto, H., Ishiyama, T., Kato, M., Uchida, K., et al. 1987. Fluorinated blasticidin S. *Agric. Biol. Chem.* 51:1183–84
66. Kawashima, A., Seto, H., Kato, M., Uchida, K., Otake, N. 1985. Preparation of flurinated antibiotics followed by ^{19}F NMR spectroscopy. I. Fluorinated vulgamycins. *J. Antibiot.* 38: 1499–1505
67. Kida, T., Ishikawa, T., Shibai, H. 1985. Isolation of two streptothricin-like antibiotics, Nos. 6241-A and B, as inhibitors of de novo starch synthesis

and their herbicidal activity. *Agric. Biol. Chem.* 49:1839–44

68. Kida, T., Shibai, H., Seto, H. 1986. Structure of new antibiotics, pereniporins A and B, from a basidiomycete. *J. Antibiot.* 39:613–15
69. Kihara, T., Kusakabe, H., Nakamura, G., Sakurai, T., Isono, K. 1981. Cytovaricin, a novel antibiotic. *J. Antibiot.* 34:1073–74
70. Kishore, G. M., Shah, D. M. 1988. Amino acid biosynthesis inhibitors as herbicides. *Annu. Rev. Biochem.* 57: 627–63
71. Kobinata, K., Sekido, S., Uramoto, M., Ubukata, M., Osada, H., et al. 1991. Isoxazole-4-carboxylic acid as a metabolite of *Streptomyces* sp. and its herbicidal activity. *Agric. Biol. Chem.* 55:1415–16
72. Kuhnt, D., Anke, T., Besl, H. 1990. Antibiotics from basidiomycetes. XXXVII. New inhibitors of cholesterol biosynthesis from cultures of *Xerula melanotricha* Dorfelt. *J. Antibiot.* 43: 1413–20
73. Kumada, Y., Anzai, H., Takano, E., Murakami, T., Hara, O., et al. 1988. The bialaphos resistance gene (*bar*) plays a role in both self-defense and bialaphos biosyntehsis in *Streptomyces hygroscopicus*. *J. Antibiot.* 41:1838–45
74. Kunze, B., Jansen, R., Pridzun, L., Jurkiewicz, E., Hunsmann, G., et al. 1992. Phenoxan, a new oxazole-pyrone from *Myxobacteria:* production, antimicrobial activity and its inhibition of the electron transport in complex I of the respiratory chain. *J. Antibiot.* 45: 1549–52
75. Kurisaki, T., Magae, J., Isono, K., Nagai, K., Yamasaki, M. 1992. Effects of tautomycin, a protein phosphatase inhibitor, on recycling of mammalian cell surface molecules. *J. Antibiot.* 45:252–57
76. Lasota, J. A., Dybas, R. A. 1991. Avermectins, a novel class of compounds: implications for use in arthropod pest control. *Annu. Rev. Entomol.* 36:91–117
77. Lauer, U., Anke, T., Sheldrick, W. S., Scherer, A., Steglich, W. 1989. Antibiotics from basidiomycetes. XXXI. Aleurodiscal: an antifungal sesterterpenoid from *Aleurodiscus mirabilis* (Berk. & Court.) Hohn. *J. Antibiot.* 42:875–82

77a. McDougall, J., Mathisen, F., Phillips, M. 1990. *Agrochemicals and Animal Health Service. Summary Report 1990.* Edinburgh: County NatWest Securities/Wood Mackenzie. 164 pp.

78. Masuda, K., Suzuki, K., Ishida-Okawara, A., Mizuno, S., Hotta, K. 1991. Pyrrolomycin group antibiotics inhibit substance P-induced release of myeloperoxidase from human polymorphonuclear leukocytes. *J. Antibiot.* 44: 533–40
79. Masuma, R., Tanaka, Y., Tanaka, H., Ōmura, S. 1986. Production of nanaomycin and other antibiotics by phosphate-depressed fermentation using phosphate-trapping agents. *J. Antibiot.* 39:1557–64
80. Matsumoto, S., Sakuda, S., Isogal, A., Suzuki, A. 1984. Search for microbial insect growth regulators; L-alanosine as an ecdysis inhibitor. *Agric. Biol. Chem.* 48:827–28
81. Miller-Wideman, M., Makkar, N., Tran, M., Isaac, B., Biest, N., et al. 1992. Herboxidiene, a new herbicidal substance from *Streptomyces chromofuscus* A7847. Taxonomy, fermentation, isolation, physico-chemical and biological properties. *J. Antibiot.* 45: 914–21
82. Mio, S., Sano, H., Shindou, M., Honma, T., Sugai, S. 1991. Synthesis and herbicidal activity of deoxy derivatives of (+)-hydantocidin. *Agric. Biol. Chem.* 55:1105–9
83. Misato, T. 1983. Recent status and future aspects of agricultural antibiotics. In *Pesticide Chemistry: Human Welfare and the Environment,* ed. J. Miyamoto, P. C. Kaerney, pp. 241–46. Oxford/ Frankfurt: Pergamon
84. Mishima, H., Ide, J., Muramatsu, S., Ono, M. 1983. Milbemycins, a new family of macrolide antibiotics. Structure determination of milbemycins D, E, F, G, H, J and K. *J. Antibiot.* 36:980–90
85. Misra, A. K. 1986. Antibiotics as crop protectants. In *ACS Symposium Series 320, Agricultural Uses of Antibiotics,* ed. W. A. Moats, pp. 49–60. Washington DC: Am. Chem. Soc.
86. Moore, R. E., Patterson, G. M. L., Mynderse, J. S., Barchi, J. Jr., Norton, T. R., et al. 1986. Toxins from cyanophytes belonging to the Scytonemataceae. *Pure Appl. Chem.* 58:263–71
87. Mrozik, H. 1985. Chemistry and biological activities of avermectin derivatives in biotechnology and its application to agriculture. In *British Crop Protection Council (BCPC) Monograph,* ed. L. Copping, 32:133–43. London: BCPC

88. Murata, H., Harada, K., Suzuki, M., Ikemoto, T., Shibuya, T. 1989. Structural elucidation of aculeximycin. II. Structures of carbohydrate moieties. *J. Antibiot.* 42:701–10
89. Naegeli, H. U., Loosli, H.-R., Nussbaumer, A. 1986. Clavamycins, new clavam antibiotics from two variants of *Streptomyces hygroscopicus.* II. Isolation and structures of clavamycins A, B and C from *Streptomyces hygroscopicus* NRRL 15846, and of clavamycins D, E and F from *Streptomyces hygroscopicus* NRRL 15879. *J. Antibiot.* 39:516–23
90. Nakagawa, A., Miura, S., Imai, H., Imamura, N., Ōmura, S. 1989. Structure and biosynthesis of a new antifungal antibiotic, phthoramycin. *J. Antibiot.* 42:1324–27
91. Nakagawa, K., Miyakoshi, S., Torikata, A., Sato, K., Tsukamoto, Y. 1991. Microbial conversion of milbemycins: hydroxylation of milbemycin A4 and related compounds by *Cunninghamella echinulata* ATCC 9244. *J. Antibiot.* 44:232–40
92. Nakagawa, K., Sato, K., Tsukamoto, Y., Torikata, A. 1992. Microbial conversion of milbemycins: 29-Hydroxylation of milbemycins by genus *Syncephalastrum. J. Antibiot.* 45:802–5
93. Nakajima, M., Itoi, K., Takamatsu, Y., Kinoshita, T., Okazaki, T., et al. 1991. Hydantocidin: a new compound with herbicidal activity from *Streptomyces hygroscopicus. J. Antibiot.* 44: 293–300
94. Nakajima, M., Itoi, K., Takamatsu, Y., Sato, S., Furukawa, Y., et al. 1991. Cornexistin: a new fungal metabolite with herbicidal activity. *J. Antibiot.* 44:1065–72
95. Nakamura, H., Shiomi, K., Iinuma, H., Naganawa, H., Obata, T., et al. 1987. Isolation and characterization of a new antibiotic, dioxapyrrolomycin, related to pyrrolomycins. *J. Antibiot.* 40:899–903
96. Nakayama, H., Takatsu, T., Abe, Y., Shimazu, A., Furihata, K., et al. 1987. Rustmicin, a new macrolide antibiotic active against wheat stem rust fungus. *Agric. Biol. Chem.* 51:853–59
97. Nishimoto, Y., Sakuda, S., Takayama, S., Yamada, Y. 1991. Isolation and characterization of new allosamidins. *J. Antibiot.* 44:716–22
98. Nishizawa, N., Kondo, Y., Koyama, M., Omoto, S., Iwata, M., et al. 1984. Studies on a new nucleoside antibiotic, dapiramicin. II. Isolation, physico-chemical and biological characterization. *J. Antibiot.* 37:1–5
99. Novak, J., Hajek, P., Rezanka, T., Vanek, Z. 1992. Nitrogen regulation of fatty acids and avermectins biosynthesis in *Streptomyces avermitilis. FEMS Microbiol. Lett.* 93:57–62
100. Ōmura, S. 1986. Philosophy of new drug discovery. *Microbiol. Rev.* 50: 259–79
101. Ōmura, S. 1988. Search for bioactive compounds from microorganisms—strategies and methods. In *Biology of Actinomycetes '88,* ed. Y. Okami, T. Beppu, H. Ogawara, pp. 26–32. Tokyo: Jpn. Sci. Soc.
102. Ōmura, S. 1992. The expanded horizon for microbial metabolites—a review. *Gene* 115:141–49
103. Ōmura, S. 1992 Trends in the search for bioactive microbial metabolites. *J. Ind. Microbiol.* 10:135–56
104. Ōmura, S., ed. 1992. *The Search for Bioactive Compounds from Microorganisms.* New York/Budapest: Springer-Verlag. 352 pp.
105. Ōmura, S., Hinotozawa, K., Imamura, N., Murata, M. 1984. The structure of phosalacine, a new herbicidal antibiotic containing phosphinothricin. *J. Antibiot.* 37:939–40
106. Ōmura, S., Ikeda, H., Tanaka, H. 1991. Selective production of specific components of avermectins in *Streptomyces avermitilis. J. Antibiot.* 44:560–63
107. Ōmura, S., Iwai, Y., Hirano, A., Nakagawa, A., Awaya, J., et al. 1977. A new alkaloid AM-2282 of *Streptomyces* origin. Taxonomy, fermentation, isolation and preliminary characterization. *J. Antibiot.* 30:275–82
108. Ōmura, S., Murata, M., Hanaki, H., Hinotozawa, K., Oiwa, R., et al. 1984. Phosalacine, a new herbicidal antibiotic containing phosphinothricin. Fermentation, isolation, biological activity and mechanism of action. *J. Antibiot.* 37: 829–35
109. Ōmura, S., Murata, M., Imamura, N., Iwai, Y., Tanaka, H. 1984. Oxetin, a new antimetabolite from an actinomycete. Fermentation, isolation, structure and biological activity. *J. Antibiot.* 37:1324–32
110. Ōmura, S., Nakagawa, A., Sadakane, N. 1979. Structure of herbimycin, a new ansamycin antibiotic. *Tetrahedron Lett.* pp. 4323–26
111. Ōmura, S., Nakagawa, A., Tanaka, Y. 1982. Structure of a new antifungal

antibiotic, irumamycin. *J. Org. Chem.* 47:5413–15

112. Ōmura, S., Otoguro, K., Imamura, N., Kuga, H., Takahashi, Y., et al. 1987. Jietacins A and B, new nematocidal antibiotics from a *Streptomyces* sp. Taxonomy, isolation, and physicochemical and biological properties. *J. Antibiot.* 40:623–29
113. Ōmura, S., Takahashi, Y., Iwai, Y., Tanaka, H. 1982. *Kitasatosporia,* a new genus of the order *Actinomycetales. J. Antibiot.* 35:1013–19
114. Ōmura, S., Takahashi, Y., Iwai, Y. 1989. Genus *Kitasatosporia.* In *Bergey's Mannual of Systemic Bacteriology, ed. S. T. Williams, M. E. Sharpe, J. G. Holt.* 4:2594–98. Baltimore: The Williams & Wilkins. 350 pp.
115. Ōmura, S., Tanaka, Y. 1985. Biosynthesis of tylosin and its regulation by ammonium and phosphate. In *Regulation of Secondary Metabolite Formation,* ed. H. Kleinkauf, H. V. Dohren, H. Dornauer, G. Nesemann, pp. 306–32. Weinheim: VCH Verlagsgesellschaft. 402 pp.
116. Ōmura, S., Tanaka, Y. 1986. Macrolide antibiotics. In *Biotechnology,* ed. H. Pape, H.-J. Rehm, 4:360–91. Weinheim: VCH Verlagsgesellschaft. 673 pp.
117. Ōmura, S., Tanaka, Y. 1991. Strategy and methods in screening of new microbial metabolites for plant protection. See Ref. 32, pp. 87–96
118. Ōmura, S., Tanaka, Y., Hisatome, K., Miura, S., Takahashi, Y., et al. 1988. Phthoramycin, a new antibiotic active against a plant pathogen, *Phytophthora* sp. *J. Antibiot.* 41:1910–12
119. Ōmura, S., Tanaka, Y., Kanaya, I., Shinose, M., Takahashi, Y. 1990. Phthoxazolin, a specific inhibitor of cellulose biosynthesis, produced by a strain of *Streptomyces* sp. *J. Antibiot.* 43:1034–36
120. Ōmura, S., Tanaka, Y., Takahashi, Y., Chia, I., Inoue, M., et al. 1984. Irumamycin, an antifungal 20-membered macrolide produced by *Streptomyces.* Taxonomy, fermentation and biological properties. *J. Antibiot.* 37:1572–78
121. Osada, H., Sonoda, T., Tsunoda, K., Isono, K. 1989. A new biological role of sangivamycin; inhibition of protein kinases. *J. Antibiot.* 42:102–5
122. Otoguro, K., Liu, Z.-X., Fukuda, K., Li, Y., Ō, S., et al. 1988. Screening for new nematocidal substances of microbial origin by a new method using the pine wood nematode. *J. Antibiot.* 41:573–75
123. Otoguro, K., Nakagawa, A., Ōmura, S. 1988. Setamycin, a 16-membered macrolide antibiotic: identification and nematocidal activity. *J. Antibiot.* 41:250–52
124. Pfefferle, W., Anke, H., Bross, M., Steffan, B., Vianden, R., et al. 1990. Asperfuran, a novel antifungal metabolite from *Aspergillus oryzae. J. Antibiot.* 43:648–53
125. Rabenhorst, R. J., Zahner, H. 1987. Biological properties and mode of action of clavams. *Arch. Microbiol.* 147: 315–20
126. Reddy, K., Jewett, G., Fatig, R. III, Brockman, M., Hatton, C., et al. 1991. New insecticidal metabolites from soil isolate W719. *J. Antibiot.* 44:962–68
127. Reichenbach, H., Gerth, K., Irschik, H., Kunze, B., Hofle, G. 1988. Myxobacteria: a source of new antibiotics. *Trends Biotechnol.* 6:115–21
128. Riddle, D. L., Georgi, L. L. 1990. Advances in research on *Caenorhabditis elegans:* applications to plant parasitic nematodes. *Annu. Rev. Phytopathol.* 28:247–70
129. Robeson, D., Strobel, G., Matusumoto, G. K., Fisher, E. L., Chen, M. H., et al. 1984. Alteichin: an unusual phytotoxin from *Alternaria eichorniae,* a fungal pathogen of water hyacinth. *Experientia* 40:1248–50

129a. Rohrer, S. P., Meinke, P. T., Hayes, E. C., Mrozik, H., Schaeffer J. M. 1992. Photoaffinity labeling of avermectin binding sites from *Caenorhabditis elegans* and *Drosophila melanogaster. Proc. Natl. Acad. Sci. USA* 89:4168–72

130. Sakuda, S., Isogai, A., Makita, T., Matsumoto, S., Koseki, K., et al. 1987. Structures of allosamidins, novel insect chitinase inhibitors, produced by Actinomycetes. *Agric. Biol. Chem.* 51: 3251–59
131. Schaeffer, J. M., Frazier, E. G., Bergstrom, A. R., Williamson, J., M., Liesch, J. M., et al. 1990. Cochlioquinone A, a nematocidal agent which competes for specific [^{3}H]ivermectin binding sites. *J. Antibiot.* 43:1179–82
132. Schuz, T. C., Fiedler, H.-P., Zahner, H., Rieck, M., Konig, W. A. 1992. Metabolic products of microorganisms. 263. Nikkomycins S_Z, S_X, S_{OZ} and S_{OX}, new intermediates associated to the nikkomycin biosynthesis of *Streptomyces tendae. J. Antibiot.* 45:199–202

133. Shephard, M. C. 1987. Screening for fungicides. *Annu. Rev. Phytopathol.* 25:189–206
134. Shibata, K., Satsumabayashi, S., Nakagawa, A., Ōmura, S. 1986. The structure and cytocidal activity of herbimycin C. *J. Antibiot.* 39:1630–33
135. Stapley, E. O., Woodruff, H. B. 1982. Avermectin, antiparasitic lactones produced by *Streptomyces avermitilis* isolated from soil in Japan. In *Trends in Antibiotic Research,* pp. 154–70. Tokyo: Jpn. Antibiotic Res. Assoc.
136. Strong, L. 1992. Avermectins: a review of their impact on insects of cattle dung. *Bull. Entomol. Res.* 82:265–74
137. Svoboda, J. A., Chitwood, D. J. 1992. Inhibition of sterol metabolism in insects and nematodes. In *ACS Symposium Series 497, Regulation of Isopentenoid Metabolism,* ed. W. D. Nes, E. J. Parish, J. M. Trzaskos, pp. 205–18. Washington, DC: Am. Chem. Soc. 270 pp.
138. Takahashi, A., Kurasawa, S., Ikeda, D., Okami, Y., Takeuchi, T. 1989. Altemicidin, a new acaricidal and antitumor substance. I. Taxonomy, fermentation, isolation and physico-chemical and biological properties. *J. Antibiot.* 42:1556–61
139. Takahashi, A., Saito, N., Hotta, K., Okami, Y., Umezawa, H. 1986. Bagougeramines A and B, new nucleoside antibiotics produced by a strain of *Bacillus circulans. J. Antibiot.* 39: 1033–46
140. Takahashi, Y., Iwai, Y., Ōmura, S. 1984. Two new species of the genus *Kitasatosporia, Kitasatosporia phosalacinea* sp. nov. and *Kitasatosporia griseola* sp. nov. *J. Gen. Appl. Microbiol.* 30:377–87
141. Takebe, H., Imai, S., Ogawa, H., Satoh, A., Tanaka, H. 1989. Breeding of bialaphos producing strains from a biochemical engineering viewpoint. *J. Ferment. Bioeng.* 67:226–32
142. Takiguchi, Y., Mishima, H., Okuda, M., Terao, M., Aoki, A., et al. 1980. Milbemycins, a new family of macrolide antibiotics: fermentation, isolation and physico-chemical properties. *J. Antibiot.* 33:1120–27
143. Tamaoki, T., Nakano, H. 1990. Potent and specific inhibitors of protein kinase C of microbial origin. *Bio/Technology* 8:732–33
144. Tanaka, Y. 1992. Fermentation processes in screening for new bioactive substances. See Ref. 104, pp. 303–26
145. Tanaka, Y., Hirata, K., Takahashi, Y., Iwai, Y., Ōmura, S. 1987. Globopeptin, a new antifungal peptide antibiotic. *J. Antibiot.* 40:242–44
146. Te Beest, D. O., Yang, X. B., Cisar, C. R. 1992. The status of biological control of weeds with fungal pathogens. *Annu. Rev. Phytopathol.* 30:637–57
147. Tombo, G. M. R., Ghisalba, O., Schar, H.-P., Frei, B., Maienfisch, P., et al. 1989. Diastereoselective microbial hydroxylation of milbemycin derivatives. *Agric. Biol. Chem.* 53:1531–35
148. Tomiya, T., Uramoto, M., Isono, K. 1990. Isolation and structure of phosphazomycin C. *J. Antibiot.* 43:118–21
149. Tomoda, H., Ōmura, S. 1990. New strategy for discovery of enzyme inhibitors: screening with intact mammalian cells or intact microorganisms having special functions. *J. Antibiot.* 43: 1207–22
150. Tsou, H.-R., Ahmed, Z. H., Fiala, R. R., Bullock, M. W., Carter, G. T., et al. 1989. Biosynthetic origin of the carbon skeleton and oxygen atoms of the LL-F28249α, a potent antiparasitic macrolide. *J. Antibiot.* 42:398–406
151. Tsukamoto, Y., Sato, K., Mio, S., Sugai, S., Yanai, T., et al. 1991. Synthesis of 5-keto-5-oxime derivatives of milbemycins and their activities against microfilariae. *Agric. Biol. Chem.* 55:2615–21
152. Tsuzuki, K., Yan, F. S., Otoguro, K., Ō, S. 1991. Synthesis and nematocidal activities of jietacin A and its analogs. *J. Antibiot.* 44:774–84
153. Ubukata, M., Uramoto, M., Uzawa, J., Isono, K. 1986. Structure and biological activity of neopeptins A, B and C, inhibitors of fungal cell wall glycan synthesis. *Agric. Biol. Chem.* 50:357–65
154. Uehara, Y., Murakami, Y., Sugimoto, Y., Mizuno, S. 1989. Mechanism of reversion of Rous sarcoma virus transformation by herbimycin A: reduction of total phosphotyrosine levels due to reduced kinase activity and increased turnover of $p^{60v\text{-}src}$. *Cancer Res.* 49: 780–85
155. Uehara, Y., Murakami, Y., Suzukae-Tsuchiya, K., Moriya, Y., Ōmura, S., et al. 1988. Effects of herbimycin derivatives on *src* oncogene function in relation to antitumor activity. *J. Antibiot.* 41:831–34
156. Uramoto, M., Itoh, Y., Sekiguchi, R., Shin-ya, K., Kusakabe, H., et al. 1988. A new antifungal antibiotic, cys-

targin: fermentation, isolation, and characterization. *J. Antibiot.* 41:1763–68
157. Watanabe, K., Umeda, K., Miyakado, M. 1989. Isolation and identification of three insecticidal principles from the red alga *Laurencia nipponica* Yamada. *Agric. Biol. Chem.* 53:2513–15
158. Yomosa, K., Hirota, A., Sasaki, H., Isogai, A. 1987. Isolation of harman and norharman from *Nocardia* sp. and their inhibitory activity against plant seedlings. *Agric. Biol. Chem.* 51:921–22
159. Yoshida, M., Ezaki, M., Hashimoto, M., Yamashita, M., Shigematsu, N., et al. 1990. A novel antifungal antibiotic, FR-900848. I. Production, isolation, physico-chemical and biological properties. *J. Antibiot.* 43:748–54

Annu. Rev. Microbiol. 1993. 47:89–115

MOLECULAR ANALYSIS OF THE PATHOGENICITY OF *STREPTOCOCCUS PNEUMONIAE*: The Role of Pneumococcal Proteins

James C. Paton
Department of Microbiology, Adelaide Children's Hospital, North Adelaide, S.A., 5006, Australia

Peter W. Andrew and Graham J. Boulnois[1]
Department of Microbiology, University of Leicester, Leicester, LE1 9HN, United Kingdom

Timothy J. Mitchell
Department of Immunology, Leicester Royal Infirmary, Leicester, LE1 5WW, United Kingdom

KEY WORDS: virulence factors, pneumolysin, neuraminidase, PspA, autolysin, genetic analysis, vaccine development

---

CONTENTS

[1]Current address: ICI Pharmaceuticals, Macclesfield, Cheshire, SK10 4TG, United Kingdom.

0066-4227/93/1001-0089$02.00

ABSTRACT

For many years the virulence of *Streptococcus pneumoniae* has largely been attributed to its antiphagocytic polysaccharide capsule. Recent evidence, however, indicates that certain pneumococcal proteins play an important part in the pathogenesis of disease, either as mediators of inflammation or by directly attacking host tissues. Pneumococci carrying defined mutations in the genes encoding any one of at least three pneumococcal proteins (the toxin pneumolysin, the major pneumococcal autolysin, and pneumococcal surface protein A) have significantly reduced virulence. Pneumococcal hydrolytic enzymes, such as neuraminidase, hyaluronidase, and IgA1 protease may also contribute to colonization and/or invasion of the host. Several of these proteins (or their detoxified derivatives) are protective immunogens in animal models and therefore warrant consideration for inclusion in human antipneumococcal vaccine formulations.

INTRODUCTION

Streptococcus pneumoniae (the pneumococcus) is a human pathogen that causes life-threatening, invasive diseases such as pneumonia, bacteremia, and meningitis with high morbidity and mortality throughout the world; young children and the elderly are particularly susceptible. In developing countries, an estimated five million children under the age of 5 years die each year from pneumonia, with *S. pneumoniae* being the single most common causative agent (81). Austrian (6) has suggested that in the United States, more than one million cases of pneumococcal pneumonia may occur each year, with a case fatality rate of 5–7%. The majority of these infections occur in the elderly. *S. pneumoniae* is also the leading cause of otitis media and sinusitis. These are less serious infections, but they are highly prevalent and have a significant impact on health-care costs in developed countries.

Many important studies (reviewed in 7) on the biology of the pneumococcus, in particular those of Avery and his many coinvestigators, were carried out during the first half of this century and established that there were numerous serotypes of *S. pneumoniae,* each producing a structurally distinct capsular polysaccharide. These substances were the first nonprotein molecules shown to be immunogenic, and antipolysaccharide antibodies were shown to provide type-specific protection against challenge with virulent pneumococci. Perhaps the most important studies of all, however, were those on the capsular transformation phenomenon, which led to the discovery that DNA was the carrier of genetic information (9).

Pneumococci are frequently isolated from the nasopharynx of healthy people, a feature that may complicate bacteriological analysis of the etiology of diseases such as pneumonia. Virtually all humans are colonized by pneumococci at some stage, and in certain populations nasopharyngeal carriage rates at any given time may exceed 70% (118). Rates of pneumococcal carriage are higher in young children and where people are living in crowded conditions, and the carriage rate in a community is related to the incidence of pneumococcal disease (118). Nasopharyngeal colonization by *S. pneumoniae* can have two consequences. In a proportion of cases [approximately 15% in one pediatric study carried out in the United States (52)], the pneumococcus becomes invasive and causes disease. This scenario is more likely in persons who have only recently become colonized. Alternatively, carriage may result in an immune response capable of eliminating the pneumococcus. This immunity is serotype specific, but does not necessarily prevent recolonization by the same serotype at a later time. Differences in the relative frequency of isolation of certain serotypes of *S. pneumoniae* from healthy persons and from cases of pneumococcal disease indicate that some capsular types are more capable of progressing from the carrier state to disease, but little is known of the bacterial and host factors involved in this process (118).

Although the advent of antimicrobial drugs, particularly penicillin, has reduced the overall mortality from pneumococcal disease, such therapy may be ineffective in high-risk groups unless given early in the course of infection (6). Moreover, the proportion of pneumococcal strains resistant to penicillin appears to be increasing. In the United States, approximately 5% of strains reportedly exhibit resistance (134), whereas in certain parts of Europe the figure exceeds 25% (10). Strains of *S. pneumoniae* with multiple resistance patterns, first detected in 1977 (62), continue to be isolated (134).

For many years the polysaccharide capsule was considered to be the *sine qua non* of pneumococcal virulence. This was based on the observation that all fresh clinical isolates of *S. pneumoniae* were encapsulated, and spontaneous nonencapsulated (rough) derivatives of such strains were almost completely avirulent. Moreover, enzymic depolymerization of the capsule of a type 3 pneumococcus increased its LD_{50} approximately 10^6-fold (8). More recently, a similar effect on virulence of type 3 *S. pneumoniae* was achieved by transposon mutagenesis of a gene essential for capsule production (149).

The precise manner in which the pneumococcal capsule contributes to virulence is not fully understood, although it does have strong antiphagocytic properties in nonimmune hosts (83, 103). Clearly, some polysaccharide serotypes are more effective than others, as certain types are far more commonly associated with human disease. Moreover, within a given serotype, virulence appears to be related to the amount of capsular polysaccharide produced (6, 83). In the immune host, however, binding of specific antibody to the capsule results in opsonization and rapid clearance of the invading pneumococci.

Notwithstanding the importance of the capsule in evading host defenses, the various capsular polysaccharides in purified form are completely nontoxic and cannot themselves account for death from pneumococcal infection. Host inflammatory responses to pneumococcal components (e.g. the cell wall) are likely to contribute to tissue injury, as reviewed by Johnston (69, 70). However, increasing evidence indicates that certain pneumococcal proteins may also play an active role. Recent advances in molecular biology have provided tools to examine the contribution of individual pneumococcal proteins (or even regions thereof) to pathogenesis, in addition to producing information applicable to the development of improved pneumococcal vaccines. In this review, we summarize the state of our knowledge of these pneumococcal products, with particular reference to the possibility of improving current preventive strategies against pneumococcal disease.

PNEUMOLYSIN

Properties and Mode of Action

Nearly 90 years have passed since the first report of production of a hemolysin by pneumococci (86). During the first half of this century, several studies were carried out on crude hemolysin preparations (30, 31, 55, 105, 106), establishing that the hemolysin was toxic, susceptible to oxidation (a process which could be reversed by thiol-reducing agents), antigenic, and irreversibly inactivated by treatment with cholesterol. The possibility that the hemolysin was somehow involved in pathogenesis was suggested by Shumway (129), who observed spherocytosis and increased osmotic fragility of erythrocytes in rabbits with pneumococcal bacteremia. Similar effects were observed in rabbits injected intravenously with a cell-free pneumococcal extract (131). However, this in vivo effect was unequivocally attributed to the toxin itself (130) only after improvements in protein chromatographic technology enabled the first purification to homogeneity of the hemolysin (by this stage referred to as pneumolysin).

Pneumolysin consists of a single 53-kDa polypeptide chain and is produced by virtually all clinical isolates of *S. pneumoniae* (73, 111). It is a member of the family of toxins referred to as the thiol-activated cytolysins (132), which are produced by four different genera of gram-positive bacteria. It differs from other members of the family, however, in that it stays within the cytoplasm rather than being secreted (64). The thiol-activated cytolysins are believed to share a common mode of action involving two steps. The first is an interaction with cholesterol in the target-cell membrane resulting in insertion of the toxin into the lipid bilayer. The second stage involves lateral diffusion and oligomerization of 20–80 toxin molecules, resulting in formation of arc and ring structures (at sufficiently high toxin concentrations) visible with electron microscopy, which are thought to be transmembrane pores (18). Probably only a small number of such pores, perhaps just one per target cell, is sufficient to induce lysis of erythrocytes (17). Although erythrocytes are particularly susceptible, thiol-activated cytolysins such as pneumolysin can interact with any animal cell that has cholesterol in its plasma membrane.

The first direct evidence for the involvement of pneumolysin in pathogenesis was the finding that immunization with highly purified toxin significantly increased the survival time of mice challenged intranasally with virulent pneumococci (111). Clues as to the possible role of pneumolysin in pathogenesis have come from studies on its biological properties. Johnson & Allen reported toxic effects when purified pneumolysin was injected into the cornea or instilled onto the conjunctiva of rabbit eyes (65). Toxicity was less apparent if rabbits were made leukopenic before challenge, which suggested the possible involvement of leukocytes in this effect (56). Johnson et al (66) had previously demonstrated that pneumolysin could lyse human polymorphonuclear leukocytes (PMNs) and platelets (albeit at much higher concentrations than that required for lysis of erythrocytes), but at sublethal doses, PMN migration was inhibited and leakage of lysozyme was enhanced. Paton & Ferrante (110) showed that very low doses (approximately 1 ng/ml) of highly purified pneumolysin significantly inhibited the respiratory burst of human PMNs, as well as their capacity to engulf and kill opsonized pneumococci. In addition, both chemotaxis and random migration of PMNs was inhibited. All these detrimental effects disappeared when the pneumolysin preparation was pretreated with cholesterol. Similar inhibitory effects of pneumolysin on the respiratory burst, degranulation, bactericidal activity, and phospholipid methylation of human monocytes have also been reported (104). Pretreatment of human lymphocytes with equally low doses of pneumolysin abrogated lymphoproliferative responses to various mitogens, as well as the capacity of stimulated lymphocytes to produce lymphokines and all three classes of immunoglobulin (44). Again these inhibitory properties of the toxin were abolished by treatment with cholesterol. No inhibitory effects on lymphocyte functions were observed when pneumolysin was added immediately after the mitogens, which is consistent with the hypothesis that the toxin exerts its effects, at least in part, by blocking membrane signal transmission (44). Interestingly, Hage-Chahine et al (54) recently reported the inhibition of both T-cell-dependent and T-cell-independent immune responses in mice infected with sublethal doses of *Listeria monocytogenes,* but this inhibition was not observed in mice infected with similar doses of a strain in which the gene encoding listeriolysin-O (a thiol-activated cytolysin closely related to pneumolysin) had been inactivated by transposon mutagenesis.

Treatment of human serum with purified pneumolysin results in activation of the classical complement pathway, in the absence of specific antibody, with concomitant depletion of serum opsonic activity (114). This phenomenon results from the capacity of the toxin to bind the Fc region of IgG (100). Activation also occurs with membrane-fixed toxin (P. W. Andrew, G. J. Boulnois & T. J. Mitchell, unpublished observation) and may result in complement attack on these membranes and generation of an inflammatory response. Interestingly, unlike the effects of pneumolysin on cells of the immune system, treatment with cholesterol did not inhibit complement activation in serum, suggesting that activation is a feature unrelated to the toxin's cytolytic activity (114). Collectively, these studies demonstrate that pneumolysin can directly interfere with the ability of the host to opsonize, phagocytose, and kill invading pneumococci, as well as block the establishment of a humoral immune response to the infection.

Recent studies have shown that pneumolysin is entirely responsible for the ability of pneumococcal culture filtrates to slow ciliary beating and disrupt the surface integrity of human respiratory epithelium in organ culture (41, 136). Pneumolysin is also cytotoxic for pulmonary endothelial cells and thus might contribute to alveolar hemorrhage during pneumococcal infections (122). Evidence for the importance of pneumolysin in pulmonary infections in vivo is provided by Feldman et al (42), who demonstrated that injection of purified pneumolysin into the apical lobe bronchus of rat lung resulted in the development of a severe lobar pneumonia, with histological changes identical to those caused by injection of live virulent pneumococci.

Two additional lines of evidence support the notion that pneumolysin is produced in vivo. Firstly, pneumolysin can be detected by Western blotting in homogenates of lung, spleen, and liver from mice challenged intranasally with virulent pneumococci (P. W. Andrew, G. J. Boulnois & T. J. Mitchell, unpublished data). Also, significant antipneumolysin antibody titers (63, 71, 72) or circulating pneumolysin immune complexes (85) are found in sera from the majority of patients with pneumococcal pneumonia. In healthy individuals, antipneumolysin levels were lowest in children under one year of age and adults over 70 years (the peak risk groups for pneumococcal pneumonia). Moreover, antipneumolysin levels in acute-phase sera from pneumonia patients were significantly lower than in age-matched healthy controls (72). This finding implies that individuals with lower pneumolysin antibody levels may be at greater risk of contracting pneumococcal pneumonia.

Cloning, Sequence Analysis, and Structure-Function Studies

The genes encoding pneumolysin from type 1 and type 2 pneumococci have been cloned into *Escherichia coli* (109, 148) and subjected to DNA sequence analysis (101, 148). Comparison of the two 471–amino acid sequences revealed that the proteins differed by only a single residue (101).

The predicted primary amino acid sequences of pneumolysin as well as six other thiol-activated toxins for which data are available (50, 53, 75, 97, 144) show no major regions of hydrophobicity. Thus, the previously observed hydrophobic nature of these toxins (68) presumably reflects the generation of hydrophobic areas during folding of the primary sequence. In contrast to the other thiol-activated toxins (which are secreted), the predicted amino acid sequence of pneumolysin lacks a typical hydrophobic N-terminal signal peptide (148), which is consistent with its cytoplasmic location in *S. pneumoniae* (64). Nevertheless, the thiol-activated toxins show extensive primary amino acid–sequence homology, which accounts for their serological cross reactivity (132).

Thiol-activation of this group of toxins was thought for many years to reflect the reduction of intramolecular disulfide bonds (132). However, in all the thiol-activated toxins so far sequenced, with the exception of ivanolysin from *Listeria ivanovii,* only one cysteine residue per molecule was predicted. This single cysteine lies near the C terminus of the protein (amino acid 428 in the case of pneumolysin) in a conserved region of 11 amino acids (Glu-Cys-Thr-Gly-Leu-Ala-Trp-Glu-Trp-Trp-Arg). In the case of seeligerolysin from *Listeria seeligerii,* this conserved amino acid sequence contains a single Ala→Phe substitution in the sixth position. The presence of only a single cysteine residue within these toxins obviously precludes the formation of intramolecular disulphide bonds. Much experimental data generated over many years suggested that cysteine residues were essential for the biological activity of these toxins, either by involvement in cholesterol binding, or by maintenance of the correct conformation of the toxin (1, 132). However, substitution of the single cysteine residue in pneumolysin with alanine by site directed mutagenesis had no detectable effect on the activity of the toxin in vitro (127). Similar results were obtained for streptolysin O (115) and listeriolysin O (98); therefore, the cysteine residue is not essential for the activity of these toxins, at least in vitro.

Despite the above, site-directed mutagenesis studies have shown that the conserved sequence around the cysteine is important for biological activity. When the cysteine residue itself was changed to either a glycine or a serine, the cytolytic and cytotoxic activities were reduced (127). The reduced activity of these mutants did not appear to result from defects in binding or the ability to form oligomers in the membrane. A similar result was obtained when other amino acids in the conserved sequence were substituted (19).

Pore-forming proteins and peptides may possess a common structural motif (107). A region of 20 amino acids (spanning the 11–amino acid conserved region around the cysteine) occur with the appropriate periodicity of hydrophobic and hydrophilic residues to form a membrane-spanning amphiphilic helix (99). This suggestion is particulary attractive when considering the comparison between pneumolysin and the antibiotic gramicidin A (34). Four tryptophan residues near the C terminus of gramicidin A play an essential role in channel formation, and the three tryptophans in the cysteine motif may play a similar role. Gramicidin A, like pneumolysin, interacts with cholesterol. It has been proposed that the ring structures of the sterol interact with tryptophan residues to orientate the antibiotic in the membrane so that the hydrophilic domains adopt a transmembrane configuration that results in the formation a pore (34). A similar mechanism may be involved in pore formation by pneumolysin, as substitution of tryptophans in the cysteine motif (particularly the one at position 433) results in a marked decrease in lytic activity of the toxin.

The importance of histidine residues in toxicity was suggested by the finding that treatment of highly purified pneumolysin with the histidine-modifying reagent diethyl pyrocarbonate resulted in loss of activity (P. W. Andrew, G. J. Boulnois & T. J. Mitchell, unpublished observation). Comparison of the amino acid sequences for the various thiol-activated toxins revealed only one conserved histidine residue (at position 367 in pneumolysin), and substitution of this with arginine resulted in a pneumolysin derivative, which had lost 99.9% of its cytolytic activity but bound to cells normally (P. W. Andrew, G. J. Boulnois & T. J. Mitchell, unpublished observation). This toxin, however, could not form oligomers in the membranes of red blood cells; hence, His367 in pneumolysin appears to be essential for the interaction that leads to pore formation.

The precise mechanism of cell binding by pneumolysin is not understood. Much of the evidence that cholesterol is the receptor is based on studies of the sterol in solution in which binding is postulated to involve interaction of the toxin with the 3-β OH group of cholesterol. However, the relative inaccessibility of cholesterol within the membrane (151) may prevent this interaction in intact cells. An N-terminal fragment of listeriolysin lacking the cysteine motif bound to cholesterol in solution (146), suggesting that the cholesterol binding site does not involve the cysteine motif. In contrast, the cell-binding site of perfringolysin may involve the cysteine (61) or the carboxyl end of the protein (145). The apparently conflicting conclusions with regard to cholesterol and cell binding may indicate that free and membrane-bound cholesterol act at different regions of the toxin. Alternatively, binding of toxin to cells may involve a receptor other than cholesterol.

The ability of pneumolysin to activate the complement pathway (114) is related to its ability to bind the Fc portion of human IgG (100). Pneumolysin is therefore another example of a bacterial protein with Fc binding activity. However, we could not detect any sequence homology between pneumolysin and these Fc binding proteins. The observation that pneumolysin has limited homology with human C-reactive protein (CRP) indicated which regions of pneumolysin may be involved in complement activation (100). A C-terminal region within CRP is partially homologous with two noncontiguous sequences of the pneumolysin molecule. The use of monoclonal antibodies raised to pneumolysin demonstrated that the toxin and CRP also share at least one cross-reactive epitope, although whether this epitope is related to the areas of sequence homology is not yet clear. CRP activates the classical complement pathway through the direct binding of C1q to CRP once it has bound to its ligand (147). Site directed mutagenesis of one of the regions of pneumolysin homologous to CRP, especially at residues Tyr384 and Asp385, reduced the ability of the toxin both to bind antibody and to activate the complement pathway (100). The cytolytic activity of these mutants was not affected. Interestingly, the amino acid substitution at Asp385 abolished the ability to activate complement but only reduced antibody binding by 70%. Therefore complement activation and antibody binding by pneumolysin are apparently related. It is not clear whether the appropriate sequence in CRP can also bind antibody, but it is known that CRP mediates complement activation by binding C1q (147). Preliminary experiments show that pneumolysin can also bind C1q (P. W. Andrew, G. J. Boulnois & T. J. Mitchell, unpublished observations). Pneumolysin may act in vivo as a CRP analogue, but one without the appropriate ligand-binding activities. In this way, pneumolysin may abrogate the effect of CRP by leading to activation of complement on autologous cells rather than at the surface of the pneumococcus via CRP bound to the cell wall.

Molecular Genetic Analysis of the Contribution of Pneumolysin to Virulence

Access to the cloned pneumolysin gene has enabled the construction by insertion-duplication mutagenesis of a defined pneumolysin-negative derivative of *S. pneumoniae* type 2 (16). The virulence of this strain was then compared with its otherwise isogenic parental type in a mouse intranasal- and intraperitoneal-challenge model. Inactivation of the pneumolysin gene increased the LD_{50} approximately 100-fold. Intravenous challenge with the wild-type strain resulted in an overwhelming bacteremia (10^8–10^9 organisms per milliliter of blood) and death within 24 hours, while the mutant pneumococci established a bacteremia (10^5–10^6 organisms per milliliter) that persisted for over a week in several cases without apparent detrimental effect. Full virulence was, however, reconstituted by back-transformation of the pneumolysin-negative pneumococci with a purified DNA fragment carrying an intact copy of the pneumolysin gene (16). Similar reduction in virulence has also been demonstrated by insertional inactivation of the pneumolysin gene of type 3 pneumococci (15). These studies confirmed that pneumolysin is directly involved in the pathogenesis of pneumococcal infections. However, the finding that inactivation of the pneumolysin gene significantly reduced, but did not abolish, the capacity of pneumococci to kill their host indicates that other pneumococcal products are also involved.

To date, the relative contributions of the structurally distinct cytotoxic and complement-activation properties of pneumolysin to pathogenesis have not been precisely determined. We have back-transformed the pneumolysin-negative type 2 pneumococcus referred to above (PLN-A) with the cloned pneumolysin gene carrying the Trp433→Phe mutation, which reduces hemolytic activity by 99.9% (A. M. Berry, J. C. Paton, P. W. Andrew, G. J. Boulnois & T. J. Mitchell, unpublished data). This resulted in the reconstitution of the pneumolysin locus, but the gene then carried the point mutation. The virulence of this strain was intermediate between that of PLN-A and the wild-type strain. Thus, unless the vestigial 0.1% of native hemolytic activity was sufficient to contribute to virulence, other activities of the toxin must also play a role in vivo. This supposition is consistent with the finding that a pneumolysin-negative (i.e. nonhemolytic) pneumococcus obtained by chemical mutagenesis (presumably a point mutation) was as virulent as the wild-type strain in a rabbit model of intracorneal infection, but a strain in which the pneumolysin gene was completely deleted had significantly reduced virulence (67). Interestingly, the inflammatory changes induced by injection of purified pneumolysin into the ligated lobes of rat lungs were reduced when pneumolysin derivatives deficient in either cytolytic activity or complement activation were administered (42).

NEURAMINIDASE

A variety of indirect evidence suggests that the pneumococccal neuraminidase also plays a role in pathogenesis. This enzyme cleaves terminal sialic acid residues from a wide variety of glycolipids, glycoproteins, and oligosaccharides on cell surfaces or in body fluids, and such activity has the potential to cause great damage to the host. Neuraminidase might also unmask potential cell-surface receptors for putative pneumococcal adhesins (80). At least two studies on fresh, clinical isolates of *S. pneumoniae* showed that all strains examined (104 in all) had neuraminidase activity (76, 108), which is consistent with our own unpublished findings. Histochemical studies of the organs from mice dying after the intraperitoneal administration of partially purified pneumococcal neuraminidase have indicated marked decreases in the sialic acid content of the kidney and liver when compared to controls (77). Also, both coma and bacteremia occur significantly more often in patients with pneumococcal meningitis when the concentration of *N*-acetyl neuraminic acid in the cerebrospinal fluid is elevated (108).

Further assessment of the contribution made by neuraminidase to pneumococcal pathogenicity was complicated by the fact that there appeared to be multiple forms of the enzyme (135, 140, 141). Lock et al (88) proposed that these forms resulted from proteolytic degradation of a parental enzyme. A single 107-kDa neuraminidase form could be purified from *S. pneumoniae* lysates treated with protease inhibitors, but in the absence of these, several smaller fully active forms were isolated. Like pneumolysin, the purified neuraminidase was toxic for mice, and immunization with the protein partially protected mice from challenge with virulent *S. pneumoniae* (89). This protection, however, was not as great as that achieved by immunization with pneumolysin, and no additive protective effect could be seen when mice were immunized with both proteins.

Two reports have described the cloning of *S. pneumoniae* genes encoding neuraminidase into *E. coli* (14, 26). Both clones were isolated on the basis of their ability to cleave a synthetic fluorogenic neuraminidase substrate, but their activity on naturally occurring substrates has not been determined. Interestingly, hybridization analysis indicated that these two neuraminidase genes are different and that individual pneumococcal isolates contain both genes (26), a fact that has complicated construction of defined neuraminidase-negative pneumococci for virulence studies. The former neuraminidase-producing clone is unstable and hence has not been further characterized. Sequence analysis of the latter clone (M. Camara, P. W. Andrew, G. J. Boulnois & T. J. Mitchell, unpublished data) revealed that the largest open reading frame (ORF) included four copies of a so-called aspartic box, Ser-X-Asp-X-Gly-X-Thr-Trp, a feature found in numerous bacterial neuraminidases (120). Moreover, the distance between certain pairs of these aspartic boxes appears to be conserved in the pneumococcal as well as other enzymes. The precise translation initiation site for the pneumococcal neuraminidase gene is somewhat ambiguous, as there are three in-frame ATG start codons, two of which are preceded by Shine-Dalgarno sequences, and there are two possible promoter sequences upstream. Thus, two translation products (110 and 114 kDa) may result, but only one of these would include a hydrophobic signal peptide at its N terminus. The C terminus of the putative neuraminidase has features typical of surface proteins of gram-positive cocci.

The C termini of these proteins are responsible for anchorage to the cell membrane and are composed of four to seven charged residues at the C terminus, which are immediately preceded by a hydrophobic domain and the sequence Leu-Pro-X-Thr-Gly-X (128). This latter sequence may be a recognition site for posttranslational modification that would release the protein from the cell surface (128).

AUTOLYSIN

Properties

The major pneumococcal autolysin is a 36-kDa *N*-acetylmuramic acid, L-alanine amidase (57), which is located in the cell envelope (35). The enzyme is thought to be bound to choline moeities of lipoteichoic acid (Forssman Antigen), which in turn is anchored to the cell membrane (21). In this form, autolysin is inactive (presumably due to lack of access to its substrate) and the association with lipoteichoic acid may be an important means of regulating its potentially suicidal activity in vivo. When cell wall biosynthesis ceases, either because of nutrient starvation or treatment with antibiotics such as penicillin, this association is disrupted and the enzyme is then able to cleave the covalent bond between the glycan chain and the peptide side chain of the choline-containing cell wall, thereby bringing about cellular autolysis (21). Treatment with detergents such as deoxycholate also releases and activates the enzyme (21).

The *S. pneumoniae* gene encoding autolysin has been cloned in *E. coli* (47), and its complete nucleotide sequence has been determined (49). Several autolysin-deficient mutants of *S. pneumoniae,* resulting from chemical mutagenesis, have also been described (48, 90, 125), all of which failed to undergo autolysis during the stationary phase of growth and were resistant to the lytic consequences of treatment with penicillin or deoxycholate. The mutants grew normally, except for the tendency to form short chains of cells rather than discrete diplococci, suggesting that autolysin might play a role in daughter-cell separation [a finding consistent with the apparent immunochemical localization of autolysin to the septal region of growing pneumococci (35)]. Transformation of one of the mutants with a recombinant plasmid carrying the wild-type gene restored the normal phenotype (121).

Possible Role in Pathogenesis

Several studies have suggested that autolysin could also (directly or indirectly) function in pneumococcal pathogenesis. Chetty & Kreger (27, 28) demonstrated that the pneumococcal purpura-producing principle was composed of high-molecular-weight peptidoglycan fragments solubilized from the cell wall by the action of autolysin. Tuomanen et al (143) demonstrated that both peptidoglycan and teichoic acid components of the pneumococcal cell wall were potent mediators of meningeal inflammation in a rabbit model. They proposed that treatment with β-lactam antibiotics, which induce autolysin-mediated release of the above components, might further contribute to inflammatory tissue injury and possibly mortality, despite effective sterilization of the cerebrospinal fluid. A similar potentiation of cell wall–induced middle ear inflammation by treatment with penicillin was also demonstrated in a chinchilla model of pneumococcal otitis media (74). However, cell wall–degradation products may not be the only harmful substances released from pneumococci by the action of autolysin. Pneumolysin, whose involvement in pathogenesis has been clearly established, is not actively secreted by *S. pneumoniae;* it is located in the cytoplasm (64). At least one other potentially deleterious pneumococcal product, neuraminidase, also appears to be cytoplasmically localized or at least strongly cell-associated (88). Thus, autolysin-induced lysis of a proportion of the invading pneumococci could directly harm the host by releasing high local concentrations of potent toxins and hydrolytic enzymes, in addition to inflammatory cell wall–degradation products.

The direct contribution of autolysin to pneumococcal virulence has been studied by using the cloned gene to construct defined autolysin-negative encapsulated type 2 or type 3 pneumococci by insertion-duplication mutagenesis (13, 15). These mutants did not spontaneously autolyse and did not release pneumolysin or neuraminidase into the culture medium, even after the addition of sodium deoxycholate. Both the type 2 and type 3 autolysin-negative strains were significantly less virulent than their otherwise isogenic parental types, as judged by intranasal and intraperitoneal LD_{50}. For both serotypes, reconstitution of the autolysin locus by back-transformation with the cloned autolysin gene restored both autolytic activity and full virulence. Interestingly, Tomasz et al (142) have also constructed autolysin-negative encapsulated pneumococci (types 3 and 6), by a slightly different procedure. After serial passage through mice "to correct for loss of virulence during in vitro growth," (142, p. 5933) both these mutants had intraperitoneal LD_{50}s similar to those of the respective encapsulated parent strain. The reason for the discordant findings is not understood, but the possibility that unknown alterations (perhaps affecting other virulence loci) occurred during animal passage cannot be excluded.

Immunization of mice with autolysin purified from recombinant *E. coli* resulted in the production of antibodies capable of inhibiting spontaneous autolysis of both rough and encapsulated pneumococcal cultures (13, 87). This experiment established that exogenous antibody could penetrate the polysaccharide capsule and interact with autolysin in the pneumcoccal cell wall. As expected, pneumococci grown in the presence of the autolysin antiserum did not release significant amounts of pneumolysin into the culture medium (87). Immunization of mice with autolysin also provided significant protection against challenge with virulent type 2 pneumococci (13, 87). The degree of protection was similar to that achieved by immunization with pneumolysin, but no additional protection was observed when mice were immunized with both antigens. Moreover, immunization of mice with autolysin did not provide any protection whatsoever against challenge with high doses of defined pneumolysin-negative pneumococci (87). Collectively, these results are consistent with the interpretation that, in the mouse intraperitoneal challenge model, the major function of autolysin in pathogenesis is to catalyze the release of pneumolysin from the cells. The relative contributions to pathogenesis of autolysin-mediated release of pneumolysin versus cell wall–degradation products is yet to be assessed in animal models of other pneumococcal diseases (e.g. meningitis and otitis media).

SURFACE PROTEINS

Pneumococcal Surface Protein A (PspA)

Monoclonal antibody studies by McDaniel et al (91) demonstrated that a protective pneumococcal protein antigen was located in the cell wall of pneumococci. This antigen, referred to as PspA, was produced by all pneumococci, but was highly variable, both immunologically and in molecular size (from 60 to 200 kDa) (33). Immunization with rough pneumococci protected mice against challenge with virulent organisms, but no protection was conferred by immunization with rough pneumococci in which the PspA gene had been inactivated by insertion-duplication mutagenesis (92). Moreover, PspA-deficient type 2 pneumococci were more readily cleared from the blood of mice during the first hour after intravenous challenge than were the otherwise isogenic wild-type organisms. This finding correlated with a longer median survival time in the former group, but the LD_{50} did not increase significantly (92). However, the relative importance of the contribution of PspA to pathogenesis appears to vary from strain to strain, as far more dramatic reductions in virulence were observed when the PspA genes of two other pneumococcal strains (belonging to serotypes 3 and 5) were inactivated (23).

The precise function of PspA in virulence is not understood. Analysis of the DNA and deduced amino acid sequences of the gene and the properties of various truncated forms of PspA (152, 153) indicated that the N-terminal half of the protein is likely to be an α-helical coiled coil. It contains the epitopes recognized by protective monoclonal antibodies, with sequence variations in this region accounting for the known antigenic variability. These features are reminiscent of the M protein of group A streptococci, which has a limited degree of amino acid sequence homology. The C-terminal portion of PspA is involved in attachment of the protein to the cell, and thus the significant degree of homology observed between this region and the C-terminal region of the pneumococcal autolysin may reflect a similar mode of attachment.

The firm attachment of PspA to the pneumococcal cell wall has frustrated purification of the mature protein for the purpose of direct assessment of its potential as a vaccine antigen. However, Talkington et al (139) have purified a truncated derivative of PspA (the 43-kDa N-terminal half) from the culture medium of a pneumococcus in which the gene had been interrupted by insertion-duplication mutagenesis. Mice immunized with this antigen were protected against challenge with a virulent strain of *S. pneumoniae* that produced a closely related PspA type. The capacity of this antigen to protect against challenge with pneumococci producing more distantly related PspA types remains to be determined but will have important implications for pneumococcal vaccine development.

37-kDa Protein

Russell et al (123), using Western blot analysis and a monoclonal antibody, demonstrated that all pneumococcal strains tested produce a 37-kDa surface antigen. This antigen was specific for *S. pneumoniae,* and they suggested that it may be a useful target for immunodiagnosis of pneumococcal disease. The gene encoding the 37-kDa antigen has also been cloned in *E. coli* (124). At present, little is known about the structure or function of this protein. However, preliminary studies have shown that immunization of mice with purified 37-kDa antigen protects them against challenge with type 3 pneumococci (20).

Pneumococcal Adhesins

The ability of bacteria to adhere to epithelial cells is an important requirement for colonization, but various studies have produced conflicting data on the adherence capacity of pneumococci (3, 4, 12, 43, 137). Some studies have suggested that pneumococci isolated from the nasopharynx of carriers or from cases of otitis media have a greater tendency to attach to nonciliated epithelial cells than isolates from blood or cerebrospinal fluid (3, 4). Also, nonciliated epithelium from otitis-prone children was more likely to have attached pneumococci than that from healthy individuals (137).

Two host-cell receptors for pneumococci have been postulated. Some studies have suggested that glycolipids containing the disaccharide

GlcNAcβ1-3Gal are the target (2), while others have reported that pneumococci adhere to glycolipids carrying GalNAcβ1-4Gal, which are found extensively in the respiratory tract (80). To date, the pneumococcal factor mediating adhesion is poorly characterized, although Andersson et al (2) have proposed that a proteinaceous, heat-extractable sandwich adhesin, with specificity for the former disaccharide receptor, exists on the pneumococcal cell surface. It has also been reported that pneumococci can bind laminin and type IV collagen (79). These proteins are components of basement membranes, which might become exposed when respiratory epithelial cells are damaged by virus infection (a known predisposing factor for pneumococcal pneumonia). In this context it is also interesting that *S. pneumoniae* is known to produce a serine protease capable of degrading these extracellular matrix proteins (32).

OTHER PROTEINS

Hyaluronidase

Hyaluronidase is an enzyme produced by a variety of wound and mucosal pathogens, including the pneumococcus (58). Its substrate is hyaluronic acid, which is found associated with mammalian connective tissue and extracellular matrix. Thus, hyaluronidase might play a role in pneumococcal pathogenesis by allowing greater microbial access to host tissue for colonization. It may also play a role in the migration of the organism between tissues, for example translocation from the lung to the vascular system.

Ninety-nine percent of clinical isolates of *S. pneumoniae* produce hyaluronidase, which is actively secreted from the pneumococcus during log-phase growth in vitro (A. M. Shepherd, P. W. Andrew, G. J. Boulnois & T. J. Mitchell, unpublished data). Two *E. coli* clones expressing the pneumococcal hyaluronidase gene were recently isolated (A. M. Shepherd, P. W. Andrew, G. J. Boulnois & T. J. Mitchell, unpublished data; A. M. Berry, R. A. Lock & J. C. Paton, unpublished data). Sequence analysis of the latter clone revealed the presence of an ORF sufficient to encode an 80-kDa protein, and an antiserum raised against the purified recombinant hyaluronidase labeled a protein of the same size in pneumococcal lysates. The hyaluronidase gene from *Streptococcus pyogenes* bacteriophage H448A has also been cloned and sequenced (59), but the amino acid homology with the putative pneumococcal sequence is negligible. The availability of the cloned pneumococcal gene will allow the study of the role of hyaluronidase in vivo via the construction of defined hyaluronidase-negative pneumococci for use in animal models of infection.

IgA1 Protease

S. pneumoniae, like several other bacterial species that colonize mucosal surfaces, produces a protease that cleaves human IgA1 at a specific point within the hinge region, yielding intact Fab and Fc fragments (94, 116). To date no conclusive evidence for the involvement of any of these proteases in pathogenesis has been presented, primarily because the enzymes are highly specific and do not cleave IgA from any animal species commonly used as models for disease. However, the Fab fragments generated by IgA1 proteases retain their specificity for antigen (95, 96), and such fragments bound to the surface of a bacterium could protect it from the immune system by blocking access to intact immunoglobulins. Also, any mechanism of IgA–immune complex elimination that depends on the Fc fragment would be abolished (78).

Genes encoding IgA1 proteases from several bacterial species have been cloned, but there is no evidence for extensive sequence homology with the respective pneumococcal gene (51). The *S. pneumoniae* gene encoding IgA1 protease has also been cloned, but instability prevented any further characterization (117). A more stable clone was recently isolated (25) but still awaits further analysis. Access to the cloned gene will permit construction of defined IgA1 protease–negative pneumococci, but assessment of the impact of such a mutation on pathogenesis will depend on the availability of transgenic animals capable of producing human IgA1.

IMPLICATIONS FOR VACCINE DEVELOPMENT

Modern polyvalent antipneumococcal vaccines for human use have all been formulated from a mixture of capsular polysaccharides selected from a predetermined number of the 84 serologically distinct serotypes. However, these polysaccharide-based vaccines have two critical shortcomings. The first is that the protection they impart is type-specific. Because of this, a formulation of serotypes that is effective for one population may not be as effective for another if serotype prevalence differs significantly. For example, studies on pneumococcal isolates from Asian populations show that only 63–73% belong to serotypes included in the current 23-valent vaccine, compared with 88–93% for the US population (82, 119).

The second shortcoming of present vaccines is that even the protection they provide against specifically included serotypes is by no means complete and may be very poor for certain high-risk groups, who have a poorer antibody response to polysaccharide vaccines than healthy adults (24, 46, 102). Young children are of particular concern. In children less than two years old, antibody response to most capsular types is generally poor (60). Also, numerous studies have shown that antibody response to several important pediatric pneumococcal types (e.g. 6A, 14, 19F and 23F) is lower in children younger than five years old (38, 84). This may explain the apparent lack of demonstrable clinical benefit in some vaccine trials in this age group (37, 93, 150), which, in view of the continuing high morbidity and mortality from pneumococcal disease, has prompted calls for the development of more effective pediatric pneumococcal vaccines.

Supplementation of vaccines with appropriate pneumococcal protein antigens could provide the answer to this problem, because young children, as a group, respond well to protein-based vaccines (29). Protective pneumococcal protein antigens would also be attractive candidates for use as carriers for polysaccharides in a conjugate vaccine. Conjugation of bacterial polysaccharides to protein carriers typically not only greatly enhances their immunogenicity, but also converts them from thymus-independent to thymus-dependent antigens capable of promoting immunological memory (11, 133). This approach has been successfully applied to the development of an effective *Haemophilus influenzae* type b vaccine (39), and studies with pneumococcal types 6B and 12F polysaccharide conjugated to either tetanus or diphtheria toxoids have also demonstrated enhanced immunogenicity of the polysaccharide moeities (40, 126). Vaccines might be improved further by the conjugation of polysaccharides to a protective pneumococcal protein (rather than a nonsense protein), thereby increasing the immunogenicity of the polysaccharide component as well as providing non-serotype-specific protection. This latter protection, conferred by the protein component, may be particularly important, as the number of polysaccharide types that will be included in conjugated formulations is likely to be limited (probably less than 10).

Pneumolysin appears to satisfy the essential criteria for inclusion as a vaccine antigen. Virtually all clinical isolates of *S. pneumoniae* produce pneumolysin, and its primary structure apparently varies little. Comparison of the deduced amino acid sequences of the pneumolysin genes from two *S. pneumoniae* strains belonging to different serotypes, isolated from opposite sides of the world 65 years apart, revealed only a single amino acid difference (101). Native pneumolysin is, of course, unsuitable for inclusion in a vaccine for humans because of its toxicity. However, as described above, site-directed mutagenesis studies have resulted in the development of pneumolysin derivatives (pneumolysoids) deficient in either cytolytic activity or complement activation, or both. These pneumolysoids are unlikely to have deleterious side effects in humans and have been shown to be at least as effective as native pneumolysin in protecting mice from challenge (112; J. C. Paton, R. A. Lock, P. W. Andrew, G. J. Boulnois & T. J. Mitchell, unpublished data).

The availability of large quantities of protein purified from recombinant *E.*

coli has permitted the synthesis of conjugates of pneumolysin derivatives with otherwise poorly immunogenic polysaccharides. Mice immunized with a conjugate of type 19F polysaccharide and the pneumolysoid containing the Trp433→Phe mutation responded with very high antibody levels to the polysaccharide (20 times higher than those immunized with type 19F polysaccharide alone) as well as significant antibody titers against the protein component (112). These conjugates provided infant mice with a high degree of protection against challenge with type 19F pneumococci (C.-J. Lee, J. C. Paton, R. A. Lock, P. W. Andrew, G. J. Boulnois & T. J. Mitchell, unpublished data). Clearly, such antigens can potentially evoke a significant anticapsular response, as well as an antivirulence-protein response in humans, including young children, thereby conferring comprehensive protection against the various forms of pneumococcal disease.

The use of the surface proteins as vaccine antigens is not complicated by toxic side effects. Immunization of mice with a purified truncated derivative of PspA provides a high degree of protection from challenge with virulent pneumococci producing a similar surface-protein type (139), but antigenic variability between strains (as revealed by studies with protective monoclonal antibodies) (33) may complicate the use of PspA in human vaccines. Polyclonal antibody studies have identified PspA epitopes that are conserved among numerous pneumococcal strains (33), but whether these are protective remains to be determined. Nevertheless, a combination of relatively few PspA types may be all that is needed for comprehensive cover (22). To date, no evidence indicates antigenic variation of the 37-kDa surface protein, and the preliminary report of its protective efficacy as an immunogen in mice renders it worthy of detailed study.

Antibodies directed against proteins protruding from the surface of pneumococci, such as anticapsular antibodies, will probably opsonize and clear the invading organisms in vivo. Antibodies directed against pneumolysin, on the other hand, probably exert their protective effect by direct neutralization of free toxin. Thus, immunization with a combination of pneumolysoid and one or more of the surface proteins could potentially provide a greater degree of protection. Collaborative studies testing the protective efficacy of immunization with combinations of pneumolysoid and PspA are currently in progress.

A final consideration relating to vaccine development is the question of mucosal immunity. Pneumococcal otitis media and sinusitis are diseases that are largely confined to mucosae. Furthermore, colonization of the nasopharyngeal mucosa by pneumococci is a recognized prerequisite for the development of these diseases as well as for the more invasive, life-threatening diseases such as pneumonia and meningitis (6). Existing purified polysaccharide vaccines induce little, if any, secretory IgA antibody response at mucosal surfaces, and these vaccines do not have a significant impact on carriage (36, 150). Conjugate vaccines may be more effective in this regard; immunization with a *H. influenzae* type b conjugate vaccine eliminated carriage of that organism in Finnish children (138). Alternatively, mucosal immunization regimens (oral or intranasal) capable of inducing secretory IgA responses against protective pneumococcal antigens might also result in blockade of this initial step in the development of a pneumococcal infection. Recently much interest has centered upon the oral vaccine potential of live attenuated *Salmonellae* species that express foreign antigens. These vaccine strains are capable of limited invasion of the gut mucosa and deliver the foreign antigens directly to gut-associated lymphoid tissue such as the Peyer's patches. This results in a strong humoral immune response and, in particular, induces production of specific antibody at mucosal surfaces, both in the gut and at other mucosal sites (5, 45). We are presently examining the possibility of using such live carriers for pneumococcal antigens. An attenuated *Salmonella* strain capable of stably expressing a toxoided form of pneumolysin has been constructed and oral immunization of mice with this organism elicits both IgG and IgA antibodies to pneumolysin (113).

CONCLUDING REMARKS

S. pneumoniae remains one of the world's foremost human pathogens. The fact that antibiotic therapy alone cannot be relied upon to prevent death from invasive pneumococcal disease, as well as the suboptimal efficacy of the existing polysaccharide vaccine, have highlighted the need to continue to improve our understanding of the mechanism by which this organism colonizes, invades, and so often kills its host. During the past decade, technological advances have permitted a far more detailed examination of this complex process. Molecular studies have resulted in the identification of potential vaccine antigens, and genetic manipulations have produced derivatives suitable for administration to humans. Preliminary studies in animal models of pneumococcal disease have been promising, but it remains to be seen whether these novel vaccine formulations can protect humans, particularly young children and other high-risk groups, from all serotypes of *S. pneumoniae*.

Literature Cited

1. Alouf, J. E. 1980. Streptococcal toxins (streptolysin O, streptolysin S, erythrogenic toxin). *Pharm. Ther.* 11:661–717
2. Andersson, B., Beachey, E. H., Tomasz, A., Tuomanen, E., Svanborg-Eden, C. 1988. A sandwich adhesin on *Streptococcus pneumoniae* attaching to human oropharyngeal epithelial cells in vitro. *Microb. Pathogen.* 4:267–78
3. Andersson, B., Gray, B. M., Dillon, H. C., Bahrmand, A., Svanborg-Eden, C. 1988. Role of adherence of *Strep-*

tococcus pneumoniae in acute otitis media. *Pediatr. Infect. Dis.* 7:476–80
4. Andersson, B., Gray, B. M., Svanborg-Eden, C. 1988. Role of attachment for the virulence of *Streptococcus pneumoniae*. *Acta Otolaryngol. Stockholm* 454:62–64 (Suppl.)
5. Attridge, S. R., Hackett, J., Morona, R., Whyte, P. 1988. Toward a live oral vaccine against enterotoxigenic *Escherichia coli* of swine. *Vaccine* 6:387–89
6. Austrian, R. 1981. Some observations on the pneumococcus and on the current status of pneumococcal disease and its prevention. *Rev. Infect. Dis.* 3:S1–17 (Suppl.)
7. Austrian, R. 1981. Pneumococcus: the first one hundred years. *Rev. Infect. Dis.* 3:183–89
8. Avery, O. T., Dubos, R. 1931. The protective action of a specific enzyme against type III pneumococcus infections in mice. *J. Exp. Med.* 54:73–89
9. Avery, O. T., MacLeod, C. M., McCarty, M. 1944. Studies on the chemical nature of the substance inducing transformation of pneumococcal types. Induction of transformation by a desoxyribonucleic acid fraction isolated from pneumococcus type III. *J. Exp. Med.* 79:137–58
10. Baquero, F., Martínez-Beltrán, J., Loza, E. 1991. A review of antibiotic resistance patterns of *Streptococcus pneumoniae* in Europe. *J. Antimicrob. Chemother.* 28(Suppl. C):31–38
11. Basten, A., Howard, J. G. 1973. Thymus independence. *Contemp. Top. Immunobiol.* 2:265–91
12. Beck, G., Puchelle, E., Plotkowski, C., Peslin, R. 1989. *Streptococcus pneumoniae* and *Staphylococcus aureus* surface properties in relation to their adherence to human buccal epithelial cells. *Res. Microbiol.* 140:563–67
13. Berry, A. M., Lock, R. A., Hansman, D., Paton, J. C. 1989. Contribution of autolysin to the virulence of *Streptococcus pneumoniae*. *Infect. Immunol.* 57:2324–30
14. Berry, A. M., Paton, J. C., Glare, E. M., Hansman, D., Catcheside, D. E. A. 1988. Cloning and expression of the pneumococcal neuraminidase gene in *Escherichia coli*. *Gene* 71:299–305
15. Berry, A. M., Paton, J. C., Hansman, D. 1992. Effect of insertional inactivation of the genes encoding pneumolysin and autolysin on the virulence of *Streptococcus pneumoniae* type 3. *Microb. Pathogen.* 12:87–93
16. Berry, A. M., Yother, J., Briles, D. E., Hansman, D., Paton, J. C. 1989. Reduced virulence of a defined pneumolysin-negative mutant of *Streptococcus pneumoniae*. *Infect. Immunol.* 57: 2037–42
17. Bhakdi, S., Roth, M., Sziegoleit, A., Tranum-Jensen, J. 1984. Isolation and identification of two hemolytic forms of streptolysin-O. *Infect. Immunol.* 46: 394–400
18. Bhakdi, S., Tranum-Jensen, J. 1986. Membrane damage by pore-forming bacterial cytolysins. *Microb. Pathogen.* 1:5–14
19. Boulnois, G. J., Mitchell, T. J., Saunders, F. K., Mendez, F. J., Andrew, P. W. 1990. Structure and function of pneumolysin, the thiol-activated toxin of *Streptococus pneumoniae*. In *Bacterial Protein Toxins,* ed. R. Rappuoli, J. E. Alouf, P. Falmagne, F. J. Fehrenbach, J. Freer, et al. Suppl. 19:43–51. Stuttgart: Gustav Fischer
20. Breiman, D. R. 1992. *Pneumococcal 37 kDa protein*. Presented at WHO/UNDP Programme for Vaccine Development Scientific Meeting on New Pneumococcal Vaccines, Geneva
21. Briese, T., Hakenbeck, R. 1985. Interaction of the pneumococcal amidase with lipoteichoic acid and choline. *Eur. J. Biochem.* 146:417–27
22. Briles, D. E. 1992. *Protection-eliciting properties of pneumococcal surface protein A (PspA)*. Presented at WHO/UNDP Programme for Vaccine Development Scientific Meeting on New Pneumococcal Vaccines, Geneva
23. Briles, D. E., Yother, J., McDaniel, L. S. 1988. Role of pneumococcal surface protein A in the virulence of *Streptococcus pneumoniae*. *Rev. Infect. Dis.* 10(Suppl. 2):S372–74
24. Broome, C. V. 1981. Efficacy of pneumococcal polysaccharide vaccines. *Rev. Infect. Dis.* 3:S82–88 (Suppl.)
25. Camara, M. 1992. *Cloning and analysis of the IgA1 protease and neuraminidase enzymes from* Streptococcus pneumoniae. PhD thesis. Univ. Leicester, United Kingdom
26. Camara, M., Mitchell, T. J., Andrew, P. W., Boulnois, G. J. 1991. *Streptococcus pneumoniae* produces at least two distinct enzymes with neuraminidase activity: cloning and expression of a second neuraminidase gene in *Escherichia coli*. *Infect. Immunol.* 59:2856–58
27. Chetty, C., Kreger, A. 1980. Characterization of the pneumococcal purpura-producing principle. *Infect. Immunol.* 29:158–64

28. Chetty, C., Kreger, A. 1981. Role of autolysin in generating the pneumococcal purpura-producing principle. *Infect. Immunol.* 31:339–44
29. Chu, C., Schneerson, R., Robbins, J. B., Rastogi, S. C. 1983. Further studies on the immunogenicity of *Haemophilus influenzae* type b and pneumococcal type 6A polysaccharide-protein conjugates. *Infect. Immunol.* 40:245–56
30. Cohen, B., Perkins, M. E., Putterman, S. 1940. The reaction between hemolysin and cholesterol. *J. Bacteriol.* 39: 59–60
31. Cole, R. 1914. Pneumococcus hemotoxin. *J. Exp. Med.* 20:346–62
32. Courtney, H. S. 1991. Degradation of connective tissue proteins by serine proteases from *Streptococcus pneumoniae*. *Biochem. Biophys. Res. Comm.* 175:1023–28
33. Crain, M. J., Waltman, W. D. II, Turner, J. S., Yother, J., Talkington, D. F., et al. 1990. Pneumococcal surface protein A (PspA) is serologically highly variable and is expressed by all clinically important capsular serotypes of *Streptococcus pneumoniae*. *Infect. Immunol.* 58:3293–99
34. de Kruiff, B. 1990. Cholesterol as a target for toxins. *Biosci. Rep.* 10:127–30
35. Díaz, E., García, E., Ascaso, C, Méndez, E., López, R., García J. L. 1989. Subcellular localization of the major pneumococcal autolysin: a peculiar mechanism of secretion in *Escherichia coli*. *J. Biol. Chem.* 264: 1238–44
36. Douglas, R. M., Hansman, D., Miles, H., Paton, J. C. 1986. Pneumococcal carriage and type-specific antibody. Failure of a 14-valent vaccine to reduce carriage in healthy children. *Am. J. Dis. Child.* 140:1183–85
37. Douglas R. M., Miles, H. B. 1984. Vaccination against *Streptococcus pneumoniae* in childhood: lack of demonstrable benefit in young Australian children. *J. Infect. Dis.* 149:861–69
38. Douglas R. M., Paton, J. C., Duncan, S. J., Hansman, D. J. 1983. Antibody response to pneumococcal vaccination in children younger than five years of age. *J. Infect. Dis.* 148:131–37
39. Eskola, J., Käyhty, H., Takala, A. K., Peltola, H., Ronnberg, P.-R., et al. 1990. A randomized, prospective field trial of a conjugate vaccine in the protection of infants and young children against invasive *Haemophilus influenzae* type b disease. *New Engl. J. Med.* 323:1381–87
40. Fattom, A., Lue, C., Szu, S. C., Mestecky, J., Schiffman, G., et al. 1990. Serum antibody response in adult volunteers elicited by injection of *Streptococcus pneumoniae* type 12F polysaccharide alone or conjugated to diphtheria toxoid. *Infect. Immunol.* 58: 2309–12
41. Feldman, C., Mitchell, T. J., Andrew, P. W., Boulnois, G. J., Read, R. C., et al. 1990. The effect of *Streptococcus pneumoniae* pneumolysin on human respiratory epithelium in vitro. *Microb. Pathogen.* 9:275–84
42. Feldman, C., Munro, N. C., Jeffery, P. K., Mitchell, T. J., Andrew, P. W., et al. 1991. Pneumolysin induces the salient histological features of pneumococcal infection in the rat lung in vivo. *Am. J. Respir. Cell Mol. Biol.* 5:416–23
43. Feldman, C., Read, R., Rutman, A., Jeffrey, P. K., Brain, A., et al. 1992. The interaction of *Streptococcus pneumoniae* with intact human respiratory mucosa in vitro. *Eur. Respir. J.* 5:576–83
44. Ferrante, A., Rowan-Kelly, B., Paton, J. C. 1984. Inhibition of in vitro human lymphocyte response by the pneumococcal toxin pneumolysin. *Infect. Immunol.* 46:585–89
45. Forrest, B. D., LaBrooy, J. T., Attridge S. R., Boehm, G., Beyer, L., et al. 1989. A candidate live oral typhoid/cholera hybrid vaccine is immunogenic in humans. *J. Infect. Dis.* 159:145–46
46. Forrester, H. L., Jahigen, D. W., LaForce, F. M. 1987. Inefficacy of pneumococcal vaccine in a high-risk population. *Am. J. Med.* 83:425–30
47. García, E., García, J. L., Ronda, C., García, P., López, R. 1985. Cloning and expression of the pneumococcal autolysin gene in *Escherichia coli*. *Mol. Gen. Genet.* 201:225–30
48. García, J. L., Sánchez-Puelles, J. M., García, P., López, R., Ronda, C. García, E. 1986. Molecular characterization of an autolysin-defective mutant of *Streptococcus pneumoniae*. *Biochem. Biophys. Res. Commun.* 137:614–19
49. García, P., García, J. L., García, E. López, R. 1986. Nucleotide sequence and expression of the pneumococcal autolysin gene from its own promoter in *Escherichia coli*. *Gene* 43:265–72
50. Geoffroy, C., Mengaud, J., Alouf, J. E., Cossart, P. 1990. Alveolysin, the thiol activated toxin of *Bacillus alvei,* is homologous to listeriolysin O, perfringolysin O, pneumolysin and strep-

tolysin O and contains a single cysteine. *J. Bacteriol.* 172:7301–5
51. Gilbert, J. V., Plaut, A. G., Fishman, Y., Wright, A. 1988. Cloning of the gene encoding streptococcal immunoglobulin A protease and its expresion in *Escherichia coli*. *Infect. Immunol.* 56:1961–66
52. Gray, B. M., Converse, G. M. III, Dillon, H. C. Jr. 1980. Epidemiological studies of *Streptococcus pneumoniae* in infants: acquisition, carriage and infection during the first 24 months of life. *J. Infect. Dis.* 142:923–33
53. Haas, A., Dumbsky, H., Kreft, J. 1992. Listeriolysin genes: complete sequence of *ilo* from *Listeria ivanovii* and *lso* from *Listeria seeligeri*. *Biochim. Biophys. Acta* 1130:81–84
54. Hage-Chahine, C. M., Del Giudice, G., Lambert, P.-H., Pechere, J.-C. 1992. Hemolysin-producing *Listeria monocytogenes* affects the immune response to T-cell-dependent and T-cell-independent antigens. *Infect. Immunol.* 60:1415–21
55. Halbert, S. P., Cohen, B., Perkins, M. E. 1946. Toxic and immunological properties of pneumococcal hemolysin. *Bull. Johns Hopkins Hosp.* 78:340–59
56. Harrison, J., Karcioglu, Z., Johnson, M. K. 1983. Response of leukopenic rabbits to pneumococcal toxin. *Curr. Eye Res.* 2:705–10
57. Höltje, J. V., Tomasz, A. 1976. Purification of the pneumococcal *N*-acetylmuramyl-L-alanine amidase to biochemical homogeneity. *J. Biol. Chem.* 251:4199–4207
58. Humphrey, J. H. 1944. Hyaluronidase production by pneumococci. *J. Pathol. Bacteriol.* 56:273–75
59. Hynes, W. L., Ferrettii, J. J. 1989. Sequence analysis and expression in *Escherichia coli* of the hyaluronidase gene of *Streptococcus pyogenes* bacteriophage H4489A. *Infect. Immunol.* 57: 533–39
60. Immunization Practices Advisory Committee. 1989. Pneumococcal polysaccharide vaccine. *Morbid. Mortal. Wkly. Rep.* 38:64–76
61. Iwamoto, M., Ohno-Iwashita, Y., Ando, S. 1987. Role of the essential thiol group in the thiol-activated toxin from *Clostridium perfringens*. *Eur. J. Biochem.* 167:425–30
62. Jacobs, M. R., Koornhof, H. J., Robins-Browne, R. M., Stevenson, C. M., Vermaak, Z. A., et al. 1978. Emergence of multiply resistant pneumococci. *New Engl. J. Med.* 299:735–40
63. Jalonen, E., Paton, J. C., Koskela, M., Kerttula, Y., Leinonen, M. 1989. Measurement of antibody responses to pneumolysin—a promising method for the etiological diagnosis of pneumococcal pneumonia. *J. Infect.* 19:127–34
64. Johnson, M. K. 1977. Cellular location of pneumolysin. *FEMS Microbiol. Lett.* 2:243–45
65. Johnson, M. K., Allen, J. H. 1975. The role of cytolysin in pneumococcal ocular infection. *Am. J. Opthalmol.* 80:518–21
66. Johnson, M. K., Boese-Marrazzo, D., Pierce, W. A. Jr. 1981. Effects of pneumolysin on human polymorphonuclear leukocytes and platelets. *Infect. Immunol.* 34:171–76
67. Johnson, M. K., Hobden, J. A., Hegenah, M., O'Callaghan, R. J., Hill, J. M., Chen, S. 1990. The role of pneumolysin in ocular infections with *Streptococcus pneumoniae*. *Curr. Eye Res.* 9:1107–14
68. Johnson, M. K., Knight, R. J., Drew G. K. 1982. The hydrophobic nature of the thiol-activated cytolysins. *Biochem. J.* 207:557–60
69. Johnston, R. B. Jr. 1981. The host response to invasion by *Streptococcus pneumoniae:* protection and pathogenesis of tissue damage. *Rev. Infect. Dis.* 3:282–88
70. Johnston, R. B. Jr. 1991. Pathogenesis of pneumococcal pneumonia. *Rev. Infect. Dis.* 13(Suppl. 6):S509–17
71. Kalin, M., Kanclerski, K., Granström, M., Möllby, R. 1987. Diagnosis of pneumococcal pneumonia by enzyme-linked immunosorbent assay of antibodies to pneumococcal hemolysin (pneumolysin). *J. Clin. Microbiol.* 25: 226–29
72. Kanclerski, K., Blomquist, S., Granström, M., Möllby, R. 1988. Serum antibodies to pneumolysin in patients with pneumonia. *J. Clin. Microbiol.* 26:96–100
73. Kanclerski, K., Möllby, R. 1987. Production and purification of *Streptococcus pneumoniae* hemolysin (pneumolysin). *J. Clin. Microbiol.* 25:222–25
74. Kawana, M., Kawana, C., Giebink, G. S. 1992. Penicillin treatment accelerates middle ear inflammation in experimental pneumococcal otitis media. *Infect. Immunol.* 60:1908–12
75. Kehoe, M. K., Miller, L., Walker, J. A., Boulnois, G. J. 1987. Nucleotide sequence of the streptolysin O (SLO) gene: structural homologies between SLO and other membrane damaging thiol-activated toxins. *Infect. Immunol.* 55:3228–32

76. Kelly, R. T., Farmer, S., Greiff, D. 1967. Neuraminidase activities of clinical isolates of *Diplococcus pneumoniae*. *J. Bacteriol.* 94:272–73
77. Kelly, R. T., Greiff, D. 1970. Toxicity of pneumococcal neuraminidase. *Infect. Immunol.* 2:115–17
78. Kilian, M., Reinholdt, J. 1987. A hypothetical model for the development of invasive infection due to IgA protease producing bacteria. *Adv. Exp. Med. Biol.* 216B:1261–69
79. Kostrzynska, M., Wadström, T. 1992. Binding of laminin, type IV collagen, and vitronectin by *Streptococcus pneumoniae*. *Zbl. Bakt.* 277:80–83
80. Krivan, H. C., Roberts, D. D., Ginsberg, V. 1988. Many pulmonary pathogenic bacteria bind specifically to the carbohydrate sequence GalNAcβ1–4Gal found in some glycolipids. *Proc. Natl. Acad. Sci. USA* 85:6157–61
81. Lancet. 1985. Acute respiratory infections in under-fives: 15 million deaths a year. *Lancet* 2:699–701
82. Lee, C.-J. 1987. Bacterial capsular polysaccharides—biochemistry, immunity and vaccine. *Mol. Immunol.* 24: 1005–19
83. Lee, C.-J., Banks, S. D., Li, J. P. 1991. Virulence, immunity and vaccine related to *Streptococcus pneumoniae*. *Crit. Rev. Microbiol.* 18:89–114
84. Leinonen, M., Sakkinen, A., Kalliokoski, R., Luotenen, J., Timonen, M., Mäkelä, P. H. 1986. Antibody response to 14-valent pneumococcal capsular polysaccharide vaccine in pre-school age children. *Pediatr. Infect. Dis.* 5:39–44
85. Leinonen, M., Syrjälä, H., Jalonen, E., Kujala, P., Herva, E. 1990. Demonstration of pneumolysin antibodies in circulating immune complexes—a new diagnostic method for pneumococcal pneumonia. *Serodiagn. Immunother. Infect. Dis.* 4:451–58
86. Libman, E. 1905. A pneumococcus producing a peculiar form of hemolysis. *Proc. N.Y. Pathol. Soc.* 5:168
87. Lock, R. A., Hansman, D., Paton, J. C. 1992. Comparative efficacy of autolysin and pneumolysin as immunogens protecting mice against infection by *Streptococcus pneumoniae*. *Microb. Pathogen.* 12:137–43
88. Lock, R. A., Paton, J. C., Hansman, D. 1988. Purification and immunological characterization of neuraminidase produced by *Streptococcus pneumoniae*. *Microb. Pathogen.* 4:33–43
89. Lock, R. A., Paton, J. C., Hansman, D. 1988. Comparative efficacy of pneumococcal neuraminidase and pneumolysin as immunogens protective against *Streptococcus pneumoniae* infection. *Microb. Pathogen.* 5:461–67
90. López, R., Sánchez-Puelles, J. M., García, E., García, J. L., Ronda, C., García, P. 1986. Isolation, characterization and physiological properties of an autolytic-deficient mutant of *Streptococcus pneumoniae*. *Mol. Gen. Genet.* 204:237–42
91. McDaniel, L. S., Scott, G., Kearney, J. F., Briles, D. E. 1984. Monoclonal antibodies against protease-sensitive pneumococcal antigens can protect mice from fatal infection with *Streptococcus pneumoniae*. *J. Exp. Med.* 160:386–97
92. McDaniel, L. S., Yother, J., Vijayakamur, M., McGarry, L., Guild, W. R., Briles, D. E. 1987. Use of insertional inactivation to facilitate studies of biological properties of pneumococcal surface protein A (PspA). *J. Exp. Med.* 165:381–94
93. Mäkelä, P. H., Leinonen, M., Pukander, J., Karma, P. 1981. A study of the pneumococcal vaccine in prevention of clinically acute attacks of recurrent otitis media. *Rev. Infect. Dis.* 3:S124–29 (Suppl.)
94. Male, C. J. 1979. Immunoglobulin A1 protease production by *Haemophilus influenzae* and *Streptococcus pneumoniae*. *Infect. Immunol.* 26:254–61
95. Mallet, C. P., Boylan, R. J., Everhart, D. L. 1984. Complement antigen-binding fragment (Fab) from secretory immunoglobulin A using *Streptococcus sanguis* immunoglobulin A protease. *Caries Res.* 18:201–8
96. Mansa, B., Kilian, M. 1986. Retained antigen-binding activity of Fabα fragments of human monoclonal immunoglobulin A1 (IgA1) cleaved by IgA1 protease. *Infect. Immunol.* 52: 171–74
97. Mengaud, J., Vicente, M.-F., Chenevert, J., Pereira, J. M., Geoffroy, C., et al. 1988. Expression in *Escherichia coli* and sequence analysis of the listeriolysin O determinant of *Listeria monocytogenes*. *Infect. Immunol.* 56:766–72
98. Michel, E., Reich, K. A., Favier, R., Berche, P., Cossart, P. 1990. Attenuated mutants of the intracellular bacterium *Listeria monocytogenes* obtained by single amino acid substitutions in listeriolysin O. *Mol. Microbiol.* 4: 2167–78
99. Mitchell, T. J., Andrew, P. W., Boulnois, G. J., Lee, C.-J., Lock, R. A., Paton, J. C. 1992. Molecular studies of pneumolysin, the thiol-activated

toxin of *Streptococcus pneumoniae,* as an aid to vaccine design. In *Bacterial Protein Toxins,* ed. B. Witholt, J. E. Alouf, G. J. Boulnois, P. Cossart, B. W. Dijkstra, et al; *Zentrabl. Bakteriol. Suppl.* 23:429–38

100. Mitchell, T. J., Andrew, P. W., Saunders, F. K., Smith, A. N., Boulnois, G. J. 1991. Complement activation and antibody binding by pneumolysin via a region of the toxin homologous to a human acute phase protein. *Mol. Microbiol.* 5:1883–88
101. Mitchell, T. J., Mendez, F., Paton, J. C., Andrew, P. W., Boulnois, G. J. 1990. Comparison of pneumolysin genes and proteins from *Streptococcus pneumoniae* types 1 and 2. *Nucleic Acids Res.* 18:4010
102. Mufson, M. A., Krause, H. E., Schiffman, G. 1983. Long term persistence of antibody following immunization with pneumococcal polysaccharide. *Proc. Soc. Exp. Biol. Med.* 173:270–75
103. Musher, D. M. 1992. Infections caused by *Streptococcus pneumoniae:* clinical spectrum, pathogenesis, immunity and treatment. *Clin. Infect. Dis.* 14:801–9
104. Nandoskar, M., Ferrante, A., Bates, E. J., Hurst, N., Paton, J. C. 1986. Inhibition of human monocyte respiratory burst, degranulation, phospholipid methylation and bactericidal activity by pneumolysin. *Immunology* 59:515–20
105. Neill, J. M. 1926. Studies on the oxidation and reduction of immunological substances. I. Pneumococcus hemotoxin. *J. Exp. Med.* 44:199–213
106. Neill, J. M. 1927. Studies on the oxidation and reduction of immunological substances. V. Production of antihemotoxin by immunization with oxidized pneumococcus hemotoxin. *J. Exp. Med.* 45:105–13
107. Ojicus, D. M., Young, J. 1991. Cytolytic pore forming proteins and peptides: is there a common structural motif? *Trends Biochem. Sci.* 16:225–28
108. O'Toole, R. D., Goode, L., Howe, C. 1971. Neuraminidase activity in bacterial meningitis. *J. Clin. Invest.* 50:979–85
109. Paton, J. C., Berry, A. M., Lock, R. A., Hansman, D., Manning, P. A. 1986. Cloning and expression in *Escherichia coli* of the *Streptococcus pneumoniae* gene encoding pneumolysin. *Infect. Immunol.* 54:50–55
110. Paton, J. C., Ferrante, A. 1983. Inhibition of human polymorphonuclear leukocyte respiratory burst, migration and bactericidal activity by the pneumococcal toxin, pneumolysin. *Infect. Immunol.* 41:1212–16
111. Paton, J. C., Lock, R. A., Hansman, D. 1983. Effect of immunization with pnuemolysin on survival time of mice challenged with *Streptococcus pneumoniae. Infect. Immunol.* 40:548–52
112. Paton, J. C., Lock, R. A., Lee, C.-J., Li, J. P., Berry, A. M., et al. 1991. Purification and immunogenicity of genetically obtained pneumolysin toxoids and their conjugation to *Streptococcus pneumoniae* type 19F polysaccharide. *Infect. Immunol.* 59:2297–2303
113. Paton, J. C., Morona, J. K., Harrer, S., Hansman, D., Morona, R. 1993. Immunization of mice with *Salmonella typhimurium* C5 *aroA* expressing a genetically toxoided derivative of the pneumococcal toxin pneumolysin. *Microb. Pathogen.* 14: In press
114. Paton, J. C., Rowan-Kelly, B., Ferrante, A. 1984. Activation of human complement by the pneumococcal toxin, pneumolysin. *Infect. Immunol.* 43:1085–87
115. Pinkney, M., Beachey, E., Kehoe, M. 1989. The thiol-activated toxin streptolysin O does not require a thiol group for cytolytic activity. *Infect. Immunol.* 57:2553–58
116. Plaut, A. G. 1983. The IgA1 proteases of pathogenic bacteria. *Annu. Rev. Microbiol.* 37:603–22
117. Pratt, S. A., Boulnois, G. J. 1987. Molecular cloning of the immunoglobulin A1 protease gene from *Streptococcus pneumoniae.* In *Streptococcal Genetics,* ed. J. J. Ferretti, R. Curtiss, III, pp. 177–80, Washington DC: Am. Soc. Microbiol.
118. Riley, I. D., Douglas, R. M. 1981. An epidemiological approach to pneumococcal disease. *Rev. Infect. Dis.* 3:233–45
119. Robbins, J. B., Austrian, R., Lee, C.-J., Rastogi, S. C., Schiffman, G., et al. 1983. Considerations for formulating the second-generation pneumococcal capsular polysaccharide vaccine with emphasis on the cross-reactive types within groups. *J. Infect. Dis.* 148:1136–58
120. Roggentin, P., Rothe, B., Kaper, J. B., Galen, J., Lawrisuk, L., et al. 1989. Conserved sequences in bacterial and viral sialidases. *Glycoconjugate* 6: 349–53
121. Ronda, C., García, J. L., García, E., Sánchez-Puelles, J. M., López, R. 1987. Biological role of the pneumococcal amidase. Cloning of the *lytA*

gene in *Streptococcus pneumoniae*. *Eur. J. Biochem.* 164:621–24
122. Rubins, J. B., Duane, P. G., Charboneau, D., Janoff, E. N. 1992. Toxicity of pneumolysin to pulmonary endothelial cells in vitro. *Infect. Immunol.* 60:1740–46
123. Russell, H., Tharpe, J. A., Wells, D. E., White, E. H., Johnson, J. E. 1990. Monoclonal antibody recognizing a species-specific protein from *Streptococcus pneumoniae*. *J. Clin. Microbiol.* 28: 2191–95
124. Sampson, J., O'Connor, S., Stinson, A., Tharpe, J., Russell, H. 1991. Molecular cloning of the gene encoding the 37-kilodalton protein of *Streptococcus pneumoniae*. *General Meeting Am. Soc. Microbiol, 91st* D-112 (Abstr.)
125. Sánchez-Puelles, J. M., Ronda, C., García, J. L., García, P., López, R., García, E. 1986. Searching for autolysin functions. Characterization of a pneumococcal mutant deleted in the *lytA* gene. *Eur. J. Biochem.* 158:289–93
126. Sarnaik, S., Kaplan, J., Schiffman, G., Bryla, D., Robbins, J. B., Schneerson, R. 1990. Studies on *Pneumococcus* vaccine alone or mixed with DTP and on *Pneumococcus* type 6B and *Haemophilus influenzae* type b capsular polysaccharide-tetanus toxoid conjugates in two- to five-year-old children with sickle cell anemia. *Pediatr. Infect. Dis. J.* 9:181–86
127. Saunders, F. K., Mitchell, T. J., Walker, J. A., Andrew, P. W., Boulnois, G. J. 1989. Pneumolysin, the thiol-activated toxin of *Streptococcus pneumoniae,* does not require a thiol group for in vitro activity. *Infect. Immunol.* 57:2547–52
128. Schneewind, O., Pancholi, V., Fischetti, V. A. 1991. Surface proteins from Gram-positive cocci have a common motif for membrane anchoring. In *Genetics and Molecular Biology of Streptococci, Lactococci and Enterococci,* ed. G. M. Dunny, P. P. Cleary, L. L. McKay, pp. 152–54. Washington DC: Am. Soc. Microbiol.
129. Shumway, C. N. 1958. Spherocytic hemolytic anemia associated with acute pneumococcal infection in rabbits. *J. Lab. Clin. Med.* 51:240–47
130. Shumway, C. N., Klebanoff, S. J. 1971. Purification of pneumolysin. *Infect. Immunol.* 4:388–92
131. Shumway, C. N., Pollock, D. 1965. The effect of a pneumococcal product upon rabbit erythrocytes in vitro and in vivo. *J. Lab. Clin. Med.* 65:432–39
132. Smyth, C. J., Duncan, J. L. 1978. Thiol-activated (oxygen-labile) cytolysins. In *Bacterial Toxins and Cell Membranes,* ed. J. Jeljaszewicz, T. Wadstrom. pp. 129–83. New York: Academic
133. Snippe, H., van Houte, A.-J., van Dam, J. E. G., de Reuver, M. J., Jansze, M., Willers, J. M. N. 1983. Immunogenic properties in mice of hexasaccharide from the capsular polysaccharide of *Streptococcus pneumoniae* type 3. *Infect. Immunol.* 40: 856–61
134. Spika, J. S., Facklam, R. R., Plikaytis, B. D., Oxtoby, M. J., Pneumococcal Surveillance Working Party. 1991. Antimicrobial resistance of *Streptococcus pneumoniae* in the United States, 1979–1987. *J. Infect. Dis.* 163:1273–78
135. Stahl, W. L., O'Toole, R. D. 1972. Pneumococcal neuraminidase: purification and properties. *Biochim. Biophys. Acta* 268:480–87
136. Steinfort, C., Wilson, R., Mitchell, T., Feldman, C., Rutman, A., et al. 1989. Effect of *Streptococcus pneumoniae* on human respiratory epithelium in vitro. *Infect. Immunol.* 57:2006–13
137. Stenfors, L.-E., Raisanen, S. 1992. Abundant attachment of bacteria to nasopharyngeal epithelium in otitis-prone children. *J. Infect. Dis.* 165: 1148–50
138. Takala, A. K., Eskola, J., Leinonen, M., Käyhty, H., Nissinen, A., et al. 1991. Reduction in oropharyngeal carriage of *Haemophilus influenzae* type b (Hib) in children immunized with an Hib conjugate vaccine. *J. Infect. Dis.* 164:982–86
139. Talkington, D. F., Crimmins, D. L., Voellinger, D. C., Yother, J., Briles, D. E. 1991. A 43-kilodalton pneumococcal surface protein, PspA: isolation, protective abilities, and structural analysis of the amino-terminal sequence. *Infect. Immunol.* 59:1285–89
140. Tanenbaum, S. W., Gulbinsky, J., Katz, M., Sun, S.-C. 1970. Separation, purification and some properties of pneumococcal neuraminidase isoenzymes. *Biochim. Biophys. Acta* 198: 242–54
141. Tanenbaum, S. W., Sun, S.-C. 1971. Some molecular properties of pneumococcal neuraminidase isoenzymes. *Biochim. Biophys. Acta* 229:824–28
142. Tomasz, A., Moreillon, P., Pozzi, G. 1988. Insertional inactivation of the major autolysin gene of *Streptococcus pneumoniae*. *J. Bacteriol.* 170:5931–34
143. Tuomanen, E., Liu, H., Hengstler, B.,

Zak, O., Tomasz, A. 1985. The induction of meningeal inflammation by components of the pneumococcal cell wall. *J. Infect. Dis.* 151:859–68

144. Tweten, R. K. 1988. Nucleotide sequence of the gene for perfringolysin O (theta toxin) from *Clostridium perfringens:* significant homology with the genes for streptolysin O and pneumolysin. *Infect. Immunol.* 56:3235–40
145. Tweten, R. K., Harris, R. W., Sims, P. J. 1991. Isolation of a tryptic fragment from *Clostridium perfringens* theta-toxin that contains sites for membrane binding and self-aggregation. *J. Biol. Chem.* 266:12449–54
146. Vasquez-Boland, J. A., Dominguez, L., Rodriguez-Feri, E. F., Fernandez-Garayzabat, J. F., Suarez, G. 1989. Preliminary evidence that different domains are involved in cytolytic activity and receptor (cholesterol) binding in listeriolysin O, the *Listeria monocytogenes* thiol-activated toxin. *FEMS Microbiol. Lett.* 65:95–100
147. Volanakis, J. E., Kaplan, M. H. 1974. Interaction of C-reactive protein complexes with the complement system. II. Consumption of guinea pig complement by CRP complexes: requirement for human C1q. *J. Immunol.* 113:9–17
148. Walker, J. A., Allen, R. L., Falmagne, P., Johnson, M. K., Boulnois, G. J. 1987. Molecular cloning, characterization and complete nucleotide sequence of the gene for pneumolysin, the sulfhydryl-activated toxin of *Streptococcus pneumoniae. Infect. Immunol.* 55:1184–89
149. Watson, D. A., Musher, D. M. 1990. Interruption of capsule production in *Streptococcus pneumoniae* serotype 3 by insertion of transposon Tn*916*. *Infect. Immunol.* 58:3135–38
150. Wright, P. F., Sell, S. H., Vaughn, W. K., Andrews, C., McConnell, K. B., Schiffman, G. 1981. Clinical studies of pneumococcal vaccines in infants. II. Efficacy and effect on nasopharyngeal carriage. *Rev. Infect. Dis.* 3:S108–12 (Suppl.)
151. Yeagle, P. 1987. In *The Membranes of Cells,* p. 292. London: Academic
152. Yother, J., Briles, D. E. 1992. Structural properties and evolutionary relationships of PspA, a surface protein of *Streptococcus pneumoniae,* as revealed by sequence analysis. *J. Bacteriol.* 174:601–9
153. Yother, J., Handsome, G. L., Briles, D. E. 1992. Truncated forms of PspA that are secreted from *Streptococcus pneumoniae* and their use in functional studies and cloning of the *pspA* gene. *J. Bacteriol.* 174:610–18

Annu. Rev. Microbiol. 1993. 47:117–38

MOLECULAR DETERMINANTS OF THE VIRULENCE AND INFECTIVITY OF CALIFORNIA SEROGROUP BUNYAVIRUSES

Christian Griot, Francisco Gonzalez-Scarano, and Neal Nathanson

Departments of Microbiology and Neurology, School of Medicine, University of Pennsylvania, Philadelphia, Pennsylvania 19104-6076

KEY WORDS: California serogroup viruses, arboviruses, virus genetics, animal models

---

CONTENTS

ABSTRACT

California bunyaviruses cause encephalitis in mammalian hosts after peripheral infection. The virulence of these viruses is determined by their ability to replicate sequentially in striated muscle, cause viremia, and invade and replicate in the central nervous system. These viruses are also able to infect vector mosquitoes following ingestion of a blood meal containing virus. Bunyaviruses are negative stranded RNA viruses with a trisegmented genome, and the large, medium, and small RNA segments encode the polymerase, the

0066-4227/93/0117$02.00

glycoproteins, and the nucleoprotein, respectively. Reassortants between virulent and avirulent virus clones have been used to map virulence determinants in mice as well as determinants of infectivity in mosquitoes. Attenuation in mice and infectivity in mosquitoes of some virus clones maps to the medium RNA segment, implying that the virus glycoproteins, which are involved in virus entry, play a role in virulence. Attenuation in mice and mosquito infectivity of other clones maps to the large RNA segment, suggesting that cell-specific differences in the function of the viral polymerase can also determine virulence and host range.

INTRODUCTION

There has been a recent resurgence of interest in the pathogenesis of viral infections. The application of molecular methods has made it fruitful to reexamine old questions and to ask new ones about the determinants of the virus-host encounter. A further stimulus has been the opportunity to take a more sophisticated approach to the prevention or treatment of viral disease.

The role of virus virulence has been one of the major questions explored during this renaissance of work on viral pathogenesis. RNA viruses are highly mutable and different clones of a single agent can vary widely in virulence. We can now study this phenomenon at the level of individual viral proteins and their functions. The genetic determinants of virus virulence and infectivity has been a central focus of our studies of California serogroup viruses and constitutes the major theme of this review.

Taxonomy of the Bunyaviruses and the California Serogroup

The family Bunyaviridae includes over 250 virus species that are classified as a single virus family because of conserved structural and genetic features. The family is divided into five genera (*Bunyavirus, Phlebovirus, Hantavirus, Nairovirus,* and *Tospovirus*) each of which has somewhat unique biological and genetic characteristics. Bunyaviruses are arthropod-borne, with the exception of the *Hantavirus* genus, and are transmitted by mosquitoes, ticks, thrips, or other insects. All genera are vertebrate viruses with the exception of the *Tospovirus* genus, which are plant viruses. Each genus contains several serogroups comprising antigenically related viruses, and the genus *Bunyavirus* includes the California serogroup. Most of the bunyaviruses are of no known medical or veterinary importance and were isolated by routine inoculation of cell cultures or mice with suspensions prepared from pooled arthropods collected in the field. Additional information on the bunyaviruses can be found in several reviews (8, 11, 18, 19, 21, 31, 40, 49, 54).

The California serogroup comprises the viruses (including several variants or subtypes) listed in Table 1. Each of these viruses has a limited geographic range, and members have been isolated from several continents including North and South America, Europe, and Africa. The viruses are grouped because they share antigenic determinants, as shown in various serological tests, including complement fixation and hemagglutination inhibition, but they can be distinguished from each other in quantitative neutralization tests (1, 7). Also, all are transmitted by mosquito vectors. Together with other members of the genus *Bunyavirus*, they share a similar genetic organization. In addition, members of the serogroup can apparently undergo reassortment among themselves but not with other viruses in the genus *Bunyavirus*, which is probably characteristic of individual serogroups within the family Bunyaviridae (11, 54).

Table 1 Members of the California serogroup of viruses[a]

Virus or strain[b]	Geographic location	Principal mosquito vector	Vertebrate host	Human infection
California encephalitis	Western United States	*Aedes melanimon*	Rodents, rabbits	Rare
La Crosse	Midwestern United States	*Aedes triseriatus*	Chipmunks, squirrels	Endemic
Snowshoe hare	Canada, northern United States	*Culiseta inornata*	Snowshoe hare	Rare
San Angelo	Western North America	*Aedes* species	Unknown	None
Tahyna	Eastern Europe	*Aedes vexans, Culiseta annulata*	Rabbits, domestic animals	Endemic
Lumbo	East Africa Fin-	?	Unknown	None
Inkoo	land, Russia	*Aedes* species	Unknown	Rare
Melao	South America	?	Unknown	Unknown
Jamestown canyon	North America	*Aedes* species, *Culiseta inornata*	White-tailed deer	Uncommon
South River	Northeast United States, Quebec	?	Unknown	None
Keystone	Eastern United States	*Aedes atlanticus*	Unknown	None
Trivittatus	North America	*Aedes trivittatus*	Unknown	Rare
Guaroa	South America	*Anopheles* species	Unknown	Unknown

[a] Adapted from ref. 21.
[b] The members of the serogroup are divided into four antigenic subtypes indicated in bold.

Epidemiology and Ecology of California Serogroup Viruses

Hammon & Reeves (32) isolated California encephalitis virus in 1950 during their studies of arbovirus encephalitis in Kern County, California. Serological tests showed that this virus caused three cases of encephalitis in humans. Subsequently, California encephalitis virus has been isolated sporadically from mosquitoes but has not been associated with encephalitis (55). In 1960 a virus, later named La Crosse virus, was isolated from a fatal case of encephalitis in La Crosse, Wisconsin (61). La Crosse virus was shown to be closely related but distinct from California encephalitis virus, and it was rapidly recognized as causing human infections over a wide area of the midwestern United States (38). Serological tests identified several other arboviruses that were related to California encephalitis virus and La Crosse virus, and together these were designated the California serogroup (Table 1).

Regular reporting of encephalitis associated with California serogroup viruses began in 1963, and reports for 1963–1991 (31) indicate that the annual incidence has remained relatively constant at about 75 cases (range 42–174) in contrast to the marked epidemicity of St. Louis encephalitis. Most cases reported as California encephalitis result from La Crosse virus, and the epidemiological patterns reflect the ecology of this virus. Over 90% of California encephalitis cases occur in the states of Minnesota, Wisconsin, Iowa, Illinois, and Indiana, with the remainder distributed over the eastern United States. Cases usually occur from July through September, and children and young adults, ages 1–19, particularly boys, are at greatest risk because of activities such as camping and hiking. Serological surveys have indicated that at least 300,000 human infections with La Crosse virus occur each year, and that more than 1000 infections occur per reported case (28, 38).

Extensive testing of mosquitoes from 1960 to the present has shown that several California serogroup viruses are enzootic in North America (Table 1). Several of these (Jamestown canyon, snowshoe hare, trivittatus viruses) infect humans and cause sporadic cases of encephalitis (9, 29, 57). However, the great majority of cases of encephalitis are caused by La Crosse virus, and little recent evidence points to California encephalitis virus itself causing human disease (55).

La Crosse virus is transmitted primarily by the woodland mosquito, *Aedes triseriatus,* and its distribution and activity explain the epidemiological patterns (1, 27). Humans probably do not transmit the virus to mosquitoes and are therefore considered dead-end hosts. *A. triseriatus* breeds in tree holes and other sites of standing water, notably tire dumps, and feeds mainly on woodland rodents, particularly squirrels and chipmunks. La Crosse virus is transmitted through this vector-vertebrate feeding cycle with an extrinsic incubation period of approximately 14 days and an intrinsic incubation period

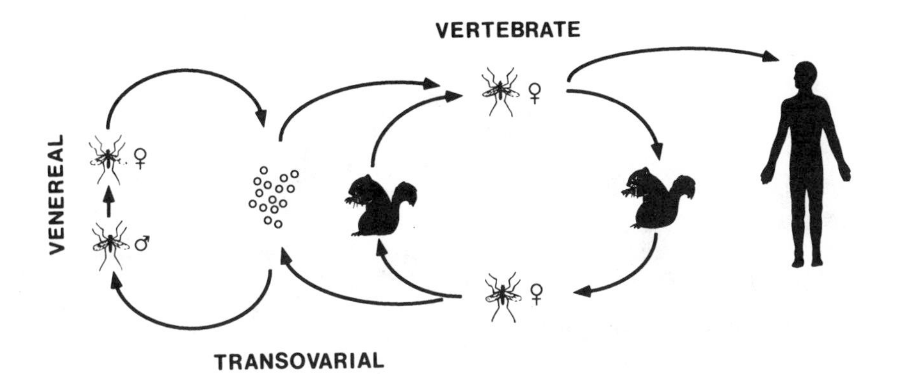

Figure 1 The maintenance cycles of La Crosse virus. In the major vector-vertebrate cycle, the principal vector is the woodland mosquito, *A. triseriatus,* and the principal vertebrate hosts are small rodents such as squirrels and chipmunks. The virus may also be transmitted by a transovarial cycle, from mosquito to mosquito, providing an overwintering mechanism for both host and parasite. Transovarial transmission also leads to infection of male mosquitoes, which can transmit during mating, thereby spreading infection within the mosquito population.

of 3–5 days. In addition, the virus can be transovarially transmitted, which may account for overwintering, and venereal transmission from transovarially infected males to females can also occur. Figure 1 summarizes these patterns.

The United States was recently invaded by the Asian tiger mosquito, *Aedes albopictus,* which has established itself across the southern United States (16, 41). This mosquito breeds extensively in standing water such as that found in tire dumps and feeds on woodland rodents and aggressively on humans. Experiments have shown that *A. albopictus* is a competent vector for La Crosse virus, and it will transmit the virus transovarially. These observations have raised the possibility that this mosquito might vector La Crosse virus, although such transmission has not yet been reported.

BUNYAVIRUS STRUCTURE, FUNCTION, AND GENETICS

Structure of the Virion and Organization of the Genome

Bunyavirions (Figure 2 and Table 2) are spherical or pleomorphic particles 80–120 nm in diameter, consisting of an outer lipid bilayer or envelope (4–5 nm thick) that surrounds the inner core (51, 59). Glycoprotein spikes extend 8–10 nm from the surface of the envelope and are composed of the G1 and G2 glycoproteins (about 600 each per virion). The core of the virion contains the RNA genome associated with about 2000 copies of the nucleocapsid (N) protein and about 25 copies of the large (L) protein per virion.

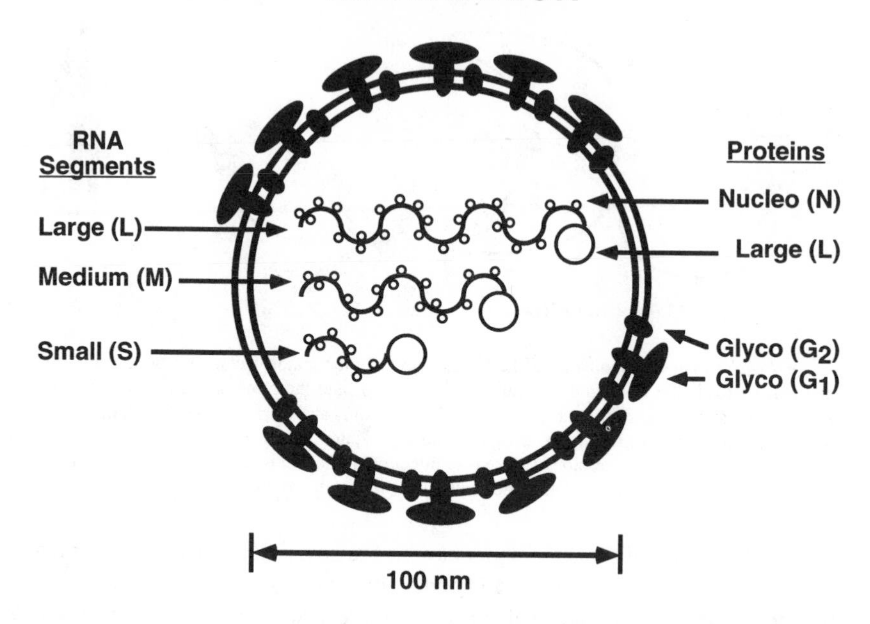

Figure 2 Structure of the bunyavirion to show its major features including the lipid envelope and its glycoproteins (600 copies per virion) surrounding the nucleocapsid core consisting of the trisegmented RNA genome that is associated with the nucleoprotein (2000 copies per virion) and the large protein (25 copies per virion). See text for details.

The bunyaviruses have a negative sense RNA genome (11) composed of three segments designated large (L), medium (M), and small (S) (Table 2). The 3′ and 5′ terminal 8–13 nucleotides are conserved among the three segments and among all viruses in the genus *Bunyavirus*. Because the 3′ and 5′ sequences are palindromic, they permit the formation of RNA-RNA duplexes, so that each segment can appear as a circle with the termini in the shape of a panhandle.

The organization of the viral genome differs somewhat for the five genera of bunyaviruses and the following description applies to the genus *Bunyavirus*. The L RNA segment contains one long open reading frame encoding the viral polymerase that is required for both transcription and replication of the genome (10–12). The M RNA segment contains a single large open reading frame that is translated as a single polyprotein and is cotranslationally translocated across the rough endoplasmic reticulum and cleaved to form the mature protein products. The polyprotein contains the G1, G2, and an NS_m protein in the order 3′ G2-NS_m-G1 5′ (11, 15, 26). The S RNA segment encodes the N

Table 2 RNA segments and viral proteins of the genus *Bunyavirus*[a]

RNA segment		Proteins encoded	
Designation	Kilobases[b]	Designation	Kilodaltons[c]
Large (L)	6.875	Large (L)	180–200
Medium (M)	4.458	Glycoprotein 1 (G1)	108–125
		Glycoprotein 2 (G2)	29–41
		Nonstructural (NS_m)	11–18
Small (S)	0.984	Nucleoprotein (N)	23–27
		Nonstructural (NS_s)	7–10

[a] Data are for La Crosse, snowshoe hare, or Bunyamwera viruses (18).
[b] Based on sequencing.
[c] Based on SDS polyacrylamide gel electrophoresis.

protein and the nonstructural (NS_S) protein in overlapping reading frames (6, 11). The N protein is an integral component of the nucleocapsid complex, elicits complement-fixing antibodies, and is the most highly conserved protein in the family Bunyaviridae, being about 40% homologous within each genus and about 80% homologous within each serogroup.

The Viral Glycoproteins and Their Functions

The glycoproteins subserve many important functions in the virus life cycle and also are important determinants of virus virulence in experimental animals. Structure and functions have been explored with the aid of monoclonal antibodies, proteolytic enzymes, manipulation of extracellular pH, monoclonal antibody resistant (MAR) mutants, and expression in recombinant vaccinia virus (15, 17, 24, 34).

The G1 protein is a type I glycoprotein, and its N terminus constitutes a large exodomain that may form a globular head bound together by multiple disulfide bridges. Its transmembrane domain is close to the C terminus, leaving a small C-terminal endodomain (18). The G2 protein has a central hydrophobic domain and is probably oriented with its N terminus external and its C terminus internal to the viral envelope, although this remains to be proven. The two glycoproteins are not covalently linked, and their precise orientation with respect to each other remains to be determined. However, we may surmise that they are assembled into heterodimers, several of which oligomerize to form a spike, although no data confirm this supposition. The open reading frame of the M RNA segment also encodes the NS_m protein, but its mature size (12–18 kDa) and nature (nonstructural or structural) remain to be determined.

VIRUS ATTACHMENT PROTEIN The G1 protein probably binds the virus to cellular receptors because a subset of anti-G1 monoclonal antibodies block attachment of virus to erythrocytes (hemagglutination inhibition) and also block binding to permissive cell lines (A. Pekosz, C. Griot, N. Nathanson & F. Gonzalez-Scarano, unpublished observations). These same anti-G1 antibodies also neutralize viral infectivity in cell culture. The G2 protein may act as the viral attachment protein in the mosquito midgut where the G1 protein on ingested virus is rapidly cleaved by proteolytic enzymes (42, 43).

CELL-TO-CELL FUSION Bunyaviruses probably enter cells via the endosomal pathway where reduced pH mediates fusion between the viral envelope and a cellular membrane, releasing the viral genome into the cytosol. The viral glycoproteins play a critical role in this process, which is mimicked by virus-mediated fusion from within or fusion from without (22). Thus, when cell cultures infected with La Crosse virus are briefly pulsed with low pH buffer, they undergo dramatic fusion (Figure 3). The G1 protein probably plays an important role in fusion, first by binding virus to cellular receptors and then by undergoing a pH-mediated conformation change (17, 23). Thus, some MAR mutants selected with anti-G1 monoclonal antibodies show a

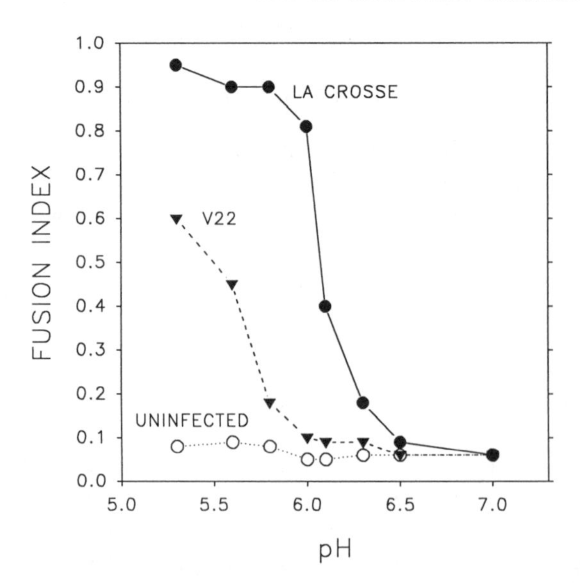

Figure 3 pH-Dependent fusion profile of wild-type La Crosse virus and monoclonal antibody resistant mutant V22, which shows that clone V22 was much less efficient in inducing fusion. An assay for fusion from within was used to determine the pH at which the virus would fuse monolayers of BHK-21 cells. The fusion index, FI, was calculated as $FI = 1 - (C/N)$, where C and N are the numbers of cells and nuclei, respectively. A fusion index of 1 would represent an infinite number of nuclei enclosed in one giant syncytium (17, 22).

markedly reduced ability to mediate fusion from within (20). The G2 protein is probably also important in fusion because liposomes containing the glycoproteins can still mediate fusion after treatment with trypsin that would normally be sufficient to remove most of the exodomain of the G1 protein (53).

IMMUNOGENICITY AND IMMUNE PROTECTION The G1 protein is considered immunodominant because hybridomas produced by immunized mice are mainly directed against the G1 protein while very few are directed against the G2 protein (24, 39). Most of the antibodies identify conformational epitopes; trypsin treatment cleaves away an N terminus fragment of the G1 protein, thus removing the domain against which many monoclonal antibodies are directed (48). This observation is consistent with the hypothesis that the G1 protein forms much of the head of the glycoprotein spikes that protrude from the virion surface.

The G1 protein is also associated with protective immunity. Many of the anti-G1 antibodies have potent neutralizing activity in cell culture because of their ability to block attachment to the putative viral receptor or to block a postbinding entry step such as fusion (52). Administered passively to newborn mice, these same antibodies provide excellent protection against a subsequent potentially lethal virus challenge administered by a peripheral route. The proteins involved in cellular immunity to the California serogroup viruses have not been studied.

Genetics, Complementation, and Reassortment

Temperature sensitive (*ts*) mutants of bunyaviruses are easily obtained through mutagenesis and can be used for genetic studies (5, 21). California serogroup viruses readily complement each other when two *ts* mutants, with *ts* lesions in different gene segments, are used to initiate dual infections at the restrictive temperature. The yield of each mutant is greater than if either one is grown independently.

Reassortants between California serogroup viruses can be constructed by coinfecting permissive cultures with two viruses. One can facilitate the selection of reassortants by using two *ts* mutants with *ts* lesions in different gene segments; the initial infection is conducted at the permissive temperature and reassortant progeny are selected by growth at the restrictive temperature (5). For California serogroup viruses, one can obtain mutants in the L and M RNA segments but not in the S RNA segment; thus, *ts* mutants cannot be used to obtain all possible reassortant genotypes (5). If wild-type parental viruses are used, all possible reassortant genotypes can be obtained, but this method requires the use of parental clones that can be distinguished by genotyping, using either a dot blot hybridization method, or by phenotyping using sodium dodecyl sulfate polyacrylamide gel electrophoresis, and involves the laborious screening of progeny to identify a complete panel of reassortants (12, 36). However, reassortants obtained in this manner have the advantage that they have not been deliberately mutagenized and are therefore more appropriate for studies in which the genotype is correlated with the biological phenotype.

The frequency of reassortment can be determined for *ts* mutants by assaying (at both restrictive and permissive temperature) the progeny of a dual infection and is usually 5–50% for California serogroup viruses. The relative frequency of reassortants of different genotypes is sufficiently unequal so that reassortment does not appear to be an entirely random event (12, 62). Reassortment can occur between many members of the California serogroup with the exception of Guaroa virus (5). However, members of the California serogroup will not reassort with viruses in other serogroups of the genus *Bunyavirus* (5, 33). Reassortment of California serogroup viruses also occurs in mosquitoes that are experimentally infected with two different viruses and may play a role in the evolution of bunyaviruses in nature (3).

PATHOGENESIS OF BUNYAVIRUS INFECTION

Experimental Murine Model of Infection

California serogroup viruses readily infect laboratory rodents, and experimental infection of mice appears to be a good model of the events that occur either in humans or in the small woodland rodents that are the vertebrate hosts in the natural maintenance cycle of La Crosse virus (50, 56). Changes in the dose and site of virus injection, the age of the host, and the virulence of the virus can cause variations in the outcome of infection from uniformly and rapidly fatal to subclinical and asymptomatic (35, 37), mimicking the spectrum of infection in chipmunks and humans. Suckling mice are highly susceptible to subcutaneous (sc) injection of virus and can be used to study the extraneural phase of infection, whereas adult mice can be used to study the central nervous system (CNS) phase because they are highly susceptible to intracerebral (ic) virus injection but do not develop viremia after sc injection (35).

Figure 4 summarizes the sequence of events that follow subcutaneous injection of virus in mice. Many of the salient steps have been well established (36, 37) and indicate that the major site of extraneural replication is striated muscle followed by a plasma viremia and invasion of the CNS, which is the target organ. Within the CNS, infection varies according to age: in suckling mice, there is an overwhelming pancellular infection whereas in adult mice infection is concentrated in neurons, in particular, in areas of the brain such

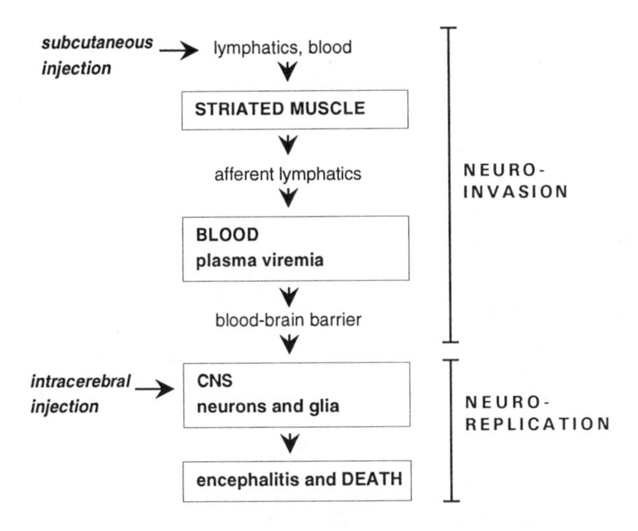

Figure 4 Pathogenesis of infection with wild-type La Crosse virus, showing the sequential events beginning with injection of the virus by mosquito bite and terminating in encephalitis. The sequence is divided into two phases, neuroinvasion followed by neuroreplication, but note that in humans most infections do not progress to invasion of the CNS. Boxes indicate those steps in infection that are well documented; unboxed steps are conjectural.

as the pyramidal cells of the hippocampus and the Purkinje cells of the cerebellum (30, 35). However, several intervening steps in the sequence are poorly documented, including (*a*) the initial site of virus replication after sc injection, (*b*) the movement of virus from the site of injection to striated muscle, (*c*) the passage of virus from the blood into the CNS, and (*d*) the possible centrifugal spread of virus from the CNS via the peripheral nervous system.

Genetic Determinants of Virulence and Infectivity

As with other animal viruses, clones of California serogroup viruses can differ in their virulence for experimental hosts. We have used two assays to measure virulence: ic injection of adult mice to assess the ability of the virus to produce lethal encephalitis (designated neurovirulence) and sc injection of suckling mice to test the ability of the virus to complete the sequential steps that carry the virus from the periphery to the CNS (designated neuroinvasiveness). In each instance, the virus was titrated in mice and the ratio plaque-forming units (PFU)/LD_{50} was computed to quantify the relative virulence of each clone (Figure 5). Experience indicated that neurovirulence and neuroinvasiveness could be distinguished in that a virus clone could be judged by both assays to be virulent, attenuated, or neurovirulent but not neuroinvasive.

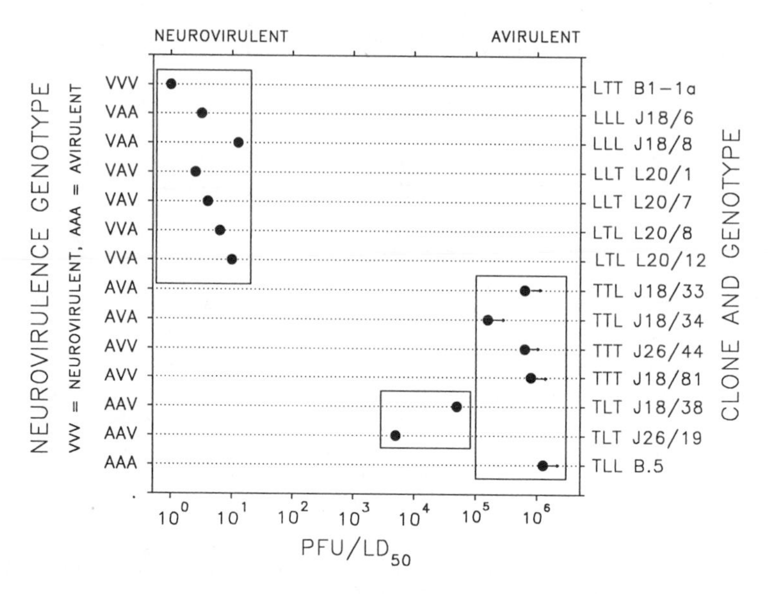

Figure 5 Neurovirulence of reassortants between a virulent clone B1-1a (genotype LTT; neurovirulence genotype VVV) and nonneurovirulent clone B.5 (genotype TLL; neurovirulence genotype AAA). Genotype indicates whether the L, M, and S RNA segments were derived from La Crosse or Tahyna virus, while the neurovirulence genotype indicates whether the segments were derived from a virulent or attenuated parental clone. Viruses were titrated by ic injection in adult mice and the ratio of PFU/LD_{50} was computed. Reassortant clones with an L RNA segment derived from the virulent parent were virulent while clones with an L RNA segment derived from the avirulent parent were attenuated. Reassortants with a virulence genotype of AAV are slightly less attenuated than expected, probably because of a modulating effect of other gene segments. Figure reproduced from data in Ref. 12.

To analyze the genetic determinants of viral virulence, we have used a three-step approach. First, virus clones were identified that exhibited different degrees of virulence. Most wild-type California serogroup isolates are both neurovirulent and neuroinvasive. To obtain markedly attenuated virus clones, wild-type virus was passaged either in mice or in cell culture, and attenuated stocks were plaque purified to yield highly attenuated clones (14, 35, 44). Preference was given to clones that were quantitatively very attenuated to permit the identification of intermediate phenotypes among the progeny of virus crosses. In addition, we used attenuated clones that replicated well in the default cell culture system (BHK-21 cells) so that attenuation was specific for the animal host and did not represent merely a global inability of the virus to replicate in all host cells. Second, each attenuated clone was crossed with a virulent clone, selected to represent a different virus within the California serogroup. This was necessary to permit genotyping of the progeny viruses derived from the cross (13). Third, a panel of reassortants representing all six possible permutations of gene segments, together with the two parental clones, was tested in mice.

NEUROVIRULENCE An attenuated clone (B.5, genotype TLL) was derived that was over one million–fold less neurovirulent than wild-type La Crosse virus; after ic injection of adult mice, the ratio PFU/LD50 was $>10^6$ for clone B.5 compared with a ratio of 1 for La Crosse virus (12, 14). When reassortants were made with a virulent clone, attenuation mapped to the L RNA segment, and pathogenesis studies showed that B.5, and reassortants with an L RNA segment derived from B.5, replicated very poorly in the brains of adult mice (Figure 6). Several other phenotypes such as inability to replicate in mouse cell lines, plaque morphology, and temperature sensitivity also mapped to the L RNA segment. Although clone B.5 was severely restricted at 39.8°C (titer reduced by 10^5), some plaques did appear, representing *ts* revertants. The *ts* revertant clones were also neurovirulence revertants, and further genetic studies localized the reversion to the L RNA segment (C. Griot, N. Nathanson & F. Gonzalez-Scarano, unpublished data). Because the L RNA segment encodes the viral polymerase, clone B.5 apparently carries a pleiotropic mutation in its polymerase that confers both temperature and host range restriction and that markedly reduces the clone's ability to replicate in neurons of the adult mouse CNS.

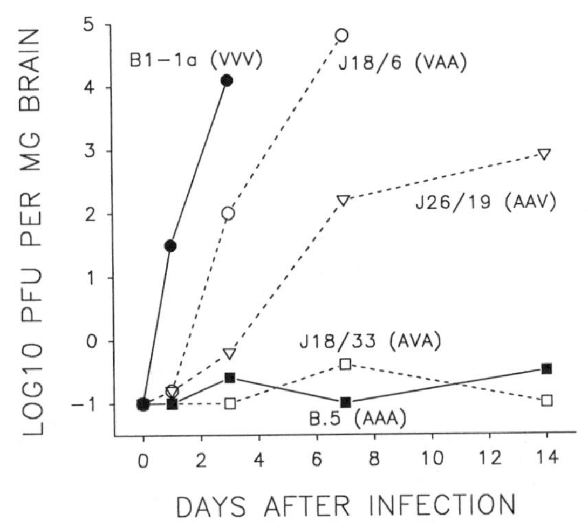

Figure 6 Brain replication after ic injection of 700 PFU of reassortants between a neurovirulent clone, B1-1a, and a nonneurovirulent clone, B.5. Reassortants with an L RNA segment derived from the virulent parent (J18/6, virulence genotype VAA) replicate vigorously, whereas reassortants with an L RNA segment derived from the avirulent parent (J18/33, virulence genotype AVA) replicate poorly. Data taken from Ref. 12.

NEUROINVASIVENESS Clone Tahyna/181-57 was derived from multiple ic passages of an isolate of Tahyna virus and was about 10,000-fold attenuated when injected sc in suckling mice, because its ratio PFU/LD_{50} is $>10^4$ compared with a ratio of 1 for wild-type La Crosse virus (35, 44). Tahyna/181-57 was neuroadapted and was highly virulent when injected ic in suckling (or adult) mice, so that its attenuation after sc injection reflected reduced neuroinvasiveness and not reduced neurovirulence. When reassortants between Tahyna/181-57 and wild-type La Crosse virus were tested in suckling mice, neuroinvasiveness mapped to the M RNA segment (Figure 7). Comparative pathogenesis studies showed that Tahyna/181-57 replicated poorly in striated muscle so that a modest virus dose (sc injection of 700 PFU) failed to cause viremia, and even a very large virus dose (sc injection of 425,000 PFU) produced a low viremia followed by death at 7 days compared with 2–3 days for wild-type La Crosse virus (Figure 8).

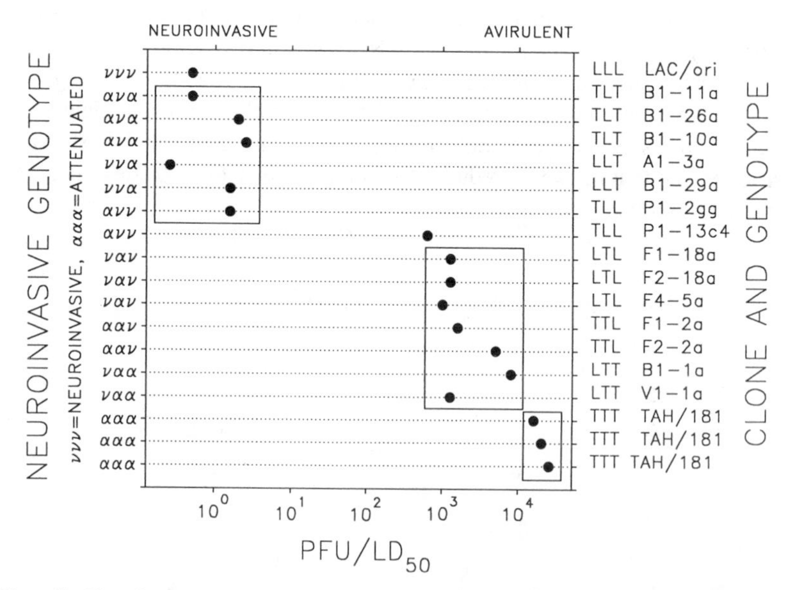

Figure 7 Neuroinvasiveness of reassortant clones derived from a cross between a neuroinvasive clone, wild-type La Crosse (genotype LLL; the neuroinvasive genotype, ννν), and a nonneuroinvasive clone, Tahyna/181-57 (genotype TTT; neuroinvasive genotype ααα). The genotype indicates whether the L, M, and S RNA segments were derived from La Crosse or from Tahyna virus, while the neuroinvasive genotype indicates whether the segments were derived from a neuroinvasive or an attenuated parental clone. Neuroinvasive indices were determined by titration of each clone by sc injection in suckling mice followed by computation of the ratio PFU/LD_{50}. Reassortants whose M RNA segment came from the neuroinvasive parent virus were neuroinvasive whereas reassortants whose M RNA segment came from the avirulent parent were nonneuroinvasive. Clone P1-13c4 is much less neuroinvasive than would be expected, perhaps reflecting a spontaneous mutation. Data taken from Ref. 36.

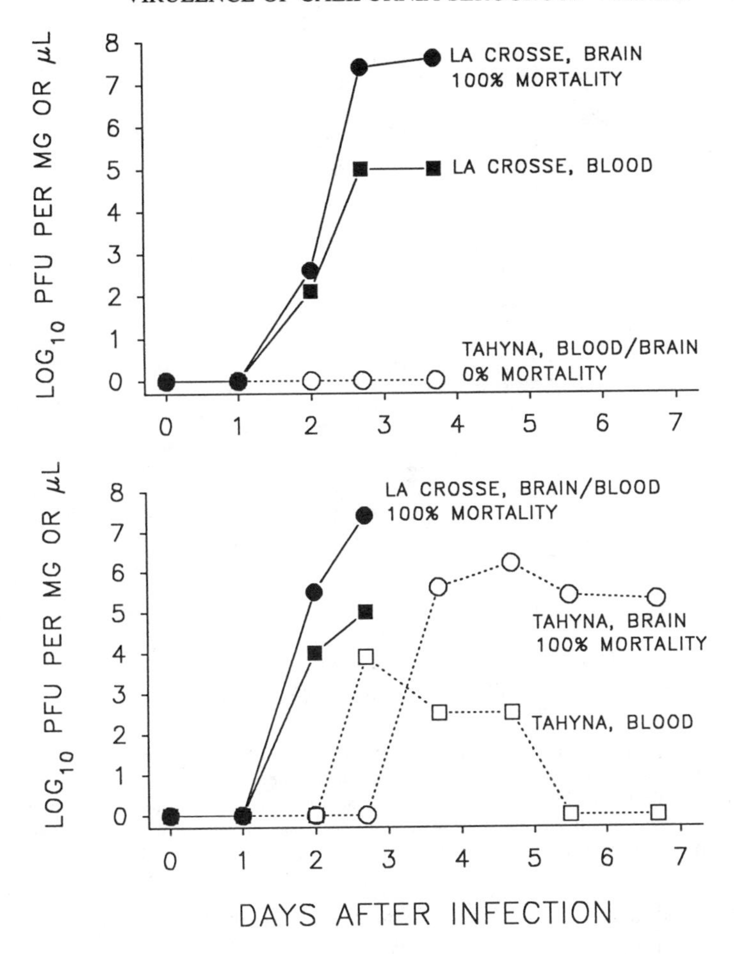

Figure 8 Replication of a neuroinvasive clone, wild-type La Crosse virus, and of a nonneuroinvasive clone, Tahyna/181-57 virus, after sc injection into suckling mice. (*Upper panel*) 700 PFU inoculum; (*lower panel*) 425,000 PFU inoculum. The nonneuroinvasive clone fails to replicate in striated muscle and produces no viremia and no mortality after injection of 700 PFU; it replicates poorly after injection of 425,000 PFU and produces mortality after a long survival time relative to the virulent clone. Data taken from Ref. 36.

Because the M RNA segment encodes the viral glycoproteins, it appeared likely that a mutation(s) in these proteins was responsible for attenuation. To further explore the putative role of the glycoproteins, a series of MAR mutants of La Crosse virus was derived using a panel of neutralizing monoclonal antibodies against the G1 protein (24, 25, 48). One of these mutants, clone V22, when injected sc into suckling mice, exhibited markedly reduced

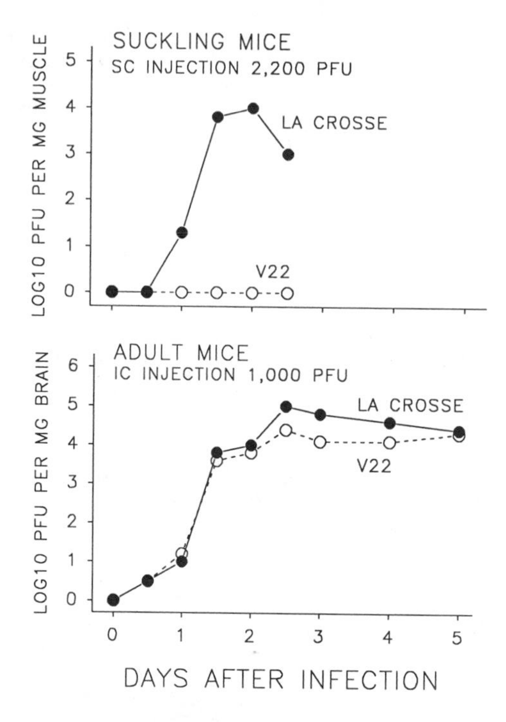

Figure 9 Replication in mice of a MAR mutant, clone V22, compared with wild-type La Crosse virus. (*Upper panel*) Injection of 2200 PFU sc in suckling mice; (*lower panel*) injection of 1000 PFU ic in adult mice. Clone V22 shows reduced neuroinvasiveness because it fails to replicate in striated muscle or kill suckling mice after sc injection, but its neurovirulence is similar to that of La Crosse virus after ic injection in adult mice. Data taken from Ref. 20.

neuroinvasiveness (20). Clone V22 also showed reduced ability to replicate in striated muscle although it replicated quite well in brain after ic injection of adult mice (Figure 9). Clone V22 also was much less effective as a fusogen (Figure 3) so that the putative mutation in the G1 glycoprotein of this clone appeared to be associated with both a reduction in fusion efficiency and in neuroinvasiveness.

Clone B.5, discussed above, showed reduced neuroinvasiveness in suckling mice (12, 14). Again, one should note that the reduced neuroinvasiveness of clone B.5 reflected its reduced ability to invade the CNS because it was highly virulent after ic injection of suckling mice in contrast to its reduced neurovirulence in adult mice. Taken together, Tahyna/181-57 and clone B.5 showed that reduced neuroinvasiveness could be produced by attenuating mutations in either the M or the L RNA segments. The availability of these clones provided us with the opportunity to examine the

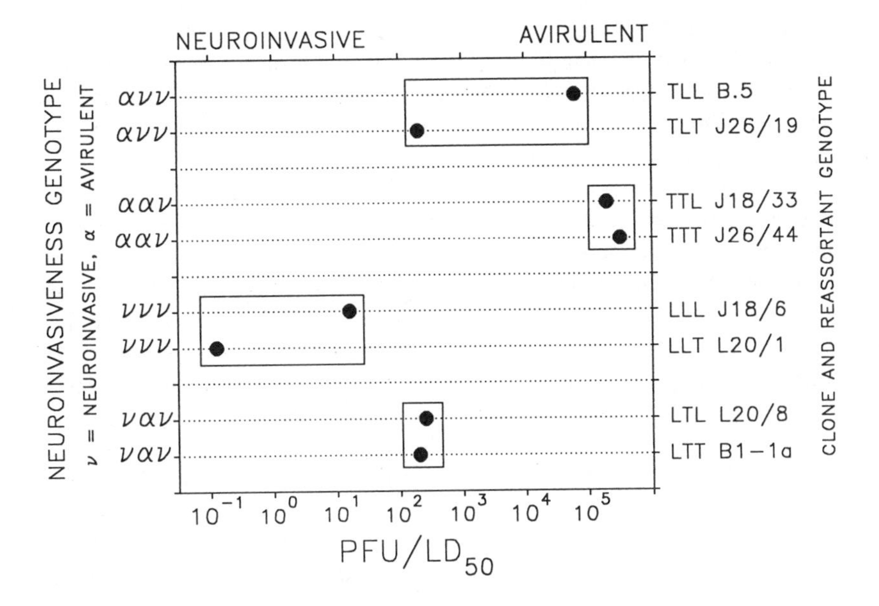

Figure 10 Neuroinvasiveness of reassortants between two nonneuroinvasive clones, B.5 (genotype TLL; neuroinvasive genotype $\alpha\nu\nu$) and B1-1a (genotype LTT, neuroinvasive genotype $\nu\alpha\nu$), of California serogroup viruses. Clones with a neuroinvasive genotype of $\nu\nu\nu$ are 100- to 1000-fold more virulent than either parental clone, and clones with a neuroinvasive genotype of $\alpha\alpha\nu$ are at least 10-fold less invasive than either parental clone (30).

interaction of two attenuated gene segments, and a panel of reassortant viruses was tested for neuroinvasiveness. [To produce this panel the attenuated M RNA segment from clone Tahyna/181-57 was represented by clone B1-1a (genotype LTT), which was reciprocal to clone B.5 (genotype TLL)]. Figure 10 shows that the effects of the two attenuating gene segments were independent and additive. Thus, clones with attenuated L and M RNA segments were 10- to 100-fold more attenuated than either parental clone, and clones with neither attenuated segment were about 1000-fold more virulent than either parental clone (30).

Experimental Infection of Mosquitoes

Like other arboviruses, California serogroup viruses replicate in insect vectors as well as in vertebrate hosts (Table 1). Infection of mosquitoes begins with a first blood meal and terminates, after an extrinsic incubation period, in the transmission of infection from vector to vertebrate host during a second blood meal. Mosquito and vertebrate infections differ in many significant ways, because the mosquito undergoes a persistent lifelong infection with little if any disease, and acquires and transmits the virus in a completely different manner from that characteristic for the vertebrate host. Virus clones cannot be considered to differ in their virulence for mosquitoes but they do differ in their ability to undergo a complete cycle of infection, and these differences can be mapped to specific gene segments.

The sequential events following ingestion of California serogroup viruses by competent mosquito hosts have been documented (45, 46, 60). The virus first enters the epithelial cells lining the midgut, and after an eclipse phase during which no infectivity can be detected, is found in the hemocoel just outside the basal lamina of the midgut. From the hemocoel, the virus has access to all body tissues and replicates in many sites including the heart, the neural ganglia, fat body, ovaries, and salivary glands. The salivary glands are one of the last tissues infected, and infected cells discharge numerous virions into the lumen of the gland whence they can be injected into the vertebrate host during feeding. The extrinsic incubation period is 7–14 days for La Crosse virus.

Infection of ovaries, including oocytes, results in widespread infection including the gonadal tissue of both female and male offspring (4, 60, 63). This provides a mechanism for transovarial transmission and for venereal spread within the mosquito population (Figure 1).

Infection of the mosquito midgut differs dramatically from infection of vertebrate tissues, since the virus is exposed to concentrated proteolytic enzymes in an environment specialized for the rapid breakdown of ingested proteins. In the midgut milieu, a rapid partial cleavage of the G1 glycoprotein of La Crosse virus probably occurs, which could remove the portion of G1 that acts as the cellular attachment domain for vertebrate cells. Ludwig et al (43) have suggested that the G2 glycoprotein, which is much less accessible to proteolytic enzymes, acts as the viral attachment protein for midgut cells, implying that the midgut cellular receptor might differ from that for vertebrate cells. However, until the receptors for California serogroup viruses have been identified, testing this proposal will be difficult.

The mosquito midgut may act as a barrier to the sequential spread of ingested virus. The concept of a midgut barrier arises from the observation that some viral isolates, which fail to be transmitted following virus feeding, may accumulate in the midgut without disseminating to other tissues (47, 60). The same isolates, when injected intrathoracically into the hemocoel, may be transmitted upon subsequent feeding.

Different California serogroup isolates vary in their ability to infect mosquitoes. When reassortants between these viruses were tested, infectivity mapped to the M RNA segment (2). Furthermore, mutations in the G1 glycoprotein may alter infectivity for mosquitoes. Clone V22, a MAR variant with reduced fusion efficiency and reduced neuroinvasiveness for the mouse (Figures 3 and 9), showed reduced infectivity for *A. triseriatus* compared with La Crosse virus, from which it was derived. A revertant clone of V22 was selected by passage in mosquitoes and had regained epitope 22 and fusion efficiency as well as mosquito infectivity (58).

FUTURE PROSPECTS

The California serogroup of viruses provide an attractive model to correlate the biological properties of viruses with their molecular and genetic determinants. The tripartite nature of the virus genome and the relatively small number of proteins it encodes facilitate detailed analysis of molecular determinants. The existence of a mouse model that mimics infection of the human and wildlife vertebrate hosts makes it relatively easy to study the pathogenesis of virus variants, and colonized *A. triseriatus* enable the characterization of replication in the natural vector. The different forms of pathogenesis of infection in vertebrate and insect hosts add another interesting dimension to the study of variant viruses.

Mutations in either the M RNA segment or in the L RNA segment can markedly alter the virulence or infectivity of California serogroup viruses. The M segment encodes the viral glycoproteins that play a key role in entry of virus into host cells, including steps such as attachment to cellular receptors, internalization, fusion, and uncoating. Present data indicate that there are several cellular receptors and that different virus clones vary in their ability to bind to these receptors, but further exploration of this entry step awaits identification of the specific receptor molecules (A. Pekosz, C. Griot, A. Tselis, N. Nathanson & F. Gonzalez-Scarano, unpublished observations). Clearly, mutations in the glycoproteins can alter the fusogenic activity of the virus, which influences pathogenesis. Further exploration of fusion requires better delineation of the role of each glycoprotein, including the identification of a fusion domain. The development of a recombinant vaccinia virus expressing the open reading frame of the M RNA segment will facilitate such studies (34). A strikingly attenuated mutant that mapped to the L RNA segment and exhibited both *ts* and cell-culture restriction implies that cell-specific differences in transcription can determine virulence and host range. The mapping of this mutation and further exploration of the polymerase functions await cloning of the L RNA segment currently underway. These ongoing studies promise to reveal new information about the functions of the viral proteins and how they influence the virulence and infectivity of California serogroup viruses.

ACKNOWLEDGMENTS

This work was supported in part by Public Health Service grants NS 20904 (Javits Award to N. N.), AI 24888 (First award to F. G.-S.), and a research fellowship from the "Schweizerische Stiftung fuer Medizinisch-Biologische Stipenden" to C. G.

Literature Cited

1. Beaty, B. J., Calisher, C. H. 1991. Bunyaviridae—natural history. *Curr. Top. Microbiol. Immunol.* 169:27–78
2. Beaty, B. J., Miller, B. R., Shope, R. E., Rohzon, E. J., Bishop, D. H. 1982. Molecular basis of bunyavirus per os infection of mosquitoes: role of the middle-sized RNA segment. *Proc. Natl. Acad. Sci. USA* 79:1295–97
3. Beaty, B. J., Sundin, D. R., Chandler, L. J., Bishop, D. H. L. 1985. Evolution of bunyaviruses by reassortment in dually infected mosquitoes (*Aedes triseriatus*). *Science* 230:548–50
4. Beaty, B. J., Thompson, W. H. 1976. Delineation of La Crosse virus in developmental stages of transovarially infected *Aedes triseriatus*. *Am. J. Trop. Med. Hyg.* 25:505–15
5. Bishop, D. H. L. 1979. Genetic potential of bunyaviruses. *Curr. Top. Microbiol. Immunol.* 86:1–33
6. Cabradilla, C. D., Holloway, B. P., Obijeski, J. F. 1983. Molecular cloning and sequencing of the La Crosse virus S RNA. *Virology* 128:463–68
7. Calisher, C. H. 1983. Taxonomy, classification, and geographic distribution of California serogroup bunyaviruses. See Ref. 8, pp. 1–18
8. Calisher, C. H., Thompson, W. H., eds. 1983. *California Serogroup Viruses*. New York: Liss. 399 pp.
9. Deibel, R., Srohingse, S., Grayson, M. 1983. Jamestown canyon virus: the etiologic agent of an emerging human disease? See Ref. 8, pp. 313–28
10. Elliot, R. M. 1989. Nucleotide analysis of the large (L) segment of Bunyamwera virus, the prototype of the family Bunyaviridae. *Virology* 173:426–36
11. Elliot, R. M. 1990. Molecular biology of the bunyaviridae. *J. Gen. Virol.* 71:501–22
12. Endres, M. J., Griot, C., Gonzalez-Scarano, F., Nathanson, N. 1991. Neuroattenuation of an avirulent bunyavirus variant maps to the L RNA segment. *J. Virol.* 65:5465–70
13. Endres, M. J., Jacoby, D. R., Janssen, R. S., Gonzalez-Scarano, F., Nathanson, N. 1989. The large RNA segment of California serogroup bunyaviruses encodes the large viral protein. *J. Gen. Virol.* 70:223–28
14. Endres, M. J., Valsamakis, A., Gonzalez-Scarano, F., Nathanson, N. 1990. A neuroattenuated bunyavirus variant: derivation, characterization, and revertant clones. *J. Virol.* 64:1927–33
15. Fazakerley, J. K., Gonzalez-Scarano, F., Strickler, J., Dietzschold, B., Karush, F., Nathanson, N. 1989. Organization of the middle RNA segment of snowshoe hare bunyavirus. *Virology* 167:422–32
16. Francy, D. B., Karabatsos, N., Wesson, D. M., Moore, C. G., Lazuick, J. S., et al. 1990. A new arbovirus from *Aedes albopictus,* an Asian mosquito established in the United States. *Science* 250:1738–40
17. Gonzalez-Scarano, F. 1985. La Crosse glycoprotein undergoes a conformational change at the pH of fusion. *Virology* 140:209–16
18. Gonzalez-Scarano, F., Endres, M., Jacoby, D. R., Griot, C., Nathanson, N. 1992. Molecular approaches to the study of bunyavirus encephalitis. In *Molecular Neurovirology,* ed. R. P. Roos, pp. 449–69. Totowa, NJ: Humana
19. Gonzalez-Scarano, F., Jacoby, D., Griot, C., Nathanson, N. 1992. Genetics, infectivity and virulence of California serogroup viruses. *Virus Res.* 24:123–35
20. Gonzalez-Scarano, F., Janssen, R., Najjar, J. A., Pobjecky, N., Nathanson, N. 1985. An avirulent G1 glycoprotein variant of La Crosse bunyavirus with a defective fusion function. *J. Virol.* 54:757–63
21. Gonzalez-Scarano, F., Nathanson, N. 1990. Bunyaviruses. In *Virology,* ed. B. N. Fields, D. M. Knipe, pp. 1195–1228. New York: Raven

22. Gonzalez-Scarano, F., Pobjecky, N., Nathanson, N. 1984. La Crosse bunyavirus can mediate fusion from without. *Virology* 132:222–25
23. Gonzalez-Scarano, F., Pobjecky, N., Nathanson, N. 1987. The fusion function of La Crosse bunyavirus is associated with the G1 glycoprotein. In *The Biology of Negative Strand Viruses*, ed. B. Mahy, D. Kolakofsky, pp. 33–39. Amsterdam: Elsevier
24. Gonzalez-Scarano, F., Shope, R. E., Calisher, C. H., Nathanson, N. 1982. Characterization of monoclonal antibodies against the G1 and N proteins of La Crosse and Tahyna, two California serogroup bunyaviruses. *Virology* 120:42–53
25. Gonzalez-Scarano, F., Shope, R. E., Calisher, C. H., Nathanson, N. 1983. Monoclonal antibodies against the G1 and N proteins of La Crosse and Tahyna viruses. See Ref. 8, pp. 145–56
26. Grady, L. J., Sanders, M. C., Campbell, W. P. 1987. The sequence of the M RNA of an isolate of La Crosse virus. *J. Gen. Virol.* 68:3057–71
27. Grimstad, P. R. 1988. California group virus disease. In *Arboviruses*, ed. T. P. Monath, pp. 99–136. Boca Raton, FL: CRC Press
28. Grimstad, P. R., Barrett, R. L., Humphrey, R. L., Sinsko, M. J. 1984. Serologic evidence for widespread infection with La Crosse and St. Louis encephalitis viruses in the Indiana human population. *Am. J. Epidemiol.* 119:913–30
29. Grimstad, P. R., Calisher, C. H., Haroff, R. N. 1984. Jamestown canyon virus (California serogroup) is the etiologic agent of widespread infection in Michigan humans. *Am. J. Trop. Med. Hyg.* 35:376–86
30. Griot, C., Endres, M. J., Gonzalez-Scarano, F., Nathanson, N. 1992. Polygenic control of neuroinvasiveness of California bunyavirus clones. *J. Cell. Biochem.* 16C:138 (Abstr.)
31. Griot, C., Tselis, A., Tsai, T. F., Gonzalez-Scarano, F., Nathanson, N. 1993. Bunyavirus diseases. See Ref. 58a. In press
32. Hammon, W. M., Reeves, W. C. 1952. California encephalitis virus—a newly described agent. I. Evidence of natural infection in man and other animals. *Calif. Med.* 77:303–9
33. Iroegbu, C. U., Pringle, C. R. 1981. Genetic interactions among viruses of the Bunyamwera complex. *J. Virol.* 37:383–94
34. Jacoby, D. R., Cooke, C., Prabakaran, I., Boland, J., Nathanson, N., Gonzalez-Scarano, F. 1993. Expression of the La Crosse M segment proteins in a recombinant vaccinia expression system mediates pH dependent cellular fusion. *Virology*. In press
35. Janssen, R., Gonzalez-Scarano, F., Nathanson, N. 1984. Mechanisms of bunyavirus virulence. Comparative pathogenesis of a virulent strain of La Crosse virus and an avirulent strain of Tahyna virus. *Lab. Invest.* 50:447–55
36. Janssen, R. S., Nathanson, N., Endres, M. J., Gonzalez-Scarano, F. 1986. Virulence of La Crosse virus is under polygenic control. *J. Virol.* 59:1–7
37. Johnson, K. P., Johnson, R. T. 1968. California encephalitis. II. Studies of experimental infection in the mouse. *J. Neuropathol. Exp. Neurol.* 27:390–400
38. Kappus, K. D., Monath, T. P., Kaminski, R. M. 1983. Reported encephalitis associated with California serogroup infections in the United States, 1963–1981. See Ref. 8, pp. 31–41
39. Kingsford, L., Ishizawa, L. D., Hill, D. W. 1983. Biological activities of monoclonal antibodies reactive with antigenic sites mapped on the G1 glycoprotein of La Crosse virus. *Virology* 129:443–55
40. Kolakofsky, D. 1991. *Bunyaviruses*. Berlin: Springer-Verlag. 380 pp.
41. Livdahl, T. P., Willey, M. S. 1991. Prospects for an invasion: competition between *Aedes albopictus* and native *Aedes triseriatus*. *Science* 253:189–91
42. Ludwig, G. V., Christensen, B. M., Yuill, T. M., Schultz, K. T. 1989. Enzyme processing of La Crosse virus glycoprotein G1: a bunyavirus-vector infection model. *Virology* 171:108–13
43. Ludwig, G. V., Israel, B. A., Christensen, B. M., Yuill, T. M., Schultz, K. T. 1991. Role of La Crosse virus glycoproteins in attachment of virus to host cells. *Virology* 181:564–71
44. Malkova, D. 1974. Comparative study of two variants of Tahyna virus. *Acta Virol.* 18:407–15
45. McLean, D. M. 1983. Yukon isolates of snowshoe hare virus. See Ref. 8, pp. 247–56
46. McLean, D. M., Grass, P. N., Judd, B. D., Stolz, K. J. 1979. Bunyavirus development in arctic and *Aedes aegypti* mosquitoes as revealed by glucose oxidase staining and immunofluorescence. *Arch. Virol.* 62:313–22
47. Miller, B. R. 1983. A variant of La Crosse virus attenuated for *Aedes*

triseriatus mosquitoes. *Am. J. Trop. Med. Hyg.* 32:1422–28

48. Najjar, J., Gentsch, J. R., Nathanson, N., Gonzalez-Scarano, F. 1985. Epitopes of the G1 glycoprotein of La Crosse virus form overlapping clusters within a single antigenic site. *Virology* 144:426–32
49. Parkin, J. S., Jacoby, D. R., Griot, C., Nathanson, N., Gonzalez-Scarano, F. 1993. Bunyaviruses. See Ref. 58a. In press
50. Patrican, L. A., DeFoliart, G. R., Yuill, T. M. 1985. La Crosse viremia in juvenile, subadult, and adult chipmunks (*Tamias striatus*) following feeding by transovarially-infected *Aedes triseriatus*. *Am. J. Trop. Med. Hyg.* 34:596–602
51. Pettersson, R. F., von Bonsdorff, C. H. 1987. Bunyaviridae. In *Animal Virus Structure*, ed. M. V. Nermut, pp. 147–57. Amsterdam: Elsevier
52. Pobjecky, N. 1988. *The fusion function of La Crosse bunyavirus*. PhD thesis, Univ. Penn., Philadelphia. 155 pp.
53. Pobjecky, N., Nathanson, N., Gonzalez-Scarano, F. 1989. Use of resonance energy transfer assay to investigate the fusion function of La Crosse virus. In *Genetics and Pathogenicity of Negative Strand Viruses*, ed. D. Kolakofsky, pp. 24–32. Amsterdam: Elsevier
54. Pringle, C. R. 1991. The Bunyaviridae and their genetics—an overview. *Curr. Top. Microbiol. Immunol.* 169:1–25
55. Reeves, W. C., Emmons, R. W., Hardy, J. L. 1983. Historical perspective on California encephalitis virus in California. See Ref. 8, pp. 19–30
56. Seymour, C., Amundsen, T. E., Yuill, T. M., Bishop, D. H. 1981. Experimental infection of chipmunks and snowshoe hares with La Crosse and snowshoe hare viruses and four of their reassortants. *Am. J. Trop. Med. Hyg.* 32:1147–53
57. Srohingse, S., Grayson, M., Diebel, R. 1984. California serogroup viruses in New York state: the role of subtypes in human infections. *Am. J. Trop. Med. Hyg.* 33:1218–27

58a. Stroop, W. G., McKendall, R. R., eds. 1993. *The Handbook of Neurovirology*. New York: Marcel Dekker. In press

58. Sundin, D. R., Beaty, B. J., Nathanson, N., Gonzalez-Scarano, F. 1987. A G1 glycoprotein epitope of La Crosse virus: a determinant of infection of *Aedes triseriatus*. *Science* 235:591–93
59. Talmon, Y., Prasad, B., Clerx, J., Wang, G., Chiu, W., Hewlett, M. 1987. Electron microscopy of vitrified-hydrated La Crosse virus. *J. Virol.* 61:2319–21
60. Tesh, R. B., Beaty, B. J. 1983. Localization of California serogroup viruses in mosquitoes. See Ref. 8, pp. 67–75
61. Thompson, W. H., Kalfayan, B., Anslow, R. O. 1965. Isolation of California encephalitis virus from a fatal human illness. *Am. J. Epidemiol.* 81: 245–53
62. Urquidi, V., Bishop, D. H. L. 1992. Non-random reassortment between the tripartite RNA genomes of La Crosse and snowshoe hare viruses. *J. Gen. Virol.* 73:2255–65
63. Watts, D. M., Pantuwatana, S., DeFoliart, G. R., Yuill, T. M., Thompson, W. H. 1973. Transovarial transmission of La Crosse virus (California encephalitis group) in the mosquito. *Science* 182:1140–43

Annu. Rev. Microbiol. 1993. 47:139–66

SUICIDAL GENETIC ELEMENTS AND THEIR USE IN BIOLOGICAL CONTAINMENT OF BACTERIA

S. Molin, L. Boe, L. B. Jensen, C. S. Kristensen, and M. Givskov
Department of Microbiology, The Technical University of Denmark, 2800 Lyngby, Denmark

J. L. Ramos
Estacion Experimental del Zaidin, CSIC, Apto 419, 18080 Granada, Spain

A. K. Bej
Department of Biology, University of Alabama at Birmingham, 1300 University Boulevard, Birmingham, Alabama 35294-1170

KEY WORDS: environmental release, stochastic killing, site-specific confinement, amplification of expression, pollution control

CONTENTS

ABSTRACT

The potential risks of unintentional releases of genetically modified organisms, and the lack of predictable behavior of these in the environment, are the subject of considerable concern. This concern is accentuated in connection

0066-4227/93/1001-0139$02.00

with the next phase of gene technology comprising deliberate releases. The possibilities of reducing such potential risks and increasing the predictability of the organisms are discussed for genetically engineered bacteria. Different approaches towards designing disabled strains without seriously reducing their beneficial effects are presented. Principally two types of strain design are discussed: actively contained bacteria based on the introduction of controlled suicide systems, and passively contained strains based on genetic interference with their survival under environmental-stress conditions.

BACKGROUND

The breakthrough in the development of genetic engineering techniques in the early 1970s was a major achievement that has since greatly impacted the biological and medical sciences. Several important proteins used in medical therapy are now being produced from engineered microorganisms; technical enzymes used in detergents are likewise produced from engineered organisms; and the engineering techniques have found their way into various diagnostic methods. Today molecular genetics is a discipline that uses not only *Escherichia coli* and a few other laboratory "pets" but is applicable to any living organism of choice, thanks to the possibilities offered by genetic-engineering methods. In recent years, interest has increased in environmental applications of genetically modified organisms of all types: plants, animals, and microorganisms. This rapid transfer of results from basic sciences to societal applications has simultaneously stimulated the basic sciences, both in choice of research projects and in governmental funding. Hence a major reason for the vast expansion of molecular biology during the past 20 years is the general awareness in society of the potential benefits from this type of research.

Both public and scientific concern grew in parallel with the fast development of the methods and their applications of the new technology. The prospect of moving genes around and constructing genetic recombinants that could not arise in nature produced a very intense debate on ethics and risks, which is still going on in most countries. The scientists were the first to point out the need for cautious behavior and for creating guidelines as prerequisites for accumulating experience and knowledge about these new hybrid organisms (10). The next obvious step was to begin a scientific investigation of the potential risks and ways to avoid or at least minimize them. In the footsteps of the technological developments came a new research area involving issues like survival in the environment, gene transfer, containment, ecological impacts, etc. A general heading for all such activities was risk assessment, and in many countries special research programs deal with these topics. An interesting observation is that many scientists with a background in fundamental molecular biology are now engaged in projects related to the risk assessment issues, even though they most likely do not think that engineered organisms represent a major threat to the environment. Instead, the many questions asked in the public debate, for which the scientific community had no precise answers, can be perceived by scientists as a challenge from a biological point of view: How far can we extrapolate our knowledge from the laboratory to the outside environment, and to what extent can we predict the fate of a constructed organism when we release it? The days of the head-shaking dismissal of the questions concerning risks, and the easy answers referring to the vast experience from the laboratory experiments are over; they have been replaced by scientific investigations of the questions and increased activity in the field of ecology. This is an interesting example of scientific choice being influenced by events in the rest of the society.

Although the risk-assessment projects, which we now see in operation in many places, are fairly recent activities, the general biological problems behind these projects have been dealt with for many years outside the area of recombinant DNA research. Growth, adaptation, and evolution of microorganisms in natural environments are topics of microbial ecology as are the participation of microorganisms in the cycle of elements (environmental metabolism) and the interactions between microorganisms and their surroundings. The principle questions concerning risks associated with the construction and use of engineered organisms are in fact directly related to ecology, and it is therefore not surprising that the present research projects within the risk-assessment programs are interdisciplinary activities involving molecular biologists along with ecologists. The methodology offered by molecular biology may in the hands of the ecologists lead to a more rapid advancement of quantitative approaches sought by the regulatory authorities as a basis for defining operational guidelines for the evaluation of proposals. Topics like survival time of introduced organisms, frequency of gene transfer to the indigenous population, displacement of original organisms, and so on, are all considered important for assessment. If these parameters can be quantified, making general guidelines and permission criteria will of course be easier.

An ideal goal often expressed is to reach a point where we can predict and control microbial life and death in a complex and uncontrollable environment. One obstacle to this is that we have difficulty defining meaningful death criteria for bacteria, let alone describing in precise terms the ecological effects of the presence of a new microbe in nature. The differentiation from active, culturable entities to viable but nonculturable dormant cells may happen to any bacterium (17, 61), making it nearly impossible to monitor survival of a bacterium after its departure from the well-controlled laboratory. Even though dormant cells can now be traced with sophisticated molecular techniques, this information tells little or nothing about the contribution from these to the overall turnover of materials or about the possibility of revival.

In the beginning of the gene technology debate, many scientists argued that bacteria used by industry can no longer survive in the environment, because the necessary functions for outside survival have been lost during the long time of laboratory cultivation; in addition, after genetic manipulation industrial bacteria would be even further disabled. Several experimental observations cast doubt on such statements (11, 14, 65), and in addition we have no easy way of interpreting reduced numbers of colony formers from environmental samples. Therefore, the wish of predicting and controlling bacterial survival may be very difficult or impossible to fulfill.

We thus seem to have only two ways out of the present dilemma (assuming that we all agree that engineered organisms are beneficial to society and should eventually become accepted tools for improving the quality of life): One is to accept the potential risk associated with the use of engineered organisms and to let the experience gained from the practical use of them provide a basis for safety and comfort. The other is to maintain as a rule that when we introduce manmade products (be it chemicals or living organisms) in the environment, we have an obligation to ensure that such products are degradable and will disappear after a limited period of time; if total elimination is not possible we should at least make the utmost effort to reduce the concentration to insignificant levels.

The present review is based on the opinion that the two options are perhaps compatible; the clean-up strategy is sound and advisable in the absence of experience, and experience may at a later point make us revise some of the precautions.

BIOLOGICAL CONFINEMENT OF BACTERIA

The first practical applications of genetic engineering were microbial productions of human and animal peptides and proteins. The enormous information about, and experience with, laboratory organisms such as *Escherichia coli, Bacillus subtilis,* and *Saccharomyces cerevisiae* made them obvious candidates as host organisms for introduced genes encoding peptide hormones or growth factors. Because many of the strains used for these genetic experiments had several mutations rendering the cells dependent on special nutritional supplies in the growth media, the generally accepted belief was that the organisms themselves would not be able to survive for a long time in the environment, let alone out-compete natural organisms.

Nevertheless, attempts were made in the early days of the technology to design particular disabled mutants for industrial and scientific use in connection with recombinant DNA techniques. The most extensive (and impressive) work was performed by Roy Curtiss III, who by sophisticated use of classical bacterial genetic techniques constructed a series of mutants of *E. coli,* of which X1776 is a famous example (20). The concept behind the design was to construct a strain that could be maintained only under laboratory conditions because of special building block requirements that are never, or only very rarely, found in nature. One example is D-amino pimelic acid, which is an essential constituent of the bacterial cell wall not naturally occurring in the environment. In addition, the strain had numerous mutations that made it highly vulnerable to environmental stress factors. Strain X1776 would certainly not have the slightest chance of surviving outside the laboratory; unfortunately, it also did not survive well in the test tube, and most likely cultivating it in large fermentors would have been very difficult.

The strain HB101, designed by Herbert Boyer (15), is another example of a disabled strain used as a host for recombinant DNA molecules. This strain is still used in many laboratories, because it is easy to grow and to transform, it is not restrictive towards foreign DNA, there are several marker mutations in the chromosome, and its *recA* character stabilizes introduced DNA. The *recA* mutation alone is probably sufficient to disable the strain, eliminating any risk of environmental establishment. Among the important features of HB101 is the absence of extrachromosomal agents (phage or plasmid DNA), which reduces the probability of transfer of the introduced genetic material from HB101 to more competitive organisms during its presumed limited, but in some cases sufficient, survival time in the environment. This feature is now considered most important in the assessment of industrial and scientific projects, and in most countries it is a rule that the host cells for foreign DNA molecules must be devoid of genetic-transfer potential.

Physically contained use of genetically engineered microorganisms involves growth of monocultures under defined and controllable conditions. In contrast, deliberate release of bacteria to the environment leaves them in competition with the natural population under conditions that constantly change. If bacteria are released to perform a task (plant protection, environmental clean-up, live vaccines, etc), a normal requirement is that they survive for some time and maintain a reasonable population size during this period. The bacterial species relevant for environmental releases are most often not as well known as e.g. *E. coli,* and consequently analogous libraries of mutant strains with markers or disabling traits do not exist. There are, therefore, two arguments against the disablement approach in case of bacterial releases: (*a*) many mutations would render the strain handicapped in the performance of its actual task in the environment, and (*b*) the proper mutations would often be difficult to obtain. (However, as discussed later in this article, certain disabling mutations may be easy to obtain without serious effects on the short-term competitive properties of the strain.)

The alternative to controlling the bacterial establishment and survival is the introduction into competitive wild-type organisms of controllable suicide functions that do not interfere with normal growth but are expressed under specific conditions defined by the physical or chemical composition of the environment. One might thereby limit survival of the strain to the exact environment where it performs its desired function so that any cell escaping this environment will be killed by induction of the suicide system. An even more specific purpose may be to eliminate gene transfer by coupling a suicide gene to the environmentally relevant gene(s), e.g. on a plasmid, and have the suicide gene controlled by a chromosomal gene in the released organism that is not found in the natural population of potential recipients. Several strategies for controlled suicide will be described in the following paragraphs; the basic idea behind them all is to couple potent killing genes to environmentally relevant expression-control circuits, which will allow the released strain to establish itself in a confined way (limited to a certain defined environment or limited in time), with severely reduced possibilities of sustaining itself.

Basic Requirements for Effective Containment Systems Based on Suicide

As discussed above, bacteria designed for a specific purpose may be eliminated through the insertion of a killing gene expressed from a controlled promoter. Many gene products are toxic to strains of bacteria if present above certain cellular concentrations, so in principle we can easily identify potential suicide functions from such observations. However, on closer examination such random phenomena often turn out to be merely growth inhibition, and expression must reach very high levels to kill. Moreover, the inhibitory effect on one species may not be reproducible in other bacterial species. In general, the best candidates are killing functions whose targets are central cellular functions, which are likely to be found in essentially all bacteria. The toxicity of the agent should be high, meaning high efficiency of killing at the lowest possible concentration. Therefore, putative killing functions must be assayed in a variety of bacteria and at a range of induction levels.

The availability of several expression systems (promoters with or without regulatory sequences) from various bacteria has made construction of fusions between these and putative killing genes quite easy. Insertion of such fusions in broad-host-range vectors (plasmids or transposons) allows rapid screening for killing efficiency in many different bacteria. Often, the expression system can be based on the *lac* promoter system active in a broad spectrum of gram-negative bacteria (4), and derivatives that function even in gram-positive bacteria have been designed (70). The major advantages of the *lac* promoter is the simple induction with isopropyl-thiogalactoside (IPTG) and the high ratio between repression and derepression. The latter characteristic is important, because in the absence of induction only a very low level of the killing function is expressed, whereas induction results in a burst of high levels of expression. Thus, screening for new killing genes and their efficiency in different bacterial strains and species is relatively simple and can be standardized once the structural gene sequences have been isolated and made ready for fusion with the expression system. In later developments of suicide systems useful for release purposes, promoter systems active under environmental conditions must be identified. Thanks to the many sophisticated reporter gene systems available today this task has become much easier.

Killing Genes

THE *gef* GENE FAMILY Studies of the maintenance functions of plasmid R1 in *E. coli* found that a locus on the plasmid, responsible in part for the high degree of plasmid stability during cell growth in the absence of selection pressure, encodes a small polypeptide (52 amino acids) that is highly toxic for the cells (30, 32). The plasmid-stability phenotype of this locus is connected to an antisense RNA regulation of the toxic polypeptide's expression, which ensures tight repression of the gene in cells harboring the plasmid due to constitutive synthesis of the antisense RNA. However, in cells to which the plasmid has not segregated, the antisense RNA rapidly decays leaving a very stable mRNA, which is then translated into the toxic protein, resulting in eradication of the plasmid-free cells. Separation of the two genes (regulator and effector) allowed a subsequent analysis of the molecular mechanisms involved in both control of expression and the mode of cell killing (28, 29, 33).

The gene encoding the small toxic polypeptide was termed *hok,* for host killing, and the antisense RNA gene was called *sok,* for suppression of killing (32). The killing of the cells after induction of the Hok protein was typically followed by a change in the bacteria's morphology as observed in the microscope; the cells became almost totally transparent except for cell material accumulating in the poles of the cells. The occurrence of such ghost cells indicated that one effect of the induction of Hok could be membrane damage, which was later confirmed by the observations that immediately after induction of the protein the transmembrane potential collapses, respiration is terminated, and the membrane contains a new small polypeptide the same size as Hok (28). Therefore, the conclusion was that the cellular membrane of *E. coli* is the target for the Hok protein, and somehow introduction of the protein into the membrane leads to cell death via interference with central components of the respiratory system.

The position of the *hok*/*sok* locus in plasmid R1 stimulated a search for similar genes in other plasmids, and in several plasmids of the enterics, homologous sequences were detected with hybridization (53). Similarly, previously described genes from several plasmids turned out to be both genetically and functionally homologous (31). In some of the hybridization experiments performed to detect plasmid homologues, more than one signal was detected, and these extra hybridization signals were detected even in plasmid-free cells. One of these was rapidly identified comparing *hok* and the so-called *relF* gene under examination by Bech et al (8). The *relF* gene is located in the *relB* gene cluster at 37 min on the chromosome map of *E. coli* (3). The functional similarity between the two genes is obvious, although no biological significance of the *relF* gene has emerged so far (28). A second *hok* homologous locus in the *E. coli* chromosome was isolated by cloning and was characterized (53); this gene, *gef,* is located at 1 min on the *E. coli* map, and the protein is functionally identical to the Hok and RelF proteins in terms of the mode of killing (56). As in the case of *hok* expression, *gef* is regulated by an antisense RNA mechanism (56) (which is not the case for *relF,* but similar to most other members of this family).

Two aspects of this gene family have been investigated in detail: regulation of gene expression and mode of action of the proteins. The former falls outside the scope of this review and has been reviewed elsewhere (31). The membrane location of the proteins was verified for the Gef protein through the analysis of its compartmentalization with antibodies raised against a synthetic peptide comprising the C-terminal part of the protein (55). The effects of expression of the proteins have been analyzed in several studies (see 31 for a review), and although all features have not been determined in all cases, they are all presumably functionally identical. A major obstacle in the analysis of these proteins is their high level toxicity, which prevents intracellular synthesis in measurable amounts. This problem was solved by the isolation of a mutant of *E. coli* that tolerates very high concentrations of the proteins (54). Immunological studies on this mutant strain showed that Gef is positioned in the cytoplasmic membrane, and construction of a gene fusion between *gef* and *phoA* that resulted in a membrane-bound enzymatically active alkaline phosphatase led to the conclusion that Gef is anchored in the membrane, with the C terminus located on the outside of the membrane (facing the periplasm). Further analysis of the protein on PAGE indicated that the membrane-bound protein is a dimer that may have been posttranslationally modified through addition of, for example, sugar moieties. From these observations a model was proposed in which the Gef protein forms a dimeric porin in the inner membrane of the cell through which transport of various substances can take place; in case of over-production of the protein this transport eventually leads to cell death (55).

The collapse of the membrane potential and the fast drop in oxygen uptake observed after induction of Hok synthesis was also observed after induction of the *gef* and the *relF* genes (28). Other membrane-connected consequences have been described for some of the plasmid genes. Induction of the small proteins was found to result in a rapid outflux of Mg^{2+} ions from the cells and a simultaneous transport of RNase from the periplasm to the cytoplasm, resulting in the degradation of ribosomal RNA (1, 51, 62). Several other compounds probably move across the membrane, thus contributing to the ghost-like appearance of the cells after a period of time. Despite these severe effects on membrane function, the cells do not lyse immediately after induction, but viability, as monitored by plating, is lost rapidly, indicating that the membrane changes after induction are irreversible (28, 32). This efficient killing of the cells by destruction of a central cellular function in the cytoplasmic membrane, as well as their small size, make the gene family an excellent source of suicide genes for the design of biological-containment systems. Although there are no precise estimates of the minimal level of protein required for killing, the fact that no Gef protein can be observed with the antibodies after induction in wild-type cells strongly indicates that low cellular concentrations are sufficient to exert the effect (A. Refn & S. Molin, unpublished data).

Three of these genes have so far been incorporated in regulated suicide model systems: *hok, gef,* and *relF* (18, 42, 47). In all cases these genes have proved to have powerful killing functions, and their killing activities ought to be no different. The finding that the apparent cellular target is the membrane suggested that the killing activity should not be restricted to *E. coli,* and transfer to other bacterial species has confirmed this (47). Insertion of the *hok* and *gef* genes fused to the *lac* promoter (or derivatives thereof) in either plasmids or chromosomes of distantly related bacteria allowed tests of killing efficiency by simple growth and viability measurements before and after IPTG addition. In all bacteria tested, growth was affected by induction of the killing genes, even though some strains were less inhibited than *E. coli* (47; S. Molin, unpublished data). Although some bacteria may be less sensitive to these proteins as a result of differences in their membranes, a series of experiments with a strain of *Bacillus thuringiensis* indicate that gene expression problems are the major cause of survival after induction of Gef (S. Molin, unpublished data). Insertion of a *lacp-gef* fusion on a plasmid in this organism showed only mild growth inhibition after IPTG addition. By treatment of the plasmid DNA with NH_2OH followed by transfer to the bacteria and screening for more severe growth inhibition after induction, a mutant with a much stronger inhibitory effect could be isolated. Mapping of the mutation showed that it was located in the *lacI* regulatory gene. Also observed in *E. coli* (L. B. Jensen, unpublished data) is that when very high levels of the Lac repressor protein are expressed, the induction with IPTG is so ineffective that no killing is observed. This observation agrees with the notion that reduced killing efficiency in other bacterial species results at least partly from suboptimal expression. The important conclusion from these results is that expression of genes in organisms that were not the source of the promoter systems used may be severely reduced for reasons other than simple promoter-strength differences; if the promoter is regulated, the balance between promoter, operator, inducer, and regulatory protein should not change dramatically.

The properties that make the *gef* gene family an interesting source of suicide functions can be summarized as follows: The small proteins are all highly toxic even in small quantities in the cell, and the effect of their synthesis is a cascade of reactions resulting from the incorporation of the polypeptides in the cytoplasmic membrane. This central target for the toxic action makes it likely that most, if not all, bacteria are susceptible to the killing activity provided sufficient protein is expressed. The results obtained so far confirm this. A bonus is that the genes are much smaller than average prokaryotic genes (less than 200 bp) in the context of genetic manipulation and analysis of sequences.

NUCLEASES The strongest argument for the potential application of the *gef*-like genes as suicide elements is that the central target, the membrane, appears to be similar in structure throughout a broad spectrum of bacteria. Alternative killing genes with common targets would of course be equally interesting. Of particular interest is the possibility of destroying the genetic material itself, because two objectives of a biological-containment system would be met: killing of the engineered bacterium as well as destruction of the source of concern, DNA.

One possible candidate for this type of activity was tested some years ago, and the idea has since frequently been mentioned, i.e. to insert a gene encoding a restriction enzyme with proper expression control and without the corresponding modification activity. In principle, induction of such an enzyme should result in degradation of the chromosomal DNA (and any other type of DNA in the cell), preventing further growth and gene transfer. This approach was tested by Stephen Cuskey with the *Eco*R1 gene (personal communication), but from a preliminary analysis he concluded that the endonuclease activity was too high; even in the absence of induction the low basal level of gene expression severely inhibited growth. Most likely one of the problems was that the enzymatic digestion of DNA results in double strand breaks that are very hard for the host cell's repair enzymes to fix. In principle, one cut may be sufficient to inactivate the cell, and the chance of that happening per generation is fairly high even if only a few enzyme molecules are present in the cell.

An alternative type of endonuclease was therefore tested. The idea was to find an enzyme with endonuclease activity producing single strand breaks with a high specific activity. The high activity was assumed to be necessary to override the repair machinery of the bacterial cell. Two such examples have been investigated, and in both cases extracellular enzymes were chosen because of their extraordinary activity: the nucleases produced and secreted by *Serratia marcescens* and *Staphylococcus aureus,* respectively. Both these enzymes are endonucleases that hydrolyze both DNA and RNA. The *S. marcescens* nuclease comprises 264 amino acids, of which 21 constitute the signal peptide that is removed during transport across the membrane (5, 6, 12, 13). There are four cysteine residues whose participation in the formation of S-S bridges could create intracellular stability problems (23). The mature *S. aureus* enzyme nuclease A consists of 149 amino acids. The precursor, nuclease B, has 19 extra amino terminal amino acids, and even further upstream is a stretch of amino acids that are removed during processing (22, 64). This enzyme has no cysteine residues.

In both biological-containment designs, the cloning strategy involved isolation of the coding sequence representing the mature enzyme (signal sequence deleted), and fusion to the *lac* promoter attached to a functional ribosome-binding site and to a start codon (ATG) (A. K. Bej & S. Molin, in preparation). In the absence of induction of nuclease expression, the cells harboring such constructs grow very well, showing that small basal levels of the enzymes create no problems. After induction, the cells stop growing, and a significant fraction of the cell population dies. Transfer of the *S. marcescens* nuclease to other bacteria showed that the effect occurred in other organisms. Agarose gel electrophoresis verified that the induced enzyme actually degrades the chromosomal DNA. The enzymes retain sufficient intracellular activity to compete with the repair enzymes in hydrolysis and religation. However, the nucleases must be present in fairly high concentrations in the cell to inhibit growth; the reason for this requirement seems to be connected to the efficiency of the repair machinery as indicated by the observation that when the nuclease genes were expressed in a *recA* mutant of *E. coli,* killing dramatically increased at much lower expression levels.

Researchers addressed the question of sulfur bridges, and their possible effect on enzymatic activity under the prevailing redox conditions inside vs outside the cell for which the enzyme is designed, by monitoring the stability of the protein after induction. The excreted, mature nuclease from *S. marcescens* is a highly stable protein, as expected for an extracellular enzyme. However, the intracellular nuclease is highly unstable; the half-life in growing cells (monitored immunologically) is approximately two minutes. In contrast, the half-life of the *S. aureus* enzymes is more than two hours. Thus the difference in the cysteine content of two enzymes may explain their very different stabilities inside the cells, but in terms of killing efficiency, the nucleases are much the same. Whether this means that the specific activity of the *S. marcescens* nuclease is much higher than that of the *S. aureus* remains to be investigated.

From the point of view of suicide function, the nucleases have an interesting potential worth pursuing. This type of killing agent has the same advantages as the Gef-like proteins with respect to the universal target found in all organisms, and it is a bonus that the genetic material itself is attacked. Also interesting is that a similar strategy has been adopted in plant cells, although suicide was not the direct objective (43). The major unsolved problems concerning nucleases as suicide agents are how much they can be optimized and to what extent the repair systems will make it difficult for the enzymes to complete their jobs. Inasmuch as *recA* mutants of bacteria represent alternative examples of contained strains (see below), the nucleases at least may become valuable components in them.

ALTERNATIVE KILLING GENES In a series of studies of extracellular enzymes from various *Serratia* species, we have worked with a phospholipase A (in addition to the nuclease), which can be excreted from both its original host organism and from *E. coli* (35, 37). The enzyme can degrade bacterial membranes, and in strains where excretion is blocked genetically, the phospholipase is detrimental to the host cell. The gene for this enzyme is included in an operon from which a second gene product, PhlB, is expressed. This protein has now been shown to be an antagonist of phospholipase; it prevents the detrimental effects of the small part of phospholipase that stays inside the cell even under normal conditions (36). Deletion of the *phlB* gene results in the conversion of the phospholipase gene, *phlA,* to an effective killing gene whose induction leads to cell lysis as a consequence of the degradation of the cytoplasmic membrane. It remains to be seen if this suicide system is easily moved to other organisms.

Early studies of the maintenance of plasmid RP4 showed that several so-called *kil/kor* gene pairs played a possible role (7, 26, 63). The *kil* genes were simply defined as genes whose expression in the absence of the corresponding *kor* gene inhibit growth, and correspondingly the *kor* (kill overriding) genes encode products that counteract the killing effect. In an attempt to design a novel biological containment system, Cuskey used the *kilA* gene from RP4 as a suicide gene after coupling it to the benzoate regulated promoter, OP2, from the TOL plasmid pWWO (21). In the tests, cells (*E. coli* or *Pseudomonas aeruginosa*) harboring plasmids with this type of suicide construct were grown and then the events following a shift to benzoate as carbon source and simultaneous induction of the *kilA* gene were monitored. The major conclusion from these studies was, however, that the KilA protein is not bacteriocidal but bacteriostatic, because growth inhibition, which did occur, was not followed by actual cell killing; i.e. if an alternative carbon source such as lactate was added to the growth-inhibited culture, the bacteria started growing again after a short lag. Thus, the usefulness of the *kilA* gene in a containment context is not obvious.

In addition to bacteria as a source of killing genes, bacterial viruses have been considered because they actually profit from killing their host cells. Lysis genes from bacteriophages have been cloned and fused with regulated gene expression systems, and the results show that this route is promising (R. Steffan, personal communication).

Expression Control Circuits

The key to the design of functional suicide containment systems lies in the regulation of expression of the suicide gene. As long as the cells are used under defined conditions in the laboratory or in a fermentation plant, molecular microbiology presents many options. However, in situations in which the contained microorganisms are aimed at functioning in the environment, very little information is available about gene expression, and even though some tests may be performed in model ecosystems, the amount one can extrapolate to the natural environment is most likely limited. The problem of identifying the optimal expression system is complex. First, the promoter for eventual transcription of the killing gene must be sufficiently active under the prevailing conditions in the environment, leading to expression of the protein in actively growing cells as well as in stationary cells. Perhaps most cells will be in a state of dormancy most of the time, and ideally the killing should happen even in this state. However, one can argue that the dormant cells do not contribute to the ecosystem activities and therefore constitute no potential risk; thus the suicide system need function only if the cells later become reactivated. The second problem is that of control. Under conditions of constant variation, predicting which outside signals are relevant for the bacteria is very difficult, and therefore it is not obvious how to choose an expression system for a specific environmental composition or condition. In general, the optimal solution will be to design the expression-control system in direct combination with the task of the organism, such that spreading of bacteria to other localities will induce their suicide system. This is possible when we have much molecular information about the particular desired function of the bacteria (see below), but this is often not the case and other regulatory alternatives must be invented. Finally, knowing whether the suicide function is active under the environmental conditions and the various physiological states of the cells is important. What will be the effect of a membrane porin in cells that have no membrane growth, and how will a nuclease survive in cells with increased proteolytic activity?

The following sections discuss some examples of gene expression control systems from the point of view that they are potentially useful in combination with suicide systems. Some have been tested already; others are being explored and tested; and still others are mentioned because it would make sense to test them.

CHEMICAL CONTROL Induction of suicide genes based on the presence or absence of a specific chemical is by far the simplest system to operate in the laboratory [e.g. IPTG addition (9)], but in a complex outside environment it is hardly realistic to water the fields with solutions of IPTG. Many chemicals that might be used in a fermentor to grow up large volumes of cells are not present in the environment, and one strategy could therefore be to let cells become derepressed for the expression of the suicide function as they were (incidentally or deliberately) introduced in the environment. Such an example was previously described, in which the *hok* gene was put under the control of the *trp* promoter from *E. coli* (47, 48). The presence of tryptophan in laboratory media repressed the promoter, but when the organism was released, the outside concentration of the amino acid was assumed to be insignificant or rapidly consumed by the bacteria, leading to derepression of Hok synthesis and killing. This approach seems suitable for reducing survival of bacteria used in contained applications (industry, research). However, if the organism is supposed to carry out a function in the environment, and should therefore stay there for some time, the stepwise change in tryptophan concentration most likely would induce rapid elimination of the cells, in conflict with the purpose of the release.

One possible solution to this problem might be to manipulate the known control systems in such a way that the growth conditions during production of the cells induced the synthesis of extraordinarily high cellular amounts of a regulatory protein repressing the suicide promoter. Such overexpression of a repressor gene could be achieved by replacing its normal promoter with a chemically inducible promoter (e.g. *lacp*). Induction of suicide will only happen in the absence of the repressor, and therefore the cells with the initial high repressor concentration would need several generations of growth before this situation occurred. In the environment, generation times are usually very long (often days or weeks), so the survival time of such a population could be long enough to achieve its goal, and yet be limited. Although this approach is somewhat speculative, the example described below concerning containment of benzoate-degrading bacteria is similar.

PHYSICAL CONTROL A classic expression system for foreign genes in *E. coli* is the set of promoters in bacteriophage λ regulated by the cI repressor. The isolation of a temperature-sensitive variant of this repressor, cI^{857} (68), has been exploited extensively as the basis for temperature-controlled expression of gene products. In addition, the two promoters, λpR and λpL, are strong, and repression at low temperatures is tight. This type of gene-expression control is very useful in laboratory tests and has also been used in connection with killing genes (32). However, an induction temperature of more than 40°C is rarely, if ever, of environmental relevance. Recently, low-temperature induction of genes was described (38, 41, 69), opening the possibility of combination with suicide genes. For soil bacteria, this type of regulation may have potential applications in biological-containment systems, but in many locations the temperature shifts between day and night lie within the temperature range of these control systems.

SITE-SPECIFIC CONTROL Bacterial degradation of aromatic compounds like benzoates has been studied in great detail for several years, and much is now known about the specific degradative pathways and their regulation (40, 46, 57). The TOL genes found in *Pseudomonas putida* are induced by the presence of very low amounts of the substrates (toluene and different benzoates), and the spectrum of inducers and substrates have been further expanded by mutation (58, 59). The control system responsible for the regulation of these genes is quite complex, but in the present context, the important feature is the substrate-dependent transcription from the p_m promoter, for which the regulatory protein is the *xylS* product. The XylS protein has no activity in the absence of substrate, e.g. 3-methylbenzoate (3-MB), but in its presence, the XylS protein is activated, and the regulator acts positively on the p_m promoter to stimulate initiation of transcription, which continues as long as 3-MB is present.

In an attempt to design a strain that is fully viable in an environment comprising 3-MB and that, due to the insertion of a regulated suicide system, is killed in the absence of the compound, a regulatory double loop was constructed: The p_m promoter was placed upstream of the *E. coli lacI* gene as a substitution for the normal promoter (18, 24). In a cell harboring this fusion and the *xylS* gene, a *lacp-gef* fusion was inserted. The resulting strain was viable in substrates containing 3-MB, because in this environment the LacI repressor is synthesized, which ensures repression of the suicide gene. When 3-MB was removed from the substrate, the cells were killed after a short growth period because no Lac repressor was expressed from the 3-MB regulated promoter. In the first demonstration of this system, the organism was *E. coli,* and no degradation of the substrate took place. The same type of construction was later inserted in *P. putida* harboring the entire TOL degradative pathway, and in this case one could follow simultaneously the degradation of 3-MB and the elimination of the bacteria after complete mineralization of the aromatic. Experiments performed in soil with and without 3-MB clearly showed that the introduction of a suicide system such as the one described here is an efficient containment strategy even in a complex environment like soil (L. B. Jensen, J. L. Ramos & S. Molin, submitted).

Similar strategies may be designed for other xenobiotic degradation pathways whose molecular regulation has been worked out. Also limiting the survival of the plant-associated bacteria to a particular growth zone may be possible. Bacterial genes responding to plant compounds such as root exudates may be identified (16), and even before the actual regulatory details have been worked out, one could insert suicide genes in a control circuit involving the plant signals so that cells escaping from the plant zone will be killed.

The availability of bacterial gene regulators allows one, by combining them with other niche-specific regulators, to design site-directed induction systems for the expression of suicide functions that will interfere with random and uncontrolled spread of manmade microorganisms from the original site of release. The increasing interest in microbial ecology and in symbiotic relationships in particular promises the identification of many such useful control systems from a broad spectrum of bacteria.

STRESS CONTROL In contrast to the life of bacteria in the nutrient-rich laboratory media, microbial life in the natural environment probably consists mostly of starvation and other types of stress (pH, extreme temperatures, toxic compounds, etc). Therefore, any organism released to the environment sooner or later will meet severe stress conditions. The bacterial responses to different forms of stress have been studied for some time, and especially for bacteria such as *E. coli* (44, 45, 50), *Salmonella typhimurium* (66), and *Vibrio* spp. (49), much is known about the genetic programs responsible for coping with these extreme growth conditions. The picture emerging from these studies is that bacteria do have types of differentiation programs in operation when conditions become harsh. The sporulation cycle of gram-positive bacteria is an excellent example of gross changes in cellular morphology and metabolism, but also many nonsporulating bacteria possess genetic programs for stress-induced morphology changes, resistance to various factors, and metabolic changes. Specific genes are activated in an ordered sequence as a response to stress, as clearly revealed by two-dimensional gel electrophoresis (39), and analogously, certain specific recovery genes are activated upon return of normal growth after a stress period (49).

Such studies of differentiation are also being carried out in different *Pseudomonas* strains, and the observations accumulated so far clearly indicate the presence of similar genetic programs in these bacteria (M. Givskov, L. Eberl & S. Molin, unpublished data). The possibility of fusing promoters of stress genes with a killing gene is an obvious containment strategy, and promoters active late after induction of a stress program are especially useful, because killing will not take place until after an extended period of stress. In this way, one can avoid an undesired rapid elimination of the population after short transient stress periods. Alternatively, fusions to specific recovery gene promoters will allow a significant period of survival followed by a subsequent reduction in the population. Among the stress genes that are particularly interesting in the context of suicide containment are those involved in starvation survival. Feast and famine will unquestionably be the changing conditions that bacteria meet in the environment, and starvation is a condition that is very easy to reproduce in the laboratory. Naturally, the more we know about such pathways and their regulation, the better opportunities will be to interfere and control survival in the environment.

STOCHASTIC CONTROL Because of the complexity and variation of natural conditions, specific regulation of gene expression by the chemical or physical constitution of the environment is problematic as a basis for controlled suicide of bacteria. In connection with the discussion of chemical induction of suicide, a particular approach based on dilution of a repressor was referred to as generation time–dependent suicide. Since the number of cell divisions taking place in the environment is unpredictable and most likely varies greatly in a population spread out in different niches, induction systems dependent on absolute time rather than doubling time may have several advantages. Thus, if killing was induced as a function of time regardless of whether the cells were growing fast, slowly, or not at all, one could predict how long it would take before the size of the population was reduced below a threshold level. Attempts to design such stochastic induction systems have been based on DNA recombination events—switches—that randomly result in activation of a suicide function. If such recombinations occur with a frequency per cell per time unit irrespective of the growth rate, a stationary cell population may be gradually reduced at a rate dependent only on the recombination frequency.

Recombinational switches are known from several operons in bacteria and bacteriophages. The switch promoter directing the synthesis of type 1 fimbriae in *E. coli* (*fimAp*) has been used as a model system for stochastic induction of suicide (25, 27, 52). The promoter is located within a 314-bp DNA sequence that inverts its orientation with a frequency of approximately 2×10^{-3} per cell per hour (measured in fast-growing cells) (25). This invertible sequence comprises a promoter that directs transcription into one of the flanking sequences, which means that any gene controlled by the promoter will be switched from on to off with the mentioned frequency. The insertion of a suicide gene next to the switch promoter, *fimAp,* was the first approach to obtain time-dependent induction of suicide (47, 48), and in a series of subsequent investigations both the *hok* and the *gef* genes have been employed (to be published). *E. coli* cells harboring such inserts on plasmids or integrated in the chromosome behave according to the expectation shown in Figure 1:

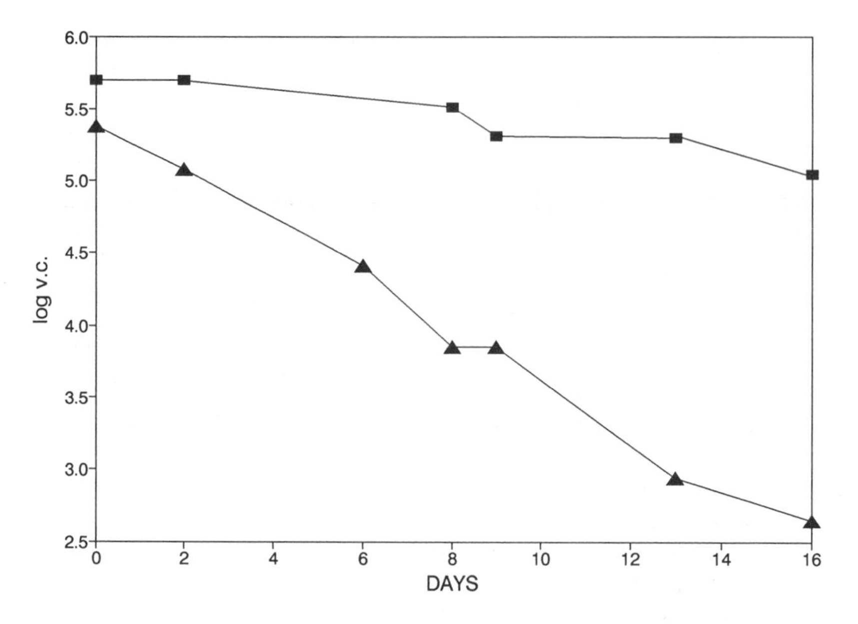

Figure 1 Stochastic killing of *E. coli*. The strain MC1000 of *E. coli* with (*triangles*) or without (*squares*) a *fimA* promoter fused to the *gef* gene inserted in a plasmid was grown up and resuspended in buffer without a carbon and energy source. From the two cultures, samples were taken at different times for determination of viable counts. The logarithm to the viable counts are plotted against time (days).

In cell suspensions where no net growth is possible (no external carbon source), viable counts constantly drop in contrast to the noncontained control strain, indicating that switching occurs in the absence of growth. In fast-growing cultures (doubling times of approximately 30 min) no effect of the suicide system on the growth rate of the culture is observed, which suggests that the killing rate is insignificant relative to the growth rate. However, if the growth condition is reduced by changing to poor carbon sources, the consequence of the presence of this type of stochastic suicide system is an apparent growth inhibition of the cell population: The slower the cells grow, the more severe is growth inhibition by the *fimAp-gef* fusion, indicating that the killing rate becomes increasingly significant relative to the decreasing obtainable growth rates under the chosen conditions.

In principle, designing functional stochastic induction systems for bacterial suicide seems possible. The particular system based on the *fimA* switch promoter unfortunately only works in the enterobacteria, and even in some of these the switch frequency is too low to be of any use. So far the best results have been obtained in *E. coli* and *S. typhimurium,* whereas no switch at all takes place in e.g. *P. putida* (L. B. Jensen & S. Molin, unpublished data). Therefore, other recombinational systems that operate in a broader spectrum of bacteria will have to be identified, and in the following section one such example is discussed.

AMPLIFICATION OF EXPRESSION The stochastic induction model system based on the *fimA* promoter was functional because a switch frequency in the range of 10^{-3} per cell per hour is below the rate where killing interferes significantly with growth of fast-growing cultures but is sufficient to eliminate stationary cultures over a period of weeks under the conditions tested in the laboratory. A weakness of this system is the inability to adjust the switch frequency to any desired value. The target for such adjustments is the recombinase activity, but so far no precise information is available about such an enzyme or its regulation.

For other recombination systems, however, we have detailed knowledge about the recombination process and the involved enzyme(s), and this knowledge allows us to interfere with the process at the level of regulation. One such example is the resolvase system of plasmids and transposons, and work has been initiated with the resolvase system of the broad host range plasmid RP4, whose normal function is to separate multimers of plasmid molecules into monomers through recombinational cross-overs (34, 60). This system also has the obvious advantage of being functional in a broad spectrum of bacteria. Many plasmids have evolved site-specific recombinases with corresponding resolution sites (*res*), providing extremely efficient intramolecular recombinations between the duplicated *res* sites in dimers and higher multimers, resulting in their resolution into monomers.

The important features of this type of recombination reaction are that a low concentration of the resolvase very efficiently catalyzes the cross-over, and that any replicon with two *res* sites will be resolved into two separate DNA molecules, each with one copy of the *res* site. In the plasmid resolution process, the two resulting DNA molecules each have a full complement of the intact plasmid molecule, and therefore both of them will be maintained in the cell. However, if one replicon molecule carries two *res* sites, the recombinational cross-over between these will result in one DNA molecule that is still a replicon but missing the DNA between the two sites, and a second DNA molecule representing the sequence between the sites but without maintenance functions. This latter molecule is abortive and will therefore be lost from the population during further growth. Because the cross-over is independent of the DNA sequences on either of the *res* sites, one can design molecules with relevant genetic information that may be deleted with an appropriate frequency.

Figure 2 shows how a recombinational system based on the RP4 resolution functions was designed for suicide purposes. In a DNA molecule, two *res*

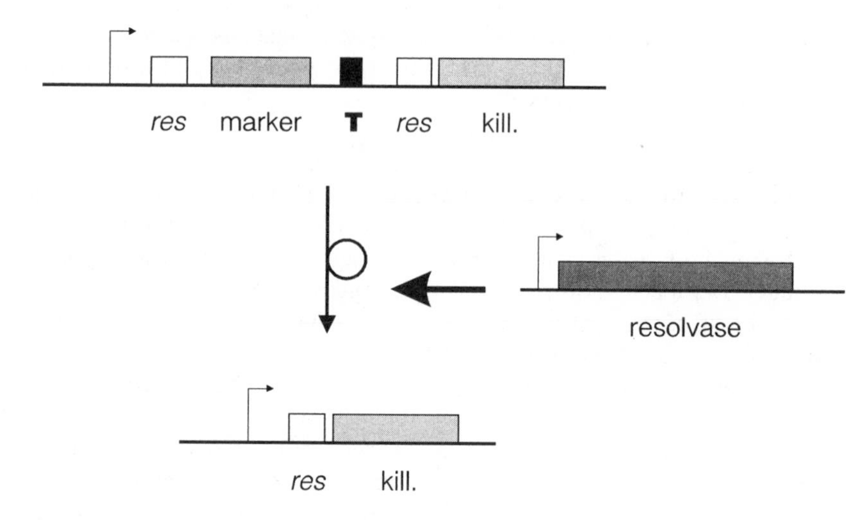

Figure 2 Schematic of the design strategy for recombinational activation of killing. See the text for further explanation.

sites are placed on either side of a strong transcription termination sequence and a marker gene (e.g. antibiotic resistance) allowing selection for maintenance of the sequence between the repeats. A strong promoter pointing towards the sites is placed on one side, and a promoterless suicide gene is placed on the other side of the sequence defined by the two *res* sites. In case of recombination mediated by the RP4 resolvase, the cross-over between the *res* sites will result in excision of the DNA between the two sites, by which the transcription terminator (and the marker gene) is deleted and later lost because of the lack of replication capacity in this fragment. As a consequence of the deletion, the strong flanking promoter will be brought in juxtaposition to the suicide gene, and cell death will result.

The resolution process requires no other components than those already mentioned (*res* sites and the resolvase). This has been clearly demonstrated in an in vitro reaction: Purified resolvase enzyme added to DNA carrying two *res* sites yields totally resolved DNA fragments of the expected sizes (C. S. Kristensen & L. Eberl, in preparation). This in vitro system could also be used to demonstrate that the efficiency of the recombination reaction depends on the amount of resolvase. Consequently one should be able to design a recombination system in which the natural resolvase promoter is replaced by an environmentally regulated promoter, thereby shifting control of the resolvase activity to the environment. This option is of significant value because promoters controlled by environmental signals may be relatively weak, making their efficacy, if combined directly with the suicide gene, questionable. However, if combined with the resolvase gene, in a context like the one described above, a weak transcription of the recombinase will be amplified due to recombination and the resulting strong transcription into the suicide gene.

Alternatively, the replacing promoter may be a very weak, constitutive promoter that will reduce the recombination frequency to levels corresponding to that observed for the *fimA* switch promoter, resulting in the possibility of a stochastic induction of suicide that is not limited to a few bacteria, as is the case for the *fimA* switch.

The Problem of Mutations

Laboratory model systems of induced suicide always contain a surviving subpopulation that can continue to grow even in the presence of inducer. Although sublethal gene expression is sometimes the cause, the main problem here is mutation; this has been studied mostly in *E. coli,* in which strains harboring the killing gene under control of various regulatable promoters are rapidly killed upon induction of transcription, thus demonstrating a high efficiency of the killing system (2, 9, 42, 47). Killing, however, is never absolute; a surviving fraction of 10^{-6}–10^{-3} is always observed, and further testing of such cells normally reveals that they no longer respond to the induction of the killing gene. This changed phenotype may be caused by two principally different genetic changes: mutations in the killing gene or its expression system that inactivate the suicide function, or mutations in other parts of the cell that make it resistant to the action of the killing function. Of course, in a nuclease-based suicide system, the target (nucleic acids) cannot be mutated to a nuclease-resistant state, but genetic changes improving the efficiency of DNA repair would most likely result in quasiresistance to the enzyme. The activity of the Gef-like proteins as porins in the cytoplasmic membrane suggest that resistant mutants with changed membrane structure might appear, and one Gef-resistant mutant has been isolated and characterized (54). One of the interesting features of this mutant is, however, that its resistance phenotype required at least two mutations, suggesting that this problem is of less significance in the context of biological containment based on a Gef killing system.

Mutations in the killing gene itself are easily identified by transferring the suicide system to a new host cell and testing for killing in a repeated induction experiment. Such mutations are by far the most frequent causes of survival after induction. If the killing gene is carried on plasmids, large deletions occur quite often, but also smaller deletions and point mutations play a significant role. In order to estimate the mutation rate of the *gef*-like genes, fluctuation test experiments have been performed (42), and the results show that these genes mutate to inactivity with rates of approximately 10^{-6} per cell per generation. In applications using many cells (large fermentors and field applications), the number of survivors escaping the suicide system would thus be quite large; therefore a major objective is to reduce survival even further.

One obvious way to reduce the mutation problem is to duplicate the killing system in the organism, either using two identical systems or combining different systems. If mutations in the two systems occur independently, the probability of simultaneous mutations in both systems during one cell cycle is, of course, the product of the two mutation rates, *in casu* 10^{-12} (per cell generation). However, inactivation of the killing systems can also occur sequentially, i.e. one killing system may mutate, after which this mutant multiplies to become a subpopulation in which the double mutant can appear as a result of a single mutation in the other killing system. Consequently, the double mutation rate (measured per cell generation) will increase with the number of cell doublings in the population, and in large populations sequential mutations will become the major route of killing escape. If the growth rates of the two single mutants are identical to that of the wild-type, the double-mutation rate will increase in a nearly linear fashion with the number of generations; if the growth rate of one or both of the single mutants is higher than that of the wild-type, the double-mutation rate will increase in an exponential fashion in the growing population.

Therefore, in a practical context involving large populations, the major concern relating to mutants is probably the ratio between the growth rates of wild-type and mutant cells. If the cell carries a regulatable killing gene that is not fully repressed under noninduction conditions, the basal level of killing gene expression may negatively affect growth. Mutations inactivating the killing genes will of course result in an increased growth rate, and the cell population will rapidly be taken over by mutants that can no longer be induced for suicide. Thus in all constructs of this type of containment system, the basal level of killing-gene expression must be as low as possible.

Alternative Strategies

The suicide strategy is not the only way to achieve containment; two other routes deserve attention. The first, attenuation of the strain, is well known from the prophylactic use of microbes (including viruses) in medical and veterinary microbiology. The basic idea is to disable the organism, either with respect to its pathogenic properties so it does not cause disease, or with respect to its survival or proliferative capacity in the environment to which it is released. The latter approach is especially of interest in the present context. The attempts to design live vaccines based on enterobacteria and, in particular, *Salmonella* spp. provide an excellent example of attenuation of bacterial proliferation. The purpose here is to use such strains as immunogens by presenting antigenic determinants to the host's immune system after oral administration. Live cells are employed because of their demonstrated efficacy, but at the same time their proliferation should be limited once they have entered the target organisms, especially because the bacteria are invasive. An efficient way to inhibit bacterial proliferation without reducing the immunogenicity or the ability to produce the cells by fermentation was to mutate the genes required for synthesis of aromatic amino acids, thereby making the strain dependent on external supplies of these essential precursors (67). In the animal gut there is a shortage of these amino acids, and therefore no growth of the bacteria is possible. Analogously, a mutation of the *crp* gene preventing the cells from exerting catabolite repression control kept the bacteria from proliferating in the gut (19). These approaches have been very useful in the design of live vaccines, and the bacteria are only able to grow and survive in environments outside the animal or human guts where they do not constitute any risk to humans or animals.

Analogously, *recA* mutations in soil bacteria have been proposed as an attenuation strategy (R. Simon, personal communication). In this case the argument is that bacteria living in soil or on plants will be exposed to irradiation frequently (sun light), and therefore mutations reducing their capacity to repair radiation damage will eventually lead to their eradication. The specific argument for the use of *recA* is that this gene is easily identified and isolated in most bacteria, and therefore also easily manipulated. The major problem of *recA* mutants is that they are often very much disabled and hence incapable of efficient competition in the environment, but for applications where long-term survival is not required this approach should be useful.

Limited survival by attenuation of strains may also be accomplished by mutations in genes whose products are essential under stress conditions. Interesting results have been obtained for *Vibrio* S14, whose stress-induced differentiation program has been studied for some years (49). If cells of this organism are starved for a carbon source they become very small, and their metabolism changes drastically. Specific proteins are synthesized at different times after onset of starvation, and the pattern of macromolecular synthesis strongly indicates the existence of a regulatory network responsible for the changes and the sequence of events. In an attempt to expose such control circuits, fusions were constructed by inserting reporter genes (β-galactosidase) into genes that are specifically induced during starvation conditions (J. Ostling & S. Kjelleberg, in preparation). Subsequently, mutagenic transposons were inserted at random in the chromosome, followed by screening for clones that showed a lack of starvation-induced gene expression. Such insertions did not interfere directly with the reporter gene—this observation led to the conclusion that the second site insertion had inactivated a regulatory gene responsible for activation of one or more genes that are normally turned on after starvation. The two described insertion mutations in genes of importance for starvation-induced gene expression had no significant effect during normal growth; however, when placed under starvation conditions in which the wild-type organism normally survives well, the double insertion mutant strain showed decreasing viability resulting in nearly total elimination after 10–15 days.

Thus, the phenotype resembles that produced by stochastic induction of suicide. This approach is very promising because the survival phenotype is one that seems ideal for releases: The cells released to the environment will face starvation conditions sooner or later; the possibility of reversion of the mutations is fairly small; and the strategy should be applicable to most bacteria. Moreover, the mutation program needed for the isolation of such strains is standardized and not too laborious, and soil bacteria, for example, apparently have similar starvation programs that may be interfered with in a similar manner (M. Givskov, L. Eberl & S. Molin, unpublished data).

Alternative Applications of Suicide Systems

In the course of construction of several suicide systems for biological containment purposes, numerous cassettes have been made in which killing genes have been fused with regulatable promoters. Many of these have been designed as simple test systems based on, for example, the IPTG inducible *lac* promoter/operator expression system, and such constructs may be useful in other contexts. One application, presently under investigation, concerns horizontal gene transfer in environmental samples from an introduced donor bacterium to the indigenous microflora of the environment (J. D. van Elsas, personal communication; S. Sørensen & T. Barkay, personal communication). In standard laboratory plasmid transfer experiments the recipient cells have a selectable marker gene (e.g. antibiotic resistance) that differs from those of the plasmid donor strain, allowing a direct selection of recipient cells harboring the plasmid. In the environmental recipient cells no such marker genes are present, and therefore donor strains must be designed that can be counterselected after plasmid transfer has been allowed. The introduction of an inducible suicide system in the donor would permit counterselection, because induction with e.g. IPTG will not affect the recipients but lead to killing of the donor. Such an application is presently being tested in experiments in which plasmid transfer from *P. putida* to sea water bacteria is monitored.

In small microcosms or biofilms where simulations of natural microbiological ecosystems are studied, the effects of introducing a certain bacterial species is a relatively easy experiment. However, the opposite experiment in which a previously introduced organism is removed is not so simple. Insertion of a regulated killing gene, like those described here, will allow studies of not only entry of strains but also sudden exits. With the modern microscopy techniques useful for studying specific bacteria in even very complex communities, such entry/exit scenarios promise a better understanding of microbial ecological principles.

DISCUSSION

The usual argument against the suicide systems as described here relates to the failure to design systems that totally eliminate a population. Even if optimized combinations of killing genes and gene expression sequences are constructed that upon induction will be 100% effective, mutations in a cell population will never be eliminated. Therefore, if an organism is considered a serious risk to the environment in connection with applications, its containment will not suffice to remove the potential risk. Even if no direct risks are foreseen, concerns remain regarding the release of engineered organisms to the environment, and we suggest that as long as such concerns prevail, attempts should be made to minimize putative problems by reducing as much as possible the exposure time and persistence of the introduced organisms. If this is the purpose of biological containment, a reduction of the population size over a limited time by a factor of a million or even less seems valuable. There are many bad examples of manmade chemicals accumulating in the environment, extensive use of antibiotics, etc, and measures are now being taken to solve those problems by encouraging the use of degradable rather than recalcitrant compounds. In the same way, the use of survival-deficient engineered organisms rather that very persistent strains, even where no obvious risks are expected, is prudent.

The important basis of these arguments is that it may be possible to design proper containment systems resulting in reduced survival under environmental conditions. In order for such systems to be useful in this context they must be easy to construct and transfer into many different bacterial species, and they should above all have a reliable efficiency, even if they do not eliminate the organism totally. Much study remains before we can conclude that all of this is possible, and it is certainly also premature to argue for one or the other strategy. Nevertheless, the combination of direct design trials and an increasing knowledge and understanding of bacterial life in nature will eventually make it possible to decide the potential of bacteriological-containment systems as a means of achieving more predictable organisms, which may be exploited in the service of both mankind and ecological systems of the environment.

ACKNOWLEDGMENTS

Much of the work presented was supported by the Danish Center for Microbiology, the Danish Center for Microbial Ecology, and the EC BAP and BRIDGE programs.

Literature Cited

1. Akimoto, S., Ono, K., Ono, T., Ohnishi, Y. 1986. Nucleotide sequence of the F plasmid gene *srnB* that promotes degradation of stable RNA in *Escherichia coli*. *FEMS Microbiol. Lett.* 33:241–45
2. Atlas, R. M., Bej, A. K., Steffan, R. J., Perlin, M. H. 1989. Approaches for monitoring and containing genetically engineered microorganisms released into the environment. *Hazard. Waste Hazard. Mater.* 6:135–44
3. Bachmann, B. J. 1990. Linkage map of *Escherichia coli* K-12, edition 8. *Microbiol. Rev.* 54:130–97
4. Bagdasarian, M. M., Amann, E., Lurz, R., Ruckert, B., Bagdasarian, M. 1983. Activity of the hybrid *trp-lac* (*tac*) promotor of *Escherichia coli* in *Pseudomonas putida*. Construction of broad-host-range, controlled-expression vectors. *Gene* 26:273–82
5. Ball, T. K., Saurugger, P. N., Benedik, M. J. 1987. The extracellular nuclease gene of *Serratia marcescens* and its secretion from *Escherichia coli*. *Gene* 57:183–92
6. Ball, T. K., Suh, Y., Benedik, M. J. 1992. Disulfide bonds are required for *Serratia marcescens* nuclease activity. *Nucleic Acids Res.* 20:4971–74
7. Barth, P. T., Ellis, K., Bechhofer, D. H., Figurski, D. H. 1984. Involvement of *kil* and *kor* genes in the phenotype of a host-range mutant of RP4. *Mol. Gen. Genet.* 197:236–43
8. Bech, F. W., Jørgensen, S. T., Diderichsen, B., Karlstrom, O. 1985. Sequence of the *relB* transcription unit from *Escherichia coli* and identification of the *relB* gene. *EMBO J* 4:1059–66
9. Bej, A. K., Perlin, M. H., Atlas, R. M., 1988. Model suicide vector for containment of genetically engineered microorganisms. *Appl. Environ. Microbiol.* 54:2472–77
10. Berg, P., Baltimore, D., Brenner, S., Roblin, R. O. III, Singer, F. 1975. Asilomar conference on recombinant DNA molecules. *Science* 188: 991–94
11. Bialey, H. 1984. Bacterial fitness and genetic engineering. *Bio/Technology* 2: 239
12. Biedermann, K., Fiedler, H., Larsen, B., Riise, E., Emborg, C., Jepsen, P. 1990. Fermentation studies and the secretion of *Serratia marcescens* nuclease by *Escherichia coli*. *Appl. Environ. Microbiol.* 56:1833–38
13. Biedermann, K., Jepsen, P. K., Riise, E., Svendsen, I. B. 1989. Purification and characterization of a *S. marcescens* nuclease produced by *E. coli*. *Carlsberg Res. Commun.* 54:17–27
14. Bouma, J. E., Lenski, R. E. 1988. Evolution of a bacteria plasmid association. *Nature* 335:351–52
15. Boyer, H. B., Roulland-Dussoiux, D. 1969. A complementation analysis of the restriction and modification of DNA in *Escherichia coli*. *J. Mol. Biol.* 41: 459–72
16. Clarke, H. R. G., Leigh, J. A., Douglas, C. J. 1992. Molecular signals in the interactions between plants and microbes. *Cell* 71:191–99
17. Colwell, R. R., Brayton, P. R., Grimes, D. J., Roszak, D. B., Huq, S. A., Palmer, L. M. 1985. Viable but non-culturable *Vibro cholerae* and related pathogens in the environment: implications for release of genetically-engineered microorganisms. *Bio/Technology* 3:817–20
18. Contreras, A., Molin. S., Ramos, J. L. 1991. Conditional-suicide containment system for bacteria which mineralize aromatics. *Appl. Environ. Microbiol.* 57:1504–8
19. Curtiss, R. III, Goldschmidt, R. M., Fletchall, N. B., Kelly, S. M. 1988. Avirulent *Salmonella typhimurium* Δ*cya* Δ*crp* oral vaccine strains expressing a streptococcal colonization and virulence antigen. *Vaccine* 6:155–60
20. Curtiss, R. III, Inoue, R. M., Pereira, D., Hsu, J. C., Alexander, L., Rock, L. 1977. Construction and use of safer bacterial host strains for recombinant DNA research. In *Molecular Cloning of Recombinant DNA*, ed. W. A. Scott, R. Werner, pp. 99–111. New York: Academic
21. Cuskey, S., Bourquin, A. 1987. Construction of bacteria with conditional lethal genetic determinants for control of released microorganisms in the environment. *Abstr. Annu. Meet. Am. Soc. Microbiol.* Q127, p. 303
22. Davis, A., Moore, B. I., Parker, D. S., Taniuchi, H. 1977. A possible precursor of nuclease A, an extracellular nuclease of *Staphylococcus aureus*. *J. Biol. Chem.* 18:6544–53
23. Doig, A. J., Williams, D. H. 1991. Is the hydrophobic effect stabilizing of destabilizing in proteins? The contribution of disulfide bonds to protein stability. *J. Mol. Biol.* 217:389–98
24. Duque, E., Ramos-Gonzalez, M.-I., Delgado, A., Contreras, A., Molin,

S., et al. 1992. Genetically engineered *Pseudomonas* strains for mineralization of aromatics: survivial, performance, gene transfer, and biological containment. In Pseudomonas: *Molecular Biology and Biotechnology,* ed. E. Galli, S. Silver, B. Witholt, 46:429–37. Washington, DC: Am. Soc. Microbiol.
25. Eisenstein, B. I. 1981 Phase variation of type 1 fimbriae in *Escherichia coli* is under transcriptional control. *Science* 214:337–38
26. Figurski, D. H., Pholman, R. D., Bechhofer, D. H., Prince, A. S., Kelton, C. A. 1982. The broad host range plasmid RK2 encodes multiple *kil* genes potentially lethal to *Escherichia coli* host cells. *Proc. Natl. Acad. Sci. USA* 79:1935–39
27. Freitag, C. S., Abraham, J. M., Clements, J. R., Eisenstein, B. I. 1985. Genetic analysis of the phase variation control of expression of type 1 fimbriae in *Escherichia coli. J. Bacteriol.* 162: 668–75
28. Gerdes, K., Bech, F. W., Jørgensen, S. T., Løbner-Olesen, A., Rasmussen, P. B., et al. 1986b. Mechanism of postsegregational killing by the *hok* gene product of the *parB* system of plasmid R1 and its homology with the *relF* gene product of the *E. coli relB* operon. *EMBO J.* 5:2023–29
29. Gerdes, K., Helin, K., Christensen, O. W., Løbner-Olesen, A. 1988. Translational control and differential RNA decay are key elements regulating postsegregational expression of the killer protein encoded by the *parB* locus of plasmid R1. *J. Mol. Biol.* 203:119–29
30. Gerdes, K., Larsen, J. E. L., Molin, S. 1985. Stable inheritance of plasmid R1 requires two different loci. *J. Bacteriol.* 161:292–98
31. Gerdes, K., Poulsen, L. K., Thisted, T., Nielsen, A. K., Martinussen, J., Andreasen, P. H. 1990. The *hok* killer gene family in Gram-negative bacteria. *New Biol.* 2:946–56
32. Gerdes, K., Rasmussen, P. B., Molin, S. 1986. Unique type of plasmid maintenance: postsegregational killing of plasmid-free cells. *Proc. Natl. Acad. Sci. USA* 83:3116–20
33. Gerdes, K., Thisted, T., Martinussen, J. 1990. Mechanism of post-segregational killing by the *hok/sok* system of plasmid R1: the *sok* anti-sense RNA regulates the formation of a *hok* mRNA species correlated with killing of plasmid free cells. *Mol. Microbiol.* 4:1807–18
34. Gerlitz, M., Hrabak, O., Schwab, H. 1990. Partitioning of broad host range plasmid RP4 is a complex system involving site-specific recombination. *J. Bacteriol.* 172:6194–6203
35. Givskov, M., Molin, S. 1992. Expression of extracellular phospholipase from *Serratia liquefaciens* is growth-phase-dependent, catabolite-repressed and regulated by anaerobiosis. *Mol. Microbiol.* 6:1363–74
36. Givskov, M., Molin, S. 1993. Secretion of *Serratia liquefaciens* phospholipase from *Escherichia coli. Mol. Microbiol.* In press
37. Givskov, M., Olsen, L., Molin, S. 1988. Cloning and expression in *Escherichia coli* of the gene for an extracellular phospholipase from *Serratia liquefaciens. J. Bacteriol.* 170:5855–62
38. Goldstein, J., Pollitt, N. S., Inouye, M. 1990. Major cold shock protein of *Escherichia coli. Proc. Natl. Acad. Sci. USA* 87:283–87
39. Hecker, M., Völker, U. 1990. General stress proteins in *Bacillus subtilis. Microbiol. Ecol.* 74:197–213
40. Inouye, M., Nakazawa S. A., Nakazawa, T. 1983. Molecular cloning of the *xylS* gene of the TOL plasmid: evidence for positive regulation of the *xylDEGF* operon by *xylS*. *J. Bacteriol.* 148:413–18
41. Jones, P. G., van Bogelen, R. A., Neidhardt, F. C. 1987. Induction of proteins in response to low temperature in *Escherichia coli. J. Bacteriol.* 169: 2092–95
42. Knudsen, S. M., Karlström, O. H. 1991. Development of efficient suicide mechanisms for biological containment of bacteria. *Appl. Environ. Microbiol.* 57:85–92
43. Mariani, C., de Beuckeleer, M., Truettner, J., Leemans, J., Goldberg, R. B. 1990. Induction of male sterility in plants by a chimaeric ribonuclease gene. *Nature* 347:737–41
44. Matin, A., 1990. Molecular analysis of the starvation stress in *Escherichia coli. Microbiol. Ecol.* 74:185–96
45. Matin, A., Auger, E. A., Blum, P. H., Schultz, J. E. 1989. Genetic basis of starvation survival in nondifferentiating bacteria. *Annu. Rev. Microbiol.* 43:293–316
46. Mermod, N., Lehrbach, P. R., Reineke, W., Timmis, K. N. 1984. Transcription of the TOL plasmid toluate catabolic pathway operon of *Pseudomonas putida* is determined by a pair of coordinately and positively regulated overlapping promoters. *EMBO J.* 3:2461–66
47. Molin, S., Klemm, P., Poulsen, L.

K., Biehl, H., Gerdes, K., Andersson, P. 1987. Conditional suicide system for containment of bacteria and plasmids. *Bio/Technology* 5:1315–18

48. Molin, S., Klemm, P., Poulsen, L. K., Biehl, H., Gerdes, K., Andersson, P. 1988. Biological containment of bacteria and plasmids to be released to the environment. In *Risk Assessment for Deliberate Releases,* ed. W. Klingmüller. pp. 127–36. Berlin/Heidelberg: Springer-Verlag
49. Nyström, T., Albertson, N. H., Flärdh, K., Kjelleberg, S. 1990. Physiological and molecular adaptation to starvation and recovery from starvation by the marine *Vibrio*. *Microbiol. Ecol.* 74: 129–40
50. Nyström, T., Neidhardt, F. C. 1992. Cloning, mapping and nucleotide sequencing of a gene encoding a universal stress protein in *Eschericha coli*. *Mol. Microbiol.* 6:3187–98
51. Ono, K., Akimoto, S., Ohnishi, Y. 1987. Nucleotide sequence of the *pnd* gene in plasmid R483 and role of the *pnd* gene product in plasmolysis. *Microbiol. Immunol.* 31:1071–83
52. Pallesen, L., Madsen, O., Klemm, P. 1989. Regulation of the phase switch controlling expression of type 1 fimbriae in *Escherichia coli*. *Mol. Microbiol.* 3:925–31
53. Poulsen, L. K., Larsen, N. W., Molin, S., Andersson, P. 1989. A family of genes encoding a cell-killing function may be conserved in all Gram-negative bacteria. *Mol. Microbiol.* 3:1463–72
54. Poulsen, L. K., Larsen, N. W., Molin, S., Andersson, P. 1992. Analysis of an *Escherichia coli* mutant strain resistant to the cell-killing function encoded by the *gef* gene family. *Mol. Microbiol.* 6:895–905
55. Poulsen, L. K., Refn, A., Molin, S., Andersson, P. 1991. Topographic analysis of the toxic Gef protein from *Escherichia coli*. *Mol. Microbiol.* 5: 1627–37
56. Poulsen, L. K., Refn, A., Molin, S., Andersson, P. 1991. The *gef* gene from *Escherichia coli* is regulated at the level of translation. *Mol. Microbiol.* 5:1639–48
57. Ramos, J. L., Michán, C., Rojo, F., Dwyer, D., Timmis, K. N. 1990. Signal-regulator interactions. Genetic analysis of the effector binding site of *xylS,* the benzoate-activated positive regulator of *Pseudomonas* TOL plasmid *meta*-cleavage pathway operon. *J. Mol. Biol.* 211:373–82
58. Ramos, J. L., Stolz, M. A., Reineke, W., Timmis, K. N. 1986. Altered effector specificities in regulators of gene expression: TOL plasmid *xylS* mutants and their use to engineer expansion of the range of aromatics degraded by bacteria. *Proc. Natl. Acad. Sci. USA* 83:8467–71
59. Ramos, J. L., Timmis, K. N. 1987. Experimental evolution of catabolic pathways of bacteria. *Microbiol. Sci.* 4:228–37
60. Roberts, R. C., Burioni, R., Helinski, D. R. 1990. Genetic charcterization of the stabilizing functions of a region of broad host range plasmid RK2. *J. Bacteriol.* 172:6204–16
61. Roszak, D. B., Colwell, R. R. 1987. Survival strategies of bacteria in the natural environment. *Microbiol. Rev.* 51:365–79
62. Sakikawa, T., Akimoto, S., Ohnishi, Y. 1989. The *pnd* gene in *E. coli* plasmid R16: nucleotide sequence and gene expression leading to cell Mg^{2+} release and stable RNA degradation. *Biochim. Biophys. Acta* 1007:158–66
63. Schreiner, H. C., Bechhofer, D. H., Pohlman, R. F., Young, C., Borden, P. A., et al. 1985. Replication control in promiscuous plasmid RK2: *kil* and *kor* functions affect expression of the essential replication gene *trfA*. *J. Bacteriol.* 163:228–37
64. Shortle, D. 1983. A genetic system for analysis of staphylococcal nuclease. *Gene* 22:181–89
65. Sobecky, P. A., Schell, M. A., Moran, M. A., Hodson, R. E. 1992. Adaptation of model genetically engineered microorganisms to lake water: growth rate enhancements and plasmid loss. *Appl. Environ. Microbiol.* 58:3630–37
66. Spector, M. P. 1990. Gene expression in response to multiple nutrient-starvation conditions in *Salmonella typhimurium*. *Microbiol. Ecol.* 74:175–84
67. Stocker, B. A. D. 1990. Aromatic-dependent *Salmonella* as live vaccine presenters of foreign inserts in flagellin. *Res. Microbiol.* 141:787–96
68. Sussman, R., Jacob, F. 1962. Sur un système de répression thermosensible chez le bactériophage λ d'Escherichia coli. *C. R. Acad. Sci.* 254:1517
69. Willimsky, G., Bang, H., Fischer, G., Marahiel, M. A. 1992. Characterization of *cspB,* a *Bacillus subtilis* inducible cold shock gene affecting cell viability at low temperatures. *J. Bacteriol.* 174: 6326–35
70. Yansura, D. G., Henner, D. J. 1984. Use of the *Escherichia coli lac* repressor and operator to control gene expression in *Bacillus subtilis*. *Proc. Natl. Acad. Sci. USA* 81:439–43

Annu. Rev. Microbiol. 1993. 47:167–97

TRANSPORT OF NUCLEIC ACIDS THROUGH MEMBRANE CHANNELS: Snaking Through Small Holes

Vitaly Citovsky
Department of Biochemistry and Cell Biology, State University of New York, Stony Brook, New York 11794

Patricia Zambryski
Department of Plant Biology, Koshland Hall, University of California, Berkeley, California 94720

KEY WORDS: nuclear transport; *Agrobacterium tumefaciens* T-DNA, snRNA, and hnRNA; plasmodesmata; cell-to-cell movement of plant viruses

---

CONTENTS

ABSTRACT

Transport of nucleic acids through cell membranes is an essential biological process that occurs in all living organisms. This review focuses on two plant systems in which nucleic acid molecules are transported through membrane channels: transport of *Agrobacterium* T-DNA through nuclear pores and

0066-4227/93/0167$02.00

movement of plant viruses through intercellular connections. To provide a broader perspective, nuclear uptake of animal viruses and nuclear import/export of small nuclear (sn) RNA and messenger (m) RNA are described. By comparing the examined cases of nucleic acid transport, the review proposes a set of general rules for this type of transport through membrane channels.

INTRODUCTION

Multicellular organisms as well as individual cells are highly compartmentalized. To maintain normal cellular functions, these compartments must constantly communicate. Intra- and intercellular communication occurs in a wide variety of biological processes ranging from the directional spread of electric impulses along neuronal axons in animals to transport of macromolecules through the intercellular connections between companion and sieve cells in plant vascular tissue. Although transport of electrolytes, hormones, and proteins is well studied, movement of nucleic acids across biological membranes is only now being addressed.

Transport of nucleic acids through cell membranes is a biological process basic to all living organisms. Nucleic acid molecules are transported through membrane channels during host-pathogen interactions (e.g. transport of viral genomes into the host cell) as well as during normal cellular processes [e.g. nuclear export/import of messenger (m) RNA]. Furthermore, transfer of genetic material has been shown to occur between evolutionarily distant organisms (9, 58, 109). A few common mechanisms potentially underlie these different events.

In this review, we focus mainly on two plant systems in which nucleic acid molecules are transported through membrane channels: (*a*) transport of *Agrobacterium* T-DNA through nuclear pores and (*b*) movement of plant viruses through intercellular connections. To provide a broader context, nuclear uptake of animal viruses and nuclear import/export of small nuclear (sn) RNA and heterogeneous nuclear (hn) RNA is also described. Then, the common structural and biochemical requirements of these transport processes are discussed.

NUCLEAR TRANSPORT

Molecular transport across the nuclear envelope involves many different proteins and nucleic acids. This transport is bidirectional and occurs exclusively through the nuclear pore complex (NPC), a 125,000-kDa structure composed of a central transporter mounted within spoke and ring protein assemblies that are integrated into the two membranes of the nuclear envelope (1) (Figure 1*a*). In the passive state, the NPC allows diffusion of small

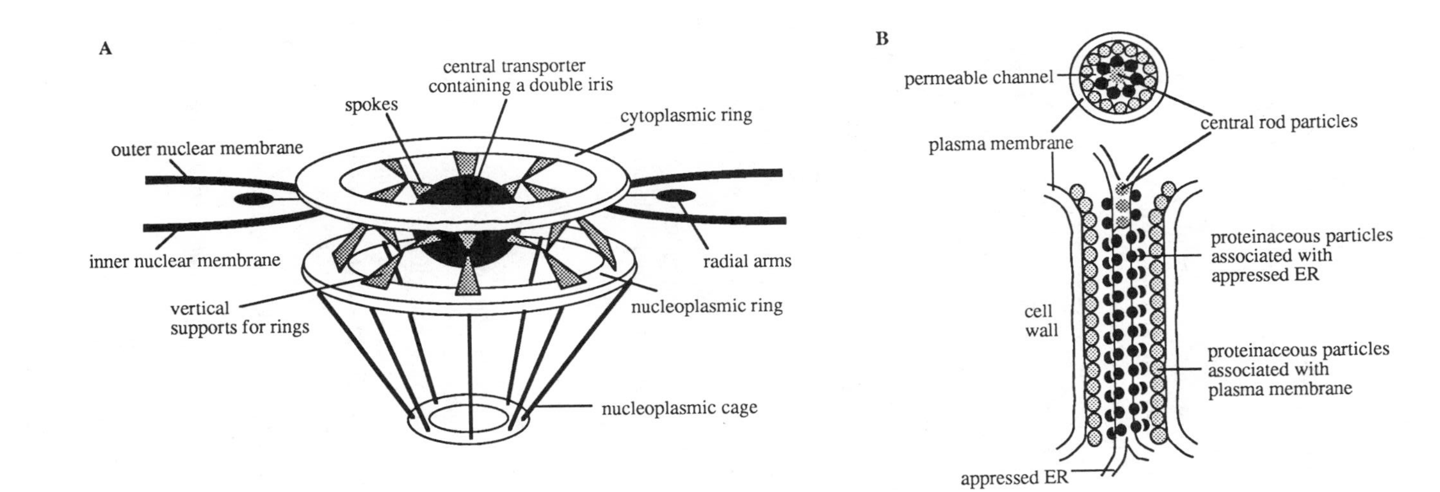

Figure 1 Schematic diagrams of the NPC and plasmodesmata. (*A*) NPC (adapted from Ref. 1). For clarity, the cytoplasmic peripheral fibrous assemblies are not shown. (*B*) A simple plasmodesmata (adapted from Refs. 22, 35). (*Top*) transverse view; (*bottom*) longitudinal view.

molecules (up to 40 kDa) (reviewed in 1, 52, 89). Transport of larger molecules occurs by an active mechanism mediated by specific nuclear localization signal (NLS) sequences contained in the transported molecule (reviewed in 46). The size limit for diffusion through the NPC was determined using microinjected foreign molecules such as dextrans (96). However, transport of small endogenous nuclear proteins such as H1 (21 kDa) also occurs by an active process (7). Recently, a model for active nuclear transport was suggested in which two symmetrical rings of protein subunits of the NPC transporter form a double iris similar to the airlock of a spacecraft (1, 2, 36); both of the irises must open sequentially to allow passage of a transported molecule. In the open conformation, the maximal size exclusion limit of the NPC is 23 nm in diameter (39).

As mentioned above, nuclear targeting of the transported molecule requires a specific NLS. With few exceptions (e.g. influenza virus nucleoprotein NLS, yeast GAL4 protein NLS), all NLSs can be classified into two groups: the SV40 large T antigen NLS (PKKKRKV) paradigm, and the bipartite motif consisting of two basic regions separated by a variable number (not less then 4) of spacer amino acids exemplified by the nucleoplasmin NLS (KR-X_{10}-KKKL) (reviewed in 37).

Nuclear transport of karyophilic molecules occurs in discrete steps. Recent data suggest that nuclear import initiates in the cytoplasm with binding of NLSs to the first cytoplasmic receptor, hsp70 (or its cognate hsc70), which may present the NLS in a locally unfolded form to the second type of cytoplasmic receptors, the NLS binding proteins (NBPs) (reviewed in 36). The NBPs then direct the transported molecule to the NPC where actual translocation across the nuclear envelope occurs (reviewed in 89).

Nuclear Import of Agrobacterium *T-DNA by the Host-Plant Cells*

The interaction of *Agrobacterium* spp. with plant cells is the only known natural example of interkingdom DNA transfer. In nature this process results in crown gall tumors, an agronomically important disease that affects most dicotyledonous plants. Most functions for *Agrobacterium*–plant cell DNA transfer are carried on a large (200 kb) Ti (tumor inducing) plasmid contained in the bacterial cell. The Ti plasmid has two important genetic components. One, the T-DNA, is copied and transferred to the plant cell. The T-DNA is delimited by two 25-bp direct repeats at its ends, the T-DNA borders. Any DNA between these borders is transported to the plant cell. Although the T-DNA is the mobile element, it does not itself encode the products that mediate its movement. Instead, a second component of the Ti plasmid, the virulence (*vir*) region, provides most of the *trans*-acting products for T-DNA transfer. Following induction of *vir* gene expression by small phenolic signal molecules excreted from wounded susceptible plant cells, a transferable copy of the T-DNA is generated. This molecule, designated the T-strand, is a single-stranded copy of the bottom strand of the T-DNA region (reviewed in 24, 126, 127).

Evidence to date suggests that the T-strand directly associates with two different protein products of the *vir* region. During T-strand synthesis, the VirD2 protein tightly (probably covalently) attaches to the 5′ end of the T-strand molecule (60, 66, 118, 125), whereas VirE2, a single-stranded (ss) DNA binding protein (SSB), is proposed to coat the T-strand along its entire length (15, 16, 28, 49). Binding of VirE2 to ssDNA is cooperative (21, 103) and results in formation of long unfolded and very thin (less than 2 nm in diameter) protein-ssDNA complexes (21). The T-strand with its associated proteins, VirD2 and VirE2, comprise the *Agrobacterium* spp. T-DNA transfer complex, designated the T-complex (63, 65).

To genetically transform the host-plant cell, the T-complex must travel from *Agrobacterium* spp. across three different biological barriers, i.e. the bacterial cell wall and cell membranes, the plant cell wall and cell membrane, and the plant nuclear envelope. Ultimately, the T-DNA is integrated into the plant genome. The relatively simple three-component structure of the T-complex (VirD2 and VirE2 proteins and the T-strand) presents a useful model system to study nuclear import of nucleic acids.

Nuclear transport of the T-complex likely occurs in a polar fashion from its 5′ end (reviewed in 126). Potentially, the VirD2 protein bound to the 5′ end of the T-strand provides a piloting function. Indeed, several recent studies demonstrate nuclear localization of VirD2 in plant cells (61, 64, 112). The amino acid sequence relevant to VirD2 mediated nuclear transport was localized to a bipartite NLS at the carboxyl terminus of the protein. Deletion of this sequence in the VirD2 protein results in reduction of *Agrobacterium* spp. tumorigenicity (106). These results together suggest that the VirD2 protein, attached to the 5′ end of the T-strand, acts to direct the T-complex to the host-cell nucleus.

However, the T-complex is a very large structure; a 20-kb T-strand (the approximate size of the nopaline-type T-DNA) would contain more than 600 molecules of VirE2 (16). This T-complex has a predicted length of 3.6 μm (21) and a combined molecular mass of about 50×10^6 Daltons. Thus, the molecular mass of the T-complex is almost 20 times larger than that of the 60S ribosomal subunit (2.8×10^6 Daltons). In addition, because the nuclear pore is approximately 60 nm thick (100), the T-complex is about 60 times longer than the dimensions of the nuclear pore. Can such a large DNA-protein complex be transported through the nuclear pore by a single molecule of VirD2? That deletion of VirD2 NLS decreased but did not completely abolish tumorigenicity suggests that VirD2 is not the sole mediator of the T-complex nuclear uptake. Since VirE2 is a major structural protein component of the T-complex, it may assist in nuclear transport.

In fact, VirE2 was identified recently as a nuclear localizing protein (23). The VirE2 nuclear localization function is mediated by two bipartite NLS sequences designated NSE 1 and 2. As opposed to the typical bipartite NLS of VirD2, the VirE2 homology to bipartite consensus NLS is not perfect. Generally, the first domain of a bipartite NLS has two adjacent basic residues, and the second domain contains at least three out of five basic amino acid residues (37). Both NSE sequences of VirE2, however, have a modified first domain in which the two basic residues are separated by one amino acid; the consensus structure of the second domain, on the other hand, is preserved in the VirE2 NLSs (23).

In addition to functioning in nuclear targeting, VirE2 acts as an SSB (15, 16, 28, 49). Mutational analysis showed that deletion of the VirE2 major ssDNA binding domain (at the carboxyl-terminus of the protein) does not alter nuclear uptake. However, site-specific mutations in NSE1 and NSE2 alter ssDNA binding (23). These latter mutations may cause conformational changes in VirE2 that inactivate its putative ssDNA binding domain(s). Conformational changes caused by deletion of short stretches of amino acid residues have been shown to inactivate single-stranded nucleic acid binding of the P30 protein of tobacco mosaic virus (20). Alternatively, both NSE sequences may reside within another VirE2 region required for ssDNA binding. For example, karyophilic signals of the progesterone and glucocorticoid receptors may overlap their DNA binding domains (53, 57). If VirE2 NLSs and the DNA binding domain(s) indeed overlap, the NLS sequences must be accessible to the nuclear transport machinery when bound to the T-strand.

That VirE2 is involved in nuclear uptake of T-DNA is further strengthened by the observation that plants transgenic for VirE2 complement the virulence of an *A. tumefaciens* strain with an inactivated *virE* locus (23). This result indicates that the VirE2 function, essential for tumor formation, is required inside the plant cell. Nuclear localization of the T-complex is one such cellular function. Also, VirE2 expressed in transgenic plants may help to protect the T-strand from cellular nucleases. These in planta functions of VirE2 also help to explain the observation that *virE* mutants can be complemented to wild-type virulence if coinoculated on plants with *A. tumefaciens* carrying a wild-type *vir* region (95). Such complementation by coinoculation is possible only if the *virE* product functions outside the bacterial cell, for example, inside the host-plant cell.

Current information on the structure of the T-complex and the properties of its protein components is integrated in the following model for the T-complex nuclear uptake (Figure 2). The T-strand with a molecule of VirD2 covalently attached to its 5′ end likely exists as a folded and collapsed structure. Following cooperative binding of VirE2, the ssDNA is unfolded to form a long and thin protein-ssDNA T-complex (Figure 2, step 1). The T-complex is composed of three structural elements: one copy of the T-strand, one VirD2 molecule, and more than 600 copies of VirE2 (16). The T-strand is not sequence specific, and any DNA sequence located between the 25-bp T-DNA border repeats can be transported to plants and function as T-DNA (reviewed in 24, 126). Thus, the T-strand probably does not possess specific nucleotide sequences for nuclear uptake; instead it likely is passively transported into the nucleus by its associated proteins [piggyback transport (89)], leaving the other two components of the T-complex, the VirD2 and VirE2 proteins, to function in nuclear transport.

Because VirE2 and VirD2 contain NLSs, both are expected to interact with the NBP receptors (Figure 2, step 2), the presumed adapter molecules between the karyophilic protein and the actual transport machinery of the NPC (reviewed in 89). VirD2 may act to initially target the T-complex to the nuclear pore in a polar fashion, facilitating linear uptake of the T-strand in a 5′-to-3′ direction. The requirement for polar transport of T-DNA has been proposed in earlier genetic and molecular studies of *A. tumefaciens*–mediated plant cell transformation (reviewed in 126); such linear transfer of the T-strand may be

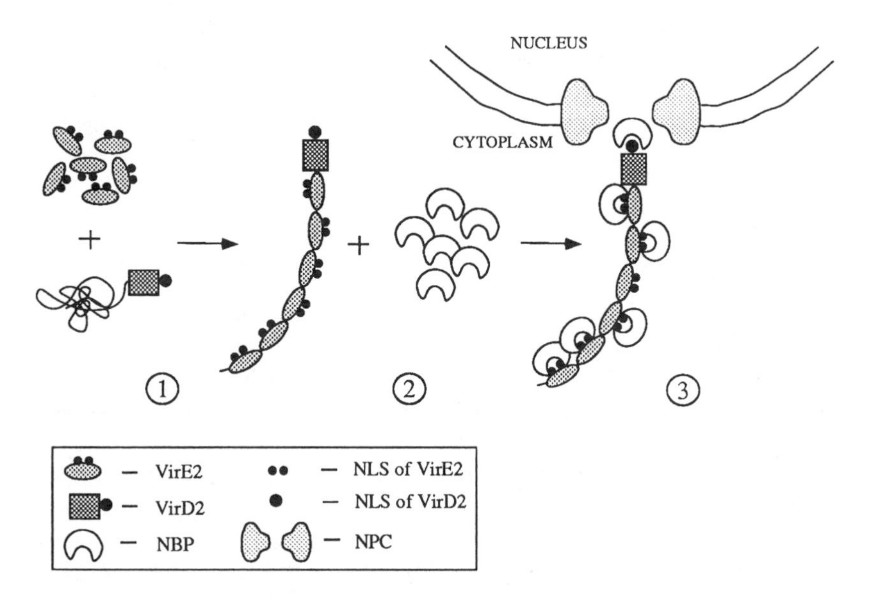

Figure 2 A model for nuclear transport of *Agrobacterium* spp. T-DNA. (*Step 1*) Formation of an unfolded T-complex. (*Step 2*) Interaction with cytoplasmic NBPs. (*Step 3*) Vectorial targeting (5′ end first) to the NPC.

necessary for its efficient integration into the plant genome. Thus, VirD2 may be critical to setting the orientation of the T-complex nuclear uptake, whereas VirE2 molecules with their NLS sequences (NSE 1 and 2) likely drive the unfolded T-complex through the nuclear pore (Figure 2, step 3). This large number of NLSs may greatly increase the probability of T-complex interaction with NBPs and, later, with NPC; multiple NLSs also may interact cooperatively with their receptors (47). Furthermore, the presence of multiple NLSs along the length of the entire T-complex may be necessary to keep open the two putative irises of the nuclear pore transporter.

There may be plant cellular proteins analogous to VirE2 that serve as molecular chaperones coating and unfolding nucleic acids and targeting them to and through nuclear pores. If true, this could explain observations that VirE2 function is not essential on some hosts (108); in this case, one can argue that another plant-cell SSB can provide the VirE2-like function. This idea supports the possibility that nuclear transport of *Agrobacterium* spp. T-complex may represent a generalized process by which ssDNA or RNA molecules move within the cell; i.e. as unfolded nucleic acid–protein complexes (see below).

Nuclear Import of Viral Genomes

Import of viral genomes into the host-cell nucleus represents another example of nucleic acid nuclear transport during host-pathogen interaction. Most animal and insect viruses as well as many plant viruses replicate in the nucleus of their host cells. Surprisingly, however, relatively little is known about mechanisms by which viral genomic nucleic acid molecules are transported into the cell nucleus. Below, we discuss possible mechanisms for nuclear localization of animal viruses as exemplified by two well-studied systems, influenza virus (an enveloped virus) and SV40 (a nonenveloped virus).

ENVELOPED VIRUSES Influenza virus is a member of the paramyxovirus group. The genome of influenza virus is composed of eight ssRNA molecules coated with viral nucleoproteins and associated with the viral polymerase. These viral RNA-protein complexes (vRNPs) directly interact with the viral matrix M1 protein that forms a shell beneath the viral envelope (59, 77 and references therein). During infection, influenza virions bind to host-cell surface receptors (sialoglycoproteins and sialoglycolipids) (reviewed in 75). The bound virions are then internalized and their envelopes fuse with endosomal membranes; this fusion is mediated by a specific viral envelope glycoprotein (HA polypeptide) and is triggered by the acidic environment of the endosome. Following fusion, the contents of the viral envelope are injected into the host-cell cytoplasm. M1 then dissociates from the vRNPs, and the latter are actively transported into the nucleus (59, 77, 121).

Nuclear transport of influenza virus is mediated by a single NLS contained in the protein component of the vRNP. The amino acid sequence of this NLS has been identified as AAFEDLRVLS (29). This signal sequence is not homologous to the most common NLSs of either the bipartite or SV40 large T-antigen types. Interestingly, influenza NLS is similar to the recently described NLS-A of a maize transcription factor, the R protein (105).

In the nucleus, influenza RNA is transcribed and replicated. Newly synthesized viral nucleoproteins, polymerase, and the M1 protein enter the nucleus, assemble with the viral RNAs, and are exported into the cytoplasm. Transport of the newly formed vRNPs out of the nucleus depends on the presence of M1, suggesting that this protein may contain a putative signal for nuclear export (77). Similar function in nuclear export of unspliced viral RNA has been described for the HIV Rev protein (reviewed in 11). Alternatively, M1 and Rev proteins may simply bind vRNPs and dislodge them from a nuclear anchor (69).

Presently, it is clear that influenza RNA is transported into the nucleus as a vRNP, and that the protein component of this complex mediates the transfer. Nuclear entry as a nucleoprotein particle also has been described for other animal (Herpes simplex viruses, reviewed in 101a) and insect enveloped viruses (Baculoviruses, reviewed in 6). Overall, however, our knowledge of nuclear transport of enveloped viruses other than influenza virus is sparse. For example, nuclear import of medically important and otherwise well-studied retroviruses is largely an unexplored subject. Retroviruses have been proposed to enter the cell nucleus as a nucleoprotein complex (reviewed in 8). One viral protein, the IN protein of avian sarcoma-leukosis virus, has been shown to contain NLS, but the role of this protein in the life cycle of the virus is unclear (86).

NONENVELOPED VIRUSES Nuclear uptake of nonenveloped viruses has been studied using SV40. SV40 is a member of the papovavirus family. Papovaviruses bind to host-cell surface receptors and are internalized in endocytotic vesicles. Virus entry into the nucleus has been proposed to occur by nonspecific fusion of the endocytotic vesicles with the outer nuclear membrane (91). This model, however, does not explain viral penetration through the inner nuclear membrane nor does it provide for the specificity and high efficiency of viral nuclear transport (for example, SV40 virions are found in the host cell nucleus as early as 15 min after infection). Recent results provide evidence that SV40 particles are karyophilic (26). Microinjection of whole SV40 virions into the cell cytoplasm resulted in substantially more rapid expression of the large T-antigen than was observed following microinjection of free SV40 DNA. Virus nuclear entry was blocked by coinjection with monoclonal antinucleoporin antibody (mAb414) or wheat germ agglutinin (WGA), ligands that bind to the nucleoporin p62 of NPC and block NLS-mediated nuclear transport of proteins. Furthermore, following SV40 microinjection, the viral structural proteins, VP1 and VP3, were detected in the cell nucleus. So far only one functional NLS of the large T-antigen type has been identified in the structural proteins of SV40 (25). This NLS may mediate nuclear uptake of the entire virion. Because the SV40 particle (50 nm in diameter) is considerably larger than the maximal diameter of the nuclear pore (23 nm as measured using microinjection of spherical gold particles coated with RNA) (39), the SV40 virion was proposed to change its shape as it passes through the fully open nuclear pore (26). Interestingly, changes in particle shape resulting from unfolding and partial uncoating were reported for nuclear export of 50-nm hnRNPs in *Chironomus tentans* (82) (see below). Alternatively, the gating potential of the nuclear pore may be higher than observed to date.

The above described observations suggest that viruses enter the host cell nucleus as either subviral particles (vRNPs) or entire virions. In all examined cases, viral proteins associated with the transported RNA or DNA molecules seem to play a critical role in the transport process. They provide specific NLS signals for targeting to and gating of the NPC; furthermore, these nucleic acid–associated proteins may be hypothesized to shape the complex into a transferable form compatible with the dimensions of the open nuclear pore.

Nuclear Traffic of snRNA and mRNA

In host-pathogen interactions, the invading microorganism generally does not invent novel metabolic pathways; instead it insinuates into the existing cellular processes and adapts them for its life cycle. Thus, nuclear uptake of *Agrobacterium* spp. T-complex and viral genomes probably follows one of the pathways for nuclear transport of cellular RNAs. To gain insight into these pathways, we review the characteristic features of RNA nuclear transport and its underlying mechanisms.

RNA molecules synthesized in the cell nucleus are divided into three classes depending on the type of the transcribing RNA polymerase (pol). Ribosomal RNAs are transcribed by pol I; both mRNA and major snRNAs are transcribed by pol II; and 5S ribosomal RNA, transfer RNAs, and other small nuclear and cytoplasmic RNAs are transcribed by pol III (reviewed in 69). Recently, significant progress has been made in our understanding of nuclear transport of pol II products, snRNA, and mRNA. Below, we focus on this field of study.

TRANSPORT OF snRNA To date, transport of uracyl-rich (U) snRNAs is the best-studied process of RNA translocation through the nuclear pore. This reflects the experimental ease with which these molecules can be handled:

they are stable, resistant to mutations, and interact with proteins, many of which are well characterized. U1-U5 snRNAs are all pol II products and have an inverted monomethyl guanosine cap (m^7GpppN); U6 (and U3 in plants), a pol III product, retains a γ-methyl triphosphate (pppG) 5′ end instead. The major nucleoplasmic U1, U2, U4, U5, and U6 snRNAs function in the processing of mRNA precursors while the nucleolar U3 snRNA is involved in ribosomal RNA maturation (reviewed in 69).

Nuclear export of snRNA Newly synthesized U snRNAs are exported into the cytoplasm. Nuclear export of these snRNAs is strongly inhibited by WGA, indicating involvement of glycoprotein components of the NPC (87). In the case of U1 snRNA, its monomethyl guanosine (mG) cap is essential for nuclear export. U1 snRNA export also is specifically inhibited by microinjection of competing amounts of the cap analog, the m^7GpppG dinucleotide (56); furthermore, U6 snRNA that lacks the mG cap is retained in the nucleus (117). These observations suggest involvement of a specific mG cap–binding receptor. Several cap-binding proteins (CABs) have been described. For example, an 80-kDa CAB found in the nuclear extracts of HeLa cells (92) has been proposed to play a direct role in U1 snRNA export from the nucleus (70). However, most known CABs (e.g. the well-studied eukaryotic translation initiation factor eIF-4E) are involved in splicing and translation rather than in nuclear export.

Nuclear import of snRNPs Following export into the cytoplasm, U snRNAs complex with their cognate proteins, the Sm polypeptides. Sm binding to U1, U2, U4, and U5 snRNAs occurs at a single-stranded consensus sequence; U3 snRNA lacks this consensus site and is associated with different protein components. In the cell cytoplasm, in addition to the association with both common (Sm) and snRNA-type specific proteins, U1-U5 snRNAs are hypermethylated at their guanosine cap to produce a 2,2,7 trimethyl guanosine (tmG) cap structure (reviewed in 78). TmG-capped U snRNA-protein complexes comprise U snRNP particles. These snRNPs move back into the nucleus to form part of a supramolecular spliceosome complex that removes intron sequences from premessenger RNA. For the assembly of the spliceosome, therefore, U snRNA molecules cross the NPC in both directions.

Because study of nuclear import is technically easier than the study of export, import of U snRNPs into the nucleus has been characterized in more detail than the U snRNA nuclear export. Nuclear localization of the major U snRNPs occurs by a complex transport pathway. The NLS for transport is likely located in the common component of most U snRNPs, the Sm proteins (79). This NLS has not yet been characterized. Efficient transport of pol II U snRNPs also requires the tmG cap (55). The mechanism by which the tmG cap operates is unknown. It may function as a binding site for specific NLS-containing shuttle proteins; alternatively, conversion of the mG into the tmG cap may simply prevent cytoplasmic anchoring of the exported U snRNPs through their mG cap structure (55). Interestingly, whereas nuclear import of U1 and U2 snRNPs is absolutely dependent on the tmG cap, nuclear uptake of U4 and U5 snRNPs has a less stringent requirement for this cap structure (44). U6 snRNA, which normally does not leave the nucleus, will migrate into the nucleus when injected into the cell cytoplasm; this nuclear import of U6 snRNA does not require the 5′ cap signal and is mediated solely by its karyophilic cognate proteins (44, 85). Recent experiments suggest that these proteins may contain the SV40 large T-antigen-type NLS signals (85).

Thus, tmG-capped U snRNPs and U6 snRNP are imported into the nucleus by two kinetically different pathways (44, 85). A third type of nuclear import pathway has been described for U3 snRNP. Recent results indicate that U3 snRNP contains a unique NLS consisting of a 5-bp 3′ terminal stem; this short structure is necessary and sufficient for U3 snRNP nuclear import (5).

Although karyophilic proteins and U snRNPs are targeted to nuclear pores by various pathways, the translocation itself likely occurs by a single mechanism. Indeed, antinucleoporin antibodies inhibit both the tmG cap–mediated and the SV40 large T-antigen type NLS-mediated import pathways, indicating involvement of the same component of the NPC (43). The three independent pathways for nuclear targeting, therefore, may represent a versatile regulatory mechanism for transport of different macromolecules through the same nuclear pore.

NUCLEAR EXPORT OF mRNA The heterogeneous nuclear (hn) RNAs, such as mRNA precursors, account for most of the sequence complexity of nuclear RNA. Immediately following transcription, hnRNAs associate with a large family of proteins to form hnRNPs; these hnRNPs are complex structures composed of more than 20 different major proteins (14, 99). Following splicing, these particles are exported into the cytoplasm through the nuclear pore. After passage through the NPC, nuclear hnRNP proteins are exchanged to form cytoplasmic mRNPs.

Although most hnRNAs that do not complete splicing are retained (and later degraded) in the nucleus, in some cases eukaryotic and viral genes (e.g. rabbit β-globin, mouse dehydrofolate reductase, SV40 late transcripts, etc) require the presence of one or more introns for nuclear export of their mRNA (reviewed in 69). In addition to splicing, mRNA export often requires formation of mature 3′ ends of mRNA molecules (40).

An important insight into the actual mechanism of hnRNA translocation through the nuclear pore has come from the electron-microscope tomography studies of the Balbiani ring (BR) granules in *Chironomus tentans* salivary glands. These granules are large 75S hnRNAs (encoding information for secretory proteins with molecular masses around 10^6 Daltons) packed into 50-nm hnRNP particles (82 and references therein). Each BR hnRNP has a ribbon-like structure folded into a ring. During transport through the nuclear pore, the hnRNP particle orients at the NPC such that the 5′ end of the RNA enters the pore first (82). This observation indicates a specific pilot function (possibly the cap structure) at the 5′ end that is recognized by the NPC apparatus.

The folded ribbon is gradually unfolded and transported through the nuclear pore. In parallel with translocation, the hnRNP is partially unpacked. Unfolding and partial uncoating of the BR particles change their diameter from 50 to 25 nm; this reduced size of the hnRNP fiber approximately corresponds to the maximal diameter of the fully open nuclear pore (23 nm) (39). Following transport, the hnRNP fiber appears in the cytoplasm in an extended conformation and is engaged in polysome formation (82). The structural features of the BR hnRNA transport, i.e. 5′-to-3′ polar translocation of the unfolded and extended RNP fiber, are remarkably similar to those proposed for the nuclear uptake of *Agrobacterium* spp. T-complex (23) (see above).

Assembly of a translation-initiation complex immediately after the nuclear export may serve as a driving force for mRNA transport through the nuclear pore. This supposition is supported by the observations that nonsense and frameshift mutations (which terminate translation prematurely) within the first two thirds of a gene result in decreased levels of cytoplasmic mRNA. However, nonsense mutations positioned in the last third of the mRNA do not affect the mRNA levels in the cytoplasm (13). These results imply that translation of mRNA may pull it out of the nucleus through the nuclear pore. Early termination of translation, therefore, may stop the transport whereas termination at later stages, when most of the mRNA has been already transported, may have a less dramatic effect (13, 52).

CELL-TO-CELL MOVEMENT OF PLANT VIRUSES

Most of our knowledge of nucleic acid transport through membrane channels has come from studies of nuclear export and import both in animal and in plant systems. In plants, however, nucleic acids also can be transported between cells. This type of transport occurs during viral infection when plant viruses spread from cell to cell through intercellular connections, the plasmodesmata. Although plasmodesmata and nuclear pores are obviously structurally different, both are complex proteinaceous pores (see Figure 1) involved in active bidirectional traffic of macomolecules. It will be interesting to see if nucleic acid transport through plasmodesmata resembles nuclear import.

One of the structural differences between animal and plant cells is that the latter are enclosed in a rigid polysaccharide cell wall. This structural feature efficiently isolates plant cells from invading virus particles, which cannot diffuse through the cell-wall matrix. Therefore, plant viruses cannot spread between the host cells by membrane fusion or endocytosis, the entry pathways used by animal viruses. Most plant viruses enter the initially infected cell following mechanical damage inflicted by a biological carrier (insect, fungus, etc) or by abrasion. After initial infection, plant viruses move into adjacent healthy cells through plasmodesmata, the only direct link between cells.

Plasmodesmata

Plant intercellular connections, the plasmodesmata, are very large and complex pore structures lined with the plasma membrane (reviewed in 101). The central region of this pore is occupied by the appressed membranes of the endoplasmic reticulum (ER) connected to the ER of the adjacent cells. The inner leaflet of the appressed ER contains a series of electron-dense (proteinaceous) particles termed the central rod (Figure 1*b*). In addition, two rows of similar proteinaceous particles (3 nm in diameter) are associated with both the inner leaflet of the plasma membrane and the outer leaflet of the appressed ER. The region between these globular subunits forms the permeable channel through which molecules pass from cell to cell (35) (Figure 1*b*); one plasmodesmata usually contains 7–10 such channels (35) (Figure 1*b*, cross-section). In some plants, plasmodesmal channel orifices are delineated by constricted neck regions proposed to act as sphincters regulating channel permeability (93). Recent observations, however, indicate that in many plants these sphincters are absent, suggesting their dispensability for regulation of the channel (35).

Ultrastructurally, plasmodesmata found in the same plant tissue can be divided into two types, single and branched. Branched plasmodesmata are composed of several single plasmodesmal structures interconnected by a large central cavity (34). Developmentally, single plasmodesmata are thought to represent primary connections formed during cell division; the branched structures may represent secondary plasmodesmata formed after cell division has occurred (34).

In contrast to the relatively well studied ultrastructure of plasmodesmata, the biochemistry of these channels is virtually unknown. Two recent studies report purification of several putative plasmodesmata-associated proteins (124) and cloning of a gene encoding one of these proteins (83). These findings, however, were based on assumed immunological homology between plasmodesmata and animal gap junctions. Although both gap junctions and plasmodesmata function as intercellular connections, they are thought to have evolved independently after animal and plant kingdoms diverged (76); furthermore, the two types of connections are completely different in their architecture (42, 101). Thus, more compelling morphological and functional evidence is needed to confirm that these isolated proteins are true plasmodesmal components rather than unrelated cell-wall or plasma-membrane constituents.

The factors that control movement of metabolites and other molecules through plasmodesmata (e.g. Ca^{2+} and phosphoinositides) all reduce flow (4, 114); no plant factors have been shown to increase plasmodesmal permeability. The size exclusion limit of plasmodesmata (1.5–2.0 nm in diameter based on the Stokes radius of diffusable dyes with the molecular mass of 700–1000 Daltons) (111, 123) is much smaller than the size of all known plant virus particles (e.g. 18 × 300 nm for TMV particles) or even free folded viral nucleic acids [10 nm for tobacco mosaic virus (TMV)] (48). To move through plasmodesmata, therefore, plant viruses have evolved to encode a specific class of nonstructural movement proteins that increase plasmodesmal permeability and transport viral nucleic acids through the enlarged plasmodesmal channels (reviewed in 3, 22, 30, 68, 80).

To date, evidence points to two different mechanisms by which viral movement proteins operate. The first mechanism involves transport of subviral nucleoprotein complexes through transiently modified plasmodesmata. TMV, one of the best-studied plant viruses, moves between cells by this mechanism. Most of our knowledge about cell-to-cell spread of plant viruses derives from studies of TMV (3, 22, 30, 68, 80); thus, this process will be described here in greater detail. Recently, a second possible mechanism for virus movement through plasmodesmata was suggested from studies with cowpea mosaic virus (CPMV) (reviewed in 67). CPMV likely moves through plasmodesmata as entire virions; this movement occurs following virus-induced irreversible modification of plasmodesmata, which removes the central appressed ER structures.

TMV-Type Cell-to-Cell Movement

TMV is a linear positive sense ssRNA tobamovirus that replicates in the host cell cytoplasm. The TMV genome encodes at least three nonstructural proteins (P126, P183, and P30) and the coat protein (P17 or CP) (50). P126 and P183 function in viral replication (97), and P30 potentiates movement of TMV through plasmodesmata. Several lines of evidence point to this role of P30: (*a*) temperature sensitive TMV mutants (LS1, Ni2519, and II27) defective in cell-to-cell spread have an altered P30 polypeptide (71, 90, 98); (*b*) in vitro mutagenesis of P30 produces phenotypes restricted in cell-to-cell movement but not in replication (84); and (*c*) P30 expressed in transgenic plants can restore cell-to-cell spread of the mutant LS1 virus at the restrictive temperature (32). Because TMV CP is dispensable for cell-to-cell spread (107), P30 is the only viral protein that mediates movement.

Until recently, the mechanism of P30 action had been a complete mystery. It began to unravel when P30 was reported to have at least two important biological activities: First, P30, expressed in transgenic plants, was shown to increase the plasmodesmal molecular size exclusion limit (123), and second, purified P30 was shown to cooperatively bind single-stranded nucleic acids (17).

P30 INTERACTION WITH PLASMODESMATA The P30 effect on plasmodesmal permeability was studied by microinjecting fluorescently labeled dextrans of increasing molecular weights (123). This study showed that the size exclusion limit of plasmodesmata in transgenic plants that express P30 is higher (5–6 nm in diameter as measured using 10- to 17-kDa dextrans) than in normal plants (1.5 nm in diameter). These observations were confirmed by coinjection of the fluorescent dyes with virus particles of tobacco rattle tobravirus (TRV) into leaf trichome cells. In trichomes, the single-file arrangement of cells greatly facilitates monitoring of dye movement. Using this system, TRV infection was shown to promote a transient increase in plasmodesmal permeability of up to 5 nm (33). The mechanism by which P30 affects plasmodesmal flow is unknown. Because no obvious ultrastructural changes can be detected in plasmodesmata in the presence of P30 (34), the increase in plasmodesmal permeability likely occurs by normal (albeit unknown) regulatory pathways rather than by physical disassembly of plasmodesmata.

The P30-induced increase in the plasmodesmal size-exclusion limit is the first experimental evidence that plasmodesmal permeability can be up-regulated. This finding has an intriguing implication for the general concept of plant intercellular communication. Traditionally, plasmodesmata have been considered to traffic only water, ions, and small metabolites. The ability of plasmodesmata to expand, however, indicates that in normal, healthy plants plasmodesmata may traffic macromolecules as well. Indeed, recent results suggest that cellular proteins are constantly transported through plasmodesmata between companion and sieve cells in the vascular bundle (45).

Presently, the study of P30 interaction with plasmodesmata has been mainly conducted in P30 transgenic plants (31, 34, 122, 123); although useful in elucidating the general role of P30 in cell-to-cell spread of TMV, these plants constitutively produce and accumulate P30, which may interfere with normal plasmodesmata function. Intercellular communication, an important biological process, may not tolerate significant or constitutive alteration. Successfully regenerated P30 transgenic plants, therefore, may have modified the movement protein to enable their own survival. In fact, the changes in plasmodesmal permeability observed in P30 transgenic plants differ from those induced when P30 (expressed in *Escherichia coli* and purified to homogeneity) (17) is microinjected into wild-type tobacco plants. When coinjected with fluorescently labeled dextrans, P30 caused larger dextrans of 20–40 kDa (with a calculated diameter of 6–9 nm) to move between cells in leaf mesophyl of wild-type tobacco; this increase in plasmodesmal permeability is twofold higher than that reported for P30 transgenic plants (10- to 17-kDa dextrans with a diameter of 5–6 nm) (123). P30-mediated movement of these large dextran molecules was detected 3–5 min after injection; 45 min after injection, the fluorescent dye was distributed between numerous cells, reaching as far as 10 to 20 cells away from the initially microinjected cell. That fluorescent dextrans could spread so far from the site of injection suggests actual movement of P30 itself between cells; such P30 movement would be necessary to affect plasmodesmal permeability in noninjected cells (E. Waigmann, W. Lucas, V. Citovsky & P. Zambryski, unpublished data).

The molecular mechanism by which P30 affects plasmodesmal permeability is unknown. For example, it is not clear whether P30 interacts directly with plasmodesmal proteins or with cytoplasmic P30-binding receptors functionally analogous to NLS-binding NBPs (89). It is well established, however, that P30 can be specifically targeted to plasmodesmata in the plant cell walls. Electron-microscope studies of virus-infected plants reported specific association of P30 with plasmodesmata (113). Furthermore, in transgenic plants, P30 was found to accumulate in the central cavity of branched (secondary) but not in single (primary) plasmodesmata (34). However, that TMV can infect young apical leaves that do not contain secondary plasmodesmata (27) implies that P30 interacts with primary plasmodesmata as well. Thus, P30 interaction with primary plasmodesmata may be transient while interaction with secondary plasmodesmata potentially results in irreversible deposition of P30 in the central cavity.

A recent study of P30 phosphorylation provided an insight into the possible mechanism of P30 accumulation in the secondary plasmodesmata (19). The carboxyl terminal part of P30 was shown to contain a major phosphorylation site, and recent results indicate that a plant cell wall–associated protein kinase phosphorylates P30 at amino acid residues Ser257, Thr261, and Ser265. This protein kinase activity is developmentally expressed, correlating the basipetal (tip-to-base) development of tobacco leaves, and appears to parallel the development of secondary plasmodesmata (34). Since P30 is specifically accumulated in secondary plasmodesmata (34), P30 phosphorylation may represent a mechanism by which the host plant deactivates this biologically potent protein by irreversible association with the cell walls.

P30 INTERACTION WITH NUCLEIC ACIDS As noted above, P30 increases the size-exclusion limit of plasmodesmata from 1.5 nm to 5–6 nm in transgenic plants (123), and to 6–9 nm in microinjected wild-type plants (E. Waigmann, W. Lucas, V. Citovsky & P. Zambryski, unpublished data). However, this increase in permeability is insufficient for transport of the 10-nm-wide free, folded TMV RNA (48). One possibility is that, prior to translocation, viral RNA is unfolded by forming a protein-RNA transport complex with P30. Thus, P30 may not only act to increase plasmodesmal permeability but may also function as a molecular chaperone for the viral nucleic acid.

Indeed, P30 has been identified as a single-stranded nucleic acid–binding protein (17). Mutational analysis of P30 revealed two independently active, single-stranded nucleic acid–binding domains at the carboxyl terminus of the protein (20). P30 binding is strong, cooperative, and sequence nonspecific (17). Preferential binding to single-stranded nucleic acids and binding cooperativity are characteristic of all known SSBs (12). Unlike other SSBs, however, P30 binds ssDNA and RNA with equal affinity (17).

Binding of P30 to viral RNA probably shapes it into a transferable form (20, 22). Free single-stranded nucleic acid molecules exist as irregular collapsed structures. Electron-microscopy observations demonstrated that P30 can unfold these structures to form long and thin single-stranded nucleic acid–protein complexes with a diameter of about 2 nm (20). This width of the complexes is compatible with the open plasmodesmal channel found in plants transgenic for or microinjected with P30 (123; E. Waigmann, W. Lucas, V. Citovsky & P. Zambryski, unpublished data) and, by implication, modified during viral infection.

VIRAL NUCLEIC ACID MOLECULES IN TRANSIT Our current knowledge of P30 biochemistry and biology is integrated in the following model for TMV RNA transport through plasmodesmata (Figure 3*a*). In this model, P30 functions as a molecular chaperone to bind and unfold TMV RNA and target it to and through the plasmodesmal channel.

The entire process begins with P30 binding to TMV RNA. Note that because replication and translation of many plant viruses are highly compartmentalized (97), movement proteins will likely associate with the viral nucleic acids rather than with host mRNAs. P30 binding results in the formation of long, thin, and unfolded ribonucleoprotein transport complexes. These complexes are then targeted to plasmodesmata. Since most processes of inter- and intracellular targeting of macromolecules require specific signal sequences, P30 may also use a putative plasmodesmata localization signal (PDL). Recent evidence indicates that movement proteins of plant viruses may specifically interact with a family of cytoplasmic 40- to 60-kDa proteins (V. Citovsky & A. Maule,

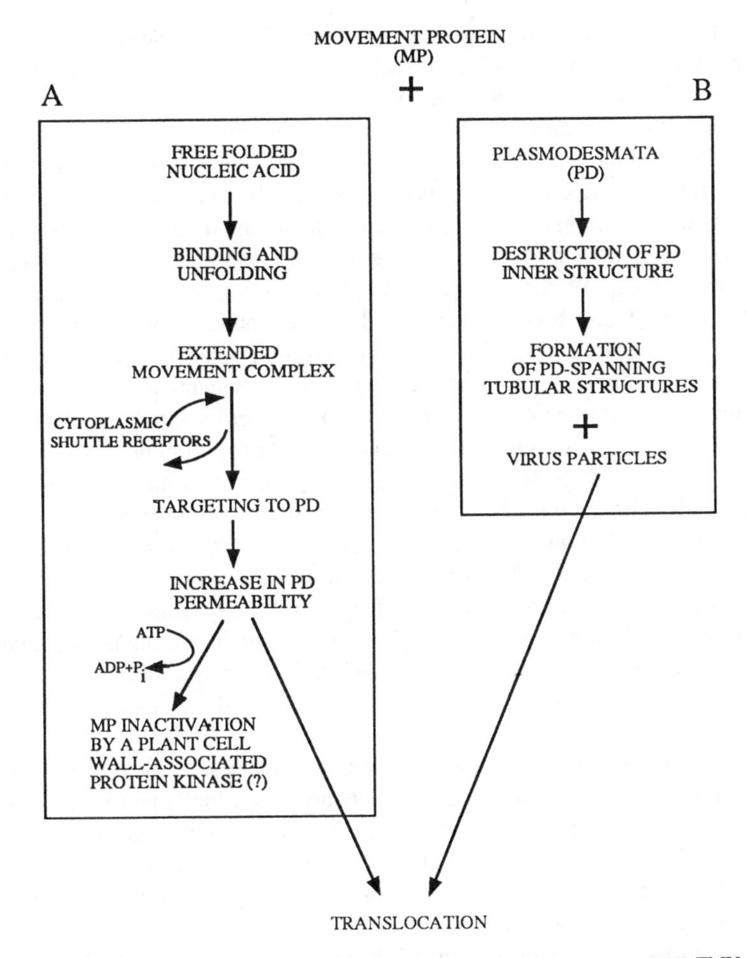

Figure 3 Two mechanisms for transport of plant viruses through plasmodesmata. (*A*) TMV-type transport as a movement protein–nucleic acid complex. (*B*) CPMV-type transport as an entire virion.

unpublished data). Potentially, these cytoplasmic receptors may function to transport ribonucleoprotein complexes to plasmodesmata; this function may be analogous to that of NBP proteins in nuclear transport. Other cellular proteins also may be involved in plasmodesmal targeting; for example, the putative movement protein of beet yellow closterovirus (BYV) has been shown to bind purified bovine brain microtubules (72). After reaching the plasmodesmal channel, P30 acts to increase its permeability. Finally, translocation through the enlarged plasmodesmal channel into the neighboring cell occurs.

Is the entire transport complex translocated? This may depend on the type of plasmodesmata in the infected tissue. In young leaves, which contain only primary plasmodesmata (34), the entire complex may be transported. In older leaves with predominantly secondary plasmodesmata (34), however, the transport complex may be partially unpacked during translocation. These possibilities are based on two lines of evidence. First, preliminary results from direct microinjection of purified P30 into plant tissue indicate that this protein itself moves between cells, suggesting that P30 also may transverse plasmodesmata as part of the transport complex (E. Waigmann, W. Lucas, V. Citovsky & P. Zambryski, unpublished). On the other hand, P30 was shown to specifically accumulate in secondary plasmodesmata found only in older leaves (34). Thus, during translocation through primary plasmodesmata in young leaves, the entire transport complex may cross into the neighboring cell. Conversely, translocation through secondary plasmodesmata in older tissue may involve partial dissociation of the transport complex so that some P30 is left behind deposited in the secondary plasmodesmata; the irreversible association of P30 with secondary plasmodesmata may be promoted by P30 phosphorylation (19). This partial uncoating of transported P30-TMV RNA complexes may resemble transport of the BR hnRNPs through the nuclear pore because these complexes shed part of their proteins during nuclear export (82) (see above). Finally, following translocation, the viral movement protein may be completely displaced to allow translation and replication of the viral RNA.

Viral transport through plasmodesmata as nucleic acid–movement protein complexes may represent the major mechanism by which plant viruses move from cell to cell. One of the criteria for such a conserved mechanism is nucleic acid binding by the movement protein. To date, at least four viruses belonging to different taxonomic groups have been shown to possess movement proteins that bind single-stranded nucleic acids: a tobamovirus (TMV) (17), a tricornavirus [alfalfa mosaic virus (AIMV)] (102), a dianthovirus [red clover necrotic mosaic virus (RCNMV)] (94), and a caulimovirus [cauliflower mosaic virus (CaMV)] (18). The most interesting finding is that the movement protein of CaMV, the gene I protein, is an SSB with a preferential binding affinity to RNA. Whereas the genomes of TMV, AIMV, and RCNMV are all single stranded, CaMV is a double-stranded DNA virus. However, CaMV is a pararetrovirus that replicates by reverse transcription of its polycistronic 35S RNA (reviewed in 104). That CaMV movement protein preferentially binds RNA suggests that this virus may move between cells as a ribonucleoprotein complex (18).

Binding of viral movement proteins to single-stranded nucleic acids regardless of nucleotide sequence specificity implies that cell-to-cell spread can be complemented between unrelated plant viruses. Indeed, the spread of TMV in resistant tomato plants is complemented by infection with an unrelated potato virus X (PVX) (110); and the phloem-limited geminivirus, bean golden mosaic virus (BGMV), can invade nonphloem leaf tissue when coinfected with the bean strain of TMV (CP-TMV) as a helper virus (10). Potentially, movement proteins of helper viruses form cell-to-cell movement complexes with genomic RNA or ssDNA molecules of transport-deficient virus strains.

CPMV-Type Cell-to-Cell Movement

Although a wide variety of plant viruses may move through plasmodesmata as subviral nucleoprotein complexes, recent evidence suggests that some plant viruses are transported as entire virions by a completely different mechanism (Figure 3*b*). This mechanism is exemplified by the comovirus CPMV. CPMV is a positive-sense ssRNA virus with a bipartite genome consisting of two RNA molecules, B-RNA and M-RNA. B-RNA encodes all the functions necessary for replication of both RNAs. M-RNA, on the other hand, is essential for CPMV spread in plants. This RNA encodes two movement proteins, of 58 and 48 kDa, and two capsid proteins, VP37 and VP23 (41, 51). Mutational analysis of M-RNA demonstrated that all four proteins are required for cell-to-cell spread of the virus (120).

The requirement of CPMV capsid proteins for movement indicates that this virus may be transported as an encapsidated viral particle. Electron-microscope studies of CPMV-infected plant tissue revealed striking tubular structures that assemble in the cell cytoplasm and extend through plasmodesmata into the neighboring cells (116). These tubular structures are densely packed with intact virions, suggesting that the tubules may function as direct conduits for transport of CPMV between cells. Immunogold labeling of the tubules detected the presence of CPMV movement proteins as well as viral capsid proteins (116). Using a CPMV mutant that does not express capsid proteins, Kasteel et al (73) showed that the latter are not required for the formation of the tubular structures; in the absence of the capsid proteins, the formed tubules are empty but otherwise structurally identical to those formed by the wild-type virus (73). The capsid proteins detected by immunostaining, then, are part of the virions found inside the tubules. Unfortunately, the antimovement protein antibodies used in these studies did not discriminate between the 49- and 58-kDa proteins (115). Thus, whether both are needed to form tubules is not known. Indirect evidence, however, suggests that only the 48-kDa protein may be present in the tubular structures (119).

What is the structure of the CPMV tubular structures? The tubules are approximately 20 μm long with 3- to 4-nm-thick walls (115). Interestingly, the formation of the tubular structures does not require the presence of plasmodesmata or even cell walls; they can be found in CPMV-infected protoplasts that lack plasmodesmata. In protoplasts, the tubular structures were enveloped by the plasma membrane and extended into the culture medium to give the protoplasts a characteristic hairy look under the light microscope (115). Because the tubules are formed regardless of plasmodesmal presence, it is not clear why in plant tissue they are always found crossing plasmodesmal channels. One possibility is that in plant tissue only tubules that perchance coincide with plasmodesmata are allowed to grow; other tubules are aborted and degraded. Alternatively, the tubular growth in plant tissue could be specifically directed toward plasmodesmata by CPMV movement protein(s); in protoplasts such targeting will not function and result in uncontrolled tubule formation. Once the CPMV tubules invade plasmodesmata, the major changes in plasmodesmal structure occur. The central appressed ER is displaced by the viral tubule, effectively destroying the functional structure of the plasmodesmal channel (116). Thus, the CPMV cell-to-cell spread through the tubules differs dramatically from the TMV-type movement, which does not impair plasmodesmal function.

The dramatic and consistent occurrence of the CPMV tubular structures likely represents a specific transport mechanism. In addition to CPMV, CaMV (74) and some nepoviruses (62) also form transplasmodesmal tubules. In the case of CaMV, which has an RNA-binding movement protein (18) (see above), two mechanisms of movement may be involved in the course of infection: a more subtle cell-to-cell movement of ribonucleoprotein complexes early in infection and aggressive transport of entire virions at later stages of infection when plasmodesmata become sufficiently modified to accommodate larger viral particles.

ARE THERE GENERAL RULES FOR NUCLEIC ACID TRANSPORT?

The great diversity seen in nucleic acid transport clearly cannot occur by a single mechanism. This review has described nine different cases of transport: nuclear uptake of *Agrobacterium* spp. T-complex, three distinct pathways for nuclear transport of snRNPs, nuclear export of hnRNPs, nuclear import of enveloped and nonenveloped viruses, and two mechanisms for cell-to-cell transport of plant viruses. Potentially, each of these processes has unique characteristics. However, one could identify major conserved structural and functional features of these pathways and propose a set of general rules for nucleic acid transport through membrane channels. A comparison of the described cases of nucleic acid transport (summarized in Table 1) identifies the following seven common features.

Table 1 Common features in nucleic acid transport through nuclear pores and plasmodesmal channels

	Transported nucleic acid				
				Viral genomic nucleic acids	
Transport characteristics	*Agrobacterium* T-DNA	snRNA	hnRNA	Animal viruses	Plant viruses
Type of channel	NPC	NPC	NPC	NPC	Plasmodesmata
Associated proteins	Yes	Yes	Yes	Yes	Yes
Localization signals	Protein NLS	Protein NLS and tmG cap	Protein NLS and tmG cap (?)	Protein NLS	Protein PDL (?)
Unfolded conformation	Yes	?	Yes	?	Yes
Polarity of transport	Yes	?	Yes	?	?
Gating of channel	Yes	Yes	Yes	Yes	Yes
Cytoplasmic shuttle receptors	Yes	Yes	Yes	Yes	Yes (?)
Transport as a single-standed or double-standard molecule	ss	ss	ss	ss/ds (?)	ss/ds (?)
Number of possible transport pathways	1	3	1 (?)	2	2

1. Association with specific proteins. No nucleic acid is known to be transported as a free molecule. Although each nucleic acid molecule is usually associated with numerous protein molecules, the number of different protein species varies considerably; for example, plant virus movement complexes may contain only one type of protein (the movement protein) (22), *Agrobacterium* spp. T-complex contains two classes of proteins (VirD2 and VirE2) (23), and vertebrate hnRNPs may contain more than 20 different proteins (14, 99).

2. Specific localization signals. Protein binding to the transported nucleic acid provides two general functions: to supply specific signals for transport and to shape the nucleic acid molecule into a transferable form (see below). In some cases, protein localization signals are sufficient for transport [*Agrobacterium* spp. T-complex (20), U6 snRNP (44, 85), and potentially, TMV cell-to-cell movement complexes (22)], whereas in others the protein signal sequences are complemented by signals contained in the nucleic acid itself [tmG cap U1–U5 snRNPs and some hnRNPs (reviewed in 69)].

3. Unfolding of the transported nucleic acid–protein complex. Unfolding of proteins plays a critical role in transport into mitochondria (88), whereas nuclear transport does not require a fully unfolded protein conformation (38). Nucleic acid–protein complexes, however, are much larger structures than individual protein molecules. The sheer bulk of these complexes may require substantial conformational changes prior to transport. For example, *Agrobacterium* spp. T-complex and cell-to-cell movement complexes of TMV may unfold prior to transport (21, 23), whereas the BR hnRNP unfolds during the translocation (82). In addition to steric considerations, unfolding of transported complexes may be required to determine the polarity of the transport, i.e. to expose the leading end of the molecules during translocation.

4. Polarity of transport. Polar transport has been proposed for nuclear uptake of *Agrobacterium* spp. T-complex (23, 126) and for nuclear export of BR hnRNPs (82). Other nucleic acid–protein complexes may also be transported in a polar fashion. For example, polarity of TmG cap signals in pol II U snRNPs (55) suggests these complexes are transported directionally. Vectorial transport of nucleic acid–protein complexes may be required for immediate processing of the emerging complex (e.g. integration, translation, etc). Potentially, transport polarity is determined by a specific signal associated with one end of the transported nucleic acid molecule (e.g. VirD2 protein in *Agrobacterium* spp. T-complex and tmG cap in snRNPs).

5. Gating of the pore. All known nucleic acid transport mechanisms involve an increase in the size of the pore to allow translocation. Presumably, these changes are mediated by the specific localization signals contained in the transported complex. The molecular mechanisms by which nuclear pores or plasmodesmal channels are gated are unknown.

6. Shuttle receptors. Nuclear transport of proteins has been shown to involve two types of cytoplasmic receptors: NBPs and, possibly, hsp70 (reviewed in 36). Because transported nucleic acids are always associated with karyophilic proteins, the nucleic acid–protein complexes may interact with cytoplasmic shuttle receptors as well. Analogous to nuclear localization, movement of at least two plant viruses (TMV and CaMV) through plasmodesmata may involve interaction with specific cytoplasmic receptors (V. Citovsky & A. Maule, unpublished data).

7. Transport of single-stranded nucleic acids. Most of the transported nucleic acids (*Agrobacterium* spp. T-strand, RNA, and genomic nucleic acids of many animal and plant viruses) are single stranded. This could be a simple coincidence; however, transport of single-stranded nucleic acids has certain advantages over transport of the double helix: single strands are polar (as linear molecules), less rigid, and easily coated by SSBs. Interestingly, all known plant viruses that move from cell to cell (presumably through plasmodesmata) possess a single-stranded genome or replicate via a single-stranded nucleic acid intermediate (3, 22, 54, 81).

FUTURE PERSPECTIVES

In recent years, the previously dormant field of transmembrane transport of nucleic acids has gained much impetus. Solution of nucleic acid transport mechanisms will profoundly affect our understanding of gene expression and host-pathogen interaction. New drug therapies may be developed to inhibit nuclear transport of pathogenic viruses. Novel approaches to production of agronomically important plants and animals resistant to infection by viruses may also be developed. Furthermore, new and efficient procedures for delivery of foreign genes into cell nuclei may be developed based on this knowledge; these procedures may contribute to the field of genetic engineering of plants and animals, which at present is still in its infancy.

ACKNOWLEDGMENTS

We thank Gail McLean, Dave Goldfarb, John Zupan, and Elisabeth Waigmann for critical reading of this manuscript. We are also grateful to Iain Mattaj, Dave Goldfarb, Bill Lucas, Christiane Stussi-Garaud, Andy Maule, Jan Van Lent, Rob Goldbach, Walt Ream, Yoshihiro Yoneda, and other colleagues for sending preprints and sharing unpublished results. The research relating to nuclear uptake of *Agrobacterium* T-complex was supported by NSF grant (89-15613) and that relating to plant virus movement by NIH grant (GM-45244-01).

Literature Cited

1. Akey, C. W. 1992. The nuclear pore complex. *Curr. Opin. Struct. Biol.* 2:258–63
2. Akey, C. W. 1990. Visualization of transport-related configurations of the nuclear pore transporter. *Biophys. J.* 58:341–55
3. Atabekov, J. G., Taliansky, M. E. 1990. Expression of a plant virus-coded transport function by different viral genomes. *Adv. Virus Res.* 38:201–48
4. Baron-Eppel, O., Hernandes, D., Jiang, L.-W., Meiners, S., Schindler, M. 1988. Dynamic continuity of cytoplasmic and membrane compartments between plant cells. *J. Cell Biol.* 106: 715–21
5. Baserga, S. J., Gilmore-Hebert, M., Yang, X. W. 1992. Distinct molecular signals for nuclear import of the nucleolar snRNA, U3. *Gen. Dev.* 6:1120–30
6. Blissard, G. W., Rohrmann, G. F. 1990. Baculovirus diversity and molecular biology. *Annu. Rev. Entomol.* 35:127–55
7. Breeuwer, M., Goldfarb, D. G. 1990. Facilitated nuclear transport of histone H1 and other small nucleophilic proteins. *Cell* 60:999–1008
8. Brown, P. O. 1989. Intergration of retroviral DNA. In *Retroviruses,* ed. R. Swanstrom, P. K. Vogt, pp. 19–48. Berlin: Springer-Verlag
9. Buchanan-Wollaston, V., Passiatore, J. E., Cannon, F. 1987. The *mob* and *oriT* mobilization functions of a bacterial plamid promote its transfer to plants. *Nature* 328:172–75
10. Carr, R. J., Kim, K. S. 1983. Evidence that bean golden mosaic virus invaded non-phloem tissue in double infections with tobacco mosaic virus. *J. Gen. Virol.* 64:2489–92
11. Chang, D. D., Sharp, P. A. 1990. Messenger RNA transport and HIV Rev regulation. *Science* 249:614–15
12. Chase, J. W., Williams, K. R. 1986. Single-stranded DNA binding proteins required for DNA replication. *Annu. Rev. Biochem.* 55:103–36
13. Cheng, J., Fogel-Petrovic, M., Maquat, L. E. 1990. Translation to near the distal end of the penultimate exon is required for normal levels of spliced triosephosphate isomerase mRNA. *Mol. Cell. Biol.* 10:5215–25
14. Choi, Y. D., Dreyfuss, G. 1984. Isolation of the heterogeneous nuclear RNA-ribonucleoprotein complex (hnRNP): a unique supramolecular assembly. *Proc. Natl. Acad. Sci. USA* 81:1997–2004
15. Christie, P. J., Ward, J. E., Winans, S. C., Nester, E. W. 1988. The *Agrobacterium tumefaciens virE2* gene product is a single-stranded-DNA-binding protein that associates with T-DNA. *J. Bacteriol.* 170:2659–67
16. Citovsky, V., De Vos, G., Zambryski, P. 1988. Single-stranded DNA binding protein encoded by the *virE* locus of *Agrobacterium tumefaciens. Science* 240:501–4
17. Citovsky, V., Knorr, D., Schuster, G., Zambryski, P. 1990. The P30 movement protein of tobacco mosaic virus is a single strand nucleic acid binding protein. *Cell* 60:637–47
18. Citovsky, V., Knorr, D., Zambryski, P. 1991. Gene I, a potential movement locus of CaMV, encodes an RNA binding protein. *Proc. Natl. Acad. Sci. USA* 88:2476–80
19. Citovsky, V., McLean, B., Zupan, J., Zambryski, P. 1993. Phosphorylation of tobacco mosaic virus cell-to-cell movement protein by a developmentally-regulated plant cell wall–associated protein kinase. *Genes Dev.* In press
20. Citovsky, V., Wong, M. L., Shaw, A., Prasad, B. V. V., Zambryski, P. 1992. Visualization and characterization of tobacco mosaic virus movement protein binding to single-stranded nucleic acids. *Plant Cell* 4:397–411
21. Citovsky, V., Wong, M. L., Zambryski, P. 1989. Cooperative interaction of *Agrobacterium* VirE2 protein with single stranded DNA: implications for the T-DNA transfer process. *Proc. Natl. Acad. Sci. USA* 86:1193–97
22. Citovsky, V., Zambryski, P. 1991. How do plant virus nucleic acids move through intercellular connections? *BioEssays* 13:373–79
23. Citovsky, V., Zupan, J., Warnick, D., Zambryski, P. 1992. Nuclear localization of *Agrobacterium* VirE2 protein in plant cells. *Science* 256:1803–5
24. Citovsky, V. C., McLean, G., Greene, E., Howard, E., Kuldau, G., et al. 1992. *Agrobacterium*–plant cell interaction: induction of *vir* genes and T-DNA transfer. In *Molecular Signals in Plant-Microbe Communications,* ed. D.

P. S. Verma, pp. 169–98. Boca Raton, FL: CRC
25. Clever, J., Kasamatsu, H. 1991. Simian virus 40 Vp2/3 small structural proteins harbor their own nuclear transport signal. *Virology* 181:78–90
26. Clever, J., Yamada, M., Kasamatsu, H. 1991. Import of SV40 virions through nuclear pore complexes. *Proc. Natl. Acad. Sci. USA* 88:7333–37
27. Culver, J. N., Lindbeck, A. G. C., Dawson, W. O. 1991. Virus-host interactions: induction of chlorotic and necrotic responses in plants by tobamoviruses. *Annu. Rev. Phytopathol.* 29:193–17
28. Das, A. 1988. *Agrobacterium tumefaciens virE* operon encodes a single-stranded DNA-binding protein. *Proc. Natl. Acad. Sci. USA* 85:2909–13
29. Davey, J., Dimmock, N. J., Colman, A. 1985. Identification of the sequence responsible for the nuclear accumulation of the influenza virus nucleoprotein in Xenopus oocytes. *Cell* 40:667–75
30. Deom, C. M., Lapidot, M., Beachy, R. N. 1992. Plant virus movement proteins. *Cell* 69:221–24
31. Deom, C. M., Schubert, K. R., Wolf, S., Holt, C. A., Lucas, W. J., et al. 1990. Molecular characterization and biological function of the movement protein of tobacco mosaic virus in transgenic plants. *Proc. Natl. Acad. Sci. USA* 87:3284–88
32. Deom, C. M., Shaw, M. J., Beachy, R. N. 1987. The 30-kilodalton gene product of tobacco mosaic virus potentiates virus movement. *Science* 327: 389–94
33. Derrick, P. M., Barker, H., Oparka, K. J. 1992. Increase in plasmodesmatal permeability during cell-to-cell spread of tobacco rattle tobravirus from individually inoculated cells. *Plant Cell* 4:1405–12
34. Ding, B., Haudsenshield, J. S., Hull, R. J., Wolf, S., Beachy, R. N., et al. 1992. Secondary plasmodesmata are specific sites of localization of the tobacco mosaic virus movement protein in transgenic tobacco plants. *Plant Cell* 4:915–28
35. Ding, B., Turgeon, R., Parthasarathy, M. V. 1992. Substructure of freeze-substituted plasmodesmata. *Protoplasma* 169:28–41
36. Dingwall, C., Laskey, R. 1992. The nuclear membrane. *Science* 258:942–47
37. Dingwall, C., Laskey, R. A. 1991. Nuclear targeting sequences—a consensus? *Trends Biochem.* 16:478–81
38. Dingwall, C., Sharnick, S. V., Laskey, R. A. 1982. A polypeptide domain that specifies migration of nucleoplasmin into the nucleus. *Cell* 30:449–58
39. Dworetzky, S. I., Feldherr, C. M. 1988. Translocation of RNA-coated gold particles through the nuclear pores of oocytes. *J. Cell Biol.* 106:575–84
40. Eckner, R., Ellmeier, W., Birnstiel, M. L. 1991. Mature mRNA 3′ end formation stimulates RNA export from the nucleus. *EMBO J.* 10:3513–22
41. Eggen, R., van Kammen, A. 1988. RNA replication in comoviruses. In *RNA Genetics,* ed. E. Domingo, J. J. Holland, P. Alquist, pp. 49–70. Boca Raton, FL: CRC
42. Evans, W. H. 1988. Gap junctions: towards a molecular structure. *BioEssays* 8:3–6
43. Featherstone, C., Darby, M. K., Gerace, L. 1988. A monoclonal antibody against the nuclear pore complex inhibits nucleocytoplasmic transport of protein and RNA in vivo. *J. Cell Biol.* 107:1289–97
44. Fischer, U., Darzynkiewicz, E., Tahara, S. M., Dathan, N. A., Luhrmann, R., et al. 1991. Diversity in signals required for nuclear accumulation of U snRNPs and variety in the pathways of nuclear transport. *J. Cell Biol.* 113:705–14
45. Fisher, D. B., Wu, Y., Ku, M. S. B. 1992. Turnover of soluble proteins in the wheat sieve tube. *Plant Physiol.* 100:1433–41
46. Garcia-Bustos, J., Heitman, J., Hall, M. N. 1991. Nuclear protein localization. *Biochim. Biophys. Acta* 1071:83–101
47. Gerace, L., Burke, B. 1988. Functional organization of the nuclear envelope. *Annu. Rev. Cell Biol.* 4:355–74
48. Gibbs, A. J. 1976. Viruses and plasmodesmata. In *Intercellular Communication in Plants: Studies on Plasmodesmata,* ed. B. E. S. Gunning, A. W. Robards, pp. 149–64. Berlin: Springer-Verlag
49. Gietl, C., Koukolikova, N. Z., Hohn, B. 1987. Mobilization of T-DNA from *Agrobacterium* to plant cells involves a protein that binds single-stranded DNA. *Proc. Natl. Acad Sci. USA* 84:9006–10
50. Goelet, P., Lomonossoff, G. P., Butler, P. J. G., Akam, M. E., Gait, M. J., et al. 1982. Nucleotide sequence of tobacco mosaic virus RNA. *Proc. Natl. Acad. Sci. USA* 79:5818–22

51. Goldbach, R., van Kammen, A. 1985. Structure, replication and expression of the bipartite genome of cowpea mosaic virus. In *Molecular Plant Virology,* ed. J. W. Davis, pp. 83–120. Boca Raton: CRC
52. Goldfarb, D., Michaud, N. 1991. Pathways for the nuclear transport of proteins and RNAs. *Trends Cell Biol.* 1:20–24
53. Guiochon-Mantel, A., Lescop, P., Christin-Maitre, S., Loosfelt, H., Perrot-Applant, M., et al. 1991. Nucleocytoplasmic shuttling of the progesterone receptor. *EMBO J.* 10:3851–59
54. Hall, M. N., Craik, C., Hiraoka, Y. 1990. Homeodomain of yeast repressor α2 contains a nuclear localization signal. *Proc. Natl. Acad. Sci. USA* 87: 6954–58
55. Hamm, J., Darzynkiewicz, E., Tahara, S. M., Mattaj, I. W. 1990. The trimethylguanosine cap structure of U1 snRNA is a component of a bipartite nuclear targeting signal. *Cell* 62:569–77
56. Hamm, J., Mattaj, I. 1990. Monomethylated cap structures facilitate RNA export from the nucleus. *Cell* 63:109–18
57. Hard, T., Kellenbach, E., Boelens, R., Maler, B. A., Dahlman, K., et al. 1990. Solution structure of the glucocorticoid receptor DNA-binding domain. *Science* 249:157–60
58. Heinemann, J. A., Sprague, J. F. Jr. 1989. Bacterial conjugative plasmids mobilize DNA transfer between bacteria and yeast. *Nature* 340:205–9
59. Helenius, A. 1992. Unpacking of the incoming influenza virus. *Cell* 69:577–78
60. Herrera-Estrella, A., Chen, Z., Van Montagu, M., Wang, K. 1988. VirD proteins of *Agrobacterium tumefaciens* are required for the formation of a covalent DNA protein complex at the 5′ terminus of T-strand molecules. *EMBO J.* 7:4055–62
61. Herrera-Estrella, A., Van Montagu, M., Wang, K. 1990. A bacterial peptide acting as a plant nuclear targeting signal: the amino-terminal portion of *Agrobacterium virD2* protein directs a β-galactosidase fusion protein into tobacco nuclei. *Proc. Natl. Acad. Sci. USA* 87:9534–37
62. Hibino, H., Tsuchizaki, T., Usugi, T., Saito, Y. 1977. Fine structures and developmental process of tubules induced by mulberry ringspot virus and satsuma dwarf virus infections. *Ann. Phytopathol. Soc. Japan* 43:255–64
63. Howard, E., Citovsky, V., Zambryski, P. 1990. The T-complex of *Agrobacterium tumefaciens. UCLA Symp. Mol. Cell Biol. New Ser.* 129:1–11
64. Howard, E., Zupan, J., Citovsky, V., Zambryski, P. 1992. The VirD2 protein of *Agrobacterium tumefaciens* contains a C-terminal bipartite nuclear localization signal: implications for nuclear uptake of DNA in plant cells. *Cell* 68:109–18
65. Howard, E. A., Citovsky, V. 1990. The emerging structure of the *Agrobacterium* T-DNA transfer complex. *BioEssays* 12:103–8
66. Howard, E. A., Winsor, B. A., De Vos, G., Zambryski, P. 1989. Activation of the T-DNA transfer process in *Agrobacterium* results in the generation of a T-strand protein complex: tight association with the 5′ ends of T-strands. *Proc. Natl. Acad. Sci. USA* 86:4017–21
67. Hull, R. 1992. Down the tube. *Curr. Opin. Biol.* 2:224–26
68. Hull, R. 1991. The movement of viruses within plants. *Semin. Virol.* 2:89–95
69. Izaurralde, E., Mattaj, I. 1992. Transport of RNA between nucleus and cytoplasm. *Semin. Cell Biol.* 3:279–88
70. Izaurralde, E., Stepinski, J., Darzynkiewicz, E., Mattaj, I. 1992. A cap binding protein that may mediate nuclear export of RNA polymerase II–transcribed RNAs. *J. Cell Biol.* 118: 1287–95
71. Jockusch, H. 1968. Two mutants of tobacco mosaic virus temperature-sensitive in two different functions. *Virology* 35:94–101
72. Karasev, A. V., Kashina, A. S., Gelfand, V. I., Dolja, V. V. 1992. HSP70-related 65 kDa protein of beet yellow closterovirus is a microtubule-binding protein. *FEBS Lett.* 304:12–14
73. Kasteel, D., Wellink, J., Verver, J., van Lent, J., van Kammen, A. 1993. Studies on the involvement of M RNA encoded proteins of cowpea mosaic virus in tubule formation. *J. Gen. Virol.* In press
74. Kitajima, E. W., Lauritis, J. A. 1969. Plant virions in plasmodesmata. *Virology* 37:681–85
75. Loyter, A., Nussbaum, O., Citovsky, V. 1988. Active function of membrane receptors in fusion of enveloped viruses with cell plasma membranes. In *Molecular Mechanisms of Membrane Fusion,* ed. S. Ohki, D. Doyle, T. D.

Flanagan, S. W. Hui, E. Mayhew, pp. 413–26. New York: Plenum
76. Lucas, W. J., Ding, B., Van der Schoot, C. 1993. Plasmodesmata: the supracellular nature of plants. *The New Phytopathol*. In press
77. Martin, K., Helenius, A. 1991. Nuclear transport of influenza virus ribonucleoproteins: the viral matrix protein (M1) promotes export and inhibits import. *Cell* 67:117–30
78. Mattaj, I. 1988. UsnRNP assembly and transport. In *Structure and Function of Major and Minor Small Nuclear Ribonucleoprotein Particles,* ed. M. L. Brinstiel, pp. 100–14. Heidelberg: Springer
79. Mattaj, I., De Robertis, E. M. 1985. Nuclear segregation of U2 snRNA requires binding of specific snRNP proteins. *Cell* 40:111–18
80. Maule, A. J. 1991. Virus movement in infected plants. *Crit. Rev. Plant Sci*. 9:457–73
81. Medberry, S. L., Lockhart, B. E. L., Olszewski, N. E. 1990. Properties of Commelina yellow mottle virus's complete DNA sequence, genomic discontinuities and transcript suggest that it is a pararetrovirus. *Nucleic Acids Res*. 18:5505–13
82. Mehlin, H., Daneholt, B., Skoglund, U. 1992. Translocation of a specific premessenger ribonucleoprotein particle through the nuclear pore studied with electron microscope tomography. *Cell* 69:605–13
83. Meiners, S., Xu, A., Schindler, M. 1991. Gap junction protein homologue from *Arabidopsis thaliana:* evidence for connexins in plants. *Proc. Natl. Acad. Sci. USA* 88:4119–22
84. Meshi, T., Watanabe, Y., Saito, T., Sugimoto, A., Maeda, T., et al. 1987. Function of the 30 kd protein of tobacco mosaic virus: involvement in cell-to-cell movement and dispensability for replication. *EMBO J*. 6:2557–63
85. Michaud, N., Goldfarb, D. 1991. Multiple pathways in nuclear transport: the import of U2 snRNP occurs by a novel kinetic pathway. *J. Cell Biol*. 112:215–23
86. Morris-Vasios, C., Kochan, J. P., Skalka, A. M. 1989. Avian sarcoma-leukosis virus pol-endo proteins expressed independently in mammalian cells accumulate in the nucleus but can be directed to other cellular compartments. *J. Virol*. 62:349–53
87. Neuman de Vegvar, H. E., Dahlberg, J. E. 1990. Nucleocytoplasmic transport and processing of small nuclear RNA precursors. *Mol. Cell. Biol*. 10: 3365–75
88. Neupert, W., Hartl, F. U., Craig, E. A., Pfanner, N. 1990. How do polypeptides cross the mitochondrial membranes? *Cell* 63:447–50
89. Nigg, E. A., Baeuerle, P. A., Luhrmann, R. 1991. Nuclear import-export: in search of signals and mechanisms. *Cell* 66:15–22
90. Nishiguchi, M., Motoyoshi, F., Oshima, M. 1978. Behavior of a temperature sensitive strain of tobacco mosaic virus in tomato leaves and protoplasts. *J. Gen. Virol*. 39:53–61
91. Nishimura, T., Kawai, N., Kawai, M., Notake, K., Ichihara, I. 1986. Fusion of SV40-induced endocytotic vacuoles with the nuclear membrane. *Cell Struct. Funct*. 11:135–41
92. Ohno, M., Kataoka, N., Shimura, Y. 1990. A nuclear cap binding protein from HeLa cells. *Nucleic Acids Res*. 18:6989–95
93. Olsen, P. 1979. The neck constriction in plasmodesmata: evidence for a peripheral sphincter-like structure revealed by fixation with tannic acid. *Planta* 144:349–58
94. Osman, T. A. M., Hayes, R. J., Buck, K. W. 1992. Cooperative binding of the red clover necrotic mosaic virus movement protein to single-stranded nucleic acids. *J. Gen. Virol*. 73:223–27
95. Otten, L., DeGreve, H., Leemans, J., Hain, R., Hooykass, P., et al. 1984. Restoration of virulence of *vir* region mutants of *A. tumefaciens* strain B6S3 by coinfection with normal and mutant *Agrobacterium* strains. *Mol. Gen. Genet*. 195:159–63
96. Paine, P. L. 1988. Nuclear protein accumulation: envelope transport or phase affinity mechanisms? In *Nucleoplasmic Transport,* ed. R. Peters, pp. 27–44. New York: Academic
97. Palikaitis, P., Zaitlin, M. 1986. Tobacco mosaic virus: infectivity and replication. In *The Rod-Shaped Viruses,* ed. M. H. V. Van Regenmortel, H. Fraenkel-Conrat, pp. 105–31. New York: Plenum
98. Peters, D. L., Murphy, T. M. 1975. Selection of temperature sensitive mutants of tobacco mosaic virus by lesion morphology. *Virology* 65:595–600
99. Pinol-Roma, S., Choi, Y. D., Matunis, M. J., Dreyfuss, G. 1988. Immunopurification of heterogeneous nuclear ribonucleoprotein particles reveals

an assortment of RNA-binding proteins. *Gen. Dev.* 2:215–27

100. Reichelt, R., Holzenburg, A., Buhle E. L. Jr., Jarnik, M., Engel, A., et al. 1990. Correlation between structure and mass distribution of the nuclear pore complex and of distinct pore complex components. *J. Cell Biol.* 110:883–94
101. Robards, A. W., Lucas, W. J. 1990. Plasmodesmata. *Annu. Rev. Plant Physiol. Plant Mol. Biol.* 41:369–419
101a. Roizman, B., Sears, A. E. 1990. Herpes simplex viruses and their replication. In *Virology,* ed. B. N. Fields, D. M. Knippe, R. M. Chanock, M. S. Hirsch, J. L. Melnick, et al. pp. 1795–1841. New York: Raven
102. Schoumacher, F., Erny, C., Berna, A., Godefroy-Colburn, T., Stussi-Garaud, C. 1992. Nucleic acid binding properties of the alfalfa mosaic virus movement protein produced in yeast. *Virology* 188:896–99
103. Sen, P., Pazour, G. J., Anderson, D., Das, A. 1989. Cooperative binding of *Agrobacterium tumefaciens* VirE2 protein to single-stranded DNA. *J. Bacteriol.* 171:2573–80
104. Shepherd, R. J. 1989. Biochemistry of DNA plant viruses. In *The Biochemistry of Plants,* ed. A. Marcus, pp. 563–616. New York: Academic
105. Shieh, M. W., Wessler, S. R., Raikhel, N. V. 1993. Nuclear targeting of the maize R protein requires two nuclear localization sequences. *Plant Physiol.* 101:353–61
106. Shruvinton, C. E., Hodges, L., Ream, W. 1992. A nuclear localization signal in the *Agrobacterium* VirD2 endonuclease is essential for tumor formation. *Proc. Natl. Acad. Sci. USA* 89:11837–41
107. Siegal, A., Zaitlin, M., Sehgal, O. P. 1962. The isolation of defective tobacco mosaic virus strains. *Proc. Natl. Acad. Sci. USA* 48:1845–51
108. Stachel, S. E., Nester, E. W. 1986. The genetic and transcriptional organization of the vir region of the A6 Ti plasmid of *Agrobacterium. EMBO J.* 5:1445–54
109. Stachel, S. E., Timmerman, B., Zambryski, P. 1986. Generation of single-stranded T-DNA molecules during the initial stages of T-DNA transfer for *Agrobacterium tumefaciens* to plant cells. *Nature* 322:706–12
110. Taliansky, M. E., Malyshenko, S. I., Pshennikova, E. S., Atabekov, J. G. 1982. Plant virus-specific transport function II. A factor controlling virus host range. *Virology* 122:327–31
111. Terry, B. R., Robards, A. W. 1987. Hydrodynamic radius alone governs the mobility of molecules through plasmodesmata. *Planta* 171:145–57
112. Tinland, B., Koukolikova-Nicola, Z., Hall, M. N., Hohn, B. 1992. The T-DNA linked VirD2 protein contains two distinct functional nuclear localization signals. *Proc. Natl. Acad. Sci. USA* 89:7442–46
113. Tomenius, K., Clapham, D., Meshi, T. 1987. Localization by immunogold cytochemistry of the virus coded 30 K protein in plasmodesmata of leaves infected with tobacco mosaic virus. *Virology* 160:363–71
114. Tucker, E. B. 1988. Inositol biphosphate and inositol triphosphate inhibit cell-to-cell passage of carboxyfluorescein in staminal hairs of *Setcreasea purpurea. Planta* 174:358–63
115. van Lent, J., Storms, M., van der Meer, F., Wellink, J., Goldbach, R. 1991. Tubular structures involved in movement of cowpea mosaic virus are also formed in infected cowpea protoplasts. *J. Gen. Virol.* 72:2615–23
116. van Lent, J., Wellink, J., Goldbach, R. 1990. Evidence for the involvement of the 58K and 48K proteins in the intercellular movement of cowpea mosaic virus. *J. Gen. Virol.* 71:219–23
117. Vankan, P., McGuigan, C., Mattaj, I. 1990. Domains of U4 and U6 snRNAs required for snRNP assembly and splicing complementation in *Xenopus* oocytes. *EMBO J.* 9:3397–3404
118. Ward, E., Barnes, W. 1988. VirD2 protein of *Agrobacterium tumefaciens* very tightly linked to the 5′ end of T-strand DNA. *Science* 242:927–30
119. Wellink, J., Jaegle, M., Prinz, H., van Kammen, A., Goldbach, R. 1987. Expression of the middle component RNA of cowpea mosaic virus in vivo. *J. Gen. Virol.* 68:2577–85
120. Wellink, J., van Kammen, A. 1989. Cell-to-cell transport of cowpea mosaic virus requires both the 58K/48K proteins and the capsid proteins. *J. Gen. Virol.* 70:2279–86
121. White, J. 1992. Membrane fusion. *Science* 258:917–24
122. Wolf, S., Deom, C. M., Beachy, R., Lucas, W. J. 1991. Plasmodesmatal function is probed using transgenic tobacco plants that express a virus movement protein. *Plant Cell* 3:593–604

123. Wolf, S., Deom, C. M., Beachy, R. N., Lucas, W. J. 1989. Movement protein of tobacco mosaic virus modifies plasmodesmatal size exclusion limit. *Science* 246:377–79
124. Yahalom, A., Warmbrodt, R. D., Laird, D. W., Traub, O., Revel, J. P., et al. 1991. Maize mesocotyl plasmodesmata proteins cross-react with connexin gap junction protein antibodies. *Plant Cell* 3:407–11
125. Young, C., Nester, E. W. 1988. Association of the VirD2 protein with the 5′ end of T-strands in *Agrobacterium tumefaciens*. *J. Bacteriol.* 170:3367–74
126. Zambryski, P. 1992. Chronicles from the *Agrobacterium*-plant cell DNA transfer story. *Annu. Rev. Plant Physiol. Plant Mol. Biol.* 43:465–90
127. Zambryski, P., Tempe, J., Schell, J. 1989. Transfer and function of T-DNA genes from *Agrobacterium* Ti and Ri plasmids in plants. *Cell* 56:193–201

Annu. Rev. Microbiol. 1993. 47:199–230

THE CELL CYCLE OF *ESCHERICHIA COLI*

William D. Donachie
Institute of Cell and Molecular Biology, University of Edinburgh, Darwin Buiding, KB, Mayfield Road, Edinburgh EH9 3JR, Scotland

KEY WORDS: control of DNA replication, genome partition, cell division, cell cycle genes and proteins

---

CONTENTS

ABSTRACT

For *E. coli* cells, growth to a critical mass leads to initiation of chromosome replication. Initiation requires ATP-bound DnaA protein, together with replication proteins. Replication is followed by decatenation and monomerization of sister-chromosome molecules. Sister chromosomes are rapidly partitioned into opposite halves of the cell, perhaps by a kinesin-like protein (MukB). Cytokinesis starts with the formation of a ring of a GTP-binding

0066-4227/93/1001-0199$02.00

protein (FtsZ), usually around the cell center. A specific enzyme (PBP3) and other proteins (FtsQ, FtsA, FtsL, FtsW) are responsible for the coordinate ingrowth of the peptidoglycan cell wall at this location. EnvA protein is required to split the resultant cross-wall to form new cell poles. The formation of the FtsZ ring is inhibited by activated MinC protein, but the MinE protein reverses this inhibition at potential division sites. A minimum cell length is probably required for partition and a minimum-size DNA-free zone for septum formation.

INTRODUCTION

The cell cycle of *Escherichia coli* appears very simple: the rod-shaped cell doubles its length and then divides in the middle. Nevertheless, this ultra-simple life cycle raises problems of such difficulty that they remain unsolved after decades of research. How is the length at division determined? How is the replication and segregation of the genome integrated with cell growth and division?

The thesis of this review is that, during log-phase growth of cultures:

1. cell growth is continual, except for three periodic events, initiation of chromosome replication, partition of sister pairs of chromosomes, and cell division;
2. only a few proteins are specifically required to carry out these cyclic events;
3. the ultimate cause of the activation of these cycle proteins is the attainment by the growing cell of a critical size (volume or length), rather than any self-sustaining oscillatory system.

Most of the cell cycle–specific proteins in *E. coli* have probably been identified, and their individual activities are rapidly being characterized (see Figure 1). I believe that the next stage of research is to determine how these functions are activated and how this activation is linked to the state of the cell (in particular, why cell dimensions are so strongly linked to cell-cycle events). The success of molecular genetics in the study of the bacterial cell cycle has been so great that we find ourselves, armed with much greater knowledge of detail, confronted once again with the same naive questions that we set out to answer in the first place.

INITIATION PROTEINS

Initiation of chromosome replication is a separate event from DNA replication itself and requires a specific set of proteins (98). Essentially, it involves the separation of complementary DNA strands at a specific site, *oriC,* that allows the entry and assembly of the replication proteins themselves (19, 20, 133).

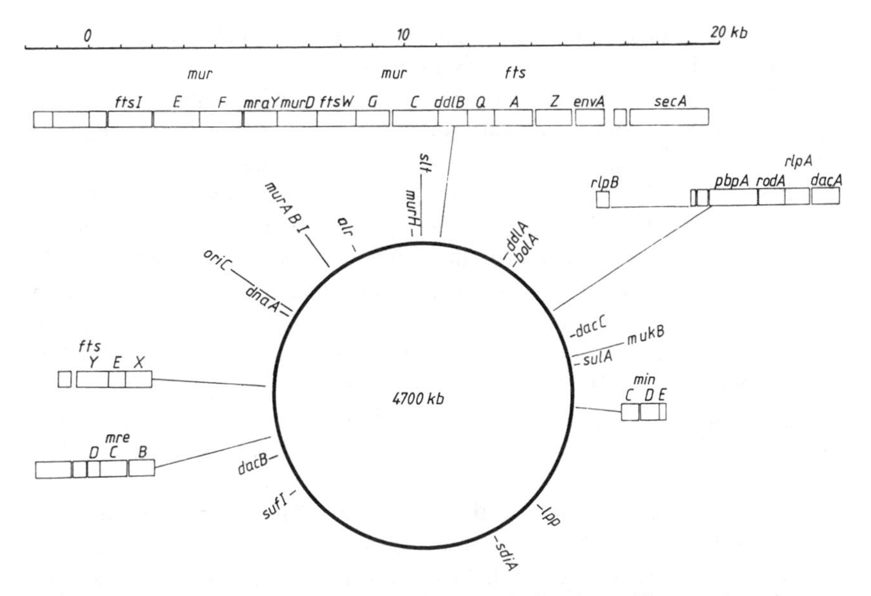

Figure 1 Cell-cycle genes of *Escherichia coli*. The loci of cycle-specific genes (genetic map: 0/100 map units at the top). Initiation of chromosome replication: *dnaA* (83.6). Chromosome partition: *mukB?* (21.1 map units). Cell division: [*ftsI*(PBP3), *ftsW?*, *ftsQ*, *ftsA*, *ftsZ*, *envA*] (2.2), (*minC*, *minD*, *minE*) (26.35), (*ftsY?*, *ftsE*, *ftsX?*) (77.6), *sulA* (22.0). Cell shape: [*pbpA*(PBP2), *rodA*] (14.5). Regulatory genes: *mreB* (73.2), *sdiA* (43.1). Genes that may contribute to cell cycle regulation: *murF* (2.2), *dacA*(PBP5) (14.4), *dacC*(PBP6) (19.05), *bolA* (10.0), *sufI* (65.3), *slt* (soluble lytic transglycosylase) (99.7). Additional loci are shown for genes required for peptidoglycan precursor synthesis and genes of still unknown function that lie within cell-cycle operons. Five gene clusters or operons are shown in more detail to larger scale (*top*). Boxes represent open reading frames, to scale. The circumference of the map represents approximately 4700 kbp of DNA.

The proteins needed to carry out this process in vitro have largely been identified.

Initiation is a cell cycle–regulated process; the frequency of initiation is linked to cell mass (45). Initiation takes place at every copy of *oriC* at each successive doubling of a fixed unit mass (or volume) (69). What is still unknown is how this link with cell size is achieved.

Modeling Initiation Control

What needs to be explained is how: (*a*) initiation takes place at every copy of *oriC* at each successive doubling of a fixed cell mass (45); (*b*) during a period of initiation, each copy of *oriC* is replicated only once (109); (*c*) if initiation is artificially blocked (e.g. by inhibition of DNA synthesis) initiation potential accumulates in proportion to cell mass, so that when DNA synthesis is allowed to resume, initiations take place successively, until the normal ratio of *oriC* to cell mass has been restored (46, 54).

A simple model to explain question *a* is that an initiator substance is made at a rate proportional to overall cell mass increase (i.e. is synthesized constitutively), that initiation takes place when the concentration of initiator reaches a critical level, and that the initiator is destroyed soon after initiation has taken place. No substance with these properties has yet been described.

In contrast, an explanation of the refractory period for reinitiation (*b*) appears to have been found. All adenines in GATC sites are methylated in mature DNA in *E. coli;* however, newly replicated DNA is methylated on the old strand only, and such hemimethylated DNA is not a substrate for further replication (129). Methylation of new strands takes only a few minutes, except for the *oriC* region, which remains in the hemimethylated state for a prolonged period [30–40% of the cell cycle, independent of growth rate (24)]. OriC DNA is associated with the cell membrane until it is fully methylated (114). A clear demonstration of the importance of methylation in initiation control is that *dam*$^-$ (*dam* methyltransferase deficient) cells have random timing of initiation (i.e. it is unlinked to cell mass) (4). Over- or underproduction of *dam* methyltransferase produces cells with aberrant copy numbers of *oriC,* presumably as the result of random initiation at individual origins (18). This finding provides a good explanation for synchronous initiation; although, as ever, it raises new questions. For example, what causes the *oriC* region to remain hemimethylated for so long?

The third requirement (*c*) for a model for initiation control presents the most problems. Why is it that successive waves of initiation take place (at intervals of much less than a generation time) after release of a prolonged block to replication? If an initiator substance simply built up to a required concentration, if all copies of the origin then replicated only once, and if the initiator was destroyed thereafter, only one round of initiation would take place and the DNA content of cells would be permanently depressed after a period of growth without DNA synthesis. We can imagine several solutions to this problem. For example, initiator is made in proportion to cell mass, but only one cell-mass equivalent is destroyed after the initiation period; initiator builds up in independent compartments (1/unit cell), and each compartment opens and closes independently as initiation takes place in it; and so on. When one remembers that chromosomes are localized within the cell (see the section on Partition) and that they move from one position to another after replication of the terminus region, the idea of independent initiation compartments may not seem so heterodox.

When initiation potential has accumulated in this way, the intervals between successive waves of initiation still depend on *dam* methylation, because increase in the level of *dam*-methylase reduces these intervals (103).

Proteins Controlling Initiation

DnaA is the only protein known to be required for initiation and not for later stages in replication. It binds to nine-basepair sequences (DnaA boxes) at *oriC* and forms a multimeric complex that causes adjacent AT-rich sequences to open and allow the entry of the replication proteins. Mutants with temperature-sensitive DnaA are unaffected in DNA synthesis per se, but cannot initiate new rounds of replication at the restrictive temperature. Investigators have often suggested that it is the level or activity of this protein that is responsible for the timing of initiation. Overproduction of DnaA stimulates DNA synthesis (3, 155) and decreases the cell mass at which initiation takes place (92). Induction of DnaA production causes synchronous initiation of replication at all copies of *oriC,* followed by a refractory period equal to about 30% of the cell cycle (i.e. the period required for methylation) (120). Therefore, the activity of DnaA seems normally to be limiting for initiation.

If DnaA is the initiator substance, then it should have the same concentration in cells that are growing at different rates (i.e. like *oriC* itself). Unfortunately, published measurements are in disagreement on this point (26, 66, 121). Nevertheless, the most recent report (66) provides evidence that the concentration of DnaA is indeed the same at different growth rates, even when the methods of previous papers were repeated.

A major problem in considering DnaA as the initiator is that transcription of *dnaA* is autoregulated (2, 21, 67, 89). Consequently, synthesis of DnaA is not constitutive, as required, and this protein ought to have a constant concentration throughout the cell cycle. A possible explanation of how DnaA levels could, nevertheless, be important in initiation timing comes from the finding that it exists in several different forms, only one of which (DnaA-ATP) is competent for initiation in vitro (132). This form is converted to DnaA-ADP, which is inactive, as is free DnaA. Conversion (rejuvenation) of DnaA-ADP, or free DnaA, to DnaA-ATP in vitro is catalyzed by acid phospholipids when the DnaA protein is complexed with *oriC* (32). The conversion of DnaA-ATP to DnaA-ADP might therefore close the initiation window, while reopening of the window might require rejuvenation of DnaA (complexed with *oriC*). Even if this were so, initiation would depend on some other periodic event in the cell and not solely on DnaA concentration.

What could be the nature of this timing signal? The activity of phospholipids in rejuvenation of DnaA suggests the possibility that a change in cell membrane–*oriC* interaction might be the signal (32), but does not specify how such a change might arise periodically, or be linked to cell size.

Despite the long period over which DnaA has been studied, essential information about it is still lacking. We do not know the proportions of the different forms of the protein, or how they change over the cell cycle. Autoregulation itself is ill understood because there are two promoters for the *dnaA* gene, and these do not respond in the same way to DnaA concentration (121).

An important observation is that cells with a mass lower than the normal mass for initiation can be made to initiate replication prematurely by treatments such as amino acid starvation (64). Thus, the potential to initiate is always present, but it must be activated in some way. Activation of DNA synthesis may lie behind the many empirical methods of synchronizing populations of cells (104).

Other Factors That Influence Initiation Timing

Initiation of DNA synthesis requires an RNA primer, and some evidence indicates that transcription into the *oriC* region from the promoters of the adjacent *mioC* gene is a factor that influences initiation (although it does not appear to be essential). Transcription from *mioC* is stringently controlled (i.e. regulated by the level of guanosine tetraphosphate) so that it is expressed in inverse proportion to growth rate (26, 27). Cyclic AMP concentration, which increases with decreasing growth rate, may also influence the activity of DnaA (78).

PARTITION

Chromosome replication is followed by partition: the process whereby sister chromosomes are relocated on opposite sides of the site of cell division. Prokaryotes such as *E. coli* carry out partition very efficiently, despite the lack of any obvious mitotic apparatus.

The Process of Partition

In growing bacterial cells, DNA is distributed throughout the cytoplasm, but inhibition of protein synthesis causes its rapid contraction into one or two tightly wound masses, or nucleoids (88). The number of nucleoids corresponds to the number of completely replicated copies of the chromosome (49, 50). Thus, partition takes place within a few minutes of the replication of the terminus region of the chromosome; i.e. there is essentially no G2 period in the *E. coli* cell cycle. Septation begins almost immediately after completion of chromosome replication, at a fixed cell length (called 2 unit lengths, or 2*L*) (53). Therefore, partition also takes place at a fixed cell length and sister-chromosome centers then separate by a fixed distance (equal to 1*L*) (50).

A few minutes of postreplication protein synthesis is required for partition to take place (50, 71, 75). This period of termination-protein synthesis is almost certainly identical to that described earlier as a prerequisite for cell division (85).

Posttermination protein synthesis may be needed for the synthesis of a specific partition protein or to allow cells to reach the critical length for partition. Evidence for a critical length comes from two sets of experiments. When protein synthesis is blocked in an asynchronous population of cells, all chromosomes can complete replication but the sisters cannot separate; if further DNA synthesis is then prevented but protein synthesis (and cell growth) allowed to resume, then the sister chromosomes will separate, but at different times in different cells over a period in which the average cell length doubles (50). This observation is consistent with a requirement for a minimum cell size for partition. The same conclusion was reached in a study of partition in cells of altered shape (49). The spherical cells of *rodA* or *pbpA* mutants grow and divide at normal rate and have normal DNA:mass ratios as well as average lengths equal to rod-shaped cells, but have volumes about six times that of the rods. A possible reason for this change in size is that a minimum linear distance is required for chromosome partition. As predicted, chloramphenicol-treated spherical cells contain only one or two nucleoids each, as in rod-shaped cells, although these nucleoids are much larger than those in the rods. The phenotype of the spherical cells is therefore exactly what is predicted if chromosome centers can be separated only when a cell has one dimension equal to $2L$ or more, and if cell division can take place only after partition has taken place (see section on Cell-Division Proteins).

Further support for a fixed partition distance comes from experiments in which both DNA synthesis and cell division were blocked while cell growth was allowed to continue (9). Under such conditions, nucleoids remain in their original positions as the cell elongates; when DNA synthesis is subsequently allowed to resume, nucleoid numbers increase rapidly but the nucleoids spread out only slowly from their initial position, as expected if sister nucleoids always separate by the same distance, $1L$.

Partition Mutants

Mutants defective in DNA partition were obtained very early (76, 77). All of these Par$^-$ mutants had similar phenotypes: at 42°C the cells continued to grow and synthesize DNA, which remained unseparated in a large mass in the center of the elongating cell. In fact, all of the *par* mutants have since proved to be defective either in DNA synthesis, in DNA gyrase, or in other topoisomerases (80, 86, 87, 113, 139). A requirement for DNA gyrase for nucleoid decatenation has been demonstrated in vivo and in vitro (138). Pairs of newly replicated sister circular chromosomes are catenated and require the action of DNA gyrase to unlink them, which probably causes the failure to partition DNA in DNA-gyrase deficient cells [*gyrA* or *gyrB* (57, 80, 87)]. The role of topoisomerase IV [*parC* and *parE* proteins (86)] may also be to unlink circles of DNA.

A further requirement for unlinking was described recently (15, 29, 90). If recombination takes place between sister DNA strands during replication of the chromosome, the chromosomes can become dimerized. Resolution of these dimers requires recombination at a specific site (*dif*) near the terminus region of the chromosome. This site-specific resolution requires the action of the XerC protein. In the absence of *dif* or XerC, partition is erratic, with some cells having unseparated DNA, while others have normal nucleoid numbers and locations.

If partition is delayed relative to cell elongation, nucleoids become crowded in the cell centers, possibly because sister chromosome centers can separate by only a fixed distance (L), no matter how long the cell itself may be (9). Therefore a mutant partly defective in DNA synthesis, so that DNA replication lags behind cell growth, also shows such nucleoid crowding, i.e. a Par-like phenotype (9, 113).

None of the mutations discussed above appear to be affected in the partition mechanism itself, and Hiraga et al (74) therefore reasoned that the phenotype of a mutant defective in partition might be quite different from the classical Par$^-$ phenotype. They argued that, even in the absence of a partition mechanism, the probability of sister chromosomes coming to occupy separate cells would be high, because the mass of DNA occupies such a large proportion of the available space within a bacterial cell. Consequently, populations of cells without any partition system would be viable and consist mainly of cells with DNA, together with a proportion of DNA-free cells. An ingenious screening system was devised in which only DNA-free cells would express β-galactosidase, and therefore only colonies containing such cells would be blue on X-gal indicator plates. Among the mutants obtained in this way were some with phenotypes similar to those predicted. These Muk mutants (from the Japanese word *mukaku,* meaning anucleate) consisted mainly of normal rods with normal nucleoids and a small percentage of DNA-free rod-shaped cells.

The defect in the *mukA1* mutant appears to be that occasionally both sister chromosomes move to the same side of the division site; thus, the chromosome-moving mechanism still appears to be operating. The defect may be in the assortment (attachment?) of sister DNA molecules at opposite ends of a chromosome-moving apparatus (note: this is my interpretation of the data and not that of the authors). This is strikingly different from the mutants that are defective in decatenation or are blocked in DNA synthesis; in such cells the DNA remains in the cell center and does not move within the cell. The *mukA1* mutation is in *tolC,* a gene coding for a membrane protein required for the lethal action of Colicin E1. Other *tolC* alleles show the same phenotype as *mukA1* (74).

A second gene, *mukB*, codes for a 177-kDa polypeptide, the largest protein known in *E. coli* (111). The sequence of this protein was used to predict a secondary structure similar to that of eukaryotic myosin heavy chains and kinesin heavy chains, and later purification and visualization of the protein in the electron microscope showed that it has the predicted structure (72). The sequence also suggests the presence of an ATP-binding site and three putative zinc fingers. The purified protein has ATP/GTP binding activity in the presence of $ZnCl_2$ and also adsorbs to DNA cellulose columns (72). The pure protein forms homodimers with pairs of globular terminal domains and two paired rod-shaped internal domains connected by a hinge region. The MukB protein is therefore just the kind of protein that might be expected to move things around in a cell and is the only such protein so far described in bacteria. Is it, however, a part of the elusive partition apparatus for chromosomes?

The *mukB* gene can be inactivated (null mutant), and the cells remain viable if grown at low temperature, although they die at higher temperatures, which do not affect *mukB*$^+$ cells. At low temperatures, the mutant forms normal-looking rod-shaped cells, of which up to 5% lack DNA. At higher temperatures, mixtures of normal and long cells are produced, and the distribution of DNA within the long cells is often abnormal, showing clumps and isolated nucleoids. Unfortunately, this phenotype is not unique to this mutant and is similar to that seen in many Par mutants, or in cells in which DNA replication has been interfered with; however, DNA replication, including the timing of initiation, is normal in *mukB* null cells (72). A nice demonstration that deletion of *mukB* affects chromosomal DNA partition specifically is that F plasmids segregate equally well into both nucleate and anucleate cells in this mutant (56). If it is true that *mukB* null cells have no partition system for chromosomes, then it is striking how normal the nucleoid numbers and positions appear, at least at low temperatures. Therefore MukB, like topoisomerases and DNA replication enzymes, may not in fact be directly involved in partition. The fact that grossly abnormal partition is seen only at high temperatures may indicate the involvement of the heat-shock response. In this respect, one should note that Δ*dnaK52* cells show a clear Par$^-$ phenotype (22). Long filaments with irregular masses of DNA, normal cells, and anucleate cells are produced in this mutant; this phenotype is indistinguishable from that of *mukB* null cells at high temperature (111). Cells without DnaK (Hsp70) presumably overproduce other heat-shock proteins, because this protein acts to turn off the regulon (143). The overproduction of some of these heat-shock proteins may therefore perturb partition. Alternatively, DnaK protein [an ATPase (158)] may be involved more directly in the functioning of the partitioning apparatus. Hence the discovery of MukB may not yet have answered the question of how chromosomes are partitioned in bacteria, but the structure of MukB is very exciting and is the closest yet found in a prokaryote to a protein that could move things around in the cell.

CELL-DIVISION PROTEINS

Cell division in gram-negative bacteria requires the coordinated ingrowth of all three layers of the cell envelope (cell membrane, peptidoglycan sacculus, and outer membrane) at a specific location. Probably only a few proteins are specifically required for this process (47). Possibly most of these division-specific proteins have now been identified (51, 52, 100). Their modes of action and interaction are now being actively investigated.

Many agents and mutations inhibit cell division in *E. coli*. The block to cell division in most cases is mediated through induced stress-response systems, such as the SOS response or the heat-shock response, which produce inhibitors of cell division. The consequent plethora of mutations and agents found to block cell division in *E. coli* once caused some doubt as to whether cell-cycle events in prokaryotes were controlled by specific sets of proteins (134), in contrast to what then seemed to be the case in yeasts (68). Nevertheless, apparently only a few proteins are specifically required to carry out cell division (47). In *E. coli* this group of primary cell division proteins certainly includes FtsA, FtsI(PBP3), FtsQ, FtsZ, and EnvA. The genes coding for these proteins are all part of a single operon (*mra*) located at 2.2 min on the genetic map. The absence of any one of these will block cell division specifically, but the stage at which division aborts depends on the protein (see below). A specific block in cell division is also found in temperature-sensitive *ftsE* mutants (123). The *ftsE* gene is part of the *ftsY-ftsE-ftsX* operon at 76 min on the genetic map (62). These three genes code for membrane-associated proteins (63). FtsE shows homology to a family of prokaryotic ATP-binding proteins (70). The role of these proteins in cell division is not yet well understood.

Cytokinesis

Cytokinesis in *E. coli* can be thought of as the separation of the cytoplasm into two compartments bounded by the cytoplasmic (inner) membrane. Normally this takes place coordinately with the ingrowth of the peptidoglycan septum, but it is probably a separate process and not simply the result of the formation of the septum itself. The evidence for this is: (*a*) the regular division of L-forms of *E. coli* despite the absence of a peptidoglycan layer (118); (*b*) the formation of gaps between cell-sized blocks of cytoplasm in plasmolyzed filamentous cells that have failed to form septa (35; K. J. Begg & W. D.

Donachie, unpublished data); (*c*) the properties of FtsZ protein and the behavior of cells that have altered FtsZ activities, as will now be discussed.

FtsZ protein is produced in large amounts [5000–20,000 molecules/cell (14)] in the cytoplasm of nondividing cells. In cells that are about to divide, the FtsZ molecules aggregate at the division site to form a circumferential ring on the inner surface of the cytoplasmic membrane (14). Cytokinesis and septation proceed together, so that the circumference of this ring decreases progressively and disappears when the septum is completed. Whether ingrowth and closure of the septum causes the cytoplasmic membrane to constrict and eventually divide or whether the FtsZ ring is contractile and pulls the cytoplasmic membrane with it as it reduces in circumference is not yet clear. A third possibility is that FtsZ-driven cytokinesis itself provides the necessary stimulus for the localized ingrowth of the peptidoglycan layer to form the septum.

Three recent papers (41, 105, 122) have shown that purified FtsZ is a GTP/GDP-binding protein in vitro. When activated (see below), FtsZ becomes a GTPase. FtsZ has a seven–amino acid sequence (GGGTGTG) similar to one of three highly conserved sequences (GGGTGSG) found in 150 eukaryotic tubulins. Two mutations that cause a block in cell division (*ftsZ.84* and *ftsZ.3*) have single amino acid substitutions in this sequence [AGGTGTG (41) and GGGAGTG (105), respectively], and the purified mutant proteins show greatly reduced GTP-binding and GTPase activities. Therefore, FtsZ may be a prokaryotic equivalent of eukaryotic cytokinetic proteins.

Cells in which FtsZ activity is inhibited continue to grow and therefore form long aseptate cylindrical cells (filaments, hence *fts*). Plasmolysis of such filaments causes the appearance of only irregularly placed plasmolysis bays (31, 35, 60), rather than the regularly spaced gaps seen when septum formation itself is directly inhibited (e.g. by β-lactam antibiotics such as furazlocillin or by mutations in other genes, such as *ftsA*). This observation may indicate that FtsZ activity is indeed required for cytokinesis per se. It also shows that septum formation depends on prior FtsZ action. Cells in which FtsZ activity is blocked are also highly resistant to β-lactam antibiotics; such antibiotics preferentially lyse cells that are in the process of septation, possibly because of localized murein hydrolase activity at the site of the ingrowing septum (131).

FtsZ protein is not by itself sufficient for cell division: FtsA, FtsQ, FtsI(PBP3), and a few other proteins (e.g. FtsW, FtsE, q.v.) are also required. Cells deficient in FtsA form constrictions at normal division sites but cannot complete division. It has therefore been suggested that FtsA is required for stages of division that take place after the formation of the FtsZ-ring (8). A specific ratio of FtsA:FtsZ protein is also essential for successful cell division (33, 42). Overproduction of FtsQ partially blocks division in minimal medium but has no effect in L-broth (25); it also causes partial inhibition of division at permissive temperatures in *ftsZ* and *ftsA* mutants (33), as well as complete inhibition of division in *ftsI* mutants (see below).

Septation

Chemical analysis of the *E. coli* sacculus has so far failed to show any striking differences between the peptidoglycan (PG) of septa or cell poles and that which forms the cylindrical portion of the cell body. Nevertheless, one enzyme is required solely for septation. PBP3 (penicillin-binding protein 3) is one of four PG-transpeptidases found in *E. coli,* but it is the only one that is required solely for septum formation (16, 130, 137). β-Lactams that preferentially bind PBP3 (e.g. furazlocillin, cephalexin, or benzyl penicillin at low concentration) block septation without inhibiting net PG synthesis, and mutations that inactivate PBP3 also block septum formation and nothing else.

PBP3 is a transmembrane protein, with its catalytic domain in the periplasmic space (17). In vitro, the protein carries out both PG-transglycosylation (elongation of new glycan strands in the PG layer) and PG-transpeptidation (the crosslinking of these chains into the complete covalently linked PG network of the sacculus) (81a). Only the different β-lactam sensitivities of the transpeptidation reaction distinguish the in vitro activity of this protein from those of the other transpeptidase/transglycosylases in the cell (i.e. PBP1a, PBP1b, and PBP2). PBP3 is present in very low abundance: about 50 molecules/cell (136).

PBP3 may act in concert with other division-specific proteins. FtsZ is clearly one of these, because inhibition of FtsZ activity leads to immediate cessation of septation at whatever stage it has reached (149). One mutation of the FtsA protein apparently alters the ability of PBP3 to bind ampicillin, although other mutations of FtsA do not (145). Overproduction of FtsQ is lethal, even at 30°C, in cells with an altered, temperature-sensitive PBP3 (105). Finally, homology between FtsW and RodA (which acts in concert with PBP2, see below) has suggested the possibility that this protein may act together with PBP3 (100). The membrane proteins (PBP3, FtsW, FtsQ) may act in a complex, perhaps with FtsA (which is associated with membrane complexes thought to be part of the septation site) (28) and with FtsZ. The numbers of the membrane proteins are so low, however, that as many as 1000 FtsZ molecules would be associated with each membrane-protein group.

The PBP3(FtsW) system is required for septum formation, but an additional requirement may be the presence of appropriate substrate molecules. The substrates for the PG-transpeptidase/transglycosylases (PBPs 1a, 1b, 2, 3) are bactoprenol-linked N-acetlyglucosaminyl:N-acetlymuramyl peptides. Figure 2 shows (in outline) the steps that lead to the formation of these

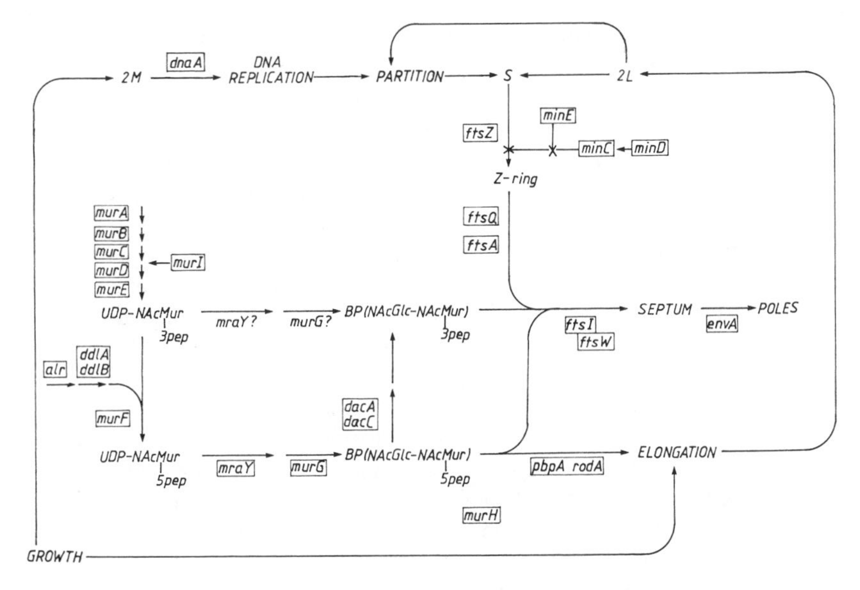

Figure 2 The morphogenetic cycle of *Escherichia coli*. A schematic view of the cell cycle—how cell growth leads to cell division. The genes coding for proteins carrying out particular steps are shown (*boxes*). The left side of the diagram shows the synthesis of peptidoglycan precursors [UDP-NAcMur-3pep or -5pep, UDP-N-acetylmuramyl tripeptide or pentapeptide, BP(NAcGlc-NAcMur-3pep or -5pep), bactoprenol-linked N-acetylglucosaminyl:N-acetylmuramyl tripeptide or pentapeptide]. About half (7/13) of the genes coding for these enzymes are located in the large operon at 2.2 min on the map (Figure 1).

Pentapeptide side chains are required as donors in all transpeptidation reactions, but tripeptide side chains may act preferentially as receptors in transpeptidation reactions carried out by septum-specific enzymes. The proportions of these two kinds of side chains depend on the opposing activities of enzymes that synthesize and add terminal D-alanine dipeptides (alanine racemase, D-alanine ligase, and D-alanine dipeptide-adding enzyme, the products of the *alr, ddlA, ddlB,* and *murF* genes) and those that remove terminal D-alanine residues (DD-carboxypeptidases PBP5 and 6, products of *dacA* and *dacC,* and DL-carboxypeptidase, gene still unknown).

Cell growth is needed for the cell to reach the required mass (2*M*) for initiation of chromosome replication by DnaA. In order for cell division to take place normally, partition of replicated sister chromosomes must occur. This can take place only in cells that have reached a minimum length (2*L*, see text and Figure 3). The action of the RodA and PBP2 proteins (*rodA* and *pbpA* genes) is required to maintain a constant cell diameter during growth, so that the critical length is reached at the same time as chromosome replication is completed (see text). This process is labeled *Elongation*. When the required cell length is reached, chromosomes separate and FtsZ protein forms a ring in the space (S) between them.

The formation of the septum requires the contraction of the Z-ring, and the peptidoglycan septum is synthesized by PBP3 (*ftsI* gene product), possibly in conjunction with FtsW. This requires a supply of tripeptide side chains. FtsQ and FtsA proteins are essential for stages that follow the formation of the Z-ring. The EnvA protein is required for the final steps of septum-splitting and cell-pole formation. The MinC protein, activated by MinD, prevents the Z-ring from forming but is prevented from doing so at S by MinE.

substrates, together with the names of the genes coding for the enzymes carrying out each reaction. The PG-specific pathway starts with uridine diphosphate (UDP)-N-acetylglucosamine, which is converted in two steps to UDP-N-acetylmuramic acid. Three steps successively add L-alanine, D-glutamic acid, and *meso*-diaminopimelic acid to form a tripeptide side chain. A final D-alanyl:D-alanine dipeptide is then added to give a pentapeptide side chain. Subsequent reactions link this to N-acetylglucosamine and then link this disaccharide unit to the bactoprenol lipid carrier, which allows it to be transported across the cell membrane into the periplasm. The bifunctional PBP enzymes then link these units to form long glycan chains (transglycosylation) and form peptide bonds between the side chains of new and preexisting chains in the sacculus (transpeptidation) (119). PBP3 may use tripeptides preferentially as acceptors in the transpeptidation reaction (6), and circumstantial evidence supports this idea. Thus, partial inhibition of the addition of the D-alanyl:D-alanine dipeptide restores the capacity of PBP3-deficient cells to form septa, as do increased levels of enzymes that remove the terminal D-alanine from completed pentapeptide side chains (10). Also, increase in MurF protein (the D-alanyl:D-alanine adding enzyme) blocks septation in cells with normal levels of PBP3, presumably by reducing the pool of tripeptide substrate (102). Increase in PBP5 (a D-alanine carboxypeptidase), in contrast, causes cells to become spherical, as though the cell-elongation system had been suppressed (97). Changes in the ratio of pentapeptide to tripeptide side chains in the peptidoglycan can therefore switch cells into and out of septation.

Septum Splitting and the Completion of Division

The product of the activity of PBP3 and its associated proteins is a double layer of peptidoglycan across the middle (usually) of the cell. These two layers are initially covalently bonded together and require the action of the EnvA protein to split them apart (154). Both over- and underproduction of EnvA are lethal (5, 140). The *envA.22* mutant was reported to produce an abnormally low level of N-acetylmuramyl-L-alanine amidase (154), but this was later denied (144). Mutants of the *amiA* gene deficient in N-acetylmuramyl-L-alanine amidase have no obvious defects (144).

Until the septal layers are split apart, the outer membrane (OM) cannot invaginate, and in the *envA.22* mutant, this membrane balloons out around the septal site (112). Normally however, the OM invaginates as the ingrowing septal layers are split, so that the OM is tightly apposed to the new cell poles. It is not known what causes the outer membrane to follow the ingrowth of the septum in this way, but this membrane is linked to the sacculus through a specific lipoprotein, which has one end embedded in the OM and the other covalently bonded to the peptidoglycan (110). Therefore, the constant formation of new bonds of this kind may keep the OM attached to the sacculus during cell division (23). Mutants of the *lpp* gene (defective in lipoprotein) growing in limiting Mg^{2+} reportedly have an OM layer that is only loosely associated with the sacculus (156), but these mutants can divide successfully. The OM therefore appears able to self-seal around new cell poles without the necessity of a direct linkage to the sacculus.

Regulating Cell-Division Proteins

A decrease in the level of any one of the essential division proteins [FtsZ, FtsA, FtsQ, FtsI(PBP3), EnvA, FtsW] will cause delayed or aborted cell division: increased levels of some of these (FtsZ, FtsA, EnvA) can also be lethal. Regulation of the levels of these proteins is therefore essential for cell survival.

THE *mra* OPERON The genes coding for all of the above cell-division proteins reside in a single operon (*mra*) (52, 100) (Figure 1). The operon contains 16 open reading frames, separated from one another by short gaps, or overlapping by a few bases. Functions and/or protein products have been identified for 14 of these. All of the known functions are steps in peptidoglycan precursor synthesis, or in cell division.

The region is considered to be a single operon because all the genes are read from the same strand, because successive open reading frames (ORFs) overlap or are separated by only a few bases, and because the only known transcription terminator is located at the downstream end (after *envA*) (5). However, this cluster of genes contains many internal promoters, so that the operon is transcribed into an overlapping set of mRNAs. Because the gaps between ORFs are so short, or nonexistent, the promoters are mostly inside ORFs. Similarly, ribosome-binding sequences in the mRNA are also generally located within protein coding sequences. Regulatory sequences, affecting either transcription or translation are also, perforce, superimposed on the protein-coding sequences.

DIFFERENTIAL EXPRESSION OF CELL-DIVISION PROTEINS The proteins of the *mra* operon are produced in widely different amounts: FtsZ, 5000–20,000 (14); FtsA, about 150 (150); and FtsQ, only about 25 molecules/cell (25). Increase or decrease in the ratio of FtsZ:FtsA blocks division (33, 42). The differential expression of these three contiguous genes appears to result mainly from differential translation of mRNA (106). The ratio of translation of FtsQ:FtsA:FtsZ from the same transcript is similar to the ratio found for these proteins in normal cells. Very low translational efficiencies are also found for FtsI(PBP3) [about 50 molecules/cell (136)] and FtsW (in vivo level unknown) (M. Khattar & G. Roberts, personal communication). Another factor in differential gene expression is probably the number and strength of upstream promoters: e.g. *ftsZ* appears to be expressed at fully normal level only if the entire upstream region from *ddlB* to *ftsA* is also present (94). At least seven promoters have been identified within this region (1, 43, 94, 125, 126, 128, 157) (Figure 2). Because four of these are within *ftsA* and one within *ftsQ*, only three of this set will transcribe *ftsA* and two transcribe *ftsQ*. The promoters also differ in strength. The exact number, location, and strengths of promoters upstream of the terminal *ddlB-envA* region are less well known, but one may assume that they also will contribute to the transcription of downstream genes.

REGULATION OF *mra* GENE TRANSCRIPTION Control of the relative levels of expression of the division proteins as described above is essentially preset, but transcription may also be regulated dynamically (1, 44, 52, 55). The major promoters of *ftsZ*, which together account for more than 90% of its transcription in normal cells (128), are all located in the proximal *ftsA* gene (Figure 1). Transcription from these promoters (expressed as a fraction of total transcription) is higher in slow-growing cells than in fast growing cells (1, 44, 52, 55; R. W. P. Smith, M. Masters & W. D. Donachie, submitted). However, because cell size increases with growth rate, the number of *ftsZ* transcripts per cell changes very little. It has been suggested that transcription of *ftsZ* is tied to the number of completed rounds of chromosome replication, so that a constant number of FtsZ molecules are produced for every division, independent of the size of the cell (1, 44, 52, 55; R. W. P. Smith, M. Masters & W. D. Donachie, submitted).

Two promoters in *ddlB* also appear to be growth-rate regulated in a similar way (1, 44, 52, 55, 151; R. W. P. Smith, M. Masters & W. D. Donachie, submitted). Thus, transcription of the *ftsQ* and *ftsA* genes may have a growth rate–dependent control similar to that of *ftsZ*.

The mechanism of this growth rate–regulated control of *ftsQAZ* transcription remains to be determined. A suggestion that at least one of the *ftsZ* promoters might be regulated (directly or indirectly) by the level of FtsA protein (44) has been challenged (124) and therefore requires reinvestigation.

A new transcription factor (SdiA) (151) has been shown to increase transcription from one of the promoters (*ftsQ2p*) located in the *ddlB* gene, but not from any other promoters in the *ftsQAZ* region. Overexpression of the *sdiA* gene results in increased production of FtsZ protein (151). Inactivation of *sdiA*, however, appears to have no phenotypic effect on the cell (151), so that role of this gene in the cell cycle remains obscure.

The *mreB* gene product is a negative regulator of transcription of *ftsI* (147). Cells that are *mreB*$^-$ overproduce PBP3 (the *ftsI* product) and are spherical in shape (147).

All the studies on growth rate–regulated transcription of the *ftsQAZ* group predict that the levels of FtsQ, FtsA, and FtsZ proteins per cell should be constant at different growth rates. Despite this prediction, recent measurement of FtsA levels (150) has shown that it has a constant concentration at different growth rates, i.e. the amount per cell increases substantially with growth rate and cell size. Bi & Lutkenhaus (14) also report that the amount of FtsZ per cell increases at higher growth rates. These observations, if confirmed, indicate that the observed growth rate–dependent regulation of transcription of the corresponding genes appears to be counterbalanced by some other control of protein levels. This conclusion is all the more surprising because an attempt to measure posttranscriptional regulation of FtsA production failed to find any regulation (106). Therefore, the reported observations on division-protein regulation contain an important paradox. Recent experiments suggest that FtsZ production may be posttranscriptionally regulated by an antisense RNA produced from the noncoding strand of the *ftsZ* gene itself (W. D. Donachie & S. J. Dewar, unpublished observations). This would act in a similar way to DicF RNA (142) (see below).

REGULATION OF FtsZ ACTIVITY FtsZ binds GTP and GDP in vitro. In the presence of Mg^{2+}, FtsZ is a GTPase. FtsZ protein requires activation before it will show GTPase activity. Activation can be brought about by K^{+} or by high concentrations of FtsZ protein itself (41, 105). Therefore, the onset of cell division might be brought about either by the achievement of a critical, self-activating, concentration of FtsZ, or by its activation by a change in the ionic composition of the cell.

REGULATION OF PBP3 ACTIVITY The activity of the PBP3(FtsW) septum-forming system is partly controlled by the supply of specific PG-substrates (see above) (10, 97). The relative activities of certain key enzymes (the products of the *murF, dacA,* and *dacC* genes; Figures 1 and 2) determine these levels. The regulation of competing enzyme activities may therefore be important in the regulation of the cell-division cycle.

Timing and Localization of Division

What is the link between chromosome replication and cell division that ensures that division follows DNA replication and that the site of division is between pairs of sister genomes? This linkage is not well understood as yet, but a specific set of proteins is required to ensure the correct placement of the septum and perhaps these may also help to ensure that septa form (usually) after chromosome replication has been completed.

Septation normally follows chromosome replication and partition, but septation does not depend absolutely on these events. If DNA replication is interrupted, then cell division does not take place because the SulA inhibitor is induced as part of the SOS response to the block in DNA replication; however, if chromosome replication is not allowed to begin at all, then the SOS response is not induced and cells will continue to grow and divide, forming a mixture of cells with and without DNA (76, 77). Similar behavior is seen in *sulA*⁻ cells if DNA replication is blocked at any stage.

In Par⁻ cells, all the DNA remains together in an ever enlarging mass in the cell center, and septa form in other parts of the cell, thereby producing DNA-less progeny cells (57, 76, 77, 79, 86, 153). The sizes of these DNA-less cells vary considerably, from near spherical minicells to very long cylindrical cells, depending on the position of the septum with respect to the cell ends (79). Certain cell lengths may be preferable, at least in some kinds of Par mutants (107), but clearly, in the absence of normal chromosome localization, septa can and do form at almost any position. This phenotype therefore suggests that septal location depends on chromosome position: a simple hypothesis is that septa are prevented from forming at positions that are too close to DNA (79, 153).

If this idea is extended to the normal cell cycle, then the signal for septation would be the appearance of a DNA-free gap in the center of the cell, an event that normally follows chromosome partition (Figure 3). What then prevents septa from forming in other DNA-free zones of similar size in the normal cell? Such zones are found at the cell ends, between the central nucleoid and the cell poles, and would normally become as large as the central zone (Figure 3). Hence, a further assumption is required, namely, that septa also cannot form near cell poles. A formal model for the timing and localization of septa is, therefore, that septa cannot form within a fixed distance of either a chromosome center or a cell pole. For this model to work, the fixed distance must be ($L/2$); i.e. half the distance between newly separated sister chromosomes.

It is not known what prevents septa from forming in the vicinity of chromosome centers, but it is known what prevents them from forming close to cell poles. The products of the *min* locus, a group of three genes, are dedicated to this function. A series of elegant papers over the past few years has revealed the main features of the way in which these genes work (37, 38). The MinC protein is an inhibitor of cell division, which is normally activated by the MinD protein. Activated MinC protein prevents division everywhere in the cell, except if the third protein, MinE, is also present; in this case, the normal situation, division is prevented only at the cell ends and not in the cell center. In mutants that lack MinC (or the whole *min* locus), division can and does take place at the cell poles, so that DNA-free minicells form. In such mutants, divisions can also occur in the cell centers, although at reduced frequency (see section on Number, below), so that *min*⁻ mutants are viable.

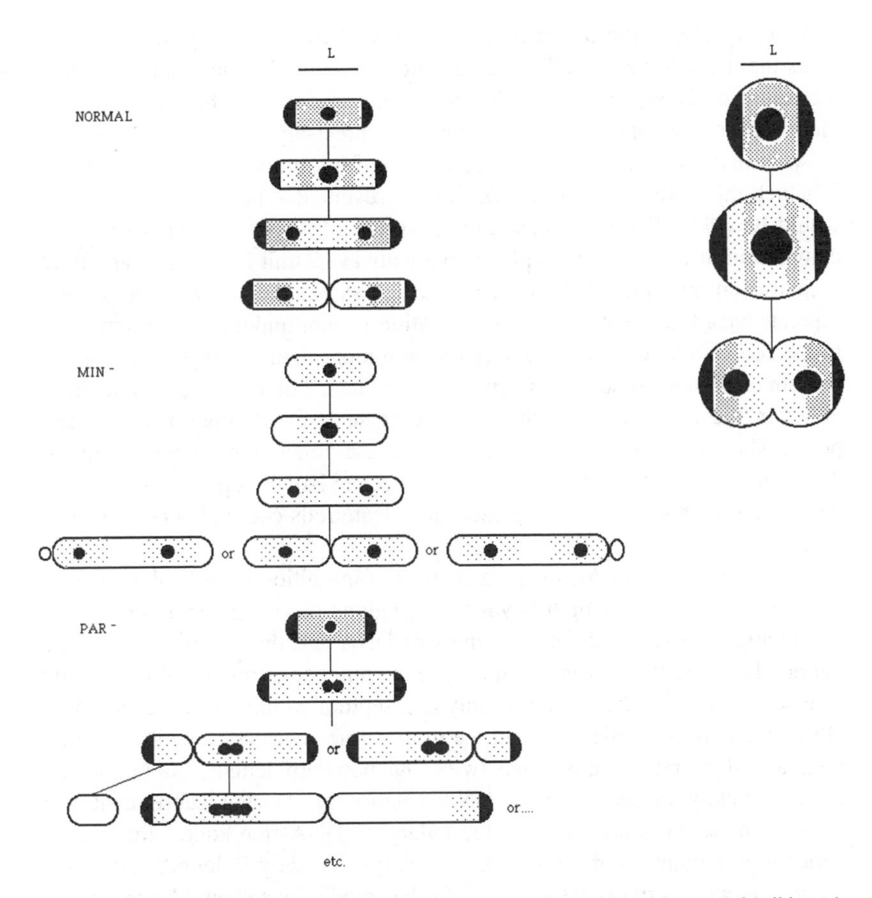

Figure 3 A model for the location of division sites. Cell cycles are shown for normal (wild-type), Min^-, Par^-, and Rod^- cells (growing at the same rate). In all cells, septation is assumed to be inhibited in zones (*shaded*) that extend a distance of *L*/2 (half the separation distance between newly completed sister chromosomes) from the centers of nucleoid masses. In all cells except Min^-, septation is also inhibited within a distance of *L*/2 of the cell poles. Nucleoids and active cell poles are shaded black. Where zones overlap, a higher density of shading is shown. Septa are assumed not to form until after chromosome partition. (A relaxation of this rule would allow septa to form at other times in both Min and Par mutants.) It is assumed that only one septum is formed per round of chromosome replication. In normal cells, septa can form only in the cell center, after partition. In Min mutants, septa can form either in the center, or near the poles (note that minicells will display a range of sizes in this model). In Par mutants, septa will form at a variety of positions, yielding a wide range of sizes of anucleate cells. In Rod mutants, partition cannot take place until the spherical cell has reached the required diameter, so that large polyploid cells are formed.

An early discussion of the role of the *min* locus (141) suggested that cell poles themselves were "old" septation sites that would remain active as sites for future division, unless inactivated by the products of the *min* genes. A modified version of this idea, in which no precisely located division sites exist, is that the Min proteins set up a zone of inhibition near the cell poles, in which the formation of the FtsZ ring is prevented. The extent of this zone would be (*L*/2). Figure 3 shows how this would work: cells at the time of completion of chromosome replication are always 2 unit length (2*L*) and have a central chromosome (49, 50, 53). At this stage, the sister chromosomes separate by a fixed distance, *L* (9); septation is then impossible, except close to the cell center, which is the only location more than *L*/2 from either a cell pole or a chromosome center. In *min*$^-$ mutants, however, septa can form either in the zone between chromosome centers, or in zones near the cell poles. The exact location of septa within these zones is not determined, so that minicells should come in a range of sizes (81b). If septation sometimes began before chromosome separation, anucleate rods could also be produced (81b).

In Par$^-$ mutants, the Min-generated zone of inhibition would still exist but, as a cell increased in length beyond 2*L*, regions in which septa could form would appear between the central mass of DNA and the cell poles. Because the number of septa is probably linked to the number of rounds of replication (see section on Number, below), only one septum would form, generating a DNA-free cell of a little less than the minimal newborn cell length, together with a cell of rather more than twice the newborn length, containing an asymmetrically placed mass of DNA (Figure 3). During subsequent cell elongation, septa would form in the enlarging DNA-free zones: this would generate a population of DNA-free cells varying widely in length (79). This model fits well with observation and further predictions should be testable.

THE MOLECULAR ACTION OF THE Min PROTEINS Overproduction of the MinC and D proteins, without a corresponding increase in MinE, is lethal, because all cell division is blocked (38, 127). The MinC protein seems to be the actual inhibitor of division, because hyperproduction of this protein can by itself block division, whereas overproduction of MinD alone does not. However, the presence of MinD causes MinC to be inhibitory at much lower concentrations (40). Recently (36), MinD was shown to be a membrane ATPase. The mechanism of MinE action is currently the most mysterious: in its absence all cell division is blocked (if MinC and MinD are present), while its overproduction creates a *min*$^-$ phenocopy (38). Thus, at normal levels, MinE protein may be considered to reverse the inhibitory effect of MinC at the cell center, but not at the cell poles; whereas at higher levels, MinE protein can prevent the action of MinC at all sites in the cell.

What is it that is inhibited by MinC? It seems probable that, like SulA, MinC interacts directly with FtsZ protein, because overproduction of FtsZ or alteration in its structure can reverse the inhibitory effect of MinC (13). Thus, both the intrinsic inhibitors of cell division in *E. coli* target FtsZ, and thereby block the very earliest stage in division.

The spatial specificity of MinCDE action raises interesting questions. Are FtsZ molecules continually attempting to form rings but being prevented by the presence of activated MinC at the poles? Is MinE, antagonizing MinCD, localized exclusively at the cell center (after partition)? Does MinE act on MinC, or on FtsZ? Does excess MinCD (or SulA) prevent the formation of FtsZ rings, or inactivate them? What is the source of spatial information in a bacterial cell: is it the positions of chromosomal centers, and if it is, how are these determined; is it distance from cell poles; is it local self-perpetuating discontinuities in the cell envelope (7, 30, 48, 101); or is it a mitotic system that separates free-floating chromosomes by a fixed distance (9)?

Housekeeping Proteins That Are Essential for Cell Division

Housekeeping proteins are also required for cell division. Many of these have no preferential effects on cell division, as opposed to cell growth and metabolism in general, but others may have more direct effects on the assembly of division-associated structures. For example, SecA is a protein required for all transport of proteins into and across membranes (83, 116, 117), but *secA*^{ts} mutants cannot divide at 42°C, although they can grow for some time (115). This may reflect a requirement for certain division proteins (PBP3, FtsQ) to be inserted into the membrane at the time of septation. Similarly, cells deficient in the molecular chaperonins GroES and GroEL (Hsp60 family) can also grow for a period, but cannot divide (61, 148). In addition, overproduction of GroE can suppress the cell-cycle block in certain *dnaA* and *ftsZ* mutants (58, 84; N. McLennan & M. Masters, unpublished data), probably because the increased chaperonin level allows the mutant proteins to fold or assemble correctly.

Intrinsic Division Inhibitors

In addition to the cycle-linked inhibitor, MinC, *E. coli* can produce other inhibitors of cell division under certain conditions. The SulA(SfiA) protein, produced as part of the SOS response to interference with DNA replication, reversibly blocks the action of FtsZ (93). Certain strains harbor a phage-relic, e-14, which produces an irreversible inhibitor of FtsZ under related circumstances (34, 95, 96). The induction of the heat-shock response also is accompanied by a block to division, which also appears to act on FtsZ (22, 146). Another phage relic (i.e. an incomplete set of genes of phage origin) produces inhibitors of division only if a regulatory gene is first inactivated (by mutation or deletion) (11). One inhibitor is a protein (DicB) that replaces MinD as activator of MinC (above) and causes inhibition of FtsZ action at all sites in the cell (39, 91, 108); another is an antisense RNA (DicF-RNA) that prevents the translation of *ftsZ* mRNA (142). Certain plasmids also carry genes for division inhibitors that act only in cells that have lost the plasmid (73, 82).

SHAPE

We have assumed here that cells must attain a minimum length before partition and cell division can take place. If initiation of replication is to take place at every doubling of the unit mass (or volume), and if division is to take place after each round of replication, then cells must attain the required length (2*L*) at the time of termination. The time taken from initiation to termination is largely independent of growth rate, and therefore, the above requirements place constraints on cell geometry. Cells are rod-shaped with a constant average length-to-width ratio (averaged over the cell cycle but independent of growth rate or cell volume), such that the required length is indeed reached at the time of completion of chromosome replication at all growth rates (49, 50, 53, 65). This cell shape is maintained by a pair of membrane proteins, RodA and PBP2 (136, 137). Together, these proteins have PG-transglycosylase and PG-transpeptidase activity (81). Inactivation of either protein causes cells to lose their rod shape and form cocci. As explained above, these cocci, probably solely because of their altered length:width ratio, are polyploid. Evidence (10, 102) indicates that cells switch between the RodA/PBP2 and PBP3(FtsW?) morphogenetic systems, perhaps because of changes in levels of preferred PG-substrates (with tripeptide side chains for the septum-forming system and with pentapeptides or tetrapeptides for the shape-determining enzymes).

NUMBER

Min$^-$ mutant cells can divide, either near a cell pole, or at the cell center, but the number of septa formed is not increased over that found in *min*$^+$ cells in which polar divisions are prevented. Consequently, a misplaced septum (near a cell pole) is formed at the expense of a centrally placed septum (and vice versa), and cells behave as if they have a limited division capacity of one septum per cycle, no matter how many potential sites for division there may be (141).

Confirmation of the limited division capacity of cells comes from the behavior of cells that are overproducing MinE protein and therefore have no inhibition of division sites by MinC (above). These cells may divide either at the cell poles (giving minicells) or between nucleoids (as in normal cells), but the number of septa formed per cell cycle appears to be restricted to one, just as in *min*$^-$ mutants (38). In a *gyrA*^{ts} (Par$^-$ phenotype at 42°C) mutant, cells continue to grow, synthesize DNA, and divide at normal frequency for an extended period at 42°C. However, because sister chromosomes cannot separate, the DNA remains in the cell center, and division produces only DNA-free cells. As in the case of the *min* mutants, the number of septa formed appears to be normal. Therefore, these cells must also have a limit of one septum per replication cycle (79).

There is now good evidence that what limits the number of divisions is the amount of FtsZ protein (12). Overproduction of FtsZ in wild-type cells causes the production of minicells (152); however, in contrast to *min* mutants, the minicells are produced in addition to a normal number of divisions at the cell center. Thus, the total number of septa formed per cycle is increased. Also, increased FtsZ will increase the number of divisions in *min* mutants, restoring the number of central divisions to the normal level without reducing the number of minicells produced (12). It has therefore been concluded that the amount of FtsZ protein is rate-limiting for cell division in normal cells (12). This conclusion was disputed recently, on the basis that cells in which translation of FtsZ mRNA is blocked by an antisense RNA (see below) do not behave as if the level of FtsZ protein is limiting for division (142). Further studies are needed to resolve this fundamental point.

The original analysis of the minicell-mutant phenotype (141) showed that the division capacity was, in a sense, quantal in nature. Thus, a cell of, say, 4 units in length (with four segregated chromosomes) would make two septa (positioned apparently at random among the two polar and three inter-chromosomal sites). The division capacity (two septal equivalents) was therefore used up in two discrete packets at two distinct sites. Recent discoveries about FtsZ protein may help in understanding how the quantal features of division capacity arise. One possibility is that the appearance of a potential division site (essentially a gap between chromosome centers, above) allows the self-assembly of FtsZ molecules into the FtsZ ring at that site, and that a fixed number of molecules are used to make each ring. If the production of FtsZ is linked to completion of rounds of chromosome replication (44), then just enough will be made to form a single ring per pair of sister chromosomes. Nucleation of a potential division site with a small number of FtsZ molecules must be assumed to initiate rapid assembly of a ring at that point, provided sufficient FtsZ molecules are present in the cell. Consequently, only a fixed number of complete rings will form when division sites become available, and this number will equal the number of newly completed rounds of chromosome replication. Alternatively, FtsZ may be present at constant concentration throughout cell growth and be activated when the cell state changes near the time of chromosome partition. Again, nucleation of a ring would rapidly deplete FtsZ and prevent the formation of a second ring at another site.

SUMMARY

E. coli cells grow until they reach the critical size (2 × unit mass, or volume) at which chromosomal DNA replication is initiated. Initiation requires a minimum concentration of active DnaA protein (probably DnaA-ATP). At this stage, replication is initiated at all copies of *oriC* that may be present. Following replication of *oriC,* this segment of DNA is hemimethylated and therefore refractory to further initiation. The replicated *oriC* DNA is sequestered in the cell membrane and remains there, in the hemimethylated state, for the next one third of the cell cycle. When this region is at last fully methylated, and therefore once again available for initiation, the initiation window in the cell cycle has closed (DnaA-ATP no longer present?) and will not reopen until the cell has once again reached the required critical mass. (DnaA synthesis is autoregulated and the protein is probably present at constant concentration throughout the cell cycle. Activation of some of this protein, enabling the initiation of DNA synthesis, may be triggered by a growth-linked event, such as a localized change in the cell membrane.)

Following initiation, DNA synthesis proceeds at a constant rate, bidirectionally around the circular chromosome, until the replication forks meet (in the *ter* region). The two newly formed sister chromosomes are interlocked at this stage, and two sets of topoisomerases (DNA gyrase and topoisomerase IV) are required to unlink them. If (an odd number of) sister-strand exchanges have taken place before replication is completed, the two sister chromosomes will form a dimer. Resolution of this dimer is carried out by the XerC protein at a unique site (*dif*) in the *ter* region.

After separation, sister chromosomes move rapidly apart by a fixed distance (unit length, *L*) within the cell (chromosome partition). Partition will take place only if the cell has reached a minimum length (2*L*), but this length is normally attained at about the same time as completion of each round of chromosome replication. The attainment of the critical length at this time results from the activity of the PBP2/RodA morphogenetic (peptidoglycan modifying) proteins, which ensure that the cell elongates as a narrow rod. A short period of posttermination protein synthesis is required for partitioning. In the absence of certain proteins (MukA, MukB), this process takes place less efficiently, so that anucleate cells are occasionally produced. (Muk proteins may carry out DNA partition directly, perhaps in combination with other undescribed proteins; in the absence of this premitotic apparatus, chromosomes may often come to be in different cells by a random process.)

Cell division begins when there is a distance of more than *L*/2 from a chromosome center (membrane attachment point?) or a cell pole. In the normal cell cycle, the first time this happens is immediately after sister-chromosome partition, and the first available site is in the cell center. The Min proteins prevent the formation of septa near cell poles. The inhibitor that blocks division near the chromosomes is unknown. When a site becomes available, activated FtsZ protein aggregates to form a ring around the inner surface of the cell membrane. The contraction of the ring beyond a certain stage also requires a fixed proportion of FtsA protein. The ring contracts and the peptidoglycan layer grows inward as a covalently bonded double layer that follows the contracting ring and eventually forms a complete double crosswall across the cell center. The formation of the septum requires a specific transpeptidase and transglycosylase (PBP3, perhaps in association with FtsW). Another transmembrane protein, FtsQ, is also required. Septum formation appears to require an adequate supply of peptidoglycan chains or precursors with tripeptide side chains; the proportion of this substrate is determined by the balance between the activities of certain enzymes. (Activation of FtsZ may be a required step in the initiation of division. This could come about by autoactivation, when the concentration of the protein reaches a critical level, or possibly by a change in ionic composition of the cell when it reaches a critical state, correlated with some morphological parameter such as a particular length, or the existence of a long DNA-free section.)

During its formation, the septum is split apart into two layers, a process that requires the EnvA protein. The outer membrane follows the separating PG layers inward, and is attached to the peptidoglycan by a specific lipoprotein.

The cell produces only enough FtsZ protein to form a single ring for each pair of newly replicated sister chromosomes. Transcription of the *ftsQAZ* group of genes may be regulated so as to produce a fixed number of transcripts per termination event. The correct ratio of FtsQ:FtsA:FtsZ is maintained by differential translation.

This is a drastically simplified view of how the cell cycle may work, on the basis of information available in October 1992. No doubt this picture will soon be modified, but the core of it is based on long-standing observations, as well as on exciting new results. The success in cataloging and characterizing the major cell-cycle proteins now opens the way to a new phase, in which the epigenetics of *E. coli* cells can be explored.

Literature Cited

1. Aldea, M., Garrido, T., Pla, J., Vicente, M. 1990. Division genes in *E. coli* are expressed coordinately to cell septum requirements by gearbox promoters. *EMBO J.* 9:3787–94
2. Atlung, T., Clausen, E. S., Hansen, F. G. 1985. Autoregulation of the *dnaA* gene of *Escherichia coli* K-12. *Mol. Gen. Genet.* 200:442–50
3. Atlung, T., Løbner-Olesen, A., Hansen, F. G. 1987. Overproduction of DnaA protein stimulates initiation of chromosome and minichromosome replication in *E. coli. Mol. Gen. Genet.* 206:51–59
4. Bakker, A., Smith, D. W. 1989. Methylation of GATC sites is required for precise timing between rounds of DNA replication in *Escherichia coli. J. Bacteriol.* 171:5738–42
5. Beall, B., Lutkenhaus, J. 1987. Sequence analysis, transcriptional organization and insertional mutagenesis of the *envA* gene of *Escherichia coli. J. Bacteriol.* 169:5408–15
6. Beck, B. D., Park, J. T. 1976. Activity of three murein hydrolases during the cell division cycle of *Escherichia coli* K-12 as measured in toluene-treated cells. *J. Bacteriol.* 167:1250–60
7. Begg, K. J., Donachie, W. D. 1977. The growth of the *E. coli* cell surface. *J. Bacteriol.* 129:1524–35
8. Begg, K. J., Donachie, W. D. 1985. Cell shape and division in *Escherichia coli:* experiments with shape and division mutants. *J. Bacteriol.* 163:615–22
9. Begg, K. J., Donachie, W. D. 1991. Experiments on chromosome separation and positioning in *Escherichia coli. New Biol.* 3:475–86
10. Begg, K. J., Takasuga, A., Edwards, D. H., Dewar, S. J., Spratt, B. G., et al. 1990. The balance between different peptidoglycan precursors determines whether *E. coli* cells will elongate or divide. *J. Bacteriol.* 172:6697–6703
11. Béjar, S., Bouché, J.-P. 1985. A new dispensible genetic locus of the terminus region involved in control of cell division in *Escherichia coli. Mol. Gen. Genet.* 201:146–50
12. Bi, E., Lutkenhaus, J. 1990. FtsZ regulates the frequency of cell division in *Escherichia coli. J. Bacteriol.* 172: 2765–68
13. Bi, E., Lutkenhaus, J. 1990. Interaction between the *min* locus and FtsZ. *J. Bacteriol.* 172:5610–16
14. Bi, E., Lutkenhaus, J. 1991. FtsZ structure associated with cell division in *Escherichia coli. Nature* 354:161–64
15. Blakely, G., Colloms, S., May, G., Burke, M., Sherratt, D. 1991. *Escherichia coli* XerC recombinase is required for chromosomal segregation at cell division. *New Biol.* 3:789–98
16. Botta, G. A., Park, J. T. 1981. Evidence for involvement of penicillin-binding protein 3 in murein synthesis during septation but not during cell elongation. *J. Bacteriol.* 145:333–40
17. Bowler, L. D., Spratt, B. G. 1989. Membrane topology of penicillin-binding protein 3 of *Escherichia coli. Mol. Microbiol.* 3:1277–86
18. Boye, E., Løbner-Olesen, A. 1990. The role of *dam* methyltransferase in the control of DNA replication in E. coli. *Cell* 62:980–89
19. Bramhill, D., Kornberg, A. 1988. Duplex opening by DnaA protein at novel sequences in initiation of replication at the origin of the E. coli chromosome. *Cell* 52:743–55
20. Bramhill, D., Kornberg, A. 1988. A model for initiation at origins of replication. *Cell* 54:915–18
21. Braun, R. E., O'Day, K., Wright, A. 1985. Autoregulation of the DNA replication gene dnaA in E. coli K-12. *Cell* 40:159–69
22. Bukau, B., Walker, G. C. 1989. Cellular defects caused by deletion of the *Escherichia coli dnaK* gene indicate roles for heat shock protein in normal metabolism. *J. Bacteriol.* 171:2337–46
23. Burdett, I. D. J., Murray, R. G. E. 1974. Electron microscope study of septum formation in *Escherichia coli* strains B and B/r during synchronous growth. *J. Bacteriol.* 119:1039–56
24. Campbell, J. L., Kleckner, N. 1990. E. coli oriC and the dnaA promoter are sequestered from dam methyltransferase following the passage of the replication fork. *Cell* 62:967–79
25. Carson, M., Barondess, J., Beckwith, J. 1991. The FtsQ protein of *Escherichia coli:* membrane topology, abundance, and cell division phenotypes due to overproduction and insertion mutations. *J. Bacteriol.* 173:2187–95
26. Chiaramello, A. E., Zyskind, J. W. 1989. Expression of *dnaA* and *mioC* as a function of growth rate. *J. Bacteriol.* 171:4272–80
27. Chiaramello, A. E., Zyskind, J. W. 1990. Coupling of DNA replication to

growth rate in *Escherichia coli:* a possible role for guanosine tetraphosphate. *J. Bacteriol.* 172:2013–19
28. Chon, Y., Gayda, R. 1988. Studies with FtsA-LacZ protein fusions reveal FtsA located at inner-outer membrane junctions. *Biochem. Biophys. Res. Commun.* 152:1023–30
29. Clerget, M. 1991. Site-specific recombination promoted by a short DNA segment of plasmid R1 and by a homologous segment in the terminus region of the *Escherichia coli* chromosome. *New Biol.* 3:780–88
30. Cook, W. R., Kepes, F., Joseleau-Petit, D., MacAlister, T. J., Rothfield, L. I. 1987. A proposed mechanism for the generation and localization of new division sites during the division cycle of *Escherichia coli. Proc. Natl. Acad. Sci. USA* 84:7144–48
31. Cook, W. R., Rothfield, L. I. 1991. Biogenesis of cell division sites in *ftsA* and *ftsZ* filaments. *Res. Microbiol.* 142:321–24
32. Crooke, E., Castuma, C. E., Kornberg, A. 1992. The chromosome origin of *Escherichia coli* stabilizes *dnaA* protein during rejuvenation by phospholipids. *J. Biol. Chem.* 267:16779–82
33. Dai, K., Lutkenhaus, J. 1992. The proper ratio of FtsZ to FtsA is required for cell division to occur in *Escherichia coli. J. Bacteriol.* 174:6145–51
34. D'Ari, R., Huisman, O. 1983. Novel mechanism of cell division inhibition assciated with the SOS response in *Escherichia coli. J. Bacteriol.* 156:243–50
35. de Boer, P. A. J., Cook, W. R., Rothfield, L. I. 1990. Bacterial cell division. *Annu. Rev. Genet.* 24:249–74
36. de Boer, P. A. J., Crossley, R. E., Hand, R. E., Rothfield, L. I. 1991. The MinD protein is a membrane ATPase required for the correct placement of the *Escherichia coli* division site. *EMBO J.* 10:4371–80
37. de Boer, P. A. J., Crossley, R. E., Rothfield, L. I. 1988. Isolation and characterization of *minB,* a complex genetic locus involved in correct placement of the division site in *Escherichia coli. J. Bacteriol.* 150:2106–12
38. de Boer, P. A. J., Crossley, R. E., Rothfield, L. I. 1989. A division inhibitor and a topological specificity factor coded for by the minicell locus determine the proper placement of the division site in Escherichia coli. *Cell* 56:641–49
39. de Boer, P. A. J., Crossley, R. E., Rothfield, L. I. 1990. Central role of the *Escherichia coli minC* gene product in two different division-inhibition systems. *Proc. Natl. Acad. Sci. USA* 87:1129–33
40. de Boer, P. A. J., Crossley, R. E., Rothfield, L. I. 1992. Roles of MinC and MinD in the site-specific septation block mediated by the MinCDE system of *Escherichia coli. J. Bacteriol.* 174: 63–70
41. de Boer, P. A. J., Crossley, R. E., Rothfield, L. I. 1992. The essential bacterial cell division protein FtsZ is a GTPase. *Nature* 359:254–56
42. Dewar, S. J., Begg, K. J., Donachie, W. D. 1992. Inhibition of cell division by an imbalance in the ratio of FtsA to FtsZ. *J. Bacteriol.* 174:6314–16
43. Dewar, S. J., Donachie, W. D. 1990. Regulation of expression of the *ftsA* cell division gene by sequences in upstream genes. *J. Bacteriol.* 172: 6611–14
44. Dewar, S. J., Kagan-Zur, V., Begg, K. J., Donachie, W. D. 1989. Transcriptional regulation of cell division genes in *Escherichia coli. Mol. Microbiol.* 3:1371–77
45. Donachie, W. D. 1968. Relationship between cell size and time of initiation of DNA replication. *Nature* 219:1077–79
46. Donachie, W. D. 1969. Control of cell division in *Escherichia coli:* experiments with thymine starvation. *J. Bacteriol.* 100:260–68
47. Donachie, W. D. 1992. What is the minimum number of dedicated functions required for a basic cell cycle? *Curr. Opin. Genet. Dev.* 2:792–98
48. Donachie, W. D., Begg, K. J. 1970. Growth of the bacterial cell. *Nature* 227:1220–24
49. Donachie, W. D., Begg, K. J. 1989. Cell length, nucleoid separation and cell division: observations on rod-shaped and spherical cells of *Escherichia coli. J. Bacteriol.* 171:4633–39
50. Donachie, W. D., Begg, K. J. 1989. Chromosome partition in *Escherichia coli* requires post-replication protein synthesis. *J. Bacteriol.* 171:5405–9
51. Donachie, W. D., Begg, K. J. 1990. Genes and the replication cycle of *Escherichia coli. Res. Microbiol.* 141: 64–75
52. Donachie, W. D., Begg, K. J., Sullivan, N. F. 1984. The morphogenes of *E. coli.* In *Microbial Development,* ed. R. Losick, L. Shapiro, pp. 27–62. New York: Cold Spring Harbor
53. Donachie, W. D., Begg, K. J., Vicente,

M. 1976. Cell length, cell growth and cell division. *Nature* 264:328–33
54. Donachie, W. D., Hobbs, D. G., Masters, M. 1968. Chromosome replication and cell division in *Escherichia coli* 15T⁻ after growth in the absence of DNA synthesis. *Nature* 219:1079–80
55. Donachie, W. D., Sullivan, N. F., Kenan, D. J., Derbyshire, S. V., Begg, K. J., Kagan-Zur, V. 1983. Genes and cell division in *E. coli*. In *Progress in Cell Cycle Controls,* ed, J. Chaloupka, A. Kotyk, E. Streiblova, pp. 28–33. Prague: Czechoslovak Acad. Sci.
56. Ezaki, B., Ogura, T., Niki, H., Hiraga, S. 1991. Partitioning of a mini-F plasmid into anucleate cells of the *mukB* null mutant. *J. Bacteriol.* 173:6643–46
57. Fairweather, N. F., Orr, E., Holland, I. B. 1980. Inhibition of DNA gyrase: effects on nucleic acid synthesis and cell division in *E. coli* K-12. *J. Bacteriol.* 142:153–61
58. Fayet, O., Louarn, J.-M., Georgopoulos, C. 1986. Suppression of the *dnaA46* mutation by amplification of the *groES* and *groEL* genes. *Mol. Gen. Genet.* 202:435–45
59. Deleted in proof
60. Foley, M., Brass, J., Birmingham, J., Cook, W., Garland, P., et al. 1989. Compartmentalization of the periplasm at cell division sites in *E. coli* as shown by fluorescence photobleaching experiments. *Mol. Microbiol.* 3:1329–36
61. Georgopolous, C. P., Eisen, H. 1974. Bacterial mutants which block phage assembly. *J. Supramol. Struct.* 2:349–59
62. Gill, D. R., Hatfull, G. F., Salmond, G. P. C. 1986. A new cell division operon in *Escherichia coli. Mol. Gen. Genet.* 205:134–45
63. Gill, D. R., Salmond, G. P. C. 1987. The *E. coli* cell division proteins, FtsY, FtsE and FtsX are inner membrane associated. *Mol. Gen. Genet.* 210:504–8
64. Grossman, N., Ron, E. Z. 1989. Apparent minimal size required for cell division in *Escherichia coli*. *J. Bacteriol.* 171:80–82
65. Grover, N. B., Woldringh, C. L., Zaritsky, A., Rosenberger, R. 1977. Elongation of rod shaped bacteria. *J. Theor. Biol.* 54:243-
66. Hansen, F. G., Atlung, T., Braun, R. E., Wright, A., Hughes, P., Kohiyama, M. 1991. Initiator (DnaA) protein concentration as a function of growth rate in *Escherichia coli* and *Salmonella typhimurium*. *J. Bacteriol.* 173:5194–99
67. Hansen, F. G., Koefoed, S., Sorensen, L., Atlung, T. 1987. Titration of DnaA protein by *oriC* DnaA-boxes increases *dnaA* gene expression in *Escherichia coli. EMBO J.* 6:255–58
68. Hartwell, L. H. 1974. *Saccharomyces cerevisiae* cell cycle. *Bacteriol. Rev.* 38:164–98
69. Helmstetter, C. E., Leonard, A. C. 1987. Coordinate initiation of chromosome and minichromosome replication in *Escherichia coli*. *J. Bacteriol.* 169: 3489–94
70. Higgins, C. F., Hiles, I. D., Salmond, G. P. C., Gill, D. R., Downie, J. A., et al. 1986. A family of related ATP-binding subunits coupled to many distinct biological processes in bacteria. *Nature* 323:448–50
71. Hiraga, S. 1990. Partitioning of nucleoids. *Res. Microbiol.* 141:50–57
72. Hiraga, S. 1992. Chromosome and plasmid partition in *Escherichia coli*. *Annu. Rev. Biochem.* 61:283–306
73. Hiraga, S., Jaffé, A., Ogura, T., Mori, H., Takahashi, H. 1986. F plasmid *ccd* mechanism in *Escherichia coli*. *J. Bacteriol.* 166:100–4
74. Hiraga, S., Niki, H., Ogura, T., Ichinose, C., Mori, H., et al. 1989. Chromosome partitioning in *Escherichia coli:* novel mutants producing anucleate cells. *J. Bacteriol.* 171:1496–1505
75. Hiraga, S., Ogura, T., Niki, H., Ichinose, C., Mori, H. 1990. Positioning of replicated chromosomes in *Escherichia coli*. *J. Bacteriol.* 172:31–39
76. Hirota, Y., Jacob, F., Ryter, A., Buttin, G., Nakai, T. 1968. On the process of cellular division in *Escherichia coli*. *J. Mol. Biol.* 35:175–92
77. Hirota, Y., Ryter, A., Jacob, F. 1968. Thermosensitive mutants of *Escherichia coli* affected in the processes of DNA synthesis and cellular division. *Cold Spring Harbor Symp. Quant. Biol.* 33: 677–93
78. Hughes, P., Landoulsi, A., Kohiyama, M. 1988. A novel role for cAMP in the control of the activity of the E. coli replication initiatior protein, DnaA. *Cell* 55:343–50
79. Hussain, K., Begg, K. J., Salmond, G. P. C., Donachie, W. D. 1987. ParD: a new gene coding for a protein required for chromosome partitioning and septum localization in *Escherichia coli*. *Mol. Microbiol.* 1:73–81
80. Hussain, K., Elliott, E. J., Salmond, G. P. C. 1987. The *parD* mutant of

Escherichia coli also carries a *gyrA.am* mutation. The complete sequence of *gyrA*. *Mol. Microbiol.* 1:259–73

81. Ishino, F., Matsuhashi, M. 1981. Peptidoglycan synthetic enzyme activities of highly purified penicillin-binding protein 3 in *E. coli:* a septum forming reaction sequence. *Biochem. Biophys. Res. Commun.* 101:905–11

81a. Ishino, F., Park, W., Tomioka, S., Tamaki, S., Takase, I., et al. 1986. Peptidoglycan synthetic activities in membranes in *Escherichia coli* caused by overproduction of penicillin-binding protein 2 and RodA protein. *J. Biol. Chem.* 261:7024–31

81b. Jaffé, A., D'Ari, R., Hiraga, S. 1988. Minicell-forming mutants of *Escherichia coli:* production of minicells and anucleate rods. *J. Bacteriol.* 170: 3094–3101

82. Jaffé, A., Ogura, T., Hiraga, S. 1985. Effects of the *ccd* function of the F plasmid on bacterial growth. *J. Bacteriol.* 163:841–49

83. Jarosik, G. P., Oliver, D. B. 1991. Isolation and analysis of dominant *secA* mutations in *Escherichia coli. J. Bacteriol.* 173:860–68

84. Jenkins, A. J., March, J. B., Oliver, I. R., Masters, M. 1986. A DNA fragment containing the *groE* genes can suppress mutations in the *Escherichia coli dnaA* gene. *Mol. Gen. Genet.* 202:446–54

85. Jones, N. C., Donachie, W. D. 1973. Chromosome replication, transcription and the control of cell division in *Escherichia coli. Nature New Biol.* 243:100–3

86. Kato, J., Nishimura, Y., Imamura, R., Niki, H., Hiraga, S., Suzuki, H. 1990. New topoisomerase essential for chromosome segregation in E. coli. *Cell* 63:393–404

87. Kato, J., Nishimura, Y., Suzuki, H. 1989. *Escherichia coli parA* is an allele of the *gyrB* gene. *Mol. Gen. Genet.* 217:178–81

88. Kellenberger, E. 1988. The bacterial chromatin. In *Chromosomes: Eukaryotic, Prokaryotic and Viral,* ed. K. W. Adolph, pp. 1–18. Boca Raton, FL: CRC Interscience

89. Kücherer, C. H., Lother, H., Kölling, R., Schauzu, A., Messer, W. 1986. Regulation of transcription of the chromosomal *dnaA* gene of *Escherichia coli. Mol. Gen. Genet.* 205:115–21

90. Kuempel, P. L., Henson, J. M., Dircks, L., Tecklenburg, M., Lim, G. F. 1991. *dif,* a *recA*-independent recombination site in the terminus region of the chromosome of *Escherichia coli. New Biol.* 3:799–811

91. Labie, C., Bouché, F., Bouché, J.-P. 1989. Isolation and mapping of *Escherichia coli* mutations conferring resistance to division inhibition protein DicB. *J. Bacteriol.* 171:4315–19

92. Løbner-Olesen, A., Skarstad, K., Hansen, F. G., von Meyenburg, K., Boye, E. 1989. The DnaA protein determines the initiation mass of Escherichia coli K-12. *Cell* 57:881–89

93. Lutkenhaus, J. F. 1983. Coupling of DNA replication and cell division: *sulB* is an allele of *ftsZ*. *J. Bacteriol.* 154: 1339–46

94. Lutkenhaus, J., Wu, H. C. 1980. Determination of transcriptional units and gene products from the *ftsA* region of *E. coli. J. Bacteriol.* 143:1281–88

95. Maguin, E., Brody, H., Hill, C. W., D'Ari, R. 1986. SOS-associated division inhibition gene *sfiC* is part of excisable element e14 in *Escherichia coli. J. Bacteriol.* 168:464–66

96. Maguin, E., Lutkenhaus, J. F., D'Ari, R. 1986. Reversibility of SOS-associated division inhibition in *E. coli. J. Bacteriol.* 166:733–38

97. Markiewicz, Z., Broome-Smith, J., Schwarz, U., Spratt, B. G. 1982. Spherical *E. coli* due to elevated levels of D-alanine carboxypeptidase. *Nature* 297:702–4

98. Masters, M. 1989. The *Escherichia coli* chromosome and its replication. *Curr. Opin. Cell Biol.* 1:241–49

99. Deleted in proof

100. Matsuhashi, M., Wachi, M., Ishino, F. 1990. Machinery for cell growth and division: penicillin-binding proteins and other proteins. *Res. Microbiol.* 141:89–103

101. McAlister, T. J., MacDonald, B., Rothfield, L. I. 1983. The periseptal annulus: an organelle associated with cell division in gram negative bacteria. *Proc. Natl. Acad. Sci. USA* 80:1372–76

102. Mengin-Lecreulx, D., Parquet, C., Desviat, L. R., Pla, J., Flouret, B., et al. 1989. Organization of the *murE-murG* region of *Escherichia coli:* identification of the *murD* gene coding for the D-glutamic-acid-adding enzyme. *J. Bacteriol.* 171:6126–34

103. Messer, W., Bellekes, V., Lother, H. 1985. Effect of *dam* methylation on the activity of the *E. coli* replication origin, *oriC*. *EMBO J.* 4:1327–32

104. Mitchison, J. M. 1971. *The Biology of the Cell Cycle.* Cambridge: Cambridge Univ. Press

105. Mukherjee, A., Dai, K., Lutkenhaus,

J. 1993. *E. coli* cell division protein FtsZ is a guanine nucleotide binding protein. *Proc. Natl. Acad. Sci. USA*. 90:1053–57

106. Mukherjee, A., Donachie, W. D. 1990. Differential translation of cell division proteins. *J. Bacteriol.* 172:6106–11

107. Mulder, E., Woldringh, C. L. 1989. Actively replicating nucleoids influence the positioning of division sites in DNA-less cell forming filaments of *Escherichia coli*. *J. Bacteriol.* 171: 4303–14

108. Mulder, E., Woldringh, C. L., Tétart, F., Bouché, J.-P. 1992. New *minC* mutations suggest different interactions of the same region of division inhibitor MinC with proteins specific for *minD* and *dicB* coinhibition pathways. *J. Bacteriol.* 174:35–39

109. Nagata, T., Meselson, M. 1968. Periodic replication of DNA in steady growing *Escherichia coli:* the localized origin of replication. *Cold Spring Harbor Symp. Quant. Biol.* 33:553–57

109a. Neidhardt, F. C., Ingraham, J. L., Magasanik, B., Low, K. B., Schaechter, M., Umbarger, H. E., eds. 1987. Escherichia coli *and* Salmonella typhimurium: *Cellular and Molecular Biology*. Washington, DC:Am. Soc. Microbiol.

110. Nikaido, H., Vaara, M. 1987. Outer membrane. See Ref. 109a, pp. 7–22

111. Niki, H., Jaffé, A., Imamura, R., Ogura, T., Hiraga, S. 1991. The new gene *mukB* codes for a 177 kDa protein with coiled-coil domains involved in chromosome partitioning in *E. coli*. *EMBO J.* 10:183–93

112. Normark, S. 1970. Genetics of a chain-forming mutant of *E. coli*. Transduction and dominance of the *envA* gene mediating increased penetration to some antibacterial agents. *Genet. Res.* 16:63–78

113. Norris, V., Alliotte, T., Jaffé, A., D'Ari, R. 1986. DNA replication termination in *Escherichia coli parB* (a *dnaG* allele), *parA*, and *gyrB* mutants affected in DNA distribution. *J. Bacteriol.* 168:494–504

114. Ogden, G. B., Pratt, M. J., Schaechter, M. 1988. The replicative origin of the *E. coli* chromosome binds to cell membranes only when hemimethylated. *Cell* 54:127–35

115. Oliver, D. B., Beckwith, J. 1981. *E. coli* mutant pleiotropically defective in the export of secreted proteins. *Cell* 25:765–72

116. Oliver, D. B., Beckwith, J. 1982. Identification of a new gene (*secA*) and gene product involved in the secretion of envelope proteins in *E. coli*. *J. Bacteriol.* 150:686–91

117. Oliver, D. B., Cabelli, R. J., Dolan, K. M., Jarosik, G. P. 1990. Azide-resistant mutants of *Escherichia coli* alter the SecA protein, an azide-sensitive component of the secretion machinery. *Proc. Natl. Acad. Sci. USA* 87:8227–31

118. Onoda, T., Oshima, A., Nakano, S., Matsuno, A. 1987. Morphology, growth and reversion in a stable L-form of *Escherichia coli* K-12. *J. Gen. Microbiol.* 133:527–34

119. Park, J. T. 1987. Murein synthesis. See Ref. 109a, pp. 663–71

120. Pierucci, O., Rickert, M., Helmstetter, C. E. 1989. DnaA protein overproduction abolishes cell cycle specificity of DNA replication from *oriC* in *Escherichia coli*. *J. Bacteriol.* 171:3760–66

121. Polaczek, P., Wright, A. 1990. Regulation of expression of the *dnaA* gene in *Escherichia coli:* role of the two promoters and the DnaA box. *New Biol.* 2:574–82

122. RayChaudhuri, D., Park, J. T. 1992. *Escherichia coli* cell division gene *ftsZ* encodes a novel GTP-binding protein. *Nature* 359:251–54

123. Ricard, M., Hirota, Y. 1973. Process of cellular division in *Escherichia coli:* physiological study on thermosensitive mutants defective in cell division. *J. Bacteriol.* 116:314–22

124. Robin, A., Joseleau-Petit, D., D'Ari, R. 1990. Transcription of the *ftsZ* gene and cell division in *Escherichia coli*. *J. Bacteriol.* 172:1392–99

125. Robinson, A. C., Kenan, D. J., Hatfull, G. F., Sullivan, N. F., Spiegelberg, R., Donachie, W. D. 1984. DNA sequence and transcriptional organization of essential cell division genes *ftsQ* and *ftsA* of *Escherichia coli:* evidence for overlapping transcriptional units. *J. Bacteriol.* 160:546–55

126. Robinson, A. C., Kenan, D. J., Sweeney, J., Donachie, W. D. 1986. Further evidence for overlapping transcriptional units in an *Escherichia coli* cell envelope-cell division gene cluster: DNA sequence and transcriptional organization of the *ddl ftsQ* region. *J. Bacteriol.* 167:809–17

127. Rothfield, L. I., de Boer, P. A. J., Cook, W. R. 1990. Localization of septation sites. *Res. Microbiol.* 141:57–63

128. Ruberti, I., Crescenzi, F., Paolozzi, L., Ghelardini, P. 1992. A class of

gyrB mutants, substantially unaffected in DNA topology, suppress the *ftsZ.84* mutation. *Mol. Microbiol.* 5:1065–72
129. Russell, D. W., Zinder, N. D. 1987. Hemimethylation prevents DNA replication in E. coli. *Cell* 50:1071–79
130. Schmidt, L. S., Botta, G., Park, J. T. 1981. Effects of furazlocillin, a β-lactam antibiotic which binds selectively to penicillin-binding protein 3, in *Escherichia coli* mutants deficient in other penicillin-binding proteins. *J. Bacteriol.* 145:623–37
131. Schwarz, U., Asmus, A., Frank, H. 1969. Autolytic enzymes and cell division of *Escherichia coli. J. Mol. Biol.* 41:419–29
132. Sekimizu, K., Bramhill, D., Kornberg, A. 1987. ATP activates dnaA protein in initiating replication of plasmids bearing the origin of the E. coli chromosome. *Cell* 50:259–65
133. Sekimizu, K., Bramhill, D., Kornberg, A. 1988. Sequential early stages in the in vitro initiation of replication at the origin of the *Escherichia coli* chromosome. *J. Biol. Chem.* 263:7124–30
134. Slater, M., Schaechter, M. 1974. Control of cell division in bacteria. *Bacteriol. Rev.* 38:199–221
135. Deleted in proof
136. Spratt, B. G. 1975. Distinct penicillin-binding proteins involved in the division, elongation and shape of *Escherichia coli* K-12. *Proc. Natl. Acad. Sci. USA* 72:2999–3003
137. Spratt, B. G. 1977. Temperature-sensitive cell division mutants of *Escherichia coli* with thermolabile penicillin-binding proteins. *J. Bacteriol.* 131:293–305
138. Steck, T. R., Drlica, K. 1985. Bacterial chromosome segregation: evidence for DNA gyrase involvement in decatenation. *Cell* 36:1081–88
139. Stewart, P. S., D'Ari, R. 1992. Genetic and morphological characterization of an *Escherichia coli* chromosome segregation mutant. *J. Bacteriol.* 174: 4513–16
140. Sullivan, N. F., Donachie, W. D. 1984. Transcriptional organization within an *Escherichia coli* cell division gene cluster: direction of transcription of the cell separation gene *envA*. *J. Bacteriol.* 160:724–32
141. Teather, R. M., Collins, J. F., Donachie, W. D. 1974. Quantal behaviour of a diffusible factor which initiates septum formation at potential division sites in *E. coli*. *J. Bacteriol.* 118:407–13
142. Tétart, F., Bouché, J.-P. 1992. Regulation of the expression of the cell cycle gene *ftsZ* by DicF antisense RNA. Division does not require a fixed number of FtsZ molecules. *Mol. Microbiol.* 6:615–20
143. Tilly, K., McKittrick, N., Zylicz, M., Georgopoulos, C. 1983. The dnaK protein modulates the heat-shock response of Escherichia coli. *Cell* 34:641–46
144. Tomioka, S., Nikaido, T., Miyakawa, T., Matsuhashi, M. 1983. Mutation of the N-acetylmuramyl-L-alanine amidase gene of *Escherichia coli* K-12. *J. Bacteriol.* 156:463–65
145. Tormo, A., Ayala, J. A., de Pedro, M. A., Vicente, M. 1986. Interaction of FtsA and PBP3 proteins in the *Escherichia coli* septum. *J. Bacteriol.* 166:985–92
146. Tsuchido, T., Van Bogelen, R. A., Neidhardt, F. C. 1986. Heat shock response in *Escherichia coli* influences cell division. *Proc. Natl. Acad. Sci. USA* 83:6959–63
147. Wachi, M., Matsuhashi, M. 1989. Negative control of cell division by *mreB*, a gene that functions in determining the rod shape of *Escherichia coli* cells. *J. Bacteriol.* 171:3123–27
148. Wada, M., Itikawa, H. 1984. Participation of *Escherichia coli* K-12 *groE* gene products in the synthesis of cellular DNA and RNA. *J. Bacteriol.* 157:694–96
149. Walker, J. R., Kovarik, A., Allan, J. S., Gustafson, R. A. 1975. Regulation of bacterial cell division: temperature-sensitive mutants of *Escherichia coli* that are defective in septum formation. *J. Bacteriol.* 123:693–703
150. Wang, H., Gayda, R. C. 1992. Quantitative determination of FtsA at different growth rates in *Escherichia coli* using monoclonal antibodies. *Mol. Microbiol.* 6:2517–24
151. Wang, X., de Boer, P. A. J., Rothfield, L. I. 1991. A factor that positively regulates cell division by activating transcription of the major cluster of essential cell division genes of *Escherichia coli*. *EMBO J.* 10:3363–72
152. Ward, J. E. Jr., Lutkenhaus, J. 1985. Overproduction of FtsZ induces minicell formation in Escherichia coli. *Cell* 42:941–49
153. Woldringh, C. L., Valkenburg, J. A. C., Pas, E., Taschner, P. E. M., Huls, P., Wientjes, F. B. 1985. Physiological and geometrical conditions for cell division in *Escherichia coli*. *Ann. Inst. Pasteur Microbiol.* 136A:131–38

154. Wolf-Watz, H., Normark, S. 1976. Evidence for a role of N-acetylmuramyl-L-alanine amidase in septum separation in *E. coli*. *J. Bacteriol.* 128:580–86
155. Xu, Y. C., Bremer, H. 1988. Chromosome replication in *Escherichia coli* induced by oversupply of DnaA. *Mol. Gen. Genet.* 211:138–42
156. Yem, D. W., Wu, H. C. 1978. Physiological characterization of an *Escherichia coli* mutant altered in the structure of murein lipoprotein. *J. Bacteriol.* 133:1419–26
157. Yi, Q.-M., Rockenbach, S., Ward, J. E. Jr., Lutkenhaus, J. 1985. Structure and expression of the cell division genes *ftsQ, ftsA,* and *ftsZ*. *J. Mol. Biol.* 184:399–412
158. Zylicz, M., LeBowitz, J. H., McMacken, R., Georgopoulos, C. P. 1983. The *dnaK* protein of *Escherichia coli* possesses an ATPase and autophosphorylating activity and is essential in an in vitro DNA replication system. *Proc. Natl. Acad. Sci. USA* 80:6431–35

Annu. Rev. Microbiol. 1993. 47:231–61

CHROMOSOME SEGREGATION IN YEAST

Barbara D. Page and Michael Snyder
Department of Biology, Yale University, New Haven, Connecticut 06511

KEY WORDS: *Saccharomyces cerevisiae, Schizosaccharomyces pombe,* microtubules, microtubule motors, spindle pole bodies, centromeres

---

CONTENTS

ABSTRACT

Because of their genetic tractability, much has been learned concerning the mechanisms of chromosome segregation in budding yeast, *Saccharomyces cerevisiae,* and fission yeast, *Schizosaccharomyces pombe.* This chapter reviews the cytology and molecular and cell biology of mitosis in both of these yeasts. Current knowledge about the components of the mitotic spindle apparatus, including spindle pole bodies, centromeres, and microtubule

0066-4227/93/1001-0231$02.00

components and motors, is summarized. Mechanisms of mitosis such as establishment and positioning of the mitotic spindle apparatus, anaphase A, and anaphase B are reviewed.

INTRODUCTION

During mitosis, replicated chromosomes are accurately segregated into two daughter cells. Chromosome segregation is mediated by the mitotic spindle apparatus, a complex structure containing spindle poles, microtubules, and chromosomes. The spindle apparatus is organized by microtubule organizing centers (MTOCs), which nucleate assembly of two types of spindle microtubules. Kinetochore (or chromosomal) microtubules link the spindle pole to the chromosome, and nonkinetochore microtubules participate in a variety of functions, such as establishment and elongation of the spindle apparatus and positioning the spindle within the cell. Although the morphology of the spindle apparatus varies among the eukaryotic taxa, its basic components are well conserved between such evolutionarily diverse cells as human and yeast.

Mitosis involves a complex series of sequential events. Microtubule organizing centers first duplicate and then separate and establish the bipolar spindle apparatus. During prometaphase, chromosomes congress at the metaphase plate. At anaphase A, sister chromatids disjoin, and the chromosomes move to opposite poles. During anaphase B, the poles separate. Finally, during telophase, the spindle apparatus disassembles and cytokinesis ensues. In many organisms, the apparatus must be specifically positioned to ensure that nuclear division occurs along the proper axis. Each of these different steps requires a number of different proteins, and many steps are coordinated with respect to one another or with other events of the cell cycle.

This article reviews chromosome segregation in the budding yeast, *Saccharomyces cerevisiae,* and the fission yeast, *Schizosaccharomyces pombe*. Chromosome segregation is extremely efficient in both these organisms; for *S. cerevisiae* chromosome loss occurs approximately once in every 10^5 cell divisions (45). Because of the ease with which these organisms can be manipulated genetically, numerous genes involved in chromosome segregation have been identified. In addition, the ability to readily construct mutations in cloned genes and to appropriately insert the mutant genes into the genome has led to a detailed functional analysis of many important mitotic components such as tubulin. Consequently, a wealth of information concerning the components and mechanisms of chromosome segregation has accumulated from studies of these organisms.

CYTOLOGY OF YEAST MITOSIS

Morphology of the Spindle Apparatus

The mitotic spindle apparatuses of both *S. cerevisiae* and *S. pombe* are relatively simple and contain few microtubules (<50) (105), as compared with the thousands of microtubules present in a mammalian spindle. Figure 1 shows an electron micrograph of an *S. cerevisiae* spindle. The spindle is enclosed by the nuclear envelope, which in fungi does not break down during mitosis. Microtubules within the apparatus are of two types. Pole-to-pole microtubules extend from one pole to a region very close to the other pole, while discontinuous microtubules terminate a short distance from the pole of origin (105; see below). In addition, microtubules extend from the pole into the cytoplasm.

In both *S. cerevisiae* and *S. pombe,* microtubules are organized by electron-dense structures called spindle pole bodies (SPBs), which are associated with the nuclear envelope (15, 17, 41, 82). In *S. cerevisiae,* SPBs consist of a trilaminar structure containing an inner plaque located next to the nucleoplasm, a central plaque in the plane of the nuclear envelope, and an outer plaque on the cytoplasmic face. Beside the SPB is a half bridge, an electron-dense, two-layered structure that is also embedded in the nuclear

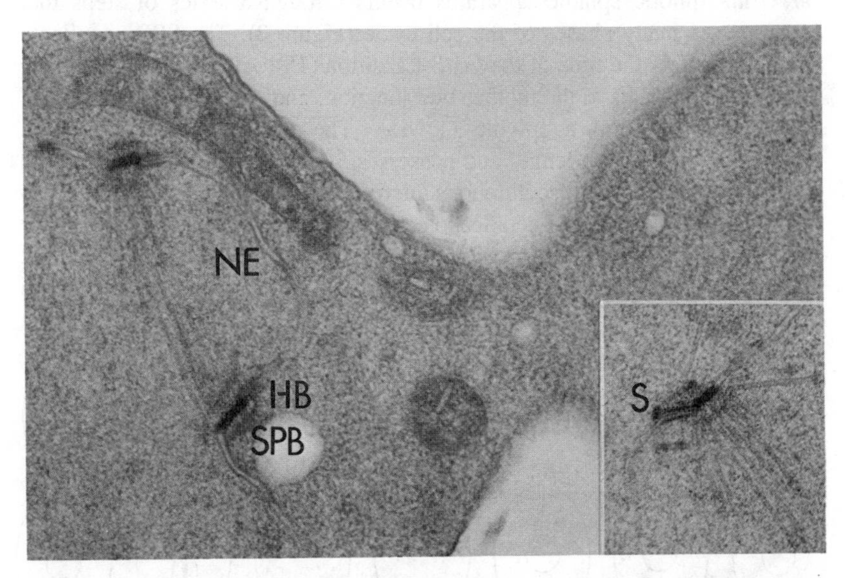

Figure 1 An electron micrograph of a *S. cerevisiae* mitotic spindle apparatus. SPB, spindle pole body; HB, half bridge; NE, nuclear envelope. The inset shows the SPB of an α-factor-treated cell with its associated satellite (S). Photographs were kindly provided by M. Winey.

envelope. Microtubules originate from a region at or adjacent to the inner plaque and extend into the nucleus to form the mitotic spindle apparatus. Microtubules also form at or next to the outer plaque to generate the cytoplasmic array (15). The orientation of microtubules emanating from the SPB is not known for either *S. cerevisiae* or *S. pombe*. The plus ends, the preferred sites of microtubule assembly, are expected to reside distal to the SPB as is the case for every eukaryotic MTOC that has been studied thus far (see below) (31, 141).

In *S. pombe,* the three chromosomes condense during mitosis and become morphologically visible (48). In *S. cerevisiae,* the position of the DNA or chromatin in the spindle can be monitored by light microscopy or electron microscopy techniques, respectively (1, 15, 105). However, the individual chromosomes are not distinguishable during mitosis. Thus, in *S. cerevisiae* some aspects of mitosis such as sister chromatid separation will be difficult to analyze using current cytological approaches.

Cytological Events of the Mitotic Cycle in Saccharomyces cerevisiae

Electron microscopy studies (15, 17) and immunofluorescence experiments with antitubulin antibodies (1) both indicate that the establishment of the *S. cerevisiae* mitotic spindle apparatus occurs through a series of steps that encompasses many phases of the cell cycle (Figure 2). The SPBs duplicate at approximately the time of the G1/S transition (15). After SPB duplication, the nucleus migrates to the mother-bud junction, and the SPBs separate from one another to form a spindle (1, 15). The newly formed spindle is approximately 1 μm in length and possesses 5 to 14 continuous pole-to-pole microtubules and many discontinuous microtubules that are approximately 0.5 μm long (105). A recent detailed electron microscopy analysis, involving serial sections of spindles from *cdc23*$^-$ arrested cells (which arrest at late G2 or early M), suggests that the pole-to-pole microtubules originate at one pole

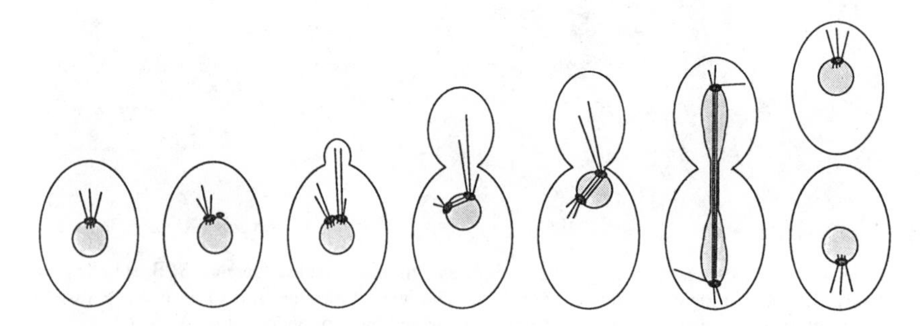

Figure 2 The mitotic cycle of *S. cerevisiae*.

and terminate very close to the other (M. Winey, personal communication). In both haploid and diploid cells, the number of discontinuous microtubules within the spindle approximates the number of chromosomes within the cell, suggesting that each *S. cerevisiae* chromosome is connected to the SPB by a single microtubule (105).

Electron microscopy analysis indicates that during mitosis the discontinuous microtubules shorten and two distinct chromatin masses become apparent (105). These fractions are thought to represent sister genomes. The decreasing length of the discontinuous microtubules is accompanied by increasing distance between the two chromatin fractions, as expected for anaphase A.

When the spindle is first established, the spindle poles are approximately 1 μm apart. Late in mitosis, the spindle rapidly extends to the length of the entire cell (~8–10 μm), and the sister genomes clearly separate (1, 15, 102). This dramatic separation may be the primary mechanism for segregating the sister genomes. At cytokinesis, the spindle breaks down and cell division occurs (1, 15).

Cytological Events of the Mitotic Cycle in Schizosaccharomyces pombe

Mitosis in *S. pombe* resembles that of *S. cerevisiae*. The cytology of the spindle apparatus and order of events, i.e. duplication of the SPB, establishment of the spindle, and separation of the genomes are similar in the two organisms (see Figures 2 and 3).

Despite the similarities, there are some important differences. First, the timing of SPB duplication differs between the two organisms: SPB duplication occurs at the G1/S transition in *S. cerevisiae* (17), but at the G2/M transition in *S. pombe* (41), suggesting that different cell cycle–control steps in the two organisms regulate SPB duplication. Second, in *S. cerevisiae,* the SPB is permanently embedded in the nuclear envelope throughout the cell cycle (15, 17), while in *S. pombe,* it resides adjacent to the envelope throughout much of the cell cycle but is inserted into the nuclear envelope during mitosis (82;

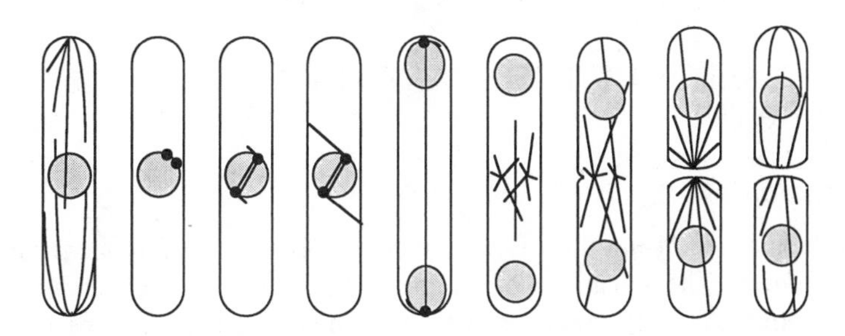

Figure 3 The mitotic cycle of *S. pombe*. Adapted from Hagan & Hyams (41).

J. R. McIntosh, personal communication). Third, *S. pombe,* unlike *S. cerevisiae,* undergoes a dramatic and complex reorganization of its microtubules during its cell cycle (41, 82). The G2/M transition is accompanied by microtubule disassembly, a feature common among eukaryotes (14).

After SPB duplication, spindle assembly occurs. Analysis of serial sections indicates that *S. pombe* discontinuous microtubules shorten during mitosis, suggestive of anaphase A (29a). The spindle then dramatically elongates during anaphase B, extending 14 μm, the length of the entire cell, and the sister genomes separate to the tips of the cell (41, 82). The nuclei then migrate to the centers of their respective incipient daughter cells, and fission occurs (41).

In *S. cerevisiae,* SPBs appear competent to organize microtubules throughout the cell cycle both in vivo and in vitro (15, 17, 19, 62). However in *S. pombe,* SPBs appear to be competent to nucleate microtubules only during mitosis. After anaphase, the SPBs apparently lose their microtubule-organizing capabilities (41, 82). Two new MTOCs appear in the center of the cell where fission will take place and organize the interphase microtubule array. Thus, in *S. pombe* the ability of the SPB to nucleate microtubules is controlled during the cell cycle, perhaps by modification of the SPB. This hypothesis is supported by in vitro microtubule assembly experiments using permeabilized cells (80). SPBs from *S. pombe* interphase cells do not nucleate microtubule assembly, but SPBs from mitotic cells do. If *S. pombe* interphase SPBs are preincubated with mitotic-state, cell-free extracts from *Xenopus* eggs, they become competent to nucleate microtubules in vitro. A likely candidate involved in this regulation of microtubule assembly is the G2 *cdc2* kinase, whose activity correlates with microtubule assembly at the SPB (35).

Based on the descriptions above, the mitotic apparatuses of the yeasts *S. cerevisiae* and *S. pombe* bear significant morphological and behavioral similarity to that of multicellular organisms. Closer examination of spindle components, including microtubules, spindle poles, and chromosomes, reveals even greater similarity between the mitotic apparatuses of diverse eukaryotic organisms.

COMPONENTS OF THE YEAST MITOTIC SPINDLE APPARATUS

Microtubules

Microtubules are 24-nm-thick fibers composed principally of α- and β-tubulin heterodimers. The dynamics of assembly differ for the two ends of the microtubule (52, 152); this asymmetry presumably results from the intrinsic polarity of the α/β heterodimers within the microtubule. The plus end is more dynamic, undergoing rapid growth and shrinkage; assembly and disassembly at the minus end is significantly slower (e.g. 52, 152).

The microtubules of both budding and fission yeast cytologically resemble those of other eukaryotes (19, 82). The amino acid sequences of yeast tubulins are highly homologous (>70%) to those of other taxa (49, 94, 121, 143). This homology likely results in conservation of function; for example, chicken-yeast chimeric β-tubulin readily incorporates into mouse microtubules in vivo with no effect upon cell morphology (11). Functional similarity of nucleation is inferred from the ability of yeast SPBs to nucleate microtubules in vitro using either bovine or porcine brain tubulin (19, 62, 80).

Both the *S. pombe* and *S. cerevisiae* genomes contain two copies of an α-tubulin gene and one essential β-tubulin gene (49, 94, 121, 143). In budding yeast, the gene products of *TUB1* and *TUB3,* the α-tubulin genes, are 90% identical to each other (121). *TUB1* is an essential gene, whereas *TUB3* is not (122). This difference probably reflects the fact that *TUB1* is more highly expressed than *TUB3*. Furthermore, increased expression of *TUB3* can suppress the lethality of a *tub1*Δ mutation, indicating functional homology between these α-tubulins (122).

Proteins that Interact with Microtubules

In multicellular organisms, numerous proteins that interact with microtubules have been identified. These proteins have various important functions such as stabilizing, binding, moving, or bundling microtubules.

MICROTUBULE MOTORS Proteins that move along microtubules include kinesin and dynein. Kinesin is a plus-end directed microtubule motor (148), and dynein is a minus-end directed motor (123). In multicellular organisms, members of the kinesin and dynein families have been implicated in chromosome segregation. Genetic analyses of two kinesin-like genes, *ncd* and *nod,* from *Drosophila melanogaster* have demonstrated that these motor molecules participate in chromosome segregation (83, 161). In mammalian cells, antibodies directed against kinesin homologs react with spindle poles (120) and microtubules (96, 160). Antibodies against cytoplasmic dynein react with the spindle poles, spindle microtubules, and kinetochores of both mammalian and avian cells (106, 137), suggesting that dynein molecules also play an important role during mitosis.

In both *S. pombe* and *S. cerevisiae,* kinesin-like genes have been identified either by isolation of mutants defective in microtubule processes (39, 54, 89) or by using the polymerase chain reaction (114, 115). The *KAR3* and *CIN8* genes from *S. cerevisiae* (54) were identified through characterization of mutants defective in karyogamy (nuclear fusion during mating) and chromosome segregation, respectively. The *cut7* gene from *S. pombe* was found through analysis of a mutant defective in mitosis (39). The *KIP1* and *KIP2* genes were isolated based on their DNA sequence similarity to other kinesin-like genes (114); yeast strains deleted for one or both of these genes have no detectable phenotype.

Sequence analysis indicates that the cut7, CIN8, and KIP1 proteins contain their motor-like domain at their amino terminus, similar to that of kinesin. The cut7 protein localizes to spindle microtubules, but it appears to be more concentrated at the SPBs (40). The *KIP1* and *CIN8* gene products both localize to spindle microtubules and are likely to participate in redundant functions. *cin8Δ* cells are defective in mitosis, but viable (54). As noted above, deletion of *KIP1* has no detectable effect upon an otherwise wild-type cell (114). An extra copy of *KIP1* suppresses $Cin8^-$ phenotypes (54), and *kip1Δ cin8Δ* cells are inviable (114).

The KAR3 kinesin-like protein contains its potential motor domain at the carboxy terminus (39, 89), a feature similar to ncd, a minus-end motor (83). In cells treated with mating pheromone (α-factor), a kar3::β-galactosidase (β-gal) fusion protein localizes to cytoplasmic microtubules and concentrates at the SPB and the microtubule tips (89). The KAR3-1 mutant protein, which is thought to cause rigor binding to microtubules, is also found along cytoplasmic microtubules (89). Also, kar3::β-gal fusion proteins localize to the SPB in vegetative cells (B. Page & M. Snyder, unpublished data). The localization of the KAR3 fusion protein and the cut7 protein indicates that the microtubule motors may play an important role at the SPB.

The cytological localization and sequence homology of motor proteins, however, must be cautiously interpreted. It is conceivable that motor proteins bind along the length of the microtubule, but then accumulate at their destination. Thus, localization of motor proteins might not represent their true site of action. Furthermore, true motor capabilities of these proteins must be proven and the directionality determined. These data will be crucial for understanding the precise role of the proteins in vivo.

In summary, at least four kinesin-like proteins have been identified in *S. cerevisiae* and one in *S. pombe*. Given the plethora of motors found in *D. melanogaster* (30), one expects to find many other motor proteins in yeast. Consistent with this hypothesis, antibodies against a conserved portion of kinesin recognize at least six polypetides by immunoblot analysis of *S. cerevisiae* proteins (114).

OTHER MICROTUBULE-INTERACTING PROTEINS Other proteins associated with microtubules have been discovered in *S. cerevisiae*. *SPO15* encodes a dynamin-related protein important for both vacuolar-protein sorting and spindle assembly (159). In vertebrate cells, dynamin bundles microtubules (128).

The BIK1 protein localizes to microtubules and contains a domain with limited sequence similarity to that of tau, a mammalian microtubule-binding protein (7, 50). The microtubule array of *bik1*$^-$ cells is reduced, and these cells are defective in both anaphase B and nuclear migration. As a result of these defects, a *bik1*$^-$ population has an increased percentage of binucleate cells. Disruptions of this gene also greatly reduce the efficiency of nuclear fusion during conjugation; in fact, this phenotype led to the identification of the *BIK1* gene (7, 145).

Finally, several proteins that associate with microtubules in vitro and in vivo have been identified by means of microtubule affinity columns (4). The discovery of yeast genes that encode microtubule motors and microtubule-binding proteins opens the investigation of these proteins, characterized traditionally with biochemistry, to genetic analysis.

Spindle Pole Bodies

γ-TUBULIN SPBs, as all MTOCs, can nucleate assembly of microtubules both in vivo and in vitro (14, 17, 80). Genetic and molecular analyses of fungal mitotic mutants have led to the identification of genes whose products compose this organelle or participate in its duplication. To date, one of the most exciting discoveries in this area has been γ-tubulin, a distantly related homolog of α- and β-tubulin. γ-Tubulin was first discovered in *Aspergillus nidulans* through the characterization of an extragenic suppressor of a β-tubulin mutation (100, 153). γ-Tubulin has since been discovered at the microtubule organizing center in organisms as diverse as humans, fruit flies, and *S. pombe* (53, 135, 162). This tubulin has not yet been found in *S. cerevisiae*. In *A. nidulans,* genetic disruption analysis indicates that this protein is required for microtubule assembly by the SPB (98), and in mammalian cells, microinjection of anti-γ-tubulin antibodies disrupts microtubule assembly (69). Thus, γ-tubulin is important for this process in a variety of organisms. As expected, the γ-tubulin gene is essential for cell growth in *S. pombe* (53).

In a proposed model for γ-tubulin function (99), γ-tubulin resides at the SPB and nucleates assembly of the 13-protofilament array that comprises a microtubule fiber. As suggested by the genetic data (153), γ-tubulin might directly interact with β-tubulin of the α/β heterodimer and thereby establish the asymmetry of the microtubule (99). γ-Tubulin itself might organize a 13-protofilament microtubule, or more likely, another component(s) organizes a 13-membered γ-tubulin nucleator.

Interestingly, the presence of γ-tubulin does not ensure microtubule assembly. In *S. pombe,* γ-tubulin is present at SPBs during interphase when they are incompetent for microtubule nucleation (80). Thus, although γ-tubulin is probably necessary for microtubule nucleation, this protein may require modifications or accessory proteins to be active.

OTHER SPB COMPONENTS In budding yeast, components of the SPB have been identified through isolation of SPBs and through various genetic screens. An immunological approach involving the large-scale isolation of SPBs and the preparation of monoclonal antibodies has provided information about their molecular composition. Four spindle components, three of which appear to be part of the SPB, have been identified (Figure 4) (117, 118). A 90-kDa protein component localizes to both the outer and inner plaques, which are associated with the cytoplasmic and nuclear microtubules, respectively. These two layers may contain the microtubule-nucleating activity of the SPB. A 42-kDa component resides within the central plaque, and a 110-kDa component localizes to a fibrous layer between the central and inner plaque. Another spindle component, an 80-kDa protein, localizes to nuclear microtubules near the inner plaque (117, 118).

The genes encoding the 42-kDa and 110-kDa polypeptides have been cloned (J. Kilmartin, personal communication), and both are essential. Sequence analysis predicts that the 42-kDa protein is unique. The 110-kDa polypeptide is predicted to contain a long coiled-coil domain and might comprise the short fibers observed in that region (J. Kilmartin, personal communication).

Another protein discovered by an immunological approach is the SPA1 protein (130). This protein is important for microtubule-associated processes, and indirect immunofluorescence experiments indicate that SPA1 localizes to the SPB and cytoplasmic sites in the cell (M. Snyder, unpublished results). Electron microscopic localization of SPA1 has not yet been performed.

Isolation of mutants defective in various SPB-mediated processes has led to the identification of genes that encode other components of this organelle. *CIK1* was found using a screen for mutants defective in both karyogamy and

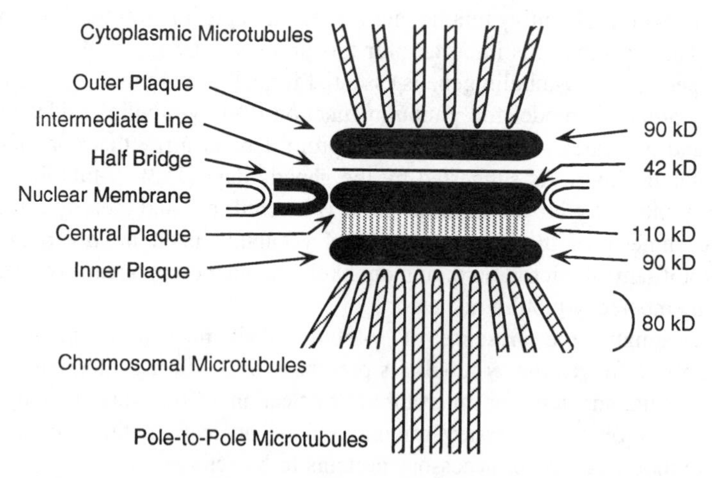

Figure 4 Diagram of the SPB of *S. cerevisiae*. Adapted from Rout & Kilmartin (117).

chromosome segregation (101). cik1::β-gal fusions localize to the SPB region at all stages of the vegetative cycle. Thus far, localization of the authentic CIK1 protein in vegetative cells has not been possible. Pheromone treatment increases the amount of the CIK1 protein by 22-fold, and anti-CIK1 antibodies stain the SPB region in α-factor-treated cells. It is not yet known where the CIK1 protein localizes within the SPB region. Affinity-purified anti-CIK1 antibodies react with the mitotic centrosome of HeLa cells, indicating that the CIK1 protein may be highly conserved (B. Page, K. Xie & M. Snyder, unpublished observation).

The *KAR1* and *CDC31* genes were identified in screens searching for mutants defective in karyogamy and progression through the cell cycle, respectively (44, 108). These genes are essential for cell growth (116, 186). Temperature-sensitive *kar1*$^-$ and *cdc31*$^-$ mutants arrest with similar phenotypes (described below), which suggests a role for these proteins in SPB duplication (6, 16, 116). Three other genes important for SPB duplication are described below.

The KAR1 protein may be a SPB component; kar1::β-gal fusions, when overproduced, localize to a region adjacent to the outer plaque of new SPBs (see below) (151). The CDC31 protein has not been localized in yeast but is homologous to a small Ca^{2+}-binding protein called centrin, which is associated with the MTOC of *Chlamydomonas reinhardtii* (6, 58, 59). The CDC31 and KAR1 proteins probably function in a similar process and interact genetically; either a dominant mutation in the *CDC31* gene or overexpression of the wild-type gene suppresses a temperature-sensitive *kar1*$^-$ mutant (149, 151). Possibly these proteins interact physically.

Chromosomes

The proper transmission of a chromosome requires *cis* DNA sequences that operate as sites for the organization of *trans*-acting factors. These DNA elements include autonomously replicating sequences (ARS) (57, 138; see also 21, 155), which act as origins of replication (13, 60); telomeres (140); and a centromere (CEN) (25), the sequence necessary for the attachment of the chromosome to the mitotic spindle. In budding yeast, these three sequences appear to be the only *cis*-acting elements required for the proper transmission of a linear chromosome (29, 91). The transmission of a circular chromosome requires only an ARS and a CEN (25). Because of the simplicity of these components in *S. cerevisiae*, artificial chromosomes are easily constructed, and the effects of mutations in a *cis*-acting element have been analyzed extensively in vivo (47, 73). Also, because these chromosome constructs are generally dispensible and can possess visual markers, strains containing a YAC (yeast artificial chromosome) are useful in the identification of *trans*-acting factors (87, 127, 132, 133). ARS and telomeres are reviewed elsewhere (8, 21, 32). This section focuses on the structure and function of the centromere.

CENTROMERE The centromeres of budding and fission yeast differ greatly (23, 25, 124). In *S. cerevisiae,* the entire sequence necessary for centromere function is only 125 base pairs in length (22, 23, 27, 33, 124). This small sequence confers proper segregation during mitosis and meiosis, and the centromere of one chromosome is interchangeable with that of another (26). The centromere is composed of three tandem elements (34, 47). CDEI (centromere DNA element one) is an 8-bp consensus sequence. CDEI is not essential for mitotic segregation, but deletion of and certain mutations within this element result in an approximately 10-fold decrease in chromosome stability (28, 37, 46, 104). The central element of the centromere, CDEII, is a 78- to 86-bp element that is highly AT-rich (>90%). Deletion of CDEII causes a 1000-fold increase in chromosome nondisjunction; this loss of function is due to the combined effects of altered spacing between CDEI and CDEIII as well as the loss of the CDEII sequence (28, 33, 37, 104). CDEIII is a 25-bp consensus element with dyad symmetry (34, 47). This element is essential for centromere function, and single base-pair substitutions within CDEIII, unlike CDEI and CDEII, can completely abolish centromere function (46, 84; see also 22, 23, 33, 124).

The *S. pombe* centromere is much larger and more complex than that of *S. cerevisiae* and contains structural features similar to that of mammalian centromeres (23, 124). *S. pombe* centromeres contain three repetitive elements [K (6.4 kb), L (4.5 kb), and B (1–2 kb); or dg (3.8 kb), dh (4.0 kb) and yn/tm (<1 kb)] that form a 9- to 14-kb repeat unit, several copies of which flank a central unique core (42, 93, 97). The central core of cen2 (the centromere of *S. pombe* chromosome 2), is 7 kb in length and relatively AT-rich; however, this core sequence is not conserved among the *S. pombe* centromeres (24). The relative arrangement of the repetitive elements also varies among the three *S. pombe* centromeres (23, 124). The smallest mitotically functional centromere contains the central core region flanked by inverted B repeats and one copy of the K repeat (42, 81). Thus, the functional *S. pombe* centromere is approximately two orders of magnitude larger than that of *S. cerevisiae* and considerably more complex. These differences are particularly interesting given that the largest *S. cerevisiae* chromosome (2 Mb) is similar in size to the smallest *S. pombe* chromosome (3.5 Mb). Thus, the existence of a more elaborate *S. pombe* centromere is probably not explained by an increase in chromatin load. The divergence between these organisms in centromere structure and size may result from different chromosomal packaging; *S. pombe* chromosomes condense extensively during mitosis and those of *S. cerevisiae* do not. As expected, given the differences just described, the centromeres of either yeast are not functional in the other (23).

CENTROMERE-BINDING PROTEINS In multicellular organisms, microtubule attachment occurs at a morphologically complex structure called the kinetochore (14, 112). This specialized structure is not cytologically detectable in budding yeast, although chromatin and microtubule contacts have been reported (105). Nevertheless, molecular and biochemical analyses of centromere-protein interactions support the existence of a specialized structure at the centromere (3, 9, 20, 71, 75, 88). Nuclease protection experiments reveal the presence of a 150- to 250-bp resistant core that includes the 125-bp conserved centromere sequence (9, 10, 33, 36).

Proteins that directly bind the centromere have been isolated based on their specific binding activity to CDEs. CBF1 (centromere binding factor) binds specifically to the conserved 8-bp centromere element CDEI (3, 20, 88). The *CBF1* gene is not essential, although deletion of this gene results in slow growth and decreased fidelity of chromosome segregation. It is not surprising that this gene is nonessential given that CDEI is not required for centromere function (33). However, centromere binding is not the only function of the CBF1 protein. CBF1 is implicated as a transcriptional regulator, not only because of its homology to known transcriptional activators, but also because CDEI sequences are found within the regulatory region of several yeast genes, including *MET2* and *MET25* (3, 20, 88). This latter feature likely explains the methionine auxotrophy of *cbf1*Δ strains. Whether the chromosome instability of *cbf1*Δ strains results only from loss of CBF1 binding at CDEI or from the combined loss of CBF1 and other mitotically important proteins whose expression is regulated by CBF1 is not known. However, these possibilities could be distinguished by evaluating the effect of *cbf1*Δ upon loss of a chromosome with a nonfunctional CDEI.

CBF3 is a 240-kDa multisubunit protein complex that binds to the essential centromere element CDEIII. This complex contains three proteins: CBF3A (110 kDa), CBF3B (64 kDa), and CBF3C (58 kDa). CBF3 binds specifically to a functional CDEIII element, and this binding activity is sensitive to phosphatase (75). Recently, the CBF3 complex was shown to contain minus-end-directed microtubule motor activity (64). Addition of the purified CBF3 complex to CDEIII oligonucleotides attached to beads is sufficient to promote movement of the beads along microtubules. No movement is observed when an olignucleotide mutated for CDEIII function is used. Furthermore, the rate of movement is comparable to anaphase chromosome movements observed in other organisms. Genes encoding two subunits, CBF3A and CBF3B, were recently cloned (W. Jiang, J. Lechner & J. Carbon, personal communication; P.-Y. Goh & J. Kilmartin, personal communication). The gene encoding CBF3A is essential. Sequence analysis predicts that the CBF3A subunit contains an ATP binding site; perhaps the CBF3A polypeptide is the catalytic subunit for motor activity.

The sequences and organization of the budding yeast centromere are very different from those of mammalian cells. However, the basic properties of this element appear to be identical. In both systems, a centromere consists of a DNA sequence that is the site of assembly for a proteinaceous structure capable of binding microtubules, and this structure possesses microtubule motor activity (64, 65, 112).

Other Components Important for Chromosome Maintenance

In both *S. cerevisiae* and *S. pombe,* numerous other components important for yeast chromosome transmission and/or mitosis have been identified through genetic screens. Mutants defective in cell cycle progression (*cdc*), chromosome maintenance (*cnd, cin, ctf,* and *chl*), and mitosis (*esp, cut,* and *nuc*) have been characterized (5, 48, 55, 74, 76, 85, 109, 133, 136, 146). For most of these genes, the manner in which they participate in chromosome segregation or mitosis is not known. Of special interest are:

1. The *CDC20* gene product, which is homologous to β-transducin (125). *cdc20*$^{-}$ temperature-sensitive mutants arrest with enhanced microtubule arrays. Possibly the CDC20 protein regulates microtubule assembly.
2. Several GTP-binding proteins (CIN1, SAR1, and others) (4, 12; M. A. Hoyt, personal communication). Because GTP-binding proteins play a role in targeting complexes within the secretory pathway, perhaps the mitotic GTP-binding proteins play a role in targeting microtubule-associated components (e.g. microtubule capture sites) at the cortex and other cellular locations.
3. The MCK1 protein kinase (95, 127). *mck1Δ* strains are defective in chromosome transmission, and *MCK1* when present on a high-copy-number plasmid suppresses mutations in CDEIII (127). Therefore, the MCK1 protein may play a role in regulating centromere function.
4. A family of proteins encoded by the *CDC23, CDC16,* and *nuc2* genes. These proteins contain a similar structural motif called the TPR repeat (48, 67, 129).

The localizations of these different gene products and the mechanisms by which they function are eagerly awaited.

MOLECULAR AND CELLULAR MECHANISMS OF MITOSIS

Mitosis in yeast can be resolved into a series of steps: (*a*) duplication of the SPB, (*b*) separation of the SPBs and establishment/maintenance of the spindle apparatus, (*c*) positioning of the yeast spindle apparatus, (*d*) attachment of the chromosome, (*e*) movement of the chromosome, and (*f*) elongation of the spindle apparatus. Many of these events involve microtubule assembly and disassembly, and/or are mediated by microtubule motors.

Duplication of the Yeast SPB

In *S. cerevisiae,* the SPB is characterized by its association with both cytoplasmic and nuclear microtubules throughout the cell cycle (1, 15, 17). In G1, the SPB acquires an electron-dense satellite on the cytoplasmic side of the half bridge (15). After the G1/S transition and by the beginning of bud formation, the SPB has clearly completed duplication. The satellite structure is no longer present and appears to have been replaced with a fully duplicated SPB that is embedded in the nuclear envelope and separated from the other SPB by a bridge (17, 18).

Several genes important for SPB duplication have been identified through various genetic screens. Temperature-sensitive *kar1*$^-$ and *cdc31*$^-$ mutants arrest at their restrictive temperature after G1 as large budded cells; the nucleus contains replicated DNA and has migrated to the mother-bud junction (16, 116). However, two SPBs are not present, which is uncharacteristic of this stage of the cell cycle. The SPB that is present is enlarged, and for *cdc31*$^-$ strains, it is approximately twice as large as a wild-type SPB and appears to nucleate twice the number of microtubules (16, 116).

Three other genes, *MPS1*, *MPS2*, and *NDC1*, important for SPB duplication have been found in *S. cerevisiae* through the characterization of thermal conditional mutants that exhibit abnormal spindles at the restrictive temperature (142, 157). As described above, only two phases of SPB duplication, the appearance of the satellite and the fully duplicated SPB, are defined by cytology. *MPS1*, *MPS2*, and *NDC1* are each required for proper assembly of the SPB, and their gene products define two additional steps necessary for SPB duplication (157; M. Winey, personal communication). At the restrictive temperature, conditional mutations in these genes, like *cdc31*$^-$ and *kar1*$^-$, result in a monopolar spindle as detected by antitubulin immunofluorescence. Electron microscopy analysis demonstrates that the monopolar spindles of *mps1*$^-$ cells contain a single enlarged SPB with an unusual half bridge that appears more electron dense than those seen in wild-type cells. Both *mps2*$^-$ and *ndc1*$^-$ cells contain one enlarged SPB embedded in the nuclear envelope and one aberrant SPB. The aberrant SPB resides on the cytoplasmic face of the nuclear envelope (156, 157), and it forms adjacent to the previous SPB at a position similar to where the satellite forms in wild-type cells. These observations suggest that in wild-type cells the satellite may be converted into the new SPB through morphogenesis and insertion into the nuclear envelope.

The defective SPB of *mps2*$^-$ and *ndc1*$^-$ strains lacks an inner plaque and its associated microtubules; however, it does organize cytoplasmic microtubules from its outer plaque. Immunofluorescence with antitubulin antibodies, of either *mps2*$^-$ or *ndc1*$^-$ cells at their restrictive temperature, reveals two foci of staining. The stronger focus corresponds to the wild-type SPB and is associated with the nuclear DNA; the weaker staining coincides with the defective SPB that migrates to the opposite side of the nucleus (142, 157).

Consistent with the cytological observations, genetic analysis has shown that the *CDC31, MPS1,* and *MPS2* gene functions are required at different points in the duplication process (157). Their functions have been ordered relative to satellite formation and to each other. *cdc31*$^{-}$ cells can be arrested at the permissive temperature in G1 after satellite formation by treatment with mating pheromone. These cells, when shifted to the restrictive temperature and transferred to a medium lacking pheromone, proceeded with SPB duplication and mitosis. This result indicates that the *CDC31* function is required prior to satellite formation. Similar experiments with *mps1*$^{-}$ and *mps2*$^{-}$ cells indicate that their gene products function after satellite formation. Analysis of *mps1*$^{-}$ *mps2*$^{-}$ double mutants has revealed that the requirement for MPS1 precedes that for MPS2. Figure 5 summarizes these results. The CDC31 protein function and satellite formation precede the requirement for MPS1 and MPS2. MPS1 acts upstream of MPS2 (157). The *MPS1* gene was recently cloned and shown to encode a protein kinase homolog (M. Winey, personal communication). Thus, the MPS1 protein probably plays a regulatory role during the duplication process.

In *S. cerevisiae,* genetic analyses of *cdc31*$^{-}$ and *ndc1*$^{-}$ mutations have also provided information about SPB duplication during meiosis. In meiosis, one spindle is assembled in meiosis I and two spindles are assembled in meiosis II; therefore, four SPBs are ultimately formed and utilized (90). At the end of meiosis, the resulting haploid SPB has half the size and nucleating capacity of the diploid SPB (62, 105); thus the SPB is reduced during this process. Sporulation of *cdc31*$^{-}$ or *ndc1*$^{-}$ cells at their restrictive temperature results in two diploid spores containing sister chromosomes, indicative of a failure of meiosis II (16, 142). These results suggest that SPB duplication only occurs at meiosis II. Thus, SPB reduction, as well as chromosome reduction, may take place during meiosis I.

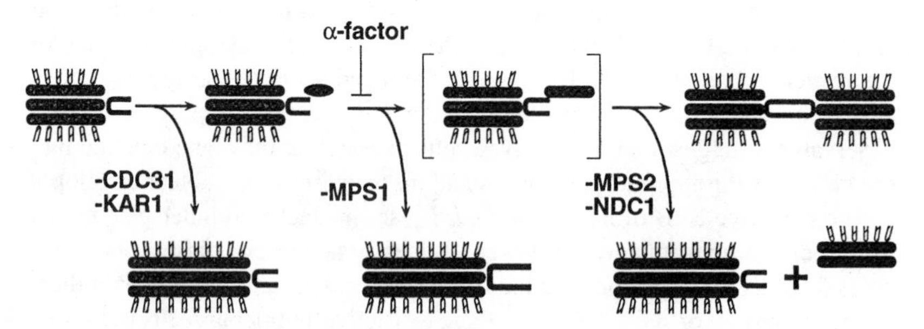

Figure 5 The SPB duplication cycle of *S. cerevisiae*. Adapted from Winey & Byers (157).

SPB Separation and Establishment of the Mitotic Spindle Apparatus

ROLE OF MICROTUBULES After SPB duplication, the SPBs and their respective half bridges migrate away from one another. Simultaneous with SPB separation is the establishment of the mitotic spindle apparatus. In *S. cerevisiae* and *S. pombe,* separation of spindle poles and assembly of the mitotic spindle are mediated by microtubules. In *S. cerevisiae,* cells treated with microtubule-depolymerizing drugs such as nocodazole and most temperature-sensitive tubulin mutants arrest with duplicated, but unseparated SPBs (61, 68). Conditional tubulin mutants of *S. pombe* also fail to form a spindle at their restrictive temperature (144).

For budding yeast, evidence indicates that both nuclear and cytoplasmic microtubules participate in SPB separation. The aberrant SPB of *mps2*$^-$ and *ndc1*$^-$ cells organizes cytoplasmic microtubules but no nuclear microtubules. In these mutants, SPB separation still occurs in the absence of interdigitating nuclear microtubules (142, 157). Recent evidence from newt lung cells also suggests that centrosome separation can occur independent of interdigitating microtubules (C. Rieder, personal communication). We speculate that SPB separation in *mps2*$^-$ and *ndc1*$^-$ mutants is mediated by attachment of microtubules emanating from the SPBs to the cortex at the distal ends of the cell. Microtubule disassembly and/or microtubule motors might then draw the SPBs apart.

Although cytoplasmic microtubules may assist in SPB separation in *S. cerevisiae,* they are not required for this process. Separation of yeast SPBs and establishment of the spindle apparatus (albeit not in the appropriate position, see below) still occurs in *tub2-104* β-tubulin mutants, which are defective in assembling cytoplasmic microtubules (139). In *S. pombe,* cytoplasmic microtubules are not detected during spindle formation and, therefore, probably do not mediate SPB separation (41).

ROLE OF MICROTUBULE MOTORS In both *S. pombe* and *S. cerevisiae,* nuclear microtubules and associated motors probably play a role in SPB separation, and strong evidence indicates that these components are required for establishment of the spindle apparatus. *cut7*$^-$, *cin8*$^-$ *kip1*Δ, and *kar3*Δ strains exhibit defects in spindle formation and/or maintenance. *cut7*$^-$ and *cin8*$^-$ *kip1*Δ temperature-sensitive strains arrest at the restrictive temperature with duplicated SPBs adjacent to one another (39, 54, 114). Deletion of *KAR3* results in accumulation of cells with a very short spindle (89).

The CIN8 and KIP1 proteins of *S. cerevisiae* reside within the center of the mitotic spindle (54, 114). These proteins might represent plus-end motors and separate SPBs by sliding microtubules in the overlap zone (Figure 6). In

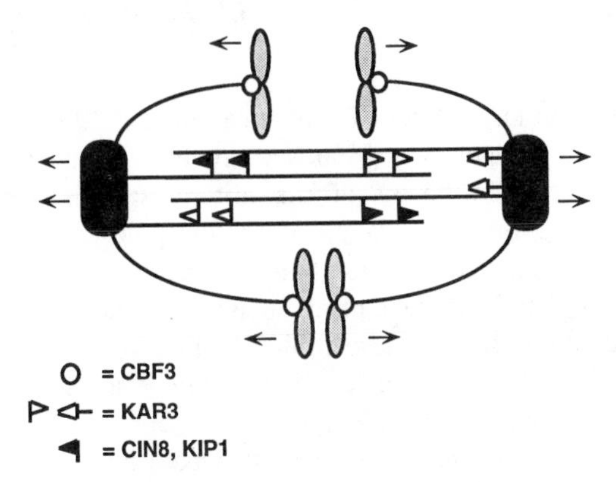

Figure 6 Possible sites of action of yeast microtubule motor proteins. The KAR3 protein might act as a minus-end motor within the spindle, or as a plus-end directed motor at the pole. In this latter case microtubule dissassembly would occur at the pole in concert with KAR3 function (89).

addition to establishing a mitotic spindle, the CIN8 and KIP1 proteins are also important for maintaining a spindle. When *cin8*$^-$ *kip1*Δ cells are first arrested with a short spindle by hydroxyurea treatment (which blocks during S phase), and then shifted to the restrictive temperature, the previously assembled spindle collapses, resulting in adjacent SPBs (119). Thus, CIN8 and KIP1 stabilize the bipolar spindle morphology and may generate the force that separates the SPBs.

The cut7 protein localizes to the SPB and is also important for establishment of the spindle apparatus. Hagan & Yanagida (40) proposed a model for how a microtubule motor associated with the SPB might reorient side-by-side SPBs to generate opposing SPBs and then separate them (see Figure 7). In this model, a cut7 protein attached to one SPB interacts with the microtubules assembled by the other SPB. [In *S. cerevisiae,* microtubules from one SPB pass near the other SPB, suggesting that this type of binding is feasible (105).] Binding of several cut7 molecules will tilt the SPBs toward one another, bringing additional cut7 proteins in contact with microtubules from the adjacent SPB. Continuation of this process will ultimately result in two SPBs that face one another. If cut7 is a plus-end-directed motor and plus ends of microtubules lie distal to the SPB, these SPB-associated motors would move to the plus end of their bound microtubule and separate the two SPBs. This movement would result in a bipolar spindle morphology with interdigitating pole-to-pole microtubules (40).

The KAR3 protein also plays a role in proper assembly or maintenance of the bipolar spindle apparatus, but its role is less clear (89). Interestingly, deletion of *KAR3* partially suppresses the *cin8*$^-$ *kip1*Δ temperature-sensitive-

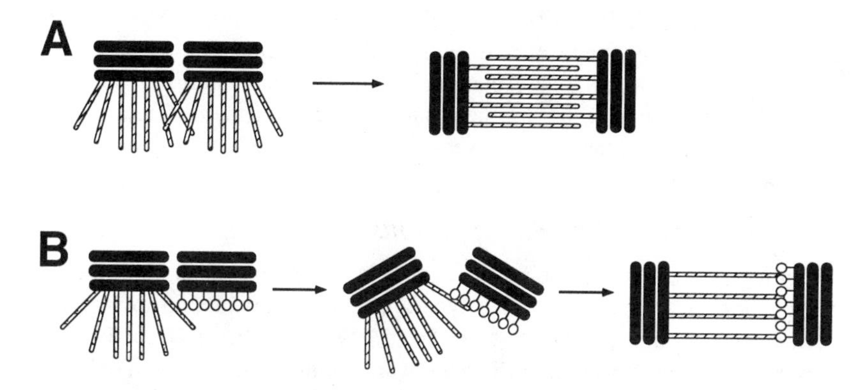

Figure 7 A model for SPB separation. (*A*) The SPB separation process. (*B*) Model for how a SPB-associated microtubule motor might mediate this process. The lollypop structures represent cut7 proteins. Adapted from Hagan & Yanagida (40).

growth and spindle-collapse phenotypes (119). The suppression of spindle collapse by loss of another microtubule motor raises the possibility that opposing forces operate in the maintenance of spindle structure. The KAR3 protein might be a minus-end motor, consistent with its ncd-like structure; this protein might slide microtubules in the direction opposite from CIN8 and KIP1 (119) (Figure 6). Alternatively, KAR3 might be a plus-end motor at the SPB that pulls microtubules toward the pole (89).

Given these two possibilities of KAR3 function, one might expect *kar3Δ* cells to contain a very long spindle; instead, they usually contain a short spindle. Possibly, *kar3Δ* cells arrest at a mitotic checkpoint. Consistent with this, the *kar3* deletion is lethal when placed in combination with *mad2-1*, *bub1*, or *bub3* (*mad2-1*, *bub1*, and *bub3* strains fail to arrest during mitotic delays; see below) (P. Meluh & M. Rose, personal communication). Furthermore, the arrest phenotypes of many of these mutants may be caused by a similar regulatory checkpoint.

The CIK1 protein is concentrated at SPBs, and *cik1Δ* cells are also defective in spindle formation and/or maintenance (101). However, CIK1 shares no homology with microtubule motor proteins. The phenotypes of *cik1Δ* cells are very similar to those of *kar3Δ* strains. CIK1 might function as a light chain that interacts with a kinesin-like heavy chain motor, such as KAR3. Consistent with a possible interaction between CIK1 and KAR3, kar3::β-gal fusion proteins no longer localize to microtubules in *cik1Δ* strains (B. Page & M. Snyder, unpublished observation). Kinesin light chains may define the localization of the heavy-chain motor or what organelles it moves along the microtubule.

Spindles from more complex eukaryotes also contain many microtubule motor proteins (86, 96, 120, 160). These motor proteins have also been implicated in many aspects of mitosis, including movement of chromosomes toward the poles or metaphase plate (65) and establishment/maintenance of the spindle apparatus (120). Further characterization of the different motors and their interacting proteins in yeast should provide a useful model for understanding mitosis in other eukaryotes.

Positioning of the Spindle Apparatus

Specific positioning of the spindle apparatus is critical for proper chromosome segregation in virtually all cell types. In yeast, orientation of spindle components occurs in several steps (1, 15, 17, 41, 70). In G1 cells of *S. cerevisiae,* the SPB becomes oriented toward the incipient bud site (131). By the completion of spindle formation, the nucleus has migrated to the neck of the bud; one SPB is oriented toward the bud and the spindle axis usually resides at an angle from the long axis of the cell (1, 17) (see Figure 2). During spindle elongation, the spindle becomes parallel to the long axis of the cell (1, 15, 17). In *S. pombe,* an equally complex series of events occurs (see Figure 3).

Analysis of tubulin mutants and nocodazole-treated cells suggests that microtubules are important for orientation and migration of spindle components (61, 68). Nocodazole-treated cells do not orient the SPB toward the bud (68). In addition, *tub2-104* mutants, which lack cytoplasmic microtubules, fail to align their spindle between the mother and daughter cells and undergo nuclear division in the mother cell (103, 139). Disruption of cytoplasmic microtubules in cells arrested with a properly positioned G2 spindle also results in frequent spindle misorientation (103). Thus, cytoplasmic microtubules are not only necessary for establishing the orientation of the spindle apparatus, they are also responsible for maintaining that orientation.

Cytological evidence also implicates cytoplasmic microtubules in spindle positioning (1, 15), and suggests models for how spindle orientation occurs. In budding yeast, analysis of synchronized cells indicates that during G1, microtubules extend from the SPB to the incipient bud site (131). This occurs even when the SPB initially resides on the side of the nucleus distal to the bud site. By the end of G1, the SPB becomes oriented toward the incipient bud site, and nearly every cell contains a microtubule bundle that extends to this region. A similar configuration is also observed in α-factor-treated cells; the SPB is oriented toward the projection tip (2, 38, 78, 116). Because pheromone-treated cells have not yet undergone SPB duplication [they do contain a satellite (15, 18)], orientation must not require two SPBs.

After bud formation, and during bud growth, one SPB remains proximal to the bud, and long cytoplasmic microtubules extend toward the tip. At anaphase, one of the SPBs will segregate into the bud. Vallen et al (150) demonstrated that segregation of the SPBs is not random. In wild-type cells overexpressing a kar1::β-gal fusion, only one of the SPBs is stained with anti-β-gal antibodies; this SPB is usually oriented toward or in the bud. Analysis of *ndc1-1* mutants indicates that the kar1::β-gal fusion is associated with the defective SPB and therefore the new SPB. Thus, these data indicate that the new SPB segregates into the daughter cell.

Figure 8 shows a model that incorporates these observations. During G1, microtubules emanating from the nascent SPB (most likely from the satellite) attach to the cortex at the incipient bud site (Figure 8, right pathway). This attachment might occur by a stochastic process of microtubule assembly and disassembly, until a cytoplasmic microtubule intersects a putative microtubule capture site. Next, the SPB orients toward the incipient bud site through

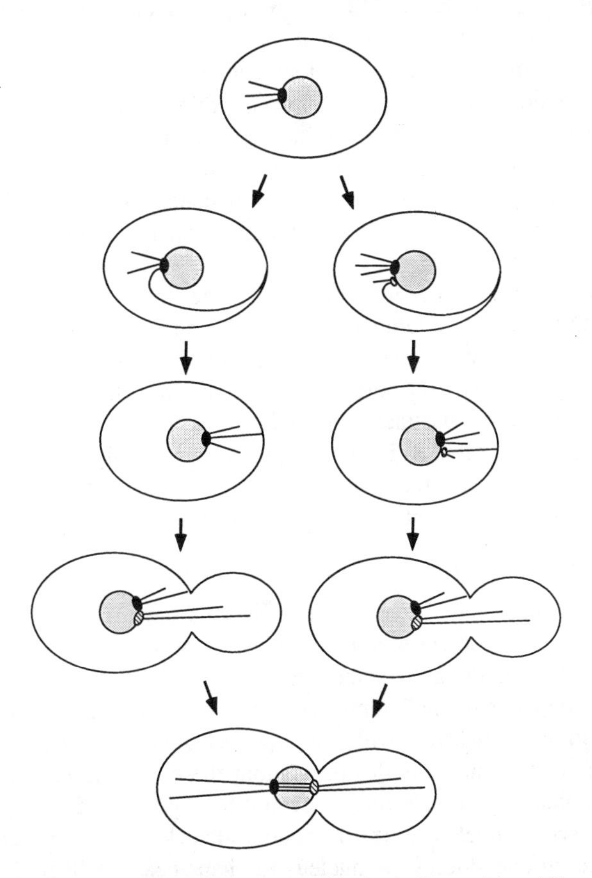

Figure 8 A model for spindle orientation and positioning in *S. cerevisiae*. Although these events are depicted as discrete steps, both nuclear and microtubule movements are expected to be very dynamic (72, 102; T. Stearns, personal communication).

processes involving microtubule disassembly and/or microtubule motors. After SPB duplication and establishment of a short spindle, the old SPB migrates to the other side of the cell and spindle elongation occurs.

Segregation of the old SPB might be mediated by old microtubule attachments to cortical sites in the mother cell that are distal to the bud. However, because the SPB becomes oriented toward the tip in late G1 and α-factor-treated cells, the old attachments must be relatively weak, if they exist at all.

An alternative hypothesis is that microtubules from the old SPB intersect the incipient bud site, and the old SPB orients toward this region (Figure 8, left pathway). After SPB duplication, microtubules extend from the new SPB toward the bud tip. During spindle formation, microtubule attachments from the old SPB release and the old SPB migrates to the opposite side of the cell, possibly through new cortical attachments.

These models predict that interactions between cytoplasmic microtubules and cortical determinants are important for orienting spindle components. The actin cytoskeleton is likely directly or indirectly involved in this process, since actin mutants are defective in proper orientation of the *S. cerevisiae* spindle (103). In other organisms, several lines of evidence indicate that interaction between astral microtubules and cortical components is important for spindle orientation (63, 66, 77).

One caveat to these simple interpretations comes from the analysis of *ndc1*$^-$ mutants. At the restrictive temperature, *ndc1-1* mutants segregate their SPBs randomly, presumably through their cytoplasmic microtubules (142, 151). Perhaps in these mutants, the cortical attachments are delayed or they become disrupted after the initial attachment. These defects might result in random attachment of SPBs after SPB duplication. Further analysis of interactions between microtubules and the cortex is necessary to elucidate the mechanisms by which spindle components are properly positioned.

Chromosome Attachment

When microtubules attach to chromosomes is not known. The CBF3A protein, a subunit of the centromere binding complex, localizes to the SPB region in nonmitotic cells and along the spindle in mitotic cells (J. Kilmartin, personal communication). Assuming that much of this protein is bound to centromeres, this localization is consistent with the hypothesis that *S. cerevisiae* chromosomes are bound to microtubules throughout the cell cycle (158). Chromosome-microtubule attachment might account for the nonrandom organization of the *S. cerevisiae* nucleus (158). In nonmitotic cells, *S. pombe* centromeres are clustered on one side of the nucleus, perhaps near the SPB (147); these also may be attached to microtubules throughout the cell cycle.

In *S. cerevisiae,* centromere attachment to microtubules has been reconstructed in vitro using crude extracts (71) and the purified motor system described above (64). In both systems, evidence indicates that the attachment occurs primarily on sides of microtubules and not simply at the ends (64; J. Kingsbury & D. Koshland, personal communication). Interestingly, in the crude extract system, microtubule-binding activity was found to be greater in nocodazole-treated cells than in α-factor-arrested cells. Thus, microtubule-binding activity appears to be greater during the time of spindle assembly. This observation is not inconsistent with chromosome attachment throughout the cell cycle, because additional microtubule binding activity is necessary after chromosome replication.

Chromosome Movement

In *S. pombe,* centromeres align at the mid-spindle early in mitosis (147). This arrangement is reminiscent of the organization of chromosomes at the metaphase plate detected in more cytologically amenable systems. It is not known whether this process occurs in *S. cerevisiae.*

The discovery that CBF3 contains minus-end microtubule motor activity suggests that CBF3 is a likely candidate for moving yeast chromosomes to the poles. Poleward chromosome movement could occur during congression of chromosomes to the metaphase plate and/or during anaphase A, when chromosomes move to their respective poles. Electron microscopy analysis indicates that anaphase A does occur in *S. cerevisiae,* as supported by the observation of shortening of the discontinuous microtubules (105).

Proper chromosome separation requires topoisomerase II (51, 126). Temperature-sensitive topoisomerase II mutants die if allowed to pass through mitosis at the restrictive temperature. Treatment of these cells with nocodazole prior to the temperature shift prevents cell death (51). Presumably, sister chromosomes become entangled during DNA replication and are disentangled during mitosis by topoisomerase II. In the absence of topoisomerase II, progression through mitosis results in chromosome breakage and/or missegregation, and therefore lethality.

Spindle Elongation

If anaphase A occurs in both budding and fission yeast, the maximum distance it generates between the two sister genomes is 1 and 2 μm, respectively. In both systems, anaphase B results in a seven- to eightfold increase in the distance between sister chromosomes (1, 15, 17, 18, 41, 82). From this viewpoint, SPB separation plays the major role in segregating the two genomes. Movement of SPBs with respect to one another is probably accomplished by interdigitating nuclear microtubules. Cytoplasmic microtubules are not required for spindle elongation, because *tub2-104* mutants, which lack cytoplasmic microtubules at the restrictive temperature, still undergo spindle elongation (139). Likely candidates for proteins mediating spindle elongation are the CIN8 and KIP1 proteins, which could potentially slide interdigitating microtubules past one another.

Anaphase B of *S. pombe* can be activated in vitro (79). This system utilizes the conditional *nuc2*$^-$ mutant, which arrests with a metaphase spindle at its restrictive temperature (48). Spindles of *nuc2*-arrested cells can be induced to undergo anaphase B by the addition of tubulin and ATP. Activation of spindle elongation is sensitive to vanadate and AMPPNP, which inhibit dynein and kinesin, respectively; thus, motor proteins are again implicated in spindle movements (79).

Regulation of Mitosis

The steps of mitosis must be tightly coordinated with one another and with respect to other events of the cell cycle. This subject has received the attention of numerous recent reviews (35, 92), and is only discussed briefly here. Considerable biochemical, molecular, and genetic evidence suggest that the onset of mitosis is controlled by the cdc2 (*S. pombe*)/CDC28 (*S. cerevisiae*) kinase. In *S. pombe,* activation of the cdc2 kinase at the onset of the G2/M transition by G2 cyclins leads to SPB duplication and the onset of mitosis (35). In *S. cerevisiae* at least two critical steps control assembly of the spindle apparatus (35, 92). Activation of the CDC28 G1 kinase controls the G1-to-S transition and SPB duplication (109, 111); the activation of the CDC28 G2 kinase is necessary for progression from G2 to M (107, 110).

Genetic analysis has identified many factors that control the onset and progression through mitosis. A subset of these regulators participate in arresting the cell cycle under damaging conditions. This arrest is termed a checkpoint and is believed to function in halting the cell cycle until conditions have improved (43, 154).

DNA damage or failure to complete DNA replication results in cell-cycle arrest prior to mitosis. In *S. cerevisiae,* several genes important for this arrest have been identified. These include *RAD9, RAD17, RAD24, MEC1, MEC2,* and *MEC3* (154; T. Weinert & L. Hartwell, personal communication). Cells containing mutations in these genes continue through mitosis and the cell cycle in the presence of damaged or unreplicated DNA. The proteins encoded by these different genes presumably function by ultimately inhibiting the activation of the CDC28 kinase and thereby preventing entry into mitosis.

S. cerevisiae and *S. pombe* cells also arrest in response to mitotic defects. Wild-type cells treated with microtubule-depolymerizing drugs and many mutants defective in mitosis undergo cell-cycle arrest. Centromeric mutations also cause a slight mitotic-arrest defect (134). The *MAD1, MAD2, MAD3, BUB1, BUB2,* and *BUB3* gene products are important for arrest at the mitotic checkpoint(s). Cells containing mutations in any of these genes fail to exhibit mitotic arrest in the presence of a microtubule-depolymerizing drug, which results in a lethal progression through the cell cycle (56, 113). The MAD and BUB gene products might function by inhibiting deactivation/destruction of the mitotic CDC28 kinase activity.

CONCLUSION

The combined genetic, biochemical, and cell biological analyses of mitosis in *S. pombe* and *S. cerevisiae* reveal structural and mechanical similarities in chromosome segregation between yeast and other eukaryotes. In the past several years, we have learned a great deal of information about mitosis from yeast. Many components important for chromosome transmission have been identified. Molecular and genetic analyses have begun to elucidate the mechanisms of SPB duplication, spindle establishment, spindle positioning, and mitotic regulation. However, these processes require further characterization and many key aspects of mitosis remain to be described. The versatility of both *S. pombe* and *S. cerevisiae* allows a variety of experimental approaches and will certainly help provide a molecular and cellular understanding of mitosis.

ACKNOWLEDGMENTS

We thank our numerous colleagues for communication of their unpublished results. J. Barrett, N. Burns, C. Costigan, J. R. McIntosh, R. Padmanabha, M. Semenov, and S. Sobel provided critical comments on the manuscript. Research from our laboratory was supported by NIH grant GM36494.

Literature Cited

1. Adams, A. E. M., Pringle, J. R. 1984. Relationship of actin and tubulin distribution to bud growth in wild-type and morphogenetic-mutant *Saccharomyces cerevisiae*. *J. Cell Biol*. 98:934–45
2. Baba, M., Baba, N., Ohsumi, Y., Kanaya, K., Osumi, M. 1989. Three-dimensional analysis of morphogenesis induced by mating pheromone α-factor in *Saccharomyces cerevisiae*. *J. Cell Sci*. 94:207–16
3. Baker, R. E., Masison, D. C. 1990. Isolation of the gene encoding the *Saccharomyces cerevisiae* centromere-binding protein CP1. *Mol. Cell. Biol.* 10:1863–72
4. Barnes, G., Louie, K. A., Botstein, D. 1992. Yeast proteins associated with microtubules in vitro and in vivo. *Mol. Biol. Cell*. 3:29–47
5. Baum, P., Yip, C., Goetsch, L., Byers, B. 1988. A yeast gene essential for regulation of spindle pole body duplication. *Mol. Cell Biol*. 8:5386–97
6. Baum, P. C., Furlong, C., Byers, B. 1986. Yeast gene required for spindle pole body duplication: homology of its product with Ca^{2+} binding proteins. *Proc. Natl. Acad. Sci. USA* 83:5512–16
7. Berlin, V., Styles, C. A., Fink, G. R. 1990. BIK1, a protein required for microtubule function during mating and mitosis in *Saccharomyces cerevisiae*, colocalizes with tubulin. *J. Cell Biol.* 111:2573–86
8. Blackburn, E., Szostak, J. 1984. The molecular structure of centromeres and telomeres. *Annu. Rev. Biochem*. 53:163–94
9. Bloom, K., Carbon, J. 1982. Yeast centromere DNA is in a unique and

highly ordered structure in chromosomes and small circular minichromosomes. *Cell* 29:305–17
10. Bloom, K. S., Amaya, E., Carbon, J., Clarke, L., Hill, A., Yeh, E. 1984. Chromatin conformation of yeast centromeres. *J. Cell Biol.* 99:1559–68
11. Bond, J., Fridovich-Keil, J., Pillus, L., Mulligan, R., Solomon, F. 1986. A chicken-yeast chimeric beta-tubulin protein is incorporated into mouse microtubules in vivo. *Cell* 44:461–68
12. Botstein, D., Segev, N., Stearns, T., Hoyt, M. A., Holden, J., Kahn, R. A. 1988. Diverse biological functions of small GTP-binding proteins in yeast. *Cold Spring Harbor Symp. Quant. Biol.* 53:629–36
13. Brewer, B., Fangman, W. 1987. The localization of replication origins on ARS plasmids in S. cerevisiae. *Cell* 51:463–71
14. Brinkley, B. R. 1985. Microtubule organizing centers. *Annu. Rev. Cell Biol.* 1:145–72
15. Byers, B. 1981. Cytology of the yeast cell cycle. See Ref. 138a, pp. 59–96
16. Byers, B. 1981. Multiple roles of the spindle pole bodies in the life cycle of *Saccharomyces cerevisiae*. In *Molecular Genetics in Yeast, Alfred Benzon Symposium,* ed. D. von Wettstein, J. Friis, M. Kielland-Brandt, A. Stenderup, 16:119–31. Copenhagen: Munksgaard
17. Byers, B., Goetsch, L. 1973. Duplication of the spindle plaques and integration of the yeast cell cycle. *Cold Spring Harbor Symp. Quant. Biol.* 38: 123–31
18. Byers, B., Goetsch, L. 1975. Behavior of the spindle plaques in the cell cycle and conjugation of *Saccharomyces cerevisiae*. *J. Bacteriol.* 124:511–23
19. Byers, B., Shriver, K., Goetsch, L. 1978. The role of spindle pole bodies and modified microtubule ends in the initiation of microtubule assembly in *Saccharomyces cerevisiae*. *J. Cell Sci.* 30:331–52
20. Cai, M., Davis, R. W. 1990. Yeast centromere binding protein CBF1, of the helix-loop-helix protein family, is required for chromosome stability and methionine prototrophy. *Cell* 61:437–46
21. Campbell, J. L., Newlon, C. S. 1991. Chromosomal DNA replication. In *The Molecular Biology and Cellular Biology of the Yeast* Saccharomyces cerevisiae: *Genome Dynamics, Protein Synthesis and Energetics,* ed. J. R. Broach, J. R. Pringle, E. W. Jones, 1:41–146. Cold Spring Harbor, NY: Cold Spring Harbor Press
22. Carbon, J. 1984. Yeast centromeres: structure and function. *Cell* 37:351–53
23. Clarke, L. 1990. Centromeres of budding and fission yeast. *Trends Genet.* 6:150–54
24. Clarke, L., Amstutz, H., Fishel, B., Carbon, J. 1986. Analysis of centromeric DNA in the fission yeast *Schizosaccharomyces pombe*. *Proc. Natl. Acad. Sci. USA* 83:8253–57
25. Clarke, L., Carbon, J. 1980. Isolation of a yeast centromere and construction of functional small circular chromosomes. *Nature* 287:504–9
26. Clarke, L., Carbon, J. 1983. Genomic substitutions of centromeres in *Saccharomyces cerevisiae*. *Nature* 305: 23–28
27. Cottarel, G., Shero, J., Hieter, P., Hegemann, J. 1989. A 125 base pair CEN6 DNA fragment is sufficient for complete meiotic and mitotic centromere functions in *Saccharomyces cerevisiae*. *Mol. Cell. Biol.* 9:3342–49
28. Cumberledge, S., Carbon, J. 1987. Mutational analysis of meiotic and mitotic centromere function in *Saccharomyces cerevisiae*. *Genetics* 117:203–12
29. Dani, G. M., Zakian, V. A. 1983. Mitotic and meiotic stability of linear plasmids in yeast. *Proc. Natl. Acad. Sci. USA* 80:3406–10
29a. Ding, R., McDonald, K. L., McIntosh, J. R. 1993. Three-dimensional reconstruction and analysis of mitotic spindles from the yeast *Schizzosaccharomyces pombe*. *J. Cell Biol.* 120:141–51
30. Endow, S. A., Hatsumi, M. 1991. A multimember kinesin gene family in *Drosophila*. *Proc. Natl. Acad. Sci. USA* 88:4424–27
31. Euteneur, U., McIntosh, J. R. 1981. Structural polarity of kinetochore microtubules in PtK1 cells. *J. Cell Biol.* 89:338–45
32. Fangman, W., Brewer, B. 1991. Activation of replication origins within yeast chromosomes. *Annu. Rev. Cell Biol.* 7:375–402
33. Fitzgerald-Hayes, M. 1987. Yeast centromeres. *Yeast* 3:187–200
34. Fitzgerald-Hayes, M., Clarke, L., Carbon, J. 1982. Nucleotide sequence comparisons and functional analysis of yeast centromere DNAs. *Cell* 29:235–44
35. Forsburg, S. L., Nurse, P. 1991. Cell cycle regulation in the yeast *Saccharomyces cerevisiae* and *Schizosaccharomyces pombe*. *Annu. Rev. Cell Biol.* 7:227–56
36. Funk, M., Hegemann, J. H.,

Philippsen, P. 1989. Chromatin digestion with restriction endonuclease reveals 150–160 bp of protected DNA in the centromere of chromosome XIV in *Saccharomyces cerevisiae*. *Mol. Gen. Genet.* 219:153–60
37. Gaudet, A., Fitzgerald-Hayes, M. 1987. Alterations in the adenine-plus-thymine-rich region of CEN3 affect centromere function in *Saccharomyces cerevisiae*. *Mol. Cell. Biol.* 7:68–75
38. Gehrung, S., Snyder, M. 1990. The SPA2 gene of *Saccharomyces cerevisae* is important for pheromone-induced morphogenesis and efficient mating. *J. Cell Biol.* 111:1451–64
39. Hagan, I., Yanagida, M. 1990. Novel potential mitotic motor protein encoded by the fission yeast *cut7*$^+$ gene. *Nature* 347:563–67
40. Hagan, I., Yanagida, M. 1992. Kinesin-related cut7 protein associates with mitotic and meiotic spindles in fission yeast. *Nature* 356:74–76
41. Hagan, I. M., Hyams, J. S. 1988. The use of cell division cycle mutants to investigate the control of microtubule distribution in the fission yeast *Schizosaccharomyces pombe*. *J. Cell Sci.* 89:343–57
42. Hahnenberger, K. M., Carbon, J., Clarke, L. 1991. Identification of DNA regions required for mitotic and meiotic functions within the centromere of *Schizosaccharomyces pombe* chromosome I. *Mol. Cell. Biol.* 11:2206–15
43. Hartwell, L., Weinert, T. 1989. Checkpoints: controls that ensure the order of cell cycle events. *Science* 246:629–34
44. Hartwell, L. H., Mortimer, R. K., Culotti, J., Culotti, M. 1973. Genetic control of the cell division cycle in yeast. V. Genetic analysis of *cdc* mutants. *Genetics* 74:267–86
45. Hartwell, L. H., Smith, D. 1985. Altered fidelity of mitotic chromosome transmission in cell cycle mutants of *S. cerevisiae*. *Genetics* 110:381–95
46. Hegemann, J. H., Shero, J. H., Cottarel, G., Phillipsen, P., Hieter, P. 1988. Mutational analysis of the centromere DNA from chromosome VI of *Saccharomyces cerevisiae*. *Mol. Cell. Biol.* 8:2523–28
47. Hieter, P., Mann, C., Snyder, M., Davis, R. W. 1985. Mitotic stability of yeast chromosomes: a colony color assay that measures nondisjunction and chromosome loss. *Cell* 40:381–92
48. Hirano, T., Hiraoka, Y., Yanagida, M. 1988. A temperature-sensitive mutation of the *Schizosaccharomyces pombe* gene *nuc2*$^+$ that encodes a nuclear scaffold-like protein blocks spindle elongation in the mitotic anaphase. *J. Cell Biol.* 106:1171–83
49. Hiraoka, Y., Toda, T., Yanagida, M. 1984. The NDA3 gene of fission yeast encodes beta-tubulin: a cold-sensitive *nda3* mutation reversibly blocks spindle formation and chromosome movement in mitosis. *Cell* 39:349–58
50. Hirokawa, N., Shiomura, Y., Okabe, S. 1988. Tau proteins: the molecular structure and mode of binding microtubules. *J. Cell Biol.* 107:1449–59
51. Holm, C., Goto, T., Wang, J. C., Botstein, D. 1985. DNA topoisomerase II required at the time of mitosis in yeast. *Cell* 41:553–63
52. Horio, T., Hotani, H. 1986. Visualization of the dynamic instability of individual microtubules by dark-field microscopy. *Nature* 321:605–7
53. Horio, T., Uzawa, S., Jung, M. K., Oakley, B. R., Tanaka, K., Yanagida, M. 1991. The fission yeast γ-tubulin is essential for mitosis and is localized at microtubule organizing centers. *J. Cell Sci.* 99:693–700
54. Hoyt, M. A., He, L., Loo, K. K., Saunders, W. S. 1992. Two *Saccharomyces cerevisiae* kinesin-related gene products required for mitotic spindle assembly. *J. Cell Biol.* 118: 109–20
55. Hoyt, M. A., Stearns, T., Botstein, D. 1990. Chromosome instability mutants of *Saccharomyces cerevisiae* that are defective in microtubule-mediated processes. *Mol. Cell. Biol.* 10:223–34
56. Hoyt, M. A., Toti, L., Roberts, B. T. 1991. *S. cerevisiae* genes required for cell cycle arrest in response to loss of microtubule function. *Cell* 66:507–17
57. Hsiao, C., Carbon, J. 1979. High frequency transformation of yeast by plasmids containing the cloned yeast *ARG4* gene. *Proc. Natl. Acad. Sci. USA* 76:3829–33
58. Huang, B., Mengersen, A., Lee, V. D. 1988. Molecular cloning of cDNA for caltractin, a basal body-associated Ca^{2+}-binding protein: homology in its protein sequence with calmodulin and the yeast *CDC31* gene product. *J. Cell Biol.* 1988:133–40
59. Huang, B., Watterson, D. M., Lee, V. D., Schibler, M. J. 1988. Purification and characterization of a basal body-associated Ca^{2+}-binding protein. *J. Cell Biol.* 107:121–31
60. Huberman, J. A., Spotila, L. D., Nawotka, K. A., El-Assouli, S. M., Davis, L. R. 1987. The in vivo rep-

lication origin of the yeast 2 μm plasmid. *Cell* 51:473–81
61. Huffaker, T. C., Thomas, J. H., Botstein, D. 1988. Diverse effects of β-tubulin mutations on microtubule formation and function. *J. Cell Biol.* 106:1997–2010
62. Hyams, J. S., Borisy, G. G. 1978. Nucleation of microtubules in vitro by isolated spindle pole bodies of the yeast *Saccharomyces cerevisiae. J. Cell Biol.* 78:401–13
63. Hyman, A. A. 1989. Centrosome movement in the early divisions of *Caenorhabditis elegans:* a cortical site determining centrosome position. *J. Cell Biol.* 109:1185–93
64. Hyman, A. A., Middleton, K., Centola, M., Mitchison, T. J., Carbon, J. 1992. Microtubule-motor activity of a yeast centromere-binding protein complex. *Nature* 359:533–36
65. Hyman, A. A., Mitchison, T. J. 1991. Two different microtubule-based motor activities with opposite polarities in kinetochores. *Nature* 351:206–11
66. Hyman, A. A., White, J. G. 1987. Determination of cell division axes in the early embryogenesis of *Caenorhabditis elegans. J. Cell Biol.* 105:2123–35
67. Icho, T., Wickner, R. 1987. Metal-binding, nucleic acid-binding finger sequences in the *CDC16* gene of *Saccharomyces cerevisiae. Nucleic Acids Res.* 15:8439–50
68. Jacobs, C. W., Adams, A. E. M., Szaniszlo, P. J., Pringle, J. R. 1988. Functions of microtubules in the *Saccharomyces cerevisiae* cell cycle. *J. Cell Biol.* 107:1409–26
69. Joshi, H. C., Palacios, M. J., McNamara, L., Cleveland, D. W. 1992. γ-Tubulin is a centrosomal protein required for cell cycle-dependent microtubule nucleation. *Nature* 356:80–83
70. Kilmartin, J. V., Adams, A. E. M. 1984. Structural rearrangements of tubulin and actin during the cell cycle of the yeast *Saccharomyces. J. Cell Biol.* 98:922–33
71. Kingsbury, J., Koshland, D. 1991. Centromere-dependent binding of yeast minichromosomes to microtubules in vitro. *Cell* 66:483–95
72. Korning, A. J., Lum, P. Y., Williams, J., Wright, R. 1993. DiOC6 staining reveals organelle structure and dynamics in living yeast cells. *Cell Mot. Cytoskel.* In press
73. Koshland, D., Kent, J., Hartwell, L. H. 1985. Genetic analysis of the mitotic transmission of minichromosomes. *Cell* 40:393–403
74. Kouprina, N., Pashina, O. B., Nikolaishwili, N., Tsouladze, A., Larionov, V. 1988. Genetic control of chromosome stability in the yeast *Saccharomyces cerevisiae. Yeast* 4: 257–69
75. Lechner, J., Carbon, J. 1991. A 240 kd multisubunit protein complex, CBF3, is a major component of the budding yeast centromere. *Cell* 64:717–25
76. Liras, P., McCusker, J., Mascioli, S., Haber, J. 1978. Characterization of a mutation in yeast causing nonradom chromosome loss during mitosis. *Genetics* 88:651–71
77. Lutz, D., Hamaguchi, Y., Inoue, S. 1988. Micromanipulation studies of the asymmetric positioning of the maturation spindle in *Chaetopterus* sp. oocytes: 1. Anchorage of the spindle to the cortex and migration of a displaced spindle. *Cell Motil. Cytol.* 11:83–96
78. Madden, K., Snyder, M. 1992. Specification of sites for polarized growth in *Saccharomyces cerevisiae* and the influence of external factors on site selection. *Mol. Biol. Cell.* 3:1025–35
79. Masuda, H., Hirano, T., Yanagida, M., Cande, W. Z. 1990. In vitro reactivation of spindle elongation in fission yeast *nuc2* mutant cells. *J. Cell Biol.* 110:417–25
80. Masuda, H., Sevik, M., Cande, W. Z. 1992. In vitro microtubule-nucleating activity of the spindle pole bodies in the fission yeast *Schizosaccharomyces pombe:* cell cycle-dependent activation in *Xenopus* cell-free extracts. *J. Cell Biol.* 117:1055–66
81. Matsumoto, T., Murakami, S., Niwa, O., Yanagida, M. 1990. Construction and characterization of centric circular and acentric linear chromosomes in fission yeast. *Curr. Genet.* 18:323–30
82. McCully, E. K., Robinow, C. F. 1971. Mitosis in the fission yeast *Schizosaccharomyces pombe:* a comparative study with light and electron microscopy. *J. Cell Sci.* 9:475–507
83. McDonald, H., Stewart, R., Goldstein, L. 1990. The kinesin-like ncd protein of Drosophila is a minus end–directed microtubule motor. *Cell* 1990:1159–65
84. McGrew, J., Diehl, B., Fitzgerald-Hayes, M. 1986. Single base-pair mutations in centromere element III cause chromosome segregation in *Saccharomyces cerevisiae. Mol. Cell. Biol.* 6: 530–38

85. McGrew, J. T., Xiao, Z., Fitzgerald-Hayes, M. 1989. *Saccharomyces cerevisiae* mutants defective in chromosome segregation. *Yeast* 5:271–84
86. McIntosh, J. R., Pfarr, C. M. 1991. Mitotic motors. *J. Cell Biol.* 115:577–85
87. Meeks-Wagner, D., Wood, J., Garvik, B., Hartwell, L. H. 1986. Isolation of two genes that affect mitotic chromosome transmission in S. cerevisiae. *Cell* 44:53–63
88. Mellor, J., Jiang, W., Rathjen, J., Barnes, C., Hinz, T., et al. 1990. CPF1, a yeast protein which functions in centromeres and promoters. *EMBO J.* 9:4017–26
89. Meluh, P. B., Rose, M. D. 1990. KAR3, a kinesin-related gene required for yeast nuclear fusion. *Cell* 60:1029–41
90. Moens, P. B., Rapport, E. 1971. Spindles, spindle plaques, and meiosis in the yeast *Saccharomyces cerevisiae*. *J. Cell Biol.* 50:344–61
91. Murray, A., Szostak, J. W. 1983. Construction of artifical chromosomes in yeast. *Nature* 305:189–93
92. Murray, A. W. 1992. Creative blocks: cell-cycle checkpoints and feedback controls. *Nature* 359:599–604
93. Nakeseko, Y., Adachi, Y., Funahashi, S., Niwa, O., Yanigida, M. 1986. Chromosome walking shows a highly homologous repetitive sequence present in all the centromere regions of fission yeast. *EMBO J.* 5:1011–21
94. Neff, N. F., Thomas, J. H., Grisafi, P., Botstein, D. 1983. Isolation of the β-tubulin gene from yeast and demonstration of its essential function in vivo. *Cell* 33:211–19
95. Neigeborn, L., Mitchell, A. P. 1991. The yeast MCK1 gene encodes a protein kinase homolog that activates early meiotic gene expression. *Genes Dev.* 5:533–48
96. Nislow, C., Lombillo, V. A., Kuriyama, R., McIntosh, J. R. 1992. A plus-end-directed motor enzyme that moves antiparallel microtubules in vitro localizes to the interzone of mitotic spindles. *Nature* 359:543–47
97. Niwa, O., Matsumoto, T., Chikashige, Y., Yanigida, M. 1989. Characterization of *Schizosaccharomyces pombe* minichromosome deletion derivatives and a functional allocation of their centromere. *EMBO J.* 8:3045–52
98. Oakley, B., Oakley, E., Yoon, Y., Jung, M. K. 1990. Gamma-tubulin is a component of the spindle pole body that is essential for microtubule function in Aspergillus nidulans. *Cell* 61:1289–1301
99. Oakley, B. R. 1992. Gamma-tubulin: the microtubule organizer? *Trends Cell Biol.* 2:1–5
100. Oakley, C. E., Oakley, B. R. 1989. Identification of γ-tubulin, a new member of the tubulin superfamily encoded by *mipA* gene of *Aspergillus nidulans*. *Nature* 338:662–64
101. Page, B., Snyder, M. 1992. CIK1: a developmentally regulated spindle pole body-associated protein important for microtubule functions in *Saccharomyces cerevisiae*. *Genes Dev.* 6:1414–29
102. Palmer, R. E., Koval, M., Koshland, D. 1989. The dynamics of chromosome movement in the budding yeast *Saccharomyces cerevisiae*. *J. Cell Biol.* 109:3355–66
103. Palmer, R. E., Sullivan, D. S., Huffaker, T., Koshland, D. 1992. Role of astral microtubules and actin in spindle orientation and migration in the budding yeast, *Saccharomyces cerevisiae*. *J. Cell. Biol.* 119:583–93
104. Panzeri, L., Landonio, L., Stotz, A., Philippsen, P. 1985. Role of conserved sequence elements in yeast centromere DNA. *EMBO J.* 4:1867–74
105. Peterson, J. B., Ris, H. 1976. Electron-microscopic study of the spindle and chromosome movement in the yeast *Saccharomyces cerevisiae*. *J. Cell Sci.* 22:219–42
106. Pfarr, C. M., Coue, M., Grissom, P. M., Hayes, T. S., Hayes, M. E. 1990. Cytoplasmic dynein is localized to kinetochores during mitosis. *Nature* 345:263–65
107. Piggott, R., Rai, R., Carter, B. 1982. A bifunctional gene product involved in two phases of the yeast cell cycle. *Nature* 298:391–93
108. Poliana, J., Conde, J. 1982. Genes involved in the control of nuclear fusion during the sexual cycle of *Saccharomyces cerevisiae*. *Mol. Gen. Genet.* 186: 253–58
109. Pringle, J., Hartwell, L. H. 1981. The *Saccharomyces cerevisiae* cell cycle. See Ref. 138a, pp. 97–142
110. Reed, S., Wittenberg, C. 1990. Mitotic role for the Cdc28 protein kinase of *Saccharomyces cerevisiae*. *Proc. Natl. Acad. Sci. USA* 87:5697–5701
111. Reed, S. I. 1980. The selection of *S. cerevisiae* mutants defective in the start event of cell division. *Genetics* 95:561–77
112. Rieder, C. L. 1982. The formation and composition of the mammalian

kinetochore and kinetochore fiber. *Int. Rev. Cytol.* 79:1–58

113. Rong, L., Murray, A. W. 1991. Feedback control of mitosis in budding yeast. *Cell* 66:519–31
114. Roof, D., Meluh, P., Rose, M. 1992. Kinesin-related proteins required for assembly of the mitotic spindle. *J. Cell Biol.* 118:95–108
115. Roof, D. M., Meluh, P., Rose, M. 1992. Multiple kinesin-related proteins in yeast mitosis. *Cold Spring Harbor Symp. Quant. Biol.* 61:693–703
116. Rose, M. D., Fink, G. R. 1987. KAR1, a gene required for function of both intranuclear and extranuclear microtubules in yeast. *Cell* 48:1047–60
117. Rout, M. P., Kilmartin, J. V. 1990. Components of the yeast spindle and spindle pole body. *J. Cell Biol.* 111: 1913–27
118. Rout, M. P., Kilmartin, J. V. 1991. Yeast spindle pole body components. *Cold Spring Harbor Symp. Quant. Biol.* 61:687–92
119. Saunders, W., Hoyt, M. A. 1992. Kinesin-related proteins required for structural integrity of the mitotic spindle. *Cell* 70:451–58
120. Sawin, K. E., LeGuellec, K., Philippe, M., Mitchison, T. J. 1992. Mitotic spindle organization by a plus-end-directed microtubule motor. *Nature* 359: 540–43
121. Schatz, P., Pillus, L., Grisafi, P., Solomon, F., Botstein, D. 1986. Two functional α-tubulin genes of the yeast *Saccharomyces cerevisae* encode divergent proteins. *Mol. Cell Biol.* 6:3711–21
122. Schatz, P., Solomon, F., Botstein, D. 1986. Genetically essential and nonessential α-tubulin genes specify functionally interchangeable proteins. *Mol. Cell. Biol.* 6:3722–33
123. Schnapp, B. J., Reese, T. S. 1989. Dynein is the motor for retrograde axonal transport of organelles. *Proc. Natl. Acad. Sci. USA* 86:1548–52
124. Schulman, I., Bloom, K. S. 1991. Centromeres: an integrated protein/DNA complex required for chromosome movement. *Annu. Rev. Cell Biol.* 7:311–36
125. Sethi, N., Monteagudo, C., Koshland, D., Hogan, E., Burke, D. 1991. The CDC20 gene product of *Saccharomyces cerevisiae,* a β-transducin homolog, is required for a subset of microtubule-dependent cellular processes. *Mol. Cell. Biol.* 11:5592–5602
126. Shamu, C. E., Murray, A. W. 1992. Sister chromatid separation in frog egg extracts requires DNA topoisomerase II activity during anaphase. *J. Cell Biol.* 117:921–34
127. Shero, J., Hieter, P. 1991. A suppressor of a centromere DNA mutation encodes a putative protein kinase (MCK1). *Genes Dev.* 5:549–60
128. Shpetner, H. S., Vallee, R. B. 1989. Identification of dynamin, a novel mechanochemical enzyme that mediates interactions between microtubules. *Cell* 59:421–32
129. Sikorski, R., Boguski, M., Goebl, M., Hieter, P. 1990. A repeating amino acid motif in CDC23 defines a family of proteins and a new relationship among genes required for mitosis and RNA synthesis. *Cell* 60:307–17
130. Snyder, M., Davis, R. W. 1988. SPA1: a gene important for chromosome segregation and other mitotic functions in S. cerevisiae. *Cell* 54:743–54
131. Snyder, M., Gehrung, S., Page, B. D. 1991. Temporal and genetic control of cell polarity in *Saccharomyces cerevisiae. J. Cell Biol.* 114: 515–32
132. Spencer, F., Connelly, C., Lee, S., Hieter, P. 1988. Isolation and cloning of conditionally lethal chromosome transmission fidelity genes in *Saccharomyces cerevisiae.* In *Cancer Cells,* ed. T. Kelly, B. Stillman, 6:441–52. Cold Spring Harbor, NY: Cold Spring Harbor Lab.
133. Spencer, F., Gerring, S. L., Connelly, C., Hieter, P. 1990. Mitotic chromosome transmission fidelity mutants in *Saccharomyces cerevisiae. Genetics* 124:237–49
134. Spencer, F., Hieter, P. 1992. Centromere mutations induce a mitotic delay in *Saccharomyces cerevisiae. Proc. Natl. Acad. Sci. USA* 89:8908–12
135. Stearns, T., Evans, L., Kirschner, M. 1991. Gamma-tubulin is a highly conserved component of the centrosome. *Cell* 65:825–36
136. Stearns, T., Hoyt, M. A., Botstein, D. 1990. Yeast mutants sensitive to antimicrotubule drugs define three genes that regulate microtubule function. *Genetics* 124:251–62
137. Steuer, E. R., Wordeman, L., Schroer, T. A., Sheetz, M. P. 1990. Localization of cytoplasmic dynein to mitotic spindles and kinetochores. *Nature* 345:266–68
138. Stinchcomb, D. T., Struhl, K., Davis, R. W. 1979. Isolation and characterization of a yeast chromosomal replicator. *Nature* 282:39–43

138a. Strathern, J., Jones, E., Broach, J.,

eds. 1981. *The Molecular Biology of the Yeast* Saccharomyces cerevisiae: *Life Cycle and Inheritance,* Vol. 1. Cold Spring Harbor, NY: Cold Spring Harbor Lab.

139. Sullivan, D. S., Huffaker, T. C. 1992. Astral microtubules are not required for anaphase B in *Saccharomyces cerevisiae. J. Cell Biol.* 119:379–88
140. Szostak, J., Blackburn, E. 1982. Cloning yeast telomeres on linear plasmid vectors. *Cell* 29:245–55
141. Telzer, B., Haimo, L. 1981. Decoration of the spindle microtubules with dynein: evidence for uniform polarity. *J. Cell Biol.* 89:373–78
142. Thomas, J. H., Botstein, D. 1986. A gene required for the separation of chromosomes on the spindle apparatus in yeast. *Cell* 44:65–76
143. Toda, T., Adachi, Y., Hiraoka, Y., Yanagida, M. 1984. Identification of the pleiotropic cell division cycle gene NDA2 as one of two different α-tubulin genes in Schizosaccharomyces pombe. *Cell* 37:233–42
144. Toda, T., Umesono, K., Hirata, A., Yanagida, M. 1983. Cold-sensitive nuclear division arrest mutants of the fission yeast *Schizosaccharomyces pombe. J. Mol. Biol.* 168:251–70
145. Trueheart, J., Boeke, J. D., Fink, G. R. 1987. Two genes required for cell fusion during yeast conjugation: evidence for a pheromone-induced surface protein. *Mol. Cell. Biol.* 7:2316–28
146. Uzawa, S., Samejima, I., Hirano, T., Tanaka, K., Yanagida, M. 1990. The fission yeast cut1$^+$ gene regulates spindle pole body duplication and has homology to the budding yeast ESP1 gene. *Cell* 62:913–25
147. Uzawa, S., Yanagida, M. 1992. Visualization of centromeric and nucleolar DNA in fission yeast by fluorescence in situ hybridization. *J. Cell Sci.* 101: 267–75
148. Vale, R. D., Reese, T. S., Sheetz, M. P. 1985. Identification of a novel force generating protein, kinesin, involved in microtubule-based motility. *Cell* 42:39–50
149. Vallen, E. A. 1992. *Genetic analysis of the yeast microtubule organizing center*. PhD thesis. Princeton Univ., Trenton, N.J. 210 pp.
150. Vallen, E. A., Hiller, M. A., Scherson, T. Y., Rose, M. D. 1992. Separate domains of KAR1 mediate distinct functions in mitosis and nuclear fusion. *J. Cell Biol.* 117:1277–87
151. Vallen, E. A., Scherson, T. Y., Roberts, T., Zee, K. V., Rose, M. D. 1992. Asymmetric mitotic segregation of the yeast spindle pole body. *Cell* 69:505–15
152. Walker, R. A., O'Brien, E., Pryer, N., Soboeiro, M., Voter, W., et al. 1988. Dynamic instability of individual microtubules analyzed by video light microscopy: rate constants and transition frequencies. *J. Cell Biol.* 107: 1437–48
153. Weil, C. F., Oakley, C. E., Oakley, B. R. 1986. Isolation of mip (microtubule-interacting protein) mutations of *Aspergillus nidulans. Mol. Cell. Biol.* 6:2963–68
154. Weinert, T., Hartwell, L. 1988. The RAD9 gene controls the cell cycle reponse to DNA damage in *Saccharomyces cerevisiae. Nature* 241:317–22
155. Williamson, D. 1985. The yeast ARS element, six years on: a progress report. *Yeast* 1:1–14
156. Winey, M., Byers, B. 1992. Spindle pole body of *Saccharomyces cerevisiae:* a model for genetic analysis of the centrosome cycle. In *The Centrosome,* ed. V. Kalnins, pp. 197–218. Orlando, FL: Academic
157. Winey, M., Goetsch, L., Byers, B. 1991. MPS1 and MPS2: novel yeast genes defining distinct steps of spindle pole body duplication. *J. Cell. Biol.* 114:745–54
158. Yang, C. H., Lambie, E. J., Hardin, J., Craft, J., Snyder, M. 1989. Higher order structure is present in the yeast nucleus: autoantibody probes demonstrate that the nucleolus lies opposite the spindle pole body. *Chromosoma* 98: 123–28
159. Yeh, E., Driscoll, R., Coltrera, M., Olins, A., Bloom, K. 1991. A dynamin-like protein encoded by the yeast sporulation gene *SPO15. Nature* 349:713–14
160. Yen, T. J., Li, G., Schaar, B. T., Szilak, I., Cleveland, D. W. 1992. CENP-E is a putative kinetochore motor that accumulates just before mitosis. *Nature* 359:536–39
161. Zhang, P., Knowles, B., Goldstein, L., Hawley, R. S. 1990. A kinesin-like protein required for distributive chromosome segregation in Drosophila. *Cell* 62:1053–62
162. Zheng, Y., Jung, M. K., Oakley, B. 1991. Gamma-tubulin is present in Drosophila melanogaster and Homo sapiens and is associated with the centrosome. *Cell* 65:817–23

Annu. Rev. Microbiol. 1993. 47:263–90

DISSIMILATORY METAL REDUCTION[1]

Derek R. Lovley
430 National Center, US Geological Survey, Reston, Virginia 22092

KEY WORDS: iron, manganese, uranium, selenium, chromium

CONTENTS

ABSTRACT

Microorganisms can enzymatically reduce a variety of metals in metabolic processes that are not related to metal assimilation. Some microorganisms can conserve energy to support growth by coupling the oxidation of simple organic acids and alcohols, H_2, or aromatic compounds to the reduction of Fe(III) or Mn(IV). This dissimilatory Fe(III) and Mn(IV) reduction influences the organic as well as the inorganic geochemistry of anaerobic aquatic sediments and ground water. Microorganisms that use U(VI) as a terminal electron acceptor play an important role in uranium geochemistry and may be a useful tool for removing uranium from contaminated environments. Se(VI) serves as a terminal electron acceptor to support anaerobic growth of some microorganisms. Reduction of Se(VI) to Se(0) is an important mechanism for the precipitation of selenium from contaminated waters. Enzymatic reduction of Cr(VI) to the less mobile and less toxic Cr(III), and reduction of soluble Hg(II) to volatile Hg(0) may affect the fate of these compounds in the environment and might be used as a remediation strategy. Microorganisms can also enzymatically reduce other metals such as technetium, vanadium, molybdenum, gold, silver, and copper, but reduction of these metals has not been studied extensively.

INTRODUCTION

Microorganisms that use metals as terminal electron acceptors, or reduce metals as a detoxification mechanism, have an important influence on the geochemistry of aquatic sediments, submerged soils, and the terrestrial subsurface. Furthermore, it is becoming increasingly apparent that microbial metal reduction may be manipulated to aid in the remediation of environments and waste streams contaminated with metals and certain organics. The purpose of this review is to give a brief overview of the physiology and ecology of microorganisms that reduce environmentally significant metals for non-assimilatory purposes.

Fe(III) REDUCTION

Microbial reduction of Fe(III) to Fe(II) has been studied not only because of its influence on iron geochemistry but also because Fe(III) is one of the most abundant potential electron acceptors for organic matter decomposition in many aquatic sediments and subsurface environments (71). Until recently, much of the Fe(III) reduction in sedimentary environments was generally considered to result from nonenzymatic reactions. However, we now know that in the anaerobic, nonsulfidogenic environments in which Fe(III) reduction is most important, dissimilatory Fe(III) reducers enzymatically catalyze nearly all of the Fe(III) reduction (71, 85).

Fe(III)-Reducing Microorganisms

In most sedimentary environments such as aquatic sediments, submerged soils, and aquifers, the oxidation of organic matter coupled to the reduction of Fe(III) requires the cooperative activity of several metabolic types of dissimilatory Fe(III) reducers (71) (Figure 1). For example, a great diversity of microorganisms that can metabolize sugars or amino acids with the reduction of Fe(III) have been described (32, 33, 36, 70). However, in all the cases examined, Fe(III) reduction is a trivial side reaction in the metabolism of these organisms (71). The primary products of the metabolism of the fermentative Fe(III)-reducing microorganisms are typical fermentation acids, alcohols, and H_2. Microorganisms that can completely oxidize sugars and amino acids to carbon dioxide with Fe(III) as the sole electron acceptor are unknown. Numerous attempts to isolate or enrich for such organisms have been unsuccessful (72). Even if microorganisms that completely oxidize glucose to carbon dioxide with Fe(III) as the sole electron acceptor exist, their

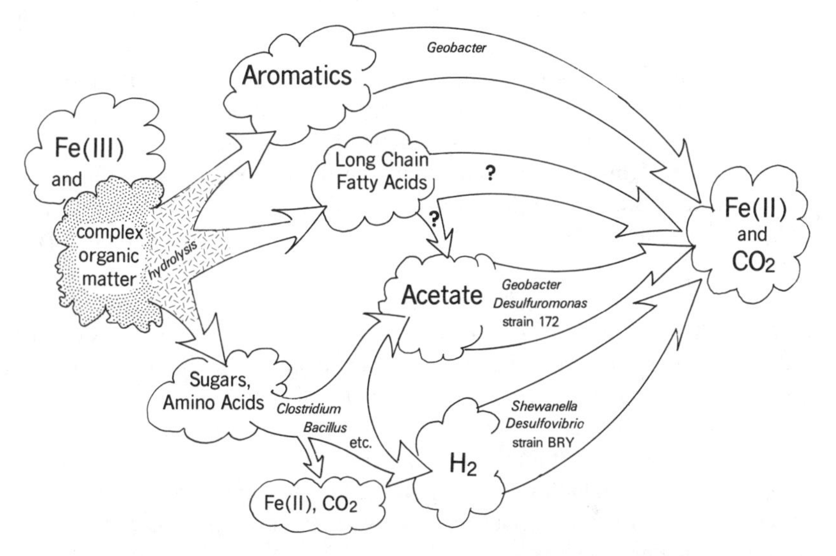

Figure 1 Model for oxidation of organic matter in sediments coupled to dissimilatory Fe(III) reduction showing examples of the microorganisms in pure culture known to catalyze the various reactions. Fermentation of sugars and amino acids has been simplified to designate the production of only the two major fermentation products, acetate and H_2. However, other short chain fatty acids are produced in lesser amounts. These include propionate and formate, which may be directly oxidized to carbon dioxide by organisms such as *Geobacter metallireducens* (propionate) or *Shewanella putrefaciens* (formate), as well as lactate, which is oxidized to acetate and carbon dioxide by organisms such as *S. putrefaciens*.

metabolism does not appear to be important in Fe(III)-reducing sediments as glucose is fermented to fatty acids rather than oxidized directly to carbon dioxide (80). Thermodynamic considerations support this idea, suggesting that microorganisms attempting to completely oxidize glucose to carbon dioxide with Fe(III) would be at a competitive disadvantage with microorganisms that convert glucose to fermentation products (80). These findings have led to the hypothesis that, in Fe(III)-reducing environments, fermentative microorganisms metabolize sugars and amino acids with relatively little Fe(III) reduction during this initial step.

Most of the electron transfer to Fe(III) during the metabolism of sugars and amino acids results from the oxidation of the fermentation products (71). Acetate is considered to be the most important fermentation product in Fe(III)-reducing sedimentary environments (80). The freshwater isolate, *Geobacter metallireducens* (formerly strain GS-15) was the first known acetate-oxidizing Fe(III) reducer (76, 79, 86). *G. metallireducens* conserves energy to support growth by the reaction:

$$\text{acetate}^- + 8\ \text{Fe(III)} + 4\ H_2O \rightarrow 2\ HCO_3^- + 8\ \text{Fe(II)} + 9\ H^+.$$

G. metallireducens oxidizes various other volatile fatty acids and simple alcohols.

G. metallireducens is in the delta proteobacteria (76). Its closest known relative is *Desulfuromonas acetoxidans*. This organism was previously known primarily for its unique ability to couple the oxidation of acetate to the reduction of S^0. However, *D. acetoxidans* can also oxidize acetate with Fe(III) as the electron acceptor (116). This finding demonstrates that some marine organisms can also effectively couple the oxidation of organic compounds to Fe(III) reduction.

Another acetate-oxidizing Fe(III) reducer, strain 172, was recovered from deep subsurface sediments of an Atlantic Coastal Plain aquifer (75). Although a detailed characterization of 172 is not yet available, the preliminary evidence suggests that it is not a *Geobacter* or *Desulfuromonas* species.

Shewanella putrefaciens (formerly *Alteromonas putrefaciens*) can also conserve energy to support growth by coupling the oxidation of organic compounds to the reduction of Fe(III) (74, 84, 96). Formate is oxidized to carbon dioxide whereas lactate and pyruvate are incompletely oxidized to carbon dioxide and acetate:

$$\text{formate}^- + 2\ \text{Fe(III)} + H_2O \rightarrow HCO_3^- + 2\ \text{Fe(II)} + 2\ H^+;$$

$$\text{lactate}^- + 4\ \text{Fe(III)} + 2\ H_2O \rightarrow \text{acetate}^- + HCO_3^- + 4\ \text{Fe(II)} + 5\ H^+;$$

$$\text{pyruvate}^- + 2\ \text{Fe(III)} + 2\ H_2O \rightarrow \text{acetate}^- + HCO_3^- + 2\ \text{Fe(II)} + 3\ H^+.$$

Formate oxidation coupled to Fe(III) reduction is a potentially important process in sediments if formate replaces H_2 as an important fermentation product in Fe(III)-reducing environments (84). However, the metabolism of lactate and pyruvate by *S. putrefaciens* is expected to be a minor pathway for carbon and electron flow in Fe(III)-reducing environments (71, 84).

The estuarine microorganism, strain BrY (22), and several *Desulfovibrio* species (D. R. Lovley, E. J. P. Phillips, J. C. Woodward & E. E. Roden, submitted) also incompletely oxidize lactate to acetate and carbon dioxide with Fe(III) as the electron acceptor. Like *S. putrefaciens,* BrY conserves energy to support growth from Fe(III) reduction. In contrast, there is no net growth during Fe(III) reduction by the *Desulfovibrio* species, but this metabolism may provide some energy.

Aquaspirillum magnetotacticum grows slowly and reduces Fe(III) in anaerobic medium that contains succinate as the sole electron donor (42). The authors of this study concluded that it is not clear whether *A. magnetotacticum* conserves energy from electron transport to Fe(III), and no data on succinate metabolism were provided.

Although the heterotrophic Fe(III)-reducing microorganisms discussed above grow best at circumneutral pH, numerous acidophilic heterotrophic bacteria reduce Fe(III) while growing in complex medium containing glucose or glycerol as well as tryptone (54). One isolate studied in detail reduces Fe(III) under both aerobic and anaerobic conditions. Evidence that energy is conserved from Fe(III) reduction is the finding that there is more cell growth as more Fe(III) is provided.

Several microorganisms can couple the oxidation of H_2 to the reduction of Fe(III). A *Pseudomonas* sp. (3), *S. putrefaciens* (84), and BrY (22) conserve energy to support growth via the reaction:

$$H_2 + 2\,Fe(III) \rightarrow 2\,H^+ + 2\,Fe(II).$$

Data (71) did not substantiate claims that another organism could grow by coupling the oxidation of H_2 to the reduction of Fe(III) (55).

Several *Desulfovibrio* species also oxidize H_2 with the reduction of Fe(III) at rates comparable to those observed with other Fe(III) reducers, but no net cell growth occurs (D. R. Lovley, E. J. P. Phillips, J. C. Woodward & E. E. Roden, submitted). The minimum threshold for H_2 uptake in *D. desulfuricans* is lower with Fe(III) serving as the electron acceptor than with sulfate, suggesting that under conditions of limiting electron donor availability, *Desulfovibrio* species will preferentially reduce Fe(III).

A wide variety of monoaromatic compounds can be completely oxidized to carbon dioxide with Fe(III) serving as the sole electron acceptor (68, 85). *G. metallireducens* is the only aromatic-oxidizing, Fe(III)-reducing microorganism as yet available in pure culture (73, 77). Aromatics oxidized by *G. metallireducens* include prevalent contaminants such as toluene, *p*-cresol, and phenol. The metabolism of *G. metallireducens* serves as a model for the oxidation of aromatic contaminants coupled to Fe(III) reduction frequently observed in polluted aquifers (73, 77).

Long-chain fatty acids are likely another important component of organic matter that is metabolized in sediments. Enrichment cultures that can oxidize long-chain fatty acids have previously been established (71), but no isolates have been purified and the pathways for oxidation of long-chain fatty acids in Fe(III)-reducing environments have not been elucidated.

In acidic environments, elemental sulfur can serve as an electron donor for Fe(III) reduction. *Thiobacillus ferrooxidans, Thiobacillus thiooxidans,* and the thermophile, *Sulfolobus acidocaldarius,* reduce Fe(III) by the following reaction (19):

$$S^0 + 6\ Fe(III) + 4\ H_2O \rightarrow HSO^-_4 + 6\ Fe(II) + 7\ H^+.$$

Initial studies indicated that neither *T. thiooxidans* nor *T. ferrooxidans* (58, 118, 130) could conserve energy to support growth from this reaction. However, subsequent studies have demonstrated that Fe(III) reduction provides energy to support amino acid transport (111) and growth (110) in *T. ferrooxidans*. Microorganisms that oxidize sulfur with the reduction of Fe(III) at circumneutral pH have not been reported.

In addition, *S. acidocaldarius* provides evidence for the existence of thermophilic Fe(III) reducers. The recovery of large quantities of ultrafine-grained magnetite from depths as great as 6.7 km below the land surface presumably indicates the presence of Fe(III)-reducing life in such environments (37).

Electron Transport to Fe(III)

Investigations into electron transport to Fe(III) in dissimilatory Fe(III) reducers are important not only for a better understanding of the mechanisms for Fe(III) reduction but also because of the potential evolutionary significance of Fe(III) reduction. The geological evidence suggests that microbial Fe(III) reduction may have evolved before other respiratory processes, such as sulfate reduction, nitrate reduction, and oxygen reduction, which can also completely oxidize multicarbon organic compounds back to carbon dioxide (71).

Much of the early work on electron transport to Fe(III) conducted with organisms grown under aerobic conditions was recently reviewed (71). The relevance of these studies to anaerobic dissimilatory Fe(III) reduction has been questioned because aerobically grown cells produce Fe(III) reductases that are not linked to energy conservation (71).

The three strains of *S. putrefaciens* that grow on nonfermentable substrates under anaerobic conditions with Fe(III) as the sole electron acceptor (28, 84, 96) must produce Fe(III) reductases that are linked to energy conservation. When cell suspensions of anaerobically grown *S. putrefaciens* are provided with Fe(III), the pH of the external medium drops, suggesting that electron transport to Fe(III) is associated with proton translocation (98). Growth under anaerobic conditions stimulates the production of *c*-type cytochromes in the outer membrane of *S. putrefaciens* (95). If these cytochromes can reduce Fe(III), this suggests a mechanism by which *S. putrefaciens* might transfer electrons to insoluble Fe(III) oxides (95). Although the Fe(III) reductase in *S. putrefaciens* has yet to be identified, it appears to be a different enzyme than the nitrate reductase (28, 98).

G. metallireducens contains a membrane-bound Fe(III) reductase also distinct from the nitrate reductase (39), as well as several membrane-bound and soluble *c*-type cytochromes (76, 99; J. E. Champine & S. Goodwin, submitted), some of which appear to be involved in electron transport to Fe(III) and other metals (76). *c*-Type cytochromes may also serve as electron carriers during Fe(III) reduction in the closely related *D. acetoxidans* (116). Menaquinone and ferredoxin are other likely components of the electron transport chain to Fe(III) in *G. metallireducens* (76; J. E. Champine & S. Goodwin, submitted). *G. metallireducens* derives electrons for Fe(III) reduction by oxidizing acetate via the citric acid cycle (24).

Thiobacillus ferrooxidans contains several enzymes that can reduce soluble Fe(III) under acidic conditions and could be involved in dissimilatory Fe(III) reduction with reduced sulfur compounds as the electron donor. A periplasmic enzyme purified to an electrophoretically homogeneous state (127) catalyzes the oxidation of sulfide to sulfite with Fe(III) as the electron acceptor (125). Another Fe(III) reductase activity located in the plasma membrane oxidizes sulfite to sulfate (126).

The first enzyme capable of reducing insoluble Fe(III) oxide at circumneutral pH was purified from the dissimilatory Fe(III) reducer, *Desulfovibrio vulgaris* (D. R. Lovley, P. K. Widman & J. C. Woodward, submitted). Attempts to find the U(VI) reductase in this organism (see below) indicated that the c_3 cytochrome was an active metal reductase. When a combination of hydrogenase and H_2 was used as a source for electrons for c_3, c_3 reduced a poorly crystalline Fe(III) oxide-producing soluble Fe(II) and the magnetic mineral magnetite. Electron transport via cytochrome c_3 may account for the ability of *D. vulgaris* and other *Desulfovibrio* species to reduce Fe(III) oxides.

Environmental Significance and Ecology of Fe(III) Reduction

Dissimilatory Fe(III) reduction has a greater overall environmental impact than microbial reduction of any other metal. Microbial Fe(III) reduction has been directly shown or implicated as an important process in the following phenomena: organic matter decomposition in a variety of freshwater, estuarine, and marine sediments; the decomposition of aromatic hydrocarbons in contaminated aquifers; the generation of undesirably high concentrations of dissolved iron in deep pristine aquifers; the oxidation of organic matter coupled to Fe(III) that resulted in the accumulation of magnetite in the Banded Iron Formations; the accumulation of magnetite around hydrocarbon seeps and formation of ultrafine-grained magnetite in aquatic sediments; the formation of other ferrous iron minerals such as siderite and vivianite; the control of the extent of methane formation in shallow freshwater environments; the release of phosphate and trace metals into water supplies; soil gleying; and corrosion of steel. A recent review examined the literature on this subject (71).

Although several microorganisms can now serve as models for the enzymatically catalyzed reduction of Fe(III), it is not clear that any of the organisms currently available in pure culture are the dominant Fe(III) reducers in sedimentary environments. For example, a study on the mechanisms for siderite ($FeCO_3$) concretion formation in salt marsh sediments found that organisms related to *G. metallireducens* or *S. putrefaciens* were not abundant in the zone of Fe(III) reduction (25a). The concretions were enriched with *Desulfovibrio*-type organisms that are now known to be active dissimilatory Fe(III) reducers (see above), though at the time they were not generally regarded as such. Further circumstantial evidence for the potential role of sulfate-reducing organisms in dissimilatory Fe(III) reduction is that sulfate reducers are abundant in the zones of deep aquifers of the Atlantic Coastal Plain in which Fe(III) reduction is the terminal electron-accepting process (25, 94). Sulfate reducers living for such long periods in these subsurface environments, in which no detectable sulfate reduction occurs, probably gain energy for survival from electron transport to Fe(III). Such findings emphasize the need for more intensive study of community structure of Fe(III)-reducing environments.

Mn(IV) REDUCTION

Mn(IV)-Reducing Microorganisms

Microbial reduction of Mn(IV) to Mn(II) greatly parallels dissimilatory Fe(III) reduction. Most of the microorganisms that reduce Mn(IV) reduce Fe(III) and vice versa. Furthermore, the fact that Fe(II) rapidly reduces Mn(IV) means that, in most environments, any microorganism that can enzymatically reduce Fe(III) will also indirectly reduce Mn(IV) (78, 97). As with Fe(III), much of the early work on dissimilatory Mn(IV) reduction was primarily conducted with microorganisms that only used Mn(IV) as a minor electron acceptor during metabolism. This early literature was reviewed previously (31, 32, 36).

In general, Fe(III)/Mn(IV) reducers oxidize the same electron donors with Mn(IV) as they do with Fe(III). The exception is the H_2-oxidizing, Fe(III)-reducing *Pseudomonas* sp., which cannot oxidize H_2 with Mn(IV) (3). All of the other well-studied microorganisms that can conserve energy from dissimilatory Fe(III) reduction such as *G. metallireducens* (79), *S. putrefaciens* (84, 96), strain BrY (22), and *D. acetoxidans* (116) can also grow with Mn(IV) as the sole electron acceptor. However, in most instances, studies on electron donor metabolism with Mn(IV) as the electron acceptor have not been as extensive as studies with Fe(III). For example, although *G. metallireducens* can reduce Mn(IV) when benzoate is provided as an electron donor (D. R. Lovley, unpublished data) the oxidation of aromatic compounds coupled to Mn(IV) reduction in this or other organisms has not been studied in detail. It seems likely, however, that organic matter could be oxidized to carbon dioxide with the reduction of Mn(IV) in a manner similar to that proposed in Figure 1 for Fe(III) reduction.

Electron Transport to Mn(IV)

Electron transport systems to Mn(IV) have been investigated in several *Bacillus* species that reduce Mn(IV) (27, 31, 36). However, the available evidence suggests that this Mn(IV) reduction is a minor side reaction that does not conserve energy to support growth for these organisms, and thus this metabolism may have little relevance to the bulk of dissimilatory Mn(IV) reduction in sedimentary environments (71).

Almost nothing is known about electron transport to Mn(IV) in organisms that conserve energy to support growth from Mn(IV) reduction. Growth of *G. metallireducens* (79), *S. putrefaciens* (84, 96), strain BrY (22), and *D. acetoxidans* (116) on nonfermentable substrates with Mn(IV) as the electron acceptor implies energy conservation through electron transport to a Mn(IV) reductase. Evidence of the potential for energy conservation as the result of electron transport to Mn(IV) was the finding that cell suspensions of *S. putrefaciens* translocated protons to the external medium when provided with lactate and Mn(IV) (98). The *c*-type cytochrome(s) in *D. acetoxidans* are oxidized by Mn(IV), suggesting that they are involved in electron transport to Mn(IV) (116).

Environmental Significance of Mn(IV) Reduction

The environmental significance of enzymatic mechanisms for Mn(IV) reduction is less clear than for Fe(III) reduction. This is because Mn(IV) is more easily reduced through nonenzymatic mechanisms than is Fe(III) (71). Most notably, no practical method has been devised for differentiating between enzymatic reduction of Mn(IV) and nonenzymatic reduction of Mn(IV) by Fe(II). However, the distinction in this case may not be meaningful because the enzymatic reduction of Fe(III) oxide followed by nonenzymatic reduction of Mn(IV) by Fe(II) results in the same end products, Mn(II) and Fe(III) oxide, as does direct enzymatic reduction of Mn(IV) (78).

In analogy to Fe(III) reduction, microbial Mn(IV) reduction has the potential to be important in the formation of reduced manganese minerals, the release of dissolved manganese into sediment pore and ground waters, and the release of trace metals bound to Mn(IV) oxides (71). Mn(IV) reduction may also serve to oxidize organic matter in aquatic sediments, deep pristine aquifers, and shallow aquifers contaminated with organics, but quantitative data are scarce (71).

U(VI) REDUCTION

Microbial reduction of soluble U(VI) to insoluble U(IV) is an important process in the global uranium cycle and may also be a useful technique for removing uranium from contaminated environments (40, 81–83). Early evidence that microorganisms might reduce U(VI) to U(IV) was the finding that cell-free extracts of *Vellionella atypica* (formerly *Micrococcus lactilyticus*) reduce U(VI) along with a variety of other metals (141). However, there is no evidence that this metal reduction is an enzymatic reaction or has any physiological significance. Subsequent studies have demonstrated that whole cells of *V. atypica* do not reduce U(VI) (D. R. Lovley, unpublished data).

U(VI)-Reducing Microorganisms

G. metallireducens was the first organism found to use U(VI) as a terminal electron acceptor. *G. metallireducens* grows by carrying out the reaction:

$$\text{acetate}^- + 4\ U(VI) + 4\ H_2O \rightarrow 2HCO_3^- + 4\ U(IV) + 9\ H^+.$$

S. putrefaciens can also grow with U(VI) as the sole electron acceptor and H_2 as the electron donor:

$$H_2 + U(VI) \rightarrow 2\ H^+ + U(IV).$$

Both organisms will grow in high-concentration (8 mM) dissolved uranium.

Several *Desulfovibrio* species can also enzymatically reduce U(VI) with either H_2 or lactate (82) (D. R. Lovley, unpublished data). However, attempts to grow *D. desulfuricans* with U(VI) as the sole electron acceptor have been unsuccessful (82).

Electron Transport to U(VI)

The enzymatic mechanisms for U(VI) reduction by *G. metallireducens* and *S. putrefaciens* are ill defined. The fact that these organisms conserve energy to support growth by oxidizing nonfermentable substrates with U(VI) as the sole electron acceptor suggests that electron transport-linked phosphorylation must be involved. Electron transport to or through *c*-type cytochrome(s) is likely based on the observation that U(VI) oxidizes the *c*-type cytochromes in whole-cell suspensions of *G. metallireducens* (76).

A U(VI) reductase has been isolated from the U(VI) reducer, *D. vulgaris* (D. R. Lovley, P. K. Widman & J. C. Woodward, submitted). The soluble fraction of *D. vulgaris* rapidly reduces U(VI) with H_2 as the electron donor. If cytochrome c_3 is removed from the soluble fraction of *D. vulgaris,* then all capacity for U(VI) reduction is lost. If cytochrome c_3 is added back, then the capacity for U(VI) reduction is restored. U(VI) rapidly oxidizes previously reduced cytochrome c_3. U(VI) is rapidly reduced when c_3 is combined with hydrogenase, the physiological electron donor for c_3, and H_2.

Environmental Significance of U(VI) Reduction

The environmental significance of microbial U(VI) reduction is that U(VI) is highly soluble in most natural waters, whereas U(IV) is highly insoluble (65). The reduction of U(VI) to U(IV) in anaerobic marine sediments is the most globally significant sink for dissolved uranium (1, 59, 136). The reductive precipitation of uranium from ground water is considered to be the mechanism for the formation of some sandstone or roll-type uranium ores (50, 53, 65). The earlier geochemical literature (50, 53, 65) suggested that U(VI) reduction in anaerobic environments results from nonenzymatic reduction of U(VI) reduction by sulfide or H_2. However, neither are effective U(VI) reductants at the temperatures and pH typical of most aquatic sediments or ground waters, and sterilization of anaerobic sediments inhibits U(VI) reduction (83). These findings suggest that U(VI)-reducing enzymes are responsible for U(VI) reduction in these environments.

Microbial U(VI) reduction may be used to remove uranium from contaminated waters and soils. In most natural surface and ground waters, U(VI) is in the form of uranyl-carbonate complexes (65). Furthermore, uranyl-carbonate complexes are the typical dissolved uranium form that results from various human activities utilizing uranium (81). Studies with *G. metallireducens* (40) and *D. desulfuricans* (82) demonstrated that U(VI)-reducing microorganisms can reduce the U(VI) in U(VI)-carbonate complexes. The U(IV) precipitates as uraninite (UO_2), all of which is extracellular. Thus, microbial uranium reduction can potentially take uranium that is dispersed in a large volume of liquid and concentrate it into a very pure, compact solid.

D. desulfuricans was chosen for detailed studies of microbial removal of U(VI) from contaminated environments because of the ease in mass culturing this organism and because its U(VI)-reducing capacity is extremely stable. For example, freeze-dried cells kept under air at room temperature lose none of their potential for U(VI) reduction even after six months of storage (81). *D. desulfuricans* effectively reduces U(VI) both at very high (24 mM) and at relatively low (< 50 nM) concentrations (81). Of the wide variety of potentially inhibiting anions and cations evaluated, only exceptionally high concentrations (> 20 μM) of copper inhibited U(VI) reduction. *D. desulfuricans* readily removed soluble U(VI) from several mine drainage waters and contaminated ground waters from a Department of Energy site.

In addition to treating uranium-contaminated waters, microbial U(VI) reduction can be used as part of a technique to concentrate uranium from contaminated soils (D. R. Lovley, E. J. P. Phillips & E. R. Landa, in preparation). In this process, uranium is leached from the soils with a bicarbonate solution and then microbial U(VI) reduction precipitates the uranium from the extract.

Microbial U(VI) reduction has several advantages over other previously proposed treatment techniques (81). These include: the ability to precipitate uranium from U(VI)-carbonate complexes; the recovery of uranium in a highly concentrated and pure form; high uranium removal per amount of biomass; the potential to simultaneously treat organic contaminants and uranium by using the organic as an electron donor for U(VI) reduction; and the potential for in situ remediation of both ground and surface waters.

Se(VI), Se(IV), AND Se(0) REDUCTION

Although selenium is classified a metalloid rather than a true metal, it is included here because of the obvious parallels between dissimilatory selenium reduction and dissimilatory metal reduction and because of its environmental significance. The discovery that toxic concentrations of selenium were adversely affecting bird populations at the Kesterson National Wildlife Refuge in California led to a surge in research on microbial metabolism of selenium. The predominant redox states of selenium in natural environments are Se(VI) (selenate, SeO_4^{-2}), Se(IV) (selenite, SeO_3^{-2}), Se(0) (elemental selenium), and Se(−II) (selenide) (29). The first three of these can potentially serve as electron acceptors for microbial metabolism.

Se(VI)-, Se(IV)-, and Se(0)-Reducing Microorganisms

A high percentage of the heterotrophic microorganisms that can be isolated from soil can reduce selenate to elemental selenium (13). *Clostridium* (56), *Citrobacter, Flavobacterium,* and *Pseudomonas* species (21) as well as an unidentified gram-negative rod (89) all reduce selenate. However, the mechanisms for selenate reduction in these organisms were not studied in detail and no evidence of selenate-dependent growth was provided.

Several organisms that can conserve energy to support growth via selenate reduction have been described. Strain SeS was purified from intertidal sediments of San Francisco Bay (105). SeS obtains energy to support growth by the reaction:

$$4\ CH_3COO^- + 3\ SeO_4^{-2} \rightarrow 3\ Se^0 + 8\ CO_2 + 4\ H_2O + 4\ H^+.$$

Pseudomonas sp. AX, which was isolated from biological reactors used for selenium removal (88) has a different mode of selenate reduction, reducing selenate to selenite:

$$CH_3COO^- + H^+ + 4\ SeO_4^{-2} \rightarrow 2\ CO_2 + 4\ SeO_3^{-2} + 2\ H_2O.$$

When *Pseudomonas* sp. AX is cocultured with a selenite-reducing microorganism, selenate is rapidly reduced to elemental selenium (88). Another selenate-reducing microorganism, designated SeS-3, grows in defined medium with lactate as the electron donor and selenate as the electron acceptor (104, 122). Selenate is reduced to selenite and elemental selenium.

The ability of microorganisms to reduce selenite to elemental selenium has been known since the turn of the century (67 and references therein) and a wide range of microorganisms are known to catalyze this reaction (29, 104). This metabolism has never been found to be associated with energy conservation and may be a detoxification mechanism (29, 104). Cell suspensions of *T. ferrooxidans* reduce red elemental selenium to selenide (2).

Environmental Significance and Ecology of Selenium Reduction

Selenate and selenite reduction to insoluble elemental selenium has been documented in a variety of soils and aquatic sediments, and the evidence suggests that this reduction is enzymatically catalyzed (30, 89, 105, 123). Because of the low concentrations of selenate and selenite in the environment, they are not important electron acceptors for organic matter oxidation (105). However, microbial selenate and selenite reduction play a central role in affecting the fate of selenium in aquatic environments and may be the principle mechanism for the removal of selenate and selenite from agricultural wastewater evaporation ponds and high-selenium ground waters (104, 106, 107).

Various lines of evidence indicate that sulfate-reducing microorganisms are not responsible for selenate reduction in sediments. Selenate and selenite are reduced at shallower depths than the zone of sulfate reduction (105, 106). At the depths where selenate is reduced, the sulfate concentrations are at levels that inhibit selenate reduction by sulfate reducers (143). Furthermore, molybdate, a specific inhibitor of sulfate reduction, did not inhibit selenate reduction (105).

Nitrate and Mn(IV) are preferred electron acceptors for selenate reducers (105, 122). Selenate reduction might be catalyzed by enzyme(s) that are also involved in dissimilatory nitrate reduction (104). The distribution of the maximum potential for denitrification and selenate reduction potentials in sediments are similar (106, 123), and the inhibition of selenate reduction in some sediments by tungstate but not molybdate might indicate the involvement of a molybdenum-containing nitrate reductase (105, 123). However, nitrate does not always inhibit selenate reduction at the low concentrations found in pore waters (106).

The immobilization of selenium that results when selenate is reduced to elemental selenium may be exploited to enhance removal of selenate from contaminated waters (89). For example, at Kesterson, flooding of previously exposed pond sediments with selenium-free water immobilized 66–110% of the dissolved selenium that had been present in the upper 1.22 m of soil (69). Immobilization is thought to result from the development of anaerobic conditions, which should stimulate selenate reduction. However, in situ stimulation of selenate reduction may not be a suitable long-term solution to the selenium-contamination problem as elemental selenium may enter the food chain through the bottom-feeding organisms (87).

Ex situ treatment processes for contaminated waters are also possible. Selenate is microbially reduced to insoluble elemental selenium when uranium-mine discharge water is passed through a soil column (56). Several investigators have suggested using combined algal-bacterial systems to remove selenium from contaminated waters (35, 106). Nitrate, which inhibits selenate reduction, is removed via the growth of algae. The water thus treated is then fed into an anaerobic digestor where a portion of the algae are oxidized and thereby become electron donors for removal of any remaining nitrate (through denitrification), and then selenate is reduced and precipitated as elemental selenium.

Cr(VI) REDUCTION

Chromium contamination of the environment is extensive (12). The reduction of highly toxic and mobile Cr(VI) to the less toxic, less mobile Cr(III) is likely to be a useful process for the remediation of contaminated waters and soils (108). This problem has stimulated interest in microorganisms that can use Cr(VI) as an electron acceptor.

Cr(VI)-Reducing Microorganisms

Early investigations demonstrated that facultative anaerobes such as *Pseudomonas dechromaticans* (117), *Pseudomonas chromatophila* (66), and *Aeromonas dechromatica* (64) remove Cr(VI) from solution by the formation of a Cr(III) precipitate, presumably $Cr(OH)_3$. Subsequent studies have demonstrated that the capacity for Cr(VI) reduction is widespread and found in such organisms as *Bacillus cereus, Bacillus subtilis, Pseudomonas aeruginosa, Achromobacter eurydice, Micrococcus roseus,* and *Escherichia coli* (43), as well as *Pseudomonas ambigua* (48), *Pseudomonas fluorescens* (17), *Enterobacter cloacae* (137), *Streptomyces* spp. (26), *Pseudomonas putida* (52), *D. desulfuricans,* and *D. vulgaris* (D. R. Lovley & E. J. P. Phillips, in preparation).

Many of these organisms reduce Cr(VI) better under aerobic conditions than under anaerobic conditions and the physiological role of this aerobic Cr(VI) reduction has not been well defined. Cr(VI) reduction is not a Cr(VI)-resistance mechanism (17), and Cr(VI) reduction in *P. putida* and possibly other Cr(VI) reducers is a side activity for enzymes that have other, as yet unidentified, natural substrates (23, 52).

Although some microorganisms reduce Cr(VI) during anaerobic growth in media in which Cr(VI) is provided as the sole electron acceptor, in no instance has Cr(VI) reduction definitely been shown to yield energy to support anaerobic growth. For example, *Pseudomonas chromatophila* purportedly uses Cr(VI) as an electron acceptor to support growth under anaerobic conditions with a variety of electron acceptors, including the nonfermentable substrate, acetate (66). However, in the only data shown, most of the Cr(VI) reduction took place after growth stopped. No studies demonstrating that anaerobic growth depended upon the presence of Cr(VI) were presented.

Pseudomonas fluorescens LB300, which reduces Cr(VI) while growing aerobically in a glucose medium, also grows in an anaerobic chamber containing oxygen-free N_2 on agar plates containing acetate as a potential electron donor and Cr(VI) as the potential electron acceptor (17). However, no Cr(VI) reduction occurs in anaerobic liquid cultures. Neither acetate oxidation nor Cr(VI) reduction under anaerobic conditions was documented. Further studies on whether electron transport to Cr(VI) can yield energy to support growth of this organism seem warranted given the difficulties in maintaining oxygen-free conditions in anaerobic chambers under a N_2 atmosphere.

Enterobacter cloacae strain HO1 reduces Cr(VI) while growing anaerobically in a medium that contains acetate and Casamino acids as potential electron donors (137). O_2 rapidly inhibits Cr(VI) reduction, but reduction resumes when O_2 is removed (62). Amino acid mixtures are the best electron donors for Cr(VI) reduction (102). The fate of the electron donors (i.e. whether they are oxidized to carbon dioxide or whether they are fermented) has not been reported. *E. cloacae* can grow anaerobically in the absence of added Cr(VI) and no evidence for Cr(VI)-dependent growth has been presented (137). In fact, in one study, the number of viable cells decreased in the initial stages after Cr(VI) was added, and ~50% of the added Cr(VI) was reduced before viable cell numbers increased over what was present prior to Cr(VI) addition (100).

Enzymatic Mechanisms for Cr(VI) Reduction

Cr(VI) reduction is an enzymatically catalyzed reaction in the Cr(VI) reducers that have been studied in detail. For example, spent medium from cultures of *P. fluorescens* LB300 does not reduce Cr(VI), and washed cell suspensions reduce Cr(VI) only if glucose or another suitable electron donor is provided (17). Cyanide (10^{-2} M) or azide (10^{-3} M) inhibit Cr(VI) reduction in crude cell extracts, and the Cr(VI)-reducing capacity is lost when the membrane fraction is removed.

Evidence for an enzymatic role in Cr(VI) reduction in *E. cloacae* is that: Cr(VI) is reduced faster with higher cell densities; no Cr(VI) reduction occurs in cell-free controls; and inhibition of growth with antibiotics inhibits Cr(VI) reduction (137). Cell-free filtrates of cultures do not reduce Cr(VI) (138). Several metabolic poisons as well as molybdate, vanadate, and tellurate inhibit Cr(VI) reduction. Cr(VI) also inhibits Cr(VI) reduction—the rate of reduction declines as the concentration rises above 1 mM (100). Temperature and pH optima for Cr(VI) reduction are characteristic of an enzymatically catalyzed reaction (63).

The Cr(VI) reductase activity in *E. cloacae* is located in the membrane fraction (138). When membrane vesicles are reduced with NADH and then exposed to Cr(VI), the *c*- and *b*-type cytochromes are oxidized (139). Further analyses have suggested that of the identifiable cytochromes in the membrane vesicles (c_{548}, c_{549}, c_{550}, b_{555} , b_{556}, and b_{558}), c_{548} might be specifically involved in Cr(VI) reduction, serving as a branch point between Cr(VI) and O_2 reduction.

In contrast to *P. fluorescens* and *E. cloacae,* the Cr(VI)-reducing activity in *P. ambigua* and *P. putida* is in the soluble fraction of the cell (48, 52). A 65-kDa protein has been purified from *P. ambigua* that can reduce Cr(VI) with NADH or NADPH serving as the electron donor (134). The enzyme initially reduces Cr(VI) to Cr(V), which is subsequently reduced to Cr(III). NADH also reduces Cr(VI) to Cr(V) in the absence of the enzyme, but at slower rates than with the enzyme.

Washed cell suspensions of *D. desulfuricans* and *D. vulgaris* rapidly reduce

Cr(VI) to Cr(III) under anaerobic conditions with H_2 as the electron donor (D. R. Lovley & E. J. P. Phillips, in preparation). H_2-dependent, Cr(VI) reductase activity in the soluble, cell-free fraction of *D. vulgaris* is lost when the soluble fraction is passed over a cation-exchange column that removes cytochrome c_3, a periplasmic protein. The capacity for Cr(VI) reduction is restored when cytochrome c_3 is added back. In the presence of H_2 and an excess of hydrogenase, cytochrome c_3 reduces Cr(VI) at a rate 50-fold faster than the maximum rate for Cr(VI) reduction by the Cr(VI) reductase purified from *P. ambigua*.

Environmental Significance of Cr(VI) Reduction

Most of the earth's chromium is tied up as insoluble Cr(III) (12). In unpolluted fresh and sea waters the concentrations of chromium are generally less than 50 nmol/liter (113). Therefore, microbial Cr(VI) reduction is of primary concern in contaminated environments or for the treatment of Cr(VI)-containing wastes. Cr(VI) is readily reduced to Cr(III) in anaerobic aquatic sediments, but nonenzymatic processes such as Cr(VI) reduction by Fe(II) are likely to be at least as important as enzymatic Cr(VI) reduction (91).

The most extensive studies on the potential use of Cr(VI)-reducing microorganisms for the removal of Cr(VI) from waste streams have been conducted with *E. cloacae* strain HO1. Early studies with this organism revealed that during Cr(VI) reduction the yellow medium turns white and turbid, presumably as the result of the formation of insoluble Cr(III) hydroxide (137). However, the Cr(III) does not readily precipitate and cannot be effectively removed with centrifugation (61). Cr(III) is removed more effectively if *E. cloacae* is placed on one side of a semipermeable membrane and the Cr(VI)-containing medium is on the other (61). Cr(VI) diffuses into culture and *E. cloacae* reduces the Cr(VI) to Cr(III), which is retained on the culture side, especially when an anion-exchange membrane that does not permit passage of Cr(III) is used (60). Despite successful removal of Cr(VI) from culture medium with *E. cloacae,* removal from industrial effluents has been problematic as heavy metals and sulfate in the effluents can inhibit Cr(VI) reduction (45, 101). Whether microbial Cr(VI) reduction would have advantages over simple chemical Cr(VI) reduction in the removal of Cr(VI) does not appear to have been evaluated.

Hg(II) REDUCTION

Many aerobic and facultative microorganisms reduce soluble ionic Hg(II) to volatile Hg(0) as a detoxification mechanism (20, 114, 132). The Hg(II) is reduced during aerobic growth and Hg(II) reduction is not linked to energy-conserving electron transport. In contrast to most types of dissimilatory metal reduction, about which very little is known, many of the mechanisms for Hg(II) reduction have been beautifully elucidated for both gram-negative and gram-positive bacteria. This includes understanding of the mechanisms for Hg(II) transport into the cell and the intracellular reduction of Hg(II), as well as mapping of the organization of the mercury-resistance (*mer*) operons and determination of the structure of the Hg(II) reductase. These topics have been reviewed frequently (for recent updates see 119, 121, 131 and references therein) and thus, because of space limitations, are not discussed further here. As recently suggested, the respiratory metal-reducing microorganism, *G. metallireducens,* may be able to reduce Hg(II) through its dissimilatory metal-reducing pathway (76).

Environmental Significance and Ecology of Microbial Hg(II) Reduction

Microbial reduction of Hg(II) may play an important role in the fate of mercury in both aquatic and terrestrial environments (11, 38). For example, microbial Hg(II) reduction in ocean waters has been proposed to be a significant source of atmospheric mercury (57). However, the study of Hg(II) reduction in the environment is complicated by the fact that the mercury cycle is much more complex than the geochemical cycles of other redox sensitive metals such as iron (51). In addition to microbial Hg(II) reduction, microbially catalyzed methylation and demethylation reactions as well as a myriad of abiotic reactions also influence mercury cycling (11, 38). Furthermore, although some of the Hg(II) reduction in natural waters clearly depends upon the presence of living microorganisms, in some instances Hg(II) reduction is primarily a nonbiological process (6, 8, 11, 112).

Mercury contamination of the environment generally enhances the capacity for microbial Hg(II) reduction. In uncontaminated waters the rate of biological Hg(II) reduction typically lags for 6–24 h until the microbial population can adapt (6, 11). This lag is in contrast to the immediate microbial Hg(II) reduction in water samples from contaminated sites or pristine waters that have been artificially preexposed to Hg(II). Mercury-contaminated environments have increased abundance of Hg(II)-resistant bacteria and/or *mer* genes as well as increased rates of Hg(II) reduction (4–6, 8–11, 103, 115). Although many of the Hg(II)-reducing microorganisms in mercury-contaminated environments do not have genes that hybridize to *mer* gene probes under highly stringent hybridization conditions (7, 8, 103, 115), current evidence indicates that these organisms also volatilize Hg(II) with a NADPH-dependent Hg(II) reductase (7).

One proposed bioremediation strategy for removing mercury from contaminated environments would be to stimulate *mer*-mediated reduction of Hg(II)

(11, 38). This could be accomplished by increasing the density of working *mer* operons, either by adding Hg(II) reducers, stimulating the growth of indigenous populations, amplifying the number of *mer* operons in indigenous microorganisms, or increasing the percentage of indigenous *mer* operons that are expressed (11, 38). However, these approaches assume that Hg(II) reduction is enzyme limited. A greater limitation in the use of microbial Hg(II) reduction to remove Hg(II) from some contaminated environments is that not all of the Hg(II) may be available for microbial reduction. For example, typically less than half and sometimes less than 10% of the Hg(II) in contaminated waters is volatilized by microbial reduction (11, 112). Much of the mercury may be bound in nonreducible forms to suspended particulate matter in the water (11). Mercury in aquatic sediments may also be unavailable for microbial reduction because sterile sediment added to water inhibits Hg(II) reduction (11) and sediment-water incubation systems volatilize less mercury than systems with water alone (112). Further evidence that mercury was in nonreducible forms in aquatic sediments was the finding that the abundance of *mer* genes in contaminated bottom sediments was two orders of magnitude lower than in the water column even though the sediment had very high mercury concentrations (11).

Preliminary experiments demonstrated that microbial Hg(II) reduction has the potential to volatilize Hg(0) from Hg(II)-containing mine- and industrial-waste waters (133) as well as residential sewage (44). However, I have been unable to find documentation of any instance in which microbial Hg(II) reduction is currently being applied for the removal of mercury from contaminated waters or waste streams.

Tc(VII) REDUCTION

Technetium is a long-lived (half-life, 2.15×10^5 years) radioactive contaminant in the environment that is a by-product of fission reactions occurring in atomic explosions and in nuclear power stations (135). Under aerobic conditions, technetium is primarily in the form of pertechnetate [Tc(VII); TcO_4^-), which is highly soluble (135). The reduced form of technetium, Tc(IV) is highly insoluble (135). Several studies have demonstrated that the development of reducing conditions in soils greatly decreases the solubility and mobility of technetium, presumably as the result of Tc(VII) reduction to Tc(IV) (120 and references therein).

One of the few studies that have directly examined the potential for microorganisms to reduce Tc(VII) found that a mixed culture of anaerobic bacteria produced more insoluble and/or adsorbed technetium from pertechnetate than mixed or pure cultures of aerobic bacteria (46). Forms of technetium other than pertechnetate were also associated with dissolved organics. However, if pertechnetate was added to bacteria-free filtrates of the mixed anaerobic cultures, no organic-associated technetium was formed. These results suggested that Tc(VII) might have served as an electron acceptor in anaerobic respiration.

Tc(VII) was reduced during the stationary growth phase of *Moraxella* and *Planococcus* spp. as oxygen was depleted from the medium (109). Studies with heat-killed cells and incubations with live cells at different temperatures suggested that Tc(VII) reduction resulted from an enzymatically catalyzed reaction.

Sulfate-reducing bacteria may also be involved in Tc(VII) reduction. The sulfate-reducing microorganisms, *Desulfovibrio gigas* and *D. vulgaris,* were even more effective than a mixed anaerobic microbial culture in converting Tc(VII) to insoluble and/or adsorbed forms (46). Molybdate, a specific inhibitor of sulfate reducers, inhibited Tc(VII) reduction in a mixed culture of anaerobic marine microorganisms (109). Both studies suggested that the ability of sulfate-reducers to remove technetium from solution was due, at least in part, to the formation of insoluble technetium sulfides and/or the release of other reducing agents. However, the only evidence for nonenzymatic reduction of Tc(VII) was the association of some technetium with dissolved organics when pertechnetate was added to bacterial-free filtrates of the sulfate-reducing cultures (46). Given the recent finding that sulfate reducers can enzymatically reduce other metals [see sections on Fe(III), U(VI), and Cr(VI) reduction, above], sulfate reducers might also be able to enzymatically reduce Tc(VII).

V(V) REDUCTION

Although microbial V(V) reduction has not been studied intensively, this metabolic capability may be widespread. Almost all of the heterotrophic bacteria and fungi isolated from a silty clay loam soil could reduce V(V) (13). However, the mechanisms for V(V) reduction were not determined.

Two V(V)-reducing microorganisms have subsequently been studied in greater detail (142). *Pseudomonas vanadium reductans* was isolated from the effluent of a metallurgical factory, and *Pseudomonas isachenkovii* was isolated from sea water. Evidence for V(V) reduction is primarily qualitative. During the anaerobic growth of these microorganisms, the V(V)-containing medium changes color from the pale yellow characteristic of dissolved V(V) to blue, which is characteristic of V(IV). With further incubation the medium turns black and opaque, presumably from the accumulation of colloidal V(III) particles. All of the V(V) is reduced within 10 days of incubation; there is no loss of V(V) in sterile controls. Several sugars, amino acids, lactate, glycerol, H_2, and CO can serve as electron donors for V(V) reduction.

Given the similarities between the geochemistry of uranium and vanadium (50), microbial vanadium reduction could be responsible for the commonly observed (47, 50) precipitation of vanadium in anaerobic environments, as seen with uranium reduction (see above), and this metabolism could potentially be used to remove vanadium from ore-processing waste streams.

Mo(VI) REDUCTION

Pseudomonas guillermondii and a *Micrococcus* sp. reduce Mo(VI) to blue reduced molybdenum [presumably Mo(V)] during growth on aerobic heterotrophic medium or in cell suspensions (13). The mechanisms for Mo(VI) reduction were not determined.

At high temperature (60°C), low pH (< 2), and with S^0 as the potential electron donor, *Sulfolobus brierleyi* and *S. acidocaldarius* reduce Mo(VI) to Mo(V) under aerobic or anaerobic conditions (18). Mo(VI) is not reduced when the organisms are grown on yeast extract instead of S^0.

Cell suspensions of the mesophile *T. ferrooxidans* also reduce Mo(VI) with S^0 (129). The rate of Mo(VI) reduction is proportional to the amount of cell protein added and the initial S^0 and Mo(VI) concentrations. The hydrogen sulfide:ferric ion oxidoreductase [see section on Fe(III) reduction] purified from *T. ferrooxidans* catalyzes S^0 oxidation to sulfite coupled to Mo(VI) reduction (129). However, Fe(III) is considered to be the physiological electron acceptor for this enzyme because the rate of Mo(VI) reduction is 20-fold lower than the rate of Fe(III) reduction.

Apparently no ecological investigations have examined the role of microbial Mo(VI) reduction in the environment. The reduction of Mo(VI) under acidic conditions could potentially influence molybdenum cycling during ore leaching. Microbial Mo(VI) reduction at circumneutral pH might be involved in the concentration of insoluble molybdenum in anaerobic marine sediments (15) and the reduction spots found in rocks (47).

Cu(II) REDUCTION

Washed cell suspensions of *T. ferrooxidans* reduce Cu(II) to Cu(I) in the presence of S^0 as a potential electron donor (128). Cu(II) is reduced under both aerobic and anaerobic conditions. However, only net reduction occurs under aerobic conditions when azide or cyanide are added to prevent the iron oxidase from oxidizing Cu(I). The hydrogen sulfide:ferric ion oxidoreductase that reduces Fe(III) and Mo(VI) (see above) also reduces Cu(II) (128). Cu(II) reduction by *T. ferrooxidans* may play a role in copper leaching (124).

Au(III), Au(I), AND Ag(I) REDUCTION

Bacillus subtilis, Aspergillus niger, Cholorella vulgaris, and *Spirulina platentis* reduce Au(III) to Au(0) (16, 34, 41). Within 5 min of exposure of isolated walls of *B. subtilis* to Au(III) chloride, granules of elemental gold form within the walls (16). Au(III) and Au(I) adsorb onto cells of *C. vulgaris* with subsequent reduction to Au(0) (41). Reduction of Au(III) to Au(I) is much faster than the reduction of Au(I) to A(0), which takes place over several days. Gold crystals form on the cell surfaces of *C. vulgaris* and within the cells (49). In some instances the gold crystals are larger than the organisms themselves (34). The mechanisms for Au(III) reduction in these organisms have not been elucidated.

In a similar manner to Au(III) and Au(I), Ag(I) adsorbs onto the cell walls of bacteria and fungi with the subsequent reduction of Ag(I) to colloidal Ag(0) (92, 93). Reduction of Ag(I) to Ag(0) through the production of unknown reducing compounds may account for silver resistance in some bacteria (14).

Dissimilatory Fe(III)-reducing microorganisms can probably also reduce Au(III) and Ag(I). Addition of Au(III) or Ag(I) resulted in the oxidation of the *c*-type cytochrome(s) in *G. metallireducens,* suggesting that the *c*-type cytochromes or a terminal reductase can transfer electrons to Au(III) (76). Washed cell suspensions of *D. vulgaris* rapidly reduce Au(III) to Au(0), and the c_3 cytochrome from *D. vulgaris* reduces Au(III) with the accumulation of Au(0) (D. R. Lovley, P. K. Widman & J. C. Woodward, submitted).

Microbial Au(III) reduction may lead to the formation of gold deposits (140) and could potentially be used for the removal of gold from waters and waste streams (34, 41, 90).

CONCLUSIONS

The intrinsic activity of dissimilatory metal-reducing microorganisms that use Fe(III), Mn(IV), U(VI), or Se(VI) as terminal electron acceptors can greatly influence the fate of these metals in aquatic sediments and groundwater. Fe(III) [and possibly Mn(IV)] reduction can also be an important mechanism for the oxidation of naturally occurring organic matter and organic contaminants in these environments. Dissimilatory reduction of U(VI), Se(VI), Cr(VI), Hg(II), Tc(VII), V(V), Au(III), and Ag(I) is a potential mechanism for removing these metals from contaminated environments or waste streams. Although microorganisms are available in pure culture that can serve as models for the reduction of each of these metals, there is no substantive information about which microorganisms are important in catalyzing metal reduction in natural environments. Furthermore, with the exception of Hg(II) reduction, very little is known about the physiology and biochemistry of dissimilatory metal reduction.

Literature Cited

1. Anderson, R. F., LeHuray, A. P., Fleisher, M. Q., Murray, J. W. 1989. Uranium deposition in Saanich Inlet sediments, Vancouver Island. *Geochim. Cosmochim. Acta* 53:2205–13
2. Bacon, M., Ingledew, W. J. 1989. The reductive reactions of *Thiobacillus ferrooxidans* on sulphur and selenium. *FEMS Microbiol. Lett.* 58:189–94
3. Balashova, V. V., Zavarzin, G. A. 1980. Anaerobic reduction of ferric iron by hydrogen bacteria. *Microbiology* 48:635–39
4. Baldi, F., Boudou, A., Ribeyre, F. 1992. Response of a freshwater bacterial community to mercury contamination ($HgCl_2$ and CH_3HgCl) in a controlled system. *Arch. Environ. Contam. Toxicol.* 22:439–44
5. Baldi, F., Semplici, F., Filippelli, M. 1991. Environmental applications of mercury resistant bacteria. *Water, Air, Soil Pollut.* 56:465–75
6. Barkay, T. 1987. Adaptation of aquatic microbial communities to Hg^{2+} stress. *Appl. Environ. Microbiol.* 53:2725–32
7. Barkay, T., Gillman, M., Liebert, C. 1990. Genes encoding mercuric reductases from selected gram-negative aquatic bacteria have a low degree of homology with *merA* of transposon Tn*501*. *Appl. Environ. Microbiol.* 56: 1695–1701
8. Barkay, T., Liebert, C., Gillman, M. 1989. Environmental significance of the potential for *mer*(Tn*21*)-mediated reduction of Hg^{2+} to Hg^0 in natural waters. *Appl. Environ. Microbiol.* 55: 1196–1202
9. Barkay, T., Liebert, C., Gillman, M. 1989. Hybridization of DNA probes with whole-community genome for detection of genes that encode microbial responses to pollutants: *mer* genes and Hg^{2+} resistance. *Appl. Environ. Microbiol.* 55:1574–77
10. Barkay, T., Olson, B. H. 1986. Phenotypic and genotypic adaptation of aerobic heterotrophic sediment bacterial communities to mercury stress. *Appl. Environ. Microbiol.* 52:403–6
11. Barkay, T., Turner, R. R., VandenBrook, A., Liebert, C. 1991. The relationships of Hg(II) volatilization from a freshwater pond to the abundance of *mer* genes in the gene pool of the indigenous microbial community. *Microb. Ecol.* 21:151–61
12. Bartlett, R. J. 1991. Chromium cycling in soils and water: links, gaps, and methods. *Environ. Health Perspect.* 92:17–24
13. Bautista, E. M., Alexander, M. 1972. Reduction of inorganic compounds by soil microorganisms. *Soil Sci. Soc. Am. Proc.* 36:918–20
14. Belly, R. T., Kydd, G. C. 1982. Silver resistance in microorganisms. *Dev. Ind. Microbiol.* 23:567–77
15. Bertine, K. K. 1972. The deposition of molybdenum in anoxic waters. *Marine Chem.* 1:43–53
16. Beveridge, T. J., Murray, R. G. E. 1976. Uptake and retention of metals by cell walls of *Bacillus subtilis*. *J. Bacteriol.* 127:1502–18
17. Bopp, L. H., Ehrlich, H. L. 1988. Chromate resistance and reduction in *Pseudomonas fluorescens* strain LB300. *Arch. Microbiol.* 150:426–31
18. Brierley, C. L., Brierley, J. A. 1982. Anaerobic reduction of molybdenum by *Sulfolobus* species. *Zentralbl. Bakteriol. Parasitenkd. Infektionskr. Hyg. Abt. 1 Orig.* C3:289–94
19. Brock, T. D., Gustafson, J. 1976. Ferric iron reduction by sulfur- and iron-oxidizing bacteria. *Appl. Environ. Microbiol.* 32:567–71
20. Brown, N. L. 1985. Bacterial resistance to mercury-reductio ad absurdum? *Trends Biochem. Sci.* 41:400–3
21. Burton, G. A. Jr., Giddings, T. H., DeBrine, P., Fall, R. 1987. High incidence of selenite-resistant bacteria from a site polluted with selenium. *Appl. Environ. Microbiol.* 53:185–88
22. Caccavo, F. Jr., Blakemore, R. P., Lovley, D. R. 1992. A hydrogen-oxidizing, Fe(III)-reducing microorganism from the Great Bay Estuary, New Hampshire. *Appl. Environ. Microbiol.* 58:3211–16
23. Cervantes, C. 1991. Bacterial interactions with chromate. *Antonie van Leeuwenhoek J. Microbiol. Serol.* 59: 229–33
24. Champine, J. E., Goodwin, S. 1991. Acetate catabolism in the dissimilatory iron-reducing isolate GS-15. *J. Bacteriol.* 173:2704–6
25. Chapelle, F. H., Lovley, D. R. 1992. Competitive exclusion of sulfate reduction by Fe(III)-reducing bacteria: a mechanism for producing discrete zones of high-iron ground water. *Ground Water* 30:29–36

25a. Coleman, M. L., Hedrick, D. B., Lovley, D. R., White, D. C., Pye, K. 1993. Reduction of Fe(III) in sediments by sulphate-reducing bacteria. *Nature* 361:436–38

26. Das, S., Chandra, A. L. 1990. Chromate reduction in *Streptomyces*. *Experientia* 46:731–33
27. De Vrind, J. P. M., Boogerd, F. C., de Vrind-de Jong, E. W. 1986. Manganese reduction by a marine *Bacillus* species. *J. Bacteriol.* 167:30–34
28. DiChristina, T. J. 1992. Effects of nitrate and nitrite on dissimilatory iron reduction by *Shewanella putrefaciens* 200. *J. Bacteriol.* 174:1891–96
29. Doran, J. W. 1982. Mircoorganisms and the biological cycling of selenium. *Adv. Microbial. Ecol.* 6:1–32
30. Doran, J. W., Alexander, M. 1977. Microbial formation of volatile Se compounds in soil. *Soil Sci. Soc. Am. J.* 40:687–90
31. Ehrlich, H. L. 1987. Manganese oxide reduction as a form of anaerobic respiration. *Geomicrobiol. J.* 5:423–31
32. Ehrlich, H. L. 1990. *Geomicrobiology*. New York: Dekker
33. Fischer, W. R. 1988. Microbiological reactions of iron in soils. In *Iron in Soils and Clay Minerals,* ed. J. W. Stucki, B. A. Goodman, U. Schwertmann, pp. 715–48. Boston: Reidel
34. Gee, A. R., Dudeney, A. W. L. 1988. Adsorption and crystallisation of gold at biological surfaces. In *Biohydrometallurgy,* ed. P. R. Norris, D. P. Kelly, pp. 437–51. Kew Surrey, UK: Rowe
35. Gerhardt, M. B., Green, F. B., Newman, D., Lundquist, T. J., Tresan, R. B., Oswald, W. J. 1991. Removal of selenium using a novel algal-bacterial process. *Res. J. Water Pollut. Control Fed.* 63:799–805
36. Ghiorse, W. C. 1988. Microbial reduction of manganese and iron. In *Biology of Anaerobic Microorganisms,* ed. A. J. B. Zehnder, pp. 305–31. New York: Wiley
37. Gold, T. 1992. The deep, hot biosphere. *Proc. Natl. Acad. Sci. USA* 89:6045–49
38. Goldstein, R. A., Olson, B. H., Porcella, D. B. 1988. Conceptual model of genetic regulation of mercury biogeochemical cycling. *Environ. Technol. Lett.* 9:957–64
39. Gorby, Y., Lovley, D. R. 1991. Electron transport in the dissimilatory iron-reducer, GS-15. *Appl. Environ. Microbiol.* 57:867–70
40. Gorby, Y. A., Lovley, D. R. 1992. Enzymatic uranium precipitation. *Environ. Sci. Technol.* 26:205–7
41. Greene, B., Hosea, M., McPherson, R., Henzi, M., Alexander, M. D., Darnall, D. W. 1986. Interaction of gold(I) and gold(III) complexes with algal biomass. *Environ. Sci. Technol.* 20:627–32
42. Guerin, W. F., Blakemore, R. P. 1992. Redox cycling of iron supports growth and magnetite synthesis by *Aquaspirillum magnetotacticum*. *Appl. Environ. Microbiol.* 58:1102–9
43. Gvozdyak, P. I., Mogilevich, N. F., Ryl'skii, A. F., Grishchenko, N. I. 1987. Reduction of hexavalent chromium by collection strains of bacteria. *Microbiology* 55:770–73
44. Hansen, C. L., Zwolinski, G., Martin, D., Williams, J. W. 1984. Bacterial removal of mercury from sewage. *Biotechnol. Bioeng.* 26:1330–33
45. Hardoyo, J. K., Ohtake, H. 1991. Effects of heavy metal cations on chromate reduction by *Enterobacter cloacae* strain HO1. *J. Gen. Appl. Microbiol.* 37:519–22
46. Henrot, J. 1989. Bioaccumulation and chemical modification of Tc by soil bateria. *Health Phys.* 57:239–45
47. Hoffman, B. A. 1990. Reduction spheroids from northern Switzerland: mineralogy, geochemistry and gentic models. *Chem. Geol.* 81:55–81
48. Horitsu, H., Futo, S., Miyazawa, Y., Ogai, S., Kawai, K. 1987. Enzymatic reduction of hexavalent chromium by hexavalent chromium tolerant *Pseudomonas ambigua* G-1. *Agric. Biol. Chem.* 51:2417–20
49. Hosea, M., Greene, B., McPherson, R., Henzl, M., Alexander, M. D., Darnall, D. W. 1986. Accumulation of elemental gold on the alga *Chlorella vulgaris*. *Inorg. Chim. Acta* 123:161–65
50. Hostetler, P. B., Garrels, R. M. 1962. Transportation and precipitation of uranium and vanadium at low temperatures with special reference to sandstone-type uranium. *Econ. Geol.* 57:137–67
51. Hurley, J. P., Watras, C. J., Bloom, N. S. 1991. Mercury cycling in a northern Wisconsin seepage lake: the role of particulate matter in vertical transport. *Water, Air, Soil Pollut.* 56: 543–51
52. Ishibashi, Y., Cervantes, C., Silver, S. 1990. Chromium reduction in *Pseudomonas putida*. *Appl. Environ. Microbiol.* 56:2268–70
53. Jensen, M. L. 1958. Sulfur isotopes and the origin of sandstone-type uranium deposits. *Econ. Geol.* 53:598–616
54. Johnson, D. B., McGinness, S. 1991. Ferric iron reduction by acidophilic

heterotrophic bacteria. *Appl. Environ. Microbiol.* 57:207–11

55. Jones, J. G., Gardener, S., Simon, B. M. 1983. Bacterial reduction of ferric iron in a stratified eutrophic lake. *J. Gen. Microbiol.* 129:131–39
56. Kauffman, J. W., Laughlin, W. C., Baldwin, R. A. 1986. Microbiological treatment of uranium mine waters. *Environ. Sci. Technol.* 20:243–48
57. Kim, J. P., Fitzgerald, W. F. 1986. Sea-water partitioning of mercury in the equatorial Pacific Ocean. *Science* 231:1131–33
58. Kino, K., Usami, S. 1982. Biological reduction of ferric iron by iron- and sulfur-oxidizing bacteria. *Agric. Biol. Chem.* 46:803–5
59. Klinkhammer, G. P., Palmer, M. R. 1991. Uranium in the oceans: where it goes and why. *Geochim. Cosmochim. Acta* 55:1799–1806
60. Komori, K., Rivas, A., Toda, K., Ohtake, H. 1990. Biological removal of toxic chromium using an *Enterobacter cloacae* strain that reduces chromate under anaerobic conditions. *Biotechnol. Bioeng.* 35:951–54
61. Komori, K., Rivas, A., Toda, K., Ohtake, H. 1990. A method for removal of toxic chromium using dialysis-sac cultures of a chromate-reducing strain of *Enterobacter cloacae*. *Appl. Microbiol. Biotechnol.* 33:117–19
62. Komori, K., Toda, K., Ohtake, H. 1990. Effects of oxygen stress on chromate reduction in *Enterobacter cloacae* strain HO1. *J. Ferment. Bioeng.* 69:67–69
63. Komori, K., Wang, P., Toda, K., Ohtake, H. 1989. Factors affecting chromate reduction in *Enterobacter cloacae* strain HO1. *Appl. Microbiol. Biotechnol.* 31:567–70
64. Kvasnikov, E. I., Stepanyuk, V. V., Klyushnikova, T. M., Serpokrylov, N. S., Simonova, G. A. et al. 1985. A new chromium-reducing, gram-variable bacterium with mixed type of flagellation. *Microbiology* 54:69–75
65. Langmuir, D. 1978. Uranium solution-mineral equilibria at low temperatures with applications to sedimentary ore deposits. *Geochim. Cosmochim. Acta* 42:547–69
66. Lebedeva, E. V., Lyalikova, N. N. 1979. Reduction of crocoite by *Pseudomonas chromatophila* sp. nov. *Microbiology* 48:517–22
67. Levine, V. E. 1925. The reducing properties of microorganisms with special reference to selenium compounds. *J. Bacteriol.* 10:217–63
68. Lonergan, D. J., Lovley, D. R. 1991. Microbial oxidation of natural and anthropogenic aromatic compounds coupled to Fe(III) reduction. In *Organic Substances and Sediments in Water*, ed. R. A. Baker, pp. 327–38. Chelsea, MI: Lewis
69. Long, R. H. B., Benson, S. M., Tokunaga, T. K., Yee, A. 1990. Selenium immobilization in a pond sediment at Kesterson Reservior. *J. Environ. Qual.* 19:302–11
70. Lovley, D. R. 1987. Organic matter mineralization with the reduction of ferric iron: a review. *Geomicrobiol. J.* 5:375–99
71. Lovley, D. R. 1991. Dissimilatory Fe(III) and Mn(IV) reduction. *Microbiol. Rev.* 55:259–87
72. Lovley, D. R. 1992. Microbial oxidation of organic matter coupled to the reduction of Fe(III) and Mn(IV) oxides. *Catena* 21:101–14
73. Lovley, D. R., Baedecker, M. J., Lonergan, D. J., Cozzarelli, I. M., Phillips, E. J. P., Siegel, D. I. 1989. Oxidation of aromatic contaminants coupled to microbial iron reduction. *Nature* 339:297–99
74. Lovley, D. R., Caccavo, F., Phillips, E. J. P. 1992. Acetate oxidation by dissimilatory Fe(III) reducers. *Appl. Environ. Microbiol.* 58:3205–6
75. Lovley, D. R., Chapelle, F. H., Phillips, E. J. P. 1990. Fe(III)-reducing bacteria in deeply buried sediments of the Atlantic Coastal Plain. *Geology* 18:954–57
76. Lovley, D. R., Giovannoni, S. J., White, D. C., Champine, J. E., Phillips, E. J. P., et al. 1993. *Geobacter metallireducens* gen. nov. sp. nov., a microorganism capable of coupling the complete oxidation of organic compounds to the reduction of iron and other metals. *Arch. Microbiol.* In press
77. Lovley, D. R., Lonergan, D. J. 1990. Anaerobic oxidation of toluene, phenol, and p-cresol by the dissimilatory iron-reducing organism, GS-15. *Appl. Environ. Microbiol.* 56:1858–64
78. Lovley, D. R., Phillips, E. J. P. 1988. Manganese inhibition of microbial iron reduction in anaerobic sediments. *Geomicrobiol. J.* 6:145–55
79. Lovley, D. R., Phillips, E. J. P. 1988. Novel mode of microbial energy metabolism: organic carbon oxidation coupled to dissimilatory reduction of iron or manganese. *Appl. Environ. Microbiol.* 54:1472–80
80. Lovley, D. R., Phillips, E. J. P. 1989. Requirement for a microbial consortium

to completely oxidize glucose in Fe(III)-reducing sediments. *Appl. Environ. Microbiol.* 55:3234–36

81. Lovley, D. R., Phillips, E. J. P. 1992. Bioremediation of uranium contamination with enzymatic uranium reduction. *Environ. Sci. Technol.* 26:2228–34
82. Lovley, D. R., Phillips, E. J. P. 1992. Reduction of uranium by *Desulfovibrio desulfuricans*. *Appl. Environ. Microbiol.* 58:850–56
83. Lovley, D. R., Phillips, E. J. P., Gorby, Y. A., Landa, E. R. 1991. Microbial reduction of uranium. *Nature* 350:413–16
84. Lovley, D. R., Phillips, E. J. P., Lonergan, D. J. 1989. Hydrogen and formate oxidation coupled to dissimilatory reduction of iron or manganese by *Alteromonas putrefaciens*. *Appl. Environ. Microbiol.* 55:700–6
85. Lovley, D. R., Phillips, E. J. P., Lonergan, D. J. 1991. Enzymatic versus nonenzymatic mechanisms for Fe(III) reduction in aquatic sediments. *Environ. Sci. Technol.* 25:1062–67
86. Lovley, D. R., Stolz, J. F., Nord, G. L., Phillips, E. J. P. 1987. Anaerobic production of magnetite by a dissimilatory iron-reducing microorganism. *Nature* 330:252–54
87. Luoma, S. N., Johns, C., Fisher, N. S., Steinberg, N. A., Oremland, R. S., Reinfelder, J. R. 1992. Determination of selenium bioavailability to a benthic bivalve from particulate and solute pathways. *Environ. Sci. Technol.* 26:485–91
88. Macy, J. M., Michel, T. A., Kirsch, D. G. 1989. Selenate reduction by a *Pseudomonas* species: a new mode of anaerobic respiration. *FEMS Microbiol. Lett.* 61:195–98
89. Maiers, D. T., Wichlacz, P. L., Thompson, D. L., Bruhn, D. F. 1988. Selenate reduction by bacteria from a selenium-rich environment. *Appl. Environ. Microbiol.* 54:2591–93
90. Mann, S. 1992. Bacteria and the Midas touch. *Nature* 357:358–60
91. Masscheleyn, P. H., Pardue, J. H., DeLaune, R. D., Patrick, W. H. Jr. 1992. Chromium redox chemistry in a lower Mississippi Valley bottomland hardwood wetland. *Environ. Sci. Technol.* 26:1217–26
92. Mullen, M. D., Wolf, D. C., Beveridge, T. J., Bailey, G. W. 1992. Sorption of heavy metals by the soil fungi *Aspergillus niger* and *Mucor rouxii*. *Soil Biol. Biochem.* 24:129–35
93. Mullen, M. D., Wolf, D. C., Ferris, F. G., Beveridge, T. J., Flemming, C. A., Bailey, G. W. 1989. Bacterial sorption of heavy metals. *Appl. Environ. Microbiol.* 55:3143–49
94. Murphy, E. M., Schramke, J. A., Fredrickson, J. K., Bledose, H. W., Francis, A. J., et al. 1992. The influence of microbial activity and sedimentary organic carbon on the isotope geochemistry of the Middendorf aquifer. *Water Resour. Res.* 28:723–40
95. Myers, C. R., Myers, J. M. 1992. Localization of cytochromes to the outer membrane of anaerobically grown *Shewanella putrefaciens* MR-1. *J. Bacteriol.* 174:3429–38
96. Myers, C. R., Nealson, K. H. 1988. Bacterial manganese reduction and growth with manganese oxide as the sole electron acceptor. *Science* 240: 1319–21
97. Myers, C. R., Nealson, K. H. 1988. Microbial reduction of manganese oxides: interactions with iron and sulfur. *Geochim. Cosmochim. Acta* 52:2727–32
98. Myers, C. R., Nealson, K. H. 1990. Respiration-linked proton translocation coupled to anaerobic reduction of manganese(IV) and iron(III) in *Shewanella putrefaciens* MR-1. *J. Bacteriol.* 172: 6232–38
99. Naik, R. R., Murillo, F. M., Stolz, J. F. 1993. Evidence for a novel nitrate reductase in the dissimilatory iron-reducing bacterium *Geobacter metallireducens*. *FEMS Microbiol. Lett.* 106: 53–58
100. Ohtake, H., Fujii, E., Toda, K. 1990. Bacterial reduction of hexavalent chromlium: kinetic aspects of chromate reduction by *Enterobacter cloacae* HO1. *Biocatalysis* 4:227–35
101. Ohtake, H., Fujii, E., Toda, K. 1990. Reduction of toxic chromate in an industrial effluent by use of a chromate-reducing strain of *Enterobacter cloacae*. *Environ. Technol.* 11:663–68
102. Ohtake, H., Fujii, E., Toda, K. 1990. A survey of effective electron donors for reduction of toxic hexavalent chromium by *Enterobacter cloacae* (strain HO1). *J. Gen. Appl. Microbiol.* 36: 203–8
103. Olson, B. H., Cayless, S. M., Ford, S., Lester, J. N. 1991. Toxic element contamination and the occurrence of mercury-resistant bacteria in Hg-contaminated soil, sediments, and sludges. *Arch. Environ. Contam. Toxicol.* 20: 226–33
104. Oremland, R. S. 1993. Biogeochemical transformations of selenium in anoxic environments. In *Selenium in the En-*

vironment, ed. W. T. Frankenberger, Jr. New York: Dekker. In press

105. Oremland, R. S., Hollibaugh, J. T., Maest, A. S., Presser, T. S., Miller, L. G., Culbertson, C. W. 1989. Selenate reduction to elemental selenium by anaerobic bacteria in sediments and culture: biogeochemical significance of a novel, sulfate-independent respiration. *Appl. Environ. Microbiol.* 55:2333–43
106. Oremland, R. S., Steinberg, N. A., Maest, A. S., Miller, L. G., Hollibaugh, J. T. 1990. Measurement of in situ rates of selenate removal by dissimilatory bacterial reduction in sediments. *Environ. Sci. Technol.* 24: 1157–64
107. Oremland, R. S., Steinberg, N. A., Presser, T. S., Miller, L. G. 1991. In situ bacterial selenate reduction in the agricultural drainage systems of western Nevada. *Appl. Environ. Microbiol.* 57:615–17
108. Palmer, C. D., Wittbrodt, P. R. 1991. Processes affecting the remediation of chromium-contaminated sites. *Environ. Health Perspect.* 92:25–40
109. Pignolet, L., Auvary, F., Fonsny, K., Capot, F., Moureau, Z. 1989. Role of various microorganisms on Tc behavior in sediments. *Health Phys.* 57: 791–800
110. Pronk, J. T., De Bruyn, J. C., Bos, P., Kuenen, J. G. 1992. Anaerobic growth of *Thiobacillus ferrooxidans. Appl. Environ. Microbiol.* 58:2227–30
111. Pronk, J. T., Liem, K., Bos, P., Kuenen, J. G. 1991. Energy transduction by anaerobic ferric iron respiration in *Thiobacillus ferrooxidans. Appl. Environ. Microbiol.* 57:2063–68
112. Regnell, O. 1990. Conversion and partitioning of radio-labelled mercury chloride in aquatic model systems. *Can. J. Fish. Aquat. Sci.* 47:548–53
113. Richard, F. C., Bourg, C. M. 1991. Aqueous geochemistry of chromium: a review. *Water Res.* 25:807–16
114. Robinson, J. B., Tuovinen, O. H. 1984. Mechanisms of microbial resistance and detoxification of mercury and organomercury compounds: physiological, biochemical, and genetic analyses. *Microbiol. Rev.* 48: 95–124
115. Rochelle, P. A., Wetherbee, M. K., Olson, B. H. 1991. Distribution of DNA sequences encoding narrow- and broad-spectrum mercury resistance. *Appl. Environ. Microbiol.* 57:1581–89
116. Roden, E. E., Lovley, D. R. 1993. Dissimilatory Fe(III) reduction by the marine microorganism, *Desulfuromonas acetoxidans. Appl. Environ. Microbiol.* 59:734–42
117. Romanenko, V. I., Koren'Ken, V. N. 1977. A pure culture of bacteria utilizing chromates and bichromates as hydrogen acceptors in growth under anaerobic conditions. *Microbiology* 46:414–17
118. Sand, W. 1989. Ferric iron reduction by *Thiobacillus ferrooxidans* at extremely low pH-values. *Biogeochemistry* 7:195–201
119. Schiering, N., Kabsch, W., Moore, M. J., Distefano, M. D., Walsh, C. T., Pai, E. F. 1991. Structure of the detoxification catalyst mercuric ion reductase from *Bacillus* sp. strain RC607. *Nature* 352:168–72
120. Sheppard, S. C., Sheppard, M. I., Evenden, W. G. 1990. A novel method used to examine variation in TcSorption among 34 soils, aerated and anoxic. *J. Environ. Radioact.* 11:215–33
121. Silver, S., Walderhaug, M. 1992. Gene regulation of plasmid- and chromosome-determined inorganic ion transport in bacteria. *Microbiol. Rev.* 56:195–228
122. Steinberg, N. A., Blum, J. S., Hochstein, L., Oremland, R. S. 1992. Nitrate is a preferred electron acceptor for growth of freshwater selenate-respiring bacteria. *Appl. Environ. Microbiol.* 58: 426–28
123. Steinberg, N. A., Oremland, R. S. 1990. Dissimilatory selenate reduction potentials in a diversity of sediment types. *Appl. Environ. Microbiol.* 56: 3550–57
124. Sugio, T., de los Santos, S. F., Hirose, T., Inagaki, K., Tano, T. 1990. The mechanism of copper leaching by intact cells of *Thiobacillus ferrooxidans. Agric. Biol. Chem.* 54:2293–98
125. Sugio, T., Katagiri, T., Inagaki, K., Tano, T. 1989. Actual substrate for elemental sulfur oxidation by sulfur: ferric ion oxidoreductase purified from *Thiobacillus ferrooxidans. Biochim. Biophys. Acta* 973:250–56
126. Sugio, T., Katagiri, T., Moriyama, M., Zhen, Y. L., Inagaki, K., Tano, T. 1988. Existence of a new type of sulfite oxidase which utilizes ferric ions as an electron acceptor in *Thiobacillus ferrooxidans. Appl. Environ. Microbiol.* 54:153–57
127. Sugio, T., Mizunashi, W., Inagaki, K., Tano, T. 1987. Purification and some properties of sulfur: ferric ion oxidoreductase from *Thiobacillus ferrooxidans. J. Bacteriol.* 169:4916–22

128. Sugio, T., Tsujita, Y., Inagaki, K., Tano, T. 1990. Reduction of cupric ions with elemental sulfur by *Thiobacillus ferrooxidans*. *Appl. Environ. Microbiol.* 56:693–96
129. Sugio, T., Tsujita, Y., Katagiri, T., Inagaki, K., Tano, T. 1988. Reduction of Mo^{6+} with elemental sulfur by *Thiobacillus ferrooxidans*. *J. Bacteriol.* 170:5956–59
130. Sugio, T., Wada, K., Mori, M., Inagaki, K., Tano, T. 1988. Synthesis of an iron-oxidizing system during growth of *Thiobacillus ferrooxidans* on sulfur-basal salts medium. *Appl. Environ. Microbiol.* 54:150–52
131. Summers, A. O. 1992. Untwist and shout: a heavy metal-responsive transcriptional regulator. *J. Bacteriol.* 174: 3097–3101
132. Summers, A. O., Barkay, T. 1989. Metal resistance genes in the environment. In *Gene Transfer in the Environment,* ed. S. Levy and R. Miller, pp. 287–308. New York: McGraw-Hill
133. Suzuki, T., Furukawa, K., Tonomura, K. 1968. Studies on the removal of inorganic mercurial compounds in waste by the cell-reused method of mercury-resistant bacterium. *J. Ferment. Technol.* 46:1048–55
134. Suzuki, T., Miyata, N., Horitsu, H., Kawai, K., Takamizawa, K., et al. 1992. NAD(P)H-dependent chromium(VI) reductase of *Pseudomonas ambigua* G-1: a Cr(V) intermediate is formed during the reduction of Cr(VI) to Cr(III). *J. Bacteriol.* 174:5340–45
135. Trabalka, J. R., Garten, C. T. 1983. Behavior of the long-lived synthetic elements and their natural analogs in food chains. *Adv. Radiat. Biol.* 10:39–104
136. Veeh, H. H. 1967. Deposition of uranium from the ocean. *Earth Plant. Sci. Lett.* 3:145–50
137. Wang, P., Mori, T., Komori, K., Sasatsu, M., Toda, K., Ohtake, H. 1989. Isolation and characterization of an *Enterobacter cloacae* strain that reduces hexavalent chromium under anaerobic conditions. *Appl. Environ. Microbiol.* 55:1665–69
138. Wang, P., Mori, T., Toda, K., Ohtake, H. 1990. Membrane-associated chromate reductase activity from *Enterobacter cloacae*. *J. Bacteriol.* 172:1670–72
139. Wang, P., Toda, K., Ohtake, H., Kusaka, I., Yabe, I. 1991. Membrane-bound respiratory system of *Enterobacter cloacae* strain HO1 grown anaerobically with chromate. *FEMS Microbiol. Lett.* 78:11–16
140. Watterson, J. R. 1992. Preliminary evidence for the involvement of budding bacteria in the origin of Alaskan placer gold. *Geology* 20:315–18
141. Woolfolk, C. A., Whiteley, H. R. 1962. Reduction of inorganic compounds with molecular hydrogen by *Micrococcus lactilyticus*. *J. Bacteriol.* 84:647–58
142. Yurkova, N. A., Lyalikova, N. N. 1991. New vandate-reducing facultative chemolithotrophic bacteria. *Microbiology* 59:672–77
143. Zehr, J. P., Oremland, R. S. 1987. Reduction of selenate to selenide by sulfate-respiring bacteria: experiments with cell suspensions and estuarine sediments. *Appl. Environ. Microbiol.* 53:1365–69

Annu. Rev. Microbiol. 1993. 47:291–319

ATP-DEPENDENT TRANSPORT SYSTEMS IN BACTERIA AND HUMANS: Relevance to Cystic Fibrosis and Multidrug Resistance

Carl A. Doige and Giovanna Ferro-Luzzi Ames
Department of Molecular and Cell Biology, University of California at Berkeley, Berkeley, California 94720

KEY WORDS: periplasmic permeases, traffic ATPases, CFTR, MDR, ATP-dependent channels, ATP-dependent transport

---

CONTENTS

ABSTRACT

The prokaryotic permeases are members of a superfamily of membrane transporters called traffic ATPases, which includes the medically important eukaryotic multidrug resistance (MDR) protein and cystic fibrosis transmembrane regulator (CFTR). Members of this superfamily have extensive sequence

0066-4227/93/1001-0291$02.00

and structural similarity, in particular in an ATP-binding motif, and are believed to use ATP to energize translocation of substrates across biological membranes. The prokaryotic histidine permease is well-characterized and serves as a convenient model system. In this review, we highlight some of the biochemical and molecular biological approaches used to study the functional and architectural organization of this permease and relate the results of these approaches to what is known about other traffic ATPases. We have identified specific regions that we believe critical for the function of the histidine permease and propose that the corresponding regions in the eukaryotic traffic ATPases are also important for their function. In light of the fact that CFTR (and possibly the MDR protein) is an ion channel, we compare the properties of channels and transporters; in addition, we discuss the possibility that other members of the traffic ATPases may also have channel-like activity.

THE SUPERFAMILY OF TRAFFIC ATPASES: OVERALL STRUCTURE

Numerous proteins involved in the transport of a variety of molecules bear strong primary sequence and secondary structure similarity, suggesting that they have similar mechanisms of action; the proteins, which include both prokaryotic and eukaryotic representatives, can be gathered into a superfamily (11, 50). Many of the prokaryotic members of the superfamily have been extensively studied and have been shown to be involved in the translocation of a variety of substrates through the cell membrane. By analogy, the eukaryotic members are probably also involved in translocation, as shown in several cases. Among the eukaryotic members of this superfamily are several medically important proteins: the P-glycoprotein (Pgp or P170) responsible for the phenomenon of multidrug resistance (MDR) in tumor cells (28, 79); a *Plasmodium falciparum* protein involved in imparting chloroquine resistance to the malaria parasite (30); the cystic fibrosis gene product (CFTR—cystic fibrosis transmembrane conductance regulator) (55, 77, 78); and protein components of the major histocompatibility complex class I (TAP-1 and TAP-2—transporter associated with antigen processing) (67). The *STE6* gene product, responsible for the export of the **a**-factor mating pheromone of *Saccharomyces cerevisiae* (58, 63), also belongs to the superfamily. We have proposed that members of this superfamily be called *traffic ATPases* (11) because: (*a*) the similarity invariably includes a putative ATP-binding site, which implies that family members hydrolyze ATP for energization; (*b*) family members translocate substrates that are extremely varied in nature and size; and (*c*) the term traffic indicates that the translocation occurs in either direction and helps distinguish this class of ATPases from other transport ATPases, such as ion-conducting ATPases. Another name suggested for these proteins is ABC (ATP-binding cassette) proteins (53). Universal features of traffic ATPases are the presence of a hydrophobic domain postulated to be embedded in the membrane and a hydrophilic domain characterized by sequence similarity that includes the above mentioned ATP-binding site.

This review summarizes the information available for the prokaryotic periplasmic transport systems (permeases), mostly from *Salmonella typhimurium* and *Escherichia coli*, which are the best-understood representatives of this superfamily, and relates it to the characteristics of the known eukaryotic systems. Periplasmic permeases are characterized by complexity of composition, their structure always including several distinct membrane-bound polypeptides that embody separate hydrophilic and hydrophobic domains; in addition, they require the presence of a soluble receptor (the periplasmic substrate-binding protein) that concentrates the substrate by binding it tightly and then transfers it to the membrane-bound complex. In contrast, eukaryotic traffic ATPases are usually composed of a single polypeptide with two homologous halves, each half comprising a hydrophobic and a hydrophilic domain. No evidence for a soluble receptor has been found.

PERIPLASMIC PERMEASES

General Features

These systems are typically composed of the soluble receptor (substrate-binding protein) located in the periplasm and a membrane-bound complex containing four subunits. Thus periplasmic permease receptors are located at the periphery of the cell and incoming substrates are exposed to them first (Figure 1). Figure 2 schematically represents a cycle of transport through a typical periplasmic permease. We refer mainly to the histidine permease as the model system (although we give references for several other well-characterized periplasmic permeases). The histidine permease comprises the histidine-binding protein, HisJ, as the receptor, and three proteins, HisQ, HisM, and HisP, that form the membrane-bound complex. Two of the membrane-bound components, HisQ and HisM, are very hydrophobic, while the third, HisP, is hydrophilic, though also firmly bound to the membrane. The receptor, HisJ, binds the substrate (histidine in this case) and changes conformation, which allows it to interact with the membrane-bound complex (14, 72), thus signaling availability of substrate in the vicinity of the membrane complex (69). Translocation occurs through a series of conformational changes concomitant with ATP hydrolysis and is hypothesized to take place either through a pore (channel?) (as shown in Figure 2), by interaction with a

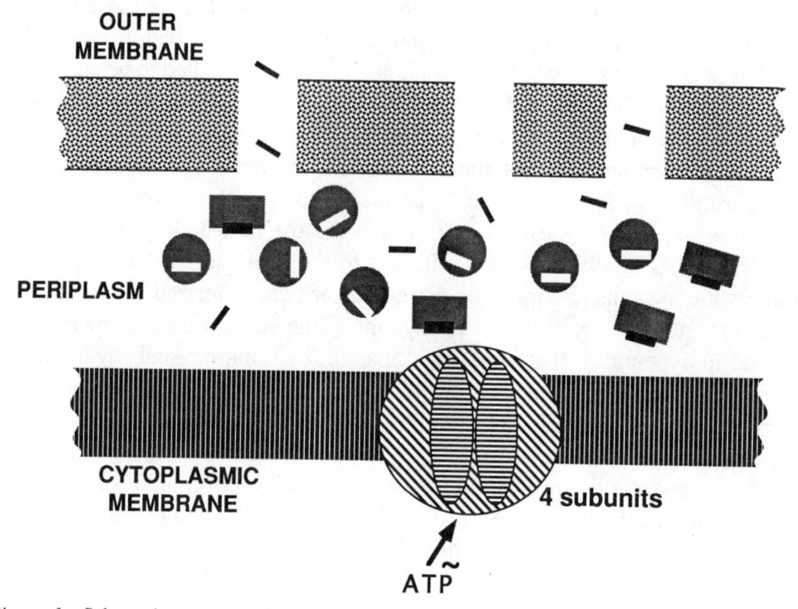

Figure 1 Schematic representation of a generic periplasmic permease. The outer membrane (*dotted*) contains proteinaceous pores that allow the substrate (*solid bars*) to enter the periplasm, where it is bound by the receptor (=periplasmic binding protein), represented by a sphere with an empty binding site (*empty rectangle*) or a rectangle when liganded to the substrate. The four membrane components (hydrophobic, *diagonal shading;* hydrophilic ATP-binding, *horizontal shading*) form a complex within the cytoplasmic membrane (*vertical shading*). When the receptor changes conformation upon binding histidine it interacts with the membrane complex. The squiggle indicates the involvement of ATP in energy coupling.

substrate-binding site(s) located in the membrane-bound complex, or by a combination of the two proposed mechanisms.

Periplasmic permeases generally accumulate substrates against large concentration gradients (10,000-fold) with apparent K_ms for uptake ranging from 0.01 to 1 μM. The concentration of receptors in the periplasm is, or can be induced to be, very high (calculated to be on the order of millimolar concentration), and the affinity for their respective substrates is usually very good, between 0.01 and 1 μM. The consequence of these two facts is that substrate is trapped and concentrated in the periplasm, thereby greatly exceeding the concentration of the free substrate in the external medium. This concentrating action, which occurs prior to the actual translocation step, may be an important distinguishing characteristic between prokaryotic and eukaryotic traffic ATPases. Such high efficiency of transport and good concentrating ability might underlie the requirements leading to such a complex structure in periplasmic permeases. For example, the very high efficiency

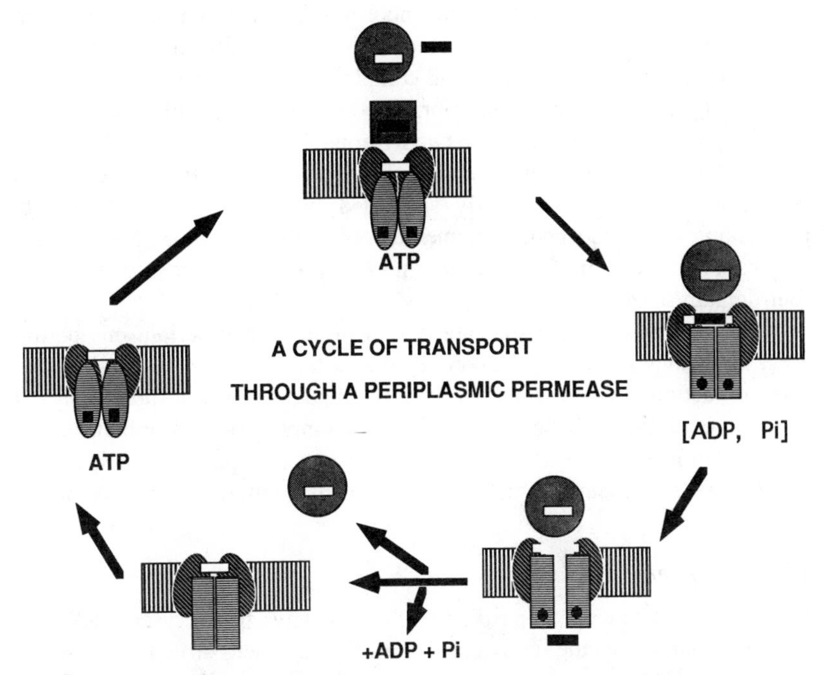

Figure 2 A cycle of transport through a periplasmic permease. The cycle is initiated by the binding of substrate (*solid rectangle*) to the periplasmic receptor, which subsequently interacts with the membrane complex. This interaction occurs probably with both hydrophobic membrane components and elicits conformational changes in the components of the membrane complex, in particular in the ATP-binding membrane protein, thus causing ATP (*solid square*) hydrolysis. Hydrolysis of ATP leads to the opening of a pore [possibly containing a substrate-specific binding site(s)] that allows the unidirectional diffusion of the substrate to the interior. After the substrate has been released to the interior, an additional conformational change [release of ADP? (*solid circle*)] closes the pore. The unliganded binding protein disassociates itself from the membrane complex, which binds ATP again and is ready to start a new cycle (modified from 11).

may constitute an evolutionary advantage because it allows recapture of biosynthetically produced amino acids that are otherwise lost from the cell (3).

The liganded form of the receptor rather than the unliganded one was shown to interact preferentially with the membrane-bound complex (71, 72), as shown in Figure 2. The existence and composition of a membrane-bound complex and the stoichiometry of its subunits have been established biochemically for the histidine permease. They include four separate polypeptides: one copy each of HisQ and HisM and two copies of the hydrophilic membrane-bound, ATP-binding protein, HisP (56, 57). The latter is the repository of the aforementioned sequence similarity; because of this extensive similarity, all the equivalent components from all periplasmic permeases are also referred to as *conserved components* (66). A similar organization was determined for the membrane-bound complex of the maltose permease (34). Thus, the basic structure of a membrane-bound complex could be visualized as being pseudosymmetrical, with two hydrophobic proteins and two subunits of a hydrophilic, ATP-binding protein. That such composition represents the general situation is also clearly supported by the analogy afforded by inspection of the sequences of numerous other permeases (53, 56, 66). The membrane-bound complex may also normally be a multimer of the basic four-domain structure.

These general characteristics are typical of all known periplasmic permeases, indicating an underlying conserved structural organization and mechanism of action, despite the vast variety of substrates transported, and a common evolutionary origin. This complex composition and organization is to be contrasted with that of shock-resistant transporters, such as the β-galactoside permease, which is composed of a single integral membrane protein.

The Receptor

Many receptors have been purified and several have been crystallized either in the liganded or the unliganded forms, and their structures resolved (reviewed in 1). A general picture has emerged of a two-domain structure in the shape of a kidney bean, in which the two lobes are well separated in the absence of substrate; when substrate finds its way to the specific substrate-binding site in the cleft between the two lobes, one lobe rotates relative to the other, trapping the substrate inside a deep pocket. Recently one receptor, the lysine-, arginine-, ornithine-binding protein (LAO, homologous to HisJ), was purified in both the liganded and unliganded forms, and the resolution of both structures has given a very clear picture of the likely sequence of events that accompany the liganding process (67a). The picture basically confirms the Venus flytrap hypothesis put forward by Quiocho and his collaborators (60). In the case of the LAO protein, because of the structure resolution in both the liganded and unliganded forms, one could postulate the existence of two forms of the unliganded receptor (open empty and closed empty) that are in dynamic equilibrium with each other, with the substrate capable of binding to one of the two lobes. In the presence of substrate, the closed form is stabilized by additional interactions between the substrate and the second lobe, resulting in a closed liganded form (67a). In effect, this model postulates that the act of binding the substrate is responsible for stabilizing the receptor in its closed conformation. Genetic analysis of the various receptor domains responsible for substrate binding and for interaction with the membrane complex are underway for several receptors and, combined with X-ray studies, will undoubtedly unravel many of the secrets of the mechanism of action of these proteins.

Because the soluble extracellular receptor has been as yet found only in prokaryotes, the question of whether it is a necessary characteristic only of the prokaryotic systems, and thus whether it reflects a basic mechanistic difference, is relevant. Receptors may be an efficient device added on to the basic membrane-bound translocating machinery by prokaryotes rather than an indispensable characteristic of traffic ATPases (see below). Therefore, the obvious similarity between eukaryotic and prokaryotic systems resides at the level of the membrane-bound complex, and it is reasonable to focus on the molecular mechanism of the membrane translocator itself as the common puzzle to be solved.

The Membrane-Bound Complex

The interaction of the liganded receptor with the membrane-bound complex has been postulated to activate the translocation mechanism by triggering a series of events resulting in the hydrolysis of ATP, the release of the receptor-bound substrate, and the opening of a specific pore in the complex through which the substrate is translocated (11, 35, 69). Clearly, in order to test such a model and to understand the mechanism of translocation, one must clarify the physical structure of the translocating complex. Ideally we would like to obtain the crystal structure of the entire complex. In the absence of such bonanza, we and others have resorted to topology studies, both biochemical and computer-assisted, and to computer modeling of the conserved components.

THE HYDROPHOBIC COMPONENTS A clear picture of the topology of the hydrophobic components of the histidine permease, HisQ and HisM, is now emerging from recent studies (56, 57). These proteins have a molecular weight around 25,000, and their sequences include several very hydrophobic stretches. Computer-assisted predictions of hydropathicity indicate that each component contains three to five membrane spanners (4). The biochemical investigation of the topology of HisQ and HisM has involved several types of impermeant reagents, such as proteolytic enzymes and antibodies, and in vitro assay systems using oriented membrane vesicles, both inside-out and right-side-out (56, 57). As shown for other integral membrane proteins, both HisQ and HisM are proteolytically digestible at either membrane surface and thus span the membrane, require strong detergent action for solubilization, and behave as expected with respect to Triton X-114 solubilization and partition analysis (25)—both partition with the detergent phase. Both are accessible to antibodies directed against their carboxyterminal peptide only in inside-out vesicles, and both are digestible with carboxypeptidase from the interior, but not exterior surfaces. HisM is digestible with aminopeptidase at the external, but not internal surface. The actual number of membrane spanners in HisQ and HisM was investigated by the use of the transposon Tn*phoA* method, which identifies the distribution of hydrophilic loops of integral membrane proteins between the external (periplasmic) and internal aspects of the cytoplasmic membrane (59). Enzymatic assays of an array of chimeric proteins fusing PhoA to HisQ or HisM clearly indicated the presence of five membrane spanners (56). Thus a combination of the above analyses established that HisQ and HisM are typical integral membrane proteins that span the membrane five times each, with their amino termini located on the external (periplasmic) side and the carboxy termini on the cytoplasmic side (56, 57). The distribution of charges between inside and outside loops within this five-spanner arrangement adheres very well to the rules established for integral membrane proteins (26, 31, 89), with positive charges preferentially inside.

How general is this type of organization for the hydrophobic components? Such analyses have not yet been fully performed for most other hydrophobic periplasmic permease components. A comparison of the predicted topology of all other known hydrophobic subunits, as predicted by hydropathicity analysis, indicates an overall striking similarity in their carboxy-terminal ends (56). In the case of the maltose permease, a topological model was deduced using Tn*phoA* fusion analysis of MalF (26, 40). Comparison of this model with that of HisQ and HisM indicates that even though sequence similarity between these proteins is limited, structural similarity is extensive. The periodicity of loops and spanners present throughout the length of HisQ and HisM matches the pattern at the carboxy-terminal two thirds of MalF, up to the long periplasmic loop between the third and fourth spanner in MalF. Because MalF is considerably larger than HisQ and HisM, its amino-terminal third forms additional membrane spanners and hydrophilic loops. In the case of the oligopeptide permease, the two hydrophobic components, OppB and OppC, have been postulated on the basis of genetic analysis procedures (namely, β-lactamase chimeric fusions) and computer-assisted predictions to have a similar organization with one additional membrane spanner at their amino-terminal ends (68). We propose that the minimum structure (as represented by HisQ and HisM) of these subunits consists of five hydrophobic helices spanning the membrane, with the amino and carboxy termini on the exterior and interior surfaces, respectively. Since HisQ and HisM are among the smallest (25,000 Daltons) of all known hydrophobic components of periplasmic permeases, it might be reasonable to assume that their organization, with a total of ten membrane spanners between the two of them, is the maximum needed to perform the basic translocation step.

Because HisQ and HisM, and equivalent hydrophobic components from several of the other permeases, show significant sequence similarity that presumably arose by duplication from an ancestral gene, they are probably also very similar structurally and form a pseudodimer within the membrane complex (4). This notion is compatible with and supported by the existence of some permeases containing only one hydrophobic component (such as the arabinose and the glutamine periplasmic permeases), in which case a true dimer of two identical subunits must be present within the complex.

A common motif in all of the hydrophobic components is a conserved region (the EAA loop) located in a large hydrophilic cytoplasmic loop that is always present between the third and fourth spanners of the minimum structure. A salient feature of this region is the triplet Glu Ala Ala (EAA) that is strongly conserved in prokaryotic subunits (4, 32, 56). At present there is no information about the function of this motif. A possible reason for its conservation is the interaction with a conserved domain of the respective ATP-binding component for signal transmission.

THE HYDROPHILIC (CONSERVED) COMPONENTS Despite the hydrophilic sequence, the conserved components of periplasmic permeases are tightly membrane bound (12, 57). However, the manner in which they are associated is controversial. In some cases they may be peripherally attached to the interior aspect of the cytoplasmic membrane (41, 83). This arrangement contrasts with what has been found for HisP, which behaves neither like a typical integral nor a peripheral membrane protein when it is in the presence of HisQ and HisM (57). Interestingly, HisP prefers to be associated with the membrane even in the absence of HisQ and HisM, but in such a case it assumes an improper association, acquiring an unusual level of sensitivity to proteases and detergents, and in general behaving like a typical peripheral membrane protein. This observation has led to the suggestion that HisP is deeply embedded in the membrane (21, 57). A possible explanation for the association of HisP with the membrane in the absence of HisQ and HisM may be found in its predicted tertiary structure, which has been modeled to include a large loop with some amphipathic characteristics (66). This loop might allow HisP's proper insertion in the membrane in the presence, and its improper association in the absence, of the hydrophobic components. The notion that the conserved component normally may be (more or less deeply) embedded in the complex, and therefore in the membrane, derives support from the recent finding that HisP is accessible to proteases and to an impermeant biotinylating reagent from the exterior surface of the membrane when the complex has been assembled in the presence of HisQ and HisM, but not in their absence (21). No analogous evidence has yet been gathered for equivalent conserved components from other permeases. Thus, a membrane-spanning disposition as a general characteristic of conserved components awaits confirmation.

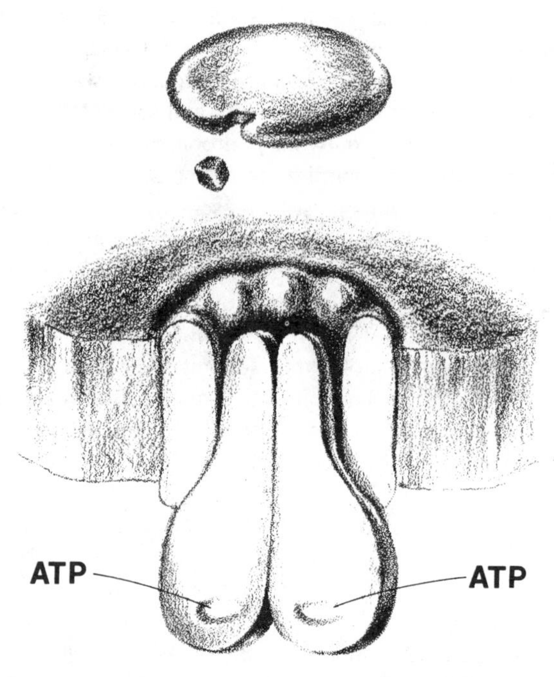

Figure 3 Three-dimensional representation of a periplasmic permease. The membrane has been sliced vertically to expose the ATP-binding domains (conserved components) as extending part-way into the membrane and encased by the hydrophobic domains. The unliganded receptor is shown above. The ATP-binding sites are located in the cytoplasmic portion of the ATP-binding domain. [Reprinted with permission from *Advances in Enzymology* copyright © 1992 (10).]

Figure 3 is an artist's rendering of a periplasmic permease that shows two conserved component subunits interacting with the membrane spanners of the hydrophobic components and with their ATP-binding sites accessible to the cytoplasm. The extent and nature of the interaction is not to be taken literally, because at present no reagents can distinguish unequivocally between partial and complete penetration of protein domains within a membrane.

The conserved components bind and/or hydrolyze ATP and GTP (35, 49, 51, 90). Hence, one of the functional domains of these components would be specifically reserved for the binding of ATP (or GTP). A definition of the portions of these molecules dedicated to this particular function would limit somewhat the task of understanding the overall function of the protein to those portions that are not included in the binding pocket. In the case of HisP, this is a particularly useful exercise because it is a relatively small protein (25,000 Daltons). Once the domain responsible for ATP binding is defined, a rather small segment of the molecule would remain that presumably is responsible for the performance of its other specific function(s), whatever these may be.

The high degree of sequence similarity in the ATP-binding motifs of the conserved components was used in an attempt at describing the ATP-binding pocket. From an extensive sequence alignment including all the known conserved components, a consensus secondary structure was predicted from which a three-dimensional structure for the pocket was derived by computer-assisted modeling. Two independent structural modeling efforts for conserved components (53, 66) have yielded somewhat different results. Possible reasons for the discrepancy have been discussed (66). Our laboratory has predicted a structure that comprises a mononucleotide-binding pocket occupying about two-thirds of the HisP molecule, with the remaining third forming the aforementioned large loop (Figure 4). The large loop has been temporarily named the *helical domain* because of its predicted secondary structure, which includes four α-helices. No experimental evidence is yet available demonstrating the existence of such a helical structure. The helical domain is presumably involved in transport-related activities other than ATP binding and hydrolysis, such as its insertion into the membrane and its interaction with HisQ and HisM in the transmission of signals to and from the site(s) of interaction with HisJ and the ATP-binding and -hydrolyzing pocket. Presumably the two-thirds of HisP engaged in forming the nucleotide-binding site are exposed on the cytoplasmic side of the membrane in order to be accessible to ATP.

Support for the model of the ATP-binding pocket was derived both from in vitro chemical modification studies using the photolabeling analogue of ATP, 8-azido ATP (65), and by a structure-function analysis of HisP mutants (84). In the latter study the ability of each of the mutant proteins to function in transport and/or to bind ATP was related to the location of the mutation within both the sequence and the predicted structure. A good correlation was found between loss of ATP-binding ability (and transport) and location of the mutation within the predicted ATP-binding pocket, while transport-negative mutations within the helical domain generally did not affect ATP binding. In addition, a set of interesting mutations that remove the dependence on signaling by the liganded receptor for ATP hydrolysis also supports the model because they are all located within the ATP-binding pocket (69). Presumably the mutations have changed the binding pocket so that it has acquired a capacity for unregulated ATP hydrolysis (69, 84, 85). Interestingly, although mutations resulting in the signal independence of ATP hydrolysis have also been found in the maltose permease (35), they are located in the hydrophobic subunits, indicating that unregulated ATP hydrolysis can arise from a disturbance in the structure of many regions of the membrane-bound complex.

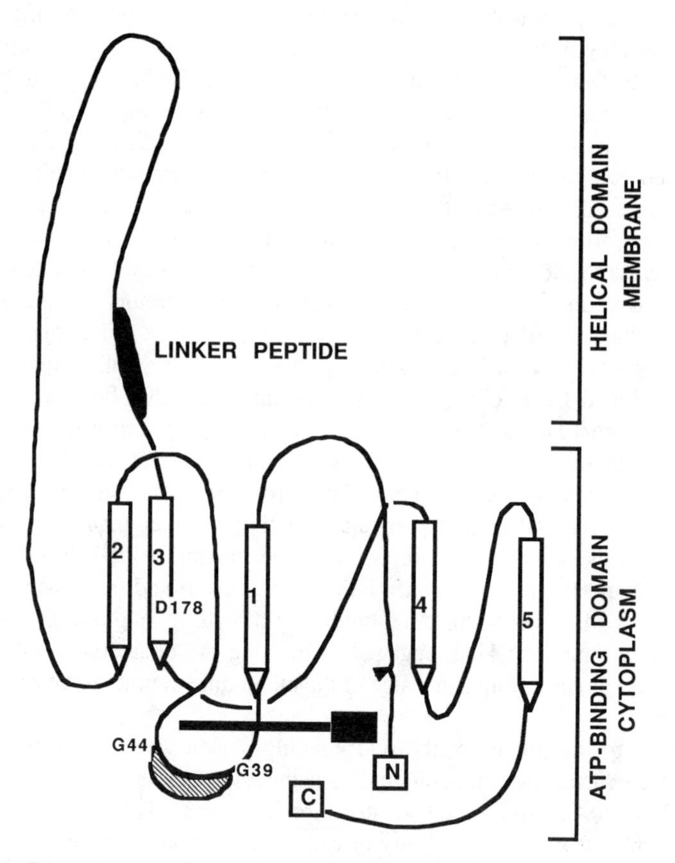

Figure 4 Schematic representation of the predicted topology of a conserved component. HisP is used as a model. Amino acid residues refer to the HisP sequence. D178, the G-rich loop, and arrows represent respectively an aspartate, a structural motif, and β-strands that are very well conserved in mononucleotide-binding proteins. The helical domain is postulated to span the membrane embedded in the hydrophobic domains, while the nucleotide-binding pocket protrudes into the cytoplasm. The linker peptide is poised to transmit signals between the two domains. The black box represents a bound mononucleotide (taken with permission from Ref. 84, with modifications).

Energy Coupling

What is the function of ATP in transport? Several in vivo and in vitro studies (13, 33, 54, 64, 71) established that ATP is the source of energy for periplasmic permease transport, thereby excluding the involvement of the proton-motive force and resolving a decade-long controversy (reviewed in 7). Reconstitution of transport in membrane vesicles (13, 71) and in proteoliposomes (24, 33) has finally enabled biochemical analysis. These in vitro systems clearly showed that ATP is hydrolyzed only concomitantly with transport. Nonhydrolyzable ATP analogues did not support transport, and a photoactivatable analogue of ATP eliminated transport (13). Attempts at deriving a stoichiometry between ATP hydrolyzed and substrate molecules transported yielded values averaging about five (24, 33). The higher-than-expected ratio is likely due to damage incurred by the complex during reconstitution resulting in uncoupling between hydrolysis and transport (24). Recently, the conserved component of the maltose permease was purified in a form free of the two hydrophobic components, and this form has been shown to hydrolyze ATP (90). This purified protein underwent a rather drastic treatment, since it was obtained by an energetic denaturation of inclusion bodies followed by renaturation; its ability to hydrolyze ATP in the absence of translocation and the other transport components may be the result of damage-induced artifacts.

It is particularly interesting that HisQ can also bind ATP (51; V. Shyamala, C. Liu & G. F.-L. Ames, in preparation). Therefore, although the conserved components are probably the units responsible for converting ATP energy into translocation work, as suggested by the fact that they harbor the well-conserved sequence similarity throughout the entire superfamily, the possibility exists that their ATP-binding sites instead have a regulatory function, with the hydrophobic subunit(s) performing the hydrolytic step.

Signaling Mechanism

The model described in Figure 2 postulates that the membrane-bound complex is in a closed conformation when "resting" and that it requires a signal in order to open up a pathway through which translocation occurs (5). The logical step at which such a signal would be initiated would be at the interaction of the liganded receptor (binding protein) with the membrane-bound complex, because at this point the substrate is brought into close proximity to the membrane-bound complex. Therefore, this interaction presumably sends a signal that triggers conformational changes resulting in ATP hydrolysis and the subsequent opening of a pore. Additional conformational changes must occur to cause: (*a*) the release of the substrate from the receptor, which involves decreasing its affinity for the substrate and therefore opening of the Venus flytrap, and (*b*) the release of the unliganded receptor. The latter may occur spontaneously once the substrate has been released as a consequence of the receptor's decreased affinity for the membrane-bound complex. Finally, possibly after the release of the products of ATP degradation (to ADP and Pi), the pathway through the complex would close again and the system would return to its initial state by binding a new molecule(s) of ATP and thus becoming ready for another cycle of transport.

In support of such a sequence of events are several facts: (*a*) the receptor interacts physically with at least one of the hydrophobic components, as shown by chemical cross-linking experiments (72); (*b*) the unliganded receptor interacts poorly or not at all with the membrane-bound complex (72; G. F.-L. Ames & K. Nikaido, unpublished data); (*c*) ATP is only hydrolyzed upon interaction of the receptor with the membrane-bound complex (24); (*d*) mutants of the membrane-bound proteins translocate the substrate independently of the presence of a receptor and hydrolyze ATP in the absence of transport and of the receptor (35, 70). Thus, ATP hydrolysis is one of the processes that must be activated following the interaction of the liganded receptor with the membrane complex; according to the scheme just presented, the energy released might be used to open the pore (or channel). Alternatively, it might rather close the pore, thus returning the system back to a state of readiness for the next cycle of transport. The signaling mechanism is clearly of great interest because it is at the heart of the molecular mechanism of action of these permeases. In addition, it is also of general interest because it will probably be useful as a model for more complex signaling mechanisms, both eukaryotic and prokaryotic.

A particularly well conserved region of the helical domain, a glycine-, glutamine-rich sequence (consensus: LSGGQQQ) located within the helical domain shortly ahead of the spot where the sequence reenters the nucleotide-binding pocket, may be involved in the postulated signaling mechanism. Argos (18) has suggested that these residues are commonly found in peptide linkers that join together separate domains within proteins. Thus this conserved sequence in traffic ATPases may be a flexible arm joining the helical domain with the nucleotide-binding domain, and it may be poised for acting as a signal transducer. This sequence has been tentatively named the *linker peptide* (8, 10, 66).

Direction of Translocation and Regulation of Transport

If the opening of the pore is an energy-utilizing mechanism dependent on the expenditure of ATP energy, then it would be reasonable to expect that periplasmic permeases translocate irreversibly. This was found to be true in reconstituted proteoliposomes: accumulated histidine did not exit from proteoliposomes, and the incorporation of ADP, inorganic phosphate, and histidine inside the proteoliposomes did not result either in the exit of histidine or the synthesis of ATP (V. Petronilli, C. Liu & G. F.-L. Ames, in preparation). Thus, translocation from the inside to the outside through the histidine permease does not occur—not surprising, considering the energetics of this reaction. In addition, both the internally accumulated histidine and the hydrolytically produced ADP inhibit transport, suggesting that the internal pool of histidine (and ADP) regulates the rate and extent of translocation and, therefore, that there is an internal substrate-recognizing site (Figure 5). This observation indicates that transport mechanisms may have evolved in which the internal concentration of substrate and the hydrolysis of ATP cause

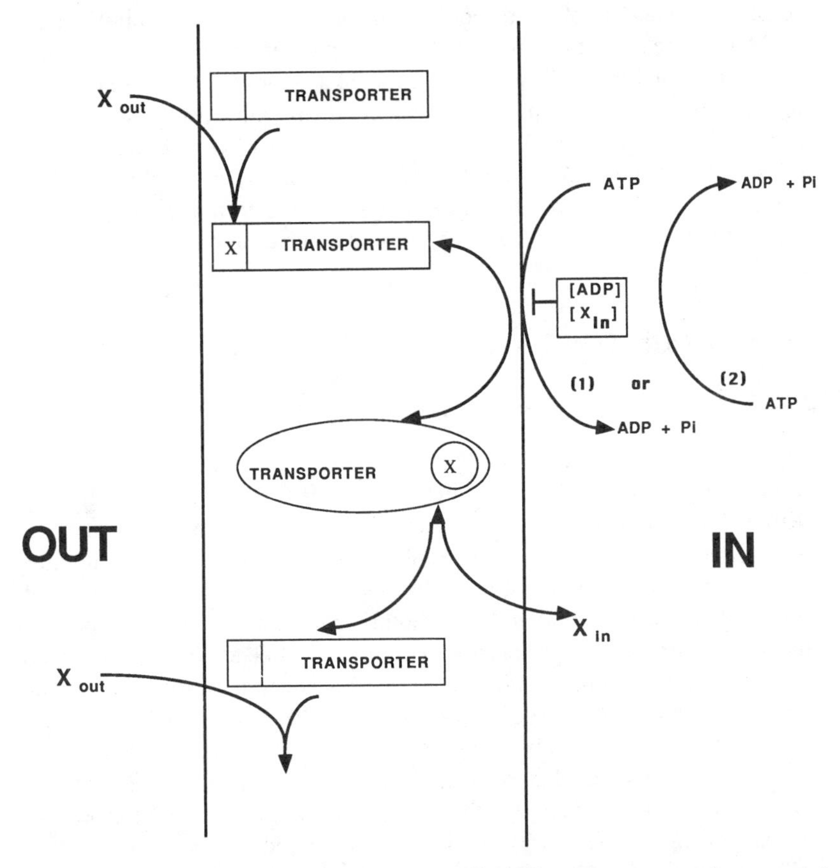

Figure 5 Directionality of transport by a generic traffic ATPase. The transporter is represented as becoming liganded with the substrate X at either the outer or inner membrane surface. An empty substrate-binding site is boxed and the transporter is shown as undergoing a conformational change upon ATP hydrolysis. Accessibility of the transporter at the inner or outer membrane is hypothesized to be mediated through the conformational changes. In mechanism 1, ATP hydrolysis causes a rearrangement that renders the transporter-bound substrate accessible at the inner surface, allowing its release. In mechanism 2, ATP hydrolysis causes the reverse rearrangement, resulting in the extrusion of the substrate through an essentially identical transporter. The presence of a binding protein is not necessary for either mechanism (modified from Ref. 8, with permission).

conformational changes resulting in translocation from the interior to the exterior by a simple reversal of the sequence of events postulated in mechanism 1 of Figure 5, i.e. by mechanism 2.

An important conclusion to be drawn from the above studies and from the existence of receptor-independent mutants is that the membrane-bound complex must be able to recognize the substrate at both surfaces, and therefore, it has the intrinsic ability to function in the absence of a soluble receptor. Elimination of the receptor from the scheme, as shown in Figure 5, would only require that the substrate be present in a sufficiently high concentration to bind to the presumed external membrane site. This is relevant for the question of extrapolation between periplasmic permeases and the (apparently) receptorless eukaryotic traffic ATPases. Therefore, it can be postulated that an original ancestral structure for traffic ATPases had the inherent ability to evolve into either a system able to translocate from the interior to the exterior, or vice versa, depending on each particular evolutionary history.

RELEVANCE TO EUKARYOTIC CARRIERS

Justification for Inclusion in the Superfamily

STRUCTURE The sequence similarity between eukaryotic and prokaryotic traffic ATPases is striking. But is inclusion of all of them in the same family justified? The sequence similarity shows that the conserved components of periplasmic permeases and the hydrophilic moieties of the eukaryotic transporters are related beyond simply the nucleotide-binding sequence motif. Indeed, the same nucleotide-binding motif is present in several other proteins that are neither involved in transport nor otherwise related to each other. For example, the three proteins that were used for modeling of the mononucleotide-binding pocket (66) bear similarity in the sequence portions relevant to the ATP-binding motif only; a comparison of their primary sequences with those of the transporters has not revealed any other significant similarity and thus excludes them from the family of traffic ATPases. Therefore, the extensive sequence similarity among the conserved components of traffic ATPases suggests a common ancestry and similar mechanism of action. This obvious evolutionary relationship has been noted and commented upon repeatedly (6, 10, 28, 79). The maintenance of the high level of sequence similarity observed between the eukaryotic and prokaryotic members of this family indicates that these proteins have additional functions in common besides binding (and hydrolyzing?) ATP. Almost certainly one of the other essential and common functions contributing to maintenance of sequence similarity is the signal-transmission mechanism from (or to) the substrate-occupied site(s) to (or from) the energy-transducing triggering site. In this scenario, the specific substrate-binding site could be located either on the conserved component, thus accounting for some of the lack of conservation within this domain, or on the hydrophobic component(s).

Additional reasons for combining all these proteins into a single superfamily are based on comparisons of the hydrophobic domains of the eukaryotic transporters (two for each molecule) with the hydrophobic prokaryotic components. Primary sequence comparisons do not reveal any significant sequence similarity. This is in agreement with the finding that the prokaryotic hydrophobic components themselves generally have relatively little sequence similarity. However, the secondary-structure prediction and membrane topology of the hydrophobic components indicates a remarkable similarity across the entire spectrum of transporters, including the hydrophobic domains of eukaryotic transporters, suggesting that they may perform similar functions. The latter are usually predicted to have six spanners per domain, which corresponds to the minimum prokaryotic structure plus one extra spanner at the amino terminal end. Thus, the hypothetical architecture of a typical eukaryotic traffic ATPase consists of two sets of membrane spanners (one in each half of the molecule), corresponding to the two hydrophobic subunits of a prokaryotic transporter, and two ATP-binding moieties. Similar to the prokaryotic systems, the entire structure may display a pseudo-two-fold symmetry, with an overall stoichiometry of two hydrophobic domains to two ATP-binding domains. A possible evolutionary history for traffic ATPases envisions an ancestral system composed of two genes coding for a single hydrophobic and a single ATP-binding protein; both genes would have undergone various duplications and fusions (Figure 6) (9). Fusion of proteins functioning individually in prokaryotes into a single polypeptide during the

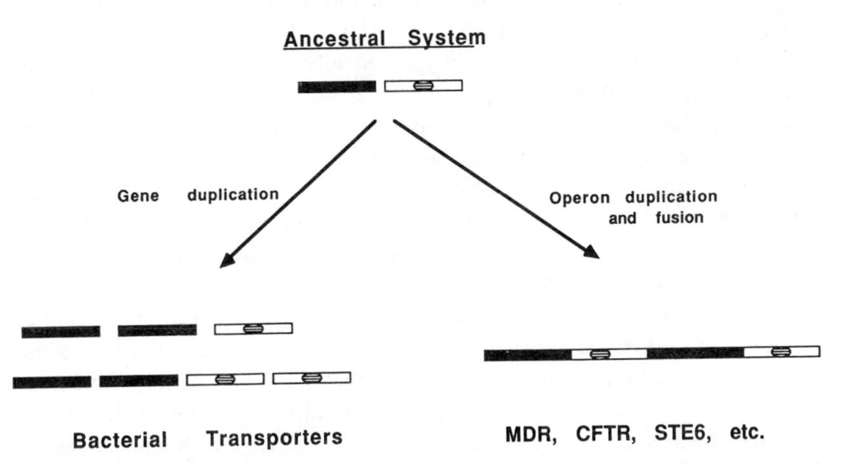

Figure 6 Evolution of traffic ATPases. The solid and empty boxes represent genes coding for the hydrophobic and conserved components, respectively; the oval shape with horizontal shading represents a mononucleotide-binding site. The eukaryotic members of the family arose by fusion of duplicated genes, while the bacterial transporters in general present a larger variety of evolutionary types, with gene fusions occurring only among duplicated conserved component genes (not shown). Not shown are genes for the receptor in prokaryotes, which also frequently undergo gene duplication (taken from 9 with permission).

evolution of eukaryotes is an event that commonly appears when eukaryotic and prokaryotic proteins are compared.

Interestingly, in the eukaryotic hydrophobic moieties a large cytoplasmic hydrophilic loop occupies a position corresponding to the prokaryotic EAA loop. Several of the eukaryotic loops also contain a conserved glutamate contiguous with residues similar to those near Glu Ala Ala in prokaryotes. This loop may have the same function as the one postulated for the prokaryotic equivalents, i.e. it interacts with the ATP-binding moiety in the transmission of energization signals. It is possible that the eukaryotic hydrophobic moieties have not conserved a complete Glu Ala Ala motif because if these sequences are necessary for holding together separate subunits in an interaction, in eukaryotes these moieties might have lost such a need after the fusion event. By analogy to the prokaryotic conserved components, the eukaryotic ATP-binding domains would be embedded partly within the hydrophobic domain with the portions carrying the ATP-binding sites accessible to the cytoplasm.

The monocomponent structure of eukaryotic traffic ATPases may also include a domain incorporating functions analogous to those of the substrate-binding protein, i.e. the high-affinity recognition and trapping of the substrate. Such a hypothetical domain could be located on either the cytoplasmic or external surface of the membrane; if a hypothetical high-affinity binding site were located on the external surface, it might function in extrusion by removing substrate from the membrane-embedded domain proper and then releasing it into the medium upon energization. From the available information, one cannot distinguish whether the affinity-labeled site(s) shown to exist in Pgp (see below) is in a domain hypothetically analogous to the substrate-binding protein or to the presumed substrate-binding sites on the membrane-bound complex of periplasmic permeases. The identification of multiple sites on a eukaryotic transporter might indicate that there are separate domains for capturing and translocating the substrate. On the other hand, the existence of mutant periplasmic permeases that function in the absence of substrate-binding proteins has clearly indicated that the latter are not absolutely necessary for transport through these transporters. In fact, teleologically speaking, these proteins may be important only in prokaryotes, which must often survive in environments where nutrients are present at extremely low concentrations and therefore need unique trapping mechanisms that allow scavenging of precious nutrients (3). Eukaryotic cells, which are normally exposed to high concentrations of substrates, may have lost the receptor in evolution because it became obsolete.

Finally, the fact that several eukaryotic traffic ATPases allow passage from the interior to the exterior, as opposed to the action of periplasmic permeases, need not be an obstacle to including them all in the same family, as already discussed above (Figure 5).

EVIDENCE THAT EUKARYOTIC TRAFFIC ATPASES ARE INVOLVED IN TRANSLOCATION The question as to whether eukaryotic proteins that are included in this superfamily are indeed involved in transport should be addressed. Studies with the Pgp protein (P-glycoprotein) provide the strongest biochemical evidence to support the role for these proteins in transport. This protein, when overexpressed in tumor cells, is believed to mediate resistance to multiple drugs by serving as an ATP-driven drug efflux pump that maintains the intracellular drug concentrations at subtoxic levels. Inside-out plasma membrane vesicles from MDR cell lines display an ATP-dependent drug transport that can be inhibited by other MDR-type drugs and that is not found in vesicles derived from drug-sensitive cell lines (37, 52). This uptake is not supported by nonhydrolyzable analogues of ATP, which suggests that the process requires ATP hydrolysis. Furthermore, Pgp is labeled by photoaffinity analogues of many of the drugs included in the MDR spectrum. The finding that azidopine and azidoprazosin, two antagonists of MDR, photoaffinity label both halves of Pgp in homologous regions—transmembrane domains 5 and 6 and 11 and 12 (27, 44)—has led to the proposal that these regions are in close proximity in the three-dimensional topology of the protein in the membrane and that together they form a drug-binding site. A mutation within the predicted transmembrane domain 11 of Pgp significantly alters the drug-resistance profile of cell lines expressing this protein (46). Because of the high level of expression of Pgp in tissues normally involved in secretory functions, e.g. the epithelia of colon, kidney, and liver, its normal function may be the secretion of physiological substrates (such as peptide hormones) and a wide range of hydrophobic molecules, including plant and microbial products. Consistent with Pgp's role as a peptide transporter, mammalian cell lines selected for resistance to a tripeptide, *N*-acetyl-leucyl-leucyl-norleucinal, overexpress Pgp (81).

Other members of the eukaryotic traffic ATPases are also implicated in transport functions: (*a*) The *S. cerevisiae* STE6 protein is necessary for the secretion of **a**-factor (58, 63). An important experiment demonstrating that yeast cells, which lack STE6 but express the cDNA for the mouse *MDR* gene, possess the ability to export **a**-factor and to mate further supports the notion that, by analogy to STE6, the physiological substrates for Pgp may include peptides (75). (*b*) The *white* and *brown* gene products of *Drosophila melanogaster* may be involved in the transfer of pigment precursors (39). And (*c*) both TAP-1 and TAP-2 proteins may be involved in the import of peptides (produced in the cytosol) into the lumen of the endoplasmic reticulum (67). Two proteins whose genes map immediately upstream of *Tap-1* and *Tap-2* are associated with proteasomes involved in peptide processing (61). These molecules may be involved in the delivery of peptide to the TAP transporter, a situation potentially analogous to the role of the binding protein in the periplasmic permeases (67).

In the case of the cystic fibrosis gene product, CFTR, the initial assertion was that this protein was unlikely to function as a chloride channel (the objections were summarized in 53). However, there is now little doubt that the CFTR serves as an ion channel. This conclusion is largely based on the fact that the ion specificity can be altered by specific mutations of charged residues predicted to lie in the transmembrane regions of the protein (16) and that purified CFTR reconstituted into planar lipid bilayers results in regulated chloride channel activity (22). The activation of the ion channel requires both cAMP-induced phosphorylation of the regulatory domain (the R domain) and the interaction of ATP with the nucleotide-binding domains. However, it is controversial as to whether ATP binding alone is sufficient for channel opening or whether ATP hydrolysis is also required (15, 73). Recent findings obtained with Pgp show that ATP binding in combination with a change in buffer osmolarity is sufficient to activate a volume-regulated chloride channel (43, 88) [although this notion is controversial (62)]. By analogy, ATP binding alone may also be sufficient for CFTR chloride-channel activity (after protein kinase A–catalyzed phosphorylation of the R domain). ATP hydrolysis would be involved in some other function—perhaps transport.

ATP Binding and ATPase Activity of Eukaryotic Traffic ATPases

Are the ATP-binding and ATP-hydrolizing properties of the eukaryotic members of this superfamily consistent with their role as traffic ATPases? Until recently very little data concerning this aspect of their function were available. Labeling with 8-azido-ATP has been demonstrated for Pgp (20, 29) and for a recombinant polypeptide corresponding to the full-length nucleotide-binding pocket of the N-terminal half of CFTR (48). In addition, a synthetic polypeptide corresponding to a portion of the N-terminal half of the nucleotide-binding domain of CFTR has also been shown to bind the ATP analogue, trinitrophenol ATP (86, 87). However, ATPase activity has only been demonstrated for Pgp with K_m values ranging between 0.3 and 1.0 mM (2, 38, 80, 82). This activity can be stimulated by MDR-type drugs and antagonists, but upon solubilization and purification, it becomes uncoupled from drug stimulation to various extents depending on the detergent used. The greatest stimulation of Pgp ATPase activity by MDR drugs and antagonists occurs in crude plasma membrane vesicles containing Pgp or in proteoliposomes, into which partially purified Pgp has been reconstituted (2, 80; F. J. Sharom, X. Yu & C. A. Doige, submitted). Presumably in these vectorial experimental systems, ATP hydrolysis occurs concomitantly with substrate transport, but this remains to be proven. The concentration of MDR drug or antagonist that gives half-maximum stimulation is roughly proportional to the relative affinities of these agents for Pgp, and the magnitude of stimulation is unique for each drug, possibly reflecting unique rates of transport.

A final point should be made concerning the ATPase activity of Pgp. A significant basal level of Pgp ATPase activity exists (i.e. is not drug stimulated) in the systems where the activity should be coupled to drug transport (i.e. plasma membrane vesicles and proteoliposomes) (2, 80; F. J. Sharom, X. Yu & C. A. Doige, submitted). The tight coupling between ATP hydrolysis and transport might be unique to the periplasmic permease members of this superfamily and might reflect the dependence of these permeases on the binding protein in transport function. In the eukaryotic systems, ATP hydrolysis may be induced by some other means, perhaps by the substrate coming into direct contact with the transporter; is it possible, evolutionarily speaking, that the tight coupling of substrate transport and ATP hydrolysis was lost in the eukaryotic proteins with the loss of the binding protein.

Are the Two Nucleotide-Binding Pockets Functionally Equivalent?

Several lines of evidence indicate that both nucleotide-binding pockets of the eukaryotic traffic ATPases are required for optimal activity. For example, in most cases, deleterious mutations in either one of these domains of Pgp (20), CFTR (17, 45), or STE6 (23) eliminates or at least compromises the function of these molecules as a whole. In addition, similar mutations in the nucleotide-binding domain of only one of the two conserved components of the *E. coli* LIV-I prokaryotic branched-chain amino acid permease significantly inhibited the function of this transporter (42). In the case of CFTR, equivalent mutations in both domains do not have identical effects: mutations in the N-terminal domain are more likely to result in defective processing of the CFTR protein rather than a specific defect in function (45). Moreover, deletion of the R domain suppresses a mutation in the C-terminal nucleotide-binding domain, but not in the N-terminal domain (76). Thus while both domains may be required for function, they may not be functionally equivalent, with one domain possibly involved in catalysis and the other serving a regulatory function. In agreement with this notion, ADP inhibits CFTR channel activity, apparently by interacting with only the C-terminal nucleotide-binding domain (17). Interestingly, in light of these observations, a recombinant fusion protein consisting of β-galactosidase and the N-terminal half of P-glycoprotein possesses an ATPase activity comparable to that of the full-length recombinant protein (82) but does not confer drug resistance when expressed in mammalian cells. No comparable experiment involving the C-terminal half of the Pgp molecule has yet been performed.

The fact that the nucleotide-binding domains of the eukaryotic traffic ATPases may be functionally distinct has direct implications for the prokaryotic permeases that possess only one gene encoding the conserved component (e.g. in the histidine permease) that presumably exists as a dimer in the transport complexes. By analogy to their eukaryotic counterparts, the two identical molecules of the conserved component may adopt distinctly unique functions through their association with the other membrane components of the transport complex (e.g. in the histidine permease, the association of one HisP molecule with HisQ and of the other with HisM).

Bacterial Permeases as Models for Investigating Mutations in the Eukaryotic Traffic ATPases

A particularly useful comparison of structure-function relationships between prokaryotic and eukaryotic traffic ATPases uses CFTR, because of the availability of numerous cystic fibrosis mutations and because it possesses a measurable activity as a chloride channel. We and others (42, 84) have compared the effect of mutations in the conserved components with similar mutations in the nucleotide-binding folds of CFTR. If the conserved component is accessible from the external surface of the membrane, thereby spanning it to some extent (21), then the nucleotide-binding domains of the eukaryotic traffic ATPases may also take the same configuration. Some support for this idea came from experiments demonstrating that one of the CFTR nucleotide-binding domains exhibited channel activity in lipid bilayers (19). If such membrane-spanning organization were correct, it also would be particularly important for the interpretation of the defect in numerous known CF mutants. For example, a large group of CF mutants is located in the linker peptide, i.e. the glutamine-, glycine-rich region postulated to be involved in signal transmission between the ATP-binding pocket and the helical domain. Recent data demonstrate that several of these mutations (e.g. Gly551→Ser and Gly1349→Asp) have no effect on the binding affinity for ATP but reduce significantly the absolute level of channel activity (17). These results are entirely consistent with the notion that these residues may be important in the transduction mechanism that couples ATP binding or hydrolysis to channel activity, and therefore that they are located within the membrane. Similarly, among the CFTR mutations located within the ATP-binding pocket, some are clearly located in the sequences characteristic of ATP-binding motifs, and many others may be equivalent to the uncoupled, receptor-independent mutations identified in the periplasmic permeases. The defect of the latter mutants may be that these mutant proteins are uncoupled and therefore constantly open, thus resulting in a continuous and improper chloride ion flux.

ANALOGIES BETWEEN CHANNELS AND TRANSPORTERS

The channel activity uncovered in CFTR raises the question as to whether periplasmic permeases, which have been referred to in the past as active transporters, can be properly used as model systems for understanding the mechanism of action of CFTR and other eukaryotic traffic ATPases. If channels were known to function in a unique fashion and had a unique biochemical structure different from that of any known transporter, the answer might be negative. However, there apparently is no inherent and insurmountable discrepancy between channels and periplasmic transporters, and much of the distinction may be semantic (9, 47).

The distinction between channels and transporters has been addressed in reviews and textbooks. In brief, both channels and transporters can be viewed as proteinaceous structures that undergo conformational changes to allow the passage of substrate. An "incipient channel lined with polar groups" (47, p. 364) is a likely ancestral form which could have evolved in the direction of a membrane-spanning pathway that, when properly activated, allows substrate access to binding sites simultaneously from both sides of the membrane, thus forming a typical channel. Within this simplified view, a periplasmic permease might be seen as a channel with one of the two outlets occluded, e.g. by way of the binding protein; alternatively, one can view the evolution of a conformational change mechanism in which channel activation would result in alternating access of a binding site, first to one side of the membrane and then to the other. The latter representation is closer to the classical view of a transporter. In either case, the distinction between channels and (at least) this kind of transporter would become blurred, depending on whether or not asymmetry of the binding sites is created by a conformational change.

A comparison of several properties considered to be characteristic of either channels or transporters also indicates that many of the presumed distinctions may be semantic (9). It seems possible that in either a periplasmic transporter or a classic channel the substrate(s) travels through a pore along which there are binding sites that may be more or less tight and that would help determine the speed of travel. In channels whose activity is routinely measured using electrophysiological techniques, low-affinity binding sites may allow a fast flow of the substrate, while the opening and closing activity is relatively slow. These are indeed the properties that render channels detectable. Channels with tighter binding sites may not be detectable using present technologies and may indeed be so slow as to be indistinguishable from transporters as far as the rate of transport is concerned. In this respect it is interesting that several translocating proteins (e.g. the H^+-ATPase of yeast) (74) exhibit channel activity under particular circumstances (36, 74). Because the CFTR channel is opened with ATP energy (15), the mechanism of energization is no longer considered an insurmountable difference between channels and transporters; this finding aligns CFTR's mechanism of action very closely with that of periplasmic permeases. At this point, the question should rather be whether periplasmic permeases could be viewed as channels that underwent more or less extensive modifications.

Periplasmic Permeases as Channels?

This possibility calls for some speculation because no attempts have been made at measuring channel activity in these systems. Such measurements would have to surmount rather large problems: Let us imagine that a periplasmic permease is indeed a channel opened only when the liganded protein signals its presence and ATP energy is consumed. In such a case, the receptor would be located right at the external mouth of the channel and would present it with a single molecule of substrate. The molecule might be transferred directly to a substrate-binding site situated somewhere within the channel, and by virtue of the fact that the channel is occluded on one side, the molecule could travel only in one direction. Alternatively, the substrate would be released from the binding protein into a small chamber where its local concentration would be quite high; it could then move down its concentration gradient towards the interior of the cell. In either case the substrate concentration, in the liganded form, would be quite high at the outer mouth of the channel because it would reflect the high concentration of the receptor to which it is liganded (see earlier); thus, reaching equal substrate concentrations on the two sides of the membrane would amount to creating a concentration gradient when the free external substrate concentration is compared to the internal concentration.

Obviously, the problem is to measure the movement of a single molecule at a time, when the channel cannot be opened without the receptor, which simultaneously blocks its opening. The solution to the problem would be to mutate a periplasmic permease so that it does not require the receptor for channel opening. Such mutants were indeed obtained in both the histidine and the maltose permeases (35, 69). They hydrolyze ATP continuously and in the total absence of a signal from the receptor, and they require high concentrations of substrate, as would be predicted for a substrate-binding site with poorer affinity located on the translocation complex. In other words, these may be channels that are uncoupled from an external signal and are continuously open, equilibrating the substrate across the membrane, but not concentrating it. These mutants offer a good chance of investigating whether periplasmic permeases function as channels by using standard channel-measuring methodologies.

CONCLUSIONS

Apparently, the numerous similarities between the prokaryotic and eukaryotic traffic ATPases justifies considering them as close relatives and extrapolating mechanistic conclusions among them, provided these extrapolations are done judiciously and supported experimentally. A reconciliation between the properties of channels and transporters would receive considerable support if periplasmic permeases were shown to function as channels on one hand, and on the other, if extensive information were obtained confirming the resemblance of the secondary and topological organizations of the eukaryotic traffic ATPases to the organizations of the prokaryotic ones. The purification and biochemical characterization of the individual components and extensive structure-function studies will doubtlessly support many of the features of various working models.

ACKNOWLEDGMENTS

Special thanks go to all the present and past members of the Ames laboratory who contributed various aspects of the knowledge accumulated over the years on periplasmic permeases. The work in this laboratory was supported by NIH grants DK12121 and DK43747 to G. F.-L. A.

Literature Cited

1. Adams, M. D., Oxender, D. L. 1989. Bacterial periplasmic binding protein tertiary structures. *J. Biol. Chem.* 256: 560–62
2. Ambudkar, S. V., Lelong, I. H., Zhang, J., Cardarelli, C. O., Gottesman, M. M., et al. 1992. Partial purification and reconstitution of the human multidrug-resistance pump: characterization of the drug-simulatable ATP hydrolysis. *Proc. Natl. Acad. Sci. USA* 89:8472–76
3. Ames, G. F.-L. 1972. Components of histidine transport. In *Biological Membranes. Proceedings of the 1972 ICN-UCLA Symposium in Molecular Biology*, ed. C. F. Fox, pp. 409–26. New York: Academic
4. Ames, G. F.-L. 1985. The histidine transport system of Salmonella typhimurium. In *Current Topics in Membranes and Transport*, pp. 103–19. New York: Academic
5. Ames, G. F.-L. 1986. Bacterial periplasmic transport systems: structure, mechanism, and evolution. *Annu. Rev. Biochem.* 55:397–425
6. Ames, G. F.-L. 1986. The basis of multidrug resistance in mammalian cells: homology with bacterial transport. *Cell* 47:323–24
7. Ames, G. F.-L. 1990. Energetics of periplasmic transport systems. In *Bacterial Energetics*, ed. T. A. Krulwich pp. 225–45. New York: Academic
8. Ames, G. F.-L. 1992. Bacterial periplasmic permeases as model systems for the superfamily of traffic ATPases including the multidrug resistance protein and the cystic fibrosis transmembrane conductance regulator. *Int. Rev. Cytol.* 137a:1–35
9. Ames, G. F.-L., Lecar, H. 1992. ATP-dependent bacterial transporters and of cystic fibrosis: Analogy between channels and transporters. *FASEB J.* 6: 2660–66
10. Ames, G. F.-L., Mimura, C., Holbrook, S., Shyamala, V. 1992. Traffic ATPases: a superfamily of transport proteins operating from *Escherichia coli* to humans. *Adv. Enzymol.* 65:1–47
11. Ames, G. F.-L., Mimura, C., Shyamala, V. 1990. Bacterial periplasmic permeases belong to a family of transport proteins operating from *E. coli* to humans: traffic ATPases. *FEMS Microbiol. Rev.* 75:429–46
12. Ames, G. F.-L., Nikaido, K. 1978. Identification of a membrane protein as a histidine transport component in *S. typhimurium*. *Proc. Natl. Acad. Sci. USA* 75:5447–51
13. Ames, G. F.-L., Nikaido, K., Groarke, J., Petithory, J. 1989. Reconstitution of periplasmic transport in inside-out membrane vesicles: energization by ATP. *J. Biol. Chem.* 264:3998–4002
14. Ames, G. F.-L., Spudich, E. N. 1976. Protein-protein interaction in transport: periplasmic histidine-binding protein J

interacts with P protein. *Proc. Natl. Acad. Sci. USA* 73:1877–81
15. Anderson, M. P., Berger, H. A., Rich, D. P., Gregory, R. J., Smith, A. E., et al. 1991. Nucleoside triphosphates are required to open the CFTR chloride channel. *Cell* 67:775–84
16. Anderson, M. P., Gregory, R. J., Thompson, S., Souza, D. W., Paul, S., et al. 1991. Demonstration that CFTR is a chloride channel by alteration of its anion selectivity. *Science* 253: 202–5
17. Anderson, M. P., Welsh, M. J. 1992. Regulation by ATP and ADP of CFTR chloride channels that contain mutant nucleotide-binding domains. *Science* 257:1701–4
18. Argos, P. 1990. An investigation of oligopeptides linking domains in protein tertiary structure and possible candidates for general gene fusion. *J. Mol. Biol.* 211:943–58
19. Arispe, N., Rojas, E., Hartman, J., Sorscher, E. J., Pollard, H. B. 1992. Intrinsic anion channel activity of the recombinant 1st nucleotide binding fold domain of the cystic fibrosis transmembrane regulator protein. *Proc. Natl. Acad. Sci. USA* 89:1539–43
20. Azzaria, M., Schurr, E., Gros, P. 1989. Discrete mutations introduced in the predicted nucleotide-binding sites of the *mdr1* gene abolish its ability to confer multidrug resistance. *Mol. Cell Biol.* 9:5289–97
21. Baichwal, V., Liu, D., Ames, G. F.-L. 1993. The ATP-binding component of a prokaryotic traffic ATPase is exposed to the periplasmic (external) surface. *Proc. Natl. Acad. Sci. USA*. 90:620–24
22. Bear, C. E., Li, C., Kartner, N., Bridges, R. J., Jensen, T. J., et al. 1992. Purification and functional reconstitution of the cystic fibrosis transmembrane conductance regulator (CFTR). *Cell* 68:809–18
23. Berkower, C., Michaelis, S. 1991. Mutational analysis of the yeast a-factor transporter STE6, a member of the ATP binding cassette (ABC) protein family. *EMBO J.* 10:3777–85
24. Bishop, L., Agbayani, R. J., Ambudkar, S. V., Maloney, P. C., Ames, G. F.-L. 1989. Reconstitution of a bacterial periplasmic permease in proteoliposomes and demonstration of ATP hydrolysis concomitant with transport. *Proc. Natl. Acad. Sci. USA* 86: 6953–57
25. Bordier, C. 1981. Phase separation of integral membrane proteins in Triton X-114 solution. *J. Biol. Chem.* 256: 1604–7
26. Boyd, D., Beckwith, J. 1989. Positively charged amino acid residues can act as topogenic determinants in membrane proteins. *Proc. Natl. Acad. Sci. USA* 86:9446–50
27. Bruggemann, E. P., Surrier, S. J., Gottesman, M. M., Pastan, I. 1992. Characterization of the azidopine and vinblastine binding site of P-glycoprotein. *J. Biol. Chem.* 267:21020–26
28. Bushman, E., Gros, P. 1993. The mouse multidrug resistance gene family: structural and functional analysis. *Int. Rev. Cytol.* 173C:169–97
29. Cornwell, M. M., Tsuruo, T., Gottesman, M. M., Pastan, I. 1987. ATP-binding properties of P-glycoprotein from multidrug resistant K Bk cells. *FASEB J.* 1:51–54
30. Cowman, A. F., Karcz, S., Galatis, D., Culvenor, J. G. 1991. A P-glycoprotein homologue of *Plasmoduium falciparum* is localized on the digestive vacuole. *J. Cell. Biol* 113:1033–42
31. Dalbey, R. E. 1990. Positively charged residues are important determinants of membrane protein topology. *Trends Biochem. Sci.* 15:253–57
32. Dassa, E., Hofnung, M. 1985. Sequence of gene *malG* in *E. coli* K-12: homologies between integral membrane components from binding protein-dependent transport systems. *EMBO J.* 4:2287–93
33. Davidson, A. L., Nikaido, H. 1990. Overproduction, solubilization, and reconstitution of the maltose transport system from *Escherichia coli*. *J. Biol. Chem.* 265:4254–60
34. Davidson, A. L., Nikaido, H. 1991. Purification and characterization of the membrane-associated components of the maltose transport system from *Escherichia coli*. *J. Biol. Chem.* 266: 8946–51
35. Davidson, A. L., Shuman, H. A., Nikaido, H. 1992. Mechanism of maltose transport in *Escherichia-coli*-transmembrane signaling by periplasmic binding proteins. *Proc. Natl. Acad. Sci. USA* 89:2360–64
36. Dierks, T., Salentin, A., Krämer, R. 1990. Pore-like and carrier-like properties of the mitochondrial aspartate/glutamate carrier after modification by SH-reagents: evidence for a preformed channel as a structural requirement of carrier-mediated transport. *Biochim. Biophys. Acta* 1028:281–88
37. Doige, C. A., Sharom, F. J. 1992. Transport properties of P-glycoprotein

in plasma membrane vesicles from multidrug-resistant Chinese hamster ovary cells. *Biochim. Biophys. Acta* 1109:161–71

38. Doige, C. A., Yu, X., Sharom, F. J. 1992. ATPase activity of partially purified P-glycoprotein from multidrug-resistant Chinese hamster ovary cells. *Biochim. Biophys. Acta* 1109:149–60
39. Dreesen, T. D., Johnson, D. J., Henikoff, S. 1988. The Brown protein of *Drosophila melanogaster* is similar to the White protein and to components of active transport complexes. *Mol. Cell Biol.* 8:5206–15
40. Froshauer, S., Green, G., Boyd, D., McGovern, K., Beckwith, J. 1988. Genetic analysis of the membrane insertion and topology of MalF, a cytoplasmic membrane protein of *Escherichia coli*. *J. Mol. Biol.* 200:501–11
41. Gallagher, M. P., Pearce, S. R., Higgins, C. F. 1989. Identification and localization of the membrane-associated, ATP-binding subunit of the oligopeptide permease of *Salmonella typhimurium*. *Eur. J. Biochem.* 180: 133–41
42. Gibson, A. L., Wagner, L. M., Collins, F. S., Oxender, D. L. 1991. A bacterial system for investigating transport effects of cystic fibrosis-associated mutations. *Science* 254:109–11
43. Gill, D. R., Hyde, S. R., Higgins, C. F., Valverde, M. A., Mintenig, G. M., et al. 1992. Separation of drug transport and chloride channel function of the human mutidrug resistance P-glycoprotein. *Cell* 71:23–32
44. Greenberger, L. M., Lisanti, C. J., Silva, J. T., Horwitz, S. B. 1991. Domain mapping of the photoaffinity drug-binding sites in P-glycoprotein encoded by mouse *mdr1b*. *J. Biol. Chem.* 266:20744–51
45. Gregory, R. J., Rich, D. P., Cheng, S. H., Souza, D. W., Paul, S., et al. 1991. Maturation and function of cystic fibrosis transmembrane conductance regulator variants bearing mutations in putative nucleotide-binding domains 1 and 2. *Mol. Cell Biol.* 11:3886–93
46. Gros, P., Croop, J., Talbot, F. 1991. A single amino acid substitution strongly modulates the activity and substrate specificity of the mouse *mdr1* and *mdr3* drug efflux pumps. *Proc. Natl. Acad. Sci. USA* 88:7289–93
47. Harold, F. M. 1986. *The Vital Force: A Study of Bioenergetics*. New York: W. H. Freeman
48. Hartman, J., Huang, Z., Rado, T. A., Peng, S., Julling, T., et al. 1992. Recombinant synthesis, purification, and nucleotide binding characteristics of the first nucleotide binding domain of the cystic fibrosis gene product. *J. Biol. Chem.* 267:6455–58
49. Higgins, C. F., Hiles, I. D., Whalley, K., Jamieson, D. K. 1985. Nucleotide binding by membrane components of bacterial periplasmic binding protein-dependent transport systems. *EMBO J.* 4:1033–40
50. Higgins, C. F., Hyde, S. C., Mimmack, M. M., Gileadi, U., Gill, D. R., et al. 1990. Binding protein-dependent transport systems. *J. Bioenerg. Biomembr.* 22:571–92
51. Hobson, A., Weatherwax, R., Ames, G. F.-L. 1984. ATP-binding sites in the membrane components of the histidine permease, a periplasmic transport system. *Proc. Natl. Acad. Sci. USA* 81:7333–37
52. Horio, M., Gottesman, M., Pastan, I. 1988. ATP-dependent transport of vinblastine in vesicles from human multidrug-resistant cells. *Proc. Natl. Acad. Sci. USA* 85:3580–84
53. Hyde, S. C., Emsley, P., Hartshorn, M. J., Mimmack, M. M., Gileadi, U., et al. 1990. Structural model of ATP-binding proteins associated with cystic fibrosis, multidrug resistance and bacterial transport. *Nature* 346:362–65
54. Joshi, A., Ahmed, S., Ames, G. F.-L. 1989. Energy coupling in bacterial periplasmic transport systems: studies in intact *Escherichia coli* cells. *J. Biol. Chem.* 264:2126–33
55. Kerem, B., Rommens, J. M., Buchanan, J. A., Markiewicz, D., Cox, T. K., et al. 1989. Identification of the cystic fibrosis gene: genetic analysis. *Science* 245:1073–80
56. Kerppola, R. E., Ames, G. F.-L. 1992. Topology of the hydrophobic membrane-bound components of the histidine periplasmic permease. Comparison with other members of the family. *J. Biol. Chem.* 267:2329–36
57. Kerppola, R. E., Shyamala, V., Klebba, P., Ames, G. F.-L. 1991. The membrane-bound proteins of periplasmic permeases form a complex: identification of the histidine permease HisQMP complex. *J. Biol. Chem.* 266: 9857–65
58. Kuchler, K., Sterne, R. E., Thorner, J. 1989. *Saccharomyes cerevisiae* STE6 gene product: a novel pathway for protein export in eukaryotic cells. *EMBO J.* 8:3973–84
59. Manoil, C., Mekalanos, J. J.,

Beckwith, J. 1990. Alkaline phosphatase fusions: sensors of subcellular location. *J. Bacteriol.* 172:515–18

60. Mao, B., Pear, M. P., McCammon, J. A., Quiocho, F. A. 1982. Hinge-bending in L-arabinose binding protein. *J. Biol. Chem.* 257:1131–33
61. Martinez, C. K., Monaco, J. J. 1991. Homology of proteasome subunits to a major histocompatibility complex-linked LMP gene. *Nature* 353:664–65
62. McEwan, G. T. A., Hunter, J., Hirst, B. H., Simmons, N. L. 1992. Volume-activated Cl^- secretion and transepithelial vinblastine secretion mediated by P-glycoprotein are not correlated in cultured human T84 intestinal epithelial layers. *FEBS* 304:233–36
63. McGrath, J. P., Varshavsky, A. 1989. The yeast STE6 gene encodes a homologue of the mammalian multidrug resistance P-glycoportein. *Nature* 340: 400–4
64. Mimmack, M. L., Gallagher, M. P., Pearce, S. R., Hyde, S. C., Booth, I. R., et al. 1989. Energy coupling to periplasmic binding protein-dependent transport systems: stoichiometry of ATP hydrolysis during transport in vivo. *Proc. Natl. Acad. Sci. USA* 86:8257–60
65. Mimura, C. S., Admon, A., Hurt, K. A., Ames, G. F.-L. 1990. The nucleotide-binding site of HisP, a membrane protein of the histidine permease. Identification of amino acid residues photoaffinity-labeled by 8-azido ATP. *J. Biol. Chem.* 265:19535–42
66. Mimura, C. S., Holbrook, S. R., Ames, G. F.-L. 1991. Structural model of the nucleotide-binding conserved component of periplasmic permeases. *Proc. Natl. Acad. Sci. USA* 88:84–88
67. Monaco, J. J. 1992. A molecular model of MHC class-I-restricted antigen processing. *J. Immunol. Today* 13:173–79

67a. Oh, B.-H., Pandit, J., Kang, C.-H., Nikado, K., Goken, S., et al. 1993. Three-dimensional structures of the periplasmic lysine-, arginine-, ornithine-binding protein with and without a ligand. *J. Biol. Chem.* In press

68. Pearce, S. R., Mimmack, M. L., Gallagher, M. P., Gileadi, U., Hyde, S. C., et al. 1992. Membrane topology of the integral membrane components, OppB and OppC, of the oligopeptide permease of *Salmonella typhimurium*. *Mol. Molbiol.* 6:47–57
69. Petronilli, V., Ames, G. F.-L. 1991. Binding protein-independent histidine permease mutants: uncoupling of ATP hydrolysis from transmembrane signaling. *J. Biol. Chem.* 266:16293–96
70. Petronilli, V., Ames, G. F.-L. 1991. Histidine transport by periplasmic permease reconstituted into proteoliposomes. *Biophys. J.* 59:322a
71. Prossnitz, E., Gee, A., Ames, G. F.-L. 1989. Reconstitution of the histidine periplasmic transport system in membrane vesicles. Energy coupling and interaction between the binding protein and the membrane complex. *J. Biol. Chem.* 264:5006–14
72. Prossnitz, E., Nikaido, K., Ulbrich, S. J., Ames, G. F.-L. 1988. Formaldehyde and photoactivatable crosslinking of the periplasmic binding protein to a membrane component of the histidine transport system of *Salmonella typhimurium*. *J. Biol. Chem.* 263: 17917–20
73. Quinton, P. M., Reddy, M. M. 1992. Control of CFTR chloride conductance by ATP levels through non-hydrolytic binding. *Nature* 360:79–81
74. Ramirez, J. A., Lecar, H., Harris, S., Haber, J. 1990. Patch-clamp studies of print and regulatory mutants of yeast plasma membrane ATP-ase (PMAI). *Biophys. J.* 57:321a
75. Raymond, M., Gros, P., Whiteway, M., Thomas, D. Y. 1992. Functional complementation of yeast *ste6* by a mammalian multidrug resistance *mdr* gene. *Science* 256:232–34
76. Rich, D. P., Gregory, R. J., Anderson, M. P., Manavalan, P., Smith, A. E., et al. 1991. Effect of deleting the R domain on CFTR-generated chloride channels. *Science* 253:205–7
77. Riordan, J. R., Rommens, J. M., Kerem, B., Alon, N., Rozmahel, R., et al. 1989. Identification of the cystic fibrosis gene: cloning and characterization of complementary DNA. *Science* 245:1066–73
78. Rommens, J. M., Iannuzzi, M. C., Kerem, B., Drumm, M. L., Melmer, G., et al. 1989. Identification of the cystic fibrosis gene: chromosome walking and jumping. *Science* 245:1059–65
79. Roninson, I. B. 1991. *Molecular and Cellular Biology of MDR in Tumor Cells*. New York: Plenum
80. Sarkadi, B., Price, E. M., Boucher, R. C., Germann, U. A., Scharborough, G. A. 1992. Expression of the human multidrug resistance cDNA in insect cells generates a high activity drug-simulated membrane ATPase. *J. Biol. Chem.* 267:4854–58
81. Sharma, R. C., Inoue, S., Roitelman, J., Schimke, R. T., Simoni, R. D. 1992. Peptide transport by the multidrug

resistance pump. *J. Biol. Chem.* 267: 5731–34
82. Shimbaku, A. M., Nishimoto, T., Ueda, K., Tohru, T. 1992. P-glycoprotein: ATP hydrolysis by the N-terminal nucleotide-binding domain. *J. Biol. Chem.* 267:4308–11
83. Shuman, H. A., Silhavy, T. J., Beckwith, J. R. 1980. Labeling of proteins with β-galactosidase by gene fusion. *J. Biol. Chem.* 255:168–74
84. Shyamala, V., Baichwal, V., Beall, E., Ames, G. F.-L. 1991. Structure-function analysis of the histidine permease and comparison with cystic fibrosis mutations. *J. Biol. Chem.* 266: 18714–19
85. Speiser, D. M., Ames, G. F.-L. 1991. Mutants of the histidine periplasmic permease of *Salmonella typhimurium* that allow transport in the absence of histidine-binding proteins. *J. Bacteriol.* 173:1444–51
86. Thomas, P. J., Shenbagamurthi, P., Sondek, J., Hullihen, J. M., Pedersøn, P. L. 1992. The cystic fibrosis transmembrane conductance regulator. *J. Biol. Chem.* 267:5727–30
87. Thomas, P. J., Shenbagamurthi, P., Ysern, X., Pedersen, P. L. 1991. Cystic fibrosis transmembrane conductance regulator: nucleotide binding to a synthetic peptide. *Science* 251:555–57
88. Valverde, M. A., Diaz, M., Sepulveda, F. V., Gill, D. R., Hyde, S. C., et al. 1992. Volume-regulated chloride channels associated with the human multidrug-resistance p-glycoprotein. *Nature* 355:830–33
89. von Heijne, G. 1986. The distribution of positively charged residues in bacterial inner membrane proteins correlates with the trans-membrane topology. *EMBO J.* 5:3021–27
90. Walter, C., Honer zu Bentrup, K., Schneider, E. 1992. Large scale purification, nucleotide binding properties, and ATPase activity of the MalK subunit of *Salmonella typhimurium* maltose transport complex. *J. Biol. Chem.* 267:8863–69

Annu. Rev. Microbiol. 1993. 47:321–50

REGULATION OF THE HEAT-SHOCK RESPONSE IN BACTERIA

T. Yura[1]
Institute for Virus Research, Kyoto University, Kyoto 606-01, Japan

H. Nagai
National Institute of Genetics, Mishima, Shizuoka 411, Japan

H. Mori
Institute for Virus Research, Kyoto University, Kyoto 606-01, Japan

KEY WORDS: heat-shock protein, stress response, σ factor, translational control, *Escherichia coli*

CONTENTS

[1]Present address: HSP Research Institute, Kyoto Research Park, Kyoto 600, Japan.

0066-4227/93/1001-0321$02.00

ABSTRACT

When bacteria cells are exposed to higher temperature, a set of heat-shock proteins (hsps) is induced rapidly and transiently to cope with increased damage in proteins. The mechanism underlying induction of hsps has been a central issue in the heat-shock response and studied intensively in *Escherichia coli*. Immediately upon temperature upshift, the cellular level of σ^{32} responsible for transcription of heat-shock genes increases rapidly and transiently. This increase in σ^{32} results from both increased synthesis and stabilization of σ^{32}, which is ordinarily very unstable. A clue to further understanding of early regulatory events came from recent analysis of translational induction and subsequent shut-off of σ^{32} synthesis. Whereas a 5′-coding region of mRNA for σ^{32} is involved in the induction meditated by the mRNA secondary structure, a distinct segment of σ^{32} polypeptide further downstream is involved in the DnaK/DnaJ-mediated shut-off and destabilization of σ^{32} that may be mutually interconnected.

INTRODUCTION

When cells of bacteria or almost any organism are exposed to high temperature, the synthesis of a set of heat-shock proteins (hsps) is rapidly induced. The primary structure of most hsps appears to be highly conserved during evolution, suggesting that they serve similar functions in all organisms (27, 62). Transient induction of hsps represents an important protective and homeostatic mechanism to cope with physiological and environmental stress at the cellular level. In recent years, both the function and regulation of hsps have undergone extensive study with bacteria, yeast, and various higher organisms. Most hsps are synthesized under nonstress conditions, albeit at reduced rates, and research indicated that hsps play fundamental roles in normal cell physiology in addition to their activity under stress conditions. These proteins are often described as molecular chaperones in that they mediate the correct folding and assembly of polypeptides but are not components of the functional assembled structures (17, 76).

Induction of hsps occurs primarily at the level of transcription: transcription of heat-shock genes is enhanced as a result of increased activity of specific transcription factors—σ^{32} in bacteria such as *Escherichia coli*, and HSF (heat-shock factor) in eukaryotes such as yeast, flies, mice, and humans (62). Moreover, some hsps, particularly HSP70 (DnaK in *E. coli*), exert a negative-feedback control on the synthesis of hsps following initial induction (14, 95), presumably by interacting directly or indirectly with the heat-shock transcription factors (12). Thus not only the structure and function of hsps but their mechanisms of regulation appear to be conserved, at least in some basic forms. Such a remarkable conservation must reflect the rapidly changing cellular requirements for hsps of most living organisms in nature and the need for rapid and fine adjustment of hsp levels to assure optimal growth and survival.

One of the outstanding problems in the field of heat-shock and stress response has been to elucidate the mechanisms regulating the expression of heat-shock genes. In this review, we concentrate on this problem by analyzing the regulation in *E. coli,* because intensive studies during the past few years have led to the accumulation of data that may provide important clues for future studies. We place special emphasis on recent progress in understanding some of the initial events of the heat-shock response. For references to earlier work on the bacterial heat-shock response and on individual hsps, previous reviews dealing with more general or comprehensive subjects (27, 28, 31, 69, 71) should be consulted. Functional aspects of hsps, one of the most widely discussed topics in biology today, are treated in a number of recent reviews (17, 27, 107, 117).

INDUCTION OF HSPS AND THE HEAT-SHOCK REGULON

In *E. coli,* as in most other organisms, transient induction of hsps is a most conspicuous response of cells when they are exposed to a modest heat shock such as a shift from 30 to 42°C: most other proteins do not change their rates of synthesis appreciably, whereas still others decrease their synthesis transiently (50, 71, 112). The induction of hsps occurs almost immediately after temperature shift, reaches its maximum at about 5 min, and declines to a new steady-state level in 20–30 min. Induction also occurs coordinately at the level of transcription (110, 111). When cells are exposed to a lethal temperature such as 50°C, most proteins cease to be synthesized and those that remain are mostly hsps (12, 71).

A breakthrough in the discovery of heat-shock regulation in *E. coli* occurred in 1975, when Cooper & Ruettinger (10) reported a nonsense mutation affecting the synthesis of a major protein essential for growth at high temperature. This mutant, Tsn-K165, carried a mutation, *rpoH* (=*htpR, hin*), that was later shown by two laboratories to affect the synthesis of several proteins now known as hsps (68, 112). Most importantly, the *rpoH* gene product was identified as a class of the σ factors (σ^{32}) that are required for transcription of the heat-shock genes (32). These findings formed a basis for subsequent analyses of regulation and function of the heat-shock response in bacteria. At least some hsps can be induced by other agents, including DNA-damaging agents (45), amino acid analogues (30), stringent response (34), starvation for carbon source (37), λ phage infection (16, 43), unfolded proteins (30, 75), alkaline shift (90), and ethanol (98), suggesting that hsps may protect cells against a variety of environmental stresses.

The heat-shock regulon under σ^{32} control includes genes for about 20 proteins as defined by two-dimensional gel electrophoresis (69, 71). Most of the major *E. coli* hsps have homologous eukaryotic counterparts, e.g. DnaK, DnaJ, GrpE, GroEL, GroES, Lon, ClpP, ClpB, and HtpG (27, 107). Many of them (DnaK, DnaJ, GrpE, GroEL, GroES, and presumably HtpG) are chaperones that assist correct folding and assembly of proteins, and consequently affect diverse cellular processes including DNA replication (see 27), RNA transcription, protein transport (e.g. 6), proteolysis (e.g. 41, 87), flagella synthesis (80), and UV mutagenesis (15). These proteins, particularly when overproduced, can protect various other proteins from heat inactivation (e.g., 24, 61, 82), consistent with the genetic evidence for key roles of GroE and DnaK proteins (among other hsps) in supporting growth at physiological temperature (46, 49, 102). DnaK, DnaJ, and GrpE proteins, known to work synergistically in λ phage and P1 plasmid DNA replication (27), are also involved in many cellular functions including feedback control of the heat-shock response (31, 89). Besides their protective roles against protein inactivation in vivo, they reactivate denatured proteins such as RNA polymerase (82), as shown by experiments using purified proteins in vitro. On the other hand, at least three hsps represent either an ATP-dependent protease (Lon, La) (29) or a catalytic (ClpP, Ti) or regulatory (ATPase) subunit (ClpB) of another protease Clp (42, 44, 59, 84). These results are thus consistent with the notion that major functions of hsps are to prevent inactivation of cellular proteins, to reactivate once inactivated proteins, and to help degrade nonreparable denatured proteins that accumulate under normal growth as well as under stress conditions (27, 49, 107).

ROLE OF σ^{32} IN TRANSCRIPTION OF HEAT-SHOCK GENES

Induction of hsps upon temperature upshift occurs through the activation of transcription from promoters specifically recognized by RNA polymerase (E) that contains σ^{32} ($E\sigma^{32}$) (11, 32, 118). Genetic experiments with an *rpoH* mutant provided first evidence that the heat induction is positively regulated by the *rpoH* gene product. Mutations in *rpoH* prevented the transient increase in hsp synthesis following a temperature upshift without appreciably affecting synthesis of other proteins (68, 112). Moreover, the extent of induction of hsps depended on the amount of *rpoH* gene product synthesized (112). The finding that an amber mutant of *rpoD* (encoding σ^{70}) producing reduced amounts (10% of wild-type) of σ^{70} exhibited markedly increased synthesis of hsps (73) suggested the possibility that the *rpoH* gene product and σ^{70} compete

Table 1 Promoters transcribed by RNA polymerase containing σ^{32}

Map Position	Promoter	−35 Region	Spacing	−10 Region	Reference
0.3'	*dnaKp1*	T C T C c C C C T T G A t	14	C C C C A T t T A	11
0.3'	*dnaKp2*	T t g g g C a g T T G A A	13	C C C C t a t T A	11
10'	*lon*	T C T C g g C g T T G A A	14	C C C C A T a T A	11
10'	*clpP*	T g T t a t g C T T G A A	14	a C C C A T a a c	59[a]
11'	*htpGp1*	g C T C t C g C T T G A A	15	C C C C A T c T c	11
28'	*topAp1*	a C a a g g g g T T G A t	16	g t C C A T a T c	52
57'	*grpE*	T g c t t C C C T T G A A	14	C C C C A T a a t	56
57'	*clpB*	c a a t a a C C T T G A A	14	C C t C A T t T A	42
67'	*rpoDphs*	T g c C a C C C T T G A A	15	g a C g A T a T A	11
83'	*ibp*	g a T a a g g C T T G A A	13	a C C C A T t t t	2[a]
90'	*rrnBp1*	a t T t c C t C T T G t c	14	C C C t A T a a t	72[a]
94'	*groE*	T t T C c C C C T T G A A	13	C C C C A T t T c	11
F	*repE*	c C T g c t t a T T G A c	17	g a C a A T c T A	103
σ^{32} consensus		T C T C − C C C T T G A A	13 ~ 17	C C C C A T − T A	
σ^{70} consensus		T T G A C A	16 ~ 18	T A T A A T	

[a] To our knowledge, direct tests using an in vitro transcription system have not been reported for these promoters.

with each other for RNA polymerase core enzyme (115). Cloning and sequencing of the *rpoH* gene revealed that regions of its predicted gene product are strikingly similar to those of σ^{70} (47, 70, 97, 114). Grossman et al (32) purified the factor that permitted transcription of the heat-shock gene as a 32-kDa σ factor (σ^{32}) that specifically recognized the heat-inducible promoters located upstream of the heat-shock genes (5, 11, 23, 93).

The distinct promoter specificity of σ^{32} provides the basis for the characteristic heat inducibility of the regulon. At least some heat-shock genes (e.g. *rpoD, groE*), however, carry an additional promoter transcribed by $E\sigma^{70}$ (93, 118). At present, 13 promoters are known to be transcribed by $E\sigma^{32}$. Table 1 lists the map positions and sequences of these promoters. In general, these promoters are specifically recognized by $E\sigma^{32}$ but not by $E\sigma^{70}$ in vitro. Such a distinct specificity was also demonstrated in vivo, at least for *dnaKp1, dnaKp2, groE, htpGp1,* and *htpGp2* promoters, through the use of *rpoH* null mutants lacking σ^{32} (118): apparently neither σ^{70} nor other σ factors endow the *E. coli* RNA polymerase core enzyme with the ability to recognize these promoters.

Recent examination of the *rrnBp1* promoter revealed an unusual case: both -35 and -10 bp sequences of a σ^{32} promoter overlap those of the previously identified σ^{70} promoter, resulting in transcription from identical start sites by $E\sigma^{32}$ and $E\sigma^{70}$ (72a). Apparently similar paradoxical cases had been found in *Bacillus subtilis* where one promoter can be utilized by two forms of RNA polymerase (38, 91, 104). A consensus sequence derived from the present list of σ^{32} promoters, which agrees well with that reported previously (11), should be compared and contrasted with consensus σ^{70} promoters (Table 1). The list also includes the *repE* promoter of F plasmid encoding a replication initiator protein; the role of σ^{32} is not restricted to the chromosomal genes. This result exemplifies the critical involvement of σ^{32} in the maintenance of conjugative plasmids such as F (103).

ROLE OF σ^{32} IN CELL GROWTH

Although the essential role of σ^{32} in cell growth at high temperature was established when an *rpoH* mutant was first isolated (10, 68, 112), its role at low temperature was assessed in later experiments; the majority of temperature-sensitive *rpoH* mutants isolated in a strain lacking known tRNA suppressors turned out to result from nonsense mutations, suggesting that σ^{32} may be dispensable at relatively low temperatures (up to 34°C) (97, 114). Indeed, *E. coli* carrying the *rpoH* null mutations (caused either by a deletion or a transposon insertion) were isolated, though they failed to grow at temperatures above 20°C (118).

The role of σ^{32} in cell growth was further assessed through the isolation and characterization of temperature-resistant revertants of the Δ*rpoH* strain

(46). This analysis found that the inability of *rpoH* null mutants to grow at high temperature is primarily caused by the lack of sufficient amounts of key heat-shock proteins, GroEL and GroES. The reason for this is that hyperproduction of GroE (but not DnaK), either by the insertion of an exogenous promoter (an IS*10*-like element) into the *groE* upstream region or by the introduction of *groE*-expressing plasmids, enabled the mutant bacteria to grow at temperatures up to 40°C (46). Moreover, the amount of GroE proteins produced among the revertants or other derivatives of the Δ*rpoH* strain tested correlated well with the ability to grow at higher temperatures. These results implied that the residual amount of GroE produced in *rpoH* null mutants, owing to transcription from a minor σ^{70} promoter, supported their growth up to 20°C (118). Furthermore, when large amounts of both GroE and DnaK were produced either by promoter insertion(s) or by plasmid(s), the Δ*rpoH* strain could grow at up to 42°C (46).

Thus, at least in the absence of σ^{32}, GroE seemed to be essential at all temperatures, whereas DnaK appeared essential under more restricted conditions. Consistent with this is the finding that deletion mutants can be isolated for *dnaK* (7, 74) but not for *groE* (27). More recently, experiments designed to assess the effects of reduced GroE contents on protein metabolism were carried out under steady-state growth conditions (40). It was found that when the amount of GroE was specifically reduced to subnormal levels, the upper limits of growth temperature dropped significantly; the amounts of other hsps produced were elevated under these conditions. The different effects of GroE and DnaK observed with the Δ*rpoH* mutant (46) may reflect different functions of the two most abundant bacterial hsps (35), as indicated by many lines of evidence (e.g. 27, 49, 107), including the recent model of sequential function of DnaK (DnaJ, GrpE) and GroEL (GroES) in assisting the folding of nascent polypeptides in *E. coli* (48). Besides their inability to grow at high temperature, the *rpoH* mutants have been shown to be defective in proteolysis (4, 29), cell division (99), growth of temperate phages (T. Yura, unpublished result), and plasmid DNA replication (103). These phenotypes should result, at least in part, from reduced levels of one or more hsps.

REGULATORY ROLES OF σ^{32} IN THE HEAT-SHOCK RESPONSE

The mechanism underlying the rapid and transient increase in the rate of transcription of hsp genes in *E. coli* was intensively studied recently (31). In addition to its catalytic role in initiating transcription from heat-shock promoters, σ^{32} is directly involved in control at three different levels: synthesis, degradation, and activity. Thus, the levels and activity of σ^{32} play an active and dynamic role in regulating and fine tuning heat-shock gene expression under a variety of conditions (31, 33, 86).

Regulation of the σ^{32} Level

Investigators initially believed that σ^{32} may be activated at high temperature, because the heat induction was so rapid (111) as to preclude any models that assume increased synthesis of σ^{32} (e.g. 32). On the other hand, the extent of heat-shock induction of hsps depended on the amounts of *rpoH* gene product produced (112). Consistent with this, increasing the rate of σ^{32} synthesis without a corresponding temperature upshift also resulted in elevated synthesis of hsps (33). In addition, σ^{32} synthesized in vitro using a purified transcription-translation system directly activated transcription of heat-shock genes (5). Immunological detection of σ^{32} provided direct evidence that the cellular

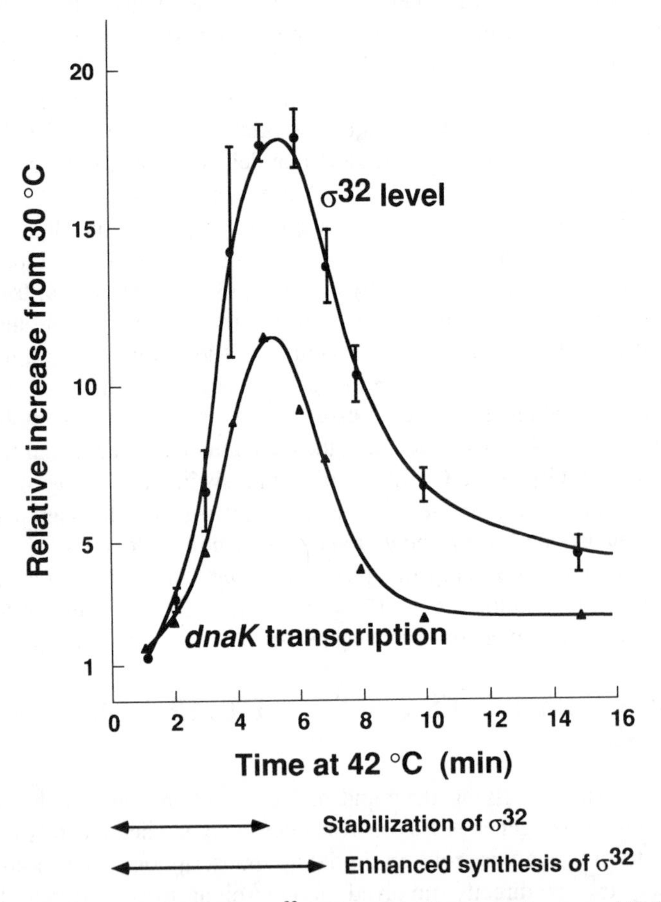

Figure 1 Time course of changes in σ^{32} level and *dnaK* transcription upon shift from 30 to 42°C. Levels of σ^{32} (immunoblotting) and relative rates of *dnaK* mRNA synthesis (pulse labeling) were determined. Reproduced with permission from Ref. 86.

levels of σ^{32} rapidly increase following a temperature upshift in both haploid cells (53, 86) and cells harboring an *rpoH*$^{+}$ multicopy plasmid (22). The ability of S-30 extracts to synthesize GroEL and DnaK proteins in vitro clearly reflected the intracellular σ^{32} levels in vivo at the time of cell harvesting: a much higher activity was obtained with cells that had been heat shocked (81).

Detailed time-course experiments using specific σ^{32} antisera were then conducted. When cells were shifted from 30 to 42°C, the intracellular concentration of σ^{32} increased 15- to 20-fold within 5 min and then declined to a new steady-state level severalfold higher than the preshift level (53, 86). The transient increase in σ^{32} levels during heat shock resulted from changes in both synthesis and stability of σ^{32} (86). Both the time course and the increase in σ^{32} levels were apparently sufficient to account for the increased transcription of heat-shock genes (Figure 1). The cellular concentration of σ^{32} during steady-state growth is very low, estimated to be only 10 to 30 molecules per cell (12). This is because σ^{32} is an extremely unstable protein (half-life = ~1 min) (33, 86, 96), and its synthesis is largely repressed at the level of translation (33, 39, 64, 86). Both the unusual instability and translational repression provide effective means of maintaining σ^{32} concentration at a minimum during steady-state growth when only low levels of hsps are needed. Following physiological or environmental stress, the levels of σ^{32} and of hsps are rapidly increased to meet the specific requirements of the given condition (33, 86).

Regulation of σ^{32} *Stability*

Perhaps one of the earliest events that occurs in response to heat-shock stress is the transient stabilization of otherwise unstable σ^{32}. As mentioned above, the half-life of σ^{32} produced from a single-copy *rpoH* gene during steady-state growth is about 1 min at both 30 and 42°C (86). During the initial studies, some variability in stability was found when σ^{32} was produced from various multicopy plasmids (3, 33, 96), presumably reflecting differences in strains, culture conditions, or other methods used. Apparently, σ^{32} is more stable when produced in larger amounts particularly at low temperatures such as 22°C (96).

Following a temperature shift from 30° to 42°C, σ^{32} is markedly though transiently stabilized for the first 4–5 min, which allows its rapid accumulation. After this initial phase, however, the marked instability of σ^{32} resumes, consistent with the transient increase in σ^{32} levels observed (Figure 1). The rapid stabilization of σ^{32} probably explains an almost instantaneous increase in the rates of hsp synthesis observed after a temperature upshift (111). Although little is known about the exact mechanisms underlying the extreme instability of σ^{32}, mutations in several genes as well as growth conditions have been shown to stabilize σ^{32}.

DnaK was first reported as a negative modulator of the heat-shock response, because a mutation in *dnaK* resulted in both elevated levels of hsps at low temperature (30°C) and prolonged synthesis of hsps upon shift to high temperature (42°C) (95). Other missense and null mutations of *dnaK* and those in two other hsp genes, *dnaJ* and *grpE,* were subsequently shown to exhibit similar phenotypes (89, 96). In all these mutants, σ^{32} was markedly stabilized, the half-life lasting 10- to 30-fold longer than that in wild-type bacteria. The increased stability but not the increased synthesis of σ^{32} accounts for the higher levels (severalfold) of hsps observed in these mutant backgrounds (89). Reduced cellular levels of GroE during steady-state growth also stabilized σ^{32} significantly and resulted in elevated synthesis of other hsps, though the effect was less striking than those observed with the *dnaK, dnaJ,* or *grpE* mutants (40). Because none of these hsps are known to be proteases, they most likely assist proteolysis directly by binding to σ^{32}, as recently shown for DnaK, DnaJ, and GrpE (25, 54), and presenting it to an unidentified protein-degradation system. Alternatively, the hsps may work indirectly through their potential effect on protease activation. *E. coli* cells infected with phage λ transiently produce high levels of hsps (16, 43) because of the stabilization of σ^{32}; the λ*c*III gene is apparently responsible for this stabilization, because overproduction of cIII protein alone enhanced hsp synthesis in uninfected bacteria (3). Recent work on the expression and stability of various σ^{32}-β-galactosidase fusion proteins using a *rpoH-lacZ* gene suggested that a specific segment of σ^{32} is required for regulation of σ^{32} degradation and feedback control of the heat-shock response (see below).

Regulation of σ^{32} Activity

Although an increase in σ^{32} level primarily causes the induction of hsps upon temperature upshift, the rate of hsp synthesis does not always reflect the intracellular levels of σ^{32} and can be reduced by inhibiting σ^{32} activity. Such a situation arises when excess hsps are synthesized following an initial overproduction of σ^{32}: whereas σ^{32} continues to be made at high levels, enhanced synthesis of hsps is observed only transiently (33). Also, the level of σ^{32} in the *dnaK, dnaJ,* or *grpE* mutants is 10- to 20-fold higher than in the wild-type, but the synthesis of hsps is only 2- to 4-fold greater, indicating that the activity of σ^{32} is inhibited (89). In addition, when *E. coli* cells grown at high temperature (42°C) are shifted to low temperature (30°C), σ^{32} levels do not appreciably decrease. However, both transcription of heat-shock genes and synthesis of hsps decrease markedly, though transiently, indicating that the activity of σ^{32} is transiently inhibited (88, 92). Furthermore, the same subset of hsps (DnaK, DnaJ, and GrpE) involved in the control of σ^{32} stability is apparently involved in the control of σ^{32} activity (C. Gross, personal communication; N. Kusukawa & T. Yura, unpublished result).

Recent experiments from two laboratories revealed that σ^{32} physically binds to DnaK, DnaJ, or GrpE. Using purified proteins, researchers showed that σ^{32} binds to wild-type DnaK, but not to DnaK756 mutant protein in vitro and that the resulting complex is disrupted by Mg^{2+}-ATP (54). Complexes between these hsps and σ^{32} are also found in vivo: histidine-tagged σ^{32} was shown to be associated with each of these hsps, apparently independently, and the complexes between σ^{32} and DnaK (or GrpE) or between σ^{32} and DnaJ behaved differently in the presence or absence of ATP (25). In further in vitro binding studies, a stable ternary complex DnaJ-σ^{32}-DnaK was formed effectively only in the presence of GrpE or ATP. Moreover, the set of three hsps was found to interfere with the ability of σ^{32} to bind to core RNA polymerase, and hence its ability to initiate transcription from heat-shock promoters (C. Georgopoulos, personal communication). These results taken together indicate that DnaK, DnaJ, and GrpE proteins work synergistically to interfere with the activity of σ^{32}. Such an apparent sequestration of σ^{32} from core RNA polymerase could presumably render σ^{32} a better substrate for one or more proteases, thus linking the control of activity to that of σ^{32} stability (27).

REGULATION OF σ^{32} SYNTHESIS

Transcription of the rpoH *Gene*

The *rpoH* gene has at least four promoters; three (P1, P4, P5) are transcribed by $E\sigma^{70}$ (19, 21, 63), whereas one (P3) is transcribed by RNA polymerase containing a novel σ factor, σ^E ($=\sigma^{24}$), that is especially active at very high temperature (18, 106). Figure 2 shows the organization of *rpoH* promoters. The most distal P1 promoter is the strongest, and together with P4, is responsible for most *rpoH* transcription under most growth conditions. The activity of P1 does not seem to be highly regulated, but the fact that its −35 region is located within the adjacent *ftsX* gene suggests a possible coupling with *fts* gene expression, perhaps mediated by the DnaA protein (13). In contrast, the activity of P4 is enhanced 2- to 3-fold upon shift from low to

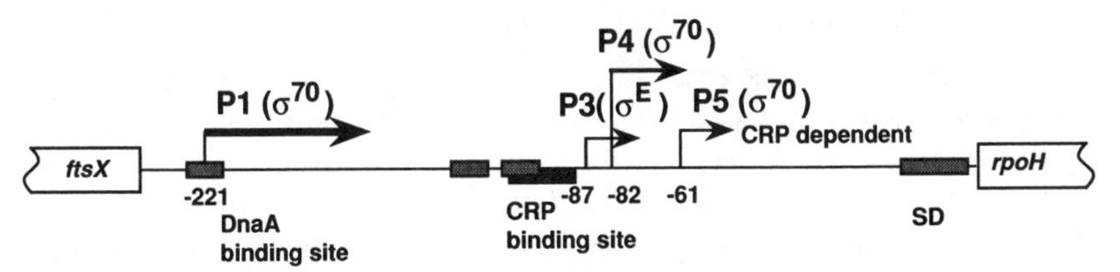

Figure 2 Organization of *rpoH* promoters and putative regulatory sites (taken from 19, 21, 63). The numbering system of promoters is that used in Refs. 19 and 63.

high temperature (19, 21). A pair of DnaA protein–binding consensus sequences found upstream of the P3 and P4 promoters were indeed shown to bind DnaA protein and specifically inhibit transcription from these promoters in vivo and in vitro (105). P5 is a weak promoter, enhanced in the absence of glucose or by addition of ethanol, and requires cAMP and cAMP-receptor protein for its function in vitro (63). However, P5 is unlikely to be involved in the *rpoH*-dependent induction of hsps during carbon starvation, because the induction is not cAMP-CRP (cAMP receptor protein) dependent (78).

The P3 promoter seems to be particularly important, because it becomes highly active under unusual conditions such as exposure to a lethal temperature (50°C) at which σ^{70} (and perhaps other σ factors) is inactivated. This result probably explains the observation that only hsps are highly and continuously expressed at 50°C (12, 71). Several other bacterial promoters carry a sequence similar to *rpoHp3*, and one of them, *htrA*, is actually transcribed in vitro by $E\sigma^{E}$ (18, 57). The *htrA* (*degP*) gene encodes a periplasmic protease that is important for survival at high temperature (55, 85). These results suggest that genes whose transcription depends on σ^{E} constitute a second heat-shock regulon of *E. coli* and that their function may strengthen or complement that of the σ^{32} regulon in coping with heat shock or other stresses (31). For example, products of these genes may be required for thermotolerance; cells that had been pretreated at 42°C acquire tolerance to a subsequent exposure to lethal temperature (55°C) (112), whereas overproduction of σ^{32} (and hsps) alone without a prior temperature upshift will not result in thermotolerance (100).

In spite of such a complex organization and potential regulatory importance of the various promoter regions in *rpoH* transcription, the bulk of the transient increase in σ^{32} synthesis observed upon temperature upshift (30° to 42°C) results from an increase in *rpoH* translation, rather than transcription. The major function of transcriptional control therefore appears to be to provide appropriate levels of *rpoH* mRNA under a variety of steady-state growth conditions, as well as to maintain the critical levels of *rpoH* message so as to tolerate some extremely stressful conditions such as lethal temperature and harmful agents, including ethanol.

Translational Induction of σ^{32} *Synthesis*

Early observations of increased levels of *rpoH* mRNA upon exposure to high temperature and of increased transcription from some *rpoH* promoters at high temperature (19, 21, 94) suggested that most of the heat-induced synthesis of σ^{32} might occur at the level of transcription. However, other results favored translational control, and the elevated levels of *rpoH* mRNA that accumulate after temperature upshift mostly result from increased mRNA stability caused by increased translation, rather than increased transcription. The following evidence indicates that the increased synthesis of σ^{32} observed after shift from 30 to 42°C occurs primarily at the translational level: First, the synthesis of σ^{32}, but not of *rpoH* mRNA, increases markedly after a temperature upshift (19). Second, heat-induced synthesis of σ^{32}-β-galactosidase fusion protein depends on the particular translational initiation region of *rpoH* used, not on the promoter region (64, 86). Third, the increase of *rpoH* mRNA level is preceded by the increase of σ^{32} synthesis: the peak rate of σ^{32} (or fusion protein) synthesis and the maximal accumulation of *rpoH* mRNA are seen at 3 and 6 min, respectively, after a temprature shift (39, 86). Finally, heat-shock induction of σ^{32}-β-galactosidase fusion protein occurs even when RNA synthesis is inhibited by rifampicin (64).

CIS-ACTING REGULATORY ELEMENTS ON *rpoH* mRNA Attempts to localize the *rpoH* region(s) that are involved in the thermoregulation of σ^{32} synthesis (39, 64) followed the studies mentioned above. Extensive analysis of 3′- and 5′-deletion derivatives of an *rpoH-lacZ* gene fusion (GF364 containing 364 nucleotides of *rpoH* fused to *lacZ*) led to the identification of two *cis*-acting

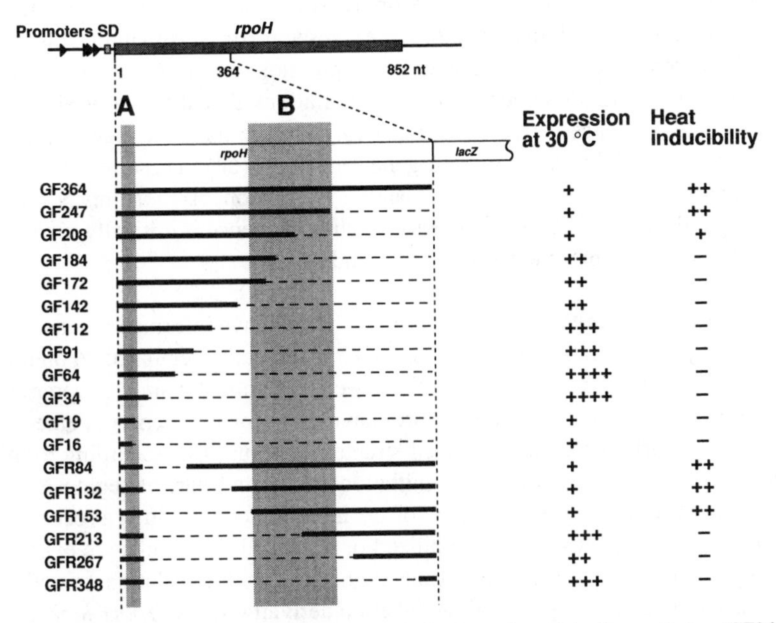

Figure 3 Structure and expression of deletion derivatives of *rpoH-lacZ* gene fusion (GF364). Cells (Δ*lac*) lysogenic for λ carrying a 3′- (GF) or 5′-deletion (GFR) derivative of GF364 were examined for expression at 30°C and at 42°C, with immunoprecipitation of pulse-labeled fusion proteins. Regions A and B represent *cis*-acting elements involved in heat induction of *rpoH* translation. Reproduced with permission from Ref. 65.

mRNA regions within the *rpoH* coding sequence. One region, designated A, corresponds to nucleotides 6–20, found immediately downstream of the initiation codon, and acts as a positive element enhancing the rate of translation: deletions extending into this region from the 3′ end exhibited drastically reduced expression (~15-fold) (Figure 3). This sequence is similar to the downstream box (83), which enhances expression of gene 0.3 of phage T7 and is complementary to nucleotides 1469–1483 of 16S rRNA. The downstream box, now found in several bacterial genes (20, 79, 83), presumably stimulates translation by stabilizing the binding of message to the 30S ribosomal subunit (79). Consistent with this expectation, base substitutions that either reduced or increased complementarity to 16S rRNA, decreased or increased the expression, respectively, in all cases tested (20, 64, 79).

The other internal *rpoH* region, designated B, is located between nucleotides ~153 and ~247 and acts as a negative element in thermoregulation, because deletions of this region from the 5′ or 3′ end resulted in partially or fully constitutive expression, i.e. increased expression at low temperature (30°C); the extent of induction upon shift to 42°C was concomitantly reduced as well (39, 64) (see Figure 3). Interestingly, some base-substitution mutations within region A exhibited constitutive expression (64, 116), indicating that besides controlling the basal level of σ^{32} expression, region A may also affect thermoregulation. Moreover, frameshift mutations that drastically alter the amino acid sequence of region B but hardly change its nucleotide sequence did not substantially interfere with gene regulation (64). These results taken together provide strong evidence that the 5′ portion of *rpoH* mRNA that corresponds to some 200 nucleotides of coding sequence is critically involved in controlling thermal induction of σ^{32} synthesis.

THE mRNA SECONDARY STRUCTURE MODEL The above information leads one to ask how translation of *rpoH* mRNA is actually regulated. Potential secondary structures of *rpoH* mRNA were examined using a computer program (36), and a plausible structure, having minimal free energy, emerged among the various alternative folded structures (Figure 4). According to this model, part of region A plus the initiation codon and part of region B can base pair, leading to a sequestration of the initiation codon. Such a secondary structure may be transiently destabilized or disrupted when cells are exposed to high temperature, thus permitting higher rates of translation initiation. Consistent with such a proposal, all deletion derivatives of an *rpoH-lacZ* gene fusion that exhibited constitutive expression were predicted to result in a loss of base pairings around the initiation region. Furthermore, random base substitution–type mutations selected on the basis of constitutive expression of an *rpoH-lacZ* gene fusion were shown to be localized within region A and a

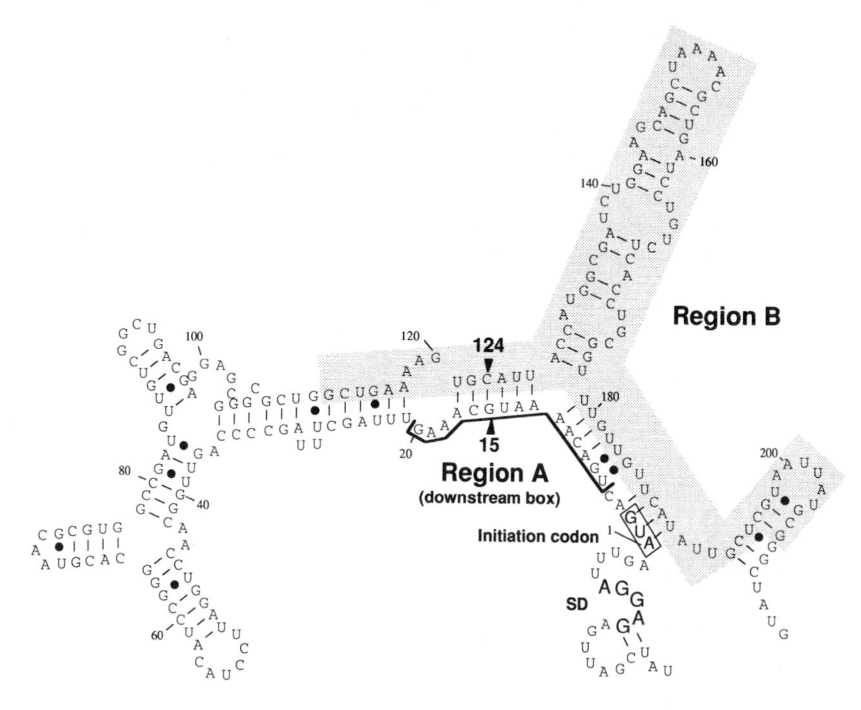

Figure 4 Predicted secondary structure of the 5′ portion of *rpoH* mRNA. The region B shown here (*shaded*) is displaced ~40 bp from that originally defined by the deletion analysis shown in Figure 3 (see text) (adapted from 64).

region spanning positions 112–208 (a shaded area in Figure 4, designated region B); the latter region, however, is shifted toward the 5′ end by ~40 nucleotides, as compared with region B originally defined by the deletion analysis (116) (cf Figures 3, 4).

Examination of additional base substitution mutations that are predicted to disrupt the secondary structure, and the putative compensatory mutations that are predicted to restore the secondary structure, further substantiated the above model (116). For example, base pairings between nucleotides 6U and 184G, 15G and 124C, 16C and 123G, and 17A and 122U appeared to be important, because single base changes predicted to disrupt a base pairing at each of these positions resulted in virtually constitutive expression. At least in some instances, compensatory changes predicted to restore the base pairing indeed resulted in an essentially parental phenotype (6G-184C; 15A-124U). Moreover, other combinations of bases at positions 15 and 124 that do not permit base pairing failed to restore the parental phenotype. The results obtained with the 15-124 pair of bases shown in Table 2a illustrate the point.

Table 2 Effects of compensatory and noncompensatory base substitution mutations on fusion protein synthesis[a]

Strain	mRNA base change	Fusion protein synthesis: Synthesis rate (30°C)	Fusion protein synthesis: Induction at 42° (x-fold)
a			
Wild-type	—	1.0[a]	5.5
15A	G → A	6.2	1.5
124U	C → U	3.8	2.3
15A 124U	G–C → A–U	1.5	5.3
124G	C → G	3.6	1.1
15A 124G	G–C → A–G	4.2	1.2
124A	C → A	2.3	1.5
15A 124A	G–C → A–A	4.6	1.1
b			
15C	G → C	3.1	2.2
124G	C → G	2.7	2.0
15C 124G	G–C → C–G	1.2	1.3

[a] See legend to Figure 3 for the methods used for determination of fusion-protein synthesis.

Interestingly, these analyses revealed other instances in which effects of single-base changes that cause constitutive expression were compensated for with respect to repression at low temperature but not compensated for with respect to heat inducibility (15C–124G; 17C–122G). The latter observation (Table 2b) suggests a possible involvement of a transacting (protein) factor that could modulate the stability of mRNA secondary structure by binding to the sequence including nucleotides 15–17, and/or 122–124. Such a hypothetical factor may be either a positive factor (activator) that would destabilize mRNA secondary structure upon heat shock, or a negative factor (repressor) that would stabilize the secondary structure under nonstress conditions and be inactivated upon heat shock.

A few additional internal deletions have been examined (116). A deletion of 12 nucleotides (from 178 to 189), predicted to disrupt a major stem structure involving the translation-initiation region, resulted in fully constitutive expression as expected from the present model. Another deletion of 51 nucleotides (127–177), predicted to delete one of the major stem structures (Figure 4) but to retain the rest of the secondary structures, exhibited significantly reduced expression at 30°C and failed to be induced at 42°C, suggesting that the deletion resulted in an overstabilization of the remaining portions of the mRNA structure (a superrepressed phenotype). These results, together with those of base substitution analysis, are consistent with the notion that thermoregulation of σ^{32} synthesis is mediated by a tight and possibly subtle adjustment of *rpoH* mRNA secondary structure, and lend further support to, though do not prove, a model such as the one depicted in Figure 4.

SEARCH FOR FACTOR(S) CONTROLLING *rpoH* TRANSLATION As to the possible transacting factor(s) involved in translational induction of σ^{32} synthesis, the heat-shock proteins DnaK, DnaJ, and GrpE that exert negative control on the heat-shock response (89, 95, 96) are not required for heat induction of σ^{32} or fusion protein synthesis: essentially normal expression at 30°C and heat induction at 42°C was observed in mutants lacking DnaK and/or DnaJ functions (64, 89). Also, autogenous regulation of either repression under steady-state growth or heat induction by σ^{32} itself is excluded, because deletion of *rpoH* had no appreciable effects on either of these processes (64). Indeed, the latter results make it unlikely that any of the heat-shock proteins under σ^{32} control are involved.

An extragenic suppressor of *rpoH,* called *suhB,* suppresses the temperature-sensitive growth of *rpoH15* missense mutant, endows the strain with a cold-sensitive growth phenotype, and enhances the synthesis of mutated σ^{32} at the translational level (113). The fact that *suhB* does not enhance σ^{32} synthesis in the *rpoH*$^{+}$ strain suggests that the wild-type SuhB protein represses *rpoH* mRNA translation when one or more hsps are under-supplied. The cold-sensitive growth of a *suhB* missense mutant, but not of a *suhB* null mutant, can be suppressed by an *rnc* mutation affecting RNase III (which specifically degrades dsRNA), suggesting an interaction between SuhB and RNase III (Y. Nakamura, personal communication). In agreement with the sequence similarity between SuhB and eukaryotic inositol monophosphate phosphatases (72), extracts of *E. coli* carrying a multicopy *suhB*$^{+}$ plasmid were recently found to contain elevated levels of phosphatase activity (K.

Shiba, personal communication). Although *suhB* could participate in modulating σ^{32} synthesis under certain conditions, the precise mechanism remains unknown.

Another temperature-sensitive mutant, *htrC*, whose product is essential for cell viability only at high temperature, shows constitutive synthesis of hsps and elevated levels of heat-shock gene transcripts (77), suggesting that it may modulate the translational control of σ^{32} synthesis. Contrary to such expectations, the *htrC* mutation must affect the stability of σ^{32}, because it exerted no effect on the rate of synthesis of the GF364 or GF807 *rpoH-lacZ* fusion proteins (H. Nagai, unpublished result).

Translational Repression of rpoH *mRNA: Feedback Control of* σ^{32} *Synthesis*

The first suggestion that the shut-off of σ^{32} synthesis, following its initial induction, occurs posttranscriptionally came from the observation that a repression of σ^{32} synthesis followed its overproduction by the λP_L promoter at 42°C, while its transcription continued at high levels for a long period (33). Subsequently, the shut-off of heat-induced synthesis of an *rpoH-lacZ* fusion protein (a fusion similar to GF807) was observed several minutes prior to the decrease in mRNA level (39). Thus, repression as well as the initial induction of σ^{32} synthesis seen after a temperature upshift occurs primarily at the level of translation, presumably reflecting the ability of the *rpoH* mRNAs to be translated.

Examination of the translational control of fusion-protein synthesis using *rpoH-lacZ* gene fusions revealed striking differences in the kinetics of expression between fusion GF807, which contained most of the *rpoH* coding sequence (807 out of 852 nucleotides), and GF364, which contained the N terminus–coding 364 nucleotides, used in most previous work. Whereas a marked induction followed by clear shut-off was observed with GF807, as with the synthesis of authentic σ^{32} (see Figure 1), similar induction but very little shut-off was observed with fusion GF364. This difference in σ^{32} synthesis is not merely a reflection of the differing stability of the fusion proteins produced (see below). When expression of fusion protein from GF807 was examined in the *ΔdnaK52* or *dnaJ259* mutant bacteria, the induction occurred normally but failed to shut-off (66), as previously found with authentic σ^{32} (89). The appreciable shut-off observed with GF364 in some earlier experiments (64) is probably explained by the assumption that the cultures used in these studies had already entered the stationary phase.

Examination of a set of in-frame 3′-deletion derivatives of GF807 led to the further localization of the *rpoH* region specifically involved in the shut-off of σ^{32} synthesis (66). Deletions retaining 433 or more nucleotides of the 5′

Table 3 Expression and stability of fusion protein with 3' deletion derivatives of gene fusion GF807[a]

Gene fusion numbers	5' *rpoH* sequence contained in fusion	Heat induction	Shut off	Stability at 30°C (half-life, min)
GF807	807	+	+	~1
GF712	712	+	+	~1
GF661	661	+	+	~1
GF533	533	+	+	~2
GF439	439	+	+	~2
GF433	433	+	+	~2
GF364	364	+	−	~20
GF247	247	+	−	~20

[a] See legend to Figure 3 for the methods used.

rpoH sequence shut off normally, whereas those carrying 364 or fewer 5′ nucleotides failed to shut off (Table 3). Thus, the small segment flanked by nucleotides 364 and 433 (coding for amino acid residues 122–144) appeared to be important for the normal shut-off of σ^{32} synthesis. Furthermore, a frameshift mutation of fusion GF807 carrying an amino acid sequence drastically altered specifically for this 23–amino acid segment showed little shut-off following initial induction. Evidently, a distinct segment of the σ^{32} polypeptide, designated C, is critically involved in the DnaK (DnaJ, GrpE)-mediated shut-off of fusion-protein synthesis, and presumably of σ^{32} synthesis, although involvement of other segment(s), particularly those located 5′ to segment C, cannot be excluded at present (66).

Possible Coupling Between Translational Repression and Degradation of σ^{32}

As previously noted, the fusion protein produced by GF807 is as unstable (half-life 1 min) as authentic σ^{32}, whereas the protein made by GF364 is quite stable (half-life 20 min) (64, 66), suggesting that a region(s) located in the C-terminal half of σ^{32} is important for its characteristic instability. A striking correlation was found between stability and shut-off of fusion proteins when the set of 3′-deletion derivatives of GF807 was examined: the same segment of σ^{32} critical for translational repression appeared to be important for its instability as well (Table 3). This rather unexpected result suggested that the shut-off of σ^{32} synthesis following heat induction and the degradation of σ^{32} may share some common mechanism or factor(s).

Several interesting possibilities raised by the above finding are that: a distinct segment of σ^{32} is critically involved in both translational repression and degradation of fusion protein and presumably of σ^{32}; this segment might represent a site of interaction with the DnaK/DnaJ proteins; and translational repression might be triggered by an interaction with DnaK/DnaJ during polypeptide chain growth, which may in turn inhibit further translation of *rpoH* mRNA.

Although it is not known how the DnaK/DnaJ-mediated shut-off of σ^{32} synthesis is related to σ^{32} degradation, a direct physical interaction between DnaK/DnaJ and the nascent σ^{32} polypeptide chain (or fusion protein) may cause both shut-off of synthesis (translational attenuation?) and degradation of σ^{32} polypeptides, as judged by the fact that these hsps and σ^{32} physically interact in vitro (54). Such an interaction could eventually lead to a repression of translation and rapid degradation of σ^{32} by an as yet unknown mechanism(s). Whatever the actual mechanism may turn out to be, there appears to be a dynamic link between the feedback controls of synthesis and degradation of σ^{32} that would play important roles in regulation of the heat-shock response.

THE FEEDBACK REGULATORY CIRCUITS

The rapid accumulation of hsps upon temperature upshift brought about by increased σ^{32} levels is shortly followed by an adaptation period during which the levels of σ^{32} and of hsps are readjusted to the new steady-state growth condition at higher temperature. The readjustment includes a shut-off of heat-induced synthesis of σ^{32} and resumed degradation of σ^{32} within several minutes after temperature shift. The central theme in this regulatory circuit is a negative feedback control exerted by a set of heat-shock proteins (DnaK, DnaJ, and GrpE) that mediate normal shut-off of heat-induced synthesis and destabilization of σ^{32} (89, 95). The same set of hsps can inhibit the activity of σ^{32} by interfering with its association with core RNA polymerase in vitro (C. Georgopoulos, personal communication), suggesting that similar modes of hsp functions may operate in vivo. Thus, the feedback control appears to be exerted at three distinct steps, namely, synthesis, degradation, and activity of σ^{32}, all contributing to a precise and prompt readjustment of the intracellular concentrations of hsps that are optimal for a given set of physiological conditions.

Such apparently multiple feedback regulatory circuits could work independently from one another. However, as we discussed above, the control of synthesis and degradation of σ^{32} during the shut-off phase is probably mediated by the distinct segment of the σ^{32} polypeptide, suggesting that the two processes may be interconnected. Segment C includes a region, spanning regions 2.4 and 3.1, that is conserved relatively little among bacterial σ factors (Figure 5*a*) (58). It is also flanked by segment 2.1, which is thought to be mainly responsible for binding to core RNA polymerase (51) and the region

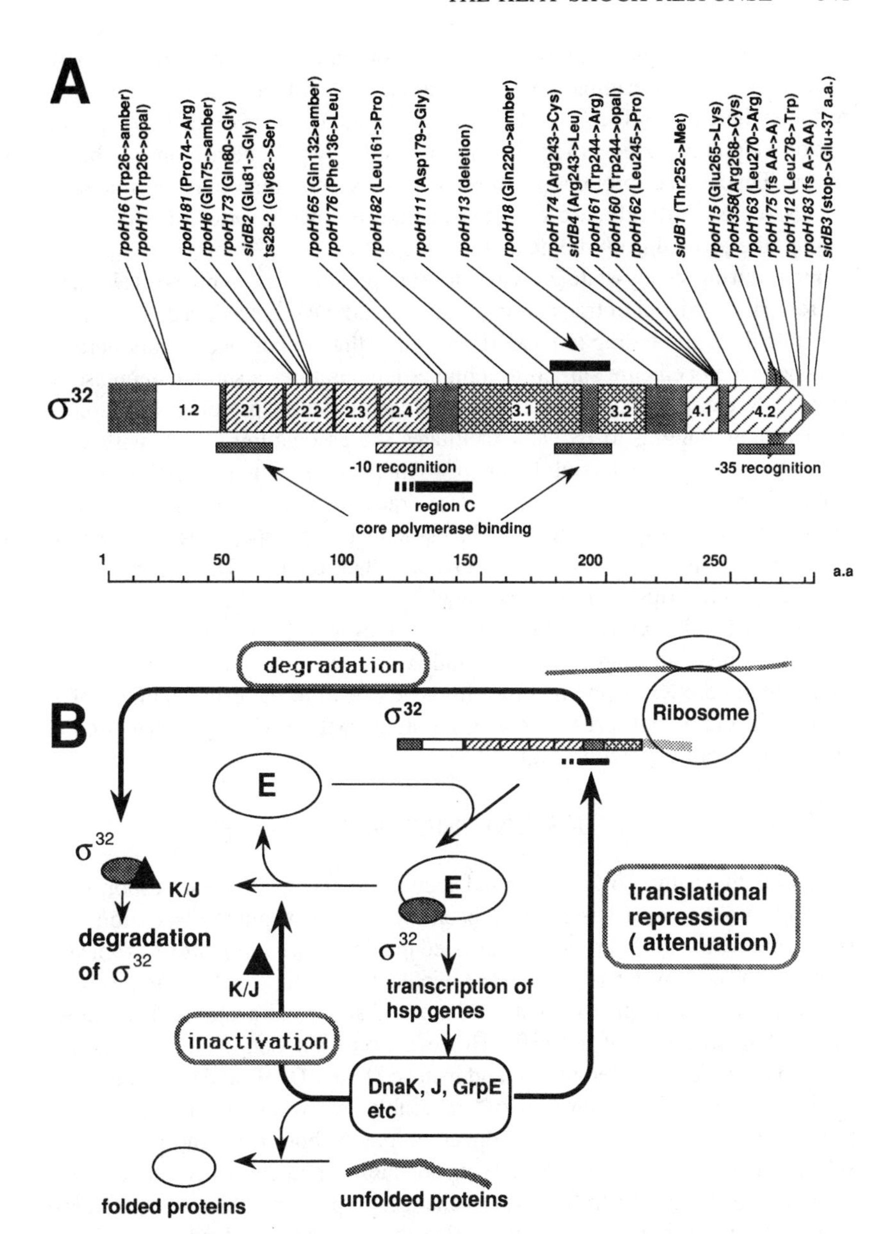

Figure 5 (*A*) Structural organization of σ^{32} protein indicating conserved regions [shown by numbers (see 58)] and mutations of known positions (taken from 8, 9, 72b, 114; R. Calendar, personal communication). (*B*) A schematic model that illustrates the DnaK (DnaJ, GrpE)-mediated feedback control circuits operating at three distinct but interconnected levels: synthesis, degradation, and activity of σ^{32}.

recently shown to play an auxiliary role in core polymerase binding (9, 119). Thus, when the N-terminal portion of nascent σ^{32} polypeptide is synthesized to somewhere beyond segment C on the ribosome, it may provide two separate but overlapping segments for binding to core RNA polymerase on one hand, and to DnaK/DnaJ proteins on the other: these bindings may be mutually exculsive. Accordingly, we can envisage the following series of events that might take place during the heat-shock response.

First, during steady-state growth at low temperature, σ^{32} synthesis is largely repressed at the level of translation initiation, and the σ^{32} produced is rendered unstable through binding to DnaK/DnaJ rather than to core RNA polymerase. Second, when cells are shifted to a higher temperature, a sudden decrease in free DnaK/DnaJ, owing to binding to unfolded/denatured proteins, will rapidly reduce their binding to σ^{32} and facilitates the association of σ^{32} with core RNA polymerase, thus stabilizing σ^{32}. In parallel with this stabilization, synthesis of σ^{32} is induced through a transient disruption of its mRNA secondary structure, and most of the nascent σ^{32} polypeptides synthesized will bind to the core RNA polymerase, rather than to DnaK/DnaJ, thus enhancing transcription from heat-shock promoters. Third, as the levels of free DnaK/DnaJ become sufficiently high, they will resume binding to σ^{32}, repress or attenuate translation, destabilize σ^{32}, and prevent the association of σ^{32} with core RNA polymerase; this would lead to the establishment of a new equilibrium appropriate for steady-state growth. Such regulatory circuits are schematically shown in Figure 5*b*.

CELLULAR SENSORS AND SIGNALS

Although the nature of sensors and signals in the heat-shock response is presently unknown, it has recently been the subject of much discussion and speculation. An earlier proposal suggested that increased amounts of abnormal proteins formed by heat-shock stress would titrate the function of protease Lon (La), an hsp required for degradation of abnormal proteins, leading to transient stabilization of σ^{32} (30). However, purified Lon protease does not seem to degrade σ^{32} (cited in 31), and instead DnaK, DnaJ, and GrpE proteins facilitate in vivo degradation of σ^{32}, probably by presenting it to unknown protease system(s). Because the concentrations of both hsps and proteolytic substrates probably do not increase appreciably upon mild heat shock, the extreme sensitivity of hsp function to changes in protein-substrate concentration could be explained by assuming that the hsps are involved primarily in polymeric interactions (31). The fact that DnaJ forms dimers (120), and recent genetic evidence suggesting that at least one form of DnaK is oligomeric (109), are consistent with such a possibility.

Another recent proposal is that the free pool of DnaK (and its eukaryotic homologue, HSP70) serves as a cellular thermometer that monitors changes in cellular concentration of unfolded or denatured proteins and regulates the expression of all hsps in prokaryotes and eukaryotes (12). Besides this indirect one, another role for DnaK as a direct thermometer was proposed on the basis of the extremely sharp temperature dependency of its autophosphorylation and ATPase activities (60). Either or both of these mechanisms may operate in sensing changes in the unfolded/denatured states of cellular proteins caused by heat and other stresses. Alternatively, DnaJ rather than DnaK was postulated to play a key regulatory function in the heat-shock response (25). The exact roles of DnaK/DnaJ and other hsps in transmitting heat-shock signals to σ^{32} at all steps of regulatory circuits would be a central issue for future studies. An entirely different model in which ribosomes are assumed to serve as the sensor for both heat shock and cold shock has been proposed (101).

A putative regulatory circuit for controlling translational induction of σ^{32} seems to offer a particular challenge because of its apparently independent nature from the rest of the regulatory circuits. Clearly, DnaK and other hsps are not directly involved in translational induction of σ^{32} observed immediately upon a temperature upshift, but instead the hypothetical regulator(s) of the *rpoH* mRNA secondary structure are likely to receive a signal generated by heat shock, rather than through titration of hsp function. Ethanol induces the synthesis of fusion protein from GF364, as well as the synthesis of authentic σ^{32} (86), suggesting that ethanol and high temperature generate similar signals in causing translational induction of σ^{32} synthesis (H. Yuzawa, unpublished result).

In contrast to the translational induction that occurs apparently independently of hsp levels, the shut-off of σ^{32} synthesis or translational repression and destabilization of σ^{32} provide major targets for negative feedback control that specifically requires the synergistic functions of DnaK, DnaJ, and GrpE. In light of what seems to be a critical involvement of a specific segment (amino acid sequence) of the σ^{32} polypeptide, direct physical interaction between one or more hsps and σ^{32} might occur during polypeptide chain elongation. In addition to, or instead of, such translational attenuation, however, direct interaction between intact σ^{32} (or σ^{32} bound to core RNA polymerase) and hsps, as suggested by experiments in vitro (C. Georgopoulos, personal communication), could also occur under certain conditions. Such an interaction would account for the inhibition of σ^{32} activity observed when σ^{32} (and hsps) is constantly overproduced by an exogenous promoter (33) or through a temperature downshift (88, 92). It might also play an auxiliary role during the shut-off phase of the heat-shock response: the slight delay in the shut-off kinetics of σ^{32} level as compared to that of *dnaK* transcription (see Figure 1) is actually in agreement with such an interpretation.

PERSPECTIVES

Intensive studies on the synthesis, stability, and activity of σ^{32} (or σ^{32}-β-galactosidase fusion protein) upon temperature upshift, coupled with a better understanding of the functional roles of hsps in recent years, have offered new insights into the mechanisms of regulation of the heat-shock response in *E. coli.* Further biochemical and genetic studies on the regulatory events occurring during the induction and shut-off phase should continue to provide basic and useful information for further understanding of early reactions related to sensors and signals of the heat-shock response. Both translational induction mediated by *rpoH* mRNA secondary structures and repression (or attenuation) modulated by DnaK and other hsps require further investigation. These studies will be important not only for elucidating the regulatory mechanism of the heat-shock response but also for further exploitation of the involvement of protein-coding sequence in translational control of gene expression in bacteria. The intriguing possibility that there are two separate pathways for the heat-shock signal transduction, one acting on translational induction of *rpoH* mRNA and the other acting on DnaK/DnaJ-mediated negative controls on translational attenuation and stabilization/destabilization of σ^{32}, should soon be investigated directly.

Among the other approaches of numerous laboratories to the regulation of the heat-shock response are biochemical studies on the interaction between σ^{32} and DnaK/DnaJ and genetic and biochemical analyses on the structure and function of σ^{32} and several important hsps such as DnaK and DnaJ. These lines of investigation should provide information essential for our ultimate understanding of the mechanisms underlying the various regulatory events along the signal-transduction pathways. However, additional factors that cannot yet be predicted from our current picture of the regulation of the heat-shock response will probably be uncovered in the near future.

Finally, in view of the recent advances in the field of heat-shock response in *E. coli,* some comparative studies with other bacteria are warranted. So far, only a few instances of *rpoH* or σ^{32} homologues have been reported in the literature (1, 26). Structural and functional analyses of heat-shock genes with *B. subtilis, Clostridium acetobutylicum,* and others led to the suggestion that the regulation of the heat-shock response is mediated by an inverted repeat consensus sequence found in front of these genes (e.g. 67, 108). It would be both interesting and rewarding to compare similarities and differences in the modes and mechanisms of heat-shock regulation within prokaryotes as well as between prokaryotes and eukaryotes.

ACKNOWLEDGMENTS

We are grateful to Carol A. Gross and her associates for invaluable advice and cooperation during past years that was instrumental to much of our recent work discussed here. We also thank C. Georgopoulos, K. Ito, T. Nagata, K. Shigesada, G. C. Walker, and other colleagues for kind advice and for sharing unpublished results with us for the writing of this review, and M. Mihara for help in preparing the manuscript.

Literature Cited

1. Allan, B., Linseman, M., McDonald, L. A., Lam, J. S., Kropinski, A. M. 1988. Heat shock response of *Pseudomonas aeruginosa*. *J. Bacteriol.* 170: 3668–74
2. Allen, S. P., Polazzi, J. O., Gierse, J, K., Easton, A. M. 1992. Two novel heat shock genes encoding proteins produced in response to heterologous protein expression in *Escherichia coli*. *J. Bacteriol.* 174:6938–47
3. Bahl, H., Echols, H., Straus, D. B., Court, D., Crowl, R., Georgopoulos, C. P. 1987. Induction of the heat shock response of *E. coli* through stabilization of σ^{32} by the phage λ cIII protein. *Genes Dev.* 1:57–64
4. Baker, T. A., Grossman, A. D., Gross, C A. 1984. A gene regulating the heat shock response in *Escherichia coli* also affects proteolysis. *Proc. Natl. Acad. Sci. USA* 81:6779–83
5. Bloom, M., Skelly, S., VanBogelen, R., Neidhardt, F., Brot, N., Weissbach, H. 1986. In vitro effect of the *Escherichia coli* heat shock regulatory protein on expression of heat shock genes. *J. Bacteriol.* 166:380–84
6. Bochkareva, E. S., Lisson, Girshovich, A. S. 1988. Transient association of newly synthesized unfolded proteins with the heat-shock GroEL protein. *Nature* 336:254–57
7. Bukau, B., Walker, G. C. 1989. Cellular defects caused by deletion of the *Escherichia coli dnaK* gene indicate roles for heat shock protein in normal metabolism. *J. Bacteriol.* 171:2337–46
8. Bukau, B., Walker, G. C. 1990. Mutations altering heat shock specific subunit of RNA polymerase suppress major cellular defects of *E. coli* mutants lacking the DnaK chaperone. *EMBO J.* 9:4027–36
9. Calendar, R., Erickson, J. W., Halling, C., Nolte, A. 1988. Deletion and insertion mutations in the *rpoH* gene of *Escherichia coli* that produce functional σ^{32}. *J. Bacteriol.* 170:3479–84
10. Cooper, S., Ruettinger, T. 1975. A temperature sensitive nonsense mutation affecting the synthesis of a major protein of *Escherichia coli* K12. *Mol. Gen. Genet.* 139:167–76
11. Cowing, D. W., Bardwell, J. C. A., Craig, E. A., Woolford, C., Hendrix, R. W., Gross, C. A. 1985. Consensus sequence for *Escherichia coli* heat shock gene promoters. *Proc. Natl. Acad. Sci. USA* 82:2679–83
12. Craig, E. A., Gross, C. A. 1991. Is hsp70 the cellular thermometer? *TIBS* 16:135–40
13. Crickmore, N., Salmond, G. P. C. 1986. The *Escherichia coli* heat shock regulatory gene is immediately downstream of a cell division operon: the *fam* mutation is allelic with *rpoH*. *Mol. Gen. Genet.* 205:535–39
14. DiDomenico, B. J., Bugaisky, G. E., Lindquist, S. 1982. The heat shock response is self-regulated at both the transcriptional and posttranscriptional levels. *Cell* 31:593–603
15. Donnelly, C., Walker, G. C. 1989. *groE* mutants of *Escherichia coli* are defective in *umuDC*-dependent UV mutagenesis. *J. Bacteriol.* 171:6117–25
16. Drahos, D. J., Hendrix, R. W. 1982. Effect of bacteriophage lambda infection on synthesis of *groE* protein and other *Escherichia coli* proteins. *J. Bacteriol.* 149:1050–63
17. Ellis, R. J., van der Vies, S. M. 1991. Molecular chaperones. *Annu. Rev. Biochem.* 60:321–47
18. Erickson, J. W., Gross, C. A. 1989. Identification of the σ^{E} subunit of *Escherichia coli* RNA polymerase: a second alternate σ factor involved in high-temperature gene expression. *Genes Dev.* 3:1462–71
19. Erickson, J. W., Vaughn, V., Walter, W. A., Neidhardt, F. C., Gross, C.

A. 1987. Regulation of the promoters and transcripts of *rpoH*, the *Escherichia coli* heat shock regulaory gene. *Genes Dev.* 1:419–32
20. Faxen, M., Plumbridge, J., Isaksson, L. A. 1991. Codon choice and potential complementarity between mRNA downstream of the initiation codon and bases 1471–1480 in 16S ribosomal RNA affects expression of *glnS*. *Nucleic Acids Res.* 19:5247–51
21. Fujita, N., Ishihama, A. 1987. Heat-shock induction of RNA polymerase sigma-32 synthesis in *Escherichia coli:* transcriptional control and multiple promoter system. *Mol. Gen. Genet.* 210: 10–15
22. Fujita, N., Ishihama, A., Nagasawa, Y., Ueda, S. 1987. RNA polymerase sigma-related proteins in *Escherichia coli:* detection by antibodies against a synthetic peptide. *Mol. Gen. Genet.* 210:5–9
23. Fujita, N., Nomura, T., Ishihama, A. 1987. Promoter selectivity of *Escherichia coli* RNA polymerase: purification and properties of holoenzyme containing the heat-shock σ subunit. *J. Biol. Chem.* 262:1855–59
24. Gaitanaris, G. A., Papavassiliou, A. G., Rubock, P., Silverstein, S. J., Gottesman, M. E. 1990. Renaturation of denatured λ repressor requires heat shock proteins. *Cell* 61:1013–20
25. Gamer, J., Bujard, H., Bukau, B. 1992. Physical interaction between heat shock proteins DnaK, DnaJ, and GrpE and the bacterial heat shock transcription factor σ^{32}. *Cell* 69:833–42
26. Garvin, L. D., Hardies, S. C. 1989. Nucleotide sequence for the *htpR* gene from *Citrobacter freundii*. *Nucleic Acids Res.* 17:4889
27. Georgopoulos, C., Ang, D., Liberek, K., Zylicz, M. 1990. Properties of the *Escherichia coli* heat shock proteins and their role in bacteriophage λ growth. See Ref. 62, pp. 191–221
28. Georgopoulos, C., Ang, D., Maddock, A., Raina, S., Lipinska, B., Zylicz, M. 1990. Heat shock response of *Escherichia coli*. In *The Bacterial Chromosome,* ed. K. Drlica, M. Riley, pp. 405–19. Washington, DC: Am. Soc. Microbiol.
29. Goff, S. A., Casson, L. P., Goldberg, A. L. 1984. Heat shock regulatory gene *htpR* influences rates of protein degradation and expression of the *lon* gene in *Escherichia coli*. *Proc. Natl. Acad. Sci. USA* 81:6647–51
30. Goff, S. A., Goldberg, A. L. 1985. Production of abnormal proteins in *E. coli* stimulates transcription of *lon* and other heat shock genes. *Cell* 41:587–95
31. Gross, C. A., Straus, D. B., Erickson, J. W., Yura, T. 1990. The function and regulation of heat shock proteins in *Escherichia coli*. See Ref. 62, pp. 167–89
32. Grossman, A. D., Erickson, J. W., Gross, C. A. 1984. The htpR gene product of E. coli is a sigma factor for heat-shock promoters. *Cell* 38:383–90
33. Grossman, A. D., Straus, D. B., Walter, W. A., Gross, C. A. 1987. σ^{32} synthesis can regulate the synthesis of heat shock proteins in *Escherichia coli*. *Genes Dev.* 1:179–84
34. Grossman, A. D., Taylor, W. E., Burton, Z. F., Burgess, R. R., Gross, C. A. 1985. Stringent response in *Escherichia coli* induces expression of heat shock proteins. *J. Mol. Biol.* 186: 357–65
35. Herendeen, S. L., VanBogelen, R. A., Neidhardt, F. C. 1979. Levels of major proteins of *Escherichia coli* during growth at different temperatures. *J. Bacteriol.* 139:185–94
36. Jaeger, J. A., Turner, D. H., Zuker, M. 1989. Improved predictions of secondary structures for RNA. *Proc. Natl. Acad. Sci. USA* 86:7706–10
37. Jenkins, D. E., Auger, E. A., Matin, A. 1991. Role of RpoH, a heat shock regulator protein, in *Escherichia coli* carbon starvation protein synthesis and survival. *J. Bacteriol.* 173:1992–96
38. Johnson, W. C., Moran, C. P. Jr., Losick, R. 1983. Two RNA polymerase sigma factors from *Bacillus subtilis* discriminate between overlapping promoters for a developmentally regulated gene. *Nature* 302:800–4
39. Kamath-Loeb, A. S., Gross, C. A. 1991. Translational regulation of σ^{32} synthesis: requirement for an internal control element. *J. Bacteriol.* 173: 3904–6
40. Kanemori, M., Yura, T. 1991. Physiological functions of the GroE heat shock proteins: effects of decreased levels of GroE. *Annu. Meet. Mol. Biol. Soc. Jpn. Abstr.* p. 196
41. Keller, J. A., Simon, L. D. 1988. Divergent effects of a *dnaK* mutation on abnormal protein degradation in *Escherichia coli*. *Mol. Micorbiol.* 2:31–41
42. Kitagawa, M., Wada, C., Yoshioka, S., Yura, T. 1991. Expression of ClpB, an analog of the ATP-dependent pro-

tease regulatory subunit in *Escherichia coli*, is controlled by a heat shock σ factor (σ^{32}). *J. Bacteriol.* 173:4247–53
43. Kochan, J., Murialdo, H. 1982. Stimulation of *groE* synthesis in *Escherichia coli* by bacteriophage lambda infection. *J. Bacteriol.* 149:1166–70
44. Kroh, H. E., Simon, L. D. 1990. The ClpP component of Clp protease is the σ^{32}-dependent heat shock protein F21.5. *J. Bacteriol.* 172:6026–34
45. Krueger, J. H., Walker, G. C. 1984. *groEL* and *dnaK* genes of *Escherichia coli* are induced by UV irradiation and nalidixic acid in an *htpR*$^+$-dependent fashion. *Proc. Natl. Acad. Sci. USA* 81:1499–1503
46. Kusukawa, N., Yura, T. 1988. Heat shock protein GroE of *Escherichia coli:* key protective roles against thermal stress. *Genes Dev.* 2:874–82
47. Landick, R., Vaughn, V., Lau, E. T., VanBogelen, R. A., Erickson, J. W., Neidhardt, F. C. 1984. Nucleotide sequence of the heat shock regulatory gene of E. coli suggests its protein product may be a transcription factor. *Cell* 38:175–82
48. Langer, T., Lu, C., Echols, H., Flanagan, J., Hayer, M. K., Hartl, F. U. 1992. Successive action of DnaK, DnaJ and GroEL along the pathway of chaperone-mediated protein folding. *Nature* 356:683–89
49. LaRossa, R. A., Van Dyk, T. K. 1991. Physiological roles of the DnaK and GroE stress proteins: catalysts of protein folding or macromolecular sponges? *Mol. Microbiol.* 5:529–34
50. Lemaux, P. G., Herendeen, S. L., Bloch, P. L., Neidhardt, F. C. 1978. Transient rates of synthesis of individual polypeptides in E. coli following temperature shifts. *Cell* 13:427–34
51. Lesley, S. A., Burgess, R. R. 1989. Characterization of the *Escherichia coli* transcription factor σ^{70}: localization of a region involved in the interaction with core RNA polymerase. *Biochemistry* 28:7728–34
52. Lesley, S. A., Jovanovich, S. B., Tse-Dinh, Y.-C., Burgess, R. R. 1990. Identification of a heat shock promoter in the *topA* gene of *Escherichia coli. J. Bacteriol.* 172:6871–74
53. Lesley, S. A., Thompson, N. E., Burgess R. R. 1987. Studies of the role of the *Escherichia coli* heat shock regulatory protein σ^{32} by the use of monoclonal antibodies. *J. Biol. Chem.* 262:5404–7
54. Liberek, K., Galitski, T. P., Zylicz, M., Georgopoulos, C. 1992. The DnaK chaperone modulates the heat shock response of *Escherichia coli* by binding to the σ^{32} transcription factor. *Proc. Natl. Acad. Sci. USA* 89:3516–20
55. Lipinska, B., Fayet, O., Baird, L., Georgopoulos, C. 1989. Identification, characterization and mapping of the *Escherichia coli htrA* gene, whose product is essential for bacterial growth only at elevated temperatures. *J. Bacteriol.* 171:1574–84
56. Lipinska, B., King, J., Ang, D., Georgopoulos, C. 1988. Sequence analysis and transcriptional regulation of the *Escherichia coli grpE* gene, encoding a heat shock protein. *Nucleic Acids Res.* 16:7545–62
57. Lipinska, B., Sharma, S., Georgopoulos, C. 1988. Sequence analysis of the *htrA* gene of *Escherichia coli:* a σ^{32}-independent mechanism of heat-inducible transcription. *Nucleic Acids Res.* 16:10053–67
58. Lonetto, M., Gribskov, M., Gross, C. A. 1992. The σ^{70} family: sequence conservation and evolutionary relationships. *J. Bacteriol.* 174:3843–49
59. Maurizi, M. R., Clark, W. P., Katayama, Y., Rudikoff, S., Pumphrey, J., Bowers, B., Gottesman, S. 1990. Sequence and structure of ClpP, the proteolytic component of the ATP-dependent Clp protease of *Escherichia coli. J. Biol. Chem.* 265: 12536–45
60. McCarty, J. S., Walker, G. C. 1991. DnaK as a thermometer: threonine-199 is site of autophosphorylation and is critical for ATPase activity. *Proc. Natl. Acad. Sci. USA* 88:9513–17
61. Mizobata, T., Akiyama, Y., Ito, K., Yumoto, N., Kawata, Y. 1992. Effects of the chaperonin GroE on the refolding of tryptophanase from *Escherichia coli:* refolding is enhanced in the presence of ADP. *J. Biol. Chem.* 267:17773–79
62. Morimoto, R., Tissières, A., Georgopoulos, C., eds. 1990. *Stress Proteins in Biology and Medicine.* Cold Spring Harbor, NY: Cold Spring Harbor Lab. 450 pp.
63. Nagai, H., Yano, R., Erickson, J. W., Yura, T. 1990. Transcriptional regulation of the heat shock regulatory gene *rpoH* in *Escherichia coli:* involvement of a novel catabolite-sensitive promoter. *J. Bacteriol.* 172:2710–15
64. Nagai, H., Yuzawa, H., Yura, T. 1991. Interplay of two *cis*-acting mRNA regions in translational control of σ^{32}

synthesis during the heat shock response of *Escherichia coli*. *Proc. Natl. Acad. Sci. USA* 88:10515–19
65. Nagai, H., Yuzawa, H., Yura, T. 1991. Regulation of the heat shock response in *E. coli:* involvement of positive and negative *cis*-acting elements in translational control of σ^{32} synthesis. *Biochimie* 73:1473–79
66. Nagai, H., Yuzawa, H., Yura, T. 1992. *DnaK-mediated negative control of the heat shock response in* E. coli: *evidence for translational attenuation in* σ^{32} *synthesis*. Presented at Meeting on Translational Control, Cold Spring Harbor, N.Y.
67. Narberhaus, F., Bahl, H. 1992. Cloning, sequencing, and molecular analysis of the *groESL* operon of *Clostridium acetobutylicum*. *J. Bacteriol.* 174: 3282–89
68. Neidhardt, F. C., VanBogelen, R. A. 1981. Positive regulatory gene for temperature-controlled proteins in *Escherichia coli*. *Biochem. Biophys. Res. Commun.* 100:894–900
69. Neidhardt, F. C., VanBogelen, R. A. 1987. Heat shock response. In Escherichia coli *and* Salmonella typhimurium: *Cellular and Molecular Biology*, ed. F. C. Neidhardt, J. L. Ingraham, K. B. Low, B. Magasanik, M. Schaechter, H. E. Umbarger, pp. 1334–45. Washington, DC: Am. Soc. Microbiol.
70. Neidhardt, F. C., VanBogelen, R. A., Lau, E. T. 1983. Molecular cloning and expression of a gene that controls the high-temperature regulon of *Escherichia coli*. *J. Bacteriol.* 153:597–603
71. Neidhardt, F. C., VanBogelen, R. A., Vaughn, V. 1984. The genetics and regulation of heat-shock proteins. *Annu. Rev. Genet.* 18:295–29
72. Neuwald, A. F., York, J. D., Majerus, P. W. 1991. Diverse proteins homologous to inositol monophosphatase. *FEBS Lett.* 294:16–18
72a. Newlands, J. T., Gaal, T., Mecsus, J., Gourse, R. L. 1993. Transcription of the *Escherichia coli rrnB* P1 promoter by the heat shock RNA polymerase (Eσ^{32}) in vitro. *J. Bacteriol.* 175:661–68
73a. Obukowicz, M. G., Slater, N. R., Krivi, G. G. 1992. Enhanced heterologous gene expression in novel *rpoH* mutants of *Escherichia coli*. *Appl. Environ. Microbiol.* 58:1511–23
73. Osawa, T., Yura. T. 1981. Effects of reduced amount of RNA polymerase sigma factor on gene expression and growth of *Escherichia coli:* studies of the *rpoD40* (amber) mutation. *Mol. Gen. Genet.* 184:166–73
74. Paek, K.-H., Walker, G. C. 1987. *Escherichia coli dnaK* null mutants are inviable at high temperature. *J. Bacteriol.* 169:283–90
75. Parsell, D. A., Sauer, R. T. 1989. Induction of a heat shock–like response by unfolded protein in *Escherichia coli:* dependence on protein level not protein degradation. *Genes Dev.* 3:1226–32
76. Pelham, H. R. B. 1986. Speculations on the functions of the major heat shock and glucose-regulated proteins. *Cell* 46:959–61
77. Raina, S., Georgopoulos, C. 1990. A new *Escherichia coli* heat shock gene, *htrC*, whose product is essential for viability only at high temperatures. *J. Bacteriol.* 172:3417–26
78. Schultz, J. E., Latter, G. I., Matin, A. 1988. Differential regulation by cyclic AMP of starvation protein synthesis in *Escherichia coli*. *J. Bacteriol.* 170:3903–9
79. Shean, C. S., Gottesman, M. E. 1992. Translation of the prophage λ *c*I transcript. *Cell* 70:513–22
80. Shi, W., Zhou, Y., Wild, J., Adler, J., Gross, C. A. 1992. DnaK, DnaJ, and GrpE are required for flagellum synthesis in *Escherichia coli*. *J. Bacteriol.* 174:6256–63
81. Skelly, S., Coleman, T., Fu, C.-F., Brot, N., Weissbach, H. 1987. Correlation between the 32-kDa σ factor levels and in vitro expression of *Escherichia coli* heat shock genes. *Proc. Natl. Acad. Sci. USA* 84:8365–69
82. Skowyra, D., Georgopoulos, C., Zylicz, M. 1990. The E. coli dnaK gene product, the hsp70 homolog, can reactivate heat-inactivated RNA polymerase in an ATP hydrolysis-dependent manner. *Cell* 62:939–44
83. Sprengart, M. L., Fatscher, H. P., Fuchs, E. 1990. The initiation of translation in *E. coli:* apparent base pairing between the 16S rRNA and downstream sequences of the mRNA. *Nucleic Acids Res.* 18:1719–23
84. Squires, C. L., Pedersen, S., Ross, B. M., Squires, C. 1991. ClpB is the *Escherichia coli* heat shock protein F84.1. *J. Bacteriol.* 173:4254–62
85. Strauch, K. L., Johnson, K., Beckwith, J. 1989. Characterization of *degP*, a gene required for proteolysis in the cell envelope and essential for growth of *Escherichia coli* at high temperature. *J. Bacteriol.* 171:2689–96

86. Straus, D. B., Walter, W. A., Gross, C. A. 1987. The heat shock response of *E. coli* is regulated by changes in the concentration of σ^{32}. *Nature* 329: 348–51
87. Straus, D. B., Walter, W. A., Gross, C. A. 1988. *Escherichia coli* heat shock gene mutants are defective in proteolysis. *Genes Dev.* 2:1851–58
88. Straus, D. B., Walter, W. A., Gross, C. A. 1989. The activity of σ^{32} is reduced under conditions of excess heat shock protein production in *Escherichia coli*. *Genes Dev.* 3:2003–10
89. Straus, D., Walter, W., Gross, C. A. 1990. DnaK, DnaJ, and GrpE heat shock proteins negatively regulate heat shock gene expression by controlling the synthesis and stability of σ^{32}. *Genes Dev.* 4:2202–9
90. Taglicht, D., Padan, E., Oppenheim, A. B., Schuldiner, S. 1987. An alkaline shift induces the heat shock response in *Escherichia coli*. *J. Bacteriol.* 169: 885–87
91. Tatti, K. M., Moran C. P. Jr. 1985. Utilization of one promoter by two forms of RNA polymerase from *Bacillus subtilis*. *Nature* 314: 190–92
92. Taura, T., Kusukawa, N., Yura, T., Ito, K. 1989. Transient shutoff of *Escherichia coli* heat shock protein synthesis upon temperature shift down. *Biochem. Biophys. Res. Commun.* 163: 438–43
93. Taylor, W. E., Straus, D. B., Grossman, A. D., Burton, Z. F., Gross, C. A., Burgess, R. 1984. Transcription from a heat-inducible promoter causes heat shock regulation of the sigma subunit of E. coli RNA polymerase. *Cell* 38:371–81
94. Tilly, K., Erickson, J., Sharma, S., Georgopoulos, C. 1986. Heat shock regulatory gene *rpoH* mRNA level increases after heat shock in *Escherichia coli*. *J. Bacteriol.* 168:1155–58
95. Tilly, K., McKittrick, N., Zylicz, M., Georgopoulos, C. 1983. The dnaK protein modulates the heat-shock response of Escherichia coli. *Cell* 34:641–46
96. Tilly, K., Spence, J., Georgopoulos, C. 1989. Modulation of stability of the *Escherichia coli* heat shock regulatory factor σ^{32}. *J. Bacteriol.* 171: 1585–89
97. Tobe, T., Ito, K., Yura, T. 1984. Isolation and physical mapping of temperature-sensitive mutants defective in heat-shock induction of proteins in *Escherichia coli*. *Mol. Gen. Genet.* 195: 10–16
98. Travers, A. A. 1982. The heat-shock phenomenon in bacteria—a protection against DNA relaxation? In *Heat Shock from Bacteria to Man,* ed. M. J. Schlesinger, M. Ashburner, A. Tissieres, pp. 127–30. Cold Spring Harbor, NY: Cold Spring Harbor Lab. 440 pp.
99. Tsuchido, T., VanBogelen, R. A., Neidhardt, F. C. 1986. Heat shock response in *Escherichia coli* influences cell division. *Proc. Natl. Acad. Sci. USA* 83:6959–63
100. VanBogelen, R. A., Acton, M. A., Neidhardt, F. C. 1987. Induction of the heat shock regulon does not produce thermotolerance in *Escherichia coli*. *Genes Dev.* 1:525–31
101. VanBogelen, R. A., Neidhardt, F. C. 1990. Ribosomes as sensors of heat and cold shock in *Escherichia coli*. *Proc. Natl. Acad. Sci. USA* 87:5589–93
102. Van Dyk, T. K., Gatenby, A. A., LaRossa, R. A. 1989. Demonstration by genetic suppression of interaction of GroE products with many proteins. *Nature* 342:451–53
103. Wada, C., Imai, M., Yura, T. 1987. Host control of plasmid replication: requirement for the σ factor σ^{32} in transcription of mini-F replication initiator gene. *Proc. Natl. Acad. Sci. USA* 84:8849–53
104. Wang, P.-Z., Doi, R. H. 1984. Overlapping promoters transcribed by *Bacillus subtilis* σ^{55} and σ^{37} RNA polymerase holoenzymes during growth and stationary phases. *J. Biol. Chem.* 259: 8619–25
105. Wang, Q., Kaguni, J. M. 1989. DnaA protein regulates transcription of the *rpoH* gene of *Escherichia coli*. *J. Biol. Chem.* 264:7338–44
106. Wang, Q., Kaguni, J. M. 1989. A novel sigma factor is involved in expression of the *rpoH* gene of *Escherichia coli*. *J. Bacteriol.* 171: 4248–53
107. Welch, W. J., Georgopoulos, C. 1993. Function and regulation of heat shock proteins. *Annu. Rev. Cell Biol.* In press
108. Wetzstein, M., Völker, U., Dedio, J., Löbau, S., Zuber, U., et al. 1992. Cloning, sequencing, and molecular analysis of the *dnaK* locus from *Bacillus subtilis*. *J. Bacteriol.* 174:3300–10
109. Wild, J., Kamath-Loeb, A., Ziegelhoffer, E., Lonetto, M., Kawasaki, Y., Gross, C. A. 1992. Partial loss of function mutations in DnaK, the *Es-*

cherichia coli homologue of the 70-kDa heat shock proteins, affect highly conserved amino acids implicated in ATP binding and hydrolysis. *Proc. Natl. Acad. Sci. USA* 89:7139–43

110. Yamamori, T., Ito, K., Nakamura, Y., Yura, T. 1978. Transient regulation of protein synthesis in *Escherichia coli* upon shift-up of growth temperature. *J. Bacteriol.* 134:1133–40
111. Yamamori, T., Yura, T. 1980. Temperature-induced synthesis of specific proteins in *Escherichia coli:* evidence for transcriptional control. *J. Bacteriol.* 142:843–51
112. Yamamori, T., Yura, T. 1982. Genetic control of heat-shock protein synthesis and its bearing on growth and thermal resistance in *Escherichia coli* K-12. *Proc. Natl. Acad. Sci. USA* 79:860–64
113. Yano, R., Nagai, H., Shiba, K., Yura, T. 1990. A mutation that enhances synthesis of σ^{32} and suppresses temperature-sensitive growth of the *rpoH15* mutant of *Escherichia coli*. *J. Bacteriol.* 172:2124–30
114. Yura, T., Tobe, T., Ito, K., Osawa, T. 1984. Heat shock regulatory gene (*htpR*) of *Escherichia coli* is required for growth at high temperature but is dispensable at low temperature. *Proc. Natl. Acad. Sci. USA* 81:6803–7
115. Yura, T., Yamamori, T., Osawa, T. 1983. Transcriptional control of heat shock-inducible operons in *Escherichia coli*. In *Microbiology—1983,* ed. D. Schlessinger, pp. 14–17. Washington, DC: Am. Soc. Microbiol.
116. Yuzawa, H., Nagai, H., Mori, H., Yura, T. 1991. *Thermoregulation of σ^{32} synthesis mediated by alternative mRNA secondary structures: a primary step of the heat shock response in* Escherichia coli. Presented at Meeting on Translational Control, Cold Spring Harbor, N.Y.
117. Zeilstra-Ryalls, J., Fayet, O., Georgopoulos, C. 1991. The universally conserved GroE (Hsp60) chaperonins. *Annu. Rev. Microbiol.* 45: 301–25
118. Zhou, Y.-N., Kusukawa, N., Erickson, J. W., Gross, C. A., Yura, T. 1988. Isolation and characterization of *Escherichia coli* mutants that lack the heat shock sigma factor σ^{32}. *J. Bacteriol.* 170:3640–49
119. Zhou, Y.-N., Walter, W. A., Gross, C. A. 1992. A mutant σ^{32} with a small deletion in conserved region 3 of σ has reduced affinity for core RNA polymerase. *J. Bacteriol.* 174:5005–12
120. Zylicz, M., Yamamoto, T., McKittrick, N., Sell, S., Georgopoulos, C. 1985. Purification and properties of the DnaJ replication protein of *Escherichia coli*. *J. Biol. Chem.* 260:7591–98

Annu. Rev. Microbiol. 1993. 47:351–83

MOLECULAR BIOLOGY OF HYDROGEN UTILIZATION IN AEROBIC CHEMOLITHOTROPHS

B. Friedrich and E. Schwartz
Institut für Pflanzenphysiologie und Mikrobiologie, Freie Universität Berlin, D-1000 Berlin 33, Germany

KEY WORDS: aerobic hydrogenotrophs, hydrogenase, nickel metalloproteins, hydrogenase genes, processing, regulation

---

CONTENTS

ABSTRACT

The aerobic bacteria capable of obtaining energy from the oxidation of H_2 form a heterogenous group that includes both facultative and obligate chemolithotrophs and representatives of both gram-negative and gram-positive

0066-4227/93/1001-0351$02.00

genera. H_2-oxidizing aerobes inhabit such diverse biotypes as soil, oceans, and hot springs. The oxidation of H_2 in these bacteria is catalyzed by [NiFe] metalloenzymes called hydrogenases. The hydrogenases studied so far belong to two families: dimeric, membrane-bound enzymes (MBH) coupled to electron transport chains and tetrameric, cytoplasmic NAD-reducing enzymes (SH). Ni^{2+} is an essential component of the active site contained in the large subunit of the MBH enzymes. The genes for the MBH enzymes are located in conserved clusters of accessory genes, some of which encode maturation functions and hydrogenase-related redox proteins. Maturation of both types of hydrogenase is apparently complex, involving specific nickel incorporation and proteolytic processing steps. In *Alcaligenes eutrophus* and *Rhodobacter capsulatus,* hydrogenase expression is regulated by transcriptional activators belonging to the response-regulator family.

INTRODUCTION

More than 60 years ago Stephenson & Stickland (152) discovered that bacteria can both evolve H_2 and utilize it to reduce artificial and physiological substrates. In the 1970s considerable effort was devoted to the purification and biochemical characterization of hydrogenase enzymes. Progress was also made toward an understanding of the complex physiological roles of hydrogenases. Three major achievements mark the development of hydrogenase research in the past decade: (*a*) the establishment of genetic systems for many H_2-oxidizing bacteria, which opened avenues for molecular studies; (*b*) the discovery that nickel is an essential constituent of many hydrogenases; and (*c*) the isolation of novel H_2 oxidizers among the aerobic hyperthermophiles, obligate lithoautotrophs, and archaebacterial species. A comprehensive treatment of the aerobic chemolithotrophs is beyond the scope of this review. Instead this survey focuses on *Alcaligenes eutrophus,* the most extensively studied representative of the true chemolithotrophs. It reviews the physiology, biochemistry, genetics, and molecular biology of H_2 oxidation in this bacterium and discusses the related H_2-oxidizing systems of representatives of the aerobic photosynthetic and diazotrophic bacteria. The reader is referred to recent reviews (1, 2, 12, 41, 42, 45, 46, 101, 113, 130, 131, 145, 146, 162, 163) on related subjects for additional information.

PHYLOGENETIC DIVERSITY OF AEROBIC H_2-OXIDIZING BACTERIA

The coupling of H_2 oxidation to the reduction of molecular O_2, catalyzed by the enzyme hydrogenase, is found in representations of taxonomically diverse groups of aerobic bacteria, in the so-called hydrogen (knallgas) bacteria (4,

5, 12), in N_2-fixing bacteria (41), and in photosynthetic microorganisms (162). Most of these bacteria are mesophiles. A few species, including *Bacillus schlegelii, Bacillus tusciae, Calderobacterium hydrogenophilum, Hydrogenobacter thermophilus* (4), and the novel isolate *Streptomyces thermoautotrophicus* (59), are thermophiles. H_2 oxidizers have been isolated from marine environments. The latter organisms include a mesophilic obligately lithoautotrophic halophile, a knallgas bacterium from trophosome tissue of the hyperthermal-vent animal *Riftia* (4), and the hyperthermophile *Aquifex pyrophilus*. 16S rRNA sequence analysis showed that the latter is the deepest phylogenetic branch off of the eubacterial domain (72).

The group of aerobic H_2-oxidizing bacteria is physiologically defined and includes species belonging to 5 gram-positive and more than 15 gram-negative genera. Most of the latter are representatives of the α and β subdivisions of the proteobacteria (5). Most H_2-oxidizing organisms have high G+C genomes (>60 mol%). The highest and lowest values are found in thermophiles (4, 59).

Phototrophic bacteria such as *Rhodobacter capsulatus* (formerly *Rhodopseudomonas capsulata*) are included in this survey because some species can utilize H_2 for chemolithoautotrophic growth in the dark at decreased O_2 tension (100). Chemolithoautotrophic growth with H_2 as the energy source has been reported for aerobic free-living N_2-fixing bacteria, some of which were initially isolated as hydrogen bacteria (5). A few strains of *Bradyrhizobium japonicum* (formerly *Rhizobium japonicum*) oxidize H_2 during symbiotic N_2 fixation, and in the free-living state they grow as chemolithoautotrophs utilizing H_2 and CO_2 as the sole sources of energy and carbon, respectively (41). H_2-oxidizing activity has also been found in symbiotic cells of certain strains of *Rhizobium*, e.g. *Rhizobium leguminosarum* (28, 41) and in the obligately aerobic N_2-fixing *Azotobacter vinelandii* (73) and *Azotobacter chroococcum* (164). In contrast to the classical H_2-oxidizing bacteria, strains of the latter two groups do not fix CO_2 autotrophically (5), and H_2 appears to be utilized as a supplementary energy source. Thus, although *Rhizobium* and *Azotobacter* spp. do not represent chemolithotrophs sensu stricto, they are included in this survey because detailed molecular analyses indicate that their H_2 metabolism is closely related to that of the chemolithoautotrophs.

ECOLOGIGAL AND PHYSIOLOGICAL SIGNIFICANCE OF H_2 OXIDATION

The geothermal habitats of the thermophilic H_2-oxidizing bacteria provide little organic substance but are often rich in H_2 and reduced sulfur compounds. Such environments offer appropriate conditions for obligate lithoautotrophs (145). *Hydrogenobacter thermophilus* was the first obligately chemolithoautotrophic aerobic H_2-oxidizing bacterium reported in the literature (81). Recently, another representative, *Streptomyces thermoautotrophicus,* was described (59). *Thiobacillus ferrooxidans* may also belong to this group, because H_2 autotrophy has been demonstrated in some strains of this acidophile (32). H_2 is not the only energy source used by obligate lithoautotrophs: *H. thermophilus* can oxidize thiosulfate (9), *S. thermoautotrophicus* uses CO (59), and *T. ferrooxidans* is well known for its ability to grow on sulfur compounds, ferrous iron, or sulfidic ores (82). The newly isolated *Thiobacillus plumbophilus* was shown to oxidize H_2 in addition to PbS and H_2S (31).

The habitat of the majority of facultatively H_2-oxidizing bacteria is less clearly defined. They are ubiquitously distributed in soil and water. The H_2 concentration in these biotopes is fairly low, because only traces of the anaerobically produced H_2 escape to oxyc environments (145). Of the total flux of H_2 formation, up to 5% is released as a byproduct of N_2 fixation (26, 41) and should be available to aerobic bacteria. The major sink of H_2 is the soil, where most of the H_2 is oxidized by microorganisms. In light of the high threshold concentration of H_2 (varying between 1.5 and 178 ppm) and the low affinity for the substrate (apparent K_m 1–60 μM H_2) (25, 146), it is not clear to what extent hydrogen bacteria participate in the H_2 cycle. Nevertheless, H_2 autotrophy does provide an alternative mode of nutrition for the facultative chemolithoautotrophs (145). In *Alcaligenes eutrophus,* the H_2-oxidizing enzyme system is derepressed during substrate limitation, supporting the notion that H_2 is utilized as a supplementary source of energy and as a reductant under starvation conditions (48, 50, 52). Because the natural environments of facultative lithotrophs are very limited in nutrients, these bacteria probably use organic and inorganic substrates concomitantly.

Adaptation to shortage of nutrients may explain the extremely versatile metabolism of facultative chemolithotrophs. They are able to utilize alternative inorganic compounds and a wide range of organic substances (12, 46). For instance, most of the CO-oxidizing species, with a few exceptions, are hydrogenotrophs using H_2 as reductant (112). Reduced sulfur compounds such as thiosulfate can be oxidized by some hydrogen bacteria, e.g. *Paracoccus denitrificans* (53).

H_2 evolution is an obligate step of the N_2-fixation process, consuming about 27% of the electron flux. The amount of H_2 released into the atmosphere depends on the total H_2-recycling capacity of the rhizobia (39). *B. japonicum* bacteroids oxidize H_2, thereby increasing the metabolic efficiency of aerobic diazotrophy (41). A similar physiological function probably accounts for the presence of H_2-oxidizing activity in *Azotobacter* sp., which recycle the H_2 evolved by nitrogenase to increase the production of ATP and to protect nitrogenase against O_2 under carbon or phosphate limitation (164, 172). Furthermore, *A. vinelandii* can grow mixotrophically with H_2 and mannose and the energy derived from H_2 oxidation facilitates the uptake of the sugar (105).

ENZYMOLOGY OF NICKEL-CONTAINING HYDROGENASES

General Characteristics

Hydrogenases (hydrogen: acceptor oxidoreductase, EC 1.18.99.1 and 1.12.2.1) catalyze reversible redox reactions with molecular H_2 according to the following equation:

$$H_2 \leftrightarrow 2H^+ + 2e^-.$$

Structure and catalytic properties of hydrogenases have been extensively reviewed in past years (1, 2, 41, 42, 46, 101, 130, 131, 146, 162, 165); therefore only a few general characteristics are summarized below as this review focuses on the enzymes from those aerobic lithotrophs that have been explored genetically.

With the exception of a novel hydrogenase found in methanogenic archaea (175), all hydrogenases characterized so far are iron-sulfur proteins that fall into two groups. One type contains only iron, and the second type contains nickel in addition to iron. Iron-only hydrogenases seem to be limited to a few anaerobic organisms (1, 2).

Bartha & Ordal (8) were the first to report nickel-dependent chemolithoautotrophic growth of *A. eutrophus*. Later studies showed that this nickel requirement was related to formation of active hydrogenase (48) and that nickel was a constituent of both hydrogenases present in this organism (54, 55). In 1980, Lancaster (94) observed novel electron paramagnetic resonance (EPR) signals in the membrane fraction of methanogens and tentatively assigned them to Ni(III) species. Graf & Thauer (63) demonstrated the presence of one Ni atom per molecule hydrogenase from *Methanobacterium thermoautotrophicum*. Since then, an increasing number of Ni-containing hydrogenases have been characterized from various sources including sulfate-reducing bacteria, enterics, phototrophic bacteria, methanogens, and hyperthermophilic archaebacterial species (1, 2, 42, 130, 131, 165). So far, all of the hydrogenases from aerobic H_2-oxidizing bacteria belong to the family of [NiFe] enzymes (23, 41, 46, 101, 146, 150). Yamazaki (171) reported the presence of selenium (in the form of selenocysteine) in addition to iron, nickel, and acid-labile sulfide in *Methanococcus vannielii*. Representatives of this subgroup of [NiFe] hydrogenases have since been found in a few methanogens and sulfate-reducing bacteria (38, 42).

The content of nonheme iron in the nickel-containing hydrogenases is reportedly in the range of 4–40 g atom/mol (130). A minimum of one [Fe-S] cluster per nickel center has been extrapolated from current data. The significance of [3Fe-4S] found in various [NiFe] hydrogenases remains unknown. The Ni center is supposed to be the catalytic site of H_2 activation (14) and has been shown to be located in the large subunit of the enzyme (70, 71).

The structure of the Ni center is unknown. EPR and X-ray absorption spectroscopy (EXAFS) studies suggest that Ni is coordinated to a varying number (one to six) of sulfur atoms (2, 130). In the presence of O_2, most of [NiFe] hydrogenases are purified in the inactive, "unready" state that correlates to a typical nickel signal (Ni-A). In contrast to the highly sensitive, irreversibly O_2-inactivated Fe-only hydrogenases, the Ni-containing enzymes can be reactivated by reduction with H_2 or other reductants, yielding the active, "ready" state that correlates with an altered nickel signal (Ni-B). An intermediate oxidation level of the hydrogenase appears to correspond to a third nickel signal (Ni-C) that may represent the form of the enzyme that interacts with H_2 (14). Although the oxidation cycle of nickel is still a matter of debate, one model suggests that the interaction of hydrogenase with H_2 involves heterolytic cleavage of the substrate to a proton, which combines with a base on the enzyme, and a hydride, which binds to a metal center (14). Three-dimensional crystal-structure analyses are urgently needed to elucidate the nature of the catalytic site. Experiments of this kind are underway, e.g. for the enzymes from sulfate-reducing bacteria (68, 116) and from the phototroph *Thiocapsa roseopersicina* (151), but the structures have not yet been solved.

Membrane-Bound Hydrogenases

Most of the [NiFe] hydrogenases are membrane-bound dimers consisting of α and β subunits with average sizes of 60 and 30 kDa, respectively (130).

The majority of the aerobic H_2-oxidizing bacteria studied so far contain single membrane-bound hydrogenases (MBH) that have been characterized to varying extents. Table 1 summarizes the biochemical data on the purified [NiFe] hydrogenases from those H_2-oxidizing bacteria that have been studied at the molecular level.

The MBH enzymes are similar in subunit composition, molecular mass, and metal content (Table 1). They are coupled to energy-generating, membrane-bound electron-transfer chains and cannot transfer electrons to NAD (146). Therefore, organisms in which H_2 oxidation is catalyzed by MBH alone need reverse electron transport for biosynthetic purposes. It has been proposed

Table 1 Biochemical data for purified hydrogenases from aerobic H_2-oxidizing bacteria

Strain	Type of enzyme[a]	Size of subunits (kDa)	Metal/cofactor content (mol/mol enzyme)	$K_m(H_2)$ (μM)	DNA sequence data[b]	Reference
Alcaligenes eutrophus	MBH	62/31	0.65 Ni; 7–9 Fe	30	+	91, 143, 144
	SH	63/56/30/26	2 Ni; 16 Fe; 1 FMN	33	+	54, 148, 159
Azotobacter vinelandii	MBH	67/31	0.68 Ni; 6.6 Fe	0.86	+	108, 150
Bradyrhizobium japonicum	MBH	65/33	0.6 Ni; 6.5 Fe; 0.56 Se	0.97	+	10, 65, 142
Nocardia opaca	SH	64/56/31/27	1.8 Ni; 16 Fe; 1 FMN	30	−	147, 173
Paracoccus denitrificans	MBH	64/34	0.6 Ni; 7.3 Fe	ND[c]	−	87
Rhodobacter capsulatus	MBH	66/34	1 Ni; 4 Fe	ND	+	95, 137
Rhodocyclus gelatinosus	MBH	69/35	ND	ND	+	137, 160

[a] MBH, membrane bound hydrogenase; SH, soluble NAD-reducing hydrogenase.
[b] +, available, −, not available.
[c] ND, not determined.

that the electrons released from H_2 oxidation by MBH enter the respiratory chain at the level of quinone (101). Recently a quinone-reactive cytochrome *b*–containing hydrogenase was isolated from the anaerobic bacterium *Wolinella succinogenes*, suggesting that cytochrome *b* is the primary electron acceptor (33). Routinely the activity of MBH is assayed by measuring the reduction of various electron acceptors such as viologen dyes, methylene blue, 2,6-dichlorophenol indophenol, phenazine methosulfate, and ferricyanide (125).

Purification of the MBH enzymes listed in Table 1 involves detergent-assisted solubilization from membranes. A few strains of *A. eutrophus* have been shown to contain MBH in the soluble fraction after cell disruption (69, 125). Recently, reconstitution of purified *B. japonicum* hydrogenase into proteoliposomes was achieved (43). The orientation of MBH in the cytoplasmic membrane is still a matter of debate; periplasmic, cytoplasmic, and integral orientations have been discussed (162). Gerberding & Mayer (60) showed that the cellular distribution of hydrogenase-specific immunochemical label depends on the growth phase and the culture conditions. A substantial degree of immunological relationship among MBH enzymes from H_2-oxidizing bacteria, including *A. eutrophus, Alcaligenes latus, A. vinelandii, B. japonicum, R. capsulatus,* and *Thiocapsa roseopersicina,* has been demonstrated using various techniques (6, 92, 144).

The NAD-Reducing Multimeric Hydrogenases

Only a few aerobic H_2-oxidizing bacteria contain a NAD-linked soluble hydrogenase (SH). This is the only H_2-oxidizing enzyme in the gram-positive actinomycete *Nocardia opaca* and in the two gram-negative species *Alcaligenes ruhlandii* and *Alcaligenes denitrificans* 4a-2 (80, 146). Most of the *A. eutrophus* strains contain both SH and MBH (12). Those [NiFe] hydrogenases that react with specific electron carriers such as the methanogenic cofactor F_{420} (8-hydroxy-5-deazaflavin) or the eubacterial cofactor NAD contain one or two extra subunits in addition to the dimeric moiety (131, 146). Whereas the multimeric F_{420}-reducing hydrogenases of methanogens are associated with the cytoplasmic membrane (7, 98, 115), the tetrameric NAD-reducing hydrogenase of *A. eutrophus* H16 resides in the cytoplasm (140, 148), where it tends to form aggregates. SH enzymes have been purified from *A. eutrophus* H16 and *N. opaca* 1b (147, 148) to maximum specific activity with the physiological substrates (Table 1) and represent two of the best-characterized hydrogenases. Despite some structural differences, the two SH enzymes from *N. opaca* and *A. eutrophus* are clearly related. The corresponding subunits of the two enzymes contain common antigenic sites and are homologous (63–80% identity over the N-terminal sequences) (173).

Studies with hydrogenase-deficient mutants of *A. eutrophus* revealed that

SH is not required for H_2-dependent autotrophic growth. Nevertheless, the loss of SH activity results in a significant increase of the doubling time from 3.6 to more than 12 h. In contrast, loss of MBH activity affects the growth rate on H_2 only slightly (69, 88), indicating that NADH production via reverse electron transport is a growth rate–limiting step. Transfer of the *A. eutrophus* SH genes into the MBH-containing *Acidovorax* (*Pseudomonas*) *facilis* reduces the doubling time of the recombinant on H_2 from 12 to 9 h (159), showing that acquisition of SH is advantageous for the host.

The NAD-linked hydrogenases are complex enzymes with a native M_r of around 175,000. They are composed of four dissimilar subunits (Table 1). Cammack et al (14) have surveyed data on the stability and deactivation/activation phenomena. The SH from *N. opaca* is particularly amenable to structural analysis because it can be separated into two dimeric moieties without loosing the individual catalytic functions (149). The large α/γ dimer (64/31 kDa) contains two [4Fe-4S] clusters, one [2Fe-2S], and one [3Fe-4S] center in addition to one FMN, and has NADH electron acceptor oxidoreductase (diaphorase) activity. The small β/δ dimer (56/27 kDa) contains one [4Fe-4S] cluster in addition to 1.3–1.7 nickel atoms and has hydrogenase activity (173). Reaggregation of the two dimers occurs in the presence of chaotropic agents, of which nickel ions are the most effective. These ions also lead to restoration of H_2-dependent NAD-reducing capacity of the holoenzyme (149). This nonspecific role of nickel in the binding of the two dimers is distinct from its essential function in hydrogenase catalysis (14, 149).

Unlike the *N. opaca* enzyme, SH from *A. eutrophus* is neither stimulated in vitro by the addition of nickel nor segregates into two distinct dimeric conformations upon nickel starvation (55, 149). Partial dissociation into smaller particles and nonspecific aggregation accompanied by inactivation has been observed in the presence of NADH. Although reassociation to the tetrameric configuration occurs in the presence of 0.5 mM $NiCl_2$, the resulting tetramer is catalytically inactive (75).

ORGANIZATION OF HYDROGENASE GENES AND FUNCTION OF THE GENE PRODUCTS

Plasmid-Linked Hydrogenase Genes

The first indication of an involvement of extrachromosomal DNA in H_2 metabolism was the pronounced genetic instability of the lithoautotrophic character of *N. opaca* and its transfer to related heterotrophic recipients such as *Rhodococcus erythropolis* and *Corynebacterium hydrocarboclastus* (132, 133). Recently, large linear plasmids, ranging in size from 180 to 510 kilobase pairs (kb) were discovered in two wild-type strains of *N. opaca* (77). DNA-DNA hybridization revealed that the genes for the NAD-reducing hydrogenase and the key enzymes of autotrophic carbon dioxide fixation form a cluster on the linear plasmid pHG201 (Table 2).

Genetic determinants of H_2 metabolism have been conclusively mapped to plasmids in various strains of *Alcaligenes* (61). One of the best characterized examples is the 450-kb megaplasmid pHG1 of *A. eutrophus* H16 (Table 2). This plasmid determines in addition to H_2-oxidizing ability (encoded by *hox* genes) (3, 49) several other metabolic activities, including autotrophic CO_2 fixation (11, 86). Conjugative transfer of plasmid pHG1 occurs at a frequency of 10^{-2} per donor among different strains of *A. eutrophus* (49) and at lower frequencies from *A. eutrophus* to *Alcaligenes hydrogenophilus* (47).

A. hydrogenophilus contains MBH and SH hydrogenase, which appear to be biochemically similar to the enzymes of *A. eutrophus* but differ significantly with respect to their regulation (47, 161). The *hox* genes of this

Table 2 Plasmids of aerobic H_2-oxidizing bacteria

Strain	Plasmid[a]	Size (kb)	Plasmid markers[b]	Reference
Nocardia opaca MR11 (formerly 1b)	pHG31-a	140	T1 resistance	
	pHG31-b	17	c	
	pHG201*	270	SH, Cfx, Tra	77, 132, 133
	pHG202*	400	c	
	pHG203*	420	c	
Alcaligenes eutrophus H16	pHG1	450	SH, MBH, Tra Cfx, Denitrification Fhp	45, 149
Alcaligenes hydrogenophilus	pHG21-a	420	SH, MBH, Tra	47, 121
	pHG21-b	360	c	
Acidovorax facilis	pHG22-a	440	MBH, Cfx	167, 168
	pHG22-b	200	c	
Pseudomonas carboxydovorans	pHCG3	128	MBH, Cox, Cfx	93, 113
Rhizobium leguminosarum	pRL6JI	300	MBH, Nod	13

[a] Asteriks mark linear plasmids.
[b] T1, thallium resistance; c, cryptic; HS, NAD-linked soluble hydrogenase; MBH, membrane-bound hydrogenase; Cfx, autotrophic CO_2 fixation; Tra, conjugal transfer functions; Fhp, flavohemoprotein; Cox, CO dehydrogenase; Nod, nodulation.

bacterium (121) are located on one of the two megaplasmids (Table 2) and could be transferred and expressed in the chemoorganoautotrophic recipient *Pseudomonas oxalaticus* (161).

Pootjes (127) first observed the frequent occurrence of hydrogenase-deficient mutants after treating *Acidovorax facilis* with plasmid-curing agents. Screening of various *A. facilis* wild-type strains for the presence of plasmid DNA revealed a rather heterogenous pattern with respect to the number and the size of the endogenous plasmids (167, 168). In one strain, *hox* genes were identified on megaplasmid pHG22-a (Table 2).

Eight out of 21 carboxydotrophic bacteria harbor plasmids ranging in size between 3.5 and 558 kb (93, 161). In *Pseudomonas carboxydovorans,* the genes for CO dehydrogenase, autotrophic CO_2 fixation, and membrane-bound hydrogenase are carried on plasmid pHCG3 (113) (Table 2).

Apparently plasmids are also widespread among N_2-fixing bacteria (13, 16, 162). In most cases, however, it is not known whether they are involved in N_2 fixation or H_2 oxidation. To our knowledge, *Rhizobium leguminosarum* (Table 2) is the only example for which hydrogenase genes have been proven to reside on plasmids (13).

Gene Nomenclature and Functional Units

Nucleotide sequences are available for the hydrogenase genes of a few chemolithotrophs. The genes for the MBH enzymes are clustered (Figure 1). The order of the genes in the different bacteria is not constant, but a conserved pattern is obviously present. While systematic investigations to define transcription units have yet to be carried out, the hydrogenase genes seem to be arranged in functional groups: structural genes encoding the enzyme subunits and a cytochrome-like membrane protein; genes for accessory proteins, some of which are involved in hydrogenase maturation; and pleiotropic genes, which encode products with a more general role in nickel metabolism. Only one set of genes for a soluble hydrogenase has been sequenced (159), obviating comparisons for this enzyme type.

Table 3 gives an overview of the various genes and gene products for the more extensively studied organisms and includes data for *Escherichia coli* for comparison. The nomenclature for the hydrogenase genes is nonuniform and therefore somewhat confusing. The designations *hox, hup,* and *hya* are used for the structural genes, *hyp* for processing functions.

MBH Structural Genes

The genes for the MBH enzymes of the aerobic H_2-oxidizing bacteria *A. eutrophus* (91), *A. chroococcum* (44), *A. vinelandii* (108), *B. japonicum* (142), *R. leguminosarum* (66), *R. capsulatus* (21, 95), and *Rhodocyclus gelatinosus* (160) have been cloned and sequenced. Comparison of these

Table 3 Hydrogenase genes nomenclature and gene products

Gene product	*Alcaligenes eutrophus*		*Azotobacter vinelandii*		*Rhizobium leguminosarum*		*Rhodobacter capsulatus*		*Escherichia coli*	
Membrane-bound hydrogenase										
Small subunit	*hoxK*	39.5[a]	*hoxK*	39.3	*hupS*	39.2	*hupS*	39.3	*hyaA*	40.6
Large subunit	*hoxG*	68.7	*hoxG*	66.8	*hupL*	66.1	*hupL*	66.0	*hyaB*	66.2
Cytochrome *b*	*hoxZ*	27.6	*hoxZ*	27.7	*hupC*	27.9	*hupM*	30.2	*hyaC*	27.6
Processing	*hoxM*	22.2	*hoxM*	22.8	*hupD*	22.1	*hupD*	22.8	*hyaD*	21.5
Unknown	—		—		*hupE*	19.0	—		—	
Metalloprotein	*hoxL*	11.9	*hoxL*	11.4	*hupF*	10.4	*hupF*	10.8	—	
Unknown	*hoxO*	17.7	*hoxO*	16.3	*hupG*	16.3	*hupG*	15.4	*hyaE*	14.5
Unknown	*hoxQ*	30.9	*hoxQ*	31.0	*hupH*	30.5	*hupH*	28.9	*hyaF*	31.5
Rubredoxin	*hoxR*	8.6	ORF1	8.1	*hupI*	8.0	—		—	
Unknown	*hoxT*	19.7	ORF2	16.3	*hupJ*	18.4	*hupJ*	30.3	—	
Unknown	*hoxV*	41.5	ORF3	36.4	*hupK*	38.7	*hupK*	30.3	—	
References	91		18, 19, 107, 108		66, 67, 134		95, 17, 22		109	
Nickel supply and processing										
Unknown	*hypA*	12.9	ORF4	12.6	*hypA*	12.4	*hypA*	12.3	*hypA*	12.8
Ni-incorporation	*hypB*	79.8	ORF5	33.2	*hypB*	32.6	*hypB*	35.4	*hypB*	31.6
Metalloprotein	*hypC*	9.8	ORF7	8.9	*hypC*	8.0	*hypC*	9.9	*hypC*	9.7
Unknown	*hypD*	41.9	ORF8	41.2	*hypD*	36.4	*hypD*	40.6	*hypD*	41.3
Membrane protein	*hypE*	36.6	—		*hypE*	36.5	*hypE*	33.0	*hypE*	35.0
Unknown	—		ORF6	80.5	*hypF*	80.7	*hypF*	77.9	*hypF* (*hydA*)	80.0
References	27		19		135		170, 22		99	

Regulation										
Activator	—		—		—		—		*fhlA*	78.5
Repressor	—		—		—		—		*hycA*	17.6
Activator	*hoxA*	53.5	—		—		*hupR1*	53.8	*hydG*	48.0
Sensor	*hoxX*	65.7	—		—		—		—	
References	36						136		22, 154	
Soluble NAD-reducing hydrogenase										
α subunit	*hoxF*	67								
δ subunit	*hoxU*	26								
γ subunit	*hoxY*	23								
β subunit	*hoxH*	55								
Unknown	*hoxW*	19								
B protein	*hoxI*	18								
References	159									

[a] Sizes of predicted gene products in kilodaltons.

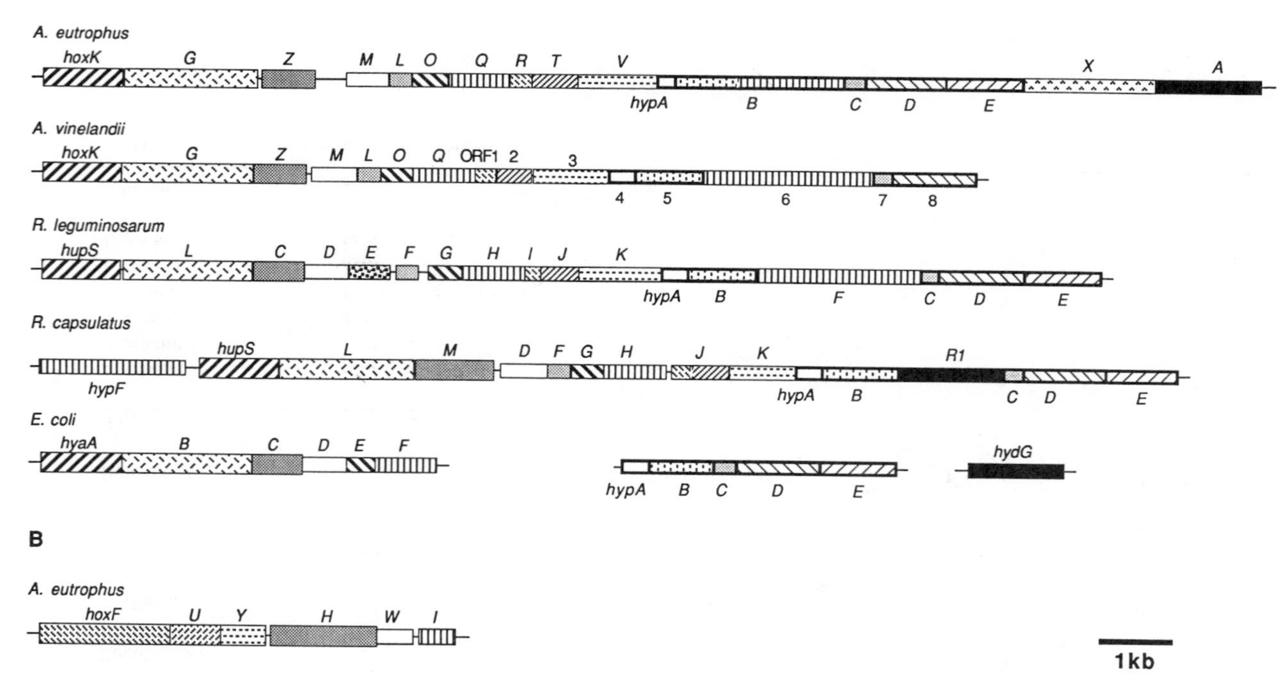

Figure 1 Organization of the hydrogenase genes of selected aerobic H_2-oxidizing bacteria. Thicker bars represent hydrogenase structural genes, thinner bars, accessory and regulatory genes. Heavy outlining emphasizes regions of homology to the *E. coli hyp* operon. Similar shading within the respective groups of genes indicates homology. *hyp* genes are labeled below the bar, all other genes, above the bar. (*A*) Genetic structure of the loci encoding the membrane-bound hydrogenases of *A. eutrophus* (27, 36, 91), *A vinelandii* (18, 19, 107, 108), *R. leguminosarum* (66, 67, 134, 135), and *R. capsulatus* (17, 22, 95, 136, 170). For comparison the *E. coli hya, hyp,* and *hyd* operons are also shown (99, 109, 154). (*B*) Genetic structure of the locus encoding the soluble hydrogenase of *A. eutrophus* (159).

nucleotide sequences reveals a constant arrangement of three conserved genes: The first and second genes determine the small and large subunits, respectively, of the dimeric enzyme. The third gene encodes an integral membrane protein with the characteristics of *b*-type cytochromes.

The small subunits are synthesized as preproteins with signal peptides that are absent from the mature proteins (91, 95, 97, 108). The exceptionally long (32–50 residues) signal sequences share the highly conserved motif Arg-Arg-X-Phe-X-Lys. These and other features (see below) suggest that a specialized secretion pathway involving hydrogenase-specific components directs the transport of hydrogenase into or through the cytoplasmic membrane (163). The small subunits also share a set of invariable cysteine, glycine, and proline residues. This array of conserved residues is found in the [NiFe] hydrogenases of other bacteria. The clustered cysteines could participate in the coordination of [Fe-S] clusters. The small subunits of the MBH enzymes have a hydrophobic C-terminal segment missing in the periplasmic hydrogenases (42). This domain may form a transmembrane helix serving to anchor the dimer to the membrane (108).

The Ni-containing large subunits share 19 strictly conserved amino acid residues located in four homologous regions of the polypeptide (element 1: Arg-Gly-X-Glu-X_{16}-Arg-X-Cys-Gly-X-Cys-X_3-His; element 2: His; element 3: Gly-X_4-Pro-Arg-Gly-X_3-His; element 4: Asp-Pro-Cys-X_2-Cys-X_2-His). The large subunits may coordinate one [Fe-S] cluster in proximity to Ni (130, 163). It is not yet known which of the conserved residues actually participate in the binding of the redox-active prosthetic groups. Preliminary results of a site-directed mutagenesis study indicate that in the large subunit of *E. coli* hydrogenase 1 an aspartate (Asp574) and a pair of cysteines (Cys576, Cys579) in the C-terminal domain, as well as arginine (Arg74) and cysteine (Cys76) residues in the N-terminal domain, are involved in Ni liganding (130). Based on these findings, spectroscopic data, and sequence comparisons, Przybyla and coworkers (130) have proposed a pseudohedral structure for the nickel active site formed by six ligands—a nitrogen ligand from arginine, an oxygen ligand from aspartate, and four thiolate ligands. Assuming a direct interaction of H_2 with Ni, we favor a structural model in which the Ni center is coordinated by only two thiolate ligands and two ligands such as nitrogen and/or oxygen. The other conserved amino acids (e.g. the four histidine residues) may serve different functions. They could operate as proton conductors (163) and/or as a signal for C-terminal processing (see below).

The third gene in each set encodes an integral membrane protein (Figure 1). Three lines of evidence indicate that these hydrophobic proteins are *b*-type cytochromes that function as intermediate electron carriers coupling the MBH to the respiratory chain: (*a*) Conserved histidine residues with potential heme-liganding capacity have been identified within hydrophobic stretches (17, 33, 67). (*b*) A hydrogenase complex consisting of three polypeptides

(HydA, HydB, HydC) was isolated from the cytoplasmic membrane of *Wolinella succinogenes* and shown to contain a cytochrome *b*. This trimeric enzyme mediated electron transfer from H_2 to a quinone, whereas the hydrogenase dimer did not (33). (*c*) Inactivation of the corresponding genes (Figure 1) of *R. capsulatus* (17) and *A. vinelandii* (141) generated mutants that failed to oxidize H_2 with O_2 as the electron acceptor but retained the ability to reduce artificial redox dyes, such as methylene blue.

MBH Accessory Genes

Nucleotide sequence analysis of the MBH loci of different H_2-oxidizing aerobes uncovered a series of conserved open reading frames immediately downstream of the hydrogenase structural genes (Figure 1). Genetic and molecular studies on *A. eutrophus* (37, 90, 91), *A. vinelandii* (18, 19, 107), and *R. capsulatus* (22) indicate that some if not all of these genes are essential for the formation of active MBH enzyme. Additional evidence comes from investigations of the *E. coli hya* operon (101). The *A. eutrophus* MBH region (91) contains seven ORFs designated *hoxM, hoxL, hoxO, hoxQ, hoxR, hoxT,* and *hoxV* downstream of the structural genes *hoxK, hoxG, and hoxZ* (Figure 1). The same basic pattern is found in the other H_2-oxidizers for which sequence data are available (Figure 1).

Little is known about the products of the individual accessory genes. These proteins could be involved in (*a*) redox reactions coupled to the hydrogenase-mediated proton translocation and/or electron transport and (*b*) the maturation of MBH, i.e. subunit processing and assembly, membrane attachment, and metal insertion. Both the *A. eutrophus hoxM* gene and its *A. vinelandii* homologue are required for normal processing of hydrogenase subunits (see below), as is the *E. coli* counterpart (91, 107, 109). The homologous genes *hoxL* (*A. eutrophus* and *A. vinelandii*) and *hupF* (*R. leguminosarum* and *R. capsulatus*) appear to encode metalloproteins involved in subunit processing (22, 91, 107, 134). The gene products are homologous to the *E. coli hypC* gene product (99) (Figure 1).

The nucleotide sequences of another group of homologous accessory genes—*hoxR* of *A. eutrophus,* ORF1 of *A. vinelandii,* and *hupI* of *R. leguminosarum* (Figure 1)—predict rubredoxins (18, 91, 134). In *R. capsulatus,* the rubredoxin gene is fused to a larger coding region (Figure 1) (22). Rubredoxins are redox-active iron metalloproteins. The role of these proteins in the context of H_2 oxidation is not known. At present there are no clues to the functions of the other accessory gene products.

SH Genes

The genes encoding the NAD-reducing hydrogenase (SH) of *A. eutrophus* H16 have been cloned and sequenced (37, 159). The SH locus, formerly designated *hoxS,* is located on megaplasmid pHG1 approximately 80 kb away from the MBH locus (90). The SH genes are flanked by two copies of an insertion element (45).

The *hoxF* and *hoxU* genes encode the α and γ subunits that form the NADH electron acceptor oxidoreductase (diaphorase) moiety of the SH. The *hoxH* and *hoxY* genes encode the β and δ subunits, components of the hydrogenase dimer. The genes belong to a transcriptional unit, *hoxFUYH* (Figure 1) (120, 159). Figure 2 presents a structural model of the SH protein based on sequence data and the results of electron microscopy (75, 159).

The β subunit, isolated from the SH^- *A. eutrophus* mutant HF14 oxidized H_2 and was shown to contain nickel and a [4Fe-4S] cluster (70, 71). This observation strongly supports the notion that the β subunit contains the catalytic site. Despite little overall homology between the primary structure of membrane-bound [NiFe] hydrogenases and the soluble SH enzyme, all 19 residues that define the consensus sequence of the putative Ni-binding subunit (130, 163) are present in the deduced sequence of the β subunit. Moreover, the β subunit shares sequence elements with the multimeric hydrogenases from methanogens, for instance an extended stretch of highly conserved residues within the C-terminal domain: Ile-Arg-Ala-Tyr-Asp-Pro-Cys-Leu-Ser-Cys-Ala-Thr-His (131, 159).

The product of the *hoxY* gene, the δ subunit, is considerably smaller than the electron transferring subunit of membrane-bound hydrogenases. The nucleotide sequence predicts a protein with nine cysteine residues that could bind at least one [Fe-S] cluster (Figure 2). On the basis of EPR and Mössbauer studies on the SH enzyme from *N. opaca*, the primary sequence of which is unknown, Zaborosch et al (173) excluded the presence of additional [Fe-S] clusters in the hydrogenase dimer. So far, the function of the δ subunit in catalysis is unclear. It may serve solely structural purposes; on the other hand it may actively participate in electron flow to the acceptor

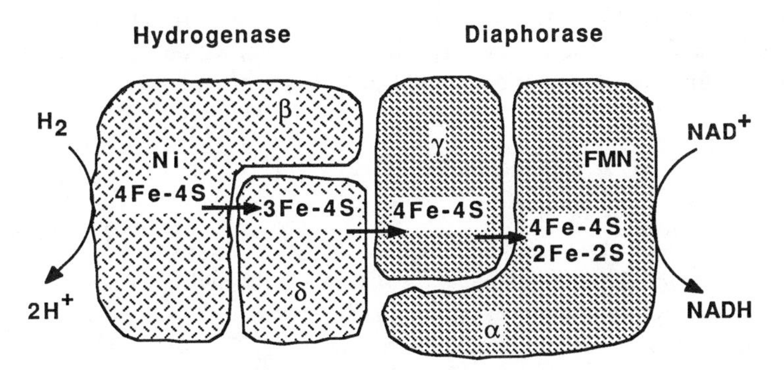

Figure 2 Structural and functional model of the NAD-dependent hydrogenase of *A. eutrophus*. The hydrogenase and diaphorase moieties are indicated by different shading. Bold arrows indicate redox processes.

site as proposed in the model shown in Figure 2. It is also possible that a redox prosthetic group in the δ subunit is involved in redox-coupled regulation of SH activity (14).

The salient feature of SH is the interaction with the cytoplasmic electron carrier NAD, and this activity is located in the diaphorase dimer of the enzyme (Figure 2) (159). Homology searches revealed striking similarities between the diaphorase dimer of the SH from *A. eutrophus* and three peripheral subunits of the mitochondrial NADH ubiquinone reductases (complex I). The α subunit of the SH appears to be a fusion of homologues of the 24- and 51-kDa subunits of complex I. Sequences highly homologous to the 51-kDa polypeptide reside in three segments of the SH α subunit and probably provide the nucleotide binding sites for NAD(H) and FMN as well as ligands for a tetranuclear [Fe-S] center (123, 124, 128, 159). Spectroscopic analysis indicates the presence of a [2Fe-2S] center in the diaphorase dimer of the SH (174). Thus, the remaining cysteine-rich motifs in the α subunit (159) could be involved in the liganding of a binuclear [Fe-S] cluster (Figure 2). The γ subunit of the SH is related to the 75-kDa subunit of complex I. Ligands for a second [4Fe-4S] center (Figure 2) appear to be present in homologous regions of both subunits (124, 159). The clear sequence similarity between the subunits of the mitochondrial complex I and the SH from *A. eutrophus* strongly suggests that the two proteins have a common ancestor.

Northern blot experiments indicate that the major SH mRNAs exceed the size predicted for a transcript encompassing the four structural genes (120). Two additional ORFs lie immediately downstream of the β subunit gene *hoxH*. One of these ORFs, designated *hoxI*, encodes an 18-kDa product called the B protein (Figure 1, Table 3) (A. Tran-Betcke, V. Popov & B. Friedrich, unpublished data). The B protein is one of five proteins expressed concomitantly with the SH enzyme and subject to control by the hydrogenase regulator HoxA (Figure 1) (78, 79). The role of the B protein is not known, but preliminary results indicate that the region downstream of the SH structural genes contains a gene or genes required for the formation of active SH (S. Thiemermann & B. Friedrich, unpublished data).

BIOSYNTHESIS AND MATURATION

Nickel Uptake

Because nickel atoms are essential constituents of the hydrogenases of aerobic H_2-oxidizing bacteria, nickel uptake can be considered an important step in the formation of active hydrogenase enzymes. The limited information available on nickel transport in these bacteria indicates that different mechanisms are involved. *A. eutrophus* is the best-studied organism with respect to nickel uptake (46). In this bacterium, two distinct systems are responsible for the accumulation of Ni^{2+} ions: (*a*) a nonspecific, low-affinity Mg^{2+} transporter (K_m for Ni^{2+}: 17 μM) and (*b*) a high-affinity system specific for Ni^{2+} (K_m 0.34 μM). Both systems are inhibited by protonophores and are susceptible to respiratory inhibitors (96). Characterization of a nickel-deficient mutant of *A. eutrophus* led to the identification of the gene for the high-affinity transporter in the hydrogenase gene cluster (35). This gene, designated *hoxN*, encodes a 33.1-kDa polypeptide with the characteristics of an integral membrane protein (40). The *hoxN* gene is subject to control by the *hoxA* locus and is expressed coordinately with the hydrogenase structural genes (T. Eitinger & B. Friedrich, unpublished data). Thus, the primary function of HoxN-mediated high-affinity Ni^{2+} transport is apparently to ensure a supply of Ni^{2+} ions for hydrogenase synthesis.

Studies on Ni^{2+} transport in *B. japonicum* have yielded disparate results for the different strains tested. An energy-independent Ni^{2+} uptake was reported for *B. japonicum* SR and its hydrogenase-constitutive derivative SR470 (155). On the other hand, an energy-dependent Mg^{2+} uptake system was shown to mediate Ni^{2+} accumulation in *B. japonicum* strain JH (56). Energy-independent Ni^{2+} uptake has also been reported for *Azotobacter chroococcum* (122). Because transport of ions is invariably energy dependent, these reports invite critical reexamination. In the phototrophic H_2 oxidizer *R. capsulatus,* Ni^{2+} uptake is mediated by a Mg^{2+} transporter (K_m 5.5 μM) (156).

Nickel Incorporation and Processing

Studies on hydrogenase-deficient mutants uncovered genetic functions that are not involved in Ni^{2+} transport but appear to have a role in hydrogenase-associated nickel metabolism. The first mutants of this type were identified in *E. coli* (169). The mutations mapping at 58–59 min on the *E. coli* chromosome curtailed formation of all three hydrogenase isoenzymes. This locus contains an operon designated *hypABCDE* (99). Mutations in one of the genes, *hypB,* are phenotypically suppressible by the addition of supplemental Ni^{2+} to the medium, suggesting that the defect perturbs some step in Ni^{2+} processing or its incorporation into the hydrogenase apoenzyme (99). Similar sets of genes have been identified in aerobic H_2-oxidizing bacteria. The members of one class of *A. eutrophus* mutants produce immunoreactive MBH and SH, but both enzymes are inactive (36). The defective enzymes lack Ni^{2+}. The locus involved contains a cluster of genes homologous to the *hypABCDE* operon of *E. coli* (27). Homologous genes have also been indentified in *R. capsulatus* (22, 170), *A. vinelandii* (19), *A. chroococcum* (34, 157), and *R. leguminosarum* (135). In *B. japonicum* strain JH, a locus involved in nickel metabolism but not related to nickel uptake has been identified (57). Mutations in this locus are accompanied by the production of inactive hydrogenase lacking nickel (57, 58). This phenotype suggests the existence of *hyp*-like genes in this organism.

There are a few clues to the functions of the individual *hyp* gene products. The deduced amino acid sequences of the products of the *hypE* homologues predict integral membrane proteins, whereas *hypC* gene products appear to be metalloproteins (Table 3). The predicted products of ORFs 5 and 6 of *A. vinelandii* (19) and of the *hypB* genes of *A. eutrophus* (27) and *R. leguminosarum* (135) contain distinctive nests of histidine residues. Such histidine-rich domains are also found in the products of accessory genes required for the synthesis of the nickel-containing ureases of *Klebsiella aerogenes* (114) and *Proteus mirabilis* (76), suggesting a Ni-related function if not Ni binding. The ORF6, *hupY,* and *hypF* gene products of *A. vinelandii, A. chroococcum,* and *R. leguminosarum,* respectively, contain a paired motif (Cys-X_2-Cys-X_{17-18}-Cys-X_2-Cys) found in zinc-finger proteins, pointing to a regulatory role for these gene products (19, 135, 157). A homologous gene, *hypF,* has been indentified outside the *hyp* cluster of *R. capsulatus* (22) and shown to encode a key regulatory component (see below). Recent studies on the *E. coli* HypB protein have shown that purified HypB protein binds and hydrolyzes GTP. A molecular model for nickel incorporation proposes that nickel-charged HypB forms a complex with the hydrogenase large subunit, transfers Ni^{2+}, and dissociates after GTP hydrolysis (106). Although experimental evidence for this model is lacking, the presence of typical GTP-binding domains in the homologous gene products (Table 3) of different organisms supports the notion that GTP hydrolysis is important for the function of these proteins.

The incorporation of nickel into nascent hydrogenases is not the only process required for the formation of the active enzymes. The maturation of MBH and SH involves mechanisms specific to the two enzyme types. *A. eutrophus* offers a unique opportunity to study both of these maturation pathways in one organism. As yet little is known about the specific steps in the maturation of the SH, but recent studies have shed some light on some of the events involved in the formation of the dimeric MBH enzyme. Each of the two subunits of the MBH exists in two electrophoretically distinct conformers (89). The faster species of both polypeptides are found predominantly in the membrane. Similar results were reported for hydrogenase 1 of *E. coli* (110). These findings suggest that the slower-moving forms are precursors that are converted into the mature forms by proteolytic processing or other modification steps, and that this conversion is intimately related to the process of membrane attachment. A nonpolar mutation in the *A. eutrophus* *hoxM* gene blocks modification of the large subunit and leads to the accumulation of the precursor form in the cytoplasm (J. Rietdorf, E. Schwartz

& B. Friedrich, unpublished data). On the other hand, immunoelectron microscopic studies link *hoxM* to the membrane attachment mechanism (91). Mutations in the homologous genes of *A. vinelandii* and *E. coli* block processing of one or both subunits (107, 110). Recently, mass spectroscopy was used to study the processing of the *A. vinelandii* MBH. The results indicate that the mature form of the large subunit had undergone C-terminal truncation (62).

The conversion of the small subunit from the slower to the faster-moving form is undoubtedly the result of proteolytic removal of a signal peptide. At present, nothing is known about the proteolytic mechanism mediating removal of the leader peptide in the organisms discussed in this review. Studies with a chimeric gene showed that the leader peptide of the *Desulfovibrio vulgaris* hydrogenase small subunit directs translocation of a β-lactamase fusion protein in *E. coli* (117).

At present nothing is known about the temporal series of events involved in the maturation of the MBH enzymes, and it is much too early to construct molecular models. In *A. eutrophus* as in *E. coli,* the insertion of nickel into the apoenzyme seems to be intimately connected with the conversion of both the small and the large subunits to their respective mature forms. In *A. eutrophus,* processing of both the small and large subunits requires functions encoded in the *hyp* region (89). In *E. coli,* defects in the *hyp* genes curtail processing of hydrogenase large subunits (74, 99). Whether inhibition of nickel incorporation is the cause or the result of defective processing is not clear.

REGULATION OF HYDROGENASE GENE EXPRESSION

This section summarizes present knowledge of the regulation of hydrogenase synthesis in *Alcaligenes eutrophus, Rhodobacter capsulatus,* and *Bradyrhizobium japonicum,* the three most extensively studied aerobic H_2 oxidizers.

The two hydrogenases of *A. eutrophus* H16 are synthesized coordinately during growth on mixtures of H_2, CO_2, and O_2 or on certain organic carbon sources. During exponential growth on preferentially utilized carbon sources such as pyruvate or succinate, the synthesis of both hydrogenases is tightly repressed. Growth on poor substrates such as glycerol derepresses the hydrogenase system to various levels, even in the absence of H_2. Thus, the production of hydrogenase in *A. eutrophus* H16 is not strictly dependent on H_2. It appears that a physiological parameter such as reductant supply is the signal triggering hydrogenase derepression (50, 52). Hydrogenase synthesis shows a pronounced temperature sensitivity: synthesis of both hydrogenases is abolished at temperatures above 33°C (51). The appearence of hydrogenase activity coincides with the appearence of immunologically detectable hydrogenase proteins and hydrogenase-specific mRNA (52, 120). Thus, the formation of hydrogenase activity upon derepression is not the result of activation of a pool of inactive precursor but entails de novo synthesis of the hydrogenase enzymes.

Genetic investigations led to the isolation of several *A. eutrophus* mutants impaired in H_2 oxidation (37, 69, 90). Hydrogenase-deficient (Hox^-) mutants revealed, in addition to defects in hydrogenase structural and accessory genes, lesions in regulatory functions. In one class of mutants, synthesis of both hydrogenases is totally eliminated (37). Another class of mutants shows a relaxed temperature control of hydrogenase synthesis: synthesis of the MBH and SH continued at temperatures above 33°C (51). Both mutations map at a locus designated *hoxA* (36). The product of *hoxA* is a response regulator–type transcriptional activator (36, 153). Experiments with promoter test constructions showed that the two hydrogenase structural gene operons and the gene encoding the nickel transporter are regulated by HoxA at the transcriptional level (E. Schwartz, T. Eitinger & B. Friedrich, unpublished data). Furthermore, MBH and SH mRNAs are reduced at least 10-fold in *hoxA* mutants (C. G. Friedrich & U. Oelmüller, personal communication). Another regulatory gene, *hoxX,* is located immediately upstream of *hoxA; hoxX* may encode a histidine protein kinase involved in signal transduction (O. Lenz, E. Schwartz, J. Dernedde, M. Eitinger & B. Friedrich, unpublished data).

The defect in another class of hydrogenase-deficient mutants maps to a chromosomal locus (69). This locus, which controls not only H_2 oxidation but also such diverse processes as denitrification, nitrate and urea assimilation, and uptake of dicarboxylic acids, contains an *rpoN* gene (138, 139, 166). Thus, expression of the genes for both hydrogenases requires the minor sigma factor 54 of the RNA polymerase.

The SH operon is expressed as three major mRNA species of 6.2, 5.9, and 0.9 kb. The latter two mRNAs are processing products of the 6.2-kb species. The soluble hydrogenase mRNA is exceptionally stable, with a functional half-life of 1 h. This unusually long-lived mRNA may, in part, account for the high-level expression of the SH. The MBH mRNA is less stable with a functional half-life of 14 min (120).

Regulation of hydrogenase synthesis in some less extensively studied chemolithotrophs superficially resembles the situation in *A. eutrophus* H16. In *Acidovorax facilis,* hydrogenase synthesis shows similar substrate and temperature dependence as well as O_2 insensitivity (167). In contrast, hydrogenase formation in the hydrogenotrophs *Paracoccus denitrificans* (strain 381) and *Alcaligenes hydrogenophilus* is subject to a different regulatory regime. In these bacteria, hydrogenase synthesis is H_2 dependent

(47, 118). Hydrogenase synthesis in *A. eutrophus* H1 (ATCC 17707) is sensitive to O_2 (15), as it is in other H_2 oxidizers including *Alcaligenes latus* (29), *Pseudomonas saccharophila* (126), *R. capsulatus,* and *B. japonicum* (see below). In *A. latus* the immunoreactive hydrogenase present under autotrophic conditions is nickel dependent, and a direct influence of nickel on hydrogenase-gene expression has been postulated (30).

Hydrogenase activity is formed in *R. capsulatus* cells both under N_2-fixing conditions and during chemolithotrophic growth. The physiological common denominator of these culture conditions is the availability of either endogenous or external H_2 (100, 162). Hydrogenase activity in *R. capsulatus* is not correlated with light, nitrogen source, or O_2 partial pressure (20). The increase in hydrogenase activity following addition of H_2 to aerobic cultures is directly correlated with an increase in immunoreactive protein and is blocked by both transcriptional and translational inhibitors. Apparently transcription and translation of the hydrogenase genes take place mainly during the initial 5-h period of induction. Thereafter posttranslational processes continue to produce active enzyme from a pool of inactive precursors (23).

Recent studies with gene fusions confirm that modulation of gene expression is in fact the basis of the observed variations in hydrogenase activity and underscore the role of H_2 in the derepression of the *hup* genes (24). H_2 is, however, not the only factor modulating the expression of the hydrogenase genes. Moreover, whether H_2 is itself the cue that triggers derepression of the hydrogenase genes or if it acts indirectly via another physiological signal is still not known.

Three hydrogenase regulatory genes have been identified in *R. capsulatus*. The gene *hupR1* encodes a response regulator of the NtrC/NifA family (136, 153). A mutation in another locus shows a characteristic intermediate phenotype, producing low but significant levels of hydrogenase. Cloning and sequencing of this locus revealed a gene homologous to the *E. coli himA* gene for the α subunit of integration host factor (IHF) (158). The finding that both a response regulator–type activator and an IHF-like protein participate in the regulation of the *R. capsulatus* hydrogenase genes suggests that the underlying mechanism involves a positive transcriptional regulation along the lines of transcriptional activation by NtrC (153). This molecular model postulates that IHF binds to a site in the upstream region between the promoter and the binding site of the response regulator, inducing or enhancing bending of the DNA and thereby facilitating interaction of regulator and RNA polymerase. Recently, Toussaint et al (158) showed that partially purified *R. capsulatus* IHF binds to DNA fragments containing the *hupSLM* promoter region in vitro. A third regulatory gene, designated *hypF,* encodes a protein with typical zinc-finger motifs. Mutations in this gene prevent stimulation of hydrogenase expression by H_2 (22).

In *A. vinelandii* and *A. chroococcum,* hydrogenase is expressed under N_2-fixing conditions (83). External H_2 stimulates hydrogenase activity in *A. vinelandii* (129).

In the root nodule *B. japonicum* bacteroids engaged in N_2 fixation can oxidize substantial amounts of H_2 produced by nitrogenase (41). In the free-living state, *B. japonicum* utilizes external H_2 for chemolithoautotrophic growth (64, 102). Because of the ease of experimental handling, most of the investigations on *B. japonicum* hydrogenase have been done with chemolithoautotrophically grown cells (101).

Physiological studies on *B. japonicum* strain 122DES have shown that O_2 and H_2 are the key factors controlling hydrogenase expression (103). Addition of H_2 to chemolithotrophic cultures triggers derepression of hydrogenase. However, derepression of hydrogenase takes place only at low O_2 partial pressures; O_2 partial pressures higher than 1% repress hydrogenase synthesis. Hydrogenase expression is independent of nitrogenase expression. The presence of organic carbon substrates (e.g. arabinose, glycerol, and sucrose) in the medium prevents hydrogenase derepression (103).

Genetic studies have yielded several *B. japonicum* mutants with altered regulatory patterns for the hydrogenase system. One class of mutants is hypersensitive to oxygen (104); another group is O_2 insensitive and H_2 independent. The mutants of both classes are also affected in substrate-mediated repression: the O_2-hypersensitive mutants are concomitantly more, and the O_2-insensitive mutants are less, susceptible to repression by organic substrates (111). Thus, in the O_2-resistant mutants, hydrogenase synthesis is constitutive. On the basis of these results, Merberg et al (111) proposed that the various mutations affect a single regulatory element responsible for both oxygen and carbon control.

The influence of O_2 on the expression of the *B. japonicum* hydrogenase genes prompted an investigation of the role of DNA supercoiling in hydrogenase regulation. It was found that gyrase inhibitors specifically block expression of the hydrogenase genes (119). The findings are, however, difficult to assess because it is not known if or to what extent the gyrase inhibitors used actually affect DNA topology in *B. japonicum*.

Experiments with plasmid-borne transcriptional fusions show that H_2 and O_2 control expression of the *B. japonicum hup* genes at the transcriptional level (85). The addition of H_2 to the gas phase leads to a dramatic activation of the *hup* promoter. Promoter activity responds in a linear fashion to increasing H_2 partial pressure over a range of 0.05–3%. Activity of the *hup* promoter shows a clear dependence on O_2 concentration in the gas phase: at O_2 partial pressures of 0.1–3%, the *hup* promoter directs high levels of transcription in the presence of H_2; transcription is sharply repressed above 3%. The *hup* promoter is also subject to control by nickel (84, 85). Analysis of deletion derivatives of the *hup* promoter region (84, 85) has pointed out the presence of a signal (or signals) required for regulation by H_2, O_2, and Ni upstream of the *hup* promoter. On the basis of these results, Kim & Maier (84) have proposed a molecular model for positive transcriptional regulation of the *hup* genes mediated by a single redox-sensitive nickel metalloprotein.

RETROSPECT AND CONCLUDING REMARKS

The aerobic H_2-oxidizing bacteria are a phylogenetically diverse group of organisms inhabiting a wide variety of biotopes from legume nodules to thermal springs. The hydrogenases, which catalyze the oxidation of H_2 in these bacteria, are synthesized under different conditions and have physiological roles as diverse as the utilization of environmental H_2 for lithoautotrophic and mixotrophic growth, and the oxidation of H_2 arising from nitrogen fixation. Nevertheless, biochemical studies have shown that all of the hydrogenases of this group of bacteria belong to two families: the dimeric, membrane-bound enzymes and the tetrameric, cytoplasmic species. All of these enzymes are nickel-containing [Fe-S] metalloproteins. Determination of three-dimensional structures will no doubt provide a key to understanding the structure and function of the hydrogenase enzymes. Recent genetic and molecular studies on the membrane-bound hydrogenases, the more widespread group, have revealed that the formation of the H_2-oxidizing apparatus is a complex process requiring the synthesis of several proteins in addition to the hydrogenase itself. Evidence indicates that some of these gene products are components of a special hydrogenase-linked electron transport chain. Formation of active hydrogenase requires multiple posttranslational steps, and sequencing has unveiled the genetic blueprint for the hydrogenase structural genes and the genes encoding the battery of accessory proteins required for maturation. Future research is faced with the dual challenge of gaining molecular insights into nickel incorporation, membrane attachment, and the other processes involved in the assembly of the H_2-oxidizing apparatus and of obtaining a detailed knowledge of the regulatory mechanisms governing hydrogenase gene expression.

Acknowledgments

We are grateful to Mrs. E. Nielsen and Mrs. D. Matzkuhn for secretarial assistance, to T. Eitinger, C. G. Friedrich, and R. A. Siddiqui for a critical reading of the manuscript, and to many collegues and coworkers who provided us with reprints and unpublished data relevant to this review. The hydrogenase research in the authors' laboratory was funded by the Deutsche Forschungsgemeinschaft, Bundesministerium für Forschung und Technologie and Fonds der Chemischen Industrie.

Literature Cited

1. Adams, M. W. W. 1990. The metabolism of hydrogen by extremely thermophilic, sulfur-dependent bacteria. *FEMS Microbiol. Rev.* 75:219–38
2. Adams, M. W. W. 1990. The structure and mechanism of iron-hydrogenases. *Biochim. Biophys. Acta* 1020:115–45
3. Andersen, K., Tait, R. C., King, W. R. 1981. Plasmids required for the utilization of molecular hydrogen by *Alcaligenes eutrophus. Arch. Microbiol.* 129:384–90
4. Aragno, M. 1992. Thermophilic, aerobic hydrogen-oxidizing (Knallgas) bacteria. See Ref. 6a, pp. 3917–33
5. Aragno, M., Schlegel, H. G. 1992. The mesophilic hydrogen-oxidizing (Knallgas) bacteria. See Ref. 6a, pp. 344–84
6. Arp, D. J., McCollum, L. C., Seefeldt, L. C . 1985. Molecular and immunological comparison of membrane-bound, H_2-oxidizing hydrogenases of *Bradyrhizobium japonicum, Alcaligenes eutrophus, Alcaligenes latus,* and *Azotobacter vinelandii. J. Bacteriol.* 163: 15–20

6a. Balows, A., Trüper, H. G., Dworkin, M., Harder, W., Schleifer, K.-H., eds. 1992. *The Prokaryotes.* New York: Springer

7. Baron, S. F., Williams, D. S., May, H. O., Patel, P. S., Aldrich, H. C., Ferry, H. G. 1989. Immunogold localization of coenzyme F_{420}-reducing formate dehydrogenase and coenzyme F_{420}-reducing hydrogenase in *Methanobacterium formicicum. Arch. Microbiol.* 151:307–13
8. Bartha, R., Ordal, E. J. 1965. Nickel-dependent chemolithoautotrophic growth of two *Hydrogenomonas strains. J. Bacteriol.* 89:1015–19
9. Bonjour, F., Aragno, M. 1986. Growth of thermophilic, obligatory chemolithoautotrophic hydrogen-oxidizing bacteria related to *Hydrogenobacter* with thiosulfate and elemental sulfur as electron and energy source. *FEMS Microbiol. Lett.* 35:11–15
10. Boursier, P., Hanus, F. J., Papen, H., Becker, M. M., Russel, S. A., Evans, H. J. 1988. Selenium increases hydrogenase expression in autotrophically cultured *Bradyrhizobium japonicum* and is a constituent of the purified enzyme. *J. Bacteriol.* 170:5594–5600
11. Bowien, B., Friedrich, B., Friedrich, C. G. 1984. Involvement of megaplasmids in heterotrophic derepression of the carbon-dioxide assimilating enzyme system in *Alcaligenes* spp. *Arch. Microbiol.* 139:305–10
12. Bowien, B., Schlegel, H. G. 1981. Physiology and biochemistry of aerobic hydrogen-oxidizing bacteria. *Annu. Rev. Microbiol.* 35:405–52
13. Brewin, N. J., DeJong, T. M., Phillips, O. A., Johnston, A. W. B. 1980. Co-transfer of determinants for hydrogenase activity and nodulation ability in *Rhizobium leguminosarum. Nature* 288:77–79
14. Cammack, R., Fernandez, V. M., Schneider, K. 1988. Nickel in hydrogenases from sulfate-reducing, photosynthetic and hydrogen oxidizing bacteria. In *The Bioinorganic Chemistry of Nickel,* ed. J. R. Lancaster, Jr., pp. 167-90. Weinheim: VCH
15. Cangelosi, G. A., Wheelis, M. L. 1984. Regulation by molecular oxygen and organic substrates of hydrogenase synthesis in *Alcaligenes eutrophus. J. Bacteriol.* 159:138–44
16. Cantrell, M. A., Hickok, R. E., Evans, H. J. 1982. Identification and characterization of plasmids in hydrogen uptake positive and hydrogen uptake negative strains of *Rhizobium japonicum. Arch. Microbiol.* 131:102–6
17. Cauvin, B., Colbeau, A., Vignais, P. M. 1991. The hydrogenase structural operon in *Rhodobacter capsulatus* contains a third gene, *hupM,* necessary for the formation of a physiologically competent hydrogenase. *Mol. Microbiol.* 5:2519–27
18. Chen, J. C., Mortenson, L. E. 1992. Two open reading frames (ORFs) identified near the hydrogenase structural genes in *Azotobacter vinelandii,* the first ORF may encode for a polypeptide similar to rubredoxins. *Biochim. Biophys. Acta* 1131:122–24
19. Chen, J. C., Mortenson, L. E. 1992. Identification of six open reading frames from a region of the *Azotobacter vinelandii* genome likely involved in dihydrogen metabolism. *Biochim. Biophys. Acta* 1131:199–202
20. Colbeau, A., Kelley, B. C., Vignais, P. M. 1980. Hydrogenase activity in *Rhodopseudomonas capsulata:* relationship with nitrogenase activity. *J. Bacteriol.* 144:141–48
21. Colbeau, A., Magnin, J.-P., Cauvin, B., Champion, T., Vignais, P. M. 1990. Genetic and physical mapping of an hydrogenase gene cluster from

Rhodobacter capsulatus. *Mol. Gen. Genet*. 220:393–99

22. Colbeau, A., Richaud, P., Toussaint, B., Caballero, J., Elster, C., et al. 1993. Organization of the genes necessary for hydrogenase expression in *Rhodobacter capsulatus*. Sequence analysis. Identification of two *hyp* regulatory mutants. *Mol. Microbiol*. 8:15–29
23. Colbeau, A., Vignais, P. M. 1983. The membrane-bound hydrogenase of *Rhodopseudomonas capsulata* is inducible and contains nickel. *Biochim. Biophys. Acta* 748:128–38
24. Colbeau, A., Vignais, P. M. 1992. Use of *hupS::lacZ* gene fusion to study regulation of hydrogenase expression in *Rhodobacter capsulatus:* stimulation by H_2. *J. Bacteriol*. 174:4258–64
25. Conrad, R., Aragno, M., Seiler, W. 1983. The inability of hydrogen bacteria to utilize atmospheric hydrogen is due to threshold and affinity for hydrogen. *FEMS Microbiol. Lett*. 18:207–10
26. Conrad, R., Seiler, W. 1980. Contribution of hydrogen production by biological nitrogen fixation to the global hydrogen budget. *J. Geophys. Res*. 85:5493–98
27. Dernedde, J., Eitinger, M., Friedrich, B. 1993. Analysis of a pleiotropic gene region involved in formation of catalytically active hydrogenase in *Alcaligenes eutrophus* H16. *Arch. Microbiol*. 159:In press
28. Dixon, R. O. D. 1972. Hydrogenase in legume root nodule bacteroids: occurrence and properties. *Arch. Microbiol*. 85:193–201
29. Doyle, C. M., Arp, D. J. 1987. Regulation of H_2 oxidation activity and hydrogenase protein levels by H_2, O_2, and carbon substrates in *Alcaligenes latus*. *J. Bacteriol*. 169:4463–68
30. Doyle, C. M., Arp, D. J. 1988. Nickel affects expression of the nickel-containing hydrogenase of *Alcaligenes latus*. *J. Bacteriol*. 170:38891–96
31. Drobner, E., Huber, H., Rachel, R., Stetter, K. O. 1992. *Thiobacillus plumbophilus* spec. nov., a novel galena and hydrogen oxidizer. *Arch. Microbiol*. 157:213–17
32. Drobner, E., Huber, H., Stetter, K. O. 1990. *Thiobacillus ferrooxidans;* a facultative hydrogen oxidizer. *Appl. Environ. Microbiol*. 56:2922–23
33. Dross, F., Geisler, V., Lenger, R., Theis, F., Krafft, T. et al. 1992. The quinone-reactive Ni/Fe-hydrogenase of *Wolinella succinogenes*. *Eur. J. Biochem*. 206:93–102
34. Du, L., Stejskal, F., Tibelius, K. H. 1992. Characterization of two genes (*hupD* and *hupE*) required for hydrogenase activity in *Azotobacter chroococcum*. *FEMS Microbiol. Lett*. 96:93–102
35. Eberz, G., Eitinger, T., Friedrich, B. 1989. Genetic determinants of a nickel-specific transport system are part of the plasmid-encoded hydrogenase gene cluster in *Alcaligenes eutrophus*. *J. Bacteriol*. 171:1340–45
36. Eberz, G., Friedrich, B. 1991. Three *trans*-acting regulatory functions control hydrogenase synthesis in *Alcaligenes eutrophus*. *J. Bacteriol*. 173: 1845–54
37. Eberz, G., Hogrefe, C., Kortlüke, C., Kamienski, A., Friedrich, B. 1986. Molecular cloning of structural and regulatory hydrogenase (*hox*) genes of *Alcaligenes eutrophus* H16. *J. Bacteriol*. 168:636–41
38. Eidsness, M. K., Scott, R. A., Pickrill, B., DerVartanian, D. V., LeGall, J., et al. 1989. Evidence for selenocysteine coordination to the active site nickel in the (NiFeSe) hydrogenase from *Desulfovibrio baculatus*. *Proc. Natl. Acad. Sci. USA* 86:147–51
39. Eisbrenner, G., Evans, H. J. 1983. Aspects of hydrogen metabolism in nitrogen-fixing legumes and other plant-microbe associations. *Annu. Rev. Plant Physiol*. 34:105–36
40. Eitinger, T., Friedrich, B. 1991. Cloning, nucleotide sequence, and heterologous expression of a high-affinity nickel transport gene from *Alcaligenes eutrophus*. *J. Biol. Chem*. 266:3222–27
41. Evans, H. J., Harker, A. R., Papen, H., Russell, S. A., Hanus, F. J., Zuber, M. 1987. Physiology, biochemistry, and genetics of the uptake hydrogenase in rhizobia. *Annu. Rev. Microbiol*. 41:335–61
42. Fauque, G., Peck, H. D. Jr., Moura, J. J. G., Huynh, B. H., Berlier, Y., et al. 1988. The three classes of hydrogenases from sulfate-reducing bacteria of the genus *Desulfovibrio*. *FEMS Microbiol. Rev*. 54:299–344
43. Ferber, D. M., Maier, R. J. 1992. Incorporation of a bacterial membrane-bound hydrogenase into proteoliposomes. *Anal. Biochem*. 203:235–44
44. Ford, C. M., Garg, N., Garg, R. P., Tibelius, K. H., Yates, M. G., et al. 1990. The identification, characterization, sequencing and mutagenesis of the genes (*hupSL*) encoding the small and large subunits of the H_2-uptake

hydrogenase of *Azotobacter chroococcum*. *Mol. Microbiol.* 4:999–1008
45. Friedrich, B., Böcker, C., Eberz, G., Eitinger, T., Horstmann, K., et al. 1990. Genes for hydrogen oxidation and denitrification form two clusters on megaplasmid pHG1 of *Alcaligenes eutrophus*. In Pseudomonas: *Biotransformations, Pathogenesis, and Evolving Biotechnology*, ed. S. Silver, A. M. Chakrabarty, B. Iglewski, S. Kaplan, pp. 408–19. Washington DC: Am. Soc. Microbiol.
46. Friedrich, B., Friedrich, C. G. 1990. Hydrogenases in lithoautotrophic bacteria. In *Advances in Autotrophic Microbiology and One-Carbon-Metabolism*, ed. G. A. Codd, L. Dijkhuizen, F. R. Tabita, pp. 55–92. Dordrecht: Kluwer
47. Friedrich, B., Friedrich, C. G., Meyer, M., Schlegel, H. G. 1984. Expression of hydrogenase in *Alcaligenes* spp. is altered by interspecific plasmid exchange. *J. Bacteriol.* 158:331–33
48. Friedrich, B., Heine, E., Finck, A., Friedrich, C. G. 1981. Nickel requirement for active hydrogenase formation in *Alcaligenes eutrophus. J. Bacteriol.* 145:1144–49
49. Friedrich, B., Hogrefe, C., Schlegel, H. G. 1981. Naturally occurring genetic transfer of hydrogen-oxidizing ability between strains of *Alcaligenes eutrophus. J. Bacteriol.* 147:198–205
50. Friedrich, C. G. 1982. Derepression of hydrogenase during limitation of electron donors and derepression of ribulosebisphosphate carboxylase during carbon limitation of *Alcaligenes eutrophus. J. Bacteriol.* 149:203–10
51. Friedrich, C. G., Friedrich, B. 1983. Regulation of hydrogenase formation is temperature-sensitive and plasmid-encoded in *Alcaligenes eutrophus. J. Bacteriol.* 153:176–81
52. Friedrich, C. G., Friedrich, B., Bowien, B. 1981. Formation of enzymes of autotrophic metabolism during heterotrophic growth of *Alcaligenes eutrophus. J. Gen. Microbiol.* 122:69–78
53. Friedrich, C. G., Mitrenga, G. 1981. Oxidation of thiosulfate by *Paracoccus denitrificans* and other hydrogen bacteria. *FEMS Microbiol. Lett.* 10:209–12
54. Friedrich, C. G., Schneider, K., Friedrich, B. 1982. Nickel in the catalytically active hydrogenase of *Alcaligenes eutrophus. J. Bacteriol.* 152: 42–48
55. Friedrich, C. G., Suetin, S., Lohmeyer, M. 1984. Nickel and iron incorporation into soluble hydrogenase of *Alcaligenes eutrophus. Arch. Microbiol.* 140:206–11
56. Fu, C., Maier, R. J. 1991. Competitive inhibition of an energy-dependent nickel transport system by divalent cations in *Bradyrhizobium japonicum* JH. *Appl. Environ. Microbiol.* 57:3511–16
57. Fu, C., Maier, R. J. 1991. Identification of a locus within the hydrogenase gene cluster involved in intracellular nickel metabolism in *Bradyrhizobium japonicum. Appl. Environ. Microbiol.* 57:3502–10
58. Fu, C, Maier, R. J. 1992. Nickel-dependent reconstitution of hydrogenase apoprotein in *Bradyrhizobium japonicum* Hup^c mutants and direct evidence for a nickel metabolism locus involved in nickel incorporation into the enzyme. *Arch. Microbiol.* 157:493–98
59. Gadkari, D., Schricker, K., Acker, G., Kroppenstedt, R. M., Meyer, O. 1990. *Streptomyces thermoautotrophicus* sp. nov., a thermophilic CO- and H_2-oxidizing obligate chemolithoautotroph. *Appl. Environ. Microbiol.* 56: 3727–34
60. Gerberding, H., Mayer, F. 1989. Localization of the membrane-bound hydrogenase in *Alcaligenes eutrophus* by electron microscopic immunocytochemistry. *FEMS Microbiol. Lett.* 50:265–70
61. Gerstenberg, C., Friedrich, B., Schlegel, H. G. 1982. Physical evidence for plasmids in autotrophic, especially hydrogen-oxidizing bacteria. *Arch. Microbiol.* 133:90–96
62. Gollin, D. J., Mortenson, L. E., Robson, R. L. 1992. Carboxyl-terminal processing may be essential for production of active NiFe hydrogenase in *Azotobacter vinelandii. FEBS Lett.* 309: 371–75
63. Graf, E. G., Thauer, R. K. 1981. Hydrogenase from *Methanobacterium thermoautotrophicum:* a nickel-containing enzyme. *FEBS Lett.* 136:165–69
64. Hanus, F., Maier, R. J., Evans, H. J. 1979. Autotrophic growth of H_2-uptake positive strains of *Rhizobium japonicum* in an atmosphere supplied with hydrogen gas. *Proc. Natl. Acad. Sci. USA* 76:1788–92
65. Harker, A. R., Lambert, G. R., Hanus, F. J., Evans, H. J. 1985. Further evidence that two unique subunits are essential for expression of hydrogenase activity in *Rhizobium japonicum. J. Bacteriol.* 164:187–91
66. Hidalgo, E., Leyva, A., Ruiz-Argüeso, T. 1990. Nucleotide sequence of the

hydrogenase structural genes from *Rhizobium leguminosarum*. *Plant Mol. Biol.* 15:367–70

67. Hidalgo, E., Palacios, J. M., Murillo, J., Ruiz-Argüeso, T. 1992. Nucleotide sequence and characterization of four additional genes of the hydrogenase structural operon from *Rhizobium leguminosarum* bv. *viciae*. *J. Bacteriol.* 174:413–39
68. Higuchi, Y., Yasuoka, N., Kakudo, M., Katsube, Y., Yagi, T., Inokuchi, H. 1987. Single crystals of hydrogenase from *Desulfovibrio vulgaris*. *J. Biol. Chem.* 262:2823–25
69. Hogrefe, C., Römermann, D., Friedrich, B. 1984. *Alcaligenes eutrophus* hydrogenase genes (*hox*). *J. Bacteriol.* 158:43–48
70. Hornhardt, S., Schneider, K., Friedrich, B., Vogt, B., Schlegel, H. G. 1990. Identification of NAD-linked hydrogenase protein species in mutant and nickel-deficient wild-type cells of *Alcaligenes eutrophus*. *Eur. J. Biochem.* 189:529–37
71. Hornhardt, S., Schneider, K., Schlegel, H. G. 1986. Characterization of a native subunit of the NAD-linked hydrogenase isolated from a mutant of *Alcaligenes eutrophus* H16. *Biochimie* 68:15–24
72. Huber, R., Wilharm, T., Huber, D., Trincone, A., Burggraf, S., et al. 1992. *Aquifex pyrophilus* gen. nov. sp. nov., represents a novel group of marine hyperthermophilic hydrogen-oxidizing bacteria. *Syst. Appl. Microbiol.* 15:340–51
73. Hyndman, L. A., Bums, R. H., Wilson, P. W. 1953. Properties of hydrogenase from *Azotobacter vinelandii*. *J. Bacteriol.* 65:522–31
74. Jacobi, A., Rossmann, R., Böck, A. 1992. The *hyp* operon gene products are required for the maturation of catalytically active hydrogenase isoenzymes in *E. coli*. *Arch. Microbiol.* 158:444–51
75. Johannssen, W., Gerberding, H., Rohde, M., Zaborosch, C., Mayer, F. 1991. Structural aspects of soluble NAD-dependent hydrogenase isolated from *Alcaligenes eutrophus* H16 and from *Nocardia opaca* 1b. *Arch. Microbiol.* 155:303–8
76. Jones, B. D., Mobley, H. L. T. 1989. *Proteus mirabilis* urease: nucleotide sequence determination and comparison with jack bean urease. *J. Bacteriol.* 171:6414–22
77. Kalkus, J., Reh, M., Schlegel, H. G. 1990. Hydrogen autotrophy of *Nocardia opaca* strains is encoded by linear megaplasmids. *J. Gen. Microbiol.* 136: 1145–51
78. Kärst, U., Friedrich, C. G. 1987. Identification of new peptides synthesized under the hydrogenase control system of *Alcaligenes eutrophus*. *Arch. Microbiol.* 147:346–53
79. Kärst, U., Suetin, S., Friedrich, C. G. 1987. Purification and properties of a protein linked to the soluble hydrogenase of hydrogen-oxidizing bacteria. *J. Bacteriol.* 169:2079–85
80. Kaur, P, Ross, K., Siddiqui, A., Schlegel, H. G. 1990. Nickel resistance of *Alcaligenes denitrificans* strain 4a-2 is chromosomally coded. *Arch. Microbiol.* 154:133–38
81. Kawasumi, T., Igarashi, Y., Kodama, T., Minoda, Y. 1984. *Hydrogenobacter thermophilus* gen. nov., sp. nov., an extremely thermophilic aerobic, hydrogen-oxidizing bacterium. *Int. J. Syst. Bacteriol.* 34:5–10
82. Kelly, D. P., Harrison, A. P. 1989. Genus *Thiobacillus*. In *Bergey's Manual of Systematic Bacteriology*, ed. J. T. Staley, M. P. Bryant, N. Pfennig, J. G. Holt, 3:1842–58. Baltimore: Williams & Wilkins
83. Kennedy, C., Toukdarian, A. 1987. Genetics of azotobacters: applications to nitrogen fixation and related aspects of metabolism. *Annu. Rev. Microbiol.* 41:227–58
84. Kim, H., Maier, R. J. 1990. Transcriptional regulation of hydrogenase synthesis by nickel in *Bradyrhizobium japonicum*. *J. Biol. Chem.* 265:18729–32
85. Kim, H., Yu, C., Maier, R. J. 1991. Common *cis*-acting region responsible for transcriptional regulation of *Bradyrhizobium japonicum* hydrogenase by nickel, oxygen, and hydrogen. *J. Bacteriol.* 173:3993–99
86. Klintworth, R., Husemann, M., Salnikow, J., Bowien, B. 1985. Chromosomal and plasmid locations for phospho-ribulose kinase genes in *Alcaligenes eutrophus*. *J. Bacteriol.* 164: 954–56
87. Knüttel, K., Schneider, K., Schlegel, H. G., Müller, A. 1989. The membrane-bound hydrogenase of *Paracoccus denitrificans*. Purification and molecular characterization. *Eur. J. Biochem.* 179:101–8
88. Kömen, R., Schmidt, K., Friedrich, B. 1992. Hydrogenase mutants of *Alcaligenes eutrophus* H16 show alterations in the electron transport system. *FEMS Microbiol Lett.* 96:173–78

89. Kortlüke, C., Friedrich, B. 1992. Maturation of membrane-bound hydrogenase of *Alcaligenes eutrophus* H16. *J. Bacteriol.* 174:6290–93
90. Kortlüke, C., Hogrefe, C., Eberz, G., Pühler, A., Friedrich, B. 1987. Genes of lithoautotrophic metabolism are clustered on megaplasmid pHG1 in *Alcaligenes eutrophus. Mol. Gen. Genet.* 210:122–28
91. Kortlüke, C., Horstmann, K., Schwartz, E., Rohde, M., Binsack, R., Friedrich, B. 1992. A gene complex coding for the membrane-bound hydrogenase of *Alcaligenes eutrophus* H16. *J. Bacteriol.* 174:6277–89
92. Kovacs, K. L., Seefeldt, L. C., Tigyi, G., Doyle, C. M., Mortenson, L. E., Arp, D. J. 1989. Immunological relationship among hydrogenases. *J. Bacteriol.* 171:430–35
93. Kraut, M., Meyer, O. 1988. Plasmids in carboxydotrophic bacteria: physical and restriction analysis. *Arch. Microbiol.* 149:540–46
94. Lancaster, J. R. Jr. 1980. Soluble and membrane-bound paramagnetic centers in *Methanobacterium bryantii. FEBS Lett.* 115:285–88
95. Leclerc, M., Colbeau, A., Cauvin, B., Vignais, P. M. 1988. Cloning and sequencing of the genes encoding the large and the small subunits of the H_2 uptake hydrogenase (*hup*) of *Rhodobacter capsulatus. Mol. Gen. Genet.* 214: 97–107
96. Lohmeyer, M., Friedrich, C. G. 1987. Nickel transport in *Alcaligenes eutrophus. Arch. Microbiol.* 149:130–35
97. Lorenz, B., Schneider, K., Kratzin, H., Schlegel, H. G. 1989. Immunological comparison of subunits isolated from various hydrogenases of aerobic hydrogen bacteria. *Biochim. Biophys. Acta* 995:1–9
98. Lünsdorf, H., Niedrig, M., Fiebig, K. 1991. Immunocytochemical localization of the coenzyme F_{420}-reducing hydrogenase in *Methanosarcina barkeri* Fusaro. *J. Bacteriol.* 173:978–84
99. Lutz, S., Jacobi, A., Schlensog, V., Böhm, R., Sawers, G., Böck, A. 1991. Molecular characterization of an operon (*hyp*) necessary for the activity of the three hydrogenase isoenzymes in *Escherichia coli. Mol. Microbiol.* 5:123–35
100. Madigan, M. T., Gest, H. 1979. Growth of the photosynthetic bacterium *Rhodopseudomonas capsulata* chemoautotrophically in darkness with H_2 as the energy source. *J. Bacteriol.* 137: 524–30
101. Maier, R. J. 1986. Biochemistry, regulation, and genetics of hydrogen oxidation in *Rhizobium. CRC Crit. Rev. Biotechnol.* 3:17–38
102. Maier, R. J., Campbell, N. E. R., Hanus, F. J., Simpson, F. B., Russell, S. A., Evans, H. J. 1978. Expression of hydrogenase in free-living *Rhizobium japonicum. Proc. Natl. Acad. Sci. USA* 75:3258–62
103. Maier, R. J., Hanus, F. J., Evans, H. J. 1979. Regulation of hydrogenase in *Rhizobium japonicum. J. Bacteriol.* 137: 824–29
104. Maier, R. J., Merberg, D. M. 1982. *Rhizobium japonicum* mutants that are hypersensitive to repression of H_2 uptake by oxygen. *J. Bacteriol.* 150:161–67
105. Maier, R. J., Prosser, J. 1988. Hydrogen-mediated mannose uptake of *Azotobacter vinelandii. J. Bacteriol.* 170:1986–89
106. Maier, T., Jacobi, A., Sauter, M., Böck, A. 1993. The product of the *hypB* gene which is required for nickel incorporation into hydrogenases is a novel guanine nucleotide binding protein. *J. Bacteriol.* 175:630–35
107. Menon, A. L., Mortenson, L. E., Robson, R. L. 1992. Nucleotide sequences and genetic analysis of hydrogen oxidation (*hox*) genes in *Azotobacter vinelandii. J. Bacteriol.* 174:4549–57
108. Menon, A. L., Stults, L. W., Robson, R. L., Mortenson, L. E. 1990. Cloning, sequencing and characterization of the (NiFe) hydrogenase-encoding structural genes (*hoxK* and *hoxG*) from *Azotobacter vinelandii. Gene* 96:67–74
109. Menon, N. K., Robbins, J., Peck, H. D. Jr., Chatelus, C. Y., Choi, E.-S., Przybyla, A. E. 1990. Cloning and sequencing of a putative *Escherichia coli* (NiFe) hydrogenase-1 operon containing six open reading frames. *J. Bacteriol.* 172:1969–77
110. Menon, N. K., Robbins, J. Wendt, J. C., Shanmugam, K. T., Przybyla, A. E. 1991. Mutational analysis and characterization of the *Escherichia coli hya* operon, which encodes (NiFe) hydrogenase 1. *J. Bacteriol.* 173:4851–61
111. Merberg, D., O'Hara, E. B., Maier, R. J. 1983. Regulation of hydrogenase in *Rhizobium japonicum:* analysis of mutants altered in regulation by carbon substrates and oxygen. *J. Bacteriol.* 156:1236–42
112. Meyer, O. 1989. Aerobic carbon monoxide oxidizing bacteria. See Ref. 145a, pp. 331–50

113. Meyer, O., Frunzke, K., Mörsdorf, G. 1993. Biochemistry of the aerobic utilization of carbon monoxide. *FEMS Microbiol. Rev.* In press
114. Mulrooney, S. B., Hausinger, R. P. 1990. Sequence of the *Klebsiella aerogenes* urease genes and evidence for accessory proteins facilitating nickel incorporation. *J. Bacteriol.* 172:5837–43
115. Muth, E. 1988. Localization of the F_{420}-reducing hydrogenase in *Methanococcus voltae* cells by immuno-gold technique. *Arch. Microbiol.* 150:205–7
116. Nivière, V., Hatchikian, C., Cambilla, C., Ferry, M. 1987. Crystallization, preliminary x-ray study and crystal activity of the hydrogenase from *Desulfovibrio gigas*. *J. Mol. Biol.* 195: 969–71
117. Nivière, V., Wong, S. L., Voordouw, G. 1992. Site-directed mutagenesis of the hydrogenase signal peptide consensus box prevents export of a β-lactamase fusion protein. *J. Gen. Microbiol.* 138:2173–83
118. Nokhal, T.-H., Schlegel, H. G. 1980. The regulation of hydrogenase formation as a differentiating character of strains of *Paracoccus denitrificans*. *Antonie van Leeuvenhoek J. Microbiol. Serol.* 46:143–55
119. Novak, P. D., Maier, R. J. 1987. Inhibition of hydrogenase synthesis by DNA gyrase inhibitors in *Bradyrhizobium japonicum*. *J. Bacteriol.* 169:2708–12
120. Oelmüller, U., Schlegel, H. G., Friedrich, C. G. 1990. Differential stability of mRNA species of *Alcaligenes eutrophus* soluble and particulate hydrogenases. *J. Bacteriol.* 172:7057–64
121. Ohi, K., Takada, V., Komemushi, S., Okazaki, M., Miura, Y. 1979. A new species of hydrogen-utilizing bacterium. *J. Gen. Appl. Microbiol.* 25:53–58
122. Partridge, C. D. P., Yates, M. G. 1982. Effects of chelating agents on hydrogenase in *Azotobacter chroococcum*. *Biochem. J.* 204:339–44
123. Patel, S. D., Aebersold, R., Attardi, G. 1991. cDNA-derived amino acid sequence of the NADH-binding 51-kDa subunit of the bovine respiratory NADH dehydrogenase reveals striking similarities to a bacterial NAD^+-reducing hydrogenase. *Proc. Natl. Acad. Sci. USA* 88:4225–29
124. Pilkington, S. J., Skehel, J. M., Gennis, R. B., Walker, J. E. 1991. Relationship between mitochondrial NADH-ubiquinone reductase and a bacterial NAD reducing hydrogenase. *Biochemistry* 30:2166–75
125. Podzuweit, H. G., Schneider, K., Knüttel, K. 1987. Comparison of the membrane-bound hydrogenases from *Alcaligenes eutrophus* H16 and *Alcaligenes eutrophus* type strain. *Biochim. Biophys. Acta* 905:435–46
126. Podzuweit, H. G., Schneider, K., Schlegel, H. G. 1983. Autotrophic growth and hydrogenase activity of *Pseudomonas saccharophila*. *FEMS Microbiol. Lett.* 19:169–73
127. Pootjes, C. F. 1977. Evidence for plasmid coding of the ability to utilize hydrogen gas by *Pseudomonas facilis*. *Biochem. Biophys. Res. Commun.* 76: 1002–6
128. Preis, D., Weidner, U., Conzen, C., Azevedo, J. E., Nehls, U., et al. 1991. Primary structures of two subunits of NADH: ubiquinone reductase from *Neurospora crassa* concerned with NADH-oxidation. Relationship to a soluble NAD-reducing hydrogenase of *Alcaligenes eutrophus*. *Biochim. Biophys. Acta* 1090:133–38
129. Prosser, J., Graham, L., Maier, R. J. 1988. Hydrogen-mediated enhancement of hydrogenase expression in *Azotobacter vinelandii*. *J. Bacteriol.* 170: 1990–93
130. Przybyla, A. E., Robbins, J., Menon, N., Peck, H. D. Jr. 1992. Structure/function relationships among the nickel-containing hydrogenases. *FEMS Microbiol. Rev.* 88:109–36
131. Reeve, J. N., Beckler, G. S. 1990. Conservation of primary structure in prokaryotic hydrogenases. *FEMS Microbiol. Rev.* 87:419–24
132. Reh, M., Schlegel, H. G. 1975. Chemolithoautotrophie als eine übertragbare, autonome Eigenschaft von *Nocardia opaca* 1b. *Nachr. Akad. Wiss. Göttingen* 12:1–10
133. Reh, M., Schlegel, H. G. 1981. Hydrogen autotrophy as a transferable genetic character of *Nocardia opaca* 1b. *J. Gen. Microbiol.* 126:327–36
134. Rey, L., Hidalgo, E., Palacios, J., Ruiz-Argüeso, T. 1992. Nucleotide sequence and organization of an H_2-uptake gene cluster from *Rhizobium leguminosarum* bv. *viciae* containing a rubredoxin-like gene and four additional open reading frames. *J. Mol. Biol.* 228:998–1002
135. Rey, L., Murillo, J., Hernando, Y., Hidalgo, E., Cabrera, E., et al. 1993. Gene organization and expression analysis of a microaerobically induced op-

eron required for hydrogenase synthesis in *Rhizobium leguminosarum* bv. *viciae*. *Mol. Microbiol.* In press

136. Richaud, P., Colbeau, A., Toussaint, B., Vignais, P. M. 1991. Identification and sequence analysis of the *hupR1* gene, which encodes a response regulator of the NtrC family required for hydrogenase expression in *Rhodobacter capsulatus*. *J. Bacteriol.* 173:5928–32
137. Richaud, P., Vignais, P. M., Colbeau, A., Uffen, L. R., Cauvin, B. 1990. Molecular biology studies of the uptake hydrogenases of *Rhodobacter capsulatus* and *Rhodocyclus gelatinosus*. *FEMS Microbiol. Rev.* 87:413–18
138. Römermann, D., Lohmeyer, M., Friedrich, C. G., Friedrich, B. 1988. Pleiotropic mutants from *Alcaligenes eutrophus* defective in the metabolism of hydrogen nitrate, urea, and fumarate. *Arch. Microbiol.* 149:471–75
139. Römermann, D., Warrelmann, J., Bender, R. A., Friedrich, B. 1989. An *rpoN* like gene of *Alcaligenes eutrophus* and *Pseudomonas facilis* controls the expression of diverse metabolic pathways, including hydrogen oxidation. *J. Bacteriol.* 171:1093–99
140. Rohde, M., Johannssen, W., Mayer, F. 1986. Immunocytochemical localization of the soluble NAD-dependent hydrogenase in cells of *Alcaligenes eutrophus*. *FEMS Microbiol. Lett.* 36: 83–86
141. Sayavedra-Soto, L. A., Arp, D. J. 1992. The *hoxZ* gene of the *Azotobacter vinelandii* hydrogenase operon is required for activation of hydrogenase. *J. Bacteriol.* 174:5295–5301
142. Sayavedra-Soto, L. A., Powell, G. K., Evans, H. J., Morris, R. O. 1988. Nucleotide sequence of the genetic loci encoding subunits of *Bradyrhizobium japonicum* uptake hydrogenase. *Proc. Natl. Acad. Sci. USA* 85:8395–99
143. Schink, B., Schlegel, H. G. 1979. The membrane-bound hydrogenase of *Alcaligenes eutrophus*. I. Solubilization, purification, and biochemical properties. *Biochim. Biophys. Acta* 567: 315–24
144. Schink, B., Schlegel, H. G. 1980. The membrane-bound hydrogenase of *Alcaligenes eutrophus*. II. Localization and immunological comparison with other hydrogenase systems. *Antonie van Leeuvenhoek J. Microbiol. Serol.* 46:1–14
145. Schlegel, H. G. 1989. Aerobic hydrogen-oxidizing (Knallgas) bacteria. See Ref. 145a, pp. 305–29

145a. Schlegel, H. G., Bowien, B., eds. 1989. *Autotrophic Bacteria. Brock/Springer Series of Contemporary Biosciences*. Madison: Science Tech.; Berlin: Springer

146. Schlegel, H. G., Schneider, K. 1985. Microbial metabolism of hydrogen. In *Comprehensive Biotechnology*, ed. M. Moo-Young, pp. 439–57. Oxford: Pergamon
147. Schneider, K., Cammack, R., Schlegel, H. G. 1984. Content and localization of FMN, Fe-S clusters and nickel in the NAD-linked hydrogenase of *Nocardia opaca* 1b. *Eur. J. Biochem.* 142:75–84
148. Schneider, K., Schlegel, H. G. 1976. Purification and properties of the soluble hydrogenase from *Alcaligenes eutrophus* H16. *Biochim. Biophys. Acta* 452:66–80
149. Schneider, K., Schlegel, H. G., Jochim, K. 1984. Effect of nickel on activity and subunit composition of purified hydrogenase from *Nocardia opaca* 1b. *Eur. J. Biochem.* 138:533–41
150. Seefeldt, L. C., Arp, D. J. 1986. Purification to homogeneity of *Azotobacter vinelandii* hydrogenase: a nickel and iron containing αβ dimer. *Biochimie* 68:25–34
151. Sherman, M. B., Orlova, E. V., Smirnova, E. A., Hovmöller, S., Zorin, N. A. 1991. Three-dimensional structure of the nickel-containing hydrogenase from *Thiocapsa roseopersicina*. *J. Bacteriol.* 173:2576–80
152. Stephenson, M., Stickland, L. H. 1931. XXVII Hydrogenase: a bacterial enzyme activating molecular hydrogen I, the properties of the enzyme. *Biochem. J.* 25:205–14
153. Stock, J. B., Ninfa, A. J., Stock, A. M. 1989. Protein phosphorylation and regulation of adaptive responses in bacteria. *Microbiol. Rev.* 53:450–90
154. Stoker, K., Reijnders, W. N. M., Oltmann, L. F., Stouthamer, A. H. 1989. Initial cloning and sequencing of *hydHG*, an operon homologous to *ntrBC* and regulating the labile hydrogenase activity in *Escherichia coli* K-12. *J. Bacteriol.* 171:4448–56
155. Stults, L. W., Mallick, S., Maier, R. J. 1987. Nickel uptake in *Bradyrhizobium japonicum*. *J. Bacteriol.* 169:1398–1402
156. Takakuwa, S. 1987. Nickel uptake in *Rhodopseudomonas capsulata*. *Arch. Microbiol.* 149:57–61
157. Tibelius, K. H., Du, L., Tito, D., Stejskal, F. 1993. The *Azotobacter chroococcum* hydrogenase gene cluster: sequences and genetic analysis of four

accessory genes, *hupA, hupB, hupY* and *hupC*. *Gene*. In press

158. Toussaint, B., Bosc, C., Richaud, P., Colbeau, A., Vignais, P. M. 1991. A mutation in a *Rhodobacter capsulatus* gene encoding an integration host factor-like protein impairs in vivo hydrogenase expression. *Proc. Natl. Acad. Sci. USA* 88:10749–53
159. Tran-Betcke, A., Warnecke, U., Böcker, C., Zaborosch, C., Friedrich, B. 1990. Cloning and nucleotide sequence of the genes for the subunits of the NAD-reducing hydrogenase of *Alcaligenes eutrophus* H16. *J. Bacteriol.* 172:2920–29
160. Uffen, R. L., Colbeau, A., Richaud, P., Vignais, P. M. 1990. Cloning and sequencing the genes encoding uptake-hydrogenase subunits of *Rhodocyclus gelatinosus*. *Mol. Gen. Genet.* 221:49–58
161. Umeda, F., Min, H., Urushira, M., Okazaki, M., Miura, Y. 1986. Conjugal transfer of hydrogen-oxidizing ability of *Alcaligenes hydrogenophilus* to *Pseudomonas oxalaticus*. *Biochem. Biophys. Res. Commun.* 137:108–13
162. Vignais, P. M., Colbeau, A., Willison, J. C., Jouanneau, Y. 1985. Hydrogenase, nitrogenase, and hydrogen metabolism in the photosynthetic bacteria. *Adv. Microb. Physiol.* 26: 155–234
163. Voordouw, G. 1992. Evolution of hydrogenase genes. *Adv. Inorg. Chem.* 38:397–422
164. Walker, C. C., Yates, M. G. 1978. The hydrogen cycle in the nitrogen-fixing *Azotobacter chroococcum*. *Biochimie* 60:225–31
165. Walsh, C. T., Orme-Johnson, W. H. 1987. Nickel enzymes. *Biochemistry* 26:4901–5
166. Warrelmann, J., Eitinger, M., Schwartz, E., Römermann, D., Friedrich, B. 1992. Nucleotide sequence of the *rpoN* (*hno*) gene region of *Alcaligenes eutrophus:* evidence for a conserved gene cluster. *Arch. Microbiol.* 158:107–14
167. Warrelmann, J., Friedrich, B. 1986. Mutants of *Pseudomonas facilis* defective in lithoautotrophy. *J. Gen. Microbiol.* 132:91–96
168. Warrelmann, J., Friedrich, B. 1989. Genetic transfer of lithoautotrophy mediated by a plasmid cointegrate from *Pseudomonas facilis*. *Arch. Microbiol.* 151:359–64
169. Waugh, R., Boxer, D. H. 1986. Pleiotropic hydrogenase mutants of *Escherichia coli* K-12: growth in the presence of nickel can restore hydrogenase activity. *Biochimie* 68:157–66
170. Xu, H.-W., Wall, J. D. 1991. Clustering of genes necessary for hydrogen oxidation in *Rhodobacter capsulatus*. *J. Bacteriol.* 173:2401–5
171. Yamazaki, S. 1982. A selenium-containing hydrogenase from *Methanococcus vannielii*. Identification of the selenium moiety as a selenocysteine residue. *J. Biol. Chem.* 257:7926–29
172. Yates, M. G., Campbell, F. O. 1989. The effect of nutrient limitation on the competition between an H_2-uptake hydrogenase positive (Hup^+) recombinant strain of *Azotobacter chroococcum* and the Hup-mutant parent in mixed populations. *J. Gen. Microbiol.* 135: 221–26
173. Zaborosch, C., Köster, M., Schneider, K., Bill, E., Schlegel, H. G., Trautwein, A. X. 1991. Electron paramagnetic resonance and Mössbauer spectroscopic studies on the *Nocardia opaca* 1b NAD-linked hydrogenase and its subunit dimers. *Int. Conf. Molec. Biol. of Hydrogenases, 3rd, Troía, Portugal,* pp. 133–34 (Abstr.)
174. Zaborosch, C., Schneider, K., Schlegel, H. G., Kratzin, H. 1989. Comparison of the NH_2-terminal amino acid sequences of the four non-identical subunits of the NAD-linked hydrogenases from *Nocardia opaca* 1b and *Alcaligenes eutrophus* H16. *Eur. J. Biochem.* 181:175–80
175. Zirngibl, C., van Dongen, W., Schwörer, B., von Bünau, R., Richter, M., et al. 1992. H_2-forming methylenetetrahydro-methanopterin dehydrogenase, a novel type of hydrogenase. *J. Biochem.* 208:511–20

Annu. Rev. Microbiol. 1993. 47:385–411

THE SURFACE *TRANS*-SIALIDASE FAMILY OF *TRYPANOSOMA CRUZI*

George A. M. Cross and Garry B. Takle[1]

Laboratory of Molecular Parasitology, The Rockefeller University, 1230 York Avenue, New York, NY 10021

KEY WORDS: lectin, glycosyltransferase, immunity, invasion, Chagas' disease, sialyltransferase

---

CONTENTS

ABSTRACT

Trypanosomes cannot synthesize sialic acids. Infectious stages of the life cycle of the human pathogen *Trypanosoma cruzi* express a cell-surface glycolipid-anchored *trans*-sialidase, which can transfer sialic acid between glyco-

[1]Present address: Innovir Laboratories, Inc., 510 East 73rd Street, New York, NY 10021.

0066-4227/93/1001-0385$02.00

conjugates. Sialic acid is transferred from host cell-surface and serum sialylglycoproteins to trypanosome cell-surface glycoconjugates. The transfer reaction is specific for donors with terminal α-2,3-linked sialic acid, and terminal β-1,4-linked galactose is the preferred acceptor. In the absence of an acceptor, the enzyme acts as a hydrolase, but cleavage is less efficient than transfer. *Trans*-sialidase activity is attributable to a few members of a large family of *T. cruzi* surface glycoproteins, many of which are simultaneously expressed. The functions of the *trans*-sialidase surface glycoprotein family are unknown but may be important for adhesion, invasion, virulence, or pathogenicity. A *trans*-sialidase is also expressed in the procyclic forms of *Trypanosoma brucei*.

INTRODUCTION

Trypanosoma cruzi is the protozoan hemoflagellate responsible for Chagas' disease, which is largely restricted to the indigent population of Central and South America, where it remains an ever-present threat of serious disease. The parasite, unlike its African relative *Trypanosoma brucei*, does not undergo antigenic variation via a variant surface glycoprotein (VSG) multigene family (20). *T. cruzi* escapes from the humoral immune response by invading and multiplying within host cells. Infection is characterized by an acute phase, with high parasitemia, followed by an asymptomatic phase, which may last for more than 10 years, leading to chronic disease, with characteristic megasyndromes of the heart and digestive tract (reviewed in 8, 44).

T. cruzi has a complex life cycle (Figure 1), with several biochemically and morphologically distinct stages in the insect and mammalian hosts. Infection of the mammalian host occurs through feces contamination of a wound or mucous membrane by blood-sucking reduviid bugs, such as *Rhodnius* and *Triatoma* species, which deposit the infective flagellated metacyclic trypomastigote form. Metacyclic trypomastigotes cannot divide and must enter host tissue cells to differentiate and multiply. *T. cruzi* is a promiscuous parasite—it can invade a wide range of cells in vitro. After invasion, the trypomastigotes escape from the parasitophorous vacuole and differentiate into nonflagellated amastigotes, which multiply extensively in the host-cell cytoplasm. Subsequently, they redifferentiate into flagellated trypomastigotes, which do not divide but are released and disseminate the infection until the immune system is able to restrict it. After the insect vector has ingested blood infected with trypomastigotes, these forms differentiate into dividing epimastigotes, which are not infective to mammals. In the later stages of the midgut infection, epimastigotes differentiate once more into infective metacyclic trypomastigotes. This transformation can be mimicked

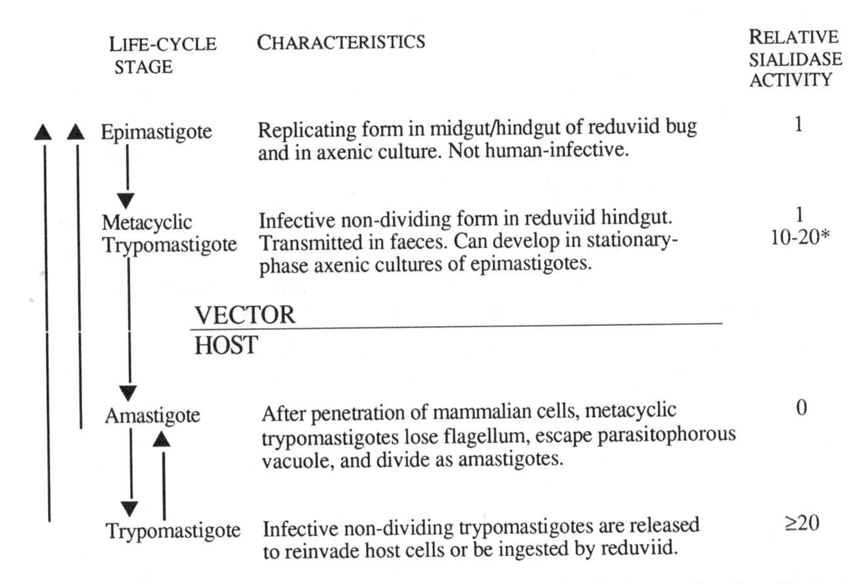

Figure 1 The life-cycle stages of *Trypanosoma cruzi*. The asterisk denotes data from Harth et al (35). Other values for relative sialidase activity are from Pereira (61).

in axenic culture (13). Blood transfusions also provide a significant route of infection in endemic and nonendemic regions (97).

Chagas' disease sensu stricto is associated exclusively with degeneration of tissues that are innervated by the autonomic nervous system, including a reduction in the number of ganglion cells, and the disease may have a strong autoimmune component. The source and significance of autoimmune responses in Chagas' disease are the subject of substantial research and debate (22, 67, 102).

The surface of *T. cruzi* has undergone intense scrutiny for glycoproteins and glycolipids that may be involved in parasite differentiation, nutrition, host cell attachment, and penetration. Recent work has identified a sialidase/*trans*-sialidase activity, which may modulate the adhesion and penetration of host cells by controlled sialylation of parasite surface glycoconjugates and/or modulation of sialic acid linkages on host cell glycoconjugates. Sialidase activity may also be involved in the exit of intracellular parasites from parasitophorous vacuoles into the cytoplasm, where they are free to divide (34). The purpose of this review is to describe background work on the *T. cruzi trans*-sialidase and its biochemical properties and structure, and to discuss possible functions of the large multigene family that encodes an array of *trans*-sialidase-related surface glycoproteins, with molecular weights ranging from approximately 80 to > 220.

SIALIDASE ACTIVITIES IN *TRYPANOSOMA CRUZI*

Sialidase activity was first recognized in *T. cruzi* in a sensitive assay in which Pereira (61) detected galactose residues, exposed after incubation of erythrocytes with live trypanosomes or lysates, by their reaction with peanut agglutinin. Subsequent studies showed that the activity was developmentally regulated and probably attributable to a polymorphic family of molecules, ranging in M_r from 160,000 to >220,000 (73, 77). Sialidase activity was absent from amastigotes, but was present in the infective-bloodstream and tissue culture–derived trypomastigotes (Figure 1). The activity in cultures of the noninfective epimastigote stage was 7- to 15-fold lower than that of tissue-culture trypomastigotes (35, 61). Recent studies suggest that epimastigotes contain an enzyme with distinct molecular characteristics (see below). There is disagreement (35, 61) about the level of sialidase activity in epimastigote-derived metacyclic trypomastigotes, which may depend on parasite strain or culture conditions.

Higher levels of sialidase activity in the infectious stages of the life cycle implicated the enzyme in processes specific to the infective forms, such as host-cell penetration or tissue pathogenesis. Although a causal relationship has not been demonstrated, certain myotropic strains of *T. cruzi* have increased sialidase activity (63). The ability of solubilized *T. cruzi* sialidase to release free sialic acid from rat myocardial and human vascular epithelial cells also suggested a possible role in host-parasite interactions (63).

Harth et al (35) purified a 60-kDa sialidase activity to apparent homogeneity. The source of their material, metacyclic trypomastigotes derived from an aged epimastigote culture subjected to separation by metrizamide gradients and anion-exchange chromatography, was different from the tissue-culture, cell-derived trypomastigotes used by Cavallesco & Pereira (11), and this variation may have contributed to differences in the findings of the two groups. Cavallesco & Pereira (11) did not purify the *T. cruzi* sialidase but cut a band showing the sialidase activity from a nondenaturing gel of enzyme released from tissue culture–derived trypomastigotes, and used this to immunize a rabbit. The resulting polyclonal antiserum showed antisialidase activity and identified a set of proteins ranging from 79 to 200 kDa. The sialidase itself could not be positively identified because the antiserum was insufficiently specific, but activity was restricted to the larger components. Monoclonal antibodies that precipitated sialidase activity from cell extracts or culture supernatants identified multiple bands in the range 121–220 kDa in tissue culture–derived trypomastigotes but recognized nothing in epimastigotes or metacyclic trypomastigotes (73).

This work suggested that antibodies to sialidase either cross-reacted with numerous proteins without sialidase activity or that several high-molecular-weight components exhibited sialidase activity. We shall see from subsequent work that the former is probably the case. At no time did Pereira report the identification of a band of 60 kDa and, in view of the purification protocol described by Harth et al (35), the 60-kDa component may have been a degradation product, containing the enzymatically active domain from a larger protein. Nevertheless, there is other evidence of a smaller epimastigote sialidase, which is not phosphatidylinositol-specific phospholipase-C (PIPLC)-susceptible and does not react with the trypomastigote sialidase monoclonal antibodies (86); this may have been the protein purified by Harth. Ultrastructural studies (72, 99) suggested that both groups were working with different proteins because antibodies to each gave quite different labeling patterns in immunoelectron microscopy. These differences might be due to antibodies to contaminants in the purified sialidase or to the extensive sialidase-related gene family now recognized in *T. cruzi* (see below).

Recent work (9) shows that epimastigotes express a sialidase that has specificity and kinetic properties that are indistinguishable from the trypomastigote enzyme, but this sialidase is structurally distinct in important ways. It lacks the immunodominant carboxy-terminal repetitive domain characteristic of the trypomastigote enzyme (see below), and it is smaller (90 kDa) and monomeric. Also, we can conclude that the epimastigote enzyme is a distinct molecular species that does not result from contamination of the cultures with metacyclic forms because its activity plateaus in early stationary phase when the numbers of metacyclic forms are increasing.

ENZYMATIC PROPERTIES OF THE *TRANS*-SIALIDASE

Sialidase and Trans-*Sialidase*

Independent studies identified sialic acid in *T. cruzi* epimastigotes, even though the parasite appeared incapable of sialic acid synthesis (12, 91). It was suggested some time ago that, because sialylation depended on the presence of serum glycoconjugates and free sialic acids could not be incorporated into the parasite, the sialic acid found on the surface of *T. cruzi* could result from a cell-surface *trans*-sialidase activity (71, 122). Sialyltransferases are normally located in the Golgi and use cytidine monophosphate (CMP)–sialic acid as donor, but evidence has been presented for the existence of cell-surface glycosyltransferases, including sialyltransferase, in mammalian cells (89, 97a, 117). That the *T. cruzi* sialidase and *trans*-sialidase activities were probably attributable to the same molecule was implied by coprecipitation of these activities with antibodies specific for a repetitive peptide motif found in some members of the *trans*-sialidase family (see below) (57, 95). These antibodies can precipitate 50–85% of the total solublizable sialidase activity of *T. cruzi*. Studies with recombinant enzymes, expressed in *Escherichia coli,* prove that sialidase and *trans*-sialidase activities can be attributed to a single molecule (106).

Donor and Acceptor Specificity

Several natural sialoglycoconjugates are susceptible to the sialidase (57, 61, 95), and 4-methylumbelliferyl *N*-acetyl neuraminic acid is an excellent synthetic substrate. Natural sialoglycoconjugates are inefficiently hydrolyzed and serve better as sialic acid donors in the presence of a suitable acceptor. To date, the biochemical properties of the *T. cruzi trans*-sialidase have not been studied with a single homogeneous molecule. Because of the diversity and extent of the *trans*-sialidase family, monocomponent enzyme could only be produced by purification of enzyme expressed from a cloned gene. Such preparations have not been studied in biochemical detail. Indeed, only two studies have reported expression of sialidase or *trans*-sialidase activity by recombinant genes (64, 106). The majority of cloned genes appear to encode proteins that are enzymatically inactive, at least in the forms in which expression has been attempted in *E. coli,* insect, or mammalian cells (e.g. G. Takle, S. Paul & A. M. Cross, unpublished observations). Thus, most of the enzymatic properties of the *T. cruzi trans*-sialidase have been studied on preparations obtained by antibody-affinity purification, the most specific protocols using monoclonal antibodies against the carboxy-terminal repetitive region that comprises up to 40% of the sequence of some, if not all, of the enzymatically active proteins of the trypomastigote stages. However, not all of the subfamily members containing the immunodominant repetitive motif have sialidase/*trans*-sialidase activity. One that has been extensively tested for activity, cloned as a dominant shed acute-phase antigen (SAPA) (70), is inactive (A. C. C. Frasch & J. E. Manning, personal communication) as, apparently, are other subfamily members (106), although failure to find activity in some studies could result from problems specific to the recombinant constructs and expression systems that were used.

The *T. cruzi trans*-sialidase specifically cleaves terminal sialic acid $\alpha(2\rightarrow3)$ glycosidic linkages (26, 57, 93, 106a). Neither $\alpha(2\rightarrow6)$-linked sialylyl glycosides nor colominic acid [an $\alpha(2\rightarrow8)$-linked polymer of sialic acid] are substrates. The viral sialidases generally cleave $\alpha(2\rightarrow3)$ linkages faster than $\alpha(2\rightarrow6)$ (reviewed in 14). In contrast, bacterial sialidases show glycosidic specificity only at low enzyme concentration, and this has led to a loose classification on the basis of $\alpha(2\rightarrow3)$ or $\alpha(2\rightarrow6)$ linkage preference (14, 16). The *Clostridium perfringens* sialidase has a very broad substrate range. Thus, the *T. cruzi trans*-sialidase seems to show the highest specificity. In *trans*-sialidase experiments in which different disaccharides were investigated as potential sialic acid acceptors (the donor was fetuin), lactose [galactosyl β(1→4) glucose] was the best acceptor, and free sialic acid was undetectable. When fetuin was incubated with either maltose [glucosyl α(1→4) glucose] or cellobiose [glucosyl β(1→4) glucose], the sialyl-disaccharide and free sialic acid were generated in roughly equal amounts (57). From 100 nmol each of [^{14}C]lactose and sialyllactose, 23 nmol of [^{14}C]sialyllactose, and only 0.5 nmol of free sialic acid, were produced. In the absence of lactose, the amount of free sialic acid produced increased only to 2.5 nmol (95). The enzyme thus appears to be more active as a transferase than as a hydrolase. A series of experiments using gangliosides as sialic acid donors and acceptors has shown that sialic acid can only be removed from terminal galactosyl residues. *N*- or *O*-linked glycans can function as sialic acid donors or acceptors (26). In mammalian cells, terminal sialic acid is most commonly, but not exclusively, linked α(2→3) to β-galactose, and α(2→6) to *N*-acetylglucosamine and *N*-acetylgalactosamine (15).

Whether the *trans*-sialidase has a functionally preferred sialic acid donor, either on host cells, in serum or tissue fluids, or inside the cell, is not known. Neither is it known whether any host glycoconjugates are incidental or essential acceptors of sialic acid during trypanosome infection. Members of the gp85 surface glycoprotein subfamily of the *trans*-sialidase family are probably sialylated (17, 18), but is not known whether the *trans*-sialidase is itself sialylated, either adventitiously or as a covalently linked reaction intermediate. When assessed with a monoclonal antibody (3C9), whose reactivity with *T. cruzi* depends on sialylation, the major consequence of *trans*-sialidase activity on tissue culture–derived trypomastigotes appears to be the sialylation of a broad range of glycoproteins, from < 50 to > 200 kDa, but monoclonal 3C9 only precipitates about 10% of radiolabeled sialylated molecules, and the *trans*-sialidase is not recognized (93). On the other hand, *T. cruzi* surface glycoconjugates of about 35 and 50 kDa, which are recognized by monoclonal antibody 10D8 in the epimastigote and metacyclic stages, but not in mouse or culture-derived trypomastigotes (119), are the major labeled glycoconjugates when epimastigotes transform to metacyclic trypomastigotes in the presence of [^{14}C]sialyllactose (92). Monoclonal antibody 10D8 inhibited the invasion of cultured cells by metacyclic trypomastigotes, and preincubation with 10D8 antibody or Fab fragments greatly reduced the infectivity of trypomastigotes to mice (119). The composition of purified 35 and 50 kDa complex is striking. Carbohydrate accounts for about 50% of the total weight and threonine for 21% of the amino acid content by weight. Judging by their composition, these molecules are glycosylphosphatidylinositol (GPI)-anchored to the cell surface (92).

Sialidase Activity in Trypanosoma rangeli

Trypanosoma rangeli is a human-infective but apparently nonpathogenic relative of *T. cruzi* (37). Because of their geographical proximity and common vectors and hosts, the comparative biochemical and genetic properties of *T. cruzi* and *T. rangeli* are of significant interest. Sialidase activity was first demonstrated in *T. rangeli* epimastigote culture supernatants (60), from which the enzyme was subsequently purified as a 70-kDa protein with broad substrate specificity and optimum activity around pH 5.5 (79). The *T. rangeli* sialidase does not show *trans*-sialidase activity (23, 24a).

The Trans-*Sialidase of* Trypanosoma brucei

A 67-kDa monomeric surface membrane–associated glycoprotein sialidase/*trans*-sialidase has been purified from the procyclic stage [the cultured analogue of the *Glossina* spp. (tsetse) midgut form] of *Trypanosoma brucei,* but this protein is absent from the mammalian-infective bloodstream forms of the parasite (23, 24, 24a, 70b). The *T. brucei* enzyme has a pH optimum very similar to the *T. cruzi trans*-sialidase. Micromolar concentrations of Hg^{2+} inhibited the *T. brucei* enzyme, suggesting an essential role for SH groups. The substrate specificity for the sialidase activity of the *T. brucei* enzyme has been studied in more detail and appears somewhat less stringent than that for the *T. cruzi trans*-sialidase. However, in the presence of 2 mM lactose, the transfer reaction exceeds the release of free sialic acid by a factor of three.

The only location of sialic acid so far identified in *T. brucei* is the GPI anchor of the major surface glycoprotein [procyclic acidic repetitive protein (PARP)/procyclin] of the insect midgut (procyclic)-stage cells grown in culture medium containing serum as a potential source of sialic acid (25a). Whether this modification occurs in genuine *Glossina* spp.–derived trypanosomes is not known. The distribution of sialic acids in insects is uncertain, although they are clearly synthesized by cells of *Drosophila melanogaster* (88). In reduviids and *Glossina* spp. the bloodmeal is also a potential source of sialic acid. An obvious question arises regarding the function of the sialidase/*trans*-sialidase in the *T. brucei* procyclic stage. Perhaps a sialic acid–binding lectin activity is involved in some aspect of trypanosome colonization or differentiation in the insect vector.

STRUCTURE OF THE *TRANS*-SIALIDASE FAMILY

Experiments with various antibodies and labeling reagents suggested that the sialidase/*trans*-sialidase proteins are members of a heterogeneous family of *T. cruzi* surface glycoproteins, ranging in size from 60 to > 200 kDa (35,

73). Biochemical and gene-cloning experiments indicate that this heterogeneity derives principally from variations in the presence and length of a carboxy-terminal repetitive domain (see below), possibly supplemented by variations in glycosylation (118). DNA cloning and hybridization studies indicate the presence of several hundred *trans*-sialidase-related genes.

A gene encoding an active *trans*-sialidase, dubbed *TCNA*, was cloned by screening a genomic expression library with polyclonal and monoclonal antisialidase antibodies (64). The published TCNA protein sequence lacks an amino-terminal secretory signal and the conserved amino-terminal sequence of the mature protein (70a). These features can be restored by a single nucleotide (G) insertion, and this correction is assumed in the following discussion. TCNA (64) and the shed acute-phase antigen (SAPA) (70) are 84% identical in amino acid sequence (their respective genes are 93% identical in nucleotide sequence) over the presumptive catalytic amino-terminal domain, and the repetitive domains are essentially identical, containing either 44 or 14 repeats, respectively, of the dodecapeptide DSSAH(S/G) TPSTP(A/V). By polyacrylamide gel electrophoresis, Parodi and coworkers (70a) separated five bands from immunoprecipitated proteins containing the carboxy-terminal repeats of SAPA and TCNA and showed that all five had identical amino-terminal sequences.

Motifs That Characterize the Trans-*Sialidase Family*

Seventeen genes of the *trans*-sialidase family have been completely or partially cloned and sequenced (Figure 2). By pairwise comparisons, the sequences can be grouped into four subfamilies, based on their overall similarities and the presence or absence of various motifs. Group I contains *TCNA* and closely related sequences. The next most closely related subfamily (group II in Figure 2) comprises genes for the so-called gp85 surface glycoproteins. The proteins of several members of this group have been completely or substantially sequenced: TSA-1 (28), Tt34c1 (101), SA85-1.1, 1.2, and 1.3 (40, 41), and pTt21 (103). Proteins of the gp85-family members lack the extensive carboxy-terminal repeats of TCNA and SAPA, but may contain short repetitive regions (28, 41, 66). None of the cloned gp85 group II members has been shown to encode sialidase activity [the activity described by Kahn et al (40) was subsequently attributed to cross-reactivity of the antiserum used (S. Kahn, personal communication)].

Pereira et al (64) distinguished several domains within the TCNA sequence that had somewhat tenuous similarities to motifs present in the mammalian LDL receptor and fibronectin or involved in protein kinase recognition and GTP binding. One motif, SXDXGXTW, seems particularly significant. It occurs, in perfect and degenerate forms, in many members of the *T. cruzi*

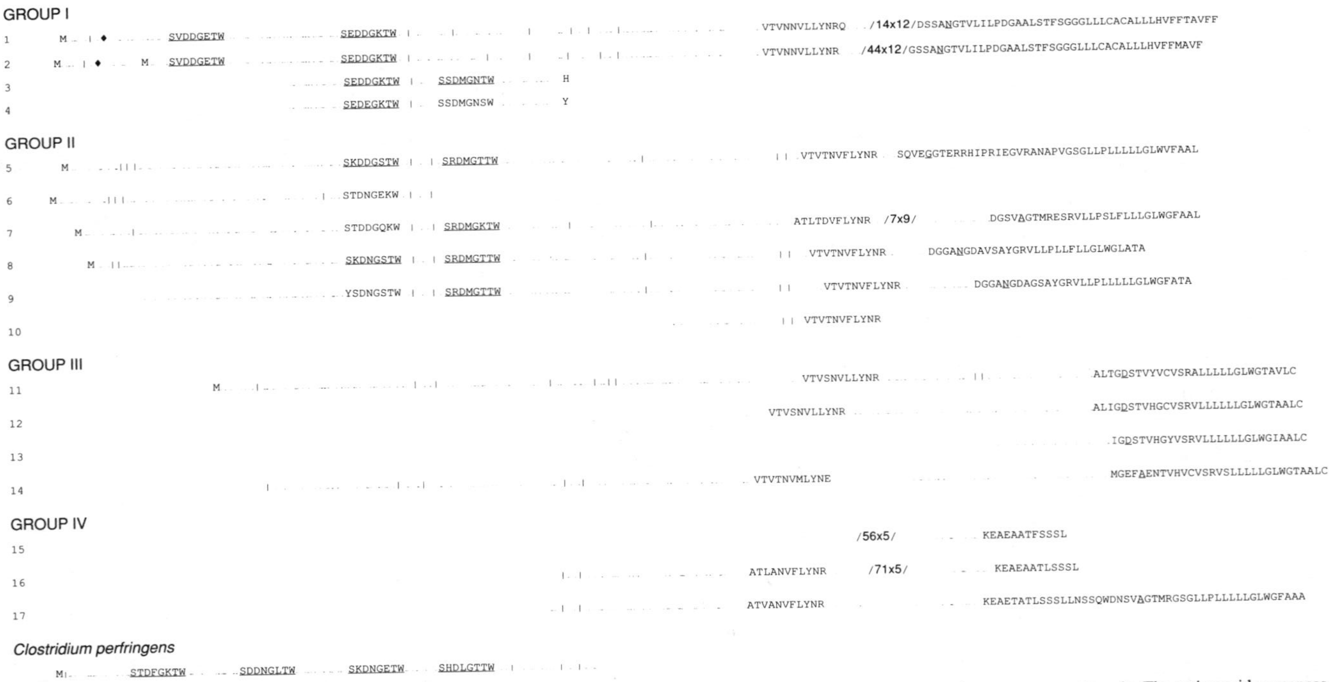

GROUP I

1 M ◆ SVDDGETW SEDDGKTW VTVNNVLLYNRQ /14x12/DSSANGTVLILPDGAALSTFSGGGLLLCACALLLHVFFTAVFF

2 M ◆ M SVDDGETW SEDDGKTW VTVNNVLLYNR /44x12/GSSANGTVLILPDGAALSTFSGGGLLLCACALLLHVFFMAVF

3 SEDDGKTW SSDMGNTW H

4 SEDEGKTW SSDMGNSW Y

GROUP II

5 M SKDDGSTW SRDMGTTW VTVTNVFLYNR SQVEGGTERRHIPRIEGVRANAPVGSGLLPLLLLLGLWVFAAL

6 M STDNGEKW

7 M STDDGQKW SRDMGKTW ATLTDVFLYNR /7x9/ DGSVAGTMRESRVLLPSLFLLLGLWGFAAL

8 M SKDNGSTW SRDMGTTW VTVTNVFLYNR DGGANGDAVSAYGRVLLPLLFLLGLWGLATA

9 YSDNGSTW SRDMGTTW VTVTNVFLYNR DGGANGDAGSAYGRVLLPLLLLLGLWGFATA

10 VTVTNVFLYNR

GROUP III

11 M VTVSNVLLYNR ALTGDSTVYVCVSRALLLLLGLWGTAVLC

12 VTVSNVLLYNR ALIGDSTVHGCVSRVLLLLLLGLWGTAALC

13 IGDSTVHGYVSRVLLLLLLGLWGIAALC

14 VTVTNVMLYNE MGEFAENTVHVCVSRVSLLLLLGLWGTAALC

GROUP IV

15 /56x5/ KEAEAATFSSSL

16 ATLANVFLYNR /71x5/ KEAEAATLSSSL

17 ATVANVFLYNR KEAETATLSSSLLNSSQWDNSVAGTMRGSGLLPLLLLLGLWGFAAA

Clostridium perfringens

M STDFGKTW SDDNGLTW SKDNGETW SHDLGTTW

Figure 2 Structures of the *trans*-sialidase-related glycoproteins of *Trypanosoma cruzi*. Subject to the following formatting constraints, the sequences are to scale. The amino acid sequences appear as a continuous line except where certain motifs are highlighted by actual letter representation. Repetitive regions, which appear largely responsible for size variation between family members, are indicated by /$x \times y$/ where x = the number and y = the amino acid periodicity of the repeats. SXDXGXTW boxes in which at least four of the five invariant residues are present are shown with perfect matches underlined. The VTVxNVfLYNR motifs are shown as and where they occur. The position of the mature amino-terminal leucine in SAPA and TCNA is indicated (*diamonds*). The initiator methionine residue (M) is indicated for sequences that appear complete at the amino terminus. The second M in TCNA (sequence 2) is the initiator deduced by the authors (64). Sequence 14 is a produced from a pseudogene: although the gene extends upstream of the region shown, it lacks an initiation codon and the upstream region contains multiple termination codons. Except for within the carboxy-terminal hydrophobic GPI anchor sequence, cysteine residues are indicated by vertical bars. Probable sites of GPI-anchor addition are indicated by the underlined residues lying about 25 amino acids upstream from the carboxy termini. A sialidase from *Clostridium perfringens* is shown for comparison. The sequences, listed in the order that they appear in the figure (common names in parentheses), are available in the GenBank™/EMBL data bases with the following accession numbers: X57235 (SAPA) (70); M61732 (TCNA) (64); D12740, D12741 (106); M64836 (Tt34c1) (101); M91469 (103); M58466 (TSA-1) (28); M62735, M62736, X53547 (SA85-1.1, 1.2, 1.3) (40, 41); M88337 (A. C. C. Frasch et al, unpublished data); M65032 (F1-160) (110); M91469, M91470 (103); M92046, M92047, M92048, (Tc13 family) (10); Y00963 (84).

trans-sialidase family (40, 101), and up to four copies are found in bacterial sialidases (83) (Figure 2).

Although the function of the repeated SXDXGXTW motif in bacterial sialidases is unknown, it is striking that a search for this motif in the protein data bases yields only five examples, excluding bacterial sialidases and members of the *T. cruzi trans*-sialidase family. The examples include a bacteriophage tail-fiber protein (81), which may bind specific carbohydrates of the bacterial host, and *Saccharomyces cerevisiae* levanase (48), which has significant homology to invertase and sucrase.

The significance of a subterminal VTVxNVfLYNR motif (amino acids that are invariant throughout the distinct branches of the *trans*-sialidase family are underlined; uppercase type indicates identity in 9 or more of the 12 sequences that extend through this region) is unknown, but it is conserved in all members of the family whose known sequence extends into the region where it resides (Figure 2). One can only speculate that this region may be involved in some receptor recognition, in interacting with other *T. cruzi* surface proteins, or as a signal in the secretory or endocytic pathways. On the other hand, the conservation of amino acid sequence could be secondary to conservation of the corresponding nucleotide sequence, which might be necessary for regulated expression or rearrangement of *trans*-sialidase genes.

Several *T. cruzi* genes that encode proteins of approximately 160 kDa, lacking *trans*-sialidase activity but homologous to the gp85/*trans*-sialidase sequences, have been described (group III sequences in Figure 2). One is a flagellar antigen (Fl-160), as defined by the localization of specific antibodies on the cell surface (110). Anti-Fl-160 antiserum also cross-reacts with specific host nervous tissues. The region of Fl-160 that is responsible for the cross-reacting antibodies is a sequence of 12 amino acids (109, 110). The published partial sequence of FL-160 does not contain any SXDXGXTW motifs, but it is included in the *trans*-sialidase family because of the presence of the VTVxNVfLYNR motif and high nucleotide homology with a proven family member (103). The large carboxy-terminal domain downstream of the VTVxNVfLYNR motif in Fl-160 is apparently nonrepetitive, and it aligns closely with a similar apparently nonrepetitive carboxy-terminal domain of the pseudogene c1821 (103). Further analysis of these regions with dot-matrix homology analysis indicated that both nucleic acid and amino acid sequences contain pseudorepeated regions, meaning that, if these sequences are aligned against themselves, certain stretches of short sequence are reiterated in a cryptic manner that is not obvious on cursory examination (103). These pseudorepetitive regions are almost certainly derived, by recombinations or replication errors involving point mutations and slippage, from family members that contain defined repeats (52, 104).

Group IV (Figure 2) contains three sequences (10) that are closely related to each other and tentatively classified with the *trans*-sialidase family by the presence of the VTVxNVfLYNR motif.

Many members of the *trans*-sialidase family contain degenerate versions of the SXDXGXTW motif. These observations suggest that the SXDXGXTW boxes are critical, in some undefined way, to the function of enzymatically active members of the *trans*-sialidase family. In TCNA and SAPA, the two perfect SXDXGXTW boxes are positioned differently from the other sequences (Figure 2). Despite the strong similarity of TCNA and SAPA sequences, SAPA does not show *trans*-sialidase or sialidase activity when expressed in either *E. coli* or insect cells (A. C. Frasch & J. E. Manning, personal comunication). Discerning the critical sequence motifs that confer *trans*-sialidase activity only by comparing linear amino acid sequences of active and inactive proteins is probably unrealistic, partly because errors in published DNA sequences can pose a critical obstacle to fastidious comparisons (49).

One study has attempted to determine the critical regions for *trans*-sialidase activity by expressing different fusions of active and inactive proteins (106). From the results of these experiments, the authors delimited a region of 101 amino acids whose origin, from an original active or inactive gene, determined the activity or inactivity of the fusion protein. By comparing the amino acid sequences in this region (Figure 2), the authors concluded that one amino acid determined activity—tyrosine in the active fusion and TCNA proteins, but histidine in the inactive protein. Unfortunately for this hypothesis, the apparently inactive SAPA also has tyrosine at the critical position. This suggests either that additional differences outside the region sequenced by Uemura et al (106) influence activity, or that the inactivity of SAPA is an artifact of the systems used to express it. Mutagenesis experiments, and the full sequences of closely related inactive and active genes (106), may clarify the situation, but complete understanding may only come through knowledge of the three-dimensional structure of both the *T. cruzi* and the bacterial sialidases.

The positions of cysteine residues in the various proteins is conserved to various extents between different subfamilies (Figure 2). Although the disposition of the cysteines is unknown, one presumes, for a cell-surface protein, that they are mostly disulfide-linked. If so, the differences in cysteine locations could indicate differences in protein folding that could be critical determinants of *trans*-sialidase activity.

Role of the Carboxy-Terminal Repetitive Motif in Trans-*Sialidase*

The 90-kDa epimastigote-specific *trans*-sialidase (9) lacks the dodecameric repetitive DSSAH(S/G)TPSTP(A/V) domain of TCNA, which strongly implies that this domain is unnecessary for activity. Although no reports have been published on the expression of variants of the *trans*-sialidase gene lacking the repeats, brief proteolysis of the mixture of native repeat-carrying molecules reduced the mixture to a single active *trans*-sialidase component on polyacrylamide gel electrophoresis or high-performance liquid chromatography. This component presumably lacks the repetitive regions (S. Schenkman, personal communication). These experiments suggest that the repetitive domain plays no role in the intrinsic catalytic activity of the *trans*-sialidase. On the other hand, some reason should explain why *trans*-sialidase activity is mainly accounted for by one or more of the dodecameric repeat-containing members of the family. The trypomastigote enzyme appears to be a trimer, based on its behavior on gel filtration (9, 64), but whether the trimerization is mediated by the dodecapeptide repetitive domain is not known. However, the epimastigote *trans*-sialidase, which is smaller and lacks the repetitive region, appears on chromatographs at the position expected for a monomeric molecule (9).

For the *trans*-sialidase repeats (57) and the immunodominant repetitive regions found in several antigenic surface molecules of malaria parasites (42), it has been suggested that the repetitive domain may somehow serve to deflect the immune responses from essential regions of a molecule or a cell surface, or otherwise unbalance the immune responses of the host. On the other hand, we believe that the immunogenicity/antigenicity of the *trans*-sialidase repeats may be incidental to the primary function of the protein or the biology of *T. cruzi*. The repeats may represent a simple solution to a functional need: to extend the enzymatically active domain of the *trans*-sialidase beyond the fuzz of other glycoproteins and glycolipids coating the parasite surface. Such a role, based on molecular modeling, has been suggested (82) for the simple Glu-Pro dipeptide repeat that comprises half of the structure of PARP/procyclin, which extends through the VSG coat during the initial stages of *T. brucei* differentiation from the bloodstream stage to the insect stage (82, 121). The sequence of the TCNA/SAPA repetitive region also indicates that it would adopt a relatively extended structure (D. Cowburn, personal communication). Macrophage sialoadhesin monomers may have a structure similar to the *T. cruzi trans*-sialidase. Although the gene for the sialoadhesin has not been reported, and hence a structure cannot be deduced, low-angle shadowing studies indicate that the protein is composed of a 9-nm-diameter globular head linked to a 35-nm-long tail (19). Low-angle shadowing of aggregates of the sialoadhesin indicate that multimerization occurs via aggregation of the head groups, and the sialic acid–binding domain was assumed to reside in the protruding tails.

Trans-*Sialidase-Family Members Are Probably GPI Anchored*

The deduced carboxy-terminal amino acid sequences of members of the *trans*-sialidase family terminate in a hydrophobic region characteristic of GPI-anchored proteins and contain a likely GPI addition site 9–12 amino acids upstream from the hydrophobic segment (21, 31, 45, 50) (Figure 2). PIPLC release of sialidase/*trans*-sialidase activity, and other enzymatically inactive members of the family, from the surface of *T. cruzi* provides experimental support for this concept (70, 86, 95), and the GPI anchor of one member of the *trans*-sialidase family (29, 118) has been structurally characterized (33). Furthermore, the *trans*-sialidase activity spontaneously released when trypomastigotes are incubated in culture medium reacts with antibodies specific for the inositol-glycan moiety generated by PIPLC treatment of GPI anchors, suggesting that release is mediated by an endogenous *T. cruzi* PIPLC (34, 95). Such an activity is present in *T. cruzi,* but has not been purified or otherwise characterized in any detail (96).

Evolution of the Trans-*Sialidase Family*

T. cruzi trans-sialidase shows no homology with the available sequences of the Golgi-associated sialyltransferases. In contrast, the extraordinary sequence conservation between a region of the *C. perfringens* sialidase, which itself shows no *trans*-sialidase activity (24a, 57), and the *T. cruzi trans*-sialidase, suggests that the *trans*-sialidase might have evolved from an ancestral bacterial gene. An evaluation of the relatedness of sialidase/*trans*-sialidase genes between different trypanosomatid species would be interesting. We could not detect activity in the promastigote stage of *Leishmania donovani* or *L. mexicana* (A. Ribeiro de Jesus & G. B. Takle, unpublished observations). The *T. cruzi trans*-sialidase seems to have spawned a large and diverse family. The most outlying members can be recognized only by the presence of the subterminal VTVxNVfLYNR motif and by the presence of probable GPI-anchor signal motifs. The sequences that are most distantly related to the *trans*-sialidase appear to have gradually diverged, although the construction of chimeric genes by segmental gene conversions may have played an important role in the evolution of this family, as has been suggested for the VSG genes of African trypanosomes (59, 105). It seems likely, therefore, that the retention of a certain level of *trans*-sialidase is strongly selected for and is crucial for the survival, pathogenicity, or virulence of *T. cruzi*. Too much *trans*-sialidase activity could be as much of a problem for the trypanosome as too little or none.

Regulation of Trans-*Sialidase-Family Gene Expression*

Serological evidence has pointed away from the possibility of antigenic variation in *T. cruzi* (69, 98). Nevertheless, we would like to determine if genes of the *T. cruzi trans*-sialidase family are expressed individually and sequentially and if genome rearrangements are associated with their expression. The first gp85/*trans*-sialidase gene to be cloned was a member of a subfamily of four genes distinguished by a short nonapeptide repetitive region

(28, 66). One of the four members of this subfamily appeared to be preferentially transcribed. This copy was telomeric (65), a situation reminiscent of VSG expression in *T. brucei* (20, 108). Subsequent work showed that other expressed members of the *trans*-sialidase family were resistant to BAL 31 digestion and therefore not situated close to telomeres (41, 101), showing that translocation to a telomeric site is not required for expression. Two reports have convincingly demonstrated, however, that clones of *T. cruzi* trypomastigotes contain transcripts from numerous *trans*-sialidase-related genes (101), and antipeptide antisera directed against different members of the family simultaneously recognize individual parasites (41). Thus there appears to be concurrent expression of multiple members of the gene family, and these observations presumably considerably underestimate the multiplicity of gene expression. However, it is still unclear what proportion of the total repertoire is expressed, what features of a particular copy specify that it will be expressed, whether the repertoire of expressed copies drifts with sequential passages through the parasite life cycle, and whether specific subfamilies are expressed at different stages in the life cycle. Cell-surface glycoprotein expression in *T. cruzi* thus seems to be very different from *T. brucei,* in which each cell appears to express a single VSG from a large and continuously evolving repertoire (reviewed in 20).

Some genes of the *trans*-sialidase family are tandemly linked (103, 106) and may be polycistronically transcribed from a single promotor, although the only experimental observation that supports this is the finding of a cDNA copy of a gp85 gene that contained sequences 5′ and 3′ to the expected mature mRNA sequence (103). Much evidence, direct and indirect, suggests that trypanosome genes are transcribed polycistronically and that the regulation of gene expression may occur subsequent to transcription initiation (107).

Some members of the gp85 family possess two in-frame methionine codons approximately 110 nucleotides apart (28, 101), the second of which best fits a consensus sequence for initiators (46). A characteristic signal sequence resides just downstream of the second methionine (115). In vitro translation/translocation experiments (G. M. Takle, unpublished observations) and examination of the mature amino-terminus of SAPA/TCNA-related proteins (70a) confirms that a signal sequence is cleaved. Either initiation site may be used in vitro (G. M. Takle, unpublished observations), and the region between the two methionines is a highly conserved region in at least one pair of sequences (101). This region contains a high net positive charge (10 positively charged amino acids out of 37) (101) and could retain the amino terminus in the cytoplasm, reversing the orientation of the hydrophobic signal sequence as it transverses the endoplasmic-reticulum membrane (36, 56). This reversed orientation may not prevent the remainder of the polypeptide from being translocated into the endoplasmic reticulum, so the issue of whether some members of the *trans*-sialidase family could become located in different cell compartments, or in different membrane orientations, is uncertain.

INTERACTIONS OF *TRYPANOSOMA CRUZI* WITH THE HOST

Biological Roles of Sialic Acids

Sialic acids are undoubtedly in an appropriate location to be involved in cellular interactions with other cells and soluble substrates, and their various interactions may derive from their great potential for structural diversity (15). Sialic acids are involved in intercellular interactions (100), substrate adhesion (111), neuronal development (120), malignancy (58), complement resistance (51) or evasion (25), biological masking (90), maintenance of red cells in circulation (38), and microbial pathology.

Numerous prokaryotic, viral, and eukaryotic pathogens use sialic acids and sialic acid receptors in their subversion of host cells. Many viruses use lectin-like hemagglutinins for attachment to host cells. The best studied of these is the influenza virus hemagglutinin (HA). Analysis of HA binding to sialic acid with X-ray crystallography has identified the binding site (116), and further work showed that viral mutants with single amino acid changes in the sialic acid binding site differ in their specificity for $\alpha(2\rightarrow3)$- or $\alpha(2\rightarrow6)$-linked sialic acids (78). This small variation in peptide sequence leading to a major difference in binding specificity indicates that it may not be easy to specify residues important for activity in the *T. cruzi* enzyme by sequence alone, and that more detailed structural analysis may be required. An important observation, relevant to all sialic acid–receptor interactions, is that HA affinity for sialic acid increases dramatically if the sialic acid ligand is dimerized with a spacer of a specified length, indicating the possible multivalency of the interaction (32).

Some bacteria contain sialic acids (reviewed in 15). Those that do are pathogenic, and many bacteria express sialidase activity (reviewed in 14). The role of host sialic acid in bacterial pathogenesis is clearly evident in some host-bacterial interactions (e.g. 80). Bacterial sialidases may function to unmask host glycoproteins, making them vulnerable to attack by bacterial proteinases. Unmasking of the T-antigen on certain host-cell types by pneumococcal sialidase leads to an autoimmune hemolytic-uremic syndrome (reviewed in 15).

The merozoite stage of the major human malaria parasite *Plasmodium falciparum* invades red blood cells via a receptor-mediated mechanism involving sialic acid moieties decorating glycophorin A (53). The receptor has been referred to as erythrocyte-binding antigen 175 (EBA175) or *P.*

falciparum sialic acid–binding protein. EBA175 binds preferentially to α(2→3)-linked sialic acids. EBA175 is a member of a family of malaria erythrocyte–binding proteins, which are also found in *Plasmodium knowlesi* and *Plasmodium vivax* (3). The erythrocyte-binding proteins of these organisms do not bind specifically to sialic acid but recognize the Duffy blood-group antigens. The structures of the genes encoding these proteins are very similar. Each gene contains four or five exons that each define a functional domain. The second exon in each case contains the Duffy antigen or sialic acid–binding domain, each with many conserved cysteines. The sequence differences between the two types of receptor are large enough to prevent prediction of the sialic acid–binding site. Both the Duffy antigen–binding proteins and the sialic acid–binding protein are found in the micronemes at the invasive apical end of the parasite and are probably components of a complex attachment/invasion process involving several receptors and enzymes working in concert.

Modulation of Cell-Surface Sialic Acid and Trans-*Sialidase Activity During the Life Cycle of* Trypanosoma cruzi

In considering the possible role of the *trans*-sialidase, one should recall that the enzyme has a broad pH optimum, centered at 7.0, and that activity declines by 50% as the pH approaches 5.5 or 8.5 (95).

Treatment of bloodstream trypomastigote forms of *T. cruzi* with proteases or, to a lesser extent, sialidase, sensitizes the cells to complement, suggesting that the presence of sialic acid is important for the aquisition of complement resistance in the infective stages (39, 43), as it is in erythrocytes (105a). Several lines of indirect evidence suggested a role for sialic acid during cell attachment and/or invasion, although the identities of the parasite and host receptors that might be involved in these processes are unknown and some of the data were in apparent conflict. Piras et al (68) showed that fetuin, but not asialofetuin or desialylated serum, stimulated infection and suggested that the effect resulted from *trans*-sialylation of *T. cruzi* surface components. In contrast, Pereira and coworkers (11, 73) showed that antibodies that inhibited the *T. cruzi* sialidase activity enhanced twofold the internalization of *T. cruzi* into cultured cells. *Vibrio cholerae* sialidase prevented this enhancement but did not, by itself, inhibit invasion. These experiments were performed in the presence of serum and before the *T. cruzi* enzyme was known to have *trans*-sialidase rather than simple hydrolase activity; hence the effects of the *V. cholerae* and *T. cruzi* enzymes are not equivalent.

In more recent experiments, the attachment of tissue culture–derived trypomastigotes to nonphagocytic L cells or Chinese hamster ovary (CHO) cells was enhanced via expression of a transfected mouse Fc receptor when the cells were incubated with small amounts of antibodies that react with sialylated *T. cruzi* surface glycoconjugates, but internalization was concomittantly inhibited. If antibodies to other *T. cruzi* surface constituents were used, both attachment and invasion were stimulated. These experiments support the notion that sialic acid–bearing surface glycoconjugates of the parasite are important for invasion of host cells rather than for attachment (94).

Early experiments showed that sialidase-specific antibodies reacted with a subpopulation of tissue culture–derived trypomastigotes (11, 73). Later work (87) showed that intracellular amastigotes lacked sialidase, but before trypomastigotes exited host cells and immediately afterwards, their surfaces reacted with antisialidase antibodies. The extracellular forms then rapidly lost reactivity with the antibody, presumably by shedding of the *trans*-sialidase (86). These experiments led the authors to suggest that sialidase activity might play a role in parasite escape from infected cells. Other authors (30) confirmed that *trans*-sialidase, identified both by antibody and enzymatic activity, reappeared in intracellular trypomastigotes as they differentiated from amastigotes. The *trans*-sialidase was shed into the cytoplasm of the host cell, but the sialic acid–dependent Ssp-3 epitope appeared only on released extracellular trypomastigotes and was never detected on intracellular trypanosomes so long as the host-cell membrane was intact. Frevert et al (30) also showed that, whereas anti-*trans*-sialidase antibodies reacted with the surface and flagellar pocket of liberated trypomastigotes, *trans*-sialidase could no longer be detected in the Golgi apparatus, in contrast to the younger intracellular trypomastigotes. In addition, they demonstrated that secretory vesicles were absent from the vicinity of the flagellar pocket, suggesting that synthesis and secretion of the *trans*-sialidase had ceased in the nondividing extracellular trypomastigotes.

Andrews et al (6, 34) suggested that sialidase activity might be involved in the escape of the initially invading trypomastigotes from the parasitophorous vacuole into the host-cell cytoplasm. Escape was faster in sialylation-defective mutant Lec-2 Chinese hamster ovary (CHO) cells (34), suggesting that removal of sialic acid from the major lysosomal glycoprotein might facilitate escape from the vacuole, a process that probably involves a parasite-encoded hemolysin, Tc-TOX, an acid-active transmembrane pore–forming protein (5, 7, 47). Once again, these experiments were interpreted before the *T. cruzi* enzyme was recognized as primarily a transferase, not a simple hydrolase. Thus, the possible function of the *trans*-sialidase within the parasitophorous vacuole needs to be reinvestigated. Also, the observations of Frevert et al (30) raise doubts about the presence of *trans*-sialidase activity at the time of escape from the parasitophorous vacuole. Within two hours of mixing tissue culture–derived trypomastigotes and LLC-MK2 cells, investigators could no longer detect *trans*-sialidase by immmunofluorescence or its activity; it appeared to have been lost from the parasite membrane during invasion or inclusion within the parasitophorous vacuole. In contrast, the Ssp-3 sialic acid-containing epitope was lost only when the parasites differentiated to amastigotes in the host-cell cytoplasm.

A Serum Sialidase Inhibitor (Cruzin)

Blood plasma inhibits *T. cruzi* sialidase activity against erythrocytes, but not against soluble glycoconjugates (76). The component responsible for this inhibitory activity was initially named *cruzin* and was later identified as high-density lipoprotein (HDL) (74). Some evidence indicated that HDL and LDL (low-density lipoprotein) at high concentration elevated epimastigote growth (74), and other evidence demonstrated that LDL and HDL, like antisialidase antibody (75), increased the infection of tissue cells in vitro. These data led Pereira and coworkers to propose a sialidase-mediated negative regulation of infection (62, 73), but more recent work (94) suggests that *trans*-sialylation is required for host-cell penetration. Furthermore, the reduction in agglutination of erythrocytes by the galactose-specific lectin peanut agglutinin, attributed to inhibition of *T. cruzi* sialidase by HDL and LDL, could very easily result from resialylation by the *trans*-sialidase, where HDL or LDL, which contain sialic acid (15, 112), are used as the donor.

Trans-*Sialidase-Related Surface Glycoproteins of* Trypanosoma cruzi

The proportion of the *trans*-sialidase-family members that show *trans*-sialidase activity is unclear, although about 80% of the cellular activity can be precipitated from extracts of *T. cruzi* by antibodies that are specific for the repetitive peptides found in TCNA and SAPA. Thus, the nonrepetitive majority of the cell-surface *trans*-sialidase-related glycoproteins of *T. cruzi* must have other functions.

Several trypomastigote stage–specific surface glycoproteins of around 83–90 kDa have been implicated in parasite attachment and/or invasion. Cloning and sequencing of genes encoding several surface glycoproteins in this size range showed that many might be members of the *trans*-sialidase family. Villalta et al (113, 114) identified a group of 83-kDa trypomastigote surface glycoproteins with differing isoelectric points that inhibited the binding of parasites to heart myoblasts in vitro. This work corroborated earlier observations of a sialylated trypomastigote-specific surface molecule referred to as Tc85 (2, 4), which was implicated in parasite invasion. Another 85-kDa trypomastigote-specific surface molecule specifically bound fibronectin, and this protein, as well as antibodies to it, inhibited the invasion of fibroblasts by trypomastigotes (54, 55). The gene for this fibronectin-binding protein has not been cloned. A major metacyclic trypomastigote-specific 90-kDa surface antigen, experimentally implicated in cell attachment and/or penetration, was identified by a monoclonal antibody (1G7) (118). Although it is glycosylated, this antigen does not seem to be sialylated, and probably contains high mannose *N*-linked oligosaccharides. The GPI-anchor structure attaching this antigen to the surface of the parasite has been investigated (33), and the sequence of a gene encoding a 1G7-reactive polypeptide showed that it is a member of the *trans*-sialidase family (29).

The functions of the *trans*-sialidase-related but enzymatically inactive surface glycoproteins are unknown, and defining their specific roles will provide an intriguing area for future research. These *trans*-sialidase homologues may have lectin-like activities, in the absence of hydrolysis or transfer. A mammalian 14-kDa β-galactoside-specific lectin contains an SXDXGTW motif, which is present in degenerate versions in several related lectins (1). A myxobacterial hemagglutinin, which recognizes a complex galactosylated receptor, contains four domains that each contain the sequence SXDXGXTL (85). A bacteriophage tail-fiber protein, gp37, which is involved in binding of the bacteriophage to sugar residues on the surface of the bacterium, contains one exact SXDXGXTW motif (81), and some bacterial glycosidases (48) contain related sequences. Thus, the SXDXGXTW motif is common to several proteins involved in carbohydrate interactions.

Members of the *T. cruzi trans*-sialidase family may also recognize a variety of monosaccharides or complex glycoprotein or glycolipid ligands, and the investigation of potential binding specificities would require considerable effort. It may be that each of the expressed copies of the gene family recognizes and binds a different saccharide structure, and in this way allows the parasite to attach and enter a diverse selection of host cells. Such promiscuity could also be facilitated by *trans*-sialidase modulation of host cell–surface sialic acids, subsequent binding by other members of the *trans*-sialidase family, or other parasite sialic acid receptors.

A sperm-surface galactosyltransferase acts in a lectin-like role by binding to an egg-surface glycoprotein, whose *O*-linked glycans it galactosylates in an interaction that suggests a possible model for the interaction of *T. cruzi trans*-sialidase with target cell glycoconjugates (48a).

One other possibility is that the nonenzymic members provide the parasite with a type of variant coat. *T. cruzi* does not rely, to the same extent as *T. brucei,* on avoiding the immune system, because *T. cruzi* divides within host cells. However, the parasite may derive some advantage by expressing a slowly varying surface, and the extensive *trans*-sialidase family may provide the genetic facility for this. The study of *trans*-sialidase-family expression is in its infancy, and the lack of probe or antibody specificity complicates the investigation of antigenic drift. The availability of methods to amplify specific sequences from single cells would provide a means for examining *trans*-sialidase-family expression in more detail.

PRESENT UNCERTAINTIES AND FUTURE DIRECTIONS

The presence of sialic acid on *T. cruzi* and/or on the target host cells, and the presence of *trans*-sialidase activity at the trypanosome surface, are important for one or more aspects of the interaction of *T. cruzi* with its host cells. Sialic acid may be important for attachment to and invasion of the cell, and for escape from the vacuole or from the cell, but experiments performed so far have not provided unequivocal evidence for specific roles, possibly because of the late recognition of the dual sialidase/*trans*-sialidase activity, together with the paucity of information on the host and trypanosome components that are specifically modified and/or required for invasion.

Our working hypothesis is that *trans*-sialidase is essential for some aspects of the parasite's ability to invade, develop, and mature in the infected host cells. This implies that the native host cell–surface sialic acid alone is insufficient for the needs of the trypanosome. The *trans*-sialidase could perform several functions. It could remodel the surface glycoconjugates of the host cell, perhaps to better prepare them as receptors for the trypanosome or to facilitate some dynamic redistribution of receptors during invasion. The *trans*-sialidase could be essential for transferring sialic acid to trypanosome surface glycoconjugates so that the proteins can fulfill their purpose during invasion or during escape from the parasitophorous vacuole or the infected cell. Either the host-cell surface or soluble sialoglycoconjugates could provide the sialic acid. There is no doubt that *trans*-sialidase transfers sialic acid to *T. cruzi* surface glycoconjugates: the 35- and 50-kDa molecules and members of the *trans*-sialidase family themselves are probably the major beneficiaries of *trans*-sialidase activity, depending on the stage of the life cycle, but their function for the trypanosome is unknown.

The ultimate test of the role of *trans*-sialidase will undoubtably involve genetic experiments, although reiteration of the relevant genes could hinder gene-disruption approaches. This problem may be lessened if the genes are arranged in simple tandem arrays. Hopefully, identification of the critical components of the *trans*-sialylation pathway will lead to the development of useful inhibitors.

ACKNOWLEDGMENTS

We thank our colleagues, particularly Simon Paul, for their insights and contributions to our work on this topic, which has been supported by the Wellcome Trust, the Carl J. Herzog Foundation, and the National Institutes of Health. We are grateful to Michael Ferguson, Carlos Frasch, Stuart Kahn, Jerry Manning, Malcolm McConville, Armando Parodi, Sergio Schenkman, and Wesley van Voorhis for permission to quote unpublished information and/or for useful discussions.

Literature Cited

1. Abbott, W. M., Feizi, T. 1991. Soluble 14-kDa beta-galactoside-specific bovine lectin. Evidence from mutagenesis and proteolysis that almost the complete polypeptide chain is necessary for integrity of the carbohydrate recognition domain. *J. Biol. Chem.* 266:5552–57
2. Abuin, G., Colli, W., Souza, W. d., Alves, M. J. M. 1989. A surface antigen of *Trypanosoma cruzi* involved in cell invasion (Tc-85) is heterogeneous in expression and molecular constitution. *Mol. Biochem. Parasitol.* 35: 229–37
3. Adams, J. H., Sim, B. K. L., Dolan, S. A., Fang, X., Kaslow, D. C., et al. 1992. A family of erythrocyte binding proteins of malaria parasites. *Proc. Natl. Acad. Sci. USA* 89:7085–89
4. Alves, M. J. M., Abuin, G., Kuwajima, V. Y., Colli, W. 1986. Partial inhibition of trypomastigote entry into cultured mammalian cells by monoclonal antibodies against a surface glycoprotein of *Trypanosoma cruzi. Mol. Biochem. Parasitol.* 21:75–83
5. Andrews, N. W., Abrams, C. K., Slatin, S. L., Griffiths, G. 1990. A T. cruzi protein immunologically related to the complement component C9: evidence for membrane pore-forming activity at low pH. *Cell* 61:1277–87
6. Andrews, N. W., Webster, P. 1991. Phagolysosomal escape by intracellular pathogens. *Parasitol. Today* 7:335–40
7. Andrews, N. W., Whitlow, M. B. 1989. Secretion by *Trypanosoma cruzi* of a hemolysin active at low pH. *Mol. Biochem. Parasitol.* 33:249–56
8. Anselmi, A., Moleiro, F. 1974. Pathogenic mechanisms in Chagas' cardiomyopathy. In *Trypanosomiasis and Leishmaniasis with Special Reference to Chagas' Disease*. 20(New series):125–36. Amsterdam: Elsevier
9. Botelho-Chaves, L., Briones, M. R. S., Schenkman, S. 1993. Trans-sialidase from *Trypanosoma cruzi* epimastigote is expressed at the stationary growth phase and is different from the enzyme expressed in trypomastigotes. *Mol. Biochem. Parasitol.* In press
10. Campetella, O., Sanchez, D., Cazzulo, J. J., Frasch, A. C. C. 1992. A superfamily of *Trypanosoma cruzi* surface antigens. *Parasitol. Today* 8:378–81
11. Cavallesco, R., Pereira, M. E. A. 1988. Antibody to *Trypanosoma cruzi* neuraminidase enhances infection in vitro and identifies a subpopulation of trypomastigotes. *J. Immunol.* 140:617–25
12. Confalonieri, A. N., Martin, N. F., Zingales, B., Colli, W., de Lederkremer, R. M. 1983. Sialoglycolipids in *Trypanosoma cruzi. Biochem. Int.* 7: 215–22
13. Contreras, V. T., Salles, J. M., Thomas, N., Morel, C. M., Goldenberg, S. 1985. In vitro differentiation of *Trypanosoma cruzi* under chemically defined conditions. *Mol. Biochem. Parasitol.* 16:315–29
14. Corfield, A. P., Schauer, R. 1982. Metabolism of sialic acids. See Ref. 89a, pp. 195–261
15. Corfield, A. P., Schauer, R. 1982. Occurrence of sialic acids. See Ref. 89a, pp. 5–50
16. Corfield, A. P., Veh, R. W., Wember, M., Michalski, J.-C., Schauer, R. 1981. The release of N-acetyl and N-glycoloyl-neuraminic acid from soluble complex carbohydrates and erythrocytes by bacterial, viral and mammalian sialidases. *Biochem. J.* 197:293–99
17. Couto, A. S., Goncalves, M. V., Colli, W., de Lederkremer, R. M. 1990. The N-linked carbohydrate chain of the 85 kilodalton glycoprotein from *Trypanosoma cruzi* trypomastigotes contains sialyl, fucosyl and galactosyl (α1-3) galactose units. *Mol. Biochem. Parasitol.* 39:101–8
18. Couto, A. S., Katzin, A. M., Colli, W., de Lederkremer, R. M. 1987. Sialic acid in a complex oligosaccharide chain of the TC-85 surface glycoprotein from the trypomastigote stage of *Trypanosoma cruzi. Mol. Biochem. Parasitol.* 26:145–54
19. Crocker, P. R., Kelm, S., Dubois, C., Martin, B., McWilliam, A. S., et al. 1991. Purification and properties of sialoadhesin, a sialic acid-binding receptor of murine tissue macrophages. *EMBO J.* 10:1661–69
20. Cross, G. A. M. 1990. Cellular and genetic aspects of antigenic variation in trypanosomes. *Annu. Rev. Immunol.* 8:83–110
21. Cross, G. A. M. 1990. Glycolipid anchoring of plasma membrane proteins. *Annu. Rev. Cell Biol.* 6:1–39
22. Eisen, H., Kahn, S. 1992. Mimicry

in *Trypanosoma cruzi:* fantasy and reality. *Curr. Opin. Immunol.* 3:507–10
23. Engstler, M., Reuter, G., Schauer, R. 1992. Identification of a trans-sialidase activity in procyclic *Trypanosoma brucei. Biol. Chem. Hoppe-Seyler* 373: 843 (Abstr.)
24. Engstler, M., Reuter, G., Schauer, R. 1992. Purification and characterization of a novel sialidase found in procyclic culture forms of *Trypanosoma brucei. Mol. Biochem. Parasitol.* 54:21–30
24a. Engstler, M., Reuter, G., Schauer, R. 1993. The developmentally regulated *trans*-sialidase from *Trypanosoma brucei* sialylates the procyclic acidic repetitive protein. *Mol. Biochem.* In press
25. Fearon, D. 1978. Regulation by membrane sialic acid of B1H-dependent decay-dissociation of amplification C3 convertase of the alternative complement pathway. *Proc. Natl. Acad. Sci. USA* 75:1971–75
25a. Ferguson, M. A. J., Murray, P., Rutherford, H., McConville, M. J. 1993. A simple purification of procyclic acidic repetitive protein and demonstration of a sialylated glycosyl-phosphatidylinositol membrane anchor. *Biochem. J.* 290: In press
26. Ferrero-Garcia, M., Trombetta, S., Sanchez, D., Frasch, A. C. C., Parodi, A. J. 1992. Specificity of *Trypanosoma cruzi* trans-sialidase. *Mem. Inst. Oswaldo Cruz* 87:58–9 (Abstr.)
27. Deleted in proof
28. Fouts, D. L., Ruef, B. J., Ridley, P. T., Wrightsman, R. A., Peterson, D. S., et al. 1991. Nucleotide sequence and transcription of a trypomastigote surface antigen gene of *Trypanosoma cruzi. Mol. Biochem. Parasitol.* 46: 189–200
29. Franco, F. R. S., Yamauchi, L. M., Cano, M. I., Carmo, M. S., Paranhos, G. S., et al. 1992. The major 90-kDa surface antigen of the metacyclic trypomastigote of *Trypanosoma cruzi* is encoded by a large gene family. *Mem. Inst. Oswaldo Cruz* 87:133 (Abstr.)
30. Frevert, U., Schenkman, S., Nussenzweig, V. 1992. Stage-specific expression and intracellular shedding of the cell surface trans-sialidase of *Trypanosoma cruzi. Infect. Immun.* 60:2349–60
31. Gerber, L. D., Kodukula, K., Udenfriend, S. 1992. Phosphatidylinositol glycan (PI-G) anchored membrane proteins. Amino acid requirements adjacent to the site of cleavage and PI-G attachment in the COOH-terminal signal peptide. *J. Biol. Chem.* 267:12168–73
32. Glick, G. D., Toogood, P. L., Wiley, D. C., Skehel, J. J., Knowles, J. R. 1991. Ligand recognition by influenza virus:the binding of bivalent sialisides. *J. Biol. Chem.* 266:23660–69
33. Guther, M. L. S., Cardoso de Almeida, M. L., Yoshida, N., Ferguson, M. A. J. 1992. Structural studies on the glycosylphosphatidylinositol membrane anchor of *Trypanosoma cruzi* 1G7-antigen. The structure of the glycan core. *J. Biol. Chem.* 267:6820–28
34. Hall, B. F., Webster, P., Ma, A. K., Joiner, K. A., Andrews, N. W. 1992. Desialylation of lysosomal membrane glycoproteins by *Trypanosoma cruzi:* into the host cell cytoplasm. *J. Exp. Med.* 176:313–25
35. Harth, G., Haidaris, C. G., So, M. 1987. Neuraminidase from *Trypanosoma cruzi:* analysis of enhanced expression of the enzyme in infectious forms. *Proc. Natl. Acad. Sci. USA* 84:8320–24
36. Hartmann, E., Rapoport, T. A., Lodish, H. F. 1989. Predicting the orientation of eukaryotic membrane-spanning proteins. *Proc. Natl. Acad. Sci. USA* 86:5786–90
37. Hoare, C. A. 1972. *The Trypanosomes of Mammals.* Oxford: Blackwell. 749 pp.
38. Jancik, J. M., Schauer, R., Andres, K. H., von During, M. 1978. Sequestration of neuraminidase-treated erythrocytes: studies on its topographic, morphologic and immunologic aspects. *Cell Tiss. Res.* 186:209–26
39. Joiner, K., Sher, A., Gaither, T., Hammer, C. 1986. Evasion of alternative complement pathway by *Trypanosoma cruzi* results from inefficient binding of factor B. *Proc. Natl. Acad. Sci. USA* 83:6593–97
40. Kahn, S., Colbert, T. G., Wallace, J. C., Hoagland, N. A., Eisen, H. 1991. The major 85-kDa surface antigen of the mammalian-stage forms of *Trypanosoma cruzi* is a family of sialidases. *Proc. Natl. Acad. Sci. USA* 88:4481–85
41. Kahn, S., Van Voorhis, W. C., Eisen, H. 1990. The major 85-kD surface antigen of the mammalian form of *Trypanosoma cruzi* is encoded by a large heterogeneous family of simultaneously expressed genes. *J. Exp. Med.* 172:589–97
42. Kemp, D. J., Coppel, R. L., Anders, R. F. 1987. Repetitive proteins and

genes of malaria. *Annu. Rev. Microbiol.* 41:181–203

43. Kipnis, T. L., David, J. R., Alper, C. A., Sher, A., Dias da Silva, W. 1981. Enzymatic treatment transforms trypomastigotes of *Trypanosoma cruzi* into activators of alternative complement pathway and potentiates their uptake by macrophages. *Proc. Natl. Acad. Sci. USA* 78:602–5
44. Koberle, F. 1974. Pathogenesis of Chagas' disease. In *Trypanosomiasis and Leishmaniasis with Special Reference to Chagas' Disease*. 20(New series):137–58. Amsterdam: Elsevier
45. Kodukula, K., Gerber, L. D., Amthauer, R., Brink, L., Udenfriend, S. 1993. Biosynthesis of glycosylphosphatidylinositol (GPI)-anchored membrane proteins in intact cells: specific amino acid requirements adjacent to the site of cleavage and GPI attachment. *J. Cell Biol.* 120:657–64
46. Kozak, M. 1991. Structural features in eukaryotic mRNAs that modulate the initiation of translation. *J. Biol. Chem.* 266:19867–70
47. Ley, V., Robbins, E. S., Nussenzweig, V., Andrews, N. W. 1990. The exit of *Trypanosoma cruzi* from the phagosome is inhibited by raising the pH of acidic compartments. *J. Exp. Med.* 171:401–13
48. Martin, I., Debarbouille, M., Ferrari, E., Klier, A., Rapoport, G. 1987. Characterization of the levanase gene of *Bacillus subtilis* which shows homology to yeast invertase. *Mol. Gen. Genet.* 208:177–84

48a. Miller, D. J., Macek, M. B., Shur, B. D. 1992. Complementarity between sperm surface β-1,4-galactosyltransferase and egg-coat ZP3 mediates sperm-egg binding. *Nature* 357:589–93

49. Miller, L. H., Roberts, T., Shahabuddin, M., McCutchan, T. F. 1993. Analysis of sequence diversity in the *Plasmodium falciparum* merozoite surface protein-1 (MSP-1). *Mol. Biochem. Parasitol.* 59:1–14
50. Moran, P., Caras, I. W. 1991. Fusion of sequence elements from non-anchored proteins to generate a fully functional signal for glycophosphatidylinositol membrane anchor attachment. *J. Cell Biol.* 115:1595–1600
51. Nairn, C. A., Cole, J. A., Patel, P. V., Parsons, N. J., Fox, J. E., et al. 1988. Cytidine 5′-monophosphate-N-acetylneuraminic acid or a related compound is the low Mr factor from human red blood cells which induces gonococcal resistance to killing by human serum. *J. Gen. Microbiol.* 134:3295–96
52. Ohno, S., Epplen, J. T. 1983. The primitive code and repeats of base oligomers as the primordial protein-coding sequence. *Proc. Natl. Acad. Sci. USA* 80:3391–95
53. Orlandi, P. A., Klotz, F. W., Haynes, J. D. 1992. A malaria invasion receptor, the 175 kDa erythrocyte binding antigen of *Plasmodium falciparum* recognises the terminal Neu 5Acα(2-3) Gal sequences of glycophorin A. *J. Cell Biol.* 116:901–9
54. Ouaissi, M. A., Afchain, D., Capron, A., Grimaud, J. A. 1984. Fibronectin receptors on *Trypanosoma cruzi* trypomastigotes and their biological function. *Nature* 308:380–82
55. Ouaissi, M. A., Cornette, J., Afchain, D., Capron, A., Gras-Masse, H., et al. 1986. *Trypanosoma cruzi* infection inhibited by peptides modeled from a fibronectin cell attachment domain. *Science* 234:603–7
56. Parks, G. D., Lamb, R. A. 1991. Topology of eukaryotic type II membrane proteins: importance of N-terminal positively charged residues flanking the hydrophobic domain. *Cell* 64:777–87
57. Parodi, A. J., Pollevick, G. D., Mautner, M., Buschiazzo, A., Sanchez, D. O., et al. 1992. Identification of the gene(s) coding for the trans-sialidase of *Trypanosoma cruzi*. *EMBO J.* 11: 1705–10
58. Passaniti, A., Hart, G. W. 1988. Cell surface sialylation and tumor metastasis. *J. Biol. Chem.* 263:7591–7603
59. Pays, E. 1989. Pseudogenes, chimaeric genes and the timing of antigen variation in African trypanosomes. *Trends Genet.* 5:389–91
60. Pereira, M. E., Moss, D. 1985. Neuraminidase activity in *Trypanosoma rangeli*. *Mol. Biochem. Parasitol.* 15: 95–104
61. Pereira, M. E. A. 1983. A developmentally regulated neuraminidase activity in *Trypanosoma cruzi*. *Science* 219:1444–46
62. Pereira, M. E. A. 1988. Does *Trypanosoma cruzi* modulate infection by inherent positive and negative control mechanisms? In *The Biology of Parasitism*, ed. P. T. Englund, A. Sher, pp. 105–9. New York: Liss
63. Pereira, M. E. A., Hoff, R. 1986. Heterogeneous distribution of neuraminidase activity in strains and clones of *Trypanosoma cruzi* and its possible

association with parasite myotropism. *Mol. Biochem. Parasitol.* 20:183–91

64. Pereira, M. E. A., Mejia, J. S., Ortega-Barria, E., Matzilevich, D., Prioli, R. P. 1991. The *Trypanosoma cruzi* neuraminidase contains sequences similar to bacterial neuraminidases, YWTD repeats of the low density lipoprotein receptor, and type III modules of fibronectin. *J. Exp. Med.* 174:179–91
65. Peterson, D. S., Fouts, D. L., Manning, J. E. 1989. The 85-kd surface antigen gene of *Trypanosoma cruzi* is telomeric and a member of a multigene family. *EMBO J.* 8:3911–16
66. Peterson, D. S., Wrightsman, R. A., Manning, J. E. 1986. Cloning of a major surface-antigen gene of *Trypanosoma cruzi* and identification of a nonapeptide repeat. *Nature* 322:566–68
67. Petry, K., Eisen, H. 1989. Chagas disease: a model for the study of autoimmune diseases. *Parasitol. Today* 5:111–16
68. Piras, M. M., Henriquez, D., Piras, R. 1987. The effect of fetuin and other sialoglycoproteins on the in vitro penetration of *Trypanosoma cruzi* trypomastigotes into fibroblastic cells. *Mol. Biochem. Parasitol.* 22:135–45
69. Plata, F., Pons, F. G., Eisen, H. 1984. Antigenic polymorphism of *Trypanosoma cruzi:* clonal analysis of trypomastigote surface antigens. *Eur J. Immunol.* 14:392–98
70. Pollevick, G. D., Affranchino, J. L., Frasch, A. C. C., Sanchez, D. O. 1991. The complete sequence of a shed acute-phase antigen of *Trypanosoma cruzi. Mol. Biochem. Parasitol.* 47: 247–50

70a. Pollevick, G. D., Sanchez, D. O., Campetella, O., Trombetta, S., Sousa, M., et al. 1993. Members of the SAPA/*trans*-sialidase protein family have identical N-terminal sequences and a putative signal peptide. *Mol. Biochem. Parasitol.* 59:169–72

70b. Pontes de Carvalho, L. C., Tomlinson, S., Vanderkerckhove, F., Bienen, E. J., Clarkson, A. B., et al. 1993. Characterization of a novel *trans*-sialidase of *Trypanosoma brucei* procyclic trypomastigotes and identification of procyclin as the main sialic acid acceptor. *J. Exp. Med.* 177:465–74

71. Previato, J. O., Andrade, A. F., Pessolani, M. C. 1985. Incorporation of sialic acid into *Trypanosoma cruzi* macromolecules. A proposal for a new metabolic route. *Mol. Biochem. Parasitol.* 16:85–96
72. Prioli, R. P., Mejia, J. S., Aji, T., Aikawa, M., Pereira, M. E. A. 1991. *Trypanosoma cruzi:* localization of neuraminidase on the surface of trypomastigotes. *Trop. Med. Parasitol.* 42:146–50
73. Prioli, R. P., Mejia, J. S., Pereira, M. E. A. 1990. Monoclonal antibodies against *Trypanosoma cruzi* neuraminidase reveal enzyme polymorphism, recognize a subset of trypomastigotes, and enhance infection in vitro. *J. Immunol.* 144:4384–91
74. Prioli, R. P., Ordovas, J. M., Rosenberg, I., Schaefer, E. J., Pereira, M. E. A. 1987. Similarity of cruzin, an inhibitor of *Trypanosoma cruzi* neuraminidase, to high-density lipoprotein. *Science* 238:1417–19
75. Prioli, R. P., Rosenberg, I., Pereira, M. E. A. 1990. High- and low-density lipoproteins enhance infection of *Trypanosoma cruzi* in vitro. *Mol. Biochem. Parasitol.* 38:191–98
76. Prioli, R. P., Rosenburg, I., Pereira, M. E. A. 1987. Specific inhibition of *Trypanosoma cruzi* neuraminidase by the human plasma glycoprotein "cruzin." *Proc. Natl. Acad. Sci. USA* 84:3097–3101
77. Prioli, R. P., Rosenberg, I., Shivakumar, S., Pereira, M. E. A. 1988. Specific binding of human plasma high density lipoprotein (cruzin) to *Trypanosoma cruzi. Mol. Biochem. Parasitol.* 28:257–63
78. Pritchett, T. J., Brossmer, R., Rose, U., Paulson, J. C. 1987. Recognition of monovalent sialosides by influenza virus H3 haemagglutinin. *Virology* 160: 502–6
79. Reuter, G., Schauer, R., Prioli, R., Pereira, M. E. A. 1987. Isolation and properties of a sialidase from *Trypanosoma rangeli. Glycoconjug. J.* 4: 339–48
80. Rfaiee, P., Leffler, H., Byrd, J. C., Cassels, F. J., Boedeker, E. C. 1991. A sialoglycoprotein complex linked to the microvillus cytoskeleton acts as a receptor for pilus (AF/R1) mediated adhesion of enteropathogenic *Escherichia coli* (RDEC-1) in rabbit small intestine. *J. Cell Biol.* 115:1021–29
81. Riede, I., Drexler, K., Eschbach, M.-L., Henning, U. 1986. DNA sequence of the tail fibre genes 37, encoding the receptor recognising part of the fibre of bacteriophages T2 and K3. *J. Mol. Biol.* 191:255–66
82. Roditi, I., Schwarz, H., Pearson, T. W., Beecroft, R. P., Liu, M. K., et

al. 1989. Procyclin gene expression and loss of the variant surface glycoprotein during differentiation of *Trypanosoma brucei*. *J. Cell Biol.* 108: 737–46

83. Roggentin, P., Rothe, B., Kaper, J. B., Galen, J., Lawrisuk, L., et al. 1989. Conserved sequences in bacterial and viral sialidases. *Glycoconjug. J.* 6:349–53
84. Roggentin, P., Rothe, B., Lottspeich, F., Schauer, R. 1988. Cloning and sequencing of a *Clostridium perfringens* sialidase gene. *FEBS Lett.* 238:31–34
85. Romeo, J. M., Esmon, B., Zusman, D. R. 1986. Nucleotide sequence of the myxobacterial haemagglutinin gene contains four homologous domains. *Proc. Natl. Acad. Sci. USA* 83:6332–36
86. Rosenberg, I., Prioli, R. P., Ortega-Barria, E., Pereira, M. E. A. 1991. Stage-specific phospholipase C-mediated release of *Trypanosoma cruzi* neuraminidase. *Mol. Biochem. Parasitol.* 46:303–6
87. Rosenberg, I. A., Prioli, R. P., Mejia, J. S., Pereira, M. E. A. 1991. Differential expression of *Trypanosoma cruzi* neuraminidase in intra- and extracellular trypomastigotes. *Infect. Immun.* 59:464–66
88. Roth, J., Kempf, A., Reuter, G., Schauer, R., Gehring, W. J. 1992. Occurrence of sialic acids in *Drosophila melanogaster*. *Science* 256:673–75
89. Roth, S., McGuire, E. J., Roseman, S. 1971. Evidence for cell-surface glycosyltransferases: their potential role in cellular recognition. *J. Cell. Biol.* 51:536–47

89a. Schauer, R., ed. 1982. *Sialic Acids: Chemistry, Metabolism and Function,* Vol. 10. New York: Springer-Verlag

90. Schauer, R. 1985. Sialic acids and their role as biological masks. *Trends Biochem. Sci.* 10:357–60
91. Schauer, R., Reuter, G., Muhlpfordt, H., Andrade, A. F. B., Pereira, M. E. A. 1983. The occurrence of N-acetyl and N-glycoloylneuraminic acid in *Trypanosoma cruzi*. *Hoppe Zeyler Z. Physiol. Chem.* 364:1053–57
92. Schenkman, S., Ferguson, M. A. J., Heise, N., Cardoso de Almeida, M. L., Mortara, R. A., et al. 1993. Mucin-like glycoproteins linked to the membrane by glycosylphosphatidylinositol anchors are the major acceptor of sialic acid in a reaction catalysed by trans-sialidase in metacyclic forms of *Trypanosoma cruzi*. *Mol. Biochem. Parasitol.* 59:293–304
93. Schenkman, S., Jiang, M. S., Hart, G. W., Nussenzweig, V. 1991. A novel cell surface trans-sialidase of Trypanosoma cruzi generates a stage-specific epitope required for invasion of mammalian cells. *Cell* 65:1117–25
94. Schenkman, S., Kurosaki, T., Ravetch, J. V., Nussenzweig, V. 1992. Evidence for the participation of the Ssp-3 antigen in the invasion of nonphagocytic mammalian cells by *Trypanosoma cruzi*. *J. Exp. Med.* 175:1635–41
95. Schenkman, S., Pontes de Carvalho, L., Nussenzweig, V. 1992. *Trypanosoma cruzi* trans-sialidase and neuraminidase activities can be mediated by the same enzymes. *J. Exp. Med.* 175:567–75
96. Schenkman, S., Yoshida, N., Cardoso de Almeida, M.-L. 1988. Glycophosphatidylinositol-anchored proteins in metacyclic *Trypanosoma cruzi*. *Mol. Biochem. Parasitol.* 29:141–51
97. Schmunis, G. A. 1991. *Trypanosoma cruzi,* the etiologic agent of Chagas' disease: status in the blood supply in endemic and nonendemic countries. *Transfusion* 31:547–57

97a. Shur, B. D. 1991. Cell surface β-1,4 galactosyltransferase: twenty years later. *Glycobiology* 1:563–75

98. Snary, D. 1980. *Trypanosoma cruzi:* antigenic invariance of the cell surface glycoprotein. *Exp. Parasitol.* 49:68–77
99. Souto-Padron, T., Harth, G., De Souza, W. 1990. Immunocytochemical localization of neuraminidase in *Trypanosoma cruzi*. *Infect. Immun.* 58:586–92
100. Stamenkovic, I., Sgroi, D., Aruffo, A., Sun Sy, M., Anderson, T. 1991. The B lymphocyte adhesion molecule CD22 interacts with leukocyte common antigen CD45RO on T-cells and α2-6 sialyltransferase, CD75, on B cells. *Cell* 66:1133–44
101. Takle, G. B., Cross, G. A. M. 1991. An 85kDa surface antigen gene family of *Trypanosoma cruzi* encodes polypeptides homologous to bacterial neuraminidases. *Mol. Biochem. Parasitol.* 48:185–98
102. Takle, G. B., Hudson, L. 1989. Autoimmunity and Chagas' Disease. In *Molecular Mimicry,* ed. M. B. A. Oldstone, 145:79–92. Berlin: Springer-Verlag
103. Takle, G. B., O'Connor, J., Young, A., Cross, G. A. M. 1992. Sequence homology and absence of mRNA defines a possible pseudogene member of the *Trypanosoma cruzi* gp85/sialidase multigene family. *Mol. Biochem. Parasitol.* 56:117–28
104. Tautz, D., Trick, M., Dover, G. A.

1986. Cryptic simplicity in DNA is a major source of genetic variation. *Nature* 322:652–56

105. Thon, G., Baltz, T., Giroud, C., Eisen, H. 1990. Trypanosome variable surface glycoproteins: composite genes and order of expression. *Genes Dev.* 4: 1374–83

105a. Tomlinson, S., Pontes de Carvalho, L., Vandekerckhove, F., Nussenzweig, V. 1992. Resialylation of sialidase-treated sheep and human erythrocytes by *Trypanosoma cruzi trans*-sialidase: Restoration of complement resistance of disialylated sheep erythrocytes. *Glycobiology* 2:549–51

106. Uemura, H., Schenkman, S., Nussenzweig, V., Eichinger, D. 1992. Only some members of a gene family in *Trypanosoma cruzi* encode proteins that express both trans-sialidase and neuraminidase activities. *EMBO J.* 11: 3837–44

106a. Vandekerckhove, F., Schenkman, S., Pontes de Carvalho, L., Tomlinson, S., Kiso, M., et al. 1992. Substrate specificity of the *Trypanosoma cruzi trans*-sialidase. *Glycobiology* 2:541–48

107. Van der Ploeg, L. H. T. 1986. Discontinuous transcription and splicing in trypanosomes. *Cell* 47:479–80

108. Van der Ploeg, L. H. T. 1991. Control of antigenic variation in African trypanosomes. *New Biol.* 3:324–30

109. Van Voorhis, W. C., Eisen, H. 1989. Fl-160, a surface antigen of *Trypanosoma cruzi* that mimics mammalian nervous tissue. *J. Exp. Med.* 169:641–52

110. Van Voorhis, W. C., Schlekewy, L., Le Trong, H. 1991. Molecular mimicry by *Trypanosoma cruzi:* the Fl-160 epitope that mimics mammalian nerve can be mapped to a 12–amino acid peptide. *Proc. Natl. Acad. Sci. USA* 88:5993–97

111. Varki, A. 1992. Selectins and other mammalian sialic acid–binding lectins. *Curr. Opin. Cell Biol.* 4:257–66

112. Vauhkonen, M. 1986. Complex-type carbohydrates of apolipoprotein-B of human plasma low-density lipoproteins. *Glycoconjug. J.* 3:35–43

113. Villalta, F., Lima, M. F., Ruiz-Ruano, A., Zhou, L. 1992. Attachment of *Trypanosoma cruzi* to host cells: a monoclonal antibody recognizes a trypomastigote stage-specific epitope on the gp83 required for parasite attachment. *Biochem. Biophys. Res. Commun.* 182:6–13

114. Villalta, F., Lima, M. F., Zhou, L. 1990. Purification of *Trypanosoma cruzi* surface proteins involved in adhesion to host cells. *Biochem. Biophys. Res. Commun.* 172:925–31

115. von Heijne, G. 1986. A new method for predicting signal sequence cleavage sites. *Nucleic Acids Res.* 14:4683–90

116. Weis, W., Brown, J. H., Cusack, S., Paulson, J. C., Skehel, J. J., et al. 1988. Structure of the influenza virus haemagglutinin complexed with its receptor, sialic acid. *Nature* 333:426–31

117. Yogeeswaran, G., Laine, R. A., Hakomori, S. 1974. Mechanism of cell contact-dependent glycolipid synthesis: further studies with glycolipid-glass complex. *Biochem. Biophys. Res. Commun.* 59:591–99

118. Yoshida, N., Blanco, S. A., Araguth, M. F., Russo, M., Gonzalez, J. 1990. The stage-specific 90-kilodalton surface antigen of metacyclic trypomastigotes of *Trypanosoma cruzi*. *Mol. Biochem. Parasitol.* 39:39–46

119. Yoshida, N., Mortara, R. A., Araguth, M. F., Gonzalez, J. C., Russo, M. 1989. Metacyclic neutralizing effect of monoclonal antibody 10D8 directed to the 35- and 50-kildalton surface glycoconjugates of *Trypanosoma cruzi*. *Infect. Immun.* 57:1663–67

120. Zhang, H., Miller, R. H., Rutishauser, U. 1992. Polysialic acid is required for optimal growth of axons on a neuronal substrate. *J. Neurosci.* 12: 3107–14

121. Ziegelbauer, K., Quinten, M., Schwarz, H., Pearson, T. W., Overath, P. 1990. Synchronous differentiation of *Trypanosoma brucei* from bloodstream to procyclic forms in vitro. *Eur. J. Biochem.* 192:373–78

122. Zingales, B., Carniol, C., de Lederkremer, R. M., Colli, W. 1987. Direct sialic acid transfer from a protein donor to glycolipids of trypomastigote forms of *Trypanosoma cruzi*. *Mol. Biochem. Parasitol.* 26:135–44

Annu. Rev. Microbiol. 1993. 47:413–40

TRANS-SPLICING OF NEMATODE PREMESSENGER RNA

Timothy W. Nilsen
Department of Molecular Biology, and Microbiology, Case Western Reserve University School of Medicine, Cleveland, Ohio 44106

KEY WORDS: RNA processing, snRNAs, *cis*- and *trans*-splicing, Sm snRNPs, *Ascaris lumbricoides, Caenorhabditis elegans,* spliced leader RNAs

---

CONTENTS

0066-4227/93/1001-0413$02.00

ABSTRACT

In nematodes, many mRNAs contain a common 5′ terminal 22-nt sequence. This sequence, the spliced leader (SL), is acquired from a small (~100 nt) SL RNA via *trans*-splicing. Parallel in vitro and in vivo experiments have begun to clarify both the mechanism and biological role of *trans*-splicing. In vitro analysis (in cell free extracts) has shown that *trans*-splicing is remarkably similar to the snRNP mediated removal of intervening sequences from pre-mRNAs (*cis*-splicing). Additionally, this analysis has suggested a mechanism that may explain how the two substrates of *trans*-splicing (the SL RNA and pre-mRNA) efficiently associate with one another in the absence of sequence complementarity. In vivo experiments suggest that a major biological function of *trans*-splicing in nematodes may be to process polycistronic transcription units. Results obtained from the study of both parasitic and free-living species are discussed, and *trans*-splicing in nematodes is compared and contrasted to the analogous process in trypanosomatid protozoans.

INTRODUCTION

Intermolecular *trans*-splicing is, by definition, an RNA processing reaction that precisely joins exons derived from separately transcribed RNAs. In one specific type of *trans*-splicing, spliced-leader (SL) addition, mRNAs acquire their 5′ terminal exon (the spliced leader) from a small RNA known as the SL RNA. Acquisition of the SL sequence occurs via a splicing reaction analogous to snRNP-mediated *cis*-splicing. This unusual mechanism of mRNA maturation has been shown to occur in a variety of lower eukaryotes including trypanosomatid protozoans, some trematodes, *Euglena,* and nematodes (28, 44, 53, 62, 64).

Trans-splicing is apparently ubiquitous in nematodes; it has been demonstrated in numerous evolutionarily diverse free-living and parasitic species. In these organisms, some but not all mRNAs receive a common 22-nucleotide SL sequence from an SL RNA that bears striking similarities to the U snRNAs required for *cis*-splicing. The same pre-mRNAs processed by *trans*-splicing contain conventional introns processed by *cis*-splicing. The existence of *trans*-splicing poses several challenging mechanistic as well as biological questions: how are specific mRNAs chosen for *trans*-splicing; how similar are *cis*- and *trans*-splicing; how can *cis*- and *trans*-splicing coexist in the same nucleus; and most intriguingly, how do the two substrates for *trans*-splicing (the SL RNA and pre-mRNA) efficiently associate with one another? Furthermore, what is the advantage of evolving or, alternatively, retaining *trans*-splicing as a mechanism of mRNA maturation?

The focus of this review is on our current understanding of the mechanism and biological role of *trans*-splicing in nematodes. Both in vitro and in vivo analyses are summarized. Also discussed are recent advances in our understanding of the mechanism of *cis*-splicing, because these studies have provided important insights into the mechanism of *trans*-splicing. Finally the process of *trans*-splicing in nematodes is compared and contrasted to the analogous process in trypanosomatid protozoans. Several reviews covering various aspects of *trans*-splicing have appeared in recent years (1, 5, 6, 11, 23a, 29, 48, 49), and the reader is referred to these for additional information and alternative perspectives.

DISCOVERY AND UNIVERSALITY OF *TRANS*-SPLICING IN NEMATODES

The first evidence for *trans*-splicing in nematodes came from the work of Hirsh and colleagues, who were studying the structure and expression of actin-encoding genes in *Caenorhabditis elegans* (28). *C. elegans,* a free living nematode, contains four actin genes; three clustered on an autosome and one located on the X chromosome. Detailed characterization of the transcripts derived from each gene showed that the mRNAs encoded by the autosomal genes shared a common 22-nucleotide (nt) 5′ terminal leader sequence; the mRNAs transcribed from the gene on the X chromosome lacked this sequence. Surprisingly, the 22-nt leader sequence was not encoded near the autosomal actin genes but instead was located in spacer regions between 5S rRNA genes in the *C. elegans* 5S gene repeat (28). Further experiments established that the leader sequence was transcribed as the first 22 nt of a small (~100 nt) nonpolyadenylated RNA. The 22-nt sequence was immediately 5′ of a potential splice donor site. Analysis of the three actin genes whose mRNAs contained the leader sequence revealed that the sites of leader addition coincided with potential splice-acceptor sites. Collectively, these observations provided strong circumstantial evidence that the 22-nt leader sequence was acquired by *trans*-splicing. Confirmation of *trans*-splicing as the mechanism of leader addition came with the in vivo characterization of intermediates predicted from a *trans*-splicing reaction (see Figure 1) (3, 66).

In addition to providing the first evidence for *trans*-splicing in nematodes, the initial studies on actin gene expression in *C. elegans* revealed two important differences between *trans*-splicing in nematodes and the analogous process in trypanosomes. First, some nematode pre-mRNAs are not *trans*-spliced (only three of four different actin mRNAs contained the SL sequence), whereas the available evidence indicates that all pre-mRNAs in trypanosomes are processed by *trans*-splicing. Second, nematode pre-mRNAs are processed by both *cis*- and *trans*-splicing (the autosomal actin genes in *C. elegans* contain

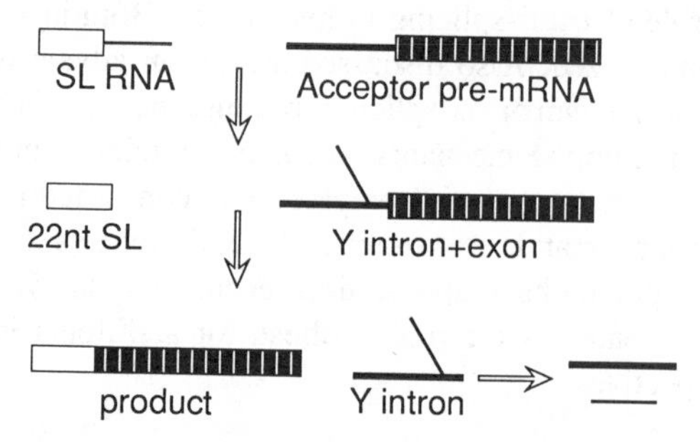

Figure 1 Schematic representation of nematode *trans*-splicing. In the first step of *trans*-splicing, cleavage at the 5′ splice site in the SL RNA yields the free 22-nt SL with a 3′ hydroxyl terminus. Concomitant with this cleavage is the formation of a new 2′5′ phosphodiester bond between the 5′ end of the intron portion of the SL RNA and an adenosine residue upstream of the *trans*-splice acceptor site. In the second step, the 3′ OH present on the free SL attacks the 3′ splice-junction phosphate, creating the *trans*-spliced product and releasing the Y intron product. Subsequently, the 2′5′ linkage in the Y intron product is hydrolyzed by debranching enzyme, producing two linear intron molecules derived from the SL RNA and acceptor pre-mRNAs.

conventional intervening sequences); in trypanosomes, no *cis* introns have been described.

Soon after the discovery of *trans*-splicing in *C. elegans,* it became apparent that the 22-nt SL sequence and SL RNAs were present in a wide variety of evolutionarily diverged nematodes. Characterization of cDNA and genomic clones showed that mRNAs encoding a tRNA synthetase in the filarial parasite, *Brugia malayi,* contained a 22-nt spliced leader sequence identical to that found on the *C. elegans* actin mRNAs. Here, as in *C. elegans,* the spliced leader was derived from an SL RNA transcribed from within the *B. malayi* 5S gene repeat (63). Similar experiments have indicated that the same 22-nt sequence is present in the parasitic nematodes *Oncocerca volvulus, Ascaris lumbricoides, Haemonchus contortus, Angiostrongylus cantonensis,* and *Anisakis* spp.; existence of the SL sequence has also been demonstrated in the free-living nematode, *Panagrellis redivivus* (4, 26, 50, 77). In all cases examined, the SL sequence itself has been perfectly conserved whereas only short sequence elements in the remainder of the SL RNA have been conserved (see below). The presence of SL sequences and SL RNAs in widely diverged nematodes strongly suggests that *trans*-splicing is used as a mechanism of mRNA maturation in all nematodes.

In most nematodes studied to date, SL RNA genes are found (as in *C. elegans*) between 5S rRNA genes in the 5S tandem repeat (48). Transcriptional orientation can be opposite to that of 5S rRNA, as in *C. elegans,* or more commonly, both 5S rRNA genes and SL RNA genes are transcribed from the same DNA strand (48). No obvious functional significance can be ascribed to the association between 5S rRNA and SL RNA genes. As discussed below, RNA polymerase II transcribes SL RNA genes, whereas RNA polymerase III synthesizes 5S rRNA. Furthermore the association of SL RNA and 5S rRNA genes is clearly not obligatory, because in some nematodes the SL RNA genes are tandemly reiterated outside of the 5S gene cluster (4, 26).

In addition to functional copies of the SL sequence (i.e. those copies transcribed as part of the SL RNA), some nematodes possess orphan copies of the 22-nt SL sequence scattered at other sites in the genome (63, 77). Similar orphan SL sequences are found in trypanosomes, and their function, if any, is obscure (45). In nematodes, the orphan copies may function as transcriptional control elements (see below).

A Second SL RNA in C. elegans

As discussed above, the demonstration of *trans*-splicing in *C. elegans* evolved from characterization of genes and mRNAs encoding actins. Similar analysis of genes and mRNAs encoding glyceraldehyde-3-phosphate dehydrogenase in *C. elegans* resulted in the surprising discovery of a second spliced leader sequence (SL2) and corresponding SL RNA (SL2 RNA) (23). SL2 RNAs are encoded by only four genes (as opposed to >100 copies of SL1 RNA), which are unlinked to 5S rRNA genes (23). SL1 RNA and SL2 RNA are not interchangeable because mRNAs that acquire the SL1 sequence do not acquire SL2 and vice versa (23). The demonstration of an alternative spliced leader in *C. elegans* introduced another layer of complexity to nematode *trans*-splicing, i.e. the question of what determines the specificity of SL1 or SL2 addition (see below).

To date, SL2 equivalents have not been described in any nematode other than *C. elegans*. It is not clear whether alternative SL RNAs are present in other nematodes, or whether additional SL RNAs will become evident in *C. elegans*.

EXPERIMENTAL SYSTEMS FOR THE STUDY OF NEMATODE *TRANS*-SPLICING

The bulk of our current understanding of the mechanism and biological significance of nematode *trans*-splicing has been derived from two organisms, *C. elegans* and *Ascaris lumbricoides* var. *suum*. As summarized briefly below, each organism possesses features that make it uniquely suitable for particular experimental approaches. Since its introduction as an experimental system 25 ago by Sydney Brenner, *C. elegans* has become the subject of investigation by an increasing number of laboratories, and it is arguably the most thoroughly understood multicellular organism in terms of anatomy, development, and genetics (73). The complete lineage and fate of each cell in the mature worm have been established, and the entire nervous system has been described in detail. Because of its simplicity, small size, short generation time, and ease of culture in the laboratory, *C. elegans* provides an exceptional system for the genetic analysis of complex questions regarding the regulation of gene expression during growth and differentiation. Well over a thousand genetic loci have been identified and ~90% of the genome has been physically mapped. The first eukaryotic genome to be completely sequenced will probably be that of *C. elegans*. Given this knowledge base and the recent advances in transformation (14), *C. elegans* has provided a valuable system for the in vivo analysis of *trans*-splicing (9, 10).

In vitro analyses are also important because cell-free systems permit the systematic biochemical dissection of macromolecular processes. However, investigators have been unable to use *C. elegans* for in vitro experiments because, to date, attempts to develop cell-free systems from this organism have been largely unsuccessful. Somewhat surprisingly, *A. lumbricoides,* an obligate parasite that cannot be propagated in the laboratory, has proven quite useful for the development of cell-free systems.

A. lumbricoides is a common intestinal parasite of pigs and humans. Because of its large size (adult females average ~27 cm in length) and ready availability, *A. lumbricoides* has for some time been a popular organism for the biochemical analysis of nematode metabolism and physiology. In addition, the unusual DNA arrangements associated with chromosome diminution have received considerable attention (67). However, only recently has *A. lumbricoides* been exploited for the biochemical analysis of RNA processing and transcription.

A. lumbricoides has an enormous reproductive capacity; a mature *A. lumbricoides* female releases ~250,000 eggs per day, and mass quantities of fertilized eggs (embryos) can be obtained from dissected uteri. *A. lumbricoides* eggs have several remarkable physical and biological properties, the most important of which is that, although fertilized, they do not initiate development until they are released from the female worm. Thus, eggs obtained by dissection constitute a homogeneous population of cells. If stored at 4°C, the embryos remain metabolically inert and can be kept for years without loss of viability. Upon transfer to 30°C, the fertilized eggs begin to develop in perfect synchrony with a cell-division time of ~24 hours. This synchrony is maintained throughout larval development, which takes ~11 days. The synchronous and slow process of *A. lumbricoides* embryogenesis enables one to obtain mass quantities of embryos at any stage of development. In addition to their obvious utility for the study of gene activation and/or repression during nematode embryogenesis, *A. lumbricoides* embryos are an excellent source material for cell-free systems. Extracts prepared from 32–64 cell embryos are remarkably versatile in their ability to catalyze various macromolecular processes, including transcription by RNA polymerases II and III, *cis*-splicing, and translation (20, 32, 33; P. A. Maroney & T. W. Nilsen, unpublished data). Furthermore, these extracts are unique in their ability to efficiently catalyze *trans*-splicing (19, 31).

Throughout the remainder of this review, information derived from in vivo analysis in *C. elegans* and in vitro analysis using *A. lumbricoides* is summarized with the implicit assumption that experimental results obtained in one system are applicable to the other and to nematode *trans*-splicing in general. However, the reader should bear in mind that individual members of the nematode phylum diverged many millions of years ago and this assumption may not be completely warranted.

NEMATODE SL RNAs—CHARACTERIZATION AND IN VITRO TRANSCRIPTION

As discussed above, SL RNAs have been characterized in several nematodes including *C. elegans* and *A. lumbricoides*. With the exception of the *C. elegans* SL2 RNA, all nematode SL RNAs have an identical 22-nt SL sequence. Furthermore, in all cases, the SL sequence is immediately 5′ to a conserved splice donor site that contains the invariant GU dinucleotide characteristic of *cis*-splice donor sites. The intron portions of the SL RNAs are of similar lengths (~80 nucleotides) and contain short sequence elements that have been stringently conserved (see below); the remainder of the intron region has diverged considerably in primary sequence (48).

Despite this divergence, all nematode SL RNAs can form nearly identical secondary structures consisting of three stem loops separated by short single-stranded regions (Figure 2) (8, 50, 77). As originally noted and discussed by Bruzik et al (8), trypanosomatid SL RNAs can adopt essentially the same secondary structure.

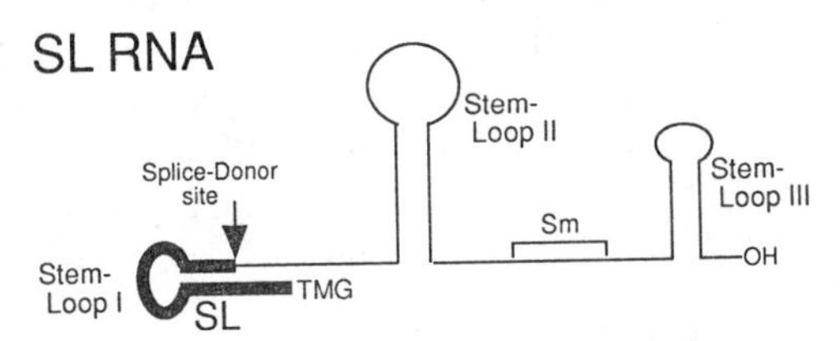

Figure 2 The nematode *trans*-spliced leader RNA drawn schematically in its predicted secondary structure. This structure contains three stem loops (see text). In stem loop I, sequences within the 22-nt SL (*thick line*) are predicted to base pair with the splice-donor site. The single-stranded region between stem loops II and III contains the Sm-binding sequence (see text).

The Nematode SL RNAs Are a Specialized Form of U snRNA

Two intriguing observations emerged from the analysis of primary sequences and potential secondary structures of nematode SL RNAs. First, in stem loop I, sequences within the SL formed part of an imperfect helix that spans the 5′ splice site (Figure 1 and see below). Second, the single-stranded region separating stem loops II and III has been rigidly conserved in all nematode SL RNAs and contains the sequence AAUUUUGGAA.

Interestingly, a portion of this conserved single-stranded region (AAUUUUGG) corresponds to the consensus Sm-binding sequence (RAU$n \geq 4$GR) found in the U snRNAs that are necessary cofactors for *cis*-splicing (36). In U snRNAs the Sm-binding sequence serves to nucleate assembly of U SnRNAs into ribonucleoprotein particles (U snRNPs) by promoting the association of a common set of core proteins with epitopes recognized by antibodies with Sm specificity (36). Concurrent analysis by several groups showed that the Sm-binding sequence in the *C. elegans* SL RNA could be functionally important. These studies demonstrated that the SL RNA (complexed with protein) was immunoprecipitable from *C. elegans* extracts with Sm antisera and that synthetic SL RNAs bound Sm proteins when incubated in HeLa cell extracts (8, 66, 71). Similar experiments showed that the Sm-binding site of *A. lumbricoides* SL RNA was functional in promoting the binding of Sm proteins (31). Furthermore, oligonucleotide affinity chromatography has been used to show that the *A. lumbricoides* SL RNP contains the common core Sm proteins present in the *A. lumbricoides* U1 and U2 snRNPs (D. LaSala & T. W. Nilsen, unpublished). As with nematode SL RNAs, trypanosomatid SL RNAs are found as RNP particles (38, 51), and these particles share common protein constituents with trypanosomatid U snRNPs (51).

In addition to Sm protein binding, nematode SL RNAs share a second feature with the U snRNAs involved in *cis*-splicing. U snRNAs (with the exception of U6 snRNA) contain a 2,2,7-trimethylguanosine-cap structure. Analysis of several nematode SL RNAs with both antibody precipitation (50, 66, 71) and direct chemical analysis (33) indicated that these RNAs contained the hypermethylated cap structure typical of U snRNAs. Collectively, these observations support the notion that nematode SL RNAs represent a specialized form of U snRNA (8, 31, 60, 66, 71). Unlike U snRNAs required for *cis*-splicing, SL RNAs are consumed during the *trans*-splicing reaction.

Transcription of SL RNAs

Given the similarity between the SL RNA and U snRNAs, it seemed likely that the SL RNA would be transcribed by RNA polymerase II (with the exception of U6 snRNA, all U snRNAs are transcribed by RNA pol II). Indeed, nuclear run-on experiments in both *C. elegans* (M. Golumb, personal communication) and *A. lumbricoides* (50) showed that SL RNA synthesis was inhibited by levels of α-amanitin that would be expected to inhibit RNA polymerase II but not RNA polymerases I and III.

Subsequent experiments showed that the *A. lumbricoides* SL RNA gene was accurately and efficiently transcribed by RNA polymerase II from cloned templates in *A. lumbricoides* embryo extracts (33). Availability of this cell-free system permitted the identification of SL RNA transcriptional control elements through mutational analysis. These experiments revealed that the SL RNA gene represents an unusual RNA polymerase II transcription unit. Efficient initiation of SL RNA synthesis required two sequence elements, one of which was centered approximately 50 nucleotides upstream from the transcriptional start site. Remarkably, the second sequence element was the 22-nt SL sequence itself; mutations in the SL sequence abolished transcription in vitro (18). DNase I footprinting showed that the SL sequence bound a protein factor; the boundaries of the footprint exactly coincided with the SL sequence. Competition experiments indicated that binding of the factor was directly correlated with transcription (18). Further analysis has indicated that the 22-nt binding factor is an ~60-kDa protein recently purified using DNA affinity chromatography (J. Denker & T. W. Nilsen, unpublished data). It will be of considerable interest to see if this factor shares any homology with known transcription factors involved in RNA polymerase II transcription.

Additional experiments characterizing SL RNA transcription showed that 3′-end formation (termination) of SL RNA synthesis in vitro was also unusual in that it depended exclusively upon gene-internal sequences (18). This contrasts to the situation in vertebrate snRNA genes in which the primary determinant of 3′-end formation is the so-called 3′-end formation box located a short distance downstream of mature 3′ ends (reviewed in 52).

The organization and expression of the SL RNA gene has some interesting implications. As noted above, the 22-nt SL sequence has been perfectly conserved in widely diverged nematodes. This conservation was commonly interpreted to mean that the 22-nt SL sequence present on *trans*-spliced mRNAs must have some important posttranscriptional function. The finding that the SL sequence is an essential promoter element for its own synthesis suggests an alternative explanation for sequence conservation, i.e. the conservation could be dictated by constraints imposed by the binding specificity of a transcription factor.

Furthermore, the finding that the SL sequence is a promoter element could explain the existence of orphan SL sequences present in some nematodes (see above). The orphan SL sequences might serve as promoter or enhancer elements for other (perhaps protein-coding) genes when uncoupled from the rest of the SL RNA.

TRANS-SPLICING IN VITRO

In addition to catalyzing transcription, *A. lumbricoides* embryo extracts catalyze both *cis*- and more importantly *trans*-splicing (19, 20, 31). The *A. lumbricoides* extract remains the only cell-free system that catalyzes leader-addition *trans*-splicing, and it has provided useful information regarding the mechanism of this unusual RNA-processing reaction.

Initial experiments showed that endogenous SL RNA, which is abundant in the extract, could serve as the SL donor to a synthetic pre-mRNA containing an authentic *trans*-splice acceptor site (19). Direct analysis of products and intermediates established that *trans*-splicing was accurate. As predicted from the characterization of processing intermediates in vivo, *trans*-splicing occurs through two successive cleavage ligation reactions (presumably transesterifications) directly analogous to the two well-characterized steps in *cis*-splicing (see Figure 1).

These analyses also identified the branch point used in *trans*-splicing. Surprisingly, two adjacent adenosines within the 5′ flanking intron were used with equal efficiency as branch acceptors. These adenosines lie within a sequence that does not share any significant identity with the consensus branch sequences used in *cis*-splicing in other organisms (27). Careful mapping of additional *trans*-splice and *cis*-splice branch sites is needed before the significance (if any) of this observation is established.

After conditions for *trans*-splicing using endogenous SL RNA were established, it was determined whether synthetic SL RNA could serve as the *trans*-splice donor. When labeled synthetic SL RNA was added to the *A. lumbricoides* extract, the RNA functioned in *trans*-splicing (31). Direct analysis showed that products and intermediates generated in these reactions were indistinguishable from those generated when endogenous SL RNA served as the *trans*-splice donor. Importantly, participation of synthetic SL RNA in *trans*-splicing did not depend upon prior depletion of endogenous SL RNA from the extract, indicating that synthetic SL RNA could effectively compete with endogenous SL RNA as the *trans*-splice donor (31).

These studies provided the necessary prelude to the biochemical dissection of nematode *trans*-splicing, because availability of a cell-free system in which both substrates of *trans*-splicing could be added exogenously permitted the use of standard mutational approaches to assess which sequences in the substrates (both SL RNA and acceptor) were necessary for function. To date relatively little has been done with the acceptor. Extensive analyses have been carried out on the SL RNA, and these are summarized in the next section.

SEQUENCES ESSENTIAL FOR SL RNA FUNCTION IN *TRANS*-SPLICING

The Nematode SL RNA Participates in Trans-*Splicing as an Sm snRNP*

As described above, the SL RNAs of nematodes share two features in common with U snRNAs: a trimethylguanosine-cap structure and a functional Sm-binding site. Point mutations introduced into the SL RNA's Sm-binding site have been used to determine if these properties were necessary for SL RNA function. Such mutations prevented both binding of Sm proteins and cap trimethylation (31). [Experiments in other systems previously showed that cap trimethylation of the U snRNAs depended upon a functional Sm-binding site (35).] These mutations also completely prevented participation of the mutant SL RNAs in *trans*-splicing. To separate the potential roles of snRNP assembly and cap trimethylation, *trans*-splicing with wild-type SL RNAs was assayed in the presence of the methylation inhibitor S-adenosyl homocysteine. In these experiments, SL RNAs with unmethylated caps functioned perfectly well in *trans*-splicing (31). Thus, to participate in *trans*-splicing, the SL RNA must assemble into an Sm snRNP particle; however, the hypermethylated cap structure is not relevant for *trans*-splicing in vitro.

Specific Exon Sequences Are not Required for SL RNA Function

SL RNAs can be considered to be chimeric molecules comprised of an exon domain (the 22-nt SL) and an snRNA-like domain (8, 60, 66, 71). It seemed likely that the exon domain would be important for SL RNA function because it has been perfectly conserved in all nematodes. Furthermore, as noted above, the SL domain is predicted to participate in an intramolecular base-pairing interaction that spans the 5′ splice site (see Figure 2). This secondary-structure feature is found in SL RNAs from all organisms studied to date (8, 50, 53, 64), and this intramolecular base pairing in the SL RNA could be functionally important because it mimics the intermolecular base pairing between U1 snRNA and *cis* 5′ splice sites (see below) (8). Thus, it was surprising that mutational disruption of this base pairing did not affect the function of the SL RNA in *trans*-splicing in vitro (34). Even more surprisingly, SL RNAs containing nearly complete truncations of the exon domain (to two nucleotides) functioned in *trans*-splicing (34). These results excluded the notion that the exon sequence contributed significantly to SL RNA function in vitro and suggested the possibility that the snRNA-like domain might be able to deliver heterologous exons to an appropriate acceptor via *trans*-splicing. Indeed, when SL RNAs were constructed with a variety of artificial exons ranging in size from 29–246 nt, each exon was used as a donor in *trans*-splicing (34). The observations described above indicated that determinants of SL RNA function resided within the snRNA-like domain of the molecule.

Sequences Within the snRNA-Like Domain Required for Function

Chemical modification interference analysis was used to define critical purine residues within the snRNA-like domain that were essential for SL RNA activity. This experiment identified nine individual purine residues that were clustered in two regions of the molecule, a short stretch on the 3′ side of stem II, and the single-stranded Sm-binding-site region between stems II and III (21). The importance of the sequence elements identified by modification interference was confirmed by mutation and by their ability, in combination, to confer SL RNA function to a fragment of an *A. lumbricoides* U1 snRNA (21).

The functional significance of the single-stranded region containing the Sm-binding site and adjacent nucleotides was examined further. Somewhat surprisingly, chemical modification of any one of the Sm-binding site purines did not inhibit assembly of the SL RNA into an Sm snRNP. This observation, coupled with the observation that a functional Sm-binding site derived from U1 snRNA (AAUUUUUGC) could not substitute for the analogous sequence in the SL RNA (AAUUUUGG), suggested that the Sm-binding region of the SL RNA had a role besides directing assembly into an Sm snRNP. To address the possibility that the significance of this sequence might (at least in part) reside in its ability to interact with another RNA, cross-linking experiments using aminomethyltrioxsalen (AMT) were performed (21). These experiments identified cross-linked species when labeled synthetic SL RNA was incubated in extract. Cross-linked molecules of identical mobility were also observed with endogenous SL RNA (21). Synthetic SL RNAs altered in critical nucleotides 3′ of the Sm-binding site efficiently assembled into Sm RNPs but failed to function in *trans*-splicing. These same mutant SL RNAs failed to cross-link, suggesting that the interaction identified by cross-linking was functionally significant.

To determine if the cross-link resulted from an interaction between the SL RNA and a known U snRNA, cross-linked species were digested with RNase H using a panel of oligodeoxynucleotides complementary to U1, U2, U4, U5, and U6 snRNAs. The cross-linked moieties were sensitive to RNase H in the presence of two separate oligodeoxynucleotides complementary to U6 snRNA and were insensitive to digestion when any of the other oligodeoxynucleotides were used (21). Inspection of the SL RNA and U6 snRNA sequences revealed a striking complementarity of 18 consecutive base pairs with one bulged nucleotide. In the SL RNA, the region of complemen-

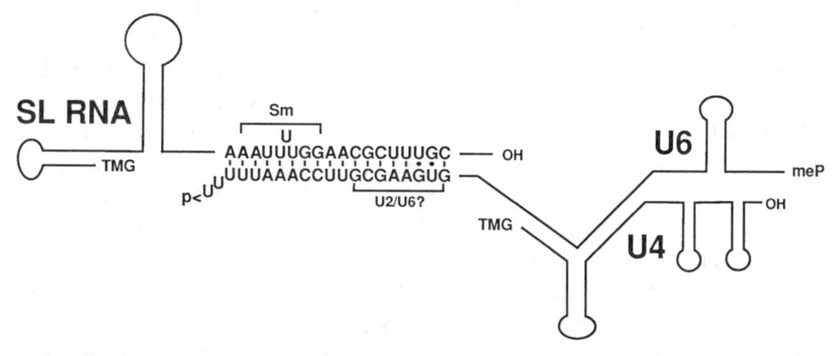

Figure 3 A potential base-pairing interaction between the *A. lumbricoides* SL RNA and *A. lumbricoides* U6 snRNA. The Sm-binding site in the SL RNA (AAUUUUGG) is bracketed, as is the region of U6 snRNA that has been shown to interact with U2 snRNA in mammalian cells (see text). Cap structures of respective RNAs (TMG, trimethylguanosine in SL RNA and U4 snRNA; meP, γ-monomethyl phosphate in U6 snRNA) are shown to indicate polarity. The AMT cross-link described in the text connects the 5′-most uridine of the SL RNA's Sm-binding site and the uridine residue (U98) five residues upstream from the 3′ end of U6 snRNA.

tarity included the Sm-binding site and extended 3′ 10 bases beyond this site (Figure 3). In U6 snRNA, the region of complementarity comprised bases 83–100. Mapping by RNA fingerprinting and partial alkaline hydrolysis placed an AMT cross-link within this region (U98 of U6 snRNA and U77 of the SL RNA) (Y. T. Yu, P. A. Maroney & T. W. Nilsen, unpublished data). U77 is the 5′-most uridine residue of the SL RNA's Sm-binding site. These results suggested that an SL RNA/U6 snRNA base pairing interaction was important for *trans*-splicing. The potential significance of this interaction is discussed below.

U snRNAs AND *TRANS*-SPLICING

Lessons from Cis-*Splicing*

Cis-splicing takes place in a large multicomponent complex known as the spliceosome. Assembly of the spliceosome on precursor mRNAs involves the ordered interaction of five U snRNPs (U1, U2, U4, U5, and U6) and a large (as yet undefined) number of protein factors (see 16, 41, 56 for recent reviews). Over the past several years a relatively clear picture of the role of each snRNP in splicing has emerged. U1 snRNP is the first snRNP to recognize the pre-mRNA, and it does so by binding to the splice donor site. This recognition is mediated by base pairing between the 5′ end of U1 snRNA and the 5′ splice site. Subsequent to U1 binding, U2 snRNP binds to the pre-mRNA at the branch site. Again this recognition involves base pairing between U2 snRNA and the pre-mRNA. Following U1 and U2 binding, U4, U5, and U6 join the presplicing complex as a triple snRNP with U4 and U6 extensively base paired to each other. In addition to its base pairing with U4 snRNA, U6 snRNA makes base-pairing contacts with U2 snRNA (22). This base pairing, initially detected with AMT cross-linking, involves bases near the 5′ end of U2 snRNA and a region near the 3′ end of U6 snRNA (22). Compensatory mutagenesis has established that this U2-U6 interaction is essential for mammalian *cis*-splicing (13, 74) but does not appear to be required for yeast *cis*-splicing. This interaction may serve to stabilize or correctly position the triple snRNP U4-U6-U5 in the spliceosome.

Recent analysis from several groups has provided new and exciting insights into additional snRNA-snRNA interactions within the fully formed complex (30, 57, 58, 72, 75). These observations have provided the basis for a model of *cis*-splicing described briefly below. Prior to the first step of splicing, exons to be spliced are juxtaposed through base-pairing interactions with both U1 and U5 snRNA. Genetic evidence in fission yeast suggests that U1 contacts conserved intron sequences at both the 5′ and 3′ splice sites (54). Perhaps simultaneously, U5 snRNP could make base pairing contacts with exon sequences immediately upstream of the 5′ splice site and downstream of the 3′ splice site (46, 47, 75). As suggested by Steitz (61), the combination of these interactions could produce a four-armed structure reminiscent of the Holliday intermediate in DNA recombination. At the same time as, or following, recognition and juxtaposition of the splice donor and acceptor sites, a dynamic reorganization occurs within the spliceosome whereupon the U4-U6 base-pairing interaction is disrupted and U6 forms a new base-pairing interaction with U2 snRNA (30). (This pairing, which is designated internal U6-U2, is distinct from the U2-U6 interaction described above.) At this point, U4 can be removed from the spliceosome without affecting the subsequent cleavage-ligation reactions (76). Thus, U4 is not an active participant in the catalysis of splicing. Although the details are at present somewhat obscure, the U1 pre-mRNA interaction may be disrupted and replaced by a U6 pre-mRNA interaction in which U6 makes base-pairing contacts with intron sequences just downstream of the 5′ exon (57, 58, 72). U5 loses contact with the 5′ exon but apparently remains associated with the pre-mRNA through base-pairing contacts with intron sequences adjacent to the 5′ splice site (72).

Following reorganization of the spliceosome, the two sequential *trans*-esterification reactions are initiated. Presumably, U6 and U2 in combination are involved in catalysis of these reactions by mechanisms analogous to those that catalyze either group I (37) or group II self-catalyzed splicing reactions (30).

Studies in Trans-*Splicing Systems*

Given the fundamental similarity between *cis*- and *trans*-splicing, *trans*-splicing is most likely catalyzed by the same mechanism used in *cis*-splicing. Soon after the discovery of *trans*-splicing in trypanosomes, studies demonstrated that these organisms possess homologues of U2, U4, and U6 snRNAs (43, 68). Using a permeabilized cell system (70) and RNase H targeted degradation, Tschudi & Ullu (69) clearly demonstrated that these snRNAs are required for trypanosome *trans*-splicing. Similar analysis using the *A. lumbricoides* extract showed that these same U snRNAs (U2, U4, and U6) are required for nematode *trans*-splicing (20). Thus, *cis*- and *trans*-splicing clearly require at least some common components. Sequence analysis of both nematode and trypanosome U snRNAs has revealed that their respective U2, U4, and U6 snRNAs can potentially form the same array of base-pairing interactions characterized in *cis*-splicing systems. Although studies have not yet shown that these interactions are necessary for *trans*-splicing, it seems relatively safe to assume that they are. The base pairing between U6 snRNA and the SL RNA described in the preceding section overlaps the region of U6 that in mammalian cells base pairs to the 5′ end of U2 snRNA. As discussed by Hannon et al (21), several considerations suggest that the U6-SL interaction is limited to the region of the SL RNA that includes the Sm-binding site and three adjacent nucleotides (see Figure 3). If this is true, U6 could, in the *trans*-spliceosome, simultaneously interact with the SL RNA and the 5′ end of U2 snRNA (see below).

Although U2, U4, and U6 snRNPs are clearly essential for *trans*-splicing, whether U1 and U5 snRNPs (known to be essential for *cis*-splicing, see above) are required for this reaction is not known. Curiously, trypanosomes (which carry out only *trans*-splicing) appear to lack homologues of U1 and U5 snRNAs (1). At face value, this observation would suggest that U1 and U5 are not required for *trans*-splicing, and their function (splice-site identification and juxtaposition) would have to be carried out by other factors. Such factors could either be proteins or other snRNAs. An obvious candidate for such an snRNA is the SL RNA itself.

Several years ago, Bruzik et al (8) suggested that the SL RNA might, in *trans*-splicing, fulfill the role of U1 snRNP in *cis*-splicing. Support for this hypothesis came from studies of the processing of chimeric *cis*-splicing substrates containing SL RNA splice-donor sites linked *in cis* to an adenoviral splice acceptor site and 3′ exon (7). Such transcripts, containing either nematode or trypanosomatid SL RNA sequences, were efficiently spliced in HeLa cell extracts in which the 5′ end of U1 snRNA was ablated by targeted RNase H digestion (7). However, subsequent analysis revealed that the SL RNA sequences did not completely obviate the need for U1 snRNP, because an additional function of U1 snRNP (independent of its base pairing to the splice site) was necessary for the splicing of the chimeric transcripts (59). It would seem that the *A. lumbricoides* extracts could provide a straight-forward system in which to address the role (or lack thereof) of U1 snRNP in *trans*-splicing. However, to date this has not proven to be the case. Because nematodes carry out both *cis*- and *trans*-splicing, they contain a full complement of snRNAs including U1 and U5 (65, 66; J. Shambaugh, G. E. Hannon & T. W. Nilsen, unpublished data). In *A. lumbricoides* extracts, oligonucleotide-directed RNase H digestion of U1 efficiently inhibited *cis*-splicing but did not affect *trans*-splicing (20). Furthermore, mutations in the 5′ splice site predicted to disrupt a U1–5′ splice site base-pairing interaction did not affect SL RNA function (34). These experiments suggest that U1 may not be involved in nematode *trans*-splicing. However, this conclusion cannot be considered definitive because, as in mammalian cells (see above) (2), U1 snRNP may perform roles in splicing independent of its ability to base pair with 5′ splice sites. We cannot rule out the possibility that U1 snRNP is performing these (or yet undiscovered) roles in *trans*-splicing. Therefore, in the absence of a totally reconstituted system, the role or lack thereof of U1 in nematode *trans*-splicing will remain an open question.

The other U snRNA implicated in splice-site recognition and juxtaposition is U5 snRNA (46, 47, 75). As stated above, U5 is apparently absent in trypanosomes but is present in nematodes. A conserved sequence within the 39-nt SL of trypanosomes might provide a U5-like function (61). Because the entire 22-nt SL can be deleted or changed in nematode SL RNAs without loss of function, the 22-nt SL sequence cannot provide this function. Indeed, *trans*-splicing in nematodes may require the participation of U5 snRNA (61). Unfortunately, as with U1, it is difficult to test this notion directly. U5 is notoriously resistant to targeted degradation and is nearly impossible to deplete from extracts by other means.

The U5–5′ splice site interaction was revealed through genetic analysis in yeast (46, 47) and more recently by site-specific cross-linking in mammalian extracts (75). Similar cross-linking experiments in *A. lumbricoides* extracts have failed to reveal any U5-SL splice-donor site interaction (J. Denker & T. W. Nilsen, unpublished data), but at present the significance of this negative finding is unclear. Thus, as with U1, the role or lack thereof of U5 in nematode *trans*-splicing remains open. The resolution of these questions might best be answered through the direct analysis of the *trans*-splicing equivalent of the *cis*-spliceosome. Though such a complex doubtless exists, to date the *trans*-spliceosome has eluded capture. Recent progress in the creation of substrate RNAs with site-specific substitutions (42) that arrest splicing (both *cis* and *trans*) after the first step (42; G. J. Hannon & T. W. Nilsen, unpublished data) should yield trapped *trans*-spliceosomes. If U5 is important for *trans*-splicing, it should be found in such complexes.

A WORKING MODEL FOR *TRANS*-SPLICING

As detailed above, extensive analysis in a variety of systems has led to an emerging picture of snRNA-snRNA interactions in *cis*-splicing. In *trans*-splicing, the evidence suggests that an additional interaction (i.e. between the SL RNA and U6 snRNA) may be functionally important (21). Collectively, these observations suggest a plausible (if somewhat speculative) model for *trans*-spliceosome assembly (Figure 4*a*). As shown, the SL RNA is base paired to U6 snRNA (interaction *a*) through the SL RNA's Sm-binding region and the 3′ end of U6. Concurrent with this base pairing, U6 is base paired to U2 snRNA [the 5′U2-3′U6 interaction (interaction *b*) (22)]. U2 snRNA is in turn base paired with the branch acceptor site of the pre-mRNA (interaction *c*). Following release of U4 snRNA (Figure 5), the additional internal base-pairing interaction between U6 and U2 [interaction *d* (30)] becomes possible. As discussed above, the role, if any, of U1 and U5 snRNPs in *trans*-splicing is unknown. The following discussion makes the assumption that these snRNPs do not participate in *trans*-splicing.

In this model, the 5′ splice site on the SL RNA is fixed relative to U6 snRNA because the two molecules are held together by base pairing. In this intimate association, U6 snRNA could conceivably interact with the 5′ splice site by base pairing (interaction *e*) without the prior intervention of U1 snRNA. Through the simultaneous interactions of U2 with U6 and U2 with the branch point, the splice acceptor site could be brought into proximity with the catalytic center of the spliceosome. If this were the case, how could the splice-acceptor site be recognized? In group II self-catalyzed splicing, a single long-range base pair has been implicated in 3′ splice-site recognition (24). Similarly, many group I introns lack recognizable 3′-exon guide sequences (40). Given these examples, 3′ splice-site recognition in *trans*-splicing may well be mediated by yet-to-be discovered secondary (or higher-order) structural interactions. Site-specific cross-linking analysis at the 3′ splice site may provide some insight into this question.

Although the foregoing model is admittedly speculative, it provides a possible scenario for a splicing reaction that would not require U1 or U5 snRNPs. More importantly, it also provides a plausible answer to one of the most important questions in *trans*-splicing, i.e. how the SL RNA and pre-mRNA efficiently associate in the absence of sequence complementarity. In this model, U2 and U6 snRNAs serve as a connecting bridge between the two substrates of *trans*-splicing through concurrent base-pairing interactions.

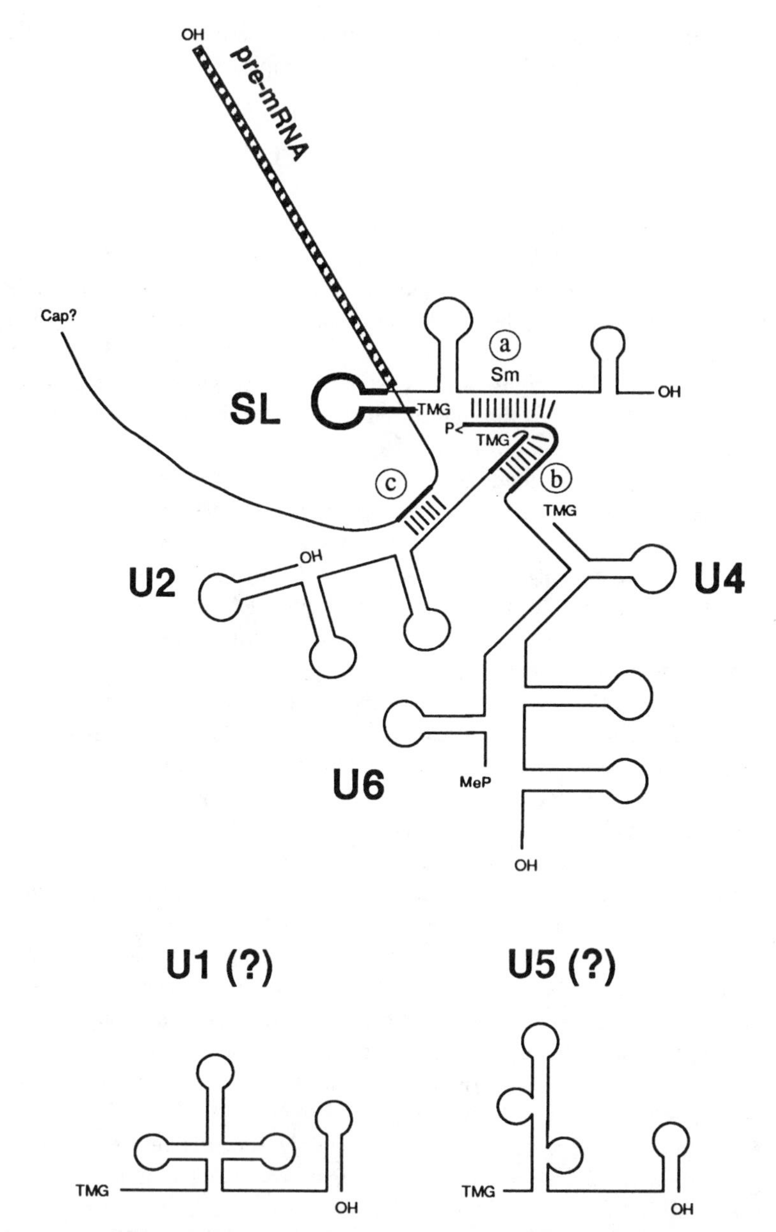

Figure 4 Model of *trans*-spliceosome organization prior to release of U4 snRNA. As discussed in the text, concurrent base-pairing interactions between the SL RNA and U6 snRNA (*a*), between U6 and U2 snRNAs (5′U2-3′U6; *b*), and between U2 snRNA and the branch-point sequence on the pre-mRNA (bold line on pre-mRNA; *c*) could link the pre-mRNA and SL RNA in appropriate geometry for *trans*-splicing. The roles, if any, of U1 and U5 snRNAs [known to juxtapose splice sites in *cis*-splicing (see text)] are unknown.

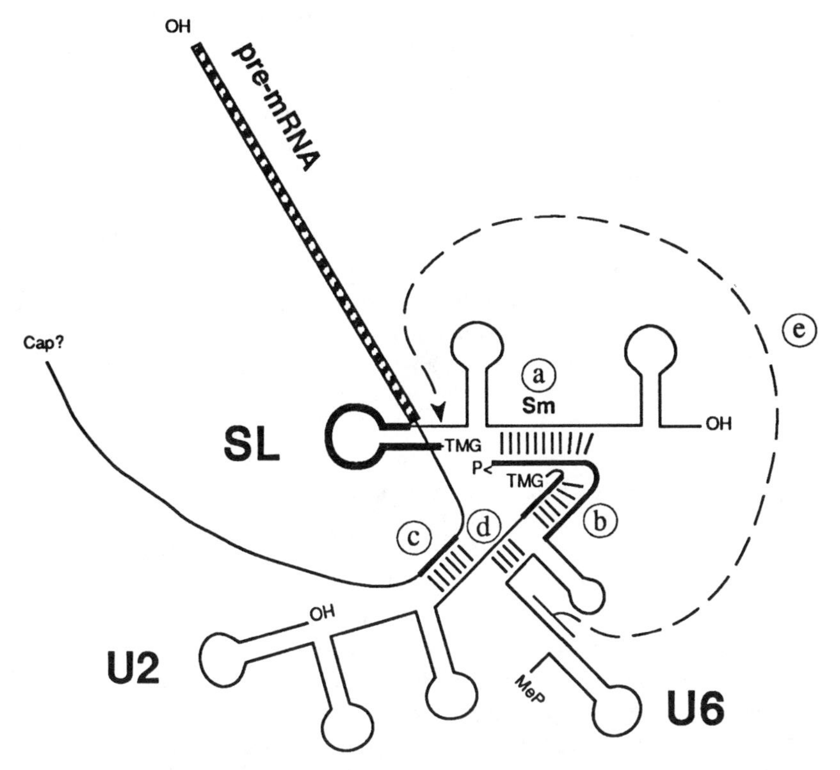

Figure 5 Possible organization of the *trans*-spliceosome following release of U4 snRNA. Following disruption of U4-U6 base pairing, two additional base-pairing interactions become possible (see text). An internal region of U6 becomes accessible for base pairing with U2 snRNA (*d*). As shown schematically, interaction *d* involves sequences in U2 that are between the 5′U2-U6 (*b*) and U2–branch point (*c*) sites. Concurrently, another internal region of U6 could interact with sequences downstream of the splice-donor site in the SL RNA (*e*).

SPLICE-SITE SELECTION IN NEMATODE *TRANS*-SPLICING

The fact that many nematode pre-mRNAs are processed by both *cis*- and *trans*-splicing (in no case has a gene been characterized that is only *trans*-spliced) presents an interesting problem regarding splice-site selection, i.e. how are internal splice-acceptor sites (which function only in *cis*) excluded as sites of SL addition? A further complication (addressed in the following section) arises in *C. elegans;* here, two distinct SL RNAs (SL1 and SL2) are used as mutually exclusive SL donors.

At least two hypotheses (either in combination or alone) could account for the exclusion of internal acceptor sites in *trans*-splicing. First, *trans*-splice acceptors could possess positive sequence elements that would promote SL addition or, conversely, *cis*-splice acceptors could contain negative sequence elements that would prevent the formation of a *trans*-splicing complex. Sites of SL addition have been determined for several pre-mRNAs in a variety of nematodes, and these sites always coincide with what appear to be conventional splice-acceptor sites. Systematic comparison of these *trans*-splice acceptor sites has not revealed any common sequence motifs that could distinguish them from *cis* acceptor sequences.

A second possible explanation for exclusion of internal acceptor sites could be that the presence of a splice donor site upstream of a splice acceptor in some way precludes *trans*-splicing. Several lines of evidence, both in vivo in *C. elegans* and in vitro in *A. lumbricoides* extracts, indicate that this latter explanation is correct. Using transformation, Blumenthal and colleagues introduced a portion of a normal *cis* intron (lacking a 5′ splice site but including a splice-acceptor site) into the 5′ untranslated sequence of a gene whose message was normally not *trans*-spliced (10). Analysis of mRNAs produced from this gene revealed that the bulk of the mRNAs contained the SL sequence. These experiments indicated that a normal *cis*-splice–acceptor site could function as a *trans*-splice acceptor and demonstrated convincingly that a non-*trans*-spliced mRNA could be converted into a *trans*-spliced mRNA in vivo. These same workers have now done the reciprocal experiment, i.e. they inserted a 5′ splice site upstream of a *trans*-splice acceptor. In these constructs, the normally *trans*-spliced mRNAs were found to be exclusively *cis*-spliced (9). Together these experiments argue strongly that an unpaired splice-acceptor site in the primary transcript is necessary and sufficient to direct *trans*-splicing. Furthermore, these results have interesting implications regarding potential biological functions of *trans*-splicing (see below).

Results paralleling those found in vivo have been observed in the *A. lumbricoides* cell-free system. Unpaired *cis*-splice–acceptor sites are used efficiently as *trans*-splice acceptors, and introduction of exonic sequences with their accompanying 5′ splice-donor site upstream of a *trans*-splice acceptor suppresses *trans*-splicing (G. J. Hannon, P. A. Maroney & T. W. Nilsen, unpublished data). The observed suppression of *trans*-splicing by an upstream splice donor site seemed as if it might result from the binding of U1 snRNP to the 5′ splice site. However, when multiple mutations were introduced at the 5′ splice site (including substitutions in the invariant GU) *trans*-splicing was still suppressed. These substrates were not processed by either *cis*- or *trans*-splicing. These experiments indicated that suppression of *trans*-splicing by an upstream exon and donor site did not depend upon a functional 5′ splice site. Elucidating the molecular basis of this phenomenon may provide fundamental information relevant to the segregation of *cis*- and *trans*-splicing. Perhaps more importantly, such experiments may also prove useful in defining sequence elements and factors involved in 5′ splice-site recognition. Although many investigators have studied 5′ splice-site identification intensively, one still cannot predict the existence of a 5′ splice site simply by sequence inspection. The suppression of *trans*-splicing may provide a powerful, if unusual, assay for determining the elements that constitute a 5′ splice site.

In addition to the problem of internal acceptor site exclusion in nematodes, *trans*-splicing (both in nematodes and trypanosomes) poses another significant question regarding splice-site selection. Even granting the notion that an unpaired acceptor site is sufficient to specify *trans*-splicing, how is the *trans*-splice–acceptor site recognized? In both yeast and in mammalian cells, U1 snRNP binding to the 5′ splice site commits a substrate to splicing (25, 39, 55), and a U1 snRNP function (independent of its pairing to the 5′ splice site) is necessary for stable association of U2 snRNP to the branch point (2). If U1 snRNP is not involved in *trans*-splicing (which is probable in nematodes and more certain in trypanosomes), what promotes U2 association with the branch-point sequence? Currently no experimental data are relevant to this point, although the SL RNP might fulfill this function in *trans*-splicing.

BIOLOGICAL ROLE OF *TRANS*-SPLICING IN NEMATODES

Although *trans*-splicing has been known to occur in nematodes for over five years, relatively little is known about its biological function(s). No evidence suggests that *trans*-splicing is either regulated or regulatory in any way. As a consequence of *trans*-splicing, specific mRNAs acquire the SL sequence and its accompanying trimethylguanosine cap. Neither the SL sequence per se nor the cap can be obligatory for any general aspect of nematode mRNA metabolism (i.e. translation or transport) because some mRNAs are not *trans*-spliced. Furthermore, SL addition cannot alter the informational content of a mRNA since the SL sequence does not contain a translation-initiation codon and SL addition cannot create such a codon. *Trans*-splicing probably does not confer any other special property to specific mRNAs because, as described above, normally *trans*-spliced mRNAs can be converted into mRNAs that lack the SL and vice versa. These conversions did not have any striking effects on the metabolism of the respective mRNAs (9, 10).

How Many Pre-mRNAs Are Trans-*Spliced?*

Any explanation for the role(s) of *trans*-splicing must take into account the prevalence of mRNAs that receive the SL sequence. Experimental data published several years ago suggested that a minority (10–15%) of pre-mRNAs in *C. elegans* were processed by *trans*-splicing (4). This estimate was obtained through hybrid arrested translation experiments in rabbit reticulocyte lysates programmed with *C. elegans* mRNA using an oligodeoxynucleotide complementary to the SL sequence (4). It now appears that these experiments underestimated the prevalence of *trans*-splicing. At the time, the prevalent hypothesis was that *trans*-spliced mRNAs did not retain the trimethylguanosine cap, and SL 2 had yet to be discovered. Because rabbit reticulocyte lysates translate trimethylguanosine-capped RNAs quite poorly (12), proteins encoded by *trans*-spliced mRNAs were probably underrepresented in the hybrid arrest experiments. Currently, the prevalence of *trans*-splicing in *C. elegans* (based simply on sequence analysis of genes and their corresponding mRNAs) is estimated to be at least 50% (T. Blumenthal, personal communication). The prevalence of *trans*-splicing in *A. lumbricoides* appears to be even higher. In hybrid arrest experiments (similar to those described above) in a homologous message-dependent translation system, greater than 80% of *A. lumbricoides* mRNAs appear to possess the SL sequence (P. A. Maroney & T. W. Nilsen, unpublished data).

These observations indicate that *trans*-splicing is much more prevalent than previously suspected. Furthermore, the data indicate that *trans*-splicing cannot be restricted to mRNAs encoding particular classes of proteins. In fact, the numerous characterized *trans*-spliced mRNAs in *C. elegans* do not encode proteins that have any particular property in common.

Translation of Trans-*Spliced mRNAs*

As discussed above, *trans*-spliced mRNAs possess both the SL sequence and a trimethylguanosine cap. The presence of this cap structure on mRNAs is unusual. In other eukaryotes, this structure is restricted to snRNAs, where it serves as a nuclear-localization signal (17, 36). Furthermore, trimethylguanosine-capped mRNAs are poorly translated in cell-free systems prepared from higher eukaryotic cells (12). Because *trans*-spliced mRNAs are both exported from the nucleus and translated, nematodes probably evolved RNA transport and translation machineries that differ from those in higher eukaryotes.

Use of the *A. lumbricoides* cell-free translation system showed that the SL sequence itself does not confer a translational advantage, i.e. synthetic mRNAs possessing or lacking the SL were translated equally. However, the trimethylguanosine cap did enhance translation efficiency; in the *A. lumbricoides* cell-free system, trimethylguanosine-capped mRNAs were translated significantly better than the same mRNAs containing a conventional monomethyl cap structure (P. A. Maroney & T. W. Nilsen, unpublished data). This observation is curious and would suggest that non-*trans*-spliced mRNAs would be at a disadvantage in translation if, in fact, they contain monomethyl caps. However, this is yet to be demonstrated. An interesting possibility is that all nematode mRNA caps may be hypermethylated.

Does Trans-*Splicing Function in the Processing of Polycistronic Transcription Units?*

In trypanosomes, many pre-mRNAs are embedded within long multicistronic primary transcripts (1, 43a, 68a). A major function of *trans*-splicing in these organisms is to mature the 5′ ends of internally located mRNAs; 3′-end maturation occurs through endonucleolytic cleavage and polyadenylation. Recent results (T. Blumenthal, personal communication) indicate that *trans*-splicing may serve a similar role in nematodes. As noted above, *C. elegans* contains two distinct SL RNAs (SL1 and SL2). Investigations designed to determine the specificity of SL1 or SL2 addition revealed that SL2-accepting pre-mRNAs were encoded by genes that were located only a short distance (~100 nt) 3′ of upstream genes transcribed in the same orientation. Transgenic analysis using a specific gene pair demonstrated that promoter elements driving expression of the SL2-accepting gene were not present in either the intergenic region or in the upstream gene itself. Instead, expression of both genes was driven by elements well 5′ of the upstream gene (J. Spieth, K. Lea, G. Brooke & T. Blumenthal, personal communication). In additional experiments, a heat-shock promoter was inserted immediately in front of the gene pair. In this case, the message made from the downstream gene contained the SL2 leader sequence, and its appearance depended upon heat shock. This experiment demonstrated that when a polycistronic RNA was created artificially, it could yield mature, correctly *trans*-spliced mRNA. From these and other analyses, Blumenthal and colleagues have concluded that polycistronic transcription units exist in *C. elegans*. Furthermore, it appears that a major (if not the only) determinant of SL2 addition in *C. elegans* is that the accepting pre-mRNAs be encoded by genes internal in such transcription units (T. Blumenthal, personal communication).

In addition to providing significant insight into the mechanism of discrimination between SL1 and SL2 addition in *C. elegans,* the studies discussed above may also bear on our general understanding of the biological role of nematode *trans*-splicing. In this regard, the demonstration of polycistronic transcription units in nematodes seems to be of fundamental importance. Common occurrence of such transcription units in nematodes would suggest that the major function of *trans*-splicing in these organisms (as it is in trypanosomes) is in the maturation of 5′ ends of mRNAs located within long poly-pre-mRNAs. In this view, SL2 *trans*-splicing in *C. elegans* would reflect a specialized form of *trans*-splicing used only when adjacent genes are in close proximity to each other. SL1 *trans*-splicing in *C. elegans,* and *trans*-splicing in general in other nematodes (which apparently lack alternative SL RNAs), would be used when adjacent pre-mRNAs are more widely spaced in the primary transcript. Although this notion has intrinsic appeal in that it would provide a functional link between *trans*-splicing in nematodes and trypanosomes, it remains largely speculative. Only two promoters of genes whose mRNAs are *trans*-spliced have been tentatively mapped in *C. elegans* (15; T. Blumenthal, personal communication), and our knowledge of promoters for genes (encoding *trans*-spliced or non-*trans*-spliced mRNAs) in other nematodes is nonexistent. Clearly, systematic transcription-unit mapping is necessary before any conclusions can be drawn regarding the role (if any) of nematode *trans*-splicing in processing polycistronic mRNAs.

CONCLUSIONS AND PERSPECTIVES

Because the majority of mRNAs are *trans*-spliced in both *A. lumbricoides* and *C. elegans, trans*-splicing apparently plays a more central role in nematode gene expression than previously suspected. Through parallel in vitro and in vivo approaches, significant progress has been made in understanding the mechanism and potential biological role of *trans*-splicing, although many fundamental questions remain to be answered.

In vitro experiments have provided a possible mechanism for the efficient association of the SL RNA and acceptor pre-mRNAs and suggest that the process of *trans*-splicing may be remarkably similar to *cis*-splicing. An important unanswered question is whether U1 and U5 snRNPs play any role in nematode *trans*-splicing. The answer should provide basic information relevant to splice-site selection and will clarify the mechanistic relationship of *trans*-splicing in nematodes and trypanosomes. It will also be important to address the role of non-snRNP-associated proteins in *trans*-splicing. Although significant progress has been made in the identification of such proteins, which are necessary for *cis*-splicing in yeast and mammalian systems (reviewed in 41), nothing is known about their presumed involvement in *trans*-splicing.

In vivo experiments have demonstrated the existence of polycistronic transcription units in *C. elegans,* suggesting that the role of *trans*-splicing in nematodes may be similar to its role in trypanosomes. It will be important to establish whether polyscistronic transcription units exist in nematodes other than *C. elegans* and if *trans*-splicing serves to process these units.

In the coming years, the combined in vitro approaches available in *A. lumbricoides* and genetic manipulations afforded by *C. elegans* should provide answers to many of the remaining questions regarding nematode *trans*-splicing. These answers may eventually shed some light on the enigmatic evolutionary origin of this unusual RNA-processing pathway.

ACKNOWLEDGMENTS

I thank members of my laboratory, especially P. Maroney, Y. T. Yu, and G. J. Hannon for helpful comments on the manuscript and J. Shambaugh for preparation of figures. I also thank T. Blumenthal and J. A. Steitz for communicating their results prior to publication. Work in the author's laboratory was supported by PHS grants GM-31528 and AI-28799, a grant from the John D. and Catherine T. MacArthur Foundation, and an award from the Burroughs Wellcome Fund.

Literature Cited

1. Agabian, N. 1990. Trans-splicing of nuclear pre-mRNAs. *Cell* 61:1157–60
2. Barabino, S. M. L., Blencowe, B. J., Ryder, U., Sproat, B. S., Lamond, A. 1990. Targeted snRNP depletion reveals an additional role for mammalian U1 snRNP in spliceosome assembly. *Cell* 63:293–302
3. Bektesh, S. L., Hirsh, D. I. 1988. *C. elegans* mRNAs acquire a spliced leader through a *trans*-splicing mechanism. *Nucleic Acids Res.* 16:5692
4. Bektesh, S. L., van Doren, K. V., Hirsh, D. 1988. Presence of the *Caenorhabditis elegans* spliced leader on different mRNAs and in different genera of nematodes. *Genes Dev.* 2: 1277–83
5. Blumenthal, T., Thomas, J. 1988. *Cis* and *trans*-mRNA splicing in *C. elegans*. *Trends Genet.* 4:305–8
6. Borst, P. 1986. Discontinuous transcription and antigenic variation in trypanosomes. *Annu. Rev. Biochem.* 55: 701–32
7. Bruzik, J. P., Steitz, J. A. 1990. Spliced leader RNA sequences can substitute for the essential 5′ end of U1 RNA during splicing in a mammalian in vitro system. *Cell* 62:889–99
8. Bruzik, J. P., van Doren, K., Hirsh, D., Steitz, J. A. 1988. *Trans*-splicing involves a novel form of small ribonucleoprotein particles. *Nature* 335: 559–62
9. Conrad, R., Liou, R. F, Blumenthal, T. 1993. Conversion of a *trans*-spliced *C. elegans* gene into a conventional gene by introduction of a splice donor site. *EMBO J.* 12:1249–56
10. Conrad, R., Thomas, J., Spieth, J., Blumenthal, T. 1991. Insertion of part of an intron into the 5′ untranslated region of a *Caenorhabditis elegans* gene converts it into a *trans*-spliced gene. *Mol. Cell Biol.* 11:1921–26
11. Cummins, C., Anderson, P. 1988. Regulatory myosin light chain genes of *Caenorhabditis elegans*. *Mol. Cell Biol.* 8:5339–49
12. Darzynkiewicz, E., Stepinski, J., Ekiel, I., Jin, Y., Sijuwade, T., Tahara, S. M. 1988. Beta-globin mRNAs capped with m7G, m2,7G or m2,2,7G differ in intrinsic translation efficiency. *Nucleic Acids Res.* 16:8953–62
13. Datta, B., Weiner, A. M. 1991. Genetic evidence for base pairing between U2 and U6 snRNAs in mammalian mRNA splicing. *Nature* 352:821–24
14. Fire, A. 1986. Integrative transformation of *Caenorhabditis elegans*. *EMBO* 5:2673–80
15. Graham, R. W., van Doren, K., Bektesh, S., Candido, E. P. M. 1988. Maturation of the major ubiquitin gene transcript in *Caenorhabditis elegans* involves the acquisition of a *trans*-spliced leader. *J. Biol. Chem.* 263:10413–19
16. Guthrie, C. 1991. Messenger RNA splicing in yeast: clues to why the spliceosome is a ribonucleoprotein. *Science* 253:157–63
17. Hamm, J., Darzynkiewicz, E., Tahara, S. M., Mattaj, I. W. 1990. The trimethyl-guanosine cap structure of U1 snRNA is a component of a bipartite nuclear targeting signal. *Cell* 62:569–77
18. Hannon, G. J., Maroney, P. A., Ayers, D. G., Shambaugh, J. D., Nilsen, T. W. 1990. Transcription of a nematode *trans*-spliced leader RNA requires internal elements for both initiation and 3′ end formation. *EMBO J.* 9:1915–21
19. Hannon, G. J., Maroney, P. A., Denker, J. A., Nilsen, T. W. 1990. *Trans*-splicing of nematode pre-messenger RNA in vitro. *Cell* 61:1247–55
20. Hannon, G. J., Maroney, P. A., Nilsen,

T. W. 1991. U small nuclear ribonucleoprotein requirements for nematode *cis-* and *trans*-splicing in vitro. *J. Biol. Chem.* 266:22792–95
21. Hannon, G. J., Maroney, P. A., Yu, Y.-T., Hannon, G. E., Nilsen, T. W. 1992. Interaction of U6 snRNA with a sequence required for function of the nematode SL RNA in trans-splicing. *Science*. 258:1775–80
22. Hausner, T.-P., Giglio, L. M., Weiner, A. M. 1990. Evidence for base-pairing between mammalian U2 and U6 small nuclear ribonucleoprotein particles. *Genes Dev*. 4:2146–56
23. Huang, X. Y., Hirsh, D. 1989. A second *trans*-spliced RNA leader sequence in the nematode *Caenorhabditis elegans*. *Proc. Natl. Acad. Sci. USA* 86:8640–44
23a. Huang, X. Y., Hirsh, D. 1992. RNA *trans*-splicing. *Genet. Eng*. 14:221–29
24. Jacquier, A. 1990. Self-splicing group II and nuclear pre-mRNA introns: how similar are they? *Trends Biochem Sci*. 15:351–54
25. Jamison, S. F., Crow, A., Garcia-Blanco, M. A. 1992. The spliceosome assembly pathway in mammalian extracts. *Mol. Cell Biol.* 10:4279–87
26. Joshua, G. W. P., Chuang, R. Y., Cheng, S. C., Lin, S. F., Tuan, R. S., Wang, C. C. 1991. The spliced leader gene of *Angiostrongylus cantonesis*. *Mol. Biochem. Parasit.* 460: 209–18
27. Keller, E. B., Noon, W. A. 1984. Intron splicing: a conserved internal signal in introns of animal pre-mRNAs. *Proc. Natl. Acad. Sci. USA* 81:7417–20
28. Krause, M., Hirsh, D. 1987. A trans-spliced leader sequence on actin mRNA in C. elegans. *Cell* 49:753–61
29. Laird, P. 1989. Trans-splicing in trypanosomes—archaism or adaptation? *Trends Genet*. 5:204–8
30. Madhani, H. D., Guthrie, C. 1992. A novel base-pairing interaction between U2 and U6 snRNAs suggests a mechanism for catalytic activation of the spliceosome. *Cell* 71:803–17
31. Maroney, P. A., Hannon, G. J., Denker, J. A., Nilsen, T. W. 1990. The nematode spliced leader RNA participates in *trans*-splicing as an Sm snRNP. *EMBO J*. 9:3667–73
32. Maroney, P. A., Hannon, G. J., Nilsen, T. W. 1989. Accurate and efficient RNA polymerase III transcription in a cell-free extract prepared from *Ascaris suum* embryos. *Mol. Biochem. Parasit.* 35:277–84
33. Maroney, P. A., Hannon, G. J., Nilsen, T. W. 1990. Transcription and cap trimethylation of a nematode spliced leader RNA in a cell free system. *Proc. Natl. Acad. Sci. USA* 87:709–13
34. Maroney, P. A., Hannon, G. J., Shambaugh, J. D., Nilsen, T. W. 1991. Intramolecular base pairing between the nematode spliced leader and its 5′ splice site is not essential for *trans*-splicing in vitro. *EMBO J.* 10: 3869–75
35. Mattaj, I. W. 1986. Cap trimethylation of U snRNA is cytoplasmic and dependent on U snRNP protein binding. *Cell* 46:905–11
36. Mattaj, I. W. 1988. U snRNP assembly and transport. In *Small Nuclear Ribonucleoprotein Particles,* ed. M. L. Birnstiel, pp. 100–14. New York: Springer-Verlag
37. McPheeters, D. S., Abelson, J. 1992. Mutational analysis of the yeast U2 snRNA suggests a structural similarity to the catalytic core of Group I introns. *Cell* 71:819–31
38. Michaeli, S., Roberts, J. G., Watkins, K. P., Agabian, N. 1990. Isolation of distinct small ribonucleoprotein particles containing the spliced leader and U2 RNAs of *Trypanosoma brucei*. *J. Biol. Chem.* 265:10582–88
39. Michaud, S., Reed, R. 1991. An ATP-independent complex commits pre-mRNA to the mammalian spliceosome assembly pathway. *Genes Dev*. 5:2534–46
40. Michel, F., Umesono, K., Ozeki, H. 1989. Comparative and functional anatomy of group II catalytic introns—a review. *Gene* 82:5–30
41. Moore, M. J., Query, C. C., Sharp, P. A. 1992. Splicing of precursors to messenger RNAs by the spliceosome. In *The RNA World,* ed. R. Gesterland, S. Adkins. Cold Spring Harbor, NY: Cold Spring Harbor Press. In press
42. Moore, M. J., Sharp, P. A. 1992. Site-specific modification of pre-mRNA: the 2′ hydroxyl groups at the splice sites. *Science* 256:992–97
43. Mottram, J., Perry, K. L., Lizardi, P. M., Luhrmann, R., Agabian, N., Nelson, R. G. 1989. Isolation and sequence of four small nuclear U RNA genes of *Trypanosoma brucei* subsp. *brucei:* identification of the U2, U4 and U6 RNA analogs. *Mol. Cell Biol*. 9:1212–23
43a. Muhich, M. L., Boothroyd, J. C. 1988. Polycistronic transcripts in trypanosomes and their accumulation during heat shock: evidence for a precursor

role in mRNA synthesis. *Mol. Cell Biol.* 8:3837–46

44. Murphy, W. J., Watkins, K. P., Agabian, N. 1986. Identification of a novel Y branch structure as an intermediate in trypanosome mRNA processing: evidence for trans-splicing. *Cell* 47:517–25
45. Nelson, R. G., Parsons, M., Barr, P. J., Stuart, K., Selkirk, M., Agabian, N. 1983. Sequences homologous to the variant antigen mRNA spliced leader are located in tandem repeats and variable orphons in *Trypanosoma brucei*. *Cell* 34:901–9
46. Newman, A., Norman, C. 1991. Mutations in yeast U5 snRNA alter the specificity of 5′ splice-site cleavage. *Cell* 65:115–23
47. Newman, A. M., Norman, C. 1992. U5 snRNA interacts with exon sequences at 5′ and 3′ splice sites. *Cell* 68:743–54
48. Nilsen, T. W. 1989. Trans-splicing in nematodes. *Exp. Parasitol.* 69:413–16
49. Nilsen, T. W. 1992. *Trans*-splicing in protozoa and helminths. *Infect. Agents Dis.* 1:212–18
50. Nilsen, T. W., Shambaugh, J., Denker, J., Chubb, G., Faser, C., et al. 1989. Characterization and expression of a spliced leader RNA in the parasitic nematode, *Ascaris lumbricoides* var *suum*. *Mol. Cell Biol.* 9:3543–47
51. Palfi, Z., Günzl, A., Cross, M., Bindereif, A. 1991. Affinity purification of *Trypanosoma brucei* snRNPs reveals common and specific protein components. *Proc. Natl. Acad. Sci. USA* 88:9097–9101
52. Parry, H. D., Scherly, D., Mattaj, I. W. 1989. 'Snurpogenesis': the transcription and assembly of U snRNP components. *Trends Biochem. Sci.* 14: 15–19
53. Rajkovic, A., Davis, R. E., Simonsen, J. N., Rottman, F. M. 1990. A spliced leader is present on a subset of mRNAs from the human parasite *Schistosoma mansoni*. *Proc. Natl. Acad. Sci. USA* 87:8879–83
54. Reich, C., Van Hoy, R., Porter, G. L., Wise, J. A. 1992. Mutations of the 3′ splice site can be suppressed by compensatory base changes in U1 snRNA in fission yeast. *Cell* 69:1159–69
55. Rosbash, M., Seraphin, B. 1991. Who's on first? *Trends Biochem. Sci.* 16:187–90
56. Ruby, S. W., Abelson, J. 1991. Pre-mRNA splicing in yeast. *Trends Genet.* 7:79–85
57. Sawa, H., Abelson, J. 1992. Evidence for a base-pairing interaction between U6 snRNA and the 5′ splice site during the splicing reaction in yeast. *Proc. Natl. Acad. Sci. USA* 89:11269–73
58. Sawa, H., Shimura, Y. 1992. Association of U6 snRNA with the 5′splice site region of pre-mRNA in the spliceosome. *Genes Dev.* 6:244–54
59. Seiwert, S. D., Steitz, J. A. 1993. Uncoupling two functions of the 5′ end of U1 snRNA during in vitro splicing. *Mol. Cell Biol.* In press
60. Sharp, P. A. 1987. Trans-splicing: variation on a familiar theme? *Cell* 50:147–48
61. Steitz, J. A. 1992. Splicing takes a holliday. *Science* 257:888–89
62. Sutton, R., Boothroyd, J. C. 1986. Evidence for trans-splicing in trypanosomes. *Cell* 47:527–35
63. Takacs, A. M., Denker, J. A., Perrine, K. G., Maroney, P. A., Nilsen, T. W. 1988. A 22-nucleotide spliced leader in the human parasitic nematode *Brugia malayi* is identical to the *trans*-spliced leader exon in *Caenorhabditis elegans*. *Proc. Natl. Acad. Sci. USA* 85:7932–36
64. Tessier, L.-H., Keller, M., Chan, R., Fournier, R., Weil, J. H., Imbault, P. 1991. Short leader sequences may be transferred from small RNAs to pre-mature mRNAs by trans-splicing in *Euglena*. *EMBO J.* 10:2621–25
65. Thomas, J., Lea, K., Zuker-Aprison, E., Blumenthal, T. 1990. The spliceosomal snRNAs of *Caenorhabditis elegans*. *Nucleic Acids Res.* 18: 2633–42
66. Thomas, J. D., Conrad, R. C., Blumenthal, T. 2988. The C. elegans trans-spliced leader RNA is bound to Sm and has a trimethylguanosine cap. *Cell* 54:533–39
67. Tobler, H., Etter, A., Mueller, F. 1992. Chromatin diminution in nematode development. *Trends Genet.* 8: 427–31
68. Tschudi, C., Richards, F. F., Ullu, E. 1986. The U2 RNA analogue of *Trypanosoma brucei* gambiense: implications for splicing mechanisms in trypanosomes. *Nucleic Acids Res.* 14: 8893–8903

68a. Tschudi, C., Ullu, E. 1988. Polygene transcripts are precursors to calmodulin mRNAs in trypanosomes. *EMBO J.* 7:455–63

69. Tschudi, C., Ullu, E. 1990. Destruction of U2, U4, or U6 small nuclear RNAs blocks trans-splicing in trypanosome cells. *Cell* 61:459–66

70. Ullu, E., Tschudi, C. 1990. Permeable trypanosome cells as a model system for transcription and *trans*-splicing. *Nucleic Acids Res*. 18:3319–26
71. van Doren, K., Hirsh, D. 1988. *Trans*-spliced leader RNAs exist as small nuclear ribonucleoprotein particles in *Caenorhabditis elegans*. *Nature* 335: 556–58
72. Wassarman, D. A., Steitz, J. A. 1992. Interactions of small nuclear RNAs with precursor messenger RNA during in vitro splicing. *Science* 257:1918–25
73. Wood, W. B., ed. 1988. *The Nematode* Caenorhabditis elegans. Cold Spring Harbor, NY: Cold Spring Harbor Press
74. Wu, J. A., Manley, J. L. 1991. Base-pairing between U2 and U6 snRNAs is necessary for splicing of mammalian pre-mRNA. *Nature* 352:818–21
75. Wyatt, J. R., Sontheimer, E. J., Steitz, J. A. 1992. Site-specific crosslinking of mammalian U5 snRNP to the 5′ splice site prior to the first step of premessenger RNA splicing. *Genes Dev*. 6:2542–53
76. Yean, S.-L., Lin, R.-J. 1991. U4 small nuclear RNA dissociates from a yeast spliceosome and does not participate in the subsequent splicing reaction. *Mol. Cell Biol*. 11:5571–77
77. Zeng, W., Alarcon, C. M., Donelson, J. E. 1990. Many transcribed regions of the *Onchocerca volvulus* genome contain the spliced leader sequence of *Caenorhabditis elegans*. *Mol. Cell Biol*. 10:2765–73

Annu. Rev. Microbiol. 1993. 47:441–65

REGULATION OF THE PHOSPHORELAY AND THE INITIATION OF SPORULATION IN *BACILLUS SUBTILIS*

James A. Hoch

Department of Molecular and Experimental Medicine, The Scripps Research Institute, 10666 N. Torrey Pines Road, La Jolla, California 92037

KEY WORDS: Spo0A, signal transduction, transcription activation, phosphorylation

---

CONTENTS

ABSTRACT

The initiation of sporulation of bacteria is a complex cellular event controlled by an extensive network of regulatory proteins that serve to ensure that a cell embarks on this differentiation process only when appropriate conditions are met. The major signal-transduction pathway for the initiation of sporulation is the phosphorelay, which responds to environmental, cell cycle, and metabolic signals, and phosphorylates the Spo0A transcription factor activating its function. Signal input into the phosphorelay occurs through activation of kinases to phosphorylate a secondary-messenger protein, Spo0F. Spo0F~P serves as a substrate for phosphoprotein phosphotransferase, Spo0B, which phosphorylates Spo0A. The pathway is regulated by transcriptional control of its component proteins and by regulating phosphate flux through the pathway. This is accomplished by several regulatory proteins, and by activated

0066-4227/93/1001-0441$02.00

Spo0A, which regulates transcription of genes for its own synthesis. Spo0A~P indirectly controls the transcription of numerous genes by regulating the level of other transcription regulators and directly activates the transcription of several regulatory proteins and sigma factors required for progression to the second stage of sporulation. Although the pathway and regulatory proteins have been identified, the signals and effectors for these regulators remain a mystery.

Introduction

Spore formation is one of the more interesting types of morphological differentiation in bacteria. This process is energy-intensive, involves the activation of numerous specific genes in a temporal sequence, and requires several hours to complete. Hence, sporulation is subject to an extensive control network that serves to ensure that a cell only embarks on this differentiation process when appropriate conditions are met. What are these conditions? What environmental and metabolic factors provide input into this network to induce sporulation? How does a cell monitor these signals and convert that information into a decision to either divide or sporulate? These are some of the questions that were asked when studies were begun of the control of the initiation process in *Bacillus subtilis*. Figure 1 graphically presents the problem.

The initiation of sporulation can only occur during a certain time period in

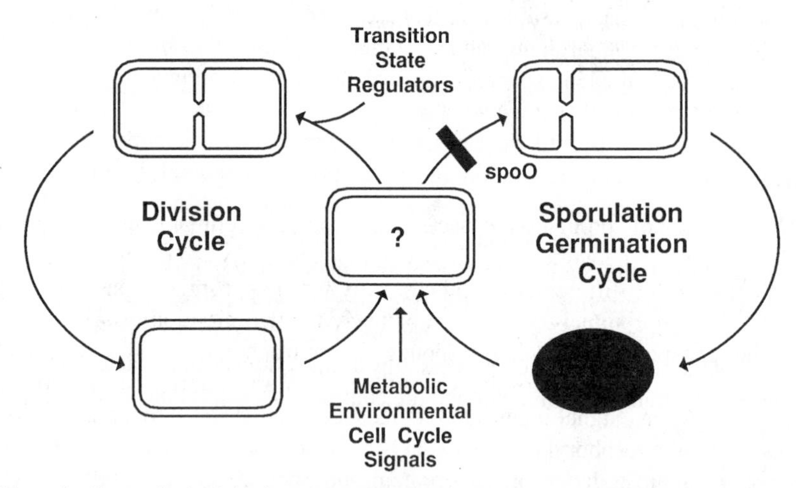

Figure 1 Factors in the initiation of sporulation. At a certain time in the cell cycle, the cell must monitor its metabolic and environmental signals and determine whether to divide or initiate sporulation. Transition-state regulators serve to direct the cell toward division rather than sporulation and respond to unknown signals. The *spo0* mutants are blocked at the very early stages of sporulation and prevent all sporulation-specific transcription.

the cell's cycle (26). In some regards, sporulation and division differ only by the location of the septum in the dividing cell. In a vegetatively growing cell, the septum is placed in the middle of the cell and two cells result, whereas in sporulation, the septum that divides the two chromosomes is placed at one end of the cell, ultimately resulting in two cells within a single cell. Regardless of the outcome, the decision to initiate sporulation cannot be made at just any time during the division cycle, but rather must be made during a distinct period or window within the cell cycle. Because sporulation normally occurs under conditions of nutritional deprivation, the cell is presumed to monitor certain environmental and metabolic signals in order to help it make this decision. Clearly, under conditions of nutritional excess and low population density, growth is favored over survival because competition depends upon increasing cell number. On the other hand, survival at high population densities, when competition for nutrients is intense, is an appropriate cause to initiate sporulation and proceed with this complex process. Thus, both nutritional signals from the environment or from the metabolism of the cell, as well as cell cycle– and density–dependent signals, might very well be involved in these initial events.

One approach to identifying the mechanism of sporulation initiation has been the isolation of mutants blocked in the initiation process (18). These mutants are termed *spo0,* or stage 0, mutants and are most commonly distinguished by their inability to produce any of the characteristic morphological structures of sporulation, as well as by defects in their expression of many enzymes, such as subtilisin, that are coincidentally produced at the very earliest stages of sporulation and/or stationary phase. The *spo0* mutants are thought to be locked in exponential growth in that they continue to grow under nutritional conditions that would normally induce sporulation, and they appear to maintain growth until the nutrients are exhausted, whereupon cell lysis occurs. Clearly, the regulation of sporulation is inexorably coupled with processes that are characteristic of stationary phase but not necessarily of sporulation. Thus, in a normal culture grown in the laboratory, sporulation occurs during the stationary phase, and many of the processes, alternate pathways, and enzymes formed during the early part of stationary phase are controlled along with sporulation because the cell controls sporulation and many of the stationary-phase processes by a single transcription factor, Spo0A. The Spo0A transcription factor and the control of its activation are the subject of this review.

spo0 *Genes and the Initiation Mechanism*

Many of the early studies of sporulation genetics dealt with the isolation of mutants deficient in the sporulation process, characterization of such mutants as to the stage of sporulation in which they stopped, and genetic mapping to differentiate and to characterize the various genetic loci involved in the sporulation process (38). Among the mutations were the *spo0* alleles thought to define the central processing unit that received the sporulation signals and transduced this information into transcriptional activation of the sporulation process (18). Genetics alone was not sufficient to characterize the products of these genes, but the cloning and identification of the size and structure of the protein product of such genes was more informative. Furthermore, molecular cloning allowed the large-scale expression of the products of the *spo0* genes. Cloning also allowed purification of the products so that in vitro biochemistry could be undertaken to investigate their functions that previously were unknown.

The first *spo0* gene to be identified as to function was the product of the *spo0H* gene (8). Sequencing studies led to a deduced structure for the *spo0H* gene product, which had some homology to bacterial sigma factors. In vitro studies of the *spo0H* gene product led to its confirmation as a sigma factor, now known as σ^H (9). This sigma factor is very important in sporulation because it is necessary, but not sufficient, for high-level transcription of the *spo0A, spo0F,* and the *kinA* genes, along with several other genes that have no relevance to the sporulation process (39). The *spo0H* gene codes for a sigma factor that regulates sporulation genes and probably other genes that are expressed and function in early stationary phase (20). These studies eliminated the *spo0H* gene product as a regulatory component of the central processing unit and focused attention onto the other *spo0* genes.

Sequence studies of the *spo0A* and *spo0F* genes were exceptionally informative as to the function of their deduced gene products. Both Spo0A and Spo0F proteins had homology to an emerging class of regulators called two-component regulatory systems (30, 46). Two-component systems are thought to be simple signal transducing switches by which an environmental signal of one kind or another activates the sensor kinase, which then activates a response-regulator protein to promote the transcription of genes specific for the particular environmental change that activated the kinase. Activation results in autophosphorylation of the kinase, which then can transfer this phosphate to the response regulator, activating its transcriptional activities. These two-component switches are widespread in bacteria, and cells are thought to contain many different pairs of these switches, enabling them to respond to a wide variety of different environmental situations. The finding of two-component switches in *spo0* genes was particularly satisfying because the environment was thought to be one of the major inducers of the sporulation process.

The *spo0A* gene coded for a protein of 29,691 M_r with the typical structure of a transcription factor of two-component systems (11, 23). The amino terminal half of the protein was homologous to response regulators, whereas the carboxyl half of the protein was unique and distinct from any other known response-regulator proteins. The product of the *spo0F* gene is a protein of 14,286 M_r without the carboxy terminal domain characteristic of transcription factors (59). At that time, the only other protein known to consist only of a response-regulator homologous domain was the CheY protein of *Salmonella typhimurium* (45). This protein acts as a vital component of the chemotaxic response of this organism. Spo0F, on the other hand, was known to not affect chemotaxis in an analogous manner, and therefore, its function remained obscure. The most curious aspect of *spo0* mutants was that none of the genes sequenced for any of the loci produced a sensor kinase molecule. Because it was thought that all response regulators were activated by phosphorylation from a sensor kinase (28), the absence of such a protein from the repertoire of genes known to be involved in sporulation initiation was particularly mystifying.

Ultimately, an important kinase for sporulation was unveiled serendipitously from the routine sequencing of a gene that gave a stage II phenotype, *spoIIJ*. The gene encoded a protein homologous to sensor kinases but without a gene for a response regulator in the same transcription unit (1, 31). Expression of this gene in *Escherichia coli* with subsequent purification of the gene product showed that the *spoIIJ* locus did indeed code for a kinase that functioned to phosphorylate both the Spo0F and the Spo0A molecules, although the relative activity on Spo0F was much higher than that observed for Spo0A (31). Therefore, a sensor kinase had been identified that phosphorylated the Spo0F protein well, yet neither the environmental stimulus for this kinase nor the function of the phosphorylated Spo0F was known.

Genetic studies had strongly indicated that the key factor in sporulation initiation was the Spo0A protein because certain mutations in Spo0A were found to suppress the need for the *spo0F* and *spo0B* gene products (22), among others. This led to the conclusion that Spo0A represented the end product that was activated by the other *spo0* gene products (19). This proposed role for Spo0A fit with its presumed transcriptional role based on homology to two-component regulatory systems.

The impasse as to how Spo0A was phosphorylated was eventually solved through in vitro biochemical studies. The first *spo0* gene to be cloned and sequenced was the *spo0B* gene (3, 10). Genetic data on *spo0B* mutants indicated that this gene product was essential for the initiation of sporulation and, therefore, must play an important role in the ultimate activation of Spo0A. The deduced amino acid sequence had no homology to any proteins of known function, and thus, its role in the process of initiation was unknown. Expression and purification of the Spo0B protein allowed its inclusion in biochemical studies of the kinase reactions with KinA, Spo0F, and Spo0A. When purified Spo0B was added to reaction mixtures with KinA, Spo0F, and

Spo0A, the Spo0B protein facilitated the phosphorylation of Spo0A. These experiments revealed a series of reactions that are now termed a *phosphorelay* (4), in which KinA phosphorylates Spo0F to yield Spo0F~P, and the phosphate group from Spo0F~P is transferred to Spo0A by the Spo0B protein, which serves here as a kinase for Spo0A that uses Spo0F~P rather than ATP as a substrate. Thus a protein that played the same role as a kinase was identified from the *spo0* genes but was simply not recognized as such because the homology between kinase proteins is in the ATP-binding domain.

The phosphorelay (Figure 2) consists of four basic reactions; the initial event is the signal transduction that causes KinA to autophosphorylate, although the nature of this signal has not been ascertained. The remainder of the phosphorelay consists of three phosphotransfer reactions in which the phosphate group is transferred first from KinA~P to Spo0F to produce Spo0F~P. Spo0B mediates the transfer of phosphate from Spo0F~P to Spo0A via a phosphorylated enzyme intermediate. Both Spo0F and Spo0A are phosphorylated as a mixed anhydride of an aspartic acid residue, which is characteristic of the phosphorylation of response regulator proteins (41). The enzyme-bound phosphorylated intermediates of KinA and Spo0B have the properties of phosphoramidates, and probably both of these are histidine-phosphate intermediates (4).

When the deduced sequences of the *spo0* gene products became available and clearly none of the genes coded for a kinase, it should have been obvious that more than one kinase could carry out this function. An experiment in which kinase genes were cloned and the effect of their mutations in combination with a *kinA*-gene mutation was tested has identified a second sporulation-specific kinase (61). The data from Table 1 indicate that *kinA*

$$\text{KinA} + \text{ATP} \longrightarrow \text{ADP} + \text{KinA{\sim}P} \quad (\text{his}_{405})$$

$$\text{KinA{\sim}P} + \text{Spo0F} \longrightarrow \text{KinA} + \text{Spo0F{\sim}P} \quad (\text{asp}_{54})$$

$$\text{Spo0F{\sim}P} + \text{Spo0B} \rightleftharpoons \text{Spo0F} + \text{Spo0B{\sim}P} \quad (\text{his}_{?})$$

$$\text{Spo0B{\sim}P} + \text{Spo0A} \rightleftharpoons \text{Spo0B} + \text{Spo0A{\sim}P} \quad (\text{asp}_{56})$$

Figure 2 Component reactions of the phosphorelay. The phosphorylated amino acid residue in each product is shown in parentheses.

Table 1 Sporulation frequencies in kinase mutants

Strain background	Relevant genotype	Cells per ml	Spores per ml	Spores (%)
168	Prototroph	1.8×10^8	1.9×10^8	100
168	*kinA::pJM8115*	4.3×10^8	1.1×10^7	2.6
168	*kinB::pJH4906*	4.7×10^8	4.8×10^8	100
168	*kinA::pJM8115 kinB::pJH4906*	6.5×10^7	<10	0

mutations depress sporulation significantly, although this is simply delayed sporulation and eventually most of the cells will sporulate (31). The mutations in *kinB,* on the other hand, have very little effect on sporulation by themselves, but in combination with *kinA* mutations they depress sporulation to almost zero. Thus, both KinA and KinB can act as initiators for the sporulation event, although in vitro studies have not absolutely proven that KinB phosphorylates Spo0F as its primary target. The *kinB* gene is encoded in an operon with another gene, *kapB,* whose gene product is required for the functioning of

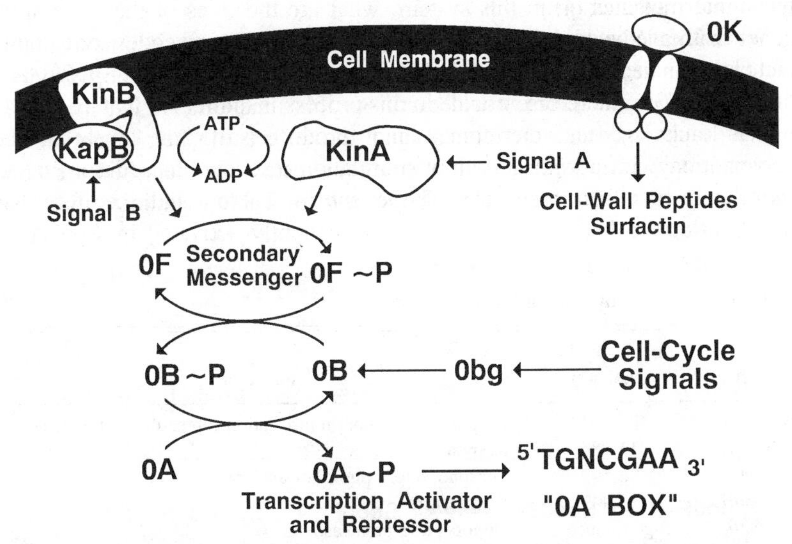

Figure 3 The role of the phosphorelay in sporulation. Two kinases, KinA and KinB-KapB, provide phosphate input to the *spo0F* secondary messenger. The Spo0F~P formed is a substrate for Spo0B and is transferred to Spo0A. Spo0A~P is a transcription factor that recognizes the 0A box in promoters it controls. Cell-cycle signals may be interpreted through Obg, a GTP-binding protein that may influence the activity of Spo0B. The Spo0K locus encodes an oligopeptide-transport system required for the transport of some peptides required for sporulation initiation.

KinB. KapB may be responsible for activation of KinB by facilitating effector ligand binding or KinB-Spo0F interaction.

The phosphorelay, as presently envisaged in Figure 3, is consistent with both the biochemical and genetic data. Mutations in the *spo0F, spo0B,* or *spo0A* genes are absolutely sporulation defective, indicating that both KinA and KinB must work through Spo0F to activate Spo0A. Additional kinases working directly on Spo0A must not exist or mutations in either *spo0F* and *spo0B* would not have the defective phenotype. The only substrate for Spo0B to phosphorylate Spo0A must be Spo0F~P. If other response regulators could serve as a substrate for this enzyme, *spo0F* mutations would not give a sporulation-defective phenotype. The ability of *sof* mutations of *spo0A* to bypass both *spo0F* and *spo0B* indicates that both *spo0F* and *spo0B* gene products modify Spo0A. This does not rule out that other kinases could function to phosphorylate Spo0A under nonphysiological conditions, or that response regulators other than Spo0F~P could serve as substrates for Spo0B under some unusual circumstances.

What Are the Functions of the Other spo0 *Genes?*

If the *kinA* or *kinB* and *spo0F, spo0B,* and *spo0A* gene products are sufficient for signal transduction in this system, what are the roles of the other *spo0* genes that have been identified (Table 2)? The *spo0H* gene codes for a sigma factor that is necessary for high-level transcription of the *spo0F, spo0A,* and *kinA* genes (39); therefore, its role in this process is indirect in that an Spo0H mutant lacks σ^H and therefore cannot produce sufficient levels of the phosphorelay components to allow sporulation to occur. In addition, σ^H is implicated in the transcription of the *spoIIA* operon and therefore has transcriptional roles in genes that are subsequently activated by Spo0A~P

Table 2 Genes involved in initiation of sporulation

Locus	Protein product (mol wt)	Function
spo0A	29,691	Response regulator, transcription repressor, and activator
spo0F	14,286	Response regulator
spo0B	22,542	Phosphoprotein phosphotransferase
spo0H	25,447	Sigma factor
spo0K	5-gene operon	Oligopeptide permease
spo0E	9,791	Negative regulator
obg	47,668	G protein, essential for growth
kinA	69,170	Cytoplasmic transmitter kinase
kinB	47,774	Membrane transmitter kinase
abrB	10,773	Multifunctional repressor or preventer
hpr	23,718	Multifunctional repressor or preventer

(64). Several other *spo0* loci were identified in the genetic studies. These loci have been cloned and characterized, but their function in at least two cases remains obscure.

The *spo0K* mutation has been found to define the oligopeptide permease system that is homologous to the *opp* operon of *S. typhimurium* (32, 40). The original *spo0K* mutant consisted of two mutations, an *opp* mutation that blocked the function of the oligopeptide permease, and a second mutation that causes strains bearing the *opp* mutation to acquire an *spo0* mutant phenotype (32). The mutation leading to sporulation deficiency, however, is clearly in the *opp* operon. The function of the oligopeptide permease is to transport small peptides, up to five amino acids, from the outside. The permease consists of five proteins: OppA, which codes for an external specificity determinant protein that recognizes peptides; OppB and OppC, which are integral membrane-spanning proteins making up the core of the permease; and OppD and OppF, which are ATP-binding proteins that are located on the cytoplasmic side of the membrane and provide the energy for peptide transport. Curiously, although OppD and OppF are both required for peptide transport in *S. typhimurium,* only OppD is required in *B. subtilis* for both peptide transport and for sporulation. Mutants in the *oppF* gene can transport at least some peptides but are deficient in competence (40). The transport of some peptides across the membrane is now believed to be the step in competence requiring the *opp* system (16), and therefore different peptides apparently utilize different ATP-binding domains for their transport. Thus, both *oppD* and *oppF* mutants are competence defective, whereas only *oppD* mutants are defective in sporulation.

What peptide is transported that is required for sporulation? This is the major underlying question that remains to be answered. It has been postulated that extracellular differentiation factors are produced that may help in the recognition of cell density, and these might be peptides in nature and transported by the oligopeptide permease system (15). Thus, lack of the *opp* system prevents such peptides from being internalized and, therefore, some aspect of sporulation is defective. The postulated role of the *opp* system in *S. typhimurium* is the recycling of cell-wall peptides that are cut from the peptidoglycan cross-linkers to allow cell-wall growth. The turnover of cell-wall peptides may play an important role in the signaling of growth and sporulation, and the inability to transport these peptides may lower their internal concentration, which could be a signal for growth rather than sporulation (32).

The *spo0E* gene codes for a small protein of 9791 Daltons, and the original mutations in this gene were found to be nonsense mutations in that portion of the gene coding for the carboxyl end of the protein (33). Investigators concluded that the *spo0E* gene product is required for the onset of sporulation and must have a positive role in this process. Subsequent studies showed that this conclusion was in error; deletions of the *spo0E* gene, rather than give rise to a sporulation-defective phenotype, in fact resulted in sporulation proficiency (35). Thus, the phenotype of the original mutations in the *spo0* gene is apparently caused by production of truncated peptides brought about by the nonsense mutations in the gene. Furthermore, overproduction of this gene product on a multicopy plasmid results in sporulation deficiency, which again suggests that this protein can interfere with sporulation, either when it is overproduced or through a modification by carboxyl truncation. Deletions of the *spo0E* gene give rise to secondary mutations in strains bearing them. These mutations are in other components of the phosphorelay, suggesting that deletion of the *spo0E* gene causes inappropriate timing of the expression of phosphorelay compensated for by secondary mutations in the pathway. Thus, the *spo0E* gene product must play some negative role in the pathway's control. An *spo0E* deletion has no effect on the relative transcription of the *kinA, spo0F, spo0B,* or *spo0A* genes, so Spo0E may control the flow of phosphate through this pathway (35). Consistent with this hypothesis is the fact that deletions of the *spo0E* gene suppress many of the missense mutations in the *spo0F* gene, which can be interpreted to mean that the pathway is much more efficient at using phosphate from low-level defective or unstable Spo0F proteins (J. A. Hoch, unpublished data). If indeed the *spo0E* gene product is a negative regulator of the pathway, then it is of some interest to determine how this works and what the target of its action is.

A third *spo0* gene, *spo0J,* has been described and recently cloned and sequenced (27, 29). The *spo0J* locus consists of two genes, resides very near the origin of replication of the chromosome, and the gene products have homology to the *korB-incC* genes of certain plasmids. These genes are somehow involved in the regulation of plasmid partition, which may indicate that the *spo0J* genes have a role in segregation of the chromosome into the two compartments that ultimately become mother cell and forespore. The forespore compartment results from an asymmetric septum produced for the initial segregation of the chromosome. This septum should certainly have special partition functions that differentiate it from a cell-division septum. Perhaps the segregation of the chromosomes is different in division and the initial sporulation event and this segregation or partition is mediated by the *spo0J* gene products. However, the exact function of *korB-incC* proteins in something as easy to study as plasmid replication and segregation is still quite murky—the functions of the *spo0J* genes may be very complex. Because *spo0J* mutations prevent subsequent spore gene expression, the chromosome segregation may have a direct effect on the control of subsequent genes in sporulation. That is, the physical act of segregating the chromosome into a forespore compartment may directly affect transcription of subsequent genes in the sporulation process.

The Transition State and Its Regulators

The transition state is roughly defined as that period of time between the end of exponential growth and the onset of stage II of sporulation. In cultures in the laboratory, this transition state lasts 1–2 h and is characterized by the initial synthesis of several proteins characteristic of early stationary phase. These proteins include enzymes such as amylase, subtilisin, and neutral proteases. Other phenomena such as the production of antibiotics also characterize stage II. In nature, the transition state may exist as an extended partially quiescent state, where nutrients are insufficient for sustained growth and division, but the insufficiency is not enough to kick off the sporulation process. Thus the cell produces extracellular carbohydrate and protein-degrading enzymes to scavenge the environment for sources of carbon and nitrogen. Antibiotics are produced to protect the ecological niche in which the organism has found itself.

In nature cells presumably spend most of their time in this type of state, only rarely having the opportunity for exponential growth when unexpected nutrient abundance occurs. The transition state is for the most part controlled by the same regulators that control sporulation. The *spo0* mutants in general never produce any of the structures characteristic of spores, and they are inhibited in the production of many of the products characteristic of the transition state, such as proteases and antibiotics. Thus, a regulatory commonality is associated with early stationary phase or transition state and sporulation. Many of these features common to both are controlled by a group of proteins termed transition-state regulators. The best characterized of these are the products of the *abrB* gene, the *hpr* gene, and the *sin* gene, although other regulators of this sort have been described. Although several important and complete reviews of the functions of transition-state regulators have recently appeared, it is worthwhile here to place their function in context with the overall control of sporulation (44, 50, 52).

The AbrB transition-state regulator is one of the most well-studied regulators of this type. The *abrB* gene codes for a protein of 10,500 Daltons that assembles into a hexameric structure for its interaction with DNA (36). AbrB prevents gene expression during exponential growth by binding to promoters of genes that are usually activated during stationary phase or the early part of sporulation (54, 68). In this regard, functioning of AbrB is a bellwether for nutritional excess and low population density where the cell has an excess of nutrients and has no desire whatsoever to enter the sporulation process. Under these conditions AbrB is maximally active, and its role is to prevent the activation of numerous genes that are associated with stationary phase or are not required under these conditions. AbrB mainly serves as a preventer of the expression of such genes and may not be the primary regulator of their expression; i.e. it prevents the expression of these genes during exponential growth even if such genes are regulated by their normal regulators. The competitive advantage that AbrB provides the cell is to minimize the expression of extraneous gene products that would interfere with the maximal rate of growth of the cell under nutrient-excess conditions. It plays a role formally but not mechanistically similar to the Cap protein of *E. coli* that prevents the expression of many unessential genes under conditions of catabolite excess. AbrB, however, does not appear to be the *B. subtilis* equivalent of Cap, but rather functions to indicate the level of Spo0A~P in the cell. Insomuch as Spo0A~P levels are controlled by catabolites and mechanisms similar to catabolite repression, AbrB is superficially similar in function to Cap.

The regulation of genes by AbrB and its mechanism of binding to sensitive promoters has been extensively studied. AbrB seems to work in a concentration-dependent manner, and no small molecule effectors have been found that either inhibit or enhance its activity toward any promoters (54). The concentration of AbrB in the cell is controlled by two factors: autoregulation by AbrB itself of its own promoter and repression of AbrB transcription by Spo0A~P (49). Thus the concentration of AbrB, and therefore its ability to prevent transcription, is controlled directly by the level of Spo0A~P in the cell. Spo0A~P binds to a region just downstream of the two promoters of *abrB* and prevents its transcription. Thus, the product of the phosphorelay, Spo0A~P, is directly responsible for controlling many genes by regulating the concentration of the transition-state regulator AbrB within the cell. Many of the genes and processes that AbrB controls are hallmarks of the stationary phase, including subtilisin and neutral protease production, motility, and competence.

A second transition-state regulator is encoded by the *hpr* gene. This gene was first identified from studies of mutants that overproduce proteases and was subsequently shown to be identical to the gene for mutations called *scoC*, for altered sporulation control, and *cat,* for catabolite resistant sporulation. Cloning and sequencing of the gene and its mutations has shown that the phenotypes result from the loss of the *hpr* gene product, and therefore the Hpr protein functions as a negative regulator of the protease genes, as well as other genes involved in catabolite repression and sporulation control (34). Overproducing the *hpr* gene product using a multicopy plasmid results in a sporulation-defective phenotype, suggesting strongly that at least some as yet unidentified genes are required for sporulation and negatively controlled by the Hpr transition-state regulator. Hpr binds to promoters for the subtilisin and neutral protease genes and most likely asserts its effects on these promoters by inhibiting transcription of their genes (21). No effector molecules are known that enhance or inhibit this DNA binding, and therefore it is not certain what environmental or metabolic stimulus controls the function of the Hpr regulator. Transcription of the *hpr* gene has been shown to be constitutive in an *spo0A* mutant and normal in an *spo0A-abrB* double mutant, suggesting that the overproduction of AbrB in an *spo0A* mutant serves to directly activate the transcription of the *hpr* gene. However, extensive experiments to confirm these observations have not been undertaken, and therefore AbrB itself is not proven to be a positive activator of this gene. If AbrB is an activator of *hpr*, then the synthesis of Hpr would be indirectly tied to the level of Spo0A~P in the cell through the level of AbrB.

One of the most interesting transition-state regulators is the product of the *sinR* gene, Sin. Sin was originally identified as the product of the gene that inhibits sporulation when present on a multicopy plasmid (12, 13). Overproduction of Sin by this means not only inhibits sporulation, but protease production and purified Sin protein can bind specifically to promoters including the subtilisin promoter (14, 21). Sin also must have some repressive effects on one or more components of the phosphorelay, because the production of Spo0A-controlled genes is curtailed by overproduction of the Sin protein. Sin appears to be constitutively expressed during growth and not subject to transcriptional controls. However, the function of the Sin protein is controlled by the product of the *sinI* gene, Isin (2). Isin functions as an inhibitor of the activity of Sin, and presumably the transcriptional regulation of genes mediated by Sin can be overcome through interaction with the Isin protein. The *sinI* gene, on the other hand, is transcriptionally controlled and its promoter is a target for binding of the Hpr protein, the AbrB protein, and Spo0A~P (21; M. A. Strauch, personal communication). Thus, not only Spo0A but also the two transition-state regulators that may be controlled by Spo0A serve to regulate transcription of *sinI*. Both Hpr and AbrB most likely act in a negative fashion to prevent production of Isin under conditions when Spo0A~P concentration is at a low level. Sin can repress sporulation genes under these conditions. When Spo0A~P accumulates through the action of the phosphorelay, AbrB and Hpr production fall and Isin can be transcribed. The fact that Spo0A~P binds to this promoter suggests that Spo0A~P positively activates the *sinI* gene to complete the formation of Isin and thereby inactivate the function of any remaining Sin.

If the above scenario is true, then the activities of all three transition-state regulators, AbrB, Hpr, and Sin, are basically controlled by the level of Spo0A~P in the cell. Whether any metabolic or environmental effectors play a role in the control of any of these transition-state regulators is still open to question. It seems unlikely that environmental input would only be affected through the kinases phosphorylating Spo0F. Unfortunately, the process of finding an effector molecule for regulatory proteins of this type is not straightforward. These could be the major regulators functioning to control the activity of the phosphorelay by preventing the flow of phosphate through the phosphorelay and lowering the production of Spo0A~P. They could do this by preventing the transcription of one or more components of the phosphorelay, or they could promote the transcription of inhibitors of the components of the phosphorelay. How any of these proteins act to prevent sporulation is not clear except that they all seem to be negative regulators of transcription.

Control of Phosphate Flux in the Phosphorelay

An article of faith in sporulation is that the environment somehow controls the onset of stationary phase and sporulation. Unfortunately, the actual effector molecules from the environment that accomplish this task are completely obscure. Environmental stimuli presumably activate sensor kinases of two-component regulatory systems to autophosphorylate and to transfer phosphate to their response-regulator proteins. Two kinases are probably responsible for all of the phosphate input into the phosphorelay, and little has been discovered of the nature of the small molecules or environmental messengers that might control their activities. However, kinase regulation may not be the only point of control of phosphate flux in the phosphorelay, and its complexity has been postulated as a means by which the cell may exert control at many levels (4).

KinA, which supplies the bulk of the phosphate to the phosphorelay, at least under laboratory conditions, appears to be a soluble enzyme with no apparent homology to any other enzymes or proteins except for the conserved homology to the kinases. Thus, the primary amino acid sequence of KinA has not helped to identify potential effectors for this kinase. Recently *cis*-unsaturated fatty acids were found to be inhibitory to KinA activity on Spo0F (51). The most inhibitory fatty acids have at least one unsaturated double bond in the *cis* configuration and a chain length of 16–20 carbon atoms. The homologous *trans*-isomers are not inhibitory, nor are saturated straight or branched-chain fatty acids. KinA may not function like a classic sensor kinase with a positive effector, but rather its activity could be simply inhibited during exponential growth by the pool of *cis*-unsaturated fatty acids. Such fatty acids, although rare in the *B. subtilis* cell, may act as specific signals, linking the initiation of sporulation to the status of membrane biosynthesis and septation, or to some other membrane-associated activity, where the free fatty acid acts as a signal. This statement reflects our ignorance of the role of fatty acids in metabolic signaling.

Through extensive kinetic analyses, C. E. Grimshaw and colleagues (personal communication) have shown that KinA is greatly stimulated by simply the presence of its substrate, Spo0F. This raises the possibility that KinA is basically inactive in its own autophosphorylation reaction in the absence of Spo0F, and any residual activity may be inhibited by the concentration of *cis*-unsaturated fatty acids in the cell. When *spo0F* transcription is induced at the end of exponential growth by a series of complicated controlling factors, the expression of Spo0F might be sufficient to activate the kinase to convert Spo0F to Spo0F~P. No other positive effector need be required. Although this is not the classical view of how such kinases are activated, it would fit the kinetic data, which show KinA to be on the order of 3% as active in the absence of Spo0F as in its presence, and Spo0F appears to induce an isomerization reaction in the enzyme to increase its activity. Thus, any environmental input into KinA activity could be indirect through the control of the transcription of *spo0F*.

KinB is a protein with six potential membrane-spanning regions, suggesting that the enzyme is an integral membrane protein with the kinase portion of the molecule in the cytoplasm of the cell (61). The *kinB* gene is present in an operon along with a gene, *kapB*, that codes for KapB, a protein of 14,600 Daltons that appears from its deduced amino acid sequence to be a moderately charged soluble protein. Inactivation of either *kinB* or *kapB* in the presence of the *kinA* mutation reduces the residual level of sporulation of the double mutant to almost zero. This result indicates that KapB is required for the activity of KinB, although one cannot rule out the possibility that KinB is not synthesized in the absence of KapB. The most likely scenario is that KapB is required to facilitate the activity of KinB, either by acting as an effector ligand–binding domain, or by facilitating the interaction of KinB with Spo0F. KapB itself appears to have no homology to response regulators and no homology to any other known protein. The reason for the membrane location of KinB is a mystery.

The *spo0E* gene product may be involved in negative control of the level of Spo0F~P, because deleting the *spo0E* gene results in suppression of poorly active mutant Spo0F proteins (J. A. Hoch, unpublished data). This sparing effect on Spo0F could be accomplished in several ways, perhaps by preventing the degradation of Spo0F~P either through an inherent phosphatase activity, or by activating the phosphatase activity of either KinA or KinB. For example, KinA has a potent phosphatase activity for Spo0F~P. Spo0E may be responding to some effector molecule, since removal of the carboxyl-terminal portion of the protein results in a sporulation-defective phenotype and presumably a constitutively active protein. What effector molecule this carboxyl portion of the protein might be interacting with is open to question.

The most likely role of Spo0E is to control the flow of phosphate through the phosphorelay in a negative fashion in response to the level of some other environmental or metabolic factor.

The *spo0B* gene resides in the same transcription unit as the gene for an essential GTP-binding protein, Obg (60). Although the juxtaposition and common control of these two genes suggests that one plays a role in the other's activity, the evidence is entirely circumstantial. Sporulation can only initiate during a certain window in a cell cycle, and the Obg protein may convey some cell-cycle control to the activity of the Spo0B phosphotransferase. Obg is essential for growth, and temperature-sensitive mutants of this protein have been isolated that do not appear to properly septate at the restricted temperature (J. Kok & J. A. Hoch, unpublished data). Whether Obg has a direct effect on septation remains to be seen, but Obg is the only candidate at present for a protein with an essential function in the cell that may convey information about the position of the cell within the cell cycle.

Transcriptional Regulation of the Phosphorelay

Regulation of the cellular concentration of the components required for the production of Spo0A~P initiates control of the flow of phosphate through the phosphorelay. This mechanism of control is readily apparent during conditions of vigorous vegetative growth in nutrient-excess conditions in which the amounts of two of the major components, Spo0F and Spo0A, are strongly controlled by transcriptional mechanisms. During the early portion of the transition state, repression of these genes is released and strong induction of both occurs (66). σ^H plays a crucial role in the induction of both *spo0A* and *spo0F* during this state, as well as being involved in the transcription of the *kinA* gene (39). Spo0A~P has some role in regulating the level of σ^H because this gene is repressed by the transition-state regulator, AbrB (63), which is subject to direct control by Spo0A~P (49). Because the σ^H required for induction is controlled by Spo0A~P, some mechanism must insure that a low level of phosphorelay components that are not under σ^H control exists in the cell. This is accomplished by differentially controlled tandem promoters for both the *spo0A* gene and the *spo0F* gene (5, 24, 58, 65).

A vegetative promoter utilizing σ^A is located upstream of the *spo0A* and *spo0F* genes and transcribes both genes at a very low level during growth in conditions of nutrient excess (Figure 4). Both of these vegetative promoters appear to be repressed as the culture moves from vegetative growth into the transition state. For the *spo0A* gene, Spo0A~P binds directly downstream of the vegetative promoter and inhibits transcription from it (55). In the *spo0F* gene, an Spo0A~P binding region occurs within the vegetative promoter, presumably again shutting off this promoter (56). The binding of Spo0A~P to both of these regions is also thought to activate transcription from the downstream σ^H promoter of both genes. The *spo0A* promoter has an additional Spo0A~P binding site just downstream of the presumed repression site of the vegetative promoter, whereas in the *spo0F* gene, Spo0A~P binding at the site of the vegetative promoter may accomplish both tasks—repressing the vegetative promoter and activating the σ^H promoter. An Spo0A~P (repression) binding site for both σ^H promoters also covers the −35 region of the *spo0A* promoter, and another is just downstream of the start site of transcription in the *spo0F* gene. Both of these sites are thought to modulate the expression of these genes in a fashion similar to autoregulation; once the Spo0A~P concentration reaches a level sufficient to accomplish its functions in the cell, Spo0A~P represses any further synthesis of both Spo0A and Spo0F (56). At this time Spo0A~P may also act as a repressor of the *kinA* gene because an Spo0A-binding site is located just downstream of this gene's start site (K. Trach, M. Strauch & J. A. Hoch, unpublished data).

This autoregulation scheme, in which the product of the phosphorelay Spo0A~P acts as a positive activator of genes not only coding for itself but also for *spo0F* required for the production of Spo0A~P, is basically an autocatalytic positive feedback loop (Figure 4). Such an arrangement might

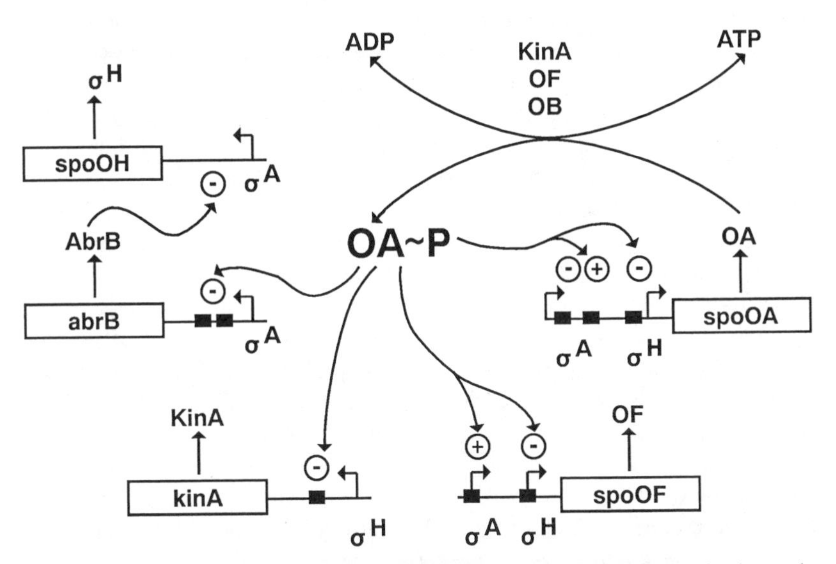

Figure 4 Autoregulation of production of Spo0A~P by Spo0A~P. The complex interactions between Spo0A~P and promoters it controls are shown. Low levels of Spo0A~P are required to repress the *abrB* promoter, which results in the relief of repression of the production of σ^H. σ^H is required for high-level expression of *kinA*, *spo0F*, and *spo0A*. Spo0A~P is a positive activator of σ^H promoters for the *spo0A* and *spo0F* genes. Spo0A~P may be a repressor for the *kinA* gene, and for *spo0F* and *spo0A* genes when Spo0A~P concentration reaches a sufficient level.

be advantageous if the cell finds itself in conditions where it needs to rapidly produce Spo0A~P in order to respond to a sudden shift from nutrient-excess to nutrient-poor conditions. In theory, the entire sporulation-induction scheme could be regulated by simply controlling the kinases that are responsible for the initial input of information into the phosphorelay. When these are off, the transcription of the genes for *spo0A* and *spo0F* is shut down except for the low-level transcription from the vegetative promoters for these genes. When the kinases become activated, a small amount of Spo0A~P is produced, which causes a rapid rise in the production of Spo0F and Spo0A through the positive feedback loop. The suggestion of such a control mechanism, although pleasing and simple, reflects our ignorance of the influence of other factors on the transcription of these genes.

Glucose strongly affects transcription of *spo0A* (65). The presence of glucose in the media prevents switching from the vegetative to the σ^H promoter of the *spo0A* gene (5). This could occur by a variety of mechanisms. For example, catabolite repression could be mediated through a DNA-binding protein with some of the properties of the Cap protein of *E. coli* that serves to prevent the transcription of the *spo0A* σ^H promoter. This hypothesis seems to be a favorite of many, especially because Chambliss and colleagues have shown that promoters controlled by catabolites including *spo0A* have a specific sequence in common, suggesting that a specific protein binds to this sequence to effect the catabolite repression of such genes (17, 62). Until such a protein is found and shown to directly bind to this promoter, however, an equally likely hypothesis is that catabolite repression is mediated through inhibition of the production of Spo0A~P, either by the inhibition of the enzymatic activity of one of the components of the pathway, by the prevention of transcription of some other gene such as the σ^H gene, or by the deactivation of the kinases by preventing effector-molecule formation.

The transition-state regulator Hpr is known to prevent sporulation when overproduced, and mutants that are deficient in this protein have catabolite-resistant sporulation. But Hpr cannot be the hypothetical catabolite-resistance protein identified by Weikert & Chambliss (62) because Hpr has no effect on the amylase gene that is presumably controlled by their regulator, and the DNA sequence of the Hpr-binding site does not correlate to the common site mentioned above (21). One of the roles of Sin, when overproduced, is to prevent the transcription of some genes controlled by Spo0A~P. Sin might transcriptionally control *spo0A* or *spo0F* genes, or both, or regulate other components of the phosphorelay scheme required for production of Spo0A~P. Finally, the possibility exists that other regulators that have not yet been identified could be controlling transcription of the genes for the phosphorelay. Clearly, the function of the transition-state regulators in this sporulation scheme is to prevent sporulation, rather than to promote it. These may be the proteins subject to the mythical nutrient regulation that we have so far been unable to explain. In the absence of transition-state regulators, presumably the production of Spo0A~P and the induction of sporulation would depend solely on activation of the kinases.

Transcriptional Properties of Spo0A~P

The Spo0A transcription regulator is clearly required for the initiation of sporulation and also plays a major role in the control of its own synthesis. This protein works both as a transcription repressor and as a transcription activator by binding at 0A boxes (5-TGNCGAA-3) in promoters (49). Several studies have shown that only the phosphorylated form of this enzyme is active in transcription (57). Site-directed mutations in the aspartic acid pocket required for phosphorylation of Spo0A give rise to a protein that is inactive in vivo, either as a repressor or an activator (55). Studies of the DNA-binding ability of Spo0A and Spo0A~P have shown that the phosphorylated form of the protein is 20–50 times more efficient at binding to its sites on the *abrB* promoter than the unphosphorylated form (49). The unphosphorylated form binds to promoters with the same specificity as the phosphorylated form, and the same is true for proteins with mutations in the aspartic acid pocket that cannot be phosphorylated. Phosphorylation may affect either the affinity of the protein for its target or the dissociation constant of the DNA-Spo0A~P complex.

The *abrB* gene appears to be extremely sensitive to the concentration of Spo0A~P in the cell. This conclusion comes from several observations, the major one being that repression of the *abrB* gene is controlled directly by Spo0A~P and occurs toward the middle of exponential growth when the Spo0A~P concentration in the cell must be quite low. A mutation in the *spo0A* gene renders the production of AbrB constitutive, subject only to its own autoregulatory controls (53). In a mutant lacking both KinA and KinB, which cannot produce enough Spo0A~P to induce a positively controlled gene such as *spoIIA*, the regulation of *abrB* is normal (61). Thus, the small amount of Spo0A~P that can be formed by crosstalk with other kinases on *spo0F* is sufficient to repress *abrB* transcription. This sensitivity may be because the *abrB* promoter contains two Spo0A~P binding sites (0A boxes) separated by one helical turn of the DNA. One possibility is that this arrangement of 0A boxes builds a high-affinity site for Spo0A~P; another is that two molecules of Spo0A~P can bind to this site and stabilize through protein-protein interactions, as well as protein-DNA interactions.

Spo0A~P is responsible for activating the transcription of three operons, *spoIIA*, *spoIIE*, and *spoIIG*, required for the transition from stage II to stage III of sporulation (42, 57, 67). It accomplishes this by activating transcription of these operons from either a σ^A promoter (in the case of *spoIIE* and *spoIIG*) or a σ^H promoter (in the case of *spoIIA*). The difference between these promoters and the *abrB* promoter, for example, is believed to be the lower concentration of Spo0A~P required for the activation of *abrB*. In addition, mutations in *spo0A*, such as *spo0A9V*, which modifies one of the very carboxyl-terminal amino acids of the protein, result in a transcription factor that can repress *abrB* efficiently, but cannot activate the transcription of *spoIIA* (37). Thus, positive activation of genes requires a certain conformation of the very carboxyl-terminal amino acids that is not required when Spo0A~P acts as a repressor. The carboxyl terminal portion of *spo0A* may need to make physical contact with the transcription complex in the case of activated promoters. The promoters for the three stage II genes have several partial 0A boxes, and in vitro studies have shown that Spo0A binds over a large portion of these promoters, extending past the –100 nucleotide of two of them (42, 67). Thus, the structure of these promoters may be such that several Spo0A~P molecules must bind in concert in order to provide the structure required for transcription activation.

The studies of Moran and colleagues on the *spoIIG* promoter and the *spoIIE* promoter have shown that these promoters are not high-affinity binding sites for the transcription complex and probably require the presence of Spo0A~P to allow the transcription complex to recognize the promoters (42, 67). Whether this occurs because of Spo0A~P–transcription complex contacts, or whether *spo0A* binding to these promoters causes a more favorable confirmation of the DNA, or both, has not been determined. The key to understanding the transcription of these stage II genes is the perception that a certain concentration barrier of Spo0A~P must be breached before these promoters turn on. No other regulatory factors have been implicated in the control of these genes, although it is certainly possible they exist.

How It Works

One can consider a bacterial culture in the laboratory as being in one of three general states: nutrient-excess exponential growth, nutrient and population density–limited transition state, and stationary phase/sporulation. Bacteria compete in nature by growing rapidly, and therefore under any conditions where nutrients are available, the first priority of a cell is to grow and divide. Under these conditions, the phosphorelay is transcribed at a maintenance level and little, if any, Spo0A~P is present in the cell. This leads to high-level production of the AbrB protein, which keeps stationary-phase genes from being transcribed and possibly interfering with the cell's growth and division. Utilization of the available nutrients leads to a more nutrient-limited situation and conversion from an exponentially dividing culture to one in which growth and division occur at a much slower rate as the cells enter the transition state. This transition-state period can be quite short in cultures in the laboratory, but may occur for extended periods in nature where limitation of nutrients may be common. The conversion to the transition state occurs with the release of transcriptional controls on the genes for the phosphorelay and the presumed accumulation of low levels of Spo0A~P. This quantity of the activated transcription factor is sufficient to cause repression of the *abrB* gene and, therefore, to release from repression those stationary-phase gene functions that are required for life in the transition state. Presumably, accumulation of Spo0A~P is not sufficient to kick off transcription of the stage II genes. Stage II gene transcription must result when nutrients become so limited that the cell cannot maintain itself in the transition state.

Several enzymes are produced during the transition state. These include amylase and other enzymes that degrade complex carbohydrates, and enzymes such as subtilisin and neutral protease that serve to produce amino acids for carbon and energy as well as nitrogen sources. During this stage, the cell is hunting through the environment for all of the possible available carbon sources it can find to continue to grow and divide, albeit, at a much reduced rate. Secondary pathways of metabolism may now be induced to take advantage of all available carbon sources that might be around. The cell produces antibiotics during this period, presumably to fend off the invasion of other bacteria or fungi that could take advantage of the substrates generated by these extracellular enzymes and to protect its space in its ecological niche. *B. subtilis* has a large armament of offensive weapons in the form of antibiotic pathways for the ongoing warfare occurring in complex habitats. *B. subtilis* and many other bacteria have developed a semiquiescent transition state called *competence,* in which the cells develop the ability to take up extraneous DNA, but that is controlled in a complex fashion by several regulators (6, 7). Thus, the transition state is a time of intensive transcription and regulation of a wide variety of genes and operons that exist solely to maintain the cell in exceedingly varied environmental states. Transition-state regulators, such as Hpr and Sin, must come into their own during these states and prevent sporulation from occurring as long as sufficient nutrients are available for the cell in the transition state. They accomplish this task by keeping Spo0A~P below the threshold level at which initiation of the transcription of stage II genes occurs and the subsequent cascade of events that ultimately leads to a spore.

It is axiomatic that sporulation is controlled by nutrient availability, but none of the research to date has been able to show what nutrients actually cause repression of sporulation and how they act. Similarly, nutrients and/or other events must be responsible for the activation of the phosphorelay, yet we have no clear picture of which nutrients these are or how they activate the kinases. We know glucose has an effect, but it is unclear what derivative form of glucose actually exerts the control, and through what proteins it works and how it works. Finally, transition-state regulators clearly exist to prevent sporulation, and they must operate by sensing the fact that sufficient nutrients exist such that sporulation should not occur, yet we have no real clue as to what nutritional compounds are controlling the activity of Hpr, Sin, or the other less-well-studied proteins that may prevent sporulation. Advances in molecular biology have allowed us to define the genes and isolate the proteins involved in this regulation but have been of limited value in determining how the proteins function and what interacts with them.

The end of the transition state is signaled by the activation of the stage II genes, *spoIIA, spoIIG,* and *spoIIE,* by Spo0A~P. At this time, the nutritional value of the environment has reached a point at which it is no longer advantageous to continue the battle, and the cell initiates sporulation and goes inert until conditions improve. The production of proteins like amylase and subtilisin shuts down, and the remaining energy available to the cell is directed toward the production of the morphological structures that ultimately characterize the quiescent spore. This is a time when form determines function (47), chromosomes condense (43), sigmas cascade (48), and compartments talk to one another (25). These exciting studies, too complex to review here, have brought forth interesting and provocative theories of gene regulation and some of them might even be right.

ACKNOWLEDGMENTS

This is manuscript #7717-MEM from the Department of Molecular and Experimental Medicine. It was supported, in part, by grant GM19416 from the National Institute of General Medical Sciences, National Institutes of Health, United States Public Health Service. The author apologizes to any of his friends whose valuable contributions to this picture of the initiation of sporulation might have been unintentionally slighted in the effort to write a readable rather than an encyclopedic review.

Literature Cited

1. Antoniewski, C., Savelli, B., Stragier, P. 1990. The *spoIIJ* gene, which regulates early developmental steps in *Bacillus subtilis,* belongs to a class of environmentally responsive genes. *J. Bacteriol.* 172:86–93
2. Bai, U., Mandic-Mulec, I., Smith, I. 1993. I-sin modulates the activity of Sin, a developmental switch protein of *Bacillus subtilis,* by protein-protein interaction. *Genes Dev.* 7:139–48
3. Bouvier, J., Stragier, P., Bonamy, C., Szulmajster, J. 1984. Nucleotide sequence of the *spo0B* gene of *Bacillus subtilis* and regulation of its expression. *Proc. Natl. Acad. Sci. USA* 81:7012–16
4. Burbulys, D., Trach, K. A., Hoch, J. A. 1991. The initiation of sporulation in *Bacillus subtilis* is controlled by a multicomponent phosphorelay. *Cell* 64: 545–52
5. Chibazakura, T., Kawamura, F., Takahashi, H. 1991. Differential regulation of *spo0A* transcription in *Bacillus subtilis:* glucose represses promoter switching at the initiation of sporulation. *J. Bacteriol.* 173:2625–32
6. Dubnau, D. 1991. The regulation of

genetic competence in *Bacillus subtilis*. *Mol. Microbiol.* 5:11–18
7. Dubnau, D. 1991. Genetic competence in *Bacillus subtilis*. *Microbiol. Rev.* 55:395–424
8. Dubnau, E., Weir, J., Nair, G., Carter, L. III, Moran, C. Jr., Smith, I. 1988. *Bacillus* sporulation gene *spo0H* codes for sigma-30 (sigma-H). *J. Bacteriol.* 170:1054–62
9. Dubnau, E., Weir, J., Nair, G., Carter, L. III, Moran, C. Jr., Smith, I. 1988. *Bacillus* sporulation gene *spo0H* codes for σ^{30} (σ^{H}). *J. Bacteriol.* 170:1054–62
10. Ferrari, F. A., Trach, K., Hoch, J. A. 1985. Sequence analysis of the *spo0B* locus reveals a polycistronic transcription unit. *J. Bacteriol.* 161: 556–62
11. Ferrari, F. A., Trach, K., LeCoq, D., Spence, J., Ferrari, E., Hoch, J. A. 1985. Characterization of the *spo0A* locus and its deduced product. *Proc. Natl. Acad. Sci. USA* 82:2647–51
12. Gaur, N. K., Cabane, K., Smith, I. 1988. Structure and expression of the *Bacillus subtilis sin* operon. *J. Bacteriol.* 170:1046–53
13. Gaur, N. K., Dubnau, E., Smith, I. 1986. Characterization of a cloned *Bacillus subtilis* gene that inhibits sporulation in multiple copies. *J. Bacteriol.* 168:860–69
14. Gaur, N. K., Oppenheim, J., Smith, I. 1991. The *Bacillus subtilis sin* gene, a regulator of alternate developmental processes, codes for a DNA-binding protein. *J. Bacteriol.* 173:678–86
15. Grossman, A. D., Losick, R. 1988. Extracellular control of spore formation in *Bacillus subtilis*. *Proc. Natl. Acad. Sci. USA* 85:4369–73
16. Hahn, J., Dubnau, D. 1991. Growth stage signal transduction and the requirements for *srfA* induction in development of competence. *J. Bacteriol.* 173:7275–82
17. Henkin, T. M., Grundy, F. J., Nicholson, W. L., Chambliss, G. H. 1991. Catabolite repression of α-amylase gene expression in *Bacillus subtilis* involves a *trans*-acting gene product homologous to the *Escherichia coli lacI* and *galR* repressors. *Mol. Microbiol.* 5:575–84
18. Hoch, J. A. 1976. Genetics of bacterial sporulation. *Adv. Genet.* 18:69–99
19. Hoch, J. A., Trach, K., Kawamura, F., Saito, H. 1985. Identification of the transcriptional suppressor *sof-1* as an alteration in the *spo0A* protein. *J. Bacteriol.* 161:552–55
20. Jaacks, K. J., Healey, J., Losick, R., Grossman, A. D. 1989. Identification and characterization of genes controlled by the sporulation-regulatory gene *spo0H* in *Bacillus subtilis*. *J. Bacteriol.* 171:4121–29
21. Kallio, P. T., Fagelson, J. E., Hoch, J. A., Strauch, M. A. 1991. The transition state regulator Hpr of *Bacillus subtilis* is a DNA-binding protein. *J. Biol. Chem.* 266:13411–17
22. Kawamura, F., Saito, H. 1983. Isolation and mapping of a new suppressor mutation of an early sporulation gene *spo0F* mutation in *Bacillus subtilis*. *Mol. Gen. Genet.* 192:330–34
23. Kudoh, J., Ikeuchi, T., Kurahashi, K. 1985. Nucleotide sequences of the sporulation gene *spo0A* and its mutant genes of *Bacillus subtilis*. *Proc. Natl. Acad. Sci. USA* 82:2665–68
24. Lewandoski, M., Dubnau, E., Smith, I. 1986. Transcriptional regulation of the *spo0F* gene of *Bacillus subtilis*. *J. Bacteriol.* 168:870–77
25. Losick, R., Stragier, P. 1992. Crisscross regulation of cell-type-specific gene expression during development in *B. subtilis*. *Nature* 355:601–4
26. Mandelstam, J., Higgs, S. A. 1974. Induction of sporulation during synchronized chromosome replication in *Bacillus subtilis*. *J. Bacteriol.* 120:38–42
27. Mysliwiec, T. H., Errington, J., Vaidya, A. B., Bramucci, M. G. 1991. The *Bacillus subtilis spo0J* gene: evidence for involvement in catabolite repression of sporulation. *J. Bacteriol.* 173:1911–19
28. Ninfa, A. J., Magasanik, B. 1986. Covalent modification of the *glnG* product, NRI, by the *glnL* product, NRII, regulates the transcription of the *glnALG* operon in *Escherichia coli*. *Proc. Natl. Acad. Sci. USA* 83:5909–13
29. Ogasawara, N., Yoshikawa, H. 1992. Genes and their organization in the replication origin region of the bacterial chromosome. *Mol. Microbiol.* 6:629–34
30. Parkinson, J. S., Kofoid, E. C. 1992. Communication modules in bacterial signaling proteins. *Annu. Rev. Genet.* 26:71–112
31. Perego, M., Cole, S. P., Burbulys, D., Trach, K., Hoch, J. A. 1989. Characterization of the gene for a protein kinase which phosphorylates the sporulation-regulatory proteins Spo0A and Spo0F of *Bacillus subtilis*. *J. Bacteriol.* 171:6187–96
32. Perego, M., Higgins, C. F., Pearce, S. R., Gallagher, M. P., Hoch, J. A. 1991. The oligopeptide transport system

of *Bacillus subtilis* plays a role in the initiation of sporulation. *Mol. Microbiol.* 5:173–85

33. Perego, M., Hoch, J. A. 1987. Isolation and sequence of the *spo0E* gene: its role in initiation of sporulation in *Bacillus subtilis*. *Mol. Microbiol.* 1:125–32
34. Perego, M., Hoch, J. A. 1988. Sequence analysis and regulation of the *hpr* locus, a regulatory gene for protease production and sporulation in *Bacillus subtilis*. *J. Bacteriol.* 170:2560–67
35. Perego, M., Hoch, J. A. 1991. Negative regulation of *Bacillus subtilis* sporulation by the *spo0E* gene product. *J. Bacteriol.* 173:2514–20
36. Perego, M., Spiegelman, G. B., Hoch, J. A. 1988. Structure of the gene for the transition state regulator, *abrB:* regulator synthesis is controlled by the *spo0A* sporulation gene in *Bacillus subtilis*. *Mol. Microbiol.* 2:689–99
37. Perego, M., Wu, J.-J., Spiegelman, G. B., Hoch, J. A. 1991. Mutational dissociation of the positive and negative regulatory properties of the Spo0A sporulation transcription of *Bacillus subtilis*. *Gene* 100:207–12
38. Piggot, P. J., Coote, J. G. 1976. Genetic aspects of bacterial endospore formation. *Bacteriol. Rev.* 40:908–62
39. Predich, M., Nair, G., Smith, I. 1992. *Bacillus subtilis* early sporulation genes *kinA*, *spo0F*, and *spo0A* are transcribed by the RNA polymerase containing σ^H. *J. Bacteriol.* 174:2771–78
40. Rudner, D. Z., Ladeaux, J. R., Breton, K., Grossman, A. D. 1991. The *spo0K* locus of *Bacillus subtilis* is homologous to the oligopeptide permease locus and is required for sporulation and competence. *J. Bacteriol.* 173:1388–98
41. Sanders, D. A., Gillece-Castro, B. L., Stock, A. M., Burlingame, A. L., Koshland, D. E. Jr. 1989. Identification of the site of phosphorylation of the chemotaxis response regulator protein, CheY. *J. Biol. Chem.* 264:21770–78
42. Satola, S. W., Baldus, J. M., Moran, C. P. Jr. 1992. Binding of Spo0A stimulates *spoIIG* promoter activity in *Bacillus subtilis*. *J. Bacteriol.* 174: 1448–53
43. Setlow, B., Magill, N., Febbroriello, P., Nakhimovsky, L., Koppel, D. E., Setlow, P. 1991. Condensation of the forespore nucleoid early in sporulation of *Bacillus* species. *J. Bacteriol.* 173: 6270–78
44. Smith, I. 1993. Regulatory proteins that control late growth development. See Ref. 44a, In press

44a. Sonenshein, A. L., Hoch, J. A., Losick, R., eds. 1993. Bacillus subtilis *and Other Gram-Positive Bacteria: Biochemistry, Physiology and Molecular Genetics*. Washington, DC: Am. Soc. Microbiol. In press

45. Stock, A., Koshland, D. E. Jr., Stock, J. 1985. Homologies between the *Salmonella typhimurium* CheY protein and proteins involved in the regulation of chemotaxis, membrane protein synthesis, and sporulation. *Proc. Natl. Acad. Sci. USA* 82:7989–93
46. Stock, J. B., Ninfa, A. J., Stock, A. M. 1989. Protein phosphorylation and regulation of adaptive response in bacteria. *Microbiol. Rev.* 53:450–90
47. Stragier, P., Bonamy, C., Karmazyn-Campelli, C. 1988. Processing of a sporulation sigma factor in Bacillus subtilis: how morphological structure could control gene expression. *Cell* 52:697–704
48. Stragier, P., Losick, R. 1990. Cascades of sigma factors revisited. *Mol. Microbiol.* 4:1801–6
49. Strauch, M., Webb, V., Spiegelman, G., Hoch, J. A. 1990. The Spo0A protein of *Bacillus subtilis* is a repressor of the *abrB* gene. *Proc. Natl. Acad. Sci. USA* 87:1801–5
50. Strauch, M. A. 1993. Control of transition state processes by the AbrB protein. See Ref. 44a, In press
51. Strauch, M. A., de Mendoza, D., Hoch, J. A. 1992. *cis*-unsaturated fatty acids specifically inhibit a signal-transducing protein kinase required for initiation of sporulation in *Bacillus subtilis*. *Mol. Microbiol.* 6:2909–17
52. Strauch, M. A., Hoch, J. A. 1992. Transition state regulators: sentinels of *Bacillus subtilis* post-exponential gene expression. *Mol. Microbiol.* 7:337–42
53. Strauch, M. A., Perego, M., Burbulys, D., Hoch, J. A. 1989. The transition state transcription regulator AbrB of *Bacillus subtilis* is autoregulated during vegetative growth. *Mol. Microbiol.* 3: 1203–9
54. Strauch, M. A., Spiegelman, G. B., Perego, M., Johnson, W. C., Burbulys, D., Hoch, J. A. 1989. The transition state transcription regulator *abrB* of *Bacillus subtilis* is a DNA binding protein. *EMBO J.* 8:1615–21
55. Strauch, M. A., Trach, K. A., Day, J., Hoch, J. A. 1992. Spo0A activates and represses its own synthesis by binding at its dual promoters. *Biochimie* 74:619–26
56. Strauch, M. A., Wu, J.-J., Jonas, R. H., Hoch, J. A. 1992. A positive

feedback loop controls transcription of the *spo0F* gene, a component of the sporulation phosphorelay in *Bacillus subtilis*. *Mol. Microbiol.* In press

57. Trach, K., Burbulys, D., Strauch, M., Wu, J.-J., Dhillon, N., et al. 1991. Control of the initiation of sporulation in *Bacillus subtilis* by a phosphorelay. *Res. Microbiol.* 142:815–23
58. Trach, K., Chapman, J. W., Piggot, P., LeCoq, D., Hoch, J. A. 1988. Complete sequence and transcriptional analysis of the *spo0F* region of the *Bacillus subtilis* chromosome. *J. Bacteriol.* 170:4194–4208
59. Trach, K., Chapman, J. W., Piggot, P. J., Hoch, J. A. 1985. Deduced product of the stage 0 sporulation gene *spo0F* shares homology with the Spo0A, OmpR and SfrA proteins. *Proc. Natl. Acad. Sci. USA* 82:7260–64
60. Trach, K., Hoch, J. A. 1989. The *Bacillus subtilis spo0B* stage 0 sporulation operon encodes an essential GTP-binding protein. *J. Bacteriol.* 171: 1362–71
61. Trach, K. A., Hoch, J. A. 1993. Multisensory activation of the phosphorelay initiating sporulation in *Bacillus subtilis:* identification and sequence of the protein kinase of the alternate pathway. *Mol. Microbiol.* 8: In press
62. Weickert, M. J., Chambliss, G. H. 1990. Site-directed mutagenesis of a catabolite repression operator sequence in *Bacillus subtilis*. *Proc. Natl. Acad. Sci. USA* 87:6238–42
63. Weir, J., Predich, M., Dubnau, E., Nair, G., Smith, I. 1991. Regulation of *spo0H*, a gene coding for the *Bacillus subtilis* σ^H factor. *J. Bacteriol.* 173: 521–29
64. Wu, J.-J., Piggot, P. J., Tatti, K. M., Moran, C. P. Jr. 1991. Transcription of the *Bacillus subtilis spoIIA* locus. *Gene* 101:113–16
65. Yamashita, S., Kawamura, F., Yoshikawa, H., Takahashi, H., Kobayashi, Y., Saito, H. 1989. Dissection of the expression signals of the *spo0A* gene of *Bacillus subtilis:* glucose represses sporulation-specific expression. *J. Gen. Microbiol.* 135:1335–45
66. Yamashita, S., Yoshikawa, H., Kawamura, F., Takahashi, H., Yamamoto, T., et al. 1986. The effect of *spo0* mutations on the expression of *spo0A*- and *spo0F-lacZ* fusions. *Mol. Gen. Genet.* 205:28–33
67. York, K., Kenney, T. J., Satola, S., Moran, C. P. Jr., Poth, H., Youngman, P. 1992. Spo0A controls the σ^A-dependent activation of *Bacillus subtilis* sporulation-specific transcription unit *spoIIE*. *J. Bacteriol.* 174:2648–58
68. Zuber, P., Losick, R. 1987. Role of AbrB in Spo0A- and Spo0B-dependent utilization of a sporulation promoter in *Bacillus subtilis*. *J. Bacteriol.* 169: 2223–30

Annu. Rev. Microbiol. 1993. 47:467–504

ADAPTIVE MUTATION: The Uses of Adversity[1]

Patricia L. Foster
Department of Environmental Health, Boston University School of Public Health, Boston University School of Medicine, Boston, Massachusetts 02118

KEY WORDS: directed mutation, spontaneous mutation, selection, evolution

---

CONTENTS

ABSTRACT

When populations of microorganisms are subjected to certain nonlethal selections, useful mutants arise among the nongrowing cells whereas useless mutants do not. This phenomenon, known as adaptive, directed, or selection-induced mutation, challenges the long-held belief that mutations only arise at random and without regard for utility. In recent years a growing number of

[1]Dedicated to John Cairns on the occasion of his seventieth birthday.

0066-4227/93/1001-0467$02.00

studies have examined adaptive mutation in both bacteria and yeast. Although conflicts and controversies remain, the weight of the evidence indicates that adaptive mutation cannot be explained by trivial artifacts and that nondividing cells accumulate mutations in the absence of genomic replication. Because this process tends to produce only useful mutations, the cells appear to have a mech- anism for preventing useless genetic changes from occurring or for eliminating them after they occur. The model that most readily explains the evidence is that cells under stress produce genetic variants continuously and at random, but these variants are immortalized as mutations only if they allow the cell to grow.

* * *

> Sweet are the uses of adversity,
> Which like the toad, ugly and venomous,
> Wears yet a precious jewel in his head.
>
> W. Shakespeare, ca 1600 (112)

INTRODUCTION

In a paper published in 1988 entitled "The Origin of Mutants," Cairns et al (16) suggested that "populations of bacteria, in stationary phase, have some way of producing (or selectively retaining) only the most appropriate mutations" (p. 144). This conclusion was based on experiments showing that when *Escherichia coli* was subjected to prolonged, nonlethal selection, advantageous mutations arose in the population, whereas nonadvantageous mutations did not. Furthermore, the mutations did not arise in the absence of specific selection for them. This phenomenon has been variously called directed, adaptive, Cairnsian, selection-induced, and stationary-phase mutation. The controversy that followed the publication of this paper has been depicted as a renewal of the battle between neo-Darwinism and Lamarckism. Even in peer-reviewed articles, scientists have exhibited a surprising fervor, verging on the religious, in debating this issue. The historical and sociological reasons for the intensity of the debate have been discussed elsewhere (57, 106). In this review, I give a brief historical perspective, attempt to evaluate the evidence, and discuss various mechanisms that could result in the production of adaptive mutations. The focus is mainly on *E. coli,* which has been the subject of the majority of studies (and is my area of expertise), but relevant results with other bacteria are included. Similar experiments with yeast are discussed in a separate section.

The scientific community has had difficulty reaching a consensus about what to name this phenomenon. "Directed mutation" implies a mechanism, whereas "stationary-phase mutation" obscures the apparent directedness of the mutations. "Selection-induced" implies that selection is inducing a process, but the role of selection is more likely to stop an ongoing process. After trying various names (15, 36, 37), I now favor "adaptive mutation," which has historical precedence (27) and is preferred by the phenomenon's discoverer (J. Cairns, personal communication). I define as adaptive those mutations that occur in nondividing cells during selection and are specific to the selection. Mutants that arise in nondividing cells but that either are not adaptive, or have not yet been shown to be adaptive, I refer to as stationary-phase mutations, again with historical precedence (101).

A HISTORICAL PERSPECTIVE

"The Last Stronghold of Lamarckism" (72)

It is difficult today to imagine the confusion and controversy that reigned in the first part of this century about the nature and variability of bacterial species. Even without the complication of impure cultures, distinguishing between genetic changes occurring among individuals in large populations and adaptation, which we now understand as enzyme induction, was nearly impossible for many bacteriologists (12, 55). The experiments of Luria & Delbrück (73), Newcombe (84), the Lederbergs (67), and Cavalli-Sforza & Lederberg (18) were of paramount importance in finally establishing that bacteria had genes, that rare individuals were the source of heritable variation, and that populations of bacteria did not have the capacity to endlessly respond to and be modified by their environment. That is, bacteria were fundamentally the same as other organisms.

Because the Luria-Delbrück experiment (73) is at the heart of the adaptive-mutation controversy, it is worth a brief review here. A mutant arising by chance during the exponential growth of a culture will give rise to a clone of identical progeny. Luria's insight was that such a mutant arising early during growth will give rise to an exceedingly large number of mutant progeny—a jackpot. Thus, among parallel cultures grown from identical initial inocula, the distribution of the numbers of mutants will be characterized by a large variance. In contrast, if mutations only arise after exposure to the selective agent, every cell so exposed has an equal probability of mutating, and the numbers of mutants obtained among parallel cultures will have the narrow variance of the Poisson distribution.

When Luria & Delbrück did this experiment, selecting for resistance to bacteriophage T1, they obtained a high-variance distribution.[2] Thus, the case was made—mutations arise at random and without regard to utility. Given

[2]In fact, because expression of T1 resistance is subject to phenotypic lag (51), Luria & Delbrück obtained a variance much larger than would be predicted (4, 73).

the importance and elegance of this experiment, criticizing it seems almost sacrilegious. But because they used a lethal selection, and dead cells cannot mutate, Luria & Delbrück's experiment did not prove that mutations cannot also arise after selection, or even in response to selection. Delbrück himself clearly understood this limitation. Commenting on a paper presented by Lwoff in 1946, Delbrück (27) drew a sharp distinction between selection for phage resistance and selection for carbohydrate utilization. He concluded: "In view of our ignorance of the causes and mechanisms of mutations, one should keep in mind the possible occurrence of specifically induced adaptive mutations" (27, p. 154).[3]

Establishment of the Mutation Paradigm

After the Luria-Delbrück experiment established that mutations can arise at random during exponential growth, it was, and still is, commonly assumed that all spontaneous mutations originate as mistakes made during genomic replication. Chemostat experiments established that spontaneous mutation rates were, in most cases, generation dependent and not time dependent (39, 88). The one notable exception to this rule, bacteria growing under tryptophan limitation (62, 87), was later shown be the result of a corresponding increase in the DNA content of the tryptophan-limited cells (61). The structure of DNA immediately suggested that spontaneous mutations arise as base mispairs, and several mispairing schemes involving rare tautomers and rotated bases were postulated (129, 134). These specific schemes have been recently and dismissively reviewed in light of current structural studies (30). Nevertheless, mispaired bases must sometimes cause mutations, as do base analogues. Similarly, mispairing can also account for mutations induced by chemical mutagens (such as alkylating agents) that create base analogues in vivo and by a number of spontaneous lesions (e.g. base deamination, alkylation, and oxidation). In addition, misaligned but correctly paired bases may cause frameshifts, deletions, complex mutations, and even base substitutions (1, 64, 98).

An entirely different aspect of mutagenesis began to be discovered in the late 1940s. When *E. coli* is exposed to certain mutagens, such as UV light and X rays, a number of apparently unrelated responses are elicited. These include mutation, filamentation, and prophage induction, as well as the increased survival and mutagenesis of irradiated bacteriophage plated on irradiated cells. It took years to understand the underlying relationship among these diverse phenomena, which are today recognized as expressions of the SOS response (32, 132, 137). Blocks to DNA replication, such as DNA lesions, result in the activation of RecA, *E. coli*'s major recombinational

[3]This comment was recently uncovered by R. Owen.

enzyme, to a state in which it facilitates the cleavage of several proteins. These include LexA (the common repressor of the SOS genes), repressors of lamboid bacteriophage, and UmuD. The UmuC protein, the carboxy-terminal fragment of the UmuD protein, and RecA promote or allow the production of mutations by a still-unknown mechanism. The simplest model for the process is that the SOS proteins modify the replication enzyme, DNA polymerase III, to allow translesion synthesis. One of the important concepts of SOS mutagenesis is that mutations of this class, unlike replication errors, are not passive but are produced only with the active participation of the cell. Furthermore, SOS-dependent mutations, like SOS-independent mutations, are created by DNA replication. The mutation is neither the DNA lesion nor the base-lesion mispair, but is the erroneous but correctly paired base pair that results from the subsequent round of replication.

Some mutagenic events may not require require DNA synthesis. These include transposition of some classes of insertional (IS) elements, excision of IS elements, and inversions, translocations, and deletions due to homologous recombination (1, 21, 38, 74, 108). However, some of these rearrangements may occur via slipped mispairing during replication (1), and most models of (nonreciprocal) recombination involve DNA synthesis (116). Clearly duplications and amplifications require at least limited DNA replication (2, 128).

Given the variety of mutagenic mechanisms, which of these is the source of spontaneous mutations in nature? That mutation rates themselves are under selection has been persuasively argued by Drake (31), and most models of evolution assume a constant mutation rate per generation. Thus, polymerase inaccuracy is widely regarded to be the source of most, if not all, spontaneous mutations. Indeed, defects in the activities that correct replication errors are powerful mutators (20). This, of course, is not at all a conclusive argument because, in wild-type cells, these error-correcting pathways may be entirely adequate for the purpose. Nor can the SOS-response be said to be a major contributor to spontaneous mutation because *umuDC* mutations are, at most, only weak antimutators (105). Considering that studies of spontaneous mutation rates typically use exponentially growing cells, whereas in nature a cell probably spends only a small part of its life in exponential growth, the origin of most natural spontaneous mutations still awaits investigation.

Evidence Contradictory to the Mutation Paradigm

Although the current controversy is a direct result of the publication of Cairns et al (16), there had been previous rumblings of dissatisfaction with the mutation doctrine. While attempting to confirm the Luria-Delbrück kinetics for reversion of a histidine auxotrophy, Ryan observed that His^+ revertants continued to arise for a period of 10 days after the cells were inoculated into medium without histidine (104). In a serious of papers published during the next nine years, he proposed and discarded many possible artifactual explanations for these results. With increasingly heroic experiments he showed that the delayed appearance of mutants did not result from slowly growing mutants, phenotypic lag, cell growth, or cell turnover (100, 101, 103). Left with no other explanation than that DNA is synthesized in the absence of cell division, Ryan attempted, but failed, to document the requisite amount of DNA replication (102). His final conclusion was that a small amount of DNA must be synthesized, but by some process that increased the error rate (102). Ryan did not investigate whether neutral mutations were likewise occurring during stationary phase, nor did he ask if the late-arising mutations would have occurred in the absence of selection for them.

Twenty years later Shapiro (113) published a study of the peculiar characteristics of fusion formation during prolonged selection. His strain, which has figured prominently in the adaptive mutation controversy, carries the regulatory region of the arabinose operon and the carboxy-terminal region of the *lacZ* gene separated by a defective Mu bacteriophage. The strain is both Lac^- and Ara^-. Cells with deletions that fuse *lacZ* in frame with *araB* can utilize lactose if arabinose is present as an inducer [these have come to be called $Lac(Ara)^+$, and the prefusion strain $Lac(Ara)^-$]. Shapiro found that after plating $Lac(Ara)^-$ cells on lactose-arabinose minimum medium at 32°C, $Lac(Ara)^+$ colonies did not appear until 5–6 days later [although if a $Lac(Ara)^+$ fusion was transduced into cells, colonies appeared on this medium within 2 days]. After 5 days, the rate of appearance of new $Lac(Ara)^+$ colonies accelerated, reached a maximum at about 20 days, and then declined. The estimated frequency of fusion production before plating was $<3 \times 10^{-11}$ per cell, whereas after plating it was 1×10^{-8} per cell. Shapiro also found that preplating culture conditions, including prolonged (static) starvation in buffer, had little effect on fusion formation. He concluded that the classical studies had been biased by the use of lethal selections, and that mutational processes would have to be reexamined in cells under nonlethal selection.

As discussed by Cairns et al (16), certain bacteria are classified as late fermenters because they develop the ability to utilize unusual (for them) sugars only after long incubation in the presence of these sugars. We now know that *E. coli* and other bacteria contain cryptic genes that can be mutationally activated (44). In some cases, more than one mutation is required (17, 42). Because the apparent frequency at which such multiple mutants arise vastly exceeds the frequency predicted if the mutations are independent, Hall (42) suggested that the mutations occur by some concerted mechanism.[4]

[4]Predating Cairns by a year, Opadia-Kadim (91) concluded that these multiple mutations resulted from a mechanism he called "postadaptive mutation."

Cairns et al (16) drew upon these discordant results and their own experiments to argue for "a non-random, possibly product-oriented form of mutation" (p. 142). First, however, they presented a critical discussion of the classical studies, emphasizing that, although these experiments proved that random mutations do occur, the use of lethal selections excluded the possibility of detecting any other sort. Thus, the occurrence of nonrandom mutations was not disproved.[5] Cairns et al presented new experimental evidence of three types. First, fluctuation analysis of reversion of a Lac^- strain to Lac^+ showed significant deviations from the Luria-Delbrück distribution toward the Poisson, suggesting that mutations were occurring after selection was applied. In addition, Lac^+ mutants continued to arise for several days after plating on lactose minimal medium. Second, mere starvation did not induce Lac^+ revertants or the Shapiro $Lac(Ara)^+$ fusion—the presence of the selective agent (lactose) was required. Third, a population accumulating Lac^+ mutants was not simultaneously accumulating mutants with a neutral phenotype (resistance to valine). These findings, together with the results of Shapiro (113) and Hall (42), led Cairns et al (16) to conclude that a process of trial and error was taking place. In the last part of the paper, they presented three molecular models that would allow cells to "subject a subset of their informational macromolecules to the forces of natural selection" (p. 145). Both the experimental evidence and the models are discussed below.

On a final historical note, Fitch, in a remarkable paper written in 1982 (34) on the occasion of the Darwinian centennial, made the theoretical argument that because mutations are advantageous during stressful times, but genome-wide mutagenesis would be deleterious, organisms probably have evolved a mechanism for selectively mutating only the genes of relevance. He then proposed a model that was essentially the same as that suggested by Davis (25) (see below).

EXPERIMENTAL EVIDENCE FOR AND AGAINST ADAPTIVE MUTATION

Table 1 summarizes various studies reporting the occurrence of spontaneous mutations during nonlethal selection. Some of these tested whether the mutations were specific to the selection, whereas others did not. Below, I focus on the types of evidence that support the idea of adaptive mutation, drawing upon these studies as appropriate.

[5]As pointed out by Keller (57), this fact has not been disputed, but most people continue to believe that "Luria & Delbrück were 'right,' even if for the 'wrong' reasons" (57, p. 294).

Table 1 Summary of studies of spontaneous mutations occurring during nonlethal selection

Organism	Reference	Selected phenotype	Mutational target	Probable mutational events	Evidence[a]
Adaptive mutations[b]					
Escherichia coli	16	Lactose utilization	*lacZ* amber	Base substitutions	A, B, C
	44	Methionine prototrophy	*metB*$^-$	Unknown	A, C
	45	Tryptophan prototrophy	*trpA* missense	Base substitutions	A, B
	45	Tryptophan prototrophy	*trpB* missense	Base substitutions	A, B, C, D
	45	Cysteine prototrophy	*cysB* missense	Base substitutions	B, C
	15	Lactose utilization	*lacI* : : *lacZ* frameshift	Frameshifts, deletions	A, B
	15	Tryptophan prototrophy	*trpE* frameshift	Frameshifts	A, B
	6	Maltodextrin uptake	*ompF*$^+$	Base substitutions	A, C
	56	Valine resistance	*ilvG* frameshift	Various	A, B
	37	Lactose utilization	*lacZ* missense	Base substitution	A, B
Pseudomonas putida plasmid	127	Halogenated alkanoic acid utilization	Poorly expressed *dehI*$^+$	Rearrangements	A, B[d]
Saccharomyces cerevisiae	48	Histidine prototrophy	*his4* missense	Base substitutions	A, C
	121	Lysine prototrophy	*lys2* frameshift	Frameshifts	A, B
Stationary-phase mutations[c]					
Escherichia coli	100	Histidine prototrophy	*his*$^-$	Unknown	A
	9	Isoleucine prototrophy	*ilv* : : Tn*3*	Excision	A
	9	Asparagine prototrophy	*asnA*$^-$	Not specified	A, D
	130	Survival in stationary phase	*surA* : : miniTn*10*	Excision	A
	47	Lactose utilization	*lacZ* missenses	Base substitutions	A
	47	Lactose utilization	*lacZ* frameshifts	Frameshifts	A
	B. Bridges (personal communication)	Tyrosine prototrophy	*tyrA* ochre	Unknown	A

Escherichia coli plasmid	11	Tetracycline resistance	*mnt*$^{+}$	Probably deletions	A, D
Salmonella typhimurium	95	Histidine prototrophy	*hisG* missense; ochre	Base substitutions; deletions	A
Clostridium thermocellum	86	Glucose, fructose utilization	Unknown	Unknown	A
Pseudomonas putida	127	Halogenated alkanoic acid resistance	*dehI*$^{+}$, *dehII*$^{+}$	Rearrangements	A
	127	Halogenated alkanoic acid utilization	*dehI*$^{-}$, *dehII*$^{-}$	Rearrangements	A, induced by starvation
Pseudomonas putida plasmid	89	Phenol utilization	Promoterless *pheAB*	Transpositions	A, D
Saccharomyces cerevisiae	48	Histidine prototrophy	*his6*	Not specified	A
Candida albicans	75	Heavy metal resistance	Unknown	Probably amplification	A
		Adenine prototrophy	*ade2/ade2*	Unknown	A
		Leucine prototrophy	*leu/leu*	Unknown	A
Disputed cases					
Escherichia coli	16, 113	Lactose utilization	Lac(Ara)$^{-}$	Mu excision, transposition	A, B
	80	Lactose utilization	Lac(Ara)$^{-}$	Mu excision, transposition	A, induced by starvation
	43	Salicin utilization	*bglF*::IS*103*	Excision	A[d], B, C
	81	Salicin utilization	*bglF*::IS*103*	Excision	A, induced by starvation
	46	Tryptophan prototrophy	*trpA trpB* missenses	Base substitutions	A[d], D

[a] The evidence that adaptive or stationary phase mutation was occurring: A. The mutants appeared late during nonlethal selection and/or the numbers of mutants in a fluctuation test had a Poisson distribution. B. The mutants did not accumulate during nonselective starvation or other appropriate nonspecific stress. C. Mutants with a neutral phenotype did not accumulate among the population during selection. D. Possible incidence of multiple mutation.

[b] The evidence supports the hypothesis that the mutations were specific for the selection imposed.

[c] The evidence supports the hypothesis that the mutations occurred in nongrowing cells during selection. Whether the mutations were also adaptive was not shown or, where indicated, the mutations were shown to be induced by nonspecific stress.

[d] The population increased under selection but possibly not enough to account for the mutations.

Do Mutations Occur After Selection?

DEVIATIONS FROM THE LURIA-DELBRUCK DISTRIBUTION If a fluctuation test gives the high variance characteristic of the Luria-Delbrück distribution, then mutations arose during prior growth of the cultures. Both Ryan (100) and Cairns et al (16) argued the obverse—that obtaining a Poisson distribution is evidence, although not proof, that mutations occur after selection is applied. However, alternative explanations for reductions in the variance of a mutant distribution have been offered (69, 106, 123). Some of these are simply wrong, and none are particularly compelling: (*a*) With a nonlethal selection, a delay in the expression of the mutant phenotype would affect only the timing, not the distribution, of the mutants. (*b*) Phenotypic lag only applies to recessive phenotypes (51), and actually increases the variance of the mutant distribution.[6] (*c*) Reduced fitness of the mutant during nonselective growth can narrow the distribution, but one cannot argue that every mutant in every case has a poor fitness relative to the nonmutant;[7] in some cases the mutant is genotypically wild-type (37, 47) or can reasonably be assumed to carry a sequence change only in the relevant gene (15). (*d*) Poor plating efficiency of the mutant on the selective medium has only a slight effect on the distribution even when the efficiency is very low (123). (*e*) If the full mutant phenotype results only after sequential mutations, these must occur at unrealistically high rates to account for mutational events occurring at normal frequencies (69).

As stated by Sarkar (106), "any observation of a deviation from the Luria-Delbrück distribution can be explained by invoking subsidiary interactions...[but]...hypothesizing such subsidiary interactions consists of making additional empirical claims and these, in turn, require experimental support" (p. 255).

THE CONTINUED APPEARANCE OF MUTANTS With the exception of the sequential mutation hypothesis, the processes discussed above cannot account for the fact that, under selection, mutants continue to appear for days or even

[6]Phenotypic lag refers to the fact that certain traits are not expressed during the first few generations after the new mutant allele has appeared (51). During the unselected growth of a population, mutants arising early will have ample generations to overcome this lag, whereas mutants that arise late will not. Thus, after selection is applied, cultures that actually contain low numbers of genotypic mutants will appear to contain none, whereas the size of jackpots will be unaffected. This will increase the variance.

[7]Even when a mutation might be deleterious, such as those that create tRNA nonsense suppressors, the effect may be slight. For example, most amber suppressors (in contrast to ochre suppressors) have little effect on growth because UAG is infrequently used as a termination codon in *E. coli* (33).

weeks (6, 15, 16, 45, 100, 113; B. Bridges, personal communication) and that the distribution of the late-arising mutants is Poisson (15, 16).[8] In some cases, e.g. the Shapiro deletion, the mutants simply never arise during exponential growth (16, 80, 113). More typically, mutants arise both before and during selection, but most of the former are detected by 1–3 days after selection is applied (6, 15, 45, 100; B. Bridges, personal communication). Late-appearing mutants could simply be slow growers, but slowly growing mutants resulting from mutations that occurred before selection was applied will have a Luria-Delbrück distribution (100). Furthermore, although all the earliest arising mutants must, necessarily, be fast growers, fast-growing mutants also arise late (15, 16, 45, 95).

Cell growth and turnover Of all the possible trivial explanations for the appearance of mutants during selection, the most important is that the cells may be dividing, allowing mutations to arise by ordinary replication-dependent processes. No experiment should be taken seriously if this artifact has not been addressed. Growth could result from: (*a*) a nonstringent selection, (*b*) utilizable contaminants in the medium, (*c*) cross-feeding by preexisting mutants, and (*d*) cell turnover. All of these phenomena may occur, and the question, in every case, is whether the amount of resulting cell growth can account for the number of mutants that appear.

Table 2 compares the mutation rates before and during selection for experiments for which these data were available. I have calculated a parameter that I call "turnover," which can be thought of as the number of DNA replications that would have had to have taken place to account for the observed number of mutations, assuming that these mutations were replication dependent and arose at the same rate as the preselection mutations. For example, with the *lacI33::lacZ* frameshift allele (15), 10^9 cells plated on lactose minimal medium will give rise to 400 Lac^+ mutants during 5 days. If these resulted from generation-dependent mutations, then 96×10^9, or 10^{11}, Lac^- cells must have existed on that plate at some time during those five days. This is about the same amount of growth that Lac^+ cells can achieve (P. L. Foster, unpublished results), whereas Lac^- cells can, at most, double (15).

Clearly in most cases the magnitude of the required amount of cell growth is so great that the mutations cannot possibly be attributed to gross population increases. Indeed, under starvation for an amino acid, the viable cell population is often declining (45, 100; B. Bridges, personal communication). Also, in our experiments, cross-feeding could not account for late-arising

[8]Phenotypic lag cannot account for the late appearance of mutants during selection if the cells must divide to overcome the lag.

Table 2 Comparison of spontaneous mutation rates during exponential growth and during nonlethal selection

		Mutation rates		Turnover required		
Organism	Mutational target	Mutations per cell per generation during nonselected growth[a]	Mutations per cell per hour during selection[b]	Total for experiment[c]	Per day[d]	Reference
Escherichia coli	*lacI* : : *lacZ* frameshift	1.4×10^{-9}	1.1×10^{-9}	96	19	15
	lacZ amber	1.7×10^{-10}	1.1×10^{-10}	75	25	16
	trpE frameshift	3.0×10^{-10}	3.1×10^{-10}	50	25	15, ud[e]
	lacZ missense	3.1×10^{-10}	7.9×10^{-11}	43	6.1	47
	trpB missense	1.5×10^{-10}	1.4×10^{-11}	26	2.2	45
	trpA missense	9.4×10^{-11}	4.5×10^{-12}	14	1.2	45
	his$^-$	3.0×10^{-8}	4.1×10^{-10}	2	0.33	100
Saccharomyces cerevisiae	*lys2* frameshift	4.9×10^{-9}	3.0×10^{-10}	7	1.5	121

[a] Calculated as $m/(2N)$ where m = average number of mutations per culture, N = cells per culture. These rates are slightly different from published rates calculated as $[m\ln(2)]/N$ (45, 47).

[b] Calculated as $M/(N_o h)$ where M = total number of mutants appearing under selection, N_o = cells originally plated, and h = hours the experiment lasted after preexisting mutations appeared. These rates are different from published rates that took into account cell death (45, 100).

[c] Calculated as M/N_o divided by the exponential-phase mutation rate.

[d] Total turnover divided by the days the experiment lasted.

[e] J. Cairns & P. L. Foster, unpublished data.

mutants. When various dilutions of revertible Lac^- cells were plated with nonrevertible Lac^- scavengers, the speed and frequency (per revertible cell) at which Lac^+ revertants arose was independent of the number of preexisting Lac^+ colonies (15).

Cryptic growth, i.e. the turnover of some of the population at the expense of the majority, is not so easy to dismiss. However, thermodynamic constraints apply—the number of cell deaths required to give rise to a new cell must be greater than one and has been estimated to be three to five (94). However, a slow input of energy (i.e. from breakdown of components of the medium) plus nutrients from dead and dying cells could allow some cell turnover. In reconstruction experiments, both Ryan (101, 103) and Hall (45) found that penicillin did not increase death rates, which means that the cells were not, apparently, undergoing cell division. However, such experiments would not detect a small subpopulation of cells that might be metabolizing but not dividing, or dividing only slowly.

DNA synthesis Cell turnover is perhaps irrelevant if what we are really interested in is DNA turnover. As discussed above, DNA synthesis is believed to be required to produce most types of mutations. If the preselection mutation rates apply, the amount of DNA synthesis needed to account for the mutations occurring during selection is the same as the turnover calculated in Table 2.

Only a few studies of DNA synthesis in cells in stationary phase have been done. After cells cease to divide, ongoing rounds of replication are completed, resulting in a 30–70% net increase in DNA, but this synthesis falls to zero within a few hours (83, 102). Ryan (102) attempted to measure DNA synthesis in histidine-starved cells with a density-shift experiment. After ^{15}N-labeled cells were starved in the presence of ^{14}N for 18 days, <5% hybrid DNA was detected, although considering the number of His^+ mutants that appeared, 67% of the DNA should have been hybrid.[9] Using ^{3}H-thymidine incorporation, Boe (9) reported 5% turnover per genome per day in cells early into (glucose-limited) stationary phase, but this rapidly declined to 0.5%. Another estimate of DNA turnover in nondividing cells, although not cells in stationary phase, was made by Grivell et al (41). After shifting a temperature-sensitive *dnaB* mutant to the nonpermissive temperature (to stop chromosomal replication), ^{3}H-thymine was incorporated into preexisting DNA at a rate equivalent to about 1% of the genome per day. However, Ryan was the only one to attempt to measure DNA turnover and mutation in the same experiment.

Taking these estimates at face value, the amount of DNA synthesis in nondividing cells, whether by genome-replication or turnover, would appear

[9]If only the *his* region was experiencing turnover, Ryan would surely not have detected it.

to be in the range of 0.5–5% per genome per day. Table 2 shows that these rates fail, in some cases by orders of magnitude, to account for the mutations that arise. We are left, therefore, with a limited number of possibilities: (*a*) the amount of DNA synthesis is being underestimated (e.g. because stationary cells might not use exogenous precursors); (*b*) the DNA turnover is restricted to a few regions of the genome that, somehow, always include the region where the mutation is being monitored; (*c*) the mutations that occur in stationary-phase cells are, in every case, independent of DNA synthesis; or (*d*) the DNA synthesis that is taking place is far more error-prone than normal DNA synthesis. These possibilities are discussed further below.

Are Mutations Stimulated by Specific Selection for Them?

The radical conclusion reached by Cairns et al (16) was not simply that mutations occur in nongrowing cells, but that the mutations appear to be directed by selective pressure. The evidence presented was, first, that starvation per se did not result in the accumulation of mutants and, second, that only mutants with the selected phenotype appeared. For convenience, these two related aspects of the phenomenon are discussed separately.

IS STRESS PER SE MUTAGENIC? *Point mutations* Cairns et al (16) introduced the delayed-overlay experiment to demonstrate that mutations did not accumulate when the cells were simply starving. Stationary-phase (glycerol-limited) Lac^- cells were plated in top agar on minimum medium without a carbon source, and at various times thereafter the plates were overlaid with top agar containing lactose. We have used this technique to study the reversion of several mutant *lacZ* alleles. In every case, no revertants accumulated during the time without lactose, and the rate at which mutant colonies appeared after the lactose was added was the same no matter how long the delay (15, 16, 37). We have also shown that Lac^+ mutants do not accumulate even in the presence of lactose if the cells are deprived of a required amino acid (15). Similar experiments, involving plating cells on filters that can then be transferred to different media, also have been used to show that starvation in the absence of specific selection is not mutagenic (45, 127).

The delayed-overlay experiment has been criticized on the grounds that mutants appearing during the starvation period might die, whereas those appearing after lactose is added would not (68). This criticism ignores that fact that the numbers of Lac^- parents and preexisting Lac^+ revertants are perfectly stable during starvation (13, 15, 37). Another claim is that the potential artifact can apply even if mutants have the same death rate as nonmutants (68). However, in our experiments (15), this would mean that during the first two days of starvation the total population on a plate would have had to decline 10-fold, and the remaining cells would have had to increase their mutation rate to exactly compensate. It seems more likely that the simple interpretation of the delayed-overlay experiments holds—starvation per se does not result in an increase in mutations.

It is true that if newly arisen mutants were somehow at a disadvantage in the absence of selection for them, the delayed-overlay experiments would fail to detect them. As stated by Cairns (13), "that, of course, is just another way of describing the very anomalies we are seeking to explain" (p. 527).

Movement of IS elements Several studies of adaptive or stationary-phase mutation have involved mutational events mediated by IS elements (9, 16, 43, 89, 113, 127). In these cases, nutritional conditions might play a role by nonspecifically triggering the movement of the elements. Complex patterns of Mu transpositions have been detected during the development of *E. coli* colonies (114), and several studies have reported increased movement of IS elements during stationary phase (3, 93, 130). In the case of the Lac(Ara)$^+$ fusion, both Shapiro (113) and Cairns et al (16) found that incubation of stationary-phase cells in the absence of lactose did not induce fusion formation. Mittler & Lenski (80) subsequently reported that after incubation of the prefusion strain with aeration in liquid (glucose-limited) minimal medium for 9 days, Lac(Ara)$^+$ cells appeared at a frequency of 2×10^{-6} after plating on selective medium. Significantly, this frequency was unchanged after several generations of growth in nonselective medium, suggesting that the mutants preexisted and were able to give rise to clones of Lac(Ara)$^+$ progeny in the absence of selection. Both I (P. L. Foster, unpublished results) and Shapiro (79) confirmed these basic results using a medium comparable to that used by Mittler & Lenski. There are two interesting questions about this controversy. First, did Mittler & Lenski actually prove that the Lac(Ara)$^+$ fusion occurred in the absence of selection for it? The answer is clearly no. Mittler & Lenski did not confirm the preexistence of Lac(Ara)$^+$ mutants with any of the methods of classical genetics invented expressly for this purpose (namely, the fluctuation test, replica-plating and sib-selection). Thus, they did not rigorously exclude the possibility that the Lac(Ara)$^+$ cells were arising from events occurring after lactose selection among cells still predisposed to fusion formation. In addition, in Mittler & Lenski's experiments, Lac(Ara)$^+$ cells had an eightfold greater survival than their Lac(Ara)$^-$ parents even in the absence of lactose. Thus, the fusion was being strongly selected whether or not lactose was present. In contrast, Cairns et al (16) starved the cells in a moderately enriched (glycerol-limited) medium that did not select against the prefusion cells (14).

The second question is, why do different experimenters get different results? Clearly, uncontrolled factors are influencing the rate at which fusions occur. Fusion formation is under control of the Mu transposition functions (113), which are responsive to the nutritional status of the cell (J. Shapiro, personal communication). In addition, Shapiro & Leach (115) provided a model by which lactose could specifically stimulate the in-frame fusions that render the cell Lac(Ara)$^+$. During the recombinational events that are required for Mu transposition, a hybrid template could form, allowing transcription of an *araB::lacZ* fusion from the *ara* promoter before rearrangements are completed. In-frame hybrids would result in β-galactosidase production, which, in the presence of lactose, could stimulate completion of the process.

Hall (43) obtained an even more controversial result involving an IS element. He found that excision of an IS*103* element from the *bglF* gene, which codes for the β-glucoside transport protein, occurred in colonies on MacConkey medium only in the presence of salicin, the substrate of the *bgl* operon. However, Mittler & Lenski (81) have reported that a low frequency of excision events can occur in cells incubated in rich broth for 15 days in the absence of salicin, although once again they did not do a fluctuation test or any form of sib-selection. Hall (personal communication) has recently disputed this finding—in his hands, no Sal$^+$ mutants were found among approximately 3×10^{12} cells during four weeks of incubation on MacConkey medium without salicin. Hall also points out that Mittler & Lenski (81) failed to demonstrate that the Sal$^+$ mutants they obtained were excision mutants.

In conclusion, starvation appears not to be universally mutagenic, but can trigger IS elements to move, depending on the nutritional status of the cell. In addition, changes in DNA topology occur in starving cells (60), and supercoiling is known to influence transposition, site-specific inversions, and other types of chromosomal rearrangements (52, 82, 107, 133). Such rearrangements may give rise to a variety of new phenotypes; for example, it is well known that cryptic genes can be activated both by insertion and by excision of IS elements (70, 92, 97). Although mutations due to these kinds of events, if they occur regardless of utility, do not meet the narrow definition of adaptive mutation, they do fall into the more general category of stationary-phase mutation.

ARE THE MUTATIONS THAT ARISE SPECIFIC FOR THE SELECTION? If mutations arise in response to selection for them, a population under selection would not accumulate mutants with a useless phenotype (16). To test this hypothesis, many cells under selection for one phenotype must be scored for a second, neutral phenotype. Because the cells are not growing, the second phenotype must be dominant and immediately expressed, making many easily scored phenotypes, like drug and phage resistances, inappropriate. Mutation to the second phenotype must be frequent enough so that the absence of mutants is meaningful, and these mutants should also be able to accumulate under selection for them. These constraints severely limit the number of mutational targets available.

Valine resistance (Val^R), which meets the above criteria, has been used as a neutral phenotype in several studies (6, 16, 43, 45, 56). All these experiments are consistent in showing that populations of cells under selection for carbohydrate utilization or reversion of an amino acid auxotrophy are not simultaneously accumulating Val^R mutants. However, it has been argued that Val^R may be a special case and its use should be viewed with caution (30, 119). *E. coli* K-12 strains are sensitive to exogenous valine because one of the three acetohydroxy acid synthase isozymes, which catalyze the first step in the biosynthesis of valine and isoleucine, is inactivated by a two–base pair deletion in the *ilvG* gene (66). Because the other two isozymes are sensitive to feedback inhibition by valine, exogenous valine deprives *E. coli* K-12 of isoleucine. The original mutation can be reverted by frameshifts, but mutations in at least eight other loci will also give a Val^R phenotype (26). Thus Val^R is a broad mutational target, but the expression of Val^R may be modified by genetic and physiological factors. Mutation rates to valine resistance vary widely (45, 56; R. Schaaper, personal communication), which suggests that under some conditions only subsets of Val^R mutants are expressed. In addition, excess valine slows the growth of even Val^R mutants (R. Schaaper, personal communication).

Hall (45) devised a reciprocal test. Colonies of a *trpB*$^-$ *cysB*$^-$ strain were grown for 11 days on filters placed on media limited for either tryptophan or cysteine. During that time, revertants appeared as papillae. Hall then switched the filters to media limited only for the other amino acid and incubated the plates for another 40 hours. Because no new papillae appeared, he argued that no Trp^+ mutants had occurred in the cysteine-limited colonies, nor had Cys^+ mutants accumulated in the tryptophan-limited colonies. One question about this experiment is whether 40 hours was sufficient time to allow a cell, starved for one amino acid but supplied with another, to switch its biosynthetic pathways in time to produce a visible papilla (36).

Thus the perfect second marker has not been found and, indeed, may never be. In addition to physiological constraints, any two mutational events may respond differently to the selective conditions (24, 68). However, a persuasive experiment was done by Steele & Jinks-Robertson (121). Using Lys^- auxotrophs of *Saccharomyces cerevisiae* (see below), they found that while mutational events giving rise to LYS prototrophs did not occur during starvation for amino acids other than lysine, recombinational events that gave the same phenotype did. Because each result was obtained under the same conditions, this experiment eliminates most artifactual explanations.

In conclusion, the few examples showing that mutations that are not selected do not arise, and the many showing that mutations do not arise when they are not useful, are consistent with the idea that, in most cases, only useful mutations accumulate during selection.

The Specificity of Mutational Events

If the mutations that occur during selection are different from those that occur during exponential growth, the two processes are likely to be different. But there is a problem of selection bias. For example, a mutant appearing during exponential growth may have several generations to express its phenotype, whereas after selection is applied, only immediately expressed phenotypes allow growth. In addition, as the cells age, the selection pressure could change to favor certain subsets of phenotypes and disfavor others.

Benson (5) reported that, after plating *lamB*$^-$ cells (missing *E. coli*'s major maltoporin) on maltodextrin medium, mutants that could use the maltodextrins, designated Dex$^+$, continued to appear for many days. All of the late-appearing mutants had mutations in *ompF,* which codes for a porin protein, although Dex$^+$ mutants in *ompC,* which codes for another porin, could be obtained if the starting strain was *ompF*$^-$. On subsequent investigation, Benson et al (6) demonstrated that OmpC(Dex$^+$) cells were at a disadvantage within a lawn of Dex$^-$ cells whereas OmpF(Dex$^+$) cells were not. Thus, the apparent mutational bias was actually a selection bias.

Cupples and colleagues (22, 23) have created a series of strains with constitutively expressed *lacZ* alleles carrying various mutations affecting the active site of β-galactosidase. Each of these can revert to Lac$^+$ by only a single, defined mutational event. In addition, all the alleles that detect base substitutions revert only to the wild-type DNA sequence; thus, all these revertants must have the same phenotype. Hall (47) compared the mutation rates of these strains during exponential growth (determined by fluctuation tests) to the frequency of Lac$^+$ papillation during lactose selection. The conclusion was that there was little correlation between the exponential mutation rate and the ability of each strain to revert during selection. For example, among the base substitutions, the event with the highest rate during exponential growth was the A:T to T:A transversion, whereas during selection it was the G:C to T:A transversion. Similarly, among the frameshift events, +A did well during nonselective growth, but −G was the winner after selection.

Using *Salmonella typhimurium,* Prival & Cebula (95) identified the mutations carried by over 1300 His$^+$ revertants isolated early (2 days) and late (5–10 days) after cells were plated on minimal medium without histidine. The two *hisG* alleles used—a missense and an ochre mutation—can revert by a variety of intragenic events as well as by the creation of extragenic suppressors. To avoid selection bias, Prival & Cebula only considered mutants that were fast-growing after isolation. The results at the two loci were strikingly different. At the missense site, there was no difference in the spectra of mutational events obtained early and late; at the ochre site, the proportions of some events were changed whereas others were the same. In both cases, extragenic suppressors were conspicuously absent among the late-arising, fast-growing revertants.

B. Bridges (personal communication) reported that the revertants of a tyrosine auxotroph that arise during prolonged selection did not result from any of the well-characterized mutations that revert this auxotrophy during growth or after treatment with mutagens. In unpublished results, we have found that base substitutions are relatively poor whereas frameshift mutations are relatively good at reverting under selection. However, factors other than the mutational event are clearly important. For example, we generated random Lac^- and Ara^- mutants with a frameshift mutagen; out of 10 of each, half of the Lac^- and only one of the Ara^- mutants reverted under selection. Finally, in general not every mutational target that could, in principle, give rise to mutants under selection does in fact do so (47, 56; B. Bridges, personal communication; P. L. Foster & J. Cairns, unpublished results).

These results, particularly those obtained with the *lacZ* and *hisG* alleles that are relatively free of selection bias, strongly support the idea that the mutational processes that give rise to mutants during selection are different from those that give rise to mutants during exponential growth. Whether the results shed any light on the mechanisms by which the mutations arise is discussed below.

Multiple Mutational Events

As mentioned above, Lenski et al (69) argued that both the deviations from the Luria-Delbrück distribution and the continued appearance of mutants observed by Cairns et al (16) could be explained by the occurrence of sequential mutations. They hypothesized that the final Lac^+ phenotype was due to two mutations. The first would give an intermediate phenotype, allowing the mutant to grow on lactose plates into an unobservable microcolony. The second mutation would then occur within this population, giving rise to a visible Lac^+ colony. This model has three parameters: the two mutation rates and the growth rate of the intermediate. To explain events occurring at frequencies of the order of 10^{-8} to 10^{-9}, as are typical of normal reversion events, and to prevent the population of the intermediate from being noticeable to the experimenter, the mutation rates must be set unreasonably high (S. Sarkar, personal communication). In fact, to model the results given in Cairns et al (16), the second mutation rate was set at 10^{-5} (69). Furthermore, the hypothesis requires additional assumptions about the nature of the mutational events. It does not seem likely that every single-step mutation that has been observed to occur during selection actually came about by a circuitous pathway. On the other hand, when two genetic events are known to be required for a particular phenotype, the frequency of the end result will be greatly increased if one of the genetic events confers some advantage. Hall (43, 46) has reported two dramatic experiments in which double mutations occurred at frequencies many orders of magnitude higher than expected; these are discussed in detail below.

THE BGL EXPERIMENT Hall's first double-mutant experiment, mentioned above, was activation of a cryptic *bgl* operon (43). In the parental strain, the regulatory gene is noninducible (designated *bglR*^{0}) and the gene, *bglF*, that encodes the β-glucoside transport protein is inactivated by an IS*103* element. When this strain was plated on MacConkey salicin plates, Hall found that up to 10% of the cells in aged colonies were *bglF*$^+$. Because *bglR*^{0} *bglF*$^+$ cells cannot utilize salicin, he concluded that each mutant represented an independent excision event, somehow driven by an anticipated need for the *bglF*$^+$ gene product. The population of *bglF*$^+$ cells could then give rise to *bglR*$^+$ mutants with the full Sal$^+$ phenotype. However, as pointed out by Symonds (125), if *bglR*^{0} *bglF*$^+$ cells had even a slight ability to grow, then the frequency of excision events could be very much lower. Mittler & Lenski (81) tested this hypothesis and indeed found that *bglR*^{0} *bglF*$^+$ cells do have a selective advantage over *bglR*^{0} *bglF*::IS*103* cells on MacConkey salicin plates. This has been confirmed by Hall (personal communication), who agrees that growth of a few *bglR*^{0} *bglF*$^+$ cells accounts for the high frequency of excision mutants that he observed. However, as mentioned above, Hall maintains that the excisions only occur when salicin is present.

THE DOUBLE-TRP EXPERIMENT Hall's second double-mutation experiment used a *trpA*$^-$ *trpB*$^-$ strain (46). Both alleles (*trpA46* and *trpB9578*) carry missense mutations, and neither the *trpA*$^+$ *trpB*$^-$ nor the *trpA*$^-$ *trpB*$^+$ cells can grow without tryptophan. When plated on medium with limiting tryptophan, which allowed colonies of about 5×10^7 cells to form, Trp$^+$ papillae began appearing on the colonies after 20 days of incubation. The frequency at which these papillae appeared was orders of magnitude greater than would be expected from the reversion rates of each single mutant. Hall then sequenced 11 revertants and found no changes other than the reverted bases themselves, a result that cannot be explained by any of the current models for adaptive mutation (46) (see below).

This, too, could be an example of growth of an intermediate. The *trpB9578* allele would allow *trpA*$^+$ *trpB*$^-$ revertants to accumulate and excrete indole, the product of the TrpA protein (C. Yanofsky, personal communication). *trpA*$^-$ *trpB*$^+$ mutants can convert indole to tryptophan (138). I observed that *trpA46 trpB*$^+$ cells could grow vigorously while being cross-fed indole by *trpA*$^+$ *trpB9578* cells, and I (36) and others (R. Kolter, personal communication) proposed that a subpopulation of *trpA*$^-$ *trpB*$^+$ cells could grow within the aging colonies and give rise to the Trp$^+$ papillae. If the time-dependent rate of reversion of *trpA46* to *trpA*$^+$ (45, 46) is applied, about 10^5 *trpA46 trpB*$^+$ cells per colony would be required to produce the number of Trp$^+$ papillae observed. Khakoo (58) independently derived the same model, incorporating a more sophisticated mathematical treatment but reaching roughly the same conclusion. Importantly, Khakoo's model reproduces the time course of appearance of Trp$^+$ papillae that Hall observed.

Hall (personal communication) has now done a reconstruction experiment with artificial colonies of 10^8 *trpA*$^-$ *trpB*$^-$ cells each seeded with an average of one *trpA*$^-$ *trpB*$^+$ and one *trpA*$^+$ *trpB*$^-$ cell. He found that the population of *trpA*$^-$ *trpB*$^+$ cells indeed increased and reached an average level of 10^5 per colony. However, he disagrees that this amount of growth can account for the number of Trp$^+$ papillae that arise. The resolution of this disagreement awaits further experimentation; however, I feel we are not obliged to account for Hall's results when considering the mechanisms by which adaptive mutations may arise.

GENE AMPLIFICATION Another type of sequential mutation can display all the attributes of adaptive mutation. If a gene has a partial activity, then successive amplifications may lead to a full phenotype. In eukaryotes, a classic example is the increased resistance to heavy metals conferred by amplification of the metallothionein genes (35). In *E. coli,* certain leaky *lacZ* alleles can produce up to 200 copies during lactose selection (128). Such events may be interesting in themselves, and they may be manifestations of a general genomic instability induced by stress (75, 76). However, because they are not stably inherited, whether amplifications per se should be considered as part of the phenomenon of adaptive mutation is questionable. Thus, it is important to check by prolonged cultivation under nonselective conditions that supposedly adaptive mutations are stable (75, 128). For example, I easily screened 500 independent Lac$^+$ revertants of the *lacI33::lacZ* frameshift allele and found that not one resulted from amplification (P. L. Foster, unpublished results).

ADAPTIVE MUTATION IN YEAST

In 1978, von Borstel reported that revertants of a tryptophan auxotroph (*trp1-1*) of *Saccharomyces cerevisiae* continued to appear for weeks during selection (131). Although von Borstel did, and still does (personal communication), attribute this to cell turnover, others have recently demonstrated adaptive mutation in *S. cerevisiae* (48, 121). Repeating his previous experiments with *E. coli,* Hall (48) found that revertants of a histidine auxotroph (a missense mutation in *his4*), scored as papillae, arose after 5 days and continued to arise for an additional 25 days after cells were plated on histidine-limited medium.

Because the histidine-starved cells did not accumulate revertants of a second defect (requirement for inositol[10]), he concluded that reversion of *his4* was specific to the selection.[11] Steele & Jinks-Robertson (121) found that revertants of a lysine auxotroph (a frameshift mutation in *lys2*) continued to appear for 8 days during lysine selection, but did not accumulate when the cells were starved for tryptophan, leucine, or both lysine and tryptophan. Both studies included penicillin-type reconstruction experiments to show that the cells were not growing. Hall used medium without inositol and Steele & Jinks-Robertson used canavanine—in both cases growing cells would have been killed, and in neither case did death rates significantly increase. Both studies demonstrated that preexisting revertants could appear within 2–3 days, and Steele & Jinks-Robertson have subsequently confirmed this result with prototrophs that have been starved for various times (S. Jinks-Robertson, personal communication).

Although the production of mutations during selection was not as vigorous as observed with bacteria, these studies are convincing. As mentioned above, a subsequent study adds even more weight to the results with *lys2*. Steele & Jinks-Robertson (122) found that LYS prototrophs due to interchromosomal recombination events also continue to arise in nondividing cells, but in this case, the production of recombinants continued whether there was selection for them or not. Thus, mutation occurred in stationary phase only when it was adaptive, but recombination occurred whether it was adaptive or not.

Delayed appearance of mutants has also been reported for *Candida albicans* (75). With long exposure to sublethal concentrations of heavy metals, colonies of resistant cells began to appear after 5–10 days and continued to appear for 1–2 weeks thereafter. These resistances could have resulted from gene amplification, although the phenotypes were stable during a short period of nonselective growth. However, revertants of two auxotrophies also appeared with similar kinetics. None of these events in *Candida albicans* have, as yet, been shown to be specific to the selection imposed.

MECHANISMS OF ADAPTIVE MUTATION

> "Is there any other point to which you would wish to draw my attention?"
> "To the curious incident of the dog in the night-time."
> "The dog did nothing in the night-time."
> "That was the curious incident," remarked Sherlock Holmes.
>
> A. C. Doyle, 1892 (29)

[10]Although this is a lethal selection, Hall (48) argued that "inositoless-death" is not exerted for one generation, and thus the INOL phenotype would have time to be expressed.

[11]Hall (48) also reported the delayed appearance of *his6* revertants, but did none of the required controls, nor did he demonstrate that the papillae were composed of HIS cells.

Two aspects of adaptive mutation need explaining. The first is, what gives rise to the mutations? If, as seems likely, DNA synthesis is required to produce adaptive mutations, this question may be equivalent to asking what forms of DNA synthesis in nongrowing cells are important for the production of mutations. The second question is, why do only useful mutations appear under selection, or (the curious incident) why do mutations not appear when they are not useful? The answer to the second question will influence the answer to the first. Additional constraints are placed by the genetic evidence, meager though it is at present. Below, I present and discuss several possible mechanisms for adaptive mutation. I use the term variant to distinguish potentially transient changes in the cell's informational molecules from mutations, which are heritable sequence changes in the DNA.

Mutational Bias

Transcription is a readily available way that an organism under stress could target mutations to useful genes (16, 25, 34). Davis (25) and Fitch (34) proposed that because transcription results in regions of single-stranded DNA, and these are particularly vulnerable to the various DNA-damaging processes that give rise to mutations, transcription per se could be mutagenic. Because one difference between a cell starving in the presence of lactose and one starving in its absence is that the *lac* operon is induced in the former case, Davis (25) argued, and gave some evidence supporting his case, that the experiments reported by Cairns et al (16) needed no other explanation.

This model in its simple form has not stood up to the evidence. First, although the addition of isopropyl-β-D-thiogalactopyranoside (IPTG), a gratuitous inducer of the *lac* operon, increases the speed at which Lac^+ mutants appear after lactose is added, it does not change the numbers of Lac^+ mutants even if a second, metabolizable, carbon source is available (16, 37). Second, constitutively expressed genes do not accumulate mutations in the absence of selection for them. This has been shown for suppressors of a *lacZ* amber mutation (16) and for two constitutively expressed *lacZ* alleles (15, 37).[12] Thus, although the transcription model might explain why mutations arise when they are useful, it does not explain why mutations do not arise when they are not useful. If transcription is inherently mutagenic, it could be a source of variant DNA sequences in nongrowing cells, but some other mechanism must account for the selectivity of adaptive mutations.

Transcription might provide at least a partial solution to the problem of the missing DNA synthesis. As discussed above, the amount of DNA synthesis

[12]Although it has been argued that a constitutively expressed gene might not be expressed in starving cells (120), the *lacI33::lacZ* fusion is not likely to be inducible, and therefore a mutant could not declare itself Lac^+ in the presence of lactose if the gene were not being constitutively expressed.

estimated to occur in nongrowing cells fails to account for the mutations that appear (Table 2). If DNA synthesis in stationary phase were confined to transcribed genes, this gap would be closed by one or two orders of magnitude, depending on how many genes are being transcribed in a nongrowing cell. Both in *E. coli* and in mammalian cells, repair of UV damage is coupled to transcription and occurs preferentially on the transcribed strand (77, 78).[13] Recently, this information was tied to a phenomenon called mutation-frequency decline (MFD). When *E. coli* strains are exposed to UV light and then held for a while under conditions that do not allow protein synthesis, mutations that would have arisen from DNA lesions in the transcribed strand are selectively lost (8, 90, 136). The *mfd* gene, which is required for MFD, encodes transcription-repair coupling factor (TRCF), which couples *E. coli*'s UvrABC excision repair enzymes to the transcribed strand (63, 111). Thus, TRCF supplies a possible mechanism by which DNA synthesis, prompted by randomly occurring DNA lesions, could be targeted to transcribed genes. Excision repair, of course, is designed to prevent errors, but it must also sometimes create errors and perhaps would be more likely to do so in starving cells.

B. Bridges (personal communication) investigated the effect of TRCF on reversion of a tyrosine auxotrophy. As would be predicted by the above model, a *mfd*$^-$ strain could not revert during selection. However, this defect was not seen if the strain was also deficient in excision repair because of a mutation in *uvrA*. The simplest interpretation of these results is that the UvrA protein (or excision repair itself) prevents mutation in stationary phase cells unless TRCF is also present (B. Bridges, personal communication).

This hypothesis is consistent with other studies showing that adaptive mutation occurs perfectly well in cells that are deficient for the UvrABC repair pathway (16, 37, 95; P. L. Foster, unpublished results). However, Grigg & Stuckey (40) showed that caffeine, which inhibits excision repair, also inhibits stationary-phase mutation. We too have found a large decrease in the appearance of Lac$^+$ revertants when caffeine is added to lactose plates; however, caffeine also decreased the number of colonies appearing on day 2, which are mostly due to mutations that occurred during growth before plating (P. L. Foster, unpublished results). Caffeine intercalates into the DNA and can cause the UvrA protein, which is the damage-recognition subunit of the UvrABC system, to bind nonspecifically to DNA (110). I suggest that in nongrowing cells, intercalated caffeine may itself, or in conjunction with the UvrA protein, prevent gene expression (e.g. by inhibiting transcription).

[13]This is not true for all tested eukaryotic genes, possibly because the coupling factor is specific for RNA polymerase II (111).

Trial And Error

Cairns et al (16) concluded that cells under selection are engaged in some "reversible process of trial and error" (p. 145). That is, variants arise continuously, but they are immortalized as mutations only if the cell achieves success. Various models have been proposed that incorporated this idea; they differ in the level at which the variants arise and in the way that natural selection is postulated to exert its effect.

VARIANT RNAs Cairns et al (16) proposed that a cell under selection might be producing highly variable mRNA molecules. If one such mutant transcript happened to code for a good protein, this might stimulate reverse transcription, immortalizing the mutation in the DNA. This process would be highly efficient if the cell had some way of associating a given transcript with its protein and targeting that specific transcript for reverse transcription (16).[14] However, the mechanism would work without this association if all transcripts that were in existence at the moment that the cell achieved success were reverse transcribed at random.

Of course this model depends on reverse transcriptase being present in *E. coli*. Subsequent to the publication of Cairns et al (16), retroviral-like elements encoding reverse transcriptase were discovered in many strains of *E. coli*, but not, so far, in K-12 strains (54). Bridges (personal communication) has shown that curing *E. coli* B of its retron has no effect on the production of mutations during selection. However, neither result precludes the possibility that reverse transcriptase from some other source might still be present in the cells.

The efficient version of this model predicts that the only sequence changes that occur in the successful cell are those that code for the good protein. Using reversion of a *lacZ* amber allele, we found that extragenic suppressors continued to arise with time during lactose selection (37). The majority of the late-arising suppressors were fast growing upon isolation, and their distribution was Poisson (P. L. Foster, unpublished results). Therefore, these suppressors had all the characteristics of adaptive mutations. Because in these suppressors the information that restored activity to the mutant protein was not present in its RNA or its DNA, there could have been no direct link of the kind proposed (37).

Although these experiments do not exclude the inefficient version of the reverse-transcription model, they do tend to discourage speculation in that direction.

[14]This was the most Lamarckian of the three models proposed by Cairns et al (16), although Sarkar points out that it only falls into a weak definition of neo-Lamarckism (106).

VARIANT DNAs *Slow repair* Stahl (118) and subsequently Boe (9) started with the assumption that in stationary phase, cells are sporadically replicating or repairing their DNA. In a nutritionally deprived cell, the error-correction enzymes that monitor newly synthesized DNA might be slow to act. This delay might allow errors on the transcribed strand to persist long enough to encode a useful protein. The cell could then replicate its DNA before the mismatch was corrected. If an active protein were not encoded, then the error would eventually be corrected. To achieve specificity, this model requires that the slow repair pathway always corrects errors in favor of the template DNA. Stahl (118) and Boe (9) both nominated the methyl-directed mismatch repair system (MMR) as the relevant pathway.

Defects in MMR greatly enhance the frequency of mutations occurring during selection (9, 37, 56). However, the critical prediction of the hypothesis is that, when MMR is not active, mutations should occur whether or not they are being selected. This proved not to be the case—when a *lac*$^-$ strain deficient in MMR was starved in the absence of lactose, no Lac$^+$ mutants accumulated (37).

If *E. coli* is defective for both of its alkyltransferases, which remove alkyl groups from DNA bases, the frequency of mutations in stationary-phase cells (but not in growing cells) is greatly increased (96). Because this repair pathway restores the original DNA sequence, it could also, potentially, reverse transient variants. However, we obtained the same result with alkyltransferase-defective cells as with MMR-defective cells—mutations did not accumulate in the absence of selection (37).

Has the slow repair model been disproved? Not conclusively. Our experiments with the MMR-defective strain were complicated by the rapid loss of viability that these mutants display during starvation (9, 37). Furthermore, DNA repair pathways are extremely abundant. Prival & Cebula (95) have postulated both increases and decreases in these various pathways to account for the differences in the mutational spectra obtained from early- and late-appearing His$^+$ revertants. Of course, the problem with this kind of analysis is that, given enough pathways and the freedom to increase or decrease activities at will, anything could be explained. Steele & Jinks-Robertson (121) argued that the Stahl model is supported by their finding that UV-light irradiation increased the frequency of adaptive mutations. However, this result is consistent with any model that assumes variants can arise at the level of DNA; if the DNA contains lesions, the mutation frequency will increase, but this tells us little about the underlying mechanism.

Extra DNA copies Adaptive mutation could be achieved if a cell made copies of its genes and allowed these to accumulate sequence changes, but kept a pristine master copy. If one of the variant DNAs allowed the cell to divide, it would be retained in the cell's successful progeny. In the absence of a useful variant, the master copy would be retained and the useless variants destroyed.

Prompted by the finding that, in at least some cases, RecA function is required to produce adaptive mutations (15, 56), we proposed a model involving gene amplification (37). Bacteria are known to duplicate regions of their genomes at frequencies of 10^{-3} to 10^{-4} per cell (2). Duplications can initiate rounds of amplification owing to RecA-dependent events (e.g. unequal crossing over) (128). Furthermore, amplified DNA sequences appear in some cases to accumulate mutations at higher rates that single-copy sequences (11, 99). Amplifications are inherently unstable, but if one DNA copy contained a useful sequence change, the cell would begin to grow, resolve the amplified region (again by a RecA-dependent process), and retain the useful copy. In the absence of success, the amplified region would also eventually be resolved, but because most copies retain the original DNA sequence, the cell would remain unmutated.

This model has been criticized by Stahl (120) on the grounds that, to achieve the required specificity, many tandem copies would have to accumulate. Otherwise, when the amplification was resolved in the absence of success, the cell would have nearly as great a chance of retaining a variant as retaining the original DNA copy. In the absence of selection for it, Stahl does not believe that "such a monstrous aberrancy" (120, p. 866) would occur. We have indeed found no evidence that amplification of the *lacI33::lacZ* allele is, of itself, advantageous (P. L. Foster, unpublished results). However, mathematical models for the evolution of arrays of hundreds of repeats, as found in eukaryotic genomes, work without any assumption of selective pressure (49). The rate-limiting step is the initial duplication, and duplications can be stimulated by activated RecA in *E. coli* (28) and by loss of cell-cycle control in eukaryotes (50). Thus, we are not ready to abandon the amplification model on the basis of Stahl's objections.

Stahl has come up with a new model that he calls "The Toe in the Water Model" (120). Stable replication is a RecA-dependent form of DNA replication that occurs in the absence of protein synthesis (65). It can be induced by DNA damage and by physiological imbalances such as shift-up from minimal to rich medium. It may also be error-prone (65). A related, but not identical, form of stable replication, which is also RecA-dependent, occurs in cells that have low RNase H levels because RNase H normally prevents random RNA transcripts from priming aberrant DNA replication (59). Thus one can imagine that stable replication could be active in starving cells and be a source of DNA variants (36, 37, 56). Stahl has proposed that in a starving cell, stable replication would initiate but not proceed past the D-loop stage and the new DNA would eventually be degraded. But if the DNA contained a useful sequence change, a full replication fork would form, the cell would replicate, and the mutation would be immortalized.

These models have clearly been great fun to devise, and they do suggest some obvious experimental approaches. However, although we (15) and Jayaraman (56) have found that the production of adaptive mutations depends on RecA function, others have not (see below). So, apparently more than one mechanism gives rise to adaptive mutations.

THE HYPERMUTABLE STATE Hall (45) proposed the most extreme form of the trial and error hypothesis. During selection, a subpopulation of cells would enter into a state of hypermutation, producing DNA-sequence changes at a high rate. If a cell achieved a successful mutation, it would exit and resume growth. If it did not succeed, it would die. Thus, unsuccessful variants would be eliminated from the population.

Recently Higgins (53) suggested a modification of this model, which might be called the altruistic model. Within a colony of *E. coli* carrying a Mu element, the majority of transpositions occur within apparently "dead" cells. These cells cannot form colonies or grow in rich medium, but yet they appear to be able to make DNA, RNA, and protein. Drawing upon examples from other bacteria, such as *Neisseria* spp., which uses DNA from dead cells to change its antigenic determinants (109), Higgins suggested that, during starvation, a subpopulation of cells that are technically dead experience a mutation rate at least 1000-fold higher than normal. If a useful mutation is achieved, the dead cell obtains enough energy to transfer its DNA (by conjugation or transduction) into colony-forming cells. Recipient cells that capture the useful sequence change would then persevere. This model has the beauty of allowing some cells to benefit from an extremely high genomic mutation rate without suffering the lethal consequences. The model might also work if all hypermutated cells transferred their DNA into the colony-forming cells whether or not the DNA had a useful sequence change. The recipient of an useful piece of DNA would then grow whereas the rest of the population might incorporate, at most, a few useless mutations.

THE MISSING NEUTRAL MUTATIONS All the trial-and-error models predict that neutral mutations should not accumulate in a population under selection. All (except the efficient reverse-transcription model) also predict that neutral mutations should be found at enhanced frequencies in the cells that achieve success. This is because, at the moment of success when the useful variant is immortalized as a mutation, other variants that existed in the cell would also be immortalized. To date, two cases of neutral mutations occurring at higher frequencies than expected among selected mutants have been published. Hall (45) reported that 2 out of 110 revertants of *trpB*$^{-}$, versus zero out of 4530 nonreverted cells, carried auxotrophies. Boe (9) found that 20 out of 2000 late-arising revertants of an asparagine auxotrophy, whereas <2 out of 2000 revertants isolated from exponentially growing cells, grew poorly on succinate (however, it is not clear that the control revertants were truly independent). These rather small numbers and can provide only tentative support for the trial-and-error models. Furthermore, if the mechanism by which variants are produced involves only small regions of the genome, as would be predicted by most of the models, then mutations should occur only within those regions. Screening for a second phenotype is unlikely to find clustered mutations.[15] Hall (46) sequenced 700 bases surrounding the *trpA* and *trpB* loci from 11 Trp$^+$ double revertants and found no sequence changes other than the reverted bases themselves. He argued that this frequency was orders of magnitude lower than would be predicted by any model. However, for the reasons discussed above, this result will have to be reevaluated.

Other Possible Contributors

A few other phenomena may be relevant to the production of mutations in stationary-phase cells. None of these supply a mechanism by which mutations would be adaptive, because all would seem to result in a general increase in mutation rates. However, they are indicative of the kinds of things that may happen to cells during prolonged nonlethal selection.

Certain stresses, such as carbon starvation and iron deficiency, change the level of modification of some tRNAs. The product of the *miaA* gene is required for modification of the adenosine 3′ to the anticodon of tRNA species that read codons beginning with U (33). Defects in *miaA*, as well as iron deficiency, increase the spontaneous rates of some mutations by 10- to 100-fold (19). Interestingly, *miaA* is part of a complex operon with *mutL*, which codes for one of the MMR enzymes (19). The immediate effect of undermodification of these tRNAs is to increase the accuracy and decrease the efficiency of translation, but the global physiological effect is dramatic, especially on the biosynthesis of amino acids that are regulated by attenuation (7). Why this should increase mutation rates is unknown, but global changes in the levels of repair enzymes and nucleotide pools could result from decreased translational efficiency (19).

Ninio (85) and subsequently Boe (10) have postulated that transcriptional and translational errors could result in error-prone DNA polymerases and defects in DNA error-correcting functions. These would then act as transient

[15]Boe (9) argued that the succinate-utilization defect resulted from mutations in the *atp* operon, which is close to *asnA;* thus, both mutations could have arisen in the same heteroduplex. However, he did not show that the mutations were in fact in *atp*, and it seems unlikely that a heteroduplex region would extend for the roughly 10 kb of DNA that separate the genes.

mutators. Ninio calculated that the contribution of such mutators to single mutational events would be insignificant but could account for up to 95% of simultaneous double mutations. However, this hypothesis seems at odds with the tRNA modification results, in which increased translational accuracy increased mutation rates.

Finally, several studies of adaptive mutation have used mutational targets carried on F′ episomes (15, 16, 37, 47). Male cells carrying F factors do conjugate, and conjugation results in single-stranded DNA (135). Furthermore, conjugal transfer among F′ cells results in elevated RecA levels (126). Thus, the mutation rates on F′ factors may be higher than on chromosomes, or the mechanisms by which the mutations arise may differ between F′ factors and chromosomes. This might explain why we found that RecA is required for production of adaptive mutations (15). However, Jayaraman (56), who used chromosomal alleles, also found that adaptive mutations (in a *mutS* background) depended on RecA function.

GENETIC CONTROLS

If the process by which mutations arise in stationary-phase cells is truly different from the process by which mutations arise during exponential growth, one should be able to find mutants altered in one pathway but not in the other. Because increases in mutation rates could result from aberrant processes that have nothing to do with the way mutations are produced in wild-type cells, the most informative genetic defects would be those that decrease mutation rates. Any defects that alter the physiology of cells in stationary phase also would be expected to affect adaptive mutation. As yet, no mutants with clear-cut phenotypes have been obtained, but the experiments that have been done at least eliminate some possible mechanisms for adaptive mutations.

The SOS response is not required to produce adaptive mutations. As mentioned above, we found that adaptive reversion of the *lacI33::lacZ* frameshift required some activity of RecA$^+$. However, the SOS functions of RecA are not necessary. Revertants arise during selection in the absence of UmuDC functions (which are required for SOS mutagenesis) and when all LexA-controlled genes other than *recA* are repressed (15). Similar results were found for Val^R mutations in a *mutS* background (56). Although we favor the hypothesis that it is RecA's recombination functions that are required, our results are also consistent with the hypotheses that RecA activity is necessary to process some protein other than LexA or UmuD, or to allow stable replication to occur (see above). At this point, it is a mystery why we found such a strong requirement for RecA$^+$, whereas others found that RecA is completely dispensable (B. Bridges, personal communication; B. Hall, personal communication; S. Benson, personal communication). These conflicting results are not due to differences in the *recA* alleles that were used; although a possible interpretation of our original experiments was that the *recA* allele we studied, *recA430,* was actually inhibiting the production of adaptive mutations (15), we have confirmed our results with a null allele of *recA* (P. L. Foster, unpublished results). The simplest explanation is that RecA is required for some, but not all, mutational events occurring in stationary-phase cells.

The UvrABC excision repair pathway could be the source of the DNA synthesis that gives rise to mutations, or it could repair spontaneous DNA lesions that, if unrepaired, give rise to mutations. Elimination of excision repair by defects in *uvrB* (16, 37, 95) or *uvrA* (B. Bridges, personal communication; P. L. Foster, unpublished results) either has no effect or increases the frequency of mutations during selection. Thus, the UvrABC pathway is certainly not required to produce adaptive mutations, but it may prevent some classes of mutations from occurring.

As discussed above, cells defective in MMR have a greatly increased mutation rate during selection (9, 37, 56), but the MMR pathway does not appear to be responsible for the reversal of useless variants (37). Nevertheless, in *mutS*$^-$ strains, the frequency of adaptive mutations increases and cell viability decreases during stationary phase (9, 37). These results suggest that DNA synthesis is occurring in stationary-phase cells (because MMR normally corrects errors only in newly synthesized DNA) and that MMR activity is required to prevent some lethal outcome. It seems unlikely that the fairly dramatic loss in viability of MMR$^-$ cells is the result of the accumulation of lethal mutations unless mutations are more likely to occur in essential genes than in nonessential ones. As Boe pointed out (9), the lethality may be a result of events occurring when cells attempt to resume replication after being plated for viability.[16]

Because cells missing all alkyltransferase activity accumulate more mutations in stationary phase than do wild-type cells, DNA bases alkylated by some endogenous source can, if unrepaired, be a significant source of mutations in nondividing cells (96). However, our results indicate that the alkyltransferases are not responsible for the specificity of adaptive mutation, because in the absence of these enzymes, mutations do not arise in the absence of selection (37). The *lacZ* allele that we used to score mutations in alkyltransferase-defective strains reverts only by a specific G:C to A:T transition (23), the mutation induced by alkylguanine (71). As the guanine is on the nontranscribed strand, variant transcripts could not arise by transcription, but could only be produced after a thymine was inserted opposite the

[16]Death by assay, i.e. that cells may be viable until their viability is tested, could be a complication in many experiments with stationary-phase cells.

alkylguanine. Thus, DNA synthesis would appear to be required to produce the variants,[17] but because mutants do not appear in the absence of selection, the newly synthesized DNA must be lost in cells that cannot benefit from it.

Finally, Cairns et al (16) demonstrated that a deletion of the *uvrB-bio* region increased adaptive reversion of a *lacZ* amber allele. Although the Δ(*uvrB-bio*) strain produces more Lac$^+$ revertants than wild-type when plated on lactose minimal medium, it fails to papillate on MacConkey lactose plates even though Lac$^+$ revertants are being produced. S. Miller & J. Cairns (personal communication) suggest that the defect interferes with entry and exit from stationary phase; thus the mutant cells may have an increased mutation rate during selection because they enter into stationary phase early but then have difficulty resuming growth in rich medium.

CONCLUSIONS

Perhaps it was inevitable that the subject of adaptive mutation would be contentious. Critics have postulated a variety of artifacts that they feel can explain the observed results. Proponents have published a variety of experiments that, in some cases, deal with those artifacts and, in some cases, do not. Thus the two sides may at times appear equivalent—an abundance of artifacts, an abundance of controls. Heated exchange may be useful to science, but nothing can ultimately be gained by a selective reading of the evidence (68) or by angry adherence to preformed beliefs (117). The experiments given in Table 1, and others that may not have been included, each have to be evaluated on their own merits.

What would constitute absolutely convincing evidence that adaptive mutations occur? It has been argued that nothing will do so except the demonstration of a molecular mechanism (106, 120). This is an entirely unwarranted burden to place upon any field of study. Traditionally, the reality of a phenomenon is established by observations of it, not necessarily by understanding its cause. Where would the science of genetics be now if classical genetics had had to wait for the discovery of DNA?

The preponderance of evidence indicates that, as Ryan observed (101), mutations can arise in stationary-phase cells. In some cases, stress and nutritional factors may trigger the movement of IS elements and other types of genomic rearrangements, but the mechanism by which point mutations arise in stationary-phase cells is entirely unclear. DNA synthesis would appear

[17]The required mutation might be created without DNA synthesis by reciprocal recombination with a small homologous sequence somewhere else in the genome (J. Cairns, personal communication).

to be required for most mutational events, but there is a vast gap between the amount of DNA synthesis that has been measured in nondividing cells and the amount that would seem to be required to produce the mutations observed (Table 2). The DNA synthesis that takes place probably is targeted to only certain regions of the genome (e.g. transcribed genes) or is unusually error prone, or both. An increase in the error rate of DNA synthesis might be an unavoidable consequence of nutritional deprivation, or it could be the result of an induced response, analogous to the SOS system.

Regardless of how mutations arise, the real mystery is why they appear to do so only when they are useful. The simplest explanation is that the role of selection is not to direct a process, but to stop a process that is creating transient variants at random. However, we still do not know the nature of the transient variants or the identity of the editing mechanism.

"Sweet are the uses of adversity." The importance of adaptive mutation is not that natural selection is being circumvented, but that natural selection is apparently being allowed to choose among a cell's population of informational macromolecules (16). Thus, individual cells not only control their phenotypes by regulating the expression of their genes, but they also seem to have access to a multitude of potential genotypes, allowing the individual to increase its variability when it would be useful to do so, while maintaining its genome more or less intact. Understanding the mechanisms underlying this phenomenon in microorganisms may also shed some light on the way mutations arise in nondividing somatic cells in mammals, leading to the success (for the cells) that we call cancer.[18]

ACKNOWLEDGMENTS

I am grateful to my colleagues for stimulating discussions, especially J. Cairns, E. Eisenstadt, M. S. Fox, F. W. Stahl, J. W. Drake, J. H. Miller, M. R. Volkert, R. Kolter, J. A. Shapiro, and B. A. Bridges. I thank the scientists who communicated results prior to publication, and I also thank the many people that I have encountered at meetings for their lively comments and ideas. I am especially grateful to J. Cairns for a patient and thoughtful critique of this manuscript. Work in my laboratory was supported by National Science Foundation grant MCB-9214137.

[18]For example, Strauss (124) has argued that because generation-dependent mutation rates cannot account for the number of genetic changes found in tumor cells, they might result from a process of adaptive mutation.

Literature Cited

1. Albertini, A. M., Hofer, M., Calos, M. P., Miller, J. H. 1982. On the formation of spontaneous deletions: the importance of short sequence homologies in the generation of large deletions. *Cell* 29:319–28
2. Anderson, R. P., Roth, J. R. 1981. Spontaneous tandem genetic duplications in *Salmonella typhimurium* arise by unequal recombination between rRNA (rrn) cistrons. *Proc. Natl. Acad. Sci. USA* 78:3113–17
3. Arber, V., Iida, S., Jutte, H., Caspers, P., Meyer, J., Hanni, C. 1978. Rearrangements of genetic material in *Escherichia coli* as observed on the bacteriophage P1 plasmid. *Cold Spring Harbor Symp. Quant. Biol.* 43:1197–1208
4. Armitage, P. 1952. The statistical theory of bacterial populations subject to mutation. *J. R. Statist. Soc. B* 14:1–40
5. Benson, S. A. 1988. Is bacterial evolution random or selective? *Nature (London)* 336:21–22
6. Benson, S. A., DeCloux, A. M., Munro, J. 1991. Mutant bias in nonlethal selections results from selective recovery of mutants. *Genetics* 129:647–58
7. Björk, G. R. 1987. Modification of stable RNA. See Ref. 83a, pp. 791–31
8. Bockrath, R., Barlow, A., Engstrom, J. 1987. Mutation frequency decline in *Escherichia coli* B/r after mutagenesis with ethyl methanesulfonate. *Mutat. Res.* 183:241–47
9. Boe, L. 1990. Mechanism for induction of adaptive mutations in *Escherichia coli. Mol. Microbiol.* 4:597–601
10. Boe, L. 1992. Translational errors as the cause of mutations in *Escherichia coli. Mol. Gen. Genet.* 231:469–71
11. Boe, L., Marinus, M. G. 1991. Role of plasmid multimers in mutation to tetracycline resistance. *Mol. Microbiol.* 5:2541–45
12. Brock, T. D. 1990. *The Emergence of Bacterial Genetics*. Cold Spring Harbor, NY: Cold Spring Harbor Lab.
13. Cairns, J. 1988. Origin of mutants disputed. *Nature (London)* 336:527–28
14. Cairns, J. 1990. Causes of mutation and Mu excision. *Nature (London)* 345: 213
15. Cairns, J., Foster, P. L. 1991. Adaptive reversion of a frameshift mutation in *Escherichia coli. Genetics* 128:695–701
16. Cairns, J., Overbaugh, J., Miller, S. 1988. The origin of mutants. *Nature (London)* 335:142–45
17. Campbell, J. J., Lengyel, J., Langridge, J. 1973. Evolution of a second gene for β-galactosidase in *Escherichia coli. Proc. Natl. Acad. Sci. USA* 70: 1841–45
18. Cavalli-Sforza, L. L., Lederberg, J. 1956. Isolation of preadaptive mutants by *sib* selection. *Genetics* 41:367–81
19. Connolly, D. M., Winkler, M. E. 1989. Genetic and physiological relationships among the *miaA* gene, 1-methylthio - N^6 - (Del^2 - isopentenyl)-adenosine tRNA modification, and spontaneous mutagenesis in *Escherichia coli* K-12. *J. Bacteriol.* 171:3233–46
20. Cox, E. C. 1976. Bacterial mutator genes and the control of spontaneous mutation. *Annu. Rev. Genet.* 10:135–56
21. Craig, N. L., Kleckner, N. 1987. Transposition and site-specific recombination. See Ref. 83a, pp. 1054–70
22. Cupples, C. G., Cabrera, M., Cruz, C., Miller, J. H. 1990. A set of *lacZ* mutations in *Escherichia coli* that allow rapid detection of specific frameshift mutations. *Genetics* 125:275–80
23. Cupples, C. G., Miller, J. H. 1989. A set of *lacZ* mutations in *Escherichia coli* that allow rapid detection of each of the 6 base substitutions. *Proc. Natl. Acad. Sci. USA* 86:5345–49
24. Danchin, A. 1988. Origin of mutants disputed. *Nature (London)* 336:527
25. Davis, B. D. 1989. Transcriptional bias: a non-Lamarckian mechanism for substrate-induced mutations. *Proc. Natl. Acad. Sci. USA* 86:5005–9
26. De Felice, M., Levinthal, M., Iaccarino, M., Guardiola, J. 1979. Growth inhibition as a consequence of antagonism between related amino acids: effect of valine in *Escherichia coli* K-12. *Microbiol. Rev.* 43:42–58
27. Delbrück, M. 1946. Heredity and variations in microorganisms. *Cold Spring Harbor Symp. Quant. Biol.* 11:154
28. Dimpfl, J., Echols, H. 1989. Duplication mutation as an SOS response in *Escherichia coli:* enhanced duplication formation by a constitutively activated RecA. *Genetics* 123:255–60
29. Doyle, A. C. 1892. The adventure of silver blaze. *The Strand Mag*. Vol. 4, No. 24
30. Drake, J. W. 1991. Spontaneous mutation. *Annu. Rev. Genet.* 25:125–46
31. Drake, J. W. 1991. A constant rate of spontaneous mutation in DNA-based

microbes. *Proc. Natl. Acad. Sci. USA* 88:7160–64
32. Echols, H., Goodman, M. F. 1991. Fidelity mechanisms in DNA replication. *Annu. Rev. Biochem.* 60:477–511
33. Eggertsson, G., Söll, D. 1988. Transfer ribonucleic acid–mediated suppression of termination codons in *Escherichia coli. Microbiol. Rev.* 52:354–74
34. Fitch, W. M. 1982. The challenges to Darwinism since the last centennial and the impact of molecular studies. *Evolution* 36(6):1133–43
35. Fogel, S., Welch, J. W. 1982. Tandem gene amplification mediates copper resistance in yeast. *Proc. Natl. Acad. Sci. USA* 79:5342–46
36. Foster, P. L. 1992. Directed mutation: between unicorns and goats. *J. Bacteriol.* 174:1711–16
37. Foster, P. L., Cairns, J. 1992. Mechanisms of directed mutation. *Genetics* 131:783–89
38. Foster, T. J., Lundblad, V., Hanley-Way, S., Halling, S. M., Kleckner, N. 1981. Three Tn10-associated excision events: relationship to transposition and role of direct and inverted repeats. *Cell* 23:215–27
39. Fox, M. 1955. Mutation rates of bacteria in steady state populations. *J. Gen. Physiol.* 39:267–78
40. Grigg, G. W., Stuckey, J. 1966. The reversible suppression of stationary phase mutation in *Escherichia coli* by caffeine. *Genetics* 53:823–34
41. Grivell, A. R., Grivell, M. B., Hanawalt, P. C. 1975. Turnover in bacterial DNA containing thymine or 5-bromouracil. *J. Mol. Biol.* 98:219–33
42. Hall, B. G. 1982. Evolution on a petri dish. In *Evolutionary Biology,* ed. M. K. Hecht, B. Wallace, G. T. Prance, 15:85–150. New York: Plenum
43. Hall, B. G. 1988. Adaptive evolution that requires multiple spontaneous mutations. I. Mutations involving an insertion sequence. *Genetics* 120:887–97
44. Hall, B. G. 1989. Selection, adaptation, and bacterial operons. *Genome* 31:265–71
45. Hall, B. G. 1990. Spontaneous point mutations that occur more often when they are advantageous than when they are neutral. *Genetics* 126:5–16
46. Hall, B. G. 1991. Adaptive evolution that requires multiple spontaneous mutations: mutations involving base substitutions. *Proc. Natl. Acad. Sci. USA* 88:5882–86
47. Hall, B. G. 1991. Spectrum of mutations that occur under selective and non-selective conditions in *E. coli. Genetica* 84:73–76
48. Hall, B. G. 1992. Selection-induced mutations occur in yeast. *Proc. Natl. Acad. Sci. USA* 89:4300–3
49. Harding, R. M., Boyce, A. J., Clegg, J. B. 1992. The evolution of tandemly repetitive DNA: recombination rules. *Genetics* 132:847–59
50. Hartwell, L. 1992. Defects in a cell cycle checkpoint may be responsible for the genomic instability of cancer cells. *Cell* 71:543–46
51. Hayes, W. 1968. *The Genetics of Bacteria and their Viruses.* New York, NY: Wiley & Sons. 2nd ed.
52. Higgins, C. F., Dorman, C. J., Stirling, D. A., Waddell, L., Booth, I. R., et al. 1988. A physiological role for DNA supercoiling in the osmotic regulation of gene expression in S. typhimurium and E. coli. *Cell* 52:569–84
53. Higgins, N. P. 1992. Death and transfiguration among bacteria. *TIBS* 17:207–11
54. Inouye, M., Inouye, S. 1991. msDNA and bacterial reverse transcriptase. *Annu. Rev. Microbiol.* 45:163–86
55. Jacob, F., Wollman, E. L. 1961. *Sexuality and the Genetics of Bacteria.* New York, NY: Academic
56. Jayaraman, R. 1992. Cairnsian mutagenesis in *Escherichia coli:* genetic evidence for two pathways regulated by *mutS* and *mutL* genes. *J. Genet.* 71:23–41
57. Keller, E. F. 1992. Between language and science: the question of directed mutation in molecular genetics. *Persp. Biol. Med.* 5:292–306
58. Khakoo, A. Y. 1992. *An investigation of the phenomenon of directed mutation.* BS thesis. Princeton Univ., Princeton, N.J.
59. Kogoma, T. 1986. RNase H-defective mutants of *Escherichia coli. J. Bacteriol.* 166:361–63
60. Kolter, R. 1993. The stationary phase of the bacterial life cycle. *Annu. Rev. Microbiol.* 47:855–74
61. Kubitschek, H. E. 1970. *Introduction to Research with Continuous Cultures.* Englewood Cliffs, NJ: Prentice-Hall
62. Kubitschek, H. E., Bendigkeit, H. E. 1964. Mutation in continuous cultures. I. Dependence of mutational response upon growth-limiting factors. *Mutat. Res.* 1:113–20
63. Kunala, S., Brash, D. 1992. Excision repair at individual bases of the *Escherichia coli lacI* gene: relation to mutation hot spots and transcription

coupling activity. *Proc. Natl. Acad. Sci. USA* 89:11031–35
64. Kunkel, T. A., Soni, A. 1988. Mutagenesis by transient misalignment. *J. Biol. Chem.* 263:14784–89
65. Lark, K. G., Lark, C. A 1979. *recA*-dependent DNA replication in the absence of protein synthesis: characteristics of a dominant lethal replication mutation, *dnaT*, and requirement for *recA*$^+$ function. *Cold Spring Harbor Symp. Quant. Biol.* 43:537–49
66. Lawther, R. P., Calhoun, D. H., Gray, J., Adams, C. W., Hauser, C. A., Hatfield, G. W. 1982. DNA sequence fine-structure analysis of *ilvG* (IlvG$^+$) mutations of *Escherichia coli* K-12. *J. Bacteriol.* 149:294–98
67. Lederberg, J., Lederberg, E. 1952. Replica plating and indirect selection of bacterial mutants. *J. Bacteriol.* 63:399–406
68. Lenski, R. E., Mittler, J. E. 1993. The directed mutation controversy and neo-Darwinism. *Science* 259:188–94
69. Lenski, R. E., Slatkin, M., Ayala, F. J. 1989. Mutation and selection in bacterial populations: alternatives to the hypothesis of directed mutation. *Proc. Natl. Acad. Sci. USA* 86:2775–78
70. Lessie, T. G., Wood, M. S., Byrne, A., Ferrante, A. 1990. Transposable gene-activating elements in *Pseudomonas cepacia*. In Pseudomonas: *Biotransformations, Pathogenesis and Evolving Biotechnology*, ed. S. Silver, A. M. Chakrabarty, B. Iglewski, S. Kaplan, pp. 279–91. Washington, DC: Am. Soc. Microbiol.
71. Loechler, E. L., Green, C. L., Essigmann, J. M. 1984. In vivo mutagenesis by O^6-methylguanine built into a unique site in a viral genome. *Proc. Natl. Acad. Sci. USA* 81:6271–75
72. Luria, S. E. 1947. Recent advances in bacterial genetics. *Bacteriol. Rev.* 11:1–40
73. Luria, S. E., Delbrück, M. 1943. Mutations of bacteria from virus sensitivity to virus resistance. *Genetics* 28:491–511
74. Mahan, M. J., Roth, J. R. 1988. Reciprocality of recombination events that rearrange the chromosome. *Genetics* 120:23–35
75. Malavasic, M. J., Cihlar, R. L. 1992. Growth response of several *Candida albicans* strains to inhibitory concentrations of heavy metals. *J. Med. Vet. Mycol.* 30:421–32
76. McClintock, B. 1978. Mechanisms that rapidly reorganize the genome. *Stadler Symp.* 10:25–47
77. Mellon, I., Hanawalt, P. C. 1989. Induction of the *Escherichia coli* lactose operon selectively increases repair of its transcribed DNA strand. *Nature (London)* 342:95–98
78. Mellon, I., Spivak, G., Hanawalt, P. C. 1987. Selective removal of transcription-blocking DNA damage from the transcribed strand of the mammalian DHFR gene. *Cell* 51:241–49
79. Mittler, J. E., Lenski, R. E. 1990. Causes of mutation and Mu excision. *Nature (London)* 345:213
80. Mittler, J. E., Lenski, R. E. 1990. New data on excisions of Mu from *E. coli* MCS2 cast doubt on directed mutation hypothesis. *Nature (London)* 344:173–75
81. Mittler, J. E., Lenski, R. E. 1992. Experimental evidence for an alternative to directed mutation in the *bgl* operon. *Nature (London)* 356:446–48
82. Mizuuchi, K., Craigie, R. 1986. Mechanism of bacteriophage Mu transposition. *Annu. Rev. Genet.* 20:385–429
83. Nakada, D., Ryan, F. J. 1961. Replication of deoxyribonucleic acid in non-dividing bacteria. *Nature (London)* 189:398–99
83a. Neidhardt, F. C., Ingraham, J. L., Low, K. B., Magasanick, B., Schaechter, M., Umbarger, H. E., eds. 1987. Escherichia coli *and* Salmonella typhimurium *Cellular and Molecular Biology*. Washington, DC: Am. Soc. Microbiol.
84. Newcombe, H. B. 1949. Origin of bacterial variants. *Nature (London)* 164:150–51
85. Ninio, J. 1991. Transient mutators: a semiquantitative analysis of the influence of translation and transcription errors on mutation rates. *Genetics* 129:957–62
86. Nochur, S. V., Roberts, M. F., Demain, A. L. 1990. Mutation of *Clostridium thermocellum* in the presence of certain carbon sources. *FEMS Microbiol. Lett.* 71:199–204
87. Novick, A., Szilard, L. 1950. Experiments with the chemostat on spontaneous mutations of bacteria. *Proc. Natl. Acad. Sci. USA* 36:708–19
88. Novick, A., Szilard, L. 1951. Experiments on spontaneous and chemically-induced mutations of bacteria growing in the chemostat. *Cold Spring Harbor Symp. Quant. Biol.* 16:337–43
89. Nurk, A., Tamm, A., Hôrak, R., Kivisaar, M. 1993. In vivo generated fusion promoters from joining of the sequences from Tn*4652* and target DNA: evidence for selection-induced

mutations in *Pseudomonas putida*. *Gene*. In press
90. Oller, A. R., Fijalkowska, I., Dunn, R. L., Schaaper, R. M. 1992. Transcription-repair coupling determines the strandedness of ultraviolet mutagenesis in *Escherichia coli*. *Proc. Natl. Acad. Sci. USA* 89:11036–40
91. Opadia-Kadima, G. Z. 1987. How the slot machine led biologists astray. *J. Theor. Biol.* 124:127–35
92. Parker, L. L., Betts, P. W., Hall, B. G. 1988. Activation of a cryptic gene by excision of a DNA fragment. *J. Bacteriol.* 170:218–22
93. Pfeifer, F., Blaseio, U. 1990. Transposition burst of the ISH27 insertion element family in *Halobacterium halobium*. *Nucleic Acids Res.* 18:6921–25
94. Postgate, J. 1967. Viability measurements and the survival of microbes under minimum stress. *Adv. Microb. Physiol.* 1:1–23
95. Prival, M. J., Cebula, T. A. 1992. Sequence analysis of mutations arising during prolonged starvation of *Salmonella typhimurium*. *Genetics* 132:303–10
96. Rebeck, G. W., Samson, L. 1991. Increased spontaneous mutation and alkylation sensitivity of *Escherichia coli* strains lacking the *ogt* O^6-methylguanine DNA repair methyltransferase. *J. Bacteriol.* 173:2068–76
97. Reynolds, A. E., Felton, J., Wright, A. 1981. Insertion of DNA activates the cryptic *bgl* operon in *E. coli* K-12. *Nature (London)* 293:625–29
98. Ripley, L. S. 1990. Frameshift mutation: determinants of specificity. *Annu. Rev. Genet.* 24:189–214
99. Roberts, J. M., Axel, R. 1982. Gene amplification and gene correction in somatic cells. *Cell* 29:109–19
100. Ryan, F. J. 1955. Spontaneous mutation in non-dividing bacteria. *Genetics* 40:726–38
101. Ryan, F. J. 1959. Bacterial mutation in a stationary phase and the question of cell turnover. *J. Gen. Microbiol.* 21:530–49
102. Ryan, F. J., Nakada, D., Schneider, M. J. 1961. Is DNA replication a necessary condition for spontaneous mutation? *Z. Vererbungsl.* 92:38–41
103. Ryan, F. J., Okada, T., Nagata, T. 1963. Spontaneous mutation in spheroplasts of *Escherichia coli*. *J. Gen. Microbiol.* 30:193–99
104. Ryan, F. J., Wainwright, L. K. 1954. Nuclear segregation and the growth of clones of spontaneous mutants of bacteria. *J. Gen. Microbiol.* 11:364–79
105. Sargentini, N. J., Smith, K. C. 1981. Much of spontaneous mutagenesis in *Escherichia coli* is due to error-prone DNA repair: implications for spontaneous carcinogenesis. *Carcinogenesis* 2:863–72
106. Sarkar, S. 1991. Lamarck contre Darwin, reduction versus statistics: conceptual issues in the controversy over directed mutagenesis in bacteria. In *Organism and the Origins of Self*, ed. A. I. Tauber, pp. 235–71. The Netherlands: Kluwer Academic
107. Schofield, M. A., Agbunag, R., Michaels, M. L., Miller, J. H. 1992. Cloning and sequencing of *Escherichia coli mutR* shows its identity to *topB*, encoding topoisomerase III. *J. Bacteriol.* 174:5168–70
108. Schofield, M. A., Agbunag, R., Miller, J. H. 1992. DNA inversions between short inverted repeats in *Escherichia coli*. *Genetics* 132:295–302
109. Seifert, H. S., Ajioka, R. S., Marchal, C., Sparling, P. F., So, M. 1988. DNA transformation leads to pilin antigenic variation in *Neisseria gonorrhoeae*. *Nature (London)* 336:392–95
110. Selby, C. P., Sancar, A. 1990. Molecular mechanisms of DNA repair inhibition by caffeine. *Proc. Natl. Acad. Sci. USA* 87:3522–25
111. Selby, C. P., Witkin, E. M., Sancar, A. 1991. *Escherichia coli mfd* mutant deficient in "mutation frequency decline" lacks strand-specific repair: "In vitro" complementation with purified coupling factor. *Proc. Natl. Acad. Sci. USA* 88:11574–78
112. Shakespeare, W. ca 1600. *As You Like It*, act 2, scene 1, lines 12–14
113. Shapiro, J. A. 1984. Observations on the formation of clones containing *araB-lacZ* cistron fusions. *Mol. Gen. Genet.* 194:79–90
114. Shapiro, J. A., Higgins, N. P. 1989. Differential activity of a transposable element in *Escherichia coli* colonies. *J. Bacteriol.* 171:5975–86
115. Shapiro, J. A., Leach, D. 1990. Action of a transposable element in coding sequence fusions. *Genetics* 126:293–99
116. Smith, G. R. 1991. Conjugational recombination in *E. coli:* myths and mechanisms. *Cell* 64:19–27
117. Smith, K. C. 1992. Spontaneous mutagenesis: experimental, genetic and other factors. *Mutat. Res.* 277:139–62
118. Stahl, F. W. 1988. A unicorn in the garden. *Nature (London)* 335:112–13
119. Stahl, F. W. 1990. If it smells like a unicorn. *Nature (London)* 346:791

120. Stahl, F. W. 1992. Unicorns revisited. *Genetics* 132:865–67
121. Steele, D. F., Jinks-Robertson, S. 1992. An examination of adaptive reversion in *Saccharomyces cerevisiae*. *Genetics* 132:9–21
122. Steele, D. F., Jinks-Robertson, S. 1993. Time-dependent mitotic recombination in *Saccharomyces cerevisiae*. *Curr. Genet.* In press
123. Stewart, F. M., Gordon, D. M., Levin, B. R. 1990. Fluctuation analysis: the probability distribution of the number of mutants under different conditions. *Genetics* 124:175–85
124. Strauss, B. S. 1992. The origin of point mutations in human tumor cells. *Cancer Res.* 52:249–53
125. Symonds, N. 1989. Anticipatory mutagenesis? *Nature (London)* 337:119–20
126. Syvanen, M., Hopkins, J. D., Griffin, T. J., Liang, T. Y., Ippen-Ihler, K., Kolodner, R. 1986. Stimulation of precise excision and recombination by conjugal proficient F′ plasmids. *Mol. Gen. Genet.* 203:1–7
127. Thomas, A. W., Lewington, J., Hope, S., Topping, A. W., Weightman, A. J., Slater, J. H. 1992. Environmentally directed mutations in the dehalogenase system of *Pseudomonas putida* strain PP3. *Arch. Microbiol.* 158:176–82
128. Tilsty, D. T., Albertini, A. M., Miller, J. H. 1984. Gene amplification in the lac region of E. coli. *Cell* 37:217–24
129. Topal, M. D., Fresco, J. R. 1976. Complementary base pairing and the origin of substitution mutations. *Nature (London)* 263:285–89
130. Tormo, A., Almiron, M., Kolter, R. 1990. *surA*, an *Escherichia coli* gene essential for survival in stationary phase. *J. Bacteriol.* 172:4339–47
131. von Borstel, R. C. 1978. Measuring spontaneous mutation rates in yeast. *Methods Cell Biol.* 20:1–24
132. Walker, G. C. 1984. Mutagenesis and inducible responses to deoxyribonucleic acid damage in *Escherichia coli*. *Microbiol. Rev.* 48:60–93
133. Wallis, J. W., Chrebet, G., Brodsky, G., Rolfe, M., Rothstein, R. 1989. A hyper-recombination mutation in S. cerevisiae identifies a novel eukaryotic topoisomerase. *Cell* 58:409–19
134. Watson, J. D., Crick, R. H. C. 1953. The structure of DNA. *Cold Spring Harbor Symp. Quant. Biol.* 18:123–31
135. Willetts, N., Skurray, R. 1987. Structure and function of the F factor and mechanism of conjugation. See Ref. 83a, pp. 1110–33
136. Witkin, E. M. 1969. Ultraviolet-induced mutations and their repair. *Annu. Rev. Genet.* 3:525–52
137. Witkin, E. M. 1976. Ultraviolet mutagenesis and inducible DNA repair in *Escherichia coli*. *Bacteriol. Rev.* 40: 869–907
138. Yanofsky, C., Crawford, I. P. 1987. The tryptophan operon. See Ref. 83a, pp. 1453–72

Annu. Rev. Microbiol. 1993. 47:505–34

GENETICS AND MOLECULAR BIOLOGY OF CHITIN SYNTHESIS IN FUNGI

Christine E. Bulawa
Myco Pharmaceuticals Inc., One Kendall Square, Suite 2200, Cambridge, Massachusetts 02139

KEY WORDS: cell wall, chitin synthase, septum, yeast, Calcofluor

---

CONTENTS

ABSTRACT

In *Saccharomyces cerevisiae,* three chitin synthases have been detected. Chitin synthases I and II, the products of the *CHS1* and *CHS2* genes, respectively, are closely related proteins that require partial proteolysis for activity in vitro. In contrast, chitin synthase III is active in vitro without protease treatment, and three genes, *CSD2* (=*CAL1*), *CSD4* (=*CAL2*), and *CAL3,* are required for its activity.

0066-4227/93/1001-0505$02.00

In the cell, the three enzymes have different functions. Chitin synthase I and II make only a small portion, <10%, of the cellular chitin. In acidic media, chitin synthase I is required for normal budding. Chitin synthase II is required for normal morphology, septation, and cell separation. Chitin synthase III is required for the synthesis of 90% of the cellular chitin, including the chitin in the bud scars and lateral wall. Mutants defective in chitin synthase III are resistant to Calcofluor and *Kluyveromyces lactis* killer toxin, they lack alkali-insoluble glucan, and under certain circumstances, they are temperature-sensitive for growth.

The available data suggest that many fungi have more than one chitin synthase and that these synthases are related to the *S. cerevisiae* CHS and CSD gene products.

INTRODUCTION

Fungal cells are enclosed by rigid walls. The cell wall (reviewed in 3, 7, 29, 31, 36, 74, 83) protects the cell from the hazards of the environment, acts as a filter permitting the passage of some molecules while excluding others, and serves as a receptor for pheromones and toxins. In addition, the wall maintains the shape of the cell. Although often viewed as a static structure, the wall frequently undergoes remodeling, thus changing structure and composition, as the cell grows and develops.

This review discusses one component of the fungal cell wall, chitin, and focuses primarily on the budding yeast *Saccharomyces cerevisiae*. The synthesis of chitin is better understood in *S. cerevisiae* than in any other fungus. Chitin synthesis occurs throughout the *S. cerevisiae* life cycle and is developmentally regulated: the amount and distribution of chitin in the cell wall change as the cell proceeds from vegetative growth to diploid formation to sporulation.

The goals of this review are to summarize and critically discuss the recent studies on chitin synthesis in *S. cerevisiae* during all phases of the life cycle, i.e. vegetative growth, mating, and sporulation; to describe the discrepancies in the published literature and, when possible, resolve these differences; and to survey studies of chitin synthesis in other fungi, concentrating on those organisms for which the most progress has been made.

STUDIES IN *SACCHAROMYCES CEREVISIAE*

Chitin Synthesis During Vegetative Growth

The cell wall of *S. cerevisiae* (reviewed in 3, 31, 36) consists of three types of structural polysaccharides: glucans, polymers of glucose containing β-(1→3) and β-(1→6) linkages (reviewed in 24, 83); mannans, heavily glycosylated proteins containing mannose as the major constituent (reviewed in 4, 51, 60, 83, 99); and chitin, a linear polymer of *N*-acetylglucosamine (GlcNAc) residues joined by β-(1→4) linkages. Glucan and mannan are major components of the cell wall; chitin, a minor component, accounts for only 1–2% of the dry weight of the wall.

Glucan and mannan are synthesized continuously during the cell cycle and are distributed uniformly over the cell surface. In contrast, chitin is made during certain portions of the cell cycle and is asymmetrically distributed in the cell wall (reviewed in 3, 31). In late G_1, a ring of chitin is made on the surface of the cell in the area where the new daughter cell, or bud, will form (for discussions of bud-site selection, see 44, 69). The bud grows by localized fusion of secretory vesicles, eventually reaching a size somewhat smaller than that of the mother cell. Following nuclear separation and cytokinesis, a thin disc of chitin is synthesized within the chitin ring to form a primary septum between mother cell and bud. This structure is reinforced by the addition of secondary septa composed of glucan and mannan. Finally, the chitin chains of the primary septum are partially hydrolyzed by an endochitinase (61), and the two cells separate. Scars are visible that mark the area where the two cells had been joined. The bud scar, a craterlike structure on the surface of the mother cell, contains the chitin ring and primary septum. A less prominent structure, the birth scar, is present on the daughter cell; it contains little or no chitin (9, 89).

Although most of the chitin in vegetative cells is present in the bud scars, some is detected in the lateral wall; the amount appears to be strain dependent (89). Recent studies suggest that the lateral wall chitin begins to appear during the final maturation of the bud, after the completion of septation (89).

Chitin synthesis poses several interesting and challenging questions. (*a*) As discussed above, chitin synthesis is regulated temporally and spatially. How is this accomplished? (*b*) Chitin is extracellular, but its precursor, UDP-*N*-acetylglucosamine (UDP-GlcNAc), is intracellular. How are chitin chains transported (29) through the plasma membrane? (*c*) Individual chains of chitin associate to form microfibrils, and some chitin is covalently attached to glucan (67, 96). What factors are required for the integration of chitin into the cell wall and how is this extracellular process regulated? Although we are still far from answering these questions, significant progress has been made in the past decade through the isolation and characterization of mutants defective in chitin synthesis. We have learned that making chitin, a seemingly simple process, is quite complicated, requiring several different chitin synthases and many different genes.

Methods for Detecting Chitin in Cells

The three most common methods for detecting cellular chitin in *S. cerevisiae* are described below, and the advantages and disadvantages of each are discussed.

STAINING WITH CHITIN-BINDING DYES Many fluorescent dyes, such as Calcofluor (77), Congo Red (103), Primulin (86), and diaethanol (64), bind to chitin. Calcofluor is most commonly used. Cells are incubated in a solution of dye, rinsed, and observed with fluorescence microscopy. This technique is easy to perform and provides excellent visualization of bud scars in vegetative cells and of mating projections in shmoos (shmoo formation is described below). However, in vitro studies show that, in addition to chitin, Calcofluor binds many β-linked polysaccharides (77), including cellulose and exopolysaccharides of *Rhizobium meliloti* (46). Thus, the detection of Calcofluor-binding material suggests, but does not prove, the presence of chitin.

LABELING WITH WHEAT GERM AGGLUTININ Wheat germ agglutinin (WGA) is a lectin that recognizes and binds GlcNAc, exhibiting an especially high affinity for β-(1→4)-linked GlcNAc oligomers consisting of three or more residues. Gold-conjugated WGA is used for the subcellular localization of chitin just as gold-conjugated antibodies are used for the subcellular localization of protein antigens. Cells are sectioned, treated with gold-conjugated WGA, and viewed using transmission electron microscopy (53, 89). The electron-dense gold particles indicate the position of the WGA, which in turn gives the position of the chitin. Although this method provides the highest resolution currently obtainable, it has two drawbacks. First, it requires considerable technical skill. Second, WGA is not absolutely specific for GlcNAc; it interacts weakly with *N*-acetylneuraminic acid, *N*-acetylgalactosamine, and Man-β-(1→4)-GlcNAc-β-(1→4)-GlcNAc-β-*N*-Asn (47).

CHEMICAL ASSAY To quantitate chitin, the amount of GlcNAc released by chitinase is measured with a colorimetric assay (21). Cells are heated in alkali to extract mannan from the cell wall and expose the chitin. The chitin chains are hydrolyzed to dimers by addition of purified chitinase and the chitobiose is cleaved to monomers by addition of a β-hexosaminidase. The amount of GlcNAc is measured with the Morgan-Elson assay, which requires a 1-aldo-2-acetamido-4-hydroxy combination for reactivity (106). Because of its specificity, this is the most reliable method for identifying chitin. However, it presents several limitations. First, it lacks sensitivity; a small change in the amount of chitin is difficult to detect. Second, it detects only alkali-insoluble chitin. Although most forms of chitin are alkali-insoluble, a few forms, such as short chains consisting of eight residues or fewer, are not. Third, it does not provide any information on the distribution of chitin in the cell.

Enzymology—a Historical Perspective

Chitin synthases have been detected in many fungi (28) and invertebrates. All catalyze the following reaction:

$$2n \text{ UDP-GlcNAc} \rightarrow [\text{GlcNAc-}\beta\text{-}(1\rightarrow4)\text{-GlcNAc}]n + 2n \text{ UDP}$$

In *S. cerevisiae,* three activities have been identified in vitro. The first to be discovered, chitin synthase I, has been extensively characterized (28) and purified to near homogeneity (56). Chitin synthase I is a zymogen that requires partial proteolysis for activity in vitro (30). Under optimal assay conditions, chitin synthase I is 10 to 20 times more active than chitin synthase II or chitin synthase III, and therefore, detection of the latter two enzymes is difficult in lysates from wild-type cells. For 15 years, chitin synthase I was thought to be the only chitin synthase in the cell. This notion was disproved by the finding that disruptants of *CHS1,* the structural gene for chitin synthase I, still synthesize apparently normal amounts of chitin in vivo (21).

Following the disruption of *CHS1,* two groups independently pursued the search for additional chitin synthases. Both were able to detect an activity in *chs1::URA3* lysates, but the two groups reached somewhat different conclusions about the properties of the new enzyme. Sburlati & Cabib (85) named the new enzyme chitin synthase 2 and found that it is a zymogen; the trypsin-treated enzyme has a pH optimum of 7.5–8.0 and requires divalent cations, with Co^{2+} giving maximal activity. Orlean (71) used the designation chitin synthase II, and in contrast to Sburlati & Cabib, he found that preincubation with trypsin lowers activity. In agreement with Sburlati & Cabib, Orlean found that the new synthase has a neutral pH optimum and that it requires divalent cations; however, in his experiments maximal activity was obtained with Mg^{2+}. The issue was resolved when we (17) demonstrated that the activities characterized by Sburlati & Cabib and by Orlean are in fact two different enzymes; in this review they are designated chitin synthase II and chitin synthase III, respectively. A critical variable for preservation of both activities is the method of cell breakage (102). When lysates are prepared

Table 1 Properties of chitin synthases in *S. cerevisiae* membrane preparations[a]

Enzyme	Trypsin	Optimal pH	Optimal Temperature (°C)	Activators	Inhibitors
Chitin synthase I	Stim.	6.5	40	Mg^{2+}, GlcNAc, digitonin	Co^{2+}, polyoxin D, nikkomycins
Chitin synthase II	Stim.	8.0	—	$Co^{2+} > Mg^{2+}$, GlcNAc	Polyoxin D, nikkomycins
Chitin synthase III[b]	Inhib.	8.0	25	$Mg^{2+} > Co^{2+}$, GlcNAc	Polyoxin D

[a] Dash, not determined; Stim., Stimulatory; Inhib., Inhibitory; GlcNAc, *N*-acetylglucosamine (data obtained from Refs. 28a, 71, 85).

[b] Designated chitin synthase II in Ref. 71.

from protoplasts, the method employed by Sburlati & Cabib, chitin synthase III activity is not consistently recovered. Both activities are recovered when cells are ruptured with glass beads, the method used by Orlean, but only chitin synthase III is detected prior to trypsin treatment. Table 1 summarizes the properties of the various enzymes.

One final comment on the assays for the three chitin synthases: the growth phase of the cells is important. Although all three synthases can be detected in lysates from exponentially growing cells (71, 85), only chitin synthase I is detectable in lysates from stationary phase cells (71; C. Bulawa, unpublished results). These data suggest that the synthesis and/or turnover of chitin synthase II and III are regulated during growth.

Genetics

Why does the cell have three enzymes for synthesizing chitin? Although far from proven, the answer seems to be that the different synthases make different pools of chitin. These pools are distinguished from one another by location in the cell wall and by size. The mutants defective in chitin synthesis can be organized into two broad groups based on the amount of the residual chitin:

Table 2 Characteristics of *S. cerevisiae* mutants defective in chitin synthesis[a]

Mutant	Enzyme[b] defect	Chitin		Chitosan spores[e]	Additional phenotypes
		Cells[c]	Shmoos[d]		
csd2	CSIII	<10%	<5%	defEM, <10%	Resistant to Calcofluor and *K. lactis* killer toxin. Ts^- under certain conditions; suppressed by *SSD1-v1*.
csd4	CSIII	<10%	<5%	wtEM	Calcofluor resistant. Ts^- under certain conditions.
cal3	CSIII	~20%	Low	—	Calcofluor resistant.
csd3	None	~10%	~20%	—	Calcofluor resistant. Ts^- under certain conditions.
chs1	CSI	>90%	<5%	wtEM	Forms refractile buds in acidic medium. Hypersensitive to polyoxin D.
chs2	CSII	>90 to 200%	—	wtEM	Defects in septation and nuclear separation.
shc1	—	>90%	>90%	—	Sporulates poorly.
csd1	—	—	—	—	Ts^-.

[a] Dash, not determined; Ts^-, temperature sensitive for growth. All data are referenced in the text.

[b] CSI means chitin synthase I; CSII, chitin synthase II; CSIII, chitin synthase III.

[c] Expressed as a percentage of the wild-type value as determined by the procedure described in the text under Chemical Assay.

[d] Amount of chitin made in response to α-factor expressed as a percentage of the wild-type values. All strains were MAT**a**. The procedure described in the text under Chemical Assay was used.

[e] defEM means defective spore wall structure as determined by transmission electron microscopy; wtEM, spore wall structure indistinguishable from that of wild-type spores as determined by transmission electron microscopy; <10%, amount of spore wall glucosamine expressed as a percentage of the wild-type value.

the zymogen-deficient mutants, which lack no more than 10% of the cellular chitin, and the chitin-deficient mutants, which lack 90% of the cellular chitin. The latter group can be subdivided into three categories: (*a*) chitin-deficient mutants lacking chitin synthase III activity, (*b*) chitin-deficient mutants with normal chitin synthase III activity, and (*c*) mutants that are conditionally defective in chitin synthesis.

All of the mutants are considered in detail in the following sections, and their properties are summarized in Table 2.

Chitin-Deficient Mutants

Chitin-deficient mutants have been isolated with two different screening procedures. Roncero et al (81) selected strains resistant to Calcofluor. When added to living cells, Calcofluor inhibits growth (80), presumably by binding

Table 3 Genes involved in chitin synthesis in *S. cerevisiae*[a]

Gene (synonym)	Essential	Chromosome (nearby gene)	Proposed function
CSD2 (*CAL1*, *CAL4*, *CAL5*, *DIT101*, *KTI2*)	No	2R[b] (*FUR4*)	Catalytic component of chitin synthase III. Required for synthesis of spore-wall chitosan.
CSD4 (*CAL2*)	No	2L[c] (*ILS1*)	Posttranslational activator of chitin synthase III. No homology to chitin synthases.
CAL3	No	12[d]	Required for chitin synthase III activity.
CSD3	No	10L[e] (*URA2*)	Unknown.
CHS1	No	—	Structural gene for chitin synthase I. Cell wall repair.
CHS2	No	2[f] (*SCO1*)	Structural gene for chitin synthase II. Synthesis of septal chitin.
SHC1	No	5R[g] (*RAD51*)	Related to *CSD4*. May be required for synthesis of spore wall chitosan.
CSD1	—	—	May be required for UDP-GlcNAc biosynthesis.

[a] In a preliminary report, we (C. Bulawa & P. W. Robbins, unpublished data) designated the chitin-deficient mutants *chs2-chs5*. Subsequently, the gene encoding chitin synthase II was designated *CHS2*, thereby assigning a second definition to this gene symbol. The chitin-deficient mutants *chs2*, *chs3*, *chs4*, and *chs5* were renamed *csd2*, *csd3*, *csd4*, and *csd1*, respectively. The nomenclature in publications that cite Ref. 19, e.g. Ref. 83, p. 110, may not be consistent with current usage.

[b] Refs. 16, 73.

[c] C. Bulawa & L. Riles, unpublished results. The position is incorrectly given as 2R in Ref. 68.

[d] B. Santos & A. Durán, personal communication.

[e] Ref. 68.

[f] Ref. 89.

[g] C. Bulawa, unpublished results.

to nascent chains and disrupting microfibril assembly (52, 80). Five Calcofluor-resistant complementation groups, *cal1*, *cal2*, *cal3*, *cal4*, and *cal5*, were originally reported (81), but subsequent analyses have shown that *cal4* (102) and *cal5* (16) are allelic to *cal1*. In a screen (16) for mutants defective in the incorporation of exogenous [^{3}H]-glucosamine into [^{3}H]-chitin, four complementation groups, *csd1*, *csd2*, *csd3*, and *csd4* (chitin synthesis defective) were identified. As judged by the ability of CEN plasmids containing the *CSD* genes to correct the Calcofluor resistance of the *cal* mutants, the following genes are allelic (16): *CSD2=CAL1* and *CSD4=CAL2*. Thus, five different genes, *CSD1*, *CSD2=CAL1*, *CSD3*, *CSD4=CAL2*, and *CAL3*, have been identified (Table 3). The gene symbol *CAL* has also been applied to genes involved in calcium metabolism (45, 68, 70). For the remainder of this review, *CAL1* and *CAL2* are referred to as *CSD2* and *CSD4*, respectively. Because *CAL3* denotes a unique complementation group, this designation is retained.

MUTANTS THAT LACK CHITIN SYNTHASE III ACTIVITY—*CSD2*, *CSD4*, AND *CAL3* Mutations in *CSD2*, *CSD4*, and *CAL3* cause a 10-fold reduction in the amount of chitin in the cell wall and a loss of chitin synthase III activity. Thus, this enzyme synthesizes most of the chitin in the cell. Another change in the composition of the cell wall is that the alkali-soluble glucan increases at the expense of the alkali-insoluble glucan. This observation is consistent with chemical data indicating a covalent linkage between chitin and glucan (67, 96) and supports the proposal that the insolubility of this glucan fraction results from its attachment to chitin. In the mutants, there is very little chitin on which to anchor the glucan, and as a result, only 6% of the glucan is alkali-insoluble compared to 34% in a wild-type strain (81).

Gene disruption experiments show that *CSD2* (16, 102), *CSD4* (18), and *CAL3* (B. Santos & A. Durán, personal communication) are not essential for growth. Under most conditions (the exception is discussed below), the mutants divide at rates comparable to those of wild-type cells and have little or no change in morphology (16, 81, 89). The multiplicity of mutants in this category is noteworthy—the synthesis of chitin synthase III is regulated and/or the enzyme consists of multiple subunits.

The CSD2 *gene* The *csd2* mutants have been studied in most detail. Much of what has been learned about these mutants may also be true for the *csd4* and *cal3* mutants because all three have the same biochemical defect, i.e. loss of chitin synthase III activity. Disruptants of *CSD2* contain a small amount of residual chitin, 5–10% of the wild-type level (16, 102), indicating that another chitin synthase is active during vegetative growth. The residual chitin has been localized by staining thin sections of disrupted cells with gold-labeled wheat germ agglutinin. A thin line of chitin is detected in the division septum; no staining is observed elsewhere in the cell (89). Thus, chitin synthase III makes the chitin of the bud scars and the lateral wall, and a different synthase makes the chitin of the primary division septum. The absence of chitin in the bud scars of *csd2* cells has also been demonstrated with Calcofluor staining (73, 81).

The conclusion that chitin synthase III makes the chitin in the lateral wall is substantiated by analyses of chitin deposition in certain cell-division cycle (*cdc*) mutants. Several *cdc* mutants, when shifted to nonpermissive conditions, are defective in polarized growth and synthesize considerable chitin in the wall, producing cells that are uniformly and brightly stained by Calcofluor (79, 94). In contrast, *cdc3 csd2* and *cdc24 csd2* double mutants shifted to nonpermissive conditions stain poorly (89), indicating that *CSD2* is required for the accumulation of lateral wall chitin in the *cdc3* and *cdc24* mutants.

Two groups have independently cloned and sequenced *CSD2*. Valdivieso et al (102) reported an open reading frame of 3297 bp, while I reported 3495 bp (16) with all of the additional nucleotides located at the beginning of the gene. The reason for this discrepancy, whether it results from allele differences or a sequencing error, is unclear.

Transcriptional studies of synchronous cells indicate that *CSD2* is expressed throughout the cell cycle. The amount of *CSD2* mRNA fluctuates severalfold over the basal level (73), peaking just prior to histone H2A expression (75), a marker of the interval between late G_1 phase and early S phase.

The deduced amino acid sequence of *CSD2* shows limited, but statistically significant, homology to known chitin synthases (e.g. the *CHS1* and *CHS2* gene products; see below), suggesting that CSD2 is a component of chitin synthase III (16, 102). Cells transformed with a 2μ plasmid containing *CSD2* do not overproduce chitin synthase III (16, 102), indicating either that amplification of the gene does not increase CSD2 or that one of the other gene products, CSD4 or CAL3, is present in limiting amount.

Other relatives of the *CSD2* gene product are two proteins (16, 22) whose exact biochemical functions are unknown: the NodC protein of *Rhizobium* species (55) and the DG42 protein of *Xenopus laevis* (82). The possible significance of these similarities is discussed below.

The CSD4 *gene* Several alleles of *CSD4* have been isolated (16, 81), and all but one, *csd4-1*, are phenotypically identical to the *csd2* mutants during vegetative growth. Although lysates from *csd4-1* strains lack chitin synthase III activity, the mutant protein appears to be partially active in vivo because cellular chitin is reduced only twofold (16). The *csd4-1* mutant is Calcofluor sensitive on rich medium, which is not surprising given the high level of residual chitin, but it is Calcofluor resistant on buffered synthetic medium (C. Bulawa, unpublished results). The reason for this difference has not yet been determined.

Overproduction of CSD4 increases chitin synthase III activity from three-to tenfold, depending on the growth medium used. Thus, in wild-type cells, CSD4 is limiting. Cooverproduction of CSD2 and CSD4 does not further elevate chitin synthase III activity (C. Bulawa, unpublished results).

CSD4 is 2088 nucleotides in length (C. Tung, C. Bulawa & P. Robbins, unpublished results), and it encodes a protein of 696 amino acids with a molecular mass of 77,000 (20). The predicted sizes are in good agreement with the experimental data—the *CSD4* mRNA is 2.5 kb, as determined by Northern analysis, and the protein is 80 kDa based on Western blotting with antibody raised against a CSD4-TrpE fusion protein. The deduced amino acid sequence terminates with a CAAX box (78), but it is not known whether the cysteine is isoprenylated. The CAAX box does not appear to be absolutely required for CSD4 function because C-terminal truncations of the open reading frame in 2μ plasmids complement *csd4-1* mutants (C. Bulawa, unpublished results).

The CAL3 *gene* Mutation of *CAL3* causes a four- to fivefold reduction in cell wall chitin and a specific loss of chitin synthase III activity. Molecular cloning and sequencing indicate that *CAL3* is 2013 bp, which is consistent with a 2-kb mRNA detected with Northern analysis. The deduced amino acid sequence predicts a protein of 74 kDa that has no similarity to any known proteins. CAL3 contains a string of 10 heptameric repeats and terminates with a tail of lysine residues (84).

POSSIBLE FUNCTIONS OF THE *CSD2, CSD4,* AND *CAL3* PROTEINS What are the biochemical functions of the *CSD2, CSD4,* and *CAL3* proteins? To answer this question, establishing the structure of chitin synthase III is critical—is it a single polypeptide or a complex of different subunits? Because the *CSD2* protein has statistically significant homology to chitin synthases, it is reasonable to postulate that *CSD2* encodes chitin synthase III or a catalytic subunit of a chitin synthase III enzyme complex. In contrast, the *CSD4* (20) and *CAL3* (B. Santos & A. Durán, personal communication) proteins show no similarity to chitin synthases. If chitin synthase III is monomeric, *CSD4* and/or *CAL3* could regulate, by transcription, translation, or posttranslational modification, the synthesis of the active *CSD2* gene product. If chitin synthase III is multimeric, *CSD4* and/or *CAL3* may encode subunits of the active enzyme. A combination of these models is also possible; for example, one gene product could control the assembly of a chitin synthase III heterodimer. Assigning a function to *CAL3* on the basis of the existing data is difficult. Regarding *CSD4,* various possibilities are discussed below:

1. *CSD4* may be required for the transcription or translation of *CSD2* or *CAL3*. It is unlikely that *CSD4* is a transcriptional activator of *CSD2* because *csd4*

null mutants contain as much *CSD2* mRNA as a wild-type strain, as judged by Northern blotting (C. Bulawa, unpublished data). Analysis of *CAL3* expression in *csd4* mutants has not been examined.

2. The CSD4 protein might activate the CSD2 or CAL3 proteins by a posttranslational modification. Many chitin synthases are activated by partial proteolysis in vitro. Although *CSD4* may encode an activating protease, the available data do not support this hypothesis: the CSD4 protein has no similarity to known proteases; it has none of the common protease motifs; and no evidence indicates that chitin synthase III is a zymogen (C. Tung & C. Bulawa, unpublished results).
3. The *CSD4* protein may be a subunit of a chitin synthase III enzyme complex. Although the deduced amino acid sequence of *CSD4* has no similarity to chitin synthases, two lines of experimental evidence (C. Bulawa, unpublished results) favor this possibility. First, when antibodies that recognize CSD4 are added to chitin synthase III assays, activity is inhibited by 50%. Second, when extra CSD4 is added to lysates prepared from wild-type cells, chitin synthase III activity is stimulated. (In this experiment, the source of CSD4 was a lysate prepared from *csd2::LEU2* disruptants containing *CSD4* on a 2μ plasmid. This preparation by itself had no detectable chitin synthase III activity.)

MUTANTS WITH NORMAL CHITIN SYNTHASE III ACTIVITY—*CSD3* Because *CSD3* is needed for chitin synthesis but not for chitin synthase III activity (16), it is the most perplexing of the genes, encoding an unanticipated component of the system. *CSD3* is 2238 nucleotides in length encoding an 86-kDa hydrophilic protein that has no obvious similarity to any known proteins. Recent gene disruption experiments indicate that *CSD3* is a nonessential gene (B. Osmond, C. Bulawa & P. Robbins, unpublished results). Therefore, CSD3 is probably not involved in the biosynthesis of UDP-GlcNAc, because this nucleotide sugar is required for protein glycosylation, an essential biosynthetic pathway (60). An intriguing possibility is that CSD3 plays a role in the spatial regulation of chitin synthesis, such as proper localization of chitin synthase III. Microvesicles that contain chitin synthases, called chitosomes, have been described and have been proposed to play a role in the transport of these enzymes (6, 62). All three chitin synthases in *S. cerevisiae* appear to lack a conventional amino terminal signal sequence. Thus, the notion that these proteins have a specific transport system, independent of the secretory pathway, deserves some consideration.

MUTANTS THAT ARE CONDITIONALLY DEFECTIVE IN CHITIN SYNTHESIS—*CSD1* When grown under permissive conditions, the *csd1-1* mutant has no obvious reduction in cellular chitin or in chitin synthase III activity. At 37°C, both chitin synthesis and growth are blocked (16). Determination of the thermal stability of chitin synthase III in the *csd1-1* mutant has not been possible because the wild-type synthase is inactive at 37°C under the assay conditions typically employed. The temperature-sensitive growth caused by *csd1-1* is not strain- or medium-dependent, in contrast to the other chitin-deficient mutants (16), which are temperature sensitive only under narrowly defined conditions (see below). These observations suggest that *csd1-1* encodes a temperature-sensitive (Ts^-) mutant protein, and therefore, *CSD1* may be an essential gene. If so, it will be important to determine whether *CSD1* is specifically involved in chitin synthesis. The *csd1-1* mutant was identified by looking for strains defective in the incorporation of exogenous [^{3}H]-glucosamine into [^{3}H]-chitin; this screen could detect mutations in many pathways, e.g. the biosynthesis of UDP-GlcNAc, acetyl-CoA, or ATP.

TEMPERATURE SENSITIVITY OF CHITIN-DEFICIENT MUTANTS When certain criteria are met, the growth of *CSD* mutants is seriously impaired. In some strains, but not others, both point and null mutations in *CSD2, CSD3,* and *CSD4* cause temperature-sensitive growth on media of low osmolarity. The temperature sensitivity is bypassed by inclusion of salts or sorbitol in the media or by a suppressor gene(s) that is present in certain strains. Neither method of suppression restores chitin synthesis (16). Strains that, in combination with *csd* mutations, confer a Ts^- or Ts^+ phenotype are designated Sup^- or SuP^+, respectively. Although *cal3* Sup^- double mutants are probably temperature sensitive on low-osmolarity media, this has not yet been tested experimentally.

Recent experiments show that CEN plasmids containing *SSD1/SRK1* (97, 108) correct the temperature-sensitive growth of *csd2* Sup^- strains but not the defect in chitin synthesis (C. Chan & C. Bulawa, unpublished results). *SSD1/SRK1* is a polymorphic gene (97); the allele that suppresses the *csd* mutations, *SSD1-v1,* was previously shown to suppress mutations in several genes involved in growth control—*sit4* (97), *slk1* (40), *pde2, bcy1,* and *ins1* (108). The SSD1/SRK1 protein may be a protein phosphatase, or it may modulate protein phosphatase activity. *SSD1-v1* can probably suppress the temperature sensitivity of *csd3* Sup^- and *csd4* Sup^- mutants, but this has not yet been tested.

ROLE OF CHITIN IN THE ACTIVITY OF *KLUYVEROMYCES LACTIS* KILLER TOXIN *Kluyveromyces lactis* killer toxin is a heterotrimer that causes sensitive cells to arrest in G_1 (95). Several lines of evidence indicate that interaction of the toxin with chitin is necessary for killing: the deduced amino acid sequence of the α-subunit of the toxin is similar to chitinases (26); the purified toxin can hydrolyze 4-methylumbelliferyl-$(GlcNAc)_2$ (26), a fluorogenic substrate for chitinases; allosamidin, an inhibitor of chitinases (58), inhibits the toxin's enzymatic activity and its ability to kill cells (26); several mutants resistant to exogenous toxin are also chitin-deficient, and it was recently determined that one of them, *kti2*, is allelic to *csd2* (25). Taken together, these data suggest that chitin acts as a receptor for the toxin. Experiments to test this model include determination of the toxin sensitivity of the *csd* and *cal* mutants and complementation analysis of the *kti*, *csd*, and *cal* mutants.

Zymogen-Deficient Mutants

The zymogen-deficient mutants lack either chitin synthase I or chitin synthase II activity and are designated *chs1* and *chs2*, respectively. These mutants were obtained in screens that monitored enzymatic activity (21, 92). Although cellular chitin is reduced by no more than 10%, these mutants have abnormalities during vegetative growth, indicating that these gene products are active. It is likely that the *CHS1* and *CHS2* gene products make a small amount of chitin, but this has not been unequivocally demonstrated. Much of the work on *CHS1* and *CHS2*, including the isolation of the mutants and the cloning, sequencing, disruption, and heterologous expression of the genes, has been reviewed (28, 33, 34, 36, 83).

CHS1 *CHS1* is the structural gene for chitin synthase I (21). Two phenotypes have been observed in *chs1* mutants during vegetative growth: they are hypersensitive to the drug polyoxin D (21), a competitive inhibitor of chitin synthases (48), and under certain culture conditions, they form small, abnormal buds that are susceptible to lysis (21).

Cabib and coworkers (32, 35) have characterized the budding defect in detail. Lysis occurs late in the cell cycle, after the separation of nuclei, and appears to be caused by a defect in cell-wall structure; a hole is visible in the birth scar of the daughter cell, and addition of sorbitol to the medium prevents lysis (32).

Many factors influence the frequency of lysed buds. In synthetic medium, 20–40% of the culture consists of lysed buds. When buffers are added that maintain the pH above 5, the number is reduced over 10-fold (35). Lysis is much less frequent in rich medium (32), but whether this is simply because of pH or because lysis frequency is also influenced by composition is not clear. Inhibition of chitinase, the enzyme that hydrolyzes chitin during cell separation, either by addition of the inhibitors allosamidin or demethylallosamidin, or by disruption of the chitinase structural gene, *CTS1*, suppresses lysis (35). Certain strains are resistant to lysis because of the presence of a recessive suppressor, *scs1;* this gene does not reduce the levels of chitinase or of chitin synthase I (93), indicating that other proteins, in addition to chitinase, play a role.

The studies on the lysis of *chs1* buds led to the proposal that chitin synthase I acts as a repair enzyme by replenishing chitin during cytokinesis (32, 35). This hypothesis implies a cause-and-effect relationship—damage to the cell wall activates chitin synthase I—that is consistent with, but not established by, the existing data. It is equally possible that chitin synthase I is activated by acidic culture conditions; if so, *chs1* mutants, when grown in acidic medium, may form a defective septum susceptible to damage during cell separation. To clearly establish the sequence of events, additional experiments are required.

The mechanism of zymogen activation is not known. Although proteinase B can activate chitin synthase I in vitro (101), mutants defective in this protease (i.e. *prb1*) grow normally in acidic synthetic medium (93), indicating that chitin synthase I is active in these strains.

CHS2 *CHS2* is the structural gene for chitin synthase II (92). Comparison of the deduced amino acid sequences of *CHS1* and *CHS2* shows that the two proteins are closely related (91); the last 750 amino acids are about 40% identical. The published alignments miss a region of significant similarity at the ends of the proteins. The optimal alignment (Figure 1) shows that, contrary to previous reports (34, 36, 91), the C termini are related to one another. The N-terminal regions are unrelated and differ in length (approximately 360 and 210 amino acids in CHS1 and CHS2, respectively).

```
        ***••••*••• *••**•••*••   *• ••••••  • *•
Ccs2    DYYAFIRSMT VLVWMFTNFV VIALVLETGG FNQFVEATDL AN........  961
Chs1    GYYANVRSLV IIFWVITNFI IVAVVLETGG IADYIAMKSI STDDTLETAK 1083
Chs2    DYYRDIRTRI VMIWMLSNLI LIMSIIQVFT PQDT...... ..........  901
Ccs1    DYAKDVRTRV VLFWMIANLV FIMTMVQVYE PGDT...... ..........  719

             •     •••** •***  ** •*•**• **•  *** • • *  * ••
Ccs2    .... LKSNR AAVFLTVILW TVAFMALFRF IGCIYYLITR LGREIKASEH 1006
Chs1    KAEIPLMTSK ASIYFNVILW LVALSALIRF IGCSIYMIVR FFK.KVTFR  1131
Chs2    .......... DNGYLIFILW SVAALAAFRV VGSMAFLFMK YLRIIVSYRN  940
Ccs1    .........G RNIYLAFILW AVAVLALVRA IGSLGYLIQT YARFFVESKS  760
```

Figure 1 Alignment of approximately 100 amino acids from the C termini of four CHS gene products. The proteins are *S. cerevisiae* chitin synthase I (Chs1) and II (Chs2) and *Candida albicans* chitin synthase I (Ccs1) and II (Ccs2). The marks above the sequences indicate the number of identical amino acids: •=2, *=3, and *=4. The numbers of the amino acid residues are shown on the right. Alignments and statistical analyses were done with the University of Wisconsin Genetics Computer Group program BESTFIT (42). Evaluation of pairwise alignments gave the percentage of identical amino acids and the number of standard deviations (SD) separating the score for the optimal alignment from the average score for 10 random alignments of the same amino acid composition. Ccs2 vs Chs1, 39% identical, 22 SD above random; Chs1 vs Chs2, 26% identical, 6 SD above random; Chs2 vs Ccs1, 42% identical, 18 SD above random.

CHS2 was originally reported to be an essential gene (92); however, subsequent work has demonstrated that it is nonessential (8, 17, 89). The viability of *chs2* spores can range from 0–90% depending on the strains and the germination media that are used (17). In certain strains, spores containing *chs2* disruptions cannot form colonies on rich medium with glucose as the carbon source (92). In some strains, this phenotype is partially suppressed, but the recovery of the mutant is low and the colonies are small. In all strains that have been tested, the recovery of the *chs2* disruptant is improved when spores are germinated on synthetic medium or on rich medium with glycerol as the carbon source (17, 89).

Disruption of *CHS2* produces several interesting phenotypes. In rich medium, *chs2* disruptants separate poorly, forming large clumps of cells, and a fraction of the cells appear to contain more than one nucleus. Unexpectedly, cellular chitin is elevated almost twofold (17, 89). All of these defects are less severe when cells are grown in synthetic medium or in rich medium with glycerol as the carbon source (17).

CHS2 is regulated transcriptionally and posttranslationally. Expression is cell-cycle regulated, peaking when 10% of the cells in a synchronous culture are in anaphase (73). Studies of cells in which *CHS2* is under the control of the *GAL1* promoter suggest that CHS2 is rapidly inactivated in vivo; when *CHS2* expression is repressed, cells lose chitin synthase II activity within about one generation (89). Chitin synthase II activity is detected only in growing cells; no activity is found in membranes prepared from stationary phase cultures (C. Bulawa, unpublished results).

Cabib and coworkers have proposed that the function of chitin synthase II is synthesis of the primary division septum. This hypothesis is based on two observations (89). First, *chs2* mutants lack the well-defined primary septa observed in wild-type cells; instead, they form thick, amorphous septa, and cell separation is impaired. Second, mutants that lack chitin synthase I and chitin synthase III (i.e. *chs1 csd2* double mutants, see above) contain a small amount of chitin located exclusively in the division septum.

Although CHS2 might synthesize the septal chitin, not all of the phenotypes of *chs2* mutants can be easily explained by this hypothesis. If the sole function of chitin synthase II is synthesis of the primary division septum, then *chs2* mutants should progress normally through the cell cycle until the time of septation, and they should either lack septa or form septa that have an abnormal structure and a deficiency of chitin. Observations of *chs2* mutants do not meet these expectations. First, although *chs2* mutants lack well-defined primary septa, they do form septa, and considerable chitin is present in the septal region, as judged by staining with WGA (89). Thus, another enzyme can synthesize chitin between mother and daughter cells; whether it contributes to the synthesis of the primary septum in wild-type cells is an important question. Second, *chs2* mutants have unexpected defects in growth. Cells are misshapen and abnormally large. Under certain conditions, chitin is overproduced as much as twofold and is abnormally distributed in the cell (17, 89). How loss of the primary septum causes these defects is not obvious. Third, a significant number of multinucleate cells are observed in *chs2* cultures (17, 89). Because nuclear segregation is normally completed before septation, cells blocked in primary septum formation are expected to be mononucleate. Of particular relevance are studies of cell division in glucosamine auxotrophs (*gfa1/gcn1*). Glucosamine starvation, which blocks both chitin and mannan biosynthesis, alters septation, but not nuclear separation; these mutants form a string of cells, each containing a single nucleus (5).

Clearly, additional data are needed to fully explain the complex phenotype of *chs2* mutants. At the present time, the following possibilities should be kept in mind: (*a*) defects in the structure of the septum may cause abnormal growth in subsequent cell cycles; (*b*) chitin synthase II may be a bifunctional protein (e.g. the conserved C-terminal domain may be catalytic whereas the unique N-terminal domain may play a role in nuclear separation); (*c*) the primary division septum may be made by a fourth, as yet unidentified synthase rather than by chitin synthase II.

MUTANTS LACKING MORE THAN ONE CHITIN SYNTHASE Loss of any single chitin synthase is not lethal, but what happens when more than one is lost? Double mutants lacking chitin synthases I and II (17, 89) or chitin synthases I and III (16, 89, 102) are viable, but triple mutants, lacking all three synthases, are not, as judged by the inability to recover such strains from the appropriate genetic crosses (89; B. Osmond & C. Bulawa, unpublished results).

What changes occur in the cell when all three chitin synthases are blocked? To answer this queston, Shaw et al (89) constructed a *chs1 chs2::LEU2 csd2* strain with a plasmid containing *CHS2* under the control of the *GAL1* promoter and then monitored chitin synthase II activity, chitin content, viability, and morphology after repression of *CHS2*. Within one generation after transfer to glucose medium, the activity of chitin synthase II fell 10-fold, and defects in cell separation and nuclear segregation were observed. Most of the cells were in groups of three; each group contained two or three nuclei located in one or two of the cells or in the junction between them. After three generations, one of the cells in each group was extremely large, with the vacuole occupying most of the intracellular space, and the number of viable cells declined by a factor of one hundred. Introduction of *CHS1* on a CEN plasmid did not rescue this strain, indicating that loss of chitin synthases II and III is lethal.

Genetics of Chitin Synthesis During Mating

Haploid cells of both mating types synthesize and respond to peptide mating pheromones. The pheromones induce a series of changes (reviewed in 65, 100) that culminates in the formation of an **a**/α diploid cell. After exposure to mating pheromone, haploid cells arrest in G_1, and the pattern of growth is altered to produce elongated, pear-shaped cells called shmoos. Chitin is diffusely deposited throughout the elongated portion of the shmoo. The amount of chitin increases in proportion to the length of exposure to pheromones.

CSD GENES AND *CAL3* All of the mutants that are chitin-deficient during vegetative growth make substantially less chitin than wild-type cells in response to mating pheromones; nonetheless, they are able to mate. Quantitative assays of pheromone-treated cultures of *csd2::LEU2, csd3-1,* and *csd4::LEU2* cells show reductions of 20-fold, 5-fold, and 20-fold, respectively (I. Jensen & P. Orlean, personal communication). The relatively high level of residual chitin in *csd3-1* shmoos is difficult to interpret because this allele might encode a partially active mutant protein; this question can be resolved by repeating the experiment with a *csd3* disruptant. Although quantitative assays have not been performed on the *cal3* mutants, the shmoos of *cal3* mutants stain poorly with Calcofluor (B. Santos & A. Durán, personal communication). In *csd* Sup^- strains, the morphology of the shmoos is abnormal (I. Jensen & P. Orlean, personal communication); the ability of *SSD1-v1* to complement this phenotype has not been examined.

The question of how the expression of the *CSD* and *CAL3* genes is regulated during mating has received little attention to date, but some information is available regarding *CSD2*. A significant amount of *CSD2* mRNA is detected in wild-type cells exposed to α-factor (73), but whether the level of *CSD2* mRNA is controlled by increased synthesis or by decreased degradation is not known. Transcriptional stimulation by α-factor requires at least two copies of the pheromone-responsive element, TGAAACA (59). In the 500-bp region upstream of the putative *CSD2* promoter, two segments match the pheromone-responsive element at six of the seven positions. One of them, however, may not be functional because it lacks the highly conserved adenine residue at position five of the TGAAACA sequence (16). The specific activity of chitin synthase III does not differ significantly before and after exposure of cells to α-factor (71).

CHS GENES In response to α-factor, the level of chitin synthase I zymogen doubles (71). Transcription of *CHS1* is stimulated by both mating pheromones

(1); the level of induction is 6- to 15-fold for MAT**a** cells treated with α-factor and 4- to 5-fold for MAT**a** cells treated with **a**-factor. Pheromones do not alter the transcription start point, –116 relative to the first ATG. Deletion analysis of 730 bp of the *CHS1* 5′ noncoding region identified a 94-bp region that confers hormone inducibility; this upstream activating sequence (UAS) contains four copies of the pheromone-responsive element.

Despite this substantial body of suggestive data, unequivocal evidence that chitin synthase I is involved in mating has not yet been obtained. Disruptants of *CHS1* have no obvious defects in shmoo formation or mating (21), and they show the same increase in chitin content in response to α-factor as wild-type cells (71). In order to detect a small increase in chitin synthesis due to chitin synthase I, MAT**a** *csd2::LEU2 CHS1* and MAT**a** *csd2::LEU2 chs1::HIS3* were exposed to α-factor; chitin was virtually undetectable in both strains as judged by chemical assays (I. Jensen & P. Orlean). Thus, chitin synthase I makes no more than a small percentage of the chitin in α-factor-treated cells.

CHS1 mRNA is almost undetectable in diploids (1); nonetheless, a significant amount of chitin synthase I activity is present in lysates from diploid cells (56). *CHS2* mRNA is undetectable in α-factor-treated cells (73).

STIMULATION OF UDP-*N*-ACETYLGLUCOSAMINE BIOSYNTHESIS DURING THE PHEROMONE RESPONSE *GFA1/GCN1* is the structural gene for L-glutamine:fructose-6-phosphate amidotransferase (105, 107); it is the first enzyme in the chitin biosynthetic pathway, catalyzing the formation of D-glucosamine-6-phosphate from D-fructose-6-phosphate and L-glutamine. Mutants defective in this gene were obtained by screening for glucosamine auxotrophs and were designated *gcn1* (107). Subsequently, the wild-type gene was cloned by complementation and the gene was renamed *GFA1* (105). α-Factor induces transcription of *GFA1* two- to threefold, and six pheromone-responsive elements are present in a 1-kb region upstream of the putative promoter. The activity of the enzyme is elevated 1.7-fold in α-factor-treated **a** cells (105). This induction is consistent with the observation that mating factors increase the de novo synthesis of amino sugars (72). Together these data suggest that substrate availability plays a role in the regulation of chitin synthesis during mating.

Genetics of Chitin Synthesis During Sporulation

Late in sporulation, after completion of the second meiotic division, the four nuclei are encircled by a double membrane, the prospore wall (reviewed in 27). Cell-wall components are deposited between the two membranes; the completed spore wall consists of four layers (14). The outer layer is a cross-linked insoluble macromolecule composed primarily of dityrosine with minor amounts of other amino acids (15). The second layer is chitosan, a deacetylated derivative of chitin (14). The two inner layers consist of glucan and mannnan.

CSD GENES AND *CAL3* The *csd2* mutations are allelic to *dit101* (73), a mutation obtained by screening for defects in the assembly of the dityrosine-rich layer of the spore wall (13). [The gene symbol *DIT* has two definitions: dityrosine and derepressed *INO1* transcription (68).] In contrast to the other *dit* mutants, which lack only the dityrosine-rich layer, *csd2* mutants lack the chitosan layer as well. This conclusion is based on ultrastructural studies, which found that the dityrosine and chitosan layers are not visible on the surface of *csd2* spores, and on quantitative assays, which found that *csd2* spores are deficient in dityrosine and glucosamine, containing about one-tenth the amount found in wild-type spores (73). Like the *csd2* mutants, glucosamine auxotrophs (i.e. *gfa1*) form spores lacking both the chitosan and dityrosine layers (107), suggesting that the chitosan layer is a prerequisite for the assembly of the dityrosine-rich layer. The mutant spores are viable but are hypersensitive to glusulase and zymolyase (13); this phenotype appears to be more severe in Sup^{-}/Sup^{-} strains (16).

The detection of chitin deacetylase activity in *Mucor rouxii* (41) and in sporulating cells of *S. cerevisiae* (P. Robbins, unpublished results) suggests that chitin is the immediate precursor of chitosan. If so, the simplest explanation for the lack of chitosan in *csd2* mutants is that they are defective in a chitin synthase active during sporulation. Because chitin synthetic activity has not yet been detected in lysates prepared from sporulating cells, direct evidence supporting this hypothesis has not been obtained. Interestingly, *CSD4*, which like *CSD2* is required for chitin synthase III activity during vegetative growth, is not required for chitosan synthesis, as judged by electron microscopy of the spore wall of *csd4* disruptants (M. Pammer, A. Ellinger, C. Bulawa & M. Breitenbach, unpublished results).

SHC1 A search of protein databases revealed that *CSD4* is very similar (49% identity over 220 amino acids) to an unidentified, partial open reading frame located downstream of *RAD51* (C. Bulawa, unpublished results).

Several lines of evidence indicate that the homologue of *CSD4, SHC1,* is required for chitin synthesis during sporulation (C. Bulawa, unpublished results). Undetectable during exponential growth, *SHC1* mRNA rises dramatically during sporulation. The pattern of *CSD4* expression is the opposite. Homozygous disruptants of *SHC1* sporulate poorly, but whether a defect in chitosan biosynthesis causes this phenotype is not known. During vegetative growth and shmoo formation (I. Jensen & P. Orlean, unpublished results), *shc1* disruptants synthesize normal amounts of chitin. The gene symbol *SHC* stands for sporulation-specific homologue of *CSD4*.

CHS GENES Electron microscopy of spores produced by homozygous *chs1 chs2* double disruptants shows no obvious defects in spore-wall ultrastructure (73). Therefore, chitin synthase I and II make little, if any, contribution to chitosan biosynthesis.

Summary of the Possible Functions of the CSD, CHS, *and* CAL3 *Gene Products*

The following summary attempts to integrate the available data into a coherent model. The model is speculative; few aspects are proven. Its purpose is to organize the current information and suggest goals of future investigations.

In vegetative cells, chitin synthase III synthesizes the chitin ring and the chitin in the lateral wall. This requires four genes, *CSD2, CSD4, CAL3,* and *CSD3;* the first three are needed for enzymatic activity. Because it is related to chitin synthases, *CSD2* is likely to encode chitin synthase III or a catalytic subunit of a chitin synthase III enzyme complex. CSD4 modulates, apparently by a posttranslational mechanism, the activity of CSD2 and may play a role in localizing synthesis to the ring (discussed further below). It is difficult to assign functions to *CAL3* and *CSD3* on the basis of the existing data (the possibilities are discussed in previous sections).

The CHS gene products make small amounts of chitin in vivo. After cytokinesis, chitin synthase II may make the primary septum, a thin disc of chitin within the chitin ring. When cells are grown in acidic medium, chitin synthase I may make a small amount of chitin in the neck between mother cell and bud.

During mating, the *CSD* and *CAL3* gene products may well have the same biochemical functions as in vegetative growth. Delocalized deposition of chitin could be achieved by changing the localizaton of chitin synthase III, which is likely to be determined in large part by the pattern of cell growth, i.e. by the factors that control shmoo formation. The observation that *CHS1* is induced by mating factors suggests that it plays a role in mating, but it is unclear what that role might be.

Chitin is the immediate biosynthetic precursor of the chitosan found in the mature spore wall. CSD4 does not play a major role in chitosan synthesis. Instead, a homologue of CSD4, SHC1, apparently modulates the activity of CSD2 during spore-wall formation. It is tempting to speculate that the region of SHC1 related to CSD4 is needed for enzymatic activity while the region unique to SHC1 plays a role in localization, directing the uniform deposition of chitin in the spore wall. We cannot determine what role, if any, *CAL3* and

CSD3 play in this process with the available data. The *CHS* genes play little or no role in chitosan synthesis.

CHITIN SYNTHESIS GENES IN OTHER FUNGI

Detection and Classification of Chitin Synthases from Diverse Fungal Species

By aligning the deduced amino acid sequences of three chitin synthase genes, *S. cerevisiae CHS1* and *CHS2* and *Candida albicans CHS1*, Bowen et al (12) identified short, completely conserved sequences suitable for designing PCR primers. Analysis of genomic DNA from 14 taxonomically diverse fungal species showed that all have at least 1 chitin synthase homologue; 13 have 2 or 3 different homologues. The amplified fragments were sequenced and their deduced amino acid sequences were aligned. With the exception of *S. cerevisiae CHS1*, the sequences fell into three distinct groups. Whether the members of each group perform similar functions in vivo is an important question for future studies. The primers designed by Bowen et al do not amplify *S. cerevisiae CSD2;* the CSD2 protein has limited similarity, 28% identity over 180 amino acids, to the CHS proteins.

Szaniszlo & Momany (98), in an analysis of three *CHS* homologues from *Wangiella dermatitidis,* identified a region of approximately 40 amino acids with similarity to the nucleotide-binding fold of bacterial periplasmic permeases and adenylate kinase. This region of the CHS proteins may be a catalytic site for binding the nucleotide portion of the substrate, UDP-GlcNAc, or it may be a regulatory site for binding of a nucleotide effector molecule.

Recent experiments indicate that the genes encoding chitin synthase III may be conserved. Sequences related to *CSD2* and *CSD4* from *S. cerevisiae* have been identified in *C. albicans* (N. Gow & C. Bulawa, unpublished results), *Aspergillus nidulans, Neurospora crassa,* and *Schizophyllum commune* (C. Specht, personal communication) by low-stringency Southern analyses.

Schizosaccharomyces pombe

A *CHS* homologue is detected in *Schizosaccharomyces pombe*. Although this result was unexpected in light of previous reports that *S. pombe* contained no chitin (21, 23, 54), it is consistent with more recent studies. Sietsma & Wessels (90) have detected a small amount of chitin, approximately 0.3% of the cell wall by dry weight, in *S. pombe* and have identified a trypsin-stimulated chitin synthase in mixed-membrane preparations. Disruption of the homologue and determination of the subcellular location of the chitin should provide important information on the role of chitin, if any, in septation. The latter experiment may be difficult; Horisberger et al (54), using gold-conjugated WGA, could not detect chitin in the wall or septum of *S. pombe*.

Candida albicans

Two genes encoding homologues of the *S. cerevisiae* CHS proteins, *CaCHS1* and *CaCHS2*, have been identified in *C. albicans*. The *CaCHS1* gene was cloned by complementation of the enzymatic defect in *S. cerevisiae chs1* mutants (2). The open reading frame consists of 2328 nucleotides and encodes a protein of 88 kDa. There is disagreement about the size of the transcript; Au-Young & Robbins (2) reported 2.8 kb while Chen-Wu et al (37) reported 3.4 kb. In the latter case the mRNA is unexpectedly large compared to the size of the open reading frame. Resolution of these discrepancies will require further investigation.

Characterization of *S. cerevisiae chs1::URA3* transformants containing *CaCHS1* on a 2μ plasmid indicates that the CaCHS1 protein requires partial proteolysis for enzymatic activity. The trypsin-stimulated activity has a pH optimum of 6.8, requires divalent cations, and is inhibited by polyoxin D (2).

Using the polymerase chain reaction (PCR), Chen-Wu et al (37) detected a second *CHS* homologue. *CaCHS2* is 3027 nucleotides encoding a protein of 116 kDa. During the yeast-hyphal phase transition, the 3.8-kb *CaCHS2* mRNA is induced severalfold. Using gene disruption technology, Gow (49) recently obtained homozygous *chs2* mutants of *C. albicans*. These strains have no obvious defects in vegetative or hyphal growth. As judged by chemical assays, there is little or no reduction in the amount of chitin in yeast cell walls; in hyphal walls, the amount of chitin is reduced by 20–50% (J. Lester & N. Gow, personal communication). Future characterization of the *chs2/chs2* mutants should include an analysis of chitin synthase activity in vitro and an evaluation of their pathogenicity.

Previous studies (2) have reported that the C termini of CaCHS1 and *S. cerevisiae* CHS1 are unique. Optimization of the alignment (Figure 1) shows that C-terminal regions of these proteins are in fact related to each other and to the corresponding regions of CaCH52 and *S. cerevisiae* CHS2.

Neurospora crassa

Degenerate oligonucleotide primers were used in the amplification of several CHS-related sequences from *N. crassa* genomic DNA by PCR. These sequences appear as nonclustered single copies in the genome. The *chs-1* gene has been mapped to linkage group V (110). The putative *chs-2* and *chs-3* genes have been mapped to linkage groups IV and V (unlinked to *chs-1*), respectively (A. Beth Din & O. Yarden, personal communication).

Repeat-induced point mutation (RIP) (88) is being used for *chs* gene inactivation experiments. Inactivation of *chs-1* by RIP produced progeny with multiple defects. The *chs-1*^{RIP} mutants grow slowly, form abnormal, swollen hyphae, and have a large (approximately 10-fold) reduction in chitin synthase activity. No defect was observed in the formation or abundance of hyphal cross-walls or in the morphology of conidia (110). Inactivation of *chs-2* did not yield a noticeable change in morphology, yet an increase in sensitivity to the antifungal compound edifenphos was observed (A. Beth Din & O. Yarden, personal communication).

Aspergillus nidulans

Three temperature-sensitive lysis mutants, *orlA1, orlB1,* and *tsE6,* have a 10-fold reduction in cell wall chitin at restrictive temperature (11). Lysis of all three mutants can be prevented by addition of glucosamine to the medium, indicating that the mutants are defective in amino sugar biosynthesis. Analysis of extracts prepared from cells grown at permissive temperature indicates that *orlA1* and *tsE6* mutants possess labile forms of L-glutamine:fructose-6-phosphate amidotransferase activity. Thus, there are apparently two structural genes for the amidotransferase in *A. nidulans*. This is somewhat surprising because this enzyme is encoded by a single gene in *S. cerevisiae* (105) and *Escherichia coli* (104). Preliminary evidence suggests that *orlA* is not the catalytic subunit of the amidotransferase; analysis of a partial sequence of this gene, approximately 80% of the total length, shows no homology to the *S. cerevisiae* amidotransferase (P. Borgia, personal communication).

Like *orlA1* and *tsE6* mutants, *bimG11* mutants lyse and are chitin deficient at restrictive temperature. The *bimG* gene encodes a type 1 protein phosphatase (43). Taken together, these data suggest that protein phosphorylation plays a role in chitin synthesis. A putative substrate for protein phosphorylation is the amidotransferase, based on the results of a series of detailed experiments (10), including the characterization of amidotransferase activity from *bimG11* mutants. Specifically, phosphorylation of the amidotransferase may increase its sensitivity to feedback inhibition by UDP-GlcNAc.

FUTURE DIRECTIONS

Saccharomyces cerevisiae

The application of genetic techniques to the study of chitin synthesis has increased our understanding of this area and provides tools for future experiments. The availability of the *CAL, CHS,* and *CSD* genes will allow the construction of gene fusions that in turn will facilitate transcriptional studies and the production of antibodies. Antibodies can be used to address the following questions.

1. What is the structure of chitin synthase III? By using immunoprecipitation experiments, investigators should be able to establish the subunit composition of chitin synthase III. One unsolved technical obstacle is the fact that chitin synthase III is particulate, and conditions for releasing it from the membrane in an active form have not been reported. Use of crosslinking reagents (109) may circumvent this problem. Interesting related experiments include mapping of the UDP-GlcNAc binding site by using substrate analogue photoprobes (50).
2. Are the CHS proteins activated by proteolysis? Antibodies raised against recombinant antigens (57) containing the unique N-terminal domains of the chitin synthases should specifically recognize CHS1 or CHS2. These antibodies should enable researchers to determine the sizes of trypsin-treated CHS1 and CHS2 and to look for the formation of truncated forms of the enzymes in vivo.
3. How is chitin synthesis regulated spatially? There is a longstanding debate over the subcellular location of the chitin synthases: are they located in the plasma membrane or are they contained in chitosomes? Localization of the *CAL, CHS,* and *CSD* gene products with immunofluorescence (76) and/or immunogold electron microscopy (38) may resolve this issue. In addition, these experiments should provide valuable information about the physiological functions of these proteins and the spatial regulation of chitin synthesis: perhaps these proteins are targeted to specific sites within the cell. If so, where are these enzymes localized in cells undergoing morphological changes, such as mating, sporulation, and pseudohyphal growth (45a)? Where are they localized in mutants with abnormalities in chitin deposition (79, 94) and in mutants defective in the secretory pathway (87)? In *csd3* mutants, which lack chitin but have chitin synthase III, is CSD2 localized correctly?

Other Organisms

In fungi, the CHS genes are highly conserved (12), and preliminary studies indicate that the CSD genes are also conserved. Determination of the functions of these genes in different taxonomic groups and different developmental stages is a major area of future research.

Chitin is a major component of the exoskeletons of arthropods (39). The regulation of chitin synthesis during insect development is not only a fascinating problem in basic science, but also an important area for the development of pesticides. The studies of chitin synthesis in fungi provide useful concepts and tools for the investigation of chitin synthesis in insects.

Recent studies suggest that chitin synthases and related proteins may not be limited to fungi and arthropods. Analysis of the deduced amino acid sequence of the *CSD2* protein (16, 22) identified two proteins, the NodC protein of *Rhizobium* spp. (55) and the DG42 (82) protein of *Xenopus laevis,* that may be distant and unanticipated relatives of chitin synthases. Both play important roles in development. The NodC protein is required for nodulation in all species of *Rhizobium* (66); the DG42 protein is expressed transiently during gastrulation, and it displays a distinct pattern of expression within the developing embryo (82). The relationships of these proteins to chitin synthases may be an important clue to their as yet unknown biochemical functions. In the case of the NodC protein, previous studies have established that *Rhizobium* spp. produce extracellular factors that trigger nodule formation in the roots of leguminous plants (66). Lerouge et al (63) have purified a nodulation factor, termed NodRm-1, from cultures of *R. meliloti*. Structural analyses showed that NodRm-1 is an *N*-acyl-tri-*N*-acetyl-β-(1→4)-D-glucosamine tetrasaccharide bearing a sulfate group on the reducing sugar moiety. The structural similarity of this molecule to chitin [poly-*N*-acetyl-β-(1→4)- D-glucosamine], together with the similarity between the *nodC* and *CSD2* gene products, suggest that *nodC* encodes an *N*-acetylglucosaminyltransferase that synthesizes the oligosaccharide backbone of NodRm-1. Consistent with this hypothesis is the observation that *nodC* mutants do not synthesize NodRm-1 (63).

Almost nothing is known of the biochemistry of the DG42 protein, and any discussion of its function is purely speculative. Nonetheless, the similarity between the NodC and DG42 proteins is noteworthy, and the elucidation of the function of the NodC protein will likely have significant implications for the understanding of the DG42 protein. If the hypothesis that *nodC* encodes an *N*-acetylglucosaminyltransferase is correct, the possibility that carbohydrates serve as signals during embryogenesis deserves consideration.

ACKNOWLEDGMENTS

I am grateful to P. W. Robbins and B. Berkowitz for their support and encouragement. I thank M. Kong, P. Orlean, B. Osmond, J. Piret, P. Robbins, and C. Specht for their critical reading of the manuscipt and P. Turner for typing. I am also grateful to the many investigators who communicated results prior to publication: S. Bartnicki-García, P. Borgia, C. Chan, A. Durán, N. Gow, P. Orlean, C. Specht, and P. Szaniszlo.

Literature Cited

1. Appeltauer, U., Achstetter, T. 1989. Hormone-induced expression of the *CHS1* gene from *Saccharomyces cerevisiae*. *Eur. J. Biochem*. 181:243–47
2. Au-Young, J., Robbins, P. W. 1990. Isolation of a chitin synthase gene (*CHS1*) from *Candida albicans* by expression in *Saccharomyces cerevisiae*. *Mol. Microbiol*. 4:197–207
3. Ballou, C. E. 1982. Yeast cell wall and cell surface. See Ref. 95b, pp. 335–60
4. Ballou, C. E. 1990. Isolation, characterization, and properties of *Saccharomyces cerevisiae mnn* mutants with

nonconditional protein glycosylation defects. *Methods Enzymol.* 185: 440–70
5. Ballou, C. E., Maitra, S. K., Walker, J. W., Whelan, W. L. 1977. Developmental defects associated with glucosamine auxotrophy in *Saccharomyces cerevisiae*. *Proc. Natl. Acad. Sci. USA* 74:4351–55
6. Bartnicki-García, S. 1989. The biochemical cytology of chitin and chitosan synthesis in fungi. In *Chitin and Chitosan,* ed. G. Skjak-Braek, T. Anthonsen, P. A. Sanford, pp. 23–35. London: Elsevier Applied Science
7. Bartnicki-García, S. 1990. Role of vesicles in apical growth and a new mathematical model of hyphal morphogenesis. In *Tip Growth in Plant and Fungal Cells,* ed. I. B. Heath, pp. 211–32. San Diego, CA: Academic
8. Baymiller, J., McCullough, J. 1990. *Analysis of a* Saccharomyces cerevisiae *strain which grows without a functional* CHS2 *gene*. Presented at 90th Annu. Meeting Soc. Microbiol., Anaheim, Calif.
9. Beran, K., Holan, Z., Baldrian, J. 1972. The chitin-glucan complex in *Saccharomyces cerevisiae*. I. IR and X-ray observations. *Folia Microbiol.* 17:322–30
10. Borgia, P. T. 1992. Roles of the *orlA, tsE,* and *bimG* genes of *Aspergillus nidulans* in chitin synthesis. *J. Bacteriol.* 174:384–89
11. Borgia, P. T., Dodge, C. L. 1992. Characterization of *Aspergillus nidulans* mutants deficient in cell wall chitin or glucan. *J. Bacteriol.* 174:377–83
12. Bowen, A. R., Chen-Wu, J. L., Momany, M., Young, R., Szaniszlo, P. J., Robbins, P. W. 1992. Classification of fungal chitin synthases. *Proc. Natl. Acad. Sci. USA* 89:519–23
13. Briza, P., Breitenbach, M., Ellinger, A., Segall, J. 1990. Isolation of two developmentally regulated genes involved in spore wall maturation in *Saccharomyces cerevisiae*. *Genes Dev.* 4:1775–89
14. Briza, P., Ellinger, A., Winkler, G., Breitenbach, M. 1988. Chemical composition of yeast ascospore wall. The second outer layer consists of chitosan. *J. Biol. Chem.* 263:11569–74
15. Briza, P., Ellinger, A., Winkler, G., Breitenbach, M. 1990. Characterization of a D,L-containing macromolecule from yeast ascospore walls. *J. Biol. Chem.* 265:15118–23
16. Bulawa, C. E. 1992. *CSD2, CSD3,* and *CSD4,* genes required for chitin synthesis in *Saccharomyces cerevisiae:* the *CSD2* gene product is related to chitin synthases and to developmentally regulated proteins in *Rhizobium* species and *Xenopus laevis*. *Mol. Cell. Biol.* 12:1764–76
17. Bulawa, C. E., Osmond, B. C. 1990. Chitin synthase I and chitin synthase II are not required for chitin synthesis in vivo in *Saccharomyces cerevisiae*. *Proc. Natl. Acad. Sci. USA* 87:7424–28
18. Bulawa, C. E., Osmond, B. C., Robbins, P. W. 1991. *Multiple genes are required for chitin synthesis in* Saccharomyces cerevisiae. Presented at Annu. Meeting Soc. Ind. Microbiol., Philadelphia, Penn.
19. Bulawa, C. E., Robbins, P. W. 1987. *Multiple genes involved in chitin synthesis in* Saccharomyces cerevisiae. Presented at Yeast Genetics and Molecular Biology, San Francisco, Calif.
20. Bulawa, C. E., Robbins, P. W. 1989. *Saccharomyces cerevisiae* mutants defective in chitin synthesis. *J. Cell Biochem. Suppl.* 13E:27
21. Bulawa, C. E., Slater, M., Cabib, E., Au-Young, J., Sburlati, A., et al. 1986. The *S. cerevisiae* structural gene for chitin synthase is not required for chitin synthesis in vivo. *Cell* 46:213–25
22. Bulawa, C. E., Wasco, W. 1991. Chitin and nodulation. *Nature* 353:710
23. Bush, D. A., Horisburger, M., Horman, I., Würsch, P. 1974. The wall structure of *Schizosaccharomyces pombe*. *J. Gen. Microbiol.* 81:199–206
24. Bussey, H. 1991. K1 killer toxin, a pore-forming protein from yeast. *Mol. Microbiol.* 5:2339–43
25. Butler, A. R., Martin, V. J., White, J. H., Stark, M. J. R. 1992. Genetic and biochemical analysis of the *Kluyveromyces lactis* toxin mode of action. *Yeast* 8:S319 (Suppl.)
26. Butler, A. R., O'Donnell, R. W., Martin, V. J., Gooday, G. W., Stark, M. J. R. 1991. *Kluyveromyces lactis* toxin has an essential chitinase activity. *Eur. J. Biochem.* 199:483–88
27. Byers, B. 1981. Cytology of the yeast life cycle. See Ref. 95a, pp. 59–96
28. Cabib, E. 1987. The synthesis and degradation of chitin. *Adv. Enzymol. Relat. Areas Mol. Biol.* 59:59–101
28a. Cabib, E. 1991. Differential inhibition of chitin synthases 1 and 2 from *Saccharomyces cerevisiae* by polyoxinD and nikkomycins. *Antimicrob. Agents Chemother.* 35:170–73
29. Cabib, E., Bowers, B., Sburlati, A., Silverman, S. J. 1988. Fungal cell wall synthesis: the construction of a

biological structure. *Microbiol. Sci.* 5: 370–75
30. Cabib, E., Farkas, V. 1971. The control of morphogenesis: an enzymatic mechanism for the initiation of septum formation in yeast. *Proc. Natl. Acad. Sci. USA* 68:2052–56
31. Cabib, E., Roberts, R., Bowers, B. 1982. Synthesis of the yeast cell wall and its regulation. *Annu. Rev. Biochem.* 51:763–93
32. Cabib, E., Sburlati, A., Bowers, B., Silverman, S. J. 1989. Chitin synthase 1, an auxiliary enzyme for chitin synthesis in *Saccharomyces cerevisiae*. *J. Cell Biol.* 108:1665–72
33. Cabib, E., Silverman, S. J., Sburlati, A., Slater, M. L. 1990. Chitin synthesis in yeast (*Saccharomyces cerevisiae*). See Ref. 59a, pp. 31–41
34. Cabib, E., Silverman, S. J., Shaw, J. A. 1991. Chitin synthetases 1 and 2 from yeast, two isoenzymes with different functions. In *NATO ASI Series,* Vol. H53, *Fungal Cell Wall and Immune Response,* ed. J. P. Latgé, D. Boucias, pp. 39–48. Berlin/Heidelberg: Springer-Verlag
35. Cabib, E., Silverman, S. J., Shaw, J. A. 1992. Chitinase and chitin synthase 1: counterbalancing activities in cell separation of *Saccharomyces cerevisiae*. *J. Gen. Microbiol.* 138:97–102
36. Cabib, E., Silverman, S. J., Shaw, J. A., Das Gupta, S., Park, H.-M., et al. 1991. Carbohydrates as structural constituents of yeast wall and septum. *Pure Appl. Chem.* 63:483–89
37. Chen-Wu, J. L., Zwicker, J., Bowen, A. R., Robbins, P. W. 1992. Expression of chitin synthase genes during yeast and hyphal growth phases of *Candida albicans*. *Mol. Microbiol.* 6: 497–502
38. Clark, M. W. 1991. Immunogold labeling of yeast ultrathin sections. See Ref. 49a, pp. 608–26
39. Cohen, E. 1987. Chitin biochemistry: synthesis and inhibition. *Annu. Rev. Entomol.* 32:71–93
40. Costigan, C., Gehrung, S., Snyder, M. 1992. A synthetic lethal screen identifies SLK1, a novel protein kinase homolog implicated in yeast cell morphogenesis and cell growth. *Mol. Cell. Biol.* 12:1162–78
41. Davis, L. L., Bartnicki-García 1984. Chitosan synthesis by the tandem action of chitin synthetase and chitin deacetylase from *Mucor rouxii*. *Biochemistry* 23:1065–73
42. Devereux, J., Haeberli, P., Smithies, O. 1984. A comprehensive set of sequence analysis programs for the VAX. *Nucleic Acids Res.* 12:387–95
43. Doonan, J. H., Morris, N. R. 1989. The bimG gene of Aspergillus nidulans, required for completion of anaphase, encodes a homolog of mammalian phosphoprotein phosphatase. *Cell* 57:987–96
44. Drubin, D. G. 1991. Development of cell polarity in budding yeast. *Cell* 65:1093–96
45. Finegold, A. A., Johnson, D. I., Farnsworth, C. C., Gelb, M. H., Judd, S. R., et al. 1991. Protein geranylgeranyltransferase of *Saccharomyces cerevisiae* is specific for cys-xaa-xaa-leu motif proteins and requires the *CDC43* gene product but not the *DPR1* gene product. *Proc. Natl. Acad. Sci. USA* 88:4448–52
45a. Gimeno, C. J., Ljungdahl, P. O., Styles, C. A., Fink, G. R. 1992. Unipolar cell divisions in the yeast S. cerevisiae lead to filamentous growth: regulation by starvation and RAS. *Cell* 68:1077–90
46. Glazebrook, J., Walker, G. C. 1989. A novel exopolysaccharide can function in place of the Calcofluor-binding exopolysaccharide in nodulation of alfalfa by Rhizobium meliloti. *Cell* 56:661–72
47. Goldstein, I. J., Poretz, R. D. 1986. Isolation and chemical properties of lectins. In *The Lectins: Properties, Functions, and Applications in Biology and Medicine,* ed. I. E. Liener, N. Sharon, I. J. Goldstein, pp. 103–15. Orlando: Academic
48. Gooday, G. W. 1990. Inhibition of chitin metabolism. See Ref. 59a, pp. 61–79
49. Gow, N. A. R. 1992. *Key genes in the regulation of the dimorphic switch of* Candida albicans. Presented at Fungal Dimorphism: 4th Symposium on Topics in Mycology, Cambridge, UK
49a. Guthrie, C., Fink, G. R., eds. 1991. *Guide to Yeast Genetics and Molecular Biology*. San Diego, CA: Academic
50. Haltiwanger, R. S., Blomberg, M. A., Hart, G. W. 1992. Glycosylation of nuclear and cytoplasmic proteins. *J. Biol. Chem.* 267:9005–13
51. Herscovics, A., Orlean, P. 1993. Glycoprotein biosynthesis in yeast. *FASEB J.* 7:540–50
52. Herth, W. 1980. Calcofluor White and Congo Red inhibit chitin microfibril assembly of *Poterioochromonas:* evidence for a gap between polymerization and microfibril formation. *J. Cell Biol.* 87:442–50
53. Horisberger, M., Vonlanthen, M. 1977.

Location of mannan and chitin on thin sections of budding yeasts with gold markers. *Arch. Microbiol.* 115:1–7

54. Horisberger, M., Vonlanthen, M., Rosset, J. 1978. Localization of α-galactomannan and of wheat germ agglutinin receptors in *Schizosaccharomyces pombe. Arch. Microbiol.* 119: 107–11
55. Jacobs, T. W., Egelhoff, T. T., Long, S. R. 1985. Physical and genetic map of a *Rhizobium meliloti* nodulation gene region and nucleotide sequence of *nodC. J. Bacteriol.* 162:469–76
56. Kang, M. S., Elango, N., Mattia, E., Au-Young, J., Robbins, P. W., Cabib, E. 1984. Isolation of chitin synthetase from *Saccharomyces cerevisiae. J. Biol. Chem.* 259:14966–72
57. Koerner, T. J., Hill, J. E., Myers, A. M., Tzagoloff, A. 1991. High-expression vectors with multiple cloning sites for construction of *trpE* fusion genes: pATH vectors. See Ref. 49a, pp. 477–90
58. Koga, D., Isogai, A., Sakuda, S., Matsumoto, S., Suzuki, A., et al. 1987. Specific inhibition of *Bombyx mori* chitinase by allosamidin. *Agric. Biol. Chem.* 51:471–76
59. Kronstad, J. W., Holly, J. A., MacKay, V. L. 1987. A yeast operator overlaps an upstream activation site. *Cell* 50: 369–77

59a. Kuhn, P. J., Trinci, A. P. J., Jung, M. J., Goosey, M. W., Copping, L. G., eds. 1990. *Biochemistry of Cell Walls and Membranes in Fungi*. New York: Springer-Verlag

60. Kukuruzinska, M. A., Bergh, M. L. E., Jackson, B. J. 1987. Protein glycosylation in yeast. *Annu. Rev. Biochem.* 56:915–44
61. Kuranda, M. J., Robbins, P. W. 1991. Chitinase is required for cell separation during growth of *Saccharomyces cerevisiae. J. Biol. Chem.* 266:19758–67
62. Leal-Morales, C. A., Bracker, C. E., Bartnicki-García, S. 1988. Localization of chitin synthetase in cell-free homogenates of *Saccharomyces cerevisiae:* chitosomes and plasma membrane. *Proc. Natl. Acad. Sci. USA* 85:8516–20
63. Lerouge, P., Roche, P., Faucher, C., Maillet, F., Truchet, G., et al. 1990. Symbiotic host-specificity of *Rhizobium meliloti* is determined by a sulphated and acylated glucosamine oligosaccharide signal. *Nature (London)* 344:781–84
64. Levitz, S. M., DiBenedetto, D. J., Diamond, R. D. 1987. A rapid fluorescent assay to distinguish attached from phagocytized yeast particles. *J. Immunol. Methods* 101:37–42
65. Lipke, P. N., Kurjan, J. 1992. Sexual agglutination in budding yeasts: structure, function, and regulation of adhesion glycoproteins. *Microbiol. Rev.* 56: 180–94
66. Long, S., Atkinson, E. M. 1990. Rhizobium sweet-talking. *Nature* 344:712–13
67. Mol, P. C., Wessels, J. G. H. 1987. Linkages between glucosaminoglycan and glucan determine alkali-insolubility of the glucan in walls of *Saccharomyces cerevisiae. FEMS Microbiol. Lett.* 41: 95–99
68. Mortimer, R. K., Contopoulou, C. R., King, J. S. 1992. Genetic and physical maps of *Saccharomyces cerevisiae,* edition 11. *Yeast* 8:817–902
69. Nelson, W. J. 1992. Regulation of cell surface polarity from bacteria to mammals. *Science* 258:948–55
70. Ohya, Y., Ohsumi, Y., Anraku, Y. 1984. Genetic study of the role of calcium ions in the cell division cycle of *Saccharomyces cerevisiae:* a calcium-dependent mutant and its trifluoperazine-dependent pseudorevertants. *Mol. Gen. Genet.* 193:389–94
71. Orlean, P. 1987. Two chitin synthases in *Saccharomyces cerevisiae. J. Biol. Chem.* 262:5732–39
72. Orlean, P., Arnold, E., Tanner, W. 1985. Apparent inhibition of glycoprotein synthesis by *S. cerevisiae* mating pheromones. *FEBS Lett.* 184:313–17
73. Pammer, M., Briza, P., Ellinger, A., Schuster, T., Stucka, R., et al. 1992. *DIT101* (*CSD2, CAL1*), a cell cycle-regulated yeast gene required for synthesis of chitin in cell walls and chitosan in spore walls. *Yeast* 9:1089–99
74. Peberdy, J. F. 1990. Fungal cell walls—a review. See Ref. 59a, pp. 5–30
75. Price, C., Nasmyth, K., Schuster, T. 1991. A general approach to the isolation of cell cycle-regulated genes in the budding yeast, *Saccharomyces cerevisiae. J. Mol. Biol.* 218:543–56
76. Pringle, J. R., Adams, A. E. M., Drubin, D. G., Haarer, B. K. 1991. Immunofluorescence methods for yeast. See Ref. 49a, pp. 565–602
77. Pringle, J. R., Preston, R. A., Adams, A. E. M., Stearns, T., Drubin, D. G., et al. 1989. Fluorescence microscopy methods for yeast. *Methods Cell Biol.* 31:357–435
78. Rine, J., Kim, S.-H. 1990. A role for isoprenoid lipids in the localization and

function of an oncoprotein. *New Biol.* 2:219–26

79. Roberts, R. L., Bowers, B., Slater, M. L., Cabib, E. 1983. Chitin synthesis and localization in cell division cycle mutants of *Saccharomyces cerevisiae*. *Mol. Cell. Biol.* 3:922–30
80. Roncero, C., Durán, A. 1985. Effect of Calcofluor White and Congo Red on fungal wall morphogenesis: in vivo activation of chitin polymerization. *J. Bacteriol.* 163:1180–85
81. Roncero, C., Valdivieso, M. H., Ribas, J. C., Durán, A. 1988. Isolation and characterization of mutants resistant to Calcofluor White. *J. Bacteriol.* 170: 1950–54
82. Rosa, F., Sargent, T. D., Rebbert, M. L., Michaels, G. S., Jamrich, M., et al. 1988. Accumulation and decay of DG42 gene products follow a gradient pattern during *Xenopus* embryogenesis. *Dev. Biol.* 129:114–23
83. Ruiz-Herrera, J. 1992. *Fungal Cell Wall: Structure, Synthesis, and Assembly*. Boca Raton, FL: CRC Press
84. Santos, B., Durán, A. 1992. Cloning of CAL3, a *Saccharomyces cerevisiae* gene involved in Calcofluor sensitivity and chitin synthesis. *Yeast* 8:S522 (Suppl.)
85. Sburlati, A., Cabib, E. 1986. Chitin synthase 2, a presumptive participant in septum formation in *Saccharomyces cerevisiae*. *J. Biol. Chem.* 261:15147–52
86. Schekman, R., Brawley, V. 1979. Localized deposition of chitin on the yeast cell surface in response to mating pheromone. *Proc. Natl. Acad. Sci. USA* 76:645–49
87. Schekman, R., Novick, P. 1982. The secretory process and yeast cell-surface assembly. See Ref. 95b, pp. 361–93
88. Selker, E. 1990. Premeiotic instability of repeated sequences in *Neurospora crassa*. *Annu. Rev. Genet.* 24:579–613
89. Shaw, J. A., Mol, P. C., Bowers, B., Silverman, S. J., Valdivieso, M. H., et al. 1991. The function of chitin synthases 2 and 3 in the *Saccharomyces cerevisiae* cell cycle. *J. Cell Biol.* 114:111–23
90. Sietsma, J. H., Wessels, J. G. H. 1990. The occurrence of glucosaminoglycan in the wall of *Schizosaccharomyces pombe*. *J. Gen. Microbiol.* 136: 2261–65
91. Silverman, S. J. 1989. Similar and different domains of chitin synthases 1 and 2 of *S. cerevisiae*. *Yeast* 5:459–67
92. Silverman, S. J., Sburlati, A., Slater, M. L., Cabib, E. 1988. Chitin synthase 2 is essential for septum formation and cell division in *Saccharomyces cerevisiae*. *Proc. Natl. Acad. Sci. USA* 85:4735–39
93. Silverman, S. J., Shaw, J. A., Cabib, E. 1991. Proteinase B is, indeed, not required for chitin synthase 1 function in *Saccharomyces cerevisiae*. *Biochem. Biophys. Res. Commun.* 174:204–10
94. Sloat, B. F., Adams, A., Pringle, J. R. 1981. Roles of the *CDC24* gene product in cellular morphogenesis during the *Saccharomyces cerevisiae* cell cycle. *J. Cell Biol.* 89:395–405
95. Stark, M. J. R., Boyd, A., Mileham, A. J., Romanos, M. A. 1990. The plasmid-encoded killer system of *Kluyveromyces lactis:* a review. *Yeast* 6:1–29

95a. Strathern, J. N., Jones, E. W., Broach, J. R., eds. 1981. *The Molecular Biology of the Yeast* Saccharomyces: *Life Cycle and Inheritance*. Cold Spring Harbor: Cold Spring Harbor Lab.

95b. Strathern, J. N., Jones, E. W., Broach, J. R., eds. 1982. *The Molecular Biology of the Yeast* Saccharomyces: *Metabolism and Gene Expression*. Cold Spring Harbor: Cold Spring Harbor Lab.

96. Surarit, R., Gopal, P. K., Shepherd, M. G. 1988. Evidence for a glycosidic linkage between chitin and glucan in the cell wall of *Candida albicans*. *J. Gen. Microbiol.* 134:1723–30
97. Sutton, A., Immanuel, D., Arndt, K. T. 1991. The SIT4 protein phosphatase functions in late G_1 for progression into S phase. *Mol. Cell. Biol.* 11:2133–48
98. Szaniszlo, P. J., Momany, M. 1993. Chitin, chitin synthase and chitin synthase conserved region homologues in *Wangiella dermatitidis*. In *NATO ASI Series, Molecular Biology and its Application to Medical Mycology,* ed. B. Maresca, G. Kobayashi, H. Yamaguchi, pp. 229–42. Berlin/Heidelberg: Springer-Verlag.
99. Tanner, W. 1990. Synthesis and function of glycosylated proteins in *Saccharomyces cerevisiae*. See Ref. 59a, pp. 109–18
100. Thorner, J. 1981. Pheromonal regulation of development in *Saccharomyces cerevisiae*. See Ref. 95a, pp. 143–80
101. Ulane, R. E., Cabib, E. 1976. The activating system of chitin synthetase from *Saccharomyces cerevisiae*. Purification and properties of the activating factor. *J. Biol. Chem.* 251:3367–74
102. Valdivieso, M. H., Mol, P. C., Shaw, J. A., Cabib, E., Durán, A. 1991. Cloning of *CAL1,* a gene required for

activity of chitin synthase 3 in *Saccharomyces cerevisiae*. *J. Cell Biol.* 114:101–9
103. Vannini, G. L., Poli, F., Donini, A., Pancaldi, S. 1983. Effects of Congo Red on wall synthesis and morphogenesis in *Saccharomyces cerevisiae*. *Plant Sci. Lett.* 31:9–17
104. Walker, J. E., Gay, N. J., Saraste, M., Eberle, A. N. 1984. DNA sequence around the *Escherichia coli unc* operon. *Biochem. J.* 224:799–815
105. Watzele, G., Tanner, W. 1989. Cloning of the glutamine:fructose-6-phosphate amidotransferase from yeast. Pheromonal regulation of its transcription. *J. Biol. Chem.* 264:8753–58
106. Wheat, R. W. 1966. Analysis of hexosamines in bacterial polysaccharides by chromatographic procedures. *Methods Enzymol.* 8:60–78
107. Whelan, W. L., Ballou, C. E. 1975. Sporulation in D-glucosamine auxotrophs of *Saccharomyces cerevisiae:* meiosis with defective ascospore wall formation. *J. Bacteriol.* 124:1545–57
108. Wilson, R. B., Brenner, A. A., White, T. B., Engler, M. J., Gaughran, J. P., Tatchell, K. 1991. The *Saccharomyces cerevisiae SRK1* gene, a suppressor of *bcy1* and *ins1,* may be involved in protein phosphatase function. *Mol. Cell. Biol.* 11:3369–73
109. Wong, S. S., Wong, L.-J. C. 1992. Chemical crosslinking and the stabilization of proteins and enzymes. *Enzyme Microb. Technol.* 14:866–74
110. Yarden, O., Yanofsky, C. 1991. Chitin synthase 1 plays a major role in cell wall biogenesis in *Neurospora crassa*. *Genes Dev.* 5:2420–30

Annu. Rev. Microbiol. 1993. 47:535–64

ANTIBIOTICS SYNTHESIZED BY POSTTRANSLATIONAL MODIFICATION

J. Norman Hansen
Department of Chemistry and Biochemistry, University of Maryland, College Park, Maryland 20742

KEY WORDS: lantibiotic, nisin, subtilin, bacteriocin, rational design

---

CONTENTS

0066-4227/93/1001-0535$02.00

ABSTRACT

Peptides that have antimicrobial activity are synthesized by many prokaryotic and eukaryotic organisms. Antimicrobial peptides commonly contain unusual amino acids that contribute to their properties and functions. Although bacteria synthesize most of these peptides by nonribosomal mechanisms, this review focuses on those that are synthesized by pathways that involve posttranslational modification of ribosomally synthesized precursor peptides. A particularly interesting class of these antimicrobial peptides is the lantibiotics, of which nisin and subtilin are the longest-known examples, although nearly a dozen new lantibiotics have been discovered in recent years. The fact that the lantibiotic structures are derived from gene-encoded peptides means that structural analogs of natural lantibiotics can be constructed by mutagenesis of their structural genes. Recent advances in our understanding of the molecular genetics of lantibiotics has made the construction of novel lantibiotics with enhanced chemical and antimicrobial properties possible. This review describes these advances and proposes future trends of research, as well as potential application of engineered lantibiotics, in the context of the general field of antimicrobial peptides.

RIBOSOMALLY SYNTHESIZED ANTIMICROBIAL PEPTIDES

The discovery and productive utilization of antibiotics has been one of the most important developments of this century, and their use for treatment of bacterial infections has been spectacularly successful. However, bacteria acquire resistance to antibiotics so easily and quickly that many have become resistant to numerous antibiotics (122, 154). Diseases such as tuberculosis that were once thought conquered forever have sprung back with drug-resistant fury (23, 65), and there is a possibility that mortality due to bacterial infections will climb to totals not seen since the past century (65, 123).

Genetic Engineering and Rational Drug Design

The discovery of penicillin in 1928 by Alexander Fleming set the stage for the antibiotic revolution. Although the effectiveness of Fleming's penicillin began to wane within the first few years of its use, the pharmaceutical industry learned how to screen the ability of a variety of microorganisms to produce new antibiotics or to improve them by making appropriate structural modifications. The concepts of rational drug design, or knowledge-based drug design, have become important. The approach involves subjecting a known antibiotic or drug to comprehensive studies of its chemical and physical properties and its mechanism of action. This information is used to hypothesize about ways to improve the drug, such as enhance its chemical stability, change its spectrum of action, or overcome resistance. These rationally inspired structures are constructed, usually by synthetic organic chemistry, and tested to see if the predicted improvements occur. This process then leads to new rounds of hypothesis, synthesis, and testing until a comprehensive understanding of the drug is attained (26, 126). For example, thousands of structural variants of penicillin have been constructed and tested (83).

A problem is that penicillin and the other common antibiotics are organic molecules that are biosynthesized by multistep enzymatic pathways, and the construction of structural analogues requires laborious organic chemistry. Synthetic organic methods for natural products are notoriously difficult; the synthesis of a single derivative often requires a year-long effort by a team of chemists. This makes the iterative process of hypothesis, synthesis, and testing required by rational drug design very time consuming and expensive. Moreover, even if a useful new derivative is discovered, its manufacture would require an expensive scale-up of what may be an exotic laboratory synthesis. Accordingly such drugs are very expensive.

The purpose of this review is to focus on antibiotics that are synthesized by a posttranslational mechanism. Such antibiotics are gene-encoded and synthesized as polypeptide precursors that undergo posttranslational modifications during their conversion to an active form. The fact that the antibiotic structure is dictated by a gene sequence means that the structure of the mature antibiotic can be modified by mutagenesis of this gene. This can be done using the powerful tools of genetic engineering, which make the construction of analogues much simpler than is possible by chemical synthesis. The availability of these analogues will allow rapid iteration of the cycles required by rational drug design, which will allow the rapid acquisition of knowledge about the structural basis of the chemical and physical properties of the antibiotic, its antimicrobial mechanism, and how to develop strategies to enhance these properties. As discussed in a later section, the properties of some structural mutants of subtilin show that this is a practical approach.

A SURVEY OF RIBOSOMALLY SYNTHESIZED ANTIMICROBIAL PEPTIDES AND PROTEINS

Proteinaceous substances with antimicrobial properties are very common. They are produced both by prokaryotes and eukaryotes, usually as biological-warfare agents against competitors. Bacteria produce many antimicrobial peptides through nonribosomal pathways (118). Many of these peptides are termed bacteriocins, which are loosely defined as bacterially produced proteinaceous antimicrobial substances that inhibit the growth of species that are closely related to the producer organism (190). Nonribosomally synthesized bacteriocins frequently contain unusual amino acids; this is not surprising, as the genetic code does not dictate their composition. However, as we shall see, many important ribosomally synthesized antimicrobial peptides also contain unusual amino acids that are introduced by posttranslational modifications.

Eukaryotic Antimicrobial Proteins and Peptides

Antibodies produced by the mammalian immune system are the most sophisticated and versatile gene-encoded antimicrobial proteins yet discovered. The immune system has an elaborate mechanism for generating astronomical numbers of structural variants and then selecting those that are highly specific for the antigens to which the organism is exposed. As far as we can tell, this mechanism can generate antibodies against any conceivable antigen or infectious agent. The key elements of the system include a method for generating variants, screening for those that are effective, and then selectively propagating those that pass the screen. Nature has chosen a combination of rational and random approaches to antibody design. They are rational in that the structural framework of all the antibodies is constant, whereas mutagenesis within the antigen-combining region is random.

Eukaryotes produce many antimicrobial peptides. For instance, magainins are present in frog skin (17, 199). These peptides have a broad spectrum of antimicrobial activity that includes many species of bacteria and fungi (200). Activity is against the cell membrane (16, 140, 141, 197), resulting in leakage (140) and formation of anion-permeable channels (16, 46). They are active against gram-positive and gram-negative bacteria, but their activity is modulated by the gram-negative lipopolysaccharide layer (134, 161, 162). Structural modification to enhance the helical structure of magainin 2 resulted in a 100-fold increase in antimicrobial activity (31).

Another class of eukaryotic antimicrobial peptides is the cecropins, first isolated from pupae of the cecropia moth (92). Subsequently, analogues were discovered in a variety of other insects (44, 129, 147, 158, 160, 193) and even mammals (128). Some of the cecropins are synthesized by a pathway involving processing of a preprocecropin sequence that undergoes several proteolytic cleavages (24). Several cecropins contain the unusual amino acid delta-hydroxylysine, which is presumably introduced by posttranslational modification of the precursor (147). A variety of cecropin analogues have been constructed, mainly by solid-phase chemical synthesis (6–8, 24, 25, 54, 145, 195), and several analogues with enhanced antimicrobial properties were designed and successfully constructed (25, 54). The mechanism of action involves binding and disrupting the lipid bilayer of the membrane (32, 186). The presence of cholesterol in the membrane inhibits this disruption, which helps explain why eukaryotic cells are insensitive to cecropins (32).

Prokaryotic Antimicrobial Peptides and Proteins

Until recently, proteinaceous antimicrobial substances produced by bacteria were generally classified as bacteriocins, whether appropriately or not. Some of the nonribosomally synthesized bacteriocins such as valinomycin and gramicidin have been thoroughly studied, both in terms of their biosynthesis and molecular mode of action (55, 79, 118, 168, 169). An unattractive feature of the bacteriocins is that they tend to have very narrow spectra of action (190), thereby limiting the ways that they can be put to practical use. Some of the bacteriocins are very large proteins and often have poorly understood molecular modes of action (190).

Microcins Produced by E. coli

The microcins are a group of low-molecular-weight antimicrobial substances produced by various strains of *Escherichia coli* (11, 170). As the study of microcins progressed, it became evident that many of them are ribosomally synthesized antimicrobial peptides. Microcins that have been identified include microcin B17 (35), H47 (127), microcin 17 (86), C7 (36, 43, 157), 15m (2), D93 (138), microcin 7 (61, 62), E492 (37–39), microcin 140 (47), and 15n (1). The most thoroughly studied microcin is B17, which is a 43–amino acid protein containing 60% glycine. B17 is an inhibitor of DNA synthesis (35). The biosynthetic genes are located in a 3.5-kb region of a 70-kb plasmid (174) that contains four open reading frames (64). Microcin B17 production occurs as the cells enter the stationary phase of growth and depends on the product of the *ompR* gene (157). The producer cell is immune to the bacteriocin because of a mechanism that pumps it out of the cytoplasm (63). Although the precursor peptide contains a 26–amino acid N-terminal region that is cleaved during maturation, it does not have the characteristics of a typical signal sequence. B17 may contain unusual amino acids (35, 183), but this has not been established with certainty.

Bacteriocins of Lactic Acid Bacteria

Many of the lactic acid bacteria are food-grade organisms used for various purposes by the food industry. Because of their long history of safe use in food, any antimicrobial substances that they produce should be excellent candidates for new food preservatives and a variety of other applications. This has prompted a search for bacteriocins among these organisms (117), and several ribosomally synthesized antimicrobial peptides have been discovered.

One is lactococcin A, isolated from *Lactococcus lactis*. It is synthesized as a 75–amino acid precursor from which a 21–amino acid N-terminal region is removed (90). The structural gene for the precursor is on a 5.6-kb restriction fragment as part of a cluster of five open reading frames. One of them encodes an immunity factor, and two others encode proteins with similarities to the HlyB and HlyD translocator proteins, suggesting the operon encodes a secretion apparatus. No unusual amino acids were found in the mature lactococcin A peptide (189). The mechanism of action of lactococcin A involves increasing the permeability of the cytoplasmic membrane, resulting in dissipation of the membrane potential (194). Lactacin B is an antimicrobial peptide produced by *Lactobacillus acidophilus* (12–14) as is lactacin F (151). A gene was cloned that contained a 75–amino acid open reading frame, which in turn contained a signal region at the N terminus. The gene also had a 57–amino acid C-terminal region that contained the lactacin F sequence. No unusual amino acids have been found (150).

Pediococcus acidilactici produces several bacteriocins called pediocins. Pediocin PA-1 is a 44–amino acid peptide that contains two disulfide bonds (84) and is synthesized as a 62–amino acid precursor from which 18 residues are removed from the N-terminal end (139). This pediocin has antilisterial activity (56). Pediocin ACH is a protease-sensitive peptide with a molecular weight of about 2700 Daltons. It also possesses antilisterial activity (198), but pediocin ACH is unusual in that its bacteriocidal effect does not seem to involve induction of membrane permeability (18, 19).

THE LANTIBIOTICS

Lantibiotics Produced by Gram-Positive Bacteria

The lantibiotics constitute a class of ribosomally synthesized antibiotic peptides that is rapidly growing in size and importance, and the rest of this review is devoted to them. Ironically, the longest-known lantibiotic, called nisin, was first observed in 1928 (171), the same year that Fleming discovered penicillin. Nisin is produced by many strains of *Streptococcus lactis*. Most of the early studies of nisin were performed by Hurst and coworkers (66, 87, 88, 93–102, 121, 142–144, 165, 192). Subtilin is a natural structural analogue of nisin produced by *Bacillus subtilis* strain ATCC 6633 and was first observed in 1948 (52). Gross and coworkers spent several years determining the complete structures of nisin and subtilin (68–77, 146). Figure 1 displays these structures.

Several decades passed before the discovery of any additional lantibiotics, until nearly a dozen were identified during the past five years or so. They have been found in several gram-positive genera, including *Bacillus, Strep-*

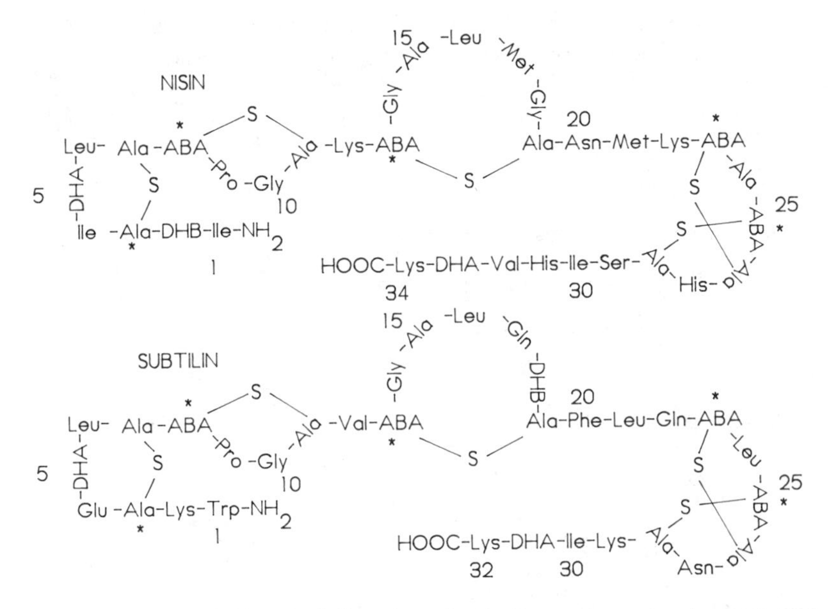

Figure 1 Structures of nisin and subtilin as determined by Gross and coworkers (see text). ABA, aminobutyric acid; DHA, dehydroalanine; DHB, dehydrobutyrine (β-methyldehydroalanine); Ala-S-Ala, lanthionine; ABA-S-Ala, β-methyllanthionine. Asterisks indicate that the amino acid has the (D)-stereo configuration at the α-carbon.

tococcus, Staphylococcus, Lactococcus, Streptomyces, and possibly others that are as yet undiscovered.

Lantibiotics Contain Unusual Amino Acids Introduced by Posttranslational Modification

Lantibiotics have an extraordinary proportion of unusual amino acids and derive their name from the fact that they contain lanthionine (42, 177). The presence of unusual amino acids led to the early assumption that they were synthesized by a nonribosomal mechanism. However, experiments to test this hypothesis for nisin supported a ribosomal mechanism (93, 103, 104), and it was demonstrated that subtilin biosynthesis involved cleavage of a precursor peptide (155). Lantibiotics also contain unusual dehydro residues called dehydroalanine (DHA) and dehydrobutyrine (DHB). Inspired solely by the structures of the unusual amino acids, Ingram hypothesized that the unusual amino acids were formed by a novel series of posttranslational modifications in which serines and threonines were dehydrated to DHA and DHB, respectively; whereupon cysteine would react with the double bond of DHA to form the thioether crosslinkage in lanthionine, or with DHB to form the

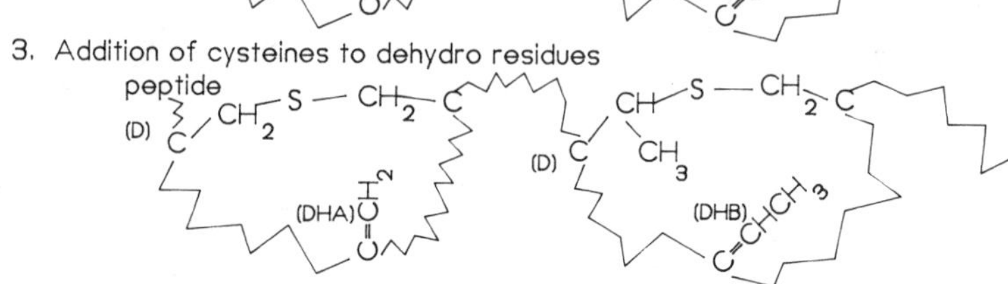

Figure 2 A hypothetical series of reactions for formation of the dehydro residues and the thioether crosslinkages in nisin. Serine and threonine are dehydrated to dehydroalanine (DHA) and dehydrobutyrine (DHB), respectively. Nucleophilic addition of the sulfhydryl of cysteine to the dehydro residues proceeds with stereo-inversion. Lanthionine and β-methyllanthionine result from the formation of thioether crosslinkages. Scheme based on that proposed by Ingram (see text).

thioether crosslinkage in β-methyllanthionine. Figure 2 shows the post-translational modification reactions Ingram hypothesized.

Although Ingram's scheme could be tested by identifying the gene for the nisin precursor, nearly 20 years passed and cloning and sequencing methodologies were developed before this was done. The nisin gene was first cloned and sequenced by Buchman et al (28), and the subtilin gene by Banerjee & Hansen (10). The epidermin (177), Pep5 (111, 177), gallidermin (176), and cinnamycin (110) genes have also been cloned and sequenced. All these gene sequences contain serines, threonines, and cysteines at exactly the locations predicted by Ingram's scheme. This is very strong circumstantial evidence in favor of Ingram's hypothesis, but the biochemical experiments demonstrating that these modifications occur as proposed have not yet been performed.

The Several Types of Lantibiotics

Lantibiotics can be categorized according to their structural features (109). The nisin-like lantibiotics include nisin (75), subtilin (71), Pep5 (114), epidermin (3), and gallidermin (112). Mersacidin (119, 156) and actagardine (116, 135, 136) can also be included in this group, although they do not quite fit within the the group's general description. In general, these molecules are linear cationic (basic) peptides up to 34 residues in size that possess sequence homology and similarities in their lanthionine bridges.

Another class is the duramycin-like lantibiotics, which includes duramycin A, duramycin B, duramycin C (57, 58, 137, 153, 181), cinnamycin (110, 137), and ancovenin (196). The duramycin-like peptides are smaller (as few as 19 residues), less cationic, or neutral, and the C-terminal residue is involved in bridge formation. In duramycin A, the C-terminal lysine forms a lysinoalanine bridge by reaction with a DHA at position 6. In Duramycin B and C, this bridge is not formed, and there is an unreacted DHA residue at position 6. Mersacidin and actagardine do not fit easily into either category (109). They are small and neutral, but their lanthionine bridges resemble those of nisin. As more lantibiotics are discovered, classifying them into a few categories will probably become increasingly difficult.

Practical Uses for Lantibiotics

Lantibiotics have already proved themselves to be of considerable practical value. Nisin is a very important food preservative (41, 97). The natural producer of nisin is *Lactococcus lactis,* which is an important organism in dairy fermentations (40, 41, 97) and has been a component of the human food supply since antiquity (99). Whereas the presence of nisin in food as a consequence of the fermentation process has always been allowed, the US Federal Drug Administration only recently acknowledged its safety and efficacy as a food additive (50, 51). Nisin is reported to be effective in treatment of bovine mastitis (180) and possibly has other therapeutic applications (22). Nisin has a very broad spectrum of activity against many species from several genera of gram-positive bacteria that are important in the dairy and food industry (80, 108). Recently nisin was found to be effective against gram-negative bacteria if their lipopolysaccharide layer is compromised by chemical treatment or genetic defects (22, 187, 188).

Epidermin and gallidermin are effective in treating a variety of skin infections (3, 112). Actagardine inhibits peptidoglycan biosynthesis (116, 135, 136). Ancovenin can inhibit the angiotensin-converting enzyme and thus has potential for use in treatment of high blood pressure (196). Cinnamycin, previously named lanthiopeptin and Ro 09-0198 (115, 182), shows antiviral activity against Herpes simplex virus, HSV1 (152). Mersacidin has shown activity as an immunosuppressor (30, 119, 156).

The range of biological effects exerted by lantibiotics is amazing. Whereas one can understand why an organism such as *L. lactis* produces an antibiotic that is targeted against other bacteria that occupy its ecological niche (99), it is hard to imagine why *Streptomyces* spp. would be under selective pressure to produce an ancovenin that inhibits mammalian angiotensin-converting enzyme, or why *Actinoplanes* spp. would be under selective pressure to produce an actagardine that inhibits HSV1, which is a human-specific virus. The answer is apparently the same as the reason why a fungus produces cyclosporin, an effective suppressant of autoimmune diseases (53, 167, 191), or why taxol, which is effective against cancerous solid tumors, can be extracted from the bark of the Pacific yew tree (4, 184). Apparently the effects of lantibiotics on mammalian viruses and enzymes are fortuitous. The lesson here is that the lantibiotics possess an enormous range of biological effects and we would benefit greatly by learning more about the relationships between their structures, properties, and functions. In this context, the susceptibility of lantibiotics to structural alteration by mutagenesis is important. Using mutagenesis, we can construct new lantibiotics and not be limited to those that Nature provides. One can envision how natural lantibiotics may be used as the foundation for designing new constructs with novel or altered functions. Some of these new forms could be directed toward infectious microorganisms that have become resistant to known antimicrobial agents or other pathogens, such as viruses, for which we currently have a limited number of antiviral agents.

Alternatively, lantibiotic peptides such as subtilin and nisin can be viewed as a kind of primitive antibody with constant and variable regions that is secreted by the bacterium in order to destroy its competitors. Unlike the mammalian immune system, which has evolved an elegant and elaborate method for generating diversity and selecting for those antibodies that are effective, the bacterium has no mechanism for generating a diverse population of lantibiotic peptides. However, such diversity is readily created in the laboratory, and ways might be devised to generate populations of lantibiotic variants that rival antibodies in their diversity of specificity. These variants could be functionally superior to antibodies in that their dehydro residues would provide a mechanism for becoming covalently attached to their hapless targets, making the destruction of these targets more certain.

The Mechanisms by which Lantibiotics Exert Their Antimicrobial Activity

The preceding section implies that the biological effects exerted by lantibiotics will not be the consequence of a single consistent mechanism. Even in the case of the nisin-like lantibiotics, which have been most extensively studied, a simple discrete mechanism of antimicrobial action has not been established. One problem appears to be that different assay systems to measure the antimicrobial effects have been employed, and the possibility that these different assay systems may detect different antimicrobial mechanisms has, until recently, been ignored.

Because nisin is highly effective in preventing the outgrowth of bacterial spores of food-borne pathogens such as *Clostridium botulinum* (179), Morris et al employed outgrowing spores of *Bacillus cereus* T to study the mechanism of nisin action (148). They obtained evidence that nisin interfered with sulfhydryl groups in the membranes of the germinated spores and speculated that nisin becomes covalently attached to membrane sulfhydryl groups by reaction with one or more dehydro residues in the nisin molecule, although they did not identify a covalent adduct. The effect of nisin action was to halt outgrowth at the early outgrowth stage without lysis. In growing cells of a variety of gram-positive species, studies of the effects of nisin, subtilin, Pep5, and other lantibiotics obtained different results. The general picture in growing cells is one in which the lantibiotic interacts with the cell membrane to form pores, which results in collapse of voltage and pH gradients, usually leading to cell lysis (20, 21, 120, 121, 172, 173, 178). We (78) have speculated that these disparate observations could be a consequence of an ability of the nisin-like lantibiotics to exert their antimicrobial effects by more than one mechanism. Experiments on subtilin structural variants, discussed below, support the idea of two distinct mechanisms of lantibiotic action.

The Problem of the Immune Factor

Virturally nothing is known about the mechanism by which a lantibiotic producer is immune to the lantibiotic it produces. Some bacteria are known to be resistant to nisin by means of proteolytic enzymes (107) or enzymes that can reduce dehydro residues (105, 106). A gene that confers nisin resistance has been cloned from a strain of *L. lactis* that is incapable of making nisin. Yet this resistance gene was absent from a nisin-producing strain, suggesting that it not the immune factor (59, 60). Although little has been published on the subject, researchers have noted that hosts that produce one type of lantibiotic are not very immune to other lantibiotics. This substrate specificity of the immune factor would seem to have an evolutionary benefit, because lantibiotics would cease to be useful weapons if immunity to one would confer immunity to all. Because this is not the case, the evolution of a new lantibiotic is worthwhile for an organism as this new lantibiotic will likely have some competitive value.

An intriguing puzzle emerges when one considers that mutation of the lantibiotic prepetide gene could give rise to a lantibiotic that is resistant to the producer's immune factor, thus killing the producer cell. Although the cell may have an elegant strategy to overcome this problem, lantibiotic evolution may have to occur as a step-by-step process in which the first step is mutation of the lantibiotic. If the immune factor is only marginally able to cope with this altered form, it will be under selective pressure to evolve and improve its immunity. Once this has occurred, the lantibiotic can mutate again, followed by selection for yet another improved immune factor. Mutation and selection would thus alternate between the lantibiotic and the immune factor until they have both become optimized in their service to the producer cell. These considerations appear to place constraints on the range of lantibiotic mutants that can be constructed and expressed, in that mutations that confer resistance to the immune factor would be lethal to the cell. Whether or not this will be important to our ability to construct and express lantibiotic mutants will depend a great deal on the mechanism of immunity, which is still completely unknown.

PROPERTIES OF KNOWN STRUCTURAL VARIANTS OF LANTIBIOTICS

Most of the structural variants of lantibiotics are naturally occurring forms. That such structural variations should occur is not surprising. After all, lantibiotic structures are gene-encoded and subject to the same effects of natural mutation and selection for improved properties as any other protein. Many, if not all, of the lantibiotics probably evolved from a common ancestor (28), and the ability to make lantibiotics has become dispersed among gram-positive bacteria by transfer of mobile genetic elements, such as the plasmid that encodes epidermin genes (9), or the transposon that encodes nisin genes (67, 163). If such a genetic element were to enter a new host, it would begin to evolve to suit the needs of the new host by the usual means of spontaneous mutation and natural selection. The principle is well illustrated by nisin and subtilin, which have clearly been evolving separately from a common ancestor for a long time (28). Although these compounds differ by 12 residues, the number and locations of the thioether bridges is conserved. They both have three dehydro residues, and the positions of two of these are conserved. One can conclude that the similarities have been conserved because they are critical for function, and the differences result from adaptation of each lantibiotic to the specific needs of its respective host. Some of the other natural variants are epidermin and gallidermin, which differ from each other by only a single residue (113). Duramycin, duramycin B and C, cinnamycin, and ancovenin all have the same organizational structure, each with 19 residues, two β-methyllanthionines and one lanthionine, and a thioether bridge between residues 1 and 18. Within this consistent structure, 13 of the amino acids vary (57). A survey of many strains of *L. lactis* has identified a strain that produces a natural nisin variant, called nisin Z, which differs by one residue—Asn instead of His at position 27 (149). An intriguing aspect of all of the different nisin-producing strains of *L. lactis* in which the gene sequences have been determined is that all of the gene sequences encoding nisin are identical, and all of the gene sequences that encode nisin Z differ from the nisin gene at the same single residue (125, 149). Many silent sites could have mutated within the nisin gene sequence without changing the nisin structure; this argues strongly that all of the nisin-producing strains of *L. lactis* have a close evolutionary relationship. What this means is not entirely clear, but it suggests that the various *L. lactis* strains available to researchers are members of a very closely related gene pool. Although this conclusion is speculative, it may reflect a tendency of the dairy industry to select certain strains with advantageous properties, which have somehow come to dominate strain collections throughout the world.

A structural variant of nisin has also been constructed by site-directed mutagenesis (124, 125), in which residue 17 of nisin Z was changed from Met to Gln, and residue 18 was changed from Gly to Thr. The mutant gene was expressed from a plasmid in a *L. lactis* strain that retained its endogenous chromosomal nisin gene. The strain produced three forms of nisin; natural nisin expressed from the chromosomal gene and two mutant forms of nisin Z that were expressed from the mutant *nisZ* gene in the plasmid. The mutant forms of nisin Z differed in that the Thr residue at position 18 had undergone dehydration in some of the molecules but not in others. It was determined that the chemical and antimicrobial properties of nisin Z and its two mutant forms did not differ significantly from nisin itself. A conservative change in position 27 (Asn changed to Gln) of nisin Z had little effect on its properties (45). Also, deletion of the the C-terminal DHA33 and Lys34 had no deleterious effect on nisin activity (29), nor did the conversion of Ile32 to Val (45).

SOME INTERESTING PROBLEMS POSED BY THE BIOSYNTHETIC PATHWAY OF LANTIBIOTICS

The Unusual Amino Acids: Processing Signals, Unusual Chemical Properties

Whereas posttranslational processing in prokaryotes is generally limited to removal of N-terminal segments, such as N-formylmethionines, or signal regions during secretion, the extent to which lantibiotic precursors undergo posttranslational modifications is extraordinary. These changes include dehydration of threonines and serines, formation of thioether crosslinkages, removal of the N-terminal leader region, and secretion from the cell. No proteins or peptides other than the lantibiotic precursors themselves have ever been observed to undergo these modifications, so one must conclude that the prelantibiotic sequence contains unique signals that are recognized by the posttranslational-modification machinery. These signals could reside solely in the leader region or solely in the mature region, or be distributed throughout the entire prepeptide. The hydropathic profiles of the leader regions of nisin, subtilin, and epidermin are highly conserved, which indicates a common function (28) and a role in signal recognition and secretion (10, 28). This posttranslational processing mechanism has the effect of circumventing the constraints of the genetic code and results in the introduction of functional groups into the peptide that confer chemical and physical properties that would otherwise be impossible (10). The mature lantibiotics accordingly have electrophilic centers that can react with a variety of nucleophilic groups (130) and display extraordinary thermal stability that is thought to be a consequence of the thioether crosslinkages (97). None of the ordinary amino acids that are defined by the genetic code is electrophilic, and the thioether crosslinkage is much more stable than the disulfide bridge in that it is insensitive to redox conditions, shorter and therefore less flexible, and less reactive toward free radicals. One might wonder if the processing-recognition signals in the prelantibiotic peptide could be incorporated into nonlantibiotic polypeptides, causing them to undergo lantibiotic-like posttranslational processing events. Suppose, for example, that the recognition signals reside solely in the leader region and that fusion of the leader region to a heterologous polypeptide would force it through the lantibiotic processing pathway, resulting in formation of dehydro residues and thioether crosslinkages within the heterologous polypeptide. Protein engineers could exploit this capability and incorporate these unusual amino acids into any arbitrary peptide or protein to confer thermal stability or unique chemical properties.

Achieving this will depend greatly on the details of the location and nature of the processing signals as well as the processing events themselves. Some information has been obtained by studying the ability of lantibiotic producers to process heterologous prepeptides. Hawkins observed that fusion of the subtilin leader region to the nisin mature region gave a chimeric prepeptide sequence that underwent the correct processing steps when expressed in subtilin-producing *B. subtilis* ATCC 6633, although the resulting nisin-like peptide was inactive despite the presence of the full complement of dehydro residues and thioether crosslinkages and the correct removal of the leader region. The result was interpreted to mean that some incorrect thioether crosslinkages had formed, and Hawkins concluded that conformational interactions between the leader and mature regions were important in guiding the formation of the correct thioether crosslinkages (82). The mutant prepeptide of Kuipers et al (124), in which Gly18 had been modified to Thr, gave rise to a mixture of products, some of which contained an unmodified Thr residue and some in which the Thr had been dehydrated to DHB. This result, together with the fact that natural nisin contains an unmodified Ser residue (Figure 1), establishes that dehydration of serines and threonines either is not inevitable, or a dehydro residue can sometimes undergo spontaneous or enzymatic rehydration.

The Leader Sequence and Secretion

Lantibiotics are produced as extracellular secretion products, so they must have a secretion pathway. The most common protein-secretion pathway in prokaryotes employs the *sec* secretion system, which recognizes a characteristic N-terminal secretion signal, typically a stretch of hydrophobic residues flanked by basic residues, that is removed during secretion (15, 27, 81, 85). Inspection of prelantibiotic peptides of nisin, subtilin, and epidermin reveals a leader region that alternates between strongly hydrophobic and hydrophilic residues, indicating that their recognition as secretion signals by the *sec* system is unlikely and that they must be secreted by an alternate pathway (28). Support for an alternate pathway is provided by the presence of genes for translocater-like proteins in the subtilin (34) and nisin operons (48). These translocator proteins have strong homology to the ATP-binding protein translocator called HlyB, which is involved with secretion of hemolysin from *E. coli* (89, 159). These same proteins have strong homology with other widely distributed ATP-binding translocator proteins, such as the multidrug-resistance proteins (5, 49, 166). A discrepancy between the systems is that the secretion signal is located at the C-terminal end of hemolysin, which is then exported without cleavage, whereas the N-terminal sequence of the prelantibiotics is removed. Apparently, lantibiotic secretion constitutes a novel system about which much remains to be learned.

THE MOLECULAR GENETICS OF LANTIBIOTICS

The Organization of Operons that Contain Genes for Subtilin, Nisin, and Epidermin Biosynthesis

Genes for subtilin biosynthesis are located in the chromosome of *B. subtilis* ATCC 6633. The subtilin structural gene is found at the 3′ end of an operon called *spa* (for small protein antibiotic). Genes for five open reading frames have been identified in the *spa* operon: *spaE, spaD, spaB, spaC,* and *spaS* (10, 33, 34). There is a temporally regulated promoter upstream from the *spaE* gene, and a ρ-independent terminator-like structure downstream from the *spaS* gene. SpaS is the subtilin prepeptide, and the SpaD and SpaB proteins possess homology to HlyD and HlyB, respectively (33, 34). Hence they presumably play a role in secretion. The only known homologues of SpaE are NisB from the nisin operon (33, 185) and EpiB from the epidermin operon (48, 175). A sequence that is characteristic of a transmembrane anchor is located at the C terminus of NisB (185), and experiments with antibodies have confirmed that NisB is associated with the membrane (48). Because this protein is encoded in operons for all three lantibiotics, it is assumed to have a functional role in lantibiotic biosynthesis. However, apart from the fact that it is membrane associated, its functional nature is unknown. The role of SpaC is also unknown. An interesting feature of the *spa* operon is that the *spaS* transcript is a long-lived (half-life of 45 min) processing product of a large polycistronic mRNA (10, 33).

The nisin gene is part of a large (68–70 kb) conjugative transposon in the chromosome of *L. lactis* (67, 91, 163). The organization of genes for nisin biosynthesis in the *nis* operon is similar to the *spa* operon, except the nisin structural gene (*nisA*) is located at the 5′ end of the *nis* operon, whereas the *spaS* gene is located at the 3′ end of the *spa* operon. The four open reading frames so far identified in the *nis* operon are *nisA, nisB, nisT,* and *nisC* (48). These are homologous to *spaS, spaE, spaB,* and *spaC,* respectively [note: *spa* gene nomenclature is that of Chung & Hansen (33), which differs from that of Engelke et al (48)]. No cognate to SpaD has been found in the *nis* operon, but it is uncertain that SpaD represents a distinct expressed protein (33). The expression signals in the *nis* operon are not yet established. The transcript of the *nisA* gene has a half-life of about 7 min (28) and appears to be a processing product, at both the 5′ and 3′ ends, of a large polycistronic mRNA (28, 185). A putative RNA processing site is located in the space between the *nisA* and the *nisB* gene (28), and there is a ρ-independent terminator-like structure downstream from the *nisB* gene (185) followed by tandem promoter-like structures that precede the *nisT* gene (185). Northern analysis has revealed a transcript with a size that corresponds to *nisB* (48). A terminator has not yet been identified downstream from the *nisC* gene. In view of the fact that the expression signals that surround the cluster of *nis* genes have not yet been identified, the conclusion that these genes constitute a complete operon is premature.

In comparison to the genes for subtilin and nisin, the epidermin genes are distinct in that they are located on a large 54-kb plasmid in *Staphylococcus epidermidis* (177) instead of in the chromosome. Regions near the epidermin structural gene were sequenced, and six open reading frames organized as two head-to-head operons were identified. One contained *epiA, epiB, epiC,* and *epiD;* the other contained *epiP* and *epiQ* (175). EpiA is the epidermin prepeptide, EpiB is homologous to NisB and SpaE (described above). Putative roles of EpiC and EpiD have not been identified. EpiP is homologous to known serine proteases (175) and thus is a candidate for the leader peptidase. EpiQ has some similarity to PhoB, a positive regulatory factor and so could play a regulatory role in expression of epidermin biosynthesis genes (175). An 8-kb restriction fragment of the original 54-kb plasmid, which contained the six complete open reading frames, was subcloned and expressed in two heterologous species, *Staphylococcus carnosus* and *Staphylococcus xylosus*. Although these species were originally not epidermin producers, they both produced epidermin after transformation with the plasmid (9).

What Genes Are Required for Lantibiotic Biosynthesis?

A central question is, what genes are necessary and sufficient for a cell to produce a lantibiotic? The biosynthetic events include formation of the prelantibiotic, dehydration and crosslinkage reactions, cleavage of the leader, and secretion. In addition, the cell must also be immune to the lantibiotic that it produces. Some of the biosynthetic reactions may be spontaneous, especially the formation of the thioether crosslinkages (78). A gene that confers immunity has not yet been discovered, although each of the lantibiotic operons has genes with unidentified functions. Whereas the *epi* operon contains a putative protease (175), proteases have been identified in neither the *spa* nor *nis* operon. Hence probably none of the identified operons contains all of the genes that participate in lantibiotic biosynthesis. They are either encoded in additional operons near those already identified, or they are host-encoded functions that could reside anywhere in the chromosome.

The Exchange of Lantibiotic Biosynthetic Genes Between Heterologous Species and Strains

One approach to solving the problem of identification of the genes required for lantibiotic biosynthesis is to carry out the transfer of the capacity for lantibiotic production from a producer strain to a heterologous nonproducer strain. The transfer of epidermin genes to heterologous nonproducer strains (described above) on an 8-kb DNA fragment in a plasmid is strong, but not conclusive, evidence that all of the genes are in the 8-kb fragment. Some could still be chromosomally encoded functions (9). The ability to produce subtilin has similarly been transferred from the natural producer (*B. subtilis* ATCC 6633) to a nonproducing strain (*B. subtilis* 168). This transfer was achieved by competence transformation involving integration of a linear DNA fragment from strain 6633 into the chromosome of strain 168, presumably by a double-crossover recombination event (131). Although the exact size of this fragment is not known, the upper limit was established to be about 40 kb

(131), which is a good deal larger than the 6.5-kb *spa* operon. Clearly, other genes associated with subtilin biosynthesis could be present outside the *spa* operon. A similar ambiguity exists for nisin. The genes for nisin production are within a 68- to 70-kb transposon that can be transferred among strains of *L. lactis* by conjugative transposition (67, 163). The *nis* operon is located within about 2 kb of the left border of this transposon (163), so more than 60 kb downstream is unexplored territory, except for a sucrose-fermentation operon that has been located near the center of the transposon (163, 164). It remains to be seen whether additional genes required for lantibiotic synthesis can be discovered in any of these systems.

SUBTILIN AS A MODEL FOR STUDIES OF LANTIBIOTIC MUTAGENESIS

The common features of lantibiotics argue that a thorough study of any particular lantibiotic will provide much information about all of them. Subtilin has become an attractive model lantibiotic because *B. subtilis* 168 has been converted to a producer of subtilin (131). *B. subtilis* 168 is second only to *E. coli* in the amount of genetic information and tools for genetic manipulation that are available. The following sections describe how a system for mutagenesis of the subtilin structural gene has permitted the design and construction of subtilin structural variants with altered, and in some instances improved, properties.

Design of a Cassette-Mutagenesis System for the Construction and Expression of Subtilin Structural Variants

The subtilin structural gene is part of the *spa* operon in the chromosome of *B. subtilis*. If a mutant copy of the subtilin gene is introduced into the cell in a manner that does not involve the removal of the endogenous natural gene, both genes could be expressed simultaneously. Liu & Hansen (132) constructed a host-vector pair consisting of a plasmid in which the mutant subtilin gene could be constructed and a host from which the endogenous gene had been deleted and replaced by an erythromycin-resistance (*erm*) gene. A chloramphenicol-resistance (*cat*) gene was placed immediately adjacent to the subtilin structural gene in the plasmid to serve as a selective marker. After mutagenesis of the subtilin gene, the plasmid was linearized and transformed into the host. Then integrants in which the subtilin-*cat* gene block had replaced the *erm* gene were selected through resistance to chloramphenicol and sensitivity to erythromycin. This not only solved the problem of the endogenous gene but resulted in placement of the mutant gene copy precisely at the normal location of the subtilin gene, at a gene dosage of one and under the same regulatory control as the natural gene (132).

Rational Design of a Subtilin Variant with Enhanced Properties

A major challenge in the design and construction of structural variants of proteins and peptides is to decide which amino acids to mutagenize. Subtilin has not been of practical use, largely because it is unstable—whereas nisin can be readily isolated and stored without loss of activity (130), subtilin undergoes spontaneous inactivation during isolation and storage. The chemical modification of DHA5 accompanies its inactivation. The disappearance of the vinyl proton resonances of DHA5 in the NMR spectrum allow one to monitor the change in DHA5 (78). The coupling of inactivation with DHA5 modification suggests that DHA5 is critical for antimicrobial activity and inactivation is a consequence of its chemical modification. Nisin and subtilin probably act by the same mechanism, yet interestingly, nisin is not subject to the same spontaneous inactivation, though it has this same DHA5 residue. DHA5 is probably a critical residue, and the environment of the DHA5 in subtilin is probably conducive to its inactivation.

Liu & Hansen proposed that the carboxyl group of the Glu4 residue (which is adjacent to DHA5) acts as a general base to catalyze the addition of a nucleophile, such as a hydroxyl group from water, to the double bond of DHA5 (132). Because the relatively stable nisin molecule has Ile at position 4, they further proposed that mutagenesis of Glu4 to Ile would eliminate this carboxyl-group participation and thus enhance the chemical and biological stability of the subtilin molecule. This mutant was constructed, and the prediction was confirmed. The mutant subtilin (called E4I-subtilin) had a chemical stability 57-fold greater than natural subtilin; the biological stability was increased by a like amount (132). These experiments further established that all the posttranslational modifications (dehydration, formation of thioether crosslinkages, cleavage of the leader sequence, and secretion) all occurred normally, so the mutation did not disrupt any processing-signal information.

These results substantiated several important principles. One is that mutagenesis can be used to functionally improve the chemical and antimicrobial properties of lantibiotics. The mutation that accomplished these improvements did not disrupt processing-signal information. The fact that the choice of mutation was determined on a rational basis dictated by knowledge of structure and chemistry, and that the effects of the change were correctly predicted, suggests that rational approaches, in general, will be useful in making decisions about how to make predictable and desirable changes in the properties of lantibiotics. It is certainly striking that such a dramatic improvement in the properties of the subtilin molecule could be made by changing just a single residue.

EVIDENCE THAT SUBTILIN CAN EXERT ITS ANTIMICROBIAL EFFECTS BY TWO DIFFERENT MECHANISMS

The results with E4I-subtilin supported the idea that an intact DHA5 is critical for subtilin antimicrobial activity. We (132) explored this further by mutating the DHA5 residue to Ala, which cannot undergo processing to a dehydro residue. This E4I/DHA5A-subtilin was devoid of activity against outgrowing *Bacillus cereus* T spores (132). However, the same mutant was unchanged, compared with E4I-subtilin, with respect to its ability to lyse vegetative cells of the same organism (133). We concluded that subtilin was exerting its antimicrobial effect against outgrowing spores by a mechanism in which DHA5 played a critical role, but lysis of vegetative cells was caused by a different mechanism in which the DHA5 residue was not important. This suggests that subtilin is an intricately designed molecule with complex chemical properties that enable it to exert its antimicrobial effects by more than one mechanism. The existence of multiple mechanisms needs to be considered when interpreting the results of future experiments.

CONSTRUCTION OF A COMMON EXPRESSION SYSTEM FOR INDUSTRIAL PRODUCTION OF LANTIBIOTICS

Lantibiotics are expected to be applied in many useful ways, e.g. as food preservatives, for treatment of infections, as antagonists to high blood pressure, and as immunosuppressives. If this promise is even partially fulfilled, production of lantibiotics on an industrial scale will be necessary. An attractive feature of lantibiotics is that they are natural secretion products of bacteria and their production will involve microbial fermentations, rather than chemical synthesis. However, several of the lantibiotics already discovered are produced by organisms that grow slowly, that are fastidious and require costly nutrients, and that secrete lantibiotics as only a minor component of their overall metabolism. These are unsatisfactory attributes for industrial organisms.

The success in transporting lantibiotic production among a variety of bacterial species and strains argues that the posttranslational-modification proteins are reasonably versatile and can function in a variety of hosts. A solution to industrial production is to transfer the production of all the lantibiotics to a common industrial strain. An obvious choice is *B. subtilis* 168, which has already been converted to the production of subtilin (131). All of the work required to optimize lantibiotic production in this strain would only have to be done once, instead of over and over again with each strain for each lantibiotic. Its use would also eliminate the need to coax countless exotic, fastidious, and parsimonious bacterial species into being efficient lantibiotic producers.

FUTURE PROSPECTS AND APPLICATIONS FOR LANTIBIOTICS

Nature has created an enormous variety of ribosomally synthesized and posttranslationally modified antimicrobial substances. So many in fact that this review does not pretend to provide comprehensive coverage and has emphasized lantibiotics for two reasons. First, a review is timely because of the wealth of new information about lantibiotics. Second, lantibiotics have characteristics that make them particularly interesting and potentially useful. Some key distinguishing features of lantibiotics are that they are synthesized by a ribosomal mechanism, are relatively small in size, and contain unusual amino acids. The advantages of these features are that the ribosomal mechanism permits the construction of structural analogues by mutagenesis; their small size makes them accessible to a comprehensive understanding of their structure-function relationships; and the posttranslational-modification system allows one to introduce into the polypeptide, by biological means, unusual amino acids that confer physical and chemical properties that are unattainable with the ordinary amino acids as defined by the genetic code. Another advantage is that lantibiotics are produced by bacteria that are convenient to work with. This particular set of features is unique to lantibiotics, which justifies our interest in them. It remains to be determined whether these features can be successfully exploited to provide new fundamental knowledge about biological processes as well as to produce new materials with useful applications in food preservation and treatment of infections and diseases. Investigation of lantibiotics may also lead to thermally stable industrial enzymes, proteins, and biological materials. As stated above, much depends on the details of the mechanism of lantibiotic synthesis and the extent to which the bacterium can be coaxed into processing heterologous substrates. If current efforts meet with even partial success, a search will likely begin for other systems that embody the features possessed by lantibiotics but that allow the introduction of a different set of unusual amino acids. Over time, discovery of such systems would provide access to methods of manufacturing biological materials with properties that are completely unknown today.

ACKNOWLEDGMENTS

The author's research on lantibiotics has been supported by NIH grant AI24454, the National Dairy Promotion and Research Board, and Applied Microbiology, Inc., New York, NY.

Literature Cited

1. Aguilar, A., Baquero, F., Martinez, J. L., Asensio, C. 1983. Microcin 15n: a second antibiotic from *Escherichia coli* LP15. *J. Antibiot. (Tokyo)* 36:325–27
2. Aguilar, A., Perez-Diaz, J. C., Baquero, F., Asensio, C. 1982. Microcin 15m from *Escherichia coli:* mechanism of antibiotic action. *Antimicrob. Agents Chemother.* 21:381–86
3. Allgaier, H., Jung, G., Werner, R. G., Schneider, U., Zahner, H. 1986. Epidermin: sequencing of a heterodetic tetracyclic 21-peptide amide antibiotic. *Eur. J. Biochem.* 160:9–22
4. Amato, I. 1992. Chemists vie to make a better taxol. *Science* 256:311; 1992. Erratum. *Science* 256(5056):428
5. Ames, G. F. L. 1986. The basis of multidrug resistance in mammalian cells: homology with bacterial transport. *Cell* 47:323–24
6. Andreu, D., Merrifield, R. B., Steiner, H., Boman, H. G. 1983. Solid-phase synthesis of cecropin A and related peptides. *Proc. Natl. Acad. Sci. USA* 80:6475–79
7. Andreu, D., Merrifield, R. B., Steiner, H., Boman, H. G. 1985. N-terminal analogues of cecropin A: synthesis, antibacterial activity, and conformational properties. *Biochemistry* 24:1683–88
8. Andreu, D., Ubach, J., Boman, A., Wahlin, B., Wade, D., et al. 1992. Shortened cecropin A-melittin hybrids. Significant size reduction retains potent antibiotic activity. *FEBS Lett.* 296:190–94
9. Augustin, J., Rosenstein, R., Wieland, B., Schneider, U., Schnell, N., et al. 1992. Genetic analysis of epidermin biosynthetic genes and epidermin-negative mutants of *Staphylococcus epidermidis. Eur. J. Biochem.* 204:1149–54
10. Banerjee, S., Hansen, J. N. 1988. Structure and expression of a gene encoding the precursor of subtilin, a small protein antibiotic. *J. Biol. Chem.* 263:9508–14
11. Baquero, F., Bouanchaud, D., Martinez-Perez, M. C., Fernandez, C. 1978. Microcin plasmids: a group of extrachromosomal elements coding for low-molecular-weight antibiotics in *Escherichia coli. J. Bacteriol.* 135:342–47
12. Barefoot, S. F., Klaenhammer, T. R. 1983. Detection and activity of lactacin B, a bacteriocin produced by *Lactobacillus acidophilus. Appl. Environ. Microbiol.* 45:1808–15
13. Barefoot, S. F., Klaenhammer, T. R. 1983. Detection and activity of lactacin B, a bacteriocin produced by *Lactococcus acidophilus. Appl. Environ. Microbiol.* 45:1808–15
14. Barefoot, S. F., Klaenhammer, T. R. 1984. Purification and characterization of the *Lactobacillus acidophilus* bacteriocin lactacin B. *Antimicrob. Agents Chemother.* 26:328–34
15. Bassford, P., Beckwith, J., Ito, K., Kumamoto, C., Mizushima, S., et al. 1991. The primary pathway of protein export in E. coli. *Cell* 65:367–68
16. Bechinger, B., Zasloff, M., Opella, S. J. 1992. Structure and interactions of magainin antibiotic peptides in lipid bilayers: a solid-state nuclear magnetic resonance investigation. *Biophys. J.* 62:12–14
17. Berkowitz, B. A., Bevins, C. L., Zasloff, M. A. 1990. Magainins: a new family of membrane-active host defense peptides. *Biochem. Pharmacol.* 39:625–29
18. Bhunia, A. K., Johnson, M. C., Ray, B. 1988. Purification, characterization and antimicrobial spectrum of a bacteriocin produced by *Pediococcus acidilactici. J. Appl. Bacteriol.* 65:261–68
19. Bhunia, A. K., Johnson, M. C., Ray, B., Belden, E. L. 1990. Antigenic property of pediocin AcH produced by *Pediococcus acidilactici* H. *J. Appl. Bacteriol.* 69:211–15
20. Bierbaum, G., Sahl, H.-G. 1985. Induction of autolysis of staphylococci by the basic peptide antibiotics Pep5 and nisin and their influence on the activity of autolytic enzymes. *Arch. Microbiol.* 141:249–54
21. Bierbaum, G., Sahl, H.-G. 1989. Influence of cationic peptides on the activity of the autolytic endo-beta-N-acetylglucosaminidase of *Staphylococcus simulans* 22. *FEMS Microbiol. Lett.* 49:223–27
22. Blackburn, P., Polak, J., Gusik, S., Rubino, S. D. 1989. Nisin compositions for use as enhanced, broad range bacteriocins. *USA Int. Patent Application No. PCT/US89/02625, 1989; International Publication Number W089/12399*
23. Bloom, B. R., Murray, C. J. L. 1992. Tuberculosis: commentary on a reemergent killer. *Science* 257:1055–64
24. Boman, H. C., Boman, I. A., Andreu,

D., Li, Z. Q., Merrifield, R. B., et al. 1989. Chemical synthesis and enzymic processing of precursor forms of cecropins A and B. *J. Biol. Chem.* 264:5852–60
25. Boman, H. G., Wade, D., Boman, I. A., Wahlin, B., Merrifield, R. B. 1989. Antibacterial and antimalarial properties of peptides that are cecropin-melittin hybrids. *FEBS Lett.* 259:103–6
26. Bruice, T. C., Mei, H. Y., He, G. X., Lopez, V. 1992. Rational design of substituted tripyrrole peptides that complex DNA by both selective minor-grove binding and electrostatic interaction with the polyphosphate backbone. *Proc. Natl. Acad. Sci. USA* 89:1700–4
27. Brundage, L., Fimmel, C. J., Mizushima, S., Wickner, W. 1992. SecY, SecE, and band 1 form the membrane-embedded domain of *Escherichia coli* preprotein translocase. *J. Biol. Chem.* 267:4166–70
28. Buchman, G. W., Banerjee, S., Hansen, J. N. 1988. Structure, expression, and evolution of a gene encoding the precursor of nisin, a small protein antibiotic. *J. Biol. Chem.* 263:16260–66
29. Chan, W. C., Bycroft, B. W., Lian, L.-Y., Roberts, C. K. 1989. Isolation and characterization of two degradation products derived from the peptide antibiotic nisin. *FEBS. Lett.* 252:29–36
30. Chatterjee, S., Chatterjee, D. K., Jani, R. H., Blumbach, J., Ganguli, B. N., et al. 1992. Mersacidin, a new antibiotic from *Bacillus*. In vitro and in vivo antibacterial activity. *J. Antibiot. (Tokyo)* 45:839–45
31. Chen, H. C., Brown, J. H., Morell, J. L., Huang, C. M. 1988. Synthetic magainin analogues with improved antimicrobial activity. *FEBS Lett.* 236: 462–66
32. Christensen, B., Fink, J., Merrifield, R. B., Mauzerall, D. 1988. Channel-forming properties of cecropins and related model compounds incorporated into planar lipid membranes. *Proc. Natl. Acad. Sci. USA* 85:5072–76
33. Chung, Y. J., Hansen, J. N. 1992. Determination of the sequence of *spaE* and identification of a promoter in the subtilin (*spa*) operon in *Bacillus subtilis*. *J. Bacteriol.* 174:6699–6702
34. Chung, Y. J., Steen, M. T., Hansen, J. N. 1992. The subtilin gene of *Bacillus subtilis* ATCC 6633 is encoded in an operon that contains a homolog of the hemolysin B transport protein. *J. Bacteriol.* 174:1417–22
35. Davagnino, J., Herrero, M., Furlong, D., Moreno, F., Kolter, R. 1986. The DNA replication inhibitor microcin B17 is a forty-three-amino-acid protein containing sixty percent glycine. *Proteins* 1:230–38
36. del Castillo, I., Gomez, J. M., Moreno, F. 1990. *mprA*, an *Escherichia coli* gene that reduces growth-phase-dependent synthesis of microcins B17 and C7 and blocks osmoinduction of *proU* when cloned on a high-copy-number plasmid. *J. Bacteriol.* 172:437–45
37. de Lorenzo, V. 1984. Isolation and characterization of microcin E492 from *Klebsiella pneumoniae*. *Arch. Microbiol.* 139:72–75
38. de Lorenzo, V. 1985. Factors affecting microcin E492 production. *J. Antibiot. (Tokyo)* 38:340–45
39. de Lorenzo, V., Pugsley, A. P. 1985. Microcin E492, a low-molecular-weight peptide antibiotic which causes depolarization of the *Escherichia coli* cytoplasmic membrane. *Antimicrob. Agents Chemother.* 27:666–69
40. Delves-Broughton, J. 1990. Nisin and its application as a food preservative. *J. Soc. Dairy Technol.* 43:73–76
41. Delves-Broughton, J. 1990. Nisin and its uses as a food preservative. *Food Technol.* 44:100–17
42. de Vos, W. M., Jung, G., Sahl, H. G. 1991. Definitions and nomenclature of lantibiotics. See Ref. 109a, pp. 457–63
43. Diaz-Guerra, L., Moreno, F., San Millan, J. L. 1989. *appR* gene product activates transcription of microcin C7 plasmid genes. *J. Bacteriol.* 171:2906–8
44. Dickinson, L., Russell, V., Dunn, P. E. 1988. A family of bacteria-regulated, cecropin D-like peptides from *Manduca sexta*. *J. Biol. Chem.* 263:19424–29
45. Dodd, H. M., Horn, N., Hao, Z., Gasson, M. J. 1992. A lactococcal expression system for engineered nisins. *Appl. Environ. Microbiol.* 58: 3683–93
46. Duclohier, H., Molle, G., Spach, G. 1989. Antimicrobial peptide magainin I from *Xenopus* skin forms anion-permeable channels in planar lipid bilayers. *Biophys. J.* 56:1017–21
47. Duro, A. F., Serrano, R., Asensio, C. 1979. Effect of the antibiotic microcin 140 on the ATP level and amino acid transport of *Escherichia coli*. *Biochem. Biophys. Res. Commun.* 88:297–304
48. Engelke, G., Gutowski-Eckel, Z., Mann, M. H., Entian, K.-D. 1992.

Biosynthesis of the lantibiotic nisin: genomic organization and membrane localization of the NisB protein. *Appl. Environ. Microbiol.* 58:3730–43

49. Fath, M. J., Skvirsky, R. C., Kolter, R. 1991. Functional complementation between bacterial MDR-like export systems: colicin V, alpha-hemolysin, and *Erwinia* protease. *J. Bacteriol.* 173: 7549–56
50. Federal Drug Administration. 1988. Nisin preparation: affirmation of GRAS status as a direct human food ingredient. *Fed. Reg.* 53:11247–51
51. Federal Drug Administration. 1989. Pasteurized process cheese spread; amendment of standard of identity. *Fed. Reg.* 54:6120–21
52. Feeney, R. E., Garibaldi, J. A., Humphreys, E. M. 1948. Nutritional studies on subtilin formation by *Bacillus subtilis*. *Arch. Biochem. Biophys.* 17:435–45
53. Feutren, G. 1992. The optimal use of cyclosporin A in autoimmune diseases. *J. Autoimmun.* 5(Suppl. A):183–95
54. Fink, J., Merrifield, R. B., Boman, A., Boman, H. G. 1989. The chemical synthesis of cecropin D and an analog with enhanced antibacterial activity. *J. Biol. Chem.* 264:6260–67
55. Fisher, R., Blumenthal, T. 1982. An interaction between gramicidin and the sigma subunit of RNA polymerase. *Proc. Natl. Acad. Sci. USA* 79:1045–48
56. Foegeding, P. M., Thomas, A. B., Pilkington, D. H., Klaenhammer, T. R. 1992. Enhanced control of *Listeria monocytogenes* by in situ-produced pediocin during dry fermented sausage production. *Appl. Environ. Microbiol.* 58:884–90; 1992. Erratum. *Appl. Environ. Microbiol.* 58(6):2102
57. Fredenhagen, A., Fendrich, G., Marki, F., Marki, W., Gruner, J., et al. 1990. Duramycins B and C, two new lanthionine containing antibiotics as inhibitors of phospholipase A2. Structural revision of duramycin and cinnamycin. *J. Antibiot. (Tokyo)* 43:1403–12
58. Fredenhagen, A., Marki, F., Fendrich, G., Marki, W., Gruner, J., et al. 1991. Duramycin B and C, two new lanthionine-containing antibiotics as inhibitors of phospholipase A2, and structural revision of duramycin and cinnamycin. See Ref. 109a, pp. 131–40
59. Froseth, B. R., Herman, R. E., McKay, L. L. 1988. Cloning of nisin resistance determinant and replication origin on 7.6-kilobase *Eco*RI fragment of pNP40 from *Streptococcus lactis* subsp. *diacetylactis* DRC3. *Appl. Environ. Microbiol.* 54:2136–39
60. Froseth, B. R., McKay, L. L. 1991. Molecular characterization of the nisin resistance region of *Lactococcus lactis* subsp. *lactis* biovar diacetylactis DRC3. *Appl. Environ. Microbiol.* 57:804–11
61. Garcia-Bustos, J. F., Pezzi, N., Asensio, C. 1984. Microcin 7: purification and properties. *Biochem. Biophys. Res. Commun.* 119:779–85
62. Garcia-Bustos, J. F., Pezzi, N., Mendez, E. 1985. Structure and mode of action of microcin 7, an antibacterial peptide produced by *Escherichia coli*. *Antimicrob. Agents Chemother.* 27: 791–97
63. Garrido, M. C., Herrero, M., Kolter, R., Moreno, F. 1988. The export of the DNA replication inhibitor microcin B17 provides immunity for the host cell. *EMBO J.* 7:1853–62
64. Genilloud, O., Moreno, F., Kolter, R. 1989. DNA sequence, products, and transcriptional pattern of the genes involved in production of the DNA replication inhibitor microcin B17. *J. Bacteriol.* 171:1126–35
65. Gibbons, A. 1992. Epidemiology of drug resistance: implications for a post-antimicrobial era. *Science* 257:1050–55
66. Gibbs, B. M., Hurst, A. 1964. Limitations of nisin as a preservative in non-dairy foods. In *Microbial Inhibitors in Foods*, ed. J. Molin, pp. 151–65. Stockholm: Almqvist and Wiksell
67. Gireesh, T., Davidson, B. E., Hillier, A. J. 1992. Conjugal transfer in *Lactococcus lactis* of a 68 kbp chromosomal fragment containing the structural gene for the peptide bacteriocin nisin. *Appl. Environ. Microbiol.* 58: 1670–76
68. Gross, E. 1977. Alpha-beta unsaturated and related amino acids in peptides and proteins. In *Protein Cross-linking*, Vol. B, ed. M. Friedman, pp. 131–53. New York: Plenum
69. Gross, E. 1978. Polypeptide antibiotics. In *Antibiotics. Isolation, Separation and Purification*, ed. M. J. Weinstein, G. H. Wagman, pp. 415–62. New York: Elsevier
70. Gross, E., Kiltz, H. H. 1973. The number and nature of alpha,beta-unsaturated amino acids in subtilin. *Biochem. Biophys. Res. Commun.* 50: 559–65
71. Gross, E., Kiltz, H. H., Nebelin, E. 1973. Subtilin part 6: the structure of subtilin. *Hoppe Seyler's Z. Physiol. Chem.* 354:810–12
72. Gross, E., Morell, J. L. 1968. The

number and nature of alpha,beta-unsaturated amino acids in nisin. *FEBS Lett.* 2:61–64
73. Gross, E., Morell, J. L. 1970. Nisin: the assignment of sulfide bridges of beta-methyllanthionine to a novel bicyclic structure of identical ring size. *J. Am. Chem. Soc.* 92:2919–20
74. Gross, E., Morell, J. L. 1971. Peptide with alpha,beta-unsaturated acids. In *Peptides 1969,* ed. E. Scoffone, pp. 356–60. Amsterdam: North-Holland
75. Gross, E., Morell, J. L. 1971. The structure of nisin. *J. Am. Chem. Soc.* 93:4634–35
76. Gross, E., Morell, J. L., Craig, L. C. 1969. Dehydroalanyllysine: identical COOH-terminal structures in the peptide antibiotics nisin and subtilin. *Proc. Natl. Acad. Sci. USA* 62:952–56
77. Gross, E., Morrell, J. L. 1967. The presence of dehydroalanine in the antibiotic nisin and its relationship to activity. *J. Am. Chem. Soc.* 89:2791–92
78. Hansen, J. N., Chung, Y. J., Liu, W., Steen, M. J. 1991. Biosynthesis and mechanism of action of nisin and subtilin. See Ref. 109a, pp. 287–302
79. Harold, F. M., Baarda, J. R. 1967. Gramicidin, valinomycin, and cation permeability of *Streptococcus faecalis. J. Bacteriol.* 94:53–60
80. Harris, L. J., Fleming, H. P., Klaenhammer, T. R. 1992. Developments in nisin research. *Food Res. Int.* 25:57–66
81. Hartl, F. U., Lecker, S., Schiebel, E., Hendrick, J. P., Wickner, W. 1990. The binding cascade of SecB to SecA to SecY/E mediates preprotein targeting to the E. coli plasma membrane. *Cell* 63:269–79
82. Hawkins, G. 1990. *Investigation of the site and mode of action of the small protein antibiotic subtilin and development and characterization of an expression system for the small protein antibiotic nisin in* Bacillus subtilis. PhD thesis. College Park, MD: Univ. Maryland
83. Hedge, P. J., Spratt, B. G. 1985. Resistance to beta-lactam antibiotics by remodelling the active site of an *E. coli* penicillin-binding protein. *Nature* 318:478–80
84. Henderson, J. T., Chopko, A. L., van Wassenaar, P. D. 1992. Purification and primary structure of pediocin PA-1 produced by *Pediococcus acidilactici* PAC-1.0. *Arch. Biochem. Biophys.* 295:5–12
85. Hendrick, J. P., Wickner, W. 1991. SecA protein needs both acidic phospholipids and SecY/E protein for functional high-affinity binding to the *Escherichia coli* plasma membrane. *J. Biol. Chem.* 266:24596–24600
86. Hernandez-Chico, C., Herrero, M., Rejas, M., San Millan, J. L., Moreno, F. 1982. Gene *ompR* and regulation of microcin 17 and colicin e2 syntheses. *J. Bacteriol.* 152:897–900
87. Hirsch, A. 1950. The assay of the antibiotic nisin. *Int. J. Food Microbiol.* 4:70–83
88. Hirsch, A., Grinsted, E. 1951. The differentiation of the lactic streptococci and their antibiotics. *J. Dairy Res.* 18:198–204
89. Holland, I. B., Kenny, B., Blight, M. 1990. Haemolysin secretion from *E. coli. Biochimie* 72:131–41
90. Holo, H., Nilssen, O., Nes, I. F. 1991. Lactococcin A, a new bacteriocin from *Lactococcus lactis* subsp. *cremoris:* isolation and characterization of the protein and its gene. *J. Bacteriol.* 173:3879–87
91. Horn, N., Swindell, S., Dodd, H. 1991. Nisin biosynthesis genes are encoded by a novel conjugative transposon. *Mol. Gen. Genet.* 228:129–35
92. Hultmark, D., Engstrom, A., Bennich, H., Kapur, R., Boman, H. G. 1982. Insect immunity: isolation and structure of cecropin D and four minor antibacterial components from *Cecropia* pupae. *Eur. J. Biochem.* 127:207–17
93. Hurst, A. 1966. Biosynthesis of the antibiotic nisin by whole *Streptococcus lactis* organisms. *J. Gen. Microbiol.* 44:209–20
94. Hurst, A. 1967. Function of nisin and nisini-like basic proteins in the growth cycle of *Streptococcus lactis. Nature* 214:1232–34
95. Hurst, A. 1972. Interactions of food-starter cultures and food-borne pathogens: the antagonism between *Streptococcus lactis* and sporeforming microbes. *J. Milk Food Technol.* 35:418–23
96. Hurst, A. 1972. Nisin: its preservative effect and function in the growth cycle of the producer organism. In *Streptococci,* ed. F. A. Skinner, L. B. Quesnel, pp. 297–314. London: Academic
97. Hurst, A. 1981. Nisin. *Adv. Appl. Microbiol.* 27:85–123
98. Hurst, A. 1983. Nisin and other inhibitory substances from lactic acid bacteria. *Food Sci.* 10:327–51
99. Hurst, A., Collins-Thompson, D. 1979. Food as a bacterial habitat. *Adv. Microb. Ecol.* 3:79–133

100. Hurst, A., Dring, G. J. 1968. The relation of the length of lag phase of growth to the synthesis of nisin and other basic proteins by *Streptococcus lactis* grown under different culture conditions. *J. Gen. Microbiol.* 50:383–90
101. Hurst, A., Kruse, H. 1972. Effect of secondary metabolites on the organisms producing them: effect of *Streptococcus lactis* and enterotoxin B on *Staphylococcus aureus*. *Antimicrobial. Agents Chemother.* 1:277–79
102. Hurst, A., Peterson, G. M. 1971. Observations on the conversion of an inactive precursor protein to the antibiotic nisin. *Can. J. Microbiol.* 17: 1379–84
103. Ingram, L. 1969. Synthesis of the antibiotic nisin: formation of lanthionine and beta-methyl-lanthionine. *Biochim. Biophys. Acta* 184:216–19
104. Ingram, L. 1970. A ribosome mechanism for synthesis of peptides related to nisin. *Biochim. Biophys. Acta* 224: 263–65
105. Jarvis, B. 1967. Resistance to nisin and production of nisin-inactivating enzymes by several species of *Bacillus*. *J. Gen. Microbiol.* 47:33–48
106. Jarvis, B., Farr, J. 1971. Partial purification, specificity, and mechanism of action of the nisin-inactivating enzyme from *Bacillus cereus*. *Biochim. Biophys. Acta* 227:232–40
107. Jarvis, B., Mahoney, R. R. 1969. Inactivation of nisin by alpha-chymotrypsin. *J. Dairy Sci.* 52:1448–50
108. Jay, J. M. 1983. Anitbiotics as food preservatives. In *Food Microbiology*, ed. A. H. Rose, 8:117–43. London: Academic
109. Jung, G. 1991. Lantibiotics: a survey. See Ref. 109a, pp. 1–34
109a. Jung, G., Sahl, H.-G., eds. 1991. *Nisin and Novel Lantibiotics*. Leiden, the Netherlands: ESCOM
110. Kaletta, C., Entian, K. D., Jung, G. 1991. Prepeptide sequence of cinnamycin (Ro 09-0198): the first structural gene of a duramycin-type lantibiotic. *Eur. J. Biochem.* 199:411–15
111. Kaletta, C., Entian, K. D., Kellner, R., Jung, G., Reis, M., et al. 1989. Pep5, a new lantibiotic: structural gene isolation and prepeptide sequence. *Arch. Microbiol.* 152:16–19
112. Kellner, R., Jung, G., Horner, T., Zahner, H., Schnell, N., et al. 1988. Gallidermin: a new lanthionine-containing polypeptide antibiotic. *Eur. J. Biochem.* 177:53–59
113. Kellner, R., Jung, G., Horner, T., Zahner, H., Schnell, N., et al. 1988. Gallidermin: a new lanthionine-containing polypeptide antibiotic. *Eur. J. Biochem.* 177:53–59
114. Kellner, R., Jung, G., Josten, M., Kaletta, C., Entian, K.-D., et al. 1989. Pep5: structure elucidation of a large lantibiotic. *Angew. Chem. Int. Ed. Engl.* 28:616–19
115. Kessler, H., Seip, S., Wein, T., Steuernagel, S., Will, M. 1991. Structure of cinnamycin (Ro 09-0198). See Ref. 109a, pp. 76–90
116. Kettenring, J. K., Malabarba, A., Vekey, K., Cavalleri, B. 1990. Sequence determination of actagardine, a novel lantibiotic, by homonuclear 2D NMR spectroscopy. *J. Antibiot. (Tokyo)* 43:1082–88
117. Klaenhammer, T. R. 1988. Bacteriocins of lactic acid bacteria. *Biochimie* 70: 337–49
118. Kleinkauf, H., von Dohren, H. 1990. Nonribosomal biosynthesis of peptide antibiotics. *Eur. J. Biochem.* 192:1–15
119. Kogler, H., Bauch, M., Fehlhaber, H.-W., Griesinger, C., Schubert, W., et al. 1991. NMR-spectroscopic investigations on mersacidin. See Ref. 109a, pp. 159–70
120. Kordel, M., Benz, R., Sahl, H. G. 1988. Mode of action of the staphylococcinlike peptide Pep 5: voltage dependent depolarization of bacterial and artificial membranes. *J. Bacteriol.* 170:84–88
121. Kordel, M., Schuller, F., Sahl, H. G. 1989. Interaction of the pore-forming peptide antibiotics Pep 5, nisin and subtilin with non-energized liposomes. *FEBS Lett.* 244:99–102
122. Koshland, D. E. J. 1992. The microbial wars. *Science* 257:1021
123. Krause, R. M. 1992. The origin of plagues: old and new. *Science* 257:1073–78
124. Kuipers, O. P., Rollema, H. S., Yap, W. M. G. J., Boot, H. J., Siezen, R. J., et al. 1992. Engineering dehydrated amino acid residues in the antimicrobial peptide nisin. *J. Biol. Chem.* 267:24340–46
125. Kuipers, O. P., Wyanda, M. G., Yap, H. S., Rollema, M. M., Beerthuyzen, R. J. S., et al. 1991. Expression of wild-type and mutant nisin genes in *Lactococcus lactis*. See Ref. 109a, pp. 250–59
126. Kuntz, I. D. 1992. Structure-based strategies for drug design and discovery. *Science* 257:1078–82
127. Lavina, M., Gaggero, C., Moreno, F.

1990. Microcin H47, a chromosome-encoded microcin antibiotic of *Escherichia coli*. *J. Bacteriol.* 172: 6585–88
128. Lee, J. Y., Boman, A., Sun, C. X., Andersson, M., Jornvall, H., et al. 1989. Antibacterial peptides from pig intestine: isolation of a mammalian cecropin. *Proc. Natl. Acad. Sci. USA* 86:9159–62
129. Li, Z. Q., Merrifield, R. B., Boman, I. A., Boman, H. G. 1988. Effects on electrophoretic mobility and antibacterial spectrum of removal of two residues from synthetic sarcotoxin IA and addition of the same residues to cecropin B. *FEBS Lett.* 231:299–302
130. Liu, W., Hansen, J. N. 1990. Some chemical and physical properties of nisin, a small protein antibiotic produced by *Lactococcus lactis*. *Appl. Environ. Microbiol.* 56:2551–58
131. Liu, W., Hansen, J. N. 1991. Conversion of *Bacillus subtilis* 168 to a subtilin producer by competence transformation. *J. Bacteriol.* 173:7387–90
132. Liu, W., Hansen, J. N. 1992. Enhancement of the chemical and antimicrobial properties of subtilin by site-directed mutagenesis. *J. Biol. Chem.* 267:25078–85
133. Liu, W., Hansen, J. N. 1993. The antimicrobial effect of mutagenized subtilin against outgrowing *Bacillus cereus* T spores and vegetative cells is caused by different mechanisms. *Appl. Environ. Microbiol.* 59:648–51
134. Macias, E. A., Rana, F., Blazyk, J., Modrzakowski, M. C. 1990. Bactericidal activity of magainin 2: use of lipopolysaccharide mutants. *Can. J. Microbiol.* 36:582–84
135. Malabarba, A., Landi, M., Pallanza, R., Cavalleri, B. 1985. Physico-chemical and biological properties of actagardine and some acid hydrolysis products. *J. Antibiot. (Tokyo)* 38:1506–11
136. Malabarba, A., Pallanza, R., Berti, M., Cavalleri, B. 1990. Synthesis and biological activity of some amide derivatives of the lantibiotic actagardine. *J. Antibiot. (Tokyo)* 43:1089–97
137. Marki, F., Hanni, E., Fredenhagen, A., van Oostrum, J. 1991. Mode of action of the lanthionine-containing peptide antibiotics duramycin, duramycin B and C, and cinnamycin as indirect inhibitors of phospholipase A2. *Biochem. Pharmacol.* 42:2027–35
138. Martinez, J. L., Perez-Diaz, J. C. 1986. Isolation, characterization, and mode of action on *Escherichia coli* strains of microcin D93. *Antimicrob. Agents Chemother.* 29:456–60
139. Marugg, J. D., Gonzalez, C. F., Kunka, B. S., Ledeboer, A. M., Pucci, M. J., et al. 1992. Cloning, expression, and nucleotide sequence of genes involved in production of pediocin PA-1, and bacteriocin from *Pediococcus acidilactici* PAC1.0. *Appl. Environ. Microbiol.* 58:2360–67
140. Matsuzaki, K., Harada, M., Funakoshi, S., Fujii, N., Miyajima, K. 1991. Physicochemical determinants for the interactions of magainins 1 and 2 with acidic lipid bilayers. *Biochim. Biophys. Acta* 1063:162–70
141. Matsuzaki, K., Harada, M., Handa, T., Funakoshi, S., Fujii, N., et al. 1989. Magainin 1-induced leakage of entrapped calcein out of negatively-charged lipid vesicles. *Biochim. Biophys. Acta* 981:130–34
142. Mattick, A. T. R., Hirsch, A. 1944. A powerful inhibitory substance produced by group N streptococci. *Nature* 154:551–52
143. Mattick, A. T. R., Hirsch, A. 1946. Sour milk and the tubercle bacillus. *Lancet* 1:417–18
144. Mattick, A. T. R., Hirsch, A. 1947. Further observations on an inhibitory substance (nisin) from lactic streptococci. *Lancet* 2:5–12
145. Merrifield, R. B., Vizioli, L. D., Boman, H. G. 1982. Synthesis of the antibacterial peptide cecropin A (1–33). *Biochemistry* 21:5020–31
146. Morell, J. L., Gross, E. 1973. Configuration of the beta-carbon atoms of the beta-methyllanthionine residues in nisin. *J. Am. Chem. Soc.* 95:6480–81
147. Morishima, I., Suginaka, S., Ueno, T., Hirano, H. 1990. Isolation and structure of cecropins, inducible antibacterial peptides, from the silkworm, *Bombyx mori*. *Comp. Biochem. Physiol. B.* 95:551–54
148. Morris, S. L., Walsh, R. C., Hansen, J. N. 1984. Identification and characterization of some bacterial membrane sulfhydryl groups which are targets of bacteriostatic and antibiotic action. *J. Biol. Chem.* 259:13590–94
149. Mulders, J. W., Boerrigter, I. J., Rollema, H. S., Siezen, R. J., de Vos, W. M. 1991. Identification and characterization of the lantibiotic nisin Z, a natural nisin variant. *Eur. J. Biochem.* 201:581–84
150. Muriana, P. M., Klaenhammer, T. R. 1991. Cloning, phenotypic expression, and DNA sequence of the gene for

lactacin F, an antimicrobial peptide produced by *Lactobacillus* spp. *J. Bacteriol.* 173:1779–88

151. Muriana, P. M., Klaenhammer, T. R. 1991. Purification and partial characterization of lactacin F, a bacteriocin produced by *Lactobacillus acidophilus* 11088. *Appl. Environ. Microbiol.* 57: 114–21
152. Naruse, N., Tenmyo, O., Tomita, K., Konishi, M., Miyaki, T., et al. 1989. Lanthiopeptin, a new peptide antibiotic. Production, isolation and properties of lanthiopeptin. *J. Antibiot. (Tokyo)* 42: 837–45
153. Navarro, J., Chabot, J., Sherrill, K., Aneja, R., Zahler, S. A., et al. 1985. Interaction of duramycin with artificial and natural membranes. *Biochemistry* 24:4645–50
154. Neu, H. C. 1992. The crisis in antibiotic resistance. *Science* 257:1064–72
155. Nishio, C., Komura, S., Kurahashi, K. 1983. Peptide antibiotic subtilin is synthesized via precursor proteins. *Biochem. Biophys. Res. Commun.* 116: 751–58
156. Niu, W. W., Neu, H. C. 1991. Activity of mersacidin, a novel peptide, compared with that of vancomycin, teicoplanin, and daptomycin. *Antimicrob. Agents Chemother.* 35:998–1000
157. Novoa, M. A., Diaz-Guerra, L., San Millan, J. L., Moreno, F. 1986. Cloning and mapping of the genetic determinants for microcin C7 production and immunity. *J. Bacteriol.* 168:1384–91
158. Okai, Y., Qu, X. M. 1989. Effects of a purified cecropin D from a Chinese silk moth on growth, function and differentiation of murine hemopoietic cells. *Immunol. Lett.* 20:127–32
159. Oropeza-Wekerle, R. L., Speth, W., Imhof, B., Gentschev, I., Goebel, W. 1990. Translocation and compartmentalization of *Escherichia coli* hemolysin (Hly A). *J. Bacteriol.* 172:3711–17
160. Qu, Z., Steiner, H., Engstrom, A., Bennich, H., Boman, H. G. 1982. Insect immunity: isolation and structure of cecropins B and D from pupae of the Chinese oak silk moth, *Antheraea pernyi*. *Eur. J. Biochem.* 127:219–24
161. Rana, F. R., Blazyk, J. 1991. Interactions between the antimicrobial peptide, magainin 2, and *Salmonella typhimurium* lipopolysaccharides. *FEBS Lett.* 293:11–15
162. Rana, F. R., Sultany, C. M., Blazyk, J. 1990. Interactions between *Salmonella typhimurium* lipopolysaccharide and the antimicrobial peptide, magainin 2 amide. *FEBS Lett.* 261:464–67
163. Rauch, P. J., de Vos, W. M. 1992. Characterization of the novel nisin-sucrose conjugative transposon Tn*5276* and its insertion in *Lactococcus lactis*. *J. Bacteriol.* 174:1280–87
164. Rauch, P. J. G., de Vos, W. M. 1992. Transcriptional regulation of the Tn*5276*-located *Lactococcus lactis* sucrose operon and characterization of the *sacA* gene encoding sucrose-6-phosphate hydrolase. *Gene* 121:55–61
165. Rayman, K., Hurst, A. 1984. Nisin: properties, biosynthesis, and fermentation. In *Biotechnology of Industrial Antibiotics*, ed. E. J. Vandamme, pp. 607–28. New York: Marcel Decker
166. Raymond, M., Gros, P., Whiteway, M., Thomas, D. Y. 1992. Functional complementation of yeast *ste6* by a mammalian multidrug resistance *mdr* gene. *Science* 256:232–34
167. Reynolds, J., Cashman, S. J., Evans, D. J., Pusey, C. D. 1991. Cyclosporin A in the prevention and treatment of experimental autoimmune glomerulonephritis in the brown Norway rat. *Clin. Exp. Immunol.* 85:28–32
168. Ristow, H., Schazschneider, B., Bauer, K., Kleikauf, H. 1975. Tyrocidine and the linear gramicidin. Do these peptide antibiotics play an antagonistic regulative role in sporulation? *Biochim. Biophys. Acta* 390:246–52
169. Ristow, H., Schazschneider, B., Vater, J., Kleinkauf, H. 1975. Some characteristics of the DNA-tyrocidine complex and a possible mechanism of the gramicidin action. *Biochim. Biophys. Acta* 414:1–8
170. Rodriguez Lemoine, V. 1984. Detection of microcin production by Enterobacteriaceae. *Acta Cient. Venez.* 35:451–52
171. Rogers, L. A. 1928. The inhibiting effect of *Streptococcus lactis* on *Lactobacillus bulgaris*. *J. Bacteriol.* 16: 321–25
172. Ruhr, E., Sahl, H.-G. 1985. Mode of action of the peptide antibiotic nisin and influence on the membrane potential of whole cells and on cytoplasmic and artificial membrane visicles. *Antimicrob. Agents Chemother.* 27:841–45
173. Sahl, H.-G. 1985. Bacteriocidal cationic peptides involved in bacterial antagonism and host defence. *Microbiol. Sci.* 2:212–17
174. San Millan, J. L., Hernandez-Chico, C., Pereda, P., Moreno, F. 1985. Cloning and mapping of the genetic determinants for microcin B17 produc-

tion and immunity. *J. Bacteriol.* 163: 275–81

175. Schnell, N., Engelke, G., Augustin, J., Rosenstein, R., Ungermann, V., et al. 1992. Analysis of genes involved in the biosynthesis of lantibiotic epidermin. *Eur. J. Biochem.* 204:57–68
176. Schnell, N., Entian, K. D., Gotz, F., Horner, T., Kellner, R., et al. 1989. Structural gene isolation and prepeptide sequence of gallidermin, a new lanthionine containing antibiotic. *FEMS Microbiol. Lett.* 49:263–67
177. Schnell, N., Entian, K. D., Schneider, U., Gotz, F., Zahner, H., et al. 1988. Prepeptide sequence of epidermin, a ribosomally synthesized antibiotic with four sulphide-rings. *Nature* 333:276–78
178. Schuller, F., Benz, R., Sahl, H. G. 1989. The peptide antibiotic subtilin acts by formation of voltage-dependent multi-state pores in bacterial and artificial membranes. *Eur. J. Biochem.* 182:181–86
179. Scott, V. N., Taylor, S. L. 1981. Effect of nisin on the outgrowth of *Clostridium botulinum* spores. *J. Food Sci.* 46:117–21
180. Sears, P. M., Smith, B. S., Rubino, S. D., Kulisek, E., Gusik, S., et al. 1991. Non-antibiotic approach to treatment of mastitis in the lactating dairy cow. *J. Dairy. Sci.* 74(Suppl. 1):203
181. Sheth, T. R., Henderson, R. M., Hladky, S. B., Cuthbert, A. W. 1992. Ion channel formation by duramycin. *Biochim. Biophys. Acta* 1107:179–85
182. Shiba, T., Wakamiya, T., Fukase, K., Ueki, Y., Teshima, T., et al. 1991. Structure of the lanthionine peptides nisin, ancovenin, and lanthiopeptin. See Ref. 109a, pp. 113–22
183. Skvirsky, R. C., Gilson, L., Kolter, R. 1991. Signal sequence-independent protein secretion in gram-negative bacteria: colicin V and microcin B17. *Methods Cell Biol.* 34:205–21
184. Song, J. I., Dumais, M. R. 1991. From yew to us: the curious development of taxol. *J. Am. Med. Assoc.* 266:1281
185. Steen, M., Chung, Y. J., Hansen, J. N. 1991. Characterization of the nisin gene as part of a polycistronic operon in chromosome of *Lactococcus lactis* 11454. *Appl. Environ. Microbiol.* 57: 1181–88
186. Steiner, H., Andreu, D., Merrifield, R. B. 1988. Binding and action of cecropin and cecropin analogues: antibacterial peptides from insects. *Biochim. Biophys. Acta* 939:260–66
187. Stevens, K. A., Klapes, N. A., Sheldon, B. W., Klaenhammer, T. R. 1992. Antimicrobial action of nisin against *Salmonella typhimurium* lipopolysaccharide mutants. *Appl. Environ. Microbiol.* 58:1786–88
188. Stevens, K. A., Sheldon, B. W., Klapes, N. A., Klaenhammer, T. R. 1991. Nisin treatment for inactivation of *Salmonella* species and other gram-negative bacteria. *Appl. Environ. Microbiol.* 57:3613–15
189. Stoddard, G. W., Petzel, J. P., van Belkum, M. J., Kok, J., McKay, L. L. 1992. Molecular analyses of the lactococcin A gene cluster from *Lactococcus lactis* subsp. *lactis* biovar diacetylactis WM4. *Appl. Environ. Microbiol.* 58:1952–61
190. Tagg, J. R., Dajani, A. S., Wannamaker, L. W. 1976. Bacteriocins of Gram-positive bacteria. *Bacteriol. Rev.* 40: 722–56
191. Tanaka, T., Kuroda, K., Sakaguchi, K. 1977. Isolation and characterization of four plasmids from *Bacillus subtilis*. *J. Bacteriol.* 129:1487–94
192. Taylor, J. I., Hirsch, A., Mattick, A. T. R. 1949. The treatment of bovine streptococcal and staphylococcal mastitis with nisin. *Vet. Rec.* 61:197–98
193. Tryselius, Y., Samakovlis, C., Kimbrell, D. A., Hultmark, D. 1992. CecC, a cecropin gene expressed during metamorphosis in *Drosophila* pupae. *Eur. J. Biochem.* 204:395–99
194. van Belkum, M. J., Kok, J., Venema, G., Holo, H., Nes, I. F., et al. 1991. The bacteriocin lactococcin A specifically increases permeability of lactococcal cytoplasmic membranes in a voltage-independent, protein-mediated manner. *J. Bacteriol.* 173:7934–41
195. van Hofsten, P., Faye, I., Kockum, K., Lee, J. Y., Xanthopoulos, K. G., et al. 1985. Molecular cloning, cDNA sequencing, and chemical synthesis of cecropin B from *Hyalophora cecropia*. *Proc. Natl. Acad. Sci. USA* 82:2240–43
196. Wakamiya, T., Ueki, Y., Shiba, T., Kido, Y., Motoki, Y. 1985. The structure of ancovenin, a new peptide inhibitor of angiotensin I converting enzyme. *Tetrahedron Lett.* 26:665–68
197. Williams, R. W., Starman, R., Taylor, K. M., Gable, K., Beeler, T., et al. 1990. Raman spectroscopy of synthetic antimicrobial frog peptides magainin 2a and PGLa. *Biochemistry* 29:4490–96
198. Yousef, A. E., Luchansky, J. B., Degnan, A. J., Doyle, M. P. 1991. Behavior of *Listeria monocytogenes* in wiener exudates in the presence of *Pediococcus acidilactici* H or pediocin

AcH during storage at 4 or 25 degrees C. *Appl. Environ. Microbiol.* 57:1461–67

199. Zasloff, M. 1987. Magainins, a class of antimicrobial peptides from *Xenopus* skin: isolation, characterization of two active forms, and partial cDNA sequence of a precursor. *Proc. Natl. Acad. Sci. USA* 84:5449–53
200. Zasloff, M., Martin, B., Chen, H. C. 1988. Antimicrobial activity of synthetic magainin peptides and several analogues. *Proc. Natl. Acad. Sci. USA* 85:910–13

Annu. Rev. Microbiol. 47:565–96

STRUCTURE-FUNCTION AND BIOGENESIS OF THE TYPE IV PILI

Mark S. Strom[1] *and Stephen Lory*
Department of Microbiology, SC-42, University of Washington, Seattle, Washington 98195

KEY WORDS: type IV pili, biogenesis, peptidase, methyltransferase, regulation

---

CONTENTS

[1]Current address: Utilization Research Division, Northwest Fisheries Science Center, National Marine Fisheries Service, 2725 Montlake Blvd. E., Seattle, WA 98112.

0066-4227/93/1001-0565$02.00

ABSTRACT

Type IV pili are adhesins expressed by a number of diverse gram-negative microorganisms. These pili are related through similarities in the primary amino acid sequences of the structural subunits, a conserved assembly machinery, and a similar mechanism of transcriptional regulation. Type IV pilus assembly is preceded by proteolytic processing and *N*-methylation of the pilin polypeptide. This process is carried out by a novel bifunctional enzyme PilD, first identified in *Pseudomonas aeruginosa*. Moreover, proteins homologous with type IV pilins have been shown to function in extracellular protein secretion in gram-negative bacteria and in transformation competence in gram-positive microorganisms. Like prepilin, these proteins are also processed and *N*-methylated by PilD. Transcription of the genes for type IV pilins is carried out by an RNA polymerase with a minor sigma factor, RpoN. In *P. aeruginosa* two other regulatory elements (PilS and PilR) are required for pilin expression. RpoN, but not PilS and PilR, is required for expression of a diverse set of bacterial genes. Therefore, regulation of synthesis and posttranslational modification and assembly of type IV pili serves as a useful model for a number of diverse biological processes in the bacterial cell.

INTRODUCTION

Many bacteria display surface appendages called pili (fimbriae) that function by mediating interaction of bacteria with surfaces of other cells. Conjugal DNA transfer is preceded by recognition of the recipient cell by the pili of the donor cell, whereas attachment of pathogenic bacteria to membranes of eukaryotic cells is mediated by the lectin component of pili. In addition to these adhesive functions, pili have been implicated in a mechanism of bacterial locomotion, called twitching motility.

Investigators have identified distinct families of pili by determining the specificity of host-receptor recognition and the seroreactivity of antibodies against pilin proteins and by comparing deduced amino acid sequences of cloned pilin structural genes. The resulting pili families in some cases include pili from taxonomically dissimilar bacteria.

This review attempts to summarize the current knowledge of pili collectively designated type IV pili. Several bacterial species produce these structures, and although they have not been implicated in conjugal DNA transfer, they are important virulence factors because of their role as bacterial adhesins. Moreover, recent work has revealed that proteins with extensive sequence similarity with the type IV pilins are essential components of biological processes such as extracellular protein secretion from gram-negative bacteria and DNA uptake by gram-positive microorganisms. Two important properties associated with expression of type IV pili, antigenic and phase variation, are not discussed here because they have been subject of several recent reviews (66, 94).

CHARACTERISTICS OF THE TYPE IV PILI FAMILY

Pili are grouped from a rather divergent collection of gram-negative microorganisms to form the type IV class according to similarities in amino acid sequence of the pilin polypeptide. The second conserved characteristic is the occurrence of *N*-methylated amino acids (phenylalanine or methionine) as the first amino acid of the mature pilin structural subunit. Alignment of amino acid sequences of the precursors of the type IV pilins (Figure 1) demonstrates the basis for the assignment of these proteins into a related family. The type IV pilin family can be further divided into two groups. Group A consists of pilins from *Pseudomonas aeruginosa* (40), *Neisseria gonorrhoeae* (65), *N. meningitidis* (82), *Moraxella bovis* (59), and *Dichelobacter* (formerly *Bacteroides*) *nodosus* (63). More recently, this class was expanded to include the pili of *Moraxella lacunata* (58), *Moraxella nonliquefaciens* (116), *Branhamella catarrhalis* (60), and *Eikenella corrodens* (89, 117). The pilins in this group are synthesized as precursors with unique short, basic, amino-terminal leader peptides. These leader sequences are removed by endoproteolytic cleavage between an invariant glycine residue and a phenylalanine residue prior to assembly of the pilin monomers into pili (Figure 1). The amino acid sequence homology of this family extends into the mature polypeptide and is most pronounced near the amino terminus. Because the amino-terminal domain of these polypeptides is highly conserved, it is often referred to as the constant domain. This domain has a strong hydrophobic character interrupted by an invariant glutamic acid residue located exactly five amino acids from the mature amino terminus. The middle of the polypeptides are considerably less homologous and contain the variable domains that comprise the antigenic epitopes of these pili. Another region of homology near the carboxy terminus contains a characteristic pair of cysteines forming a disulfide loop.

The second group of the type IV pilin family, group B, currently has two members, the subunits of the toxin-coregulated pili (TCP) of *Vibrio cholerae* (26, 95), and the subunits of bundle-forming pili (BFP) (19) of enteropathogenic *Escherichia coli*. These proteins are homologous with each other and with the group A members near their amino termini, including the location of the invariant glutamic acid within the hydrophobic region. Although the carboxy-terminal regions of both TcpA (the TCP subunit) and BfpA (the BFP subunit) differ, they do contain a pair of cysteines that presumably form an intrachain disulfide bond.

```
                                              ▼
A  Pa PilA   ------------------MKAQKGFTLIELMIVVAIIGILAAIAIPQYQNYVAR   36
   Ek EcpA   m-----------------KQVQKGFTLIELMIVIAIIGILAAIALPLYQDYISK   37
   Mb TfpQ   ------------------MNAQKGFTLIELMIVIAIIGILAAIALPAYQDYISK   36
   Mn TfpA   ------------------MNAQKGFTLIELMIVIAIIGILAAIALPAYQDYIAR   36
   Dn FimA   m-----------------KSLQKGFTLIELMIVVAIIGILAAFAIPAYNDYIAR   37
   Ng PilE   m-----------------NTLQKGFTLIELMIVIAIVGILAAVALPAYQDYTAR   37
   Nm PilE   m-----------------NTLQKGFTLIELMIVIAIVGILAAVALPAYQDYTAR   37

B  Vc TcpA   mqllkqlfkkkfvkeehdkKTGQEGMTLLEVIIVLGIMGVVSAGVVTLAQRAIDS   55
   Ec BfpA   mvskimn-----------KKYEKGLSLIESAMVLALAATVTAGVMFYYQSASDS   43

   Pa PilA   SEGASALASVNPLKTTVEEALSRGWSVKSGTGTedatkkevplgvaadanklgti   91
   Ek EcpA   SQVTRAYGEMAGTKTAIEAALFEGRTPVLAATAaagaaatppnewvgmldnpt-S   91
   Mb TfpQ   SQTTRVVGELAAGKTAVDAALFEGKTPKLGKAAndteediglttggtarsnlms   91
   Mn TfpA   AQVSEAFTLADGLKTSISTnrqngrcfadgkdtaadgvdiitgkygkatileenp   91
   Dn FimA   SQAAEGVSLADGLKVRIAEnlqdgeckgpdadpasgvvgnkdtgkyalaeidgty   92
   Ng PilE   AQVSEAILLAEGQKSAVTEYYLNHGKWPENNTSAGVASPPsDIKGKYVKEVEVKN   92
   Nm PilE   AQVSEAILLAEGQKSAVTEYYLNHGEWPGNNTSAGVATSS-EIKGKYVKSVEVKN   91

   Vc TcpA   QIMTKAAQSLNSIQVALTQtyrglgnypatadataaskltsglvslgkissdeak   110
   Ec BfpA   NKSQNAISEVMSATSAINGlyigqtsysgldstillntsaipdnykdttnkkitn   98

   Pa PilA   alkpdpadgtadit-----LTFTMGGAGPKNKGKIITLtrtaadglwkc------   135
   Ek EcpA   NLLSAATLTPGANAGDVTFVGTLGENANSSIHGATITLTCTASGEWTCAVAA-GT   145
   Mb TfpQ   svnigggaf---ATGAGTLEATLGNRANKDIAGAVITQSRDAEGVWTCTING-SA   142
   Mn TfpA   ntadglicgiyye--------FNTTGVSDKLIGKTIALkadekagklvletvnSK   138
   Dn FimA   dasktaagdpngckvnitygqgtaadkisklitgkklvldqlvngsfiq----GD   143
   Ng PilE   GVVTATMLSSGVN--------------NEIKGKKLSLWARRENGSVKWFCGQPV   132
   Nm PilE   GVVTATMLSSGVN--------------KEIKGKKLSLWAKRQNGSVKWFCGQPV   131

   Vc TcpA   npfngtnmnifsfprnaaankafaisvdgltqaqcktlitsvgdmfpyiaikagg   165
   Ec BfpA   pfggelnvgpannntafgyyltltrldkaacvslatlnlgtsakgygvnisgenn   153

   Pa PilA   TSDQDEQFIPKGCSr----------------------------------------   150
   Ek EcpA   ATGWKTKFVPSGCN-----------------------------------------   159
   Mb TfpQ   APGWKSKFVPTGCKe----------------------------------------   157
   Mn TfpA   TTNVENKYLPSAFKkp---------------------------------------   154
   Dn FimA   GTDLADKFIPNAVKakk--------------------------------------   160
   Ng PilE   TRT---DDDTVADAKDGKEIDTKHLPSTCRDkasdak------------------   166
   Nm PilE   TRNdtdDTVAAVAADNTGNINTKHLPSTCRDasdas-------------------   167

   Vc TcpA   avaladlgdfensaaaaetgvgviksiapasknldltnithveklckgtapfgva   220
   Ec BfpA   itsfgnsadqaakstaitpaeaatackntdstnkvtyfmk---------------   193

   Vc TcpA   fqns                                                      224
```

Figure 1 Comparison of the amino acid sequences of representative group A and group B type IV pilins. Species abbreviations with the structural subunit designation are: Pa, *P. aeruginosa,* strain PAK (41); Ek, *E. corrodens,* strain ATCC 23834 (89); Mb, *M. bovis,* strain EPP63 (Q pilin) (28); Mn, *M. nonliquefaciens,* strain NCTC 7784 (116); Dn, *D. nodosus,* strain AC6 (7); Ng, *N. gonorrhoeae,* strain MS11 (6); Nm, *N. meningitidis,* strain C311 (82); Vc, *Vibrio cholerae,* strain Z17561, classical (26); Ec, *Escherichia coli,* E2348/69 (19). This alignment was done using MACAW (Multiple Alignment Construction and Analysis Workbench from the National Center for Biotechnology Information), which searches for and aligns regions of functional similarity using a strategy of pairwise comparison of multiple sequences (93). The shaded regions with the single-letter amino acid code in capital letters indicates the regions of highest similarity. White lettering on black indicates the presence of a particular block of similarity in more than 67% of the input sequences, white on gray between 37 and 66%, and black capital letters on white representing a homologous block found in at least two of the sequences. The cleavage site of the type IV pilins is denoted by an upside-down triangle.

The precursors of TcpA and BfpA are synthesized with longer leader peptides than those of group A: 25 and 13 amino acids, respectively, with a net basic character. The signal peptides of TcpA and BfpA end with glycine, as is the case with group A members. The first amino acid of mature TcpA is methionine, and that of BfpA is leucine, which contrasts with the presence of invariant phenylalanine found at the mature amino termini of group A members. The amino-terminal methionine of mature TcpA is *N*-methylated, but the modification of the amino-terminal leucine of BfpA has not been determined. The conservation of sequence near the amino terminus surrounding the cleavage site places these proteins into the type IV pilin family, and a further conservation of the machinery involved in biogenesis is expected. This machinery includes the leader peptidase and *N*-methyltransferase for removal of the leader peptide and methylation of the amino terminus, a common mechanism of secretion across the membrane and possibly a conserved mechanism for assembly of subunits into mature pili.

STRUCTURE OF THE TYPE IV PILUS ORGANELLE

The pili belonging to group A are flexible, approximately 6 nm in diameter, and have a length ranging from 1000 to 2500 nm. Each type IV pilus is estimated to contain 500–1000 subunits, based on their average length and assuming that they are indeed composed of a polymer of a single polypeptide. The pilin subunits of *P. aeruginosa* appear to be arranged in a helical manner, with approximately five subunits per turn around the long axis of the filament (76). More recently, the crystallization of *N. gonorrhoeae* pilin and subsequent X-ray diffraction analysis (77) has led to the proposal that each individual pilin subunit folds into an antiparallel 4–α-helix bundle similar to tobacco mosaic virus coat protein and myohemerythrin (29, 78). Dissociation of pili from either *P. aeruginosa* or *N. gonorrhoeae* with nonionic detergents yielded a stable dimeric form of the pilin subunit that appears to be the initial building block leading to the polymerized helical structure. The assembly therefore involves interaction of subunits to form dimers, followed by assembly of higher-order structures from the dimerized subunits. However, dimerized subunits have been isolated only from assembled pili, and dimerized pilin has not been shown to be an obligatory precursor of organelle assembly. A detailed biochemical analysis of *P. aeruginosa* pilin, involving alkaline pH titrations, solvent perturbation, quenching of tryptophan fluorescence with acrylamide, and circular dichroism, demonstrated that tyrosine residues at positions 24 and 27 in the hydrophobic domain are at a dimer/dimer interface in both native pili and in reassembled pilin filaments (122, 123). Dissociation of pili by octyl glucoside resulted in exposure of these tyrosines. These two residues are conserved in all type IV pilins of group A , but are not present in TcpA or BfpA.

The type IV pilins of group B (TcpA and BfpA) assemble into straight fibers approximately 7 nm wide and of variable length (30, 114). Little structural information is available on the arrangement of the subunits within the filament. One of the striking features of these pili is their propensity to aggregate laterally when expressed on the bacterial surface or when purified. Conceivably, aggregation of filaments between individual bacteria may allow for formation of microcolonies in their natural environment or in infected tissues.

ROLE OF TYPE IV PILI IN VIRULENCE

Adherence to Epithelial Cells

Type IV pili have been found on gram-negative pathogens that cause a variety of diseases in animals and humans. The contribution of pili to virulence lies primarily in their ability to promote attachment to various types of receptors during tissue colonization. In contrast, bacterial attachment to phagocytic cells, mediated by pili or through antipilin opsonic antibody, is an important host-defense mechanism. Evidence for the role of pili in pathogenesis is examined here for selected microorganisms.

P. AERUGINOSA *P. aeruginosa* possesses several adhesins that allow binding to a variety of epithelial cell types and mucins, which are amino sugars found in the secretions that bathe respiratory-tract cells. *P. aeruginosa* pili form one class of these adhesins and have been extensively studied from this perspective. Initial observations by Woods et al (126), who first suggested a role for pili in bacterial adherence, have been confirmed in several tissue culture systems. While the contribution of pili to virulence of *P. aeruginosa* has not been extensively investigated, piliated organisms were 10-fold more virulent than nonpiliated strains in a burned mouse model. The susceptibility of healthy mice to different strains did not vary regardless of the strain's piliation (92).

Pili can function in bacterial adhesion by direct interactions of the assembled pilin subunit with a tissue receptor. Alternatively, pili can incorporate minor pilin-like subunits that are responsible for receptor recognition and binding. Cumulative evidence suggests that pilin subunits do recognize certain cellular receptors. Many bacteria expressing type IV pili, however, do express other adhesins on their surface (79, 97).

Two approaches demonstrated the involvement of pili in bacterial adhesion. Monoclonal antibodies specific for defined regions of the pilin monomer, purified pili, and synthetic peptides corresponding to specific peptide sequences within the pilin subunit inhibited bacterial attachment to buccal and tracheal epithelial cells (16–18, 37, 52, 53). Moreover, peptides containing the cysteine-cysteine bridge region were more efficient in blocking bacterial adherence than peptides specific to other epitopes recognized by monoclonal antibodies. Also, antisera raised against oxidized peptides containing the cysteine residues inhibited adherence of both homologous and heterologous strains (37, 53). These studies suggested that the binding domain on the pilus for epithelial cells contained the cysteine-cysteine bridge, and that this region was also a conserved antigenic determinant.

Evidence from several laboratories has raised the possibility that additional adhesins, not associated with pili, function as *P. aeruginosa* adhesins. Loss of functional pili reduces adherence to cultured epithelial cells (13, 97) and bovine tracheal cells (91), but does not abolish it, while adhesion of *P. aeruginosa* to mucins, or carbohydrate components of mucin, is not affected by mutations in the pilin structural gene (87, 88). Interestingly, certain regulatory mutations that abolish expression of pili result in complete loss of bacterial adherence to most cells and to mucin (13, 87). This suggests the possibility that expression of multiple adhesins by *P. aeruginosa* is regulated by one or several common regulatory elements.

In addition to mediating adherence to epithelial cells or mucin, pili of *P. aeruginosa* may be the targets recognized by phagocytic cells. Speert and coworkers have demonstrated that the presence of pili on the surface of nonmucoid *P. aeruginosa* was required for nonopsonic phagocytosis by human neutrophils and monocyte-derived macrophages (98, 99). Furthermore, fibronectin-stimulated macrophages phagocytize *P. aeruginosa* cells grown in vivo, on nutrient agar plates, or in static broth (46–48), conditions known to allow bacteria to achieve maximal piliation. The same strains grown in agitated cultures were not taken up, and electron microscope examination showed that surface pili were missing, presumably because of the shearing action of the growth conditions. In addition, a pilin gene transposon mutant also was not phagocytized, leading to the conclusion that pili serve as ligands for fibronectin-stimulated macrophages and subsequent phagocytosis.

N. GONORRHOEAE Pili appear to play an essential role in infections of human hosts by *N. gonorrhoeae,* the organism responsible for the sexually transmitted disease gonorrhea (10). The organism adheres to a variety of cultured epithelial cells and, as with all such adhesins, it is postulated that the gonococcal pili facilitate the initial attachment to and colonization of human mucosal tissues prior to invasion and development of overt disease (108, 120). Evidence for involvement of pili in interaction of pathogen with host cells is based primarily on studies of in vitro attachment using organ and cell-culture systems. Piliated gonococci mediate adherence to cultured vaginal epithelial cells (57), human fallopian tubes (62), and erythrocytes (11). Monoclonal antibodies against peptides corresponding to constant and variable regions of the pilus prevent adherence to a human endometrial carcinoma cell line (90), suggesting that pilin polypeptides themselves mediate adhesion to cells. However, more recent studies have demonstrated that surface proteins other than pili mediate adherence of *N. gonorrhoeae* to specific glycolipids, such as lactosylceramide and gangliotriaosylceramide derived from a human cervical carcinoma (14, 107). A nonpilus adhesin has been isolated and cloned and has been shown to be a 36-kDa protein on the surface of gonococci that is not associated with pili (79).

DICHELOBACTER NODOSUS *D. nodosus* is responsible for ovine footrot, a contagious, debilitating disease of sheep (24). The common type IV pilus of *D. nodosus* has been established as the major host-protective immunogen (101), and vaccines prepared either from purified pili or piliated whole cells offer protection against homologous challenge (100). In fact, this is the only vaccine generated against a type IV pilin that has proven effective in prevention of disease. There are no reports addressing the role of pili in adherence. Moreover, it is not clear whether the efficacy of the pilus-based vaccine results from its ability to interfere with adherence or from increased opsonophagocytosis of the bacteria.

MORAXELLA BOVIS *M. bovis* is the causative agent of infectious bovine keratoconjunctivitis, the most common ocular disease of cattle worldwide. Pili are considered to be a virulence factor in the disease because only piliated organisms cause infections by mediating attachment (2, 12, 39, 68), under experimental conditions (55). Vaccination of cattle with pili of *M. bovis* has been shown to be protective against homologous challenge (54).

NEISSERIA MENINGITIDIS *N. meningitidis* is a causative agent of bacterial meningitis and sepsis, especially in children. Pili expressed by this pathogen fall into two distinct classes. Class I pili consist of pilin subunits that share serological and amino acid–sequence similarity to the members of the type IV family, whereas class II pili appear to be unique (82, 119). Interaction of piliated organisms bearing class I or class II pili with a variety of cultured cells was investigated (118). Class I pili were the primary determinant of binding to three different human epithelial cell lines and an endothelial cell line. To further investigate the correlation between piliation and adherence, Nassif et al (69) isolated variants of a nonpiliated *N. meningitidis* that express class I pili but differ in their ability to bind to human endometrial carcinoma cells. All adhering piliated meningococci expressed one antigenic variant while the nonadhering variants expressed another. Transformation of nonadhering meningococci with the pilin gene from adhering variants conferred adherence on the recipient. These data suggest that only certain pilin serotypes can function as adhesins.

PILI OF *V. CHOLERAE* AND ENTEROPATHOGENIC *E. COLI* Colonization of the small intestine by *V. cholerae* and subsequent production of toxins causes the clinical symptoms of cholera. The disease is characterized by a profuse diarrhea caused primarily by electrolyte imbalance following the actions of the cholera enterotoxin on intestinal epithelial cells. Because of the peristaltic action of the intestinal tract, attachment of the pathogen to the intestinal lining is an important event during the host-pathogen interaction. The direct involvement of *V. cholerae* TCP in the pathogenesis of cholera was demonstrated in an infant mouse model, in which a dramatic increase in the infectious dose occurred with a strain bearing a mutation in TcpA, the pilus structural subunit (114). The same mutant was unable to colonize the mouse intestinal tract, establishing the role of TCP as a colonization factor. The function of TCP as a potential adhesin was confirmed by demonstrating the inability of *tcpA* mutants to agglutinate erythrocytes. The precise nature of the receptor for TCP is not known.

Enteropathogenic *E. coli* (EPEC) are important agents of diarrheal disease of children in developing countries. The EPEC isolates can adhere to a variety of tissue-culture cells in a pattern called localized adherence, in which bacteria form clusters on the epithelial cell surface. The bundle-forming pili mediate the localized adherence of EPEC because this action can be blocked by antibodies to BFP (30). Moreover, EPEC strains carrying transposon insertions in *bfpA* have lost the ability to adhere to epithelial cells. Purified BfpA can mediate hemagglutination of human or mouse red blood cells. Although BFP and TCP have many structural similarities, there is no evidence that the receptors for the pili of these intestinal pathogens are similar.

BIOGENESIS OF TYPE IV PILI

Pilus biogenesis includes the concerted synthesis of major and minor subunits, various posttranslational modifications, and assembly into a functional organelle. This process likely involves one or more accessory proteins. Genes responsible for biogenesis of *E. coli* pili are often clustered near the structural subunit genes, presumably to allow for coordinate transcriptional control (36, 50, 75). Type IV pilin genes are all located in the bacterial chromosome, with the exception of the BfpA pilin gene of enteropathogenic *E. coli,* which is located on the 92-kilobase virulence plasmid (30). However, the structural subunit genes are usually surrounded by genes that do not encode biogenesis

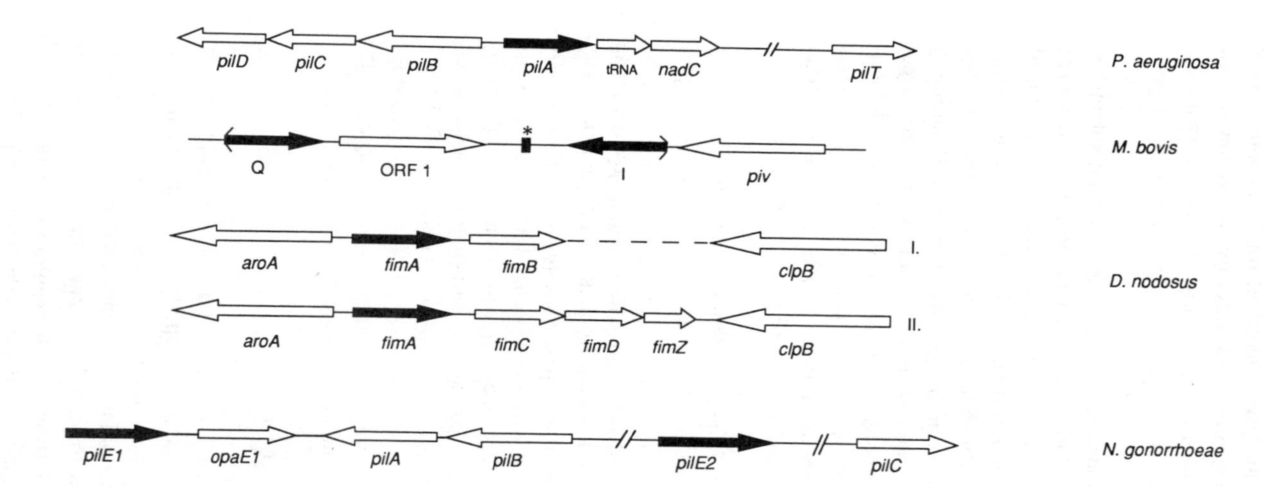

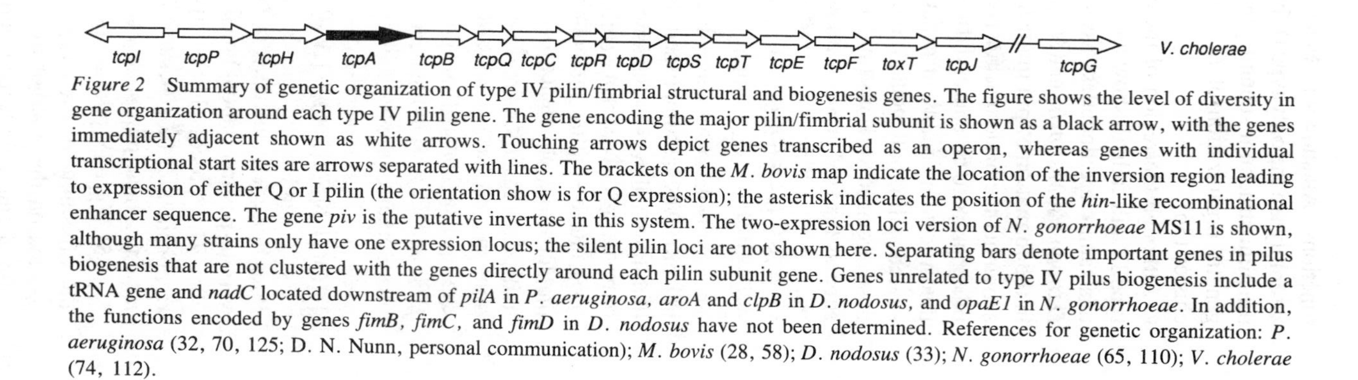

Figure 2 Summary of genetic organization of type IV pilin/fimbrial structural and biogenesis genes. The figure shows the level of diversity in gene organization around each type IV pilin gene. The gene encoding the major pilin/fimbrial subunit is shown as a black arrow, with the genes immediately adjacent shown as white arrows. Touching arrows depict genes transcribed as an operon, whereas genes with individual transcriptional start sites are arrows separated with lines. The brackets on the *M. bovis* map indicate the location of the inversion region leading to expression of either Q or I pilin (the orientation show is for Q expression); the asterisk indicates the position of the *hin*-like recombinational enhancer sequence. The gene *piv* is the putative invertase in this system. The two-expression loci version of *N. gonorrhoeae* MS11 is shown, although many strains only have one expression locus; the silent pilin loci are not shown here. Separating bars denote important genes in pilus biogenesis that are not clustered with the genes directly around each pilin subunit gene. Genes unrelated to type IV pilus biogenesis include a tRNA gene and *nadC* located downstream of *pilA* in *P. aeruginosa, aroA* and *clpB* in *D. nodosus*, and *opaE1* in *N. gonorrhoeae*. In addition, the functions encoded by genes *fimB, fimC*, and *fimD* in *D. nodosus* have not been determined. References for genetic organization: *P. aeruginosa* (32, 70, 125; D. N. Nunn, personal communication); *M. bovis* (28, 58); *D. nodosus* (33); *N. gonorrhoeae* (65, 110); *V. cholerae* (74, 112).

functions. As shown in Figure 2, the genes required for *V. cholerae* TCP assembly are the only type IV biogenesis determinants extensively linked with the pilin structural subunit gene. Therefore, TCP most closely resembles the genetic organization of *E. coli* pili, as exemplified by the *pap* pilus gene cluster (36).

Although the region of the chromosome flanking the structural subunit of type IV pili fails to show a conserved biogenesis machinery, the members of this family are doubtless assembled by a similar mechanism. Expression of *D. nodosus, M. bovis,* and *N. gonorrhoeae* pilin genes in *P. aeruginosa* results in the formation of pili in the heterologous host (5, 25, 34, 61). These studies show that the basic machinery involved in biogenesis of the type IV group A pili appears to be conserved. The various posttranslational modifications and the ordered assembly of subunits involves recognition of structural features of the pilin polypeptide found in all members of this group. Similar heterologous expression and assembly of TCP and BFP by those bacteria that express type IV group A pilins has not been reported.

Genetic Determinants of Type IV Pilus Biogenesis in P. aeruginosa

The first isolation of genes directly involved in biogenesis of prototypical type IV pili was accomplished by localized transposon mutagenesis of the chromosomal region flanking the pilin structural gene (*pilA*) by Nunn et al (70). The effects of these chromosomal transposon insertions on the assembly of pili were monitored by electron microscope examination and by observing mutant sensitivity to killing by the pilus-specific bacteriophage PO4. The transposon insertions revealed that a 4.0-kilobase (kb) region adjacent to the pilin structural gene was required for the formation of mature pili. DNA sequencing revealed the presence of three open reading frames designated *pilB, pilC,* and *pilD* that encode 62-, 38-, and 32-kDa proteins, respectively. The predicted sizes of these proteins were confirmed by expression in *E. coli* using a T7 promoter-polymerase expression system. Although mutations in either *pilB, pilC,* or *pilD* resulted in the absence of pili on the bacterial surface, they all synthesized pilin antigen in the cell at a level comparable with that produced by the wild-type bacteria. The functions of PilB and PilC in biogenesis is not clear, but the precise role of PilD has been elucidated.

P. aeruginosa *PilB and PilC*

P. aeruginosa containing mutations in either *pilB* or *pilC* are pilus-deficient but express pilin that is localized predominately in the cytoplasmic membrane (70). The pilin subunits expressed by *pilB* or *pilC* mutants appear to be in the fully processed form (i.e. the six–amino acid leader peptide is removed); hence the defect in these mutants occurs during assembly of pilin following interaction of the subunits with the membrane. PilB lacks an identifiable secre- tion-signal sequence, and hydrophobicity analysis of the deduced polypeptide sequence suggests that it is a cytoplasmic protein. PilC, however, may be an integral cytoplasmic membrane protein resulting from the presence of four po- tential membrane-spanning domains and absence of a typical signal sequence.

The predicted amino acid sequences of PilB share significant similarity with several other proteins, two of which are expressed by *P. aeruginosa*. One of these is PilT, a protein required for twitching motility in *P. aeruginosa* (125). *P. aeruginosa* XcpR (4) is also similar to PilB, suggesting that these proteins arose by gene duplication from an ancestral gene. This similarity extends to the predicted amino acid sequences of ORF1 from the *Bacillus subtilis comG* operon, to PulE in *Klebsiella oxytoca,* and to a lesser extent to VirB-11 of *Agrobacterium tumefaciens* (125). Although the similarity extends across the entire sequences of these proteins, there are two notable conserved regions. The first conserved domain contains a consensus nucleotide binding sequence, GXXXXGK(T), which is common to many nucleotide-binding proteins from both prokaryotes and eukaryotes (121). The other prominent homologous region contains the less strongly conserved secondary domain (R/KXXXGXXXL-4 hydrophobic-D), which is present in only some nucleotide-binding proteins (27, 121). Therefore, PilB (and PilT) are probably cytoplasmic nucleotide-binding proteins. One possible function for PilB is to supply energy for subunit translocation or assembly. Because many of the PilB homologues are involved in movement of macromolecules across biological membranes, they may also provide energy for the translocation processes as well.

P. aeruginosa *PilD Is a Leader Peptidase*

The transposon insertions identified a third gene, *pilD,* located distal from the pilin structural gene (70). The predicted amino acid sequence of PilD showed a protein of high hydrophobicity, with five to six transmembrane-spanning helices and without a typical signal peptide. This suggested the possibility that PilD is an integral cytoplasmic membrane protein, which was confirmed by localization studies using antibodies to PilD.

Studies of *P. aeruginosa pilD* mutants demonstrated a complete loss of pili with an accumulation of precursor, unprocessed prepilin in the bacterial cell envelope. The product of the *pilD* gene, therefore, could represent a leader peptidase or a component essential for the prepilin-processing reaction, catalyzed by another protease. To directly examine the role of PilD in leader-peptide cleavage, PilD was purified to homogeneity by immunoaffinity chromatography using antibodies raised against a peptide deduced from the *pilD*-coding sequence. Purified PilD, in the presence of nonionic detergents and phospholipids, could cleave the leader sequence from purified prepilin as well as from prepilin of *N. gonorrhoeae,* another member of the type IV group A family (71).

N-*Methyltransferase Activity of PilD*

One of the characteristics of the type IV pilin family is that the first amino acid of the mature protein (Phe or Met) is *N*-methylated following cleavage of the leader peptide. The type IV prepilin peptidase, PilD, in addition to cleaving the six–amino acid leader peptide, is the same enzyme that catalyzes the *N*-methylation of the amino-terminal residue in *P. aeruginosa* pilin (106). The bifunctionality of the enzyme has been clearly demonstrated using purified PilD, purified prepilin, and *S*-adenosyl-L-methionine (Ado-Met) as the methyl donor. In addition, PilD can recognize and efficiently methylate pilin that has been first cleaved in the absence of the methyl donor, thus suggesting that cleavage and methylation are independent reactions, perhaps involving two different active sites. The latter was confirmed by demonstrating that the methyltransferase activity of PilD can be competitively inhibited with Ado-Met analogues without affecting peptidase activity.

Other bacterial prepilin leader peptidases that are homologues of PilD very likely have *N*-methyltransferase activity as well. PulO, a protein similar to PilD that is involved in maturation of components of the extracellular secretion apparatus of pullulanase in *Klebsiella oxytoca* (see below), cleaves the gonococcal prepilin when both are expressed in *E. coli* (22). Moreover, the amino-terminal residue of this mature pilin could not be identified by standard amino-terminal sequencing, very likely because of a coupled modification, leading Dupuy et al (22) to speculate that either *E. coli* has a similar *N*-methyltransferase activity or that PulO is also the *N*-methyltransferase. The bifunctionality of PilD may very well be conserved among all of the type IV peptidases. Using a *P. aeruginosa pilD* probe, Dupuy et al identified a clone that very likely encodes the gene for *N. gonorrhoeae pilD*. Another group has recently sequenced the *N. gonorrhoeae pilD* homologue (51a). A comparison of the deduced amino acid sequence of PilD^{GC} to *P. aeruginosa* PilD demonstrated a 48% identity and 65% similarity between the two proteins. TcpJ, the prepilin leader peptidase of *V. cholerae* (44), may also be a similar bifunctional enzyme, because the methionine of the TcpA structural subunit is completely methylated (95).

Structure-Function of PilD

Hydropathy analysis of PilD shows a typical membrane protein with at least five putative transmembrane regions. A single large, relatively hydrophilic domain contains a region that is most conserved among PilD homologues (TcpJ, PulO, OutO, ComC, and PilD^{GC}) (44, 51a, 56, 67, 86). The location of this domain in the cytoplasm was confirmed by fusing alkaline phosphatase and β-galactosidase to PilD in the middle of this hydrophilic stretch (101a). The former hybrid protein lacked alkaline phosphatase activity, while the latter showed high levels of β-galactosidase activity, thus confirming the cytoplasmic location of this segment of the PilD polypeptide. This region is also notable for the presence of four cysteine residues, arranged in a pairwise fashion with the cysteines of each pair either adjacent or separated by 2 amino acids with approximately 22 amino acids between pairs (44, 51a, 101a, 125).

Examination of the effects of various inhibitors showed that the cysteine-specific alkylating agents, *N*-ethylmaleimide (NEM) and iodoacetamide, as well as reversible agents such as *p*-chloromercuribenzoate, inhibit both the prepilin peptidase and *N*-methyltransferase activities of PilD (101a). The role of cysteines in catalysis was confirmed by site-directed mutagenesis of the codons for all five cysteines in *P. aeruginosa* PilD. Substitutions of individual cysteines with serine or glycine within the conserved hydrophilic domain of PilD substantially reduced both leader peptidase and *N*-methyltransferase activity. Similar substitutions of the single amino-terminal cysteine found within the amino-terminal hydrophobic segment had no effect on either activity. Therefore, the active sites of the enzyme are within adjacent domains of the cytoplasmic segment of PilD. Other parts of PilD, however, could participate in proteolysis and methylation, including segments within the membrane. Interestingly, none of the mutations completely abolished the activity of the enzyme, and even substitutions that lowered the enzymatic activity of PilD to 5% of the wild-type allowed bacteria to assemble sufficient pili to make the mutant bacteria susceptible to killing by the pilus-specific phage PO4.

The Specificity and Kinetics of Leader-Peptide Cleavage and N-*Methylation of Mature Pilin*

The cleavage site for the type IV pilin precursors is between glycine and phenylalanine residues (methionine in TcpA) in the sequence Gly-Phe(Met)-Thr-Leu-Ile(Leu)-Glu, which is conserved among all members of the type IV pilin family (Figure 1). The precise role of the leader sequence is not known. Although the sequence cannot function as an export signal, membrane insertion of pilin is very likely directed by the hydrophobic portion of the mature polypeptide, based on the cellular localizations of pilin-alkaline phosphatase (PhoA) fusions (102).

Attempts to define the minimal sequence of prepilin that still functions during the initial stages of pilus biogenesis relied on engineering of gene fusions between the deleted pilin gene from *P. aeruginosa* and the *E. coli phoA* gene. The hybrid proteins expressed from the gene fusions contained the entire six–amino acid leader sequence, along with 3, 26, 41, and 46 residues of the mature protein fused to PhoA at the carboxy terminus. The hybrid proteins containing the 26-, 41-, and 46–amino acid portion of the pilin were translocated across the cytoplasmic membrane, as indicated by the high levels of PhoA activity in *P. aeruginosa* (102; M. S. Strom, unpublished results). It was not possible to determine leader-sequence cleavage of the PilA-PhoA fusions. However, when purified, only the largest hybrid was efficiently methylated by PilD, suggesting that the 46–amino acid amino-terminal portion of pilin, together with the leader sequence, contains the necessary information for translocation across the cytoplasmic membrane, with concomitant leader-peptide cleavage and *N*-methylation. The smallest fusion did not direct translocation of PhoA across the inner membrane, and neither it nor the hybrids containing 26 and 41 amino acid residues of mature pilin were *N*-methylated in vitro in the presence of PilD and Ado-Met (102; M. S. Strom, unpublished results). Similarly engineered hybrid proteins between *N. gonorrhoeae* pilin and PhoA that contained a 20- or 34–amino acid segment of the gonococcal prepilin were also not processed in vivo (23).

Oligonucleotide-directed site-specific mutagenesis of *pilA* was used to construct a series of amino acid substitutions in the pilin precursor to identify the specific amino acids within the amino terminus of prepilin of *P. aeruginosa* that are important for processing, posttranslational modification, and assembly of the pilin subunits into mature pili (103). This mutational analysis showed that the glycine residue at the −1 position relative to the leader peptide cleavage site is absolutely required for prepilin processing prior to pilus biogenesis. Substitution of alanine for glycine at this position resulted in partial processing of prepilin, suggesting that an amino acid with a small side group was required at this position for cleavage by PilD.

A second class of mutations engineered in the pilin protein of *P. aeruginosa* centered around the phenylalanine at the +1 position. This residue can be substituted by a variety of other residues with little effect on formation of pilin, as assessed by electron microscopy and sensitivity to killing by the pilus-specific phage PO4. Substitutions within the leader peptide that increased the net charge to +3 from +2, or lowered it to +1, 0, or −1, had no effect on either pilin processing or pilus biogenesis. However, the leader sequence itself is necessary, because a deletion mutation that removed this sequence resulted in unstable pilin expression and no pilus assembly.

Additional mutations in the mature portion of *P. aeruginosa* prepilin were also examined for processing and pilus assembly (80). Prepilin with a deletion of eight residues (from Ile+4 to Ala+11 of the mature protein) resulted in a form that was not proteolytically processed, whereas a four–amino acid deletion (from Ile+4 to Met+7) was partially processed. The construct with the four–amino acid deletion, but not the eight, was still incorporated into the membrane. Apparently, these deletions altered the overall hydrophobic character of the mature portion, which then affected recognition of the prepilin substrate by PilD. Interestingly, the majority of point mutations in the hydrophobic domain near the amino terminus of pilin have little or no effect on processing of prepilin or its assembly into pili (103). The only exception is an invariant glutamic acid, found at the +5 position from the cleavage site. Substitutions at this position with either hydrophobic or basic amino acids have no effect on prepilin processing, but the pilin is not *N*-methylated and subunits are not assembled into pili (80, 103). Furthermore, the same mutant pilin was not a substrate for *N*-methylation in vitro by PilD (M. S. Strom, unpublished results). Although the presence of an acidic residue within the hydrophobic amino-terminal domain is not essential for cleavage of the leader sequence by PilD, it may be part of the recognition sequence for the methyltransferase activity catalyzed by the same bifunctional enzyme.

To assess whether phenylalanine at the amino terminus is the only residue that can be *N*-methylated, several mutations at the +1 position were examined for the presence of *N*-methylated amino acids (103). Substitutions of methionine, alanine, and tyrosine resulted in processed pilin with the substitute residue almost fully *N*-methylated. However, two mutations gave opposing results. In the first, a serine residue at this position was not methylated although pili were produced. In the second, a glycine substitution resulted in fully processed pilin that was approximately 50% *N*-methylglycine at the amino terminus, but no functional pili were detected by phage PO4 sensitivity or by electron microscopy. These results suggest that PilD, which catalyzes the *N*-methylation of pilin, can modify a wide range of amino acids. Although in some mutants no obvious defects result from lack of methylation, the pili assembled from unmethylated subunits may differ in some subtle way from methylated pili. This difference would not be obvious when piliated bacteria are examined by electron microscopy.

The kinetics of cleavage of the leader peptide from the *P. aeruginosa* pilin precursor was also determined (104). A K_m of approximately 650 μM and a turnover rate, or K_{cat}, of 180 min^{-1} were measured. Similar rates were obtained when PilD-mediated processing of the *N. gonorrhoeae* prepilin was measured. Therefore, the differences in length and net charge of the *N. gonorrhoeae* prepilin leader sequence does not affect the overall rate of cleavage by PilD. The rates of processing of prepilins obtained by site-directed mutagenesis of the codon for the phenylalanine at the +1 position were lower than those of the wild-type substrate. Substitution of a methionine at this position resulted in substrate that was processed at a rate closest to the phenylalanine-containing wild-type prepilin. Interestingly, substitution of an asparagine at this position resulted in a turnover rate approximately one-tenth that of wild-type substrate but with a higher affinity for PilD. This finding could form the basis for the design of peptide inhibitors or pseudosubstrates of PilD that may aid in the study of the cleavage reaction.

Prepilin Processing and Biogenesis of N. gonorrhoeae *Pili*

The prepilin of *N. gonorrhoeae* can be efficiently processed by *P. aeruginosa* PilD, and gonococcal membranes contain an enzyme with a prepilin leader-peptidase activity (71). The structural requirements for processing within prepilins of all group A type IV pilins are therefore very likely identical.

Gonococci can produce a truncated form of pilin that is exported from the cell in a soluble form. These S-pilins are processed at a different site within the amino terminus of the mature protein and are smaller than the pilin proteins assembled into pili. Two types of genetic events can lead to truncation of pilin and its export. Mutations that alter the leader-peptide cleavage site, specifically changing the invariant glycine at the -1 position to serine, result in cleavage of ~40 amino acids from the amino terminus and export of pilin from the cell. Alternate cleavage and export of pilin also often accompanies mutations in *pilC,* which encodes a minor protein component of pili (see below). In addition to these mutations, several changes occur in the mature portion of pilin, presumably facilitating cleavage at an alternative site.

Gonococcal outer membranes contain a minor protein of M_r 110,000 that is also present in purified pili (42). The gene for this protein, designated *pilC,* was cloned and sequenced and found to exist as two nonlinked copies, *pilC1* and *pilC2,* in most strains of *N. gonorrhoeae*. However, in the piliated strain MS11 (P+), only the *pilC2* copy is expressed. Insertional inactivation of *pilC2,* but not *pilC1,* prevented the formation of pili but did not abolish expression of the pilin subunits; hence PilC functions in assembly of pili. One class of spontaneous nonpiliated variants is a result of the absence of PilC. PilC is not expressed because of a frameshift mutation in a cluster of G residues located in the region encoding the PilC signal peptide. Reversion back to the piliated phenotype results from the restoration of the reading frame within this G-track. The alteration of piliation by modulating expression of PilC is therefore another mechanism of phase variation. Interestingly, many of the mutants unable to express PilC synthesize a secreted form of pilin (S-pilin) that is generated by alternative processing within the amino-terminal region of the mature polypeptide (43). The reversible PilC-dependent expression also leads to sequence alterations in the expression locus for pilin, and therefore represents another mechanism of antigenic variation. The exact role of PilC in biogenesis of gonococcal pili has not been determined, and a protein that structurally or functionally resembles PilC in other bacteria expressing type IV pili has not been identified as yet.

Biogenesis of Toxin-Coregulated Pili of V. cholerae

The ToxR-coregulated pilin gene (*tcpA*) of *V. cholerae* was originally identified using Tn*phoA* insertions that require ToxR, the transcriptional regulator of the cholera-toxin operon, for expression (114). The gene encoding TcpA is part of a gene cluster involved in regulation and assembly of the TCP pilus. Other genes in the cluster include *tcpB, tcpQ, tcpC, tcpR, tcpD, tcpS, tcpT, tcpE, tcpF,* and *tcpJ,* located immediately downstream of *tcpA,* and *tcpH, tcpP,* and *tcpI,* located upstream of *tcpA*. These genes are also regulated by the product of the *toxR* gene (Figure 2) (74, 112). Another gene located outside of this cluster, *tcpG,* encodes a product required for efficient biogenesis of TCP pili (81). Although the organization of the *tcp* genes originally suggested some resemblance to the biogenesis functions of the *E. coli* Pap pili operon, the complete sequence of the *tcp* genes revealed little similarity with the Pap pili operon. Therefore, the mechanism of biogenesis of the group B members of the type IV pilin family may be unique.

Two genes required for assembly of TCP encode proteins of known biochemical function. TcpJ is a prepilin peptidase, responsible for the cleavage of the 25–amino acid leader peptide from the precursor of TcpA (44). It is also highly similar to PilD, especially in the cytoplasmic region that contains the active-site cysteines. Because mature TcpA undergoes the same postcleavage *N*-methylation as the other type IV pilins, TcpJ, like PilD, may be a bifunctional enzyme with both protease and methyltransferase activities.

The only gene not part of the *tcp* gene cluster is *tcpG,* which encodes a 24-kDa periplasmic protein. The sequence of TcpG shows striking similarity to several bacterial thioredoxins, *E. coli* DsbA (a periplasmic thiol-disulfide interchange protein), and eukaryotic protein disulfide isomerase (81, 127). All of these proteins have been implicated in mediating thio-disulfide exchange reactions, and the homology of these proteins with TcpG is in the cysteine-containing active centers. Purified TcpG has oxidoreductase activity and probably participates in some chaperone-like function, perhaps controlling folding of TcpA by formation of transient disulfide-linked intermediates or by catalyzing formation of correct disulfide bonds. Interestingly, mutations in TcpG also cause a pleiotropic defect in the extracellular secretion of cholera toxin A subunit and a major protease. These results indicate that the function of TcpG is not restricted to biogenesis of pili, but is also involved in some steps during extracellular protein secretion.

Two additional products of the *tcp* operon share sequence similarity with the biogenesis proteins of *P. aeruginosa*. TcpT is a homologue of PilB (and PilT) as well as other related proteins involved in extracellular protein secretion (see below) and displays a well-conserved nucleotide-binding region. TcpE reportedly has homology to *P. aeruginosa* PilC (45). These similarities with the biogenesis genes of *P. aeruginosa* further reinforce the overall relationship among the type IV pili not only in the primary sequence of the major subunits but in the overall mechanism of organelle biogenesis. Whether conservation of the export or assembly functions can be extrapolated to the point that they are interchangeable remains to be experimentally determined.

TWITCHING MOTILITY

Most bacteria expressing type IV pili can move on solid surfaces; this is called twitching motility. The ability of type IV pili to undergo reversible depolymerization/assembly of the filament presumably mediates this action. Recently, a gene necessary for twitching motility in *P. aeruginosa* was isolated by restoring twitching motility to a nonretractable, hyperpiliated strain, PAK/2pfs, with a DNA expression library derived from another *P. aeruginosa* prototype wild-type strain, PAO1 (125). This gene (*pilT*) was sequenced, and its 344–amino acid product was shown to share extensive homology with two other *P. aeruginosa* proteins, PilB and XcpR, that are required for biogenesis of pili and extracellular protein secretion, respectively. One of the striking features of these proteins is the presence of two highly conserved domains that contain consensus sequences found in several nucleotide-binding proteins in both prokaryotes and eukaryotes (as discussed above) (115, 121). Further analysis of the PilT amino acid sequence shows that it is largely hydrophilic and lacks both a leader sequence common to exported proteins and stretches of hydrophobic residues that could act as membrane-spanning domains. Thus, PilT is probably a cytoplasmic nucleotide-binding protein, although its role in twitching motility is as yet undefined.

EXPRESSION OF TYPE IV PILIN: CONTROL AT THE TRANSCRIPTIONAL LEVEL

The ability to control expression of the genes encoding structural components as well as proteins involved in assembly of pili is important for several reasons. Formation of functional pili requires interactions between various structural and nonstructural proteins in vastly different ratios, ranging from a few molecules of an enzyme to several thousand copies of the structural subunits. Differential expression of genes assures accumulation of the various proteins in precise relative proportions as the pilus filament is assembled. Expression of pili specifying adhesins and additional bacterial virulence factors results in the evolution of regulatory networks that control gene expression in response to the environmental signals of the host. Finally, because pili are prominent surface structures, they also are occasionally targets of host defenses, and alternating induction and shutoff of pilus expression allows for successful evasion of the immune system. This section discusses selective examples of the mechanism of transcriptional control of expression of type IV pilins.

P. aeruginosa *Pilin*

Johnson et al (40) were first to recognize that the promoter of *pilA,* the pilin structural gene, resembles the promoter for genes transcribed by RNA polymerase containing an alternative sigma factor, σ^{54}, encoded by the enterobacterial *rpoN* gene. The promoters of all members of the type IV group A pilin family contain a conserved motif found in RpoN-dependent promoters, $G_{-24}G...(N_{10})...GC_{-12}$, found upstream of the transcriptional start site (51). The group B pilins lack this sequence and hence are transcribed by the RNA polymerase with the major sigma factor, σ^{70}. The requirement for RpoN in transcription of the pilin gene in *P. aeruginosa* was verified by constructing a strain with an insertionally inactivated RpoN that could not transcribe the pilin gene (38, 38a).

Two additional genes have been identified that encode positive regulators of pilin transcription. A screen of a Tn*5* library of *P. aeruginosa* using a pilin-*lacZ* transcriptional fusion yielded two mutations in linked genes. These genes were cloned by genetic complementation of the transposon mutants. The deduced sequence of one of these regulatory genes, designated *pilR,* belongs to the response regulator class and is homologous in several regions to the prototype of this family, NtrC (38). The most extensive homology between PilR and NtrC occurs at the amino terminus and in a central region. The amino-terminal domain contains conserved aspartic acid residues at positions 11 and 54 and a lysine at position 104 of the 446–amino acid protein (~50 kDa). These conserved residues are found in all members of this class. The central domain of PilR contains two motifs found in a superfamily of ATP-binding proteins (27). In NtrC, this nucleotide-binding domain is required for the endogenous ATPase activity of its phosphorylated form, which is essential for formation of the open complex during initiation of transcription (124). If analogous to other members of the NtrC-family, PilR requires phosphorylation of one of the aspartic acid residues for it to activate transcription of the pilin gene. Mutant forms of PilR, with substitutions at Asp54, cannot function in transcription of the pilin gene (J. Boyd & S. Lory, in preparation).

The second *P. aeruginosa* pilin transcriptional activator gene has been designated *pilS,* and according to sequence homology, it is a sensor or histidine kinase member of the two-component class (J. Boyd & S. Lory, in preparation). The autophosphorylation of a single histidine in the sensor protein, followed by the transfer of the phosphate to the aspartic acid residue of the regulatory protein, relays an environmental signal to the level of gene expression. Mutants of *P. aeruginosa,* in which the histidine at position 115 of PilS was changed to arginine, leucine, or proline, did not synthesize pili. The mechanism of transcriptional regulation mediated by the PilR/PilS pair seems to resemble the other members of the two-component regulatory family. However, the specific environmental or nutritional conditions that influence the regulation of pilus biosynthesis have yet to be determined for *P. aeruginosa* or any of the other organisms expressing type IV pili.

N. gonorrhoeae *Pilin*

As predicted by the presence of a consensus RpoN promoter sequence upstream of the pilin gene in *N. gonorrhoeae* (*pilE*), the expression of this type IV pilin gene very likely involves RpoN and the cognate regulatory factors. Additional regulatory proteins, which do not resemble the *P. aeruginosa* transcriptional factors, have been described. Two closely linked genes, designated *pilA* and *pilB,* were isolated and shown to act *in trans* to regulate pilin gene expression (111). Initial experiments showed that the *pilB* gene product, in conjunction with the product of *pilA,* decreases pilin expression in *N. gonorrhoeae,* and hence, gonococcal mutants in PilB are hyperpiliated. In *E. coli,* the *pilA* gene product stimulated pilin-promoter activity. However, a *N. gonorrhoeae pilA* mutant could not be obtained, suggesting that this gene has an essential regulatory function in gonococci. Attempts to construct null mutants in both *pilA* or *pilB* by transposon mutagenesis were also unsuccessful. Construction of $pilA^{-}/pilA^{+}$ merodiploid gonococci led to a drastic reduction in the amount of pilin production; taken with the stimulatory effect of PilA on the pilin promoter in *E. coli,* these results suggested that this gene product is also a regulatory protein.

Analysis of the amino acid sequences of PilA and PilB showed similarities to members of the two-component regulatory-system family (109). PilB is homologous to the histidine kinase sensor component, and work with gene fusions to alkaline phosphatase and cellular localization studies show it to be a cytoplasmic membrane protein with both periplasmic and cytoplasmic domains. As with several members of this class of regulatory proteins, the periplasmic domain is very likely to be the sensor, whereas the cytoplasmic domain is the transmitter in an as-yet uncharacterized signaling pathway. The cellular localization of PilB in gonococci is therefore different than that of PilS, its counterpart in *P. aeruginosa.* However, as with PilS, the environmental signals that induce PilB to transmit an activation signal to initiate pilin transcription are unknown.

PilA is homologous to the response-regulator proteins of the two-component family (109). As with PilR of *P. aeruginosa,* gonococcal PilA appears to have DNA-binding capability and also has a carboxy-terminal ATP-binding consensus sequence. However, two characteristics of PilA distinguish it from PilR as well as from most other response regulators. First, the phospho-acceptor residue in all regulatory elements of the two-component family is an aspartic acid, while PilA contains a glutamic acid at the corresponding position. Second, PilA appears to be an essential gene for *N. gonorrhoeae*, whereas PilR is not necessary in *P. aeruginosa*. PilA therefore most likely controls expression of an as-yet unidentified gene that is essential for gonococcal growth under most laboratory conditions.

V. cholerae *TCP*

The molecular mechanism controlling expression of toxin-coregulated pili of *V. cholerae* involves activities of regulatory elements that coordinately control expression of several other virulence factors. *V. cholerae* mutants in *toxR*, the gene encoding the regulator of the cholera toxin operon, are unable to express pili. Expression of a fusion between *tcpA*, which is the gene for the structural subunit of TCP, and alkaline phosphatase also depends on functional ToxR (114). Genes under *toxR* control include not only the structural subunit gene, but also the remaining linked genes of the TCP operon, as well as the unlinked gene *tcpG*. The genetic organization of the *tcp* accessory genes suggests that a single promoter upstream from *tcpT* directs transcription of all of the remaining *tcp* genes, whereas the gene for *tcpI* is transcribed in the opposite direction. Interestingly, a site of transcriptional initiation, or mRNA processing, has been also identified upstream of *tcpA* mRNA (R. Taylor, personal communication).

Among all of the ToxR regulated genes, only transcription of the *ctx* operon is activated in *E. coli* in the presence of ToxR. This observation led to a search for other transcriptional elements that may act together with ToxR in transcriptional initiation. ToxT was identified as a regulatory factor required for expression of many toxin-coregulated genes, including the genes of the *tcp* operon. ToxT is a 32-kDa transmembrane protein that shares similarity with the members of the AraC family of transcriptional activators (15, 73). It may act as a site-specific DNA-binding protein with several unusual features. Transcription of *toxT* depends on ToxR, and hence ToxR exerts its activity on many toxin-coregulated genes, including the *tcp* locus indirectly, by regulating *toxT*. Interestingly, the gene for ToxT is within the *tcp* operon, and it may therefore be under autoregulatory control from regulatory sites that precede this operon. The products of *tcpH* and *tcpI* apparently encode additional regulators of the *tcp* operon, although their mechanism of action is not known (113).

The *tcp* genes respond to environmental signals transmitted to the various structural and biogenesis genes of *V. cholerae* pili via the ToxT-ToxR cascade. The intramembrane disposition of ToxR and its ability to recognize promoters of regulated genes suggests that this protein transmits signals from the periplasmic side of the inner membrane and effects transcription via its cytoplasmic domain by binding regulatory sites near promoters, including the one near *toxT*. The environmental conditions that control expression of *tcp* genes include osmolarity, temperature, and pH. That these signals control coordinate expression of various cell-associated and extracellular virulence factors may reflect the different environments that *V. cholerae* experiences prior to and during colonization of the lower intestinal tract (64).

RELATIONSHIP OF TYPE IV PILI AND PROTEIN EXPORT

Phenotypic characterization of *P. aeruginosa* mutants in the pilus biogenesis determinants *pilB, pilC,* and *pilD* revealed that PilD functions not only in processing of prepilin, but that it has a second role in extracellular protein secretion. Mutations in *pilD* lead to periplasmic accumulation of exotoxin A, elastase, alkaline phosphatase, and phospholipase C—enzymes that are actively exported from wild-type *P. aeruginosa* (105). The secretion signal of exotoxin A, found in the periplasm of *pilD* mutants, is cleaved. These findings suggested that PilD has additional substrates in the bacterial cell, and these substrates are essential components of the extracellular-secretion apparatus. Independently, Bally et al (3) isolated a gene essential for extracellular protein export from *P. aeruginosa*. The product of this gene, XcpA, is identical to PilD, hence confirming the central role of PilD in protein export as well as biogenesis of pili.

In order to identify substrates of PilD, a series of degenerate oligonucleotides corresponding to possible codons spanning the prepilin cleavage site were designed and used as probes to screen cloned DNA from sites of transposon insertions in pleiotropic export-deficient *P. aeruginosa* mutants. The probes identified a region of DNA containing four genes that encode proteins with a high degree of similarity to prepilin (72) (Figure 3). These proteins, termed PddABCD (for PilD-dependent proteins), are required for extracellular secretion and are substrates of PilD (106; D. N. Nunn & S. Lory, submitted). Analysis of a cluster of genes originally isolated in export-deficient *P. aeruginosa* mutants and designated *xcp* led to identification of four genes (*xcpTUVW*) (4) that are identical to *pddABCD*.

The similarity between type IV pilin and proteins of the extracellular secretion apparatus is not restricted to *P. aeruginosa*. Several genes have been identified during the past two years that are homologues of type IV prepilins, and with a single exception, they are part of a protein-export machinery. All of these bacteria also express a protein similar to PilD. These include the pilin homologues PulGHIJ and the leader peptidase/methylase PulO in *K. oxytoca* (85). Extracellular secretion of degradative enzymes in the plant pathogens *Erwinia chrysanthemi, Erwinia carotovora,* and *Xanthomonas campestris,* and in *Aeromonas hydrophila,* involve similar genes, organized in clusters,

		▼
Type IV pilin		
***P. aeruginosa* PAK**	MetLysAlaGlnLysGly	PheThrLeuIleGluLeuMetIleValVal-
Extracellular protein secretion		
P. aeruginosa		
PddA (XcpT)	LeuGlnArgArgGlnGlnSerGly	PheThrLeuIleGluIleMetValValVal-
PddB (XcpU)	MetArgAlaSerArgGly	PheThrLeuIleGluLeuMetValValMet-
PddC (XcpV)	MetLysArgAlaArgGly	PheThrLeuLeuGluValLeuValAlaLeu-
PddD (XcpW)	MetArgLeuGlnArgGly	PheThrLeuLeuGluLeuLeuIleAlaIle-
K. oxytoca		
PulG	MetGlnArgGlnArgGly	PheThrLeuLeuGluIleMetValValIle-
PulH	ValArgGlnArgGly	PheThrLeuLeuGluMetMetLeuIleLeu-
PulI	MetLysLysGlnSerGly	MetThrLeuIleGluValMetValAlaLeu-
PulJ	MetIleArgArgSerSerGly	PheThrLeuValGluMetLeuLeuAlaLeu-
E. chrysanthemi		
OutG	MetGluArgArgGlnArgGly	PheThrLeuLeuGluIleMetValValIle-
OutH	ValArgGlnArgGly	PheThrLeuLeuGluIleMetLeuValVal-
OutI	MetLysGlnGlnGly	MetThrLeuLeuGluValMetValAlaLeu-
OutJ	ValLysGlnProGluArgGly	PheThrLeuLeuGluValMetLeuAlaLeu-
X. campestris		
XpsG	MetIleLysArgSer..*6*..ArgAlaGlyGlnAlaGly	MetSerLeuLeuGluIleIleIleValIle-
XpsH	MetArgValAlaArg.*13*..ArgArgGlnLeuArgGly	SerSerLeuLeuGluMetLeuLeuValIle-
XpsI	MetLysHisGlnArgGly	TyrSerLeuIleGluValIleValAlaPhe-
XpsJ	MetArgProArgAlaAlaGly	PheThrLeuIleGluValLeuLeuAlaThr-
A. hydrophila		
ExeG	MetGlnLysArgArgGlnSerGly	PheThrLeuLeuGluValMetValValIle-
DNA uptake		
B. subtilis		
ComG ORF3	MetAsnGluLysGly	PheThrLeuValGluMetLeuIleValLeu-
ComG ORF4	LeuAsnIleLysLeuAsnGluGluLysGly	PheThrLeuLeuGluSerLeuLeuValLeu-
ComG ORF5	MetTrpArgGluAsnLysGly	PheSerThrIleGluThrMetSerAlaLeu-
Consensus	-Gly	Phe or Met ThrLeu*Pho*Glu-(*Pho*16-18)-

Figure 3 Comparison of the amino termini of the precursors of type IV pilin–like proteins involved in extracellular protein secretion or DNA uptake. These are compared with the type IV pilin from *P. aeruginosa* PAK (41). The upside-down triangle denotes the putative leader peptide-cleavage site. The sequences for these proteins were first described in references: *P. aeruginosa* PddABCD/XcpTUVW (4, 72); *K. oxytoca* PulGHIJ (85); *Erwinia chrysanthemi* OutGHIJ (31, 56); *Xanthomonas campestris* XpsGHIJ (21, 35); *Aeromonas hydrophila* ExeG (8); *B. subtilis* ComG (ORF3,4,5) (1, 9).

several of which encode type IV pilin–like proteins (Figure 3) (8, 20, 21, 31, 35; G. P. Salmond, personal communication).

The genes responsible for extracellular protein export are usually clustered and in some instances organized into a single operon (84). This region contains not only genes homologous to type IV pilin, but other functions related to biogenesis of type IV pilins. For example, in *P. aeruginosa,* significant homologies can be found among products of *xcpR* and *pilB* and *pilT* (4, 125). Also, PilC of *P. aeruginosa* is similar to XcpS (4). Corresponding genes can be found in *K. oxytoca, Erwinia* sp., and *X. campestris*. These findings suggest that the genes found in the export cluster specify determinants that facilitate assembly of organelles resembling pili by ordered assembly of the four heterologous subunits encoded by adjacent genes. The existence of a complex composed of the four pilin homologues has not been identified in the bacterial envelope.

The only homologues of type IV pilins not associated with extracellular protein secretion are encoded by genes in *B. subtilis* required for the competence state and DNA uptake. Clearly, the bifunctional enzyme PilD and its homologues have a central role not only in type IV pilus biogenesis, but also in macromolecular transport across cell membranes.

CONCLUDING REMARKS

The discovery during the past decade that a diverse group of pathogenic bacteria express pili related by their structure and their mechanisms of regulation and assembly has provided a very useful model to study expression of virulence factors. The role of type IV pili in virulence is now firmly established, and studies of pilus biogenesis and function are aimed not only at understanding the mechanism of bacterial pathogenesis better but also at exploring the possibility of novel approaches to prevention and therapy of infectious disease.

Numerous questions remain to be answered regarding the role of pili in pathogenesis. Specifically, the precise molecular nature of the interaction of an adhesive component of pili with host receptors is not completely clear. Complicating these studies is the observation that the surfaces of virtually all bacteria that express type IV pili also possess other adhesins. The receptor-binding contribution of the pilin subunits and the range of receptors recognized by the different adhesins is therefore difficult to measure. Promising developments in bacterial genetics of many bacteria with type IV pili should allow construction of isogenic mutants that cannot express individual adhesins. Such strains will be extremely useful in dissecting the complex interactions of bacteria with their hosts.

The processes of secretion and assembly of pili have been studied primarily using genetic methods. In the future, efforts must be directed towards understanding the mechanisms of biogenesis and identifying the precise function of products of individual biogenesis genes in the formation of pili. Although some of the proteins have known biological functions, such as PilD of *P. aeruginosa* and TcpJ and TcpG of *V. cholerae,* the products of the majority of genes identified by their nonpiliated phenotypes have no known functions. No information is available at this time on the mechanisms that allow subunits of type IV pilin to cross the inner and outer membrane and how the pili are anchored in the bacterial envelope. Based on the unique properties of the type IV pili, answers to these questions will very likely provide insights on previously undescribed basic biological processes of membrane targeting and macromolecular assembly.

ACKNOWLEDGMENTS

We thank Jessica Boyd and David Simpson for a careful and critical reading of this manuscript. Work performed in S. L.'s laboratory has been supported by Public Health Service grants AI21451 and AI32624 from the National Institutes of Health. We also gratefully acknowledge the following for generously sharing their results and manuscripts prior to publication: Michael Donnenberg, Ann Progulske-Fox, Peter Hoyne, Michael Koomey, Paul Manning, Thomas Meyer, Staffan Normark, Anthony Pugsley, Maggie So, Ron Taylor, and Tone Tønjum.

Literature Cited

1. Albano, M., Breitling, R., Dubnau, D. A. 1989. Nucleotide sequence and genetic organization of the *Bacillus subtilis comG* operon. *J. Bacteriol.* 171:5386–5404
2. Annuar, B. O., Wilcox, G. E. 1985. Adherence of *Moraxella bovis* to cell cultures of bovine origin. *Res. Vet. Sci.* 39:241–46
3. Bally, M., Ball, G., Badere, A., Lazdunski, A. 1991. Protein secretion in *Pseudomonas aeruginosa:* the *xcpA* gene encodes an integral inner membrane protein homologous to *Klebsiella pneumoniae* secretion function protein PulO. *J. Bacteriol.* 173:479–86
4. Bally, M., Filloux, A., Akrim, M., Ball, G., Lazdunski, A., Tommassen, J. 1992. Protein secretion in *Pseudomonas aeruginosa:* characterization of seven *xcp* genes and processing of secretory apparatus components by prepilin peptidase. *Mol. Microbiol.* 6: 1121–32
5. Beard, M. K., Mattick, J. S., Moore, L. J., Mott, M. R., Marrs, C. F., Egerton, J. R. 1990. Morphogenetic expression of *Moraxella bovis* fimbriae (pili) in *Pseudomonas aeruginosa. J. Bacteriol.* 172:2601–7
6. Bergstroem, S., Robbins, K., Koomey, J. M., Swanson, J. 1986. Piliation control mechanisms in *Neisseria gonorrhoeae. Proc. Natl. Acad. Sci. USA* 83:3890–4
7. Billington, S. J., Rood, J. I. 1991. Sequence of fimbrial subunit-encoding genes from virulent and benign isolates of *Dichelobacter* (*Bacteroides*) *nodosus. Gene* 99:115–9
8. Bo, J., Howard, S. P. 1992. The *Aeromonas hydrophila exeE* gene, required both for protein secretion and normal outer membrane biogenesis, is a member of a general secretion pathway. *Mol. Microbiol.* 6:1351–61
9. Breitling, R., Dubnau, D. 1990. A membrane protein with similarity to N-methylphenylalanine pilins is essential for DNA binding by competent *Bacillus subtilis. J. Bacteriol.* 172: 1499–1508
10. Britigan, B. E., Cohen, M. S., Sparling, P. F. 1985. Gonococcal infection: a model of molecular pathogenesis. *New Engl. J. Med.* 312:1683–94
11. Buchanan, T. M., Pearce, W. A. 1976. Pili as a mediator of the attachment of gonococci to human erythrocytes. *Infect. Immun.* 13:1483–89
12. Chandler, R. L., Smith, K., Turfrey, B. A. 1985. Exposure of bovine cornea to different strains of *Moraxella bovis*

and to other bacterial species in vitro. *J. Comp. Pathol.* 95:415–23

13. Chi, E., Mehl, T., Nunn, D., Lory, S. 1991. Interaction of *Pseudomonas aeruginosa* with A549 pneumocyte cells. *Infect. Immun.* 59:822–28
14. Deal, C. D., Stromberg, N., Nyberg, G., Normark, S., Karlsson, K. A., So, M. 1987. Pilin independent binding of *Neisseria gonorrhoeae* to immobilized glycolipids. *Antonie Van Leeuwenhoek J. Microbiol. Serol.* 53: 425–30
15. DiRita, V. J., Parsot, C., Jander, G., Mekalanos, J. J. 1991. Regulatory cascade controls virulence in *Vibrio cholerae*. *Proc. Natl. Acad. Sci. USA* 88:5403–7
16. Doig, P., Paranchych, W., Sastry, P. A., Irvin, R. T. 1989. Human buccal epithelial cell receptors of *Pseudomonas aeruginosa:* identification of glycoproteins with pilus binding activity. *Can. J. Microbiol.* 35:1141–45
17. Doig, P., Sastry, P. A., Hodges, R. S., Lee, K. K., Paranchych, W., Irvin, R. T. 1990. Inhibition of pilus-mediated adhesion of *Pseudomonas aeruginosa* to human buccal epithelial cells by monoclonal antibodies directed against pili. *Infect. Immun.* 58:124–30
18. Doig, P., Todd, T., Sastry, P. A., Lee, K. K., Hodges, R. S., et al. 1988. Role of pili in adhesion of *Pseudomonas aeruginosa* to human respiratory epithelial cells. *Infect. Immun.* 56:1641–6
19. Donnenberg, M. S., Giron, J. A., Nataro, J. P., Kaper, J. B. 1992. A plasmid-encoded type IV fimbrial gene of enteropathogenic *Escherichia coli* associated with localized adherence. *Mol. Microbiol.* 6:3427–37
20. Dow, J. M., Daniels, M. J., Dums, F., Turner, P. C., Gough, C. 1989. Genetic and biochemical analysis of protein export from *Xanthomonas campestris*. *J. Cell Sci. Suppl.* 11:59–72
21. Dums, F., Dow, J. M., Daniels, M. J. 1991. Structural characterization of protein secretion genes of the bacterial phytopathogen *Xanthomonas campestris* pathovar *campestris:* relatedness to secretion systems of other gram-negative bacteria. *Mol. Gen. Genet.* 229:357–64
22. Dupuy, B., Taha, M. K., Possot, O., Marchal, C., Pugsley, A. P. 1992. PulO, a component of the pullulanase secretion pathway of *Klebsiella oxytoca,* correctly and efficiently processes gonococcal type IV prepilin in *Escherichia coli*. *Mol. Microbiol.* 6:1887–94
23. Dupuy, B., Taha, M. K., Pugsley, A. P., Marchal, C. 1991. *Neisseria gonorrhoeae* prepilin export studied in *Escherichia coli*. *J. Bacteriol.* 173: 7589–98
24. Elleman, T. C. 1988. Pilins of *Bacteroides nodosus:* molecular basis of serotypic variation and relationships to other bacterial pilins. *Microbiol. Rev.* 52:233–47
25. Elleman, T. C., Hoyne, P. A., Stewart, D. J., McKern, N. M., Peterson, J. E. 1986. Expression of pili from *Bacteroides nodosus* in *Pseudomonas aeruginosa*. *J. Bacteriol.* 168:574–80
26. Faast, R., Ogierman, M. A., Stroeher, U. H., Manning, P. A. 1989. Nucleotide sequence of the structural gene, *tcpA,* for a major pilin subunit of *Vibrio cholerae*. *Gene* 85:227–31
27. Fry, D. C., Kuby, S. A., Mildvan, A. S. 1986. ATP-binding site of adenylate kinase: mechanistic implications of its homology with *ras*-encoded p21, F1-ATPase, and other nucleotide-binding proteins. *Proc. Natl. Acad. Sci. USA* 83:907–11
28. Fulks, K. A., Marrs, C. F., Stevens, S. P., Green, M. R. 1990. Sequence analysis of the inversion region containing the pilin genes of *Moraxella bovis*. *J. Bacteriol.* 172:310–16
29. Getzoff, E. D., Parge, H. E., McRee, D. E., Tainer, J. A. 1988. Understanding the structure and antigenicity of gonococcal pili. *Rev. Infect. Dis.* 10(S2):296–99
30. Gir'on, J. A., Ho, A. S., Schoolnik, G. K. 1991. An inducible bundle-forming pilus of enteropathogenic *Escherichia coli*. *Science* 254:710–13
31. He, S. Y., Lindeberg, M., Chatterjee, A. K., Collmer, A. 1991. Cloned *Erwinia chrysanthemi* out genes enable *Escherichia coli* to selectively secrete a diverse family of heterologous proteins to its milieu. *Proc. Natl. Acad. Sci. USA* 88:1079–83
32. Hobbs, M., Dalrymple, B., Delaney, S. F., Mattick, J. S. 1988. Transcription of the fimbrial subunit gene and an associated transfer RNA gene of *Pseudomonas aeruginosa*. *Gene* 62: 219–27
33. Hobbs, M., Dalrymple, B. P., Cox, P. T., Livingstone, S. P., Delaney, S. F., Mattick, J. S. 1991. Organization of the fimbrial gene region of *Bacteroides nodosus:* class I and class II strains. *Mol. Microbiol.* 5:543–60
34. Hoyne, P. A., Haas, R., Meyer, T.

F., Davies, J. K., Elleman, T. C. 1992. Production of *Neisseria gonorrhoeae* pili (fimbriae) in *Pseudomonas aeruginosa*. *J. Bacteriol.* 174: 7321–27

35. Hu, N. T., Hung, M. N., Chiou, S. J., Tang, F., Chiang, D. C., et al. 1992. Cloning and characterization of a gene required for the secretion of extracellular enzymes across the outer membrane by *Xanthomonas campestris* pv. *campestris*. *J. Bacteriol.* 174:2679–87
36. Hultgren, S. J., Normark, S., Abraham, S. N. 1991. Chaperone-assisted assembly and molecular architecture of adhesive pili. *Annu. Rev. Microbiol.* 45: 383–415
37. Irvin, R. T., Doig, P., Lee, K. K., Sastry, P. A., Paranchych, W., et al. 1989. Characterization of the *Pseudomonas aeruginosa* pilus adhesin: confirmation that the pilin structural protein subunit contains a human epithelial cell-binding domain. *Infect. Immun.* 57:3720–26
38. Ishimoto, K. S., Lory, S. 1992. Identification of *pilR* which encodes a transcriptional activator of the pilin gene in *Pseudomonas aeruginosa*. *J. Bacteriol.* 174:3514–21

38a. Ishimoto, K. S., Lory, S. 1989. Formation of pilin in *Pseuodomonas aeruginosa* requires the alternative σ factor (RpoN) of RNA polymerase. *Proc. Natl. Acad. Sci. USA* 86:1954–57

39. Jackman, S. H., Rosenbusch, R. F. 1984. In vitro adherence of *Moraxella bovis* to intact corneal epithelium. *Curr. Eye Res.* 3:1107–12
40. Johnson, K., Parker, M. L., Lory, S. 1986. Nucleotide sequence and transcriptional initiation site of two *Pseudomonas aeruginosa* pilin genes. *J. Biol. Chem.* 261:15703–8
41. Johnson, K., Parker, M. L., Lory, S. 1986. Nucleotide sequence and transcriptional initiation site of two *Pseudomonas aeruginosa* pilin genes. *J. Biol. Chem.* 261:15703–8
42. Jonsson, A. B., Nyberg, G., Normark, S. 1991. Phase variation of gonococcal pili by frameshift mutation in *pilC*, a novel gene for pilus assembly. *EMBO J.* 10:477–88
43. Jonsson, A. B., Pfeifer, J., Normark, S. 1992. *Neisseria gonorrhoeae* PilC expression provides a selective mechanism for structural diversity of pili. *Proc. Natl. Acad. Sci. USA* 89:3204–8
44. Kaufman, M. R., Seyer, J. M., Taylor, R. K. 1991. Processing of TCP pilin by TcpJ typifies a common step intrinsic to a newly recognized pathway of extracellular protein secretion by gram-negative bacteria. *Genes Dev.* 5:1834–46
45. Kaufman, M. R., Shaw, C. E., Jones, I. D., Taylor, R. K. 1993. Biogenesis and regulation of the *Vibrio cholerae* toxin-coregulated pilus: analogies to other virulence factor secretory systems. *Gene* 126:43–49
46. Kelly, N. M., Kluftinger, J. L., Pasloske, B. L., Paranchych, W., Hancock, R. E. 1989. *Pseudomonas aeruginosa* pili as ligands for nonopsonic phagocytosis by fibronectin-stimulated macrophages. *Infect. Immun.* 57:3841–45
47. Kluftinger, J. L., Kelly, N. M., Hancock, R. E. 1989. Stimulation by fibronectin of macrophage-mediated phagocytosis of *Pseudomonas aeruginosa*. *Infect. Immun.* 57:817–22
48. Kluftinger, J. L., Kelly, N. M., Jost, B. H., Hancock, R. E. 1989. Fibronectin as an enhancer of nonopsonic phagocytosis of *Pseudomonas aeruginosa* by macrophages. *Infect. Immun.* 57:2782–85
49. Koga, T., Ishimoto, K., Lory, S. 1993. Genetic and functional characterization of the gene cluster specifying expression of *Pseudomonas aeruginosa* pili. *Infect. Immun.* 61:1371–77
50. Krogfelt, K. A. 1991. Bacterial adhesion: genetics, biogenesis, and role in pathogenesis of fimbrial adhesins of *Escherichia coli*. *Rev. Infect. Dis.* 13: 721–35
51. Kustu, S., Santero, E., Keener, J., Popham, D., Weiss, D. 1989. Expression of sigma 54 (*ntrA*)-dependent genes is probably united by a common mechanism. *Microbiol. Rev.* 53:367–76

51a. Lauer, P., Albertson, N. H., Koomey, M. 1993. Conservation of genes encoding components of a type IV pilus assembly/two-step protein export pathway in *Neisseria gonorrhoeae*. *Mol. Microbiol.* 8:357–68

52. Lee, K. K., Doig, P., Irvin, R. T., Paranchych, W., Hodges, R. S. 1989. Mapping the surface regions of *Pseudomonas aeruginosa* PAK pilin: the importance of the C-terminal region for adherence to human buccal epithelial cells. *Mol. Microbiol.* 3:1493–99
53. Lee, K. K., Sastry, P. A., Paranchych, W., Hodges, R. S. 1989. Immunological studies of the disulfide bridge region of *Pseudomonas aeruginosa* PAK and PAO pilins, using anti-PAK pilus and antipeptide antibodies. *Infect. Immun.* 57:520–26

54. Lepper, A. W. 1988. Vaccination against infectious bovine keratoconjunctivitis: protective efficacy and antibody response induced by pili of homologous and heterologous strains of *Moraxella bovis*. *Aust. Vet. J.* 65:310–16
55. Lepper, A. W., Power, B. E. 1988. Infectivity and virulence of Australian strains of *Moraxella bovis* for the murine and bovine eye in relation to pilus serogroup sub-unit size and degree of piliation. *Aust. Vet. J.* 65:305–9
56. Lindberg, M., Collmer, A. 1992. Analysis of eight out genes in a cluster required for pectic enzyme secretion by *Erwinia chrysanthemi:* sequence comparison with secretion genes from other gram-negative bacteria. *J. Bacteriol.* 174:7385–97
57. Mardh, P. A., Westrom, L. 1976. Adherence of bacteria to vaginal epithelial cells. *Infect. Immun.* 13:661–66
58. Marrs, C. F., Rozsa, F. W., Hackel, M., Stevens, S. P., Glasgow, A. C. 1990. Identification, cloning, and sequencing of *piv,* a new gene involved in inverting the pilin genes of *Moraxella lacunata*. *J. Bacteriol.* 172:4370–77
59. Marrs, C. F., Schoolnik, G., Koomey, J. M., Hardy, J., Rothbard, J., Falkow, S. 1985. Cloning and sequencing of a *Moraxella bovis* pilin gene. *J. Bacteriol.* 163:132–39
60. Marrs, C. F., Weir, S. 1990. Pili (fimbriae) of *Branhamella* species. *Am. J. Med.* 88:36s–40s
61. Mattick, J. S., Bills, M. M., Anderson, B. J., Dalrymple, B., Mott, M. R., Egerton, J. R. 1987. Morphogenetic expression of *Bacteroides nodosus* fimbriae in *Pseudomonas aeruginosa*. *J. Bacteriol.* 169:33–41
62. McGee, Z. A., Johnson, A. P., Taylor-Robinson, D. 1981. Pathogenic mechanisms of *Neisseria gonorrhoeae:* observations on damage to human fallopian tubes in organ culture by gonococci of colony type 1 or type 4. *J. Infect. Dis.* 143:413–22
63. McKern, N. M., Stewart, D. J., Strike, P. M. 1988. Amino acid sequences of pilins from serologically distinct strains of *Bacteroides nodosus*. *J. Protein Chem.* 7:157–64
64. Mekalanos, J. J. 1992. Environmental signals controlling expression of virulence determinants in bacteria. *J. Bacteriol.* 174:1–7
65. Meyer, T. F., Billyard, E., Haas, R., Storzbach, S., So, M. 1984. Pilus genes of *Neisseria gonorrhoeae:* chromosomal organization and DNA sequence. *Proc. Natl. Acad. Sci. USA* 81:6110–14
66. Meyer, T. F., Gibbs, C. P., Haas, R. 1990. Variation and control of protein expression in *Neisseria*. *Annu. Rev. Microbiol.* 44:451–77
67. Mohan, S., Aghion, J., Guillen, N., Dubnau, D. 1989. Molecular cloning and characterization of *comC,* a late competence gene of *Bacillus subtilis*. *J. Bacteriol.* 171:6043–51
68. Moore, L. J., Rutter, J. M. 1989. Attachment of *Moraxella bovis* to calf corneal cells and inhibition by antiserum. *Aust. Vet. J.* 66:39–42
69. Nassif, X., Lowy, J., Stenberg, P., O'Gaora, P., Ganji, A., So, M. 1993. Antigenic variation of pilin regulates the adhesion of *Neisseria meningitidis* to a human epithelial cell line. *Mol. Microbiol.* 8:719–26
70. Nunn, D., Bergman, S., Lory, S. 1990. Products of three accessory genes, *pilB, pilC,* and *pilD,* are required for biogenesis of *Pseudomonas aeruginosa* pili. *J. Bacteriol.* 172:2911–19
71. Nunn, D. N., Lory, S. 1991. Product of the *Pseudomonas aeruginosa* gene *pilD* is a prepilin leader peptidase. *Proc. Natl. Acad. Sci. USA* 88:3281–85
72. Nunn, D. N., Lory, S. 1992. Components of the protein-excretion apparatus of *Pseudomonas aeruginosa* are processed by the type IV prepilin peptidase. *Proc. Natl. Acad. Sci. USA* 89:47–51
73. Ogierman, M. A., Manning, P. A. 1992. Homology of TcpN, a putative regulatory protein of *Vibrio cholerae,* to the AraC family of transcriptional activators. *Gene* 116:93–97
74. Ogierman, M. A., Zabihi, S., Mourtzios, L., Manning, P. A. 1993. Genetic organization and nucleotide sequence of the promoter distal region of the *tcp* gene cluster of *Vibrio cholerae*. *Gene*. 126:51–60
75. Oudega, B., De, G. F. K. 1988. Genetic organization and biogenesis of adhesive fimbriae of *Escherichia coli*. *Antonie van Leeuwenhoek J. Microbiol. Serol.* 54:285–99
76. Paranchych, W., Frost, L. 1988. The physiology and biochemistry of pili. *Adv. Microbiol. Physiol.* 29:53–114
77. Parge, H. E., Bernstein, S. L., Deal, C. D., McRee, D. E., Christensen, D., et al. 1990. Biochemical purification and crystallographic characterization of the fiber-forming protein pilin from *Neisseria gonorrhoeae*. *J. Biol. Chem.* 265:2278–85
78. Parge, H. E., McRee, D. E., Capozza,

M. A., Bernstein, S. L., Getzoff, E. D., Tainer, J. A. 1987. Three dimensional structure of bacterial pili. *Antonie Van Leeuwenhoek J. Microbiol. Serol.* 53:447–53
79. Paruchuri, D. K., Seifert, H. S., Ajioka, R. S., Karlsson, K. A., So, M. 1990. Identification and characterization of a *Neisseria gonorrhoeae* gene encoding a glycolipid-binding adhesin. *Proc. Natl. Acad. Sci. USA* 87:333–37
80. Pasloske, B. L., Paranchych, W. 1988. The expression of mutant pilins in *Pseudomonas aeruginosa:* fifth position glutamate affects pilin methylation. *Mol. Microbiol.* 2:489–95
81. Peek, J. A., Taylor, R. K. 1992. Characterization of a periplasmic thiol:disulfide interchange protein required for the functional maturation of secreted virulence factors of *Vibrio cholerae*. *Proc. Natl. Acad. Sci. USA* 89:6210–14
82. Potts, W. J., Saunders, J. R. 1988. Nucleotide sequence of the structural gene for class I pilin from *Neisseria meningitidis:* homologies with the *pilE* locus of *Neisseria gonorrhoeae*. *Mol. Microbiol.* 2:647–53
83. Deleted in proof
84. Pugsley, A. P. 1993. The complete general protein secretory pathway in gram-negative bacteria. *Microbiol. Rev.* 57:50–108
85. Pugsley, A. P., Dupuy, B. 1992. An enzyme with type IV prepilin peptidase activity is required to process components of the general extracellular protein secretion pathway of *Klebsiella oxytoca*. *Mol. Microbiol.* 6:751–60
86. Pugsley, A. P., Reyss, I. 1990. Five genes at the 3 end of the *Klebsiella pneumoniae pulC* operon are required for pullulanase secretion. *Mol. Microbiol.* 4:365–79
87. Ramphal, R., Carnoy, C., Fievre, S., Michalski, J. C., Houdret, N., et al. 1991. *Pseudomonas aeruginosa* recognizes carbohydrate chains containing type 1 (Gal beta 1–3GlcNAc) or type 2 (Gal beta 1–4GlcNAc) disaccharide units. *Infect. Immun.* 59:700–4
88. Ramphal, R., Koo, L., Ishimoto, K. S., Totten, P. A., Lara, J. C., Lory, S. 1991. Adhesion of *Pseudomonas aeruginosa* pilin-deficient mutants to mucin. *Infect. Immun.* 59:1307–11
89. Rao, V. K., Progulske-Fox, A. 1993. Identification of a novel type 4 pilin gene arrangement: cloning and sequence of two N-methylphenylalanine pilin genes from *Eikenella corrodens*. *J. Gen. Microbiol.* 139:651–60
90. Rothbard, J. B., Fernandez, R., Wang, L., Teng, N. N., Schoolnik, G. K. 1985. Antibodies to peptides corresponding to a conserved sequence of gonococcal pilins block bacterial adhesion. *Proc. Natl. Acad. Sci. USA* 82:915–19
91. Saiman, L., Ishimoto, K., Lory, S., Prince, A. 1990. The effect of piliation and exoproduct expression on the adherence of *Pseudomonas aeruginosa* to respiratory epithelial monolayers. *J. Infect. Dis.* 161:541–48
92. Sato, H., Okinaga, K., Saito, H. 1988. Role of pili in the pathogenesis of *Pseudomonas aeruginosa* burn infection. *Microbiol. Immunol.* 32:131–39
93. Schuler, G. D., Altschul, S. F., Lipman, D. J. 1991. A workbench for multiple alignment construction and analysis. *Proteins* 9:180–90
94. Seifert, H. S., So, M. 1988. Genetic mechanisms of bacterial antigenic variation. *Microbiol Rev.* 52:327–36
95. Shaw, C. E., Taylor, R. K. 1990. *Vibrio cholerae* O395 *tcpA* pilin gene sequence and comparison of predicted protein structural features to those of type 4 pilins. *Infect. Immun.* 58:3042–49
96. Deleted in proof
97. Simpson, D. A., Ramphal, R., Lory, S. 1992. Genetic analysis of *Pseudomonas aeruginosa* adherence: distinct genetic loci control attachment to epithelial cells and mucins. *Infect. Immun.* 60:3771–79
98. Speert, D. P., Loh, B. A., Cabral, D. A., Eftekhar, F. 1985. Opsonin-independent phagocytosis of *Pseudomonas aeruginosa*. *Antibiot. Chemother.* 36:88–94
99. Speert, D. P., Loh, B. A., Cabral, D. A., Salit, I. E. 1986. Nonopsonic phagocytosis of nonmucoid *Pseudomonas aeruginosa* by human neutrophils and monocyte-derived macrophages is correlated with bacterial piliation and hydrophobicity. *Infect. Immun.* 53:207–12
100. Stewart, D. J., Clark, B. L., Peterson, J. E., Emery, D. L., Smith, E. F., et al. 1985. The protection given by pilus and whole cell vaccines of *Bacteroides nodosus* strain 198 against ovine foot-rot induced by strains of different serogroups. *Aust. Vet. J.* 62:153–59
101. Stewart, D. J., Clark, B. L., Peterson, J. E., Griffiths, D. A., Smith, E. F. 1982. Importance of pilus-associated antigen in *Bacteroides nodosus* vaccines. *Res. Vet. Sci.* 32:140–47

101a. Strom, M. S., Bergman, P., Lory, S. 1993. Identification of active site cysteines in the conserved domain of PilD, the bifunctional type IV pilin leader peptidase/N-methyltransferase of *Pseudomonas aeruginosa*. *J. Biol. Chem.* In press

102. Strom, M. S., Lory, S. 1987. Mapping of export signals of *Pseudomonas aeruginosa* pilin with alkaline phosphatase fusions. *J. Bacteriol.* 169:3181–8

103. Strom, M. S., Lory, S. 1991. Amino acid substitutions in pilin of *Pseudomonas aeruginosa*. Effect on leader peptide cleavage, amino-terminal methylation, and pilus assembly. *J. Biol. Chem.* 266:1656–64

104. Strom, M. S., Lory, S. 1992. Kinetics and sequence specificity of processing of prepilin by PilD, the type IV leader peptidase of *Pseudomonas aeruginosa*. *J. Bacteriol.* 174:7345–51

105. Strom, M. S., Nunn, D., Lory, S. 1991. Multiple roles of the pilus biogenesis protein *pilD:* involvement of *pilD* in excretion of enzymes from *Pseudomonas aeruginosa*. *J. Bacteriol.* 173:1175–80

106. Strom, M. S., Nunn, D. N., Lory, S. 1993. A single bifunctional enzyme, PilD, catalyzes cleavage and N-methylation of proteins belonging to the type IV pilin family. *Proc. Natl. Acad. Sci. USA* 90:2404–8

107. Stromberg, N., Deal, C., Nyberg, G., Normark, S., So, M., Karlsson, K. A. 1988. Identification of carbohydrate structures that are possible receptors for *Neisseria gonorrhoeae*. *Proc. Natl. Acad. Sci. USA* 85:4902–6

108. Swanson, J. 1983. Gonococcal adherence: selected topics. *Rev. Infect. Dis.* 5(Suppl. 4):S678–84

109. Taha, M. K., Dupuy, B., Saurin, W., So, M., Marchal, C. 1991. Control of pilus expression in *Neisseria gonorrhoeae* as an original system in the family of two-component regulators. *Mol. Microbiol.* 5:137–48

110. Taha, M. K., So, M., Seifert, H. S., Billyard, E., Marchal, C. 1988. Pilin expression in *Neisseria gonorrhoeae* is under both positive and negative transcriptional control. *EMBO J.* 7:4367–78

111. Taha, M. K., So, M., Seifert, H. S., Billyard, E., Marchal, C. 1988. Pilin expression in *Neisseria gonorrhoeae* is under both positive and negative transcriptional control. *EMBO J.* 7:4367–78

112. Taylor, R., Shaw, C., Peterson, K., Spears, P., Mekalanos, J. 1988. Safe, live *Vibrio cholerae* vaccines? *Vaccine* 6:151–4

113. Taylor, R. K. 1989. Genetic studies of enterotoxin and other potential virulence factors of *Vibrio cholerae*. In *Genetics of Bacterial Diversity*, ed. D. A. Hopwood, K. F. Chater, pp. 310–29. London: Academic

114. Taylor, R. K., Miller, V. L., Furlong, D. B., Mekalanos, J. J. 1987. Use of *phoA* gene fusions to identify a pilus colonization factor coordinately regulated with cholera toxin. *Proc. Natl. Acad. Sci. USA* 84:2833–37

115. Thompson, D. V., Melchers, L. S., Idler, K. B., Schilperoort, R. A., Hooykaas, P. J. 1988. Analysis of the complete nucleotide sequence of the *Agrobacterium tumefaciens virB* operon. *Nucleic Acids Res.* 16:4621–36

116. Tønjum, T., Marrs, C. F., Rozsa, F., Bovre, K. 1991. The type 4 pilin of *Moraxella nonliquefaciens* exhibits unique similarities with the pilins of *Neisseria gonorrhoeae* and *Dichelobacter (Bacteroides) nodosus*. *J. Gen. Microbiol.* 137:2483–90

117. Tønjum, T., Weir, S., Bovre, K., Progulske-Fox, A., Marrs, C. F. 1993. Sequence divergence in two tandemly located pilin genes of *Eikenella corrodens*. *Infect. Immun.* 61:1909–16

118. Virji, M., Alexandrescu, C., Ferguson, D. J., Saunders, J. R., Moxon, E. R. 1992. Variations in the expression of pili: the effect on adherence of *Neisseria meningitidis* to human epithelial and endothelial cells. *Mol. Microbiol.* 6: 1271–79

119. Virji, M., Heckels, J. E. 1983. Antigenic cross-reactivity of *Neisseria* pili: investigations with type- and species-specific monoclonal antibodies. *J. Gen. Microbiol.* 129:2761–68

120. Virji, M., Heckels, J. E. 1984. The role of common and type-specific pilus antigenic domains in adhesion and virulence of gonococci for human epithelial cells. *J. Gen. Microbiol.* 130: 1089–95

121. Walker, J. E., Saraste, M., Runswick, M. J., Gay, N. J. 1982. Distantly related sequences in the alpha- and beta-subunits of ATP synthase, myosin, kinases and other ATP-requiring enzymes and a common nucleotide binding fold. *EMBO J.* 1:945–51

122. Watts, T. H., Kay, C. M., Paranchych, W. 1982. Dissociation and characterization of pilin isolated from *Pseudomonas aeruginosa* strains PAK and PAO. *Can. J. Biochem.* 60:867–72

123. Watts, T. H., Kay, C. M., Paranchych, W. 1983. Spectral properties of three

quaternary arrangements of *Pseudomonas* pilin. *Biochemistry* 22:3640–46
124. Weiss, D. S., Batut, J., Klose, K. E., Keener, J., Kustu, S. 1991. The phosphorylated form of the enhancer-binding protein NTRC has an ATPase activity that is essential for activation of transcription. *Cell* 67:155–67
125. Whitchurch, C. B., Hobbs, M., Livingston, S. P., Krishnapillai, V., Mattick, J. S. 1991. Characterisation of a *Pseudomonas aeruginosa* twitching motility gene and evidence for a specialised protein export system widespread in eubacteria. *Gene* 101:33–44
126. Woods, D. E., Straus, D. C., Johanson, W. J., Berry, V. K., Bass, J. A. 1980. Role of pili in adherence of *Pseudomonas aeruginosa* to mammalian buccal epithelial cells. *Infect. Immun.* 29:1146–51
127. Yu, J., Webb, H., Hirst, T. R. 1992. A homologue of the *Escherichia coli* DsbA protein involved in disulphide bond formation is required for enterotoxin biogenesis in *Vibrio cholerae*. *Mol. Microbiol.* 6:1949–58

Annu. Rev. Microbiol. 1993. 47:597–626

MOLECULAR BIOLOGY OF THE LysR FAMILY OF TRANSCRIPTIONAL REGULATORS

Mark A. Schell
Department of Microbiology, University of Georgia, Athens, Georgia 30602

KEY WORDS: transcription activation, DNA binding, helix-turn-helix, protein domains

---

CONTENTS

ABSTRACT

The LysR family is composed of >50 similar-sized, autoregulatory transcriptional regulators (LTTRs) that apparently evolved from a distant ancestor into subfamilies found in diverse prokaryotic genera. In response to different

0066-4227/93/1001-0597$02.00

coinducers, LTTRs activate divergent transcription of linked target genes or unlinked regulons encoding extremely diverse functions. Mutational studies and amino acid sequence similarities of LTTRs identify: (*a*) a DNA-binding domain employing a helix-turn-helix motif (residues 1–65), (*b*) domains involved in coinducer recognition and/or response (residues 100–173 and 196–206), (*c*) a domain required for both DNA binding and coinducer response (residues 227–253). DNA footprinting studies suggest that in the absence of coinducer many LTTRs bind to regulated promoters via a 15-bp dyadic sequence with a common structure and position (near −65). Coinducer causes additional interactions of LTTRs with sequences near the −35 RNA polymerase binding site and/or DNA bending that results in transcription activation.

INTRODUCTION

The LysR family of transcriptional regulators was first reported by Henikoff et al in 1988 (48), although others independently noticed an incipient family (22, 29, 104). Since then the LysR family has grown from 9 to over 50 members that regulate diverse genes and complex regulons in many prokaryotic genera. New members are discovered almost monthly. With the exception of the two-component systems (111), LysR-type transcriptional regulators (LTTRs) may be the most common type of positive regulators in prokaryotes. This review summarizes the general distinguishing characteristics of the family, the structure-function organization of LTTR polypeptides, and the biochemical and molecular aspects of their mode(s) of action. I also discuss the phylogeny and distribution of LTTRs to give insight into their origins, evolution, and diversity of their distribution and action.

GENERAL CHARACTERISTICS OF LysR FAMILY MEMBERS

Table 1 lists the currently known (March 1993) LysR family members; a few others (e.g. CitR, ORF2) have been detected from incomplete or unpublished DNA sequences (49, 107a; S. Jin & A. Sonenshein, personal communication). Usually, a >20% amino acid sequence identity with another LTTR family member or a consensus sequence for the highly conserved amino terminus (N terminus) indicates the presence of an LTTR (48, 49; A. Bairoch, personal communication). In the N-terminal halves of the majority of LTTRs, >20% of the residues at each aligned position are identical and 53% show a similarity coefficient above a log-odds threshold of 10 (128). Some, but not all, of this similarity derives from conservation of a helix-turn-helix DNA-binding motif. In comparison, most aligned residues in the carboxy-terminal (C-terminal) halves show only 4% identity, and 27% have a log-odds threshold >10 (128).

Table 1 Properties of LysR family members

Member	Origin	Function of regulated target genes	Coinducer	Reference
AlsR	*Bacillus subtilis*	Acetoin synthesis	Acetate or pH	89a
AmpR	*Rhodobacter capsulatus, Enterobacter cloacae, Citrobacter freundii*	β-Lactamase	β-Lactam or derivative	18, 51, 65
BlaA	*Streptomyces* (several spp.)	β-Lactamase	Unknown	123
CatM	*Acinetobacter calcolaceticus*	Catechol catabolism	*cis-cis*-muconate	77
CatR	*Pseudomonas putida*	Catechol catabolism	*cis-cis*-muconate	92
CfxR	*Alcaligenes eutrophus*	Carbon dioxide fixation (rubisco)	Formate?	134
CfxO[a] (OrfD)	*Xanthobacter flavus*	Carbon dioxide fixation?	Unknown	73
ChvO[a] (GbpR)	*Agrobacterium tumefaciens*	ChvE virulence regulator?	Unknown	55, 64
ClcR	Plasmid pAC25	Chlorocatechol catabolism	3-Chlorobenzoate?	24, 35
CynR	*Escherichia coli*	Detoxification of cyanate	Cyanate	116
CysB[b]	*Salmonella typhimurium, Escherichia coli*	Cysteine biosynthesis	*O*-acetyl serine or *N*-acetylserine	79
DgdR	*Pseudomonas cepacia*	Dialkylglycine decarboxylase	2-Methylalanine	57a
GltC	*Bacillus subtilis*	Glutamate synthase	Glutamine?	7
IciA[c]	*Escherichia coli*	Inhibition of initiation of DNA replication in vitro	Unknown	121
IlvY	*Escherichia coli*	Isoleucine/valine biosynthesis	Acetolactate or acetohydroxybutyrate	132
IrgB	*Vibrio cholerae*	Iron-regulated virulence factors	None?	42
LeuO[c]	*Escherichia coli*	Unknown	Unknown	48
LysR	*Escherichia coli*	Lysine biosynthesis	Diaminopimelate	114
MetR	*S. typhimurium, E. coli*	Methionine biosynthesis	Homocysteine	86
MleR[b]	*Lactococcus lactis*	Malolactic enzyme	L-malate	89
MprR	*Streptomyces* (several spp.)	Metalloprotease	Unknown	28
Nac[b]	*Klebsiella aerogenes*	Use of poor nitrogen sources	None	5, 107a
NahR	NAH7 plasmid of *P. putida*	Naphthalene/salicylate catabolism	Salicylate	104

Table 1 *(Continued)*

Member	Origin	Function of regulated target genes	Coinducer	Reference
NhaR[b] (AntO)	*Escherichia coli, Salmonella enteriditis*	Na^+-H^+ antiporter	Na^+ or Li^+	68, 85, 88
NocR	*A. tumefaciens* Ti plasmids	Nopaline catabolism	Nopaline	129
NodDs >14 species	*Rhizobium, Bradyrhizobium, Azorhizobium*	Development of nitrogen fixation symbiosis (*nod* genes)	Various flavonoids	1, 94
OccR	*A. tumefaciens* Ti plasmid	Octopine catabolism	Octopine	46
ORF294 (XapR?)[c]	*Escherichia coli*	Xanthosine phosphorylase?	Unknown	13
OxyR[b] (MomR, Mor)	*Escherichia coli, Salmonella typhimurium*	Oxidative stress response; phage restriction; aggregation	Direct oxidation of OxyR; H_2O_2?	8, 23, 131
PhcA[b]	*Pseudomonas solanacearum*	Regulators of virulence factors	Unknown	12, 101
RbcR	*Chromatium vinosum, Thiobacillus ferroxidans*	Carbon dioxide fixation?	Photoautotrophy	63a, 128
SdsB	*Pseudomonas* sp.	Alkyl sulfatase	Na-dodecyl-SO_4	29
SpvR[b] (VsdA, MkaC)	*Salmonella* virulence plasmids	Synthesis of virulence factors	Unknown	17, 63
SyrM[c]	*Rhizobium meliloti*	Exopolysaccharide synthesis; nodulation genes	None	2
TcbR	*Pseudomonas* plasmid J51	Chlorocatechol metabolism	3-Chlorobenzoate	127
TdcO	*Escherichia coli*	Threonine degradation?	Anaerobic metabolite?	43
TfdS[a] (TfdO)	*A. eutrophus* plasmid pJP4	2,4-Dichlorophenol hydroxylase	Chloromaleyl acetate or chlorodiene lactone	57, 115
TrpI	*P. putida, P. aeruginosa, P. syringae*	Tryptophan biosynthesis	Indoleglycerol phosphate	22

[a] Only partial DNA sequence available. Alternate designations are in parentheses.
[b] Divergent promoter structure absent.
[c] Divergent promoter structure not known.

Consensus sequences for LTTRs are available (49, 128). Some LTTRs have anomalously low Lys-to-Arg ratios (128), possibly as a result of the high G+C content of some of their genes: Lys codons contain only A+T, whereas Arg codons contain only G+C. This possibly leads to the selection of Arg over Lys in proteins encoded by high G+C DNA.

Most family members share four other characteristics: (*a*) They encode coinducer-responsive transcriptional activator proteins with sizes ranging from 276 to 324 residues. (*b*) Independent of the presence of coinducer they bind at regulated targets to DNA sequences that have a similar position and structural motif. (*c*) Each is divergently transcribed from a promoter that is very close to and often overlaps a promoter of a regulated target gene. (*d*) Because an overlapping divergent promoter allows simultaneous bidirectional control of transcription (4), most LTTRs repress their own transcription by 3- to 10-fold (i.e. negatively autoregulate), possibly to self-maintain constant levels. Several LTTRs appear to lack this divergent promoter structure (Table 1). Of these, OxyR, Nac, and CysB still negatively autoregulate. MleR (89) and AlsR (89a) apparently do not autoregulate. MkaC (SpvR) (118), some NodDs (107, 108), and PhcA (12) show more than sixfold positive autoregulation (i.e. increase their own transcription). Usually LTTRs activate transcription of target promoters between 6- and 200-fold, but only in the presence of a small signal molecule (coinducer). Exceptions are Nac (5) and NodD3 (76), which do not require a coinducer, and CatM (77), AmpR of *Enterobacter cloacae* (51), and IciA (121), which may act mainly as repressors. CysB, OccR, Nac, MomR (OxyR), and MetR show repressor-like effects at some promoters. TfdS is purportedly a repressor-activator (57), but its 40% maximum repression is much lower than that expected for a repressor.

The following LTTRs have been purified to near homogeneity: AmpR (3), TrpI (21), MetR (70), CatR (84, 93), IlvY (W. Hatfield, unpublished results), OxyR (113), OccR (130), IciA (121), Nac (5), and CysB (74). In general ammonium sulfate precipitation (<30% saturation) and chromatography on heparin-agarose, DNA cellulose, or methyl agarose were used to fractionate extracts of *Escherichia coli* overexpressing an LTTR. Sizes of active LTTRs suggest that in vivo they are multimeric: TrpI (21), CysB (74), and NahR (100) likely are tetramers, but MetR (70), CatR (84), IlvY, NodD3 (34), Nac (5), and IciA (121) are likely dimers. This variation may signify a fundamental difference between these LTTRs, with implications for their mode of action or could be an experimental artifact. In vivo NodD of *Rhizobium leguminosarum* appears membrane associated, which may help it respond to its hydrophobic coinducers (flavonoids) dissolved in the membrane (106). Preliminary data suggest NhaR may be membrane associated (88); however, all other purified LTTRs appear to be soluble cytoplasmic proteins. Binding of a coinducer to a purified LTTR has not been demonstrated, but coinducer effects on interactions of several LTTRs with their target promoters are documented (see below).

DIVERSITY OF TARGET GENES

As seen in Table 1, LTTRs regulate very diverse genes and functions. Some LTTRs (e.g. NodDs, CysB, OxyR, PhcA) control multiple, unlinked target genes (regulons), while others (MetR, IrgB, SyrM, Nac) are themselves regulated by other genes forming complex regulatory networks. For example, IrgB (42) is required for expression of the divergently transcribed *irgA* gene, which likely encodes the vibriobactin receptor, a virulence factor of *Vibrio cholerae*. Iron strongly represses transcription of *irgB* and *irgA* from start sites overlapping by 5 bp, likely because of binding of a Fur-like repressor to the overlapping sites. Another example is Nac of *Klebsiella aerogenes* (5), which in concert with CAP-cAMP, transcriptionally activates *hut* (encoding histidase), *put* (encoding proline utilization), and *ure* (encoding urease), but represses *gdh* (encoding glutamate dehydrogenase) in response to nitrogen limitation (107a). Nac activation of *put* does not require a coinducer; rather, the NTR system and Nac itself control *nac* transcription by σ^{54} RNA polymerase. SpvR positively regulates three linked genes required for spleen invasion by *Salmonella typhimurium* and that are expressed only during stationary phase (17, 26, 118). This involves the alternate sigma factor, KatF, which could interact with or control expression of SpvR (78). Alternatively, SpvR may require a coinducer produced only during stationary phase. A final example is PhcA, which regulates transcription of unlinked genes encoding virulence factors of the phytopathogen *P. solanacearum* (12). PhcA control appears indirect, because it appears to regulate expression of several transcriptional activators that in turn regulate the virulence genes; PhcA may also be part of a cell density–sensitive regulatory network (101).

DISTRIBUTION, EVOLUTION, AND PHYLOGENY

LTTRs are widely distributed in diverse genera of prokaryotes (Table 1). The G+C contents of genes encoding LTTRs vary between 75% (BlaA) and 28% (MleR). The majority are in the genomes of *Proteobacteria* (purple bacteria) of the α and γ subgroups. A few have been found in the β subgroup, but none in the δ subgroup. MleR, AlsR, BlaA, MprR, and GltC are in gram-positive bacteria, but none have been found in *Archaebacteria* or eukaryotes. However, because many prokaryotic genera have not been subjected to extensive genetic characterization, the observed distribution of LTTRs may be nonrepresentative. The large genetic distances between prokaryotes with LTTRs and vast differences in G+C content suggest that a progenitor LTTR arose early in prokaryotic evolution. The structural diversity of the different coinducers that stimulate various LTTRs to activate transcription (e.g. aromatics, various aliphatics, ions, amino acid derivatives) also suggests a long-term divergence of signal recognition by LTTRs. However the many LTTRs on self-transmissible plasmids (which probably move freely throughout the prokaryotic community) may have promoted a more recent and rapid dissemination and evolution.

Schlaman et al presented a phylogenetic tree for 42 LTTRs (107). This tree and functional similarities clearly point to two subfamilies in addition to the NodDs. The first consists of ClcR, TcbR, TfdS (TfdO), and possibly CatR and CatM. ClcR shows ~60% amino acid sequence identity with TcbR and the first 177 available residues of TfdS (24). Moreover, both ClcR and TcbR activate by 15-fold homologous divergent promoters of highly homologous genes encoding for oxidation of chlorocatechols and likely respond to the same coinducer (a metabolite of 3-chlorobenzoate). TfdS, like ClcR, is plasmid-borne and is required for induction of very similar chlorocatechol degradation genes. ClcR shows 32% amino acid sequence identity with CatR or CatM, which regulate divergently transcribed genes encoding similar enzymes for oxidation of nonchlorinated catechols. ClcR and TcbR may have diverged from CatR to recognize chlorocatechols and activate genes for their degradation.

Another possible subfamily is comprised of RbcR, CfxR, and CfxO, each of which is required for expression of divergently transcribed genes encoding for CO_2-fixation enzymes (e.g. Rubisco). The amino acid sequence of RbcR is 45% identical to CfxR, which is 40% identical to the 150 residues of the available sequence of CfxO (63a, 73, 128). That these three very similar LTTRs are each from a different branch of *Proteobacteria* (and *rbcR*-like genes are absent from the cyanobacterium *Anacystis nidulans*) suggests that they diverged from an LTTR in an ancestral phototroph (128). Sequence similarities and the phylogenetic tree (107) suggest other possible subfamilies: (*a*) NahR, SyrM, LeuO; (*b*) AmpR, TrpI, LysR, OccR/NocR; (*c*) CysB, GltC, CfxR/RbcR. However, members of these putative subfamilies do not share any obvious common patterns of function, coinducer, target genes, or origin, indicating substantial divergence. The conserved size, organization, and diversity of LTTRs suggest they are likely old but useful and efficient regulators.

DOMAIN ORGANIZATION OF LTTR POLYPEPTIDES

Insight into important structural features, functional domains, and mode of action can be derived by looking at the locations of amino acid substitutions that cause altered function of LTTRs and the extent to which the residue substituted for is conserved in all LTTRs. Although the mutational studies identifying important functional regions are limited to a few LTTRs—NahR, NodD, OxyR, CysB, AlsR, and AmpR (3, 14, 25, 54, 89a, 100, 113)—the extensive amino acid sequence similarity between LTTRs makes extrapolation reasonable. An alignment of the amino acid sequences of 22 LTTRs was recently published (128); for the discussion below, residue position 1 of a generic LTTR is defined as aligning with the Met translational initiator of Nod, NahR, AmpR, and TrpI (whose N termini align exactly without gaps to residue 55 in this alignment); the Met initiator of most other LTTRs is between positions 3 and 6.

Amino-Terminal DNA-Binding Domain

The region of greatest amino acid sequence identity between LTTRs is clearly the 66 N-terminal residues. The central portion of this highly conserved region (residues 23–42) is nearly 40% identical in all LTTRs; secondary structure predictions (40) and other methods (30) predict it contains a helix-turn-helix DNA-binding motif (HTH). The structure and functions of the 20-residue HTH have been well studied (11, 56, 83). The HTH of LTTRs (see Figure 1) varies somewhat from classic HTHs, mostly in poor conservation of Gly at HTH position 9 and presence of Pro at HTH position 13 (LTTR positions 31 and 35, respectively). The highly conserved Ala at HTH position 5 (LTTR position 27) and Val/Leu/Ile at HTH position 15 (LTTR position 37) are usually present. Overall the most highly conserved residues (identical in 70% of LTTRs) are Ala27, Thr(Ser)33, Gln34, Pro35, Ser(Thr)38, Leu44, and Glu45.

Mutagenesis experiments largely confirm the presence of an HTH between residues 23–42, which along with ~15 residues on either side mediates specific binding of LTTRs to their regulated promoters. However, as discussed later, other regions also are important for DNA binding. In a study with NahR (100), 40% of the 13 independent, single–amino acid substitutions affecting DNA-binding activity mapped between residues 26 and 56—replacing either Ala27 (universally conserved at HTH position 5) or the adjacent Thr26 destroys NahR's ability to specifically bind to and activate transcription of its targets. Substitutions for Arg43 or Arg45 in NahR (100), or Ala40 or Ser42 of NodD, a related LTTR (14, 104), caused a 15-fold decrease in coinducer stimulated transcription activation, likely due to loss of DNA binding ability. Substitutions for Leu23 or Arg43 of NodD destroyed autoregulatory ability, which likely depends on specific DNA binding (14). Null phenotypes caused by substitutions at Arg58 or Pro62 of NodD (14) and Thr56 of NahR (100) also probably result from loss of DNA-binding ability. Amino acid substitution for the highly conserved Pro35 of NahR rendered it unable to activate transcription (100) and changed its interactions with a target promoter (54).

A substitution at the same position of AmpR also caused loss of in vivo function (3). Substitution for the highly conserved Ser38 in CysB destroyed its DNA binding activity (25). A substitution of Pro for Leu33 (LTTR position 37) inactivated MetR, whereas replacing Leu19 (LTTR position 23) reduced its ability to activate transcription by sixfold (71).

Thus the vast majority of amino acid substitutions reported to affect DNA binding of LTTRs are clustered between residues 23 and 62, in or near the putative HTH; a few others cluster in the C terminus between 240–290 (see below). Circumstantial evidence supports the primary importance of these two regions in DNA binding (14, 100), because mutations outside these two regions are much less likely to affect DNA binding. Maxon et al (71) suggested a region of MetR overlapping the LTTR HTH region was a leucine zipper, a motif usually involved in multimerization. However, substitutions for two of the four key Leu residues of the putative zipper had no effect on in vivo activity or the multimerization state of MetR. Most evidence is more consistent with the conclusion that the N terminus of LTTRs forms an HTH, not a leucine zipper. Domains involved in multimerization of LTTRs have yet to be defined.

Coinducer Recognition/Response Domains

Residues 100–150 of different LTTRs show much less similarity to one another than residues 1–66; outside these regions only residues 236–246 show any significant conservation between LTTRs (Figure 1). Even the closely related NodDs from different species of *Rhizobium* show extensive C-terminal sequence divergence (1). Exchanging the C-terminal portions of several NodDs with different coinducer specificities showed that recognition function lies between residues 122 and 270 (52, 109), leading to the assumption that the C-terminal halves of LTTRs are involved in coinducer recognition, an assumption supported by many subsequent experiments. Studies of mutant LTTRs with altered coinducer response, all mapping between residues 102 and 253 of NodD (14, 72), NahR (54, 100), AmpR (3), OxyR (113), AlsR (89a), and CysB (25), has further refined the location of coinducer recognition/response functions into subdomains.

Gyorgypal et al (45) noticed that residues 109–171 and 189–216 of NodD had 45% sequence similarity with the steroid-binding domain consensus sequence of the human estrogen receptor. In vivo, NodD and the receptor likely recognize chemically similar signal ligands (i.e. flavanones); in fact, estradiol was recognized by NodD because it caused a 25-fold activation of a NodD-regulated promoter (45). The presence of a putative β-turn-β ligand-binding crevice motif (10, 45) (residues 172–188 and 222–245) also supports the hypothesis that residues 109–245 of NodD are involved in coinducer recognition. The position of amino acid substitutions altering ligand-responsive transcription activation by LTTRs without affecting DNA-

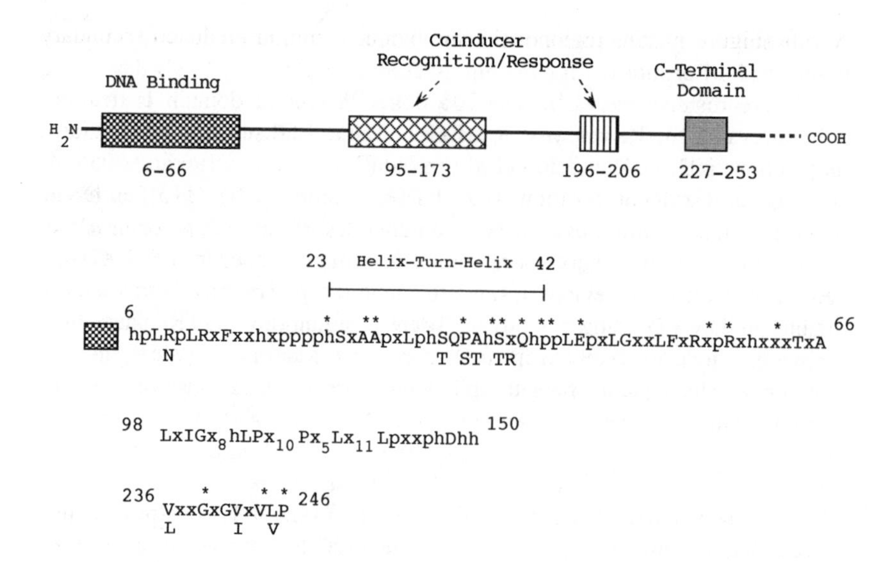

Figure 1 Model for domain organization of an LTTR. Residue numbers and domains (*boxes*) are as defined in text. At the bottom are 70% consensus sequences for highly conserved regions in single-letter code; alternates are below. x, any residue; h, hydrophobic residue (V, I, L, M); p, hydrophylic residue (T, S, N, Q, D, E, K, R, H). Positions where substitution destroyed DNA binding activity are marked (*). Substitution positions altering coinducer response but not DNA binding are: 102, 107, 126, 130, 135, 152, 173, 196, 199, 203, 204, 205, 206, 227, 253 (coinducer-insensitive); and 95, 116, 123, 149, 154, 165, 201, 206, 231, 234, 252, 284 (coinducer-independent).

binding activity support this conclusion. McIver et al (72) isolated coinducer-independent *nodD* alleles caused by substitutions at positions 95, 123, or 154 that activated transcription of regulated promoters by >15-fold without coinducer. Substitutions at positions 116 or 154 of NahR (54) or positions 149 or 165 of CysB (25) also give a coinducer-independent phenotype. Residues 102 and 135 of AmpR were also implicated in coinducer-responsive transcription activation (3). Such mutations may alter the conformation of LTTRs in a way that mimics the change caused by binding of coinducer.

Several coinducer-insensitive NahR mutants, which bind normally to their target promoters but fail to activate them in response to coinducer (100), map in the region of NahR (at positions 107, 126, 130, 152, and 173) that aligns with the NodD segment with predicted homology to the estrogen receptor. Position 622 of the receptor [which takes part in ligand binding (19)] aligns with residue 150 of NodD and NahR, close to the position of many substitutions that alter coinducer response of LTTRs. LTTR residues 102–154 are surely important for coinducer-responsive transcription activation, perhaps in ligand binding/interaction. Although there is some minor sequence conservation (Figure 1), this region lacks an obvious common predicted secondary structure (40) or motif.

The 11 residues between 196–206 define a second domain critical for coinducer recognition/response in LTTRs. Amino acid substitutions in NahR at positions 203, 204, and 206 (100), in NodD at positions 196 and 205 (14, 15), or in OxyR at position 199 (LTTR position 204) (113) eliminate transcription activation, but not DNA binding. Substitution at position 201 of NahR (54) or AlsR (89a) confers coinducer-independence. In the NodD–estrogen receptor alignment (45), Cys530 of the receptor, which is involved in ligand binding (47), corresponds to NodD position 205. LTTRs share little sequence similarity here except for a somewhat conserved basic residue at position 206. Secondary-structure predictions for several LTTRs suggest this region is a random coil.

Conserved Carboxy-Terminal Domain

Only one region in the C-terminal halves of LTTRs shows concentrated amino acid sequence similarity: residues 236–246 containing part of the β-turn-β ligand-binding crevice predicted for NodD (45) (Figure 1). The importance of this and small flanking regions in the multiple functions of LTTRs is supported by the fact that mutations here cause either altered response to coinducer or loss of DNA binding. Substitution mutations at positions 231 and 252 in NahR (54) or position 234 of OxyR (23) caused a coinducer-independent phenotype, while substitutions at NahR positions 227 or 253 made it coinducer insensitive (100). Positions 252–254 of the coinducer-independent NodD3 are the only positions that significantly vary from those of related coinducer-responsive NodDs (94). Substitutions at positions 239, 244, or 246 (highly conserved in all LTTRs) in NahR (100) or position 264 of AmpR (3) lead to loss of DNA binding; in contrast to other DNA-binding mutants, these were not *trans*-dominant, leading to speculation that this region could be involved in multimerization. Deletion of the last eight residues of NahR causes a loss of function (100) and substitution at NodD position 284 caused a coinducer-independent phenotype (14), suggesting even the last residues of LTTRs are important for function; however, adding 10 residues to the end of NodD3 did not affect its function (94).

Summary

The data discussed above suggest LTTR polypeptides have conserved and similarly organized functional domains (Figure 1). Secondary-structure predictions for 11 LTTRs suggest a three- or four-domain model (89). Such compartmentalization of functional domains for LTTRs, although oversimplified, provides a working model for testing and revision. The highly conserved 65 N-terminal residues, of which the central 20 are likely a helix-turn-helix motif, probably play an important role in recognition/binding of LTTRs (most likely as dimers) to cognate promoters via classical interactions with the nucleotide bases of the major groove (56). The conserved C-terminal domain (residues 227–253) is also generally important for DNA interactions. It may play a role in interactions between LTTR subunits and/or protein-DNA interactions occurring during transcription activation. X-ray crystallography of an LTTR-DNA complex will help to confirm which LTTR sequences interact with DNA and how. Recognition (possibly binding) of coinducer by LTTRs involves residues 95–173 (especially 149–154), which may function as a general ligand-binding pocket; ligand specificity may be conferred by less-conserved residues in or near this region. Coinducer recognition/response also involves the 196–206 region, which may also contain more specific ligand recognition functions or could be a hinge structure involved in the hypothesized conformational change of LTTRs caused by binding coinducer. Future experiments on coinducer response should focus on these two regions.

MECHANISMS OF ACTION OF INDIVIDUAL FAMILY MEMBERS

TrpI

TrpI (22), one of the earliest and best-characterized LTTRs, is found only in *Pseudomonas aeruginosa, P. putida,* and *P. syringae,* where it positively regulates the divergently transcribed *trpBA* genes in response to the coinducer indoleglycerol phosphate (INGP). TrpBA encodes tryptophan synthetase, which catalyzes the last step in tryptophan biosynthesis, converting INGP and serine to tryptophan and glyceraldehyde-3-phosphate (27). Gel-retardation (gel-shift) assays (36) and DNase I footprinting (37) showed that TrpI (in extracts of TrpI-overproducing *E. coli*) specifically binds to the 54-bp *trpI-trpBA* intergenic region (20, 21). Without INGP, TrpI-enriched extracts protected a sequence (site I) between −77 and −52 (relative to the *trpBA* transcription start site) from DNase I digestion; with 10^{-3} M INGP, a larger region (−77 to −32) was protected, extending into the −51 to −32 region (site II). With the cleavage reagent hydroxylradical (122), which had greater penetration, TrpI appeared to protect 3-bp regions around −76, −66, and −55; with INGP, additional 3-bp regions near −45 and −34 were also protected, again suggesting INGP caused a conformational change the TrpI-DNA complex (21). The 10-bp spacing of protections implies TrpI binds on one helical face of the DNA.

In gel-shift DNA binding assays with purified TrpI and the *trpI-trpBA* promoter region, two retarded species (i.e. two different protein-DNA complexes) appeared (20). The faster-migrating species (complex 1) was proposed to contain one TrpI (tetramer?) bound to the fragment, and the slower species (complex 2) was proposed to have multiple bound TrpIs (20, 21). DNA bending can also cause slower mobility. INGP increased complex 2 formation by 15-fold, implying that INGP stimulates binding of more TrpIs at site II. Deletion of site II prevented formation of low amounts of complex 2 but did not affect binding of TrpI in complex 1. Thus it was proposed (20) that without INGP, TrpI primarily occupies site I (giving complex 1), whereas with INGP additional TrpIs bind, and both sites I and II become occupied (giving complex 2). This hypothesis was supported by DNase I footprint analysis of gel-purified complex 1 and complex 2 (38). Because site I and II DNA sequences are very different and Trp occupancy of site II requires site I, INGP may foster cooperative binding of additional TrpIs, mostly via protein-protein interactions with TrpIs already on site I. The observation that INGP stimulates formation of low amounts of complex 2 in the absence of an intact site II (20, 38) supports this idea. Less likely, INGP may cause a conformational change or bending of the TrpI–site I complex, extending contacts of bound TrpI into site II.

In vitro transcription analysis showed that high levels of purified TrpI alone stimulates *Pseudomonas* RNA polymerase to form open complexes and activate transcription at the *trpBA* promoter and that INGP greatly stimulated this activation (39). Simultaneous mutation of two site II nucleotides did not affect TrpI binding to site I or II, but did reduce activation threefold, indicating that binding to sites I and II alone is not sufficient to activate transcription (39). An 8-bp alteration in site II caused nearly complete loss of TrpI+INGP-induced transcription activation; although it did not affect the TrpI footprinting at site I, it abolished INGP-induced interaction with site II (38), correlating protection of site II with transcription activation. These studies showed that by binding at site I TrpI represses its own transcription 10-fold and that RNA polymerase cannot form open complexes simultaneously at the overlapping *trpI-trpAB* promoters.

In summary, TrpI usually occupies site I, apparently via specific binding to an interrupted dyad at its core (-74 to -59: GTgAG-N_5-CTgAC), even in the absence of INGP, and as a result causes autoregulation. INGP stimulates cooperative binding of a second TrpI to the first via protein-protein interactions, after which it contacts nucleotides in site II near -35 and somehow increases open complex formation and hence transcription initiation. Alternatively, INGP may alter the conformation of the TrpI-DNA complex (possibly via bending) to increase transcription initiation.

IlvY

Another LTTR regulating amino acid biosynthetic genes is IlvY, which controls the divergently transcribed *ilvC* gene encoding acetohydroxy acid isomeroreductase, responsible for the second step in the common pathway for biosynthesis of isoleucine, valine, and leucine in *E. coli* (132). Expression of *ilvC* is regulated in vivo by its substrates, acetolactate and acetohydroxybutyrate (ACHB). Using transcriptional fusions of the *ilvC* and *ilvY* promoters to *galK*, Wek & Hatfield (133) showed that IlvY caused a 15-fold increase in *ilvC* transcription in response to ACHB and a twofold negative autoregulation of *ilvY* transcription independent of ACHB. Using IlvY-enriched cell extracts (~20% IlvY) and gel-shift DNA binding assays, they showed that IlvY specifically bound to the *ilvY-ilvC* intergenic region with a K_{app} of 1 nM; ACHB increased affinity threefold (133).

DNase I footprinting showed IlvY protected two sites upstream of the *ilvC* transcription start: O1 (−76 to −51) and O2 (−44 to −18); ACHB did not dramatically affect protection. Addition of RNA polymerase to the footprinting reactions showed that it protected the −18 to +24 region of the *ilvC* promoter only when both IlvY and ACHB were present (132). DNase I footprinting and in vitro transcription experiments with purified IlvY protein (W. Hatfield, personal communication) showed that the amount of ACHB required to stimulate *ilvC* transcription in vitro is the same as that required to give a strong IlvY-dependent DNase I hypersensitivity at −37. Dimethylsulfate footprinting analysis showed IlvY strongly protected four guanines of O1 (−70, −69, −61, and −60) from methylation and weakly protected three others (−38, −30, −29) in O2; ACHB did not affect protection. The IlvY-protected guanines (positions underlined) are contained in two interrupted dyads in O1 (−72 to −58: TTGCAAaaaTTGCAA) and O2 (−43 to −25: TATatCaatttccGcaATA (here and below, upper case = dyadic positions; lower case = nondyadic).

Gel-shift DNA binding assays (133) showed the K_{app} of IlvY for O1 was reduced 14-fold by deletion of O2, while the affinity of IlvY for O2 was reduced over 200-fold by deletion of O1, strongly suggesting cooperative binding between IlvYs at O1 and O2. Although ACHB did not affect this cooperative binding, it did enhance formation of a slower-migrating IlvY-O1 complex, possibly through binding of additional IlvYs to the one already bound at O1 or through an ACHB-induced conformational change in the complex. DNase I footprinting supported the latter hypothesis. Thus, irrespective of coinducer, IlvY binds simultaneously and cooperatively on one helical face to different dyadic sequences in O1 and O2, likely resulting in *ilvY* autoregulation, because O1 includes its start site. How ACHB stimulates transcription of *ilvC* is unclear. That ACHB altered DNase I sensitivity of nucleotides in the complex near the RNA polymerase–binding site indicates that it probably causes a conformational change in the IlvY-DNA complex.

OccR

OccR (46) controls genes for metabolism of octopine, which is used by *Agrobacterium tumefaciens* for growth in crown gall tumors (110). Genes for conjugal transfer of the Ti plasmid (60) and possibly others (130) may also be under OccR control. Catabolism of octopine (a rare pyruvate-arginine condensate synthesized in T-DNA-transformed plant cells) as carbon and nitrogen source is mediated by the *occ* operon of plasmid pTIA6, composed of *occQMPJ* (encoding octopine permease), *ooxAB* (encoding octopine oxidase), and *ocd* (encoding ornithine cyclodeaminase) (59, 110). Induction of *occ* transcription by octopine requires OccR (110), an LTTR divergently transcribed from a site 46 bp upstream of the *occ* transcription-start site (46). Experiments with *occR::lacZ* fusions and plasmid-borne *occR* showed a 10-fold negative autoregulation; analysis of *occR* mRNA in wild-type and *occR* mutants showed less autoregulation (46). Plasmid pTiC58 of *A. tumefaciens* has (in addition to OccR) another LTTR, NocR, which is 35% identical in amino acid sequence to OccR; it independently regulates transcription of nopaline catabolic genes in response to nopaline but has not been further characterized (129).

In vitro transcription experiments with *E. coli* RNA polymerase and purified OccR showed that transcription from the *occQ* promoter required OccR and octopine as coinducer and that OccR strongly repressed its own transcription (130). Gel-retardation assays showed specific binding of OccR to the *occR-occQ* divergent promoter with a K_{app} of 1 nM. Opposite to IlvY, coinducer decreased the OccR K_{app} twofold. Octopine caused a decrease in the mobility of the OccR-DNA complex; position-dependent mobility analysis (58) suggested this decrease resulted from a 62° OccR-induced bend located near −50 of the *occ* promoter DNA (130). Octopine relaxed bending by 16°, presumably shifting the complex toward a conformation that promotes productive interactions with RNA polymerase. DNase I footprinting with purified OccR showed an ~55-bp protection zone (−80 to −28 relative to the *occQ* transcription-start site) with hypersensitivities appearing near −70, −40, and −52; octopine caused a slight decrease in footprint length and hypersensitivities. Upstream of the putative bend is an interrupted dyad (−71 to −57: ATAA-N_7-TTAT) with a LysR motif (41), making it a likely site for interaction with OccR and its autoregulation.

OxyR

OxyR controls a regulon that mediates the oxidative-stress response by *E. coli* or *S. typhimurium* (112, 113). All aerobic organisms must defend against toxic oxidants resulting from incomplete reduction of oxygen. Exposure of

E. coli to 0.05 mM H_2O_2 induces production of 30 proteins that protect it from exposure to 200-fold higher levels of H_2O_2 (112). Several of these H_2O_2-induced proteins and their genes have been characterized [e.g. KatG (catalase), AhpCF (alkyl peroxide reductase), and GorA (glutathione reductase)] and shown to be positively regulated by OxyR (112, 120). Independently Bolker & Kahmann (8) discovered that *momR,* which negatively regulates the Mu phage *mom* gene encoding a modification enzyme for protection of Mu DNA from restriction, is the same gene as *oxyR* (8). MomR binds to sequences of *mom* between −92 and −50 to repress its expression only if the three GATC *dam* sites there are not methylated (8). In addition, the *mor* gene, controlling switching between different colony morphologies, aggregation behaviors, and piliation states of *E. coli* (131), also appears to be the same as *oxyR*.

OxyR is unlinked to any of its target genes and thus lacks the characteristic divergent promoter. However, recent evidence (G. Storz, unpublished data) suggests an untranslated RNA transcribed divergently from *oxyR* may be involved in regulation. Nonetheless OxyR, negatively autoregulates fivefold by binding to the −27 to +21 region of its own promoter (23). Unlike other LTTRs, OxyR may activate transcription after it has been directly oxidized rather than in response to binding of a specific coinducer. This hypothesis is based on in vitro transcription assays that showed that while purified OxyR specifically activated transcription of *katG* or *ahpC* 100-fold, addition of 100 mM dithiothreitol (DTT, a reductant) specifically eliminated the OxyR-dependent transcription activation; removal of DTT reversed the effect (113). Thus, the oxidation state of OxyR affects its ability to activate transcription. The mechanism and/or oxidant controlling OxyR activity in vivo is unknown; direct, reversible binding of an oxidant-like signal ligand is still possible. Examination of the role of OxyR's five cysteines in its oxidation showed that only Cys199 is required for activity (113).

Deletion of *ahpC* promoter sequences upstream of −46 (but not upstream of −68) eliminated the fivefold OxyR-dependent activation observed in vivo (120). Extracts of OxyR-overproducing cells protected the *ahp* or *katG* promoters between approximately −79 to −33 from DNase I (120). The oxidation state of the OxyR (and hence activity) used was unknown. Subsequent DNase I footprinting with purified OxyR under oxidized (active) or reduced (inactive) conditions (113) showed that both forms still bound to the −79 to −33 regions, but the *katG* footprint was slightly longer with active OxyR, while the *oxyR* footprint was shorter. A DNase I hypersensitivity near −55 caused by inactive OxyR was drastically reduced when the active form was used; this was taken as evidence of a conformational change in the preformed OxyR-DNA complex caused by oxidation of OxyR into a transcriptionally active form (113).

Gel-retardation binding assays showed that oxidized or reduced OxyR has the same high affinity and specificity for its target sites upstream of *katG*, *ahpC*, and *oxyR*, implying that binding is not highly regulated by oxidation state (119). These sites show little conservation of sequence or features (e.g dyads, repeats). Methylation interference analysis (which detects guanines required for protein-DNA binding) of *katG* and *oxyR* showed that OxyR interacts with nucleotides throughout the DNase I–protected regions, most strongly with guanines in the −68 to −54 region, and less so with adenines around −40 (119). A weak, 40-bp degenerate dyadic consensus sequence was proposed for the OxyR binding site (119). The binding sites and mechanism of action of OxyR appear somewhat different from those of other LTTRs, perhaps so OxyR can recognize many different promoters in response to oxidative stress.

CatR and CatM

CatR of *Pseudomonas putida* regulates *catBC*, encoding *cis,cis*-muconate lactonizing enzyme and muconolactone isomerase, respectively, which are part of a pathway for use of benzoate as a carbon and energy source (92). Induction of expression of *catBC* in vivo by 20-fold requires *cis-cis*-muconate (CCM) and *catR*, which is divergently transcribed from a site 48 bp upstream of the *catBC* transcription start (92). Indirect qualitative evidence of autoregulation of *catR* has been reported (92). Gel retardation assays showed that purified CatR bound to the *catR-catBC* promoter region irrespective of the presence of the coinducer CCM. Hydroxylradical footprinting (122) showed that CatR protected three regions on one helical face between −78 and −54 (relative to the *catBC* transcription start site) and caused hypersensitivity of two sites near −50 and −80; CCM did not appear to affect the footprint (93). However, recent experiments (84) showed that CCM caused CatR to protect new additional sequences of the *catBC* promoter [activation binding site (ABS): −48 to −22] from DNase I; without CCM only sequences between −79 and −53 [repression binding site (RBS)] were protected. Methylation interference assays implicated guanines at −69, −68, −63, −62, and −61 in CatR binding. These residues (positions underlined) are symmetrically located in an interrupted dyad (−73 to −56: CAgACC-N_4-gGGTaTG). The hypersensitivity near −50 was again observed irrespective of CCM. Gel retardation assays showed CCM increased formation of a slower-moving CatR-DNA complex; methylation of a guanine at −41 in ABS blocked formation of this complex (84), suggesting this guanine is important for CCM-induced interaction of CatR with the ABS. Deletion of the RBS eliminated binding of CatR; deletion of the ABS lowered binding only twofold. From these observations came the proposal that CatR binds as a dimer to the RBS independent of the ABS and that binding of a second CatR dimer to the ABS occurs through CCM-induced cooperative interactions between CatR molecules that cause increased RNA polymerase binding (84). Studies of AmpR (3, 66; E. Bartowsky, personal communication) suggest this protein uses a similar two-site mechanism, where coinducer also causes interactions with a downstream site near −45 that are required for transcription activation.

CatM of *Acinetobacter calcoaceticus,* similar to CatR, is divergently transcribed from the *catBC* genes encoding enzymes for catabolism of ring-cleavage products from benzoate (77). CatBC of *A. calcoaceticus* and *P. putida* are isofunctional, highly homologous, and induced by CCM. However unlike most other LTTRs, CatM appears to predominantly act as a negative regulator of *catBC* and *catA* (77). This conclusion is based on two observations: (*a*) in a *catM* deletion strain *catA* is constitutively expressed, and (*b*) in a strain with a mutant CatM caused by an amino acid substitution at residue 156, *catBC* is constitutively expressed (77). However, similarly located substitutions in LTTRs that are activators can also give a constitutive phenotype (54, 72). CatM shows 34% amino acid sequence identity to CatR, but as the shortest LTTR (250 residues), CatM lacks C-terminal domains possibly involved in transcription activation. In *E. coli,* CatM did not repress a *catBCDEF::lacZ* fusion but did threefold negatively autoregulate a *catM::lacZ* fusion; gel retardation assays showed CatM binds to an 85-bp fragment of the *catM-catBC* divergent promoter region, irrespective of CCM (90). In vivo methylation protection assays in *E. coli* showed, like LTTRs that are activators, CatM protected symmetrical guanines at −69 and −61 upstream of the *catBC* transcription start site in an interrupted dyad (−72 to −58: ATA$\underline{C}$-N_7-$\underline{G}$TAT). CatM also protected a guanine at −9, but not as strongly in the presence of CCM (90). Although CatM binding to the −72 to −58 sequence may be for autoregulation, and the apparent CCM-sensitive binding at −9 may act in repression of *catBC,* further studies of this atypical mechanism and more proof that CatM functions only as a repressor are necessary.

NahR

The *nahR* gene is on the NAH7 plasmid of *P. putida* (135). Insertional inactivation of *nahR* causes loss of expression of two plasmid-borne operons, *nah* and *sal,* which encode the 14 enzymes for metabolism of naphthalene or salicylate as a carbon source (99, 135, 136). High-level transcription of *nah* and *sal* is induced by salicylate and requires *nahR,* which is divergently transcribed from a site 60 bp upstream of the *sal* operon transcription start site; the *nah* promoter is 17 kb upstream (97, 98). Gel-retardation DNA-binding assays showed that NahR specifically bound to the *nahR-sal* divergent promoter region, irrespective of coinducer (103). This binding is likely the cause of the threefold negative autoregulation of *nahR* (102). Deletion of the *sal* or *nah* promoter sequences upstream of −45 (but not upstream of −83) caused loss of *nahR*+salicylate-induced transcriptional activation (102, 105). DNase I footprinting with partially purified NahR showed that it protected 70% identical sequences between −82 and −47 of the *nah* or *sal* promoters; coinducer did not affect footprint (103). Mutational analysis of this conserved region of *sal* showed that nucleotides at positions −74, −73, and −61 were critical for transcription activation, while those at −69, −67, −64, and −59 were not (103).

Most studies of LTTR-DNA interactions have used purified proteins of uncertain specific activity with nonsupercoiled DNA in vitro. Absence of supercoiling and other in vivo conditions may alter results. Thus in vivo dimethylsulfate methylation protection (96) studies with NahR provide a unique perspective (54). In vivo, NahR strongly protected two guanines at −71 and −62 of the *nah* and *sal* promoters; both were on the same helical face and were part of an interrupted dyad (positions underlined) within the conserved NahR-binding sites (−73 to −60: tTCA-N_6-TGAt). Coinducer caused NahR to additionally protect new guanines (−58, −45, and −35 at *nah*; −42 and −40 at *sal*) and also caused hypersensitivity at −52 of *nah*. In vivo footprinting with NahR mutants having altered coinducer response showed that salicylate-induced DNA contacts and hypersensitivity were always present when NahR was activating transcription (54). Two mutants (Gly203→Asp; Pro35→Ser) made these contacts but did not activate transcription, indicating that contacts are required but not sufficient for increased transcription. Thus NahR, like CatR and TrpI, always binds to a dyadic sequence around −65, but when coinducer is present, further interactions with promoter DNA near the −35 region occur that somehow stimulate RNA polymerase to increase transcription initiation.

MetR

The disperse genes encoding methionine biosynthesis in *E. coli* and *S. typhimurium* are regulated by interacting regulatory gene products. MetJ, along with its corepressor *S*-adenosyl methionine, negatively regulates all the genes (except *metH*). MetR positively regulates *metE* and *metH*, encoding enzymes for the transmethylation of homocysteine (HC) to give methionine (95). In vivo and in vitro *metE* transcription is induced more than fivefold by HC but only with MetR (70, 124). MetR also positively regulates by threefold transcription of *metA* (69) and *glyA* (87), encoding two pathway enzymes: homoserine succinyl transferase and serine hydroxymethyl transferase.

MetR is divergently transcribed from a site 27 bp upstream of the *metE* transcription start site (87). MetR negatively autoregulates fivefold, but unlike most other LTTRs, this requires a corepressor, HC. In vivo and in vitro the MetJ repressor also negatively regulates *metR* 50-fold (70, 126), likely by binding to the *metR-metE* intergenic region. Gel-retardation assays with MetR-enriched extracts demonstrated specific binding of MetR to the intergenic region that was unaffected by HC; two retarded species were observed (125). DNase I footprinting showed that MetR protected the 24-bp region between −72 and −48 upstream of the *metE* transcription start and caused hypersensitivity at −56; 6-bp deletions here caused a 40-fold decrease in MetR-activation of *metE* transcription (125). Further footprinting with purified MetR revealed a dyadic consensus sequence (−62 to −50: TGAA-N_5-TTCA) centered near −58 in the DNase I protected regions of *metA* (69), *metE* (125), and *metH* (16).

Substitution mutations at −62, −61, −51, or −50 in the conserved dyad arms of the consensus sequence at *metH* (16) caused >80% reduction in MetR binding and transcription activation. Mutations at −56 reduced activation fourfold without affecting binding, suggesting position −56 is exclusively involved in transcription activation. Mutations at −59 or −53 had no effect. The mechanism for MetR+HC-mediated transcription activation may be similar to that of NahR, TrpI, or CatR because the size, structure, and position of their binding sites are similar. However, the effect of HC on MetR interactions with individual nucleotides is unknown, and further comparative studies are needed.

CysB

Biosynthesis of cysteine requires more than 16 genes, some in operons dispersed around the chromosome of *E. coli* and *S. typhimurium: cysJIH* encoding for NADPH-sulfite reductase and 3′-phosphoadenosine 5′-phosphosulfate sulfotransferase, *cysPTWA* encoding a periplasmic sulfate/thiosulfate transport system, and *cysK* encoding *O*-acetylserine (thiol)-lyaseA. All these are in the *cys* regulon under the positive control of the LTTR CysB (79). CysB, like OxyR, is not linked to any of its many target genes and thus lacks the typical divergent promoter of LTTRs. Nonetheless, in vitro transcription assays showed that purified CysB negatively autoregulates itself by fourfold, but only in the absence of its coinducers *N*-acetylserine (NAS) or *O*-acetylserine (82); gel-shift assays and DNase I footprinting showed that NAS reduced by more than fourfold the binding of CysB to the −10 to +36 region of its own promoter (82).

In vitro transcription assays showed that CysB greatly stimulated transcription from the *cysJIH* (80, 81), *cysK* (75), and *cysP* (53) promoters, but only with NAS. Thiosulfate (and less so sulfide) blocked this NAS+CysB-mediated activation; thus these compounds are probably antiinducers of the *cys* regulon, and possibly inhibit binding of NAS to CysB (81). DNase I footprinting identified CysB binding sites at several promoters, but the sites differed in number, sequence, arrangement, and reactivity (53, 75). The *cysJIH, cysK,* and *cysP* promoters have one CysB-protected site between −76 and −35, −78 to −39, and −85 to −41, respectively. At *cysK* (and likely the others), this site is required for in vivo function (75). Both the *cysP* and *cysK* promoters have one additional CysB binding site (−19 to +25 and −115 to −79, respectively). Deletion of this extra site at *cysK* did not affect expression, implying the extra site sequesters CysB for action at the downstream site (75). NAS inhibits binding of CysB to this extra site but stimulates binding to the *cysP* site near +25. CysB-protected sequences are weakly dyadic but do not show a strong consensus, implying that CysB may have degenerate recognition sequences like OxyR. As with OccR, binding of CysB to the *cysK* promoter may induce DNA bending that is relaxed when coinducer is present (75). Many details of CysB-mediated regulation remain unclear, but the system appears more complex than many other LTTR systems.

NodD and SyrM

Some rhizobia establish a species-specific symbiotic relationship with plants, in which differentiated bacteria in root nodules reduce N_2 into NH_3. Different flavonoids in root exudates of the appropriate host plant are recognized by members of the NodD subfamily of LTTRs (reviewed in 44, 67, 107). NodDs activate by 30-fold transcription of *nod* operons, whose products initiate and control establishment of the symbiosis (67). Some rhizobia have one or two NodDs, whereas others like *Rhizobium meliloti* have three. Although the *R. meliloti* NodDs are >80% identical in amino acid sequence, they are functionally distinct and interact in a complex fashion (76).

R. meliloti has another LTTR, SyrM, that controls *nod* genes and genes encoding for exopolysaccharide synthesis (2). Transcription of *syrM* is activated by *nodD3* and vice versa, resulting in a self-amplifying regulatory circuit (117), which may also involve NolR, a repressor of *nodD* expression (61, 62). Like other LTTRs, some NodDs are divergently transcribed from regulated *nod* genes and negatively autoregulate (15), whereas others are not (76) or are positively autoregulated (107, 108). Most NodDs activate transcription in response to different flavonoids, while NodD3 apparently activates without a coinducer (76).

Gel-retardation assays (32, 50) show that NodDs bind to a 47-bp sequence (−75 to −28: *nod* boxes) that is highly conserved in promoters of *nod* genes of many rhizobia (91). DNase I footprinting of gel-purified protein-DNA complexes showed that partially purified NodD1 or NodD3 of *R. meliloti* identically protected the *nod* box at *nodABC, nodFE,* and *nodH,* except that NodD3 alone caused a strong hypersensitivity near −47 (33). Protection patterns were not altered by coinducer or use of the more-penetrating nuclease

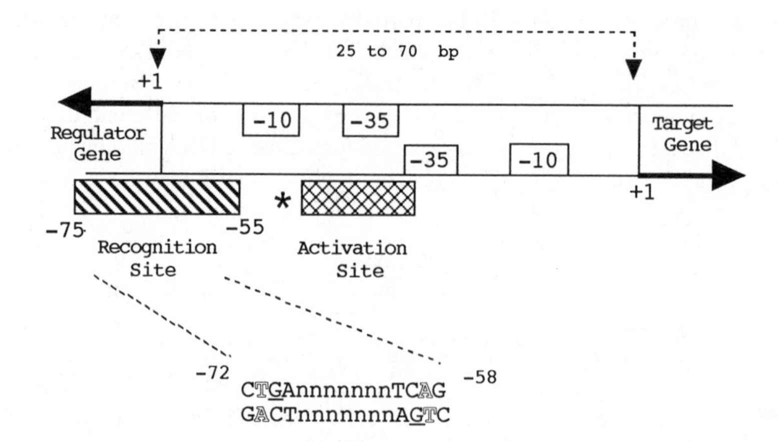

Figure 2 Typical structure of a divergent promoter regulated by an LTTR. +1, transcription start site; −35 and −10 consensus sequences are boxed. An asterisk marks the location of the common LTTR-induced hypersensitivity (i.e. possible bending site). A hypothetical recognition-site sequence is shown below to illustrate the interrupted dyadic sequence and T-N_{11}-A-motif (*outlined letters*); guanines expected to be directly involved in binding are underlined.

Cu-phenanthroline. Recent methylation-interference studies (34) suggested that specific guanines in two regions of the *nod* boxes of *nodA, nodH,* and *nodF* (−75 to −58 and −45 to −27) interacted with the inducer-independent NodD3. Position-dependent mobility-shift analysis (58) implied that NodD3 bends the *nod* box DNA between two sites near the −47 hypersensitivity (34). In the upstream site the guanines critical for binding (positions underlined) are located in or near a highly conserved interrupted dyad (−75 to −59: CATCc-N_7-aGATG), where substitutions at positions 3 and 15 (41) or deletion of the GATG (34) destroyed function. Nucleotides 3 and 15 purportedly define a motif common to the binding site of many LTTRs (41). The downstream site has a different interrupted dyad (−40 to −28: GATT-N_5-AATC) but is on the same helical face; insertion of 2 or 10 bp into the GATT destroys binding and activation by NodD3, whereas a 4-bp (but not 10-bp) insertion at −58 caused loss of activation and binding (34). Ethylation interference patterns and other data suggested that NodD3 binds to these two different dyadic sites in a similar manner and that they can be separated by one integral helical turn without affecting transcription activation (34).

Summary and Model

The highly conserved features of LTTR polypeptides and the limited data on promoter interactions presented above suggest many LTTRs may function in very similar ways. For example, most characterized LTTRs bind to their regulated promoters at an ~15-bp, partially dyadic sequence centered near −65 (recognition site; Figure 2), which has the conserved T-N_{11}-A motif proven critical for binding of NodD, NahR, and MetR (41). Binding often involves symmetrical guanines in the two dyad arms (on different strands, i.e. in adjacent major grooves), as expected for a dimeric DNA binding protein using a helix-turn-helix motif (56). The T-N_{11}-A motif may be a general anchor or guide for LTTR binding, while adjacent nucleotides of the dyad (varying in position and sequence) confer recognition specificity. These recognition sites usually overlap the LTTR's promoter (Figure 2) and binding here (which is usually insensitive to coinducer) is likely responsible for autoregulation. The conservation of the structure and position of the site indicates it may have coevolved with each LTTR. Unpublished analyses suggest many other LTTR systems (e.g. AmpR, Nac, GltC) have similarly organized and located binding sites (K. Goethals; E. Bartowsky; T. Goss; A. Sonenshein, personal communications).

All LTTRs also interact with a dissimilar sequence (site) downstream on the same side of the DNA helix (activation site; Figure 2). For many LTTRs, interaction with this site (near the −35 RNA polymerase binding site) often requires coinducer and is a prerequisite for transcription activation. The recognition site, however, is often the primary (or only) determinant of promoter binding, while coinducer-dependent cooperative protein-protein interactions seem to be required for occupancy of the activation site. In vitro studies showed that some LTTRs (e.g. OccR, IlvY, OxyR) apparently occupy both sites even without coinducer. However, the activation state of these LTTRs was not totally certain, and if coinducer stimulates cooperative binding between LTTR molecules, the in vitro conditions may artificially force complete occupancy. Alternatively, this difference may define two mechanistic classes of LTTRs. Nonetheless, in the presence of coinducer the interactions of all LTTRs with target promoters are very similar. In most cases, coinducer appears to alter the structure of the LTTR-DNA complex, as evidenced by changes in cleavage sensitivity of nucleotides in or near the activation site (e.g. the usually observed hypersensitivity near −50) or complex mobility. These LTTR+coinducer-mediated changes may signify DNA bending or new LTTR-DNA interactions that increase the affinity for RNA polymerase, causing increased transcription initiation (9).

CONCLUDING REMARKS

I have tried to succinctly summarize current knowledge about the rapidly expanding LysR family and have presented speculative models for domain organization and function of LTTRs and their mechanism(s) of action. These models should serve as a basis for further analysis and as guides for beginning characterization of new family members. Further studies along these lines will hopefully lead to more-refined models and subclassification of LTTRs into smaller functional or evolutionary groups.

ACKNOWLEDGMENTS

I thank all those who provided published and unpublished information, especially: K. Goethals, H. Schlaman, S. Long, R. Fisher, N. Kredich, A. Chakrabarty, T. Goss, R. Bender, S. Henikoff, F. Norel, and E. Bartowsky. I thank P. Bates and V. Burke for help in manuscript preparation and T. Hoover, S. Winans, and T. Denny for critical review. Support from NSF grant IBN 91-17544 is also acknowledged.

Literature Cited

1. Appelbaum, E. R., Thompson, D. V., Idler, K., Chartrain, N. 1988. *Rhizobium japonicum* USDA 191 has two *nodD* genes that differ in primary structure and function. *J. Bacteriol.* 170:12–20
2. Barnett, M. J., Long, S. R. 1990. DNA sequence and translational product of a new nodulation-regulatory locus: SyrM has sequence similarity to NodD proteins. *J. Bacteriol.* 172:3695–3700
3. Bartowsky, E., Normark, S. 1991. Purification and mutant analysis of *Citrobacter freundii* AmpR, the regulator for chromosomal AmpC β-lactamase. *Mol. Microbiol.* 5:1715–25
4. Beck, C. F., Warren, R. A. J. 1988. Divergent promoters, a common form of gene organization. *Microbiol. Rev.* 52:318–26
5. Bender, R. A. 1991. The role of the NAC protein in the nitrogen regulation of *Klebsiella aerogenes*. *Mol. Microbiol.* 5:2575–80
6. Deleted in proof
7. Bohannon, D. E., Sonenshein, A. L. 1989. Positive regulation of glutamate biosynthesis in *Bacillus subtilis*. *J. Bacteriol.* 171:4718–27
8. Bolker, M., Kahmann, R. 1989. The *Escherichia coli* regulatory protein OxyR discriminates between methylated and unmethylated states of the phage Mu *mom* promoter. *EMBO J.* 8:2403–10
9. Bracco, L., Kotlarz, D., Kolb, A., Dieckermann, S., Buc, H. 1989. Synthetic curved DNA sequences can act as transcriptional activators in *Escherichia coli*. *EMBO J.* 8:4289–96
10. Branden, C. A. 1980. Relation between structure and function of α/β proteins. *Q. Rev. Biophys.* 13:317–38
11. Brennan, R. G., Matthews, B. W. 1989. The helix-turn-helix DNA binding motif. *J. Biol. Chem.* 264:1903–6
12. Brumbley, S., Carney, B., Denny, T. P. 1993. Phenotype conversion in *Pseudomonas solanacearum* due to spontaneous inactivation of PhcA, a putative LysR transcriptional regulator. *J. Bacteriol.* In press
13. Brun, Y. V., Breton, R., Lanouette, P., Lapointe, J. 1990. Precise mapping and comparison of two evolutionarily related regions of the *Escherichia coli* K-12 chromosome. *J. Mol. Biol.* 214: 825–43
14. Burn, J. E., Hamilton, J. D., Wootton, J. C., Johnston, A. W. B. 1989. Single and multiple mutations affecting properties of the regulatory gene *nodD* of *Rhizobium*. *Mol. Microbiol.* 3:1567–77
15. Burn, J., Rossen, L., Johnston, A. W. B. 1987. Four classes of mutations in the *nodD* gene of *Rhizobium leguminosarum* biovar. *viciae* that affect its ability to autoregulate and/or activate other *nod* genes in the presence of flavonoid inducers. *Genes Dev.* 1:456–64
16. Byerly, K. A., Urbanowski, M. L., Stauffer, G. V. 1991. The MetR binding site in the *Salmonella typhimurium metH* gene: DNA sequence constraints on activation. *J. Bacteriol.* 173:3547–53
17. Caldwell, A. L., Gulig, P. A. 1991. The *Salmonella typhimurium* virulence plasmid encodes a positive regulator of a plasmid-encoded virulence genes. *J. Bacteriol.* 173:7176–85

18. Campbell, J. I. A., Scahill, S., Gibson, T., Ambler, R. P. 1989. The phototrophic bacterium *Rhodopseudomonas capsulata* sp108 encodes an indigenous class A β-lactamase. *Biochem. J.* 260: 803–12
19. Carlstedt-Duke, J., Stromstedt, P. E., Persson, B., Cederlund, E., Gustafsson, J. A., Jornvall, H. J. 1988. Identification of hormone-interacting amino acid residues within the steroid-binding domain of the glucocorticoid receptor in relation to other steroid hormone receptors. *J. Biol. Chem.* 263: 6842–46
19a. Chakrabarty, A. M., Kaplan, S., Iglewski, B., Silver, S., eds. 1990. Pseudomonas: *Biotransformations, Pathogenesis, and Evolving Biotechnology*. Washington, DC: Am. Soc. Microbiol.
20. Chang, M., Crawford, I. P. 1991. In vitro determination of the effect of indoleglycerol phosphate on the interaction of purified TrpI with its DNA-binding sites. *J. Bacteriol.* 173:1590–97
21. Chang, M., Crawford, I. P. 1990. The roles of indoleglycerol phosphate and the TrpI protein in the expression of *trpBA* from *Pseudomonas aeruginosa*. *Nucleic Acids Res.* 18:979–88
22. Chang, M., Hadero, A., Crawford, I. P. 1989. Sequence of the *Pseudomonas aeruginosa trpI* activator gene and relatedness of *trpI* to other procaryotic regulatory genes. *J. Bacteriol.* 171: 172–83
23. Christman, M. F., Storz, G., Ames, B. N. 1989. OxyR, a positive regulator of hydrogen peroxide-inducible genes in *Escherichia coli* and *Salmonella typhimurium*, is homologous to a family of bacterial regulatory proteins. *Proc. Natl. Acad. Sci. USA* 86:3484–88
24. Coco, W., Rothmel, R., Henikoff, S., Chakrabarty, A. M. 1993. Nucleotide sequence and initial functional characterization of *clcR*, a LysR family activator of the *clcABC* chlorocatechol operon in *Pseudomonas putida*. *J. Bacteriol.* 175:417–27
25. Colyer, T. E., Kredich, N. M. 1991. Characterization of constitutive CysB protein interactions with *cys* promoters of *S. typhimurium*. *Ann. Meeting Am. Soc. Microbiol.* p. 164. Abstr. H-57
26. Coynault, C., Robbe-Saule, V., Popoff, M., Norel, F. 1992. Growth phase and SpvR regulation of transcription of *Salmonella typhimurium spvABC* virulence genes. *Microb. Path.* 13:133–45
27. Crawford, I. P. 1989. Evolution of a biosynthetic pathway: the tryptophan paradigm. *Annu. Rev. Microbiol.* 43: 567–600
28. Dammann, T., Wohlleben, W. 1992. A metalloprotease gene from *Streptomyces coelicolor* Müller and its transcriptional activator, a member of the LysR family. *Mol. Microbiol.* 6:2267–78
29. Davison, J., Brunel, F., Phanopoulos, A., Prozzi, D., Terpstra, P. 1992. Cloning and sequencing of *Pseudomonas* genes determining sodium dodecyl-sulfate biodegradation. *Gene 114:19–24*
30. Dodd, I. A., Egan, J. B. 1990. Improved detection of helix-turn-helix DNA binding motifs in protein sequences. *Nucleic Acids Res.* 18:5019–26
31. Deleted in proof
32. Fisher, R. F., Egelhoff, T. T., Mulligan, J. T., Long, S. R. 1988. Specific binding of proteins from *Rhizobium meliloti* cell-free extracts containing NodD to DNA sequences upstream of inducible nodulation genes. *Genes Dev.* 2:282–93
33. Fisher, R. F., Long, S. R. 1989. DNA footprint analysis of the transcriptional activator proteins NodD1 and NodD3 on inducible *nod* gene promoters. *J. Bacteriol.* 171:5492–5502
34. Fisher, R. F., Long, S. R. 1993. Interactions of NodD at the *nod*-box: NodD binds to two distinct sites on the same face of the helix and induces a bend in the DNA. *J. Mol. Biol.* In press
35. Frantz, B., Chakrabarty, A. M. 1987. Organization and nucleotide sequence determination of a gene cluster involved in 3-chlorocatechol degradation. *Proc. Natl. Acad. Sci. USA* 84:4460–64
36. Fried, M., Crothers, D. 1981. Equilibria and kinetics of *lac*-repressor-operator interactions by polyacrylamide electrophoresis. *Nucleic Acids Res.* 9: 6505–29
37. Galas, D. J., Eggert, M., Waterman, M. S. 1985. Rigorous pattern-recognition methods for DNA sequences: analysis of promoter sequences from *Escherichia coli*. *J. Mol. Biol.* 186: 117–28
38. Gao, J., Gussin, G. N. 1991. Mutations in TrpI binding site II that differentially affect activation of the *trpBA* promoter of *Pseudomonas aeruginosa*. *EMBO. J.* 10:4137–44
39. Gao, J., Gussin, G. N. 1991. Activation of the *trpBA* promoter of *Pseudomonas aeruginosa* by TrpI protein in vitro. *J. Bacteriol.* 173:3763–69

40. Garnier, J., Osguthorpe, D. J., Robson, B. 1978. Analysis of the accuracy and implications of simple methods for predicting the secondary structure of globular proteins. *J. Mol. Biol.* 120:97–120
41. Goethals, K., VanMontagu, M., Holsters, M. 1992. Conserved motifs in a divergent *nod* box of *Azorhizobium caulinodans* ORS571 reveal a common structure in promoters regulated by LysR-type proteins. *Proc. Natl. Acad. Sci. USA* 89:1646–50
42. Goldberg, M. B., Boyko, S. A., Calderwood, S. B. 1991. Positive transcriptional regulation of an iron-regulated virulence gene in *Vibrio cholerae*. *Proc. Natl. Acad. Sci. USA* 88:1125–29
43. Goss, T. J., Datta, P. 1985. Molecular cloning and expression of the biodegradative threonine dehydratase gene (*tdc*) of *Escherichia coli* K12. *Mol. Gen. Genet.* 201:308–14
44. Györgypal, Z., Kiss, G. B., Kondorosi, A. 1991. Transduction of plant signal molecules by the *Rhizobium* NodD proteins. *Bioessays* 13:575–81
45. Györgypal, Z., Kondorosi, A. 1991. Homology of the ligand-binding regions of *Rhizobium* symbiotic regulatory protein NodD and vertebrate nuclear receptors. *Mol. Gen. Genet.* 226:337–40
46. Habeeb, L. F., Wang, L., Winans, S. C. 1991. Transcription of the octopine catabolism operon of the *Agrobacterium* tumor-inducing plasmid pTiA6 is activated by a LysR-type regulatory protein. *Mol. Plant-Microbe Interact.* 4: 379–85
47. Harlow, K. W., Smith, D. N., Katzenellenbogen, J. A., Greene, G. L., Katzenellenbogen, B. S. 1989. Identification of cysteine 530 as the covalent attachment site of an affinity-labeling estrogen (ketononestrol aziridine) and antiestrogen (tamoxifen aziridine) in the human estrogen receptor. *J. Biol. Chem.* 264:17476–85
48. Henikoff, S., Haughn, G. W., Calvo, J. M., Wallace, J. C. 1988. A large family of bacterial activator proteins. *Proc. Natl. Acad. Sci. USA* 85:6602–6
49. Henikoff, S., Wallace, J. C., Brown, J. P. 1990. Finding protein similarities with nucleotide sequence databases. *Methods Enzymol.* 183:111–32
50. Hong, G. F., Burn, J. E., Johnston, A. W. B. 1987. Evidence that DNA involved in the expression of nodulation (*nod*) genes in *Rhizobium* binds to the product of the regulatory gene *nodD*. *Nucleic Acids Res.* 15:9677–90
51. Honore, N., Nicolas, M. H., Cole, S. T. 1986. Inducible cephalosporinase production in clinical isolates of *Enterobacter cloacae* is controlled by a regulatory gene that has been deleted from *Escherichia coli*. *EMBO J.* 5: 3709–14
52. Horvath, B., Bachem, C. W. B., Schell, J., Kondorosi, A. 1987. Host-specific regulation of nodulation genes in *Rhizobium* is mediated by a plant-signal, interacting with the *nodD* gene product. *EMBO J.* 6:841–48
53. Hryniewicz, M. M., Kredich, N. M. 1991. The *cysP* promoter of *Salmonella typhimurium:* characterization of two binding sites for CysB protein, studies of in vivo transcription initiation, and demonstration of the anti-inducer effects of thiosulfate. *J. Bacteriol.* 173:5876–86
54. Huang, J., Schell, M. A. 1991. In vivo interactions of the NahR transcriptional activator with its target sequences: inducer mediated changes resulting in transcription activation. *J. Biol. Chem.* 266:10830–38
55. Huang, M. W., Cangelosi, G. A., Halperin, W., Nester, E. W. 1990. A chromosomal *Agrobacterium tumefaciens* gene required for effective plant signal transduction. *J. Bacteriol.* 172: 1814–22
56. Jordon, S. R., Pabo, C. 1988. Structure of the lambda complex at 2.5 Å resolution: details of repressor-operator interactions. *Science* 242:893–99
57. Kaphammer, B. K., Olsen, R. H. 1990. Cloning and characterization of *tfdS,* the repressor-activator gene of *tfdB,* from the 2,4-dichlorophenoxyacetic acid catabolic plasmid pJP4. *J. Bacteriol.* 172:5856–62
57a. Keller, J. W., Allen-Daley, E. S., Wang, X. 1992. *Pseudomonas cepacia* DgdR, a new DNA binding protein in the LysR family. Revised sequence, expression in *E. coli,* and gel shift assay. *FASEB J.* 6:A357 (Abstr. 2048)
58. Kim, J., Zwieb, C., Wu, C., Adhya, S. 1989. Bending of DNA by regulatory proteins: construction and use of a DNA bending vector. *Gene* 85:15–23
59. Klapwijk, P. M., Oudshoorn, M., Schilperoot, R. A. 1977. Inducible permease involved in the uptake of octopine, lysopine and octopinic acid by *Agrobacterium tumefaciens* strains carrying virulence-associated plasmids. *J. Gen. Microbiol.* 102:1–11
60. Klapwijk, P. M., Schilperoot, R. A. 1979. Negative control of octopine degradation and transfer genes of octopine Ti plasmids in *Agrobacterium tumefaciens*. *J. Bacteriol.* 139:424–31

61. Kondorosi, E., Buire, M., Cren, M., Iyer, N., Hoffmann, B., Kondorosi, A. 1991. Involvement of the *syrM* and *nodD3* genes of *Rhizobium meliloti* in *nod* gene activation and in optimal nodulation of the plant host. *Mol. Microbiol.* 5:3035–48
62. Kondorosi, E., Pierre, M., Cren, M., Haumann, U., Buire, M., et al. 1991. Identification of NolR, a negative trans-acting factor controlling the *nod* regulon in *Rhizobium meliloti. J. Mol. Biol.* 222:885–96
63. Krause, C., Roudier, C., Fierer, J., Harwood, J., Guiney, D. 1991. Molecular analysis of the virulence locus of the *Salmonella dublin* plasmid pSDL2. *Mol. Microbiol.* 5:307–16

63a. Kusano, T., Sugawara, K. 1993. Specific binding of *Thiobacillus ferroxidans* RbcR to the intergenic sequence between the *rbc* operons and the *rbcR* gene. *J. Bacteriol.* 175:1019–25

64. Lafferty, S., Nester, E. 1992. The chromosomal virulence gene *chvE* of *A. tumefaciens* is regulated by a LysR family member. *Proc. 6th Int. Symp. on Molecular Plant Microbe Interactions, Seattle, Wash.* Abstr. #21
65. Lindberg, F., Westman, L., Normark, S. 1985. Regulatory components in *Citrobacter freundii ampC* β-lactamase induction. *Proc. Natl. Acad. Sci. USA* 82:4620–24
66. Lindquist, S., Lindberg, F., Normark, S. 1989. Binding of the *Citrobacter freundii* AmpR regulator to a single DNA site provides both autoregulation and activation of inducible AmpC-β-lactamase gene. *J. Bacteriol.* 171:3746–53
67. Long, S. R. 1989. Rhizobium-legume nodulation: life together in the underground. *Cell* 56:203–14
68. Mackie, G. A. 1986. Structure of the DNA distal to the gene for ribosomal protein S20 in *Escherichia coli* K12: presence of a strong terminator and an IS1 element. *Nucleic Acids Res.* 14: 6965–81
69. Mares, R., Urbanowski, M. L., Stauffer, G. V. 1992. Regulation of the *Salmonella typhimurium metA* gene by the MetR protein and homocysteine. *J. Bacteriol.* 174:390–97
70. Maxon, M. E., Redfield, B., Cai, X., Shoeman, R., Fujita, K., et al. 1989. Regulation of methionine synthesis in *Escherichia coli:* effect of the MetR protein on the expression of the *metE* and *metR* genes. *Proc. Natl. Acad. Sci. USA* 86:85–89
71. Maxon, M. E., Wigboldus, J., Brot, N., Weissbach, H. 1990. Structure-function studies on *Escherichia coli* MetR protein, a putative prokaryotic leucine zipper protein. *Proc. Natl. Acad. Sci. USA* 87:7076–79
72. McIver, J., Djordjevic, M. A., Weinman, J. J., Bender, G. L., Rolfe, B. G. 1989. Extension of host range of *Rhizobium leguminosarum* caused by point mutations in *nodD* that result in alterations in regulatory function and recognition of inducer molecules. *Mol. Plant-Microbe Interact.* 2:97–106
73. Meijer, W. G., Arnberg, A. C., Enequist, H. G., Terpstra, P., Lidstrom, M. E., Dijkhuizen, L. 1991. Identification and organization of carbon dioxide fixation genes in *Xanthobacter flavus* H4–14. *Mol. Gen. Genet.* 225: 320–30
74. Miller, B. E., Kredich, N. M. 1987. Purification of the CysB protein from *Salmonella typhimurium. J. Biol. Chem.* 262:6006–8
75. Monroe, R. S., Ostrowski, J., Hryniewicz, M. M., Kredich, N. M. 1990. In vitro interactions of CysB protein with the *cysK* and *cysJIH* promoter regions of *Salmonella typhimurium. J. Bacteriol.* 172:6919–29
76. Mulligan, J. T., Long, S. R. 1989. A family of activator genes regulates expression of *Rhizobium meliloti* nodulation genes. *Genetics* 122:7–18
77. Neidle, E. L., Hartnett, C., Ornston, L. N. 1989. Characterization of *Acinetobacter calcoaceticus catM,* a repressor gene homologous in sequence to transcriptional activator genes. *J. Bacteriol.* 171:5410–21
78. Norel, F., Robbe-Saule, V., Popoff, M. Y., Coynault, C. 1993. The putative sigma factor KatF (RpoS) is required for the transcription of the *Salmonella typhimurium* virulence gene SpvB in *E. coli. FEMS Microbiol. Lett.* 99:271–76
79. Ostrowski, J., Jagura-Burdzy, G., Kredich, N. M. 1987. DNA sequences of the *cysB* regions of *Salmonella typhimurium* and *Escherichia coli. J. Biol. Chem.* 262:5999–6005
80. Ostrowski, J., Kredich, N. M. 1989. Molecular characterization of the *cysJIH* promoters of *Salmonella typhimurium* and *Escherichia coli:* regulation by *cysB* protein and *N*-acetyl-L-serine. *J. Bacteriol.* 171:130–40
81. Ostrowski, J., Kredich, N. M. 1990. In vitro interactions of CysB protein with the *cysJIH* promoter of *Salmonella typhimurium:* inhibitory effects of sulfide. *J. Bacteriol.* 172:779–85

82. Ostrowski, J., Kredich, N. M. 1991. Negative autoregulation of *cysB* in *Salmonella typhimurium:* in vitro interactions of CysB protein with *cysB* promoter. *J. Bacteriol.* 173:2212–18
83. Pabo, C. O., Sauer, R. J. 1984. Protein DNA recognition. *Annu. Rev. Biochem.* 53:293–321
84. Parsek, M., Shinabarger, D. L., Rothmel, R. K., Chakrabarty, A. M. 1992. The role of *catR* and *cis-cis*-muconate in the activation of the *catBC* operon involved in benzoate degradation by *Pseudomonas putida. J. Bacteriol.* 174:7807–18
85. Pinner, E., Carmel, O., Bercovier, H., Sela, S., Padan, E. Schuldiner, S. 1992. Cloning, sequencing and expression of the *nhaA* and *nhaR* genes of *Salmonella enteriditis. Arch. Microbiol.* 157:323–28
86. Plaman, L. S., Stauffer, G. V. 1987. Nucleotide sequence of the *Salmonella typhimurium metR* gene and the *metR-metE* control region. *J. Bacteriol.* 169: 3932–37
87. Plaman, M., Stauffer, G. V. 1989. Regulation of the *Escherichia coli glyA* gene by the *metR* gene product and homocysteine. *J. Bacteriol.* 171:4958–62
88. Rahav-Manor, O., Carmel, O., Karpel, R., Taglicht, D., Glaser, G., et al. 1992. NhaR, a protein homologous to a family of bacterial regulatory proteins (LysR), regulates *nhaA,* the sodium proton antiporter gene in *Escherichia coli. J. Biol. Chem.* 267:10433–38
89. Renault, P., Gaillardin, C., Heslot, H. 1989. Product of the *Lactococcus lactis* gene required for malolactic fermentation is homologous to a family of positive regulators. *J. Bacteriol.* 171: 3108–14
89a. Renna, M. C., Najimudin, N., Winik, L. R., Zahler, S. A. 1993. Regulation of the *Bacillus subtilis alsS, alsD,* and *alsR* genes involved in post-exponential phase production of acetoin. *J. Bacteriol. In press*
90. Romero-Arroyo, C. E. 1992. *In vitro studies of transcription regulation of the* catBCEFD *operon in* Acinetobacter calcoaceticus *by CatM.* MS thesis, Univ. Ga., Athens, Ga. 62 pp.
91. Rostas, K., Kondorosi, E., Horvath, B., Simoncsits, A., Kondorosi, A. 1986. Conservation of extended promoter regions of nodulation genes in *Rhizobium. Proc. Natl. Acad. Sci. USA* 83:1757–61
92. Rothmel, R. K., Aldrich, T. L., Houghton, J. E., Coco, W. M., Ornston, L. N., Chakrabarty, A. M. 1990. Nucleotide sequencing and characterization of *Pseudomonas putida catR:* a positive regulator of the *catBC* operon is a member of the LysR family. *J. Bacteriol.* 172:922–31
93. Rothmel, R. K., Shinabarger, D. L., Parsek, M. R., Aldrich, T. L., Chakrabarty, A. M. 1991. Functional analysis of the *Pseudomonas putida* regulatory protein CatR: transcriptional studies and determination of the CatR DNA-binding site by hydroxyl-radical footprinting. *J. Bacteriol.* 173:4717–24
94. Rushing, B. G., Yelton, M. M., Long, S. R. 1991. Genetic and physical analysis of the *nodD3* region of *Rhizobium meliloti. Nucleic Acids Res.* 19:921–27
95. Saint-Girons, I. C., Parsot, M., Zaikin, O., Barzu, O., Cohen, G. N. 1988. Methionine biosynthesis in *Enterobacteriaceae:* biochemical, regulatory, and evolutionary aspects. *Crit. Rev. Biochem.* 23:S1–S42
96. Sasse-Dwight, S., Gralla, J. D. 1988. Probing the *Escherichia coli glnALN* upstream activation mechanism in vivo. *Proc. Natl. Acad. Sci. USA* 85:8934–38
97. Schell, M. A. 1985. Transcriptional control of the *nah* and *sal* hydrocarbon degradation operons by the *nahR* gene product. *Gene* 36:301–9
98. Schell, M. A. 1986. Homology between nucleotide sequences of the promoter regions of the *nah* and *sal* operons of the NAH7 plasmid of *Pseudomonas putida. Proc. Natl. Acad. Sci. USA* 83:369–73
99. Schell, M. A. 1990. Regulation of naphthalene degradation genes of plasmid NAH7: example of a generalized positive control system in *Pseudomonas* and related bacteria. See ref. 19a, pp. 165–77
100. Schell, M. A., Brown, P. H., Raju, S. 1990. Use of saturation mutagenesis to localize probable functional domains in the NahR protein, a LysR-type transcription activator. *J. Biol. Chem.* 265: 3844–50
101. Schell, M. A., Denny, T. P., Clough, S. J., Huang, J. 1993. Further characterization of genes encoding extracellular polysaccharide of *Pseudomonas solanacearum* and their regulation. In *Advances in Molecular Genetics of Plant-Microbe Interactions,* ed. E. W. Nester, D. P. S. Verma, 2:231–39. Dordrecht: Kluwer
102. Schell, M. A., Faris, E. 1987. Transcriptional regulation of the *nah* and *sal* naphthalene degradation operons of plasmid NAH7 of *P. putida.* In *RNA*

Polymerase and the Regulation of Transcription, ed. W. S. Reznikoff, R. Burgess, J. Dahlberg, C. Gross, M. Record, pp. 455–58. New York: Elsevier

103. Schell, M. A., Poser, E. 1989. Demonstration, characterization, and mutational analysis of NahR protein binding to the *nah* and *sal* promoters. *J. Bacteriol.* 171:837–46
104. Schell, M. A., Sukordhaman, M. 1989. Evidence that the transcription activator encoded by the *Pseudomonas putida nahR* gene is evolutionarily related to transcriptional activators encoded by the *Rhizobium nodD* genes. *J. Bacteriol.* 171:1952–59
105. Schell, M. A., Wender, P. E. 1986. Identification of the *nahR* gene product and nucleotide sequences required for its activation of the *sal* operon. *J. Bacteriol.* 166:9–14
106. Schlaman, H. R., Spaink, H. P., Okker, R. J., Lugtenberg, B. J. 1989. Subcellular localization of the *nodD* gene product in *Rhizobium leguminosarum. J. Bacteriol.* 171:4686–93
107. Schlaman, H. R., Okker, R. J., Lugtenberg, B. J. 1992. Regulation of nodulation gene expression by NodD in *Rhizobia. J. Bacteriol.* 174:5177–82

107a. Schwacha, A., Bender, R. A. 1993. The *nac* (nitrogen assimilation control) gene from *Klebsiella aerogenes. J. Bacteriol.* 175:2107–24

108. Smit, G. V., Puvanesarajah, V., Carlson, R. W., Barbour, W. M., Stacey, G. 1992. *Bradyrhizobium japonicum nodD1* can be specifically induced by soybean flavonoids that do not induce the *nodYABCSUIJ* operon. *J. Biol. Chem.* 267:310–18
109. Spaink, H. P., Wijffelman, C. A., Okker, R. J., Lugtenberg, B. J. 1989. Localization of functional regions of the *Rhizobium nodD* product using hybrid *nodD* genes. *Plant Mol. Biol.* 12:59–73
110. Stachel, S. E., An, G., Flores, C., Nester, E. W. 1985. A Tn*3 lacZ* transposon for the random generation of β-galactosidase gene fusions: application to the analysis of gene expression in *Agrobacterium. EMBO J.* 4:891–98
111. Stock, J. B., Ninfa, A. J., Stock, A. M. 1989. Protein phosphorylation and regulation of adaptive responses in bacteria. *Microb. Rev.* 53:450–90
112. Storz, G., Tartaglia, L. A. 1992. OxyR: a regulator of antioxidant genes. *J. Nutrit.* 122:627–30
113. Storz, G., Tartaglia, L. A., Ames, B. N. 1990. Transcriptional regulator of oxidative stress-inducible genes: direct activation by oxidation. *Science* 248: 189–94
114. Stragier, P., Danos, O., Patte, J. 1983. Regulation of diaminopimelate decarboxylase synthesis in *Escherichia coli. J. Mol. Biol.* 168:307–50
115. Streber, W. R., Timmis, K. N., Zenk, M. H. 1987. Analysis, cloning, and high-level expression of 2,4-dichlorophenoxyacetate monooxygenase gene *tfdA* of *Alcaligenes eutrophus* JMP134. *J. Bacteriol.* 169:2950–55
116. Sung, Y., Fuchs, J. A. 1992. The *Escherichia coli* K-12 *cyn* operon is positively regulated by a member of the LysR family. *J. Bacteriol.* 174:3645–50
117. Swanson, J., Mulligan, J., Long, S. R. 1993. Regulation of SyrM and *nodD3* in *Rhizobium meliloti. Genetics.* In press
118. Taira, S., Riikonen, P., Saarilati, H., Sukupolvi, S., Rhen, M. 1991. The *mkaC* virulence gene of *Salmonella* serovar *typhimurium* 96 Kb plasmid encodes a transcriptional activator. *Mol. Gen. Genet.* 228:381–84
119. Tartaglia, L. A., Gimenos, C. J., Storz, G., Ames, B. N. 1991. Multidegenerate DNA recognition by the OxyR transcriptional regulator. *J. Biol. Chem.* 267:2038–45
120. Tartaglia, L. A., Storz, G., Ames, B. N. 1989. Identification and molecular analysis of *oxyR*-regulated promoters important for the bacterial adaption to oxidative stress. *J. Mol. Biol.* 210:709–19
121. Thony, B., Hwang, D. S., Fradkin, L., Kornberg, A. 1991. *iciA* an *Escherichia coli* gene encoding a specific inhibitor of chromosomal initiation of replication in vitro. *Proc. Natl. Acad. Sci. USA* 88:4066–70
122. Tullius, T. D., Dombroski, B. A., Churchill, M. E. A., Kam, L. 1987. Hydroxylradical footprinting: a high resolution method for mapping protein-DNA contacts. *Methods Enzymol.* 155: 537–58
123. Urabe, H., Ogawara, H. 1992. Nucleotide sequence and transcriptional analysis of activator-regulator proteins for β-lactamase in *Streptomyces cacaoi. J. Bacteriol.* 174:2834–42
124. Urbanowski, M. L., Stauffer, G. V. 1989. Role of homocysteine in *metR*-mediated activation of the *metE* and *metH* genes in *Salmonella typhimurium* and *Escherichia coli. J. Bacteriol.* 171: 3277–81
125. Urbanowski, M. L., Stauffer, G. V.

1989. Genetic and biochemical analysis of the MetR activator-binding site in the *metE-metR* control region of *Salmonella typhimurium*. *J. Bacteriol.* 171: 5620–29
126. Urbanowski, M. L., Stauffer, G. V. 1987. Regulation of the *metR* gene of *Salmonella typhimurium*. *J. Bacteriol.* 169:5841–44
127. Van Der Meer, J. R., Frijters, A. C., Leveau, J. H., Eggen, R. I., Zehnder, A. J., De Vos, W. M. 1991. Characterization of the *Pseudomonas* sp. strain P51 gene *tcbR*, a LysR-type transcriptional activator of the *tcbCDEF* chlorocatechol oxidative operon, and analysis of the regulatory region. *J. Bacteriol.* 173:3700–8
128. Viale, A. M., Kobayashi, H., Akazawa, T., Henikoff, S. 1991. RcbR, a gene coding for a member of the LysR family of transcriptional regulators, is located upstream of the expressed set of ribulose 1,5-bisphosphate carboxylase/oxygenase genes in the photosynthetic bacterium *Chromatium vinosum*. *J. Bacteriol.* 173:5224–29
129. Von Lintig, J., Zanker, H., Schröder, J. 1991. Positive regulators of opine-inducible promoters in the nopaline and octopine catabolism regions of Ti plasmids. *Mol. Plant-Microbe Interact.* 4:370–78
130. Wang, L., Helmann, J. D., Winans, S. C. 1992. The A. tumefaciens transcriptional activator OccR causes a bend at a target promoter, which is partially relaxed by a plant tumor metabolite. *Cell* 69:659–67
131. Warne, S. R., Varley, J. M., Boulnois, G. J., Norton, M. G. 1990. Identification and characterization of a gene that controls colony morphology and auto-aggregation in *E. coli* K12. *J. Gen. Microbiol.* 136:455–62
132. Wek, R. C., Hatfield, G. W. 1986. Nucleotide sequence and in vivo expression of the *ilvY* and *ilvC* genes in *Escherichia coli* K12. *J. Biol. Chem.* 261:2441–50
133. Wek, R. C., Hatfield, G. W. 1988. Transcriptional activation at adjacent operators in the divergent-overlapping *ilvY* and *ilvC* promoters of *Escherichia coli*. *J. Mol. Biol.* 203:643–63
134. Windhövel, U., Bowien, B. 1991. Identification of *cfxR*, an activator gene of autotrophic CO_2 fixation in *Alcaligenes eutrophus*. *Mol. Microbiol.* 5:2695–2705
135. Yen, K.-M., Gunsalus, I. C. 1982. Plasmid gene organization: naphthalene/salicylate oxidation. *Proc. Natl. Acad. Sci USA* 79:874–78
136. Yen, K.-M., Gunsalus, I. C. 1985. Regulation of naphthalene catabolic plasmid NAH7. *J. Bacteriol.* 162:1008–13

Annu. Rev. Microbiol. 1993. 47:627–58

ENZYMES AND PROTEINS FROM ORGANISMS THAT GROW NEAR AND ABOVE 100°C

Michael W. W. Adams
Department of Biochemistry, University of Georgia, Athens, Georgia 30602

KEY WORDS: hyperthermophiles, *Archaea*, anaerobes, sulfur-dependent organisms

---

CONTENTS

ABSTRACT

Microorganisms that can grow at and above 100°C were discovered a decade ago, and about 20 different genera are now known. These so-called hyperthermophiles are the most ancient of all extant life; all but two genera are

0066-4227/93/1001-0627$02.00

classified as *Archaea*. All have been isolated from geothermal heated environments including deep-sea hydrothermal vents. This group includes some methanogenic and sulfate-reducing species, but the majority are strictly anaerobic heterotrophs that utilize complex peptide mixtures as sources of energy, carbon, and nitrogen. Only a few species are saccharolytic. Most of the hyperthermophiles absolutely depend on the reduction of elemental sulfur (S^0) to H_2S for significant growth, a property that severely limits their large-scale culture in conventional fermentation systems. Consequently, most physiological and metabolic studies have focused on those that can also grow in the absence of S^0, including species of the *Archaea, Pyrococcus* and *Thermococcus,* and the bacterium *Thermotoga*. The fermentative pathways for the metabolism of both peptides and carbohydrates in the *Archaea* appear to depend upon enzymes that contain tungsten, an element seldom used in biological systems. The mechanisms of S^0 reduction and energy conservation remain unclear. Enzymes purified from the S^0-reducing hyperthermophiles include proteases, amylolytic-type enzymes, hydrogenases, redox proteins, various ferredoxin-linked oxidoreductases, dehydrogenases, and DNA polymerases, some of which are active up to 140°C. However, complete amino acid sequences are known for only a handful of these proteins, and the three-dimensional structure of only one hyperthermophilic protein has been determined. Potential mechanisms by which proteins and various biological cofactors and organic intermediates are stabilized at extreme temperatures are only now beginning to emerge.

INTRODUCTION

This review focuses on enzymes and proteins that function in vivo at temperatures near 100°C, i.e. molecules isolated from organisms that grow at temperatures of 90°C and above. Microorganisms that thrive near the normal boiling point of water were first isolated in 1982 by Stetter from shallow marine volcanic vents (136). Since then about 20 different genera of organisms have been discovered that can grow near 100°C, several of which were recovered from the proximity of deep-sea hydrothermal vents (for reviews, see 2, 5, 73, 78, 137, 139). Because these life forms with their extraordinary characteristics were isolated in just the past few years, we know little about the novel biochemistry that must be required to sustain them under such conditions. This review aims to summarize the current knowledge about hyperthermophilic organisms and the enzymes that have been characterized from them, with an emphasis on the sulfur-dependent species, the predominant type of hyperthermophile.

HYPERTHERMOPHILIC ORGANISMS

Definition and Evolutionary Aspects

Hyperthermophiles are defined here as organisms able to grow at 90°C and above with an optimal growth temperature of at least 80°C. Such organisms, which were isolated only in the past few years, form a fairly cohesive group (2, 4), in distinct contrast to the numerous moderately thermophilic organisms that grow optimally in the range 50–70°C (150). All but two of the hyperthermophilic genera are classified as *Archaea* (formerly *Archaebacteria*) (151, 153, 154). The *Archaea* and *Eukarya* likely had a common ancestor not shared by the *Bacteria,* and the first organisms to have diverged from the *Eukarya/Archaea* lineage were the hyperthermophiles (148, 153). Moreover, the two currently known hyperthermophilic genera are the most ancient of all *Bacteria* (31, 153). Hyperthermophilic organisms therefore appear more closely related than all other organisms to the ancestor of all extant life, having evolved when the earth was much hotter than it is at present (151, 153). Indeed, phylogenetic analysis suggests that the rest of biology results from evolutionary pressures to adapt to low (less than 100°C) temperatures (see 6).

Physiology

Table 1 lists the hyperthermophilic genera currently known. The majority are from marine environments including hot sediments at depths down to 100 m or so and deep-sea sediments and hydrothermal vents located up to 4000 m below sea level (see 11, 61, 65). All but two of the organisms are obligate anaerobes. The exceptions are the archaeon *Acidianus infernus* and the bacterium *Aquifex pyrophilus*. The former is an obligate autotroph that obtains energy for growth either by oxidizing S^0 with O_2 to produce sulfuric acid or by reducing S^0 to H_2S (128). *A. infernus* is classified within the *Sulfolobales,* which also includes some less thermophilic genera (129, 137, 139). *Aquifex pyrophilus* is a microaerophilic chemolithotroph that uses H_2, thiosulfate, and S^0 as electron donors and O_2 (<0.5%, v/v) or nitrate as the electron acceptor (62). It is the first microaerophilic (or nitrate-reducing) hyperthermophile.

The ability of organisms to use O_2 is therefore very limited at temperatures above 90°C, and the reduction of S^0 rather than O_2 appears to be the predominant means of energy conservation by hyperthermophiles. Hence, all the hyperthermophiles shown in Table 1 are anaerobic S^0-reducing organisms, except for the methanogens, which utilize only H_2 and CO_2 as an energy source, and the related (152) sulfate-reducing genus *Archaeglobus* (141). The type strain, *Archaeoglobus fulgidus,* grows on sulfate plus lactate, pyruvate, or formate as the sole carbon and energy source to produce CO_2 and H_2S

Table 1 Hyperthermophilic genera

Classification	T_{max}	(T_{opt})	%GC	Habitat[a]	First isolated	Reference
S^0-dependent *Archaea*						
Thermoproteales						
Thermoproteus (T. tenax, T. neutrophilus, T. uzoniensis)	92°	(88°)	56	c	1981	20, 143, 164
Staphylothermus marinus	98°	(92°)	35	d/m	1986	51, 137
Desulfurococcus (D. amylolyticus, D. mucosus)	90°	(87°)	51	d/m/c	1982	21, 66, 163
Thermofilum (T. pendens, T. librum)	100°	(88°)	57	c	1983	159
Pyrobaculum (P. organotrophum, P. islandicum)	102°	(100°)	46	c	1987	58
Sulfolobales						
Acidianus infernus	96°	(90°)	31	m/c	1986	128
Pyrodictiales						
Pyrodictium (P. brockii, P. occultum, P. abysi)	110°	(105°)	62	d/m	1983	105, 139, 140
Thermodiscus maritimus	98°	(90°)	49	m	1986	137
Thermococcales						
Pyrococcus (P. furiosus, P. woesii)	105°	(100°)	38	d/m	1986	50, 162
Thermococcus (T. litoralis, T. celer, T. stetteri)	97°	(88°)	57	d/m	1983	88, 95, 161
Hyperthermus butylicus	110°	(100°)	56	m	1990	160
ES-1	91°	(82°)	59	d/m	1989	103
ES-4	108°	(100°)	55	d/m	1991	104
GB-D	103°	(95°)	39	d/m	1992	67
GE-5	102°	(94°)	44	d/m	1992	49

Sulfate-reducing *Archaea*						
Archaeoglobales						
Archaeoglobus (A. profundis, A. fulgidus)	95°	(83°)	46	d/m	1987	30, 138, 158
Methanogenic *Archaea*						
Methanococcales						
Methanococcus igneus	91°	(88°)	31	d/m	1983	29
Methanobacteriales						
Methanothermus (M. sociabilis, M. fervidus)	97°	(83°)	33	c	1981	80, 142
Methanopyrales						
Methanopyrus kanderi	110°	(100°)	60	d/m	1990	32, 59, 79
Bacteria						
Thermotogales						
Thermotoga (T. maritima, T. neapolitana)	90°	(80°)	46	m	1986	60, 64
Aquificales						
Aquifex pyrophilus	95°	(85°)	40	m	1992	31, 62

[a] Species have been isolated from shallow marine vents (m), deep sea hydrothermal vents (d), and/or from continental hot springs (c). The sulfur-dependent genera are grouped in separate orders, except for *Hyperthermus butylicus*, ES-1, ES-4, GB-D, and GE-5, as these have yet to be formally classified. Modified from Ref. 6.

(138). When thiosulfate replaces sulfate, some growth is also obtained using H_2/CO_2 or fumarate as the carbon source. Sulfate-reducing organisms growing above 100°C were recently identified in a deep sea environment (70), but these have so far resisted isolation.

Various genera carry out the predominant metabolism of the known hyperthermophiles—anaerobic S^0 reduction—but virtually all are strict organotrophs and most obtain energy by S^0 respiration. Only a few species, such as *Thermoproteus tenax* (143), *Thermoproteus neutrophilus* (143), *Pyrobaculum islandicum* (58), *Pyrodictium brockii* (140), and *Pyrodictium occultum* (139), are facultative autotrophs and can use H_2 as the electron donor for S^0 reduction and CO_2 as the sole carbon source. Species of *Desulfurococcus* (21, 66, 163), *Thermofilum* (159), *Thermodiscus* (137), and *Staphylothermus* (51), and *Pyrodictium abyssi* (105), *Pyrobaculum organotrophum* (58), *Thermococcus celer* (161), *Thermococcus stetteri* (88), *Pyrococcus woesei* (162), and *Hyperthermus butylicus* (160) are strict heterotrophs that use organic substrates to reduce S^0. The same is true for the deep-sea hyperthermophiles ES-1 (103), ES-4 (104), GB-D (thought to be a *Pyrococcus* sp.) (67), and GE-5 (49), which have yet to be formally classified. Complex organic mixtures, including yeast and meat extracts, tryptone, peptone, and casein, are the source of reducing equivalents for S^0 reduction, as well as the sources of carbon and nitrogen. Carbohydrates are utilized by only a few species, including *Thermoproteus tenax* (143), *Pyrodictium abyssi* (105), *Pyrococcus furiosus* (50), *Pyrococcus woesei* (162), and *Thermococcus litoralis* (92). All of the heterotrophic hyperthermophiles, however, including the saccharolytic ones, appear to require peptides as a nitrogen source, and with the exceptions noted below, optimal growth depends upon S^0 and thus the production of H_2S.

Only three of the S^0-dependent archaea, *P. furiosus* (50), *P. woesei* (162), and *T. litoralis* (95), grow well both with and without S^0. *T. litoralis* preferentially grows on peptides whereas *Pyrococcus* spp. are saccharolytic, utilizing maltose, starch, and glycogen. They are obligate heterotrophs that obtain energy for growth by fermentation and dispose of reductant either as H_2 or, if S^0 is present, as H_2S. These properties are also shared by the hyperthermophilic bacteria *Thermotoga maritima* (60) and *Thermotoga neapolitana* (64), although unlike the saccharolytic archaea, they utilize a much larger range of substrates, including monosaccharides.

Practical Aspects

The obligate requirement of S^0 for the majority of the hyperthermophiles has severely hindered attempts to grow them in sufficient quantities for enzyme purification. The production of large amounts of H_2S prohibits the use of conventional stainless steel fermentors because of corrosion, a situation made worse by the relatively high salt concentrations required by many of the marine isolates. For small-scale studies, glass vessels can be used (see 100), and a continuous culture system for growing *P. furiosus* has been described that can operate near 100°C using S^0-containing media (27). The use of a 2-liter foil-lined fermentor for the growth of *P. woesei* was also recently reported (113). Corrosion-resistant, enamel-lined fermentation systems for the large-scale culture (300 liters and above) of H_2S-producing hyperthermophiles are present in Stetter's laboratory at the University of Regensburg and at the author's institution at the University of Georgia. Although high pressure ought to be relevant to growing deep-sea isolates, hydrostatic pressures comparable to those in the natural environments only marginally affect the isolate's growth rate and the upper temperature limit for growth (67, 93, 94, 111). An obligately barophilic (or extremely halophilic) hyperthermophile has yet to be isolated.

NOVEL METABOLIC PATHWAYS OF HYPERTHERMOPHILES

Primary Pathways of Carbon Metabolism

To date *P. furiosus* is the best studied of the hyperthermophilic S^0-dependent species. It grows optimally near 100°C by a fermentative-type metabolism (50) in which certain carbohydrates and pyruvate (116) are oxidized to organic acids, H_2, and CO_2. Growth also depends upon a source of peptides (yeast extract, tryptone, or peptone), and these can serve as the sole carbon and energy source, although growth yields are lower (28, 72, 85, 96). S^0 is not required for growth, but if added to the medium, it is reduced to H_2S (50, 85). *P. furiosus* produces an extracellular amylase (77) when grown on starch, and the products of starch hydrolysis are further converted to glucose by an intracellular α-glucosidase (40). Bryant & Adams (28) found that tungsten greatly stimulates the growth of this organism, which gave some insight into the pathway for glucose metabolism in *P. furiosus*. Tungsten (W) is an element seldom used in biological systems—indeed, before the 1989 study (28), only one other W-containing enzyme was known (157). A tungsten-containing enzyme was subsequently purified from *P. furiosus* (90) and was shown to be an aldehyde ferredoxin oxidoreductase (AOR) (91). AOR catalyzed the oxidation of a variety of aldehydes to the corresponding acid, but it did not utilize aldehyde phosphates, coenzyme A (CoA), or NAP(P). Ferredoxin (8), which is the physiological electron donor for the H_2-evolving hydrogenase of this organism (28), served as an electron carrier for AOR (90).

Because the conventional pathways for glucose oxidation do not require an aldehyde oxidation step, we (91) proposed that *P. furiosus* contains a novel pyroglycolytic pathway. This pathway was based on the partially non-phosphorylated, NAD(P)-dependent pathway elucidated by Danson and co-workers in some aerobic archaea (43). The pathway in *P. furiosus* differs in that the three oxidation steps between glucose and acetate are linked to the reduction of ferredoxin and thus to H_2 production via the hydrogenase. The first of these steps is catalyzed by a new enzyme, glucose ferredoxin oxidoreductase, whereas W-containing AOR is proposed to catalyze the second step, the oxidation of glyceraldehyde to glycerate (91). A more conventional pyruvate ferredoxin oxidoreductase (POR) (17) generates acetyl CoA from pyruvate. A unique consequence of this pathway is that *P. furiosus* could convert glucose to acetate without the participation of the thermolabile cofactors, NAD(P)(H) (91). Schönheit and coworkers (117) recently measured all but one of the enzyme activities of the pyroglycolytic pathway.

W-containing AORs with properties similar to those of the *P. furiosus* enzyme have been purified from *T. litoralis* and ES-4 (69, 92), suggesting that these organisms also utilize the pyroglycolytic pathway. The activity of *P. furiosus* AOR is increased threefold in cells grown on pyruvate rather than maltose (117), and because the organism contains low activities of both fructose-1, 6-bisphosphate aldolase, and glyceraldehyde-3-phosphate dehydrogenase, gluconeogenesis might also occur via the pyroglycolytic pathway. Here AOR reduces glycerate rather than oxidizes glyceraldehyde (117). Net ATP synthesis does not occur in the production of pyruvate from glucose, and the only apparent site of energy conservation is in the production of acetate. Schönheit and coworkers showed that *P. furiosus* (116) and other acetate-producing hyperthermophiles (117) contain an acetyl CoA synthetase that catalyzes ATP production in a single step from acetyl CoA, ADP, and inorganic phosphate. Previously, such an enzyme was found only in a eukaryotic parasite and in a mesophilic anaerobe (see 116).

Tungsten also appears to play a key regulatory role in *P. furiosus* (119). The addition of W to energy-limited continuous cultures growing on maltose and peptides not only stimulated growth, it dramatically increased AOR activity and decreased total protease activity. This suggests a preference for use of maltose over peptides as the carbon and energy source in the presence of W. Conversely, cells do grow, albeit poorly, without the addtion of W. Whether growth-limiting amounts of this element are contaminants in the complex media or whether *P. furiosus* can ferment sugars independent of W-containing enzymes is not known. Although cells contain low levels of glyceraldehyde-3-phosphate dehydrogenase (166), triose phosphate isomerase, and phosphoglycerate kinase (see 91), they apparently lack hexokinase, glucose-6-phosphate dehydrogenase, 2-keto-3-deoxy-6-phosphogluconate aldolase, and phosphofructokinase (117).

The hyperthermophilic bacterium *Thermotoga maritima* also obtains energy for growth by the fermentation of a range of carbohydrates to H_2, CO_2, acetate, and lactate (60). Growth does not depend upon S^0, although if present this element is reduced to H_2S (68). Amylase, xylanase, β-glucosidase, 4-α-glucanotransferase, cellobiohydrolase, and cellulase activities have been detected in various *Thermotoga* spp. (24, 82, 114, 126, 130). The growth of *Thermotoga maritima* is also stimulated by tungsten (71), but it lacks AOR activity (91). Apparently, glucose is oxidized by a conventional Embden-Meyerhof-Parnas pathway, as demonstrated by the presence of glyceraldehyde-3-phosphate dehydrogenase (156) and lactate dehydrogenase (155) and the measurement of key enzyme activities in cell-free extracts (68). Although the role of W in *Thermotoga maritima* is unknown, the pathways of sugar fermentation found in hyperthermophilic bacteria and in hyperthermophilic archaea seem to differ greatly.

Of the hyperthermophilic S^0-dependent archaea, the only other primary pathway for carbon metabolism that has been studied concerns the autotroph *Thermoproteus neutrophilus*. Fuchs and coworkers have shown with elegant ^{13}C NMR studies (144) that this organism utilizes an unusual reductive citric acid cycle for CO_2 fixation. Acetyl CoA is carboxylated via pyruvate to yield oxaloacetate, which via the reductive carboxylations of succinyl CoA and 2-ketoglutarate ultimately yields citrate. The pathway differs from previous versions in that part of the cycle is reversible. Also, a new type of ATP citrate lyase reaction was postulated that reversibly generates acetyl CoA and oxaloacetate from citrate in the presence of CoASH and ATP.

The hyperthermophilic sulfate-reducing organism *Archaeoglobus fulgidus* grows on lactate and sulfate as the sole carbon and energy sources, producing CO_2 and H_2S (30, 141). This organism contains methanogenic cofactors such as methanofuran, methanopterin, and coenzyme F_{420}, but not coenzymes M or F_{430} (1) nor methyl CoM reductase (138, 141). The pathway of lactate oxidation to CO_2 has been elucidated (89). The organism uses tetrahydromethanopterin as a C-1 carrier rather than the tetrahydrofolate used by mesophilic sulfate-reducing bacteria in the carbon monoxide dehydrogenase pathway of acetyl CoA oxidation. The other major difference is that in *Archaeoglobus fulgidus* CO_2 is generated from formylmethanofuran whereas in the mesophiles CO_2 is derived from free formate (via formyltetrahydrofolate) (121). The pathway of hyperthermophilic sulfate reduction appears to be identical to that found in the mesophilic bacteria (see 42). Similarly, the pathway of CO_2 reduction to CH_4 in the hyperthermophilic methanogens is the same as in the mesophilic species (25, 127).

Primary Pathways of Nitrogen Metabolism

Proteins and peptides serve as nitrogen sources for all of the heterotrophic, S^0-reducing hyperthermophiles, and the majority also use proteins and peptides as their sole carbon sources. However, little is known about how these organisms utilize the peptide nitrogen and obtain energy and carbon compounds for biosynthesis when grown on peptides. High intracellular proteolytic activities have been measured in several hyperthermophilic archaea (19, 35, 41, 47, 75, 133), and high intracellular concentrations of glutamate dehydrogenase (GDH) are found in *P. furiosus* (39, 112) and in *T. litoralis* and ES-4 (K. Ma, F. T. Robb, & M. W. W. Adams, unpublished data). Kinetic analyses indicate that the latter enzymes function in vivo to convert glutamate produced from aminotransferase reactions to 2-ketoglutarate. Both *T. litoralis* and *P. furiosus* also contain distinct 2-ketoglutarate ferredoxin oxidoreductases (KGOR) (X. Mai & M. W. W. Adams, unpublished data). KGORs in these organisms catalyze the conversion of 2-ketoglutarate to succinyl CoA and CO_2, using ferredoxin as the electron acceptor. Excess reductant from 2-ketoglutarate oxidation can therefore be channeled to hydrogenase to evolve H_2. KGOR and GDH may be key enzymes in the link between peptide utilization and energy and carbon metabolism. *P. furiosus* also contains an indolepyruvate ferredoxin oxidoreductase (IOR) that catalyzes the CoASH-dependent reduction of *P. furiosus* ferredoxin using indole pyruvate, phenyl pyruvate, and *p*-hydoxyphenyl pyruvate, but not pyruvate or 2-ketoglutarate, as substrates (X. Mai & M. W. W. Adams, unpublished data). Presumably, these intermediates are generated by the transamination of aromatic amino acids and may serve as significant nitrogen and energy sources in *P. furiosus*.

The recent observation (92) that tungsten stimulates the growth of *T. litoralis* (95) suggests that the hyperthermophilic metabolism of peptides involves some unusual biochemistry. *T. litoralis* harbors another type of W-containing enzyme, formaldehyde ferredoxin oxidoreductase (FOR). This is present at high concentrations, comparable to AOR in *P. furiosus* (91), and it is proposed to be involved in peptide fermentation (92). *T. litoralis* produces H_2 during growth; hence FOR could dispose of the excess reductant generated by amino acid oxidation via ferredoxin and the hydrogenase, both of which have been purified (Z. T. Zhou, J.-B. Park, & M. W. W. Adams, unpublished data). Moreover, *P. furiosus* will grow with peptides as sole carbon and energy sources. It also contains FOR, but at about 10% of the concentration of AOR (92). Although *T. litoralis* reportedly cannot metabolize sugars (95), maltose does stimulate its growth in peptide-limited cultures (92). Finally, an AOR-type enzyme has been purified from *T. litoralis* (69, 92), suggesting that this organism also uses the pyroglycolytic pathway. Thus, *T. litoralis* and *P. furiosus* contain reciprocal amounts of two different types of aldehyde-oxidizing tungstoenzymes, one (AOR) involved in sugar fermentation and one (FOR) potentially involved in peptide fermentation.

The only report dealing with biosynthetic reactions in the hyperthermophilic organisms concerns the pathways of arginine synthesis in *Thermotoga maritima* and *P. furiosus* (146). The former generates arginine from glutamate via N-acetylated intermediates in an eight-step pathway identical to that found in mesophilic bacteria. However, none of the enzymes catalyzing the first five steps, from glutamate to ornithine, could be detected in *P. furiosus*. Because this organism is obligately dependent upon a source of peptides for growth, it is not known if arginine is required.

Mechanisms of S^0 Reduction and Energy Conservation

Although the reduction of S^0 to H_2S is extremely limited in the mesophilic world (81), it virtually characterizes the known hyperthermophiles. They can be subdivided into those that obligately utilize S^0 as a terminal electron acceptor and those that will reduce S^0 to H_2S but appear to obtain energy by fermentation rather than by S^0 respiration. The form of sulfur reduced by all hyperthermophiles is probably polysulfide, which is produced by the nucleophilic attack of sulfide (S^{2-}) on the S_8 ring of elemental sulfur (18). The mechanism of S^0 reduction has been investigated in only one autotrophic hyperthermophile, *Pyrodictium brockii* (102). Maier and coworkers demonstrated that *Pyrodictium brockii* appears to contain a primitive membrane-bound electron transport chain for energy conservation during the reduction of S^0 by H_2 (100). It contains a Ni hydrogenase (101), a *c*-type cytochrome, and a novel quinone, and resembles the membrane system of mesophilic, H_2-oxidizing aerobes (100). The S^0-reducing activity of *Pyrodictium brockii* has not been characterized. An enzyme capable of reducing S^0 to H_2S has been purified from the facultatively anaerobic chemolithotroph *Desulfurolobus ambivalens,* which grows at temperatures up to 87°C (see 74). However, the enzyme functions as a sulfur oxygenase and was not present in cells grown anaerobically on H_2/S^0.

In the heterotrophic hyperthermophiles such as *P. furiosus* that do not absolutely require S^0, its reduction is thought to be a detoxification mechanism to remove fermentatively produced H_2, as H_2 inhibits growth (50). Reductant is channeled by unknown mechanisms to S^0 rather than to protons (2). However, a more recent study showed that, when maltose is the carbon source, the yield of *P. furiosus* cells (dry wt/mol of maltose) grown in the presence of S^0 was almost twice that of cells grown without S^0, and that S^0 reduction was equivalent to an ATP yield of 0.5 mol ATP/mol S^0 reduced (118). Furthermore, the pattern of fermentation products, which were consistent with the postulated pyroglycolytic pathway (92), were unchanged when S^0 was present. Thus, S^0 reduction appears to play a role in energy conservation by this organism, although the mechanism of coupling ATP synthesis to S^0 reduction is not known.

An investigation (84) into the sulfur reductase activity of *P. furiosus* (measured by the production of sulfide from S^0) using reduced ferredoxin as the electron donor showed that it was cytoplasmic and polysulfide was also used as a substrate. Moreover, virtually all of the activity was associated with the hydrogenase that had been previously purified from this organism (28). Thus, this bifunctional enzyme, tentatively termed sulfhydrogenase, can dispose of the excess reductant generated during fermentation using either protons, to produce H_2, or S^0 (polysulfide), to produce H_2S. In addition, several hydrogenases purified from both hyperthermophilic and mesophilic representatives of the *Archaea* and *Bacteria* domains reduce S^0 to H_2S, indicating that the function of some form of ancestral hydrogenase was S^0 reduction rather than, or in addition to, the reduction of protons (84). The question still remains, however, as to how the intracellular production of sulfide can lead to the conservation of energy in hyperthermophiles such as *P. furiosus*.

Role of Tungsten

Apparently, W-containing enzymes may play key roles in the primary metabolic pathways of some of the heterotrophic hyperthermophiles. Yet enzymes containing the analogous element molybdenum (Mo) are ubiquitous in the rest of the biological world and are involved in many fundamental biological processes (see 134). Only certain moderately thermophilic species of acetogenic clostridia and methanogens are also known to utilize tungsten, and the element was recently shown to be a component of carboxylic acid reductase (see 149) and formylmethanofuran dehydrogenase (see 120). These anaerobic organisms are not obligately dependent upon W because analogous molybdoenzymes are synthesized if W is omitted and Mo is added to a culture medium (120, 149). Whether the heterotrophic hyperthermophiles are the first organisms to absolutely require W remains to be established. Indeed, since hyperthermophily is regarded as an ancient phenotype (151, 153), one could speculate that the earliest life forms were not only hyperthermophilic, they were also W dependent (4). Presumably, they then evolved into the mesophilic, Mo-dependent species that now overwhelmingly predominate. Possibly, W is required rather than Mo because W (but not Mo) may be present in the environments in which hyperthermophilic organisms are found, such as near deep-sea hydrothermal vents. Alternatively, W might be more readily stabilized in the oxidation states and at the redox potentials required to catalyze biologically relevant reactions at extreme temperatures (4). As yet no experimental evidence supports either of these possibilities.

HYPERTHERMOPHILIC ENZYMES AND PROTEINS

Because of the problems in culturing the H_2S-producing hyperthermophiles, most enzymological studies have focused on species of *Pyrococcus, Thermococcus,* and *Thermotoga,* organisms that can be grown in the absence of S^0 (see 6). Table 2 lists the enzymes that have been reasonably well characterized so far, together with some of their molecular, physiological, and temperature-dependent properties.

Proteases

P. furiosus contains at least 13 distinct polypeptides with proteolytic activity (35). These appear to exist as active aggregates, and they may be related to the multicatalytic proteinase complex found in eukaryotes and moderately thermophilic archaea (133). Kelly and coworkers (19) discovered that upon boiling cell-free extracts of *P. furiosus* with SDS (1%) for 24 h, only four proteins remained. Two of these had proteolytic activity and were termed S66 and S102, reflecting their apparent M_r values on SDS gels. The S66 protein is a serine-type protease and is the hydrolysis product of a 200-kDa nonproteolytic precursor. It appears to be the same as one of the components of a membrane-associated protease complex from *P. furiosus* (47). Cowan and coworkers (41) have purified an extracellular protease from *Desulfurococcus mucosus*. It was active up to 125°C and substrate-specificity studies indicated a preference for hydrophobic residues.

Carbohydrate-Metabolizing Enzymes

Pyrococcus spp. produce several inducible amylolytic (starch-degrading) enzymes both intracellularly and extracellularly (26, 40, 76, 77). More than 80% of the α-amylase activity (hydrolysis of α-1,4-glucosidic linkages) of *P. furiosus* was extracellular, and this enzyme consisted of a high-molecular-weight complex of two different subunits with M_r values of 137,000 and 96,000 (77). The complex was active up to 140°C, and unlike the analogous enzyme from aerobic bacteria, it did not require metal ions for activity. On the other hand, the extracellular amylase of *P. woesei* is a monomeric enzyme of M_r 70,000 and is active up to 130°C (76). Its activity was also independent of metal ions. Neither of the pyrococcal enzymes produced glucose from starch, which also contrasts with the bacterial enzymes.

P. furiosus also produces an intracellular α-glucosidase to further hydrolyze the products of the amylase reaction to glucose (40). The enzyme is more than twice the size of the corresponding enzyme from moderately thermophilic bacteria. It has a high specific activity for the hydrolysis of maltose and is active up to 130°C (40). *P. furiosus* also produces an intracellular sucrose α-glucohydrolase (or invertase). It is active up to 117°C, but its molecular properties are similar to those of mesophilic invertases (10). The function of this enzyme is unknown because *P. furiosus* does not utilize sucrose as a carbon source. The novel hyperthermophile ES-4 contains an unusual extracellular amylolytic enzyme (123) that hydrolyzes both α-1,4 (amylase-like) and α-1,6 (pullulanase-like) glucosidic linkages, similar to amylopullulanases isolated from moderately thermophilic bacteria. The ES-4 enzyme is glycosylated and was stabilized by Ca^{2+} ions, which increased the maximum temperature at which activity could be detected from 130 to 140°C (123).

Table 2 Molecular, temperature-dependent, and physiological properties of enzymes purified from S^0-reducing hyperthermophiles

Enzyme	M_r[a]	T_{opt}[b] (°C)	t_{50}[c] (h/°C)	Function	Reference
Pyrococcus furiosus					
Protease	66 (α)	115	33/98	Peptide hydrolysis	19, 47
Amylase	?	100	2/120	Starch hydrolysis	77
α-Glucosidase	125 (α)	110	48/98	Maltose hydrolysis	40
Sucrose α-glucohydrolase	114 (α)	105	48/95	Unknown	10
Hydrogenase	160 (αβγδ)	>95	2/100	H_2 production	2, 4, 28
Ferredoxin	7 (α)	>95	>24/95	Electron transfer	4, 8, 36
Rubredoxin	5 (α)	>95	>24/95	Unknown	14
Aldehyde oxidoreductase	85 (α)	>95	6/80	Glycolytic enzyme	90, 91
Pyruvate oxidoreductase	100 (αβγ)	>95	0.3/90	Pyruvate oxidation	17
Glutamate dehydrogenase	270 (α_6)	95	10/100	Glutamate oxidation	39, 112
DNA polymerase	93 (α)	>75	20/95	Replication	83, 86
Thermococcus litoralis					
Ferredoxin	7 (α)	>95	>24/95	Electron transfer	33
Formaldehyde oxidoreductase	280 (α_4)	95	2/80	Unknown	92
DNA polymerase	93 (α_4)	75	7/95	Replication	98
Pyrococcus woesei					
Amylase	70 (α)	100	6/100	Starch hydrolysis	76
GAPDH[d]	150 (α_4)	nd[e]	0.7/100	Glycolysis?	166
Thermoproteus tenax					
GAPDH (NAD)	196 (α_4)	nd	0.3/100	Unknown	55
GAPDH (NADP)	156 (α_4)	nd	0.5/100	Unknown	55

ES-4					
Amylopullulanase	140 (α)	118	20/98	Starch degradation	123
Pyrodictium brockii					
Hydrogenase	118 ($\alpha\beta$)	>90	1/98	H_2 oxidation	101
Desulfurococcus mucosus					
Protease	52 (α)	100	1.5/95	Peptide hydrolysis	41
Thermotoga maritima					
4-α-Glucanotransferase	53 (α)	70	3/80	Starch degradation	82
Hydrogenase	280 (α_4)	>95	1/90	H_2 production	4, 71
GAPDH	148 (α_4)	>90	2/95	Glycolysis	156
Lactate dehydrogenase	144 (α_4)	>90	1.5/90	Pyruvate reduction	155
Thermotoga sp. strain FjSS3-B.1					
Xylanase	31 (α)	105	1.5/95	Xylan hydrolysis	130
Cellobiohydrolase	36 (α)	105	1.1/108	Unknown	114

[a] Molecular weight in kilodaltons (the number and/or type of subunits is given in parentheses).
[b] Optimum temperature for catalytic activity in vitro.
[c] Time required to lose 50% of catalytic activity after incubation at the indicated temperature.
[d] GAPDH: glyceraldehyde-3-phosphate dehydrogenase.
[e] Not determined due to the instability of the substrates at high temperature.

Thermotoga spp. also contain several saccharolytic enzymes. Two distinct amylases have been located in the toga, or outer sheath, of *Thermotoga maritima* (126), and its 4-α-glucosyl transferase (82), which disproportionates 1,4-α-linked maltodextrins by glucosyl transfer reactions, was purified from an *Escherichia coli* clone expressing the corresponding *Thermotoga maritima* gene (which was not sequenced). The recombinant enzyme showed a temperature optimum of approximately 70°C, suggesting differences in stability between it and the native form. Unlike the saccharolytic archaea, *Thermotoga* spp. utilize xylan as a carbon source and secrete an endo-1,4-β-xylanase (130). Immobilization of the purified enzyme on glass beads served to increase its temperature optimum from 105 to 112°C, and the immobilized enzyme retained 25% of its activity after 1 h at 130°C in molten sorbitol (130). In addition, although *Thermotoga* spp. do not utilize cellulose, at least one strain (FjSS-B.1) produces an extracellular cellobiohydrolase. This is the most stable cellulase reported so far and is active up to 105°C (114).

Hydrogenases

Many of the S^0-dependent hyperthermophiles use H_2 as a source of energy and reductant, and some produce H_2 from fermentative metabolisms (2). Hydrogenase, the enzyme responsible for catalyzing reversible activation of H_2, has been purified and characterized from the heterotrophs *P. furiosus* (2, 4, 28) and *Thermotoga maritima* (4, 6, 71) and from the autotroph *Pyrodictium brockii* (101). Hydrogenases from mesophilic organisms that function in H_2 oxidation are typically membrane-bound dimeric enzymes that contain a nickel atom at their active sites (2, 107), whereas the H_2-evolving enzymes are soluble monomeric proteins and H_2 catalysis takes place at a novel iron-sulfur cluster (2, 3). Thus, the dimeric Ni-containing hydrogenase of *Pyrodictium brockii* is membrane-bound and appears similar to its mesophilic counterparts, except in its thermostability (101). Surprisingly, the *P. furiosus* enzyme, which is located in the cytoplasm and functions to evolve H_2, is also a Ni hydrogenase (28). It differs from the mesophilic enzymes in its high H_2 evolution activity, the use of ferredoxin as its physiological electron donor, and its insensitivity to the classical hydrogenase inhibitors (2, 28). In contrast, the H_2-evolving enzyme of *Thermotoga maritima* is a cytoplasmic Fe hydrogenase (71). Like the corresponding mesophilic enzymes, it is extremely O_2 sensitive (t_{50} in air ~10 s), but it differs in its low H_2 evolution activity in vitro and its inability to use *Thermotoga maritima* ferredoxin as an electron carrier. Moreover, it lacks the novel H_2-activating FeS cluster and has a different site and/or mechanism for catalyzing H_2 production (71). H_2 evolution by *T. litoralis* is accomplished by a cytoplasmic, dimeric Ni hydrogenase that uses ferredoxin as an electron donor (Z. H. Zhou, J.-B. Park & M. W. W. Adams, unpublished data). *Thermococcus stetteri* H_2

production is also catalyzed by a soluble hydrogenase, although its molecular properties were not reported (108).

Redox Proteins

The ferredoxin from *P. furiosus* is a soluble low-molecular-weight protein containing a single $[Fe_4S_4]$ cluster (8). It serves as an electron carrier for the hydrogenase and for several oxidoreductase enzymes. Although ferredoxins of this type have been purified from a wide range of mesophilic organisms (see 4), the *P. furiosus* protein is by far the most stable (8) and it is the only example of a 4Fe-ferredoxin in which one of the four cysteinyl residues that normally coordinate the $[Fe_4S_4]$ cluster has been replaced, in this case with an aspartyl residue (33). An OH^- molecule appears to be bound to the unique iron site (97), which gives the cluster unusual spectroscopic properties (36) and enables it to bind exogeneous ligands in vitro (38). Incomplete cysteinyl ligation also facilitates both the removal of a single Fe atom to yield a $[Fe_3S_4]$ cluster (36) and its reconstitution with other metal ions to give mixed-metal clusters, such as the first example of a $[NiFe_3S_4]$ cluster in a biological system (37). The electronic and magnetic properties of the mixed-metal sites and the interactions between the cluster and the protein have also been investigated (33, 135). Mixed-metal FeS clusters and FeS clusters lacking complete sulfur coordination are thought to be part of the active sites of various metalloenzymes, and extremely thermostable models of such sites might be generated using *P. furiosus* ferredoxin (see 4).

Ferredoxins have also been purified from ES-4 and *Thermotoga maritima* (J. M. Blamey & M. W. W. Adams, unpublished data) and from *T. litoralis* (see 33). They all contain a single $[Fe_4S_4]$ cluster and resemble *P. furiosus* ferredoxin in molecular size and thermal stability. In spite of this, the *T. litoralis* protein is seven amino acids shorter than the *P. furiosus* protein, and sequence alignment reveals identity in only 39 of 59 residues (66%). Both proteins show about 40% sequence identity with the prototypical mesophilic ferredoxin from *Desulfovibrio gigas* (see 33). In further contrast to the *P. furiosus* protein, the $[Fe_4S_4]$ cluster in *T. litoralis* ferredoxin has complete cysteinyl coordination. Even with these relatively small proteins, however, sequence comparisons of the hyperthermophilic ferredoxins with their mesophilic relatives give no clue as to potential mechanisms of hyperthermostability.

The rubredoxin from *P. furiosus* has also been purified (14) and extensively characterized. It is a small protein (M_r 5400) containing a single Fe atom coordinated by four cysteinyl residues. Mesophilic rubredoxins have been isolated from a range of anaerobic bacteria, and although their physiological function is unclear, they have been well studied [12 amino acid sequences and 4 crystal structures are known (see 4)]. As discussed below, the small size and extreme stability of *P. furiosus* rubredoxin (it is unaffected after 24 h at 95°C), make it an ideal protein with which to study mechanisms of protein hyperthermostability (12, 14, 15, 23, 45, 147). In addition, the effect of temperature on its redox properties has been reported (4, 132), and the ability to replace its iron atom with other metal ions has enabled the study of metal-protein interactions (13, 16), the electronic properties of its metal site (53, 56), and the potential of the Ni-substituted protein to function as a model of the active site of Ni hydrogenases (57).

Ferredoxin-Dependent Oxidoreductases

The putative pyroglycolytic enzyme, aldehyde ferredoxin oxidoreductase (AOR) of *P. furiosus,* is an extremely O_2-sensitive, monomeric, tungsten-containing iron-sulfur protein that catalyzes the oxidation of numerous aliphatic and aromatic aldehydes in vitro (91). It is unstable at ambient temperatures and must be purified rapidly in buffer containing glycerol (90). The enzyme contains two FeS centers in addition to tungsten and exhibits remarkable temperature-dependent redox and spectroscopic properties (90). The tungsten is present as a mononuclear site, the coordination structure of which has been determined (52). It is ligated to a unique pterin species known as molybdopterin, which also serves to coordinate the molybdenum in all known molybdenum-containing enzymes (with the exception of nitrogenase) (109). Interestingly, the pterin in AOR is the type usually found in eukaryotic molybdoenzymes (69). AOR has also been purified from ES-4 and *T. litoralis* (S. Mukund & M. W. W. Adams, unpublished data). Like the *P. furiosus* enzyme, AORs from these organisms are very O_2-sensitive, WFeS-containing proteins that contain the same unmodified pterin (69).

Formaldehyde ferredoxin oxidoreductase (FOR) from *T. litoralis* contains a mononuclear W site and an FeS cluster (92). It differs from *P. furiosus* AOR in molecular weight and subunit composition and in its sensitivity to inactivation by O_2 and various inhibitors. Furthermore, FOR catalyzes only the oxidation of C1-C3 aliphatic aldehydes and does not utilize aromatic aldehydes, aldehyde phosphates, CoA, nor NAD(P). Nevertheless, FOR and *P. furiosus* AORs appear to be related as their N-terminal amino acid sequences show significant homology (92). FOR has also been purified from *P. furiosus* and resembles the *T. litoralis* enzyme in its molecular and catalytic properties (S. Mukund & M. W. W. Adams, unpublished data).

In the fermentative hyperthermophiles, pyruvate ferredoxin oxidoreductase (POR) catalyzes the final oxidation step in the conversion of sugars to acetate—the production of acetyl CoA and CO_2 from pyruvate. *P. furiosus* POR is an extremely O_2-sensitive Fe/S-containing protein and is quite thermolabile with a half-life at 80°C of only 23 min (17). The molecular weight of the enzyme is less than half that of the mesophilic PORs, which have been obtained from anaerobic bacteria, protozoa, and extremely halophilic archaea (see 17). The PORs of *T. litoralis* and *Thermotoga maritima* have also been purified, and these have molecular compositions almost identical to that of *P. furiosus* POR (J. M. Blamey & M. W. W. Adams, unpublished data). All three enzymes also show no activity towards other potential substrates such as 2-ketoglutarate, phenyl pyruvate, or indole pyruvate. We (17) have speculated that the hyperthermophilic PORs may represent some ancient form of a pyruvate-oxidizing enzyme. Their three dissimilar subunits may be analogous to the E1, E2, and E3 enzymes of the pyruvate dehydrogenase multienzyme complex found in aerobic organisms; these may also have evolved into the single subunit of the PORs found in anaerobic mesophiles (17).

Ferredoxin-dependent oxidoreductases that catalyze the oxidative decarboxylation of 2-ketoglutarate to succinyl CoA (KGOR) have been purified from both *P. furiosus* and *T. litoralis* (X. Mai & M. W. W. Adams, unpublished data). These enzymes do not utilize pyruvate (nor aromatic keto acids) and, like the hyperthermophilic PORs, are O_2-sensitive, Fe/S-containing, trimeric proteins of M_r approximately 100,000. This is less than half the size of the one KGOR purified from a mesophilic organism, the extreme halophile *Halobacterium halobium* (see 17), suggesting that both KGOR and POR evolved from an ancestral, trimeric keto-acid oxidoreductase. However, the indolepyruvate ferredoxin oxidoreductase (IOR) recently purified from *P. furiosus* is distinct from POR and KGOR. Although it too is an O_2-sensitive, FeS-containing protein, it has an M_r value of 160,000 and a tetrameric structure ($\alpha_2\beta_2$) (X. Mai & M. W. W. Adams, unpublished data).

Dehydrogenases

P. furiosus glutamate dehydrogenase (GDH), which catalyzes the reversible interconverion of glutamate and 2-ketoglutarate, is the most thermostable dehydrogenase currently known (39, 112). It represents about 20% of the total cytoplasmic protein, which suggests it plays a major role in the metabolism of this organism. It differs from the GDHs of mesophilic bacteria in its ability to utilize both NADH and NADPH with high affinity, a property shared with eukaryotic GDHs. *P. furiosus* GDH is inactive at ambient temperature and undergoes activation above 40°C. Differential scanning microcalorimetry (DSC) showed that activation occurs with a reversible increase in heat capacity (ΔH = 180 kcal mol^{-1}; T_m = 60°C) and an increase in UV absorption (46). This suggests that activation involves some exposure of hydrophobic groups to solvent. On the other hand, thermal inactivation occurs with T_m = 113°C, an irreversible increase in heat capacity (ΔH = 410 kcal mol^{-1}), and a large increase in A_{280}, more than eight times that observed during activation (46).

The enzyme therefore appears well adapted to the growth temperature range of *P. furiosus* (75–105°C).

Glyceraldehyde-3-phosphate dehydrogenase (GAPDH) from *P. woesei* also utilizes both NAD and NADP (166). It was insensitive to the antibiotic pentalenolactone, a potent inhibitor of bacterial and eukaryotic GAPDHs, and was stabilized by a variety of salts at high temperature. For example, its half-life at 104°C was 9 min, but no activity was lost after 30 min in the presence of 250 mM potassium citrate. The gene for GAPDH has been cloned, sequenced, and expressed in *E. coli* (166). The thermal stability and kinetic properties of the recombinant enzyme were identical to those of the native protein. The enzyme showed 50% sequence identity with the GAPDHs from mesophilic methanogens. Amino acid sequence comparisons with mesophilic and thermophilic GAPDHs from both archaea and bacteria indicated an increase in average hydrophobicity and decrease in average chain flexibility for the hyperthermophilic enzyme (166). Some of the amino acid changes previously identified between homologous mesophilic and thermophilic proteins were observed with the *P. woesei* GAPDH, e.g. preference for alanine, but others were not, e.g. preference for threonine (166). Two distinct GAPDHs have been purified from *Thermoproteus tenax* (55). One was specific for NAD and the other for NADP, and they differed in molecular weight and thermal stability. The presence of two types of GAPDHs in the same organism was previously found only in some eukaryotes (55).

GAPDH from *Thermotoga maritima* is present at high concentrations within the cytoplasm (yield 1.5 mg/g of cells), is specific for NAD, and has been obtained in a crystalline form (156). The melting temperature (T_m) for the pure protein measured with DSC was 109°C, comparable to that for *P. furiosus* GDH (46). The amino acid sequence of the native protein showed 60% identity with the GAPDHs from moderately thermophilic bacteria and 33% identity with all known GAPDHs (124). However, detailed comparisons with the sequences of homologous GAPDHs from various sources led to the conclusion that enhanced intrinsic thermal stability cannot be attributed to specific amino acid exchanges in terms of the traffic rules of thermophily (110). Extensive analysis of the effects of temperature, pH, and chaotropic agents on the stability of the *Thermotoga maritima* enzyme were recently reported (110, 125). Compared with its quantities of GAPDH, *Thermotoga maritima* contains extremely low amounts of lactate dehydrogenase (LDH). Although insufficient amounts of the pure enzyme were obtained for detailed characterization (155), studies showed that it is specific for NAD, and as with other bacterial LDHs, fructose-1,6-bisphosphate stimulated its activity. Moreover, when NAD was also present, the half-life of the enzyme at 90°C increased from 2 to 150 min, suggesting that these reagents also play a role in stabilizing the enzyme in vivo (155).

DNA Polymerases

With the advent of the polymerase chain reaction (PCR) as a routine tool in molecular biology and its requirement for a thermostable DNA polymerase, interest in this enzyme from hyperthermophiles has become substantial. Three are commercially available: Vent and Deep Vent DNA polymerases (both from New England Biolabs, Mass.) and Pfu DNA polymerase (from Stratagene, Calif.), purified from *T. litoralis,* strain GB-D, and *P. furiosus,* respectively. Details on the high fidelity of the *P. furiosus* DNA polymerase in the PCR reaction have been reported (83) and both the *P. furiosus* and *T. litoralis* enzymes have been cloned and sequenced (86, 98). Analysis of the deduced amino acid sequence for *P. furiosus* DNA polymerase revealed that it had considerable homology with highly conserved regions in α-like polymerases, a group that includes human, yeast, and other eukaryotic DNA polymerases, but it was much less homologous with Pol1-like DNA polymerases, which include the bacterial and bacteriophage enzymes (86). Characterization of the gene for *T. litoralis* DNA polymerase showed that it was split by two intervening sequences that form one contiguous open reading frame. One of the introns codes for an endonuclease, while the other self-splices in *E. coli* to generate the active enzyme (98). This is the first report of introns in a gene that codes for a protein in any archaeon or bacterium.

Miscellaneous Proteins

The gene for elongation factor Tu of *Thermotoga maritima* has been cloned and sequenced and expressed in *E. coli* (145), and the same was accomplished with the structural gene for glutamine synthetase (GS) (115). GS activity was measured in cell-free extracts of *Thermotoga maritima,* and its sequence was deduced from the cloned gene, which was identified using the corresponding gene from a mesophile. This report (115) also gave the deduced amino acid sequence of GS from *P. woesei*. An oligomeric outer membrane protein termed Ompα has also been isolated from *Thermotoga maritima,* and its gene was cloned and sequenced (48). A novel enzyme known as reverse gyrase, which introduces positive superturns in DNA, has been detected in a wide variety of hyperthermophiles including *Thermotoga maritima* (see 22), and the enzyme has been purified from *Desulfurococcus amylolyticus* (131). It is not found in mesophilic organisms (22), and its activity may afford DNA some protection against thermal denaturation. However, the first plasmid to be identified in a hyperthermophilic organism (49) is apparently relaxed at physiological temperatures (34). In addition, Baumeister, Stetter, and coworkers (99) purified a large cylindrical protein complex from *Pyrodictium occultum* that exhibits ATPase activity at 100°C and becomes highly enriched as a result of heat shock (from 102 to 108°C). Immunoanalysis showed an analogous protein is present in several other hyperthermophilic archaea but not in *Thermotoga maritima*. The complex was proposed to represent a novel type of chaperonin related to the *groEL/hsp60* family.

THE STABILITY OF BIOMOLECULES NEAR AND ABOVE 100°C

Understanding the mechanisms that stabilize globular proteins, especially at extreme temperatures, has been and continues to be one of the most challenging problems in both biotechnology and biochemistry (see e.g. 63, 87, 106, 165). The hyperthermophilic proteins reveal a new aspect of protein stability, particularly because proteins in vitro may be destroyed at 80–100°C by the hydrolysis of certain peptide bonds, deamination of asparaginyl residues, and cleavage of disulfide bonds (7). Significant insight into stabilizing mechanisms should be obtained by comparisons of homologous hyperthermophilic and mesophilic proteins at the molecular level. However, several studies have now shown that amino acid sequence information is not sufficient and detailed structural information is required (14, 110, 115, 124, 125, 156, 166).

The only three-dimensional structure of a hyperthermophilic protein determined so far is that of *P. furiosus* rubredoxin (12, 15, 23, 45, 147). Comparisons of its sequence with those of over a dozen mesophilic rubredoxins gave no indication of potential stabilizing mechanisms, in spite of the fact that these proteins contain only about 50 residues (14). The structure of the *P. furiosus* protein was determined independently using two-dimensional NMR (15), X-ray crystallography (45), and molecular modeling (147). All three methods, which were complementary in many ways, gave virtually identical structures (12, 23, 147). Comparisons with the crystal structures of mesophilic rubredoxins showed that the hydrophobic core of the hyperthermophilic protein exhibited significant sequence and structural homology to the mesophilic proteins, and that hyperthermostability appeared to be conferred by relatively minor changes mainly in the surface residues (15, 45, 147).

The most striking difference between the structures of *P. furiosus* rubredoxin and the mesophilic proteins was in the N-terminal region. Because the *P. furiosus* protein lacks an N-terminal methionine and a glutamyl residue at position 15 (which is proline in the mesophilic proteins), its N terminus is part of the hydrogen-bonding network of a three-stranded β-sheet. In contrast, the N-terminal residues of the mesophilic proteins are not incorporated into the β-sheet and are highly disordered. The N terminus is probably the first part of the protein to unzip at high temperatures, and the enhanced stability of the *P. furiosus* protein likely results in part from a reduction in the lability of its N terminus (14, 15). These conclusions suggest that the elucidation of the factors that stabilize much larger hyperthermophilic enzymes may be exceedingly complex, even if high-resolution structures are available.

Although most of the enzymes and proteins that have been purified from hyperthermophiles are remarkably thermostable in vitro (Table 2), some are unstable at ambient temperature (71, 91). In addition, the question arises as to how these organisms stabilize various biological cofactors and intermediates, many of which are extremely thermolabile, at least in vitro (see 44). For example, NADPH and NADH have half-lives of a few minutes above 90°C (112), and glyceraldehyde-3-phosphate and glyceraldehyde cannot be used in in vitro assays above 70°C (55, 91). Some form of stabilization of labile enzymes, cofactors, and simple organic compounds must therefore occur in vivo to enable cellular function at temperatures near 100°C. One explanation is that the hyperthermophiles have cytoplasmic thermoprotectants, solutes that minimize denaturation. For example, *P. woesei* contains K^+ ions and the novel sugar di-inositol-1,1′-phosphate at concentrations of approximately 0.8 M (122). Similarly, thermophilic and hyperthermophilic methanogens contain intracellular concentrations of 1.0–2.3 M K^+ ions and up to 1.2 M trianionic cyclic 2,3-diphosphoglycerate (54, 59, 79). Both of these carbon compounds stabilize enzymes purified from the respective organisms (see also 25). However, the intracellular concentration of K^+ ions (and Na^+) in the hyperthermophile *Thermoproteus tenax* is less than 100 mM (55). Hence, if cytoplasmic solutes do serve a role in stabilizing biological molecules in the hyperthermophiles, there is apparently no universal mechanism. Clearly, much remains to be learned about hyperthermophilic biochemistry.

SUMMARY

Numerous enzymes and proteins have recently been purified and characterized from hyperthermophilic S^0-reducing organisms. From the *Archaea,* complete amino acid sequences for five proteins have been reported: three derived from proteins (ferredoxin and rubredoxin from *P. furiosus* and ferredoxin from *T. litoralis*) (14, 33) and two from gene sequences (*P. woesei* GAPDH and *T. litoralis* DNA polymerase) (69, 98). In addition, five complete amino acid sequences have been reported for *Thermotoga maritima* proteins: lactate dehydrogenase (from the protein) (155), GAPDH, glutamine synthetase, Ompα, and EF-Tu (from gene sequences) (9, 48, 115, 124). So far the genes for three hyperthermophilic proteins have been expressed in *E. coli* (GAPDH from *P. woesei* and glutamine synthetase and EFTu from *Thermotoga maritima*) (69, 115, 145). Finally, the three-dimensional structure of only one hyperthermophilic protein has been determined, that of *P. furiosus* rubredoxin (12, 15, 23, 45, 147). The list of purified hyperthermophilic enzymes will expand considerably in the very near future, although the elucidation of the mechanisms that enable some of these proteins to retain catalytic activity in vitro at temperatures in excess of 140°C may not be so forthcoming.

ACKNOWLEDGMENTS

Research carried out in the author's laboratory was supported by grants from the Department of Energy, the National Science Foundation, the Office of Naval Research, and the National Institutes of Health.

Literature Cited

1. Achenbach-Richter, L., Stetter, K. O., Woese, C. R. 1987. A possible missing link among archaebacteria. *Nature* 327: 348–49
2. Adams, M. W. W. 1990. The metabolism of hydrogen by extremely thermophilic, sulfur-dependent bacteria. *FEMS Microbiol. Rev.* 75:219–38
3. Adams, M. W. W. 1990. The structure and mechanism of iron-hydrogenases. *Biochim. Biophys. Acta* 1020:115–45
4. Adams, M. W. W. 1992. Novel iron sulfur clusters in metalloenzymes and redox proteins from extremely thermophilic bacteria. *Adv. Inorg. Chem.* 38: 341–96
5. Adams, M. W. W., Kelly, R. M., eds. 1992. *Biocatalysis at Extreme Temperatures: Enzyme Systems Near and Above 100°C.* Washington, DC: Am. Chem. Soc., Series No. 498. 215 pp.
6. Adams, M. W. W., Park, J.-B., Mukund, S., Blamey, J., Kelly, R. M. 1992. See Ref. 5, pp. 4–23
7. Ahern, T. J., Klibanov, A. M. 1985. The mechanism of irreversible enzyme inactivation at 100°C. *Science* 228: 1280–84
8. Aono, S., Bryant, F. O., Adams, M. W. W. 1989. A novel and remarkably thermostable ferredoxin from the hyperthermophilic archaebacterium *Pyrococcus furiosus*. *J. Bacteriol.* 171: 3433–39
9. Bachleitner, M., Ludwig, W., Stetter, K. O., Schleifer, K. H. 1989. Nucleotide sequence of the gene coding for the elongation factor Tu from the extremely thermophilic eubacterium *Thermotoga maritima*. *FEMS Microbiol. Lett.* 57:115–20
10. Badr, H. R., Sims, K. A., Adams, M. W. W. 1993. Purification and characterization of a sucrose α-glucohydrolase from *Pyrococcus furiosus* exhibiting a temperature optimum above 100°C. *J. Bacteriol.* Submitted
11. Baross, J. A., Deming, J. W. 1993. Growth at high temperatures: isolation and taxonomy, physiology and ecology. In *Microbiology of Deep Sea Hydrothermal Vent Environments,* Vol. 1, ed. D. M. Karl. Caldwell, NJ: Telford. In press
12. Blake, P. R., Day, M. W., Hsu, B. T., Joshua-Tor, L., Park, J.-B., et al. 1992. Comparison of the X-ray structure of native rubredoxin from *Pyrococcus furiosus* with the NMR structure of the zinc-substituted protein. *Protein Sci.* 1:1522–25
13. Blake, P. R., Park, J.-B., Adams, M. W. W., Summers, M. F. 1992. Novel observation of NH-S(Cys) hydrogen bond-mediated scalar coupling in ^{113}Cd-substituted rubredoxin from *Pyrococcus furiosus*. *J. Am. Chem. Soc.* 114:4931–33
14. Blake, P. R., Park, J. B., Bryant, F. O., Aono, S., Magnuson, J. K., et al. 1991. Determinants of protein hyperthermostability. 1. Purification, amino acid sequence, and secondary structure from NMR of the rubredoxin from the hyperthermophilic archaebacterium, *Pyrococcus furiosus*. *Biochemistry* 30:10885–91
15. Blake, P. R., Park, J.-B., Zhou, Z. H., Hare, D. R., Adams, M. W. W., Summers, M. F. 1992. Solution state structure by NMR of zinc-substituted rubredoxin from the marine hyperthermophilic archaebacterium, *Pyrococcus furiosus*. *Protein Sci.* 1:1508–21

16. Blake, P. R., Summers, M. F., Adams, M. W. W., Park, J.-B., Zhou, Z. H., Bax, A. 1992. Quantitative measurement of small through-hydrogen-bond and through-space ^{1}H-^{113}Cd and ^{1}H-^{199}Hg couplings in metal-substituted rubredoxin from *Pytrococcus furiosus*. *J. Biomol. NMR* 2:527–33
17. Blamey, J. M., Adams, M. W. W. 1993. Purification and characterization of pyruvate ferredoxin oxidoreductase from the hyperthermophilic archaeon *Pyrococcus furiosus*. *Biochim. Biophys. Acta*. 1161:19–27
18. Blumentals, I. I., Itoh, M., Olson, G. J., Kelly, R. M. 1990. Role of polysulfides in reduction of elemental sulfur by the hyperthermophilic archaebacterium *Pyrococcus furiosus*. *Appl. Environ. Microbiol*. 56:1255–62
19. Blumentals, I. L., Robinson, A. S., Kelly, R. M. 1990. Characterization of sodium dodecyl sulfate-resistant proteolytic activity in the hyperthermophilic archaebacterium *Pyrococcus furiosus*. *Appl. Environ. Microbiol*. 56: 1992–98
20. Bonch-Osmolovskaya, E. A., Miroschnichenko, M. L., Kostrikina, N. A., Chernych, N. A., Zavarzin, G. A. 1990. *Thermoproteus uzoniensis* sp. nov., a new extremely thermophilic archaebacterium from Kamchatka continental hot springs. *Arch. Microbiol*. 154:556–59
21. Bonch-Osmolovskaya, E. A., Slesarev, A. I., Miroshnichenko, M. L., Svetlichnaya, T. P., Alekseev, V. A. 1988. Characteristics of *Desulfurococcus amylolyticus*—a new extremely thermophilic archaebacterium isolated from thermal springs of Kamchatka and Kunashir island. *Mikrobiologiya* 57:78–85
22. Bouthier de la Tour, C., Portemer, C., Nadal, M., Huber, R., Forterre, P., Duguet, M. 1991. Reverse gyrase in thermophilic eubacteria. *J. Bacteriol*. 173:3921–23
23. Bradley, E. A., Stewart, D. E., Adams, M. W. W., Wampler, J. E. 1993. Investigations of the thermostability of rubredoxin models using molecular dynamics simulations. *Protein Sci*. 2:650–65
24. Bragger, J. M., Daniel, R. M., Coolbear, T., Morgan, H. W. 1989. Very stable enzymes from extremely thermophilic archaebacteria and eubacteria. *Appl. Microbiol. Biotechnol*. 31:556–61
25. Breitung, J., Schmitz, R. A., Stetter, K. O., Thauer, R. K. 1991. N^5,N^{10}-methylenetetrahydromethanopterin cyclohydrolase from the extreme thermophile *Methanopyrus kanderli:* increase of catalytic efficiency (k_{cat}/K_m) and thermostability in the presence of salts. *Arch. Microbiol*. 156:517–24
26. Brown, S. H., Costantino, H. R., Kelly, R. M. 1990. Characterization of amylolytic enzyme activities associated with the hyperthermophilic archaebacterium *Pyrococcus furiosus*. *Appl. Environ. Microbiol*. 56:1985–91
27. Brown, S. H., Kelly, R. M. 1989. Cultivation techniques for hyperthermophilic archaebacteria: continuous culture of *Pyrococcus furiosus* at temperatures near 100°C. *Appl. Environ. Microbiol*. 55:2086–88
28. Bryant, F. O., Adams, M. W. W. 1989. Characterization of hydrogenase from the hyperthermophilic archaebacterium, *Pyrococcus furiosus*. *J. Biol. Chem*. 264:5070–79
29. Burggraf, S., Fricke, H., Neuner, A., Kristjansson, J. K., Rouvier, P., et al. 1990. *Mechanococcus igneus* sp. nov., a novel hyperthermophilic methanogen from a shallow submarine hydrothermal system. *Syst. Appl. Microbiol*. 13:263–69
30. Burggraf, S., Jannasch, H. W., Nicolaus, B., Stetter, K. O. 1990. *Archaeglobus profundus* sp. nov. represents a new species within the sulfate-reducing archaebacteria. *Syst. Appl. Microbiol*. 13:24–28
31. Burggraf, S., Olsen, G. J., Stetter, K. O., Woese, C. R. 1992. A phylogenetic analysis of *Aquifex pyrophilus*. *Arch. Microbiol*. 15:352–56
32. Burggraf, S., Stetter, K. O., Rouviere, P., Woese, C. R. 1991. *Methanopyrus kandleri:* an archaeal methanogen unrelated to all other known methanogens. *Syst. Appl. Microbiol*. 14:346–51
33. Busse, S. A., La Mar, G. N., Yu, L. P., Howard, J. B., Smith, E. T., et al. 1992. Proton NMR investigation of the oxidized three-iron clusters in the ferredoxins from the hyperthermophilic archaea, *Pyrococcus furiosus* and *Thermococcus litoralis*. *Biochemistry* 31:11952–62
34. Charbonnier, F., Eraauso, G., Barbeyron, T., Prieur, D., Forterre, P. 1992. Evidence that a plasmid from a hyperthermophilic archaebacterium is relaxed at physiological temperatures. *J. Bacteriol*. 174:6103–8
35. Connaris, H., Cowan, D. A., Sharp, R. J. 1991. Heterogeneity of proteinases from the hyperthermophilic archaeo-

bacterium *Pyrococcus furiosus*. *J. Gen. Microbiol.* 137:1193–99
36. Conover, R. C., Kowal, A. T., Fu, W., Park, J.-B., Aono, S., et al. 1990. Spectroscopic characterization of the novel iron-sulfur cluster in *Pyrococcus furiosus* ferredoxin. *J. Biol. Chem.* 265:8533–41
37. Conover, R. C., Park, J.-B., Adams, M. W. W., Johnson, M. K. 1990. The formation and properties of a $NiFe_3S_4$ cluster in *Pyrococcus furiosus* ferredoxin. *J. Am. Chem. Soc.* 112: 4562–64
38. Conover, R. C., Park, J.-B., Adams, M. W. W., Johnson, M. K. 1991. Exogenous ligand binding to the $[Fe_4S_4]$ cluster in *Pyrococcus furiosus* ferredoxin. *J. Am. Chem. Soc.* 113:2799–2800
39. Consalvi, V., Chiaraluce, R., Politi, L., Vaccaro, R., De Rosa, M., Scandurra, R. 1991. Extremely thermostable glutamate dehydrogenase from the hyperthermophilic archaebacterium *Pyrococcus furiosus*. *Eur. J. Biochem.* 202:1189–96
40. Constantino, H. R., Brown, S. H., Kelly, R. M. 1990. Purification and characterization of an α-glucosidase from a hyperthermophilic archaebacterium, *Pyrococcus furiosus*, exhibiting a temperature optimum of 105 to 115°C. *J. Bacteriol.* 172:3654–60
41. Cowan, D. A., Smolenski, K. A., Daniel, R. M., Morgan, H. W. 1987. An extremely thermostable extracellular proteinase from a strain of the archaebacterium *Desulfurococcus* growing at 88°C. *Biochem. J.* 247:121–33
42. Dahl, C., Koch, H.-G., Keuken, O., Trüper, H. G. 1990. Purification and characterization of ATP sulfurylase from the extremely thermophilic archaebacterial sulfate-reducer, *Archaeoglobus fulgidus*. *FEMS Microbiol. Lett.* 67:27–32
43. Danson, M. J. 1989. Central metabolism of the archaebacteria: an overview. *Can. J. Microbiol.* 35:58–64
44. Dawson, R. M. C., Elliott, D. C., Elliott, W. H., Jones, K. M., eds. 1986. *Data for Biochemical Research*. Oxford: Clarendon. 580 pp.
45. Day, M. W., Hsu, B. T., Joshua-Tor, L., Park, J.-B., Zhou, Z. H., et al. 1992. X-ray crystal structure of the oxidized and reduced forms of the rubredoxin from the marine hyperthermophilic archaebacterium, *Pyrococcus furiosus*. *Protein Sci.* 1: 1494–1507
46. Di Ruggiero, J., Klump, H., Kessel, M., Park, J.-B., Adams, M. W. W., Robb, F. T. 1992. Glutamate dehydrogenase from the hyperthermophile *Pyrococcus furiosus:* thermal denaturation and activation. *J. Biol. Chem.* 267:22681–85
47. Eggen, R., Geerling, A., Watts, J., de Vos, W. M. 1990. Characterization of pyrolysin, a hyperthermoactive serine protease from the archaebacterium *Pyrococcus furiosus*. *FEMS Microbiol. Lett.* 71:17–20
48. Engel, A. M., Cejka, Z., Lupas, A., Lottspeich, F., Baumeister, W. 1992. Isolation and cloning of Ompα, a coiled-coil protein spanning the periplasmic space of the ancestral eubacterium *Thermotoga maritima*. *EMBO J.* 11:4369–78
49. Erauso, G., Charbonnier, F., Barbeyron, T., Forterre, P., Prieur, D. 1992. Preliminary characterization of a hyperthermophilic archaebacterium with a plasmid, isolated from a North Fiji basin hydrothermal vent. *C. R. Acad. Sci. Paris* 314:387–93
50. Fiala, G., Stetter, K. O. 1986. *Pyrococcus furiosus* sp. nov. represents a novel genus of marine heterotrophic archaebacteria growing optimally at 100°C. *Arch. Microbiol.* 145:56–61
51. Fiala, G., Stetter, K. O., Jannasch, H. W., Langworthy, T. A., Madon, J. 1986. *Staphylothermus marinus* sp. nov. represents a novel genus of extremely thermophilic submarine heterotrophic archaebacteria growing up to 98°C. *Syst. Appl. Microbiol.* 8:106–13
52. George, G., Prince, R. C., Mukund, S., Adams, M. W. W. 1992. Aldehyde ferredoxin oxidoreductase from the hyperthermophilic archaebacterium *Pyrococcus furiosus* contains a tungsten oxo-thiolate center. *J. Am. Chem. Soc.* 114:3521–23
53. George, S. J., van Elp, J., Chen, J., Chen, C. T., Ma, Y., et al. 1992. L-edge X-ray absorption spectroscopy of *Pyrococcus furiosus* rubredoxin. *J. Am. Chem. Soc.* 114:4426–27
54. Hensel, R., König, H. 1988. Thermoadaptation of methanogenic bacteria by intracellular ion concentration. *FEMS Microbiol. Lett.* 49:75–79
55. Hensel, R., Laumann, S., Lang, J., Heumann, H., Lottspeich, F. 1987. Characterization of two glyceraldehyde-3-phosphate dehydrogenases from the extremely thermophilic archaebacterium *Thermoproteus tenax*. *Eur. J. Biochem.* 170:325–33
56. Huang, Y.-H., Moura, I., Moura, J. J. G., LeGall, J., Park, J.-B., et al.

1992. Resonance Raman studies of nickel tetrathiolates and nickel-substituted rubredoxins and desulforedoxin. *Inorg. Chem.* 32:406–12

57. Huang, Y.-H., Park, J.-B., Zhou, Z. H., Adams, M. W. W., Johnson, M. K. 1993. Oxidized nickel-substituted rubredoxin as a model for the Ni-C EPR signal of NiFe-hydrogenases. *Inorg. Chem.* 32:375–76
58. Huber, R., Kristjansson, J. K., Stetter, K. O. 1987. *Pyrobaculum* gen. nov., a new genus of neutrophilic, rod-shaped archaebacteria from continental solfataras growing optimally at 100°C. *Arch. Microbiol.* 149:95–101
59. Huber, R., Kurr, M., Jannasch, H. W., Stetter, K. O. 1989. A novel group of abyssal methanogenic archaebactreia (*Methanopyrus*) growing at 110°C. *Nature* 342:833–34
60. Huber, R., Langworthy, T. A., König, H., Thomm, M., Woese, C. R., et al. 1986. *Thermotoga maritima* sp. nov. represent a new genus of unique extremely thermophilic eubacteria growing up to 90°C. *Arch. Microbiol.* 144:324–33
61. Huber, R., Stoffers, P., Cheminee, J. L., Richnow, H. H., Stetter, K. O. 1990. Hyperthermophilic archaebacteria occurring within the crater and open sea plume of erupting Macdonald seamount. *Nature* 345:179–81
62. Huber, R., Wilharm, T., Huber, D., Trincone, A., Burggraf, S., et al. 1992. *Aquifex pyrophilus* gen. nov. sp. nov., represents a novel group of marine hyperthermophilic hydrogen oxidizing bacteria. *Arch. Microbiol.* 15:340–51
63. Jaenicke, R. 1991. Protein stability and molecular adaption to extreme conditions. *Eur. J. Biochem.* 202:715–28
64. Jannasch, H. W., Huber, R., Belkin, S., Stetter, K. O. 1988. *Thermotoga neapolitana* sp. nov. of the extremely thermophilic, eubacterial genus *Thermotoga*. *Arch. Microbiol.* 150:103–4
65. Jannasch, H. W., Mottl, M. J. 1985. Geomicrobiology of deep-sea hydrothermal vents. *Science* 229:717–25
66. Jannasch, H. W., Wirsen, C. O., Molyneaux, S. J., Langworthy, T. A. 1988. Extremely thermophilic fermentative archaebacteia of the genus *Desulfurococcus* from deep-sea hydrothermal vents. *Appl. Environ. Microbiol.* 54:1203–9
67. Jannasch, H. W., Wirsen, C. O., Molyneaux, S. J., Langworthy, T. A. 1992. Comparative physiological studies on hyperthermophilic archaea isolated from deep sea hot vents with emphasis on *Pyrococcus* strain GB-D. *Appl. Environ. Microbiol.* 58:3472–81
68. Janssen, P. H., Morgan, H. W. 1992. Heterotrophic sulfur reduction by *Thermotoga* sp. strain FjSS3.B1. *FEMS Microbiol. Letts.* 96:213–18
69. Johnson, J. L., Rajagopalan, K. V., Mukund, S., Adams, M. W. W. 1993. Identification of molybdopterin as the organic component of the tungsten cofactor in four enzymes from hyperthermophilic archaea. *J. Biol. Chem.* 268:4848–53
70. Jorgensen, B. B., Isaksen, M. F., Jannasch, H. W. 1992. Bacterial sulfate reduction above 100°C in deep-sea hydrothermal vent sediments. *Science* 258:1756–57
71. Juszczak, A., Aono, S., Adams, M. W. W. 1991. The extremely thermophilic eubacterium, *Thermotoga maritima,* contains a novel iron-hydrogenase whose cellular activity is dependent upon tungsten. *J. Biol. Chem.* 266: 13834–41
72. Kelly, R. M., Blumentals, I. I., Snowden, L. J., Adams, M. W. W. 1992. Physiological and biochemical characteristics of *Pyrococcus furiosus,* a hyperthermophilic archaebacterium. *Ann. N. Y. Acad. Sci.* 665:309–20
73. Kelly, R. M., Deming, J. W. 1988. Extremely thermophilic bacteria: biological and engineering considerations. *Biotech. Prog.* 4:47–62
74. Kletzin, A. 1992. Molecular characterization of the *sor* gene, which encodes the sulfur oxygenase/reductase of the thermoacidophilic archaeum *Desulfurolobus ambivalens*. *J. Bacteriol.* 174:5854–69
75. Klingeberg, M., Hashwa, F., Antranikian, G. 1991. Properties of extremely thermophilic proteases from anaerobic hyperthermophilic bacteria. *App. Microbiol. Biotechnol.* 34:715–19
76. Koch, R., Spreinat, A., Lemke, K., Antranikian, G. 1991. Purification and properties of a hyperthermoactive α-amylase from the archaebacterium *Pyrococcus woesei*. *Arch. Microbiol.* 155:572–78
77. Koch, R., Zablowski, P., Spreinat, A., Antranikian, G. 1990. Extremely thermostable amylolytic enzyme from the archaebacterium *Pyrococcus furiosus*. *FEMS Microbiol. Lett.* 71:21–26
78. Kristjansson, J. K., Stetter, K. O. 1991. Thermophilic bacteria. In *Thermophilic Bacteria,* ed. J. K. Kristjansson, pp. 1–19. Boca Raton, FL: CRC
79. Kurr, M., Huber, R., König, H.,

Jannsch, H. W., Fricke, H., et al. 1991. *Methanopyrus kanderli,* gen. and sp. nov. represents a novel group of hyperthermophilic methanogens, growing at 110°C. *Arch. Microbiol.* 156: 239–47
80. Lauerer, G., Kristjansson, J. K., Langworthy, T. A., König, H., Stetter, K. O. 1986. *Methanothermus sociabilis* sp. nov., a second species within the *Methanothermaceae* growing at 97°C. *Syst. Appl. Microbiol.* 8:100–5
81. Le Faou, A., Rajagopal, B. S., Daniels, L., Fauque, G. 1990. Thiosulfate, polythionates and elemental sulfur assimilation and reduction in the bacterial world. *FEMS Microbiol. Rev.* 75:351–82
82. Liebl, W., Feil, R., Gabelsberger, J., Kellerman, J., Schleifer, K. H. 1992. Purification and characterization of a novel thermostable 4-α-glucanotrasferase of *Thermotoga maritima* cloned in *Escherichia coli. Eur. J. Biochem.* 207:81–88
83. Lundberg, K. S., Shoemaker, D. D., Short, J. M., Sorge, J. A., Adams, M. W. W., Mathur, E. 1991. High fidelity amplification with a thermostable DNA polymerase isolated from *Pyrococcus furiosus. Gene* 108:1–6
84. Ma, K., Schicho, R. N., Kelly, R. M., Adams, M. W. W. 1993. Hydrogenase of the hyperthermophile, *Pyrococcus furiosus,* is an elemental sulfur reductase or sulfhydrogenase: evidence for a sulfur-reducing hydrogenase ancestor. *Proc. Natl. Acad. Sci. USA* In press
85. Malik, B., Su, W.-w., Wald, H. L., Blumentals, I. I., Kelly, R. M. 1989. Growth and gas production for hyperthermophilic archaebacterium, *Pyrococcus furiosus. Biotechnol. Bioeng.* 34: 1050–57
86. Mathur, E. J., Adams, M. W. W., Callen, W. N., Cline, J. M. 1991. The DNA polymerase gene from the hyperthermophilic archaebacterium, *Pyrococcus furiosus,* shows sequence homology with α-like DNA polymerases. *Nucleic Acid Res.* 19:6952–53
87. Menendez-Arias, L., Argos, P. 1989. Engineering protein thermal stability. *J. Mol. Biol.* 206:397–406
88. Miroshnichenko, M. L., Bonch-Osmolovskaya, E. A., Neuner, A., Kostrikina, N. A., Chernych, N. A., Alikseev, V. A. 1989. *Thermococcus stetteri* sp. nov., a new extremely thermophilic marine sulfur-metabolizing archaebacterium. *Syst. Appl. Microbiol.* 12:257–62
89. Möller-Zinkhan, D., Börner, G., Thauer, R. K. 1989. Function of methanofuran, tetrahydromethanopterin and coenzyme F_{420} in *Archaeglobus fulgidus. Arch. Microbiol.* 152:362–68
90. Mukund, S., Adams, M. W. W. 1990. Characterization of a tungsten-iron-sulfur protein exhibiting novel spectroscopic and redox properties from the hyperthermophilic archaebacterium, *Pyrococcus furiosus. J. Biol. Chem.* 265:11508–16
91. Mukund, S., Adams, M. W. W. 1991. The novel tungsten-iron-sulfur protein of the hyperthermophilic archaebacterium, *Pyrococcus furiosus,* is an aldehyde ferredoxin oxidoreductase: evidence for its participation in a unique glycolytic pathway. *J. Biol. Chem.* 266:14208–16
92. Mukund, S., Adams, M. W. W. 1993. Characterization of a novel tungsten-containing formaldehyde ferredoxin oxidoreductase from the hyperthermophilic archaeon, *Thermococcus litoralis.* A role for tungsten in peptide catabolism. *J. Biol. Chem.* In press
93. Nelson, C. M., Schuppenhauer, M. R., Clark, D. S. 1991. Effects of hyperbaric pressure on a deep-sea archaebacterium in stainless steel and glass-lined vessels. *Appl. Environ. Microbiol.* 57:3576–80
94. Nelson, C. M., Schuppenhauer, M. R., Clark, D. S. 1992. High-pressure, high-temperature bioreactor for comparing effects of hyperbaric and hydrostatic pressure on bacterial growth. *Appl. Environ. Microbiol.* 58:1789–93
95. Neuner, A., Jannasch, H., Belkin, S., Stetter, K. O. 1990. *Thermococcus litoralis* sp. nov.: a new species of extremely thermophilic marine archaebacteria. *Arch. Microbiol.* 153: 205–7
96. Parameswaran, A. K., Su, W.-w., Schicho, R. N., Provan, C. N., Malik, B., Kelly, R. M. 1988. Engineering considerations for growth of bacteria at temperatures around 100°C. *Appl. Biochem. Biotechnol.* 18:53–73
97. Park, J.-B., Fan, C., Hoffman, B. M., Adams, M. W. W. 1991. Potentiometric and electron nuclear double resonance properties of the two spin forms of the $[4Fe\text{-}4S]^{1+}$ cluster in the novel ferredoxin from the hyperthermophilic archaebacterium, *Pyrococcus furiosus. J. Biol. Chem.* 266:19351–56
98. Perler, F. B., Comb, D. G., Jack, W. E., Moran, L. S., Qiang, B., et al. 1992. Intervening sequences in an

archaea DNA polymerase gene. *Proc. Natl. Acad. Sci. USA* 89:5577–81

99. Phipps, B. M., Hoffmann, A., Stetter, K. O., Baumeister, W. 1991. A novel ATP-ase complex selectively accumulated upon heat shock is a major cellular component of thermophilic archaebacteria. *EMBO J.* 10:1711–22
100. Pihl, T. D., Black, L. K., Schulman, B. A., Maier, R. J. 1992. Hydrogen-oxidizing electron transport components in the hyperthermophilic archaebacterium *Pyrodictium brockii*. *J. Bacteriol.* 174:137–43
101. Pihl, T. D., Maier, R. J. 1991. Purification and characterization of the hydrogen uptake hydrogenase from the hyperthermophilic archaebacterium *Pyrodictium brockii*. *J. Bacteriol.* 173: 1839–44
102. Pihl, T. D., Schicho, R. N., Black, L. K., Schulman, B. A., Maier, R. J., Kelly, R. M. 1990. Hydrogen sulfur autotrophy in the hyperthermophilic archaebacterium *Pyrodictium brockii*. *Biotech. Genet. Eng. Rev.* 8:345–77
103. Pledger, R. J., Baross, J. A. 1989. Characterization of an extremely thermophilic archaebacterium isolated from a black smoker polychaete (*Paralvinella* sp.) at the Juan de Fuca Ridge. *Syst. Appl. Microbiol.* 12:249–56
104. Pledger, R. J., Baross, J. A. 1991. Preliminary description and nutritional characterization of a chemoorganotrophic archaeobacterium growing at temperatures up to 110°C isolated from a submarine hydrothermal vent environment. *J. Gen. Microbiol.* 137:203–13
105. Pley, U., Schipka, J., Gambacorta, A., Jannasch, H. W., Fricke, H., et al. 1991. *Pyrodictium abyssi* sp. nov. represents a novel heterotrophic marine archaeal hyperthermophile growing at 110°C. *Syst. Appl. Microbiol.* 14:245–53
106. Privalov, P. L., Gill, S. J. 1988. Stability of protein structure and hydrophobic interaction. *Adv. Protein Chem.* 39:191–234
107. Przybyla, A. E., Robbins, J., Menon, N., Peck, H. D. Jr. 1992. Structure/function relationships among nickel-containing hydrogenases. *FEMS Microbiol. Rev.* 88:109–36
108. Pucheva, M. A., Slobodkin, A. I., Bonch-Osmolovskaya, E. A. 1992. Investigation of hydrogenase activity of the extremely thermophilic archaebacterium *Thermococcus stetteri*. *Mikrobiologiya* 60:661–66
109. Rajagopalan, K. V., Johnson, J. L. 1992. The pterin molybdenum cofactors. *J. Biol. Chem.* 267:10199–10202
110. Rehaber, V., Jaenicke, R. 1992. Stability and reconstitution of D-glyceraldehyde-3-phosphate dehydrogenase from the hyperthermophilic eubacterium *Thermotoga maritima*. *J. Biol. Chem.* 267:10999–11006
111. Reysenbach, A. L., Deming, J. W. 1991. Effects of hydrostatic pressure on growth of hyperthermophilic archaebacteria from the Juan de Fuca Ridge. *Appl. Environ. Microbiol.* 57: 1271–74
112. Robb, F. T., Park, J.-B., Adams, M. W. W. 1992. Characterization of an extremely thermostable glutamate dehydrogenase: a key enzyme in the primary metabolism of the hyperthermophilic archaebacterium, *Pyrococcus furiosus*. *Biochim. Biophys. Acta* 1120:267–72
113. Rüdiger, A., Ogbonna, J. C., Märkl, H., Antranikian, G. 1992. Effect of gassing, agitation, substrate supplementation and dialysis on the growth of an extremely thermophilic archaeon *Pyrococcus woesei*. *Appl. Microbiol. Biotechnol.* 37:501–4
114. Ruttersmith, L. D., Daniel, R. M. 1991. Thermostable cellobiohydrolase from the thermophilic eubacterium *Thermotoga* strain FjSS3-B. 1—purification and properties. *Biochem. J.* 277: 887–91
115. Sanangelantoni, A. M., Forlani, G., Ambroselli, F., Cammarano, P., Tiboni, O. 1992. The *glnA* gene of the extremely thermophilic eubacterium *Thermotoga maritima:* cloning, primary structure and expression in *Escherichia coli*. *J. Gen. Microbiol.* 138:383–93
116. Schäfer, T., Schönheit, P. 1991. Pyruvate metabolism of the hyperthermophilic archaebacterium *Pyrococcus furiosus*. *Arch. Microbiol.* 155:366–77
117. Schäfer, T., Schönheit, P. 1992. Maltose fermentation to acetate, CO_2 and H_2 in the anaerobic hyperthermophilic archaeon *Pyrococcus furiosus:* evidence for the operation of a novel sugar fermentation pathway. *Arch. Microbiol.* 158:188–202
118. Schicho, R. N., Ma, K., Adams, M. W. W., Kelly, R. M. 1993. Bioenergetics of sulfur reduction in the hyperthermophilic archaeon, *Pyrococcus furiosus*. *J. Bacteriol.* 175:1823–30
119. Schicho, R. N., Snowden, L. J., Mukund, S., Park. J.-B., Adams, M. W. W., Kelly, R. M. 1993. Influence of tungsten on metabolic patterns in

Pyrococcus furiosus, a hyperthermophilic archaeon. *Arch. Microbiol.* 159: 380–85

120. Schmitz, R. A., Albracht, S. P. J., Thauer, R. K. 1992. A molybdenum and a tungsten isoenzyme of formylmethanofuran dehydrogenase in the thermophilic archaeon *Methanobacterium wolfei. Eur. J. Biochem.* 209:1013–18

121. Schmitz, R. A., Linder, D., Stetter, K. O., Thauer, R. K. 1991. N^5,N^{10}-methylenetetrahydromethanopterin reductase (coenzyme F_{420}-dependent) and formylmethanofuran dehydrogenase from the hyperthermophile *Archaeoglobus fulgidus. Arch. Microbiol.* 156: 427–34

122. Scholz, S., Sonnenbichler, J., Schäfer, W., Hensel, R. 1992. Di-myo-inositol-1,1′-phosphate: a new inositol phosphate isolated from *Pyrococcus woesei. FEBS Lett.* 306:239–42

123. Schuliger, J. W., Brown, S. H., Baross, J. A., Kelly, R. M. 1993. Purification and characterization of a novel amylolytic enzyme from ES4, a marine hyperthermophilic archaeon. *Mol. Mar. Biol. Biotechnol.* In press

124. Schultes, V., Deutzmann, R., Jaenicke, R. 1990. Complete amino acid sequence of glyceraldehyde-3-phosphate dehydrogenase from the hyperthermophilic eubacterium *Thermotoga maritima. Eur. J. Biochem.* 192:25–31

125. Schultes, V., Jaenicke, R. 1991. Folding intermediates of hyperthermophilic D-glyceraldehyde-3-phosphate dehydrogenase from *Thermotoga maritima* are trapped at low temperatures. *FEBS Lett.* 290:235–38

126. Schumann, J., Wrba, A., Jaenicke, R., Stetter, K. O. 1991. Topographical and enzymatic characterization of amylases from the extremely thermophilic eubacterium *Thermotoga maritima. FEBS Letts.* 282:122–26

127. Schwörer, B., Thauer, R. K. 1991. Activities of formylmethanofuran dehydrogenase, methylenetetrahydromethanopterin dehydrogenase, methylenetetrahydromethanopterin reductase, and heterodisulfide reductase in methanogenic bacteria. *Arch. Microbiol.* 155:459–65

128. Segerer, A., Neuner, A., Kristjansson, J. K., Stetter, K. O. 1986. *Acidianus infernus* gen. nov. sp. nov., and *Acidianus brierleyi* comb. nov.: facultatively aerobic, extremely acidophilic thermophilic sulfur-metabolizing archaebacteria. *Int. J. Syst. Bacteriol.* 36: 559–64

129. Segerer, A. H., Trincone, A., Gahrtz, M., Stetter, K. O. 1991. *Stygiolobus azoricus* gen. nov., sp. nov. represents a novel genus of anaerobic, extremely thermoacidophilic archaebacteria of the Order *Sulfolobales. Int. J. Syst. Bacteriol.* 41:495–501

130. Simpson, H., Haufler, U. R., Daniel, R. M. 1991. An extremely thermostable xylanase from the thermophilic eubacterium *Thermotoga. Biochem. J.* 277: 413–17

131. Slesarev, A. I. 1988. Positive supercoiling catalyzed in vitro by ATP-dependent topoisomerase from *Desulfurococcus amylolyticus. Eur. J. Biochem.* 173:395–99

132. Smith, E., Adams, M. W. W. 1992. A temperature-controlled, anaerobic cell for direct electrochemical studies. *Anal. Biochem.* 207:94–99

133. Snowden, L. J., Blumentals, I. I., Kelly, R. M. 1992. Regulation of proteolysis in *Pyrococcus furiosus,* a hyperthermophilic archaebacterium. *Appl. Microbiol. Environ.* 58:1134–41

134. Spiro, T. G., ed. 1985. *Molybdenum Enzymes, Metals in Biology Series,* Vol. 7. New York: Wiley. 611 pp.

135. Srivastava, K. K. P., Surerus, K. K., Conover, R. C., Johnson, M. K., Park, J.-B., et al. 1992. Mössbauer study of the $ZnFe_3S_4$ and $NiFe_3S_4$ clusters in *Pyrococcus furiosus* ferredoxin. *Inorg. Chem.* 32:927–36

136. Stetter, K. O. 1982. Ultrathin mycelia forming organisms from submarine volcano areas having an optimum growth temperature of 105°C. *Nature* 300:258–60

137. Stetter, K. O. 1986. Diversity of extremely thermophilic archaebacteria. In *The Thermophiles, General, Molecular and Applied Microbiology,* ed. T. D. Brock, pp. 39–74. New York: Wiley

138. Stetter, K. O. 1988. *Archaeoglobus fulgidus* gen. nov., sp. nov.: a new taxon of extremely thermophilic archaebacteria. *Syst. Appl. Microbiol.* 10:172–73

139. Stetter, K. O., Fiala, G., Huber, G., Huber, R., Segerer, G. 1990. Hyperthermophilic microorganisms. *FEMS Microbiol. Rev.* 75:117–24

140. Stetter, K. O., König, H., Stackebrandt, E. 1983. *Pyrodictium* gen. nov., a new genus of submarine disc-shaped sulfur reducing archaebacteria growing optimally at 105°C. *Syst. Appl. Microbiol.* 4:535–51

141. Stetter, K. O., Lauerer, G., Thomm, M., Neuner, A. 1987. Isolation of extremely thermophilic sulfate reducers:

evidence for a novel branch of archaebacteria. *Science* 236:822–24

142. Stetter, K. O., Thomm, M., Winter, J., Wildegruber, G., Huber, et al. 1981. *Methanothermus fervidus,* sp. nov., a novel extremely thermophilic methanogen isolated from an Icelandic hot spring. *Zentrabl. Bakteriol. Parasitenkd. Infektionskr. Hyg. Abt. 1: Orig. Reihe C* 2:166–78
143. Stetter, K. O., Zillig, W. 1985. *Thermoplasma* and the thermophilic sulfur-dependent archaebacteria. In *The Bacteria,* ed. C. R. Woese, R. S. Wolfe, 8:85–170. New York: Academic
144. Strauss, A., Eisenreich, W., Bacher, A., Fuchs, G. 1992. ^{13}C-NMR study of autotrophic CO_2 fixation pathways in the sulfur-reducing archaebacterium *Thermoproteus neutrophilus* and in the phototrophic eubacterium *Chloroflexus aurantiacus. Eur. J. Biochem.* 205: 853–66
145. Tiboni, O., Sanangelantoni, A. M., Cammarano, P., Cimino, L., Di Pasquale, G., Sora, S. 1989. Expression in *Escherichia coli* of the *tuf* gene from the extremely thermophilic eubacterium *Thermotoga maritima:* purification of the *Thermotoga* elongation factor Tu by thermal denaturation of the mesophile host cell protein. *Syst. Appl. Microbiol.* 12:127–33
146. Van de Casteele, M., Demarez, M., Legrain, C., Glansdorff, N., Pierard, A. 1990. Pathways of arginine biosynthesis in extreme thermophilic archaeo- and eubacteria. *J. Gen. Microbiol.* 136: 1177–83
147. Wampler, J. E., Bradley, E. A., Stewart, D. E., Adams, M. W. W. 1993. Modelling the structure of *Pyrococcus furiosus* rubredoxin by homology to other X-ray structures. *Protein Sci.* 2:640–49
148. Wheelis, M. L., Kandler, O., Woese, C. R. 1992. On the nature of global classification. *Proc. Natl. Acad. Sci. USA* 89:2930–34
149. White, H., Simon, H. 1992. The role of tungstate and/or molybdate in the formation of aldehyde oxidoreductase in *Clostridium thermoaceticum* and other acetogens; immunological distances of such enzymes. *Arch. Microbiol.* 158:81–84
150. Wiegel, J. K. W., Ljungdahl, L. G. 1986. The importance of thermophilic bacteria in biotechnology. *CRC Crit. Rev. Biotechnol.* 3:39–107
151. Woese, C. R. 1987. Bacterial evolution. *Microbiol. Rev.* 51:221–71
152. Woese, C. R., Achenbach, L., Rouviere, P., Mandelco, L. 1991. Archael phylogeny: reexamination of the phylogenetic position of *Archaeglobus fulgidus* in light of certain composition-induced artifacts. *Syst. Appl. Microbiol.* 14:364–71
153. Woese, C. R., Kandler, O., Wheelis, M. L. 1990. Towards a natural system of organisms: proposal for the domains of Archaea, Bacteria and Eucarya. *Proc. Natl. Acad. Sci. USA* 87:4576–79
154. Woese, C. R., Magrum, L. J., Fox, G. E. 1978. Archaebacteria. *J. Mol. Evol.* 11:245–52
155. Wrba, A., Jaenicke, R., Huber, R., Stetter, K. O. 1990. Lactate dehydrogenase from the extreme thermophile *Thermotoga maritima. Eur. J. Biochem.* 188:195–201
156. Wrba, A., Schweiger, A., Schultes, V., Jaenicke, R., Zavodszky, P. 1990. Extremely thermophlic D-glyceraldehyde-3-phosphate dehydrogenase from the eubacterium *Thermotoga maritima. Biochemistry* 29:7584–92
157. Yamamoto, I., Saiki, T., Liu, S.-M., Ljungdahl, L. G. 1983. Purification and properties of NADP-dependent formate dehydrogenase from *Clostridium thermoaceticum,* a tungsten-selenium protein. *J. Biol. Chem.* 258:1826–32
158. Zellner, G., Stackebrandt, E., Kneifel, H., Messner, P., Sleytr, U. B., et al. 1989. Isolation and characterization of a thermophilic, sulfate-reducing archaebacterium, *Archaeoglobus fuldidus* strain Z. *Syst. Appl. Microbiol.* 11:151–60
159. Zillig, W., Gierl, G., Schreiber, G., Wunderl, S., Janekovic, D., et al. 1983. The archaebacterium *Thermofilum pendens* represents a novel genus of the thermophilic, anaerobic sulfur-respiring Thermoproteales. *Syst. Appl. Microbiol.* 4:79–87
160. Zillig, W., Holz, I., Janekovic, D., Klenk, H.-P., Imsel, E., et al. 1991. *Hyperthermus butylicus,* a hyperthermophilic sulfur-reducing archaebacterium that ferments peptides. *J. Bacteriol.* 172:3959–65
161. Zillig, W., Holtz, I., Janekovic, D., Schafer, W., Reiter, W. D. 1983. The archaebacterium *Thermococcus celer* represents a novel genus within the thermophilic branch of the archaebacteria. *Syst. Appl. Microbiol.* 4:88–94
162. Zillig, W., Holz, I., Klenk, H.-P., Trent, J., Wunderl, S., et al. 1987. *Pyrococcus woesei,* sp. nov., an ultrathermophile marine archaebacterium, representing a novel order, *Ther-*

mococcales. *Syst. Appl. Microbiol.* 9: 62–70

163. Zillig, W., Stetter, K. O., Prangishvilli, D., Schafer, W., Wunderl, S., et al. 1982. Desulfurococcaceae, the second family of extremely thermophilic, anaerobic, sulfur-respiring Thermoproteales. *Zentrabl. Bakteriol. Parasitenkd. Infektionskr. Hyg. Abt. 1: Orig. Reihe C* 3:304–17
164. Zillig, W., Stetter, K. O., Schäfer, W., Janekovic, D., Wunderl, S., et al. 1981. Thermoproteales: a novel type of extremely thermoacidophilic anaerobic archaebacterium isolated from Icelandic solfataras. *Zentrabl. Bakteriol. Parasitenkd. Infektionskr. Hyg. Abt. 1: Orig. Reihe C* 2:200–27
165. Zuber, H. 1988. Temperature adaptation of lactate dehydrogenase: structural, functional and genetic aspects. *Biophys. Chem.* 29:171–79
166. Zwickl, P., Fabry, S., Bogedain, C., Haas, A., Hensel, R. 1990. Glyceraldehyde-3-phosphate dehydrogenase from the hyperthermophilic archaebacterium *Pyrococcus woesei:* characterization of the enzyme, cloning and sequencing the gene, and expression in *Escherichia coli. J. Bacteriol.* 172: 4329–38

Annu. Rev. Microbiol. 1993. 47:659–84

GENETICS FOR ALL BACTERIA

B. W. Holloway
Department of Genetics and Developmental Biology, Monash University, Clayton, Victoria 3168, Australia

KEY WORDS: conjugation, transduction, transformation, electroporation, mapping, whole genome analysis, physical and genetic maps

---

CONTENTS

ABSTRACT

The availability of genetic analysis has now been extended to a wide variety of bacteria. While the traditional methods of conjugation, transduction, and transformation have made major contributions to microbiology and genetics, new recombinant DNA techniques and the development of new equipment for characterization and isolation of DNA fragments have enabled genome analysis of many bacteria for which no genetic information was previously available. These new procedures have enabled the construction of detailed physical/genetic maps as well as precise measurements of genome size, and provided new data on functional arrangements of genes in the bacterial genome. Such information is proving increasingly valuable for many aspects

0066-4227/93/1001-0659$02.00

of microbiology as well as for the genetic manipulation of bacteria important in human disease, agriculture, and biotechnology.

INTRODUCTION AND HISTORICAL PERSPECTIVE

For many years the literature on bacterial genetics was dominated by work on *Escherichia coli* K12, *Salmonella typhimurium* LT2, *Bacillus subtilis* 168, and *Streptomyces coelicolor* A3(2). The accumulation of an impressive body of information has influenced aspects of biological research not restricted to microbiology. The most notable example is the development of recombinant DNA techniques that have, and will continue to have, a pervasive influence on the understanding of all living organisms. Microbiologists soon recognized the value of genetic analysis for understanding and investigating microbial phenomena. Workers studying organisms other than those listed above sought to use these newly developed techniques to investigate other genetic characteristics of their bacteria of interest. The predominance of conjugation in genetic analysis during those early days meant that the F factor was almost the only agent known to promote conjugal exchange in bacteria. As it turned out, the F factor of *E. coli* is very organism specific in its action. In addition, the host range of bacteriophages is very strain specific and limits the range of organisms for which transduction is available. Finally, the techniques for obtaining transformation competent cells were found to be highly fastidious and needed to be precisely defined in each organism examined. Development of systems of genetic analysis for individual strains of other bacteria was very labor intensive, and investigators devoted considerable effort to searching for new conjugative plasmids with chromosome mobilizing ability (cma), and bacteriophages with transducing activity, and to devising more general techniques for establishing competence to enable transformation analysis in a wider range of bacteria.

These efforts are described below. The overall conclusion is that genetic analysis using one or another of these three systems is now possible in a wide range of different bacteria. Two factors have encouraged the extension of genetic analysis to a wider range of bacteria. The first is the use of newly isolated organisms in biotechnology applications and the need to analyze and manipulate their genome. To this activity has been added a greater interest in genetics by those bacteriologists studying plant bacterial interactions, notably nitrogen fixation and plant disease caused by bacteria. The second factor is the development and use of recombinant DNA techniques that have enabled genomic analysis of bacteria without the classical mechanisms of gene exchange, the need to have selective markers to obtain recombinants, or even the need to isolate mutants, although, in the immortal words of Salvadore Luria, "When you have a mutant you are better off than when you don't."

This review does not aim to provide an encyclopedic coverage of all the bacteria for which genetic analysis is now possible, nor does it attempt to describe genetic analysis of the four organisms identified above, except where such work is important for specific aspects of genetic analysis for other bacteria. Numerous reviews cover the genetic analysis of these organisms including those by Neidhardt et al (124), Hopwood & Chater (83), and Lovett (108). The focus here is to understand genomic analysis, mapping, and those general approaches that are or will be important for an even wider understanding of bacterial genome structure in a greater number of organisms than presently exists. This review does not include any of the numerous papers describing the cloning and characterization of individual bacterial genes or chromosomal segments, the procedures by which this has been achieved, or the significance of this large body of data. Previous reviews on genetic exchange processes in bacteria include those by Low & Porter (109), Ball (6), Scaife et al (142), Birge (11), and Holloway (74). A recent compendium of papers (120) provides an excellent outline of methods for use in both in vivo and molecular genetic systems of bacterial genetics for a variety of bacteria.

THE AIMS OF GENETIC ANALYSIS

Any traditional system of genetic analysis for bacteria must include the ability to (*a*) produce and select desired mutants, (*b*) introduce DNA into bacterial cells, and (*c*) provide for the selection and characterization of recombinants or mutants that have undergone complementation. Given that these requirements have been achieved, a number of common aims for the genetic analysis of bacteria can be identified:

1. The understanding of genome structure, the measurement of genome size, the number and size of individual replicons, and the distinction between chromosomally located genes and plasmid-borne genes.
2. The identification of particular gene arrangements on both chromosomes and plasmids that have significance for biological properties.
3. The use of genetic data to explain biological characteristics and phenomena, particularly in regard to disease, natural habitat, regulation, evolution, and taxonomy.
4. The manipulation of bacterial strains to produce novel recombinant forms with special properties, which can be used to solve problems in biotechnology, medical research, or agriculture.

The extent and success of these aims will depend upon the focus of the investigator, the nature of the organism, and the genetic data sought. Clearly it has been much easier to obtain more sophisticated and extensive data with

E. coli than with some newly isolated autotroph. Such limitations have encouraged the development of methods of genetic analysis that are applicable to a wider range of organisms. These methods have exploited particular recombinant DNA techniques and new equipment that together enable one to obtain data on genome structures and linkage relationships for all bacteria, and these data are not restricted by biological factors dependent on in vivo genetic transfer techniques.

The Historical Approach to Bacterial Genetic Analysis and Its Problems

Traditionally, mutants have been isolated by the use of chemical mutagens, but as shown by the example of *N*-methyl-*N*-nitrosoguanidine, inherent sources of error can result in multiple genetic changes. Nevertheless, numerous excellent mutants have been produced by one or another chemical mutagen. The availability of transposon mutagenesis has made possible the production of mutants that have inbuilt selectivity and are likely to be at a single site; and should physical genetic analysis be available, the mutation can be located with the utmost precision. Problems that arise include the type of selection inherent in the transposon. This is commonly antibiotic resistance, but other types, e.g. substrate utilization, are available, and in some situations the transposon insertion may be unstable (59). Techniques of transposon mutagenesis and related topics have been comprehensively reviewed by Berg et al (9).

In Vivo Systems of Genetic Exchange

In what must now be called classical mechanisms, the three principal ways by which bacteria can exchange genetic information are conjugation, transduction, and transformation. The critical feature of these mechanisms for any individual bacteria is the frequency of exchange, which must be at least 10^{-6} recombinants per donor cell. The second critical feature is a selection mechanism to identify these rare recombinants in the presence of an excess of parental bacterial cells. The elegance of the use of auxotrophs in *E. coli* by Lederberg & Tatum (100) created high expectations for other bacteria, but this system is not universally available because not all bacteria grow on a defined medium. The innovative skills of bacterial geneticists are best displayed in seeking new ways of selecting recombinants. Antibiotic resistance, use of carbon and nitrogen sources for growth, and conditional lethal mutants have all been used, but the lack of an effective recombinant selection procedure for individual species has often limited the effectiveness of a gene exchange system.

Conjugation systems of genetic exchange have in practice been the most likely to produce extensive knowledge of chromosomal gene arrangement in bacteria. The prime requirement of a such system is usually a plasmid that will promote chromosome transfer. The first such plasmid was F, a native plasmid found in *E. coli* K12, but conjugative plasmids capable of promoting chromosome transfer have been difficult to find in most organisms. One exception is *P. aeruginosa,* whose clinical strains frequently carry plasmids capable of acting as sex factors (37). Fortunately, the ubiquitous drug resistance plasmids have proved to be the principle means by which conjugation can be used as a genetic tool in various bacteria, and as early as 1962, investigators showed that antibiotic resistance plasmids could promote chromosome transfer in *E. coli* (167). The next key discovery was that of Lowbury et al (110), who demonstrated the first plasmids that could replicate in a range of unrelated gram-negative bacteria and be transferred freely among different genera. These IncP1 plasmids, as they are now described (174), have a host range that extends to most gram-negative organisms including *Acinetobacter, Agrobacterium, Azospirillum, Azobacteria, Erwinia, Escherichia, Pseudomonas, Rhizobium, Rhodobacter, Vibrio,* and *Zymomonas* (73, 174). IncP1 plasmids can promote chromosome transfer (161), but their effectiveness to do so can be increased by the construction of variants with enhanced frequencies of gene transfer, such as R68.45 (64, 65) or pMO514 (129). Wide host–range plasmids do not necessarily display conjugal competence in a variety of species. The IncP10 plasmid R91-5 has a wide host range of transfer among many gram-negative genera but has only a narrow host range of replication, namely *P. aeruginosa* (22). R91-5 is excellent for mapping in *P. putida* PPN (36, 166), but less effective in other strains of *P. putida* (M. I. Sinclair, unpublished data) and in *P. syringae* (126).

Transposable elements that were mainly derived from plasmids carrying antibiotic resistance markers have also been used to construct chromosome mobilizing systems for a range of bacteria. The concept has been to mimic the native *E. coli* system in which chromosome transfer takes place because of limited homology between the conjugative F plasmid and regions of the chromosome where common insertion sequences (IS) are located. Using transposable elements, an artificial homology can be constructed by the presence of a transposable element on both a plasmid and the chromosome. This has been achieved using well-known transposons such as Tn*1*, Tn*3*, Tn*5*, Tn*10*, Tn*501*, and Mu to create chromosome mobilizing systems in *Salmonella, Pseudomonas, Vibrio, Klebsiella, Rhizobium,* and *Staphylococcus,* to name some of the genera for which this approach has been successful (9, 27, 36, 71, 85, 88, 96, 165, 166). Construction of IncP1-Mu hybrids has enabled the high transposition frequency of Mu to be combined with the wide bacterial host range of the IncP1 plasmids (9).

The development of conjugation systems of gene exchange in gram-positive organisms has been slower than in the gram-negative bacteria. The study of antibiotic resistance plasmids was less extensive in these organisms than in the gram-negative bacteria, and, in addition, systems of genetic exchange have proved to be more elusive and application of molecular genetics has been more difficult (28). All this has been redressed over the past ten years, and new insights into mechanisms of gene exchange in bacteria have resulted. The variety of plasmids and transposable elements in gram-positive organisms is less than in gram-negative organisms so that there has been a greater reliance on recombinant DNA procedures for genetic analysis and genome mapping. The exception, of course, has been in the case of *Streptomyces*.

The best studied conjugative exchange systems in other gram-positive bacteria are those for *Enterococcus faecalis*. These include pheromone-inducible plasmids (30, 183), broad host-range plasmids (143), and conjugative transposons (29). A comprehensive review has recently been provided by Dunny et al (46). Scott (146) has stressed the role of conjugative transposition in the basic biology of bacterial mating. While studies of these systems have been particularly revealing in terms of mechanisms of exchange of antibiotic resistance determinants and new insights into the conjugal DNA transfer process in gram-positive organisms (45), they have not contributed to the knowledge of gene arrangement or genome structure of gram-positive organisms.

In lactic acid bacteria, both transduction and conjugation have played an important role in genetic analysis. Given the widespread use of these bacteria in food production and biotechnology, gene transfer studies have shown that most of the economically important characteristics are plasmid coded, including ability to ferment lactose, use of proteins, and resistance to bacteriophage. A transposon mutagenesis technique involving Tn*916* and its relatives has been developed for chromosomal gene identification (50, 54).

In organisms such as *Staphylococcus aureus,* transduction, transformation, conjugation, and "phage mediated conjugation" have all been identified as mechanisms of genetic transfer, which has been mainly studied to increase understanding of the spread of antibiotic resistance in this medically important organism (113). The development of a chromosome map for *S. aureus* of about 80 markers has been achieved by transformation and protoplast fusion (131). In the case of *Bacillus subtilis,* the most extensively mapped gram-positive organism, a combination of transduction and transformation has been used to locate more than 700 genes on the chromosome (132).

Despite the immense amount of work currently conducted on a wide range of bacteria using recombinant DNA techniques, work also continues to extend the use of classical genetic exchange mechanisms to a wider range of bacteria. In addition to the organisms for which such mechanisms exist, lists of which already have been published (6, 7, 104, 120, 142), new systems for such exchange continue to be discovered and additional limited chromosome mapping achieved as shown by the following examples.

1. Conjugation: *Agrobacterium tumefaciens* (41, 133); *Ancylobacter aquaticus* (13); *Azotobacter vinelandii* (14, 91); *Bacteroides denticola* (63); *Bordetella pertussis* (154); *Erwinia chrysanthemi* (84); *Erwinia stewartii* (115); *Halobacterium volcanii* (118); *Lactococcus lactis* (19, 54); *Methylobacillus flagellatum* (176); *Mycobacterium* (87); *Myxococcus xanthus* (56, 150); *Paracoccus denitrificans* (55); *Pseudomonas putida* (166); *Pseudomonas solanacearum* (15); *Pseudomonas syringae* (126).
2. Transduction: *Desulfovibrio desulfuricans* (135); *Paracoccus denitrificans* (55); *Pseudomonas cepacia* (114); *Rhodococcus erythropolis* (35); *Vibrio cholerae* (62); *Vibrio fischeri* (103); *Xanthobacter autotrophicus* (186).
3. Transformation: *Agmenellum quadruplicatum* (48); *Azotobacter* (91); *Corynebacterium* (148); *Halobacterium halobium, H. volcanii* (31); *Lactobacillus acidophilus* (105); *Mycobacterium* (159); *Neisseria gonorrhoea* (21).

There are now substantial chromosome maps for a number of bacteria generated by these in vivo techniques, and every three years these are updated (128).

That bacteria have evolved mechanisms of gene exchange which do not fit neatly into the major compartments of conjugation, transduction, and transformation will not be a surprise to microbiologists. For example, in *Rhodobacter capsulatus* there takes place the process known as capsduction, in which transfer of packaged fragments of DNA occurs. The packages known as the gene transfer agent (GTA) have been shown to contain fragments of bacterial chromosome that have a constant length. Such fragments cannot replicate in the recipient parent unless integrated into the chromosome (145).

OVERCOMING THE PROBLEMS OF GENETIC ANALYSIS

Despite this extensive and impressive array of organisms for which genetic analysis has been achieved, some problems are evident. For example, the number of organisms with extensive chromosome maps is limited; for many organisms of importance in microbiology in vivo genetic analysis is either not available or of limited use. The reasons are varied. Mutants may be hard to find, selection for recombinants is ineffective, plasmids with effective chromosome mobilizing ability do not function outside a limited range of bacteria, or the biology of the organism is unfriendly to geneticists. As a result, new technology has been adapted to solve these problems. This innovation has come primarily from recombinant DNA technology, combined with improvements in instrumentation for handling macromolecules.

The Role of New Technology

Transformation clearly is an area that would benefit from new technology, given the highly precise and vexatious reproducibility of conditions that is required to obtain competent recipient cells. Examples of transformation of plasmid DNA in many bacteria are too numerous to list here, but this work has helped to develop new technologies for the introduction of DNA, both that of linear chromosomal fragments and circular plasmids, into bacteria. The most important advance has been the development of electroporation. In this procedure, high-voltage, high-current exponential pulses of controlled characteristics are passed through bacterial cells, resulting in the enhanced entry of DNA into bacteria for which conditions of competence have not been established, or at higher frequency in those for which transformation is an established technique. The process is technically simple and commercially produced equipment is readily available. It has general applicability for many bacteria, and it is a procedure that is likely to undergo further improvement (93). Bacteria for which electroporation has resulted in more effective uptake of DNA for genetic analysis include *Bacillus, Bordetella, Brevibacterium, Campylobacter, Clostridium, Corynebacterium,* cyanoabacteria, *Enterococcus, Erwinia, Escherichia, Haemophilus, Klebsiella, Listeria, Lactobacillus, Leuconostoc, Mycobacterium, Myxococcus, Pediococcus, Propionbacterium, Pseudomonas, Rhizobium, Salmonella, Staphylococcus, Streptococcus,* and *Vibrio* (24, 120, 121, 149). Much remains to be understood about the phenomenon of competence. Stewart & Carlson (163) have provided an excellent summary on the problems associated with transformation, and they point out the ways in which this phenomenon could be used for a better understanding of gene exchange among bacteria in natural environments. Clearly, there is a need for greater understanding of the mechanism of DNA uptake by bacterial cells. An example of the benefits of ready uptake of DNA is shown by *Acinetobacter calcoaceticus,* which is unusual in showing exceptionally high competence without the need for special cultural conditions; this property has been used to advantage in chromosomal mapping. Averhoff et al (4) using *A. calcoaceticus* AP1 (89) mapped genes of the *p*-hydroxybenzoate (pob) pathway. *E. coli* colonies carrying cloned *A. calcoaceticus* genes contain DNA that will transform *A. calcoaceticus* so that it is only necessary to replicate the colonies onto a layer of the recipient. This technique has been used to demonstrate supraoperonic clustering of *pob* genes (185), and this natural transformation makes *A. calcoaceticus* an excellent subject to study the relative location, regulation, and evolution of genes with physiologically interdependent functions. In addition, it would be interesting to understand the mechanism of competence in *A. calcoaceticus* with the aim of extending this property to other bacteria. It has been possible to identify genes that affect the transformability of *Haemophilus influenzae* (98).

Transformation by uptake of DNA by competent cells or by electroporation has assumed a greater importance in genetic analysis because it is often the only way in which isolated DNA can be returned to a living cell, particularly in the case of plasmids or other vectors carrying cloned genetic material. Genetic manipulation of any bacterium is only possible when the manipulated DNA can be returned to the living cell, and in the absence of conjugative plasmids and bacteriophages, transformation is the key technique. Hence, electroporation will play a more essential role in the genetic manipulation and genetic analysis of many different bacteria in which uptake of DNA has been the limiting factor.

Complementation Mapping

A general feature of bacterial genetic exchange is that only fragments of the entire genome are transferred from a donor to a recipient. While the fraction of the whole genome transferred differs markedly between conjugation, transduction, and transformation, the genetic implication is that a number of separate crosses are needed to analyze and to map the whole bacterial chromosome. New techniques have been devised that can be generally applied to a wider variety of bacteria but that still have in common the fragmentation of the genome and its subsequent analysis. The approaches differ in the size of the fragment, how it is constructed, and the recombinant DNA techniques used for its characterization. The most difficult aspect of these procedures is identification of the natural arrangement of the fragments that exist in the whole chromosome from which they are derived. There are a number of ways in which this ordering process can be made easier. For short regions of the chromosome (usually up to 100 kb), one can use a biological precloning step. This was first described with a derivative of the F plasmid in *E. coli,* the F prime plasmid (80). A hybrid was formed consisting of a segment of bacterial chromosome and the intact F plasmid; this hybrid retained all the transfer properties of the F plasmid. The use of such plasmid-bacterial chromosome hybrids has been extended to other bacteria through the use of primes derived from R plasmids. For example, the IncP1 derivative R68.45 generates R prime plasmids in a variety of genera including *Acinetobacter, Escherichia, Klebsiella, Methylobacillus, Methylophilus, Pseudomonas, Rhizobium, Rhodobacter,* and *Salmonella* (71, 176). R prime analysis has been particularly valuable in *Rhizobium* because genes affecting a function of interest may not be clustered or contiguous on the chromosome and the size of R prime chromosome inserts may be more than 100 kb (123). Undoubtedly, prime plasmids occur in natural populations of bacteria and could be an important method of horizontal transfer of genes between unrelated bacteria. While the wide host-range feature of IncP1 plasmids was unusual when they were first discovered, clearly this ability is shared by other groups of plasmids, thus reinforcing the role of plasmid transfer for gene reassortment between unrelated bacteria. Coplin (34) has pointed out the potential role of such gene exchange in bacteria associated with plant disease with respect to host range of phytopathogens and the nature of the disease process.

A second approach is to use cosmid cloning to fragment the genome. A variety of cosmid vectors are available, but those of particular use for mapping include the genotype for wide bacterial host range; examples are pLA2917 (2) or pLAFR1 (52). Cosmids can clone up to 40 kb of inserted DNA, although for complementation mapping in bacteria, 25 kb is more suitable.

An example of the use of cosmid genome libraries for mapping and genetic analysis is provided by the methylotrophs, which are widely recognized as intransigent to genetic analysis. They are biologically diverse, few bacteriophages are known for them, conjugative plasmids are inefficient in transferring chromosome, and mutants that can be used to select rare recombinants are not easily found. Despite these collective disadvantages, the desire to understand the biology of these organisms is considerable, because of their fascinating ability to use unusual substrates, their role in the carbon cycle, and hence their importance for the environment, as well as their industrial potential (43, 72, 104). Some methylotrophs are more amenable to traditional approaches: for example, *Methylobacillus flagellatum* has been successfully analyzed using conjugation and prime plasmids (147, 176).

The essential steps for cosmid library complementation mapping are as follows. (*a*) Construction of a cosmid library of the organism in question using a wide host-range cosmid, for example, one derived from an IncP1 plasmid, e.g. pLA2917 (2). Each cosmid clone should be kept as an individual culture and for most organisms at least 1000 such clones will be required to give a comprehensive genome library. Choice of a host for the cosmids is important. *E. coli* S17-1 (152) is highly effective because it has an IncP plasmid integrated into the chromosome, so that two factor matings can be used to transfer the cloned DNA fragments to any gram-negative recipient. (*b*) Individual genes on individual cosmids can be identified by patching the cosmids on characterized mutants of a suitable recipient using a medium that will enable growth of that mutant, or will produce a color change as a result of enzyme activity, if the mutant allele is complemented. (*c*) A suitable recipient organism must be selected. *Pseudomonas aeruginosa* is a particularly effective host for complementation because it expresses heterologous genetic material well (60), an extensive range of mutants is available, and *P. aeruginosa* can be made phenotypically restriction deficient by growth at 43°C (70), thus enabling effective expression of incoming heterologous genetic material. *E. coli* is also used for this purpose, given the wide range of available mutants.

Using a combination of R prime plasmid transfer and transfer of cosmids, mapping of methylotroph chromosomes was accomplished by complementation of both *P. aeruginosa* and *E. coli* markers, chiefly auxotrophs. Moore et al (122) mapped 19 markers into four linkage groups in *M. methylotrophus*. Linkage was established where any individual prime plasmid complemented more than one *P. aeruginosa* marker. This map was extended using pLA2918 derived cosmids; 28 markers were mapped in *M. methylotrophus* and 25 in *M. viscogenes* (112). Intergeneric complementation has also been used to map chromosomal genes in *Agrobacterium tumefaciens* (182), *Rhodopseudomonas viridis* (107), and *Methylobacillus flagellatum* (176).

R primes can also be used as the first step to prepare a cosmid bank of a selected limited region of a bacterial chromosome. This was done by Zhang & Holloway (190) for the *catA* region of *P. aeruginosa* PAO. Using an R68.45 derivative, an R prime was isolated carrying a 120-kb segment of the *P. aeruginosa* PAO chromosome, which included the *catA* gene. A cosmid bank was constructed from this prime plasmid, and then overlapping cosmids were ordered into a contig covering 120 kb. This procedure enabled the precise ordering of the *cat, ben,* and *ant* genes and the identification of two new genes affecting glycine utilization and the permeability of carbon substrates.

The development of a variety of recombinant DNA techniques has led to many highly significant advances in bacterial genetic analysis. Firstly, the techniques are equally applicable to all bacteria. Secondly, by combining particular techniques, whole genome mapping can be achieved. Finally, these techniques enable genes to be located, which would be impossible by any of the traditional gene exchange mechanisms in which rare recombinants are selected. Given all these advantages, it will now be possible to compare chromosome structure and gene arrangement in a wide variety of bacteria, an achievement not possible when mapping data could only be obtained by the techniques of conjugation, transduction, and transformation in a limited number of strains and species. A survey of bacterial chromosome organization by Krawiec & Riley (95) updates an earlier, but more extensive publication (44).

Physical Analysis of Genomes

Two key techniques that have made possible these achievements are pulsed field gel electrophoresis (PFGE) and the identification and use of restriction endonucleases that cut DNA infrequently. Smith & Condemine (155) have given a practical account of how these techniques can be applied generally to bacterial systems. A variety of equipment can be used in PFGE; the key common feature is that double-stranded DNA fragments can be separated in a size range from about 5 megabases down to 15 kb. PFGE was first described by Schwartz & Cantor (144), and since that time a number of variations in the method have been introduced. These include orthogonal field alternation gel electrophoresis (OFAGE) (144); transverse alternating field electrophoresis (TAFE) (53); contour clamped homogenous electric field (CHEF) (26); and programmable, autonomously controlled electrode (PACE) (12). Despite these variations and the extensive use of the technique, the precise mechanism by which the separation of DNA molecules of different sizes is achieved is still not completely understood. The enzymes that must be used to achieve a suitable number of fragments for any given bacterium will vary according to the base composition of the bacterial DNA. For example, for *E. coli* there are 27 *Not*I and 22 SfiT sites resulting in fragments that can be aggregated to a genome size of 4,700 kb (156). These two enzymes are not suited for organisms with a high G+C content such as *Pseudomonas*. McClelland et al (116) have shown that the nucleotide sequence CTAG is found infrequently in most prokaryotes, and restriction endonucleases that include this sequence in their recognition sequence cut bacterial genomes infrequently. The enzymes most commonly used in a range of bacteria for whole genome physical analysis include *Spe*I, *Avr*II, *Sal*I, *Xba*I, *Nhe*I, *Dpn*I, *Asa*I, *Sma*I, *Rsr*II, *Nal*I and *Sac*II (116).

An important bonus of this technique is the ease with which genome size can be measured, a parameter that was previously subject to considerable error when measured by other techniques. Krawiec & Riley (95) list 40 bacterial species for which PFGE analysis has been carried out, with genome measurements ranging from 585 kb (*Mycoplasma genitalium)* to 9454 kb *Myxococcus xanthus*.

One important outcome of the use of PFGE and restriction endonuclease digestion is the construction of a physical map. In most cases it is possible to map the whole genome and the relative positions of the sites for one or more restriction enzymes that have been located. Unlike many genetic maps of bacteria, the entire genome can be mapped by this procedure and such a map can provide an essential framework on to which to build a genetic map. Landmarks of any physical map include restriction enzyme sites, insertion sequence sites, transposon insertion sites, repetitive extragenic sequences, and any base sequences that have been ordered by sequencing techniques.

Physical and Genetic Maps

The physical examination of the genome obtained by these techniques is only the first step in genome analysis. It does not necessarily distinguish between chromosome and plasmid components, although this has been accomplished in *Rhodobacter sphaeroides* (168, 169). Establishment of the physical map enables the construction of a combined physical and genetic map. This is achieved by probing the fragments of the restriction-digested genome with cloned material. This material can have a variety of origins: either chromosomal segments cloned in plasmids, bacteriophage or cosmids, individual cloned genes, oligonucleotides synthesized from sequence information obtained from either a protein or DNA data base, or polymerase chain reaction (PCR) generated sequences.

Improvements of the basic techniques are overcoming some of the difficulties encountered. Two-dimensional PFGE with sequential digestion by two restriction enzymes and the use of end labeling to detect small fragments can overcome problems such as repetitive DNA sequences, regions of the genome not amenable to cloning, unfavorable locations of restriction endonuclease sites, and cloning artifacts (139).

Physical/genetic maps are being generated in a wide variety of bacteria including *Anabaena* (7), *Bacillus subtilis* (179); *Bordetella pertussis* (164), *Brucella melitensis* (1), *Campylobacter coli* (187), *Campylobacter jejuni* (127), *Caulobacter crescentus* (47), *Clostridium perfringens* (20, 140), *E. coli* (156), *Haemophilus influenzae* (8, 90), *Haloferax vulcanii* (23), *Helicobacter pylori* (173), *Lactococcus lactis* (177), *Listeria monocytogenes* (119); *Methanobacterium thermoautotrophicum* (162), *Methanococcus voltae* (151), *Mycoplasma pneumoniae* (184), *Myxococcus xanthus* (25), *Neisseria gonorrhoea* (10, 38), *Pseudomonas aeruginosa* (78, 82, 136–139, 178), *Pseudomonas putida* (75, 78), *Pseudomonas solanacearum* (77), *Rhodobacter sphaeroides* (168–171), *Salmonella typhimurium* (106), *Streptococcus mutans* (66), *Streptomyces coelicolor* (92), and *Thermococcus celer* (125).

In a limited number of bacteria, the combination of physical analysis, probing, and data obtained from conjugational and transductional mapping has resulted in extreme precision of mapping and extensive characterization of the chromosome. This is true of *E. coli,* which has a genetic map with 1403 loci (5), a *Not*I, *Sfi*I physical map (156), a physical map constructed by sorting of a large genomic library (94), an ordered cosmid collection (172), an alignment of DNA sequences to a genomic restriction map (141), and a polymerase chain reaction (PCR)–generated physical map of some *E. coli* genes (181). Itaya et al (86) have used a gene-directed mutagenesis method to construct a highly precise physical and genetic map of *B. subtilis*. The method allows the introduction of nucleotide sequences into a selected region of the chromosome without affecting other regions. Restriction enzyme sites can be eliminated or converted to different restriction enzyme sites, and this enables precise alignment of designated restriction sites. By creation of enzyme sites in gene loci of known location, a highly precise correlation of the physical and genetic maps is achieved with a method that is generally applicable to all bacteria. Kuspa et al (97) used yeast artificial chromosomes

(YAC) to construct a physical genetic map of *Myxococcus xanthus*. The large genome size of this organism (9454 kb) is particularly suited to the larger insert fragments (averaging 111 kb) that can be cloned by the YAC technique. Such varied approaches provide models by which more detailed genome analysis can be achieved in a wider variety of organisms. For example, in *Rhodobacter sphaeroides,* the presence of two chromosomal linkage groups has been confirmed both by physical analysis and by the development of a new conjugation system. Suwanto & Kaplan (170, 171) identified and characterized a self-transmissible, endogenous plasmid, the S factor in *R. sphaeroides*. A 600-bp *oriT*-containing fragment of the S factor can promote chromosomal transfer, and Hfr strains have been constructed (171). Mapping by interrupted mating has confirmed the presence of two linkage groups demonstrated previously (168, 169), and both chromosomes have been shown to carry essential housekeeping genes. In some organisms, it is important to be able to determine whether a function maps on the chromosome or on a plasmid. A variety of methods can be used. For example, if the plasmid is conjugative, then the function can be transferred by mating and the physical presence of the plasmid in the recipient can be correlated with the acquisition of the phenotype being studied. For some organisms, e.g. *Rhizobium meliloti,* special methods have been developed using transposon insertions into the plasmids and a *recA* recipient (58).

The precision of mapping available by these techniques and their applicability in locating markers for which classical approaches would be unsuccessful is shown by the following examples. The site of origin of chromosomal replication, *oriC* of *P. aeruginosa* PAO and *P. putida* PPN was cloned and sequenced by Yee & Smith (188). Using this cloned material, Smith et al (157) mapped the *oriC* locus and a second *ori* site of as yet unknown function to the base pair level of precision, as *oriC* contains a *Dra*I site and hence could be precisely located on the physical map. The second example concerns the insertion of plasmid DNA into the bacterial chromosome. One of the issues in understanding the structure and evolution of the pseudomonad chromosome is the manner by which new genetic material, and hence new functions, have been acquired during evolution. The genetic map of the *P. aeruginosa* chromosome now includes nearly 400 genes (76), and Holloway & Morgan (81) have proposed an accretion process to explain chromosome organization in pseudomonads in which new genetic material was acquired by integration of DNA segments from plasmids, bacteriophages, and transposable elements into a specific segment of the chromosome. Evidence supporting this view has been obtained by combined physical and genetic mapping. Transposons have been demonstrated on the TOL plasmid pWW0 (175), and Sinclair & Holloway (153) showed that these transposons, Tn*4651* and Tn*4653*, could insert into the chromosome of *P. aeruginosa* PAO in a region consistent with the accretion proposal of Holloway & Morgan. *Spe*I digestion and PFGE showed that 16 independent insertions occurred within a 334-kb region of the chromosome, with nine inserts limited to a 10-kb area. Clearly, this mapping has precisely identified a region of the *P. aeruginosa* PAO chromosome that possesses special properties of acquisition of new genetic material, as wild type *P. aeruginosa* does not possess the genetic information to degrade toluene.

Whole Genome Structure

The availability of whole genome analysis procedures has enabled a totally new dimension of comparative genome structure to be developed, which will have important implications for bacterial evolution, taxonomy, and patterns of regulatory behavior. The data currently available from the physical/genetic maps of bacteria, as listed above, indicates that most bacteria have a single circular chromosome. In one case, which will probably not be unique, two chromosomes have been identified, in addition to a number of autonomous plasmids (168–171). This has led investigators to define, in more precise terms than was previously possible, the difference between a chromosome and a plasmid. Key features of this classification are the presence of essential housekeeping genes, the stoichometric relationships of the various classes of replicons, the constant presence of the chromosomes in other strains of the same species but the variability in plasmid composition in different isolates, and the nontransferability of the chromosomal elements. Undoubtedly, other examples will be found that differ from the usual single circular chromosome structure found in most bacteria. In one case, *Borrelia burgdorfii,* linear chromosomes have been observed (49). *Azotobacter vinelandii* has been shown to have multiple copies of chromosome-like plasmids with multiple copies of genes and up to 40 times the DNA content per cell that is found in *E. coli* (134).

Genomic analysis has provided much improved data on genome size of bacteria. Some caution in interpretation is necessary because PFGE of restriction enzyme fragments does not necessarily distinguish between chromosomal and plasmid components and gives a whole genome measurement of DNA content. Differences in genome size of isolates of the same species may reflect differences in plasmid content rather than chromosome size, but the technique is sufficiently flexible so that it can be used to examine independent isolates for differences in genomic content (68).

At the sequence level on the chromosome, there is clear evidence of extragenic structures. Insertion sequences (IS) have been well characterized in the *E. coli* chromosome (39), and there is a need to identify the occurrence of similar structures in other bacteria. Attempts to find IS in the *P. aeruginosa* chromosome have been quite unsuccessful (V. Krishnapillai, unpublished data), but a detailed description of IS in *P. cepacia* has been provided by Lessie and his colleagues (101, 102), and their role in the catabolic potential of that species has been identified. One interesting feature of IS in *P. cepacia* is that they are found on plasmids and do not have a stable life when inserted into the chromosome. This is in contrast to the situation in *E. coli* where stable location of IS at identified sites on the chromosome is an important aspect of Hfr formation with the F plasmid. These differences in the genomic location of IS in different bacteria may affect their ability to survive in varying environmental niches.

A variety of other components of the prokaryotic genome will need to be mapped, characterized, and have their role in the genetic economy established. These include repetitive extragenic palindromic sequences (57, 69) and short interspersed repetitive sequence elements (111). Such elements are thought to influence bacterial virulence, gene regulation, transcription, chromosome structure, and chromosomal rearrangements. Genome analysis and mapping should help answer the question of why bacteria possess these genomic components.

An additional important aspect of genome structure is the pattern of distribution of genes on plasmids relative to the chromosome and the potential for integration of all or part of plasmids into the chromosome. This topic has been addressed to some extent in the pseudomonads in which genes for utilization of substrates are commonly located on plasmids (51). Plasmids are significant for pathogenicity in *P. solanacearum* in which megaplasmids up to and exceeding 1000 kb have been demonstrated in almost all isolates examined and have been shown to carry virulence and hypersensitive response (hrp) genes (3, 16, 17). Other genera where the chromosomal and plasmid components of the genome need to be better defined include *Staphylococcus,* particularly with respect to antibiotic resistance genes, and *Rhizobium,* whose nitrogen-fixing functions are distributed between the chromosome and plasmids. Sobral et al (160) have demonstrated physical identification and separation of plasmids and chromosome in *R. meliloti* by PFGE using the TAFE system. Given the range of interchanges between unrelated bacteria documented by Mazodier & Davies (117), techniques of genome analysis now provide the means to acquire much better data to establish how present day organisms have diverged or converged with respect to related species, and the extent of lateral genetic exchange can now be defined in more precise detail. Grothues & Tümmler (61) have applied PFGE techniques to the taxonomy of pseudomonads. In general, dendrograms deduced from PFGE data corresponded to other systems used for classification, but some exceptions were found.

At the level of gene organization, little is known of how present day arrangements have arisen. In *B. subtilis* Zeigler & Dean (189) have examined the orientation of genes on the *B. subtilis* chromosome using published sequence data. Of 96 genes, 91 were oriented so that the promoters were proximal to the chromosomal origin of replication, and hence gene transcription is codirectional with replication, confirming previous proposals of Brewer (18). As yet, the generality of this arrangement remains to be determined. An analogy can be drawn between analysis of the complexity of bacterial genomes and the results of the complete sequencing of the single strand bacteriophage ϕX174. The latter showed that different genes overlapped and that the same DNA sequence was used for two different informational purposes. Likewise, a comprehensive knowledge of gene arrangement and the identification of multipurpose regulatory units will yield new insights on how the bacterial genome interacts with and responds to the environment. An indication of such interactions is provided by the genetic analysis of bacterial virulence genes, as shown by the control of alginate synthesis of isolates of *P. aeruginosa* isolated from cystic fibrosis patients. (40, 42).

USES OF GENOME ANALYSIS OF BACTERIA

Genetics is now a prime force in understanding most aspects of microbial activities. The practical aspects of the role of bacteria in human, animal, and plant disease and the role of bacteria in modern commercial biology have stimulated genetic studies to address practical ends. Specifically, the need to solve problems of bacterial pathogenicity in humans, animals, and plants and the multiple uses of bacteria for biotechnology are driving the expansion of genetic analysis to more diverse bacteria. Smith (158) has highlighted the role that genetics must play in searching for hitherto unknown virulence determinants of bacteria. Cloning of individual genes is a vital tool in identifying individual genes and gene products, but multigene unlinked genetic determinants can only be evaluated using whole genome analysis, and this is exemplified by situations where the environment is a component of pathogenicity.

The acquisition of specific genomic regions through lateral evolution may be a mechanism by which virulence characteristics are acquired. This can occur by the process of lysogenic conversion in which the prophage confers specific changes in the bacterial phenotype. The best known case is that of *Corynebacteria diphtheriae,* (130) but similar cases are those of *Pseudomonas aeruginosa* (67), *Staphylococcus aureus,* (32) and *Shigella flexneri* (180). It would be interesting to establish whether lysogeny plays any role in bacterial diseases of plants. Lysogeny in phytopathogenic bacteria has been reported (79), and soil is a recognized source of bacteriophage for many bacteria. Cook & Sequeira (33) have suggested that acquisition of a genomic fragment of a bacteriophage may have been responsible in part for the host specificity of isolates of *Pseudomonas solanacearum,* which can cause bacterial wilt in potatoes.

While cloning procedures are essential for construction of modified strains for specific purposes, in some commercial situations the use of such strains could inhibit public acceptance of products due to regulatory situations. This applies particularly to food products, and while the recent acceptance by the US Government of recombinant DNA procedures for food products makes acceptance of such products easier, such an attitude is not prevalent worldwide. Gasson (54) has indicated the importance of in vivo techniques of genetic recombination for the more general adoption of bacteriophage-resistant dairy starter cultures of *Lactococcus lactis* by the dairy industry and consumers of dairy products.

Likewise, there are considerably fewer regulatory hurdles encountered in the release of microorganisms for agricultural control purposes if they have been constructed by means other than the use of recombinant DNA techniques. Lecadet et al (99) have used a combination of transduction and electroporation to construct strains of *Bacillus thuringiensis* with novel insecticidal properties and enhanced expression of toxin, thus providing a more effective pesticide and one better able to be effective against the emergence of toxin-resistant insect variants.

SUMMARY

The evolution of techniques and approaches to the whole genome analysis of bacteria has been enthusiastically and effectively embraced by many bacteriologists working with many different bacteria as shown by the range of physical/genetic maps now available. The next five years will be a period of data gathering as gene distribution patterns and other indices of chromosome structure become better known. At present, it is not possible to predict the implications of genome structure analysis. One early benefit will undoubtedly be a better understanding of ribosomal gene patterns, given the high level of conservation of the sequences involved. And given the use of ribosomal DNA in bacterial taxonomy, there should be a more immediate nexus between this criterion of bacterial relationship and its genetic basis. PFGE data have already contributed to the clarification of the taxonomy of the pseudomonads (61). Of particular interest will be the definition of the fine lacework of structural genes, their regulatory elements, and noncoding sequences and interactions of these elements with the extracellular environment. An understanding of this aspect of genomic structure must surely help to solve the basic nature of disease production and interactions involving the environment and bacterial communities. Nearly fifty years of microbial genetics has already made a major contribution to microbiology. It will continue to do so in the future.

ACKNOWLEDGMENTS

I thank V. Krishnapillai for his critical reading of the text and J. Elliston for preparation of the manuscript. Work in the author's laboratory is supported by the Australian Research Council, the Australian Centre for International Agricultural Research, and the Celgene Corporation.

Literature Cited

1. Allardet-Servent, A., Carles-Nurit, M-J., Bourg, G., Michaux, S., Ramuz, M. 1991. Physical map of the *Brucella melitensis* 16 M chromosome. *J. Bacteriol.* 173:2219–24
2. Allen, L. N., Hanson, R. S. 1985. Construction of broad-host-range cosmid cloning vectors: identification of genes necessary for growth of *Methylobacterium organophilum* on methanol. *J. Bacteriol.* 161:955–62
3. Arlat, M., Barberis, P., Trigalet, A., Boucher, C. 1990. Organization and expression of *hrp* genes in *Pseudomonas solanacearum*. In *Proceedings of the 7th International Conference on Plant Pathogenic Bacteria,* ed. Z. Klement, pp. 419–24. Budapest: Akadémici Kiadó
4. Averhoff, B., Gregg-Jolly, L., Elsemore, D., Ornston, L. N. 1992. Genetic analysis of supraoperonic clustering by use of natural transformation in *Acinetobacter calcoaceticus*. *J. Bacteriol.* 174:200–4
5. Bachman, B. J. 1990. Linkage map of *Escherichia coli* K12, Edition 8. *Microbiol. Rev.* 54:130–97
6. Ball, C., ed. 1984. *Genetics and Breeding of Industrial Microorganisms*. Boca Raton, Fla: CRC Press. 203 pp.
7. Bancroft, C., Wolk, C. P., Oren, E. V. 1989. Physical and genetic maps of the genome of the heterocyst forming cyanobacterium *Anabaena* sp. strain PCC7120. *J. Bacteriol.* 171:5940–48
8. Barcak, G. J., Chandler, M. S., Redfield, R. J., Tomb, J-F. 19. Genetic systems in *Haemophilus influenzae*. *Methods Enzymol.* 204:321–42
9. Berg, C. M., Berg, D. E., Groisman, E. A. 1989. Transposable elements and the genetic engineering of bacteria. In *Mobile DNA,* ed. D. E. Berg, M. M. Howe, pp. 879–926. Washington, DC: Am. Soc. Microbiol. 972 pp.
10. Bihlmaier, A., Römling, U., Meyer, T. F., Tümmler, B., Gibbs, C. P. 1991. Physical and genetic map of the *Neisseria gonorrhoea* strain MS11-N198 chromosome. *Mol. Microbiol.* 5:2529–40
11. Birge, E. A. 1988. *Bacterial and Bacteriophage Genetics*. Berlin: Springer-Verlag. 414 pp. 2nd ed.
12. Birren, B. W., Hood, L., Lai, E. 1989. Pulsed field gel electrophoresis: studies of DNA migration made with the programmable autonomously-controlled electrode electrophoresis system. *Electrophoresis* 10:302–9
13. Bittle, C., Konopka, A. 1990. IncP-1 mediated transfer of loci involved with gas vesicle production in *Ancylobacter aquaticus*. *J. Gen. Microbiol.* 136: 1259–63
14. Blanco, G., Ramos, F., Medina, J. R., Tortolero, M. 1990. A chromosomal linkage map of *Azotobacter vinelandii*. *Mol. Gen. Genet.* 224:241–47
15. Boucher, C., Arlat, M., Zischek, C., Boistard, P. 1988. Genetic organization of pathogenicity determinants of *Pseudomonas solanacearum*. In *Physiology and Biochemistry of Plant-Microbial Interactions,* ed. N. T. Keen, T. Kosuge, L. L. Walling, pp. 83–95. Rockville, Md: Am. Soc. Plant Physiol.
16. Boucher, C. A., Gijsegem, F. V., Barberis, P. A., Arlat, M., Zischek, C. 1987. *Pseudomonas solanacearum* genes controlling both pathogenicity on tomato and hypersensitivity are clustered. *J. Bacteriol.* 169:5626–32
17. Boucher, C. A., Martinet, A., Barberis, P. A., Alloing, G., Zischek, C. 1986. Virulence genes are carried by a megaplasmid of the plant pathogen *Pseudomonas solanacearum*. *Mol. Gen. Genet.* 205:270–75
18. Brewer, B. J. 1988. When polymerases collide: replication and the transcriptional organization of the *E. coli* chromosome. *Cell* 53:679–86
19. Bringel, F., van Alstine, G. L., Scott, J. R. 1992. Transfer of Tn*916* between *Lactococcus lactis* subs. *lactis* strains

is nontranspositional: evidence for a chromosome fertility function in strain MG1363. *J. Bacteriol.* 174:5840–47
20. Canard, B., Cole, S. T. 1989. Genome organization of the anaerobic pathogen *Clostridium perfringens. Proc. Natl. Acad. Sci. USA* 86:6676–80
21. Cannon, J. G., Sparling, P. F. 1984. The genetics of the gonococcus. *Annu. Rev. Microbiol.* 38:111–33
22. Chandler, P. M., Krishnapillai, V. 1974. Phenotypic characteristics of R factors of *Pseudomonas aeruginosa:* R factors transferable only in *Pseudomonas aeruginosa. Genet. Res.* 23:251–57
23. Charlebois, R. L., Schalkwyk, L. C., Hofman, D. C., Doolittle, W. F. 1991. Detailed physical map and set of overlapping clones covering the genome of the archaebacterium *Haloferax volcanii. J. Mol. Biol.* 222:509–24
24. Chassy, B. M., Mercenier, A., Flickinger, J. 1988. Transformation of bacteria by electroporation. *Trends Biotechnol.* 6:303–9
25. Chen, H., Kuspa, A., Keseler, I., Shimkets, L. J. 1991. Physical map of the *Myxococcus xanthus* chromosome. *J. Bacteriol.* 173:2109–15
26. Chu, G. 1989. Pulsed field electrophoresis in contour clamped homogenous electric fields for the resolution of DNA by size or topology. *Electrophoresis* 10:290–95
27. Chumley, F. G., Henzel, R., Roth, J. R. 1979. Hfr formation directed by Tn*10*. *Genetics* 91:639–55
28. Clewell, D. B. 1981. Plasmids, drug resistance and gene transfer in the genus *Streptococcus. Microbiol. Rev.* 45:409–36
29. Clewell, D. B., Gawron-Burke, M. C. 1986. Conjugative transposons and the dissemination of antibiotic resistance in streptococci. *Annu. Rev. Microbiol.* 40:635–59
30. Clewell, D. B., Weaver, K. E. 1989. Sex pheromones and plasmid transfer in *Enterococcus faecalis,* a review. *Plasmid* 21:175–84
31. Cline, S. W., Lam, W. L., Charlebois, R. L., Schalkwyk, L. C., Doolittle, N. F. 1989. Transformation methods with halophilic archaebacteria. *Can. J. Microbiol.* 35:148–52
32. Coleman, D. C., Sullivan, D. J., Russell, R. J., Arbuthnot, J. P., Carey, B. F., Pomeroy, H. M. 1989. *Staphylococcus aureus* bacteriophages mediating the simultaneous lysogenic conversion of β-lysin, staphylokinase and enterotoxin A: molecular mechanism of triple conversion. *J. Gen. Microbiol.* 135:1679-97
33. Cook, D., Sequeira, L. 1991. The use of subtractive hybridization to obtain a DNA probe specific for *Pseudomonas solanacearum* race 3. *Mol. Gen. Genet.* 227:401–10
34. Coplin, D. L. 1989. Plasmids and their role in the evolution of plant pathogenic bacteria. *Annu. Rev. Phytopathol.* 27: 187–212
35. Dabbs, E. R. 1987. A generalised transducing bacteriophage for *Rhodococcus erthropolis. Mol. Gen. Genet.* 206:116–20
36. Dean, H. F., Morgan, A. F. 1983. Integration of R91–5::Tn*501* into the *Pseudomonas putida* PPN chromosome and genetic circularity of the chromosomal map. *J. Bacteriol.* 153:485–97
37. Dean, H. F., Royle, P., Morgan, A. F. 1979. Detection of FP plasmids in hospital isolates of *Pseudomonas aeruginosa. J. Bacteriol.* 138:249–50
38. Dempsey, Jo Ann F., Lilaker, W., Madhure, A., Snodgrass, T. L., Cannon, J. G. 1991. Physical map of the chromosome of *Neisseria gonorrhoea* FA1090 with locations of genetic markers, including *opa* and *pil* genes. *J. Bacteriol.* 173:5476–86
39. Deonier, J. 1987. Locations of native insertion elements. See Ref. 124, pp. 982–89
40. Deretic, V., Konyecsni, W. M., Mohr, C. D., Martin, D. W., Hibler, N. S. 1989. Common denominators of promoter control in *Pseudomonas* and other bacteria. *Bio/Technology.* 7:1249–54
41. Dessaux, Y., Petit, A., Ellis, J. G., Legrain, C., Demarez, M., et al. 1989. Ti-plasmid controlled chromosome transfer in *Agrobacterium tumefaciens. J. Bacteriol.* 171:6363–66
42. DeVault, J. D., Berry, A., Misra, T. K., Darzins, A., Chakrabarty, A. M. 1989. Environmental sensory signals and microbial pathogenesis: *Pseudomonas aeruginosa* infection in cystic fibrosis. *Bio/Technology* 7:352–58
43. de Vries, G. E., Kues, U., Stahl, U. 1990. Physiology and genetics of methylotrophic bacteria. *FEMS Microbiol. Rev.* 75:57–102
44. Drlica, K., Riley, M., eds. 1990. *The Bacterial Chromosome*. Washington, DC: Am. Soc. Microbiol. 469 pp.
45. Dunny, G. M. 1991. Mating interactions in gram positive bacteria. In *Cell-Cell Interactions,* ed. M. Dworkin. Washington, DC: Am. Soc. Microbiol. 374 pp.
46. Dunny, G. M., Cleary, P. P., McKay,

L. L., eds. 1991. *Genetics and Molecular Biology of Streptococci, Lactococci and Enterococci*. Washington, DC: Am. Soc. Microbiol. 310 pp.

47. Ely, B., Ely, T. W., Geradot, C. J., Dingwall, A. 1990. Circularity of the *Caulobacter crescentus* chromosome determined by pulsed-field gel electrophoresis. *J. Bacteriol.* 172:1262–66
48. Essich, E., Stevens, S. E., Porter, R. D. 1990. Chromosomal transformation in the cyanobacterium *Agmenellum quadruplicatum*. *J. Bacteriol.* 172: 1916–22
49. Ferdows, M. S., Barbour, A. G. 1989. Megabase-sized linear DNA in the bacterium *Borrelia burgdorferi*, the lyme disease agent. *Proc. Natl. Acad. Sci. USA* 86:5969–73
50. Fitzgerald, G. F., Gasson, M. J. 1988. In vivo gene transfer systems and transposons. *Biochimie* 70:489–502
51. Frantz, B., Chakrabarty, A. M. 1986. Degradative plasmids in *Pseudomonas*. In *The Bacteria. The Biology of* Pseudomonas, ed. J. R. Sokatch, 10:295–324. Orlando: Academic. 617 pp.
52. Friedman, A. M., Long, S. R., Bowen, S. E., Buikema, W. J., Ausubel, F. M. 1982. Construction of a broad host range cloning vector and its use in the genetic analysis of *Rhizobium* mutants. *Gene* 18:289–98
53. Gardiner, K., Patterson, D. 1989. Transverse alternating field electrophoresis and applications to mammalian genome mapping. *Electrophoresis* 10: 296–302
54. Gasson, M. J. 1990. In vivo genetic systems in lactic acid bacteriology. *FEMS Microbiol. Rev.* 87:43–60
55. Ghozlan, H. A., Ahmadian, R. M., Frohlich, M., Soby, S., Kleiner, D. 1991. Genetic tools for *Paracoccus denitrificans*. *FEMS Microbiol. Lett.* 82:303–6
56. Gill, R. E., Cull, M. G., Fly, S. 1988. Genetic identification and cloning of a gene required for developmental cell interactions in *Myxococcus xanthus*. *J. Bacteriol.* 170:5279–88
57. Gilson, E., Clément, J-M., Brutlag, D., Hofnung, M. 1984. A family of dispersed repetitive extragenic palindromic DNA sequences in *E. coli*. *EMBO J.* 3:1417–21
58. Glazebrook, J., Walker, G. C. 1991. Genetic techniques in *Rhizobium meliloti*. *Methods. Enzymol.* 204:398–418
59. Goldberg, J. B., Won, J., Ohman, D. E. 1990. Precise excision and instability of the transposon Tn*5* in *Pseudomonas aeruginosa*. *J. Gen. Microbiol.* 136: 789–96
60. Gray, G. L., McKeown, K. A., Jones, A. J. S., Seeburg, P. H., Heyneker, H. L. 1984. *Pseudomonas aeruginosa* secretes and correctly processes human growth hormone. *Bio/Technology* 2: 161–65
61. Grothues, D., Tümmler, B. 1991. New approaches in gene analysis by pulsed-field gel electrophoresis: application to the analysis of *Pseudomonas* species. *Mol. Microbiol.* 5:2763–76
62. Guidolin, A., Manning, P. A. 1987. Genetics of *Vibrio cholerae* and its bacteriophages. *Microbiol. Rev.* 51: 285–98
63. Guiney, D. G., Hasegawa, P. 1992. Transfer of conjugal elements in oral black pigmented *Bacterioides (Prevotella)* spp. involves DNA rearrangements. *J. Bacteriol.* 174:4853–55
64. Haas, D., Holloway, B. W. 1976. R factor variants with enhanced sex factor activity in *Pseudomonas aeruginosa*. *Mol. Gen. Genet.* 144:243–51
65. Haas, D., Holloway, B. W. 1978. Chromosome mobilization by the R plasmid R68.45: a tool in *Pseudomonas* genetics. *Mol. Gen. Genet.* 158:229–37
66. Hantman, M. J., Tudor, J. J., Sun, S., Marri, L., Piggott, P. J., Daneo-Moore, L. 1991. Physical and genetic mapping of the *Streptococcus mutans* GS-5 genome. See Ref. 46, pp. 309–11
67. Hayashi, T., Baba, T., Matsumoto, H., Terawaki, Y. 1990. Phage conversion of cytotoxin production in *Pseudomonas aeruginosa*. *Mol. Microbiol.* 4:1703–9
68. Hector, J. S. R., Johnson, A. R. 1990. Determination of genome size of *Pseudomonas aeruginosa* by PFGE: analysis of restriction fragments. *Nucleic Acids Res.* 18:3171–74
69. Higgins, C. F., McLaren, R. S., Newbury, S. F. 1988. Repetitive extragenic palindromic sequences, mRNA stability and gene expression: evolution by gene conversion—a review. *Gene* 72:3-14
70. Holloway, B. W. 1965. Variations in restriction and modification following increase of growth temperature of *Pseudomonas aeruginosa*. *Virology* 25:634–42
71. Holloway, B. W. 1979. Plasmids that mobilize bacterial chromosome. *Plasmid* 2:1–19
72. Holloway, B. W. 1984. Genetics of methylotrophs. In *Methylotrophs: Microbiology, Biochemistry and Genetics*, ed. C. T. Hou, pp. 87–106. Boca Raton, Fla: CRC Press. 180 pp.

73. Holloway, B. W. 1986. Chromosome mobilization and genomic organization in *Pseudomonas*. See Ref. 51, pp. 230–86
74. Holloway, B. W. 1992. Genetic exchange processes in prokaryotes. In *Biotechnology. Genetic Fundamentals and Genetic Engineering*, ed. H-J. Rehm, G. Reed, A. Pühler, P. Stadler, 2:48–72. Weinheim: VCH
75. Holloway, B. W., Bowen, A., Dharmsthiti, S., Krishnapillai, V., Morgan, A., et al. 1990. Genetic tools for the manipulation of metabolic pathways. In *Proceedings of the 6th International Symposium on Genetics of Industrial Microorganisms. GIM 90*, ed. H. Heslot, J. Davies, J. Florent, L. Bobichon, G. Durand, L. Penasse, pp. 227–38. Paris: Soc. Fr. Microbiol. 554 pp.
76. Holloway, B. W., Carey, E. 1993. *Pseudomonas aeruginosa* PAO. See Ref. 128. In press
77. Holloway, B. W., Escuadra, M. D., Krishnapillai, V. 1993. Whole genome analysis of pseudomonads and its application to *Pseudomonas solanacearum*. In *Bacterial Wilt: The Disease and Its Causative Agent* Pseudomonas solanacearum, ed. A. C. Hayward, G. L. Hartman. Wallingford: C. A. B. Int. In press
78. Holloway, B. W., Escuadra, M. D., Morgan, A. F., Saffery, R., Krishnapillai, V. 1992. The new approaches to whole genome analysis of bacteria. *FEMS Microbiol. Lett.* 100: 101–6
79. Holloway, B. W., Krishnapillai, V. 1975. Bacteriophages and bacteriocins. In *Genetics and Biochemistry of* Pseudomonas, ed. P. H. Clarke, M. H. Richmond, pp. 99–132. London: Wiley. 366 pp.
80. Holloway, B. W., Low, K. B. 1987. F-prime and R-prime factors. See Ref. 124, pp. 1145–53
81. Holloway, B. W., Morgan, A. F. 1986. Genome organization in *Pseudomonas*. *Annu. Rev. Microbiol.* 40: 79–105
82. Holloway, B. W., Ratnaningsih, E., Krishnapillai, V., Tummler, B., Römling, U. 1993. A physical and genetic map of *Pseudomonas aeruginosa*. See Ref. 128. In press
83. Hopwood, D. A., Chater, K. E. 1989. *Genetics of Bacterial Diversity*. London: Academic. 449 pp.
84. Hugouvieux-Cotte-Pattat, N., Reverchon, S., Robert-Baudouy, J. 1989. Expanded linkage map of *Erwinia chrysanthemi* strain 3937. *Mol. Microbiol.* 3:573–82
85. Ichige, A., Matsutani, S., Oishi, K., Mizushima, S. 1989. Establishment of gene transfer systems for and construction of the genetic map of a marine *Vibrio* strain. *J. Bacteriol.* 171:1825-34
86. Itaya, M., Tanaka, T. 1991. Complete physical map of the *Bacillus subtilis* 168 chromosome constructed by a gene directed mutagenesis method. *J. Mol. Biol.* 220:631–34
87. Jekkel, A., Csajági, Eac., Ilköy, Eac., Ambrus, G. 1989. Genetic recombination by sphaeroplast fusion of steroltransforming *Mycobacterium* strains. *J. Gen. Microbiol.* 135:1727–33
88. Johnson, S. R., Romig, W. R. 1979. Transposon facilitated recombination in *Vibrio cholerae*. *Mol. Gen. Genet.* 170: 93-101
89. Juni, E. 1978. Genetics and physiology of *Acinetobacter*. *Annu. Rev. Microbiol.* 32:349–71
90. Kauc, L., Mitchell, M., Goodgal, S. H. 1989. Size and physical map of the chromosome of *Haemophilus influenzae*. *J. Bacteriol.* 171:2474–79
91. Kennedy, C., Toukdarian, A. 1987. Genetics of *Azotobacters:* applications to nitrogen fixation and related aspects of metabolism. *Annu. Rev. Microbiol.* 41:227–58
92. Kieser, M. H., Kieser, T., Hopwood, D. A. 1992. A combined genetic and physical map of the *Streptomyces coelicolor* A3(2) chromosome. *J. Bacteriol.* 174:5496–507
93. Kilbane, J. J., Bielaga, B. A. 1991. Instantaneous gene transfer from donor to recipient microorganisms via electroporation. *Biotechniques* 10:354–65
94. Kohara, Y., Akiyama, K., Isono, K. 1987. The physical map of the whole *E. coli* chromosome: application of a new strategy for rapid analysis and sorting of a large genomic library. *Cell* 50:495–508
95. Krawiec, S., Riley, M. 1990. Organization of the bacterial chromosome. *Microbiol. Rev.* 54:502–39
96. Krishnapillai, V., Royle, P., Lehrer, J. 1981. Insertions of the transposon Tn*1* into the *Pseudomonas aeruginosa* chromosome. *Genetics* 97:495–511
97. Kuspa, A., Vallbrath, D., Cheng, Y., Kaiser, D. 1989. Physical mapping of the *Myxococcus xanthus* genome. *Proc. Natl. Acad. Sci. USA* 86:8917–21
98. Larson, T. G., Roszczyk, E., Goodgal, S. 1991. Molecular cloning of two linked loci that increase the trans-

formability of transformation-deficient mutants of *Haemophilus influenzae*. *J. Bacteriol*. 173:4675–82
99. Lecadet, M. M., Chaufaux, J., Ribier, J., Lereclus, D. 1992. Construction of novel *Bacillus thuringiensis* strains with different insecticidal activities by transduction and transformation. *Appl. Environ. Microbiol*. 58:840–49
100. Lederberg, J., Tatum, E. L. 1946. Novel genotypes in mixed cultures of biochemical mutants of bacteria. *Cold Spring Harbor Symp. Quant. Biol*. 11:113–14
101. Lessie, T. G., Gaffney, T. 1986. Catabolic potential of *Pseudomonas cepacia*. See Ref. 51, pp. 439–81
102. Lessie, T. G., Wood, M. S., Byrne, A., Ferrante, A. 1990. Transposable gene inactivating elements in *Pseudomonas cepacia*. In *Pseudomonas: Biotransformations, Pathogenesis and Evolving Biotechnology*, ed. S. Silver, A. M. Chakrabarty, B. Iglewski, S. Kaplan, pp. 279–91. Washington, DC: Am. Soc. Microbiol. 423 pp.
103. Levisohn, R., Moreland, J., Nealson, K. H. 1987. Isolation and characterization of a generalized transducing phage for the marine luminous bacterium *Vibrio fischeri* MJ-1. *J. Gen. Microbiol*. 133:1577–82
104. Lidstrom, M., Stirling, D. I. 1990. Methylotrophs: genetics and commercial applications. *Annu. Rev. Microbiol*. 44:27–58
105. Lin, J. H-C., Savage, D. C. 1986. Genetic transformation in *Lactobacillus acidophilus*. *J. Gen. Microbiol*. 132: 2107–11
106. Liu, S. L., Sanderson, K. E. 1992. A physical map of the *Salmonella typhimurium* LT2 genome made by using *Xba*I analysis. *J. Bacteriol*. 174: 1662–72
107. Long, F. S., Oesterhelt, D. 1989. A gene transfer system for *Rhodopseudomonas viridis*. *J. Bacteriol*. 171: 4425–35
108. Lovett, P. S. 1984. *Bacillus*. See Ref. 6, pp. 44–62
109. Low, K. B., Porter, D. D. 1978. Modes of gene transfer and recombination in bacteria. *Annu. Rev. Genet*. 12:249–87
110. Lowbury, E. J., Kidson, A., Lilly, H. A., Ayliffe, G. A. J., Jones, R. J. 1969. Sensitivity of *Pseudomonas aeruginosa* to antibiotics: emergence of strains highly resistant to carbenicillin. *Lancet* 2:448–52
111. Lupski, J. R., Weinstock, G. M. 1992. Short interspersed repetitive DNA sequences in prokaryotic genomes. *J. Bacteriol*. 174:4525–29
112. Lyon, B. R., Kearney, P. P., Sinclair, M. I., Holloway, B. W. 1988. Comparative complementation mapping of *Methylophilus* spp. using cosmid clone libraries and prime plasmids. *J. Gen. Microbiol*. 134:123–32
113. Lyon, B. R., Skurray, R. 1987. Antimicrobial resistance of *Staphylococcus aureus:* genetic basis. *Microbiol. Rev*. 51:88-134
114. Matsumoto, H., Itoh, Y., Ohta, S., Terawaki, Y. 1986. A generalized transducing phage of *Pseudomonas cepacia*. *J. Gen. Microbiol*. 132:2583–86
115. McCammon, S., Coplin, D. L. 1982. Chromosome mobilization in *Erwinia stewartii* by plasmid pDC252.1 and bacteriophage Mu *cts*62. *Phytopathology* 72:1001 (Abstr.)
116. McClelland, M., Jones, R., Patel, Y., Nelson, M. 1987. Restriction endonucleases for pulsed field mapping of bacterial genomes. *Nucleic Acids Res*. 15:5989–6005
117. Mazodier, P., Davies, J. 1991. Gene transfer between distantly related bacteria. *Annu. Rev. Genet*. 25:147–71
118. Mevarech, M., Werczberger, R. 1985. Gene transfer in *Halobacterium volcanii*. *J. Bacteriol*. 162:461–62
119. Michel, E., Cossart, P. 1992. Physical map of the *Listeria monocytogenes* chromosome. *J. Bacteriol*. 174:7098–103
120. Miller, J. H., ed. 1991. Bacterial genetic systems. *Methods Enzymol*. 204: 1–669
121. Minton, N. P., Brehm, J. K., Oultram, J., Swinfield, T. J., Schimming, S., et al. 1990. Development of genetic systems for *Clostridium acetobutylicum*. See ref. 75, pp. 759–70
122. Moore, A. T., Nayudu, M., Holloway, B. W. 1983. Genetic mapping in *Methylophilus methylotrophus* AS1. *J. Gen. Microbiol*. 129:785–99
123. Nayudu, M., Rolfe, B. G. 1987. Analysis of R-primes demonstrates that genes for broad host nodulation of *Rhizobium* strain NGR-234 are dispersed on the Sym plasmid. *Mol. Gen. Genet*. 206:326–37
124. Neidhardt, F. C., Ingraham, J. L., Low, K. B., Magasanik, B., Schaechter, M., Umbarger, H. E., eds. 1987. Escherichia coli *and* Salmonella typhimurium *Cellular and Molecular Biology*, 2:807–1649. Washington, DC: Am. Soc. Microbiol.
125. Noll, K. M. 1989. Chromosome map

of the thermophilic archaebacterium *Thermococcus celer*. *J. Bacteriol.* 171: 6720-25
126. Nordeen, R. O., Holloway, B. W. 1990. Chromosome mapping in *Pseudomonas syringae* pv *syringae* strain PS224. *J. Gen. Microbiol.* 136:1231–39
127. Nuijten, P. J. M., Bartels, C., Bleumink-Pluym, N. M. C., Gaastra, W., van der Zeijst, B. A. M. 1990. Size and physical map of the *Campylobacter jejuni* chromosome. *Nucleic Acids Res.* 18:6211-14
128. O'Brien, S., ed. 1993. *Genetic Maps. Locus Maps of Complex Genomes*. Cold Spring Harbor, NY: Cold Spring Harbor Lab. Press. In press
129. O'Hoy, K., Krishnapillai, V. 1987. Recalibration of the *Pseudomonas aeruginosa* PAO chromosome map in time units using high-frequency-of recombination donors. *Genetics* 115:611–18
130. Pappenheimer, A. M. 1984. The diphtheria bacillus and its toxin: a model system. *J. Hyg.* 93:397–404
131. Pattee, P. A. 1990. Genetic and physical mapping of the chromosome of *Staphylococcus aureus* NCTC8325. See Ref. 44, pp. 163–96
132. Piggot, P. J. 1990. Genetic map of *Bacillus subtilis* 168. See Ref. 44, pp. 107–46
133. Pischl, D. L., Farrand, S. K. 1984. Characterization of transposon Tn*5*-facilitated donor strains and development of a chromosomal linkage map for *Agrobacterium tumefaciens*. *J. Bacteriol.* 159:1-8
134. Punita, Jafri, S., Reddy, M. A., Das, H. K. 1989. Multiple chromosomes of *Azotobacter vinelandii*. *J. Bacteriol.* 171:3133-38
135. Rapp, B. J., Wall, J. D. 1987. Genetic transfer in *Desulfovibrio desulfuricans*. *Proc. Natl. Acad. Sci. USA* 84:9128–30
136. Ratnaningsih, E., Dharmsthiti, S., Krishnapillai, V., Morgan, A., Sinclair, M., Holloway, B. W. 1990. A combined physical and genetic map of *Pseudomonas aeruginosa* PAO. *J. Gen. Microbiol.* 136:2351–57
137. Römling, U., Duchêne, M., Essar, D. W., Galloway, D., Guidi-Rontani, C., et al. 1992. Localization of *alg, opr, phn, pho,* 4.5S RNA, 6S RNA, *tox, trp* and *xcp* genes, *rrn* operons and the chromosomal origin on the physical genome map of *Pseudomonas aeruginosa* PAO. *J. Bacteriol.* 174:327–30
138. Römling, U., Grothues, D., Bautsch, W., Tümmler, B. 1989. A physical genome map of *Pseudomonas aeruginosa* PAO. *EMBO J.* 8:4081–89
139. Römling, U., Tümmler, B. 1991. The impact of two-dimensional pulsed field gel electrophoresis techniques for the consistent and complete mapping of bacterial chromosomes: refined physical map of *Pseudomonas aeruginosa* PAO. *Nucleic Acids Res.* 19:3199–206
140. Rood, J. I., Cole, S. T. 1991. Molecular genetics and pathogenesis of *Clostridium perfringens*. *Microbiol. Rev.* 55:621–48
141. Rudd, K. E., Miller, W., Ostell, J., Benson, D. A. 1990. Alignment of *Escherichia coli* K12 DNA sequences to a genomic restriction map. *Nucleic Acids Res.* 18:313–21
142. Scaife, J., Leach, D., Galizzi, A., eds. 1985. *Genetics of Bacteria*. London: Academic. 286 pp.
143. Schaberg, D. R., Zervos, M. J. 1986. Intergeneric and interspecies gene exchange in gram-positive cocci. *Antimicrob. Agents Chemother.* 30:817–22
144. Schwartz, D. C., Cantor, C. R. 1984. Separation of yeast chromosome sized DNAs by pulsed field gradient gel electrophoresis. *Cell* 37:67–75
145. Scolnik, P. A., Marrs, B. L. 1987. Genetic research with photosynthetic bacteria. *Annu. Rev. Microbiol.* 41: 701–26
146. Scott, J. R. 1992. Sex and the single circle: conjugative transposition. *J. Bacteriol.* 174:6005–10
147. Serebrijski, I. G., Kazakova, S. M., Tsygankov, Y. D. 1989. Construction of Hfr-like donors of the obligate methanol-oxidizing bacterium *Methylobacillus flagellatum* KT. *FEMS Microbiol. Lett.* 59:203–6
148. Serwold-Davis, T. M., Groman, N., Rabin, M. 1987. Transformation of *Corynebacterium diphtheriae, Corynebacterium ulcerans, Corynebacterium glutamicum* and *Escherichia coli* with the *C. diphtheriae* plasmid pNG2. *Proc. Natl. Acad. Sci. USA* 84:4964–68
149. Shigekawa, K., Dower, W. J. 1988. Electroporation of eukaryotes and prokaryotes: a general approach to the introduction of macromolecules into cells. *Biotechniques* 6:742–51
150. Shimkets, L. J., Gill, R. E., Kaiser, D. 1983. Developmental cell interactions in *Myxococcus xanthus* and the *spoC* locus. *Proc. Natl. Acad. Sci. USA* 80:1406–10
151. Silzman, J., Klein, A. 1991. Physical and genetic map of the *Methanococcus voltae* chromosome. *Mol. Microbiol.* 5:505-13

152. Simon, R., Priefer, U., Pühler, A. 1983. A broad host range mobilization system for in vivo genetic engineering: transposon mutagenesis in gram negative bacteria. *Bio/Technology* 1:784-91
153. Sinclair, M. I., Holloway, B. W. 1991. Chromosomal insertion of TOL transposons in *Pseudomonas aeruginosa* PAO. *J. Gen. Microbiol.* 137:1111–20
154. Smith, C. J., Coote, J. G., Parton, R. 1986. R plasmid-mediated chromosome mobilization in *Bordetella pertussis*. *J. Gen. Microbiol.* 132:2685–92
155. Smith, C. L., Condemine, G. 1990. New approaches for physical mapping of small genomes. *J. Bacteriol.* 172: 1167–72
156. Smith, C. L., Econome, E., Schutt, A., Klco, S., Cantor, C. R. 1987. A physical map of the *E. coli* genome. *Science* 236:1448-53
157. Smith, D. W., Yee, T. W., Baird, C., Krishnapillai, V. 1991. Pseudomonad replication origins: a paradigm for bacterial origins? *Mol. Microbiol.* 5:2581–87
158. Smith, H. 1989. The mounting interest in bacterial and viral pathogenicity. *Annu. Rev. Microbiol.* 43:1–22
159. Snapper, S. B., Lugosi, L., Jekkel, A., Melton, R. E., Kieser, T., et al. 1988. Lysogeny and transformation in mycobacteria: stable expression of foreign genes. *Proc. Natl. Acad. Sci. USA* 85:6987–91
160. Sobral, B. W. S., Honeycutt, R. J., Atherley, A. G., McLelland, M. 1991. Electrophoretic separation of the three *Rhizobium meliloti* replicons. *J. Bacteriol.* 173:5173–80
161. Stanisich, V. A., Holloway, B. W. 1971. Chromosome transfer in *Pseudomonas aeruginosa* mediated by R factors. *Genet. Res.* 17:169–72
162. Stettler, R., Leisinger, T. 1992. Physical map of the *Methanobacterium thermoautotrophicum* Marburg chromosome. *J. Bacteriol.* 174:7227–34
163. Stewart, G. J., Carlson, C. A. 1986. The biology of natural transformations. *Annu. Rev. Microbiol.* 40:211–35
164. Stibitz, S., Garletts, T. L. 1992. Derivation of a physical map of the chromosome of *Bordetella pertussis* Tohama I. *J. Bacteriol.* 174:7770–77
165. Stout, V. G., Iandolo, J. J. 1990. Chromosome gene transfer during conjugation by *Staphylococcus aureus* is mediated by transposon-mediated mobilization. *J. Bacteriol.* 172:6148–50
166. Strom, A. D., Hirst, R., Petering, J., Morgan, A. 1990. Isolation of high frequency of recombination donors from Tn*5* chromosomal mutants of *Pseudomonas putida* PPN and recalibration of the genetic map. *Genetics* 126:497–503
167. Sugino, Y., Hirota, Y. 1962. Conjugal fertility associated with resistance factor R in *Escherichia coli*. *J. Bacteriol.* 84:902–10
168. Suwanto, A., Kaplan, S. 1989. Physical and genetic mapping of the *Rhodobacter sphaeroides* 2.4.1 genome: genome size, fragment identification and gene localization. *J. Bacteriol.* 171:5840–49
169. Suwanto, A., Kaplan, S. 1989. Physical and genetic mapping of the *Rhodobacter sphaeroides* 2.4.1 genome: presence of two unique circular chromosomes. *J. Bacteriol.* 171:5850–59
170. Suwanto, A., Kaplan, S. 1992. A self transmissible narrow-host-range endogenous plasmid of *Rhodobacter sphaeroides* 2.4.1: physical structure, incompatibility determinants, origin of replication and transfer function. *J. Bacteriol.* 174:1124–34
171. Suwanto, A., Kaplan, S. 1992. Chromosome transfer in *Rhodobacter sphaeroides:* Hfr formation and genetic evidence for two unique circular chromosomes. *J. Bacteriol.* 174:1135-45
172. Tabata, S., Higashitani, A., Takanami, M., Akiyama, K., Kohara, Y., et al. 1989. Construction of an ordered cosmid collection of the *Escherichia coli* K12 W3110 chromosome. *J. Bacteriol.* 171:1214–18
173. Taylor, D. E., Eaton, M., Cheng, N., Salama, S. M. 1992. Construction of a *Helicobacter pylori* genome map and demonstration at the genome level. *J. Bacteriol.* 174:6800–6
174. Thomas, C. T., Smith, C. A. 1987. Incompatibility group P plasmids: genetics, evolution and use in genetic manipulation. *Annu. Rev. Microbiol.* 41:77–101
175. Tsuda, M., Iino, T. 1987. Genetic analysis of a transposon carrying toluene degrading genes on a TOL plasmid pWW0. *Mol. Gen. Genet.* 210:270–76
176. Tsygankov, Y. D., Kazakova, S. M., Serebrijski, 1990. Genetic mapping of the obligate methylotroph *Methylobacillus flagellatum:* characteristics of prime plasmids and mapping of the chromosome in time of entry units. *J. Bacteriol.* 172:2747-54
177. Tulloch, D. L., Finch, L. R., Hillier, A. J., Davidson, B. E. 1991. Physical map of the chromosome of *Lactococcus lactis* subsp. *lactis* DL11 and localization of six putative rDNA operons. *J. Bacteriol.* 173:2768–75
178. Tümmler, B., Römling, U., Ratna-

ningsih, E., Morgan, A. F., Krishnapillai, V., Holloway, B. W. 1992. A common system of nomenclature for the physical map of the chromosome of *Pseudomonas aeruginosa* PAO. In Pseudomonas *Molecular Biology and Biotechnology,* ed. E. Galli, S. Silver, B. Witholt, pp. 9–11. Washington, DC: Am. Soc. Microbiol. 443 pp.

179. Ventra, L., Weiss, A. S. 1989. Transposon mediated restriction mapping of the *Bacillus subtilis* chromosome. *Gene* 78:29–36
180. Verma, N. K., Brandt, J. M., Verma, D. J., Lindberg, A. A. 1991. Molecular characterization of the O-acetyl transferase gene of converting bacteriophage SF6 that adds group antigen 6 to *Shigella flexneri. Mol. Microbiol.* 5:71–75
181. Versalovic, J., Koeuth, T., McCabe, E. R. B., Lupski, J. R. 1991. Use of the polymerase chain reaction for physical mapping of *Escherichia coli* genes. *J. Bacteriol.* 13:5253–55
182. Waekens, F., Verdickt, K., Vanduffel, L., Vanderlyden, J., Van Gool, A., Mergeay, M. 1987. Intergeneric complementation of *Agrobacterium tumefasciens* chromosomal genes and its potential use for linkage mapping. *FEMS Microbiol. Lett.* 44:329-34
183. Weaver, K. E., Clewell, D. B. 1990. Regulation of the PAD1 sex pheromone response in *Enterococcus faecalis:* effects of host strain and *traA, traB* and *C* region mutants on expression of an *E* region pheromone-inducible *lacZ* fusion. *J. Bacteriol.* 172:2633-41
184. Wenzel, R., Pirkl, E., Herrmann, R. 1992. Construction of an *Eco*RI restriction map of *Mycoplasma pneumoniae* and localization of selected genes. *J. Bacteriol.* 174:7289–96
185. Wheelis, M. L., Stanier, R. Y. 1970. The genetic control of dissimilatory pathways in *Pseudomonas putida. Genetics* 66:245–66
186. Wilke, D., Schlegel, H. G. 1979. A defective generalized transducing bacteriophage in *Xanthobacter autotrophicus* GZ29. *J. Gen. Microbiol.* 115: 403–10
187. Yan, W., Taylor, D. E. 1991. Sizing and mapping of the genome of *Campylobacter coli* strain UA417R using pulsed-field gel electrophoresis. *Gene* 101:117–20
188. Yee, T. W., Smith, D. W. 1990. *Pseudomonas* chromosomal replication origins: a bacterial class distinct from *Escherichia coli*-type origins. *Proc. Natl. Acad. Sci. USA* 87:1278–82
189. Zeigler, D. R., Dean, D. H. 1990. Orientation of genes in the *Bacillus subtilis* chromosome. *Genetics* 125: 703–8
190. Zhang, C., Holloway, B. W. 1992. Physical and genetic mapping of the *catA* region of *Pseudomonas aeruginosa. J. Gen. Microbiol.* 138:1097–107

Annu. Rev. Microbiol. 1993. 47:685–713

GENETICS OF DIFFERENTIATION IN *STREPTOMYCES*

Keith F. Chater
John Innes Institute, Colney Lane, Norwich NR4 7UH, United Kingdom

KEY WORDS: development, sporulation, regulatory network, aerial mycelium, antibiotics

---

CONTENTS

0066-4227/93/1001-0685$02.00

ABSTRACT

The use of mutants and molecular genetics has begun to reveal how differentiation is brought about in multicellular, mycelial *Streptomyces* spp. Alternative pathways to aerial mycelium formation may be activated on different media. In some cases, extracellular signals are transmitted or exchanged. Extracellular proteins may act as morphogens in the erection of aerial hyphae. A rare codon, UUA, is apparently confined to mRNAs from a few genes important only for early stages of differentiation; it is absent from vegetative or developmentally late mRNAs. Sporulation of aerial hyphae involves at least four specific regulatory proteins, including a sigma factor and an unusual small protein. A complex interplay between the sporulation regulatory genes, rather than a simple linear dependence cascade, is emerging.

INTRODUCTION: DIFFERENTIATION AS A NORMAL ASPECT OF PROKARYOTES

When the general physiologies of the progeny resulting from cell division differ in a way that is more than a flexible short-term response to changing environmental conditions, the progeny may be said to have differentiated. Differentiation is often associated with morphological change, so in the past the morphological uniformity of many bacterial cell populations, especially in the exponential cultures usually used for metabolic and physiological studies, was taken as tacit evidence of their inability to differentiate. More recently, this misconception was corrected: rapid growth is only one of the differentiated states of simple bacteria such as *Escherichia coli,* and this state is probably not the one most often adopted in natural environments. Stationary-phase *E. coli* cells differ from rapidly growing ones in many positive attributes, and there may be more than one kind of stationary-phase *E. coli* cell (75). *E. coli* colonies even exhibit multicellular differentiation, inasmuch as they display heritable pattern formation (74). The detailed analysis of these phenomena promises to illuminate our understanding of some of the more readily recognizable examples of differentiation in prokaryotes. Nevertheless, morphologically more obvious differentiation, such as sporulation, does offer significant experimental benefits, particularly in recognizing or separating different cell types and in relating differentiation to function.

In *Streptomyces* spp., multicellular differentiation is indeed obvious, because colonies grow as a branching vegetative mycelium that forms dispersive units by sending spore-bearing branches into the air. The attraction of *Streptomyces* spp. as subjects for research in developmental biology is augmented by their most famous attribute—they produce an extraordinary range of antibiotics and other secondary metabolites, an ability termed physiological differentiation (17) because production typically occurs after the main period of rapid vegetative growth and assimilative metabolism (29). The regulation of secondary metabolism and its interplay with morphological differentiation in *Streptomyces* have been reviewed elsewhere (14, 21, 22). This chapter focuses on genes involved in the formation of aerial hyphae and spores.

AN OVERVIEW OF GROWTH AND DEVELOPMENT IN STREPTOMYCETES

Growth and a First Stationary Phase

For the purposes of this article, only an outline of the main features of vegetative growth of *Streptomyces* spp., which are reviewed in more detail elsewhere (14, 41), is necessary. Growth occurs mainly by cell-wall extension at hyphal tips; hyphal branching allows quasiexponential growth kinetics. Hyphal compartments (especially the tip cells) contain many copies of the genome. Vegetative septa are thus infrequent, and they are not usually the sites of cell separation, so the vegetative (or substrate) mycelium forms a coherent mat. As colonies grow, the parts farthest from the advancing edge may accumulate various kinds of storage materials such as glycogen, lipids, and polyphosphate (10, 58; K. A. Plaskitt & K. F. Chater, unpublished data), which are made in many bacteria when their growth is limited by one kind of nutrient and other nutrients are present in excess. These parts of the colony probably grow more slowly, either because of nutrient limitation or because conditions have become growth inhibitory. Secondary metabolism also takes place as colonies age, though the exact profile of the secondary metabolites produced also depends on such external parameters as the availability of potentially repressing nutrients like glucose, phosphate, or readily assimilable nitrogen (29). It is not known whether the cellular compartments that contain storage compounds also produce secondary metabolites.

A New Lease of Life: the Emergence and Growth of Aerial Hyphae

Under normal laboratory conditions, a white growth of aerial hyphae emerges from the surface of colonies after one or two days, at about the same time as storage compounds and secondary metabolites become detectable. Aerial growth follows a short period of reduced macromolecular synthesis (33) and seems to involve the reuse of material first assimilated into the substrate mycelium either as macromolecules such as DNA and proteins (63) or as storage compounds (10). Thus, as aerial hyphae grow, many cells in the substrate mycelium die (82). Aerial hyphae presumably also need both internal osmotic pressure sufficient to drive growth and a sense of direction. Osmotic potential may be generated through the solubilization of macromolecules such as glycogen (18, 19). Directionality may perhaps be supplied by the condensation of surface molecules peculiar to aerial hyphae only on surfaces that are not exposed to water (84). Candidates for this role are the so-called Saps (spore-associated proteins) (34). The Saps may also fulfill other functions, including the breaking of surface tension at the air-water interface (84), the prevention of aerial hyphal desiccation, the maintenance of an aqueous compartment outside the cell membrane, or the provision of a hydrophobic surface to spores (analogous to the action of fungal hydrophobins) (20).

A Second Stationary Phase—Sporulation

Eventually, extension growth of aerial hyphae stops and, in contrast to vegetative growth, regularly spaced and synchronously formed crosswalls subdivide the hyphal tips into many unigenomic compartments, each destined to become a spore (reviewed in 58). The sporulation septa are morphologically different from the crosswalls in vegetative hyphae, typically consisting of two membrane layers separated by a double layer of cell-wall material—an arrangement that permits the eventual separation of adjacent spores (58). These structural differences suggest that some genes should be specifically involved in sporulation septation, perhaps in concert with genes involved in vegetative crosswall formation. Indeed, mutants lacking sporulation septa but with normal vegetative septa have been characterized (15, 62). Spore chains may consist of many tens of spores, and a single colony grown for 4–6 days on minimal agar medium may yield 10^7 spores. It is not known what defines the boundary between the sporulating hyphal tip and the nonsporulating stem—perhaps a single crosswall forms first, defining the compartment that will then become subdivided into spore compartments. New glycogen deposits form in the sporulating compartment (10), glycogen being otherwise absent from aerial hyphae. In some species, glycogen is detectable before sporulation septa are fully formed (10), whereas in *Streptomyces coelicolor* A3(2), the main organism discussed in this review, the septa are completed first (K. A. Plaskitt & K. F. Chater, unpublished data). This observation may reflect the differing morphology, especially thickness, of the ingrowing crosswalls in different species (58), thicker crosswalls perhaps taking longer to form. Probably no more than a few minutes are needed for sporulation septation in *S. coelicolor* (83; K. A. Plaskitt & K. F. Chater, unpublished data).

The glycogen that forms in young spore chains may act as a temporary sink for carbon metabolites, facilitating their transport from the lower part of the aerial hyphae by steepening the concentration gradient. As in lower parts of the colony, glycogen may also influence osmotic pressure; reduced osmotic pressure associated with the conversion of glucose-1-phosphate into glycogen could perhaps contribute to the growth cessation associated with sporulation. Later, as glycogen is degraded again, the associated osmotic-pressure increase might help to bring about the shape change by which ellipsoidal spores are generated from cylindrical spore compartments (19). Other storage deposits also appear in the sporulating parts of aerial hyphae. In particular, the trehalose contribution to dry weight increases from 2% in vegetative hyphae to 5% in aerial hyphae and 12% in spores (10), and spherical electron-transparent bodies thought to contain lipids, and electron-dense granules thought to contain polyphosphate, are often seen with electron microscopy in sporulating parts of hyphae (58; K. A. Plaskit & K. F. Chater, unpublished data). During rounding up of spores, the spore wall thickens, and in most species, a pigment is deposited in the wall. At the same time, glycogen deposits disappear from the cytoplasm (10), though other storage materials may be retained in the mature spores.

THE IMPORTANCE OF DEVELOPMENTAL MUTANTS

It is difficult to study the complex developmental biology of *Streptomyces* colonies using only biochemical techniques, because of the heterogeneity of cell types and the asynchrony of the developmental processes. The use of genetics allows important genes, and hence processes, to be identified; their genetic regulation to be analyzed; and developmentally regulated—as opposed to developmental regulatory—genes to be isolated. Probably the most important starting point is the isolation of mutants with developmental abnormalities. Mutants relevant to this review are of two general types: those lacking an aerial mycelium (*bld* mutants, most of which turn out to be conditional—see below) and those that produce an aerial mycelium but fail to develop the pigmentation associated with mature spores (*whi* mutants). The characterization of such mutants in *Streptomyces* spp. has been done mostly in *S. coelicolor* A3(2), because of the early choice of this strain for genetic studies (42). These studies have recently culminated in a combined genetic and physical map of the chromosome (50). Figure 1 shows the locations of developmentally relevant genes on the map. Important and complementary developmental work has also been done with mutants of *Streptomyces griseus* because of two special attributes—its ability to sporulate in submerged culture and its use of an extracellular autoregulator, A-factor, as part of the mechanism activating secondary metabolism and sporulation (66).

Mutants that Are Pleiotropically Defective in Secondary Metabolism and Aerial Mycelium Formation

From the outset we should emphasize that *bld* mutants seldom simply and unconditionally lack an aerial mycelium. Nearly all of them can be made to produce an aerial mycelium and spores either by changing the medium

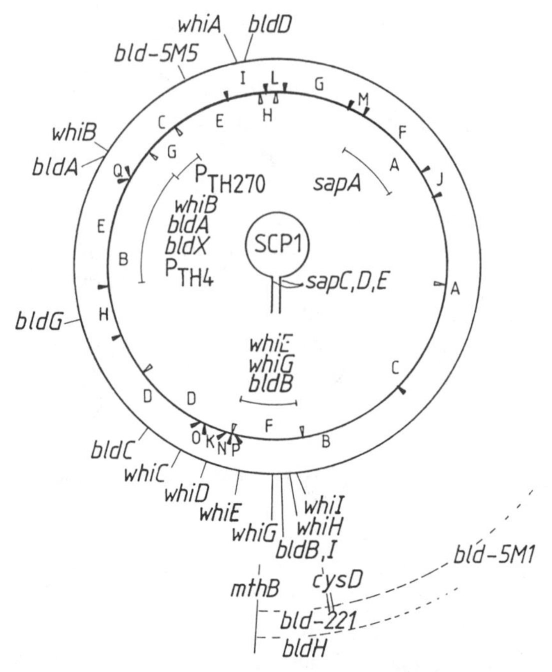

Figure 1 Locations of developmental genes on the combined genetic and physical map of the *Streptomyces coelicolor* A3(2) chromosome. (*Outer circle*) Genetic linkage map of developmental genes. (*Inner circle*) Restriction map for *draI* (*open arrowheads*) and *aseI* (*filled arrowheads*), with the locations of physically mapped developmental genes indicated inside the circle. Also included is the linear plasmid SCP1, shown with its terminal inverted repeats aligned. The map is based on information in Ref. 50.

composition (especially by replacing glucose in minimal medium with mannitol) or by growing them near the wild-type parent.

Only four classes of *S. coelicolor bld* mutants (*bldA,B,G,* and *H*) are represented by more than one mutant allele. However, the list of other *bld* mutants of this strain that genetically and/or phenotypically differ from these four classes is always growing. The following groups can now be added to a fairly comprehensive recent tabulation of 12 classes (41): *bld-5M1* and *bld-5M5*, the most studied of five *bld* mutations induced by Tn*4563* mutagenesis (73); 17 *bld* mutations identified after Tn*5353* mutagenesis, many of which appear to involve transcriptional fusions of the *bld* gene to the *luxAB* cassette in the transposon (76); and *bld-261* (J. Willey & R. Losick, personal communication) (see below). A proper assessment of any of these mutants will become possible only when more representatives are available, or when the relevant genes have been cloned: the relationships of many of the new mutants to each other or to previously identified mutants are still not clear, though both their phenotypes and Southern analyses indicate the diversity of the transposition-induced mutants (76).

Early observations showed that *bld* mutants of *S. coelicolor* were nearly all defective in secondary metabolism (64; see 41 for a review) though not always completely so: undecylprodigiosin, one of the four known antibiotics of this strain [actinorhodin, methylenomycin, and calcium-dependent antibiotic are the others (43)] is abundant in *bldE* (40) and *bldF* (70) mutants and is even produced in *bldA* mutants grown on low phosphate medium (35); blue pigment—presumably actinorhodin—is seen when *bldB* mutants are incubated for a long time; secondary metabolism is restored to *bldH* mutants when mannitol replaces glucose as carbon source in minimal medium (13); and methylenomycin and a low level of pigment (actinorhodin and/or undecylprodigiosin) are produced by *bldC* mutants (64; K. F. Chater, unpublished data). All this suggests that the regulatory mechanisms of individual biosynthetic pathways for secondary metabolism differ, despite elements common to their regulation, in the ways in which they integrate and respond to sensory input from the plethora of physiological situations that can be encountered (21). More remarkable, however, is the virtual absence of descriptions of *S. coelicolor bld* mutants that are entirely unaffected in their secondary metabolism. This may mean that all those genes involved in aerial mycelium but not secondary-metabolite formation are also essential for growth and so cannot readily be identified by mutations; that such genes are functionally redundant, so that mutations in more than one gene would be needed for a mutant phenotype to be apparent; or even that there is positive-feedback regulation of secondary metabolism by some aspect of aerial mycelium formation. Whatever the reason, clearly all sufficiently described *bld* mutants of *S. coelicolor* are pleiotropic.

This pleiotropy seems in many cases to extend to effects on the aerial mycelium– and spore-associated Sap proteins, which are absent from most *bld* mutants (see below), and on storage-compound metabolism. Electron microscopy (K. A. Plaskitt & K. F. Chater, unpublished) has shown that *bldA* mutants growing on glucose-containing minimal medium—conditions under which they form no aerial mycelium—accumulate no glycogen but contain a superabundance of lipid bodies: when mannitol replaces glucose, lipid bodies are greatly reduced in abundance, glycogen remains absent from the substrate mycelium and aerial hyphae, and spores are formed. Possibly, glucose represses the utilization of the lipid for aerial growth, whereas mannitol is permissive. Despite its absence from the substrate mycelium, glycogen is seen in developing spore chains of *bldA* mutants growing on mannitol. It seems that the intact *bldA* gene is needed for glycogen deposition in one location, but not in another.

Four of the *S. coelicolor bld* genes (*bldA,B,D,* and *G*) have been cloned by complementing mutants, in each case using φC31 bacteriophage-based vectors (36, 71; R. Passsantino, B. K. Leskiw & K. F. Chater, unpublished data). But only in the case of *bldA* do we have an idea of what the gene product does: it is the only tRNA that efficiently recognizes the rare leucine codon UUA (53, 55). The entire *bldA* gene can be deleted (56), strongly indicating that no genes essential for vegetative growth contain TTA codons (54). In fact, TTA codons have been found almost exclusively in pathway-specific regulatory and resistance genes associated with the biosynthesis of secondary metabolites, and it is certainly tempting to consider that the level of leucyl-tRNA$^{Leu}_{TTA}$ may, at least in some circumstances, be an important regulator (54). A more sceptical view would suggest that TTA is merely a rare, probably inefficiently translated codon in the generally GC-rich *Streptomyces* genome and that it has been eliminated from genes whose expression needs to be efficient (i.e. many vegetative genes) (32, 54). Because phenotypically comparable *bldA* mutants exist not only in the closely related *S. coelicolor* A3(2) and *Streptomyces lividans* 66, but also in *S. griseus* NRRL B-2682 (2, 61), the range of genes that have retained TTA codons is remarkably conserved. This conservation suggests adaptive benefits. Some recent experiments have addressed the potential regulatory role of *bldA* by analyzing the timing of transcription and processing of *bldA* RNA, together with the timing of transcription and translation of genes containing TTA codons, in relation to the life cycle (32, 56). Points of agreement and disagreement between the results are dealt with in later sections.

The occurrence of *bldA* mutants in *S. griseus* has made it possible to investigate how the mutants differ physiologically from the wild-type when subjected to the nutritional downshift used to set off sporulation of this species in submerged culture. Interestingly, spore-like bodies are formed much more rapidly in the mutant than in the wild-type under these conditions (61), in agreement with earlier observations that aberrant sporulation septation took place in hyphae on the bald surface of *bldA* colonies growing on agar (24).

The study of *S. griseus bld* mutants also led to the identification of a possible target for *bldA* action. Certain mutants phenotypically resembling *bldA* mutants are not complemented by *bldA* DNA (61); some of those can be complemented by DNA including a gene called *orf1590* that contains a TTA codon (2, 3). An *S. coelicolor* homologue, which would specify a protein with 92% identity to the ORF1590 product, but lacking a 56–amino acid segment, also contains the TTA codon (61). The phenotype of mutations in the *S. coelicolor* gene has not yet been reported. Mutants of another group of *S. griseus bldA*-like mutants are complemented by neither *bldA* nor *orf1590* (2, 61).

Growth near the wild-type or any of the three groups of *bldA*-like mutants (collectively called class III) can cause another phenotypic class of *bld* mutants (class II) of *S. griseus* to sporulate, providing evidence for a diffusible extracellular factor (2, 61). It is not clear how this relates to other diffusible factors described below, such as SapB or the signal cascade leading to its production, or A-factor, which is secreted by other strains of *S. griseus*—notably those that produce streptomycin—and in them is required for normal sporulation and secondary metabolism.

Two other classes of *S. griseus bld* mutants are phenotypically unlike known *S. coelicolor* mutants: those unable to sporulate or produce secondary metabolites in any conditions tested (class I) and those affected only in morphology (class IV), a defect phenotypically suppressible (as is that of class III mutants) by growth on minimal medium containing glucose and an ammonium salt (2, 61).

A-Factor and Its Role in Differentiation in Some Strains of Streptomyces griseus

Reference has already been made to factor-stimulated differentiation in *Streptomyces* spp. The most celebrated example concerns the γ-butyrolactone compound A-factor, which was discovered through the study of mutants (49). This factor and its action are extensively reviewed elsewhere (47) and are discussed only briefly here. Streptomycin-producing strains of *S. griseus* generate nonproducing *bld* mutant colonies at a very high frequency, and many of these recover the wild-type phenotype when grown near wild-type colonies. The extracellular compound responsible, A-factor, probably diffuses freely into cells, and at $\sim 10^{-9}$ M binds to a cytoplasmic binding protein and inactivates the protein's ability to repress sporulation and secondary metabolism (65). Repression of sporulation possibly occurs via intermediate components in some kind of regulatory cascade, as indicated by the isolation of cloned genes that allow A-factor deficient mutants to sporulate (47). Many *Streptomyces* spp. produce compounds related to A-factor (30), while not necessarily using them in the same way as in *S. griseus*. For example, certain mutants (*afsA*) of *S. coelicolor* that cannot produce A-factor-like compounds do not show any obvious morphological or physiological deficiencies (35a).

Some A-factor-like compounds do not interact with the A-factor-binding protein of *S. griseus*. Notably, *Streptomyces virginiae* produces related compounds that have no effect on *S. griseus* but are needed for production of virginiamycin antibiotics in *S. virginiae*. The binding protein for these *S. virginiae* butanolides has been identified biochemically and genetically, and remarkably, it is homologous with the *E. coli* NusG protein. Its genetic determinant in *S. virginiae* is part of a ribosomal-protein operon similar to that containing *nusG* in *E. coli* (69). In *E. coli,* NusG is believed to form part of the RNA polymerase quaternary complex during transcript elongation and plays an ill-defined role in modulating transcription termination (57).

In summary, compounds like A-factor are widespread among *Streptomyces* spp. and other bacteria (including many gram-negative species) (4), and their biological activities may be very specific. Although a clearcut role for them in secondary metabolism and/or differentiation has been demonstrated in published accounts in only a few cases, other unpublished examples are known, so it seems likely that similar situations will eventually prove to be widespread.

SapB, an Unusual Small Protein, Is Implicated in Aerial Mycelium Formation and Is Absent from Many bld *Mutants*

Sap proteins associated with the surface of aerial hyphae and spores were referred to above. Attempts at partial sequencing were successful with SapA,C,D, and E (see below), but the tiny SapB protein (estimated to be 3 kDa by electrophoresis) proved refractory to this approach. The theoretical mass of the protein, 1982 Daltons, which was calculated from its overall composition (Ala, 1; Arg, 2; Asp, 3; Gly, 2; Ile, 1; Leu, 5; Ser, 1; Thr, 3), is smaller by 99 Daltons than the mass of protein actually observed using mass spectrometry, suggesting that SapB is modified (84). Indeed, a positive reaction of SapB with Schiff's reagent revealed the presence of vicinal hydroxyl groups normally associated with the sugar residues of glycoproteins (84), though glycosylation would add more than 99 Daltons to the molecular mass. Circumstantial evidence now points to the nonribosomal synthesis of SapB, because it is made by *S. coelicolor* during exposure to chloramphenicol (J. Willey & R. Losick, personal communication).

Although these complications have prevented the easy cloning of the genetic determinant(s) of SapB, intriguing progress has been made with the help of anti-SapB antibodies (84). The use of colony blots, made by pressing an Immobilon membrane against an agar surface on which colonies are growing and then treating it with antiserum as if it were a Western blot, revealed that SapB diffuses for several millimeters around colonies. Two mutants that produce no SapB (when grown on rich medium—see below) were identified using this technique (84) and both lacked an aerial mycelium. In addition, previously identified *bldA,B,C,D,G,H,* and *I* mutants as well as another new mutant, *bld-261* (J. Willey & R. Losick, personal communication), were all found to be SapB$^-$. Moreover, colonies of these *bld* mutants produced aerial mycelium when grown very close to a strong producer of SapB on a suitable rich medium, and a transient aerial mycelium could also be induced by the addition of purified SapB (84; J. Willey & R. Losick, personal communication). Thus SapB may be both a diffusible signal molecule for aerial mycelium development and a surface component of aerial hyphae. SapB is essential for differentiation only on rich medium—production of SapB even by morphologically wild-type cultures is undetectable on minimal medium, even when mannitol is the carbon source (84), i.e. the conditions in which nearly all the SapB$^-$ *bld* mutants form sporulating aerial mycelium. Clearly, all workers in the field of aerial mycelium development should note the culture media used in each other's experiments—nearly all the work done on the initial genetics and phenotypes of *whi* and *bld* mutants has used minimal medium, whereas the work on Saps has almost wholly used the rich medium R2YE.

A further surprise in the fascinating SapB story is the discovery that in some pairwise combinations, different *bld* mutants grown near each other on rich medium exhibit a kind of extracellular complementation, such that one of the pair produces both aerial mycelium and SapB. Four extracellular complementation groups have been found (Figure 2) and interpreted in terms of a cascade of extracellular signals reminiscent of those exchanged during fruiting body development in *Myxococcus xanthus* (48). On this hypothesis, the *bld* mutants might conceivably be primarily defective in elements of this cascade (84; J. Willey & R. Losick, personal communication).

These discoveries and models are unexpected in relation to the following other aspects of *bld* mutant phenotypes and genetics. (*a*) The *bld* mutants are defective in secondary metabolism and these defects are not overcome by extracellular complementation. This might imply that the signaling and morphological defects are an indirect consequence of defects that principally affect secondary metabolism. (*b*) In one case, *bldA*, the gene function is known—it specifies a tRNA (see above). Presumably, at least one gene needed in signal B production contains a TTA codon. (*c*) Because SapB is probably made nonribosomally, its synthesis would be expected to require a substantial enzymatic apparatus, because the nonribosomal synthesis of peptide antibiot-

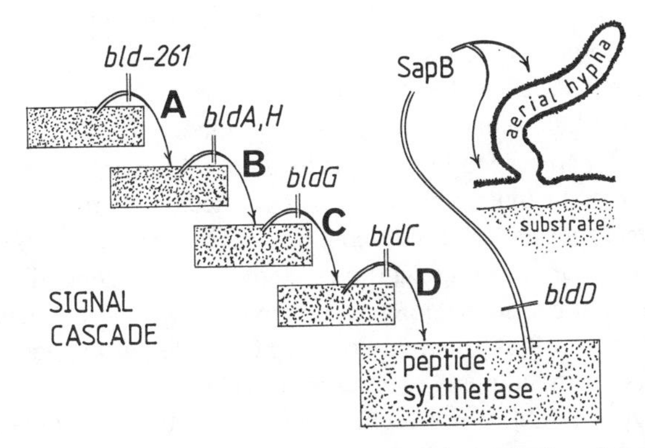

Figure 2 Proposed hierarchical signal cascade leading to the production of SapB. The scheme was deduced by J. Willey & R. Losick (personal communication) from their unpublished results.

ics involves some of the largest known proteins, typically specified by more than 10 kb of genetic information (60). These genes should present a target for mutagenesis that is about 100-fold larger than that provided by *bldA*, and such mutants should therefore be abundant. They should also extracellularly complement all the mutants that lack SapB because of deficiencies in the signaling pathway. As yet, a *bld* mutant with this phenotype, the *bldD* mutant, is represented only once among the many *S. coelicolor bld* mutants so far studied in sufficient detail (J. Willey & R. Losick, personal communication). Genetic studies with genes encoding other Saps are reviewed in a later section.

Mutants Unable to Produce Spores

Formation of the gray spore pigment of *S. coelicolor* depends on genes that determine earlier stages of sporulation. This is shown by the morphological phenotypes of mutants whose colonies stay white upon prolonged incubation (45). Linkage analysis of 52 such mutants identified nine *whi* loci (Figure 1) (15, 24). Among these mutants only five formed abundant spores. These were thin-walled in the single *whiD* mutant, rod-shaped in two mutants (these are termed *whiF99* and *whi79* but may be unusual *whiG* mutants), and apparently morphologically normal in two *whiE* mutants. Three other mutants (*whiC193*, *whi-53*, and *whi-77*) that were genetically distinct from other *whi* mutants produced spores at greatly reduced frequency. The remaining 44 mutants were all severely defective in sporulation and defined the loci *whiA* (6 alleles), *whiB* (2 alleles), *whiG* (14 alleles), *whiH* (7 alleles), and *whiI* (15 alleles) (15). The morphological phenotypes of constructed double mutants for representative alleles of these five loci suggested the following simple pattern of epistatic interactions: *whiG* → *whiH* → *whiA,B* → *whiI* (16). This pattern was presumed to follow the normal sequence of morphological events during sporulation of the wild-type, in which initially straight aerial hyphae (the *whiG* phenotype) became curled (the *whiH* phenotype), then more curled (the *whiA* and *whiB* phenotypes), before sporulation septa formed (in *whiI* mutants, these are unusually widely spaced in tightly curled aerial hyphae) (16). However, results discussed below have now shown that the epistatic sequence is not a simple result of a corresponding linear regulatory cascade of gene expression and moreover that the mutant phenotypes do not accurately reflect intermediate stages in normal sporulation. Phase-contrast microscopy and scanning electron microscopy show that young wild-type aerial hyphae generally grow in a somewhat curled form, so that long straight hyphae like those of *whiG* mutants are not typical. The loose or tight coils typical of aerial hyphae of *whiH,A*, and *B* mutants, or the almost knotted aerial hyphae of *whiI* mutants, likewise do not closely resemble the irregular twisting of the wild-type structures, suggesting that the relationship between gene function and morphology is somewhat complex.

Several of the *whi* genes have been studied in some detail. The nature of their products and details of their transcription and dependencies are given in later sections.

MOLECULAR GENETICS OF GENES INVOLVED IN AERIAL MYCELIUM FORMATION

Transcription of bldA *and Processing of the Transcript to a Mature tRNA*

The primary *bldA* RNA appears to be transcribed from a single promoter with a sequence at −10 resembling that of the major class of bacterial promoters (56). Promoter-probing and S1 nuclease protection experiments show that the promoter is active even in very young cultures in minimal or rich liquid medium or during surface growth (32, 56). In older cultures, activity of this promoter becomes difficult to detect in S1 experiments, but a form of *bldA* RNA that has been processed to give the mature 5′ end remains abundant (32, 56). It thus seems that the relative rates of primary transcription, processing, and degradation change with culture age. A marked increase in the abundance of the 5′-processed form of the *bldA* transcript was observed late in growth in liquid culture, or apparently coinciding with early aerial growth on surface culture (56), supporting early dot-blot results (53) [though independent experiments showed only a slight indication of such an increase (32)]. This contrasted with observations made with a different tRNA (for a frequent lysine codon). This tRNA was equally abundant in young or old cultures (56). Thus, *bldA* has atypical regulation compared with the general pattern of stable RNA accumulation observed in *E. coli,* which is broadly positively—rather than inversely—correlated with growth rate. Leskiw et al (56) also reported evidence (from promoter-probing and in vitro transcription) that an antisense transcript originates from a promoter within the tRNA-encoding sequence of *bldA*. This RNA might conceivably influence processing or stability of the primary *bldA* transcript in vivo. (It could also provide a technical obstacle in S1 studies, by competing with the probe for in vitro hybridization with *bldA* sense RNA.) The complexity of *bldA* transcription and processing will make further detailed investigations necessary.

Is Translation of UUA Codons Limited in Rapidly Growing Cultures?

Using a prototrophic strain of *S. coelicolor* A3(2) and cultures grown in liquid minimal medium containing Casamino acids and polyethylene glycol (after a transfer of germinated spores from a rich germination medium), Gramajo et al (32) found little or no difference in the ratio of a UUA-containing mRNA to its encoded protein (as assessed by western blotting) at different stages of growth. This was also true when the reporter system was modified by site-directed mutagenesis to eliminate the UUA codon or insert two more in tandem. The reporter system was especially relevant to the potential regulatory role of *bldA*, because it consisted of the TTA-containing 5′ end of *actII*-ORF4, the pathway-specific regulatory gene for biosynthesis of the antibiotic actinorhodin (31), fused in frame to a gene, *ermE*, that encodes a ribosomal RNA methylase for which antiserum was available. In the same sets of samples, the level of mature *bldA* tRNA increased only slightly during culture (32).

On the other hand, Leskiw et al (56) used a reporter gene, *ampC* from *E. coli*, containing seven TTA codons. Their host strain differed from the one used by Gramajo et al (32) in several ways (auxotrophic and resistance mutations, and changes in phage sensitivity and the profile of transposable elements), and their experiments were done with cultures grown either in liquid yeast extract–malt extract medium containing 34% sucrose or on the surface of the rich medium R2YE. In these experiments, the ratio of *ampC*-encoded β-lactamase activity to its mRNA increased significantly in older cultures, an increase potentially attributable to the increased *bldA* tRNA in the same sets of samples. Overall, these results suggest that the ability to translate UUA codons is a major regulatory influence in one set of conditions, but not in another.

Unexpected Complexity in the Expression of the S. griseus bld *Gene* orf1590

Figure 3 summarizes some of the major features of *orf1590*. Evidence indicates the existence of two *orf1590* promoters, one within the coding region for the N terminus of a presumptive protein (3). Use of this downstream promoter would allow translation from a plausible second translation start codon. Thus, *orf1590* appears to specify two related gene products. The N-terminus unique to the larger protein contains a potential DNA-binding motif. It has been proposed that a homodimer of the larger protein might activate sporulation, and heterodimers between the large and small proteins might be inactive, providing a posttranslational regulatory device superimposed on *bldA*-dependent translational regulation via the TTA codon present in the coding region for both proteins (3, 61). In formulating this model, the investigators noted that the amount of the smaller transcript relative to the larger one decreased under conditions used to initiate sporulation. Those sequence features thought to be important for *orf1590* expression are all conserved in the *S. coelicolor* homologue (61). McCue et al (61) noted a few other cases of nested promoters giving rise to C-terminally coextensive proteins with different N termini.

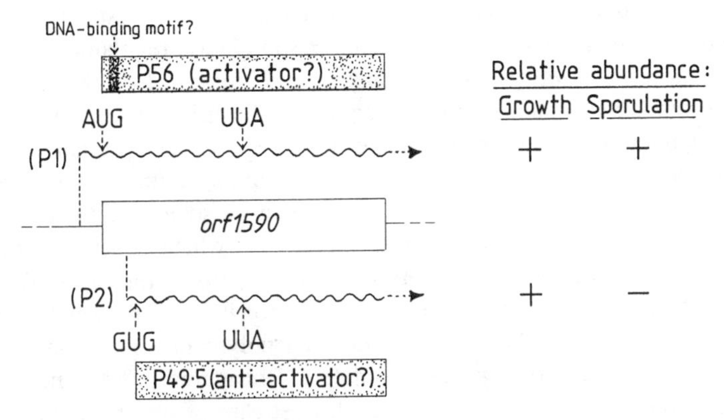

Figure 3 Features of *orf1590* of *Streptomyces griseus*. The diagram shows two alternative mRNA species (*wavy arrows*) that are transcribed from different promoters (P1 and P2). The probable translational start codons (AUG and GUG) for translation into the respective proteins P56 and P49.5 are indicated. Both transcripts include a potentially *bldA*-dependent UUA codon. Both mRNA species are detectable during vegetative growth, but the shorter is absent during sporulation. Information from Ref. 61.

Genes Encoding Spore-Associated Proteins SapA,C,D, and E

The SapA,C,D, and E proteins that can be washed from *S. coelicolor* spores with alkaline SDS-dithiothreitol solution are also produced by various *whi* mutants and are therefore probably present on presporulation aerial hyphae (34; J. McCormick, R. Santamaria & R. Losick, personal communication). Partial sequencing of these four proteins has led to the cloning of the structural genes (34; J. McCormick, R. Santamaria & R. Losick, unpublished data cited in 84). Remarkably, the SapC,D, and E determinants are located near each other on the long terminal inverted repeats of a large linear plasmid, SCP1, present in *S. coelicolor* (Figure 1) (J. McCormick, personal communication). Because SCP1 can be lost without affecting the morphological phenotype, SapC,D, and E are not essential for normal morphogenesis. However, $SCP1^+$ strains inhibit aerial mycelium development of $SCP1^-$ strains, a phenomenon involving the SCP1-specified antibiotic methylenomycin (81; K. F. Chater, unpublished data), so SapC,D, and E may well be associated with self-protection against this effect.

Attempts to disrupt SapA production have not been reported, and its role remains unknown. S1 nuclease studies and transcriptional fusions to the *luxAB* reporter system (34) showed that the *sapA* gene is transcribed only in older cultures, including liquid-grown cultures in which morphological differentiation is not detectable. Thus, SapA is probably first expressed at or before the earliest stages of aerial mycelium formation. Transcription from the *sapA* promoter is not eliminated by mutations in those *bld* genes tested (*bldA,B* and *C*), nor by mutations in the *whiB,G,* or *H* sporulation-regulatory genes (see below) or in the *whiE* genes (encoding spore pigment; also see below). The *sapA* promoter is extraordinary among prokaryotic promoters in that it appears to overlap into the transcribed region of *sapA*. Sequences upstream of −8 are not needed for correctly positioned transcription initiation, nor for its developmental regulation (H. Im & A. Schauer, personal communication). A similar situation was observed for promoter P1a, which is active during the stationary phase of the actinomycete *Micromonospora echinospora* (5). It may be significant that the transcription start sites for P1a and the *sapA* promoter fall in the sequence GTGGC. A novel RNA polymerase form may perhaps be involved in transcribing these promoters. [Intriguingly, RNA polymerase III of eukaryotes depends on features downstream of the transcription start site (72).]

MOLECULAR GENETICS AND REGULATORY INTERDEPENDENCE OF GENES INVOLVED IN THE SPORULATION OF AERIAL HYPHAE

The Sigma Factor-Like whiG *Gene Product and Two Potential Target Promoters for Its Action*

The amino acid sequence of the deduced product of *whiG* (23) shows nearly 40% identity with each of three sigma factors [σ^D of *Bacillus subtilis* (39) and σ^F of *Salmonella typhimurium* (67) and *Pseudomonas aeruginosa* (78)] that direct RNA polymerase to transcribe many of the genes in the later part of the transcriptional cascades leading to motility and chemotaxis. The four proteins are identical at 57 of ~278 residues, most markedly in the conserved regions 2.4 and 4.2 that are known from studies of other sigma factors to interact with the −10 and −35 regions, respectively, of cognate promoters (38). It is therefore not surprising that the promoters recognized by the σ^D and σ^F forms of RNA polymerase resemble each other, with characteristic consensus sequences at −10 and, to a lesser extent, at −35 (37). Two further *whiG* homologues have been cloned and sequenced, from *Streptomyces aureofaciens* (J. Kormanec, personal communication) and from a representative of a genus closely related to *Streptomyces, Streptoverticillum griseocarneum* (J. Soliveri & K. F. Chater, unpublished). Both gene products are very similar to the *whiG* gene product (92 and 90% identity, respectively). The *S. griseocarneum* gene, when introduced into *S. coelicolor* at high copy number, resembled *whiG* itself in eliciting premature and ectopic sporulation (see below). It also caused a *whiG* mutant to sporulate. Close *whiG* homologues were present in all *Streptomyces* spp. tested (8), but absent from representatives of some other genera of actinomycetes, at least as detected by exposure of Southern blots to a probe comprising the highly conserved region 2 (J. Soliveri, C. Granozzi & K. F. Chater, unpublished data). The genera containing an apparent *whiG* homologue were *Streptoverticillum, Streptosporangium, Saccharopolyspora,* and *Amycolatopsis,* all of which produce a sporulating aerial mycelium. No homologues were detected in representatives of genera that do not form an aerial mycelium: *Corynebacterium, Mycobacterium, Nocardia,* or *Actinoplanes*. The last-named case was particularly surprising because the spores of *Actinoplanes* spp. are motile, so one might have expected that some of the motility-associated genes should be transcribed by a σ^{WhiG} form of RNA polymerase holoenzyme.

The introduction into *S. coelicolor* of multiple copies of a small fragment of *B. subtilis* DNA that included a σ^{D}-dependent promoter gave a white colony phenotype and reduced sporulation (23). Thus the foreign promoter was probably sequestering the σ^{WhiG} form of RNA polymerase, an observation that was exploited in the cloning of two apparently σ^{WhiG}-dependent promoters of *S. coelicolor,* both of which conferred a white-colony phenotype at high copy number and caused expression of a reporter gene, *xylE,* present in the vector (80). S1 mapping verified that each promoter-containing fragment contained a transcription start point and that the chromosomal copies of both promoters were active. Transcripts detectable during growth on solid medium appeared at the same time as the aerial mycelium, and neither transcript was present in a *whiG* mutant. The two promoters (P_{TH4} and P_{TH270}) both contained appropriately placed regions of similarity (6/8 bases) to the consensus sequences for the −10 regions of promoters recognized by σ^{D}-containing RNA polymerase in *B. subtilis* (or by σ^{F}-containing enzyme in gram-negative bacteria), and weak (2/4) similarity to the consensus −35 regions of these promoters (80) [the latter region is absent from some apparently σ^{D}- or σ^{F}-dependent promoters (37)] (Figure 4). Taken together, these data strongly suggest that the two newly identified promoters are direct targets for σ^{WhiG}-containing RNA polymerase, though rigorous proof will require in vitro transcription and evidence from compensatory mutations in the promoters and *whiG*.

The two promoters are from different parts of the *S. coelicolor* chromosome (Figure 1) and are not from known *whi* genes (50). The sequences surrounding the promoters seem to come from within transcription units containing more than one gene, because each promoter is not only upstream of a potential translation initiation signal followed by apparently protein-coding DNA but is within another potentially protein-coding sequence whose stop codon overlaps the downstream coding sequence in a way suggesting translational coupling (80) (Figure 4). Probably, therefore, the 3′ parts of these DNA fragments can be transcribed from more than one promoter, allowing the

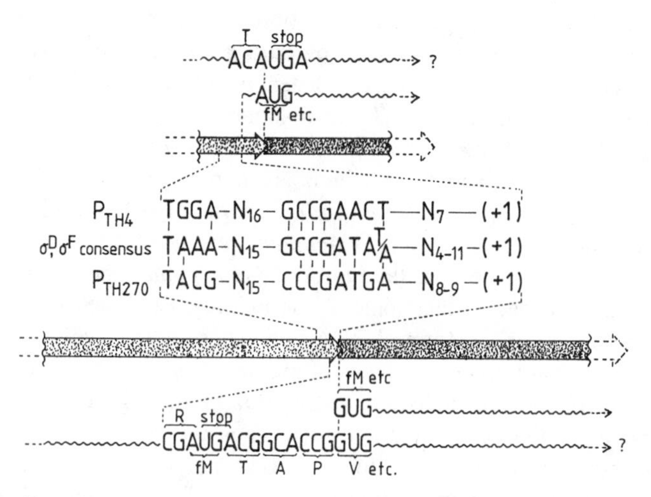

Figure 4 Two promoters from *S. coelicolor* that depend on *whiG*. Consensus sequences for σ^D-dependent promoters of *Bacillus subtilis* and σ^F-dependent promoters of *Escherichia coli* and *Salmonella typhimurium* are combined from Refs. 1, 37, 52. The data are from Ref. 80.

alternatives of coordinate or uncoupled expression of the upstream and downstream genes. The small amounts of DNA sequence available so far (corresponding to *Sau*3AI fragments of 233 and 759 bp) do not resemble any sequences present in data bases.

The apparent success of this strategy for the isolation of *whiG*-dependent promoters implies that σ^{WhiG} may be present in limiting amounts. This was also suggested by the dramatic effect of introducing extra copies of the wild-type *whiG* sequence, which caused early and spatially inappropriate (ectopic) sporulation—an effect that was absent when a small in-frame deletion was introduced into the cloned gene (23). Thus, σ^{WhiG} appears to be a crucial determinant of the spatial localization of sporulation. In view of the importance of σ^{WhiG} concentration, investigations of whether there is an anti-σ^{WhiG} protein similar to the anti-σ^F protein recently described in *S. typhimurium* are needed (68).

The whiB *Gene Encodes a Small, Highly Charged Protein and Has Apparent Homologues Even in Nonsporulating Actinomycetes*

Sequencing of a DNA fragment that restored sporulation to a *whiB* mutant revealed a single small open reading frame encoding a WhiB polypeptide of 87 amino acids (26). This protein contains a high proportion (35%) of charged residues, with a net excess of acidic over basic residues giving a predicted P_i of 4.5. Most of the basic residues are near the C terminus. A probe consisting of little more than the *whiB* open reading frame detected a hybridizing band in DNA from all *Streptomyces* spp. tested and from representatives of 10 different genera of *Actinomycetales,* including several that appeared to lack a *whiG* homologue (J. Soliveri & K. F. Chater, unpublished data). The deduced WhiB proteins of *S. coelicolor* and its close relative *S. lividans* are identical (26) and differ from the WhiB protein of *Streptoverticillum griseocarneum* by only three highly conservative substitutions (L5F, D11E, and D12E) (J. Soliveri & K. F. Chater, unpublished data). The cloned *S. griseocarneum* gene restores sporulation—albeit somewhat inefficiently—to *whiB* mutants of *S. coelicolor* (J. Soliveri & K. F. Chater, unpublished data). The *Streptomyces aureofaciens whiB* gene encodes a product differing from that of *S. griseocarneum* only at one position (A61G) (J. Kormanec, personal communication). Unlike the situation with *whiG,* high copy numbers of *whiB* do not appear to cause ectopic sporulation (26).

Only two *whiB* mutant alleles were identified in the original genetic analysis of 52 *whi* mutants (15). One of these, *whiB218,* is a deletion of at least 3.5 kb including *whiB* and a large part of an unidentified gene downstream (but well separated) from it; the other, *whiB70,* contains a base transition resulting in the replacement of a leucine with a rigid proline residue at a predicted flexible turn region next to the potentially α-helical C terminus (26).

The amino acid sequences of these proteins do not closely resemble those of any characterized proteins in data bases, though the possession of a predicted α-helical acidic region near the N terminus and a predicted basic α-helix at the C terminus is reminiscent of some eukaryotic transcription factors (26). However, recent data-base entries have revealed two uncharacterized apparent WhiB homologues, each showing about 25% amino acid identity to WhiB and to each other [for comparison, a similar degree of identity is found among functionally distinct members of the σ^{70} family of sigma factors to which σ^{WhiG} belongs (23, 59)]. Both homologues were derived by translation of DNA sequences from actinomycetes (N. K. Davies, personal communication). In one case, the genetic determinant converges with the *groES-groEL* operon of *Mycobacterium tuberculosis* (51), and in the other it resides on the antisense strand of the *saf* gene from *S. griseus* (27), which had been identified by its ability to stimulate extracellular enzyme production when introduced into *S. lividans* on a high-copy-number plasmid. The latter situation is extraordinary and requires further analysis because, although the arguments for the original choice of sense strand for *saf* are persuasive (27), the GC bias at the three codon positions characteristic of *Streptomyces* genes (7) would suggest that the opposite strand is the coding one. The alignments themselves are also rather persuasive, because all three proteins would be of similar size and all would share the following properties: acidic N termini, basic regions near their C termini, four similarly positioned cysteine residues, and two similarly positioned tryptophan residues. Further analysis of what seems to be a novel protein family will be of some interest.

The Complex whiE *Locus Determines Synthesis of Spore Pigment by the Polyketide Route*

Cloned DNA from the *whiE* locus, which caused the normally white spores of a *whiE* mutant to be somewhat more strongly pigmented than the wild-type spores, contained seven consecutive open reading frames, most encoding recognizable homologues of proteins involved in the biosynthesis of polyketide antibiotics, such as actinorhodin, tetracenomycin C, oxytetracycline, and granaticin (44), that contain multiple aromatic rings (25). Disruption of the wild-type chromosomal cluster with the cloned DNA generated the *whiE* mutant phenotype (11), and disruption of similar clusters in *Streptomyces halstedii* and *Streptomyces avermitilis* also gave unpigmented spores (9; T. MacNeil & D. MacNeil, personal communication). Southern blotting revealed *whiE*-like clusters in about half of a random collection of *Streptomyces* spp. (8).

Horinouchi & Beppu (46) cloned a piece of DNA that was unidentified at the time but is now known to include most of the published sequence in the *whiE* cluster except part of ORFI. Insertion of promoters upstream of this piece of DNA caused the production of a diffusible brown pigment whose preliminary characterization indicated that it was a 16-carbon oxidized naphthalene (possibly a naphthaquinone) with a β-hydroxy-β-keto acid side chain (46). This compound is unlikely to be identical to the spore pigment, because of the incomplete *whiE* ORF1, and because gene disruption experiments have shown that a diverging gene upstream of the *S. halstedii whiE*-like cluster (the *sch* genes) (8), and also present in *S. coelicolor* (11), is required for normal pigmentation (G. Blanco, personal communication).

Comparison of the products of conserved genes for biosynthesis of polyketide antibiotics and spore pigments in various *Streptomyces* spp. suggests that an ancient evolutionary split occurred between the two classes of genes before the antibiotic clusters diverged from each other (8). The regulatory mechanisms probably also diverged at about the same time. Certainly *whiE* and *sch* promoters resemble each other in their dependence on three genes quite distinct from those known to be needed in *S. coelicolor* for transcription of the related structural genes for biosynthesis of actinorhodin (21, 22) (see below).

Maximal Transcription of whiB, whiG, *and* whiE *Coincides with Aerial Mycelium Development*

Transcription of *whiB, whiG,* and *whiE* has been studied mainly with S1 experiments and the use of promoter-probe vectors. In the case of the *whiB* promoter region, two start sites were detectable by S1 mapping (77). The more upstream start site indicated a weak constitutive promoter (P1), while the more downstream site revealed a promoter (P2) that was almost inactive in young surface cultures, becoming much stronger when the aerial mycelium was developing. However, fusions of these promoters to the reporter gene *xylE* did not yield any activity, suggesting that the promoter region is either very weak, only transiently active, or sensitive to its context (77). Weak or transient activity is also consistent with the finding (E. Vijgenboom & K. F. Chater, unpublished data) that antiserum raised against WhiB protein purified from an *E. coli* expression system failed to detect WhiB protein in western blots of *S. coelicolor* cell extracts.

The *whiG* gene appears from S1 mapping to be transcribed from only a single start point, with little or no activity detectable in young surface cultures before aerial mycelium formation (G. L. Brown & K. F. Chater, unpublished data). A similar overall time course, but with evidence of greater promoter complexity, was seen for a *whiG* homologue of *Streptomyces aurofaciens* (J. Kormanec, personal communication). Results with *whiG-xylE* fusions in *S. coelicolor* also gave a similar time course (G. L. Brown & K. F. Chater, unpublished data).

A single promoter (*whiEP1*) seems to be responsible for transcription of much of that part of the *whiE* cluster for which DNA sequence has been published, though evidence from promoter probing suggests that a promoter internal to the cluster may augment transcription of ORFs V, VI, and VII (11). The pattern of *xylE* reporter gene activation by *whiEP1*, or of *whiE* transcript accumulation during growth on solid medium, again shows that the onset of *whiEP1* activity coincides with the visible growth of aerial hyphae.

The temporal resolution of these experiments (and of those with the two *whiG*-dependent promoters described earlier) is very limited, so it is not known whether promoters *whiB* P2, *whiG* P, *whiE* P1, P_{TH4}, and P_{TH270} are expressed sequentially or at essentially the same time, nor whether their expression is actually localized to the aerial mycelium.

Regulation of Transcription During Sporulation Involves a Network Rather than a Simple Linear Cascade

The availability of both a small collection of promoters presumptively expressed during aerial mycelium development, and several mutants blocked in that process, has made it possible to investigate transcriptional dependence relationships during sporulation. Despite the crucial importance of σ^{WhiG} in the initiation of sporulation, it is not needed for transcription from the known promoters of *whiE* or of *whiG* itself (G. L. Brown & K. F. Chater, unpublished data)—an observation compatible with the absence from any of these promoters of appropriately positioned sequences resembling the consensus sequences recognized by σ^D in *Bacillus subtilis* or σ^F in enteric bacteria. Thus, P_{TH4} and P_{TH270} are the only *whiG*-dependent promoters known so far. The wider dependence pattern of P_{TH4} and P_{TH270} was studied during surface growth by introducing fusions with the reporter gene *xylE* in a high-copy-number plasmid into various mutants. The two promoters proved to be relatively inactive in a *whiH119* mutant but showed no obvious reduction in activity in *whiA85,B70,* and *I17* mutants, or in a *bldA39* mutant grown in conditions suppressing aerial mycelium formation (79).

With the use of analogous fusions to *xylE,* it was found that several promoters from developmental genes had very low activity in *whiA85,B70,* or *H119* mutants while having strong activity in *whiG71, whiI17,* and *bldA39* mutants. The promoters showing this dependence pattern comprised two isolated from the *whiE* (spore pigment) cluster (11); their two homologues, together with one additional promoter, isolated from the *sch* (spore pigment) cluster of *S. halstedii* (G. Blanco, personal communication) and, remarkably, the *whiG* promoter (G. L. Brown & K. F. Chater, unpublished data). These results tend to support the tentative deduction (from its sequence) that WhiB is a transcriptional activator (26) and suggest that *whiA* and *whiH* may also specify transcriptional activators (especially because *whiA,B,* and *H* are the only classes of *whi* mutants that affect *whiE* expression).

Apparent transcriptional start sites have been determined for only two of the *whiA,B,H*-dependent promoters (*whiEP1* and *whiGP*) (11; G. L. Brown & K. F. Chater, unpublished data). Fairly extensive similarity between sequences immediately upstream of the start sites may indicate targets for the action of the *whiA,B,* or *H* gene products. These sequences are also conserved in the equivalent regions upstream of the *whiG* homologues of *S. aureofaciens* (J. Kormanec, personal communication) and *S. griseocarneum* (J. Soliveri & K. F. Chater, unpublished data) and of the *whiE* homologues of *S. halstedii* (*sch* genes) (8) and *Streptomyces curacoi* (*cur* genes) (6).

What regulates expression of the putative regulatory genes *whiA,B,* and *H*? Transcriptional dependence has so far been studied only in the case of *whiB*. In S1 experiments with RNA isolated from surface cultures of various mutants during aerial mycelium development, both *whiB* promoters were found to be active in representative *bldA, whiA, whiB, whiG,* and *whiH* mutants (77). Thus, *whiB* expression does not seem to be strongly dependent on any of these genes, so the mechanism of its developmental regulation is mysterious. Its developmentally regulated, stronger promoter (P2) contains appropriately located sequences resembling the consensus -10 and -35 sequences for the major class of bacterial promoters, so *whiBP2* may be recognized by RNA polymerase containing σ^{HrdB}, which seems to be the principal sigma factor in *S. coelicolor* (12).

These results appear at first sight to be compatible with a simple linear regulatory cascade, in which WhiB levels (perhaps together with levels of the as yet entirely uncharacterized WhiA and WhiH proteins) increase until they are sufficient to activate both a set of structural genes encoding particular morphological or metabolic functions (including the *whiE* genes) and the regulatory gene(s) needed for the next step in the cascade, represented by *whiG*. A threshold level of σ^{WhiG} would then activate a further set of morphological/metabolic genes for late stages of sporulation, including the genes controlled by P_{TH4} and P_{TH270}. However, this model is too simple. First, it predicts that the morphology of *whiA,B,* or *H* single mutants should determine that of *whiAG, whiBG,* or *whiGH* double mutants, when in fact the *whiG* mutation is morphologically epistatic to *whiA,B,* and *H* mutations (16). Second, it does not account for the apparently crucial role of σ^{WhiG} levels in determining whether sporulation will take place anywhere in the colony (23). Third, though it predicts that P_{TH4} and P_{TH270} should depend on all the genes needed for *whiG* expression, these two promoters do not seem to depend on *whiA* or *whiB* (79). Fourth, it does not explain why *whiG* mutants fail to produce gray-pigmented aerial hyphae (see below).

We currently do not have enough information to clarify these inconsistencies. Nevertheless, plausible, if imprecise, models can be erected that are more consistent with the available observations (Figure 5). A particularly convenient assumption would be that the effect of WhiA and WhiB is to increase *whiG* expression from an earlier, lower, WhiA- and WhiB-independent level, and that this hypothetical earlier, low-level expression is needed to kick-start sporulation and to express P_{TH4} and P_{TH270}. The *whiH* dependence of P_{TH4} and P_{TH270} probably does not directly result from *whiH* dependence of *whiG,* because this would imply morphological epistasis of *whiH* over *whiG*—the reverse of what was observed (16). Probably, therefore, WhiH protein and σ^{WhiG} are both directly needed for full expression of P_{TH4} and P_{TH270}.

Because a *whiG* mutation is morphologically epistatic to representative *whiA,B,* and *H* mutations, some effect of *whiG* expression must be needed for *whiA,B,* or *H* expression to be phenotypically manifested: the hyphae must be conditioned in some way. The early morphological events observed during sporulation—hyphal curling, nucleoid segregation, and sporulation septation—are likely to involve complex interactions among proteins involved in cell-wall biosynthesis and cell division. The expression of some of the relevant genes, and/or the posttranscriptional activation of the gene products, is presumably orchestrated by the major known candidates for sporulation regulatory genes, i.e. *whiA,B,G,H,* and *I*. The precise dependence of the synthesis or activity of particular proteins on particular *whi* genes, and the nature of the interactions between these proteins, can be expected to determine the pattern of epistasis. Therefore, an evaluation of the occurrence and roles

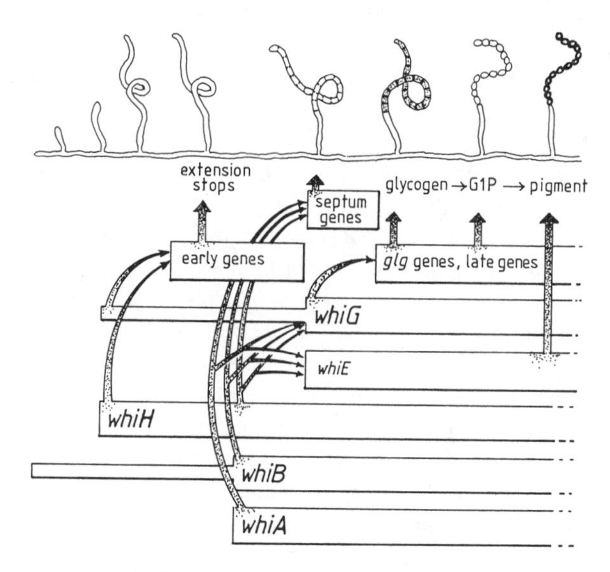

Figure 5 A tentative model for the regulatory network underlying sporulation in aerial hyphae of *Streptomyces coelicolor* A3(2). The upper part of the diagram illustrates the progressive metamorphosis of an aerial hypha into a pigmented spore chain. The points during development at which genes are thought to become active are indicated by the alignment of the left-hand ends of the boxes with the developmental stages. Increased expression of genes is indicated by vertical widening of the boxes. There is no information about when genes are switched off. Arrows indicate dependence. Among the early genes would be those transcribed from P_{TH4} and P_{TH270}. Among the late genes might be those for spore-wall thickening. Presumptive regulatory genes are depicted by larger lettering. Other abbreviations: *glg* genes, genes concerned with the biosynthesis and degradation of glycogen; G1P, glucose-1-phosphate.

of enzymes and proteins involved in growth and division, and their regulatory dependence on *whi* genes, is critical. The identification of penicillin-binding proteins and homologues of cell-division proteins known from other bacteria will be an important next stage in understanding *Streptomyces* sporulation.

Reference was made in an earlier section to the transient deposition of glycogen during sporulation. This appears to be regulated developmentally, under the direct or indirect control of σ^{WhiG}, rather than as a metabolic response to nutrient limitation at hyphal tips, because it is also observed during ectopic sporulation induced in the substrate mycelium or in liquid medium by multiple copies of *whiG* (K. A. Plaskitt & K. F. Chater, unpublished data). Because the disappearance of glycogen from developing spore chains shows some correlation with the formation of visible spore pigment, glycogen might be a metabolic precursor of the pigment (see above)—a view supported by the absence of glycogen from the hyperpigmented sporulating parts of aerial hyphae of a strain carrying extra copies of the *whiE* cluster and by the increased persistence of sporulation-associated glycogen in a *whiE* mutant (25; K. A.

Plaskitt & K. F. Chater, unpublished data). Thus, *whiG* mutations may prevent pigmentation by interference with sporulation-associated glycogen metabolism, rather than solely by a genetic regulatory cascade.

CONCLUDING REMARKS

As outlined in this review, several significant advances have been made in our understanding of morphological differentiation in streptomycetes.

The genetic complexity of aerial mycelium formation has become more obvious as more *bld* genes have been identified, and the fascinating involvement of extracellular signaling in this process has taken wing with the recent studies of SapB (84; J. Willey & R. Losick, personal communication). The significance of the extraordinary confinement of TTA codons to genes needed around the time of aerial mycelium emergence remains enigmatic. Sporulation-associated genes show no sign of transcriptional dependence on *bld* genes, which tends to rule out a coherent genetic regulatory cascade linking aerial-mycelium formation to sporulation. Hence the activation of sporulation only in aerial hyphae may depend instead on sensing of some structural, metabolic, or environmental signal(s) normally peculiar to aerial hyphae. The first information addressing the transcriptional regulatory interactions of sporulation genes has emerged and tends to indicate a surprisingly nonhierarchical interplay, considering that the sporulation process takes place over a period of several hours. This observation may indicate that many physiological processes are simultaneously activated and that the biochemical interplay of these processes itself contributes significantly to the organized series of events. Possibly, the switch-off timing of genes may also contribute to the orchestration of events. On the other hand, the apparent lack of genetic complexity (compared with sporulation in *Bacillus subtilis,* for example) may be misleading—possibly, many important functions are also necessary for normal growth, can be fulfilled by more than one mechanism (e.g. isoenzymes, alternative pathways, etc), or are not needed for spore pigment to form. Any such circumstances would have resulted in the relevant genes remaining undiscovered. The evidence supports alternative pathways at an early developmental stage: growth of aerial mycelia depends on SapB only in rich medium (84), and many *bld* mutants can produce sporulating aerial hyphae when mannitol replaces glucose as the carbon source in minimal medium.

The critical mass of knowledge about regulatory interactions that is now available will lead exponentially to new experimental opportunities, perhaps particularly in the biochemical in vitro reconstruction of situations deduced from genetic studies. Increased emphasis will be laid on the genes that, when activated by regulatory genes, bring about the observed morphological or physiological changes. Attention will be paid not only to how genes are activated in a spatially specific manner, but also to how they are kept unexpressed in different hyphal cell types. Homologues of *S. coelicolor* genes will be studied from or in other streptomycetes and actinomycetes, as well as in more diverse bacteria.

ACKNOWLEDGMENTS

I am grateful to Mervyn Bibb, Mark Buttner, David Hopwood, and Tobias Kieser for constructive comments on the manuscript and Meredyth Limberg and Anne Williams for their patience and skill in typing succesive versions. Thanks are also due to Hannah Im, Jan Kormanec, Richard Losick, Joseph McCormick, Ramon Santamaria, Alan Schauer, and Joanne Willey for generously providing unpublished information. Unpublished work from my laboratory referred to in this review was supported by the Agriculture and Food Research Council, the John Innes Foundation, the European Community BRIDGE program (BIOT-CT91-0255), and European Community Research Training Fellowships to J. Soliveri and E. Vijgenboom.

Literature Cited

1. Arnosti, D. N., Chamberlin, M. 1989. Secondary sigma factor controls transcription of flagellar and chemotaxis genes in *Escherichia coli*. *Proc. Natl. Acad. Sci. USA* 86:830–34
2. Babcock, M. J., Kendrick, K. E. 1988. Cloning of DNA involved in sporulation of *Streptomyces griseus*. *J. Bacteriol.* 170:2802–8
3. Babcock, M. J., Kendrick, K. E. 1990. Unusual transcriptional and translational features of a developmental gene of *Streptomyces griseus*. *Gene* 95:57–63
4. Bainton, N. J., Bycroft, B. W., Chhabra, S. R., Stead, P., Gledhill, L., et al. 1992. A general role for the *lux* autoinducer in bacterial cell signalling: control of antibiotic biosynthesis in *Erwinia*. *Gene* 116:87–91
5. Baum, E. Z., Buttner, M. J., Lin, L.-S., Rothstein, D. M. 1989. Transcription from the P1 promoters of *Micromonospora echinospora* in the absence of native upstream DNA sequences. *J. Bacteriol.* 171:6503–10
6. Bergh, S., Uhlen, M. 1992. Analysis of a polyketide synthesis-encoding gene cluster of *Streptomyces curacoi*. *Gene* 117:131–36
7. Bibb, M. J., Findlay, P. R., Johnson, M. W. 1984. The relationship between base composition and codon usage in bacterial genes and its use for the simple and reliable identification of protein-coding sequences. *Gene* 30: 157–66
8. Blanco, G., Brian, P., Pereda, A., Méndez, C., Salas, J. A., Chater, K. F. 1993. Hybridization and DNA sequence analyses suggest an early evolutionary divergence of related biosynthetic gene sets for polyketide antibiotics and spore pigments in *Streptomyces* spp. *Gene*. In press
9. Blanco, G., Pereda, A., Méndez, C., Salas, J. A. 1992. Cloning and disruption of a DNA fragment of *Streptomyces halstedii* involved in the biosynthesis of a spore pigment. *Gene* 112:59–65
10. Braña, A. F., Méndez, C., Diaz, L. A., Manzanal, M. B., Hardisson, C. 1986. Glycogen and trehalose accumulation during colony development in *Streptomyces antibioticus*. *J. Gen. Microbiol.* 132:1319–26
11. Brian, P. 1992. *A developmentally regulated spore pigment locus from* Streptomyces coelicolor *A3(2)*. PhD thesis. Univ. East Anglia, Norwich, UK
12. Brown, K. L., Wood, S., Buttner, M. J. 1992. Isolation and characterization of the major vegetative RNA polymerase of *Streptomyces coelicolor* A3(2): renaturation of a sigma subunit using GroEL. *Mol. Microbiol.* 6:1133–39
13. Champness, W. C. 1988. New loci required for *Streptomyces coelicolor* morphological and physiological differentiation. *J. Bacteriol.* 170:1168–74
14. Champness, W. C., Chater, K. F. 1993. The regulation and integration

of antibiotic production and morphological differentiation in *Streptomyces* spp. In *Regulation of Bacterial Differentiation,* ed. P. Piggot, P. Youngman, C. Moran. Washington, DC: Am. Soc. Microbiol. In press

15. Chater, K. F. 1972. A morphological and genetic mapping study of white colony mutants of *Streptomyces coelicolor*. *J. Gen. Microbiol.* 72:9–28
16. Chater, K. F. 1975. Construction and phenotypes of double sporulation deficient mutants in *Streptomyces coelicolor* A3(2). *J. Gen. Microbiol.* 87:312–25
17. Chater, K. F. 1984. Morphological and physiological differentiation in *Streptomyces*. In *Microbial Development,* ed. R. Losick, L. Shapiro, pp. 89–115. Cold Spring Harbor: Cold Spring Harbor Lab.
18. Chater, K. F. 1989. Sporulation in *Streptomyces*. In *Regulation of Procaryotic Development,* ed. I. Smith, R. Slepecky, P. Setlow, pp. 277–99. Washington, DC: Am. Soc. Microbiol.
19. Chater, K. F. 1989. Multilevel regulation of *Streptomyces* differentiation. *Trends Genet.* 5:372–76
20. Chater, K. F. 1991. Saps, hydrophobins, and aerial growth. *Curr. Biol.* 1:318–20
21. Chater, K. F. 1992. Genetic regulation of secondary metabolic pathways in *Streptomyces*. In *Secondary Metabolites: Their Function and Evolution, Ciba Foundation Symposium 171,* ed. D. J. Chadwick, J. Whelan, pp. 144–56. Chichester: Wiley & Sons
22. Chater, K. F., Brian, P., Brown, G. L., Plaskitt, K. A., Soliveri, J., et al. 1993. Problems and progress in the interactions between morphological and physiological differentiation in *Streptomyces coelicolor*. In *Genetics and Molecular Biology of Industrial Microorganisms,* ed. C. L. Hershberger, P. Skatrud, G. Hegeman. Washington, DC: Am. Soc. Microbiol. In press
23. Chater, K. F., Bruton, C. J., Plaskitt, K. A., Buttner, M. J., Méndez, C., Helmann, J. 1989. The developmental fate of *S. coelicolor* hyphae depends crucially on a gene product homologous with the motility σ factor of B. subtilis. *Cell* 59:133–43
24. Chater, K. F., Merrick, M. J. 1976. Approaches to the study of differentiation in *Streptomyces coelicolor* A3(2). In *Second International Symposium on the Genetics of Industrial Microorganisms,* ed. K. D. MacDonald, pp. 583–93. London: Academic
25. Davis, N. K., Chater, K. F. 1990. Spore colour in *Streptomyces coelicolor* A3(2) involves the developmentally regulated synthesis of a compound biosynthetically related to polyketide antibiotics. *Mol. Microbiol.* 4:1679–91
26. Davis, N. K., Chater, K. F. 1992. The *Streptomyces coelicolor whiB* gene encodes a small transcription factor-like protein dispensable for growth but essential for sporulation. *Mol. Gen. Genet.* 232:351–58
27. Daza, A., Gil, J. A., Vigal, T., Martin, J. F. 1990. Cloning and characterization of a gene of *Streptomyces griseus* that increases production of extracellular enzymes in several species of *Streptomyces*. *Mol. Gen. Genet.* 222:384–92
28. Deleted in proof
29. Demain, A. L., Aharonowitz, Y., Martin, J.-F. 1983. Metabolic control of secondary biosynthetic pathways. In *Biochemistry and Genetic Regulation of Commercially Important Antibiotics,* ed. L. C. Vining, pp. 49–72. Reading, MA: Addison-Wesley
30. Efremenkova, O. V., Anisova, L. N., Bartoshevich, Y. E. 1985. Regulators of differentiation in actinomycetes. *Antibiot. Medit. Biotekhnol.* 9:687–707
31. Fernández-Moreno, M. A., Caballero, J. L., Hopwood, D. A., Malpartida, F. 1991. The act cluster contains regulatory and antibiotic export genes, direct targets for translational control by the bldA tRNA gene of Streptomyces. *Cell* 66:769–80
32. Gramajo, H., Takano, E., Bibb, M. J. 1993. Stationary phase production of the antibiotic actinorhodin is transcriptionally regulated. *Mol. Microbiol.* 7:837–45
33. Granozzi, C., Billeta, R., Passantino, R., Sollazzo, M., Puglia, A. M. 1990. A breakdown in macromolecular synthesis preceding differentiation in *Streptomyces coelicolor* A3(2). *J. Gen. Microbiol.* 136:713–18
34. Guijarro, J., Santamaria, R., Schauer, A., Losick, R. 1988. Promoter determining the timing and spatial localization of transcription of a cloned *Streptomyces coelicolor* gene encoding a spore-associated polypeptide. *J. Bacteriol.* 170:1895–1901
35. Guthrie, E. P., Chater, K. F. 1990. The level of a transcript required for production of a *Streptomyces coelicolor* antibiotic is conditionally dependent on a tRNA gene. *J. Bacteriol.* 172:6189–93

35a. Hara, O, Horinouchi, S., Uozumi, T., Beppu, T. 1983. Genetic analysis of A-factor synthesis in *Streptomyces coelicolor* A3(2) and *Streptomyces*

griseus. *J. Gen. Microbiol.* 129:2939–44
36. Harasym, M., Zhang, L.-H., Chater, K., Piret, J. 1990. The *Streptomyces coelicolor* A3(2) *bldB* region contains at least two genes involved in morphological development. *J. Gen. Microbiol.* 136:1543–50
37. Helmann, J. D. 1991. Alternative sigma factors and the control of flagellar gene expression. *Mol. Microbiol.* 5:2875–82
38. Helmann, J. D., Chamberlin, M. J. 1988. Structure and function of bacterial sigma factors. *Annu. Rev. Biochem.* 42:839–72
39. Helmann, J. D., Marquez, L. M., Chamberlin, M. J. 1988. Cloning, sequencing and disruption of the *Bacillus subtilis* σ^{28} gene. *J. Bacteriol.* 170: 1568–74
40. Hodgson, D. A. 1980. *Carbohydrate utilization in* Streptomyces coelicolor *A3(2)*. PhD thesis. Univ. East Anglia, Norwich, UK
41. Hodgson, D. A. 1992. Differentiation in actinomycetes. In *Prokaryotic Structure and Function: A New Perspective,* ed. S. Mohan, C. Dow, J. A. Cole, pp. 407–40. Cambridge: Cambridge Univ. Press
42. Hopwood, D. A. 1967. Genetic analysis and genome structure in *Streptomyces coelicolor. Bacteriol. Rev.* 31:373–403
43. Hopwood, D. A. 1988. Towards an understanding of gene switching in *Streptomyces,* the basis of sporulation and antibiotic production. The Leeuwenhoek lecture, 1987. *Proc. R. Soc. London Ser. B* 235:121–38
44. Hopwood, D. A., Sherman, D. H. 1990. Molecular genetics of polyketides and its comparison to fatty acid biosynthesis. *Annu. Rev. Genet.* 24:37–66
45. Hopwood, D. A., Wildermuth, H., Palmer, H. M. 1970. Mutants of *Streptomyces coelicolor* defective in sporulation. *J. Gen. Microbiol.* 61:397–408
46. Horinouchi, S., Beppu, T. 1985. Construction and application of a promoter-probe plasmid that allows chromogenic identification in *Streptomyces lividans. J. Bacteriol.* 162:406–12
47. Horinouchi, S., Beppu, T. 1992. Regulation of secondary metabolism and cell differentiation in *Streptomyces:* A-factor as a microbial hormone and the AfsR protein as a component of a two-component regulatory system. *Gene* 115:167–72
48. Kaiser, D. 1989. Multicellular development in *Myxobacteria*. In *Genetics of Bacterial Diversity,* ed. D. A. Hopwood, K. F. Chater, pp. 243–63. London: Academic
49. Khokhlov, A. S., Tovarova, I. I., Borisova, L. N., Pliner, S. A., Schevchenko, L. A., et al. 1967. A-factor responsible for the biosynthesis of streptomycin by a mutant strain of *Actinomyces streptomycini. Dokl. Akad. Nauk. SSSR* 177:232–35
50. Kieser, H. M., Kieser, T., Hopwood, D. A. 1992. A combined genetic and physical map of the *Streptomyces coelicolor* A3(2) chromosome. *J. Bacteriol.* 174:5496–5507
51. Kong, T. H., Coates, A. R. M., Hickman, C. J., Shinnick, T. M. 1991. *Genbank/EMBL Accession Number X60350*
52. Kutsukake, K., Ohta, Y., Iino, T. 1990. Transcriptional analysis of the flagellar regulon of *Salmonella typhimurium. J. Bacteriol.* 172:741–47
53. Lawlor, E. J., Baylis, H. A., Chater, K. F. 1987. Pleiotropic morphological and antibiotic deficiencies result from mutations in a gene encoding a tRNA-like product in *Streptomyces coelicolor* A3(2). *Genes & Dev.* 1:1305–10
54. Leskiw, B. K., Bibb, M. J., Chater, K. F. 1991. The use of a rare codon specifically during development? *Mol. Microbiol.* 5:2861–67
55. Leskiw, B. K., Lawlor, E. J., Fernandez-Abalos, J. M., Chater, K. F. 1991. TTA codons in some genes prevent their expression in a class of developmental, antibiotic-negative *Streptomyces* mutants. *Proc. Natl. Acad. Sci. USA* 88:2461–65
56. Leskiw, B. K., Mah, R., Lawlor, E. J., Chater, K. F. 1993. Accumulation of *bldA*-specified tRNA is temporally regulated in *Streptomyces coelicolor* A3(2). *J. Bacteriol.* 175:1995–2005
57. Li, J., Horwitz, R., McCracken, S., Greenblatt, J. 1992. NusG, a new *Escherichia coli* elongation factor involved in transcriptional antitermination by the N protein of phage λ. *J. Biol. Chem.* 267:6012–19
58. Locci, R., Sharples, G. P. 1983. Morphology. In *The Biology of the Actinomycetes,* ed. M. Goodfellow, M. Mordarski, S. T. Williams, pp. 165–99. London: Academic
59. Lonetto, M., Gribskov, M., Gross, C. A. 1992. The σ^{70} family: sequence conservation and evolutionary relationships. *J. Bacteriol.* 174:3843–49
60. Marahiel, M. A. 1992. Multidomain enzymes involved in peptide synthesis. *FEBS Lett.* 307:40–43
61. McCue, L. A., Kwak, J., Babcock,

M. J., Kendrick, K. E. 1992. Molecular analysis of sporulation in *Streptomyces griseus*. *Gene* 115:173–79
62. McVittie, A. M. 1974. Ultrastructural studies on sporulation in wild-type and white colony mutants of *Streptomyces coelicolor*. *J. Gen. Microbiol.* 81:291–302
63. Méndez, C., Braña, A. F., Manzanal, M. B., Hardisson, C. 1985. Role of substrate mycelium in colony development in *Streptomyces*. *Can. J. Microbiol.* 31:446–50
64. Merrick, M. J. 1976. A morphological and genetic mapping study of bald colony mutants of *Streptomyces coelicolor*. *J. Gen. Microbiol.* 96:299–315
65. Miyake, K., Kuzuyama, T., Horinouchi, S., Beppu, T. 1990. The A-factor-binding protein of *Streptomyces griseus* negatively controls streptomycin production and sporulation. *J. Bacteriol.* 172:3003–8
66. Ochi, K. 1990. *Streptomyces griseus* as an excellent object for studying microbial differentiation. *Actinomycetologica* 4:23–30
67. Ohnishi, K., Kutsukake, K., Suzuki, H., Iino, T. 1990. Gene *fliA* encodes an alternative sigma factor specific for flagellar operons in *Salmonella typhimurium*. *Mol. Gen. Genet.* 221:139–47
68. Ohnishi, K., Kutsukake, K., Suzuki, H., Iino, T. 1992. A novel transcriptional regulation mechanism in the flagellar regulon of *Salmonella typhimurium:* an anti-sigma factor inhibits the activity of the flagellum-specific sigma factor, σ^F. *Mol. Microbiol.* 6: 3149–57
69. Okamoto, S., Nihira, T., Kataoka, H., Suzuki, A., Yamada, Y. 1992. Purification and molecular cloning of a butyrolactone autoregulator receptor from *Streptomyces virginiae*. *J. Biol. Chem.* 267:1093–98
70. Passantino, R., Puglia, A. M., Chater, K. F. 1991. Additional copies of the *actII* regulatory gene induce actinorhodin production in pleiotropic *bld* mutants of *Streptomyces coelicolor* A3(2). *J. Gen. Microbiol.* 137:2059–64
71. Piret, J. M., Chater, K. F. 1985. Phage-mediated cloning of *bldA,* a region involved in *Streptomyces coelicolor* morphological development, and its analysis by genetic complementation. *J. Bacteriol.* 163:965–72
72. Sakonju, S., Bogenhagen, D. F., Brown, D. D. 1980. A control region in the center of the 5S RNA gene directs specific initiation of transcription: I. The 5′ border of the region. *Cell* 19:13–25
73. Schauer, A. T., Nelson, A. D., Daniel, J. B. 1991. Tn*4563* transposition in *Streptomyces coelicolor* and its application to isolation of new morphological mutants. *J. Bacteriol.* 173:5060–67
74. Shapiro, J. A. 1988. Bacteria as multicellular organisms. *Sci. Am.* 258:62–69
75. Siegele, D. A., Kolter, R. 1992. Life after log. *J. Bacteriol.* 174:345–48
76. Sohaskey, C. D., Im, H., Schauer, A. T. 1992. Construction and application of plasmid- and transposon-based promoter-probe vectors for *Streptomyces* spp. that employ a *Vibrio harveyi* luciferase reporter cassette. *J. Bacteriol.* 174:367–76
77. Soliveri, J., Brown, K. L., Buttner, M. J., Chater, K. F. 1992. Two promoters for the *whiB* sporulation gene of *Streptomyces coelicolor* A3(2), and their activities in relation to development. *J. Bacteriol.* 174:6215–20
78. Starnbach, M. N., Lory, S. 1992. The *fliA* (*rpoF*) gene of *Pseudomonas aeruginosa* encodes an alternative sigma factor required for flagellin synthesis. *Mol. Microbiol.* 6:459–69
79. Tan, H. 1991. *Molecular genetics of developmentally regulated promoters in* Streptomyces coelicolor *A3(2)*. PhD thesis. Univ. East Anglia, Norwich, UK
80. Tan, H., Chater, K. F. 1993. Two developmentally controlled promoters of *Streptomyces coelicolor* A3(2) that resemble the major class of motility-related promoters in other bacteria. *J. Bacteriol.* 175:933–40
81. Vivian, A. 1971. Genetic control of fertility in *Streptomyces coelicolor* A3(2): plasmid involvement in the interconversion of UF and IF strains. *J. Gen. Microbiol.* 69:353–64
82. Wildermuth, H. 1970. Development and organisation of the aerial mycelium in *Streptomyces coelicolor*. *J. Gen. Microbiol.* 60:43–50
83. Wildermuth, H., Hopwood, D. A. 1970. Septation during sporulation in *Streptomyces coelicolor*. *J. Gen. Microbiol.* 60:51–59
84. Willey, J., Santamaria, R., Guijarro, J., Geistlich, M., Losick, R. 1991. Extracellular complementation of a developmental mutation implicates a small sporulation protein in aerial mycelium formation by Streptomyces coelicolor. *Cell* 65:641–50

Annu. Rev. Microbiol. 1993. 47:715–38

EVALUATING BIOREMEDIATION: Distinguishing Fact from Fiction

Michael J. R. Shannon and Ronald Unterman
Envirogen, Inc., Lawrenceville, New Jersey 08648

KEY WORDS: environmental biotechnology, biorestoration, pollution control, hazardous waste, biodegradation, detoxification, abiotic losses

---

CONTENTS

ABSTRACT

Bioremediation options encompass diverse types of biochemical mechanisms that may lead to a target's mineralization, partial transformation, humification, or altered redox state (e.g. for metallic elements). Because these various mechanisms produce alternative fates of the targeted pollutants, it is often necessary to use diverse evaluation criteria to qualify a successful bioremediation. Too often target depletion from a treated matrix can be mistakenly

0066-4227/93/1001-0715$02.00

ascribed to biological activity when in fact the depletion is caused by abiotic losses (e.g. volatilization, leaching, and stripping). Thus, effective, and therefore convincing, evaluation requires that experimental and engineering designs anticipate all possible routes of target depletion and that these routes be carefully monitored.

INTRODUCTION

Biological remedies for pollution reduction (the application of which is referred to as bioremediation) have received increasing attention since the 1980s and have attained widespread recognition in the wake of certain environmental catastrophes such as the Exxon Valdez and Mega Borg oil spills and the Iraq-Kuwait war and its aftermath. Although biological processes have been in common use since the early 1900s to treat nonrefractory wastes (e.g. sewage), their use for treating hazardous and refractory chemical wastes (associated with chemical/industrial activities) is more recent, and many applications are still under development. This increase in bioremediation applications has been fostered, in part, by our expanding knowledge of how these chemicals are metabolized by existing microbes (3, 24, 30), the isolation and utilization of new microbes (74), and our ability to rationally design novel metabolic capabilities using genetic engineering (90). Thus we have increased our repertoire of useful biocatalytic reactions for biotreatment applications. In addition, the biological remedies for waste treatment, which have generally used the metabolic activities of whole microorganisms to consume waste chemicals, are now also using cell-free enzymes, cell components, and cell products.

Because bioremediation is a developing industry, a learning curve is expected for the potential clients, those who provide the technology, and those who regulate the cleanups. All three participants are faced with the task of evaluating this relatively new technology, both technically and economically, and ultimately of assuring that defined cleanup goals are achieved. This assurance typically unfolds during initial laboratory-scale biotreatability studies, extends through pilot studies, and concludes during full-scale implementation of the bioremediation. During this process, it is in the interest of all parties to properly evaluate the technology to ensure that the treatment goal is being achieved and also, where necessary, to prove the biological nature of the process.

Proper evaluation of bioremediation options begins by determining what constitutes an acceptable cleanup goal (which is not the subject of this review); for example, one must determine whether destruction, detoxification, or physical removal of the chemical target(s) is the goal of the remediation. Only after making this decision can one choose, design, and conduct an appropriate means of bioremediation. Monitoring and evaluating the bioremediation process during implementation, however, is not always straightforward and is often confounded by any of several factors. Bioremediation encompasses an array of biotreatment processes that can vary significantly in their mechanisms of action; thus each option may require a unique monitoring approach. For example, for biodegradation processes in which the chemical targets are destroyed or transformed, especially biotreatments in open systems, attaining a complete mass balance of the target is often impossible. In such cases, determining the success of the bioremediation may rely on indirect measurements that are known to correspond directly with target degradation. Furthermore, bioremediation is frequently intended for use in large, open systems and is directed at chemical targets that often have the propensity to migrate, intact, away from the treatment zone without being detected (i.e. they disappear), thereby confounding effective analysis.

Although the applicability of bioremediation has been cited accurately in trade and technical journals (36, 58, 68), claims are too often overblown. In fact, the intellectual and commercial integrity of bioremediation has been compromised by some reports in magazines, trade journals, and product literature. For example, reports have claimed the aerobic degradation of tetrachloroethylene (PCE) and the degradation of dioxin (implying 2,3,7,8-tetrachlorodibenzodioxin), when in fact these transformations are beyond the present capabilities of bioremediation and for that matter known biochemistry. These and other exaggerated claims for bioremediation have created a climate of skepticism for this industry by misleading those searching for realistic treatment alternatives for site remediation and pollution prevention.

A rudimentary understanding of biological processes will generally aid those assessing any true biotreatment option, and help them to identify false claims. The person performing the evaluation can become sufficiently familiar with both the theoretical potential and actual capabilities of biodegradation by referring to technical literature and even general microbial textbooks. This understanding of realistic biological capabilities forms the basis for knowledgeable decision making. Therefore, the purpose of this review is to: (*a*) outline the diversity of biotreatment options that are applicable to polluted sites and process effluents, (*b*) describe the abiotic processes that are sometimes confused with true bioremediation, and (*c*) provide an overview of evaluation methodology spanning laboratory feasibility studies to field implementation. This discussion includes both in situ and reactor based technologies.

THE DIVERSITY OF TRUE BIOREMEDIATION OPTIONS

The diverse bioremediation options include many different types of attenuation mechanisms that can, depending on the nature of the target compound and matrix, yield various end results. Understanding these differences is important because a proper evaluation approach must anticipate the possible, alternative fates of the targets and relate these possibilities back to the stated goal of the cleanup. The mechanisms characteristic of bioremediation can result in the physical removal, sequestering, detoxification, or destruction of target pollutants.

Bioremediation options that result in the collection and subsequent removal of an untransformed target compound from a polluted site (e.g. bioadsorption) are conceptually easy to evaluate and are analogous to any physical extraction process. In this type of remediation, determination of success is performed by simply monitoring the quantity of target collected and the amount remaining in the contaminated matrix; in other words a mass balance is easy to determine. For other options that aim solely to reduce the toxicity associated with a target compound, one must evaluate the process with reference to a toxicity study. Proving remediation is more complex when the target compounds are destroyed to form daughter products (such as CO_2 or organic metabolites) or if the targets are humified, as may happen if the bioremediation is conducted in soil or sediment environments. In these cases, the parent compound cannot always be monitored directly and a mass balance that includes metabolites can be difficult to achieve. Evaluation is especially problematic for in situ treatments because the biological process may take weeks, and perhaps months, and in that time span abiotic losses are increasingly likely. Understanding these and other related issues associated with the various bioremediation options is critical to designing and assessing proper evaluation methodologies.

Biodegradation of Organic Compounds

The biodegradation of organic compounds encompasses most of the current commercially available bioremediation options. Perhaps the best known application of biodegradation is for the in situ cleanup of fuel spills. In contrast to these treatments, compounds such as pentachlorophenol (23), phenol, and 2,4-dichlorophenoxyacetic acid (2,4-D) are often biodegraded in contained and well-controlled facilities. These cases exemplify biodegradation processes in which microbial species grow on and consume the target pollutants. In addition, cometabolic (nongrowth) biodegradation processes have also been used to biotreat recalcitrant halogenated compounds, for example, trichloroethylene (TCE) and polychlorinated biphenyls (PCB) at the pilot scale (1, 28). The classes of target compounds amenable to biodegradation processes include, but by no means are limited to, complex mixtures of petroleum hydrocarbons, pesticides, herbicides, organic solvents, and haloorganic compounds.

Biodegradation is a type of biotransformation that destroys the backbone structure of the parent compound, ultimately yielding daughter products (metabolites) of less complexity or lower mass. This may be contrasted with processes by which the target is destroyed or attenuated via synthetic reactions. The degree of target destruction can range from a simple displacement of atoms or functional groups (partial biodegradation) to the complete oxidation of the compound to yield inorganic forms of its constituent atoms (e.g. CO_2, H_2O, and inorganic salts). This latter process (mineralization) results in a successful cleanup and suggests a priori that a detoxification has also been achieved. However, in the case of a partial biodegradation, the remediation is deemed successful only if the metabolites are determined to be nontoxic or, alternatively, susceptible to subsequent reactions that lead to their eventual detoxification. These secondary reactions may be abiotic, resulting in polymerization or spontaneous breakdown, or biotic if resident microorganisms can complete the degradation. In terms of an evaluation process, one should note that because the physico-chemical behavior of the products can differ from that of the parent compound, a chosen analytical procedure must be valid for these as well (77).

Biodegradation may alter the microbial populations and chemical characteristics of the biotreated matrix, especially if the targets promote cell growth, and these secondary effects can be used during the evaluation process. For example, if the organisms responsible for the biodegradation (individuals, or synergistic and commensal communities) derive nutrients or energy from a compound, their populations will likely expand (52). One can sometimes observe bacterial growth in cases where the target compound serves as an electron-accepting species (26). Also a shift in the relative proportion of cell types (27, 29, 85) may occur, or the microbial predators that have been enriched by the species growing on the targets may increase (57). In other cases, a transient increase in the toxicity of the materials under treatment may be observed, thereby supporting the conclusion that a chemical target has been oxidized, albeit to a transient toxic daughter product that will undergo subsequent degradation (96). If the biodegradation occurs via cometabolism, however, the secondary effects resulting from microbial growth are, at least, less predictable, because cometabolism is by definition a fortuitous, non-growth supporting process.

Humification and Polymerization Reactions

In numerous cases, biotransformation without biodegradation, can ameliorate the toxicity associated with organic pollutants via humification or polymerization mechanisms. Although this is not a classic example of a bioremediation, it is nevertheless a common reaction in soil that results in the effective removal of toxic compounds from the soluble fraction of the soil environment. Research into the fate of pesticide residues in humus (4, 13) has shown that microbial populations (and abiotic mechanisms) in soil often transform parent compounds to intermediates that are subsequently incorporated into soil organic components, although this phenomenon is not limited to pesticides. The biodegradation of petroleum hydrocarbons by using biotreatments resembling land-farming methods can lead to the humification of polycyclic aromatic hydrocarbons (16), and other studies have indicated that even relatively refractory haloorganic compounds such as PCBs and pentachlorophenol can be incorporated into humic material (64). Although the mechanism for PCB humification was unclear, other research has suggested that it may be mediated in part by an enzymatic polymerization (48). Indeed the contribution of extracellular microbial enzymes to the process of humification has been recognized as an important contribution to determining the fate of soil pollutants (11) and has been explored as a potential bioremediation option for decontaminating soils (76).

In some cases the polymerization of organic compounds can lead to increased toxicity as when aromatic amines are polymerized to form azo compounds (44). Nevertheless, many polymerized products are less likely to leach from soil (81) and are in many cases less toxic than their parent compounds. For example, Bollag et al demonstrated that various phenolic compounds were less acutely toxic after their copolymerization with natural soil components such as syringic acid (12). While it is true that the parent compound remains in the soil environment after humification, its subsequent release is unlikely to occur at a significant rate (80). An evaluation of this biodetoxification mechanism, however, does require prolonged fate and toxicity studies to alleviate concerns for rerelease and reexposure.

Abiotic Transformations as a Component of Bioremediation

Abiotic transformations can also influence the fate of target pollutants and are of particular importance in soils and sediments (57, 61). Abiotic transformations, such as hydrolysis, oxidation-reduction reactions, and photodegradation, are distinct from physical abiotic losses such as sorption, leaching, and stripping, which result in the migration of the intact target compound (e.g. from soil to air). Abiotic transformation reactions have been thoroughly explored in the context of pesticide fate studies (62). Understanding these reactions can be useful for certain types of bioremediation because they can serve to prime otherwise nonbiodegradable targets for biodegradation, or they can complete the detoxification and degradation of otherwise incomplete biotransformation reactions.

Hydrolytic reactions are quite prominent in determining the fate of certain organic compounds in soils. With many pesticides, an initial abiotic hydrolysis can be the first step of degradation to yield products that are subsequently mineralized by soil microorganisms in what can be considered a two-stage abiotic-biotic degradation. These reactions are less significant for the bioremediation of more refractory compounds that are not easily hydrolyzed.

The effect of solar irradiation on the dissipation or attenuation of chemical pollutants has been extensively addressed, primarily with pesticides, and is another example of the potential role abiotic transformation can play in bioremediation. These abiotic reactions are also known to affect the transformation of compounds other than pesticides; for example, solar irradiation contributes to the degradation of partially biodegraded PCBs in vitro (9, 50). The possibility of applying a photodegradation process step has been investigated for surface soils that have been contaminated with other similar aromatic compounds such as dioxin (61). Other biologically recalcitrant compounds can also be primed for a subsequent biodegradation step, as was shown with benzo[*a*]pyrene in an irradiated aqueous system (63).

Conversely, abiotic transformations can complete degradations that were initiated by microbial metabolism. For example, herbicides such as the acylanalides are partially biodegraded in soils to an intermediate that is subsequently incorporated into humic material via abiotic mechanisms (39). The released chloroanaline moiety exhibits considerable persistence in the soil environment because it is humified via abiotic processes. Another example is the cometabolic biodegradation of TCE by microorganisms such as *Methylosinus trichosporium,* which results in the formation of a chemically labile epoxide that spontaneously degrades abiotically in the aqueous environment of soils (66).

Thus, abiotic reactions contribute to the biotransformation of chemical targets and should be considered in the evaluation of bioremediation processes.

Biotransformation of Metals and Metal Compounds

Another type of bacterial transformation that can lead to detoxification, and therefore true bioremediation, is the oxidation or reduction of heavy and other potentially toxic metals. Here, obviously no degradation (change in the nuclear structure of the element) occurs; however, the metal's oxidation state is altered so that it is either: (*a*) more water soluble and is removed by leaching, (*b*) inherently less toxic, (*c*) less water soluble so that it precipitates and is less bioavailable or removed from the matrix, or (*d*) volatilized and removed from the polluted matrix.

Metals that are potentially toxic (such as arsenic, chromium, mercury, or uranium) can often be removed from polluted water environments by chemical reduction. For example, the extensive use of uranium has led to the pollution of surface waters, ground waters, and industrial-waste streams. In response, the US Geological Survey (56) has investigated the enzymatic detoxification of dissolved U(VI) to a precipitated form, U(IV). The reduced form is insoluble in aqueous media thus decreasing its toxicity by reducing its bioavailability and potentially permitting the development of a biologically mediated separations process. Industrial processes that use chromate can also pollute water environments with the more toxic form of the chromium anion, Cr(VI), which can also be detoxified by bacterially mediated reduction. The enzymatic mechanism responsible for the reduction of Cr(VI) to Cr(III) is being studied (42) and may ultimately lead to a commercial bioremediation process. (For a review on metal reduction in this volume, please see 55a.)

Alternatively, materials that are contaminated by metal sulfides can be decontaminated via oxidation reactions (15). This process, known as bioleaching, is mediated by bacteria such as *Thiobacillus ferrooxidans,* whose oxidation of insoluble metal sulfides to water-soluble sulfates allows for leaching to remove the metal compound. This approach may be useful for selectively removing metal-containing complexes from bulky waste materials so they can be inexpensively disposed of as nonhazardous waste.

Some metals (e.g. As, Se, Hg) become readily volatile when methylated or when transformed to their metallic forms. As reported by several labs (43, 89), some bacterial strains can reduce mercuric ion [Hg(II)] to volatile metallic mercury [Hg(0)]. This natural reduction process is thought to be responsible for the environmental detoxification of mercury and is now being developed for future commercial applications.

Biologically Mediated Accumulation/Stabilization

In contrast to the transformation reactions described thus far, biological components or processes can also be used to collect or sequester target compounds in their intact state without chemical modifications. For example, heavy and potentially toxic metals can be concentrated and collected by either specific or nonspecific binding of the metal ions to cells, cell components, or proteins. After their physical removal from the environment, they can be either recovered or stabilized and buried. In this approach, a biological material adsorbs the contaminant (biosorption), and the mechanism is a purely physical/chemical one, the end result of which yields an accumulation of the target pollutant. The biological components can be either living or dead and the process does not necessarily involve active cell metabolism (84). Currently results with algae have been the most promising; however, bacterial and fungal cells have also been found to bind metals from solution (51, 53).

The nonspecific mechanism for this binding is analogous to an ion exchange resin, in that the cell-wall components are comprised of many ionic functional groups (e.g. amino, carboxyl, hydroxyl). The specific binding of metals such as copper and cadmium is mediated by specific sulfhydryl groups and binding-pocket dimensions (53). The potential of certain microorganisms to bioaccumulate radioactive metal compounds is also well known, as described in a recent study conducted to isolate a cesium-accumulating bacterium (91).

These and other bioaccumulation phenomena are being developed for the treatment of hazardous waste in the environment.

Another treatment that does not chemically transform the target compound uses biodegradation processes as an indirect means to reduce the risks associated with the leaching of organics from soil. Compounds that themselves are partially or completely soluble in water (e.g. nitrobenzene and methanol) are often responsible for increasing the solubility (hence leaching) of relatively insoluble compounds when both are present as cocontaminants (72). One company (Groundwater Technology, Inc.) has demonstrated that by biodegrading the more soluble cocontaminant, one can retard the rate of leaching of target compounds (in their case, PCBs) even without degrading the target compound. They accomplish this by stimulating the in situ degradation of the more soluble component, thereby rendering the PCBs less leachable and therefore "biostabilized."

FALSE BIOREMEDIATION: ABIOTIC LOSS OF TARGET COMPOUNDS

Abiotic losses of target compounds occur by physical means including stripping, leaching, or sorption. The inability to detect and quantify target losses through these routes results from inadequate monitoring of surrounding environments (air, water, or soil) or incomplete sampling (e.g. of bioreactors). In other cases, the monitoring and sampling may be adequate, but inappropriate analytical techniques may miss the target's presence within the sample (77).

A notorious (albeit nonbiological) case that exemplifies some of these possibilities was the false claim for PCB destruction resulting from a particular solidification process. As reported in the popular press, field observations from several sites in the late 1980s showed that when PCB-containing sludge was solidified (in a legitimate process requiring quicklime), the subsequent concentration of PCBs in the solidified material was substantially reduced and in some cases could not be detected. It was hypothesized that the quicklime perhaps contributed to the in situ chemical destruction of PCBs. Subsequent laboratory experiments showed that these results were repeatable under "controlled" laboratory conditions. This laboratory verification was not submitted to a reviewed journal but was instead released directly to the popular press before alternative explanations were investigated.

The alternative explanations for the apparent removal of PCB from the treated material included: (*a*) loss by volatilization, (*b*) increased physical sorption into the treated matrix, and (*c*) in the case of the original field observations, leaching from the treated site. Subsequent studies by others showed that the PCB "destruction" could in fact be attributed to both physical encapsulation and perhaps volatilization. One study addressed the possibility that PCB would volatilize because of the heat generated from the hydration of quicklime (an exothermic reaction) (78). The apparatus used in that experiment allowed for a complete mass balance of the PCB isomer used, and particular care was taken to distinguish the quantity of PCB volatilized through the reactor's outlet and that quantity volatilized and subsequently redeposited to the surface of the reactor's headspace. Volatilization of the PCB isomer did occur, but to such a small extent that this alone could not account for the loss of PCBs as originally observed. In a conclusive study, Hughes et al (41) showed that the quicklime treatment encapsulated the PCB-spiked sand and dramatically reduced the extractability of PCBs. The PCBs were recoverable, however, after acidification to dissociate the lime matrix. This example clearly demonstrated the danger in relying solely on target disappearance as proof of target destruction (93) and showed that the experimental apparatus and analytical techniques used must be designed so as to anticipate and detect potential abiotic losses.

Abiotic losses of a target compound under field conditions are monitored effectively only in enclosed systems. These include bioreactors or land-based treatment facilities that have been constructed with liners for leachate collection and a sealed overhead cover to trap volatiles. Such enclosures allow for the accounting of all materials passing through the system. For larger open treatment systems and most in situ treatments, however, expecting a good quantitative mass balance is almost futile, and more indirect methods of evaluation may become necessary (58). The following sections discuss issues related to abiotic losses during biodegradation processes.

Stripping

Bioremediation of materials containing readily volatile targets needs to be carefully monitored to ensure that target removal is due to microbial metabolism and not simply to stripping. Problems associated with the sampling and analysis of such materials are innumerable; nevertheless appropriate procedures are critical and are available (46). One biodegradation study brought to our attention was an assessment of the biodegradability of two volatile compounds (chlorobenzene and nitrobenzene) in a lagoon sample. This biotreatability study was performed in open, aerated tanks, and the report concluded that depletion of the target compounds resulted primarily from biodegradation. In the study the sludge had been amended with appropriate nutrients, the pH was maintained near neutrality, and the tank was aerated for almost two months to provide an oxygen source. Because of the inherent volatility of the target compounds, such an open-reactor configuration was inappropriate—because of the reactor's construction, the degree of target stripping was never quantified. This study might have been acceptable if the tank had been sealed and configured with one outlet port to serve as an exhaust. The rate of target stripping through this port could then have been quantified by using on-line analysis or in-line carbon traps.

An alternative treatability approach to conduct such a test is the use of a series of serum vials sealed with Teflon®-lined rubber septa, which precludes the possibility of stripping. By periodically introducing oxygen into the vials or by using a large head space, the vial contents can be maintained under aerobic conditions. Any tendency for the volatile pollutants to migrate through the Teflon® liner and into the butyl-rubber stopper can be monitored with controls that are sacrificed over the time course of the experiment. This experimental design also allows for the entire vial contents to be extracted to assure a complete recovery of the target compound and possible metabolites.

Large reactor processes for pilot-scale testing or full-scale biotreatment generally cannot be completely sealed, so for these systems a mass balance is possible only if all routes to and from the reactor are carefully monitored during the process. The rate of target-compound introduction into a reactor can be monitored as well as the breakthrough of the target compound and its metabolites as they leave the reactor. Such an approach was successfully demonstrated by Sullivan (88) using a submerged fixed-film bioreactor. The application for that particular bioreactor presented a problem because the target compounds were present in an industrial wastewater at extremely low concentrations. Although the target compounds were readily biodegradable (aerobically) and served as growth substrates, the air flow through the bioreactor meant that stripping might outcompete the rate of biodegradation. Careful monitoring of the system and calculations to account for dilution effects at high airstream flows demonstrated that biodegradation was the primary source of target-compound attenuation and that in fact air stripping accounted for less than 4% of the compound's loss.

Air stripping can likewise be an issue during the enrichment phase for the in situ stimulation of aerobic microbial consortia (18). For example, column experiments have shown that biotreatments utilizing enrichment of methane-oxidizing bacteria for the cometabolism of volatile, chlorinated aliphatics (e.g. TCE) are confounded by air stripping. Initially, the enrichment for methane-oxidizing bacteria is so slow that the passage of air through the system can result in significant stripping of the target compound. However, after the methane oxidizers are established, the rate of TCE biodegradation can proceed at a sufficiently high rate to overcome any potential for air stripping. Thus during start-up of in situ bioremediation for readily volatile targets the initial air stream may need to be recirculated or, alternatively, treated above ground. As these examples suggest, air stripping during remediation can be minimized or avoided by relying on properly engineered systems or well-designed bioreactors. Indeed, a significant research effort is underway to address how various reactor designs affect air stripping (37) and demonstrates the crucial role afforded by design and systems engineering in the bioremediation industry.

Leaching

Leaching of target pollutants from an in situ treatment zone is an abiotic loss mechanism that can pose particularly difficult monitoring problems. Unlike stripping, which can be monitored by constructing overhead enclosures or by well-designed air monitoring protocols, leaching is difficult to detect and quantify because of the complexities of soils, sediments, and subsurface geology (e.g. channeling).

Leaching of a compound may occur because the compound is inherently water soluble or, if it is not soluble, because it binds to other soil components that are. For example, compounds regarded as essentially nonwater soluble (e.g. PCBs) can bind to the hydrophobic surfaces of soluble humic and fulvic acids and thereby be mobilized (21). Alternatively, bioremediation processes themselves may stimulate the leaching rate of target compounds. For example, the aerobic biodegradation of trinitrotoluene (TNT) proceeds via nitro group reduction to amino intermediates that, in the presence of oxygen, dimerize to form diazo compounds. In addition to being more toxic than the parent compound, these dimers are also leachable (35).

Biotreatment of contaminated soils in situ often involves the application of nutrient media to the soil, and some constituents in these media can cause solubilization of otherwise nonsoluble soil organic fractions. For example, the ability of phosphate ions to solubilize humic materials has been demonstrated and may be problematic because hydrophobic contaminants bind to these soil components (21). The potential for chemicals to leach from soil or sediment matrices can be addressed if the biotreatment test systems are constructed to allow for leachate collection. Lysimeters are one such tool and have been used in laboratory studies, in pilot field tests, and full-scale remediations. Mueller has shown that one can predict target leaching using a laboratory-scale "land farming chamber" (analogous to a field lysimeter) (65). The material to be treated is placed into a funnel held inside a small beaker. Periodically, as the biotreatment proceeds, subsamples of soil material and leachate are collected and analyzed for the appearance of the chemical targets or their metabolites, thereby allowing analysis of the entire biotreatment system. Lysimeters have also been used in field studies to follow the fate of diesel-fuel contamination during biotreatment (96), and laboratory and pilot scale studies have been used to characterize the fate of leachable pollutants (73).

Thus by carefully designing closed treatability and pilot systems, one can assess the contribution that leaching has on target-chemical loss in biotreatment systems; however, full-scale in situ bioremediation is still problematic. One approach used to aid in site monitoring is to distribute wells in and around the treatment zone when there is sufficient reason to believe the targets may leach. Alternatively, the bioremediation can be conducted within an enclosed system that allows the collection of leachate.

Sorption

The sorption of a target compound, especially an insoluble compound to a matrix component, or to surfaces within a bioreactor, can be of sufficient magnitude that the target effectively disappears from subsampled materials, resulting in an overestimate of the target's degradation. Assuming the reactor materials are appropriate and do not adsorb the target compound, other phenomenon related to sorption should be considered in evaluating bioremediation.

For example, despite a high degree of mixing, uneven distribution of an insoluble target compound within a bioreactor may occur. This latter anomaly was demonstrated in a laboratory study conducted to mimic the behavior of PCB soils in a bioslurry reactor (94). In an attempt to evaluate mechanisms of abiotic loss, Unterman et al (94) set up and evaluated a mock biodegradation system following the design of a highly publicized "EPA-approved" PCB bioremediation system. These researchers suspected that no biodegradation was occurring in the process because an analysis of the company's treated vs control samples showed proportional loss of all PCB congeners. This type of loss is characteristic of abiotic and not biological processes (93).

In the mock biodegradation study, soil contaminated with Aroclor 1260 was placed in a laboratory-scale, slurry bioreactor constructed as a closed system. The reactor was sparged with an inert gas (argon) to preclude aerobic biodegradation, and volatilization was monitored by incorporating a Florisil sampling tube in the outlet port. Small subsamples of the slurry were periodically removed during the course of the 19-day experiment. The subsamples indicated that, as in the highly publicized process, the concentration of PCB in the suspended slurry phase did decrease significantly (from 7800 ppm to 620 ppm). However, after extracting the entire contents of the reactor and off-gas, almost all of the PCB (86%) could be recovered. When only the amount of PCB present in the sampled slurry fraction was monitored, the results indicated that greater than 90% reduction in starting PCB concentration had been achieved. A careful analysis of the separate fractions remaining within the reactor showed that about 23% of the recovered material was bound to the surface of the reactor and impeller; about 25% was associated with tar balls (i.e. coalesced PCB oil droplets) at the bottom of the reactor; 15% was associated with the soil and water remaining in the bottom of the reactor; and 0.1% was volatilized (23% was removed in the experimental sampling).

This mock biodegradation study points out a significant pitfall for assessing biodegradation of insoluble, hydrophobic substrates such as PCBs. Although the time-course analysis indicated a progressive loss of PCB as would be expected from a biotreatment process, the mass balance calculations showed that the aeration and stirring of the soil resulted in a redistribution of dispersed PCBs from the soil to difficult-to-sample locations in the reactor (i.e. glassware, stirrer, and coalesced droplets of PCB).

A similar example is the bioremediation of a pentachlorophenol (PCP)-contaminated site (23). One batch of contaminated soil was biotreated in a soil slurry bioreactor, and it was observed that the PCP concentration in the aqueous phase diminished by 40% after only 5 days of treatment, which was an atypically high rate of biodegradation. Because the group performing the remediation had carefully monitored both the aqueous and soil phases in the bioreactor, this substantial rate of PCP depletion was easily explained and not mistakenly attributed to biodegradation. Apparently, the reactor had been charged with a soil sample that was cocontaminated with a volatile organic solvent that increased the capacity of the aqueous phase for PCP. After stripping removed the solvent during the first 5 days of treatment, the PCP repartitioned to the soil phase and was subsequently biodegraded but at slower rates corresponding well to the authors' previous experiences with PCP degradation.

These latter cases have shown that one must consider sorption as an abiotic loss mechanism in addition to volatilization or leaching. Target compounds that are relatively insoluble in aqueous systems (e.g. PCBs) tend to partition from the aqueous phase and accumulate or sequester around surfaces (including soil particles) within a bioreactor or, if liquid, to coalesce as a free phase. This sequestering can be detected by extracting the entire reactor contents or by comparing test results to abiotic controls. However, such controls are impractical for larger systems, and in those cases it may be prudent to selectively sample regions within a bioreactor that may act as a collection surface.

THE SCOPE OF EFFECTIVE EVALUATION

An evaluation of bioremediation requires one or more of the following: (*a*) appraising the biodegradability of the target compound by referring to prior literature (including laboratory feasibility studies and environmental data), (*b*) demonstrating biotreatability using actual site samples or surrogate matrices (i.e. simulation testing in the laboratory using bench-scale bioreactors or microcosms), and (*c*) demonstrating the bioremediation during pilot testing or full-scale remediation. This listing is not necessary for all cases but suggests the possible evaluation steps that can be considered in a formal step-by-step evaluation process.

Literature and Data Base Review

Reference to technical literature is a simple first step in determining the potential biodegradability of a particular target compound. Although in some cases the microbes or the environmental conditions conducive to their use in bioremediation are well characterized and published (5), care should be taken in evaluating these reports. The literature, especially trade journals and marketing brochures, contains a lot of misinformation.

Periodically the scientific literature contains reports indicating the biodegradability of compounds that are not in fact biodegradable in the manner cited. This can be attributed to the fact that alternative loss mechanisms were not addressed (i.e. insufficient care with controls, sampling, and/or analysis). One example was a report that Mirex could serve as a sole carbon and energy source for the growth of a bacterial culture. This conclusion was corrected in a subsequent letter to the journal, which pointed out that this process was theoretically impossible aerobically (6). Other reports have indicated growth on compounds that do not, to our knowledge, support growth aerobically, such as chloroform (17). Although aerobic growth on chloroform may be possible, to date this has never been demonstrated. (Methanotrophs operating under cometabolic conditions can degrade chloroform, but they derive their energy from the cosubstrate.) Other published experiments have included suggestions that were beyond the scope of the experimental data. One study, for example, demonstrated that a particular white rot fungus could biodegrade several relatively recalcitrant compounds, shown through the use of radiolabeled compounds and the monitoring of the evolution of $^{14}CO_2$. The same report also concluded that 2,3,7,8-TCDD (dioxin) was biodegradable by the fungus because trace amounts of $^{14}CO_2$ product were detected. However, no reference was made to the radiochemical purity of the test compound (19).

To aid in the interpretation of some of these results, several papers have been published that discuss the thermodynamic limitations that determine which target chemicals may serve as possible growth substrates (25, 26, 95). However, remember that thermodynamic limitations can be overcome within the context of cometabolism where the target compound is biodegradable only because the required energy for the reaction is supplied by oxidation of a cosubstrate. Ultimately one must rely on empirical studies to discover and elucidate these degradative processes.

In contrast to the aforementioned false positives, the technical literature (including published treatability databases) can also contain false-negative results. Remembering this is important if one is conducting an on-line computer search for the biotreatability of a specific compound X, because a significant portion of the biotreatability literature summarizes results from standardized tests developed for wastewater treatment. These test protocols are limited for predicting the biodegradability of many chemical targets of interest because they mimic the conditions typical of aerobic and anaerobic wastewater treatment facilities (7, 67). Tests under these conditions can also miss important cometabolic processes or alternatively can yield false-negative results if the target or an intermediate product is toxic (54). An excellent overview of how treatability tests should be conducted and the pitfalls associated with these general approaches has been published (34).

Laboratory Biotreatability Testing

Laboratory studies are useful and often critical because they can predict biotreatability potential using actual site samples, or they can be run in parallel to pilot or full-scale remediations for aiding evaluation. A third use, although not germain here, is that they may help establish critical parameters for system operation.

From a fundamental point of view, claims that biological processes are responsible for the transformation of a compound can be tested by noting specific features unique to biological phenomena. First, biochemical transformations behave in a manner predicted by Arrhenius analysis, namely, as temperature increases by 10°C there is a corresponding doubling of the biochemical reaction rate (49). This is also true with chemical reactions, except that biochemical reactions will generally be irreversibly inactivated at higher temperatures (e.g. 40–45°C). Second, unlike chemical catalysts, which do not reproduce themselves, living cells that catalyze biochemical transformations are transferable, although in some cases transfer is difficult (70). Other means of proving that microbial activity is responsible for a transformation include pasteurizing the soil or other matrix (98) or employing metabolic inhibitors known to inactivate specific metabolic reactions (22).

Biotreatability studies should be conducted by applying the selected biological process to actual site samples to ensure that actual environmental conditions are simulated. This is an important step because the characteristics (chemical, physical, and biological) of site soils, sediments, groundwater, or industrial effluents can increase or attenuate microbial degradation rates that may have been predictably modeled under cleaner laboratory conditions. Studies using cationic and anionic surfactants have clearly demonstrated that although the fate of these targets can be modeled relatively well in laboratory systems, the field correlation is not easy to establish (31, 83).

In some cases, the biodegradation potential can be retarded under field conditions because the target compound may adsorb to organic compounds in soil, thereby decreasing the rate of degradation (33). Alternatively, some target compounds are readily biodegradable in soil but not necessarily in a clean system without the soil, as has been demonstrated for compounds whose degradation proceeds owing to commensal interactions among a consortia of microbes (40).

Thus, proper biodegradability assessment of a target should generally be performed using representative site samples. A good review of the experimental approach for aerobic biodegradation studies is available (92). In general, if the test compounds are expected to support microbial growth, biodegradation can be monitored by using a biometer flask that allows for the collection of CO_2 from the metabolized target compound. Alternatively, a modified fernbach flask system developed by Marinucci & Bartha (60) allows one to quantify radiolabeled metabolites when the target compounds are present in low concentrations or when it is necessary to distinguish between CO_2 and volatile organic products. The hardware needed to construct these systems is available commercially. Similarly, standard biodegradability testing under anaerobic conditions has also been developed (8, 82).

When biotreatability studies are performed in simulated laboratory systems and run in parallel to pilot or full-scale biotreatments, they can be very effective for aiding the field evaluations. However, care in interpreting these parallel studies is needed because samples displaced from the environment and studied in the laboratory do not necessarily mimic the metabolic activities occurring in the field. For example, in vitro studies with sediment samples have shown that the methane production rates in the laboratory can differ significantly depending on where in the vertical profile the sample was obtained (47). Deep sediments produced methane at higher rates in the laboratory compared with production rates of the same sediments in the field, unlike the surface sediments whose methanogenic activity was reproduced in the laboratory. Difficulties associated with predicting the rates of a target compound decomposition have even been noted with relatively nonrecalcitrant materials such as wheat residue (87).

An interesting study by Acton & Barker (2) demonstrated that biodegradation was observed in situ but was not demonstrated in laboratory microcosms constructed to model and predict behavior in situ. This is perhaps an example of a microbial disturbance effect in which the environmental sample was disrupted when moved to the laboratory. This result is disquieting because typically the proof of in situ bioremediation needs to be buttressed with reference to laboratory experiments run in parallel.

Thus, site samples obtained and studied in the laboratory do not necessarily represent what one might see in the natural environment and may produce misleading results and poor extrapolation from laboratory to field data. Ultimately field studies may be required to solve this dilemma.

FIELD EVALUATION

Successful bioremediation can be monitored either directly or indirectly. Direct measurements require that the chemical targets themselves be monitored for removal or detoxification. Alternatively, if the biochemical or microbiological process is well characterized, the progression of bioremediation can be evaluated indirectly by monitoring: (*a*) metabolites derived from the degradative activity (including CO_2), (*b*) cosubstrate depletion (including oxygen), or (*c*) known changes corresponding to the microbial ecology of the treated system. Examples of the latter would include, for example, increased frequency of specific microbial populations or relevant genetic elements and changes in the chemical composition of the treated site (e.g. pH).

Direct Measurement of Target Depletion

Reference to control plots or control reactors can convincingly demonstrate the removal of a chemical. The most effective controls would differ from the biotreated plot or reactor by only one variable, such as the absence of the biocatalyst or of conditions conducive to microbial growth. Good use of controls to monitor or demonstrate the biologically mediated depletion of target compounds has been demonstrated for field biotreatments of PCBs (1), cationic surfactants (27), PCP (23), diesel oil (96), *para*-nitrophenol (86), and oil spills (69).

Often, however, well-designed controls are not available for reference. For instance, an accidental spill may need quick and decisive action (29), or direct monitoring may be elusive (e.g. deep subsurface). Nevertheless, internal standards can help to overcome some of these difficulties if they can serve as a conservative reference point from which the target's concentration can be reliably calculated. The physico-chemical behavior of the standards must be predictable and similar to that of the target compound. Internal standards can be organic compounds or inorganic ions (such as chloride or bromide); some tests have incorporated both (14).

Internal standards may be added to the treated matrix at some point during sample preparation or extraction (97), or they may be added before or during the bioremediation to provide estimates of target migration (14, 79). Alternatively, a site may already contain appropriate internal references if it is polluted with complex chemical mixtures. Biodegradation in such a case can be monitored by noting the differential depletion (extent or rate) of the constituent compounds. This example has been particularly well documented for PCB biodegradation, which results in patterns of congener depletion that are distinctly characteristic of the organism accomplishing the biodegradation (10, 93). Studies have used this phenomenon to verify or characterize the involvement of microbial metabolism in the transformations of creosote and crude oils in the environment (32, 75). The successful use of internal standards relies on the ability to demonstrate that the standard is not biodegraded within the time span of the biotreatment. For example, two internal standards used for assessing the cleanup of the Exxon Valdez oil spill were reported to be biodegradable, and the use of hopane has been proposed as a better conservative reference (20).

Indirect Measurements of Target Depletion

Direct measurements that are difficult to perform (for example, for in situ bioremediation), or that only provide incomplete evidence for biotransformations, may need to be supplemented by indirect measurements of biochemical transformation. Some indirect measures were discussed above (in the section on biodegradation of organic compounds); additionally, Madsen (58) recently presented an excellent overview of approaches to indirect measurements.

Aerobic biodegradation reactions will typically be accompanied by the uptake of cosubstrates (e.g. O_2) or the generation of metabolites (e.g. CO_2). However, such changes, are not easily measured during full-scale remediations except for in enclosed environments (e.g. bioreactors) or perhaps in some bioventing operations (38). Gasoline-contaminated sites, for example, have been treated by in situ bioventing, and during the process O_2 uptake and CO_2 release were detected and quantified. While these measurements can allow for a reasonable mass balance, their general applicability may be limited for practical use. Instead, monitoring for the appearance of metabolites that are indicative or unique to microbial metabolism may be more useful.

An excellent example showed that the biological degradation of TCE leads to the production of *cis*-1,2-dichloroethylene and that abiotic reactions lead instead to 1,1-dichloroethylene (45). Thus, some biotransformation reactions may lead to the production of unique biotic products. Some types of microbial metabolism can lead to the production of toxic intermediates (3). One field demonstration, for example, showed that the biotreatment of a soil contaminated with hydrocarbons resulted in a transient increase in the soil's acute toxicity. Presumably this increase resulted from the production of short-chain fatty acids. The transient nature of the toxicity is easily measured by using commercial tests (e.g. Microtox®) and may be useful as an indirect, qualitative measure of a biodegradation process (96).

Other indirect measurements, in the case of growth processes, may rely on the fact that target compounds can induce acclimation within a microbial community. This was demonstrated using *para*-nitrophenol (PNP) as a growth substrate in pond waters (86). The control area of a test pond received no PNP compared with the remaining test area, which received enough PNP to promote bacterial growth. After depletion of the test compound by biodegradation, both the test and control areas received fresh PNP, with the result that the PNP was depleted rapidly from the acclimated but not from the nonacclimated areas. Thus, another indirect measure is a differential lag time between acclimated and nonacclimated test systems.

It may also be possible to correlate increased cell members with bioremediation as was shown at a gasoline-contaminated site (55). In another very interesting approach, Madsen et al (59) correlated the degradation of a target compound with an increase in predator populations that were feeding on stimulated bacterial populations. Alternatively, one can quantify or characterize the genetic make-up of the microbial communities involved in a remediation, for example, by using catabolic gene markers to detect the occurrence of specific catabolic cell types. Finally, a new analytical technique is proving useful for characterizing community structure by reference to phospholipid ester-linked fatty acids as a qualitative marker of specific microbial communities (71).

CONCLUSION

This review aimed to demonstrate that bioremediation, even of difficult chemicals, is real, but also that there are numerous pitfalls in evaluating its efficacy for any particular soil, sediment, groundwater, or industrial wastestream. The applications of biotreatment for both remedial cleanup as well as industrial effluent control are growing rapidly and, where applicable, provide a safe, natural, and cost-effective process. These technologies have their basis in fundamental as well as applied research from environmental laboratories throughout the world that have developed a wealth of excellent scientific information. Therefore, one must be suspicious of unsubstantiated commercial claims that fly in the face of scientific knowledge. If the best peer-reviewed scientists in the world can't degrade pollutant Q, what is the probability that Dr. X from Superbug, Inc. has miraculously succeeded? One must keep an open mind, but be skeptical. The challenge we face is to continue developing the scientific and engineering base that provides the underpinnings for both the technology and its evaluation. If bioremediation is to fully reach its potential commercial applications, it will be necessary for vendors, users, and regulators to understand the various processes and mechanisms of bioremediation as well as develop and apply accurate measures of evaluation so as to distinguish fact from fiction.

Acknowledgments

We wish to thank Debbie Fugill for excellent assistance in the preparation of this manuscript.

Literature Cited

1. Abramowicz, D. A., Harkness, M. R., McDermott, J. B., Salvo, J. J. 1992. 1991 In situ Hudson River research study: a field study on biodegradation of PCBs in Hudson River sediments. *Final Rep.*, General Electric Co.
2. Acton, D. W., Barker, J. F. 1992. In situ biodegradation potential of aromatic hydrocarbons in anaerobic groundwaters. *J. Contam. Hydrol.* 9:325–52
3. Alexander, M. 1981. Biodegradation of chemicals of environmental concern. *Science* 211:132–38
4. Bartha, R. 1980. Pesticide residues in humus. *ASM News* 46:356–60
5. Bartha, R. 1986. Biotechnology of petroleum pollutant biodegradation. *Microb. Ecol.* 12:155–72
6. Bartha, R. 1986. Comments concerning "Search for mirex-degrading soil microorganisms" by J. Ashlanzadeh and H. G. Hedrick. *Soil Sci.* 142:241
7. Battersby, N. S. 1990. A review of biodegradation kinetics in the aquatic environment. *Chemosphere* 21:1243–84
8. Battersby, N. S., Wilson, V. 1988. Evaluation of a serum bottle technique for assessing the anaerobic biodegradability of organic chemicals under methanogenic conditions. *Chemosphere* 17:2441–60
9. Baxter, R. M., Sutherland, D. A. 1984. Biochemical and photochemical processes in the degradation of chlorinated biphenyls. *Environ. Sci. Technol.* 18:608–10
10. Bedard, D. L., Unterman, R., Bopp, L. H., Brennan, M. J., Haberl, M. L., Johnson, C. 1986. Rapid assay for screening and characterizing microorganisms for the ability to degrade polychlorinated biphenyls. *Appl. Environ. Microbiol.* 51:761–68
11. Bollag, J.-M., Loll, M. J. 1983. Incorporation of xenobiotics into soil humus. *Experientia* 39:1221–31
12. Bollag, J.-M., Shuttleworth, K. L., Anderson, D. H. 1988. Laccase-mediated detoxification of phenolic compounds. *Appl. Environ. Microbiol.* 54:3086–91
13. Bollag, J.-M., Sjoblad, R. D., Minard, R. D. 1977. Polymerization of phenolic intermediates of pesticides by a fungal enzyme. *Experientia* 33:1564–66
14. Borden, R. C., Bedient, P. B. 1987. In situ measurement of adsorption and biotransformation at a hazardous waste site. *Water Resources Bulletin* 23:629–36
15. Bosecker, K. 1986. Bacterial metal recovery and detoxification of industrial waste. In *Workshop on Biotechnology for the Mining, Metal-Refining and Fossil Fuel Processing Industries, Biotech. Bioeng. Symp. No. 16,* ed. H. L. Ehrlich, D. S. Holmes, pp. 105–20. New York: Wiley
16. Bossert, I., Kachel, W. M., Bartha, R. 1984. Fate of hydrocarbons during oily sludge disposal in soil. *Appl. Environ. Microbiol.* 47:763–67
17. Bouwer, E. J., Rittman, B. E., McCarty, P. L. 1982. Correspondence on: anaerobic degradation of halogenated 1- and 2-carbon organic compounds. *Environ. Sci. Technol.* 16:130
18. Broholm, K., Christensen, T. H., Jensen, B. K. 1991. Laboratory feasibility studies on biological in situ treatment of a sandy soil contaminated with chlorinated aliphatics. *Environ. Technol. Lett.* 12:279–89
19. Bumpus, J. A., Tien, M., Wright, D., Aust, S. D. 1985. Oxidation of persistent environmental pollutants by a white rot fungus. *Science* 228:1434–36
20. Butler, E. L., Douglas, G. S., Steinhauer, W. G., Prince, R. C., Aczel, T., et al. 1991. Hopane, a new chemical tool for measuring oil biodegradation. In *On-Site Bioreclamation, Processes for Xenobiotic and Hydrocarbon Treatment,* eds. R.E. Hinchee and R.F. Olfenbuttel, pp. 515–21. Stoneham: Butterworth-Heinemann
21. Chin, Y.-P., Weber, W. J., Chiou, C. T. 1991. A thermodynamic partition model for binding of nonpolar organic compounds by organic colloids and implications for their sorption to soils and sediment. In *Organic Substances and Sediments in Water,* ed. R. A. Baker, 1:251–73. Chelsea, MI: Lewis. 392 pp.
22. Compeau, G. C., Bartha, R. 1985. Sulfate-reducing bacteria: principal methylators of mercury in anoxic estuarine sediment. *Appl. Environ. Microbiol.* 50:498–502
23. Compeau, G. C., Mahaffey, W. D., Patras, L. 1991. Full-scale bioremediation of contaminated soil and water. See Ref. 77a, pp. 91–109
24. Dagley, S. 1975. A biochemical approach to some problems of environmental pollution. In *Essays in Biochemistry,* ed. P. N. Campbell, W. N. Aldridge, 2:81–130. London: Academic

25. Dolfing, J., Harrison, B. K. 1992. Gibbs free energy of formation of halogenated aromatic compounds and their potential role as electron acceptors in anaerobic environments. *Environ. Sci. Technol.* 26:2213–18
26. Dolfing, J., Tiedje, J. M. 1987. Growth yield increase linked to reductive dechlorination in a defined 3-chlorobenzoate degrading methanogenic coculture. *Arch. Microbiol.* 149:102–5
27. Federle, T. W., Pastwa, G. M. 1988. Biodegradation of surfactants in saturated subsurface sediments: a field study. *Ground Water* 26:761–70
28. Flathman, P. E. 1992. Bioremediation technology advances via broad research and applications. *Gen. Eng. News* 12:4–11
29. Frankenberger, W. T. 1989. In situ bioremediation of an underground diesel fuel spill: a case history. *Environ. Manage.* 13:325–32
30. Gibson, D. T. 1984. *Microbial Degradation of Organic Compounds*. New York: Marcel Dekker. 535 pp.
31. Gledhill, W. E., Trehy, M. L., Carson, D. B. 1991. Comparative biodegradability of anionic surfactants in synthetic and natural test systems. *Chemosphere* 22:873–80
32. Godsy, E. M., Goerlitz, D. F., Grbic-Galic, D. 1987. Anaerobic biodegradation of creosote contaminants in natural and simulated ground-water ecosystems. In *Geological Survey Program on Toxic Waste—Groundwater Contamination, 3rd Tech. Meeting,* Pensacola, pp. A:17–19. Menlo Park: US Geological Survey
33. Gordon, A. S., Millero, F. J. 1985. Adsorption mediated decrease in the biodegradation rate of organic compounds. *Microb. Ecol.* 11:289–98
34. Grady, C. P. L. Jr. 1985. Biodegradation: its measurement and microbiological basis. *Biotech. Bioeng.* 27: 660–74
35. Griest, W. H., Stewart, A. J., Tyndal, R. L., Ho, C. H., Tan, E. 1990. Characterization of explosives processing waste decomposition due to composting. *Report No. DOE IAC 1016-B123-A1*
36. Hardman, D. J. 1991. Microbial pollution control: a technology in its infancy. *Chem. Ind.* 7:244–46
37. Hill, G. A., Tomusiak, M. E., Quail, B., Van Cleave, K. M. 1991. Bioreactor design effects on biodegradation capabilities of VOCs in wastewater. *Environ. Progress* 10:147–53
38. Hoeppel, R. E., Hinchee, R. E., Arthur, M. F. 1991. Bioventing soils contaminated with petroleum hydrocarbons. *J. Ind. Microbiol.* 8:141–46
39. Hsu, T.-S., Bartha, R. 1974. Interaction of pesticide-derived chloroaniline residues with soil organic matter. *Soil Sci.* 116:444–52
40. Hsu, T.-S., Bartha, R. 1979. Accelerated mineralization of two organophosphate insecticides in the rhizosphere. *Appl. Environ. Microbiol.* 37: 36–41
41. Hughes, B. M., McKenzie, D. E., Mappes, G. W. 1991. *Proc. Electric Power Research Institute PCB Seminar, Baltimore,* pp. 6:1–5. Palo Alto, CA: Electric Power Research Inst.
42. Ishibashi, Y., Cervantes, C., Silver, S. 1990. Chromium reduction in *Pseudomonas putida*. *Appl. Environ. Microbiol.* 56:2268–70
43. Ji, G., Salzberg, S. P., Silver, S. 1989. Cell-free mercury volatilization activity from three marine *Caulobacter* strains. *Appl. Environ. Microbiol.* 55: 523–25
44. Kaake, R. H., Roberts, D. J., Stevens, T. O., Crawford, R. L., Crawford, D. L. 1992. Bioremediation of soils contaminated with the herbicide 2-*sec*-butyl-4,6-dinitrophenol (Dinoseb). *Appl. Environ. Microbiol.* 58:1683–89
45. Kästner, M. 1991. Reductive dechlorination of tri- and tetrachloroethylenes depends on transition from aerobic to anaerobic conditions. *Appl. Environ. Microbiol.* 57:2039–46
46. Keith, L. H. 1991. *Environmental Sampling and Analysis: A Practical Guide*. Chelsea, MI: Lewis. 143 pp.
47. Kelly, C. A., Chynoweth, D. P. 1980. Comparison of *in situ* and *in vitro* rates of methane release in freshwater sediments. *Appl. Environ. Microbiol.* 40:287–93
48. Klibanov, A. M., Tu, T.-S., Scott, K. P. 1983. Peroxidase-catalyzed removal of phenols from coal-conversion waste waters. *Science* 221:259–61
49. Kohring, G.-W., Zhang, X., Wiegel, J. 1989. Anaerobic dechlorination of 2,4-dichlorophenol in freshwater sediments in the presence of sulfate. *Appl. Environ. Microbiol.* 55:2735–37
50. Kong, H.-L., Sayler, G. S. 1983. Degradation and total mineralization of monohalogenated biphenyls in natural sediment and mixed bacterial culture. *Appl. Environ. Microbiol.* 46:666–72
51. Kuhn, S. P., Pfister, R. M. 1990. Accumulation of cadmium by immobilized *Zoogloea ramigera* 115. *J. Ind. Microbiol.* 6:123–28

52. Leahy, J. G., Colwell, R. R. 1990. Microbial degradation of hydrocarbons in the environment. *Microbiol. Rev.* 54:305–15
53. Lerch, K. 1980. Copper metallothionein, a copper-binding protein from *Neurospora crassa*. *Nature* 284:368–70
54. Lindgaard-Jørgensen, P., Rieman, B. 1989. ^{3}H-thymidine incorporation preliminary investigations of a method to forecast the toxicity of chemicals in biodegradability tests. *Chemosphere* 19:1447–55
55. Litchfield, C. D., Erkenbrecher, C. W. Jr., Matson, C. E., Fish, L. S., Levine, A. 1988. Evaluation of microbial detection methods and interlaboratory comparisons during a peroxide-nutrient enhanced in situ bioreclamation. *Water Sci. Technol.* 20:11–12
55a. Lovley, D. R. 1993. Dissimilatory metal reduction. *Annu. Rev. Microbiol.* 47:263–90
56. Lovely, D. R., Phillips, E. J. P. 1992. Bioremediation of uranium contamination with enzymatic uranium reduction. *Environ. Sci. Technol.* 26:2228–34
57. Macalady, D. L., Tratnyek, P. G., Grundl, T. J. 1986. Abiotic reduction reactions of anthropogenic organic chemicals in anaerobic systems: a critical review. *J. Contam. Hydrol.* 1:1–28
58. Madsen, E. L. 1991. Determining in situ biodegradation: facts and challenges. *Environ. Sci. Technol.* 25: 1663–73
59. Madsen, E. L., Sinclair, J. L., Ghiorse, W. C. 1991. *In situ* biodegradation: microbiological patterns in a contaminated aquifer. *Science* 252:830–33
60. Marinucci, A. C., Bartha, R. 1979. Apparatus for monitoring the mineralization of volatile ^{14}C-labeled compounds. *Appl. Environ. Microbiol.* 38: 1020–22
61. Miller, G. C., Hebert, V. R., Miller, W. W. 1989. Effect of sunlight on organic contaminants at the atmosphere-soil interface. In *Reactions and Movements of Organic Chemicals in Soils*. pp. 99–110. Madison, WI: Soil Sci. Soc. Am. 474 pp.
62. Miller, G. C., Zepp, R. G. 1983. Extrapolating photolysis rates from the laboratory to the environment. *Res. Rev.* 85:89–110
63. Miller, R. M., Singer, G. M., Rosen, J. D., Bartha, R. 1988. Photolysis primes biodegradation of benzo[*a*]pyrene. *Appl. Environ. Microbiol.* 54: 1724–30
64. Moza, P., Shuenert, I., Klein, W., Korte, F. 1979. Studies with 2,4′,5-trichlorobiphenyl-^{14}C and 2,2′,4,4′,6-pentachlorobiphenyl-^{14}C in carrots, sugar beets, and soil. *J. Agric. Food Chem.* 27:1120–24
65. Mueller, J. G., Lantz, S. E., Blattmann, B. O., Chapman, P. J. 1991. Bench-scale evaluation of alternative biological treatment processes for the remediation of pentachlorophenol- and creosote-contaminated materials: solid-phase bioremediation. *Environ. Sci. Technol.* 25:1045–55
66. Newman, L. M., Wackett, L. P. 1991. Fate of 2,2,2-trichloroacetaldehyde (chloral hydrate) produced during trichloroethylene oxidation by methanotrophs. *Appl. Environ. Microbiol.* 57:2399–2402
67. Nyholm, N. 1992. The European system of standardized legal tests for assessing the biodegradability of chemicals. *Environ. Tech. Chem.* 10:1237–46
68. Pope, T. 1990. Increasing role for “bugs” in waste cleanup. *Waste Age* 21:86–88
69. Pritchard, P. H., Costa, C. F. 1991. EPA’s Alaska oil spill bioremediation project. *Environ. Sci. Technol.* 25:372–79
70. Quensen, J. F. III, Tiedje, J. M., Boyd, S. A. 1988. Reductive dechlorination of polychlorinated biphenyls by anaerobic microorganisms from sediments. *Science* 242:752–54
71. Rajendran, N., Matsuda, O., Imamura, N., Urushigawa, Y. 1992. Variation in microbial biomass and community structure in sediments of eutrophic bays as determined by phospholipid ester-linked fatty acids. *Appl. Environ. Microbiol.* 58:562–71
72. Rao, P. S. C., Lee, L. S., Pinal, R. 1990. Cosolvency and sorption of hydrophobic organic chemicals. *Environ. Sci. Technol.* 24:647–54
73. Reinhart, D. R., Pohland, F. G. 1991. The assimilation of organic hazardous wastes by municipal solid waste landfills. *J. Ind. Microbiol.* 8:193–200
74. Roberts, L. 1987. Discovering microbes with a taste for PCBs. *Science* 237:975–77
75. Rowland, S. J., Alexander, R., Kagi, R. I., Jones, D. M., Douglas, A. G. 1986. Microbial degradation of aromatic components of crude oils: a comparison of laboratory and field observations. *Org. Geochem.* 9:153–61
76. Ruggiero, P., Sarkar, J. M., Bollag, J.-M. 1989. Detoxification of 2,4-dichlorophenol by a laccase im-

mobilized on soil or clay. *Soil Sci.* 147:361–70

77. Sayler, G. S., Fox, R. 1991. Environmental biotechnology: perceptions, reality, and applications. See Ref. 77a, pp. 1–13

77a. Sayler, G. S., Fox, R., Blackburn, J. W., eds. 1991. *Environmental Biotechnology for Waste Treatment*. New York: Plenum

78. Sedlak, D. L., Dean, K. E., Armstrong, D. E., Andren, A. W. 1991. Interaction of quicklime with polychlorobiphenyl-contaminated solids. *Environ. Sci. Technol.* 25:1936–40

79. Semprini, L., Hopkins, G. D., Roberts, P. V., Grbic-Galic, D., McCarty, P. L. 1991. A field evaluation of in situ biodegradation of chlorinated ethenes: Part 3, studies of competitive inhibition. *Ground Water* 29:239–50

80. Shannon, M. J. R. 1990. *Use of a fungal laccase for immobilization and detoxification of leachable phenolic pollutants in soil*. PhD thesis. Rutgers University, New Brunswick, N.J. 136 pp.

81. Shannon, M. J. R., Bartha, R. 1988. Immobilization of leachable toxic soil pollutants by using oxidative enzymes. *Appl. Environ. Microbiol.* 54:1719–23

82. Shelton, D. R., Tiedje, J. M. 1984. General method for determining anaerobic biodegradation potential. *Appl. Environ. Microbiol.* 47:850–57

83. Shimp, R. J., Schwab, B. S. 1991. Use of a flow-through in situ environmental chamber to study microbial adaptation processes in riverine sediments and periphyton. *Environ. Tox. Chem.* 10:159–67

84. Sneddon, J., Pappas, C. P. 1991. Binding and removal of metal ions in solution by an algae biomass. *Am. Environ. Lab.* 10:9–13

85. Song, H.-G., Bartha, R. 1990. Effects of jet fuel spills on the microbial community of soil. *Appl. Environ. Microbiol.* 56:646–51

86. Spain, J. C., Van Veld, P. A., Monti, C. A., Pritchard, P. H., Cripe, C. R. 1984. Comparison of *p*-nitrophenol biodegradation in field and laboratory test systems. *Appl. Environ. Microbiol.* 48: 944–50

87. Stroo, H. F., Bristow, K. L., Elliot, L. F., Papendick, R. I., Campbell, G. S. 1989. Predicting rates of wheat residue decomposition. *Soil Sci. Soc. Am. J.* 53:91–99

88. Sullivan, K. M., Skladany, G. J. 1987. *Decay theory biological treatment for low-level organic contaminated groundwater and industrial waste*. Presented at Superfund '87: Proceedings of the 8th National Conference, Silver Spring, MD

89. Summers, A. O. 1986. Organization, expression, and evolution of genes for mercury resistance. *Annu. Rev. Microbiol.* 40:607–34

90. Timmis, K. N., Rojo, F., Ramos, J. L. 1988. Prospects for laboratory engineering of bacteria to degrade pollutants. See Ref. 90a, pp. 61–79

90a. Omenn, G. S., ed. 1988. *Environmental Biotechnology: Reducing Risks from Environmental Chemicals Through Biotechnology*. New York: Plenum

91. Tomioka, N., Uchiyama, H., Yagi, O. 1992. Isolation and characterization of cesium-accumulating bacteria. *Appl. Environ. Microbiol.* 58:1019–23

92. United States Food and Drug Administration. 1987. *Environmental Assessment Technical Assistance Handbook: Aerobic Biodegradation in Soil. Report No. PB87–175345*

93. Unterman, R. 1991. What is the K_m of disappearase? See Ref. 77a, pp. 159–62

94. Unterman, R., Bedard, D. L., Brennan, M. J., Bopp, L. H., Mondello, F. J., et al. 1988. Biological approaches for polychlorinated biphenyl degradation. See Ref. 90a, pp. 253–69

95. Vogel, T. M., Criddle, C. S., McCarty, P. L. 1987. Transformations of halogenated aliphatic compounds. *Environ. Sci. Technol.* 21:722–36

96. Wang, X., Bartha, R. 1990. Effects of bioremediation on residues, activity, and toxicity in soil contaminated by fuel spills. *Soil Biol. Biochem.* 22:501–5

97. Wang, X., Yu, X., Bartha, R. 1990. Effect of bioremediation on polycyclic aromatic hydrocarbon residues in soil. *Environ. Sci. Technol.* 24:1086–89

98. Ye, D., Quenson, J. F. III, Tiedje, J. M., Boyd, S. A. 1992. Anaerobic dechlorination of polychlorobiphenyls (Aroclor 1242) by pasteurized and ethanol-treated microorganisms from sediments. *Appl. Environ. Microbiol.* 58: 1110–14

Annu. Rev. Microbiol. 1993. 47:739–63

GENETICALLY ENGINEERED PROTECTION AGAINST VIRUSES IN TRANSGENIC PLANTS

John H. Fitchen and Roger N. Beachy
The Scripps Research Institute, La Jolla, California 92037

KEY WORDS: virus resistance, coat protein–mediated resistance, replicase-mediated resistance, pathogen-derived resistance

---

CONTENTS

ABSTRACT

Transgenic plants carrying nucleotide sequences derived from plant viruses can exhibit increased resistance to viral disease. Many viral sequences confer some level of either resistance to infection or suppression of disease symptoms

0066-4227/93/1001-0739$02.00

(tolerance). These include segments of viral genomes encoding capsid or coat proteins, sequences encoding proteins that are or may be subunits of the viral replicase, sequences incapable of encoding proteins, entire genomes of defective interfering viruses and satellite viruses, and complete genomes of mild strains of virus. The transgene may act on initiation of infection, replication of virus, spread of the infection throughout the plant, and symptom development. More than one of these processes can be impaired by a single transgene derived from a single viral gene. The level of protection ranges from very low to high, while the breadth of protection ranges from very narrow, where protection is only observed against closely related strains of the virus from which the transgene was derived, to moderately broad, extending to other viruses. Data are insufficient to establish a molecular mechanism of resistance for most of the described examples. In addition, although the use of a particular segment of the viral genome confers resistance in one virus-host system, analogous sequences from a different virus in another host may be ineffective.

INTRODUCTION

The presence of one virus in a plant can interfere with infection by another virus or strain, resulting in cross protection (23). The idea that individual viral components contained in plants might also interfere with virus infection predated the development of stable gene-transfer techniques for transgenic plants (28). A broad and formal statement of the hypothesis that genes conferring resistance against a virus might be derived from the viral genome was made in 1985 (85). It stated that certain key gene products of a pathogen present in the plant in a dysfunctional form, in excess, or at an inappropriate stage during the viral replication cycle could disrupt infection by the invading pathogen. Although initially termed parasite-derived resistance, the concept is now widely referred to as pathogen-derived resistance.

This review summarizes the progress made in applying and understanding pathogen-derived resistance to viruses in transgenic plants. Examples of transgenic plants in which the transgene is not derived from the pathogen per se, but from viral satellites associated with some viruses, are included. However, this chapter does not describe many of the details of individual examples in which resistance has been demonstrated, even though some of these examples may be of great agronomic importance. Rather, those experiments are emphasized that best help to understand how resistance is conferred at the cellular and molecular levels. The examples have been divided into categories according to the nature of the sequences used as transgenes.

PROTECTION CONFERRED BY SEQUENCES ENCODING VIRAL COAT PROTEINS

In 1986, Powell et al (70) described experiments in which transgenic tobacco were produced that had a chimeric gene containing a cDNA copy of the coat protein (*CP*) gene of tobacco mosaic virus (TMV). The plants were found to have an increased level of resistance to TMV infection. The resistance phenomenon was given the name coat protein–mediated resistance. Since the initial demonstration, numerous examples of resistance in plants transgenic for viral *CP* gene constructs have been reported. Table 1 lists reports in the refereed literature, as of late 1992.

Individual Examples of Resistance

The conceptual simplicity of the approach and the availability of cDNA clones of *CP* genes of many plant viruses have facilitated the broad and rapid implementation of the coat-protein strategy. The technical barriers to its application to new viruses, however, are substantial. In many cases, limitations of gene transfer and plant regeneration technologies have necessitated the use of plant hosts (often *Nicotiana* species) that are model systems. Furthermore, the transgenic plants are often challenged by mechanical

Table 1 Examples of transgenic plants in which protection is conferred by viral CP sequences

Plant	Viral CP	References
Tobacco, tomato	Tobacco mosaic virus (TMV)	1, 62–64, 70, 84
Tomato	Tomato mosaic virus (ToMV)	84
Tobacco, *Nicotiana benthamiana*	Tobacco rattle virus (TRV)	95, 96
Potato, tobacco	Potato virus X (PVX)	31, 33, 41, 42, 45
Potato, *Nicotiana debneyii*	Potato virus S (PVS)	53, 54
Tobacco	Soybean mosaic virus (SMV)	88
Potato	Potato virus Y (PVY)	42, 45, 93
Papaya, tobacco	Papaya ringspot virus (PRV)	22, 48
Tobacco	Tobacco etch virus (TEV)	46, 47
N. benthamiana	Watermelon mosaic virus II (WMVII)	60
N. benthamiana	Zucchini yellow mosaic virus (ZYMV)	60
N. benthamiana, N. clevelandii	Plum pox virus (PPV)	79
Rice	Rice stripe virus (RSV)	30
Tobacco, alfalfa, tomato	Alfalfa mosaic virus (ALMV)	1, 32, 49, 92, 96, 97
Tobacco	Tobacco streak virus (TSV)	97
Tobacco, cucumber	Cucumber mosaic virus (CMV)	18, 27, 59, 74
Tobacco	Arabis mosaic virus (ArMV)	10
Potato	Potato leafroll virus (PLRV)	5, 43, 44, 94
Tobacco	Tomato spotted wilt virus (TSWV)	19, 25, 52, 68

inoculation with virus-containing solutions, whereas many plant viruses are spread naturally by insect vectors.

Constraints on the length of this review do not allow discussion of most of the examples in which resistance conferred by *CP* sequences has been demonstrated. Seven examples have been chosen on the basis of the extent to which the resistance mechanism has been explored and the degree to which the example helps to illustrate the diversity of probable mechanisms.

TOBACCO MOSAIC VIRUS Tobacco mosaic virus (TMV) is the most extensively characterized plant virus. The virion is a rigid rod 18 nm in diameter and 300 nm long containing a single-stranded (+)-sense RNA genome that is 6395 nucleotides in length. The capsid consists of approximately 2130 identical 17.5-kDa subunits of virus-encoded coat protein. In the virus particle, the coat protein forms a helical array with the RNA located at the axial interfaces of the subunits. The structure of the coat protein has been determined to atomic resolution. For additional information about the viruses described in this chapter, the reader is referred to *Plant Virology* by Matthews (55).

A chimeric gene was constructed to express TMV *CP* in plants using a cloned cDNA derived from the 3′ end of the TMV genome (70). Because cDNA copies of TMV promoters do not function in plants, a promoter from the plant DNA virus cauliflower mosaic virus, which can drive high levels of expression of foreign genes in many plant tissues, was used for the chimeric gene (65). Sequences from the 3′ end of the *Agrobacterium tumefaciens nos* gene provide transcription termination and polyadenylation signals.

The chimeric gene was transferred to tobacco (*Nicotiana tabacum*) tissue by *A. tumefaciens*–mediated transfer (98). Plants regenerated from the tissue were allowed to self-fertilize and their progeny were analyzed. Several plant lines derived from independent transformation events were used in the analysis.

After mechanical inoculation with solutions containing purified TMV, the transgenic plants were found to lack symptoms or to exhibit a delay in symptom development compared with similarly inoculated control plants. The proportion of plants that remained without symptoms and the duration of the delay in symptom development in those that did become infected were reduced when the concentration of virus in the challenge inoculum was increased (70). Plants that did not develop symptoms did not contain detectable virus in uninoculated leaves (64). Transgenic plants that express the *CP* gene (CP^+ plants) had fewer initial sites of infection than did nontransgenic (CP^-) plants, and virus concentrations in the inoculum had to be 1000- to 10,000-fold higher for CP^+ plants than for CP^- plants to yield the same numbers of infection sites (62, 64). However, when TMV RNA rather than TMV was included in the inoculum, infections were initiated at roughly the same frequency on CP^+ and CP^- plants, and similar proportions of CP^+ and CP^- plants developed systemic disease (64).

Spread of the virus infection from the initial site was also delayed, although the effect was marginal (100). Delays in virus accumulation were observed only in regions of the inoculated leaf more than 2.5 mm distant from the initially infected cells, in the stem adjacent to or above the inoculated leaf, or in the upper leaves of the plant. The delay in spread of infection was not overcome by inoculation with TMV RNA.

The level of coat protein accumulation and the degree of resistance were positively correlated when resistance was scored across several plant lines accumulating different levels of coat protein (71). Furthermore, when transgenic plants were grown at higher temperatures, drastic reductions in coat protein accumulation were observed and resistance was overcome (61).

Protoplasts prepared from leaf tissue of transgenic TMV CP^+ plants and from control plants were used to study resistance in single cells (75). Inoculation of CP^+ protoplasts resulted in fewer infected protoplasts than inoculation of CP^- protoplasts. In the few infected CP^+ protoplasts, however, newly synthesized virus accumulated to approximately the same levels as in infected CP^- protoplasts. When protoplasts were inoculated with TMV RNA, approximately the same proportion of both CP^+ and CP^- protoplasts became infected and virus accumulated to similar levels.

The transgenic plants were tested for resistance to infection by another strain of TMV, by other viruses in the tobamovirus group, and by viruses belonging to other groups (1, 62). Protection against a closely related strain was essentially as effective as homologous protection. Protection against other tobamoviruses depended upon which tobamovirus was included in the inoculum and its apparent degree of relatedness to TMV according to *CP* sequence. The coat protein of tomato mosaic virus (ToMV) shares 82% amino acid identity with the TMV coat protein, and tobacco mild green mosaic virus (TMGMV), shares 72% identity. These viruses were similar to TMV in the extent to which the plants were able to resist infection. In contrast, ribgrass mosaic virus (RMV), the coat protein of which is only 45% identical to TMV coat protein, was much more able to infect and cause disease in CP^+ plants. There was, therefore, a rough correlation between the degree of similarity between the coat proteins of the challenge virus and TMV and the extent of resistance conferred by the TMV *CP* transgene. Protection against viruses belonging to other groups was very weak or was nonexistent (1). Symptoms were assessed after inoculation with potato virus X (PVX), the type member of the potexvirus group of plant viruses; potato virus Y (PVY), the type member of the potyvirus group; cucumber mosaic virus (CMV), the type member of the cucumoviruses; and alfalfa mosaic virus (AlMV). After inoculation with very low concentrations of PVX, PVY, or CMV, slight delays were observed in systemic spread of the virus and in development of disease symptoms in transgenic TMV CP^{+} plants compared with control plants. However, these delays were not observed when virus concentrations in the inoculum were increased. Because of the limited and ephemeral nature of the effect, it was not possible to determine if the delays resulted directly from the expression of the transgene or whether an altered physiological status in the transgenic plants was responsible (80).

POTATO VIRUS X Potato virus X (PVX) is a flexuous rod-shaped virus. The 25-kDa coat protein subunits are arranged in a helix with an outside diameter of 12 nm in the virus particle. Transgenic tobacco plants accumulating PVX coat protein to 0.02–0.10% of total extractable protein were produced (31). Plants that expressed a gene construct with the *CP* coding region in the inverted orientation, thus making an antisense transcript, were also produced. The transgenic plants with the gene construct directing the synthesis of coat protein were significantly less susceptible to infection than were control plants. CP^{+} plants exhibited reduced numbers of infection sites on the inoculated leaves, a delay or absence of systemic disease symptoms, and reduced virus accumulation in both inoculated and upper leaves. The degree of protection was correlated with the level of accumulation of transgene-derived coat protein. In marked contrast to the TMV CP^{+} plants described earlier, the transgenic PVX CP^{+} plants were as resistant to inoculation with PVX RNA as they were to inoculation with whole virus. The plant lines carrying the antisense gene construct showed only a low degree of protection.

Transgenic potato plants accumulating PVX coat protein to as much as 0.1% of soluble leaf protein were recovered and tested by mechanical inoculation with PVX (33, 41, 42, 45). The transgenic plants showed a dramatic reduction in the accumulation of virus compared with CP^{-} plants. In CP^{+} plants that did develop infections, symptom development was delayed. In one study a correlation was observed between expression level of the *CP* gene and the reduction in virus accumulation, as in tobacco. In contrast, the most resistant line in another study accumulated lower levels of PVX coat protein than less-resistant lines. Field testing of the transgenic potato plants demonstrated effective resistance under highly variable conditions.

VIRUSES OF THE POTYVIRUS GROUP The virus particles of viruses in this group are long, flexuous rods with a helical arrangement of coat-protein subunits. In contrast to the coat proteins described so far, the coat protein of potyviruses is not translated in its final form of about 30 kDa. Rather, nearly the entire viral genome is translated into a single polyprotein that is cleaved into the functional viral proteins. Levels of sequence identity between the coat proteins of different potyviruses range upward from about 50% over the whole protein, but the central region of the coat protein sequence tends to be the most similar, whereas the NH_3 and COOH termini tend to be more divergent.

Potato virus Y (PVY) is the type member of the potyvirus group. Potato lines expressing a *CP* gene of strain PVY^0 accumulated coat protein to 0.01–0.05% of total protein and exhibited a high degree of resistance to infection when challenged by mechanical inoculation (42, 45). Many of the plants lacked detectable virus accumulation at 35 days postinoculation. Some lines were also resistant to challenge by the viruliferous aphids that are the natural vector of the disease.

In contrast to the protection conferred against TMV by TMV *CP*, resistance was greatest in plants accumulating lower levels of coat protein. This was also true for transgenic plants accumulating the coat protein of strain PVY^N (93). Although highly effective protection was observed in one line, no transgene-derived coat protein could be detected (21). This protein could be detected only in those plants that became infected with another PVY strain and only by using detection with strain-specific monoclonal antibodies. Presumably, heterologous incorporation of the transgenic coat protein into mixed virions stabilized the coat protein, allowing accumulation to higher levels. To further demonstrate a lack of a positive correlation between coat protein levels and protection, a gene lacking an initiating methionine codon was constructed (93). Transgenic plants carrying this gene were as resistant to PVY infection as were plants carrying the construct with the initiation signals added.

Tobacco etch virus (TEV) is another member of the potyvirus group. Gene constructs to direct the synthesis of full-length as well as truncated forms of the coat protein were made, as were constructs to produce an antisense transcript and a transcript without the added translational initiation site (46, 47). Transgenic plants carrying the full-length gene construct accumulated TEV coat protein to as much as 0.01% of total extractable protein.

The transgenic plants were challenged by mechanical inoculation with relatively high levels of purified virus or virus-containing plant sap (46). No delay in the appearance of symptoms was observed, but there was a slight attenuation of symptoms and an eventual emergence of younger leaf tissue devoid of symptoms and virus. Transgenic plant lines expressing truncated genes appeared to have more effective resistance than lines expressing the entire gene, although only a few lines were analyzed and the variation between lines may be great. Much more striking was the delay in the appearance of symptoms, and the amelioration of symptom severity in plant lines accumulating only *CP*-related RNA sequences (47). Many individual plants of both an antisense plant line and a plant line expressing the gene construct without the initiating methionine codon remained entirely asymptomatic and did not accumulate virus. The resistance was independent of inoculum concentration over the range used; resistance was not conferred against infection by other potyviruses.

The effects of the various transgenes at the cellular level were examined in protoplasts (47). TEV replication was normal in protoplasts prepared from plants accumulating full-length or truncated coat proteins. Protoplasts from the plants carrying the construct without the initiating methionine codon did not support TEV replication at a detectable level. Protoplasts carrying the antisense construct supported replication at reduced levels.

Soybean mosaic virus (SMV) is also a member of the potyvirus group. The coat protein of SMV shares 61 and 58% sequence identity with PVY and TEV coat proteins, respectively. Transgenic SMV CP^{+} plants inoculated with PVY exhibited a 3- to 4-day delay in symptom development compared with inoculated control plants (88). Some percentage of the plants completely escaped infection in each experiment. Protection was observed at virus concentrations in the inoculum 100-fold higher than those required to completely infect nontransgenic plants. Transgenic SMV CP^{+} plants inoculated with TEV developed symptoms more slowly than did similarly inoculated nontransgenic plants, and the final symptoms were much less severe. Furthermore, symptoms on transgenic plants tended to fade with time and newly expanded parts of the plants were often symptom free. Plant lines exhibiting the greatest protection against infection did not contain the highest SMV coat protein levels.

Virus-resistant transgenic plants have also been produced with *CP* sequences of the potyviruses papaya ringspot virus (PRV), watermelon mosaic virus II (WMVII), zucchini yellow mosaic virus (ZYMV), and plum pox virus (PPV) (22, 48, 60, 79).

ALFALFA MOSAIC VIRUS Alfalfa mosaic virus (AlMV) has a tetrapartite genome of single-stranded, (+)-sense RNA. Each segment of the genome is encapsidated independently, and four isodiametric particle sizes are observed—the smallest is nearly isometric while the larger particles are bacilliform. A single species of 24-kDa coat protein forms the protein component of all of the particles.

Transgenic tobacco, tomato, and alfalfa expressing an AlMV *CP* gene accumulate AlMV coat protein up to 0.05–0.80% of total soluble protein (1, 32, 49, 91, 92, 96, 97). In general, symptoms on coat protein–accumulating lines were reduced in the inoculated leaves and were less severe or nonexistent in their upper leaves. Furthermore, resistant plants accumulated several hundred–fold less virus (49). The level of protection decreased with increasing concentration of virus in the inoculum. Plants that accumulated the highest levels of coat protein developed fewer primary infections and were slower in developing systemic infections. Resistance was overcome in some plant lines by inoculation with AlMV RNA while other lines were protected against RNA (49, 91). Resistance was also observed after inoculation of CP^+ protoplasts (32). A low but significant degree of protection was reported against infection by the unrelated viruses PVX and cucumber mosaic virus (CMV) (1).

TOMATO SPOTTED WILT VIRUS Tomato spotted wilt virus (TSWV), the type member of the tospovirus group, is an enveloped virus 80–110 nm in diameter. The 29-kDa nucleocapsid protein is not exposed on the surface of the virions.

Transgenic tobacco lines carrying three different TSWV *CP*-based chimeric genes differing by a few residues at the amino terminus of the encoded product were produced (19, 25, 52, 68). Plants that expressed one of the gene constructs accumulated TSWV nucleocapsid protein to 1.5% of the total soluble protein (25). Virus challenge was by mechanical inoculation or by viruliferous thrips. Resistance was observed as a delay in symptom development and a reduction in the proportion of CP^+ plants that eventually became infected. The number of sites of infection on transgenic leaves was also reduced on CP^+ plants. In some cases, including after challenge by thrips, protection was complete or nearly complete and virus could not be recovered from the inoculated plants.

The levels of protection observed differed in plants transformed with the different gene constructs, but the influence of environmental and other factors, rather than the differences between the constructs, may have been responsible. Furthermore, the levels of protection in different plant lines transgenic for the same construct varied widely. Therefore, comparisons of one or a few lines with one construct to one or a few lines with another are not valid. In each case there was no correlation between the degree of protection and the level of coat protein in the plants. However, when the plants were challenged with a closely related virus, those plant lines accumulating the highest levels of coat protein appeared to be the most effectively protected (68). Protection was not observed in any line against more distantly related strains. Gene constructs lacking a translational initiation sequence conferred resistance equivalent to that conferred by translationally competent constructs (19).

Mechanisms of Resistance

The examples of resistance described in Table 1 all result from the introduction into plants of chimeric genes designed to express viral *CP* sequences. However, it is highly unlikely that a single mechanism accounts for resistance in all of the examples. Indeed, the resistance derived from a single *CP* gene in a single plant species may act by more than one mechanism and may inhibit several different stages in the infection or replication process.

RESISTANCE TO TMV The system most thoroughly studied among the above examples is resistance to TMV in transgenic TMV CP^+ tobacco. In this system, several mechanisms contribute to the resistance.

A major component of coat protein–mediated resistance must result from an action of the transgene-derived coat protein on an early event in the infection process because the number of infection sites is lower on CP^+ plants than on control plants (64). This impaired step probably occurs after entry of the virus and prior to completion of the normal process of uncoating (removal of the protein coat). The interference must follow viral entry because the characteristics of resistance against virus introduced by electroporation into protoplasts derived from CP^+ leaves were fundamentally very similar to the characteristics of resistance in inoculated leaves (75). It must precede uncoating because the major block in CP^+ plants or protoplasts is overcome when TMV RNA is used in the inoculum (64, 75).

In this window between TMV virion entry and the complete uncoating of the RNA, three stages have been identified in the normal infection process. First, a few dozen coat-protein subunits are removed from the 5′ end of the virion (58). Whether the removal is host-assisted or whether it depends solely upon intracellular conditions such as Ca^{2+} concentration is not known. Structures believed to correspond to virions in this stage of the infection process can be created in vitro (99). Second, ribosomal subunits bind to the RNA exposed by removal of coat protein. Structures corresponding to this stage of infection, referred to as striposomes, have been observed in vitro and in vivo (86, 99). Third, translation by the ribosomes proceeds along the genome with cotranslational displacement of the remaining coat protein (99).

Several lines of evidence suggest that the major block to infection in TMV CP^+ plants occurs during or prior to the removal of the first few dozen coat-protein subunits. That the restriction does not follow the removal of the coat-protein subunits was established by demonstrating that purified virions exposed briefly to pH 8.0, a treatment known to strip at most 60–70 coat-protein subunits from the 5′ end, were able to overcome much of the resistance in CP^+ protoplasts (75). An analysis of the virus-related particles isolated from protoplasts at short time intervals after inoculation showed that the restriction occurs prior to the formation of striposomes (101). In experiments with protoplasts from control plants, rapidly sedimenting particles believed to be striposome complexes were observed by 7 min postinoculation. In protoplasts from CP^+ plants, relatively few such particles were observed, even up to 1 h postinoculation. Additional evidence that the blocked step in CP^+ plants precedes the completion of uncoating comes from experiments with artificial virus particles (66). Messenger RNA for a marker enzyme was coated with TMV coat protein in vitro, and the resulting particles were introduced into CP^+ and CP^- protoplasts by electroporation. Synthesis of the marker enzyme was 100-fold less efficient in the CP^+ protoplasts than in the CP^- protoplasts.

At least two alternative models have been proposed based upon the described results (78). Both are derived from models originally proposed to explain classical cross-protection between viruses (7, 20). In one model, a subcellular component acts as a receptor or uncoating site for invading virions and is responsible for initiating the uncoating process. In transgenic plants that are accumulating coat protein, coat protein would bind to the receptor, preventing the association of virions with the receptor and thus rendering them unable to initiate infection. In the other model, the initiation of uncoating is triggered by a change in physiological conditions upon entry into the cell, and initiation of uncoating is a reversible, dynamic process. In cytoplasm that contains transgene-derived coat protein, the equilibrium between uncoating and recoating of the end of the virion would be shifted in favor of recoating. Ribosomes would be unable to bind to the end of the genome and infection would not ensue. The essential difference between the two models lies in the extent of participation of specific host factors in the initial uncoating of the 5′ terminus of the virus. Ribosomes are probably not directly involved in the resistance reaction because resistance is specific to related viruses.

Both models predict that the transgene-derived protein is the resistance-conferring entity. This is probably the case for TMV resistance in transgenic tobacco, as protection was not observed in transgenic plants expressing a nontranslatable gene (71). Furthermore, both models predict that plants accumulating higher levels of coat protein would exhibit a greater degree of resistance, at least over some range, which has also been observed.

If the first model is correct, multimeric aggregates of coat protein might be expected to confer more effective resistance than coat-protein monomers if they could block a receptor more effectively. To evaluate the ability of various structures to confer resistance, exogenous coat protein in different aggregation states was coinoculated with virus into protoplasts prepared from nontransgenic plants (76). Significant protection was observed, but the most active form appeared to consist of smaller, rather than larger, aggregates. However, in these experiments and in experiments to determine the aggregation state of the coat protein in transgenic plants, the possibility remained that the aggregates observed are not the active form per se but are dynamically interconverted with the active form (70).

According to both models and according to the extent that resistance results from a block in an early event in the infection process, transgene expression would be required for resistance in the cell types in which infection normally initiates. Several tests of this model have been made. In one approach, a plant-derived promoter weakly active in the epidermal cells of the leaf but quite active in the mesophyll cells of the interior of the leaf was used to drive transgene expression (16). A much lower degree of resistance was observed in transgenic plants carrying this gene construct compared with transgenic plants carrying the more uniformly expressed construct. Conversely, plants carrying a construct with a promoter active in the epidermis but not in the underlying cells were resistant, although only against very low virus concentrations in the inoculum (U. Reimann-Philipp & R. N. Beachy, submitted). The low level of resistance was attributed to very low levels of coat-protein accumulation in epidermal cells.

The mechanism(s) of the interference with steps in TMV infection after the initial events is much less certain. Although spread from infection sites in CP^+ compared with CP^- plants may be reduced, it has not been possible to determine whether the slower spread resulted from interference with a movement process or from a slightly reduced rate of replication, compounded as each adjacent cell became infected (100).

RESISTANCES TO VIRUSES OTHER THAN TMV Although all of the examples described in Table 1 represent demonstrations of resistance conferred by expression of genes based upon *CP* sequences, they differ in important ways. From a mechanistic standpoint, the most important difference is the likelihood that the accumulation of RNA, rather than of protein, is responsible for resistance observed in some of the examples. In several cases, resistant plants have been generated by transformation with *CP* gene constructs designed to be transcribed into translationally incompetent mRNA. The mRNA resistance was in some cases as effective as that conferred by translationally competent constructs, suggesting that RNA may be the active entity, even when protein does accumulate. The examples of resistance to potato leafroll virus and to tomato spotted wilt virus appear to be of this type (19, 44). In other cases, RNA-mediated resistance was much less effective or was much more easily overcome than protein-mediated resistance, suggesting at most a lesser contribution by RNA-mediated resistance. The examples of resistance to cucumber mosaic virus and to potato virus X are probably of this type (18, 31). In resistance to potyviruses, the role played by RNA is less certain (46, 93).

Differences in the reactions of resistant plants to challenge inoculation by virus may also reflect differences in mechanisms of resistance. In general, resistance is manifested as a reduction in the number of initial infections and reduced or delayed systemic spread of virus and symptoms. However, examples vary in the relative magnitude of the two effects, and additional signs of resistance are observed in some cases. For example, while in many systems very few infections are initiated, most AlMV-resistant tomato plants become infected but the infection does not spread through the plant (92). In plant lines resistant to TEV or SMV or some other potyviruses, many of the plants become infected but exhibit a delay in symptom formation and virus accumulation and eventually produce new, virus-free growth (46, 88).

Differences in the frequency of resistant plants in a population of transgenic plants all carrying the same gene construct may also reflect differences in the mechanism(s) of protection. In the production of TMV-resistant plants, almost all transformed lines that accumulated coat protein had some degree of resistance (71). In the production of lines resistant to TEV, only a low percentage of the recovered lines showed a resistant phenotype (46). When the degree of protection does not correlate with levels of gene expression, the low frequency may indicate that some other, unidentified criterion plays an essential role in resistance.

Resistance to inoculation with viral RNA was demonstrated in plants transgenic for *CP* sequences from alfalfa mosaic virus, arabis mosaic virus, potato virus X, and potato virus S, indicating that resistance to these viruses may occur at a stage later than virus uncoating (10, 31, 53, 91). Several possible mechanisms have been proposed. In the case of alfalfa mosaic virus, the coat protein has many functions, not all of which are essential for conferring resistance (35, 36, 73, 91, 96). During infection, the accumulation of high levels of coat protein may function as a switch from replication to translation and encapsidation, much as nucleocapsid protein of (−)-strand animal viruses controls the balance between replicative and transcriptional RNA synthesis (3, 12). In transgenic plants that accumulate coat protein, the switch may be triggered prematurely, preventing infection. In contrast, in the case of resistance against PVX, the ability to protect against inoculation with viral RNA is still compatible with a model in which uncoating is inhibited by terminal recoating. The key difference is that on PVX RNA, encapsidation may initiate near the 5′ end, thereby enabling recoating by a few subunits to block transcription (31, 50). On TMV RNA, nucleation of assembly is far removed from the 5′ end. In resistance to PVX, a partially uncoated virion would not overcome protection; in resistance to TMV, partially coated virions do.

The lack of a correlation between levels of coat-protein accumulation and degree of protection may suggest that the mechanisms proposed so far are inadequate to explain the resistance against some viruses. However, the measurements of the levels of coat-protein accumulation are relatively imprecise and may not reflect relevant concentrations. For example, potato leafroll virus (PLRV) invades only the phloem cells of the plant. Only coat protein in the phloem would be expected to be active, but samples taken to measure coat-protein levels have included only a small proportion of phloem cells. Similarly, many viruses initiate infection in epidermal cells and might not be influenced by overall coat-protein levels in the leaf. Establishing a relationship between bulk coat-protein concentrations and coat protein in these individual tissues is difficult because expression of foreign genes by the cauliflower mosaic virus (CaMV) 35S promoter, used in most of the resistance examples reviewed here, is not uniform throughout plant tissues (6, 9). Indeed, the patterns of expression in different tissues can vary widely between individual transgenic plants transformed with the same gene construct (6). Furthermore, what is reported as the level of coat protein accumulation does not reflect the subcellular location of the coat protein. Coat protein may accumulate in compartments of the cell inaccessible to invading virions (4). In addition, coat protein may form multimeric structures, and the most abundant structures are not necessarily active in conferring resistance, but may only represent the most stable form of the protein (11, 76).

Application to Virology and Agriculture

Resistance conferred by expression of genes encoding virus coat protein has stimulated research in basic plant virology. Obviously, transgenic plants that block one or more stages in the infection process provide useful tools for study of the process. Analysis is simplified in comparison to naturally occurring forms of resistance because the block is conferred by a single gene of known sequence. The contributions of coat protein–mediated resistance to our understanding of the early events in TMV infection have been reviewed (77). The use of mutant viruses in conjunction with coat protein–mediated resistance, and the combination of coat protein–mediated resistance with other types of engineered resistance, will allow more thorough dissection of virus infection.

Virus protection in transgenic plants conferred by expression of *CP* genes will almost certainly prove useful for commercial agriculture. Field tests with transgenic tomatoes resistant to ToMV or TMV, transgenic potatoes resistant to PVX or PVY, and transgenic cucumbers resistant to CMV have been very encouraging (27, 41, 42, 63, 84).

In addition, the coat-protein strategy of disease resistance has many specific advantages. Perhaps foremost among them is that even in the absence of a mechanistic understanding of the phenomenon, the strategy can generally be extended to other plant species and resistance against other viruses. In examples in which model plant systems were tested prior to application to plant species that were difficult to transform, the results from the model systems have predictive value (22, 32). Furthermore, cloned cDNA copies of *CP* genes can be obtained with relative ease. In cases where broad-spectrum resistance is expected, it may not be necessary to isolate the *CP* gene of the pathogenic virus if the *CP* gene of a related virus or strain is available (88).

PROTECTION CONFERRED BY SEQUENCES ENCODING REPLICASE-RELATED PROTEINS

The first example of resistance in transgenic plants conferred by a nonstructural viral gene was not planned as a test of pathogen-derived resistance. Transgenic tobacco plants carrying a chimeric gene directing the expression of a replication-related protein encoded by the TMV genome were found to have a high level of resistance to TMV infection (26). Subsequent demonstrations of resistance have been made using plants transformed with related sequences from pea early browning virus, cucumber mosaic virus, and potato virus X. In these examples, the transgenes encode defective or functional forms of components of viral replicase complexes. As a group they have been referred to as replicase-mediated resistances.

Resistance to Tobacco Mosaic Virus

A cDNA copy of an open reading frame from the middle of the TMV genome, cloned between promoter sequences of the CaMV 35S promoter and transcription-termination and polyadenylation sequences from the *Agrobacterium tumefaciens nos* gene, was used for transformation of tobacco (26). The 54-kDa protein encoded by the open reading frame may have a role in replication of the virus during infection in nontransgenic plants. Translation of a subgenomic RNA that has been found associated with polyribosomes and that accumulates during infection would be expected to yield the 54-kDa protein. The deduced amino acid sequence of the protein contains several motifs characteristic of viral RNA-dependent RNA polymerases; however, the 54-kDa protein has never been detected in infected tissue. Alternatively, the polypeptide may naturally exist only as part of a 183-kDa protein. This protein is translated from the genomic RNA by infrequent readthrough of a leaky termination codon at the end of the sequence encoding a 126-kDa protein (69). The 126-kDa protein is the major virus-encoded component of the viral replicase. The 183-kDa protein is also required for replication (39).

The transgenic plants were resistant to infection when challenged with either TMV virions or TMV RNA, even at very high concentrations. Inoculated plants remained symptom free through maturity and did not accumulate virus. Resistance was also observed against closely related strains of TMV but not against more distantly related strains or against other viruses. Protoplasts prepared from the transgenic plants were also resistant to infection, but low levels of replicative forms of TMV were detected (15). mRNA transcripts of the 54-kDa ORF were identified in the transgenic plants, but no 54-kDa protein could be detected. However, an intact initiation codon for the 54-kDa ORF was required for resistance (14).

Resistance to Pea Early Browning Virus

Pea early browning virus (PEBV) is a member of the tobravirus group. Translational readthrough from the ORF encoding a 141-kDa protein would generate a 201-kDa protein. These proteins contain regions of sequence similar to the 126- and 183-kDa proteins of TMV. Initiation of translation just beyond the termination codon of the ORF encoding the 141-kDa protein would result in the synthesis of a 54-kDa protein of the same sequence as the COOH-terminal domain of the 201-kDa protein.

Nicotiana benthamiana, a host for PEBV, was transformed with a chimeric gene for expression of the 54-kDa ORF (51). Three fourths of the plant lines were completely resistant to infection by mechanical inoculation with purified virus. Resistance was also demonstrated against other strains of the virus. Plant lines transgenic for sequences encoding truncated forms of the 54-kDa protein were not resistant.

Resistance to Potato Virus X

The PVX genome encodes a 165-kDa protein that is presumed to be the viral replicase. Chimeric genes that direct the expression of full-length or truncated versions of the 165-kDa protein were used for transformation of tobacco (13). Many of the lines recovered after transformation with the construct for expression of the full-length protein exhibited some level of disease-symptom reduction compared with nontransgenic controls. Three lines had reduced numbers of infection sites and considerably reduced virus accumulation (up to 25,000-fold) in both the inoculated and upper leaves. These plants were equally resistant to PVX and PVX RNA. A single transformant with the gene construct encoding the NH_3-terminal third of the protein was resistant to PVX infection. The resistance was similar to that of the most resistant line transformed with the gene encoding the entire protein.

Resistance to Cucumber Mosaic Virus

The genome of cucumber mosaic virus (CMV) encodes 110- and 97-kDa components of the viral replicase. A chimeric gene capable of encoding a truncated protein derived from the NH_3-terminal two thirds of the 97-kDa protein was used to generate transgenic tobacco (2). The domain of the protein removed in the truncation included the Gly-Asp-Asp motif common to replicases of RNA viruses (82).

In some plant lines, resistance was manifested only as a delay in symptom development and only against low virus concentrations in the inoculum. In two lines, however, the plants did not develop symptoms even after inoculation with virus concentrations >1000-fold higher than that needed to infect control plants. These lines were also resistant to inoculation with viral RNA. Chlorotic lesions and associated virus replication were sometimes observed in the inoculated leaves, which indicated cell-to-cell spread of the infection. In uninoculated leaves of these plants, neither virus nor viral RNA could be detected.

Mechanisms of Replicase-Mediated Resistance

The studies on replicase-mediated resistance against TMV have shed some light on its mechanism (14, 15). The results suggest that the number of initially infected cells in transgenic and nontransgenic plants are the same, but that virus replication is markedly reduced in cells of the transgenic plants. Although replication is not halted absolutely, it is so severely impeded that little or no systemic spread can occur within the plant during its lifetime (15). That the active entity was the 54-kDa protein, rather than its message, was clearly established, but the identity of the protein, as a naturally occurring independent entity or as a severely truncated form of the 186-kDa protein, remains unclear (14).

The replicase-mediated resistance to CMV described above is closely analogous to an engineered resistance to bacteriophage Qβ in *Escherichia coli* (37, 38). The phage-encoded β-subunit of the RNA-dependent RNA polymerase was modified by mutation of the consensus Gly-Asp-Asp sequence and by truncation of the COOH-terminal portion of the protein. Proteins with truncations of up to 25% or substitutions of the Gly residue could not support replication and conferred immunity. The experiments demonstrated interference with translation, and it was hypothesized that defective replicase inhibits protein synthesis by competing with ribosomes for access to the genome. The defective CMV protein could act by preventing the accumulation of virus-encoded replicase or by inhibiting the function of the accumulated replicase.

Inhibition of virus replication in mammalian cell lines has been demonstrated with wild-type (90) and mutant (34, 67) forms of replication-associated proteins encoded by animal viruses. Resistance conferred by at least two alternative modes of action has been demonstrated. *Trans*-dominant forms of the adeno-associated virus Rep78 protein, resulting from a mutation in a consensus purine nucleotide–binding site, bind to the origin of viral replication, preventing the binding of wild-type protein (67). In contrast, *trans*-dominant mutant forms of the HIV Rev protein associate with wild-type Rev protein to form nonfunctional complexes (34).

PROTECTION CONFERRED BY THE ACCUMULATION OF RNA

Most of the examples of RNA-mediated resistance to virus infection in transgenic plants were discovered in control plant lines in studies of coat protein–mediated resistance. In some cases, plants with RNA-mediated resisitance were protected to a degree comparable to the plants that accumulated coat protein, but in others, the degree of resistance was much less. In experiments with sequences from a few viruses, both anitsense and translationally impaired sense RNAs conferred protection. Some of the experiments with other viruses did not test for RNA protection. As more experiments are completed, more examples of RNA-mediated resistance will likely be discovered.

Transgenic tobacco plants carrying a translationally defective TSWV nucleocapsid protein gene exhibited levels of resistance similar to those reported for experiments with translationally competent gene constructs (19). In addition, the phenotype of the resistance was identical to that reported in the plants carrying the translationally competent genes, suggesting that the previously reported *CP*-mediated resistance had actually been an example of RNA-mediated resistance. Protection was observed against various strains and isolates of TSWV, but not against two related viruses with about 80% nucleotide sequence identity to TSWV.

Resistance to PLRV infection was observed in transgenic potato plants accumulating high levels of either sense or antisense transcripts of the PLRV *CP* gene (44). The sense transcript was predicted to be translatable, but no coat protein could be detected. The degree and phenotype of resistance were similar for the different transgenic lines.

Resistance to infection by the potyvirus PVY^N was observed in tobacco plants transgenic for a *CP* gene construct lacking a start codon (93). The level and characteristics of the resistance were similar to those observed for plant lines carrying a construct with a start codon. Among the transgenic lines, the amounts of accumulated *CP* RNA and the level of resistance were not correlated. Resistance to TEV, another potyvirus, was observed in transgenic tobacco accumulating antisense *CP* transcripts or sense transcripts containing a frameshift mutation (46, 47). Only low levels of resistance were observed in transgenic tobacco accumulating antisense transcripts of *CP* genes of CMV or PVX (31, 81).

The use of sense or antisense transcripts of regions of viral genomes other than *CP* has also been investigated. Resistance to TMV infection was observed in tobacco plants accumulating an antisense RNA complementary to the 3′ end of the TMV genome (72). The degree of protection was considerably less than in plants accumulating coat protein. Resistance to infection by the plant DNA geminivirus tomato golden mosaic virus (TGMV) was conferred by RNA complementary to an ORF encoding a protein required for replication (8). Fewer transgenic than control plants developed symptoms, and the symptoms were of reduced severity. A broad correlation was observed between RNA levels and resistance, and virus replication was reduced in transgenic tissue. Three different regions of the viral genome of CMV were tested as antisense transcripts (81). One plant line, accumulating an RNA complementary to a region encoding a protein required for replication, was resistant.

Other lines, including some that accumulated similar or higher levels of the same antisense transcript, were not resistant.

A mechanistic understanding of resistance is not available for any of the RNA-based examples described. Potential mechanisms include binding to the genome to inhibit replication or translation and competition for viral or host factors needed for replication. Inhibition of in vitro translation of TMV by antisense oligodeoxynucleotides complementary to a region near the 5′ end of the virus has been reported (17). Replication of turnip yellow mosaic virus (TYMV) was inhibited, in vitro, by small sense RNAs that contain the 3′-terminal region of the genome (57).

The application of RNA-mediated protection to agriculture seems promising, especially if used in combination with other types of resistance. Although the spectrum of RNA-mediated resistance against related viruses may be less broad than equivalent spectra for protein-mediated resistance, the degree of protection in some cases is quite high.

PROTECTION CONFERRED BY TRANSGENE COPIES OF MILD STRAINS, SATELLITES AND SATELLITE RNAS, AND DEFECTIVE INTERFERING VIRUSES

As described over half a century ago, infection by mild strains of a virus can often prevent or reduce the symptoms of subsequent infections by more severe strains of the virus. In 1988, tobacco plants carrying a cDNA copy of the genome of a mild TMV strain, under the transcriptional control of the CaMV 35S promoter, were produced (102). The transgenic plants are constitutively infected with the mild strain of TMV and, as expected, they develop only mild symptoms when challenged with severe strains of TMV.

Virus satellites and satellite RNAs depend on the helper virus for replication and are not related by sequence similarity to the virus. Satellites capable of enhancing or ameliorating symptom severity have been identified (83, 89). Transgenic plants expressing cloned copies of several different satellites or satellite RNAs have been produced. In experiments with a satellite of tobacco ringspot virus (TRV), the accumulation of satellite increased drastically following infection with TRV, presumably because of replication of the transcriptionally derived RNAs (24). The transgenic plants exhibited a marked delay of symptoms, and newly produced upper parts of the plants lacked symptoms for several weeks and then developed only mild symptoms. Virus accumulation was markedly reduced in the inoculated leaves and was not detectable in the upper leaves. In experiments with satellite RNAs of cucumber mosaic virus, the transgenic tobacco plants inoculated with CMV accumulated large amounts of satellite RNA (29, 40, 56, 103). Replication of CMV was greatly decreased, and symptom development was largely suppressed. Inoculation of the transgenic plants with the related virus tomato aspermy virus

(TAV) induced satellite RNA accumulation, and symptoms of the infection were attenuated, but little or no decrease in TAV replication was observed (29). In tests with tomato plants that express the CMV satellite RNA, fruit yields observed for transgenic plants were 50% higher than those for nontransgenic control plants (89).

A fragment of the genome of the DNA geminivirus African cassava mosaic virus (ACMV) also confers protection (87). In transgenic plants inoculated with ACMV, the DNA fragment was replicated and amplified to high levels. In these plants, the ACMV accumulation was reduced 70%, suggesting that the subgenomic DNA of the transgene was amplified at the expense of the virus. Systemic infection took longer to become established in the transgenic plants, and the symptoms that developed were less severe than in nontransgenic plants. No reduction in severity of symptoms or level of viral DNA was observed when the plants were challenged with related geminiviruses.

The methods described in this section have practical disadvantages. In some cases, active pathogens or active components of a pathogenic mixture are produced. In addition, the transgene-derived component may recombine with another invading virus, extending its host range or virulence. Such modifications are much more likely in virus-tolerant plants than in virus-resistant plants.

CONCLUDING REMARKS

The number of examples of pathogen-derived resistance utilizing a variety of virus or virus-associated sequences is growing rapidly. Coat protein–mediated resistance is more widely applied than other approaches, perhaps in part because it was the first described, but also because it appears to provide a broader type of resistance. However, other sequences can provide extremely high levels of resistance against challenge by a narrow range of viruses. Alternative strategies and the combining of several strategies may provide broad protection as well as high levels of resistance. This will in turn lead to the development of crop plants that in the field have heightened levels of resistance to viruses that are both mechanically transmitted and vector borne.

For the virologist, the phenotype of disease resistance offers substantial opportunity to explore the molecular mechanisms of virus infection, replication, and spread while studying cellular and molecular bases of resistance. Clearly, additional experimental approaches are needed before pathogen-derived resistance is fully understood and exploited in agriculture.

Acknowledgment

We thank Amelia T. Briones for her help in the preparation of this chapter.

Literature Cited

1. Anderson, E. J., Stark, D. M., Nelson, R. S., Powell, P. A., Tumer, N. E., Beachy, R. N. 1989. Transgenic plants that express the coat protein genes of tobacco mosaic virus or alfalfa mosaic virus interfere with disease development of some nonrelated viruses. *Phytopathology*. 79:1284–90
2. Anderson, J. M., Palukaitis, P., Zaitlin, M. 1992. A defective replicase gene induces resistance to cucumber mosaic virus in transgenic tobacco plants. *Proc. Natl. Acad. Sci. USA* 89:8759–63
3. Arnheiter, H., Davis, N. L., Wertz, G., Schubert, M., Lazzarini, R. A. 1985. Role of the nucleocapsid protein in regulating vesicular stomatitis virus RNA synthesis. *Cell* 41:259–67
4. Banerjee, N., Zaitlin, M. 1992. Import of tobacco mosaic virus coat protein into intact chloroplasts in vitro. *Mol. Plant-Microbe Interact*. 5:466–71
5. Barker, H., Reavy, B., Kumar, A., Webster, K. D., Mayo, M. A. 1992. Restricted virus multiplication in potatoes transformed with the coat protein gene of potato leafroll luteovirus: similarities with a type of host gene-mediated resistance. *Ann. Appl. Biol*. 120: 55–64
6. Barnes, W. M. 1990. Variable patterns of expression of luciferase in transgenic tobacco leaves. *Proc. Natl. Acad. Sci. USA* 87:9183–87
7. Bawden, F. C., Kassanis, B. 1944. The suppression of one plant virus by another. *Ann. Appl. Biol*. 32:52–57
8. Bejarano, E. R., Lichtenstein, C. P. 1992. Prospects for engineering virus resistance in plants with antisense RNA. *Trends Biotechnol*. 10:383–87
9. Benfey, P. N., Chua, N. H. 1989. Regulated genes in transgenic plants. *Science* 244:174–81
10. Bertioli, D. J., Cooper, J. I., Edwards, M. L., Hawes, W. S. 1992. Arabis mosaic nepovirus coat protein in transgenic tobacco lessens disease severity and virus replication. *Ann. Appl. Biol*. 120:47–54
11. Bertioli, D. J., Harris, R. D., Edwards, M. L., Cooper, J. I. 1991. Transgenic plants and insect cells expressing the coat protein of arabis mosaic virus produce empty virus-like particles. *J. Gen. Virol*. 72:1801–9
12. Blumberg, B. M., Leppert, M., Kolakofsky, D. 1981. Interaction of VSV leader RNA and nucleocapsid protein may control VSV genome replication. *Cell* 23:837–45
13. Braun, C. J., Hemenway, C. L. 1992. Expression of amino-terminal portions or full-length viral replicase genes in transgenic plants confers resistance to potato virus X infection. *Plant Cell* 4:735–44
14. Carr, J. P., Marsh, L. E., Lomonossoff, G. P., Sekiya, M. E., Zaitlin, M. 1992. Resistance to tobacco mosaic virus induced by the 54-kDa gene sequence requires expression of the 54-kDa protein. *Mol. Plant-Microbe Interact*. 5:397–404
15. Carr, J. P., Zaitlin, M. 1991. Resistance in transgenic tobacco plants expressing a nonstructural gene sequence of tobacco mosaic virus is a consequence of markedly reduced virus replication. *Mol. Plant-Microbe Interact*. 4:579–85
16. Clark, W. G., Register, J. C. III, Nejidat, A., Eichholtz, D. A., Sanders, P. R., et al. 1990. Tissue-specific expression of the TMV coat protein in transgenic tobacco plants affects the level of coat protein-mediated virus protection. *Virology* 179:640–47
17. Crum, C., Johnson, J. D., Nelson, A., Roth, D. 1988. Complementary oligodeoxyribonucleotide mediated inhibition of tobacco mosaic virus RNA translation in vitro. *Nucleic Acids Res*. 16:4569–81
18. Cuozzo, M., O'Connell, K. M., Kaniewski, W., Fang, R. X., Chua, N. H., Tumer, N. E. 1988. Viral protection in transgenic tobacco plants expressing the cucumber mosaic virus coat protein or its antisense RNA. *Bio/Technology* 6:549–57
19. De Haan, P., Gielen, J. J. L., Prins, M., Wijkamp, I. G., Van Schepen, A., et al. 1992. Characterization of RNA-mediated resistance to tomato spotted wilt virus in transgenic tobacco plants. *Bio/Technology* 10:1133–37
20. De Zoeten, G. A., Fulton, R. W. 1975. Understanding generates possibilities. *Phytopathology* 65:221–22
21. Farinelli, L., Malnoe, P., Collet, G. F. 1992. Heterologous encapsidation of potato virus Y strain O (PVYO) with the transgenic coat protein of PVY strain N (PVYN) in *Solanum tuberosum* CV. *bintje*. *Bio/Technology* 10:1020–25
22. Fitch, M. M. M., Manshardt, R. M., Gonsalves, D., Slightom, J. L.,

Sandford, J. C. 1992. Virus resistant papaya plants derived from tissues bombarded with the coat protein gene of papaya ringspot virus. *Bio/Technology* 10:1466–72
23. Fulton, R. W. 1986. Practices and precautions in the use of cross protection for plant virus disease control. *Annu. Rev. Phytopathol.* 24:67–81
24. Gerlach, W. L., Llewellyn, D., Haseloff, J. 1987. Construction of a plant disease resistance gene from the satellite RNA of tobacco ringspot virus. *Nature* 328:802–5
25. Gielen, J. J. L., De Haan, P., Kool, A. J., Peters, D., Van Grinsven, M. W. J. M., Goldbach, R. W. 1991. Engineered resistance to tomato spotted wilt virus, a negative-strand RNA virus. *Bio/Technology* 9:1363–67
26. Golemboski, D. B., Lomonossoff, G. P., Zaitlin, M. 1990. Plants transformed with a tobacco mosaic virus nonstructural gene sequence are resistant to the virus. *Proc. Natl. Acad. Sci. USA* 87:6311–15
27. Gonsalves, D., Chee, P., Provvidenti, R., Seem, R., Slightom, J. L. 1992. Comparison of coat protein-mediated and genetically-derived resistance in cucumbers to infection by cucumber mosaic virus under field conditions with natural challenge inoculations by vectors. *Bio/Technology* 10:1562–70
28. Hamilton, R. I. 1980. Defenses triggered by previous invaders: viruses. In *Plant Disease,* pp. 279–303. New York: Academic. 5th ed.
29. Harrison, B. D., Mayo, M. A., Baulcombe, D. C. 1987. Virus resistance in transgenic plants that express cucumber mosaic virus satellite RNA. *Nature* 328:799–801
30. Hayakawa, T., Zhu, Y., Itoh, K., Kimura, Y., Izawa, T. 1992. Genetically engineered rice resistant to rice stripe virus, an insect-transmitted virus. *Proc. Natl. Acad. Sci. USA* 89:9865–69
31. Hemenway, C., Fang, R. F., Kaniewski, W. K., Chua, N. H., Tumer, N. E. 1988. Analysis of the mechanism of protection in transgenic plants expressing the potato virus X coat protein or its antisense RNA. *EMBO J.* 7:1273–80
32. Hill, K. K., Jarvis-Eagan, N., Halk, E. L., Krahn, K. J., Liao, L. W., et al. 1991. The development of virus-resistant alfalfa, *Medicago sativa* L. *Bio/Technology* 9:373–77
33. Hoekema, A., Huisman, M. J., Molendijk, L., van den Elzen, P. J. M., Cornelissen, B. J. C. 1989. The genetic engineering of two commercial potato cultivars for resistance to potato virus X. *Bio/Technology* 7:273–78
34. Hope, T. J., Klein, N. P., Elder, M. E., Parslow, T. G. 1992. Trans-dominant inhibition of human immunodeficiency virus type 1 *rev* occurs through formation of inactive protein complexes. *J. Virol.* 66:1849–55
35. Houwing, C. J., Jaspars, E. M. J. 1986. Coat protein blocks the in vitro transcription of the virion RNAs of alfalfa mosaic virus. *FEBS Lett.* 209: 284–88
36. Houwing, C. J., Jaspars, E. M. J. 1987. In vitro evidence that the coat protein is the programming factor in alfalfa mosaic virus-induced RNA synthesis. *FEBS Lett.* 221:337–42
37. Inokuchi, Y., Hirashima, A. 1987. Interference with viral infection by defective RNA replicase. *J. Virol.* 61: 3946–49
38. Inokuchi, Y., Hirashima, A. 1990. Interference with viral infection by RNA replicase deleted at the carboxy-terminal region. *J. Biochem.* 108:53–58
39. Ishikawa, M., Meshi, T., Motoyoshi, F., Takamatsu, N., Okada, Y. 1986. In vitro mutagenesis of the putative replicase genes of tobacco mosaic virus. *Nucleic Acids Res.* 14:8291–8305
40. Jacquemond, M., Amselem, J., Tepfer, M. 1988. A gene coding for a monomeric form of cucumber mosaic virus satellite RNA confers tolerance to CMV. *Mol. Plant-Microbe Interact.* 1:311–16
41. Jongedijk, E., de Schutter, A. A. J. M., Stolte, T., van den Elzen, P. J. M., Cornelissen, B. J. C. 1992. Increased resistance to potato virus X and preservation of cultivar properties in transgenic potato under field conditions. *Bio/Technology* 10:422–29
42. Kaniewski, W., Lawson, C., Sammons, B., Haley, L., Hart, J., et al. 1990. Field resistance of transgenic Russet Burbank potato to effects of infection by potato virus X and potato virus Y. *Bio/Technology* 8:750–54
43. Kawchuk, L. M., Martin, R. R., McPherson, J. 1990. Resistance in transgenic potato expressing the potato leafroll virus coat protein gene. *Mol. Plant-Microbe Interact.* 3:301–7
44. Kawchuk, L. M., Martin, R. R., McPherson, J. 1991. Sense and antisense RNA-mediated resistance to potato leafroll virus in Russet Burbank potato plants. *Mol. Plant-Microbe Interact.* 4:227–53
45. Lawson, C., Kaniewski, W., Haley,

L., Rozman, R., Newell, C., et al. 1990. Engineering resistance to mixed virus infection in a commercial potato cultivar: resistance to potato virus X and potato virus Y in transgenic Russet Burbank. *Bio/Technology* 8:127–34

46. Lindbo, J. A., Dougherty, W. G. 1992. Pathogen-derived resistance to a potyvirus: immune and resistant phenotypes in transgenic tobacco expressing altered forms of a potyvirus coat protein nucleotide sequence. *Mol. Plant-Microbe Interact.* 5:144–53
47. Lindbo, J. A., Dougherty, W. G. 1992. Untranslatable transcripts of the tobacco etch virus coat protein gene sequence can interfere with tobacco etch virus replication in transgenic plants and protoplasts. *Virology* 189:725–33
48. Ling, K., Namba, S., Gonsalves, C., Slightom, J. L., Gonsalves, D. 1991. Protection against detrimental effects of potyvirus infection in transgenic tobacco plants expressing the papaya ringspot virus coat protein gene. *Bio/Technology* 9:752–58
49. Loesch-Fries, L. S., Merlo, D., Zinnen, T., Burhop, L., Hill, K., et al. 1987. Expression of alfalfa mosaic virus RNA 4 in transgenic plants confers virus resistance. *EMBO J.* 6:1845–51
50. Lok, S., Abouhaidar, M. G. 1986. The nucleotide sequence of the 5′ end of papaya mosaic virus RNA: site of in vitro assembly initiation. *Virology* 153:289–96
51. MacFarlane, S. A., Davies, J. W. 1992. Plants transformed with a region of the 201-kilodalton replicase gene from pea early browning virus RNA 1 are resistant to virus infection. *Proc. Natl. Acad. Sci. USA* 89:5829–33
52. MacKenzie, D. J., Ellis, P. J. 1992. Resistance to tomato spotted wilt virus infection in transgenic tobacco expressing the viral nucleocapsid gene. *Mol. Plant-Microbe Interact.* 5:34–40
53. MacKenzie, D. J., Tremaine, J. H. 1990. Transgenic *Nicotiana debneyii* expressing viral coat protein are resistant to potato virus S infection. *J. Gen. Virol.* 71:2167–70
54. MacKenzie, D. J., Tremaine, J. H., McPherson, J. 1991. Genetically engineered resistance to potato virus S in potato cultivar Russet Burbank. *Mol. Plant-Microbe Interact.* 4:95–102
55. Matthews, R. E. F. 1991. *Plant Virology*. San Diego: Academic. 835 pp. 3rd ed.
56. McGarvey, P. B., Kaper, J. M., Avila-Rincon, M. J., Pena, L., Diaz-Ruiz, J. R. 1990. Transformed tomato plants express a satellite RNA of cucumber mosaic virus and produce lethal necrosis upon infection with viral RNA. *Biochem. Biophys. Res. Commun.* 170: 548–55
57. Morch, M. D., Joshi, R. L., Denial, T. M., Haenni, A. L. 1987. A new "sense" RNA approach to block viral RNA replication in vitro. *Nucleic Acids Res.* 15:4123–30
58. Mundry, K. W., Watkins, P. A. C., Ashfield, T., Plaskitt, K. A., Eisele-Walter, S., Wilson, T. M. A. 1991. Complete uncoating of the 5′ leader sequence of tobacco mosaic virus RNA occurs rapidly and is required to initiate cotranslational virus disassembly in vitro. *J. Gen. Virol.* 72:769–77
59. Namba, S., Ling, K., Gonsalves, C., Gonsalves, D., Slightom, J. L. 1991. Expression of the gene encoding the coat protein of cucumber mosaic virus (CMV) strain-WL appears to provide protection to tobacco plants against infection by several different CMV strains. *Gene* 107:181–88
60. Namba, S., Ling, K., Gonsalves, C., Slightom, J. L., Gonsalves, D. 1992. Protection of transgenic plants expressing the coat protein gene of watermelon mosaic virus II or zucchini yellow mosaic virus against six potyviruses. *Phytopathology* 82:940–46
61. Nejidat, A., Beachy, R. N. 1989. Decreased levels of TMV coat protein in transgenic tobacco plants at elevated temperatures reduce resistance to TMV infection. *Virology* 173:531–38
62. Nejidat, A., Beachy, R. N. 1990. Transgenic tobacco plants expressing a coat protein gene of tobacco mosaic virus are resistant to some other tobamoviruses. *Mol. Plant-Microbe Interact.* 3:247–51
63. Nelson, R. S., McCormick, S. M., Delannay, X., Dube, P., Layton, J., et al. 1988. Virus tolerance, plant growth, and field performance of transgenic tomato plants expressing coat protein from tobacco mosaic virus. *Bio/Technology* 6:403–9
64. Nelson, R. S., Powell Abel, P., Beachy, R. N. 1987. Lesions and virus accumulation in inoculated transgenic tobacco plants expressing the coat protein gene of tobacco mosaic virus. *Virology* 158:126–32
65. Odell, J. T., Nagy, F., Chua, N.-H. 1985. Identification of DNA sequences required for activity of the cauliflower mosaic virus 35S promoter. *Nature* 313:810–12
66. Osbourn, J. K., Watts, J. W., Beachy,

R. N., Wilson, T. M. A. 1989. Evidence that nucleocapsid disassembly and a later step in virus replication are inhibited in transgenic tobacco protoplasts expressing TMV coat protein. *Virology* 172:370–73

67. Owens, R. A., Trempe, J. P., Chejanovsky, N., Carter, B. J. 1991. Adeno-associated virus rep proteins produced in insect and mammalian expression systems: wild-type and dominant-negative mutant proteins bind to the viral replication origin. *Virology* 184:14–22
68. Pang, S.-Z., Nagpala, P., Wang, M., Slightom, J. L., Gonsalves, D. 1992. Resistance to heterologous isolates of tomato spotted wilt virus in transgenic tobacco expressing its nucleocapsid protein gene. *Phytopathology* 82:1223–29
69. Pelham, H. R. B. 1978. Leaky UAG termination codon in tobacco mosaic virus RNA. *Nature* 272:469–71
70. Powell, P. A., Nelson, R. S., De, B., Hoffmann, N., Rogers, S. G., et al. 1986. Delay of disease development in transgenic plants that express the tobacco mosaic virus coat protein gene. *Science* 232:738–43
71. Powell, P. A., Sanders, P. R., Tumer, N., Fraley, R. T., Beachy, R. N. 1990. Protection against tobacco mosaic virus infection in transgenic plants requires accumulation of coat protein rather than coat protein RNA sequences. *Virology* 175:124–30
72. Powell, P. A., Stark, D. M., Sanders, P. R., Beachy, R. N. 1989. Protection against tobacco mosaic virus in transgenic plants that express tobacco mosaic virus antisense RNA. *Proc. Natl. Acad. Sci. USA* 86:6949–52
73. Quadt, R., Rosdorff, H. J. M., Hunt, T. W., Jaspars, E. M. J. 1991. Analysis of the protein composition of alfalfa mosaic virus RNA-dependent RNA polymerase. *Virology* 182:309–15
74. Quemada, H. D., Gonsalves, D., Slightom, J. L. 1991. Expression of coat protein gene from cucumber mosaic virus strain C in tobacco: protection against infections by CMV strains transmitted mechanically or by aphids. *Phytopathology* 81:794–802
75. Register, J. C. III, Beachy, R. N. 1988. Resistance to TMV in transgenic plants results from interference with an early event in infection. *Virology* 166:524–32
76. Register, J. C. III, Beachy, R. N. 1989. Effect of protein aggregation state on coat protein-mediated protection against tobacco mosaic virus using a transient protoplast assay. *Virology* 173:656–63
77. Register, J. C. III, Nelson, R. S. 1992. Early events in plant virus infection: relationships with genetically engineered protection and host gene resistance. *Semin. Virol.* 3:441–51
78. Register, J. C. III, Powell, P. A., Nelson, R. S., Beachy, R. N. 1989. Genetically engineered cross protection against TMV interferes with initial infection and long distance spread of the virus. In *Molecular Biology of Plant-Pathogen Interactions,* ed. B. Staskawicz, P. Ahlquist, O. Yoder, pp. 269–81. New York: Liss
79. Regner, F., da Camara Machado, A., da Camara Machado, M. L., Steinkellner, H., Mattanovich, D., et al. 1992. Coat protein mediated resistance to plum pox virus in *Nicotiana clevelandii* and *N. benthamiana*. *Plant Cell Rep.* 11:30–33
80. Reinero, A., Beachy, R. N. 1989. Reduced photosystem II activity and accumulation of viral coat protein in chloroplasts of leaves infected with tobacco mosaic virus. *Plant Physiol.* 89:111–16
81. Rezaian, M. A., Skene, K. G. M., Ellis, J. G. 1988. Anti-sense RNAs of cucumber mosaic virus in transgenic plants assessed for control of the virus. *Plant Mol. Biol.* 11:463–71
82. Rezaian, M. A., Williams, R. H. V., Gordon, K. H. J., Gould, A. R., Symons, R. H. 1984. Nucleotide sequence of cucumber-mosaic-virus RNA 2 reveals a translation product significantly homologous to corresponding proteins of other viruses. *Eur. J. Biochem.* 143:277–84
83. Roux, L., Simon, A. E., Holland, J. J. 1991. Effects of defective interfering viruses on virus replication and pathogenesis. *Adv. Virus Res.* 40:181–210
84. Sanders, P., Sammons, B., Kaniewski, W., Haley, L., Layton, J., et al. 1992. Field resistance of transgenic tomatoes expressing the tobacco mosaic virus or tomato mosaic virus coat protein genes. *Phytopathology* 82:683–90
85. Sanford, J. C., Johnston, S. A. 1985. The concept of parasite-derived resistance—deriving resistance genes from the parasite's own genome. *J. Theor. Biol.* 113:395–405
86. Shaw, J. G., Plaskitt, K. A., Wilson, T. M. A. 1986. Evidence that tobacco mosaic virus particles disassemble cotranslationally in vivo. *Virology* 148: 326–36
87. Stanley, J., Frischmuth, T., Ellwood,

S. 1990. Defective viral DNA ameliorates symptoms of geminivirus infection in transgenic plants. *Proc. Natl. Acad. Sci. USA* 87:1–5

88. Stark, D. M., Beachy, R. N. 1989. Protection against potyvirus infection in transgenic plants: evidence for broad spectrum resistance. *Bio/Technology* 7: 1257–62
89. Tien, P., Wu, G. 1991. Satellite RNA for the biocontrol of plant disease. *Adv. Virus Res.* 39:321–39
90. Tsunetsugu-Yokota, Y., Matsuda, S., Maekawa, M., Saito, T., Takemori, T., Takebe, Y. 1992. Constitutive expression of the *nef* gene suppresses human immunodeficiency virus type 1 (HIV-1) replication in monocytic cell lines. *Virology* 191:960–63
91. Tumer, N. E., Kaniewski, W., Haley, L., Gehrke, L., Lodge, J. K., Sanders, P. 1991. The second amino acid of alfalfa mosaic virus coat protein is critical for coat protein-mediated protection. *Proc. Natl. Acad. Sci. USA* 88:2331–35
92. Tumer, N. E., O'Connell, K. M., Nelson, R. S., Sanders, P. R., Beachy, R. N., et al. 1987. Expression of alfalfa mosaic virus coat protein gene confers crossprotection in transgenic tobacco and tomato plants. *EMBO J.* 6:1181–88
93. Van der Vlugt, R. A. A., Ruiter, R. K., Goldbach, R. 1992. Evidence for sense RNA-mediated protection to PVYN in tobacco plants transformed with the viral coat protein cistron. *Plant Mol. Biol.* 20:631–39
94. Van der Wilk, F., Willink, D. P. L., Huisman, M. J., Huttinga, H. 1991. Expression of the potato leafroll luteovirus coat protein gene in transgenic potato plants inhibits viral infection. *Plant Mol. Biol.* 17:431–39
95. van Dun, C. M., Bol, J. F. 1988. Transgenic tobacco plants accumulating tobacco rattle virus coat protein resist infection with tobacco rattle virus and pea early browning virus. *Virology* 167:649–52
96. van Dun, C. M. P., Bol, J. F., Van Vloten-Doting, L. 1987. Expression of alfalfa mosaic virus and tobacco rattle virus coat protein genes in transgenic tobacco plants. *Virology* 159:299–305
97. van Dun, C. M. P., Overduin, B., Van Vloten-Doting, L., Bol, J. F. 1988. Transgenic tobacco expressing tobacco streak virus or mutated alfalfa mosaic virus coat protein does not cross-protect against alfalfa mosaic virus infection. *Virology* 164:383–89
98. Weising, K., Schell, J., Kahl, G. 1988. Foreign genes in plants: transfer, structure, expression, and applications. *Annu. Rev. Genet.* 22:421–77
99. Wilson, T. M. A. 1984. Cotranslational disassembly increases the efficiency of expression of TMV RNA in wheat germ cell-free extracts. *Virology* 138: 353–56
100. Wisniewski, L. A., Powell, P. A., Nelson, R. S., Beachy, R. N. 1990. Local and systemic spread of tobacco mosaic virus in transgenic tobacco. *Plant Cell* 2:559–67
101. Wu, X., Beachy, R. N., Wilson, T. M. A., Shaw, J. G. 1990. Inhibition of uncoating of tobacco mosaic virus particles in protoplasts from transgenic tobacco plants that express the viral coat protein gene. *Virology* 179:893–95
102. Yamaya, J., Yoshioka, M., Meshi, T., Okada, Y., Ohno, T. 1988. Cross protection in transgenic tobacco plants expressing a mild strain of tobacco mosaic virus. *Mol. Gen. Genet.* 215: 173–75
103. Yie, Y., Zhao, F., Zhao, Z., Liu, Y. Z., Liu, Y. L., Tien, P. 1992. High resistance to cucumber mosaic virus conferred by satellite RNA and coat protein in transgenic commercial tobacco cultivar G-140. *Mol. Plant-Microbe Interact.* 5:460–65

Annu. Rev. Microbiol. 1993. 47:765–90

GENETIC MANIPULATION OF NEGATIVE-STRAND RNA VIRUS GENOMES

Adolfo García-Sastre[1] *and Peter Palese*
Department of Microbiology, Mount Sinai School of Medicine, One Gustave L. Levy Place, New York, New York 10029

KEY WORDS: RNA transfection, reverse genetics, genetic manipulation, viral vectors, virus replication

---

CONTENTS

ABSTRACT

Negative-strand RNA viruses have been refractory to genetic manipulation using recombinant DNA techniques. Recently, new techniques were developed that allowed the rescue of synthetic RNA molecules into influenza A viruses and, subsequently, into other negative-strand RNA viruses. These techniques are presently being used to study the molecular biology of these

[1]Permanent address: Department of Biochemistry and Molecular Biology, Faculty of Biology, University of Salamanca, Spain.

0066-4227/93/1001-0765$02.00

viruses. Questions concerning *cis*- and *trans*-acting elements that are involved in transcription and replication of negative-sense RNA viral genomes can now be addressed with reverse genetic approaches. Further development of this methodology has enabled the construction—by recombinant DNA techniques—of influenza A viruses that contain altered genomes. The phenotypic characteristics and possible applications of these novel transfectant viruses are also discussed.

INTRODUCTION

Negative-strand RNA virus genomes consist of single-stranded RNA (ssRNA) of negative polarity. Consequently, this genomic RNA needs to be transcribed into mRNA in order to direct the synthesis of viral proteins in the host cell. This function is controlled by a virally coded RNA-dependent RNA polymerase, which is also responsible for the replication of the genome; thus, no DNA of viral origin is involved in virus replication. The viral RNA polymerase forms a tight ribonucleoprotein (RNP) complex with the genomic RNA both in virions and in infected cells. In addition, negative-strand RNA viruses are enveloped viruses whose ribonucleocapsids are surrounded by a lipid bilayer derived from the host cell. Viral glycoproteins involved in receptor-binding and in entry into the host cell are inserted in this envelope.

The negative-strand RNA viruses belong to six families: *Rhabdoviridae, Filoviridae, Paramyxoviridae, Orthomyxoviridae, Bunyaviridae,* and *Arenaviridae*. Members of the first three virus families have nonsegmented RNA genomes, whereas genomes of viruses belonging to the other three families are segmented and consist of either two (*Arenaviridae*), three (*Bunyaviridae*), or seven or eight (*Orthomyxoviridae*) different RNA molecules. These viruses include several important human and animal pathogens: parainfluenza viruses, mumps virus, measles virus, respiratory syncytial virus (RSV) (*Paramyxoviridae*); vesicular stomatitis virus (VSV), rabies virus (*Rhabdoviridae*); influenza virus types A, B, and C (*Orthomyxoviridae*); lymphocytic choriomeningitis virus (LMCV) (*Arenaviridae*); and several encephalitis and hemorrhagic fever viruses (*Arenaviridae, Bunyaviridae,* and *Filoviridae*) (78). The genome organization and the structural similarities of these viruses suggest that they have common viral ancestors (107, 108). In addition, the nonclassified Borna viruses appear to contain RNA genomes of negative polarity (74). Neither these viruses, nor hepatitis delta virus, a nonconventional negative-strand RNA virus, are considered in this review.

Recombinant DNA techniques have been used extensively to genetically manipulate DNA viruses and many RNA viruses. These studies have contributed to a better understanding of the molecular biology of virus infection, replication, and pathogenicity and have enabled the engineering of viral vectors suitable for the expression of foreign proteins in host cells. Several DNA viruses, such as vaccinia viruses (22, 23), herpes simplex viruses (32), cytomegaloviruses (106), baculoviruses (76), adenoviruses (34), and parvoviruses (99), have been used to express foreign proteins in cell culture. Some of these may be used in the future as vaccine vectors for the delivery of heterologous immunogens, or they may allow the transfer of novel genes of interest into target cells (77, 98, 99). Genetically altered retrovirus genomes are also being used for gene transfer and for stable expression of foreign sequences in mammalian cells (75).

Recombinant DNA technology has also led to the construction of infectious full-length cDNA clones of positive-strand RNA viruses, such as bacteriophages (111), picornaviruses (93), plant viruses (1), insect viruses (19), alphaviruses (95), and flaviviruses (53, 94). Specific mutagenesis of these cDNA clones has provided important insights into the molecular biology of the viruses. In addition, positive-strand RNA viral vectors carrying foreign epitopes or expressing heterologous polypeptides have been constructed using polioviruses (2, 89), Sindbis virus (36, 58, 65), Semliki Forest virus (64), or plant viruses (28, 110). RNA transcription in vitro of a linearized full-length cDNA copy of the genomes of these viruses yields viral RNA molecules that can be transfected into appropriate host cells and initiate productive virus infections. This is possible because the RNA of positive-strand RNA viruses is used directly as mRNA for translation of the viral proteins responsible for the amplification of input RNA as well as for the packaging of the RNA into infectious virus particles. Coronaviruses, another family of positive-strand RNA viruses, can also be genetically manipulated. In this case, one can introduce novel or altered genes by taking advantage of the high rate of homologous RNA recombination (51, 63, 114).

In contrast to DNA and positive-strand RNA viruses, negative-strand RNA viruses have been refractory to genetic manipulation. The genomic RNA of these viruses is not in itself infectious. In vitro synthesized viral RNA introduced into permissive cells cannot initiate infection, because in the transfected cell the negative-sense viral RNA is not translated. The present review analyzes recent developments toward circumventing this problem and describes new reverse genetic methodologies used to study *cis*- and *trans*-acting elements involved in viral RNA replication. The review emphasizes transfectant influenza viruses that exhibit novel phenotypic characteristics. These viruses have one or more of their wild-type genomic RNAs replaced with in vitro transcribed synthetic RNAs. The reader may also wish to consult other recent reviews concerning the genetic manipulation of negative-strand RNA viruses (9, 30, 71).

REVERSE GENETICS OF NEGATIVE-STRAND RNA VIRUSES

The Ribonucleoprotein Transfection Method

Amplification, transcription, and rescue of synthetic RNA molecules derived from a negative-strand RNA virus was achieved for the first time in 1989 (72). A biologically active influenza virus RNP complex was reconstituted using synthetic RNA and purified viral proteins, and amplification and expression of the reporter gene was driven by an influenza helper virus (Figure 1).

Influenza A viruses are negative-strand RNA viruses that contain a segmented genome of eight different RNA molecules (55, 81). Coding sequences (in negative polarity) are flanked in each influenza virus RNA segment by short strings of noncoding nucleotides. Interestingly, the first 12 and 13 nucleotides of the 3′ and 5′ ends of each RNA segment, respectively, are highly conserved among different RNA segments of the same virus and also among different influenza A virus strains. In addition, these ends are partially complementary and are responsible for the panhandle structure the viral RNAs adopt in virions and in infected cells (42). These features lead to the hypothesis that the *cis*-acting sequences responsible for replication, transcription, and packaging of the viral RNA are contained within the noncoding regions. Each RNA segment has a coding capacity for one or two viral proteins (56). The three largest RNA segments (approximately 2.3 kb each) code for the PB1, PB2, and PA proteins (P proteins), which form the RNA-dependent RNA polymerase, and, together with the viral RNAs and the NP protein, are found in the RNP complexes (45, 46, 113). In addition, there are six other viral proteins: the hemagglutinin (HA) and the neuraminidase (NA), which are glycoproteins inserted into the viral envelope; the M1 protein, which forms a layer between the viral envelope and the RNPs, and the M2 protein, which is a transmembrane protein with ion-channel activity (91); and the nonstructural proteins, NS1 and NS2, which are both encoded by the shortest RNA segment.

Viral RNP cores containing transcriptional activity can be isolated from disrupted viruses or from infected cells (4, 21, 46, 52, 66, 92, 96, 104). The templates for this activity are endogenous viral RNA molecules that are tightly associated with the purified RNPs. In order to use synthetic RNAs as templates, a procedure to separate the viral proteins and RNA from the RNP cores was needed. Szewczyk et al (109) developed a method for renaturing the NP and P proteins following their isolation from SDS-polyacrylamide gels. RNPs reconstituted with viral RNAs purified from virions were able to transcribe template RNAs, at least at low levels. However, this method was

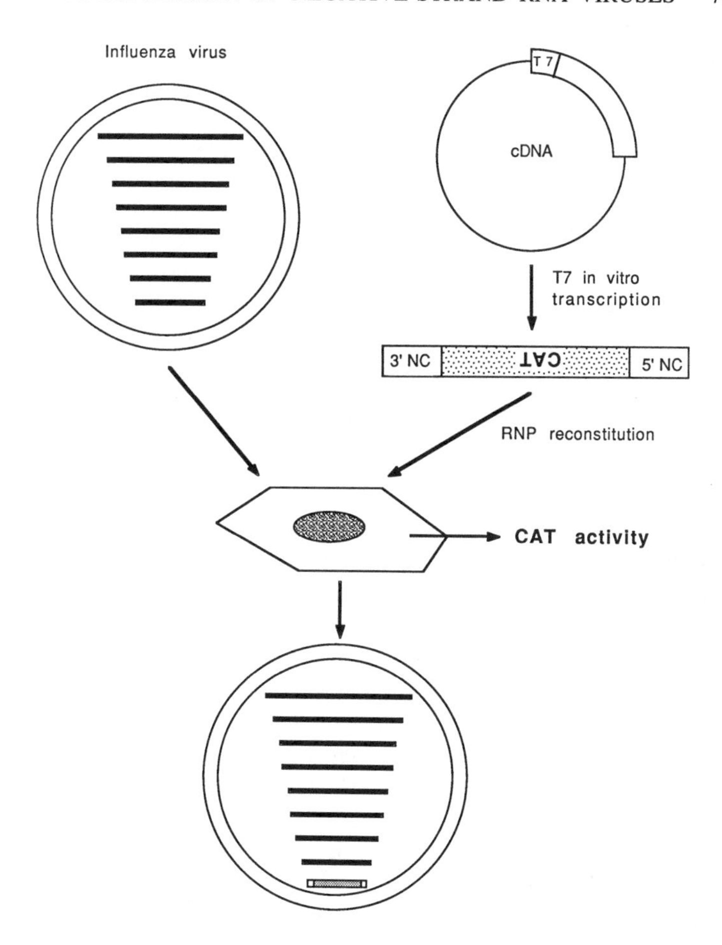

Figure 1 RNP transfection of a synthetic RNA molecule into influenza virus–infected cells (72). A plasmid (cDNA) was constructed containing the open reading frame of the CAT gene in negative polarity flanked by the 3′ and 5′ noncoding (NC) termini of an influenza virus genomic segment. In vitro transcription from this plasmid DNA by T7 RNA polymerase yields RNA that can be used for RNP reconstitution by adding purified viral polymerase-nucleoprotein complexes. The obtained RNPs are transfected into influenza virus–infected cells. CAT activity is detected as a result of replication and transcription of the synthetic RNA, which is also rescued into the new budding viruses. Black boxes represent the eight different influenza virus RNA segments. Stippled boxes represent the RNA containing the CAT gene. The components in the figure are not to scale.

not developed further for the reconstitution of RNP using synthetic RNAs derived from plasmid cDNA.

Honda et al (39) reported a method for separating the NP protein from the RNP complexes. Isolated RNP cores obtained from purified virions were centrifuged through a discontinuous CsCl-glycerol gradient. The NP protein was found in the upper fractions whereas the P proteins and viral RNA (P-RNA complexes) were in the fractions close to the bottom of the gradient. These authors also showed that functional RNP complexes could be reconstituted by adding back the purified NP to the P-RNA complexes (40). However, they could not show RNP reconstitution using exogenous or synthetic RNA.

Parvin et al (84) first demonstrated functional reconstitution using synthetic RNA and purified NP-P protein fractions. The latter were isolated following centrifugation of purified native RNPs on CsCl gradients. These reconstituted RNP complexes were able to copy synthetic RNA molecules whose 15 3′-terminal nucleotides were identical to those present in the viral RNAs. The studies suggested that the RNA promoter sequence for influenza virus RNA synthesis is located in the 3′-terminal nucleotides.

The above-described reconstitution system was then used by Luytjes et al (72) for introducing synthetic RNA sequences into the genomes of influenza virus particles (Figure 1). A recombinant RNA molecule (IVACAT1) was constructed containing—in antisense orientation—the open reading frame of the chloramphenicol acetyltransferase (CAT) gene flanked by the 3′ and 5′ terminal noncoding sequences of the NS gene of influenza A/PR/8/34 virus. Transfection of reconstituted RNP containing this gene into influenza virus–infected cells revealed the replication, transcription, and translation of IVACAT1. The helper virus provided the viral proteins needed for viral RNA synthesis. The foreign gene was also packaged into progeny virions, and CAT activity was detected after several passages in tissue culture of the virus containing the recombinant RNA. However, the activity was greatly decreased, indicating that the CAT gene probably functioned as a defective interfering RNA. The experiment also demonstrated that the noncoding sequences of influenza A virus genes contain the signals required for replication, transcription, and packaging of the viral RNA. The RNP transfection method was further improved by coupling the in vitro transcription from plasmid cDNA with the RNP reconstitution (26). Yamanaka et al (118), instead of using the helper influenza virus, added native RNP cores themselves to drive the replication of the introduced RNPs. After complexing synthetic viral RNA with CsCl-glycerol gradient purified viral NP (117), the resulting RNA-protein complexes were incubated with whole purified native RNP cores in the presence of 1 M NaCl. This procedure most likely resulted in a loosening of the native RNP complex. Next, the salt concentration was reduced to allow an association of the synthetic RNA with the viral polymerase, and the whole mixture was then transfected into cells. Replication and transcription of the input RNA could be detected in the absence of influenza virus.

Recently, Seong & Brownlee reported yet another method for in vitro RNP reconstitution (101). In this case, the isolated RNPs were freed from endogenous RNA molecules by digestion with micrococcal nuclease (17, 47).

The addition of ethylene glycol-*bis*(β-aminoethyl ether)N,N,N′,N′-tetraacetic acid (EGTA) then inactivated the micrococcal nuclease enzyme. The reconstituted RNPs obtained after incubation of nuclease-treated viral cores with synthetic viral RNA supported RNA transcription in vitro and allowed rescue of the gene in vivo. Moreover, CAT expression of IVACAT1 was also detected, although at low levels, when the RNP transfection was done in the absence of helper virus, suggesting that the reconstituted RNPs alone could initiate transcription of the synthetic RNA in vivo. Martín et al (73) reported that NP-P complexes can also be isolated from influenza virus–infected cells rather than from purified virus. Using this polymerase preparation for reconstitution, synthetic RNA templates were transcribed in vitro and successfully rescued into influenza virus in vivo.

In all these systems, replication and transcription of the input RNP is driven either by superinfection with a helper virus or by the proteins contained in the purified RNP cores. However, Huang et al (43) showed that the helper virus can be substituted by coexpressing the P and NP proteins in the same cell using vaccinia virus recombinants (22). Reconstituted RNPs were transfected into vaccinia virus–infected cells expressing different combina-

Table 1 Transfection methods for amplification of synthetic RNAs of negative-strand RNA viruses in vivo

Viruses	Transfected sample	Helper functions	References
Influenza A virus	Reconstituted RNPs: RNA plus CsCl-purified NP and P proteins	Influenza helper virus	26, 72
		Coexpression of NP and P proteins from DNA recombinant viruses	20,[a] 43
	Reconstituted RNPs: RNA plus nuclease digested RNP cores	Influenza helper virus[b]	101
	Reconstituted RNPs: RNA plus CsCl-purified NP plus RNP cores	RNP cores	118
	Reconstituted RNPs:[a] RNA plus *E. coli*–expressed NP protein	Coexpression of P and NP proteins in cells using inducible promoters	50
Sendai virus	RNA plus cellular extracts[a]	Sendai helper virus	82
	Linearized cDNA[c]	T7-responsive cDNA plasmids expressing NP, P/C, and L proteins	10
RSV	Naked RNA	RSV helper virus	14
VSV	Plasmid cDNA[c,d]	T7-responsive cDNA plasmids expressing VSV proteins	85

[a] Naked RNA can also be transfected.
[b] Although influenza virus infection of transfected cells enhances efficiency, it is not absolutely required.
[c] DI RNA is obtained in vivo by T7 polymerase transcription from the transfected plasmid.
[d] Correct 3′ end of the DI RNA is assured by rybozyme cleavage.

tions of influenza virus proteins. This vector-driven replication system demonstrated that only the influenza virus NP, PB1, PB2, and PA proteins are required for the replication and transcription of the viral RNPs in vivo. In addition, such a system can be used to study the *trans*-acting domains of the polymerase proteins involved in RNA replication. Kimura et al (50) used a similar method in which the NP and P proteins were expressed in a mouse cell line in response to dexamethasone (80). These cells supported replication of a synthetic RNA containing the 3′ and 5′ ends of the NS gene of influenza A virus. Prior to transfection, the synthetic viral RNA was covered with partially purified NP obtained from an NP gene–expressing *Escherichia coli* strain. RNA transcription was also demonstrated after transfection of naked synthetic RNA, although at low levels. Expression of functional NP and P proteins, which allowed the amplification and expression of transfected reconstituted RNPs, has also been achieved using SV40 vectors (20). In this case, naked RNA is also used for the transfection. Table 1 summarizes all transfection methods used for the replication and transcription of synthetic RNAs.

Studies on the Replicative Cycle of Influenza Virus

The RNP-transfection systems described above have been successfully used for the study of the molecular biology of influenza virus transcription and replication. When influenza virus enters the host cell, the RNPs are quickly transported into the nucleus (for a recent review, see 38), where viral RNA synthesis takes place. Viral RNAs (vRNAs) are used as templates by the viral RNA polymerase to produce two different species of positive-strand RNAs (52) (Figure 2). Complementary RNAs (cRNAs) are full-length copies of the vRNAs, whereas mRNAs are capped and polyadenylated. The addition of the poly(A) tail occurs at a stretch of uridine residues close to the 5′ end of the vRNAs. Thus, the mRNAs do not contain the genetic information present at the very 5′ end of the vRNAs. Amplification of the vRNA is achieved via copying of the full-length cRNAs into new full-length vRNA molecules.

Promoter sequences recognized by the viral polymerase in both vRNA and cRNA have been studied by a reverse genetics approach (62, 84, 90, 101, 102, 118). One can mutate different nucleotide positions in both vRNA and cRNA 3′ ends and study the effects of these changes on promoter activity in vitro and in vivo. Although discrepancies emerge according to the polymerase preparation and template used (103), promoter sequences have been defined within the first 12–14 nucleotides at the 3′ end of vRNAs (84, 90, 101, 102, 118), and within the first 11–13 nucleotides at the 3′ end of cRNAs (62, 101). As expected, these nucleotide positions are highly conserved in influenza virus RNA segments. Some positions in the promoters appear to be more critical than others (62, 84, 90, 102, 118). In addition, both vRNA and cRNA

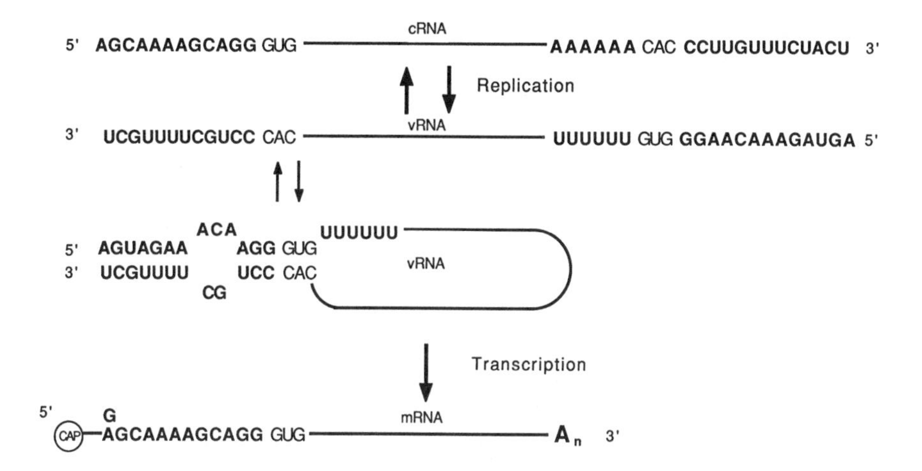

Figure 2 Replication and transcription of influenza virus RNA. An NS gene (vRNA) is represented both in the panhandle and linear configurations. Nucleotide positions conserved in different RNA segments are shown in bold letters. mRNA synthesis is primed by short capped RNA molecules that are cleaved from cellular mRNAs by the PB2 protein. Transcription starts at the penultimate nucleotide of the vRNA and terminates at a stretch of uridine residues where polyadenylation takes place. This stretch is next to a double-stranded RNA structure that is formed by the partially complementary 3′ and 5′ ends of the vRNA. vRNAs also direct the synthesis of full-length complementary RNAs (cRNAs), which are not capped and are used as templates for the generation of new vRNA molecules. Although studies have shown that the vRNAs in virions adopt a panhandle structure, there is no direct experimental evidence for the existence of linear configurations of the replicative vRNAs and cRNAs.

promoters seem to require a free 3′ end for efficient activity (62, 90), although not all experimenters agree on this matter (118). Furthermore, expression in vivo of a reporter gene (CAT) under the control of different mutated vRNA or cRNA promoters was more sensitive to changes than transcription in vitro. This finding suggests that the nucleotides that are contained in the promoters are also involved in other *cis*-acting signals required for CAT expression (62, 90). For example, inefficient transcription, polyadenylation, transport, and/or translation of the mRNA would affect the level of CAT activity. In this context, the reader should note that the available in vitro systems do not allow the synthesis of poly(A)-containing mRNA from synthetic vRNA templates. Thus, the in vivo and in vitro systems are not strictly comparable.

Luo et al (70) studied the requirements for mRNA polyadenylation. Synthetic analogs of the IVACAT1 construct were mutated and studied in vivo. A stretch of 5–6 uridine residues close to the 5′ end of the vRNA and juxtaposed to the RNA panhandle structure is required for efficient polyadenylation of the mRNA. Accordingly, mutations affecting either the uridine stretch or the panhandle structure reduced mRNA formation (70). These results support a stuttering mechanism for poly(A) addition of viral mRNA. The RNA panhandle structure by itself or in association with protein(s) could function as a physical barrier to the viral polymerase (Figure 2). However, it is still not clear what the signal responsible for the switch from mRNA to cRNA synthesis is. The NP protein may play a role in this step (5, 104), perhaps by melting the panhandle vRNA structure to form a linear configuration and allowing readthrough of the polymerase at the mRNA termination site. Further experimentation is needed to determine the conditions under which the template RNAs assume a panhandle or a linear configuration. Another important question is whether regulatory sequences exist that are specific for individual influenza virus RNA segments. Reverse genetics studies should help in answering this crucial question (24, 69, 112).

The possible functions of the NS1 and NS2 proteins in viral RNA replication have also been studied (43). IVACAT1 was RNP transfected into cells expressing PB1, PB2, PA, and NP proteins. The additional expression of NS1 and NS2 proteins did not alter the levels of CAT activity in this system. However, other experiments have suggested regulatory functions for the NS proteins as well as for the M1 protein (3, 24).

Reverse genetics has also been used to study the mechanism of inhibition of influenza virus by the murine Mx1 protein (44). Previous studies showed that the interferon-inducible Mx1 protein is associated with specific resistance to influenza virus infection (37), probably by inhibiting viral primary transcription (88). Huang et al (44) have implicated the PB2 protein as a target for the inhibition of influenza virus replication by the Mx1 protein. However, at this time no direct interaction between the Mx1 and the PB2 proteins has been demonstrated.

Reverse Genetics of Other Negative-Strand RNA Viruses

Synthetic analogs of viral RNAs have also been successfully rescued into paramyxoviruses and vesicular stomatitis virus (VSV) (Figure 3, Table 1). These viruses differ from influenza viruses in that they possess a non-segmented genomic RNA and replicate and transcribe the RNA in the cytoplasm of the host cell. Park et al (82) modified the RNP-transfection method developed for influenza virus for use with Sendai virus. The latter is a prototype paramyxovirus whose RNA contains a leader sequence and six transcriptional units (NP, P/C, M, F, HN, and L) (Figure 3). The F and HN proteins are glycoproteins inserted into the viral envelope; the M or matrix protein surrounds the ribonucleoprotein, which is composed of RNA associated with the NP, P, and L proteins. Leader RNA and mRNAs are transcribed by the viral RNA polymerase due to the presence of conserved transcription start and termination, polyadenylation, and restart sequences. Replication of the genomic RNA occurs via a full-length replicative intermediate of positive polarity that is synthesized by readthrough of transcription termination signals (for review, see 29). A recombinant RNA was first engineered that contained the CAT gene in negative polarity flanked by the 3′ and 5′ ends of the Sendai virus genomic RNA (Figure 3). When this recombinant RNA was transfected into Sendai virus–infected cells, it was replicated, transcribed, and packaged into new Sendai virus particles. In contrast to the influenza virus system, in vitro RNP reconstitution was not needed, although addition of cellular extracts to the synthetic RNA prior to transfection enhanced the expression level of the CAT gene. This reverse genetics system was further used to study the requirements for RNA editing in Sendai virus (83). Most paramyxoviruses can synthesize from the P/C gene mRNAs containing nontemplated G-nucleotide inserts (for a review, see 57). In Sendai virus, edited mRNAs encode a cysteine-rich V protein, which has been proposed to be responsible for down regulation of viral RNA synthesis (18). A 24-nucleotide sequence derived from the P/C gene and containing the consensus paramyxovirus editing sequence was introduced into a synthetic Sendai virus model RNA. This sequence was sufficient to promote mRNA editing when the recombinant RNA was transfected into Sendai virus–infected cells.

Calain et al (10) showed that encapsidation and replication of a synthetic RNA molecule by the Sendai virus RNA polymerase can also be driven in cells that express the L, P/C, and NP proteins of Sendai virus. These proteins were previously shown to be required for RNA polymerase activity and for the encapsidation of the genomic RNA (41). The authors transfected into cells a linearized cDNA plasmid containing a cloned Sendai virus defective interfering (DI) RNA under the control of the T7 polymerase (Figure 3). Infection of these cells by a vaccinia virus recombinant expressing T7 polymerase led to the synthesis of the DI RNA. When these cells were cotransfected with plasmids expressing NP, P/C, and L proteins via T7 polymerase, the DI RNA was encapsidated and replicated. Because the DI RNA does not direct the synthesis of mRNA, this system can be used for studying *cis*- and *trans*-acting signals in Sendai virus replication.

Foreign synthetic RNA could also be rescued into respiratory syncytial virus (RSV), another member of the paramyxovirus family (Figure 3). Collins et al (14) reported that a foreign gene flanked by the noncoding 3′ and 5′ ends of the RSV genomic RNA is replicated, transcribed, and packaged into virions after transfection into RSV infected cells. These authors also used the system for studying the *cis*-acting sequences in the RNA that are required for amplification and rescue. They showed that the first five 3′-terminal nucleotides do not appear to be essential for RNA transcription and packaging, although they are highly conserved. However, the RNA reporter gene did not tolerate the addition of extra nucleotides at the 3′ end. These authors also found that the amplification, expression, and packaging of the synthetic RNA

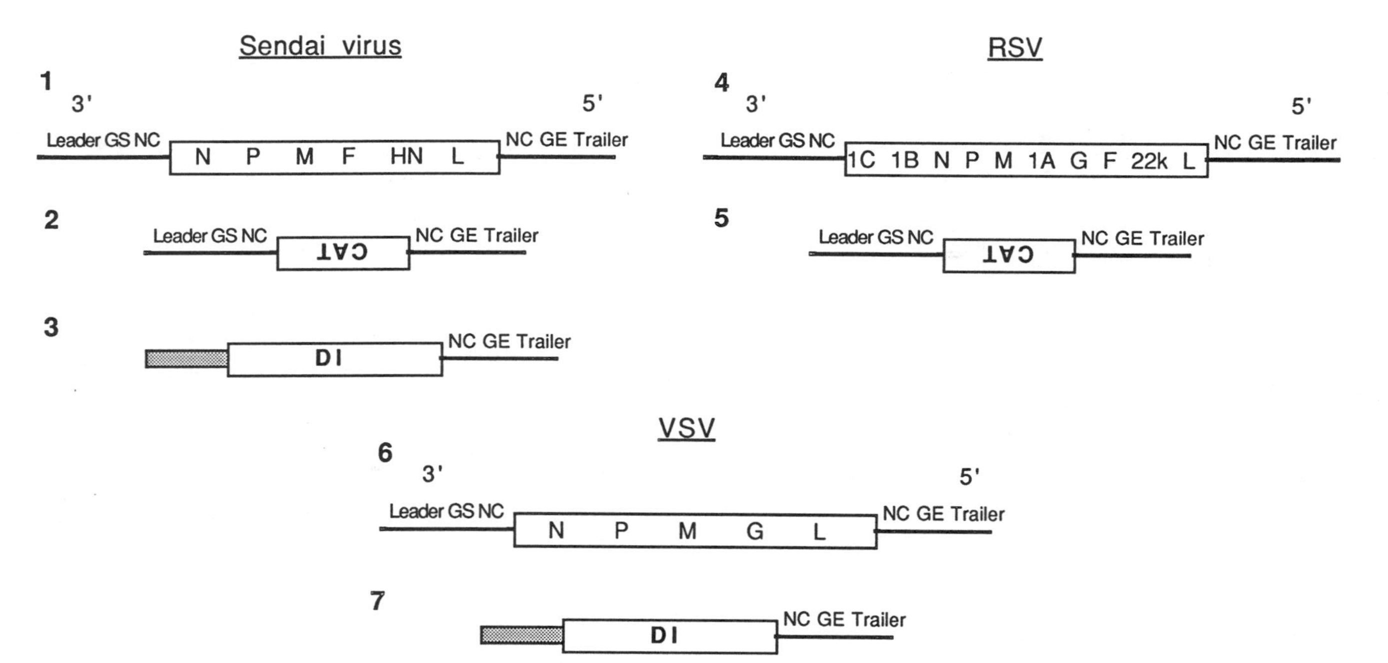

Figure 3 Schematic representation of genomic RNAs of Sendai virus, respiratory syncytial virus (RSV), and vesicular stomatitis virus (VSV), and of the synthetic RNAs rescued into these viruses. Leader, trailer, gene start (GS), gene end (GE), and noncoding (NC) sequences located at the 3' and 5' ends of the RNA molecules are represented by a line. 1, 4, 6. Transcriptional units in the genomic RNAs are identified by letters or combination of letters and numbers. 2, 5. Synthetic RNAs expressing CAT have been rescued into Sendai virus and RSV infected cells (14, 82). These RNAs contain the CAT gene in negative polarity flanked by the 3' and 5' ends of the genomic RNA. 3, 7. Synthetic defective interfering (DI) RNAs have also been used in replicative systems established in cells that express Sendai virus or VSV polymerases and N proteins (10, 85). These DI RNAs are based on naturally occurring copy-back DI molecules, and they consist of the 5' ends of the viral genomes, including a portion of the L open reading frame, and 3' ends (*stippled boxes*) that are complementary to the 5' terminal regions of the DI RNAs.

could be accomplished with 53 and 52 nucleotides from the 3′ and 5′ vRNA ends, respectively. In addition, substitution of the vRNA promoter by a cRNA promoter, as found in DI RNAs, did not affect the rescue and expression of the gene.

Finally, a reverse genetics system has been developed for vesicular stomatitis virus (VSV), which is a prototype rhabdovirus. The replicative cycle of VSV is very similar to that of paramyxoviruses. The genome has the coding capacity for five proteins (Figure 3): the N (nucleocapsid) protein is tightly associated with the genomic RNA; the P (phosphoprotein) and L (large) proteins constitute the viral RNA polymerase; the M protein covers the nucleocapsid and is surrounded by a bullet-shape envelope; and the G glycoprotein is involved in virus attachment and internalization into the host cell (115). Pattnaik & Wertz (86) showed that the coexpression of N, P, and L proteins driven by a vaccinia vector supports RNA replication of naturally occurring infectious DI particles of VSV. When the M and G proteins were also expressed, new infectious DI particles budded from the cell (87). This system can now be used for a structure-function analysis of all VSV proteins. In order to investigate the *cis*-acting signals in the VSV genome responsible for replication, transcription, and packaging, Pattnaik et al (85) substituted the natural DI genome with a cDNA DI clone. After T7 transcription of this cDNA clone in cells expressing T7 polymerase, ribozyme cleavage generated an exact 3′ terminus of the DI RNA (Figure 3). This system allows the analysis of both *cis*- and *trans*-acting elements required for VSV RNA replication and packaging, assembly, budding, and infectivity.

GENETIC MANIPULATION OF INFLUENZA VIRUS GENOMES

Selection Systems

As described above, the RNP transfection method has allowed the rescue by influenza virus of a foreign gene that expresses CAT. The first influenza virus gene exchanged by an in vitro synthesized RNA was the NA gene (25). The combination of RNP transfection of an engineered NA gene with a selection system for the novel (transfectant) virus enabled the virus gene exchange (Figure 4). Influenza A virus containing the NA gene of influenza A/WSN/33 virus can form plaques in MDBK cells in the absence of trypsin. On the other hand, WSN-HK virus does not form plaques in MDBK cells without trypsin (100). WSN-HK is identical to influenza A/WSN/33 virus except for the NA gene, which is derived from influenza A/HK/8/68 virus. When an in vitro synthesized NA gene of influenza A/WSN/33 virus was RNP-transfected into

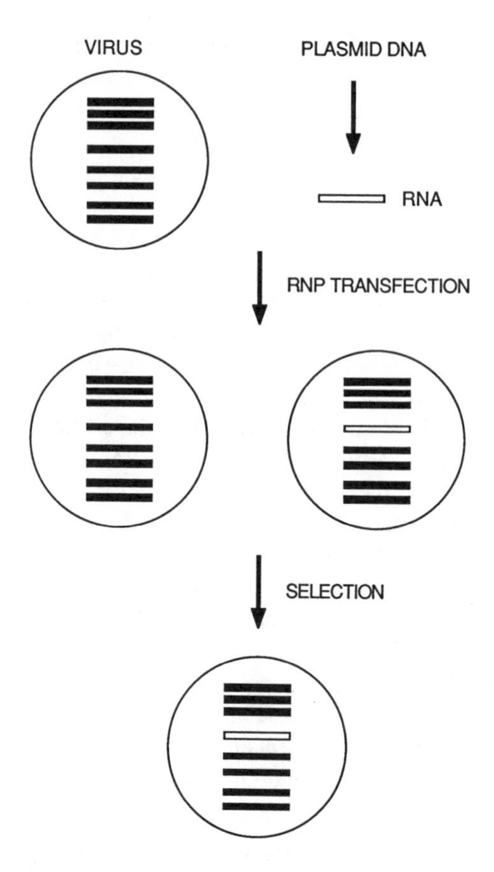

Figure 4 Generation of influenza virus transfectants. A synthetic RNA (*open box*) is transcribed in vitro from plasmid cDNA. This RNA molecule is complexed into ribonucleoproteins (RNPs) by incubation with purified NP and viral polymerases. The RNPs are transfected into cells infected with an influenza helper virus. As a result, some of the new budding viruses contain the transfected gene. Finally, a selection system is used to isolate transfectant viruses. Full boxes depict RNAs derived from the helper virus, and the open box identifies the rescued RNA.

helper WSN-HK virus–infected cells, viruses containing the transfected gene were selected by plaquing on MDBK cells without trypsin (25).

Recent developments have enabled the exchange of the HA and NS genes of influenza A virus by cDNA-derived RNAs (26). Selection of viruses containing the transfected HA gene is achieved by using neutralizing antibodies that are directed against the HA protein of the helper virus and suppress its replication. NS gene transfections have been performed using a temperature-sensitive influenza virus strain with a defect in the NS gene (105).

Viruses containing a transfected wild-type NS gene can be selected at the nonpermissive temperature.

RNP transfection will likely allow the manipulation of other influenza virus genes if selection systems for the transfectants become available. In the past, the only method for exchanging genes in influenza A viruses was to coinfect cells with different strains. Because of the segmented nature of the influenza virus genome, the progeny virions might receive RNA segments from either parental strain. Such reassortment, however, only leads to the exchange of naturally occurring genes in influenza viruses. The RNP-transfection system allows the replacement of influenza A virus genes with synthetic recombinant RNA molecules. This opens the possibility of introducing specific mutations into the genome of influenza A viruses. The new viruses formed as a result of RNP transfection have been called transfectants (26).

Genetic Manipulation of the Noncoding Sequences of Influenza Virus Genes

The noncoding sequences of influenza virus genes contain the signals responsible for replication, transcription, and packaging of the genes. The characterization of important nucleotide positions in the vRNA and cRNA promoters (see above) can provide guidance in constructing transfectant viruses with novel phenotypic properties. Further analysis of the rescued viruses will enhance our knowledge about the regulation of influenza virus transcription, replication, and translation (69).

Influenza viruses containing a gene with altered noncoding sequences were first generated by Muster et al (79). A transfectant virus (NA/B-NS) was obtained in which the noncoding region of the neuraminidase gene was derived from the NS gene of influenza B virus. The fact that this chimeric virus is viable suggests that influenza B virus promoter sequences can be recognized by the influenza A virus RNA polymerase. However, the NA gene of the NA/B-NS virus was not as efficiently replicated as the other seven genes (67). The reduction in the synthesis of the NA gene in infected cells was also reflected in a lower representation of this gene in packaged virions. Consequently, the NA/B-NS virus preparation contains a higher proportion of defective particles lacking the NA gene than does a wild-type preparation. Interestingly, this observation is correlated with an attenuated phenotype of the virus both in tissue culture and in mice (79). In addition, the NA/B-NS virus was able to elicit a protective immune response in animals against a challenge with wild-type virus. The attenuation characteristic makes this virus a potential prototype for a live virus vaccine (for review, see 13). Attenuation by down-regulation of a viral gene, leading to an increase in noninfectious particles, could thus represent a new principle for generating live virus vaccines that would be applicable to other viruses as well. Attempts have been made to identify the minimum number of B-NS gene-derived sequences in the NA RNA of the NA/B-NS virus that could confer this novel phenotype (7).

Changes in the Hemagglutinin Protein

The HA protein of influenza A virus is the most abundant glycoprotein of the viral envelope. This protein contains the receptor-binding site of the virus, mediates the membrane fusion required for internalization of the virus into the host cell, and is the major antigenic component on the surface. Fourteen different HA subtypes have been identified and, for the H1 and H3 subtypes, five B-cell antigenic regions (A, B, C, D, and E) have been delineated (for a review, see 116). Li et al (61) obtained different transfectant influenza A viruses with chimeric HAs. Portions of the antigenic site B of subtype H1 were replaced with the corresponding regions of subtype H2 or H3. The introduced epitopes retained their immunological properties and were recognized by antibodies raised against their own HA subtype. Mice immunized with the chimeric transfectant virus W (H1)-H3, which possesses part of the H3-specific B site epitope in its H1 HA, developed neutralizing antibodies

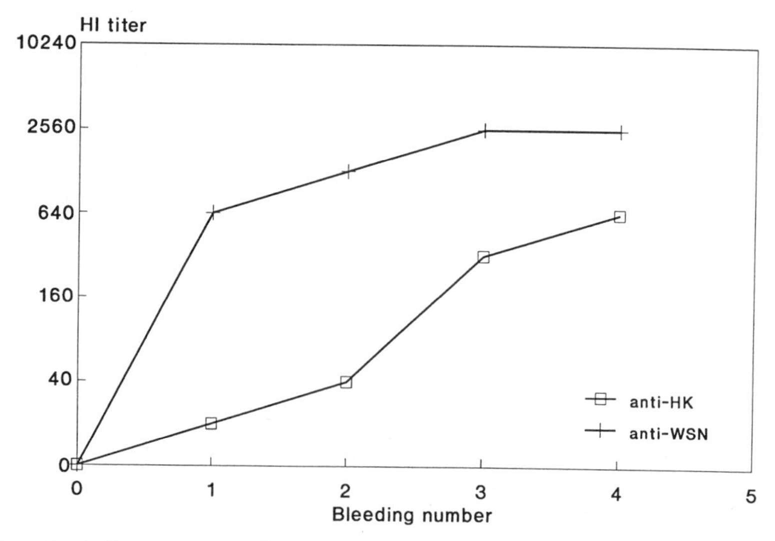

Figure 5 Antibody response in mice immunized with W(H1)-H3 transfectant virus (61). Ten micrograms of purified W(H1)-H3 virus were administered intraperitoneally into mice on days 0, 7, 21, and 35. Sera were collected on days 0, 7, 14, 35, and 42 (bleeding numbers 0, 1, 2, 3, and 4, respectively). Hemagglutination inhibition (HI) titers against A/WSN/33 (H1N1) virus and A/Hong Kong/8/68 (H3N2) virus are indicative of the presence in serum of antibodies that recognize H1 and H3 hemagglutinins, respectively.

against influenza A viruses of both subtypes (Figure 5). These results led to attempts to present foreign T-cell and B-cell epitopes in the HA protein. A malaria sporozoite T-cell epitope has been inserted into site E of the HA (60), and different B-cell epitopes from several pathogens have been inserted into site B of the HA (35, 59). The properties of the novel transfectant viruses and their ability to induce specific immune responses in mice are currently under investigation.

Changes in the Neuraminidase Protein

The NA protein of influenza A virus is the second major glycoprotein of the viral envelope. The NA protein plays a role in the release of virions from infected cells by removing sialic acids involved in HA binding. Thus, this viral enzyme contributes to the spread of the virus during infection. The NA molecule consists—from its amino terminal to its carboxy terminal—of a short cytoplasmic domain, a transmembrane region, a stalk of variable length, and a globular head where the active site is located (for review, see 15). The coding region of the NA gene of influenza A virus has also been a target for genetic manipulation. Castrucci et al (11) reported the construction of an influenza virus (FL79) containing a foreign sequence of eight amino acids inserted into the stalk of its neuraminidase. This sequence in the NA was recognized by monoclonal antibodies, although infection of mice by the chimeric virus did not induce detectable levels of antibodies against the foreign sequence. Interestingly, in mice the chimeric virus showed attenuation characteristics similar to those of the NA/B-NS virus. Thus, FL79 virus was attenuated in mice and could confer immunity against wild-type virus infections. The attenuation might result from slightly reduced neuraminidase activity of the chimeric virus as compared to that of the wild-type virus.

In addition, the length of the stalk region of the NA protein can be altered, and mutations in the stalk may modulate virulence (12, 68). For example, a transfectant influenza A/WSN/33 virus that contained a deletion of 28 amino acids in the stalk region of the NA protein showed an altered host-range phenotype in tissue culture (68). A similar deletion of the stalk region of the NA resulted in attenuated virus phenotypes in mice (12). These studies demonstrate that the stalk region tolerates insertions and replacements of its amino acid sequence and that it therefore might be a convenient site for the introduction of foreign epitopes. Manipulation of the stalk could also result in the generation of viruses with different levels of attenuation. Modulation of viral pathogenicity can also be achieved by mutating the cytoplasmic domain of the NA protein (8; M. Doymaz & P. Palese, unpublished observations).

Genetic Manipulation of the NS Gene

The NS gene of influenza A virus codes for the two nonstructural proteins, NS1 and NS2. The NS1 protein is translated from the unspliced form of the NS mRNA, whereas the NS2 protein is translated from the spliced mRNA. Enami et al (27) engineered a viral RNA that encodes only the NS1 protein. This RNA was RNP transfected into cells infected with an influenza helper virus containing a temperature-sensitive (*ts*) defect in the NS1 protein. Transfectant viruses rescued at the nonpermissive temperature possessed nine different RNA segments instead of the usual eight segments: viable transfectant viruses required both the helper virus NS gene and the transfected NS1 gene in order to produce the wild-type NS1 and NS2 proteins. The ability of influenza virus to package more than eight RNA segments is in agreement with the theory of random RNA packaging (16). According to this theory, vRNA segments are randomly chosen among a pool of eight different vRNAs that carry the same specific packaging signal. Since 5–10% of virus particles are infectious, this model requires each particle to have more than ten RNA segments in order to generate the correct percentage of viruses containing a full complement of the eight different vRNAs. This theory also predicts that an influenza virus with nine different RNA segments would produce fewer infectious particles than a wild-type virus, because the probability of generating such infectious particles would be lower. The transfectant virus preparations containing nine different RNA segments showed a lower ratio of infectious particles to physical particles than did the wild-type virus preparations, thus supporting the random RNA packaging hypothesis.

Further support for this model came from experiments allowing quantification of vRNA levels in infected cells. If the RNAs were packaged randomly, one would expect that the ratio among all the viral RNA segments would be the same within the cell and after being packaged into virions. This correlation between cellular vRNA and virion vRNA levels was evident in the case of the transfectant NA/B-NS virus (67). As discussed in the previous section, the level of the NA gene in infected cells was reduced compared with that of the other seven genes; the same reduction of the NA gene was also seen in the purified RNA of progeny virions. Nevertheless, these experiments provide only indirect evidence for the random-packaging model, and further experimentation is needed to elucidate the precise mechanism of RNA packaging and to define the proteins involved in RNP morphogenesis.

Other Transfectant Influenza Viruses

In order to further investigate the potential of influenza virus to accommodate additional genes, several transfectant viruses have been constructed with extra sequences in their NA gene downstream of the open reading frame. Viruses containing more than 900 extra nucleotides in the NA have been rescued (31).

Bergmann et al (6) reported the rescue of influenza virus NA genes generated via nonhomologous RNA recombination. The recombinant RNAs were derived from the transfected NA gene and viral RNA fragments present in the polymerase preparation used for the RNP reconstitution. Although RNA recombination has been extensively documented for positive-strand RNA viruses (for review, see 54), only a few cases of RNA recombination have been reported for negative-strand RNA viruses such as influenza viruses (48, 49). The experiments described above indicate that recombination between influenza virus RNA genes occurs, albeit infrequently, and that this mechanism may contribute to the evolutionary changes of the virus in nature.

Future Directions and Perspectives

The RNP transfection method has been used for the introduction of changes in the NA, HA, and NS genes of influenza A virus. These studies have already provided new insights into the molecular biology of influenza viruses. We believe that this technique will continue to teach us more about the structure-function relationships of viral proteins, as well as about their roles in viral replication and pathogenesis. Specifically, new selection systems should become available that will allow the genetic manipulation of other influenza virus genes in order to construct additional novel transfectant viruses with specific biological and molecular characteristics. To avoid cumbersome selection methods, which always require the presence of an undesired helper virus, systems will be developed that provide all the required viral proteins from cDNA clones. Because of the absence of helper virus in such complementation systems, biochemical studies can be done on transfectant viruses that undergo only a single replicative cycle.

Some of the changes introduced into the genomes of influenza viruses have been responsible for the attenuation phenotypes of the transfectant viruses (11, 12, 79). Attenuation of influenza viruses is a prerequisite to developing safe and effective live virus vaccines against influenza. Genetic stability of the attenuated phenotype could be assured by the introduction of additional attenuation markers. Genetic manipulation of influenza viruses by DNA recombinant technologies is one of the routes by which this goal can be achieved. Furthermore, the RNP-transfection system should also facilitate the use of influenza virus as a viral vector. Until now, only foreign epitopes have been stably expressed in the viral glycoproteins (11, 35, 59–61). However, preliminary evidence suggests that we will be able to engineer influenza viruses containing bicistronic genes that express an entire foreign protein in addition to an essential viral protein (Figure 6). If such constructs are stable and express

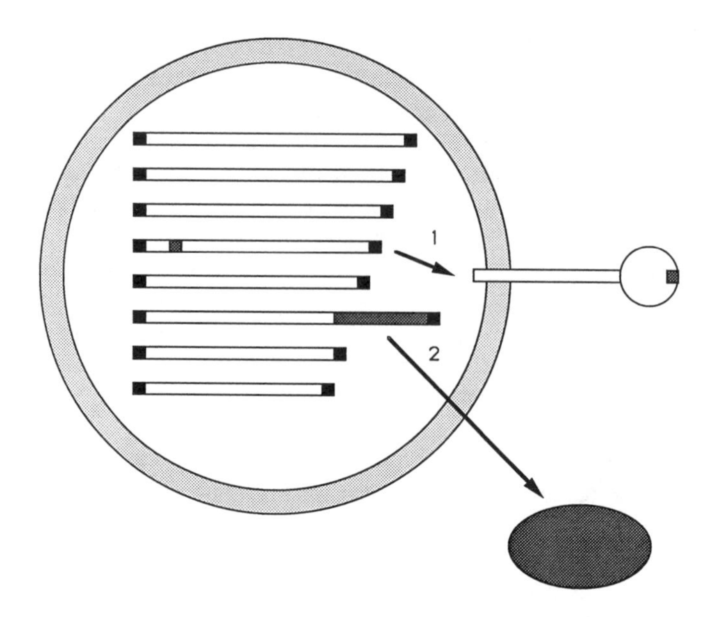

Figure 6 Possibilities of using influenza virus as a vector. 1. Epitope grafting. A viral protein is used for the presentation of a foreign epitope that has been inserted into the protein's amino acid sequence. The insert in the gene and in the surface protein is cross hatched. 2. Bicistronic genes. An influenza virus gene is engineered that encodes a foreign protein in addition to an essential viral protein. The foreign protein (*stipled oval*) is indicated.

foreign proteins at high levels, the possibility of using influenza virus as a vector for vaccination against other diseases would be greatly increased.

GENETIC MANIPULATION OF OTHER NEGATIVE-STRAND RNA VIRUSES: PERSPECTIVES

Reverse genetics studies of influenza viruses have increased our understanding of the fundamental nature of replication, transcription, and packaging of influenza virus genomes. Recently, this approach was also applied to other negative-strand RNA viruses (10, 14, 18, 82, 85–87). *Cis*- and *trans*-acting signals important for viral replication can now be studied in the case of Sendai virus, RSV, and VSV. Furthermore, systems have been developed to study the replication and expression of transfected RNA of *Reoviridae,* a family of double-stranded RNA viruses (33, 97). The genome manipulation of this group of viruses was subject to constraints similar to those experienced with negative-strand RNA viruses. In time reverse genetics methods will become available for the study of all negative-strand and double-stranded RNA viruses. Site-directed mutagenesis of the genomes of these viruses will become possible with the availability of full-length genomic cDNA clones and the use of appropriate selection systems to rescue infectious virions. This process will facilitate the generation of novel RNA viruses for possible use as therapeutic and scientific tools.

NOTE. See note added in proof, page 790.

ACKNOWLEDGMENTS

We are grateful to Dr J. Ortín for helpful suggestions and discussions on this review and for generously providing us with relevant in-press manuscripts.

Literature Cited

1. Ahlquist, P., French, R., Janda, M., Loesch-Fries, L. S. 1984. Multicomponent RNA plant virus infection derived from cloned viral cDNA. *Proc. Natl. Acad. Sci. USA* 81:7066–70
2. Almond, J. W., Burke, K. L. 1990. Poliovirus as a vector for the presentation of foreign antigens. *Semin. Virol.* 1:11–20
3. Alonso-Caplen, F. V., Nemeroff, M. E., Qiu, Y., Krug, R. M. 1992. Nucleocytoplasmic transport: the influenza virus NS1 protein regulates the transport of spliced NS2 mRNA and its precursor NS1 mRNA. *Genes Dev.* 3:1534–44

3a. Ausubel, F. M., Brent, R., Kingston, R. E., Moore, D. D., Seidman, J. G., et al, eds. 1991. *Current Protocols in Molecular Biology*. New York: Greene/Wiley Interscience

4. Beaton, A. R., Krug, R. M. 1984. Synthesis of the templates for influenza virion RNA replication in vitro. *Proc. Natl. Acad. Sci. USA* 81:4682–86
5. Beaton, A. R., Krug, R. M. 1986. Transcription antitermination during influenza viral template RNA synthesis requires the nucleocapsid protein and the absence of a 5′ capped end. *Proc. Natl. Acad. Sci. USA* 83:6282–86
6. Bergmann, M., García-Sastre, A., Palese, P. 1992. Transfection-mediated recombination of influenza A virus. *J. Virol.* 66:7576–80
7. Bergmann, M., Luo, G., Muster, T., García-Sastre, A., Palese, P. 1992. Characterization of the attenuated influenza A NA/B-NS transfectant virus. *Options for the Control of Influenza Int. Sci. Conf., 2nd, Courchevel,* p. 69
8. Bilsel, P., Castrucci, M. R., Kawaoka, Y. 1992. *Rescue of influenza A neuraminidase mutants*. Presented at Annu. Meet. Am. Soc. Virol., Cornell Univ., Ithaca
9. Bredenbeek, P. J., Rice, C. M. 1992. Animal RNA virus expression systems. *Semin. Virol.* 3:297–310
10. Calain, P., Curran, J., Kolakofsky, D., Roux, L. 1992. Molecular cloning of natural paramyxovirus copy-back defective interfering RNAs and their expression from DNA. *Virology* 191: 62–71
11. Castrucci, M. R., Bilsel, P., Kawaoka, Y. 1992. Attenuation of influenza A virus by insertion of a foreign epitope into the neuraminidase. *J. Virol.* 66: 4647–53
12. Castrucci, M. R., Kawaoka, Y. 1993. Biologic importance of neuraminidase stalk length in influenza A virus. *J. Virol.* 67:759–64
13. Clements, M. L. 1992. Influenza vaccines. In *Vaccines: New Approaches to Immunological Problems,* ed. R. W. Ellis, pp. 129–50. Boston: Butterworth-Heinemann
14. Collins, P. L., Mink, M. A., Stec, D. S. 1991. Rescue of synthetic analogs of respiratory syncytial virus genomic RNA and effect of truncations and mutations on the expression of a foreign reporter gene. *Proc. Natl. Acad. Sci. USA* 88:9663–67
15. Colman, P. M. 1989. Neuraminidase: enzyme and antigen. See Ref. 51a, pp. 175–218
16. Compans, R. W., Choppin, P. W.

1975. Reproduction of myxoviruses. In *Comprehensive Virology,* ed. H. Fraenkel-Conrat, R. R. Wagner, 4:179–252. New York: Plenum

17. Cuatrecasas, P., Fuchs, S., Anfinsen, C. B. 1967. Catalytic properties and specificity of the extracellular nuclease of *Staphylococcus aureus. J. Biol. Chem.* 242:1541–47
18. Curran, J., Boeck, R., Kolakofsky, D. 1991. The Sendai virus P gene expresses both an essential protein and an inhibitor of RNA synthesis by shuffling modules via mRNA editing. *EMBO J.* 10:3079–85
19. Dasmahapatra, B., Dasgupta, R., Saunders, K., Selling, B., Gallagher, T., Kaesberg, P. 1986. Infectious RNA derived by transcription from cloned cDNA copies of the genomic RNA of an insect virus. *Proc. Natl. Acad. Sci. USA* 83:63–66
20. De la Luna, S., Martín, J., Portela, A., Ortín, J. 1993. Influenza virus naked RNA can be expressed upon transfection into cells co-expressing the three subunits of the polymerase and the nucleoprotein from simian virus 40 recombinant viruses. *J. Gen. Virol.* 74:535–39
21. Del Río, L., Martínez, C., Domingo, E., Ortín, J. 1985. In vitro synthesis of full-length influenza virus complementary RNA. *EMBO J.* 4:243–47
22. Earl, P. L., Moss, B. 1991. Generation of recombinant vaccinia viruses. See Ref. 3a, pp. 16.17.1–16.17.16
23. Earl, P. L., Moss, B. 1991. Characterization of recombinant vaccinia viruses and their products. See Ref. 3a, pp. 16.18.1–16.18.10
24. Enami, M. 1992. Influenza virus NS1, NS2 and M1 proteins as regulatory proteins for the viral gene expression. *Options for the Control of Influenza Int. Sci. Conf., 2nd, Courchevel,* p. 79
25. Enami, M., Luytjes, W., Krystal, M., Palese, P. 1990. Introduction of site-specific mutations into the genome of influenza virus. *Proc. Natl. Acad. Sci. USA* 87:3802–5
26. Enami, M., Palese, P. 1991. High-efficiency formation of influenza virus transfectants. *J. Virol.* 65:2711–13
27. Enami, M., Sharma, G., Benham, C., Palese, P. 1991. An influenza virus containing nine different RNA segments. *Virology* 185:291–98

27a. Fields, B. N., Knipe, D. M., Chanock, R. M., Hirsch, M. S., Melnick, J. L., et al, eds. 1990. *Virology*. New York: Raven. 2nd ed.

28. French, R., Janda, M., Ahlquist, P. 1986. Bacterial gene inserted in an engineered RNA virus: efficient expression in monocotyledonous plant cells. *Science* 231:1294–97
29. Galinski, M. S., Wechsler, S. L. 1991. The molecular biology of the *Paramyxovirus* genus. See Ref. 50a, pp. 41–82
30. García-Sastre, A., Palese, P. 1993. Infectious influenza viruses from cDNA-derived RNA: reverse genetics. In *Regulation of Gene Expression in Animal Viruses,* ed. L. Carrasco, N. Sonenberg, E. Wimmer, pp. 107–14. New York: Plenum
31. García-Sastre, A., Percy, N., Barclay, W., Palese, P. 1992. An influenza virus carrying the internal ribosome entry site of the encephalomyocarditis virus in its neuraminidase gene. *Abstract Book NATO Advanced Study Institute and EEC Course on the Regulation of Gene Expression by Animal Viruses, Mallorca,* p. S19
32. Glorioso, J. C., Goins, W. F., Fink, D. J. 1992. Herpes simplex virus-based vectors. *Semin. Virol.* 3:265–76
33. Gorziglia, M. I., Collins, P. L. 1992. Intracellular amplification and expression of a synthetic analog of rotavirus genomic RNA bearing a foreign marker gene: mapping *cis*-acting nucleotides in the 3′-noncoding region. *Proc. Natl. Acad. Sci. USA* 89:5784–88
34. Grunhaus, A., Horwitz, M. S. 1992. Adenoviruses as cloning vectors. *Semin. Virol.* 3:237–52
35. Guinea, R., Racaniello, V. R., Palese, P. 1992. Influenza virus as a vector for the presentation of foreign antigens. *Abstract Book NATO Advanced Study Institute and EEC Course on the Regulation of Gene Expression by Animal Viruses, Mallorca,* p. S18
36. Hahn, C. S., Hahn, Y. S., Braciale, T. J., Rice, C. M. 1992. Infectious Sindbis virus transient expression vectors for studying antigen processing and presentation. *Proc. Natl. Acad. Sci. USA* 89:2679–83
37. Haller, O., Arnheiter, H., Lindenmann, J., Gresser, I. 1980. Host gene influences sensitivity to interferon action selectively for influenza virus. *Nature (London)* 283:660–62
38. Helenius, A. 1992. Unpacking the incoming influenza virus. *Cell* 69:577–78
39. Honda, A., Uéda, K., Nagata, K., Ishihama, A. 1987. Identification of

the RNA polymerase-binding site on genome RNA of influenza virus. *J. Biochem.* 102:1241–49

40. Honda, A., Uéda, K., Nagata, K., Ishihama, A. 1988. RNA polymerase of influenza virus: role of NP in RNA chain elongation. *J. Biochem.* 104: 1021–26
41. Horikami, S. M., Curran, J., Kolakofsky, D., Moyer, S. A. 1992. Complexes of Sendai virus NP-P and P-L proteins are required for defective interfering particle genome replication in vitro. *J. Virol.* 66:4901–8
42. Hsu, M.-T., Parvin, J. D., Gupta, S., Krystal, M., Palese, P. 1987. Genomic RNAs of influenza viruses are held in a circular conformation in virions and in infected cells by a terminal panhandle. *Proc. Natl. Acad. Sci. USA* 84: 8140–44
43. Huang, T.-S., Palese, P., Krystal, M. 1990. Determination of influenza virus proteins required for genome replication. *J. Virol.* 64:5669–73
44. Huang, T., Pavlovic, J., Staeheli, P., Krystal, M. 1992. Overexpression of the influenza virus polymerase can titrate out inhibition by the murine Mx1 protein. *J. Virol.* 66:4154–60
45. Inglis, S. C., Carroll, A. R., Lamb, R. A., Mahy, B. W. J. 1976. Polypeptides specified by the influenza virus genome. I. Evidence for eight distinct gene products specified by fowl plague virus. *Virology* 74:489–503
46. Ishihama, A., Nagata, K. 1988. Viral RNA polymerases. *CRC Crit. Rev. Biochem.* 23:27–76
47. Jackson, R. J., Hunt, T. 1983. Preparation and use of nuclease-treated rabbit reticulocyte lysates for the translation of eukaryotic messenger RNA. *Methods Enzymol.* 96:50–74
48. Jennings, P. A., Finch, J. T., Winter, G., Robertson, J. S. 1983. Does the higher order structure of the influenza virus ribonucleoprotein guide sequence rearrangements in influenza viral RNA? *Cell* 34:619–27
49. Khatchikian, D., Orlich, M., Rott, R. 1989. Increased viral pathogenicity after insertion of a 28S ribosomal RNA sequence into the haemagglutinin gene of an influenza virus. *Nature (London)* 340:156–57
50. Kimura, N., Nishida, M., Nagata, K., Ishihama, A., Oda, K., Nakada, S. 1992. Transcription of a recombinant influenza virus RNA in cells that can express the influenza virus RNA polymerase and nucleoprotein genes. *J. Gen. Virol.* 73:1321–28

50a. Kingsbury, D. W., ed. 1991. *The Paramyxoviruses*. New York: Plenum

51. Koetzner, C. A., Parker, M. M., Ricard, C. S., Sturman, L. S., Masters, P. S. 1992. Repair and mutagenesis of the genome of a deletion mutant of the coronavirus mouse hepatitis virus by targeted RNA recombination. *J. Virol.* 66:1841–48

51a. Krug, R. M., ed. 1989. *The Influenza Viruses*. New York: Plenum

52. Krug, R. M., Alonso-Caplen, F. V., Julkunen, I., Katze, M. G. 1989. Expression and replication of the influenza virus genome. See Ref. 51a, pp. 89–152
53. Lai, C.-J., Zhao, B., Hori, H., Bray, M. 1991. Infectious RNA transcribed from stably cloned full-length cDNA of dengue type 4 virus. *Proc. Natl. Acad. Sci. USA* 88:5139–43
54. Lai, M. M. C. 1992 RNA recombination in animal and plant viruses. *Microbiol. Rev.* 56:61–79
55. Lamb, R. A. 1989. Genes and proteins of the influenza viruses. See Ref. 51a, pp. 1–87
56. Lamb, R. A., Horvath, C. M. 1991. Diversity of coding strategies in influenza viruses. *Trends Genet.* 7:261–66
57. Lamb, R. A., Paterson, R. G. 1991. The nonstructural proteins of paramyxoviruses. See Ref. 50a, pp. 181–214
58. Levis, R., Huang, H., Schlesinger, S. 1987. Engineered defective interfering RNAs of Sindbis virus express bacterial chloramphenicol acetyltransferase in avian cells. *Proc. Natl. Acad. Sci. USA* 84:4811–15
59. Li, S., Palese, P. 1992. Expression of foreign epitopes on the surface of influenza viruses. *Options for the Control of Influenza Int. Sci. Conf., 2nd, Courchevel,* p. 93
60. Li, S., Rodrigues, M., Rodriguez, D., Rodriguez, J. R., Esteban, M., et al. 1993. Priming with recombinant influenza virus followed by administration of recombinant vaccinia virus induces CD8+ T cell–mediated protective immunity against malaria. *Proc. Natl. Acad. Sci. USA.* 90:5214–18
61. Li, S., Schulman, J. L., Moran, T., Bona, C., Palese, P. 1992. Influenza A virus transfectants with chimeric hemagglutinins containing epitopes from different subtypes. *J. Virol.* 66: 399–404
62. Li, X., Palese, P. 1992. Mutational analysis of the promoter required for influenza virus virion RNA synthesis. *J. Virol.* 66:4331–38
63. Liao, C.-L., Lai, M. M. C. 1992.

RNA recombination in a coronavirus: recombination between viral genomic RNA and transfected RNA fragments. *J. Virol.* 66:6117–24
64. Liljeström, P., Garoff, H. 1991. A new generation of animal cell expression vectors based on the Semliki Forest virus replicon. *Bio/Technology* 9:1356–61
65. London, S. D., Schmaljohn, A. L., Dalrymple, J. M., Rice, C. M. 1992. Infectious enveloped RNA virus antigenic chimeras. *Proc. Natl. Acad. Sci. USA* 89:207–11
66. López-Turiso, J. A., Martínez, C., Tanaka, T., Ortin, J. 1990. The synthesis of influenza virus negative-strand RNA takes place in insoluble complexes present in the nuclear matrix fraction. *Virus Res.* 16:325–38
67. Luo, G., Bergmann, M., García-Sastre, A., Palese, P. 1992. Mechanism of attenuation of a chimeric influenza A/B transfectant virus. *J. Virol.* 66:4679–85
68. Luo, G., Chang, J., Palese, P. 1993. Alterations of the stalk of the influenza virus neuraminidase: deletions and insertions. *Virus Res.* In press
69. Luo, G., Doymaz, M., Bergmann, M., Palese, P. 1992. Regulation of influenza A virus RNA expression. *EMBO Annu. Symp., 18th, Frankfurt,* p. 173
70. Luo, G., Luytjes, W., Enami, M., Palese, P. 1991. The polyadenylation signal of influenza virus RNA involves a stretch of uridines followed by the RNA duplex of the panhandle structure. *J. Virol.* 65:2861–67
71. Luo, G., Palese, P. 1992. Genetic analysis of influenza virus. *Curr. Opin. Genet. Dev.* 2:77–81
72. Luytjes, W., Krystal, M., Enami, M., Parvin, J. D., Palese, P. 1989. Amplification, expression, and packaging of a foreign gene by influenza virus. *Cell* 59:1107–13
73. Martín, J., Albo, C., Ortín, J., Melero, J. A., Portela, A. 1992. In vitro reconstitution of active influenza virus ribonucleoprotein complexes using viral proteins purified from infected cells. *J. Gen. Virol.* 73:1855–59
74. McClure, M. A., Thibault, K. J., Hatalski, C. G., Lipkin, W. I. 1992. Sequence similarity between Borna disease virus p40 and a duplicated domain within the paramyxovirus and rhabdovirus polymerase proteins. *J. Virol.* 66:6572–77
75. Miller, A. D., Rosman, G. J. 1989. Improved retroviral vectors for gene transfer and expression. *BioTechniques* 7:980–90
76. Miller, L. K. 1988. Baculoviruses as gene expression vectors. *Annu. Rev. Microbiol.* 42:177–99
77. Moss, B. 1991. Vaccinia virus: a tool for research and vaccine development. *Science* 252:1662–67
78. Murphy, F. A., Kingsbury, D. W. 1990. Virus taxonomy. See Ref. 27a, pp. 9–35
79. Muster, T., Subbarao, E. K., Enami, M., Murphy, B. R., Palese, P. 1991. An influenza A virus containing influenza B virus 5′ and 3′ noncoding regions on the neuraminidase gene is attenuated in mice. *Proc. Natl. Acad. Sci. USA* 88:5177–81
80. Nakamura, Y., Oda, K., Nakada, S. 1991. Growth complementation of influenza virus temperature-sensitive mutants in mouse cells which express the RNA polymerase and nucleoprotein genes. *J. Biochem.* 110:395–401
81. Palese, P. 1977. The genes of influenza virus. *Cell* 10:1–10
82. Park, K. H., Huang, T., Correia, F. F., Krystal, M. 1991. Rescue of a foreign gene by Sendai virus. *Proc. Natl. Acad. Sci. USA* 88:5537–41
83. Park, K. H., Krystal, M. 1992. In vivo model for pseudo-templated transcription in Sendai virus. *J. Virol.* 66:7033–39
84. Parvin, J. D., Palese, P., Honda, A., Ishihama, A., Krystal, M. 1989. Promoter analysis of influenza virus RNA polymerase. *J. Virol.* 63:5142–52
85. Pattnaik, A. K., Ball, L. A., LeGrone, A. W., Wertz, G. W. 1992. Infectious defective interfering particles of VSV from transcripts of a cDNA clone. *Cell* 69:1011–20
86. Pattnaik, A. K., Wertz, G. W. 1990. Replication and amplification of defective interfering particle RNAs of vesicular stomatitis virus in cells expressing viral proteins from vectors containing cloned cDNAs. *J. Virol.* 64:2948–57
87. Pattnaik, A. K., Wertz, G. W. 1991. Cells that express all five proteins of vesicular stomatitis virus from cloned cDNAs support replication, assembly, and budding of defective interfering particles. *Proc. Natl. Acad. Sci. USA* 88:1379–83
88. Pavlovic, J., Haller, O., Staeheli, P. 1992. Human and mouse Mx proteins inhibit different steps of the influenza virus multiplication cycle. *J. Virol.* 66:2564–69
89. Percy, N., Barclay, W. S., Sullivan, M., Almond, J. W. 1992. A poliovirus replicon containing the chloramphenicol

acetyltransferase gene can be used to study the replication and encapsidation of poliovirus RNA. *J. Virol.* 66:5040–46
90. Piccone, M. E., Fernández-Sesma, A., Palese, P. 1993. Mutational analysis of the influenza virus vRNA promoter. *Virus Res.* 28:99–112
91. Pinto, L. H., Holsinger, L. J., Lamb, R. A. 1992. Influenza virus M_2 protein has ion channel activity. *Cell* 69:517–28
92. Plotch, S. J., Bouloy, M., Ulmanen, I., Krug, R. M. 1981. A unique cap (m^7GpppXm)-dependent influenza virion endonuclease cleaves capped RNAs to generate the primers that initiate viral RNA transcription. *Cell* 23:847–58
93. Racaniello, V. R., Baltimore, D. 1981. Cloned poliovirus complementary DNA is infectious in mammalian cells. *Science* 214:916–19
94. Rice, C. M., Grakoui, A., Galler, R., Chambers, T. J. 1989. Transcription of infectious yellow fever virus RNA from full-length cDNA templates produced by in vitro ligation. *New Biol.* 1:285–96
95. Rice, C. M., Levis, R., Strauss, J. H., Huang, H. V. 1987. Production of infectious RNA transcripts from Sindbis virus cDNA clones: mapping of lethal mutations, rescue of a temperature-sensitive marker, and in vitro mutagenesis to generate defined mutants. *J. Virol.* 61:3809–19
96. Rochovansky, O. M. 1976. RNA synthesis by ribonucleoprotein-polymerase complexes isolated from influenza virus. *Virology* 73:327–38
97. Roner, M. R., Sutphin, L. A., Joklik, W. K. 1990. Reovirus RNA is infectious. *Virology* 179:845–52
98. Rosenfeld, M. A., Yoshimura, K., Trapnell, B. C., Yoneyama, K., Rosenthal, E. R., et al. 1992. In vivo transfer of the human cystic fibrosis transmembrane conductance regulator gene to the airway epithelium. *Cell* 68:143–55
99. Russell, S. J., Brandenburger, A., Flemming, C. L., Collins, M. K. L., Rommelaere, J. 1992. Transformation-dependent expression of interleukin genes delivered by a recombinant parvovirus. *J. Virol.* 66:2821–28
100. Schulman, J. L., Palese, P. 1977. Virulence factors of influenza A viruses: WSN virus neuraminidase required for plaque production in MDBK cells. *J. Virol.* 24:170–76
101. Seong, B. L., Brownlee, G. G. 1992. A new method for reconstituting influenza polymerase and RNA in vitro: a study of the promoter elements for cRNA and vRNA synthesis in vitro and viral rescue in vivo. *Virology* 186: 247–60
102. Seong, B. L., Brownlee, G. G. 1992. Nucleotides 9 to 11 of the influenza A virion RNA promoter are crucial for activity in vitro. *J. Gen. Virol.* 73: 3115–24
103. Seong, B. L., Kobayashi, M., Nagata, K., Brownlee, G. G., Ishihama, A. 1992. Comparison of two reconstituted systems for in vitro transcription and replication of influenza virus. *J. Biochem.* 111:496–99
104. Shapiro, G. I., Krug, R. M. 1988. Influenza virus RNA replication in vitro: synthesis of viral template RNAs and virion RNAs in the absence of an added primer. *J. Virol.* 62:2285–90
105. Snyder, M. H., London, W. T., Maassab, H. F., Chanock, R. M., Murphy, B. R. 1990. A 36 nucleotide deletion mutation in the coding region of the NS1 gene of an influenza A virus RNA segment 8 specifies a temperature-dependent host range phenotype. *Virus Res.* 15:69–84
106. Spaete, R. R., Mocarski, E. S. 1987. Insertion and deletion mutagenesis of the human cytomegalovirus genome. *Proc. Natl. Acad. Sci. USA* 84:7213–17
107. Strauss, E. G., Strauss, J. H., Levine, A. J. 1990. Virus evolution. See Ref. 27a, pp. 167–90
108. Strauss, J. H., Strauss, E. G. 1988. Evolution of RNA viruses. *Annu. Rev. Microbiol.* 42:657–83
109. Szewczyk, B., Laver, W. G., Summers, D. F. 1988. Purification, thioredoxin renaturation, and reconstituted activity of the three subunits of the influenza A virus RNA polymerase. *Proc. Natl. Acad. Sci. USA* 85:7907–11
110. Takamatsu, N., Ishikawa, M., Meshi, T., Okada, Y. 1987. Expression of bacterial chloramphenicol acetyltransferase gene in tobacco plants mediated by TMV-RNA. *EMBO J.* 6:307–11
111. Taniguchi, T., Palmieri, M., Weissmann, C. 1978. Qβ DNA-containing hybrid plasmids giving rise to Qβ phage formation in the bacterial host. *Nature (London)* 274:223–28
112. Tsai, I.-K., Maassab, H. F., Shaw, M. W. 1992. Investigating the function of segment-specific noncoding sequences in influenza virus genes. *Options for the Control of Influenza Int. Sci. Conf., 2nd, Courchevel.* p. 113
113. Ulmanen, I., Broni, B. A., Krug, R. M. 1981. Role of two of the influenza

virus core P proteins in recognizing cap I structures ($m^7GpppNm$) on RNAs and in initiating viral RNA transcription. *Proc. Natl. Acad. Sci. USA* 78: 7355–59

114. Van der Most, R. G., Heijnen, L., Spaan, W. J. M., De Groot, R. J. 1992. Homologous RNA recombination allows efficient introduction of site-specific mutations into the genome of coronavirus MHV-A59 via synthetic co-replicating RNAs. *Nucleic Acids Res.* 20:3375–81
115. Wagner, R. R. 1990. Rhabdoviridae and their replication. See Ref. 27a, pp. 867–81
116. Wiley, D. C., Skehel, J. J. 1987. The structure and function of the hemagglutinin membrane glycoprotein of influenza virus. *Annu. Rev. Biochem.* 56:365–94
117. Yamanaka, K., Ishihama, A., Nagata, K. 1990. Reconstitution of influenza virus RNA-nucleoprotein complexes structurally resembling native viral ribonucleoprotein cores. *J. Biol. Chem.* 265:11151–55
118. Yamanaka, K., Ogasawara, N., Yoshikawa, H., Ishihama, A., Nagata, K. 1991. In vivo analysis of the promoter structure of the influenza virus RNA genome using a transfection system with an engineered RNA. *Proc. Natl. Acad. Sci. USA* 88:5369–73

Note Added in Proof

Since the completion of this review, a new system for the in vitro transcription of synthetic VSV RNA molecules has been published (4). Also, a neuraminidase-minus influenza A virus mutant has been described that allows the efficient rescue of cloned neuraminidase genes into infectious virus (3). In addition, the rescue of synthetic RNA by human parainfluenza virus type 3—a paramyxovirus—has been achieved (2), and the rescue by RSV of a synthetic RNA that is almost half the size of the viral genomic RNA has been reported (1). The latter experiments suggests that RSV will also be amenable to genetic manipulation once a full length cDNA clone becomes available.

Literature Cited

1. Collins, P. L., Mink, M. A., Hill, M. G. III. Camargo, E., Grosfeld, H., Stec, D. S. 1993. Rescue of a 7502-nucleotide (49.3% of full-length) synthetic analog of respiratory syncytial virus genomic RNA. *Virology* 195:252–56
2. Dimock, K., Collins, P. L. 1993. Rescue of synthetic analogs of genomic RNA and replicative-intermediate RNA of human parainfluenza virus type 3. *J. Virol.* 67:2772–78
3. Liu, C., Air, G. M. 1993. Selection and characterization of a neuraminidase-minus mutant of influenza virus and its rescue by cloned neuraminidase genes. *Virology* 194:403–7
4. Smallwood, S., Moyer, S. A. 1993. Promoter analysis of the vesicular stomatitis virus RNA polymerase. *Virology* 192:254–63

Annu. Rev. Microbiol. 1993. 47:791–819

THE CELLULOSOME: The Exocellular Organelle of *Clostridium*

Carlos R. Felix[1] *and Lars G. Ljungdahl*
Center for Biological Resource Recovery and Department of Biochemistry, University of Georgia, Life Sciences Building, Athens, Georgia 30602-7229

KEY WORDS: cellulases, endoglucanases, multienzyme complex, cellulolytic, cellulose, cellulose-binding protein

---

CONTENTS

ABSTRACT

The cellulolytic enzyme complex of the anaerobic thermophile *Clostridium thermocellum* is reviewed. This complex, called the cellulosome, is cell associated and has a mass of from 2×10^6 to 6.5×10^6 Daltons. It consists of from 14 to 26 different polypeptides. Cellulosomes form larger complexes,

[1]Permanent Address: Departamento de Biologia Celular, 70.910 Universidade de Brasília, Brasília, D.F. Brasil.

0066-4227/93/1001-0791$02.00

polycellulosomes, with masses from 50×10^6 to 80×10^6 Daltons. The cellulosome efficiently hydrolyzes crystalline cellulose whereas individual polypeptides alone or in mixtures do not. Many of the polypeptides are catalytically active and can be characterized as endoglucanases, xylanases, and cellodextrinases. Several of the polypeptides have been sequenced including the largest subunit, CipA, that is a glycoprotein with a mass of 210 kDa. CipA has a cellulose-binding domain and nine internal repeated sequences postulated to bind eight catalytic subunits and a special peptide (ORF3p). The ORF3p anchors the CipA to the cell surface. CipA can be characterized as a scaffold holding the catalytic subunits that line up with the cellulose fiber. This arrangement allows a multiple cutting of the cellulose glucan chain. A similar system has been observed for other cellulosome-like complexes, notably *Clostridium cellulovorans*.

INTRODUCTION

The cellulosome, which was first described by Lamed et al (81), is a multicomponent cellulolytic complex. It, like other true cellulolytic enzyme systems, hydrolyzes crystalline cellulose to its building block, the disaccharide cellobiose (85). The complex was first described as an antigenically active, cellulose-binding factor (CBF) located on the cell surface and in the culture medium of the thermophilic, gram-positive, anaerobic cellulolytic bacterium *Clostridium thermocellum* (7). CBF was reported to have a mass of 2.1×10^6 Daltons and to contain 14 different polypeptides (80).

The cellulosome has emerged as a specialized exocellular structure (79, 90). As a single multicomponent topologically oriented catalyst, the cellulosome effects hydrolysis of cellulose and hemicellulose. Its multifunctionality represents the efficiency of the evolutionary processes that provided clostridia and other bacteria with a mechanism that enables cells to obtain energy from the two most abundant but intrinsically intractable substrates. Adorning the cell, cellulosomes have a significant physiological role. On the cell surface they appear as polycellulosomal aggregates promoting adherence of the bacterium to cellulose (8, 79, 89, 90, 132).

Several *C. thermocellum* strains have been isolated and described. These strains produce cellulosomes that apparently vary in the number of polypeptide components and in size (74, 80, 90). Ultrastructural observations indicate that cellulosomes are composed of a set of polypeptides arranged in an ordered chain-like array (90). Such architecture apparently renders the complex resistant to the most unfriendly conditions. Its behavior as a discrete entity has facilitated the isolation of cellulosomes as homogenous preparations from residual cellulose by desorption at low ionic strength (60, 61). However, the very durability of the complex (61, 80, 86) has bedeviled attempts to isolate native individual components. Recombinant DNA technology has then been used as an alternative means of examining the properties of individual polypeptides. Although this work has been very successful (10, 28) and has yielded much information on the genes and structures of individual polypeptides with cellulolytic activities, notably endoglucanases, it has not revealed until very recently information about the structure and action of the cellulosome complex. Reconstitution of the cellulosome from engineered proteins as well as from components isolated directly from it is still a challenge for biochemists.

The major functional characteristic of the cellulosome is its complete or true cellulolytic activity, i.e. it hydrolyzes both amorphous, crystalline, and highly ordered cellulose. The mechanism of cellulose hydrolysis by the cellulosome is still a vexing question. The goal of most research in the cellulose field has been to find a superior cellulase for applied use. Clues as to the structure/function relationship and to the possible mechanism whereby cellulose is hydrolyzed by the cellulosome have come mostly from electron-microscope observations (76, 90). A multicutting event mechanism (90) and a contact corridors theory (75, 76) have been proposed as results of these observations. However, neither hypothesis has so far been proven. The juxtaposition of polypeptides forming the complex, the mode of action, and the properties of each individual component alone or in cooperative action (synergism) are the keys to the efficient hydrolysis of highly ordered cellulose by the cellulosome. This efficient hydrolysis is also determined by the rule: better adsorption—better catalysis (71). Whether any of the proposals is correct is still a matter for discussion, and in fact, our knowledge of the cellulosome lags behind that of fungal cellulolytic systems. Nevertheless, several aspects are common to both bacterial and fungal cellulolytic systems. Persistent and productive efforts of researchers working in both areas have helped eliminate the disparities between fields. These differences are due much more to the properties of the cellulosome itself than to any lack of interest on the part of the researchers in the magnificent architecture of this complex, or in the possible use of clostridia as a source of catalyst for bioconversion. Here we limit our discussion to the cellulosome discovered first as a discrete exocellular organelle of *C. thermocellum*. For general knowledge of cellulolytic systems, the reader is referred to symposium volumes edited by Aubert et al (5), Wood & Kellog (135), Coughlan (24), Akin et al (3), and Haigler & Weimer (49). These books cover properties and genetics of fungal and bacterial cellulases and cellulolytic systems. Other more specialized reviews are those on ultrastructure by Mayer (89) and cloning and genetics by Béguin and coworkers (10, 13, 14) and Gilkes et al (41).

CELLULOSOMES AND CELLULOSOME-LIKE COMPLEXES AMONG CLOSTRIDIA AND OTHER MICROORGANISMS

One of the best demonstrations of the binding of *C. thermocellum* to its substrate, cellulose, is that by Wiegel & Dykstra (132). This binding is clearly mediated by the cellulosome, which was first described as a cellulose-binding factor (CBF) present on the cell surface of *C. thermocellum* strain YS and in its culture medium (7, 80). Antibodies prepared against the surface of wild-type *C. thermocellum* confirmed this association. These antibodies were rendered specific to the CBF by the removal of nonspecific antibodies by adsorption with an adherence-defective mutant strain. The antibodies reacted positively with the wild-type cells but not with the mutant cells (7). Further ultrastructural analyses by specific immunolabeling and by general staining procedures using cationized ferritin showed the cellulosome as protuberant structures on the surface of the cells (8, 9, 79). These protuberances protract to form fibrous contact corridors after adhesion of the cells to cellulose (8). Actually, the protuberances represent extremely large and complex structures consisting of polycellulosomes comprised of discrete entities, i.e. cellulosomes.

Cellulosomes and polycellulosomes have been purified from cultures of *C. thermocellum* strain JW20 growing on cellulose. In young cultures, almost all of the cellulolytic complexes are bound to cellulose. However, extraction with distilled water releases them from the cellulose (60). Electron-microscope examination of the water extract revealed two major particles with diameters of 21 and 61 nm. The masses for the particles were calculated to be 4.2×10^6 and 100×10^6 Daltons, respectively, assuming perfect spheres (26). When the particles were observed to be flattened instead of perfect spheres, these figures were adjusted to range from 2.0×10^6 to 2.5×10^6 Daltons for the cellulosome, and from 50×10^6 to 80×10^6 Daltons for the polycellulosome (90). The two types of particles were separated by gel-filtration and ultracentrifugation into originally bound large (OBL) cellulose complexes (polycellulosomes) and originally bound small (OBS) cellulose complexes (cellulosomes). The complexes had almost identical specific activities hydrolyzing crystalline cellulose and carboxymethyl cellulose (CMC) (endoglucanase activity). SDS-polyacrylamide gel electrophoresis demonstrated that the two complexes had similar if not identical polypeptide compositions (60, 61).

Mayer et al (90), examining the cellulolytic complexes with electron microscopy, found that polycellulosomal complexes were covered by a skin that may be the same fibrous material described as contact corridors (8). Both polycellulosomes and cellulosomes remain attached to the cellulose after the bacterial cells leave. As the cellulolytic process proceeds and the culture ages, tightly packed cellulosomes decompose to loosely packed cellulosomes and ultimately to free polypeptides (90). This explains the presence of cellulosomal components and cellulolytic activity in the culture medium mainly in late stages. A more detailed discussion of the organization of the cellulosome is presented below.

Several strains of *C. thermocellum* have been described including ADZ, NCIB 10682, ATCC 27405, JW20, YM4, and F7 (30, 37, 70, 92, 97). They all have cellulosomes, but cellulosomes from different strains do vary. The reported masses range from 2.0×10^6 to 2.5×10^6 Daltons for strains YS and JW20 (77, 90), whereas the masses are 3.5×10^6 Daltons for strain YM4 (90) and 6.5×10^6 Daltons for strain ATCC 27405 (137). These data are rough estimates based on electron-microscope observations and results of gel filtrations. Polycellulosomes have not been observed for strain YM4 (90). The numbers of polypeptides constituting the cellulosomes seem also to differ. Lamed et al (80) reported that the cellulosome from *C. thermocellum* strain YS contains 14 different polypeptides of masses ranging from 45 kDa to 210 kDa, and strain JW20 has 26 polypeptides with apparent masses ranging from 37.5 to 185 kDa (74). The number of different polypeptides in the cellulosome has been determined by SDS-gel electrophoresis. It should be noted that the cellulosome is quite resistant to SDS treatment, and separation of the polypeptides after such treatment is not always achieved. Remarkably, after SDS-gel electrophoresis, several of the polypeptides are still enzymatically active, hydrolyzing carboxymethyl cellulose or xylan, but not crystalline cellulose (74).

The occurrence of cellulosomes is not restricted to *C. thermocellum*. The anaerobic cellulolytic bacteria *Acetivibrio cellulolyticus, Bacteroides cellulosolvens, Clostridium cellobioparum, Clostridium cellulovorans,* and *Ruminococus albus* have cell-surface localized cellulosome-like structures (79). Significantly, these bacteria share all the properties common to cellulolysis by *C. thermocellum*. All produce cellulolytic complexes containing a high-molecular-weight polypeptide (210 kDa) that has been characterized as the cellulose-binding polypeptide. Their cellulases require Ca^{2+} and thiol for activity against crystalline cellulose. These bacteria also interact specifically with the *Griffonia simplicifolia* isolectin B4 and with the antibody against *C. thermocellum* YS cellulosome (79). *Clostridium* strain C7, a mesophilic cellulolytic bacterium isolated from fresh water sediment (83), lacks cellulosome clusters on its surface (19). The cellulolytic system of this bacterium, which displays true cellulolytic activity, is found in the medium from which it can be isolated as a multicomponent cellulosome-like complex with a mass of 700 kDa consisting of at least 15 polypeptides (19, 20).

The cellulosome of *C. cellulovorans* is smaller than that of *C. thermocellum,* but the properties are very similar. It is released from the cellulose with deionized water, has a mass of 900 kDa, and consists of at least seven polypeptides ranging in mass from 40 to 170 kDa (121). The 170-kDa subunit, which lacks enzymatic activity, has high affinity for cellulose and apparently serves as the cellulose-binding protein (CbpA) in the same way as the 210-kDa subunit from *C. thermocellum*. The primary sequence of the CbpA from *C. cellulovorans* was recently reported (123). The cellulolytic complex in the mesophile *C. cellulolyticum* apparently is also somewhat smaller and less complex than cellulosome of *C. thermocellum* (15). Substantial genetic work on the *C. cellulolyticum* cellulolytic complex has been performed, and five genes coding for its cellulolytic components have been identified (34).

A high-molecular-mass (670 kDa) complex composed of subunits is present in the anaerobic cellulolytic fungus *Neocallimastix frontalis* (133). The authors refer to this complex as a cellulosome-type enzyme fraction. However, they recognize important differences between the *N. frontalis* complex and the true cellulosome from anaerobic bacteria that include size, amount of extracellular enzyme, and association with β-glucosidase (not present in cellulosome). A cellulolytic complex may be formed by another anaerobic fungus, *Neocallimastix patriciarum*. The genes for three cellulolytic enzymes have been cloned and the enzymes expressed in *Escherichia coli;* two of these, CelB and CelC, appear to be of the type endoglucanase, whereas the third, CelA, seems to be a cellobiohydrolase (138).

At least 20 different clostridial bacteria are known to hydrolyze cellulose. They and many other anaerobic cellulolytic bacteria may produce cellulosome-like cellulolytic enzyme complexes. However, *Clostridium stercorarium,* a thermophile, produces two enzymes designated avicelase I and II. One of these, avicelase I, with a mass of 100 kDa, has been purified and found to hydrolyze crystalline cellulose without requiring additional proteins and cofactors. It produces cellobiose and glucose with cellobiose and cellotetraose as intermediates. The enzyme can be designated an endoglucanase, which is distinguished from *C. thermocellum* endoglucanases by its ability to hydrolyze crystalline cellulose (17).

Considering the cellulosome starts as a surface protuberance (9, 90) and eventually forms polycellulosomes, it is worth noting that noncellulolytic amylolytic anaerobes also form surface protuberances, which have been named amylosomes (77). Hence, Lamed & Bayer (76) have proposed the term hydrolysomes to describe polymer-degrading multienzyme complexes.

Analyses of the *N. frontalis* complex (133) and the so-called amylosome (77) would be very helpful in reassessing the previous statements that cellulosomes are a feature of cellulolytic bacteria only (28, 79).

PROPERTIES OF THE CELLULOSOME

General Properties

In the previous section, we pointed out that the number of polypeptides in the cellulosome may vary with the strain, and clearly the size of cellulosomes must reflect the total number of polypeptides they contain. The only estimate of the total number of polypeptides is for the cellulosome from *C. thermocellum* strain YM4. Electron-microscope examination revealed that it contains 45–50 polypeptides ranging from 20 to about 200 kDa (90). Many of the polypeptides were of the same size. This observation, together with SDS-gel electrophoresis patterns of cellulosomes from different sources, clearly demonstrated that some of the polypeptides are present in larger numbers than others.

One common feature seems to be that cellulosomes contain a large polypeptide that serves as the cellulose-binding protein. This polypeptide has no known enzymatic activity. Its mass in the *C. thermocellum* cellulosome is 210 kDa and in the *C. cellulovorans* cellulosome 170 kDa (80, 81, 122). The work by Wu et al (137) illustrates the importance of this subunit. These authors isolated two major proteins from the cellulosome of *C. thermocellum* strain ATCC 27405. One, designated S_S, had a mass of 82 kDa and hydrolyzed carboxymethyl cellulose but not avicel. The second fraction, designated S_L and assumed to correspond to the cellulose-binding subunit, had a mass of 250 kDa and exhibited no enzymatic activity. It, together with the S_s fraction, could hydrolyze crystalline cellulose (avicel), albeit at a low rate compared with the intact cellulosome. The integrity of the cellulosome is needed for an efficient hydrolysis of crystalline cellulose. None of the isolated subunits alone has such activity, although, as should be discussed, several of the subunits have cellulolytic activities using carboxymethyl cellulose as a substrate.

The cellulosome is remarkably stable. It can be stored in deionized water or 50% (w/v) ethanol for months at room temperature without loss of activity (84). Cellulosomes can be concentrated by acetone (66% v/v), and the presence of 8 M urea during gel filtration does not change the elution pattern, indicating that urea is not an effective agent for the dissociation of the cellulosome (7). Treatment of cellulosomes with sodium dodecyl sulfate (SDS) slowly leads to complete loss of activity against crystalline cellulose, but 86% of original CMCase activity remains after incubation for 10 days at room temperature with 2% SDS.

Enzymatic Activity of the Cellulosome

Hydrolysis of cellulose to its monomer glucose is accomplished by cellulolytic enzyme systems that traditionally include three types of enzymes: endo-1,4-β-glucanase (EC 3.2.1.4), exo-1,4-β-glucanase or cellobiohydrolase (EC 2.3.1.91), and cellobiase or β-glucosidase (EC 3.2.1.21). In brief, endoglucanases attack amorphous cellulose, carboxymethylcellulose, and phosphoric acid–swollen cellulose, producing water-soluble cellooligosaccharides that in turn may be hydrolyzed to cellobiose and glucose. Purified endoglucanases have little apparent capacity to degrade crystalline cellulose. Cellobiohydrolases degrade cellulose by splitting cellobiose units from the nonreducing end of the chain of swollen partially degraded amorphous cellulose and soluble cellooligosaccharides (cellotriose to cellohexaose). β-Glucosidase hydrolyzes cellobiose and soluble cellooligosaccharides to glucose, but cellulose is not degraded. The above concept of cellulose hydrolysis has been developed mostly in studies of cellulolytic fungi, notably *Trichoderma reesei* and *Phanerochaete chrysosporium* (*Sporotrichum pulverulentum*). When separated, the enzymes do not hydrolyze crystalline cellulose, but when combined, they interact in a synergistic way to catalyze this hydrolysis. Several reviews cover the cellulolytic enzymes from fungi (25, 28, 31, 32).

The work on the enzymatic cellulolytic systems of anaerobic bacteria was originally influenced by the thinking developed in studies of fungi. However, the discovery of the cellulosome produced by *C. thermocellum* changed the research approach. As stated above, the cellulosome displays true cellulolytic activity, i.e. the complex effects hydrolysis of crystalline cellulose, amorphous or substituted cellulose, and cellooligosaccharides (51, 59–61, 65, 78, 80, 81, 88). Clostridial cellulosomes require Ca^{2+} and thiols for hydrolysis of crystalline cellulose but catalyze the hydrolysis of carboxymethyl cellulose (CMC) and amorphous cellulose just as effectively in the absence of these reagents (60, 65, 78). The cellulosomal systems of other cellulolytic bacteria, e.g. *A. cellulolyticus* (88) and *Ruminococcus flavefaciens* (38) appear to have the same properties.

Hydrolysis of ordered cellulose by clostridial enzyme systems seems to be a major property of the intact cellulosome and polycellulosome. Any dissociation leads to lower specific activity or total loss of activity, although the activity against amorphous or substituted cellulose is maintained (29, 33, 52, 60, 61, 80, 110, 137).

Cellobiose is the major product of hydrolysis of cellulose by cellulosomes and is also a competitive inhibitor of such hydrolysis (60, 61, 64, 78). These findings have led to the suggestion that endoglucanases and cellobiohydrolases are components of the cellulosome.

Endoglucanases

Endoglucanases cleave internal cellulosic glycosidic bonds of cellulose in a random fashion. This results in a rapid decrease in chain length of the β-glucan together with a slow increase in reducing groups (134). The review by Rapp & Beermann (108) discusses several properties of endoglucanases, which have been classified in families according to their hydrophobic properties (58) and catalytic domains (41).

The presence of endoglucanases among the polypeptides constituting the cellulosome was an early observation made by eluting separated protein bands from SDS-polyacrylamide gel electrophoresis and assaying the activity, or by overlay activity staining techniques. The first technique showed that the cellulosome of *C. thermocellum* has as many as 10 proteins displaying endoglucanase activity (61, 80, 81). Activity staining with the substrate carboxymethyl cellulose added to the SDS-polyacrylamide gels showed that, out of 26 different protein bands detected on the gels, 16 exhibited endoglucanase activity (74, 104). Nonetheless, very few endoglucanases have been isolated from *C. thermocellum* culture fluids. A monomeric 56-kDa endoglucanase (105) and an 83- to 94-kDa endoglucanase containing 11–12% carbohydrate (101) were the first examples of cellulolytic enzymes isolated from the culture medium. Additional endoglucanases have now been purified from culture fluids. They include three from strain ATCC 27405 with masses 60 kDa, 110 kDa (52), and 76 kDa (110). The activity of the latter enzyme is stimulated by Mg^{2+} and Ca^{2+} and inhibited by thiol-blocking agents, and it shares these properties with the intact cellulosome. These endoglucanases quite possibly could be constituents of the cellulosome, but this has not been conclusively demonstrated, except for the 56-kDa enzyme (105).

Endoglucanases from isolated cellulosomes from *C. thermocellum* have been purified. As discussed above, an endoglucanase termed S_s (small subunit) with a molecular mass of 82 kDa was isolated concurrently with the cellulose binding protein S_L after the cellulosome had been dissociated by mild SDS treatment (73, 137). The same group later reported the isolation of the S_s endoglucanase from the extracellular fluid (33). This endoglucanase was identified as the original S_s by the use of specific antibodies—it was of the same size and had similar enzymatic properties. The gene coding for the S_s (CelS) endoglucanase has recently been sequenced (130a). CelS is a major component of the cellulosome. A second endoglucanase was also isolated after mild SDS treatment of the cellulosome obtained from strain YM4 (98). It has a mass of 51 kDa. This endoglucanase shares several properties including the same NH_2-terminal sequence with the endoglucanase isolated from culture fluid by Pétré et al (105) and the endoglucanase A produced by an *E. coli* clone (11). It may be a first example of a *C. thermocellum* endoglucanase obtained through three entirely different procedures: extracted from the culture fluid, isolated as a constituent of the cellulosome, and cloned from recombinant *E. coli*. In the last procedure, the enzyme produced contains a signal sequence of 32 amino acids. The enzyme isolated from the cellulosome was glycosylated and contained galactose and *N*-acetylglucosamine. Treatment of this enzyme with SDS generated three different bands of slightly different molecular weights, as shown by SDS-PAGE (98). It was deduced that the enzyme was deglycosylated during the SDS treatment and eventually converted to an unique 48-kDa protein, which was stable on further treatment. This observation indicates one difficulty in establishing the correct number of polypeptides of cellulosomes: the number of bands might not correlate to the number of different polypeptides because the cellulosome may be only partially denaturated, or because the same polypeptide may yield multiple bands on SDS-PAGE owing to the presence of nonproteinaceous material such as carbohydrates.

The presence of many endoglucanases in *C. thermocellum* and the complex nature of its cellulosome have been substantiated by gene cloning and DNA sequencing studies. At least 31 endoglucanase genes have been cloned so far from *C. thermocellum* (10, 18, 56, 71, 72, 93, 114). In addition a *Clostridium* sp., strain F1, which closely resembles *C. thermocellum,* contains four endoglucanase genes that on restriction maps do not match those of the *C. thermocellum* genes cloned previously (114), and strain F7 seems to have seven genes differing from those previously cloned (127). As some of the endoglucanase genes have been cloned at least twice independently (56), a necessary question is whether the quite large number of protein components displaying endoglucanase activity (61, 74, 80, 81, 104) does indeed represent multiplicity of endoglucanase genes, or does it represent polypeptides modified by posttranscriptional processes, thus yielding multiple bands during SDS-PAGE procedures and resulting in an overestimation of the number of polypeptide components of the cellulosome.

Nevertheless, eight different endoglucanase genes from *C. thermocellum* have been sequenced, i.e. *celA* (11), *celB* (45), *celC* (118), *celD* (66), *celE* (50), *celF* (99), *celH* (139), and *celX* (50). The corresponding endoglucanases CelA (117), CelB (12), CelC (106, 119), CelD (67), and CelE (50) have been purified from *E. coli* clones and characterized.

Most cloning and sequencing work has been performed using *E. coli* as host, but other microorganisms have been used successfully. The *celA* gene has been cloned into *Bacillus stearothermophilus, Bacillus subtilis* (68, 124), *Thermus thermophilus* (82), and yeast (16, 112). Interestingly, in the yeast endoglucanase A is formed as a glycoprotein of mass 97 kDa containing about

50% carbohydrate. The *celE* gene is expressed in *Lactobacillus plantarum,* and the gene product is excreted (6).

The endoglucanases of *C. thermocellum* differ in physical and enzymic properties. The *celA* gene encodes a polypeptide of 488 amino acids with a mass of 52,503 Daltons (11), which includes a signal sequence of 32 amino acids. This sequence is not found in the enzyme isolated from the cellulosome (98). The native enzyme isolated from *C. thermocellum* is glycosylated, which may explain why it has been found as multiple bands on SDS-PAGE of masses ranging from 49 to 60 kDa. The enzyme hydrolyzes CMC and glucans with alternating β-1,4 and β-1,3 linkages (lichenan) but not crystalline cellulose or laminarin, a β-1,3 linked glucan (117). It does not hydrolyze cellodextrins.

The *celB* gene encodes endoglucanase B, which apparently consists of 536 amino acids, including a signal peptide, and has a mass of 63,857 Daltons (12, 45). Antibodies prepared against the enzyme produced by *E. coli* were used to screen supernatant fluids from *C. thermocellum* cultures. A polypeptide with a mass of 66,000 Daltons was identified as endoglucanase B. The enzyme has not been isolated from the intact cellulosome. Endoglucanase B has a lower specific activity against CMC as compared with endoglucanase A. However, it hydrolyzes cellotetraose and cellopentaose but not cellotriose or cellobiose (15). Perhaps it should be considered a cellodextrinase, a type of enzyme isolated from *Fibrobacter* (*Bacteroides*) *succinogenes* and described by Huang & Forsberg (62).

The *celC* genes from *C. thermocellum* strain ATCC27405 and strain F1 have been cloned into *E. coli* and sequenced (115, 118). The genes are very similar; both have an open reading frame of 1029 nucleotides and encode polypeptides with predicted masses of 40,439 (118) and 40,905 Da (115). Only six nucleotide substitutions were found in the coding region. They led to three replacements at the amino acid level: Cys156→Tyr, Val157→Ile, Arg167→Met. These substitutions led to a difference in mass of 49 and not 466 as stated by the authors. The polypeptide expressed in the recombinant *E. coli* strains may have an atypical signal peptide sequence, and it is not processed when the enzyme is exported into the periplasm. The enzyme has been purified from recombinant strains of *E. coli* (106, 119). It has high activity toward barley-β-glucan and lichenan, which are mixed linkage β-glucans with β-1,3, and β-1,4 bonds. It acts slowly as an endoglucanase on CMC and hydrolyzes cellotriose, cellotetraose, and cellopentaose, but not cellobiose. In fact, cellobiose seems to be the main product, and the enzyme has features resembling a cellobiohydrolase or a cellodextrinase.

Endoglucanase D is a typical endoglucanase. It has a specific activity toward CMC of 428 IU mg^{-1}, which is one of the highest reported activities of an endoglucanase (15). It also is part of the cellulosome of *C. thermocellum* (75). The *celD* gene and its product endoglucanase D have been studied intensely. The gene encodes a polypeptide of 649 amino acids, which has a calculated mass of 72,334. Included in the amino acid sequence is a typical signal peptide of 41 or 42 amino acid residues (66). The mature protein has a mass of 65 kDa. It is produced at a high level (15% of total protein), and it forms cytoplasmic granules in an *E. coli* strain containing the plasmid carrying the DNA fragment encoding endoglucanase D (67). The purified enzyme has been crystallized. Its three-dimensional structure was recently determined with X-ray crystallography (69). Three forms of the enzyme were crystallized from three different solutions containing ammonium sulfate, $CaCl_2$, and the inhibitor 0-iodobenzyl-1-thio-β-cellobioside. The enzyme apparently contains one zinc binding site and three calcium binding sites. Previously work showed that Ca^{2+} binds to the endoglucanase D of *C. thermocellum* with an association constant of $2.03 \times 10^6\ M^{-1}$ (22) and that the presence of Ca^{2+} decreases the K_d for CMC and increases the thermostability of the enzyme. The enzyme has five cysteine residues, none of which form disulfide bridges. As mentioned above, the cellulosome requires Ca^{2+} and thiols for the hydrolysis of crystalline cellulose (63). The substrate-binding site is in a large barrel-like structure consisting of residues 136–574. The barrel is formed by six inner and six outer helices running in opposite directions that are connected by loops. On one side of the barrel, these loops form the active site, which includes His516 (126) and Glu555 (21). In addition, Asp198 and Asp201 apparently participate in the catalytic process.

Endoglucanase E is one of the constituents of the cellulosome of *C. thermocellum*. It has a mass of 85 kDa as determined with SDS-PAGE (55). The *celE* gene has been cloned in *E. coli*, and this clone produces an immature polypeptide that has 814 amino acids. The first 34 amino acids, which apparently constitute a signal peptide, are removed to form the mature protein of 780 amino acids. The calculated mass of this protein is 84,016 Daltons. Like several endoglucanases from *C. thermocellum*, endoglucanase E forms three separate domains: The catalytic domain comprised of the N-terminal 334 amino acids, a cellulose-binding domain separate from the catalytic site and located between residues 432 and 671, and a reiterated domain at the C terminus (50). This latter domain seems highly conserved among several but not all endoglucanases. It is present in endoglucanases A, B, D, and H of *C. thermocellum*. The function of this domain has not yet been established. It and the cellulose-binding domain are not needed for activity toward CMC (50).

Endoglucanase F of *C. thermocellum* has not been investigated much, although it has been sequenced (99). The transcription of its *celF* gene has been studied together with the genes of *celA*, *celC*, and *celD*. The size of endoglucanase F is about 730 amino acids as calculated from its structural gene, which contains 2.2 kb (94).

The *celH* gene of *C. thermocellum* encodes a protein of 900 amino acid with a mass of 102,301 Daltons. The protein contains a typical signal peptide of 44 amino acids (139). The C terminal has the reiterated 24–amino acid stretches of other endoglucanases, and the active site is located in a central 328–amino acid region. The enzyme is active toward CMC, barley β-glucan, larchwood xylan, *p*-nitrophenyl-β-D-cellobioside, and methylbelliferyl-β-D-cellobioside.

A *celX* gene has been identified in *C. thermocellum*. It encodes a protein of 673 amino acids and it has the 23–amino acid reiterated sequences (50). The protein has not been isolated and at present it must be characterized as a putative endoglucanase.

Two endoglucanases designated endoglucanase-1 and -2 of *C. thermocellum* have been isolated from a recombinant strain of *E. coli* (42). They have similar masses, 41 and 42 kDa respectively, and are considered isoforms. They also have high activity toward CMC and cellodextrins. The high activity toward CMC seems to distinguish these endoglucanases from endoglucanase C described above.

The cellulosome of *C. cellulovorans* has been well characterized (121). Genes encoding four endoglucanases (*engA, engB, engC,* and *engD*) have been cloned and expressed in *E. coli* (54, 122). The nucleotide sequences have been determined for *engB* and *engD* (36, 53). The enzymes purified from *E. coli* have been characterized extensively and have been shown to be part of the cellulosome (35). The enzymes exhibit rather broad substrate specificity and act on CMC, xylan, lichenin, laminarin, and mannan. They slowly hydrolyze microgranular cellulose, and avicel, and they exhibit some activity toward *p*-nitrophenyl β-D-cellobioside and similar substrates. Like the endoglucanases from *C. thermocellum,* EngB and EngD have several distinct regions. The catalytic activity of the two enzymes appears to reside in their N-terminal regions, although with EngD the C-terminal region is needed for the slow hydrolysis of crystalline cellulose. It has been suggested that the N and C termini of EngD act synergistically to hydrolyze cellulose (53). The EngD C terminus also seems to contain a cellulose-binding domain. EngB has the highly conserved reiterated region found in endoglucanases A, B, D, E, and H as well as in the xylanase coded by *xynZ* of *C. thermocellum* (35).

Cellobiohydrolases

Individual endoglucanases of *C. thermocellum* or mixtures of them do not hydrolyze crystalline cellulose. The main product of the hydrolysis of cellulose by the cellulosome is cellobiose. Exoglucanases or cellobiohydrolases degrade amorphous cellulose by consecutive removal of cellobiose from the nonreducing end of the glucan chain. Thus, by analogy with fungal cellulolytic enzyme systems that degrade crystalline cellulose through the synergetic action of endo- and exoglucanases, the cellulosome may have a cellobiohydrolase component (27, 71). Evidence for this is that genes have been found in *C. thermocellum* encoding proteins with activity toward methylumbelliferyl-β-D-cellobioside (93). In addition, four of the polypeptides separated from the cellulosome of *C. thermocellum* JW20, the 58-, 72.5-, 94-, and 110-kDa polypeptides, catalyze the hydrolysis of *p*-nitrophenyl-β-D-cellobioside (74). However, although artificial substrates have been useful for studing exocellobiohydrolases, one should remember that the activity of exo- and endoglucanases and β-glucosidases overlaps considerably (128, 129). Such artificial substrates are cleaved by endoglucanases, β-glucosidases, and xylanase (15). Therefore, one cannot conclude that the above proteins are cellobiohydrolases on account of the reported results (74, 104). In fact, the presence of cellobiohydrolases in anaerobic bacteria has been questioned.

Nevertheless, an exo-β-1,4-glucanase was isolated from the culture medium of *Ruminococcus flavefaciens* FD-1 (38, 131). The enzyme, purified to homogeneity as determined by silver staining of both denaturing and non-denaturing polyacrylamide gels, is a dimer of 118 kDa subunits. It hydrolyzes *p*-nitrophenyl-β-D-cellobioside and filter paper to cellobiose as the end product. The first example of a *C. thermocellum* cellobiohydrolase was provided recently by cloning the corresponding gene into *E. coli,* purifying the recombinant protein, and demonstrating that this protein was precisely a cellobiohydrolase (71). The enzyme reportedly acted on both amorphous and crystalline cellulose with cellobiose as virtually the only product. That the enzyme was a cellobiohydrolase was demonstrated by its hydrolysis of 3-fluoromethylumbelliferylcellooligosacharides (3F-MUF-G*n,* where n = 1–4). 3F-MUF-G was not a substrate, and no colored fluorophore was formed on hydrolysis of 3F-MUF-G3, but one did form on digestion of 3F-MUF-G2 and 3F-MUF-G4. The enzyme binds cellobiose 2000 times tighter than glucose (71).

A defined cellobiohydrolase was reported to be present in the cellulosome from *C. thermocellum* (96). The cellobiohydrolase acting protein, termed S8-tr, was isolated after proteinase digestion of purified cellulosomes. The 68-kDa polypeptide S8-tr was traced to be a truncated product of the 75-kDa S8 subunit of the cellulosome. It displayed high activity toward amorphous cellulose and xylan, moderate activity toward avicel, low activity toward CMC, and no activity toward chromogenic or fluorogenic derivatives of cellobiose, glucose, and xylose. The product of hydrolysis of amorphous cellulose and avicel was cellobiose. A comparison of enzymatic activities between the intact S8 subunit and its truncated form S8-tr would be interesting, but unfortunately, this has not been done. However, Morag et al (96) stated that the S8 subunit obtained from *C. thermocellum* YS and the subunit termed S_s from *C. thermocellum* ATTC 27405 are coincidental. As described above, the S_S subunit has been isolated both from the cellulosome and the extracellular fluid (33). It has been characterized as a rather typical endoglucanase. Thus, the relationship between the S8 subunit and its truncated form S8-tr with the S_S subunit seems unclear. The question still remains: is this enzyme a cellobiohydrolase?

β-*Glucosidases and Cellobiose Phosphorylases*

A third enzyme generally associated with cellulolytic enzyme systems is β-glucosidase or cellobiase, which hydrolyzes cellobiose to glucose. Weak β-glucosidase, β-xylosidase, β-galactosidase, and β-mannosidase activities have been demonstrated in cellulosomes from *C. thermocellum* JW20 (74). But given that cellobiose is the main product of hydrolysis of cellulose by the cellulosome, β-glucosidase is not present in the cellulosome nor is it normally found in the culture fluid. Utilization of cellobiose by *C. thermocellum* follows internalization of this disaccharide (102). *C. thermocellum* (1, 2), *Ruminococcus flavefaciens* (107), *Fibrobacter (Bacteroides) succinogenes* (48), and *Acetivibrio celluloyticus* (113) are among several anaerobic bacteria that have cell-associated or cytoplasmic β-glucosidases. One should note that *C. thermocellum* has at least two β-glucosidases, the genes of which have been cloned into *E. coli* and characterized (43, 44, 109).

The β-glucosidases may certainly be involved in the metabolism of cellobiose by hydrolyzing it to glucose. However, in *C. thermocellum* and some other bacteria, cellobiose and also cellodextrins are more likely metabolized by cellobiose phosphorylase (4, 103) and cellodextrin phosphorylase (120). Because the phosphorylase and the β-glucosidases are cytoplasmic enzymes, cellobiose and cellodextrins must be taken up by the cell. This uptake has not been investigated much with *C. thermocellum*. However, evidence has been presented that the cellobiose uptake occurs by active transport (102). This observation has been well documented for *R. flavefaciens* (57) and *F. succinogenes* (87).

Xylanases

We noted when discussing individual endoglucanases that some of them exhibit activity toward xylan. The presence of xylanase activity among the polypeptides of the cellulosome from *C. thermocellum* has also been demonstrated by staining of SDS-PAGE gels using xylan as substrate. With this technique, 13 polypeptide bands displayed xylanase activity (74, 104). Of these, eight also exhibited endoglucanase activity. Two xylanase genes have been discovered in recombinant *E. coli* clones carrying *C. thermocellum* DNA (56). One of these, *xynZ,* has been sequenced. It encodes a protein composed of 837 amino acids yielding a mass of 92,159 (46). It carries a putative signal peptide of 28 amino acids and has the reiterated segment of 24 amino acids also present in endoglucanases A, B, D, E, and H. This segment was located between residues 429 and 488. The active site was determined by deletion analysis of *xynZ*. After removal of 508 codons from the 5′ end, the gene still encoded a polypeptide with enzyme activity demonstrating that the active site was located in the C-terminal part of the enzyme. The shorter enzyme had in fact a specific activity 220-fold that of the complete protein. The truncated protein was purified from the recombinant *E. coli,* and antiserum raised against this truncated protein reacted with a polypeptide of 90 kDa present in the *C. thermocellum* cellulosome (47).

The Cellulose-Binding Proteins

A major component of the cellulosome of *C. thermocellum* is a large polypeptide that does not display any enzymatic activity (60, 74, 80, 81, 104, 136). The apparent molecular mass of this subunit as determined by SDS-PAGE ranges from 185 to 250 kDa, and the designations S1 (80, 81) and S_L (137) have been used, although it has recently been renamed CipA (for cellulosome-integrating protein) (see 37b). This polypeptide may act as a cellulose-binding factor and/or a scaffolding protein binding and supporting the enzymatically active subunits of the cellulosome. The S1 subunit contains perhaps as much as 40% of covalently bound carbohydrate, and a tetrasaccharide of the following structure has been isolated from the cellulosome (40):

$$
\begin{array}{l}
\text{3-0Me-D-Glc}p\text{NAc-}\alpha(1\rightarrow 2) \\
\qquad\qquad\qquad\qquad\qquad \downarrow \\
\qquad\qquad\qquad\qquad\quad \text{D-Gal}f\text{-}\alpha(1\rightarrow 2)\text{-D-GalOH} \\
\qquad\qquad\qquad\qquad\qquad \uparrow \\
\qquad\qquad \text{D-gal}p\text{-}\alpha(1\rightarrow 3)
\end{array}
$$

Evidence indicates that the tetrasaccharide is associated with the 210-kDa polypeptide. The presence of a terminal α-galactopyranose in the tetrasaccharide and the fact that the *Griffonia simplicifolia* GSI-B4 isolectin reacts with the cellulosome (79) have led to the speculation that the 210-kDa polypeptide occupies an outer position on the cellulosome and that the tetrasaccharide may be involved in the cellulosome's association with cellulose (28). Thus, the cellulose-binding subunit may have three functions: binding the cellulosome to the bacterial cell, attaching the cellulosome to the substrate, and supporting the enzymatically active subunits. In fact, in *C. thermocellum* a polypeptide with 447 residues termed ORF3p, encoded by an open reading frame (ORF) located downstream from CipA, might serve as an anchoring factor for the cellulosome on the cell surface by binding the duplicated segment that is present at the COOH end of the CipA (37b) (see Figure 2, below).

Until recently, the cellulose-binding subunit, based on SDS-PAGE analyses, was considered to consist of a single polypeptide, but some findings indicated it may consist of subunits. First, polycellulosomes (OBL) have a higher level of the cellulose-binding subunit than the cellulosome (OBS) (61). Second, simple dialysis of the cellulosome against double-distilled water or lowering of the pH to below 5 caused the S1 polypeptide to migrate on SDS-PAGE faster than normally observed, suggesting that the polypeptide separated into smaller components (95). Addition of 50 mM Tris-HCl, pH 7.5, to the dialyzed preparation restored the SDS-PAGE profile to its appearance prior to the dialysis, demonstrating that the ionic strength and pH are important for the integrity of the cellulosome. That ionic strength effects the adsorption of the cellulosome to cellulose is evidenced by the fact that distilled water or 50% (v/v) ethanol, but not 5 mM salt solutions, releases the cellulosome bound to cellulose (61, 86).

The possibility that the cellulose-binding protein was composed of subunits was temporarily strengthened when a cloned DNA fragment from *C. thermocellum* directed in *E. coli* the expression of a 37-kDa peptide that crossreacted with antibodies prepared against the S_L subunit (111). This peptide, which has low levels of endoglucanase and avicelase activities, binds to cellulose and interacts positively with the *C. thermocellum* S_s endoglucanase (137). However, it has now been found that the above-mentioned DNA fragment encoded only part of the S_L (S1) protein. The entire gene encodes a protein with a mass of 196,800 Daltons (111). The sequence reportedly contains eight homologous domains of about 165 residues that may bind the cellulolytic enzymes. In addition, it has separate domain(s) for binding to cellulose (39, 116).

The cellulose-binding protein (CbpA) of the cellulosome from *C. cellulovorans* has been cloned and sequenced (122). The gene encodes a protein containing 1848 amino acids and has a mass of 189,036 Daltons. It has a signal peptide sequence of 28 amino acids, two types of cellulose binding domains of about 100 amino acids, and eight conserved hydrophobic regions of about 140 amino acids. As was suggested for the S1 protein of *C. thermocellum,* these eight regions are proposed to bind the catalytic subunits of the cellulosome.

Nonproteinaceous Components of or Interacting with the Cellulosome

Carbohydrate accounts for as much as 6–12% by weight of the cellulosomal mass (84). Sugars identified in the cellulosomes from *C. thermocellum* strain JW20 and YM4 are xylose, mannose, galactose, glucose, and *N*-acetylglucosamine. The amounts of xylose, mannose, and glucose vary considerably, and these sugars can be removed, to a large extent, by extended dialysis. Galactose (60–90 mg/mg protein) and *N*-acetylglucosamine (5–10 mg/mg protein) on the other hand are apparently integral parts of the cellulosome. Several subunits of the cellulosome separated by SDS-PAGE react positively upon glycoprotein staining (61, 74, 104). The major glycopolypeptide is S1, the 210-kDa subunit identified as the cellulose-binding protein (75). As was discussed in the previous section, S1 contains a tetrasaccharide, the structure of which has been determined (40). It presumably plays an important role in the functions of S1 in binding to the cellulose and the microbial cell and as a scaffold of the cellulolytic subunits. Electron-microscope analysis revealed cellulosomal particles as rows of equidistantly spaced polypeptide subunits attached to each other by ultrathin fibrils, arranged parallel to the main axis of the cellulosome (90). The ultrathin fibrils connecting the proteins to one another and to the central mass may be comprised of carbohydrate. Connecting the fibrils may be the function of the tetrasaccharide (40) associated with the 210-kDa polypeptide. Polycellulosomes, when examined with electron microscopy, appear to be enclosed in a thin skin layer (90), which may form protuberances (8, 75). This skin layer may be composed of large glycoconjugates identified in the cell surface of *C. thermocellum* (75).

In addition to the tetrasaccharide of the cellulose-binding protein S1, a disaccharide D-Gal*p*-β(1→4)-D-GalOH has been isolated from the cellulosome (40). With which cellulosomal subunit(s) this disaccharide is associated is not known. However, we have noted when discussing endoglucanses that some of them are glycosylated (101) and that endoglucanase A of the cellulosome of strain YM4 contains galactose and *N*-acetylglucosamine (98).

The hydrolysis of crystalline cellulose by the cellulosome requires Ca^{2+} (60, 65, 78). Endoglucanase D has three binding sites for calcium and one binding site for zinc (69). The presence of calcium in this enzyme lowers the K_d for CMC and affects the thermal stability of the enzyme (22). Plasma emission spectroscopy of several cellulosome preparations of *C. thermocellum* strain JW20 have shown that calcium is present at a level of 43 nmol mg^{-1} of protein (23). Other metals including zinc are present in much smaller quantities (less than 3 nmol mg^{-1}). Calcium probably has a structural role. Thus, treatment of the cellulosome with EDTA causes a slow disintegration with production of at least three subcellulosomal particles, a pattern similar to the treatment of the cellulosome with SDS under cold conditions (81, 98). The EDTA treatment also changes the SDS-PAGE pattern of the cellulosome in that several of the subunits move faster in the gel. The changes are not observed when Ca^{2+} is present during the treatment.

One feature of *C. thermocellum* is that it produces a yellow substance (91, 100, 130, 132) that coats the cellulose. This substance is produced shortly before the cellulase complex forms and in substantial amounts only when the bacterium is grown on cellulose (59). It adheres to cellulose and seems to facilitate binding of the cellulosome to the insoluble substrate (78, 86). Thus it was termed yellow affinity substance (YAS) (86). YAS cellulose binds the cellulosome more tightly than regular cellulose, and the cellulosome complexes can easily be recovered from the YAS cellulose by extraction with distilled water (61, 86). The residual cellulose-YAS complex has proven useful for purification of the *C. thermocellum* cellulosomes (60, 61). The YAS is easily extracted from YAS cellulose with acetone or 50% alcohol (60, 84). Its UV/visible spectrum in this solvent shows a major peak at 440–445 nm, which disappears on oxidation and exposure to sunlight. This water-insoluble substance has been postulated to be a carotenoid-like pigment with the elemental composition $C_{52}H_{94}O_{19}N$, which corresponds to a M_r of 1036. The YAS has been speculated to function as a signal for synthesis/secretion of cellulose by *C. thermocellum* (84). However, the possible physiological role of the YAS is not defined and so far has been overlooked.

MACROMOLECULAR ORGANIZATION OF THE CELLULOSOME AND POSSIBLE MECHANISMS OF ACTION

The cellulosome and polycellulosome are among the most complex enzyme systems being studied. One must wonder why nature has created such a system to catalyze the seemingly very simple hydrolysis of β-glycosidic bonds. Clearly, one cannot ignore the properties of the highly ordered substrate cellulose, which is in nature associated with hemicellulose, other polysaccharides, and lignin. This association may explain the presence in the cellulosome of polypeptides that hydrolyze xylan, lichenan, laminarin, etc. However, the composition of the cellulosomes, and the mode of action of their individual components, are the main factors that deeply influence the hydrolytic process. Research has now advanced to a state at which we can propose some reasonable models of action by the cellulosome. Some earlier models based mainly on observations made with electron microscopy (8, 90) seem to have held up and actually are being supported by new knowledge obtained using the powerful technique of modern molecular biology (10, 13).

In considering the life cycle of the cellulosome, we first note that it consists of between 14–26 different polypeptides (74, 80). The exact number of polypeptides is not known because some of them apparently yield multiple bands on SDS-PAGE depending on association with other polypeptides and the presence of carbohydrates or other nonproteinaceous materials. The large number of polypeptides constituting the cellulosome has been confirmed by genetic analyses of *C. thermocellum,* which have demonstrated several different genes encoding endoglucanases, β-glucosidases, xylanases, and perhaps cellobiohydrolases (15, 71, 108). The genes seem to be scattered throughout the bacterial genome and each apparently belongs to a monocistronic transcription unit (15). This raises questions not yet answered regarding the control and coordination of the synthesis of the different polypeptides of the cellulosome.

The cellulosomes are probably assembled on the bacterial cell surface where they can be seen with the electron microscope as tightly packed clusters of cellulosomes forming polycellulosomes (8, 79, 90). The evidence that the individual polypeptides have signal peptides needed for transport through the cell membrane supports this observation (see previous section on endoglucanases, xylanases, and cellulose-binding protein). The polypeptides are likely bagged inside a skin-like cover extending over the polycellulosomes (90). This cover may consist of peptidoglycan remnants or large glycoconjugates (75).

Upon contact with the substrate, cellulose, the skin cover appears to hook up with the cellulose, forming fibrous contact corridors and thus attaching the bacterial cells and the cellulosomes to the cellulose (8, 75, 132). Conceivably at this stage, cellobiose, cellodextrins, and glucose, products of the cellulolysis, pass through the contact corridors directly to the cell. This is a seemingly very efficient process benefiting the cell. Eventually the cell, because it is well fed or because of the accumulation of fermentation products and lowering of the pH of the culture medium, sporulates, leaving the cellulosomes attached to the cellulose (132). The cellulolytic process continues, but the polycellulosomes and cellulosomes decompose and ultimately form free polypeptides in the culture fluid (59, 61). The polypeptides maintain endoglucanase activity or other activities involving soluble substrates, but crystalline cellulose is not degraded further.

Ultrastructural observations with the electron microscope have given clues about a possible mechanism by which the cellulosome hydrolyzes cellulose (89, 90). The first-formed cellulosomes are tightly packed particles, but as they age, they seem to loosen. The loosened cellulosomes appear as globular particles (polypeptides) arranged in ordered chain-like arrays bound together by fibrils. Each cellulosome appears to have four or more of these chains, each chain composed of five to eight identical polypeptides. Remarkably, this is exactly the structure one might imagine if the noncatalytic cellulose-binding (scaffolding) subunit (S1 or S_L) of the *C. thermocellum* cellulosome binds at each of its eight homologous domains to a polypeptide, e.g. the endoglucanase D subunit. As noted above, the X-ray crystallography of this subunit indicated a barrel-like structure with the catalytic site on one side (top) of the barrel (69). Binding to the S1 cellulose-binding subunit is likely mediated by the reiterated conserved C-terminal domain (125), which in X-ray crystallography appeared as disordered and outside the barrel (69). Additional observations made with electron microscopy (90) involved an estimation of the average

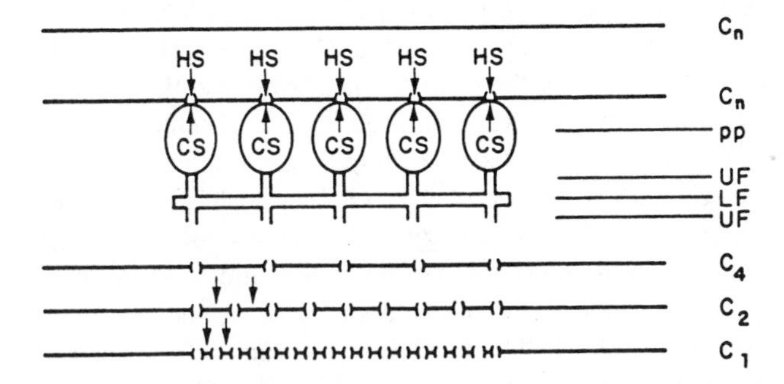

Figure 1 Dramatic view of the main aspect of the structural concept of the *C. thermocellum* cellulosome from strains JW20 and YM4 and the inherent structure-function relationships. Abbreviations: CS, catalytic site; HS, hydrolysis site, i.e. the site on the cellulose chain at which a gycosidic bond is cleaved; UF, ultrathin fibrils; *Cn*, C4, C2, C1, cellodextrins of various lengths (C_n is cellulose and C_1 is cellobiose). Reproduced from Mayer et al (90), with permission.

center-to-center distance between catalytic polypeptides in a single chain of 4 nm. Assuming each polypeptide is an individual enzyme, this represents the distance between catalytic sites of neighboring subunits. This distance is also the length of a cellooligosaccharide of four cellobiose units (C_4). Thus, a multicutting event may take place along a cellulose fiber aligned beside a row of subunits to release C_4-cellooligosaccharides. These may then be cut

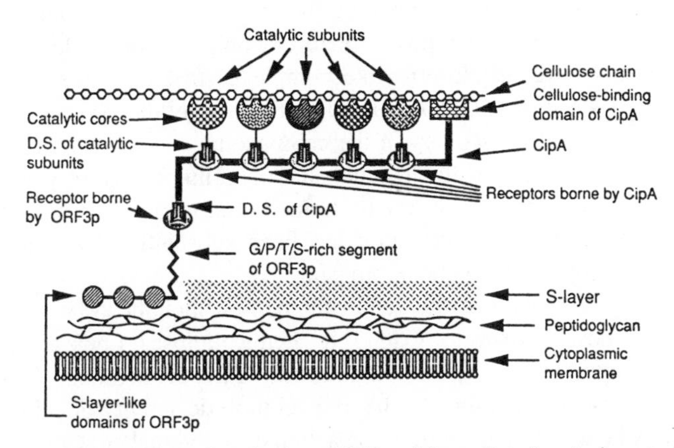

Figure 2 Hypothetical model showing how ORF3p might mediate the attachment of the cellulosome to the cell surface of *C. thermocellum*. The scheme for the organization of the cellulosome itself is derived from evidence showing that CipA mediates binding to cellulose of the catalytic S_s subunit (136), that catalytic subunits bind to the repeated domains (receptors) of CipA by means of their duplicated segments (37a, 125), and that CipA comprises a cellulose- binding domain (95a, 116). The cores of the catalytic subunits are asssumed to be poised along a cellulose chain, enabling quasisimultaneous, multiple cutting events, as proposed by Mayer et al (90). D.S., 22-amino-acid duplicated segment. Reproduced from Fujino et al (37b), with permission.

by other catalytic subunits, e.g. dextrinases, similarly arranged within the cellulosome to yield mainly cellobiose (C_1), the main product of the degradation of cellulose by the cellulosome. Some additional support for the above mechanism is the observation that desorption of cellulosomes with water from residual cellulose is always accompanied by release of fairly soluble cellooligosaccharides (C. R. Felix & L. G. Ljungdahl, unpublished observation). Furthermore, molecular biology studies have substantiated the proposed organization and the possible mechanism of action of cellulosomes. The S1, or S_L, or CipA protein, with the catalytic subunits bound to its repeated domains, is possibly attached to the *C. thermocellum* cell by a surface protein, the ORF3p (37b). This is further illustrated by Figures 1 and 2.

Other bacteria may also use the mechanisms outlined here. For example, the *C. cellulovorans* cellulosome has a cellulose-binding protein (CbpA) almost identical to the S1 or S_L subunit of *C. thermocellum*. It contains eight repeating domains that may be binding sites for catalytic polypeptides (122). The endoglucanase, EngB, of *C. cellulovorans* contains the reiterated region found in endoglucanases of *C. thermocellum* (36) that is postulated to be the binding site between the catalytic subunits and the cellulose-binding protein (125).

CONCLUDING REMARKS

In this review, we have concentrated on the cellulosome of *C. thermocellum* because we felt a more composed picture would emerge. Clearly, other microorganisms have cellulosome-like complexes that resemble in varying degrees the cellulosome of *C. thermocellum*. Several cellulolytic systems have evolved in nature. The discovery of the cellulosome was a breakthrough and began the real search for the mechanism of cellulose hydrolysis by *C. thermocellum*. Since this discovery by Lamed et al (80), much progress has been made. Biochemical information has been substantiated by the use of electron-microscope and genetic approaches.

The hydrolytic activity of the cellulosome, first thought to be unique for cellulose, is now known to extend to hemicellulose. Endo-actions are traditionally reported for the multiple subunits of the cellulosome. However, the exo-action is still questioned. The true cellulolytic activity of cellulosome is commonly believed to result from the action of individual enzymes working in a cooperative fashion. A multicutting-event mechanism was postulated based on electron-microscope observations and has now been essentially substantiated by biochemical and genetic information. Nevertheless, plausible as this model is, the mechanism of cellulose hydrolysis by the cellulosome is still far beyond our understanding. A long list of questions remains to be answered: Is the number of subunits overestimated, or do amorphogenesis and cellulolysis by the cellulosome indeed require such a wide diversity of subunits? Do the subunits serve other functions besides being hydrolytic? What are the events involved in synthesis of subunits and assembly of the cellulosome? Also overlooked are the roles of the yellow affinity substance (YAS) and of the carbohydrates present in the cellulosome. Additional information on these compounds may lead to a clear picture of the cellulosome architecture and its mechanism of action. In fact, the literature concerned with the cellulosome is accumulating very fast, and providing all the information needed to support the topics discussed would be extremely difficult. We hope this review has given a rather accurate account of the present knowledge of the cellulosome. Perhaps it will stimulate additional research and open up new avenues leading to a better understanding of an intriguing and fascinating enzyme complex.

ACKNOWLEDGMENTS

Work on cellulolytic enzyme systems in the authors' laboratory is funded by contract #DE-FG09-86ER13614 from the US Department of Energy. A Georgia Power Distinguished Professorship to LGL is gratefully acknowledged. CRF is on an extended sabbatical and acknowledges financial support from the Universidade de Brasília and the scholarship from the Conselho Nacional de Desenvolvimento Científico e Tecnológico do Brasil (proc. no. 260 192/91.0 RHAE/CNPq). We would like to express our thanks to Professor P. Béguin and Professor A. L. Demain for supplying reprints of several publications and to Mr. Sang-ki Choi, a graduate student in our group, for allowing incorporation of recent results. We would like to dedicate this paper to Michael Coughlan, who contributed much to our understanding of cellulolytic enzyme systems. Michael died February 3, 1993.

Literature Cited

1. Aït, N., Creuzet, N., Cattaneo, J. 1979. Characterization and purification of thermostable β-glucosidase from *Clostridium thermocellum*. *Biochem. Biophys. Res. Commun.* 90: 537–46
2. Aït, N., Creuzet, N., Cattaneo, J. 1982. Properties of β-glucosidase purified from *Clostridium thermocellum*. *J. Gen. Microbiol.* 128:569–77
3. Akin, D. E., Ljungdahl, L. G., Wilson, J. R., Harris, J. J., eds. 1990. *Microbial and Plant Opportunities to Improve Lignocellulose Utilization by Ruminants*. New York: Elsevier
4. Alexander, J. K. 1968. Purification and specificity of cellobiose phosphorylase from *Clostridium thermocellum*. *J. Biol. Chem.* 243:2899–2904
5. Aubert, J.-P., Béguin, P., Millet, J., eds. 1988. *Biochemistry and Genetics of Cellulose Degradation. FEMS Symp. No. 43*. London: Academic
6. Bates, E. E. M., Gilbert, H. J., Hazlewood, G. P., Huckle, J., Laurie, J. I., Mann, S. P. 1989. Expression of a *Clostridium thermocellum* endoglucanase gene in *Lactobacillus plantarum*. *Appl. Environ. Microbiol.* 55:2095–97
7. Bayer, E. A., Kenig, R., Lamed, R. 1983. Adherence of *Clostridium thermocellum* to cellulose. *J. Bacteriol.* 156:818–27

8. Bayer, E. A., Lamed, R. J. 1986. Ultrastructure of the cell surface cellulosome of *Clostridium thermocellum* and its interaction with cellulose. *J. Bacteriol.* 167:828–36
9. Bayer, E. A., Setter, E., Lamed, R. 1985. Organization and distribution of the cellulosome in *Clostridium thermocellum. J. Bacteriol.* 163:552–59
10. Béguin, P. 1990. Molecular biology of cellulose degradation. *Annu. Rev. Microbiol.* 44:219–48
11. Béguin, P., Cornet, P., Aubert, J.-P. 1985. Sequence of a cellulose gene of the thermophilic bacterium *Clostridium thermocellum. J. Bacteriol.* 162:102–5
12. Béguin, P., Cornet, P., Millet, J. 1983. Identification of the endoglucanase encoded by *celB* gene of *Clostridium thermocellum. Biochimie* 65:495–500
13. Béguin, P., Gilkes, N. R., Kilburn, D. G., Miller, R. C. Jr., O'Neill, G. P., Warren, R. A. J. 1987. Cloning of cellulose genes. *CRC Crit. Rev. Biotechnol.* 6:129–62
14. Béguin, P., Millet, J., Chavaux, S., Yague, E., Tomme, P., et al. 1989. Genetics of bacterial celluloses. See Ref. 24, pp. 57–72
15. Béguin, P., Millet, J., Grepinent, O., Navarro, A., Juy, M., et al. 1988. The *cel* genes of *Clostridium thermocellum.* See Ref. 5, pp. 267–84
16. Benitez J., Silva, A., Vazquez, R., Noa, M. D., Hollenberg, C. P. 1989. Secretion and glycosylation of *Clostridium thermocellum* endoglucanase A encoded by the *celA* gene in *Saccharomyces cerevisiae. Yeast* 5: 299–306
17. Bronnenmeier, K., Staudenbauer, W. L. 1990. Cellulose hydrolysis by a highly thermostable endo-1,4-β-glucanase (Avicelase I) from *Clostridium stercorarium. Enzyme Microb. Technol.* 12:431–36
18. Bumazkin, B. K., Velikodvorskaya, G. A., Tuka, K., Mogutov, M. A., Strongin, A. Y. A. 1990. Cloning of *Clostridium thermocellum* endoglucanase genes in *Escherichia coli. Biochem. Biophys. Res. Commun.* 167: 1057–64
19. Cavedon, K., Leschine, S. B., Canale-Parola, E. 1990. Cellulose system of a free-living mesophilic *Clostridium* (strain C7). *J. Bacteriol.* 172:4222–30
20. Cavedon, K., Leschine, S. B., Canale-Parola, E. 1990. Characterization of the extracellular cellulose from a mesophilic *Clostridium* (strain C7). *J. Bacteriol.* 172:4231–37
21. Chauvaux, S., Béguin, P., Aubert, J.-P. 1992. Site-directed mutagenesis of essential carboxylic residues in *Clostridium thermocellum* endoglucanase *celD. J. Biol. Chem.* 267:4472–78
22. Chauvaux, S., Béguin, P., Aubert, J.-P., Bhat, K. M., Gow, L. A., et al. 1990. Calcium-binding affinity and calcium-enhanced activity of *Clostridium thermocellum* endoglucanase D. *Biochem. J.* 265:261–65
23. Choi, S.-K., Ljungdahl, L. G. 1993. Effects of Ca^{2+} and EDTA on the dissociation of the cellulosome complex from *Clostridium thermocellum. Abstr. Annu. Meeting Am. Soc. Microbiol.* 93:283
24. Coughlan, M. P., ed. 1989. *Enzyme Systems for Lignocellulose Degradation.* New York: Elsevier
25. Coughlan, M. P. 1991. Cellulose degradation by fungi. In *Microbial Enzymes and Biotechnology,* ed. W. M. Fogarty, C. T. Kelly, pp. 1–36. Amsterdam: Elsevier
26. Coughlan, M. P., Hon-nami, K., Honnami, H., Ljungdahl, L. G., Paulin, J. J., et al. 1985. The cellulolytic enzyme complex of *Clostridium thermocellum* is very large. *Biochem. Biophys. Res. Commun.* 130: 904–9
27. Coughlan, M. P., Ljungdahl, L. G. 1988. Comparative biochemistry of fungal and bacterial cellulolytic enzyme systems. See Ref. 5, pp. 11–30
28. Coughlan, M. P., Mayer, F. 1990. Cellulose-decomposing bacteria and their enzyme systems. In *The Prokaryotes,* ed. A. B. Balows, Berlin: Springer-Verlag
29. Creuzet, M., Frixon, C. 1983. Purification and characterization of an endoglucanase from a newly isolated thermophilic anaerobic bacterium. *Biochimie* 65:149–56
30. Duong, T.-V. C., Johnson, E. A., Demain, A. L 1983. Thermophilic anaerobic and cellulolytic bacteria. *Top. Enzyme Ferment. Biotechnol.* 7:156–95
31. Erikson, K.-E. L., Blanchette, R. A., Ander, P. 1990. *Microbial and Enzymtic Degradation of Wood and Wood Components.* Berlin: Springer-Verlag
32. Erikson, K.-E., Wood, T. M. 1985. Biodegradation of cellulose. In *Biosynthesis and Biodegradation of Wood Components,* ed. T. Higuchi, pp. 469–503. New York: Academic
33. Fauth, U. Romaniec, M. P. M., Kobayashi, T., Demain, A. L. 1991. Purification and characterization of endoglucanase S_s from *Clostridium thermocellum. Biochem. J.* 279:67–73
34. Fierobe, H.-P, Gaudin, C., Belaich,

A., Loutfi, M., Bagnara, C., Belaich, J.-P. 1991. Characterization of endoglucanase A from *Clostridium cellulolyticum*. *J. Bacteriol.* 173:7956–62

35. Foong, F., Doi, R. H. 1992. Characterization and comparison of *Clostridium cellulovorans* endoglucanases-xylanases *engB* and *engD* hyper- expressed in *Escherichia coli*. *J. Bacteriol.* 174:1403–9

36. Foong, F., Hamamoto, T., Shoseyov, O., Doi, R. H. 1991. Nucleotide sequence and characteristics of endoglucanase gene *engB* from *Clostridium cellulovorans*. *J. Gen. Microbiol.* 137: 1729–36

37. Freier D., Mothershed, C. P., Wiegel, J. 1988. Characterization of *Clostridium thermocellum* JW20. *J. Appl. Environ. Microbiol.* 54:204–11

37a. Fujino, T., Béguin, P., Aubert, J.-P. 1992. Cloning of a *Clostridium thermocellum* DNA fragment encoding polypeptides that bind the catalytic components of the cellulosome. *FEMS Microbiol. Lett.* 94:165–70

37b. Fujino, T., Béguin, P., Aubert, J.-P. 1993. Organization of a *Clostridium thermocellum* gene cluster encoding the cellulosomal scaffolding protein CipA and protein possibly involved in attachment of the cellulosome to the cell surface. *J. Bacteriol.* 175:1891–99

38. Gardner, R. M., Doerner, K C., White, B. A. 1987. Purification and characterization of an exo-β-1,4-glucanase from *Ruminococcus flavefaciens* FD-1. *J. Bacteriol.* 169:4581–88

39. Gerngross, V. T., Romaniac, P. M., Huskisson, N. S., Demain, A. L. 1993. Sequencing of a *Clostridium thermocellum* gene (CipA) encoding the cellulosomal SL protein reveals an unusual degree of internal homology. *J. Mol. Biol.* In press

40. Gerwig, G. J., de Waard, P., Kamerling, J. P., Vliegenthart, J. F. G., Morgenstern, E., et al. 1989. Novel *O*-linked carbohydrate chains in the cellulose complex (cellulosome) of *Clostridium thermocellum*. 3-*O*-methyl-*N*-acetylglucosamine as a constituent of a glycoprotein. *J. Biol. Chem.* 264:1027–35

41. Gilkes, N. R., Henrissat, B., Kilburn, D. G., Miller, R. C. Jr., Warren, R. A. J. 1991. Domains in microbial β-1,4-glucanases: sequence conservation, function, and enzyme families. *Microbiol. Rev.* 55:303–15

42. Golovchenko, N. P., Katayeva, I. A., Bukhtiyarova, M. G., Aminov, R. I., Tsoi, T. V., Akimenko, V. K. 1991. Isolation and certain properties of a thermostable endoglucanase from *E. coli* C 600 (pKNE-102), encoded by a gene from *Clostridium thermocellum*. *Biokhimiya* 56:49–54. Engl. transl. 56: 34–38

43. Gräbnitz, F., Rücknagel, K. P., Seiss, M., Staudenbauer, W. L. 1989. Nucleotide sequence of the *Clostridium thermocellum bglB* gene encoding thermostable β-glucosidase B: homology to fungal β-glucosidase. *Mol. Gen. Genet.* 217:70–76

44. Gräbnitz, F., Staudenbauer, W. L. 1988. Characterization of two β-glucosidase genes from *Clostridium thermocellum*. *Biotechnol. Lett.* 10:73– 78

45. Grépinet, O., Béguin, P. 1986. Sequence of the cellulose gene of *Clostridium thermocellum* coding for endoglucananse B. *Nucleic Acids Res.* 14:1791–99

46. Grépinet, O., Chebrou, M.-C., Béguin, P. 1988. Nucleotide sequence and deletion analysis of the xylanase gene (*xynZ*) of *Clostridium thermocellum*. *J. Bacteriol.* 170:4582–88

47. Grépinet, O., Chebrou, M.-C., Béguin, P. 1988. Purification of *Clostridium thermocellum* xylanase Z expressed in *Escherichia coli* and identification of the corresponding product in the culture medium of *Clostridium thermocellum*. *J. Bacteriol.* 170:4576–81

48. Groleau, D., Forsberg, C. W. 1981. Cellulolytic activity of the rumen bacterium *Bacterioides succinogenes*. *Can. J. Microbiol.* 27:517–30

49. Haigler, C. H., Weimer, P. J., eds. 1991. *Biosynthesis and Biodegradation of Cellulose*. New York: Marcel Dekker

50. Hall, J., Hazlewood, G. P., Barker, P. J., Gilbert, H. J. 1988. Conserved reiterated domains in *Clostridium thermocellum* endoglucanases are not essential for catalytic activity. *Gene* 69: 29–38

51. Halliwell, G., Bryant, M. P. 1963. The cellulolytic activity of pure strains of bacteria from the rumen of cattle. *J. Gen. Microbiol.* 32:441–48

52. Halliwell, G., Halliwell, N. 1989. Cellulolytic enzyme components of cellulase complex of *Clostridium thermocellum*. *Biochim. Biophys. Acta.* 992: 223–29

53. Hamamoto, T., Foong, F., Shoseyov, O., Doi, R. H. 1992. Analysis of functional domains of endoglucanases from *Clostridium cellulovorans* by gene cloning, nucleotide sequencing and

chimeric protein construction. *Mol. Gen. Genet.* 231:472–79
54. Hamamoto, T., Shoseyov, O., Foong, F., Doi, R. H. 1990. A *Clostridium cellulovorans* gene, *engD*, codes for both endo-β-1,r-glucanase and cellobiosidase activities. *FEMS Microbiol. Lett.* 72:285–88
55. Hazlewood, G. P., Davidson, K., Clarke, J. H., Durrant, A. J., Hall, J., Gilbert, H. J. 1990. Endoglucanase E, produced at high level in *Escherichia coli* as a *lacZ'* fusion protein, is part of the *Clostridium thermocellum* cellulosome. *Enzyme Microb. Technol.* 12:656–62
56. Hazlewood, G. P., Romaniec, M. P. M., Davidson, K., Grépinet, O., Béguin, J., et al. 1988. A catalogue of *Clostridium thermocellum* endoglucanase, β-glucosidase and xylanase genes cloned in *Escherichia coli*. *FEMS Microbiol. Lett.* 51:231–36
57. Helaszek, C. T., White, B. A. 1991. Cellobiose uptake and metabolism by *Ruminococcus flavefaciens*. *Appl. Environ. Microbiol.* 57:64–68
58. Henrissat, B., Claeyssens, M., Tomme, P., Lemesle, L., Mornon, J.-P. 1989. Cellulose families revealed by hydrophobic cluster analysis. *Gene* 81:83–85
59. Hon-nami, H., Coughlan, M. P., Hon-nami, K., Ljungdahl, L. G. 1987. The time course production of the yellow affinity substance and of the bound and free cellulose complexes by *Clostridium thermocellum* JW20. *Proc. R. Irish Acad.* 87B:83–92
60. Hon-nami, K., Coughlan, M. P., Hon-nami, H., Carreira, L. H., Ljungdahl, L. G. 1985. Properties of the cellulolytic enzyme system of *Clostridium thermocellum*. *Biotechnol. Bioeng. Symp.* 15:191–205
61. Hon-nami, K., Coughlan, M. P., Hon-nami, H., Ljungdahl, L. G. 1986. Separation and characterization of the complexes constituting the cellulolytic enzyme system of *Clostridium thermocellum*. *Arch. Microbiol.* 145:13–19
62. Huang, L., Forsberg, C. W. 1988 Purification and comparision of the periplasmic and extracellular forms of the cellodextrinase from *Bacteroides succinogenes*. *Appl. Environ. Microbiol.* 54:1488–93
63. Johnson, E. A., Demain, A. L. 1984. Probable involvement of sulphydryl groups and a metal as essential components of the cellulosome of the cellulase of *Clostridium thermocellum*. *Arch. Microbiol.* 137:135–38
64. Johnson, E. A., Reese, E. T., Demain, A. L. 1982. Inhibition of *Clostridium thermocellum* cellulase by end products of cellulolysis. *J. Appl. Biochem.* 4:64–71
65. Johnson, E. A., Sakajoh, M., Halliwell, G., Madia, A., Demain, A. L. 1982. Saccharification of complex cellulosic substrates by the cellulase system from *Clostridium thermocellum*. *Appl. Environ. Microbiol.* 43:1125–32
66. Joliff, G., Béguin, P., Aubert, J.-P. 1986. Nucleotide sequence of the cellulase gene *celD* encoding endoglucanase D of *Clostridium thermocellum*. *Nucleic Acids Res.* 14:8605–13
67. Joliff, G., Béguin, P., Juy, M., Millet, J., Ryter, A., et al. 1986. Isolation, crystallization and properties of a new cellulase of *Clostridium thermocellum* overproduced in *Escherichia coli*. *Bio/Technology* 4:896–900
68. Joliff, G., Edelman, A., Klier, A., Rapoport, G. 1989. Inducible secretion of a cellulase from *Clostridium thermocellum* in *Bacillus subtilis*. *Appl. Environ. Microbiol.* 55:2739–44
69. Juy, M., Amit, A. G., Alzari, P. M., Poljak, R. J., Claeyssens, M., Béguin, P. 1992. Three-dimensional structure of a thermostable bacterial cellulase. *Nature* 357:89–91
70. Khraptsova, G. I., Kostrikina, N. A., Biryuzova, V. I., Loginova, L. G. 1986. Isolation of *Clostridium thermocellum* from hot springs in Buryataiya. *Mikrobiologiya* 55:496–501. Engl. transl. 55:385–91
71. Klesov, A. A. 1990. Biochemistry and enzymology of cellulose hydrolysis. *Biokhimiya* 55:1731–65. Engl. transl. 55:1295–1318
72. Kobayashi, T., Romaniec, M. P. M., Fauth, V., Barker, P. J., Demain, A. L. 1992. Cloning and expression in *Escherichia coli* of *Clostridium thermocellum* DNA encoding subcellulosomal proteins. *Enzyme Microb. Technol.* 14:447–53
73. Kobayashi, T., Romaniec, M. P. M., Fauth, V., Demain, A. L. 1990. Subcellulosome preparation with high cellulase activity from *Clostridium thermocellum*. *Appl. Environ. Microbiol.* 56: 3040–46
74. Kohring, S., Wiegel, J., Mayer, F. 1990. Subunit composition and glycosidic activities of the cellulase complex from *Clostridium thermocellum* JW 20. *Appl. Environ. Microbiol.* 56:3798–3804
75. Lamed, R., Bayer, E. A. 1988. The cellulosome concept: exocellular/extracellular enzyme reactor centers for ef-

ficient binding and cellulolysis. See Ref. 5, pp. 101–16
76. Lamed, R., Bayer, E. A. 1991. Cellulose degradation by thermophilic anaerobic bacteria. See Ref. 49, pp. 377–410
77. Lamed, R., Bayer, E. A., Saha, B. C., Zeikus, J. G. 1988. Biotechnological potential of enzymes from unique thermophiles. In *Proc. Int. Biotechnol. Symp. 8th, Soc. Fr. Microbiol.* ed. G. Duran, L. Bobichon, pp. 371–83
78. Lamed, R., Kenig, R., Setter, E., Bayer, 1985. Major characteristics of the cellulolytic system of *Clostridium thermocellum* coincide with those of the purified cellulosome. *Enzyme Microbiol. Technol.* 7:37–41
79. Lamed, R., Naimark, J., Morgernstern, E., Bayer, E. A. 1987. Specialized cell surface structures in cellulolytic bacteria. *J. Bacteriol.* 169:3792–3800
80. Lamed, R., Setter, E., Bayer, E. A. 1983. Characterization of a cellulose-binding, cellulase-containing complex in *Clostridium thermocellum. J. Bacteriol.* 156:828–36
81. Lamed, R., Setter, E., Kenig, R., Bayer, E. A. 1984. The cellulosome—a discrete cell surface organelle of *Clostridium thermocellum* which exhibits separate antigenic, cellulose-binding and various catalytic activities. *Biotechnol. Bioeng. Symp.* 13:163–81
82. Lasa, I., De Grado, M., De Pedro, M. A., Berenguer, J. 1992. Development of *Thermus-Escherichia* shuttle vectors and their use for expression of the *Clostridium thermocellum celA* gene in *Thermos thermophilus. J. Bacteriol.* 174:6424–31
83. Leschine, S. B., Canale-Parola, E. 1983. Mesophilic cellulolytic clostridia from freshwater enviroments. *Appl. Environ. Microbiol.* 46:728–37
84. Ljungdahl, L. G., Coughlan, M. P., Mayer, F., Mori, Y., Hon-nami, H., Hon-nami, K. 1988. Macrocellulase complexes and yellow affinity substance from *Clostridium thermocellum. Methods Enzymol.* 160:483–500
85. Ljungdahl, L. G., Eriksson, K.-E. 1985. Ecology of microbial cellulose degradation. *Adv. Microbiol. Ecol.* 8: 237–99
86. Ljungdahl, L. G., Petterson, B., Erikson, K.-E., Wiegel, J. 1983. A yellow affinity substance involved in the cellulolytic system of *Clostridium thermocellum. Curr. Microbiol.* 9:195–200
87. Maas, L. K., Glass, T. L. 1991. Cellobiose uptake by the cellulolytic ruminal anaerobe *Fibrobacter (Bacteroides) succinogenes. Can. J. Microbiol.* 37:141–47
88. MacKenzie, C. R., Patel, G. B., Bilous, D. 1987. Factors involved in hydrolysis of microcrystalline cellulose by *Acetivibrio cellulolyticus. Appl. Environ. Microbiol.* 53:304–8
89. Mayer, F. 1988. Cellulolysis: ultrastructural aspects of bacterial systems. *Electron Microsc. Rev.* 1:69–85
90. Mayer, F., Coughlan, M. P. Mori, Y., Ljungdahl, L. G. 1987. Macromolecular organization of the cellulolytic complex of *Clostridium thermocellum* as revealed by electron microscopy. *Appl. Environ. Microbiol.* 53:2785–92
91. McBee, R. H. 1948. The culture and physiology of a thermophylic cellulose-fermenting bacterium. *J. Bacteriol.* 56: 653–63
92. McBee, R. H. 1950. The anerobic thermophilic cellulolytic bacteria. *Bacteriol. Rev.* 14:51–63
93. Millet, J., Pétré, D., Béguin, P., Raynaud, O., Aubert, J.-P. 1985. Cloning of ten distinct DNA fragments of *Clostridium thermocellum* coding for cellulases. *FEMS Lett.* 29:145–49
94. Mishra, S., Béguin, P., Aubert, J.-P. 1991. Transcription of *Clostridium thermocellum* genes *celF* and *celD. J. Bacteriol.* 173:80–85
95. Morag (Morgenstern), E., Bayer, E. A., Lamed, R. 1991. Anomalous dissociative behavior of the major glycosylated component of the cellulosome of *Clostridium thermocellum. Appl. Biochem. Biotech.* 30:129–36
95a. Morag, E., Bayer, E. A., Lamed, R. 1992. Unorthodox intrasubunit interactions in the cellulosome of *Clostridium thermocellum:* identification of structural transitions induced in the S1 subunit. *Appl. Biochem. Biotechnol.* 33: 205–17
96. Morag (Morgenstern), E., Halevy, I., Bayer, E. A., Lamed, L 1991. Isolation and properties of a major cellobiohydrolase from the cellulosome of *Clostridium thermocellum. J. Bacteriol.* 173:4155–62
97. Mori, Y. 1990. Characterization of a symbiotic coculture of *Clostridium thermohydrosulfuricum* YM3 and *Clostridium thermocellum* YM4. *Appl. Environ. Microbiol.* 56:37–42
98. Mori, Y. 1992. Purification and characterization of an endoglucanase from the cellulosome (multicomponent cellulose complex) of *Clostridium thermocellum. Biosci. Biotechnol. Biochem.* 56:1199–1203

99. Navarro, A., Chebrou, M.-C, Béguin, P., Aubert, J.-F. 1992. Nucleotide sequence of the cellulase gene *celF* of *Clostridium thermocellum*. *Res. Microbiol.* In press
100. Ng, T. K., Weimer, P. J., Zeikus, J. G. 1977. *Arch. Microbiol.* 114:1–7
101. Ng, T. K., Zeikus, J. G. 1981. Purification and characterization of an endoglucanase (1,4,β-D-glucan-glucanohydrolase) from *Clostridium thermocellum*. *Biochem. J.* 199:341–50
102. Ng, T. K., Zeikus, J. G. 1982. Differential metabolism of cellobiose and glucose by *Clostridium thermocellum* and *Clostridium thermohydrosulfuricum*. *J. Bacteriol.* 150:1391–99
103. Ng, T. K., Zeikus, J. G. 1986. Synthesis of (^{14}C) cellobiose with *Clostridium thermocellum* cellobiose phosphorylase. *Appl. Environ. Microbiol.* 52:902–4
104. Pape, S. 1989. *Enzymatische charakterisierung des Cellulase-Komplexes von* Clostridium thermocellum *JW 20*. PhD dissertation, Univ. Gottingen
105. Pétré, J., Longin, R., Millet, J. 1981. Purification and properties of an endo-β-1,4-glucanase from *Clostridium thermocellum*. *Biochimie* 63:629–39
106. Pétré, D., Millet, J., Longuin, R., Béguin, P., Girard, H., Aubert, J.-P. 1986. Purification and properties of the endoglucanase C of *Clostridium thermocellum* produced in *Escherichia coli*. *Biochimie* 68:687–95
107. Pettipher, G. L., Latham, M. J. 1979. Production of enzymes degrading plant cell walls and fermentation of cellobiose by *Ruminococcus flavefaciens* in batch and continuous culture. *J. Gen. Microbiol.* 118:29–38
108. Rapp, P., Beermann, A. 1991. *Bacterial cellulases*. See Ref. 49, pp. 535–97
109. Romaniec, M. P. M., Davidson, K., Hazlewood, G. P. 1987. Cloning and expression in *Escherichia coli* of *Clostridium thermocellum* DNA encoding β-glucosidase activity. *Enzyme Microb. Technol.* 9:474–78
110. Romaniec, M. P. M., Fauth, U., Kobayashi, T., Huskisson, N. S., Barker, P. J., Demain, A. L. 1992. Purification and characterization of a new endoglucananse from *Clostridium thermocellum*. *Biochem. J.* 283:69–73
111. Romaniec, M. P. M., Kobayashi, T., Fauth, U., Gerngross, U. T., Demain, A. L. 1991. Cloning and expression of a *Clostridium thermocellum* DNA fragment that encodes a protein related to cellulosome component SL. *Appl. Biochem. Biotech.* 31:119–34
112. Sacco, M., Millet, J., Aubert, J.-P. 1984. Cloning and expression in *Saccharomyces cerevisiae* of a cellulase gene from *Clostridium thermocellum*. *Ann. Microbiol. Inst. Pasteur* 135A: 485–88
113. Saddler, J. N., Khan, A. W. 1981. Cellulolytic enzyme system of *Acetivibrio cellulolyticus*. *Can. J. Microbiol.* 27:228–94
114. Sakka, K., Furuse, S., Shimada, K. 1989. Cloning and expression in *Escherichia coli* of the thermophilic *Clostridium* sp. F1 genes related to cellulose hydrolysis. *Agric. Biol. Chem.* 53:905–11
115. Sakka, K., Shimanuki, T., Shimada, K 1991. Nucleotide sequence of *celC*307 encoding endoglucanase C307 of *Clostridium* sp. strain F1. *Agric. Biol. Chem.* 55:347–50
116. Salamitou, S., Tokatlidis, K., Béguin, P., Aubert, J.-P. 1992. Involvement of separate domains of the cellulosomal protein S1 of *Clostridium thermocellum* in binding to cellulose and in anchoring of catalytic subunits to the cellulosome. *FEBS Lett.* 304:89–92
117. Schwarz, W. H., Gräbnitz, F., Staudenbauer, W. L. 1986. Properties of a *Clostridium thermocellum* endoglucanase produced in *Escherichia coli*. *Appl. Environ. Microbiol.* 51: 1293–99
118. Schwarz, W. H., Schimming, S., Rücknagel, K. P., Burgschwaiger, S., Kreil, G., Staudenbauer, W. L. 1988. Nucleotide sequence of the *celC* gene encoding endoglucanase C of *Clostridium thermocellum*. *Gene* 63:23–30
119. Schwarz, W. H., Schimming, S., Staudenbauer, W. L. 1988. Degradation of barley β-glucan by endoglucanase C of *Clostridium thermocellum*. *Appl. Microbiol. Technol.* 29:25–31
120. Sheth, K., Alexander, J. K. 1969. Purification and properties of β-1,4-oligoglucan: orthophosphate glucosyltransferase from *Clostridium thermocellum*. *J. Biol. Chem.* 224:457–64
121. Shoseyov, O., Doi, R. H. 1990. Essential 170-kDa subunit for degradation of crystalline cellulose by *Clostridium cellulovorans*. *Proc. Natl. Acad. Sci. USA* 87:2192–95
122. Shoseyov, O., Hamamoto, T., Foong, F., Doi, R. H. 1990. Cloning of *Clostridium cellulovorans* endo-1,r-β-glucanase genes. *Biochem. Biophys. Res. Commun.* 169:667–72
123. Shoseyov, O., Takagi, M., Goldstein,

M. A., Doi, R. H. 1992. Primary sequence analysis of *Clostridium cellulovorans* cellulose binding protein A. *Proc. Natl. Acad. Sci. USA* 89: 3483–87

124. Soutschek-Bauer, E., Staudenbauer, W. L. 1987. Synthesis and secretion of a heat-stable carboxymethylcellulase from *Clostridium thermocellum* in *Bacillus subtilis* and *Bacillus stearothermophilus*. *Mol. Gen. Genet.* 208: 537–41
125. Tokatlidis, K., Salamitou, S., Béguin, P., Dhurjati, P., Aubert, J.-P. 1991. Interaction of the duplicated segment carried by *Clostridium thermocellum* cellulases with cellulosome components. *FEBS Lett.* 291:185–88
126. Tomme, P., Chauvaux, S., Béguin, P., Millet, J., Aubert, J.-P., Claeyssens, M. 1991. Identification of a histidyl residue in the active center of endoglucanase D from *Clostridium thermocellum*. *J. Biol. Chem.* 266: 10313–18
127. Tsoi, T. V., Bukbtiyarova, M. G., Aminov, R. I., Golovchenko, N. P., Kataeva, I. A., Kosheleva, I. A. 1990. Cloning and expression of *Clostridium thermocellum* F7 cellulase genes in *Escherichia coli* and *Bacillus subtilis*. *Genetika* 26:1349–60. English translation *Soviet Genet.* 26:877–85
128. van Tilbeurgh, H., Claeyssens, M., DeBruyne, C. K. 1982. The use of 4-methylumbelliferyl and other chromophoric glycosides in the study of cellulolytic enzymes. *FEBS Lett.* 149: 152–56
129. van Tilbeurgh, H., Loontiens, F. G., De Bruyne, C. K., Claeyssens, M. 1988. Fluorogenic and chromogenic glycosides as substrates and ligands of carbohydrases. *Methods Enzymol.* 160: 45–49
130. Viljoen, S. A., Fred, E. B., Peterson, W. H. 1926. The fermentation of cellulose by thermophilic bacteria. *J. Agric. Sci.* 16:1–17

130a. Wang, W. K., Kruus, K., Wu, J. H. D. 1993. Cloning and DNA sequencing of the gene coding for *Clostridium thermocellum* cellulase S_S (CelS), a major cellulosome component. *J. Bacteriol.* 175:1293–1302

131. White, B. A., Rasmussen, M. A., Gardner, R. M. 1988. Methyl-cellulose inhibition of exo-β-1,4-glucanase A from *Ruminococcus flavefaciens* FD-1. *Appl. Environ. Microbiol.* 54:1634–36
132. Wiegel, J., Dykstra, M. 1984. *Clostridium thermocellum:* adhesion and sporulation while adhered to cellulose and hemicellulose. *Appl. Microbiol. Biotechnol.* 20:59–65
133. Wilson, C. A., Wood, T. M. 1992. The anaerobic fungus *Neocallimastix frontalis:* isolation and properties of a cellulosome-type enzyme fraction with the capacity to solubilize hydrogen-bond-ordered cellulose. *Appl. Microbiol. Biotechnol.* 37:125–9
134. Wood, T. M., McCray, S. I. 1979. Synergism between enzymes involved in the solubilization of native cellulose. *Adv. Chem. Ser.* 181:181–209
135. Wood, W. A., Kellog, S. T., eds. 1988. Biomass, cellulose and hemicellulose. *Methods Enzymol.* 160:1–774
136. Wu, J. H. D., Demain, A. L 1988. See Ref. 5, pp. 117–31
137. Wu, J. H. D., Orme-Johnson, W. H., Demain, A. L. 1988. Two components of an extracellular protein aggregate of *Clostridium thermocellum* together degrade crystalline cellulose. *Biochemistry* 27:1703–9
138. Xue, G.-P., Orpin, C. G., Gobius, K S., Aylward, J. H., Simpson, G. D. 1992. Cloning and expression of multiple cellulase cDNAs from the anaerobic rumen fungus *Neocallimastix patriciarum* in *Escherichia coli*. *J. Gen. Microbiol.* 138:1413–20
139. Yague, E., Béguin, P., Aubert, J.-P. 1990. Nucleotide sequence and deletions analysis of the cellulase-encoding gene *celH* of *Clostridium thermocellum*. *Gene* 89:61–67

Annu. Rev. Microbiol. 1993. 47:821–53

THE PROTEASES AND PATHOGENICITY OF PARASITIC PROTOZOA[1]

James H. McKerrow
Department of Veterans Affairs Medical Center, San Francisco, California 94121

Eugene Sun and Philip J. Rosenthal
Departments of Pathology and Medicine, University of California, San Francisco, California 94143

Jacques Bouvier
Institut de Biochimie, Universite de Lausanne, Epalinges, Suisse

KEY WORDS: parasites, proteinase, inhibitors, pathogenesis, life cycle

---

CONTENTS

ABSTRACT

Protozoan parasites are among the most prevalent pathogens worldwide. Diseases like malaria, leishmaniasis, amebiasis, and trypanosomiasis affect hundreds of millions of people. Recent advances in our understanding of the biochemistry and molecular biology of these organisms has focused attention on specific parasite molecules that are key to the parasite life cycle or the pathogenesis of the diseases they produce. One group of enzymes that plays myriad roles in these processes are the parasite-derived proteases. Different types of proteases are frequently expressed at different stages of the parasite life cycle to support parasite replication and metamorphosis. Intracellular parasites such as those that produce malaria and Chagas' disease express high levels of protease activity to efficiently degrade host proteins like hemoglobin. In other instances, such as infection with *Entamoeba histolytica,* the causative agent of amebiasis, proteases released by the parasite can damage host cells and tissues, contributing to host tissue damage and parasite invasion. Detailed studies of these enzymes have led to model systems for the study of parasite gene regulation, parasite metabolism, and the host-parasite interplay. In some instances, proteases appear to be promising targets for the development of new antiparasitic chemotherapy.

INTRODUCTION TO PROTEASES AND THEIR INHIBITORS

Proteolytic enzymes, or proteases that catalyze the degradation of peptide bonds in both proteins and peptides, are found throughout the plant and animal kingdoms. Proteases that catalyze the cleavage of an internal peptide bond within a protein are often referred to as proteinases or, less commonly, endoproteases. Proteases that catalyze the cleavage of one or two amino acids from the extreme amino terminus or carboxy terminus of a protein or peptide are referred to as exopeptidases or, more specifically, aminopeptidases and carboxypeptidases, depending upon from which end of the substrate molecule the amino acids are liberated (14, 149). These variations in substrate specificity of proteases are important, because they often give clues to protease function.

While some issues remain unresolved regarding the kinetics of protease-catalyzed reactions and some aspects of protease-substrate interactions at the molecular level, researchers generally agree that a key feature of protease catalysis is the ability of the enzyme to bind the so-called transition state, a structural intermediate in the pathway to peptide-bond cleavage. Favoring binding of the transition state over the native substrate conformation helps overcome the activation-energy barrier; this is necessary for peptide-bond cleavage to occur under physiologic conditions.

Many natural and synthetic protease inhibitors have been identified (14,

21). Natural inhibitors play important roles in posttranslational regulation of proteolytic activity in both parasites and their hosts (141). Synthetic inhibitors are important tools in subclassifying proteases into the four major protease groups (14). A pragmatic corollary of the transition-state theory is that development of protease inhibitors as pharmaceuticals has often required the design of inhibitors to mimic the transition state. For further details on the kinetics of protease-catalyzed reactions and transition-state theory, the reader is referred to comprehensive reviews (19, 58, 115).

Peptide substrates are used to map active-site specificity of proteases and provide specific and sensitive assays. By convention (16) the amino acid residues on the amino-terminal side of the peptide bond being cleaved are designated P1, P2, P3...P*n*. On the carboxy terminal side, they are P1′, P2′, P3′...P*n*′ Many substrates have been designed with a leaving group that has a high extinction coefficient that can be detected following cleavage of the peptide, either spectrophotometrically or spectrofluorometrically (190). In other cases, these more convenient substrates are not suitable or available for the enzyme in question and an underivatized synthetic peptide substrate is used and its cleavage monitored by HPLC.

Serine proteases are the largest family of proteolytic enzymes. They are named for the amino acid serine, whose hydroxyl group is involved in the binding and catalysis of substrate at the active site. In cysteine or thiol proteases, the thiol group of the amino acid cysteine plays a similar role. Metalloproteases contain a metal ion in the active site that plays a key role in substrate binding, and aspartyl proteases utilize two aspartic acid side chains in catalysis. Subclassifying a protease into one of these classes, like recognizing whether it is a protease or exopeptidase, is important because this classification provides clues to the enzyme's function in vivo. For example, aspartyl proteases have pH optima between 2.5 and 4.5. It would be unlikely that an aspartyl protease functions in host invasion by a parasite unless the specific locale in which the organism is invading is one with an unusually low pH. On the other hand, both serine and metalloproteases have pH optima in the neutral to slightly alkaline range. Cysteine proteases can have very broad pH optima, from 5.5 to 7.5, and therefore function in a variety of environments as long as sufficient reducing agent is present to keep the active-site thiol group from being oxidized.

FUNCTION OF PROTEASES IN PARASITIC PROTOZOA AND THE DISEASES THEY PRODUCE

Protease activity in extracts or excretory/secretory products of protozoa can be measured either by utilizing protein substrates or synthetic peptide substrates. Ideally, if a natural target substrate for a protease is known, then that substrate, or a peptide mimic of it, should be used in assays. Unfortunately, proteases are often discovered long before their function is known, by the use of less specific protein substrates such as hemoglobin, immunoglobulins, denatured collagen, or fibrinogen. Although these proteins may play important roles in other host functions, they can be degraded by a variety of proteases, most of which never come in contact with the substrates in a biological setting.

Before reviewing work on proteases of specific organisms, an overview of documented and theoretical functions for parasite proteases focuses attention on the most active research directions in the field. A current, but by no means exhaustive, list of potential functions includes:

1. Invasion of the host facilitated by catalyzing degradation of connective tissues.
2. Metabolism within the host taking advantage of specific, abundant host proteins in the environment where the parasite resides.
3. Immune evasion or modulation by degradation or activation of host immune molecules.
4. Interaction with host blood coagulation or fibrinolytic systems that themselves are composed of protease-catalyzed reaction cascades.
5. Parasite remodeling during transition of one morphologic stage to another.
6. Activation or turnover of parasite peptide hormones, enzymes, or regulatory proteins.
7. Degradation of host cytoskeletal proteins during invasion or rupture of host cells.

Although the host-parasite relationship is very intricate and highly evolved, one must be cautious in attributing functions such as those listed above to proteases identified in extracts of parasitic organisms. For example, proteases are often identified that will cleave immunoglobulin molecules in an in vitro assay. While this is an intriguing theoretical possibility, it by no means confirms that immunoglobulins are the natural substrate for the enzyme. Enzymes such as pepsin and papain can cleave collagen, fibrinogen, immunoglobulins, and hemoglobin, but undoubtedly did not evolve to catalyze cleavage of these molecules in vivo. How, then, can protease function be documented? In the case of parasitic protozoa, rigorous genetic analysis of function has only recently become possible. The development of vectors that allow gene transfer into protozoa has enabled investigators to analyze function by transformation with a protease gene, or knockout of a protease gene (60, 247). To be convincing, such studies would have to analyze genetically altered organisms in an animal model of infection. The limitations of such a genetic approach might include the copy number of the target gene and the possibility that induced mutations might be lethal, and therefore provide less direct information on function.

The development of specific and irreversible protease inhibitors for at least some of the classes of proteases allows design of experiments in which inhibition of the enzyme can be correlated with a specific function in vitro or in vivo. Limitations to this approach are the availability of specific and effective inhibitors for the protease in question and the pharmacokinetics of these inhibitors in an animal model of infection.

The following discussion of individual protozoa emphasizes recent studies. Several excellent previous reviews on the proteases of protozoa should be consulted for more detail and historical information (55, 137, 171, 175).

PROTEASES AS VIRULENCE FACTORS IN *ENTAMOEBA HISTOLYTICA*

Entamoeba histolytica, the causative agent of amebiasis, has been the object of intense investigation and controversy concerning whether pathogenicity is genetically predetermined (152). Only 10% of the hundreds of millions of infected individuals manifest signs and symptoms of invasive amebiasis—i.e. colitis, dysentery, and amebic abscess (200). With the development of axenic laboratory cultures of *E. histolytica,* several factors have been identified as contributing to the virulence of isolated strains (201), including proteases of the thiol (cysteine) protease and metalloprotease classes. Detailed historical reviews of these proteases are available (108, 200).

The Major Protease Activity of E. histolytica: *a Thiol Protease*

Two independent lines of investigation identified a thiol protease with broad pH optima and cathepsin B–like substrate specificity as the major intracellular and extracellular protease of *E. histolytica.* Several investigators had hypothesized that invasion and tissue destruction by virulent strains of *Entamoeba* required expression and release of a histolytic protease (153, 166). In independent investigations, several groups attempted to identify factors responsible for the cytotoxic effect of trophozoites on tissue-culture monolayers (84, 138, 140, 150). By 1980 it was clear that two different mechanisms were operating to produce cytotoxicity. One was a contact-dependent cytolysis (188, 199, 202, 203) that has recently been attributed, at least in part, to an ion channel (amebapore) inserted in the plasma membrane of target cells (87, 119, 259). The second mechanism of cytotoxicity involved a secreted factor from trophozoites. This activity was inhibited by protease inhibitors (84, 138, 140, 150). Gadasi and coworkers, and Lushbaugh and coworkers, subsequently demonstrated that this cytotoxin copurified with cysteine protease activity and that the greater cytotoxic effect of HM-1 strain amebae versus the HK-9 strain was associated with greater protease activity (84, 140). While some overlap of terminology still occurs, most investigators now reserve the term cytotoxic effect for the contact-dependent cytolysis of target cells, and cytopathic effect for the effect of the extracellular (secreted) cysteine protease that causes host cells to round up and detach without necessarily being lysed.

When the major cysteine protease of *E. histolytica* was purified to homogeneity, Keene et al (110) and Luaces & Barrett (136) independently confirmed that the purified protease could produce a cytopathic effect. Keene and collaborators extended these observations to show that a specific, irreversible inhibitor of the cysteine protease could prevent destruction of tissue culture monolayers by live trophozoites (109).

The strong correlation between the level of protease activity and the virulence of laboratory strains underscored the importance of cysteine protease activity in *E. histolytica* infections (84, 110, 136, 138). This correlation was extended to actual clinical isolates by Reed et al (205), who correlated the level of protease activity with the clinical severity of disease in 20 patients. They also noted that patients with invasive disease produced antibodies to the cysteine protease, while those with noninvasive infections did not (205).

E. histolytica cysteine protease activity has a pH optimum between 6 and 7, depends on reducing agents, is inhibited by thiol protease class inhibitors, and exhibits multiple isoforms ranging in M_r from 16,000 to 66,000 (110, 136, 139, 181, 185, 230). To date, the etiology of the multiple cysteine protease species is still obscure. There are three distinct cysteine protease genes (see below), but each codes for a protease of subunit M_r 23,000–27,000. None of the gene products are glycosylated. Luaces & Barrett (136) suggested the higher M_r species may be multimers.

The substrate specificity of the major cysteine protease activity of *E. histolytica* is cathepsin B–like, with a preference for dipeptides containing arginine at both the P1 and P2 positions (110, 136, 182, 229). Like the lysosomal cysteine proteases of mammalian cells, the ameba cysteine protease had a broad substrate specificity, with activity against casein, gelatin, insulin, and—of more biologic import—type I collagen, fibronectin, and laminin (110, 136, 228, 231, 234). Several investigators have noted the fascinating parallels between release of a cathepsin B–like protease from invasive amebae and similar enzymes found in the extracellular milieu of invasive tumor cells (137, 204).

Recently, genes coding for cysteine proteases of *E. histolytica* were isolated in two different laboratories (204, 244, 245). Three genes have been identified. Two are 80–85% identical and occur on the same chromosome-sized DNA fragment on Southern blot analysis of field inversion gel electrophoresis of *E. histolytica* DNA (204). A third gene is only 40% identical to the other two and occurs on two different chromosome-sized DNA fragments. While the two research groups reached different conclusions about which genes are associated with pathogenic versus nonpathogenic isolates of *E. histolytica,* both groups agreed that mRNA levels for cysteine proteases are significantly higher in pathogenic strains of amebae. Data from our group suggests that this is a result of gene dosage (204).

None of the gene products appear to be glycosylated. Nevertheless, they all appear to be exported from the endoplasmic reticulum to endosome-like vesicles in the ameba cytoplasm. Targeting to these endosome structures cannot be by the same pathways followed in higher eukaryotes where a mannose-6 phosphate carbohydrate signal is used. From the endosome, cysteine protease can be released into the extracellular milieu by exocytosis. The rate of release of cysteine protease from trophozoites of pathogenic *Entamoeba* is virtually identical to that of exocytosis of fluoroscein-conjugated dextran (M. Brown & J. McKerrow, unpublished data). These observations also support the hypothesis of Orozco that enhanced phagocytosis and virulence are correlated (180). The higher the phagocytic rate, the more cysteine protease would be released by exocytosis. Pathogenic amebae may have evolved because of enhanced phagocytosis, giving them a competitive edge in intraluminal replication. While tissue destruction and invasion are the consequence of this activity, invasive organisms are lost because they never encyst and leave the body like their intraluminal counterparts.

The E. histolytica *Metallocollagenase*

A second proteolytic enzyme proposed to play an important role in the invasion of virulent *E. histolytica* trophozoites is a surface-associated or membrane-associated metallocollagenase. Collagenase activity was first detected in assays with whole live trophozoites by Muñoz et al (163). More recently, electron microscopy showed that incubation of trophozoites with collagen in vitro induces the formation of electron-dense granules. These granules accumulate at the trophozoite plasma membrane and are released into the extracellular milieu (164). The collagenase activity was inhibited by EDTA, an inhibitor of metalloproteases, but not soybean trypsin inhibitor, phenyl methyl sulfonyl fluoride (PMSF), or N-ethyl maleimide (NEM). However, EDTA also had a profound effect on viability of organisms in the in vitro experiments. To date, the structural and biochemical analysis of the metallocollagenase remains very preliminary. A confounding problem in analysis of this enzyme is that it appears to be present in relatively low concentrations compared with the cysteine protease discussed above, which also has collagenolytic activity. Nevertheless, degradation of collagen by either or both enzymes strongly correlates with virulence (248).

THE CYSTEINE PROTEASE OF *TRYPANOSOMA CRUZI*

Trypanosoma cruzi is the causative agent of Chagas' disease (American trypanosomiasis). Chagasic heart disease is estimated to affect over 16 million persons in Central and South America, and over 90 million are at risk (88). *T. cruzi* is the leading cause of heart disease in South America. As with other protozoan parasites, numerous protease species have been detected (6, 22, 24, 25, 91, 198), but the major protease activity of *T. cruzi* is a cysteine protease (25). Unlike that of *E. histolytica,* the cysteine protease of *T. cruzi* (cruzain, cruzipain) is not, strictly speaking, a virulence factor but rather is required for intracellular replication and differentiation of the parasite.

Historically, this enzyme became the focus of three independent lines of research. In the first, Cazzulo and colleagues in Buenos Aires were studying the proteases of *T. cruzi* for their possible role in parasite metabolism and cellular invasion. They identified and subsequently purified the major protease, a cysteine protease with a cathepsin L–like preference for dipeptide substrates containing phenylalanine and arginine (25, 42–44, 194).

In parallel, Scharfstein and colleagues at FIOCRUZ in Rio de Janeiro were studying the major parasite antigens recognized during a human infection. They identified a glycoprotein designated GP57/51 as a major antigen and promising serodiagnostic reagent (221, 222).

We were developing techniques to allow cloning of cysteine protease genes by homology-based PCR. In the course of this work, we included a sample of *T. cruzi* genomic DNA and cloned a cysteine protease gene, the coding sequence of which contained an amino terminal sequence identical to that identified by Cazzulo and coworkers as the major protease and later by Scharfstein and coworkers as GP57/51 (43, 69, 144, 165).

More recently, several studies have elucidated structural information about cruzain, clarifying some issues regarding isoforms (72, 145), but at the same time raising intriguing questions about processing, localization, and function. The cruzain gene codes for a protein that includes the expected preproenzyme form as well as an unusual carboxy-terminal extension of approximately 145 amino acids (8, 72). The prepro form is similar to that found in other members of the papain superfamily, but the carboxy-terminal extension is relatively rare, having been identified only in other trypanosomatid cysteine proteases, including those of *Trypanosoma brucei* and *Leishmania* spp., as well as in a cysteine protease gene from the tomato (151, 173). The C-terminal domain is characterized by an unusual repeat of six threonines occurring shortly after the amino terminus of the mature protein. A series of studies by Eakin et al utilizing expression of recombinant protease in *Escherichia coli* to examine mutants with and without the C-terminal domain, showed that the C-terminal domain is not required for protein folding or activity (71, 72). Cazzulo and coworkers presented evidence that the C-terminal domain may be autocatalytically removed by the protease (8). Eakin has also suggested that dimerization may be occurring between two C-terminal domains to produce a covalently linked dimer. To date, the most likely function for the C-terminal domain is to target the protease within the parasite cell and/or anchor it to a cell membrane. However, the location of the protease in different stages of the parasite is still somewhat controversial. Evidence strongly supports targeting to an endosome-like vesicle in some stages of the parasite and possibly to the surface of the organism in other stages (26, 240).

Several functions for the cysteine protease have been proposed, including intracellular metabolism, parasite-macrophage interaction, host cell invasion, and adhesion to target cells (41, 189, 240). Use of specific, irreversible inhibitors of the protease has provided some clues to the function of cruzain. Although all stages of the parasite life cycle are to some extent susceptible to inhibition of the cysteine protease, transformation events appear to be particularly vulnerable. Low micromolar concentrations of fluoromethyl ketone-derivatized or diazomethane-derivatized peptides can block metacyclogenesis, trypomastigote-to-amastigote transformation, and subsequent amastigote-to-trypomastigote transformation (7, 22, 99, 155). These events mark the development of infective forms, initial replication of the parasite within the host cell, and release of new infective forms, respectively. The efficacy of these inhibitors against parasite replication in vitro has raised optimism that the protease may be a target for new chemotherapy. Preliminary studies have indicated that the inhibitors that are effective against the parasite in vitro are relatively nontoxic in animals in concentrations necessary to affect parasite replication (J. Engel, unpublished data).

Further studies with fluoromethyl ketone-derivatized peptides, as well as prediction of new nonpeptide inhibitors by computer modeling, are ongoing. An X-ray crystallographic data set for cruzain has been collected that should facilitate structure-function analyses (71). Because of the seeming ubiquitous importance of cysteine proteases in protozoan parasites, the development of inhibitors with a potential in vivo use may have far-ranging applications. Aside from the issue of new chemotherapy, the use of specific protease inhibitors in vivo should provide new clues to the function of proteases in the host-parasite relationship.

PROTEASES FROM *LEISHMANIA*

Leishmaniasis is classified by the World Health Organization (WHO) as one of the six major parasitic diseases. Over 12 million people worldwide are infected by the different species of the protozoan parasite *Leishmania*. The disease has a wide spectrum of clinical manifestations, varying from self-healing cutaneous ulcers to nonresolving mucocutaneous lesions and hepatosplenomegaly leading to death.

The Promastigote Surface Protease

Promastigotes of Old and New World species of *Leishmania* express at high density (5 × 10^5 molecules/cell) a surface membrane glycoprotein, known as gp63 (27, 30, 47, 49, 75, 120, 177). This protein is bound to the membrane by a glycosyl-phosphatidylinositol (GPI) anchor (28, 76, 224) similar to that which attaches a wide variety of proteins to membranes (59, 79, 135). The protein is present on the promastigotes residing in the midgut of the phlebotomine sandfly vector (62, 93) and has been detected at the surface of all *Leishmania* species examined so far (31, 73). Recently, this abundant glycoprotein was shown to be a zinc metalloprotease (29, 50, 107) and is now also named by some investigators the promastigote surface protease (PSP). Its predominant β-structure, determined by Raman spectroscopy (105) and circular dichroism (29), distinguishes it from the well-characterized metalloendopeptidase thermolysin (121). With the exception of 1,10-phenanthroline and zinc, no inhibitors, including α_2-macroglobulin, have any effect on the activity of the membrane-bound enzyme (77, 100). Surface metalloprotease activity is not only a highly conserved feature of the genus *Leishmania* but also occurs at the surface of the monogenetic trypanosomatids *Crithidia* and *Herpetomonas* spp. (73, 226). Peptide-mapping analyses showed that these related proteases are structurally conserved among the recognized species of *Leishmania* (52, 78).

The complete nucleotide sequence for the protease has been deduced (38, 39, 156, 242, 254) and the enzyme has been shown to be encoded by a family of tandemly linked genes, all of which map to a single chromosome (40). Although messenger RNA was reported to be constitutively transcribed in both stages of the parasite (40, 255) and a polypeptide crossreacting with PSP was detected in amastigotes of several *Leishmania* species (50, 82, 154), information about the proteolytic activity, amount, and properties of the amastigote PSP is still controversial. Distinct RNA populations for the enzyme have been shown to be differentially expressed during development of promastigotes to an infectious form (195). PSP is down-regulated 300-fold in the amastigote of *Leishmania major,* the intracellular form infecting the host macrophages (227). This observation could confirm previous reports that PSP is a promastigote stage-specific enzyme (80). Its involvement in the early phases of infection as a ligand for the mannosyl-fucosyl receptor, as an acceptor for C3b deposition, or as the major surface antigen (47, 157, 192, 219) has led to considerable effort to use the surface metalloprotease as a molecularly defined vaccine. Indeed, intraperitoneal vaccination of inbred mice with purified *Leishmania mexicana* PSP with Freund's complete adjuvant confers significant protection to CBA mice upon challenge infection (218); however, attempts to protect BALB/c mice with recombinant *L. major* PSP were unsuccessful (97). More recently, synthetic peptides of predicted T-cell epitopes of *L. major* PSP were shown to protect mice from subsequent low-dose challenge infection with both *L. major* and *L. mexicana* promastigotes (106). However, T-cells from human cutaneous leishmaniasis patients do not recognize purified PSP in vitro (104), and the immune response to PSP does not correlate with susceptibility of inbred mouse strains to cutaneous infection with *L. mexicana* (134).

Despite PSP being one of the best characterized surface molecules of *Leishmania* (74), its precise role in the life cycle of the parasite remains under speculation. The peptide substrate specificity of PSP was reported to be essentially defined by the P′ subsites of the substrate (32, 103), as shown for other metalloendopeptidases (143, 179, 191). The kinetic parameters obtained with the model peptide substrate, H_2N-L-I-A-Y-L-K-K-A-T-COOH, constituted the first step toward the rational design of a fluorogenic substrate for PSP based on the radiationless energy transfer (RET) methodology (33). This approach, which has been applied successfully to the design of substrates for several enzymes (113, 168, 243) to explore their subsite specificities, provides a highly sensitive assay for the identification of novel inhibitors for the enzyme. These could be used to investigate the precise role of PSP in the life cycle of *Leishmania* and may eventually serve as model compounds for the chemotherapy of leishmaniasis, should the enzymatic activity be shown to be crucial for survival of the parasite in the vertebrate host.

Cysteine Proteases

Cysteine proteases have been identified in several species of *Leishmania* (175, 207). Several studies showed that the leishmanicidal activity of peptide esters results from their hydrolysis by cysteine proteases (3, 4, 85, 193, 196, 197), and specific inhibitors of this class of enzymes arrest growth of the parasites (56). These inhibitors and esters, however, are toxic for monocytes and lymphoid cells, thus precluding their therapeutic use. The cysteine protease inhibitor, Z-Phe-Ala-fluoromethyl ketone (Z-F-A-FMK), has been shown to arrest the cytopathic effect of *E. histolytica* (109, 110) and the development of *Plasmodium falciparum* (212, 216) and *T. cruzi* (72, 99). Several fluoromethyl-ketone peptide inhibitors were tested on the promastigotes and amastigotes of *L. major* in vitro. The effectiveness of these inhibitors in killing both stages of the parasites was determined, and results showed that two fluoromethyl ketone-derivatized peptides were effective in killing the parasites but not host cells (J. Bouvier & J. A. Sakanari, unpublished data).

In contrast to the promastigote surface protease, several biochemically distinct cysteine proteases are expressed in the amastigote form (207, 208,

227). At least two different genes encoding cysteine proteases have been cloned and analyzed (160, 241), but biochemical analyses suggest that there are at least three distinct proteases (J. Bouvier & J. A. Sakanari, unpublished data). Recently, two soluble exopeptidases, an aminopeptidase and a carboxypeptidase, were characterized in *L. major;* they have been shown to be present in both amastigote and promastigote stages of the parasite (225).

MALARIAL PROTEASES

Malaria is the most important protozoan infection worldwide, with hundreds of millions of cases and approximately one million deaths reported annually (253). The erythrocytic malarial life cycle, which is responsible for all of the clinical manifestations of malaria, is initiated by the invasion of host erythrocytes by free merozoites. The intraerythrocytic parasites then develop from small ring-stage organisms to larger, more metabolically active trophozoites and then to multinucleated schizonts. The erythrocytic cycle is completed when mature schizonts rupture erythrocytes, releasing numerous invasive merozoites. Malarial proteases are likely required during both the invasion of erythrocytes by merozoites and the rupture of erythrocytes by mature schizonts, as during both events erythrocyte cytoskeletal proteins must be hydrolyzed and several parasite proteins are proteolytically processed. Parasite proteases also appear to mediate the hydrolysis of globin into free amino acids by erythrocytic parasites. Nonerythrocytic stages of malaria parasites (sporozoites, which are injected by the bite of an infected mosquito; exoerythrocytic liver stages; sexual gametocyte stages; and all mosquito stages) are difficult to study and no specific information regarding proteases of these stages is yet available.

Numerous reports have described neutral and acidic proteases of human (*Plasmodium falciparium*), monkey (*Plasmodium knowlesi*), rodent (*Plasmodium berghei, P. chabaudi, P. yoelii, P. vinckei*) and avian (*Plasmodium lophurae*) malaria parasites. Older studies (46, 53, 54, 122–125, 161, reviewed in 237) are often difficult to interpret, because of inconsistencies with older proteolytic assays and the difficulty of distinguishing host and parasite enzyme activities. Most studies reported since 1980 have attempted to control for the activities of host enzymes, and this discussion concentrates on these more recent studies (Table 1), with emphasis on the proposed functions of the malarial proteases that have been identified (Table 2).

Proteases Mediating Erythrocyte Invasion and Rupture

The process of erythrocyte invasion includes merozoite binding to erythrocyte receptors, the alignment of the apical end of the merozoite (containing the rhoptry and microneme organelles) toward the erythrocyte membrane, the

Table 1 Properties of malarial proteases[a]

Protease class	*Plasmodium* species	Size (M_r)	pH Optimum	Life cycle stage[b]	Reference
Serine	*falciparum*	75,000	7.0	M	211
	falciparum	76,000	7.4–8.3	S/M	36, 37
	chabaudi	68,000	7.5	S/M	35
Cysteine	*berghei, chabaudi*	68,000	7.4	S/M	17, 18, 232
	falciparum	68,000	7.4	S/M	92, 146, 233
	falciparum	35,000–40,000	7.0	S	211
	falciparum	28,000	5.5–6.0	T	211, 213–216
	vinckei	27,000	4.0–6.0	NR	212
Aspartic	*lophurae*	37,000	3.5	NR	237, 238
	falciparum	148,000	3.5	NR	95
	yoelii	50,000	3.0	NR	2
	falciparum	<10,000	4.5	T/S	250
	falciparum	36,000–38,000	4.5–5.0	T	251, 252
	berghei	18,000–20,000	3.2	NR	220
	falciparum	40,000	5.0	T	89
	falciparum	55,000	3.9	NR	10
Aminopeptidase	*yoelii, chabaudi*	90,000	7.0	NR	48
	falciparum	186,000	7.5	NR	95
	falciparum	63,000	7.5	T/S	248, 251
Unknown	*falciparum, berghei*	37,000	5.0	NR	64

[a] Malarial proteases reported since 1980. Values for M_r and pH optimum may not always be equivalent, as assay conditions have varied among the different studies. When the pH optimum for proteolytic activity was not determined, the pH at which the enzyme was studied is reported.
[b] T, trophozoite; S, schizont; M, merozoite; NR, not reported.

secretion of rhoptry and microneme contents, and the entrance of the parasite (within a parasitophorous vacuole) into the erythrocyte cytoplasm (186). The mechanism of rupture of the erythrocyte by mature schizonts is poorly understood, but the secretion of a portion of rhoptry and microneme contents has been observed at the time of rupture in *P. knowlesi* (12). During both erythrocyte invasion and rupture, the erythrocyte cytoskeleton serves as an important barrier to malaria parasites in addition to the erythrocyte plasma membrane. Most likely, the rhoptry and microneme contents that are released during erythrocyte invasion and rupture include proteases that degrade cytoskeletal proteins.

Several investigators have studied the effects of peptide protease inhibitors on the invasion and rupture of erythrocytes by malaria parasites. Available evidence suggests roles for both serine and cysteine proteases in these processes, though distinguishing the two processes is difficult with *P. falciparum*, as invasive merozoites have not been successfully isolated from

Table 2 Proteolytic functions of malaria parasites[a]

Function	Protease class	Life-cycle stage	Candidate proteases (M_r)[b]	Reference
Erythrocyte invasion	Serine	Schizont/Merozoite	75,000	211
Erythrocyte rupture			76,000	35–37
	Cysteine	Schizont/Merozoite	68,000	17, 18, 92, 146, 232, 233
			35,000–40,000	211
Hemoglobin degradation	Aspartic	Trophozoite	37,000	237, 238
			148,000	95
			50,000	2
			<10,000	250
			18,000–20,000	220
			40,000	89
			55,000	10
			36,000–38,000	251, 252
	Cysteine	Trophozoite	28,000	211–216

[a] Malarial proteases with potential functions suggested by their catalytic class and the life cycle stage at which they are active.

[b] When equivalent proteases have been isolated from different species, the size of the *P. falciparum* protease is shown.

this species. Studies with isolated viable merozoites of other malaria species have been helpful in this regard. Chymostatin (an inhibitor of serine proteases), but not leupeptin (an inhibitor of cysteine proteases and trypsin-like serine proteases) (96) in *P. knowlesi,* and several serine protease inhibitors (35) in *P. chabaudi,* specifically blocked the invasion of erythrocytes by isolated merozoites. Pretreatment of human erythrocytes with the serine protease chymotrypsin (68) or pretreatment of murine erythrocytes with a *P. chabaudi* serine protease of M_r 68,000 (35) reversed the effects on erythrocyte invasion of serine protease inhibitors, further suggesting that serine protease activity is required for invasion. Cysteine protease activity may also be required, as peptide inhibitors of a *P. falciparum* M_r 68,000 cysteine protease (146) and specific inhibitors of calpain (178) appeared to inhibit invasion in *P. falciparum*. Other investigators also showed that in *P. knowlesi* (13) and *P. falciparum* (65, 68, 142) chymostatin and leupeptin blocked erythrocyte rupture and/or invasion. Considering the process of erythrocyte rupture in both *P. knowlesi* (96) and *P. falciparum* (142), the incubation of parasites with either chymostatin or leupeptin was followed by the accumulation of erythrocytes containing unreleased merozoites. In these erythrocytes schizonts apparently completed development into merozoites, but the rupture of the erythrocyte cytoskeleton and/or membrane required to free the merozoites was blocked by the serine and cysteine protease inhibitors.

Multiple proteins of mature schizonts and merozoites are proteolytically processed immediately before or during erythrocyte rupture and invasion (reviewed in 36), suggesting that proteolytic fragments have roles in these processes. Serine protease inhibitors inhibited the processing of both the M_r 230,000 major surface antigen of *P. knowlesi* schizonts and the M_r 190,000 *P. falciparum* analogue of this protein (also known as merozoite surface protein 1) (20, 61). Leupeptin altered the processing of P126, a parasitophorous vacuole protein of *P. falciparum* schizonts (63). These results suggest that serine and/or cysteine proteases may be involved in the processing of schizont/merozoite proteins.

Several proteases have been identified as potentially involved in erythrocyte invasion and rupture by malaria parasites. A 68-kDa cysteine protease was identified in schizonts and merozoites of *P. berghei, P. chabaudi,* and *P. falciparum* (17, 18, 92, 146, 232, 233). Antisera directed against this protease localized primarily to the merozoite apex (18) suggest that the protease might be released from the rhoptry organelle during invasion. A cysteine protease (M_r 35,000–40,000) of mature schizonts and a serine protease (M_r 75,000) of merozoites were identified in highly synchronized *P. falciparum* parasites (211). The stage-specific activities of these proteases suggest that they may also be involved in erythrocyte invasion or rupture. A 76-kDa serine protease of *P. falciparum* schizonts and merozoites was shown to be bound in an inactive form to the schizont/merozoite membrane by a glycosyl-phosphatidyl inositol anchor and to be activated by phosphatidylinositol-specific phospholipase C during the merozoite stage (37), suggesting a role in erythrocyte invasion. A 68-kDa *P. chabaudi* analogue of the schizont/merozoite serine protease that was also activated by phosphatidylinositol-specific phospholipase C cleaves the erythrocyte cytoskeletal protein band 3 (35). Also, a protease of *P. falciparum* and *P. berghei* (M_r 37,000) that was inhibited by both chymostatin and leupeptin, but was not easily categorized as to catalytic class (64), hydrolyzed erythrocyte cytoskeletal proteins (spectrin and band 4.1), suggesting potential roles in both erythrocyte invasion and rupture.

The gene encoding a serine-rich *P. falciparum* schizont antigen (referred to independently as SERA, SERP, P126, P113, and Pf140) (reviewed in 66) that is located on the parasitophorous vacuole (67) and induces partial protective immunity in Saimiri monkeys (187) was recently sequenced (111, 126). This gene encodes a 111- to 113-kDa protein, a portion of which was subsequently shown to have limited sequence homology to cysteine proteases (101), particularly near highly conserved active-site residues. However, a serine residue replaced the predicted active-site cysteine (70, 158). Subsequently, a related gene [*SERP H* (112)] was characterized; this gene encodes a 130-kDa parasitophorous vacuole protein with sequence similarity and antigenic crossreactivity with SERA. Interestingly in *SERP H,* the catalytic cysteine is conserved. Whether SERA or *SERP H* have protease activity is unknown, but the identification of these proteins on the parasitophorous vacuole of schizonts suggests potential roles for these perhaps multifunctional proteins in the proteolysis of erythrocyte cytoskeletal or parasite proteins. SERA is itself processed in a manner that is in part inhibited by leupeptin (63), suggesting that hydrolysis (perhaps autohydrolysis) may be required for protease activity.

Proteases Mediating Hemoglobin Degradation

Extensive evidence suggests that the degradation of host erythrocyte hemoglobin is necessary for the growth of intraerythrocytic malaria parasites. These parasites apparently require hemoglobin as a source of free amino acids, as they have a limited capacity for synthesizing amino acids and the quantity of free amino acids within erythrocytes is not sufficient for parasite needs (223). The following evidence suggests that malaria parasites degrade hemoglobin into free amino acids: (*a*) the hemoglobin content of infected erythrocytes decreases 25–75% during the life cycle of erythrocytic parasites (11, 94, 217), (*b*) the concentration of free amino acids is greater in infected than uninfected erythrocytes (235), (*c*) the composition of the amino acid pool of infected erythrocytes is similar to the amino acid composition of hemoglobin (15, 45, 260), and (*d*) the infection of erythrocytes containing radiolabeled hemoglobin is followed by the appearance of labeled amino acids in parasite proteins (83, 236, 246).

In *P. falciparum,* hemoglobin degradation occurs predominantly in trophozoites and early schizonts, the stages at which the parasites are most metabolically active (148, 258). Trophozoites ingest erythrocyte cytoplasm via a specialized organelle known as the cytostome (and possibly also via pinocytosis), and then transport the cytoplasm within vesicles to a large central food vacuole (1). In the food vacuole hemoglobin is broken down into heme, which is a major component of malarial pigment (239), and globin, which is hydrolyzed to its constituent free amino acids. The food vacuole is an acidic organelle (117, 257) that appears to be analogous to lysosomes (116). Several lysosomal proteases are well characterized, including cysteine (cathepsins B, H, L) and aspartic (cathepsin D) proteases (23), and malaria parasites apparently contain analogous food vacuole proteases that degrade hemoglobin.

Investigations have identified soluble aspartic proteases of M_rs 148,000 (95), <10,000 (251), 40,000 (89), and 55,000 (10) of *P. falciparum*; 37,000 of *P. lophurae* (238); 50,000 of *P. yoelii* (2); and 18,000–20,000 of *P. berghei* (220). In addition, three related membrane-associated aspartic proteases of M_r 36,000–38,000 were recently isolated from *P. falciparum* food vacuole preparations (252, 253). The aspartic proteases all had acid pH optima and degraded denatured hemoglobin. In a more biologically relevant assay, the

40-kDa *P. falciparum* aspartic protease cleaved purified hemoglobin of apparently native conformation at a specific site on the α-globin chain (89), suggesting that this proteolytic cleavage initiates the hydrolysis of hemoglobin. The gene encoding the M_r 40,000 aspartic protease has recently been identified and partially sequenced; the sequence predicts a protein with ~40% amino acid identity with human renin (D. E. Goldberg, personal communication).

In several studies, incubation with leupeptin caused *P. falciparum* trophozoite food vacuoles to fill with apparently undegraded erythrocyte cytoplasm (68, 213, 250), suggesting that the inhibitor allowed normal transport of erythrocyte cytoplasm to the food vacuole but blocked the degradation of cytoplasmic components. Analysis of the leupeptin-treated parasites showed that they contained large quantities of undegraded globin, while globin was undetectable in control parasites (213). Leupeptin inhibits both cysteine and some serine proteases, but the highly specific cysteine protease inhibitor E-64 also caused the accumulation of undegraded globin (9, 213), and globin accumulation was not seen after incubation with inhibitors of other classes of proteases including chymostatin and pepstatin (9, 68, 213, 250). These results suggest that a cysteine protease is required for an initial step in hemoglobin degradation by *P. falciparum*.

A *P. falciparum* trophozoite cysteine protease that has the biochemical features expected for a food vacuole hemoglobinase has been identified (211). This M_r 28,000 protease degraded hemoglobin in vitro (213), and its size, acid pH optimum, substrate specificity, and inhibitor sensitivity were all very similar to those of cathepsin L (213, 214). Specific inhibitors of cathepsin L also inhibited the M_r 28,000 protease, blocked hemoglobin degradation, and prevented parasite development. The degree of inhibition of the cysteine protease correlated with the degree of inhibition of hemoglobin degradation and parasite development, supporting the hypothesis that the M_r 28,000 protease is the cysteine protease required for hemoglobin degradation (214, 216). A *P. falciparum* cysteine protease gene was recently characterized (215), and the deduced gene product predicted to represent the active enzyme (after cleavage of a proform) has 37% amino acid identity to cathepsin L and appears to be the gene encoding the 28-kDa trophozoite cysteine protease. In an animal model system, a *P. vinckei* cysteine protease (M_r 27,000) that has similar biochemical properties to the *P. falciparum* cysteine protease has been identified (212), and a *P. vinckei* gene that predicts a mature cysteine protease that is 56% identical to the *P. falciparum* cysteine protease has been characterized (P. J. Rosenthal, unpublished observation).

Two groups have reported the purification of food vacuoles of *P. falciparum* by density gradient centrifugation (51, 90). In one study, the food vacuoles had ATPase activity, as expected for an acidic organelle (51). In the other study, extracts of purified food vacuoles contained acid hemoglobinase activity that was inhibited by both aspartic and cysteine protease inhibitors (90). The degradation of hemoglobin by food vacuole extracts was blocked 67% by the aspartic protease inhibitor pepstatin and 32% by the cysteine protease inhibitor E-64. These results were similar to those obtained when the degradation of hemoglobin by soluble trophozoite extracts was studied [71% inhibition by pepstatin; 50% inhibition by E-64 (213)]. With food vacuole extracts, however, pepstatin most effectively blocked initial cleavages of hemoglobin, suggesting that an aspartic protease is responsible for the initial cleavage of hemoglobin. These results appear to conflict with the studies noted above showing that cysteine protease inhibitors completely block the hydrolysis of globin (213). The conflicting conclusions might be explained by the requirement for one class of proteases to be processed by the other for activation. The activation of the lysosomal cysteine proteases cathepsin B and cathepsin L has, in fact, been shown to be blocked by pepstatin (170). In any event, though the precise steps involved in hemoglobin degradation in the food vacuole remain unclear, as in lysosomes, both aspartic and cysteine proteases are probably active and both classes of enzymes are probably required for the complete hydrolysis of hemoglobin.

The complete hydrolysis of globin to free amino acids most likely requires exopeptidase activity in addition to the aspartic and cysteine proteases described above. Knowledge of malarial exopeptidases is limited, but neutral aminopeptidases have been reported in murine parasites (48) and *P. falciparum* (95, 249, 252).

Malarial Proteases as Potential Chemotherapeutic Targets

The cysteine protease inhibitor E-64 and the aspartic protease inhibitor pepstatin have both been shown to block *P. falciparum* development (9, 68, 213, 250). In a recent study, the two inhibitors acted synergistically, though E-64 appeared to act primarily on trophozoites and schizonts, and pepstatin primarily on rings and schizonts (9). Only E-64 specifically blocked globin hydrolysis (9, 213, 250). Numerous peptide fluoromethyl ketone inhibitors of cysteine proteases inhibited *P. falciparum* growth at nanomolar-to-micromolar concentrations (210, 214, 216). In a comparison of seven fluoromethyl ketone inhibitors, Z-Phe-Arg-CH_2F was the most effective inhibitor of the M_r 28,000 *P. falciparum* cysteine protease (IC_{50} = 0.36 nM); it also most effectively blocked trophozoite hemoglobin degradation (blocked at 100 nM) and most effectively killed cultured parasites (IC_{50} = 64 nM) (216). Z-Phe-Arg-CH_2F was also nontoxic to cultured mammalian cells at concentrations up to 100 μM (216). In an animal-model system, a related fluoromethyl ketone, morpholine urea-Phe-HPhe-CH_2F, strongly inhibited the M_r 27,000 *P. vinckei* cysteine protease (IC_{50} = 5.1 nM), blocked *P. vinckei* protease activity in vivo after a single subcutaneous dose, and when administered for four days, cured 80% of murine malaria infections (212). Thus, despite the theoretical limitations of potentially rapid degradation in vivo and inhibition of host proteases, peptide protease inhibitors appear to be promising candidates as antimalarial drugs.

In order to identify nonpeptide inhibitors, the structure of the *P. falciparum* trophozoite cysteine protease was recently modeled based on the predicted protease sequence (215), and the known structures of other papain-family proteases (206) and potential inhibitors were identified from the Fine Chemicals Directory with computer modeling techniques (118). Screening of potential nonpeptide inhibitors has identified a low micromolar lead compound (206). Ultimately, a better understanding of the biochemical properties and biological roles of malarial proteases will foster the development of protease inhibitors that specifically inhibit parasite enzymes and thus are the most suitable candidates for chemotherapy.

PROTEASES OF *TRYPANOSOMA BRUCEI*

Trypanosoma brucei is the etiologic agent of African sleeping sickness, a disease transmitted by the tsetse and characterized by waves of parasitemia and progressive meningoencephalitis that is ultimately fatal if untreated. An estimated 20,000 new cases are reported each year (256); control of the disease is largely hampered by the toxicity of current drugs. The organism differs from the related protozoans *T. cruzi* and *Leishmania* species in that it remains extracellular throughout its life cycle, replicating and persisting in the bloodstream and lymphatic system.

The major proteolytic activity in *T. brucei* results from a cysteine protease (131, 133, 174, 209). On SDS-polyacrylamide gels copolymerized with fibrinogen or gelatin, the predominant activity is observed at 28 kDa, with optimal activity at pH 5–6. Activity is enhanced by thiol reducing agents and is inhibited by the cysteine protease inhibitors E-64 and leupeptin. The substrate specificity of the enzyme resembles that of cathepsin L, with preference for basic residues (arginine > lysine) at the P1 position, and large, hydrophobic residues (phenylalanine) at P2; the dipeptide substrate Z-Phe-Arg-AMC (Z, N-benzyloxycarbonyl; AMC, 7-amino-4-methylcoumarin) is customarily used in the study of this enzyme (132, 133, 174, 209).

A cysteine protease gene has been cloned and sequenced (159) and predicts a full-length protein of 450 amino acids. A 217–amino acid segment of the sequence has significant similarity to the active protease domains of the *Dictyostelium discoideum* cysteine protease, human cathepsin L, actinidin, and papain. The upstream sequence encodes a putative signal peptide and a 105-residue pro region that also has homology to the corresponding regions of other cysteine proteases. A 108-residue C-terminal extension contains a segment of nine consecutive prolines, as well as one of two potential *N*-glycosylation sites.

The biological role of the cysteine protease in the life cycle of *T. brucei* is unknown. Density gradient experiments (132) and immunolabeling studies (131) localize the protease to lysosomal organelles, suggesting a role as a catabolic enzyme analogous to the mammalian cathepsins. Proteolytic activity in the short, stumpy form of the parasite infective for the tsetse vector is increased compared with that of the long, slender form predominant in the human blood stream (147, 183), suggesting a possible role for this activity in differentiation or adaptation to the insect host.

A second, more recently characterized protease appears to be a cytosolic enzyme with optimal activity at pH 8, which is not inhibited by the cysteine protease-specific compounds E-64 and cystatin (102, 114, 132, 147, 209). Diisopropylfluorophosphate (DFP), leupeptin, and *N-a-p*-tosyl-L-lysine-chloromethylketone (TLCK) inhibit the enzyme, indicating that it is probably a serine protease. However, the enzyme is not inhibited by phenylmethylsulfonylfluoride, a specific serine protease inhibitor; thus, some ambiguity remains about its precise classification (102, 114, 132, 209). The enzyme has a molecular mass of 80,000 as determined using SDS-PAGE and 110,000 using gel filtration (114). The substrate preference for arginine at P1 resembles that of the cysteine protease but is broader at P2, where it is active against arginine, phenylalanine, and glycine substrates (114, 132). In contrast to the cysteine protease, activity on gelatin substrate gels is not detectable (102). The activity of this protease is two to three times higher in procyclic forms compared with bloodstream forms (147), but the significance of this and its physiologic role remains to be elucidated.

TRICHOMONAS VAGINALIS

Trichomonas vaginalis is a flagellated protozoan that is a major sexually transmitted cause of vaginitis and urethritis. The organism can infect the urogenital tract of both women and men, though in the latter it is usually asymptomatic. Adherence to the vaginal squamous epithelium elicits an acute inflammatory response with a subepithelial neutrophilic infiltrate and results in a characteristic purulent vaginal discharge (169). The pathogenesis of this disease process is unknown.

Compared with other protozoans, trichomonads have very high levels of proteolytic activity (172). These are primarily cysteine proteases, but non-cysteine proteases have been recently described as well (34). Many distinct protease activities can be resolved by substrate gel electrophoresis (57, 130, 176); as many as 23 different species appear on two-dimensional gels (167).

To what extent these actually represent separate gene products, as opposed to multimers or different modifications of a few gene products, is unclear, especially as the substrate specificities are quite similar (176). Two intracellular cysteine proteases have been partially purified, protease D and protease H with M_rs of 18,000 and 64,000, respectively (172). Two distinct secreted cysteine proteases have also been identified; one has a M_r of 60,000 with apparent subunits of 23,000 and 43,000, while the other has a M_r of 30,000 (86). In general, three groups of cysteine proteases can be identified on the basis of peptide substrate specificity: one that hydrolyzes Z-Arg-Arg-AMC, another that hydrolyzes Z-Phe-Arg-AMC, and a third group that hydrolyzes Boc-Val-Leu-Lys-AMC (176).

The role of proteases in the physiology of the organism or its pathogenicity is unknown. They may be involved in catabolizing exogenous proteins, given their association with subcellular compartments (128, 129) and the observation that the organism ingests particulate matter and bacteria, which are then digested in phagolysosomes (81). An interesting observation has been made relating protease activity to cytadherence. Pretreatment of parasites with the cysteine protease inhibitor TLCK results in loss of cytadherence, which can be restored by the addition of protease-containing parasite extract (5). A surface protease may be involved in the modification of adhesins, rendering them competent to bind their epithelial cell targets. A 43-kDa protease has been localized to the parasite surface based on protease K susceptibility, though its function has not been delineated (167).

GIARDIA LAMBLIA

Giardia lamblia is the most prevalent intestinal parasite in the United States and worldwide. It is a major cause of traveler's diarrhea, waterborne outbreaks of diarrhea, and diarrhea in institutional settings. Cysteine protease activity in the trophozoite stage has been identified and characterized (98, 127, 184). The enzyme hydrolyzes the fluorogenic substrate *a*-*N*-benzoyl-DL-arginine-2-naphthylamide (BANA), as well as urea-denatured hemoglobin. Enzymatic activity is enhanced by dithiothreitol and inhibited by several cysteine protease inhibitors (98, 184). On gelatin substrate gels, forms of M_rs 40,000 and 105,000 appear (98), whereas gel filtration elutes a major activity peak at 38,000 and a minor peak at 66,000 (184). The enzyme appears to be lysosomal, based on its cosedimentation with acid phosphatase-containing particles after isopycnic centrifugation (127). Although the enzyme proteolyses IgA (184), the cleavage pattern differs significantly from that of bacterial IgA proteases, which are metalloenzymes specific for the hinge region of the IgA1 molecule (162). The significance of this activity is unknown, as is the role of the protease in general; speculated functions include protein degradation and involvement in the transformation between cyst and trophozoite forms (excystation).

CONCLUDING REMARKS

While by no means an exhaustive survey of the proteases of parasitic protozoa, the examples that have been discussed highlight the remarkable diversity of function these enzymes have. Even a structurally related group of enzymes like the cysteine proteases can play roles from facilitating parasite invasion to mediating transformation from one stage to the next in the life cycle. It is also clear that in many instances, like that of gp63 in *Leishmania,* the exact function of the protease remains, to date, a mystery. This is true despite the fact that hundreds of thousands of protease molecules are present on the surface of the organism. Elucidating the structural relationships between parasite proteases provides information not only on their specific function, but on structure-function relationships for the entire protease family to which they belong. Primary sequence data on parasite proteases are being accumulated at such a rapid rate that data sets of natural variations of amino acids in the active site can be studied for their effect on substrate specificity as an alternative to investigator-derived data sets from site-directed mutagenesis. Studies of the molecular evolution of a family of proteases, or phylogenetic trees for the organisms, benefit from this sequence data as well. Finally, there is an increasing optimism that some of these enzymes may be targets for new chemotherapeutic attempts to solve a number of major health problems worldwide.

Literature Cited

1. Aikawa, M. 1988. Fine structure of malaria parasites in the various stages of development. See Ref. 254a, pp. 97–129
2. Aissi, E., Charet, P., Bouquelet, S., Biguet, J. 1983. Endoprotease in *Plasmodium yoelii nigeriensis*. *Comp. Biochem. Physiol.* 74B:559–66
3. Alfieri, S. C., Pral, E. M. F., Shaw, E., Ramazeilles, C., Rabinovitch, M. 1991. *Leishmania amazonensis:* specific labeling of amastigote cysteine proteinases by radioiodinated N-benzyloxycarbonyl-tyrosyl-alanyl diazomethane. *Exp. Parasitol.* 73:424–32
4. Alfieri, S. C., Shaw, E., Zilberfarb, V., Rabinovitch, M. 1989. *Leishmania amazonensis:* involvement of cysteine proteinases in the killing of isolated amastigotes by L-leucine methyl ester. *Exp. Parasitol.* 68:423–31
5. Arroyo, R., Alderete, J. F. 1989. *Trichomonas vaginalis* surface proteinase activity is necessary for parasite adherence to epithelial cells. *Infect. Immun.* 57:2991–97
6. Ashall, F. 1990. Characterisation of an alkaline peptidase of *Trypanosoma cruzi* and other trypanosomatids. *Mol. Biochem. Parasitol.* 38:77–88
7. Ashall, F., Angliker, H., Shaw, E. 1990. Lysis of trypanosomes by peptidyl fluoromethyl ketones. *Biochem. Biophys. Res. Commun.* 170:923–29
8. Åslund, L., Henriksson, J., Campetella, O., Frasch, A. C. C., Pettersson, U., Cazzulo, J. J. 1991. The C-terminal extension of the major cysteine proteinase (cruzipain) from *Trypanosoma cruzi*. *Mol. Biochem. Parasitol.* 45:345–48
9. Bailly, E., Jambou, R., Savel, J.,

Jaureguiberry, G. 1992. *Plasmodium falciparum:* differential sensitivity in vitro to E-64 (cysteine protease inhibitor) and pepstatin A (aspartyl protease inhibitor). *J. Protozool.* 39:593–99

10. Bailly, E., Savel, J., Mahouy, G., Jaureguiberry, G. 1991. *Plasmodium falciparum:* isolation and characterization of a 55-kDa protease with a cathepsin D-like activity from *P. falciparum*. *Exp. Parasitol.* 72:278–84
11. Ball, E. G., McKee, R. W., Anfinsen, C. B., Cruz, W. O., Geiman, Q. M. 1948. Studies on malarial parasites. IX. Chemical and metabolic changes during growth and multiplication in vivo and in vitro. *J. Biol. Chem.* 175:547–71
12. Bannister, L. H., Mitchell, G. H. 1989. The fine structure of secretion by *Plasmodium knowlesi* merozoites during red cell invasion. *J. Protozool.* 36:362–67
13. Banyal, H. S., Misra, G. C., Gupta, C. M., Dutta, G. P. 1981. Involvement of malarial proteases in the interaction between the parasite and host erythrocyte in *Plasmodium knowlesi* infections. *J. Parasitol.* 67:623–26
14. Barrett, A. J. 1986. An introduction to the proteinases. In *Proteinase Inhibitors,* ed. A. J. Barrett, G. Salvesen, pp. 3–22. Amsterdam: Elsevier Science
15. Barry, D. N. 1982. Metabolism of *Babesia* parasites in vitro: amino acid production by *Babesia rodhaini* compared to *Plasmodium berghei*. *Aust. J. Exp. Biol. Med. Sci.* 60:175–80
16. Berger, A., Schecter, I. 1970. Mapping the active site of papain with the aid of peptide substrates and inhibitors. *Philos. Trans. R. Soc. London Ser. B*. 257:249–64
17. Bernard, F., Mayer, R., Picard, I., Deguercy, A., Monsigny, M., Schrevel, J. 1987. *Plasmodium berghei* and *Plasmodium chabaudi:* a neutral endopeptidase in parasite extracts and plasma of infected animals. *Exp. Parasitol.* 64:95–103
18. Bernard, F., Schrevel, J. 1987. Purification of a *Plasmodium berghei* neutral endopeptidase and its localization in merozoite. *Mol. Biochem. Parasitol.* 26:167–74
19. Beynon, R. J., Bond, J. S. 1989. *Proteolytic Enzymes: A Practical Approach*. Oxford/New York: IRL/Oxford Univ. Press
20. Blackman, M. J., Holder, A. A. 1992. Secondary processing of the *Plasmodium falciparum* merozoite surface protein-1 (MSP1) by a calcium-dependent membrane-bound serine protease: shedding of $MSP1_{33}$ as a noncovalently associated complex with other fragments of the MSP1. *Mol. Biochem. Parasitol.* 50:307–16
21. Bode, W., Huber, R. 1991. Ligand binding: proteinase-protein inhibitor interactions. *Curr. Opin. Struc. Biol.* 1:45–52
22. Bonaldo, M. C., D'Escoffier, L. N., Salles, J. M., Goldenberg, S. 1991. Characterization and expression of proteases during *Trypanosoma cruzi* metacyclogenesis. *Exp. Parasitol.* 73:44–51
23. Bond, J. S., Butler, P. E. 1987. Intracellular proteases. *Annu. Rev. Biochem.* 56:333–64
24. Bongertz, V., Hungerer, K. D. 1978. *Trypanosoma cruzi:* isolation and characterization of a protease. *Exp. Parasitol.* 45:8–18
25. Bontempi, E., Franke de Cazzulo, B. M., Ruiz, A. M., Cazzulo, J. J. 1984. Purification and some properties of an acidic protease from epimastigotes of *Trypanosoma cruzi*. *Comp. Biochem. Physiol.* 77B:599–604
26. Bontempi, E., Martinez, J., Cazzulo, J. J. 1989. Subcellular localization of a cysteine proteinase from *Trypanosoma cruzi*. *Mol. Biochem. Parasitol.* 33:43–48
27. Bordier, C. 1987. The promastigote surface protease of *Leishmania*. *Parasitol. Today* 3:151–53
28. Bordier, C., Etges, R. J., Ward, J., Turner, M. J., Cardoso de Almeida, M.-L. 1986. *Leishmania* and *Trypanosoma* surface glycoproteins have a common glycophospholipid membrane anchor. *Proc. Natl. Acad. Sci. USA* 83:5988–91
29. Bouvier, J., Bordier, C., Vogel, H., Reichelt, R., Etges, R. 1989. Characterization of the promastigote surface protease of *Leishmania* as a membrane-bound zinc endopeptidase. *Mol. Biochem. Parasitol.* 37:235–45
30. Bouvier, J., Etges, R. J., Bordier, C. 1985. Identification and purification of membrane and soluble forms of the major surface protein of *Leishmania promastigotes*. *J. Biol. Chem.* 260: 15504–9
31. Bouvier, J., Etges, R. J., Bordier, C. 1987. Identification of the promastigote surface protease in seven species of *Leishmania*. *Mol. Biochem. Parasitol.* 24:73–79
32. Bouvier, J., Schneider, P., Etges, R. J., Bordier, C. 1990. Peptide substrate specificity of the membrane-bound

metalloprotease of *Leishmania*. *Biochemistry* 29:10113–19

33. Bouvier, J., Schneider, P., Malcolm, B. 1992. A fluorescent peptide substrate for the surface metalloprotease of *Leishmania*. *Exp. Parasitol.* In press
34. Bózner, P., Demes, P. 1991. Proteinases in *Trichomonas vaginalis* and *Tritrichomonas mobilensis* are not exclusively of the cysteine type. *Parasitology* 102:113–15
35. Braun-Breton, C., Blisnick, T., Jouin, H., Barale, J. C., Rabilloud, T., et al. 1992. *Plasmodium chabaudi* p68 serine protease activity required for merozoite entry into mouse erythrocytes. *Proc. Natl. Acad. Sci. USA* In press
36. Braun-Breton, C., Pereira da Silva, L. 1988. Activation of a *Plasmodium falciparum* protease correlated with merozoite maturation and erythrocyte invasion. *Biol. Cell* 64:223–31
37. Braun-Breton, C., Rosenberry, T. L., Pereira da Silva, L. 1988. Induction of the proteolytic activity of a membrane protein in *Plasmodium falciparum* by phosphatidyl inositol-specific phospholipase C. *Nature* 332:457–59
38. Button, L. L., McMaster, W. R. 1988. Molecular cloning of the major surface antigen of *Leishmania*. *J. Exp. Med.* 167:724–29
39. Button, L. L., McMaster, W. R. 1990. Correction. *J. Exp. Med.* 171:589
40. Button, L. L., Russell, D. G., Klein, H. L., Medina-Acosta, E., Karess, R., McMaster, W. R. 1989. Genes encoding the major surface glycoprotein in *Leishmania* are tandemly linked at a single chromosomal locus and are constitutively transcribed. *Mol. Biochem. Parasitol.* 32:271–84
41. Calderon, R. O., Lujan, H. D., Aguerri, A. M., Bronia, D. H. 1989. *Trypanosoma cruzi:* involvement of proteolytic activity during cell fusion induced by epimastigote form. *Mol. Cell Biochem.* 86:189–200
42. Cazzulo, J. J., Cazzulo Franke, M. C., Martínez, J., Franke de Cazzulo, B. M. 1990. Some kinetic properties of a cysteine proteinase (cruzipain) from *Trypanosoma cruzi*. *Biochim. Biophys. Acta* 1037:186–91
43. Cazzulo, J. J., Couso, R., Raimondi, A., Wernstedt, C., Hellman, U. 1989. Further characterization and partial amino acid sequence of a cysteine proteinase from *Trypanosoma cruzi*. *Mol. Biochem. Parasitol.* 33:33–42
44. Cazzulo, J. J., Hellman, U., Couso, R., Parodi, A. J. A. 1990. Amino acid and carbohydrate composition of a lysosomal cysteine proteinase from *Trypanosoma cruzi*. Absence of phosphorylated mannose residues. *Mol. Biochem. Parasitol.* 38:41–48
45. Cenedella, R. J., Rosen, H., Angel, C. R., Saxe, L. H. 1968. Free amino-acid production in vitro by *Plasmodium berghei*. *Am. J. Trop. Med. Hyg.* 17:800–3
46. Chan, V. L., Lee, P. Y. 1974. Host cell specific proteolytic enzymes in *Plasmodium berghei*-infected erythrocytes. *Southeast Asian J. Trop. Med. Public Health* 5:447–49
47. Chang, K.-P., Chaudhuri, G., Fong, D. 1990. Molecular determinants of *Leishmania* virulence. *Annu. Rev. Microbiol.* 44:499–529
48. Charet, P., Aissi, E., Maurois, P., Bouquelet, S., Biguet, J. 1980. Aminopeptidase in rodent *Plasmodium*. *Comp. Biochem. Physiol.* 65B:519–24
49. Chaudhuri, G., Chang, K.-P. 1988. Acid protease activity of a major surface membrane glycoprotein (gp63) from *Leishmania mexicana* promastigotes. *Mol. Biochem. Parasitol.* 27:43–52
50. Chaudhuri, G., Chaudhuri, M., Pan, A., Chang, K.-P. 1989. Surface acid proteinase (gp63) of *Leishmania mexicana*. *J. Biol. Chem.* 264:7483–89
51. Choi, I., Mego, J. L. 1988. Purification of *Plasmodium falciparum* digestive vacuoles and partial characterization of the vacuolar membrane ATPase. *Mol. Biochem. Parasitol.* 31:71–78
52. Colomer-Gould, V., Galvao-Quintao, L., Keithly, J., Noguiera, N. 1985. A common major surface antigen on amastigotes and promastigotes of *Leishmania* species. *J. Exp. Med.* 162: 902–16
53. Cook, L., Grant, P. T., Kermack, W. O. 1961. Proteolytic enzymes of the erythrocytic forms of rodent and simian species of malarial plasmodia. *Exp. Parasitol.* 11:372–79
54. Cook, R. T., Aikawa, M., Rock, R. C., Little, W., Sprinz, H. 1969. The isolation and fractionation of *Plasmodium knowlesi*. *Mil. Med.* 134:866–83
55. Coombs, G., North, M., eds. 1991. *Biochemical Protozoology*. London/ Washington, DC: Taylor & Francis. 635 pp.
56. Coombs, G. H., Baxter, J. 1984. Inhibition of *Leishmania* amastigote growth by antipain and leupeptin. *Ann. Trop. Med. Parasitol.* 78:21–24
57. Coombs, G. H., North, M. J. 1983. An analysis of the proteinases of *Trichomonas vaginalis* by polyacryl-

amide gel electrophoresis. *Parasitology* 86:1–6

58. Craik, C. S., Largman, C., Fletcher, T., Roczniak, S., Barr, P. J., et al 1985. Redesigning trypsin: alteration of substrate specificity. *Science* 228: 291–97
59. Cross, G. A. M. 1990. Glycolipid anchoring of plasma membrane proteins. *Annu. Rev. Cell Biol.* 6:1–39
60. Cruz, A., Coburn, C. M., Beverley, S. M. 1991. Double targeted gene replacement for creating null mutants. *Proc. Natl. Acad. Sci. USA* 88:7170–74
61. David, P. H., Hadley, T. J., Aikawa, M., Miller, L. H. 1984. Processing of a major parasite surface glycoprotein during the ultimate stages of differentiation in *Plasmodium knowlesi*. *Mol. Biochem. Parasitol.* 11:267–82
62. Davies, C. R., Cooper, A. M., Peacock, C., Lane, R. P., Blackwell, J. M. 1990. Expression of LPG and gp63 by different developmental stages of *Leishmania major* in the sandfly *Phlebotomus papatasi*. *Parasitology* 101:337–43
63. Debrabant, A., Delplace, P. 1989. Leupeptin alters the proteolytic processing of P126, the major parasitophorous vacuole antigen of *Plasmodium falciparum*. *Mol. Biochem. Parasitol.* 33:151–58
64. Deguercy, A., Hommel, M., Schrevel, J. 1990. Purification and characterization of 37-kilodalton proteases from *Plasmodium falciparum* and *Plasmodium berghei* which cleave erythrocyte cytoskeletal components. *Mol. Biochem. Parasitol.* 38:233–44
65. Dejkriengkraikhul, P., Wilairat, P. 1983. Requirement of malarial protease in the invasion of human red cells by merozoites of *Plasmodium falciparum*. *Z. Parasitenkd.* 69:313–17
66. Delplace, P., Bhatia, A., Cagnard, M., Camus, D., Colombet, G., et al. 1988. Protein p126: a parasitophorous vacuole antigen associated with the release of *Plasmodium falciparum* merozoites. *Biol. Cell* 64:215–21
67. Delplace, P., Fortier, B., Tronchin, G., Dubremetz, J., Vernes, A. 1987. Localization, biosynthesis, processing and isolation of a major 126 kDa antigen of the parasitophorous vacuole of *Plasmodium falciparum*. *Mol. Biochem. Parasitol.* 23:193–201
68. Dluzewski, A. R., Rangachari, K., Wilson, R. J. M., Gratzer, W. B. 1986. *Plasmodium falciparum:* protease inhibitors and inhibition of erythrocyte invasion. *Exp. Parasitol.* 62:416–22
69. Eakin, A. E., Bouvier, J., Sakanari, J. A., Craik, C. S., McKerrow, J. H. 1990. Amplification and sequencing of genomic DNA fragments encoding cysteine proteases from protozoan parasites. *Mol. Biochem. Parasitol.* 39:1–8
70. Eakin, A. E., Higaki, J. N., McKerrow, J. H., Craik, C. S. 1989. Cysteine or serine proteinase? *Nature* 342:132
71. Eakin, A. E., McGrath, M. E., McKerrow, J. H., Fletterick, R. J., Craik, C. S. 1993. Production of crystallizable cruzain, the major cysteine protease from *Trypanosoma cruzi*. *J. Biol. Chem.* In press
72. Eakin, A. E., Mills, A. A., Harth, G., McKerrow, J. H., Craik, C. S. 1992. The sequence, organization, and expression of the major cysteine protease (cruzain) from *Trypanosoma cruzi*. *J. Biol. Chem.* 267:7411–20
73. Etges, R. J. 1992. Identification of a surface metalloproteinase on 13 species of *Leishmania* isolated from humans, *Crithidia fasciculata,* and *Herpetomonas samuelpessoai*. *Acta Trop.* 50: 205–17
74. Etges, R. J., Bouvier, J. 1991. The promastigote surface proteinase of *Leishmania*. See Ref. 55, pp. 221–33
75. Etges, R. J., Bouvier, J., Bordier, C. 1986. The major surface protein of *Leishmania* promastigotes is anchored in the membrane by a myristic acid-labeled phospholipid. *EMBO J.* 5:597–602
76. Etges, R. J., Bouvier, J., Bordier, C. 1986. The major surface protein of *Leishmania* promastigotes is a protease. *J. Biol. Chem.* 261:9099–9101
77. Etges, R. J., Bouvier, J., Bordier, C. 1989. The promastigote surface protease of *Leishmania:* pH optimum and effects of protease inhibitors. In *Leishmaniasis, the First Century (1885–1985). The Current Status and New Strategies for Control,* ed. D. T. Hart, NATO ASI Ser. A, 163:627–33. New York: Plenum
78. Etges, R. J., Bouvier, J., Hoffman, R., Bordier, C. 1985. Evidence that the major surface proteins of three *Leishmania* species are structurally related. *Mol. Biochem. Parasitol.* 14: 141–49
79. Ferguson, M. A. J., Williams, A. F. 1988. Cell-surface anchoring of proteins via glycosyl-phosphatidylinositol structures. *Annu. Rev. Biochem.* 57:285–320
80. Fong, D., Chang, K.-P. 1982. Surface antigenic change during differentiation of a parasitic protozoan, *Leishmania*

mexicana: identification by monoclonal antibodies. *Proc. Natl. Acad. Sci. USA* 79:7366–70

81. Francioli, P., Shio, H., Roberts, R. B., Muller, M. 1983. Phagocytosis and killing of *Neisseria gonorrhoeae* by *Trichomonas vaginalis*. *J. Infect. Dis.* 47:87
82. Frommel, T. O., Button, L. L., Fujikura, Y., McMaster, W. R. 1990. The major surface glycoprotein (gp63) is present in both life stages of *Leishmania*. *Mol. Biochem. Parasitol.* 38:25–32
83. Fulton, J. D., Grant, P. T. 1956. The sulphur requirements of the erythrocytic form of *Plasmodium knowlesi*. *Biochem. J.* 63:274–82
84. Gadasi, H., Kobiler, D. 1983. *Entamoeba histolytica:* correlation between virulence and content of proteolytic enzymes. *Exp. Parasitol.* 55:105–10
85. Galvao-Quintao, L., Alfieri, S. C., Ryter, A., Rabinovitch, M. 1990. Intracellular differentiation of *Leishmania amazonensis* promastigotes to amastigotes: presence of megasomes, cysteine proteinase activity and susceptibility to leucine-methyl ester. *Parasitology* 101: 7–13
86. Garber, G., Lemchuk-Favel, L. T. 1989. Characterization and purification of extracellular proteases of *Trichomonas vaginalis*. *Can. J. Microbiol.* 35:903–9
87. Gitler, C., Calef, E., Rosenberg, I. 1984. Cytopathogenecity of *Entamoeba histolytica*. *Philos. Trans. R. Soc. London Ser. B* 307:73–85
88. Godal, T., Najera, J. 1990. Tropical diseases. In *WHO Division of Control of Tropical Diseases,* pp. 12–13. Geneva, Switzerland: World Health Organization
89. Goldberg, D. E., Slater, A. F. G., Beavis, R., Chait, B., Cerami, A., Henderson, G. B. 1991. Hemoglobin degradation in the human malaria pathogen *Plasmodium falciparum:* a catabolic pathway initiated by a specific aspartic protease. *J. Exp. Med.* 173:961–69
90. Goldberg, D. E., Slater, A. F. G., Cerami, A., Henderson, G. B. 1990. Hemoglobin degradation in the malaria parasite *Plasmodium falciparum:* an ordered process in a unique organelle. *Proc. Natl. Acad. Sci. USA* 87:2931–35
91. Greig, S., Ashall, F. 1990. Electrophoretic detection of *Trypanosoma cruzi* peptidases. *Mol. Biochem. Parasitol.* 39:31–38
92. Grellier, P., Picard, I., Bernard, F., Mayer, R., Heidrich, H.-G., et al. 1989. Purification and identification of a neutral endopeptidase in *Plasmodium falciparum* schizonts and merozoites. *Parasitol. Res.* 75:455–60
93. Grimm, F., Jenni, L., Bouvier, J., Etges, R. J., Bordier, C. 1987. The promastigote surface protease of *Leishmania donovani infantum* in the midgut of *Phlebotomus perniciosus*. *Acta Trop.* 44:375–77
94. Groman, N. B. 1951. Dynamic aspects of the nitrogen metabolism of *Plasmodium gallinaceum* in vivo and in vitro. *J. Infect. Dis.* 88:126–50
95. Gyang, F. N., Poole, B., Trager, W. 1982. Peptidases from *Plasmodium falciparum* cultured in vitro. *Mol. Biochem. Parasitol.* 5:263–73
96. Hadley, T., Aikawa, M., Miller, L. H. 1983. *Plasmodium knowlesi:* studies on invasion of rhesus erythrocytes by merozoites in the presence of protease inhibitors. *Exp. Parasitol.* 55:306–11
97. Handman, E., Button, L. L., McMaster, R. W. 1990. *Leishmania major:* production of recombinant gp63, its antigenicity and immunogenicity in mice. *Exp. Parasitol.* 70:427–35
98. Hare, D. F., Jarroll, E. L., Lindmark, D. G. 1989. *Giardia lamblia:* characterization of proteinase activity in trophozoites. *Exp. Parasitol.* 68:168–75
99. Harth, G., Andrews, N., Mills, A. A., Engel, J., Smith, R., McKerrow, J. H. 1992. Peptide-fluoromethyl ketones arrest intracellular replication and intercellular transmission of *Trypanosoma cruzi*. *Mol. Biochem. Parasitol.* In press
100. Heumann, D., Burger, D., Vischer, T., de Colmenares, M., Bouvier, J., Bordier, C. 1989. Molecular interactions of *Leishmania* promastigote surface protease with human α2- macroglobulin. *Mol. Biochem. Parasitol.* 33: 67–72
101. Higgins, D. G., McConnell, D. J., Sharp, P. M. 1989. Malarial proteinase? *Nature* 340:604
102. Huet, G., Richet, C., Demeyer, D., Bisiau, H., Soudan, B., et al. 1992. Characterization of different proteolytic activities in *Trypanosoma brucei brucei*. *Biochim. Biophys. Acta* 1138: 213–21
103. Ip, H. S., Russell, D. G., Cross, G. A. M. 1990. *Leishmania mexicana mexicana* gp63 is a site-specific neutral endopeptidase. *Mol. Biochem. Parasitol.* 40:163–72
104. Jaffe, C. L., Shor, R., Trau, H., Passwell, J. H. 1990. Parasite antigens

recognized by patients with cutaneous leishmaniasis. *Clin. Exp. Immunol.* 80: 77–82

105. Jähnig, F., Etges, R. 1988. Secondary structure of the promastigote surface protease of *Leishmania*. *FEBS Lett.* 241:79–82
106. Jardim, A., Alexander, J., Teh, H. S., Ou, D., Olafson, R. 1990. Immunoprotective *Leishmania major* synthetic T cell epitopes. *J. Exp. Parasitol.* 172:645–48
107. Jongeneel, C. V., Bouvier, J., Bairoch, A. 1989. A unique signature identifies a family of zinc-dependent metallopeptidases. *FEBS Lett.* 242:211–14
108. Keene, W. E. 1990. *Purification and characterization of the major neutral proteinase of* Entamoeba histolytica: *a virulence factor for invasive amebiasis.* PhD thesis. Univ. Calif., Berkeley, 110 pp.
109. Keene, W. E., Hidalgo, M. E., Orozco, E., McKerrow, J. H. 1990. *Entamoeba histolytica:* correlation of the cytopathic effect of virulent trophozoites with secretion of a cysteine proteinase. *Exp. Parasitol.* 71:199–206
110. Keene, W. E., Petitt, M. G., Allen, S., McKerrow, J. H. 1986. The major neutral proteinase of *Entamoeba histolytica*. *J. Exp. Med.* 163:536–49
111. Knapp, B., Hundt, E., Nau, U., Küpper, H. A. 1989. Molecular cloning, genomic structure and localization in a blood stage antigen of *Plasmodium falciparum* characterized by a serine stretch. *Mol. Biochem. Parasitol.* 32: 73–84
112. Knapp, B., Nau, U., Hundt, E., Küpper, H. A. 1991. A new blood stage antigen of *Plasmodium falciparum* highly homologous to the serine-stretch protein SERP. *Mol. Biochem. Parasitol.* 44:1–14
113. Knight, C. G., Willenbrock, F., Murphy, G. 1992. A novel coumarin-labelled peptide for sensitive continuous assays of the matrix metalloproteinases. *FEBS Lett.* 296:263–66
114. Kornblatt, M. J., Mpimbaza, G. W. N., Lonsdale-Eccles, J. D. 1992. Characterization of an endopeptidase of *Trypanosoma brucei brucei*. *Arch. Biochem. Biophys.* 293:25–31
115. Kraut, J. 1988. How do enzymes work? *Science* 242:533–40
116. Krogstad, D. J., Schlesinger, P. H. 1987. Acid-vesicle function, intracellular pathogens, and the action of chloroquine against *Plasmodium falciparum*. *New Engl. J. Med.* 317: 542–49
117. Krogstad, D. J., Schlesinger, P. H., Gluzman, I. Y. 1985. Antimalarials increase vesicle pH in *Plasmodium falciparum*. *J. Cell Biol.* 101:2302–9
118. Kuntz, I. D., Blaney, J. M., Oatley, S. J., Langridge, R., Ferrin, T. E. 1982. A geometric approach to macromolecule-ligand interactions. *J. Mol. Biol.* 161:269–88
119. Leippe, M., Sebastian, E., Schoenberger, O. L., Horstmann, R. D., Muller-Eberhard, H. J. 1991. Pore-forming peptide of pathogenic *Entamoeba histolytica*. *Proc. Natl. Acad. Sci. USA* 88:7659–63
120. Lepay, D. A., Nogueira, N., Cohn, Z. 1985. Surface antigens of *Leishmania donovani* promastigotes. *J. Exp. Med.* 157:1562–72
121. Levitt, M., Greer, J. 1977. Automatic identification of secondary structure in globular proteins. *J. Mol. Biol.* 114: 181–293
122. Levy, M. R., Chou, S. C. 1973. Activities and some properties of an acid proteinase from normal and *Plasmodium berghei*-infected red cells. *J. Parasitol.* 59:1064–70
123. Levy, M. R., Chou, S. C. 1974. Some properties and susceptibility to inhibitors of partially purified acid proteases from *Plasmodium berghei* and from ghosts of mouse red cells. *Biochim. Biophys. Acta* 334:423–30
124. Levy, M. R., Chou, S. C. 1975. Inhibition of macromolecular synthesis in the malarial parasites by inhibitors of proteolytic enzymes. *Experimentia* 31:52–54
125. Levy, M. R., Siddiqui, W. A., Chou, S. C. 1974. Acid protease activity in *Plasmodium falciparum* and *P. knowlesi* and ghosts of their respective host red cells. *Nature* 247:546–49
126. Li, W.-B., Bzik, D. J., Horii, T., Inselburg, J. 1989. Structure and expression of the *Plasmodium falciparum* SERA gene. *Mol. Biochem. Parasitol.* 33:13–26
127. Lindmark, D. G. 1988. *Giardia lamblia:* localization of hydrolase activities in lysosome-like organelles of trophozoites. *Exp. Parasitol.* 65:141–47
128. Lindmark, D. G., Müller, M., Shio, H. 1975. Hydrogenosomes in *Trichomonas vaginalis*. *Parasitology* 63:552–54
129. Lockwood, B. C., North, M. J., Coombs, G. H. 1988. The release of hydrolases from *Trichomonas vaginalis* and *Tritrichomonas foetus*. *Mol. Biochem. Parasitol.* 30:135–34
130. Lockwood, B. C., North, M. J., Scott,

K. I., Bremner, A. F., Coombs, G. H. 1987. The use of a highly sensitive electrophoretic method to compare the proteinases of trichomonads. *Mol. Biochem. Parasitol.* 24:89–95

131. Lonsdale-Eccles, J. D. 1991. Proteinases of African trypanosomes. See Ref. 55, pp. 200–7
132. Lonsdale-Eccles, J. D., Grab, D. J. 1987. Lysosomal and non-lysosomal peptidyl hydrolases of the bloodstream forms of *Trypanosoma brucei brucei*. *Eur. J. Biochem.* 169:467–75
133. Lonsdale-Eccles, J. D., Mpimbaza, G. W. N. 1986. Thiol-dependent proteases of African trypanosomes. *Eur. J. Biochem.* 155:469–73
134. Lopez, J. A., Reims, H.-A., Etges, R. J., Button, L. L., McMaster, W. R., et al. 1991. Genetic control of the immune response in mice to *Leishmania mexicana* surface protease. *J. Immunol.* 146:1328–34
135. Low, M. G., Saltiel, A. R. 1988. Structural and functional roles of glycosyl-phosphatidylinositol in membranes. *Science* 239:268–75
136. Luaces, A. L., Barrett, A. J. 1988. Affinity purification and biochemical characterization of histolysin, the major cysteine proteinase of *Entamoeba histolytica*. *Biochem. J.* 250:903–9
137. Lushbaugh, W. B. 1988. Proteinases of *Entamoeba histolytica*. See Ref. 200, pp. 219–31
138. Lushbaugh, W. B., Hofbauer, A. F., Pittman, F. E. 1984. Proteinase activities of *Entamoeba histolytica* cytotoxin. *Gastroenterology* 87:17–27
139. Lushbaugh, W. B., Hofbauer, A. F., Pittman, F. E. 1985. *Entamoeba histolytica:* purification of cathepsin B. *Exp. Parasitol.* 59:328–36
140. Lushbaugh, W. B., Kairalla, A. B., Cantey, J. R., Hofbauer, A. F., Pittman, F. E. 1979. Isolation of a cytotoxin-enterotoxin from *Entamoeba histolytica*. *J. Inf. Dis.* 139:9–17
141. Lustigman, S., Brotman, B., Huima, T., Prince, A. M., McKerrow, J. H. 1992. Molecular cloning and characterization of onchocystatin, a cysteine proteinase inhibitor of *Onchocerca volvulus*. *J. Biol. Chem.* 267:17339–46
142. Lyon, J. A., Haynes, J. D. 1986. *Plasmodium falciparum* antigens synthesized by schizonts and stabilized at the merozoite surface when schizonts mature in the presence of protease inhibitors. *J. Immunol.* 136:2245–51
143. Mäkinen, P. L., Clewell, D. B., An, F., Mäkinen, K. K. 1989. Purification and substrate specificity of a strongly hydrophobic extracellular metalloendopeptidase ("gelatinase") from *Streptococcus faecalis* (strain OG1–10). *J. Biol. Chem.* 264:3325–34
144. Martinez, J., Campetella, O., Frasch, A. C. C., Cazzulo, J. J. 1991. The major cysteine proteinase (cruzipain) from *Trypanosoma cruzi* is antigenic in human infections. *Infect. Immun.* 59:4275–77
145. Martinez, J., Cazzulo, J. J. 1992. Anomalous electrophoretic behaviour of the major cysteine proteinase (cruzipain) from *Trypanosoma cruzi* in relation to its apparent molecular mass. *FEMS Microbiol. Lett.* 74:225–29
146. Mayer, R., Picard, I., Lawton, P., Grellier, P., Barrault, C., et al. 1991. Peptide derivatives specific for a *Plasmodium falciparum* proteinase inhibit the human erythrocyte invasion by merozoites. *J. Med. Chem.* 34:3029–35
147. Mbawa, Z. R., Gumm, I. D., Fish, W. R., Lonsdale-Eccles, J. D. 1991. Endopeptidase variations among different life-cycle stages of African trypanosomes. *Eur. J. Biochem.* 195: 183–90
148. McColm, A. A., Shakespeare, P. G., Trigg, P. I. 1980. Analysis of proteins synthesized in vitro by the erythrocytic stages of *Plasmodium knowlesi*. *Parasitology* 81:177–98
149. McDonald, J. K. 1985. An overview of protease specificity and catalytic mechanisms: aspects related to nomenclature and classification. *Histochem. J.* 17:773–85
150. McGowan, K., Deneke, C. F., Thorne, G. M., Gorbach, S. L. 1982. *Entamoeba histolytica:* purification, characterization, strain virulence, and protease activity. *J. Infect. Dis.* 146: 616–25
151. McKerrow, J. H. 1991. New insights into the structure of a *Trypanosoma cruzi* protease. *Parasitol. Today* 7:132–33
152. McKerrow, J. H. 1992. Pathogenesis in amebiasis: is it genetic or acquired? *Infect. Agent. Dis.* 1:11–14
153. McLaughlin, J., Faubert, G. 1977. Partial purification and some properties of a neutral sulfhydryl and an acid proteinase from *Entamoeba histolytica*. *Can. J. Microbiol.* 23:420–25
154. Medina-Acosta, E., Karess, R. E., Schwarz, H., Russell, D. G. 1989. The promastigote surface protease (gp63) of *Leishmania* is expressed but differentially processed and localized in the amastigote stage. *Mol. Biochem. Parasitol.* 37:263–74

155. Meirelles, M. N., Juliano, L., Carmona, E., Silva, S. G. Costa, E. M., et al. 1992. Inhibitors of the major cysteinyl proteinase (GP57/51) impair host cell invasion and arrest the intracellular development of *Trypanosoma cruzi* in vitro. *Mol. Biochem. Parasitol.* 52:175–84
156. Miller, R. A., Reed, S. G., Parsons, M. 1990. *Leishmania* gp63 molecule implicated in cellular adhesion lacks an Arg-Gly-Asp sequence. *Mol. Biochem. Parasitol.* 39:267–74
157. Mosser, D. M., Edelson, P. J. 1987. The third component of complement (C3) is responsible for the intracellular survival of *Leishmania major*. *Nature* 327:329–31
158. Mottram, J. C., Coombs, G. H., North, M. J. 1989. Cysteine or serine proteinase? *Nature* 342:132
159. Mottram, J. C., North, M. J., Barry, J. D., Coombs, G. H. 1989. A cysteine proteinase cDNA from *Trypanosoma brucei* predicts an enzyme with an unusual C-terminal extension. *FEBS Lett* 258:211–15
160. Mottram, J. C., Robertson, C. D., Coombs, G. H., Barry, J. D. 1992. A developmentally regulated cysteine proteinase gene of *Leishmania mexicana*. *Mol. Microbiol.* 6:1925–32
161. Moulder, J. W., Evans, E. A. 1946. The biochemistry of the malaria parasite. VI. Studies on the nitrogen metabolism of the malaria parasite. *J. Biol. Chem.* 164:145–57
162. Mulks, M. H., Knapp, J. S. 1987. Immunoglobulin A1 protease types of *Neisseria gonorrhoeae* and their relationship to auxotype and serovar. *Infect. Immun.* 55:931–36
163. Muñoz, M. L., Calderon, J., Rojkind M. 1982. The collagenase of *Entamoeba histolytica*. *J. Exp. Med.* 155: 42–51
164. Muñoz, M. L., Lamoyi, E., León, G., Tovar, R., Pérez-García, J., et al. 1990. Antigens in electron-dense granules from *Entamoeba histolytica* as possible markers for pathogenicity. *J. Clin. Microbiol.* 28:2418–24
165. Murta, A. C. M., Persechini, P. M., Padron, T. dS., de Souza, W., Guimaraes, J. A., Scharfstein, J. 1990. Structural and functional identification of GP57/51 antigen of *Trypanosoma cruzi* as a cysteine proteinase. *Mol. Biochem. Parasitol.* 43:27–38
166. Neal, R. A. 1960. Enzymic proteolysis by *Entamoeba histolytica:* biochemical characteristics and relationship with invasiveness. *Parasitology* 50:531–50
167. Neale, K. A., Alderete, J. F. 1990. Analysis of the proteinases of representative *Trichomonas vaginalis* isoloates. *Infect. Immun.* 58:157–62
168. Ng, M., Auld, D. 1989. A fluorescent oligopeptide energy transfer assay with broad applications for neutral proteases. *Anal. Biochem.* 183:50–56
169. Nielsen, M. H., Nielsen, R. 1975. Electron microscopy of *Trichomonas vaginalis* Donné. Interaction with vaginal epithelium in human trichomoniasis. *Acta Pathol. Microbiol. Scand.* 83:305
170. Nishimura, Y., Kawabata, T., Kato, K. 1988. Identification of latent procathepsins B and L in microsomal lumen: characterization of enzymatic activation and proteolytic processing in vitro. *Arch. Biochem. Biophys.* 261:64–71
171. North, M. J. 1991. Proteinases of parasitic protozoa: an overview. See Ref. 55, pp. 180–85
172. North, M. J. 1991. Proteinases of trichomonads and *Giardia*. See Ref. 55, pp. 234–44
173. North, M. J. 1991. Trypanosomes and tomatoes. *Parasitol. Today* 7:249
174. North, M. J., Coombs, G. H., Barry, J. D. 1983. A comparative study of the proteolytic enzymes of *Trypanosoma brucei, T. equiperdum, T. evansi, T. vivax, Leishmania tarentolae,* and *Crithidia fasiculata*. *Mol. Biochem. Parasitol.* 9:161–80
175. North, M. J., Mottram, J. C., Coombs, G. H. 1990. Cysteine proteinases of parasitic protozoa. *Parasitol. Today* 6:270–75
176. North, M. J., Robertson, C. D., Coombs, G. H. 1990. The specificity of trichomonad cysteine proteinases analysed using fluorogenic substrates and specific inhibitors. *Mol. Biochem. Parasitol.* 39:183–94
177. Olafson, R. W., Thomas, J. R., Ferguson, M. A. J., Dwek, R. A., Chaudhuri, M., et al 1990. Structures of the N-linked oligosaccharides of gp63, the major surface glycoprotein from *Leishmania mexicana amazonensis*. *J. Biol. Chem.* 265:12240–47
178. Olaya, P., Wasserman, M. 1991. Effect of calpain inhibitors on the invasion of human erythrocytes by the parasite *Plasmodium falciparum*. *Biochem. Biophys. Acta* 1096:217–21
179. Orlowski, M., Michaux, C., Molineaux, C. J. 1988. Substrate-related potent inhibitors of brain metalloendopeptidase. *Biochemistry* 27: 597–602

180. Orozco, E., Rodriguez, M. A., Hernandez, F. D. 1988. The role of phagocytosis in the pathogenic mechanism of *Entamoeba histolytica*. See Ref. 200, pp. 326–38

181. Ostoa-Saloma, P., Cabrera, N., Becker, I., Perez-Montfort, R. 1989. Proteinases of *Entamoeba histolytica* associated with different subcellular fractions. *Mol. Biochem. Parasitol.* 32:133–44

182. Otte, J., Werries, E. 1989. Specificity of a cysteine proteinase of *Entamoeba histolytica* against various unblocked synthetic peptides. *Mol. Biochem. Parasitol.* 33:257–64

183. Pamer, E. G., So, M., Davis, C. E. 1989. Identification of a developmentally regulated cysteine protease of *Trypanosoma brucei*. *Mol. Biochem. Parasitol.* 33:27–32

184. Parenti D. 1989. Characterization of a thiol proteinase in *Giardia lamblia*. *J. Infect. Dis.* 160:1076–80

185. Perez-Montfort, R., Ostoa-Saloma, P., Velaszquez-Medina, L., Montfort, I., Becker, I. 1987. Catalytic classes of proteinases of *Entamoeba histolytica*. *Mol. Biochem. Parasitol.* 26:87–98

186. Perkins, M. E. 1989. Erythrocyte invasion by the malarial merozoite: recent advances. *Exp. Parasitol.* 69:94–99

187. Perrin, L. H., Merkli, B., Loche, M., Chizzolini, C., Smart, J., Richle, R. 1984. Antimalarial immunity in Saimiri monkeys: immunization with surface components of asexual blood stages. *J. Exp. Med.* 160:441–51

188. Petri, W. A., Smith, R. D., Schlesinger, P. H., Murphy, C. F., Ravdin, J. I. 1987. Isolation of the galactose-binding lectin that mediates the in vitro adherence of *Entamoeba histolytica*. *J. Clin. Invest.* 80:1238–44

189. Piras, M. M., Henriquez, D., Piras, R. 1985. The effect of proteolytic enzymes and protease inhibitors on the interaction *Trypanosoma cruzi* fibroblasts. *Mol. Biochem. Parasitol.* 14: 151–63

190. Powers, J. C. 1986. Serine proteases of leukocyte and mast cell origin: substrate specificity and inhibition of elastase, chymases, and tryptases. In *Advances Inflammation Research,* ed. I. Otterness. New York: Raven

191. Pozsgay, M., Michaux, C., Liebman, M., Orlowski, M. 1986. Substrate and inhibitor studies of thermolysin-like neutral metalloendopeptidase from kidney membrane fractions. Comparison with thermolysin. *Biochemistry* 25: 1292–99

192. Puentes, S. M., Dwyer, D. M., Bates, P. A., Joiner, K. A. 1989. Binding and release of C3 from *Leishmania donovani* promastigotes during incubation in normal human serum. *J. Immunol.* 143:3743–49

193. Rabinovitch, M. 1989. Leishmanicidal activity of amino acid and peptide esters. *Parasitol. Today* 5:299–301

194. Raimondi, A., Wernstedt, C., Hellman, U., Cazzulo, J. J. 1991. Degradation of oxidised insulin A and B chains by the major cysteine proteinase (cruzipain) from *Trypanosoma cruzi* epimastigotes. *Mol. Biochem. Parasitol.* 49:341–44

195. Ramamoorthy, R., Donelson, J. E., Paetz, K. E., Maybodi, M., Roberts, S. C., Wilson, M. E. 1992. Three distinct RNAs for the surface protease gp63 are differentially expressed during development of *Leishmania donovani chagasi* promastigotes to an infectious form. *J. Biol. Chem.* 267:1888–95

196. Ramazeilles, C., Juliano, L., Chagas, J. R., Rabinovitch, M. 1990. The anti-leishmanial activity of dipeptide esters on *Leishmania amazonensis* amastigotes. *Parasitology* 100:201–7

197. Ramazeilles, C., Rabinovitch, M. 1989. *Leishmania amazonensis:* uptake and hydrolysis of [^{3}H]-amino acid methyl esters by isolated amastigotes. *Exp. Parasitol.* 68:135–43

198. Rangel, H. A., Arauacjo, P. M. F., Camargo, I. J. B., Bonfitto, M., Repka, D., et al. 1981. Detection of a proteinase common to epimastigote, trypomastigote and amastigote of different strains of *Trypanosoma cruzi*. *Tropenmed. Parasit.* 32:87–92

199. Ravdin, J. 1986. Pathogenesis of disease caused by *Entamoeba histolytica:* studies of adherence, secreted toxins, and contact-dependent cytolysis. *Rev. Infect. Dis.* 8:247–60

200. Ravdin, J. I., ed. 1988. *Amebiasis: Human Infection by* Entamoeba histolytica. New York: Wiley & Sons. 838 pp.

201. Ravdin, J. I. 1988 Pathogenesis of amebiasis: an overview. See Ref. 200, pp. 166–76

202. Ravdin, J. I., Croft, B. Y., Guerrant, R. L. 1980. Cytopathogenic mechanisms of *Entamoeba histolytica*. *J. Exp. Med.* 152:377–90

203. Ravdin, J. I., Guerrant, R. L. 1982. A review of the parasite cellular mechanisms involved in the pathogenesis of amebiasis. *Rev. Infect. Dis.* 4:1185–1207

204. Reed, S., Bouvier, J., Pollack, A. S., Engel, J. C., Brown, M., et al. 1992.

Cloning of a virulence factor of *Entamoeba histolytica:* pathogenic strains possess a unique cysteine proteinase gene. *J. Clin. Invest.* In press

205. Reed, S., Keene, W. E., McKerrow, J. H. 1989. Thiol proteinase expression correlates with pathogenicity of *Entamoeba histolytica. J. Clin. Microbiol.* 27:2772–77
206. Ring, C. S., Sun, E., McKerrow, J. H., Lee, G. K., Rosenthal, P. J., et al. 1993. Structure-based inhibitor design using model built structures. *Proc. Natl. Acad. Sci. USA.* In press
207. Robertson, C. D., Coombs, G. H. 1990. Characterisation of three groups of cysteine proteinases in the amastigotes of *Leishmania mexicana mexicana. Mol. Biochem. Parasitol.* 42:269–76
208. Robertson, C. D., Coombs, G. H. 1992. Stage-specific proteinases of *Leishmania mexicana mexicana* promastigotes. *FEMS Microbiol. Lett.* 94:127–32
209. Robertson, C. D., North, M. J., Lockwood, B. C., Coombs, G. H. 1990. Analysis of the proteinases of *Trypanosoma brucei. J. Gen. Microbiol.* 136:921–25
210. Rockett, K. A., Playfair, J. H. L., Ashall, F., Targett, G. A. T., Angliker, H., Shaw, E. 1990. Inhibition of intraerythrocytic development of *Plasmodium falciparum* by proteinase inhibitors. *FEBS Lett.* 259:257–59
211. Rosenthal, P. J., Kim, K., McKerrow, J. H., Leech, J. H. 1987. Identification of three stage-specific proteinases of *Plasmodium falciparum. J. Exp. Med.* 166:816–21
212. Rosenthal, P. J., Lee, G. K., Smith, R. E. 1993. Inhibition of a *Plasmodium vinckei* cysteine proteinase cures murine malaria. *J. Clin. Invest.* In press
213. Rosenthal, P. J., McKerrow, J. H., Aikawa, M., Nagasawa, H., Leech, J. H. 1988. A malarial cysteine proteinase is necessary for hemoglobin degradation by *Plasmodium falciparum. J. Clin. Invest.* 82:1560–66
214. Rosenthal, P. J., McKerrow, J. H., Rasnick, D., Leech, J. H. 1989. *Plasmodium falciparum:* inhibitors of lysosomal cysteine proteinases inhibit a trophozoite proteinase and block parasite development. *Mol. Biochem. Parasitol.* 35:177–84
215. Rosenthal, P. J., Nelson, R. G. 1992. Isolation and characterization of a cysteine proteinase gene of *Plasmodium falciparum. Mol. Biochem. Parasitol.* 51:143–52
216. Rosenthal, P. J., Wollish, W. S., Palmer, J. T., Rasnick, D. 1991. Antimalarial effects of peptide inhibitors of a *Plasmodium falciparum* cysteine proteinase. *J. Clin. Invest.* 88:1467–72
217. Roth, E. F., Brotman, D. S., Vanderberg, J. P., Schulman, S. 1986. Malarial pigment-dependent error in the estimation of hemoglobin content in *Plasmodium falciparum*-infected red cells: implications for metabolic and biochemical studies of the erythrocytic phases of malaria. *Am. J. Trop. Med. Hyg.* 35:906–11
218. Russell, D. G., Alexander, J. 1988. Effective immunization against cutaneous leishmaniasis with defined membrane antigens reconstituted into liposomes. *J. Immunol.* 140:1274–79
219. Russell, D. G., Talamas-Rohana, P. 1989. *Leishmania* and the macrophage: a marriage of inconvenience. *Immunol. Today* 10:328–33
220. Sato, K., Fukabori, Y., Suzuki, M. 1987. *Plasmodium berghei:* a study of globinolytic enzyme in erythrocytic parasite. *Zentrabl. Bakteriol. Parasitenkd. Infektionskr. Hyg. Orig. Reihe A* 264:487–95
221. Scharfstein, J., Luquetti, A., Murta, A. C. M., Senna, M., Rezende, J. M., et al. 1985. Chagas' disease: serodiagnosis with purified Gp25 antigen. *Am. J. Trop. Med. Hyg.* 34:1153–60
222. Scharfstein, J., Schechter, M., Senna, M., Peralta, J. M., Mendonça-Previato, L., Miles, M. A. 1986. *Trypanosoma cruzi:* characterization and isolation of a 57/51,000 m.w. surface glycoprotein (GP57/51) expressed by epimastigotes and bloodstream trypomastigotes. *J. Immunol.* 137:1336–41
223. Scheibel, L. W., Sherman, I. W. 1988. Plasmodial metabolism and related organellar function during various stages of the life-cycle: proteins, lipids, nucleic acids and vitamins. See Ref. 254a, pp. 219–52
224. Schneider, P., Ferguson, M. A. J., McConville, M. J., Mehlert, A., Homans, S. W., Bordier, C. 1990. Structure of the glycosyl-phosphatidylinositol membrane anchor of the *Leishmania major* promastigote surface protease. *J. Biol. Chem.* 265:16955–64
225. Schneider, P., Glaser, T. A. 1992. Characterisation of two soluble exopeptidases in the protozoan parasite *Leishmania major. Biochem. J.* Submitted
226. Schneider, P., Glaser, T. A. 1992. Purification and characterisation of a surface metalloproteinase from

Herpetomonas samuelpessoai and comparison with the promastigote surface metalloprotease (PSP) of *Leishmania major*. *Mol. Biochem. Parasitol.* In press

227. Schneider, P., Rosat, J.-P., Bouvier, J., Louis, J., Bordier, C. 1992. *Leishmania major:* differential regulation of the surface metalloprotease in amastigote and promastigote stages. *Exp. Parasitol.* 75:196–206
228. Scholze, H., Schulte, W. 1988. On the specificity of a cysteine proteinase from *Entamoeba histolytica*. *Biomed. Biochim. Acta* 47:115–23
229. Scholze, H., Schulte, W. 1990. Purification and partial characterization of the major cysteine protease from *Entamoeba histolytica*. *Biomed. Biochim. Acta* 49:455–63
230. Scholze, H., Werries, E. 1984. A weakly acidic protease has a powerful proteolytic activity in *Entamoeba histolytica*. *Mol. Biochem. Parasitol.* 11:293–300
231. Scholze, H., Werries, E. 1986. Cysteine proteinase of *Entamoeba histolytica*. I. Partial purification and action on different enzymes. *Mol. Biochem. Parasitol.* 18:103–12
232. Schrevel, J., Bernard, F., Maintier, C., Mayer, R., Monsigny, M. 1984. Detection and characterization of a selective endopeptidase from *Plasmodium berghei* by using fluorogenic peptidyl substrates. *Biochem. Biophys. Res. Commun.* 124:703–10
233. Schrevel, J., Grellier, P., Mayer, R., Monsigny, M. 1988. Neutral proteases involved in the reinvasion of erythrocytes by *Plasmodium merozoites*. *Biol. Cell* 64:233–44
234. Schulte, W., Scholze, H., Werries, E. 1987. Specificity of a cysteine proteinase of *Entamoeba histolytica* towards the α1-CB2 peptide of bovine collagen type 1. *Mol. Biochem. Parasitol.* 25: 39–43
235. Sherman, I. W., Mudd, J. B. 1966. Malaria infection (*Plasmodium lophurae*): changes in free amino acids. *Science* 154:287–89
236. Sherman, I. W., Tanigoshi, L. 1970. Incorporation of ^{14}C-amino acids by malaria (*Plasmodium lophurae*) IV. In vivo utilization of host cell hemoglobin. *Int. J. Biochem.* 1:635–37
237. Sherman, I. W., Tanigoshi, L. 1981. The proteases of *Plasmodium:* a cathepsin D–like enzyme from *Plasmodium lophurae*. In *Biochemistry of Parasites*, ed. G. Slutzky, pp. 137–49. Oxford: Pergamon
238. Sherman, I. W., Tanigoshi, L. 1983. Purification of *Plasmodium lophurae* cathepsin D and its effects on erythrocyte membrane proteins. *Mol. Biochem. Parasitol.* 8:207–26
239. Slater, A. F. G. 1992. Malaria pigment. *Exp. Parasitol.* 74:362–65
240. Souto-Padrón, T., Campetella, O., Cazzulo, J. J., de Souza, W. 1990. Cysteine proteinase in *Trypanosoma cruzi:* immunocytochemical localization and involvement in parasite-host cell interaction. *J. Cell. Sci.* 96:485–90
241. Souza, A. E., Waugh, S., Coombs, G. H., Mottram, J. C. 1992. Characterization of a multi-copy gene for a major stage-specific cysteine proteinase of *Leishmania mexicana*. *FEBS Lett.* In press
242. Steinkraus, H. B., Langer, P. J. 1992. The protein sequence predicted from a *Leishmania guyanensis* gp63 major surface glycoprotein gene is divergent as compared with other *Leishmania* species. *Mol. Biochem. Parasitol.* 52: 141–44
243. Stöcker, W., Ng, M., Auld, D. S. 1990. Fluorescent oligopeptide substrates for kinetic characterization of the specificity of *Astacus* protease. *Biochemistry* 29:10418–25
244. Tannich, E., Nickel, R., Buss H., Horstmann, R. D. 1992. Mapping and partial sequencing of the genes coding for two different cysteine proteinases in pathogenic *Entamoeba histolytica*. *Mol. Biochem. Parasitol.* 54:109–11
245. Tannich, E., Scholze, H., Nickel, R., Horstmann, R. D. 1991. Homologous cysteine proteases of pathogenic and nonpathogenic *Entamoeba histolytica*. *J. Biol. Chem* 266:4798–4803
246. Theakston, R. D. G., Fletcher, S. A., Maegraith, B. G. 1970. The use of electron microscope autoradiography for examining the uptake and degradation of haemoglobin by *Plasmodium berghei*. *Ann. Trop. Med. Parasitol.* 64:63–71
247. Tobin, J. F., Wirth, D. F. 1992. A sequence insertion targeting vector for *Leishmania enriettii*. *J. Biol. Chem.* 267:4752–58
248. Vander Jagt, D. L., Baack, B. R., Hunsaker, L. A. 1984. Purification and characterization of an aminopeptidase from *Plasmodium falciparum*. *Mol. Biochem. Parasitol.* 10:45–54
249. Vander Jagt, D. L., Caughey, W. S., Campos, N. M., Hunsaker, L. A., Zanner, M. A. 1989. Parasite proteases and antimalarial activities of protease

inhibitors. *Prog. Clin. Biol. Res.* 313: 105–18

250. Vander Jagt, D. L., Hunsaker, L. A., Campos, N. M. 1986. Characterization of a hemoglobin-degrading, low molecular weight protease from *Plasmodium falciparum. Mol. Biochem. Parasitol.* 18:389–400

251. Vander Jagt, D. L., Hunsaker, L. A., Campos, N. M. 1987. Comparison of proteases from chloroquine-sensitive and chloroquine-resistant strains of *Plasmodium falciparum. Biochem. Pharmacol.* 36:3285–91

252. Vander Jagt, D. L., Hunsaker, L. A., Campos, N. M., Scaletti, J. V. 1992. Localization and characterization of hemoglobin-degrading aspartic proteinases from the malarial parasite *Plasmodium falciparum. Biochim. Biophys. Acta* 1122:256–64

253. Walsh, J. A. 1989. Disease problems in the Third World. *Ann. N.Y. Acad. Sci.* 569:1–16

254. Webb, J. R., Button, L. L., McMaster, W. R. 1991. Heterogeneity of the genes encoding the major surface glycoprotein of *Leishmania donovani. Mol. Biochem. Parasitol.* 48:173–84

254a. Wernsdorfer, W. H., McGregor, I., eds. 1988. *Malaria: Principles and Practice of Malariology.* Edinburgh: Churchill Livingstone

255. Wilson, M. E., Hardin, K. K. 1990. The major *Leishmania donovani chagasi* surface glycoprotein in tunicamycin-resistant promastigotes. *J. Immunol.* 143:678–84

256. World Health Organization. 1986. Epidemiology and control of African trypanosomiasis. *WHO Tech. Rep. Ser.* 739:36–58

257. Yayon, A., Cabantchik, Z. I., Ginsburg, H. 1984. Identification of the acidic compartment of *Plasmodium falciparum*-infected human erythrocytes as the target of the antimalarial drug chloroquine. *EMBO J.* 3:2695–2700

258. Yayon, A., Vande Waa, J. A., Yayon, M., Geary, T. G., Jensen, J. B. 1983. Stage-dependent effects of chloroquine on *Plasmodium falciparum* in vitro. *J. Protozool.* 30:642–47

259. Young, J. D., Young, T. M., Lu, L. P., Unkeless, J. C., Cohn, Z. A. 1982. Characterization of a membrane pore-forming protein from *Entamoeba histolytica. J. Exp. Med.* 156:1677–90

260. Zarchin, S., Krugliak, M., Ginsburg, H. 1986. Digestion of host erythrocyte by malaria parasites is the primary target for quinoline-containing antimalarials. *Biochem. Pharmacol.* 35:2435–42

Annu. Rev. Microbiol. 1993. 47:855–74

THE STATIONARY PHASE OF THE BACTERIAL LIFE CYCLE

Roberto Kolter
Department of Microbiology and Molecular Genetics, Harvard Medical School, 200 Longwood Avenue, Boston, Massachusetts 02115

Deborah A. Siegele
Department of Biology, Texas A & M University, College Station, Texas 77843

Antonio Tormo
Departamento de Bioquímica y Biología Molecular, Facultad de Ciencias Químicas, Universidad Complutense de Madrid, Madrid, Spain

KEY WORDS: stationary phase, survival, starvation, gram-negative bacteria

---

CONTENTS

ABSTRACT

In the natural environment bacteria seldom encounter conditions that permit periods of exponential growth. Rather, bacterial growth is characterized by long periods of nutritional deprivation punctuated by short periods that allow

0066-4227/93/1001-0855$02.00

fast growth, a feature that is commonly referred to as the feast-or-famine lifestyle. In this chapter we review the recent advances made in our understanding of the molecular events that allow some gram-negative bacteria to survive prolonged periods of starvation. After an introductory description of the properties of starved gram-negative bacteria, the review presents three aspects of stationary phase: entry into stationary phase, responses during prolonged starvation, and reentry into the growth cycle.

INTRODUCTION

In the natural environment bacteria seldom encounter conditions that permit continuous balanced growth. When nutrients are plentiful, bacteria can sustain relatively fast growth rates. But the very fact that bacterial populations can use nutrients efficiently to generate rapid increases in their biomass means that they are nutritionally starved most of the time. Still, these organisms can survive for extremely long periods in the absence of nutrients.

Laboratory conditions do not exactly reflect what bacteria find in nature. However, one can simulate short periods of nutrient availability and prolonged periods of starvation by growing cultures in synthetic media. In most media, exponentially growing cells quickly use up the available nutrients and cease their exponential increase in biomass, thus entering a phase of the culture referred to as stationary phase. This part of the bacterial life cycle has always attracted the attention of investigators, and in recent years, through the application of modern genetic and biochemical approaches, exciting discoveries have been made with regards to the molecular mechanisms that bacteria utilize to survive during stationary phase. The ability of many bacteria to form dormant spores or multicellular aggregates in response to starvation has been extensively studied, and the reader is referred to several recent reviews on the subject (46, 52, 82, 83), two of which appear in this volume (17, 38). This review focuses on the responses that gram-negative organisms, in particular *Escherichia coli,* mount when confronted with starvation conditions in the laboratory. Several review articles that treat the same subject from different perspectives have appeared recently (32, 47, 54, 55, 80).

Stages of the Escherichia coli *Life Cycle*

During normal exponential growth, *E. coli* cells undergo cycles of cell growth and division in which daughter cells are virtually identical to the mother cell. Theoretically, the cessation of growth in response to starvation could result simply from the arrest of metabolic activity anywhere along this growth cycle. Growth could then be reinitiated by restarting the cycle from its point of arrest once nutrients were again available. Whereas this scheme might be possible, the sudden arrest of growth could halt key metabolic processes, DNA

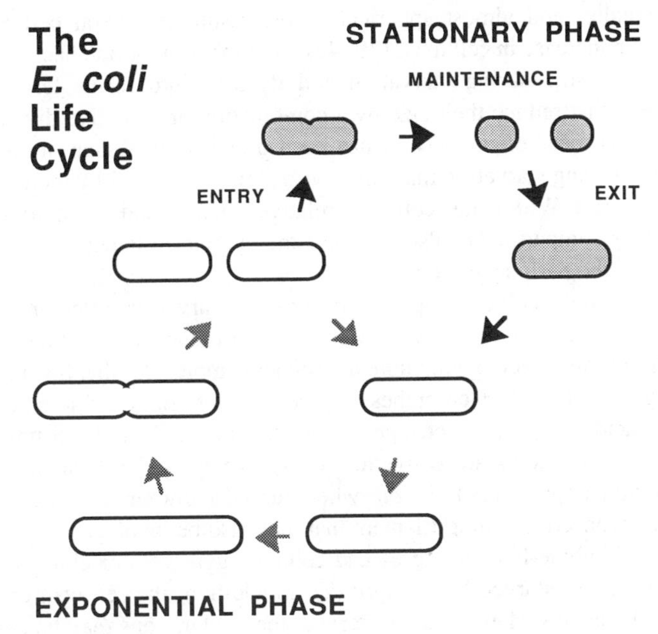

Figure 1 The three stages of stationary phase as part the *E. coli* life cycle.

replication in particular, at stages at which severe and irreparable damage could occur. In order to insure their survival, bacteria should be able to make an orderly transition into stationary phase such that the cell cycle is not arrested randomly. In addition, bacteria must also be able to remain viable during prolonged periods of starvation and to exit stationary phase and return to the exponential cell cycle when starvation is relieved. Thus the physiology of starved bacteria can be divided into three stages: entry into stationary phase, maintenance of viability, and exit from stationary phase (Figure 1). The results reviewed here show that, upon being starved, gram-negative bacteria can enter a developmental program that results in metabolically less active and more resistant cells. However, in contrast to other microbial developmental programs such as sporulation, this starvation-induced differentiation does not appear to be an all-or-none process involving an irreversible commitment to a program. Rather, the differentiation appears more gradual: the slower the growth rate of the culture, the more growing cells resemble starved cells.

Properties of Starved Gram-Negative Bacteria

Morphological changes that are brought about by starvation are apparent through both light and electron microscopic examination. The familiar rod shape of growing *E. coli* is lost in stationary phase because cells become much smaller and almost spherical as the result of several cell divisions without an increase in cell mass (39, 48). A number of marine bacteria greatly decrease in size during starvation and develop into forms termed ultramicrocells that reduce their size by a combination of reductive divisions and endogenous metabolism (47). It has been proposed that such dramatic size reduction during starvation may improve a strain's survival by increasing cell numbers (58). Within the cell, the relative volume and disposition of the subcellular compartments also changes; the cytoplasm is condensed and the volume of the periplasm increases (72).

Changes in the cell envelope that result from starvation reflect the need for protection and insulation from stressful environments. On their surface, starved cells are covered with more hydrophobic molecules that favor adhesion and aggregation (47). Membranes may become less fluid and less permeable as fatty acid composition changes. For example, in *E. coli* all unsaturated membrane fatty acids are converted to the cyclopropyl derivatives as cells enter stationary phase (19). *E. coli,* when starved at low temperatures, produce curli, a fibronectin-binding filament that may also be involved in aggregation (10, 69). Fimbrae-like structures and cellular aggregates or clumps are also characteristic of starved *Vibrio* spp. (2). In addition, the cell wall undergoes structural changes when cells are starved; these alterations may be correlated with increased resistance to autolysis (67, 92).

The chromosome undergoes topological changes consistent with the reduction in gene expression observed in starved cells. After several hours in stationary phase, changes in the negative superhelical density of plasmids become apparent (13) and the nucleoid condenses (12, 59). This could in part result from the production of large amounts of the histone-like proteins H-NS (84) and Dps (8) during starvation.

Bacterial genera such as *Escherichia, Salmonella,* and *Vibrio* are not generally considered to form differentiated cells as a result of starvation. Clearly these species do not form classical spores, but the morphological changes described above make it evident that major structural changes do occur when they are starved. These changes, in combination with changes in metabolism and physiology, confer on starved gram-negative cells many of the properties of classical spores.

The spores that result from starvation of many gram-positive bacteria are characterized by their extreme resistance to different environmental stresses. In nonsporulating gram-negative bacteria, starvation also induces the development of a more resistant state (54). *E. coli* cells that have been starved are more resistant to heat shock, oxidative stress, and osmotic challenge than exponential-phase cells (42, 43, 49). Although resistance to these stresses can be induced during growth by exposure to nonlethal levels of heat, H_2O_2, or salt, the resistance produced by starvation is even more protective.

Many of the properties identified as stationary-phase induced may also be important for growth under conditions of limiting or poor nutrient availability. Hence for some physiological processes, stationary phase may represent a maximally slow growth rate (18). Many functions induced during stationary phase are also induced when the cells are growing with a long doubling time; several promoters induced during stationary phase also show an inverse proportionality of expression as a function of growth rate (6, 7, 18). In some cases, this inverse proportionality results in the presence of a constant number of gene-product molecules per cell, which may be important for the organization of processes such as cell division (94).

ENTRY INTO STATIONARY PHASE

Definition

Bacteria growing in batch culture will inevitably reach a point when the growth rate decreases, indicating the onset of stationary phase. But what is really meant by stationary phase? Growth might be prevented by the exhaustion of any one of several essential nutrients, but does stationary phase represent a homogeneous physiological state? No—*stationary-phase cultures* is only a descriptive term: cultures in which the number of bacterial cells ceases to increase are said to be in stationary phase. This description does not distinguish whether or not the cells are metabolically active or even if they are undergoing cell division or not. It simply refers to a culture that shows no further increase in the number of cells, which points out some of the difficulties with the term. The time of the onset of stationary phase will differ depending on what criterion is used to define entry into this phase. Because of the reduction in cell size upon entry into stationary phase, cells will stop increasing in size while they continue to increase in number. If cell growth is measured by optical density, the defined onset of stationary phase will be much earlier than if the cessation of growth is defined as the moment when cell numbers cease increasing. Therefore, the investigator must recognize that entry into stationary phase is a transition period beginning at the point in the exponential phase when all cellular parameters cease increasing at equal rates, i.e. DNA, protein, and total cell mass no longer increase together, and continuing until the time when no further increase in cell number is detected.

Just as one must recognize that the entry into stationary phase is a transition period for the cell, one should understand that the cell physiology during this time will vary enormously depending on the composition of the medium in the which cells are growing and on the conditions in which starvation came about. Experimentally, a culture can enter stationary phase in many ways. The most clearly defined ones involve starvation for a single nutrient, for which two methods can be used. One can take a culture in midexponential phase, spin out the cells, and resuspend them in identical medium that lacks one specific nutrient, e.g. a source of carbon. Alternatively, cultures can be incubated in medium containing one nutrient at a concentration low enough such that it will be exhausted before all other nutrients. Although the resuspension method is useful in defining a quick transition between conditions that permit growth and those that do not, it suffers from the fact that it artificially places the cells in a new medium. If any extracellular signaling molecules accumulate during the entry into stationary phase, these will be lacking in resuspension experiments.

Protein Synthesis During Entry into Stationary Phase

The patterns of proteins synthesized during the entry into stationary phase have been analyzed extensively using two-dimensional SDS polyacrylamide gel electrophoresis and gene fusions (31, 49, 65, 66, 85–87). The results obtained have provided an initial picture of the molecular events that occur in the cell as it senses starvation. Each nutritional starvation condition that leads to the cessation of growth results in the induction of a characteristic set of proteins that accompanies the inevitable decrease in the overall rate of protein synthesis. While the proteins that are induced vary widely depending on the conditions of starvation, a core set of 15–30 proteins is always induced in *E. coli* and has been designated the Pex (postexponential) proteins (55).

A kinetic analysis of the proteins induced during the onset of stationary phase has revealed that not all proteins are induced with the same kinetics. *E. coli, Salmonella typhimurium,* and *Vibrio* spp. have temporal classes of gene expression. The expression of some genes is induced very early while others are not induced for many hours. The length of time that their expression is on also varies, but little is known about the molecular mechanisms responsible for the different kinetics.

Proteins synthesized by starved cell during entry into stationary phase are involved in maintaining viability during prolonged starvation. Inhibiting protein synthesis with chloramphenicol during the first few hours in stationary phase greatly increases the rate at which cultures lose viability, while there is only a small decrease in viability when protein synthesis is inhibited after cells have been in stationary phase for several hours (66, 72). Some of the functions needed for maintaining maximal viability may function specifically in stationary phase and be dispensable during growth. This idea is supported by the isolation of *E. coli* mutants that, while appearing normal during logarithmic growth, fail to survive during stationary phase (49, 56, 90). Given the importance of the proteins made during stationary phase, the regulation of their synthesis must be critical for the survival of the cell.

A Stationary Phase–Specific Sigma Factor

The starvation-induced expression of many genes is controlled by an alternative sigma factor known as σ^S or σ^{38}. Several investigators independently identified the gene encoding σ^S, and thus it initially received several different designations. Researchers found several regulatory mutations, without realizing they were in the same locus, that mapped near 59 min and affected a variety of processes: near ultraviolet light resistance (*nur*) (93), acid phosphatase production (*appR*) (91), and HPII catalase production (*katF*) (50). The first indication that this locus encoded an alternative sigma factor came from the nucleotide sequence of *katF* (60). Subsequently, a search for carbon starvation–inducible gene fusions identified *csi2::lacZ*, an insertion at 59 min with a pleiotropic phenotype with respect to acid phosphatase, HPII catalase production, and overall stationary-phase gene expression (49). This finding led to the recognition that *appR* and *katF* were different names for the same locus. Based on the gene's nucleotide sequence and its role in activating and repressing the synthesis of many proteins at the onset of starvation, the gene designation *rpoS* was proposed (49). Subsequent studies further demonstrated the central role of *rpoS* in the development of increased resistance at the onset of starvation (56). More recently, the purified protein was shown to have sigma-factor function in vitro (89). Hence we refer to the gene as *rpoS* and its product as σ^S.

At least 30 proteins require *rpoS* for their expression during starvation (56). Over a dozen of these proteins have been identified and many of them are important for the development of the resistant state seen in stationary phase (32). A partial list of *rpoS*-dependent genes and their products or function includes: *katE*, HPII catalase (50); *xthA*, exonuclease III (75); *appA*, acid phosphatase (91); *mcc*, microcin C7 (25); *bolA*, cell-shape determination (16, 48); *osmB*, lipoprotein (45); *treA*, periplasmic trehalase (34); *otsAB*, trehalose synthesis (34); *cyxAB*, third cytochrome oxidase (20); *glgS*, glycogen primer (33); *dps*, DNA protection (8); and *csgA*, curli fibronectin binding fibers (68).

With all of these genes identified, one might expect that a consensus sequence for *rpoS*-dependent promoters would have been derived. Although a possible −10 and −35 consensus sequence that differed from the σ^{70} consensus was proposed (48), the consensus broke down as more *rpoS*-dependent genes were identified (32). The analysis of *rpoS*-dependent promoter sequences suffers from the fact that it is still not known how many of these promoters are directly recognized by a σ^S-containing RNA polymerase holoenzyme and how many are regulated by *rpoS* indirectly. The lack of a consensus could indicate a regulatory cascade, with some genes directly transcribed by σ^S holoenzyme, while others are further down in the cascade. In addition, the little biochemical evidence available for the σ^S holoenzyme indicates that this form of RNA polymerase can recognize, in vitro, many σ^{70} promoters (89). Thus, although σ^S is clearly a sigma factor, how σ^S-dependent promoters are recognized in vivo remains unknown.

Several findings relevant to this lead us to speculate about a possible mechanism of promoter recognition by σ^S holoenzyme. First, the *rpoS* dependence of *mcc* and *csgA* promoters can be suppressed by mutations in *osmZ* (57, 68), the gene encoding the histone-like protein H-NS (or H1), which is known to affect DNA topology (84). Second, these promoters are stimulated by positive activators (10, 57). Third, genetic evidence suggests that the *dps* promoter can be transcribed by σ^S holoenzyme during starvation and by σ^{70} holoenzyme in the presence of the activator OxyR (M. Almirón, G. Storz & R. Kolter, unpublished results). Fourth, as mentioned above, σ^S holoenzyme can recognize many σ^{70} promoters in vitro, including *lacUV5*p and *trp*p (89). Based on these observations, we speculate that several promoters transcribed by σ^{70} in vivo that depend on an activator may be transcribed by σ^S holoenzyme, independent of activators. This would define a subset of promoters that respond to specific induction during growth by σ^{70} and an activator but that can also be induced by starvation via recognition by σ^S. Most questions regarding promoter recognition by σ^S holoenzyme remain unanswered, and future experimentation will require combined biochemical and physiological analyses if we are to understand the mechanisms by which a promoter can be differentially recognized by two forms of RNA polymerase under different physiological conditions.

The σ^S-dependent induction of many genes in stationary phase suggests that σ^S activity is regulated in response to starvation. Evidence from genetic analyses using transcriptional and translational fusions to *rpoS* and σ^S-dependent genes indicates transcriptional and posttranscriptional regulation of *rpoS* expression (49, 51, 61). In addition, posttranslational modification of the *rpoS* product may occur to modulate its activity (35). This multiple-level regulation is reminiscent of the regulation of the activity of the heat-shock sigma, σ^{32} (95).

The induction of the σ^S regulon at the onset of stationary phase requires that a variety of starvation conditions be recognized as such and that these signals be transduced to produce active σ^S (Figure 2). It is not known whether all nutrient limitations activate σ^S via the same or by different pathways. Several candidates have been suggested as signals for activation of the σ^S regulon. For example, subtle fluctuations in the ΔpH or $\Delta\psi$ may provide a common signal that regulatory sensors can transduce to the cell's machinery, signifying starvation (80). The concentrations of both ppGpp and AppppA drastically change as a result of a variety of stresses including starvation, and mutants defective in their synthesis display lowered viability in stationary phase (44). Another metabolite that accumulates towards the end of the

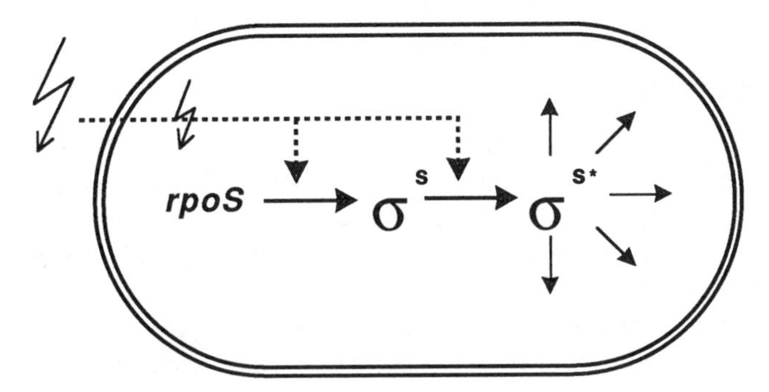

Figure 2 Schematic of the possible activation of the *rpoS* regulon by starvation signals.

logarithmic growth phase in some media and is subsequently metabolized is acetate. Although they are controversial, some reports suggest that acetate activates the σ^S regulon (76).

Other Regulators of Stationary-Phase Gene Expression

The product of *rpoS* is not the only regulatory molecule controlling the induction of transcription during stationary phase. Induction of approximately 20 of the proteins induced by carbon starvation is *rpoS* independent, as judged with two-dimensional gel electrophoresis (56). These proteins include three heat-shock proteins (DnaK, GroEL, and HtpG) whose induction during starvation depends on σ^{32} rather than on σ^S. Many genes that are induced in response to glucose starvation depend on cAMP and are induced in an *rpoS*-independent fashion (77). For example, *cstA* appears to encode a peptide-transport system (78). In addition, the stationary-phase induction of the glycogen biosynthetic genes *glgB* and *glgC* is under the control of a negative regulatory molecule, the product of the *csrA* gene (73).

Stationary-phase induction of the promoter of the microcin B17 (*mcb*) operon is independent of *rpoS,* and transcription of the *mcb* promoter in vitro requires σ^{70} (16, 48). The microcin promoter is also under the control of the positive activator OmpR (36, 37) and the negative regulator, MprA (23, 24). The −10 and −35 regions of the *mcb* promoter are nearly identical to those of the *bolA* promoter, whose stationary-phase induction is *rpoS* dependent (16, 48). This observation suggests that sequences outside the −10 and −35 regions are important for discriminating between σ^S- and σ^{70}-dependent promoters. Perhaps sequences or structures such as bends in the DNA could be important in this process, which is consistent with the hypothesis of promoter recognition discussed above.

RESPONSES DURING PROLONGED STARVATION

Protein Synthesis

After several hours in stationary phase *E. coli, Vibrio* spp., or *S. typhimurium* cells complete the developmental process that results in the observed increased resistances. Afterwards their metabolic activity is greatly reduced. A priori, there is no reason to expect that any metabolic activity remains in these cells in the subsequent days of starvation. However, starved gram-negative cells are not truly dormant—they remain metabolically active even after many weeks of starvation. This sharply contrasts with the dormant state of the spores produced by many gram-positive organisms.

E. coli cells in stationary-phase cultures continue to change in many ways. The pattern of proteins synthesized during prolonged incubations in stationary phase changes over time (8). Just as there are temporal classes of proteins expressed during the first few hours of starvation, there are temporal classes of proteins whose expression increases and decreases over many days of starvation. At the onset of stationary phase, the overall rate of protein synthesis is reduced to approximately 20% of the rate observed for a culture growing exponentially in minimal glucose medium (4). This rate drops precipitously between 10 and 20 h after the onset of stationary phase, and after 11 days in stationary phase the rate is about 0.05% that of a growing culture (M. Almirón & R. Kolter, unpublished results). Yet even at this slow overall rate, the pattern of proteins synthesized changes almost daily. The environmental changes leading to these variations in gene expression have not been investigated.

What is the role of proteins synthesized in cells that have been starved for many days? One of the proteins that continues to be synthesized at the highest relative rates for many days is Dps, a DNA-binding protein with regulatory and protective roles in starved cells (8). Although this protein is important for resistance to oxidative damage, presumably because it protects the DNA directly, is does not appear essential for the prolonged survival of glucose-starved cells. In fact, protein synthesis during prolonged starvation does not appear to be as important for survival as protein synthesis that occurs as cells enter stationary phase. When chloramphenicol is added as cells enter stationary phase, survival decreases markedly during the following days (72). However, if chloramphenicol is added after cells have been in stationary phase for many days, it still inhibits protein synthesis but has no effect on viability during, at least, the following three days (M. Almirón & R. Kolter, unpublished results). The changes in gene expression that occur after many days may be related to extremely long-term survival. Although this is a testable hypothesis, such tests remain to be done.

Survival Kinetics

What is the rate of survival during prolonged starvation? *E. coli* cells can remain viable during many days of starvation. However, the fraction of the original viable counts that survives depends on the conditions of starvation (79). For example, in M63- or MOPS-glucose–starved cultures, some investigators observe no significant decrease in colony-forming units (CFU)/ml for 7 days (49, 79), and the evidence indicates little or no cell turnover under these conditions (72). Yet other groups report a significant loss of viability during the first week of incubation (21, 72). At present the reasons for these discrepancies are not understood, but they could result from differences in strain backgrounds and the specific culture conditions. In a variety of minimal media, starvation for nutrients other than carbon (especially phosphate) always results in much faster kinetics of death (21, 79).

The survival of cultures incubated in rich medium is very different than in minimal glucose medium. In Luria broth the number of CFU/ml have decreased by one or two orders of magnitude between 2 and 5 days of incubation (79, 96). After that time, the number of viable cells (usually 10^8 CFU/ml) remains remarkably constant for many weeks and even months. These results are in agreement with those reported much earlier for *Serratia* and *Sarcinia* spp. (88). Nutrients derived from the 90–99% of the population that dies could allow some cells to grow, a phenomenon often referred to as cryptic growth (70, 71, 74). We have evidence that cryptic growth does occur; however, our data indicate that the cells that can grow under these conditions are mutants that have a competitive advantage enabling them to take over stationary-phase cultures (96).

Mutants with a Competitive Advantage in Stationary Phase

The ability of a fraction of the cells to survive prolonged periods in spent rich medium raised the possibility that surviving cells in aged cultures might be better adapted to those incubation conditions than freshly starved cells. Initially, we formulated two alternative hypotheses to explain the ability of a fraction of the cells to survive. One hypothesis proposed that after one day, a fraction of the population entered a developmental program that led to a death-resistant state. Alternatively, the second hypothesis proposed that the survivors are no different from those cells that died. But since cellular death is a stochastic process, the fraction of cells that did not die early remained viable because of nutrients released by the dying cells. To distinguish between these two hypotheses, we carried out a series of experiments in which cells from an overnight or "young" culture were mixed with cells from a 10-day-old or "aged" culture. The number of viable cells of each type was determined over the span of several days after mixing. In order to distinguish the cells from the aged and young cultures, they were differentially marked with chromosomally encoded antibiotic-resistance markers. We reasoned that if surviving cells in an aged culture resulted from a death-resistant program, they should not suffer the two-order-of-magnitude drop in viable counts when added, as a minority population, to a young culture. If, on the other hand, cells from the aged cultures were physiologically identical to cells from the young culture, then the viable counts of both types of cells should be roughly parallel, with both populations undergoing the two-order-of-magnitude drop in their viable counts.

The results obtained from such mixing experiments were quite unexpected (96). In the mixed cultures, cells from the aged cultures neither suffer a drop in viability nor remain at a constant titer as predicted by the naive hypotheses. Instead, they grow and take over the culture, resulting in the death of the young cells (Figure 3). This "competitive advantage in stationary-phase" phenotype results from mutations, not from a physiological adaptation. Although mutations at different loci appear able to confer this stationary-phase phenotype, the best-characterized mutations so far have been mapped to the *rpoS* locus (96). Several different alleles at that locus have been found to confer the phenotype, including a 46-bp duplication near the end of the gene that replaces the last 4 codons of the gene with 39 new codons. All of the *rpoS* alleles that confer this phenotype display a reduced ability to induce the expression of several σ^S-dependent genes. However, some expression of the σ^S regulon is required because null mutations in *rpoS* do not confer this phenotype.

The growth advantage of *rpoS* mutants over wild-type cells probably results

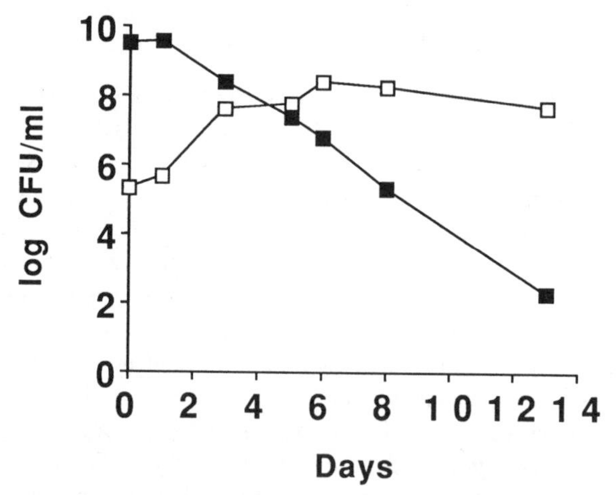

Figure 3 The ability of cells from aged cultures to grow and take over young cultures. Cells from a 10-day-old culture (*open boxes*) were introduced into a 1-day-old culture (*closed boxes*).

from changes in gene expression that affect the overall cell physiology in stationary phase. Full induction of the *rpoS*-stimulated genes results in a cell with increased resistance to environmental assaults. While better able to survive prolonged starvation, such cells may be less suited for growth. Mutations in *rpoS* that affect expression of this regulon may lead to cells that fail to develop such high resistances but are better suited for growth on scarce nutrients. Interestingly, when the *rpoS* genes from many different laboratory strains were sequenced, many different alleles were found (40). In fact, many laboratory strains of *E. coli* harbor *rpoS* alleles that express the regulon at lower levels and that confer a competitive advantage in stationary phase. This suggests that, over the years of being maintained in laboratories, these strains have been subjected to stationary-phase selections similar to those imposed on the aged cultures described here.

The selection of mutants that can grow and take over cultures is not limited to one cycle of aging. When a strain already containing an *rpoS* allele that confers a competitive advantage in stationary phase is incubated for 10 days, new mutants more fit for those conditions once again grow and take over. The mutations that result in the competitive advantage in the second cycle are found at loci other than *rpoS* (96). And recently, we selected mutants with a competitive advantage in stationary phase in a third cycle of aging (M. M. Zambrano & R. Kolter, unpublished observation).

These results show that stationary-phase cultures are dynamic and can undergo population shifts. The mixed-culture experiments revealed the presence of mutants in aged cultures that have a growth advantage under certain stationary-phase conditions. The composition of the population in the culture changes as fitter cells grow and take over the population, reminiscent of the population changes that take place during continuous growth of *E. coli* cultures (26). The periodic appearance of mutants with a growth advantage over the parent strain has also been reported in cultures growing over long periods of time in chemostats or by serial transfer techniques (11, 64). A more recent study showed that prolonged growth at higher temperature also selects for *E. coli* cells better able to grow under high-temperature conditions (15). However, in the mixed stationary-phase cultures, cells were not provided with a continuous supply of nutrients but were incubated in the same medium for several days after exponential growth had stopped. Clearly cell growth can take place under these conditions in the absence of fresh nutrients, possibly because of the release of nutrients from the dead cells.

These findings have important implications for the study of the origin of mutations in starved microorganisms. Several reports, reviewed elsewhere in this volume (28), suggest that in stationary-phase cultures mutations occur more often when they are advantageous and thus are a direct response to particular environmental challenges. The observation that mutants can grow in a stationary-phase culture raises the possibility that many of the observed postselection mutations could arise from a minority of mutant cells growing slowly in the presence of starved cells without changing the overall bacterial counts.

REENTERING THE GROWTH CYCLE FROM STATIONARY PHASE

Starved cells respond rapidly to the addition of fresh nutrients. The quickness of the initiation of the recovery process is similar to what is observed in germinating spores (29). However, germination is very different from starvation recovery or outgrowth. Bacterial spores are metabolically dormant, but as discussed above, even after prolonged starvation, many bacterial species remain metabolically active and capable of RNA and protein synthesis. Nonsporulating starved bacteria only require the addition of nutrients to resume growth, whereas spore germination is stimulated by activating agents such as heat, low or high pH, or reducing agents (29). Germination does not depend on de novo protein synthesis and is not sensitive to a variety of metabolic inhibitors. Further growth of the germinated spore is known as outgrowth. The events occurring when starved *E. coli* cells are given nutrients may be more analogous to the events occurring during spore outgrowth than to the events of germination per se.

Because starved cells can respond quickly to the availability of fresh nutrients, they must possess the pertinent nutrient-uptake systems during starvation. Indeed, starved cells are primed for nutrient uptake. For example in several marine *Vibrio* spp., the mannitol uptake system is maintained during growth and starvation while glucose- and amino acid–transport systems are induced during starvation (5, 22, 27, 30, 53).

RNA synthesis restarts almost immediately after the addition of fresh nutrients while protein synthesis lags a short time. When *Vibrio* sp. S14 is starved for carbon, nitrogen, and phosphate by resuspension, RNA synthesis increases within 20 s of addition of nutrients and protein synthesis increases after 2 min (3). The initial increase in protein synthesis does not depend on new RNA synthesis; it is not inhibited by the addition of rifampicin. The protein synthesis that occurs in the absence of new RNA synthesis must result from translation of messages already present in the cell. This observation suggests that long-lived transcripts are present during starvation. In *E. coli* starved for glucose by resuspension or by exhaustion, RNA and protein synthesis also increases very rapidly (41; D. A. Siegele & R. Kolter, submitted).

Increases in cell mass, the rate of DNA synthesis, and cell number follow the reinitiation of RNA and protein synthesis. Cells start to increase in size relatively soon after the addition of nutrients while the initiation of DNA synthesis lags. The lag time seen before DNA synthesis is initiated presumably reflects the time needed for the mass of cells to reach a critical size (4, 9).

At some point after resuming growth, cells lose the enhanced resistance to various environmental challenges that is characteristic of stationary phase cells. Once expression of σ^S-dependent genes ceases, loss of resistance could occur gradually as the molecules involved are diluted out during growth. Alternatively, active mechanisms could turn over such molecules. In the case of Dps, its concentrations are reduced by dilution rather than by degradation (M. Almirón & R. Kolter, unpublished results). In contrast, starved *Vibrio* sp. S14 cells lose their heat resistance abruptly after the addition of nutrients (4).

The expression of some genes is limited to the recovery period. Two-dimensional polyacrylamide gel analysis of protein synthesis identified different kinetic classes of proteins specific for the recovery period in *Vibrio* sp. S14 (4). The synthesis of approximately 20 starvation-inducible proteins is rapidly repressed after the addition of nutrients to starved cells. Proteins whose synthesis is induced during recovery are temporally regulated, and at least three kinetic classes have been identified. Group I proteins were synthesized during the first 20 min, while group II and III proteins were induced 20 and 80 min, respectively, after nutrient addition. Eleven of the group I and II proteins are not synthesized in starved cells or in exponentially growing cells. This observation suggests not only that regulatory mechanisms exist to turn on and off expression of these groups of genes, but that the gene products are only needed during this particular growth transition.

In *E. coli,* the only protein currently known to be induced during recovery is Fis, an abundant DNA-binding protein with little sequence specificity. Little or no detectable Fis protein is detected in stationary-phase cells, but a burst of Fis synthesis occurs when stationary-phase cells are transferred to fresh medium (14, 62, 63). Mutants lacking Fis protein have an increased lag period and cannot respond quickly to nutrient upshifts during growth (14, 62). Cells in which Fis protein production continues in stationary phase because of its expression from a heterologous promoter have reduced viability in stationary phase, and surviving cells also have a prolonged lag period, indicating that both excessive and deficient Fis levels are deleterious and that the level of Fis protein must be carefully balanced for optimal growth in environments where the nutrient supply fluctuates (14).

Are there cellular functions specific for the recovery period? Although many such functions are likely, only one has been identified in *E. coli:* the product of the *surB* gene (D. A. Siegele & R. Kolter, submitted). Mutants lacking *surB* cannot resume aerobic growth at high temperature after glucose starvation. These mutant cells do not have an overall temperature-sensitive phenotype. Once growth has started at low temperature they can be shifted to high temperature and grow normally; the temperature-sensitive period for these mutants is specific during stationary-phase exit. Mutants analogous to *surB* and temperature sensitive for vegetative outgrowth have been isolated in *Bacillus subtilis* (1). When germinated at the nonpermissive temperature, the mutants arrest growth at a specific stage characteristic for each mutant strain. The mutants germinate normally, and vegetative growth is normal at high temperature, indicating that the gene products are needed only during the initiation of growth after spore germination.

SUMMARY AND PERSPECTIVES

We feel the reader should take home two main points from this review. First, gram-negative bacteria can enter and exit starvation conditions successfully as a result of elaborate physiological and morphological differentiation. Second, in contrast to other differentiation pathways, the starvation response in these organisms represents the extreme end of a spectrum; many of the starvation-inducible genes are expressed in such a way that their basal level during growth increases as the growth rate decreases. The adaptations that bacteria undergo during starvation are not limited to a few aspects of the cell but rather involve global changes in cell physiology. All the physiological processes that we know so well during growth likely undergo interesting and varied alterations in the stationary phase. Stationary-phase cultures hold unexplored terrain awaiting those interested in almost any area of bacterial physiology.

ACKNOWLEDGMENTS

We gratefully acknowledge the many colleagues that sent us published and unpublished material to help in the writing of this review. Work on various aspects of stationary phase has been supported by a grant from the National Science Foundation to R. K. D. S. was the recipient of a US Public Health Service Postdoctoral Fellowship. R. K. is the recipient of an American Cancer Society Faculty Research Award.

Literature Cited

1. Albertini, A. M., Baldi, M. L., Ferrari, E., Isnenghi, E., Zambelli, M. T., Galizzi, A. 1979. Mutants of *Bacillus subtilis* affected in spore outgrowth. *J. Gen. Microbiol.* 110:351–63
2. Albertson, N. H. 1990. *On the adaptation of the marine* Vibrio *S14 to starvation and recovery*. PhD thesis. Univ. Göteborg, Göteborg, Sweden
3. Albertson, N. H., Nyström, T., Kjelleberg, S. 1990. Functional m-RNA half-lives in the marine *Vibrio* sp. S14 during starvation and recovery. *J. Gen. Microbiol.* 136:2195–99

4. Albertson, N. H., Nyström, T., Kjelleberg, S. 1990. Macromolecular synthesis during recovery of the marine *Vibrio* sp. S14 from starvation. *J. Gen. Microbiol.* 136:2201–7
5. Albertson, N. H., Nyström, T., Kjelleberg, S. 1990. Starvation-induced modulations in binding protein-dependent glucose transport by the marine *Vibrio* sp. S14. *FEMS Microbiol. Lett.* 70:205–10
6. Aldea, M., Garrido, T., Hernandez-Chico, C., Vicente, M., Kushner, S. R. 1989. Induction of a growth-phase-dependent promoter triggers transcription of *bolA*, an *E. coli* morphogene. *EMBO J.* 8:3923–31
7. Aldea, M., Garrido, T., Pla, J., Vicente, M. 1990. Division genes in *Escherichia coli* are expressed coordinately to cell septum requirements by gearbox promoters. *EMBO J.* 9:3787–94
8. Almirón, M., Link, A., Furlong, D., Kolter, R. 1992. A novel DNA binding protein with regulatory and protective roles in starved *E. coli*. *Genes Dev.* 6:2646–54
9. Amy, P. S., Pauling, C., Morita, R. Y. 1983. Recovery from nutrient starvation by a marine *Vibrio* sp. *Appl. Environ. Microbiol.* 45:1685–90
10. Arnqvist, A., Olsén, A., Pfeifer, J., Russel, D. G., Normark, S. 1992. The Crl protein activates cryptic genes for curli formation and fibronectin binding in *Escherichia coli*. *Mol. Microbiol.* 6:2443–52
11. Atwood, K. C., Schneider, L. K., Ryan, F. J. 1951. Periodic selection in *Escherichia coli*. *Proc. Natl. Acad. Sci. USA* 37:146–55
12. Baker, R. M., Singleton, F. L., Hood, M. A. 1983. Effects of nutrient deprivation on *Vibrio cholerae*. *Appl. Environ. Microbiol.* 46:930–40
13. Balke, V. L., Gralla, J. D. 1987. Changes in the linking number of supercoiled DNA accompany growth transitions. *J. Bacteriol.* 169:4499–4506
14. Ball, C. A., Osuna, R., Ferguson, K. C., Johnson, R. C. 1992. Dramatic changes in Fis levels upon nutrient upshift in *Escherichia coli*. *J. Bacteriol.* 174:8043–56
15. Bennet, A. F., Dao, K. M., Lenski, R. E. 1990. Rapid evolution in response to high temperature selection. *Nature* 346:79–81
16. Bohannon, D. E., Connell, N., Keener, J., Tormo, A., Espinosa-Urgel, M., et al. 1991. Stationary-phase-inducible "gearbox" promoters: differential effects of *katF* mutations and the role of σ^{70}. *J. Bacteriol* 173:4482–92
17. Chater, K. 1993. Genetics of differentiation in *Streptomyces*. *Annu. Rev. Microbiol.* 47:685–713
18. Connell, N., Han, Z., Moreno, F., Kolter, R. 1987. An *E. coli* promoter induced by the cessation of growth. *Mol. Microbiol.* 1:195–201
19. Cronan, J. E. 1968. Phospholipid alterations during growth of *Escherichia coli*. *J. Bacteriol.* 95:2054–61
20. Dassa, J., Fsihi, H., Marck, C., Dion, M., Kieffer-Bontemps, M., Boquet, P. L. 1992. A new oxygen-regulated operon in *Escherichia coli* comprises the genes for a putative third cytochrome oxidase and for pH 2.5 acid phosphatase (*appA*). *Mol. Gen. Genet.* 229:342–52
21. Davis, B. D., Luger, S. J., Tai, P. C. 1986. Role of ribosome degradation in the death of starved *E. coli* cells. *J. Bacteriol.* 166:439–45
22. Davis, C. L., Robb, F. T. 1985. Maintenance of different mannitol uptake systems during starvation of oxidative and fermentative marine bacteria. *Appl. Environ. Microbiol.* 50:743–48
23. del Castillo, I., Gómez, J. M., Moreno, F. 1990. *mprA*, an *Escherichia coli* gene that reduces growth-phase-dependent synthesis of microcins B17 and C7 and blocks osmoinduction of *proU* when cloned on a high-copy-number plasmid. *J. Bacteriol.* 172:437–45
24. del Castillo, I., González-Pastor, J. E., SanMillán, J. L., Moreno, F. 1991. Nucleotide sequence of the *Escherichia coli* regulatory gene *mprA* and construction and characterization of *mprA*-deficient mutants. *J. Bacteriol.* 173: 3924–29
25. Díaz-Guerra, L., Moreno, F., Millan, J. L. S. 1989. *appR* gene product activates transcription of microcin C7 plasmid genes. *J. Bacteriol.* 171:2906–8
26. Dykhuizen, D. E., Hartl, D. L. 1983. Selection in chemostats. *Microbiol. Rev.* 47:150–68
27. Faquin, W. C., Oliver, J. D. 1984. Arginine uptake by a psychrophilic marine *Vibrio* sp. during starvation-induced morphogenesis. *J. Gen. Microbiol.* 130:1331–35
28. Foster, P. 1993. Adaptive mutation: the uses of adversity. *Annu. Rev. Microbiol.* 47:467–504
29. Foster, S. J., Johnstone, K. 1989. The trigger mechanism of bacterial spore germination. See Ref. 82a, pp. 89–108
30. Geesey, G. G., Morita, R. Y. 1979.

Capture of low arginine at low concentrations by a marine psychrophilic bacterium. *Appl. Environ. Microbiol.* 38:1092–97

31. Groat, R. G., Schultz, J. E., Zychlinsky, E., Bockman, A., Matin, A. 1986. Starvation proteins in *Escherichia coli:* kinetics of synthesis and role in starvation survival. *J. Bacteriol.* 168:486–93
32. Hengge-Aronis, R. 1993. Survival of hunger and stress: the role of *rpoS* in early stationary phase gene regulation in Escherichia coli. *Cell* 72:165–68
33. Hengge-Aronis, R., Fischer, D. 1992. Identification and molecular analysis of *glgS,* a novel growth-phase-regulated and *rpoS*-dependent gene involved in glycogen synthesis in *Escherichia coli. Mol. Microbiol.* 6:1877–86
34. Hengge-Aronis, R., Klein, W., Lange, R., Rimmele, M., Boos, W. 1991. Trehalose synthesis genes are controlled by the putative sigma factor encoded by *rpoS* and are involved in stationary phase thermotolerance in *Escherichia coli. J. Bacteriol.* 173:7918–24
35. Hengge-Aronis, R., Lange, R., Henneberg, N., Fischer, D. 1993. Osmotic regulation of *rpoS*-controlled genes in *Escherichia coli. J. Bacteriol.* 175:259–65
36. Hernández-Chico, C., Herrero, M., Rejas, M., SanMillán, J. L., Moreno, F. 1982. Gene *ompR* and regulation of microcin B17 and colicin E2 synthesis. *J. Bacteriol.* 152:897–900
37. Hernández-Chico, C., San Millán, J. L., Kolter, R., Moreno, F. 1986. Growth phase and OmpR regulation of transcription of the Microcin B17 genes. *J. Bacteriol.* 167:1058–65
38. Hoch, J. A. 1993. Regulation of the phosphorelay and the initiation of sporulation in *Bacillus subtilis. Annu. Rev. Microbiol.* 47:441–65
39. Ingraham, J. L., Maaløe, O., Neidhardt, F. C. 1983. *Growth of the Bacterial Cell.* Sunderland, MA: Sinauer
40. Ivanova, A., Renshaw, M., Guntaka, R. V., Eisenstark, A. 1993. DNA base sequence variability in *katF* (putative sigma factor) gene of *Escherichia coli. Nucleic Acids Res.* 20:5479–80
41. Jacobson, A., Gillespie, D. 1968. Metabolic events occurring during recovery from prolonged glucose starvation in *Escherichia coli. J. Bacteriol.* 95:1030–39
42. Jenkins, D. E., Auger, E. A., Matin, A. 1991. Role of RpoH, a heat shock regulator protein, in *Escherichia coli* carbon starvation protein synthesis and survival. *J. Bacteriol.* 173:1992–96
43. Jenkins, D. E., Chaisson, S. A., Matin, A. 1990. Starvation-induced cross protection against osmotic challenge in *Escherichia coli. J. Bacteriol.* 172: 2779–81
44. Johnstone, D. B., Farr, S. B. 1991. AppppA binds to several proteins in *Escherichia coli,* including the heat shock and oxidative stress proteins DnaK, GroEL, E89, C45, C40. *EMBO J.* 10:3897–3904
45. Jung, J. U., Gutierrez, C., Martin, F., Ardourel, M., Villarejo, M. 1990. Transcription of *osmB,* a gene encoding an *Escherichia coli* lipoprotein, is regulated by dual signals. *J. Biol. Chem.* 265:10574–81
46. Kim, S. K., Kaiser, D., Kuspa, A. 1992. Control of cell density and pattern by intercellular signalling in *Myxococcus* development. *Annu. Rev. Microbiol.* 46:117–39
47. Kjelleberg, S., Hermansson, M., Mårdén, P., Jones, G. W. 1987. The transient phase between growth and nongrowth of heterotrophic bacteria, with emphasis on the marine environment. *Annu. Rev. Microbiol.* 41:25–49
48. Lange, R., Hengge-Aronis, R. 1991. Growth phase-regulated expression of *bolA* and morphology of stationary-phase *Escherichia coli* cells are controlled by the novel sigma factor σ^S. *J. Bacteriol.* 173:4474–81
49. Lange, R., Hengge-Aronis, R. 1991. Identification of a central regulator of stationary phase gene expession in *E. coli. Mol. Microbiol.* 5:49–59
50. Loewen, P. C., Triggs, B. L. 1984. Genetic mapping of *katF,* a locus that with *katE* affects the synthesis of a second catalase species in *Escherichia coli. J. Bacteriol.* 160:668–75
51. Loewen, P. C., von Ossowski, I., Switala, J., Mulvey, M. R. 1993. KatF (σ^S) synthesis in *Escherichia coli* is subject to posttranscriptional regulation. *J. Bacteriol.* 175:2150–53
52. Losick, R., Stragier, P. 1992. Crisscross regulation of cell-type-specific gene expression during development in *B. subtilis. Nature* 355:601–4
53. Mårdén, P., Nyström, T., Kjelleberg, S. 1987. Uptake of leucine by a gram-negative heterotrophic bacterium during exposure to starvation conditions. *FEMS Microbiol. Ecol.* 45:223–41
54. Matin, A. 1991. The molecular basis of carbon-starvation-induced general resistance in *Escherichia coli. Mol. Microbiol.* 5:3–10

55. Matin, A., Auger, E. A., Blum, P. H., Schultz, J. E. 1989. Genetic basis of starvation survival in nondifferentiating bacteria. *Annu. Rev. of Microbiol.* 43:293–316
56. McCann, M. P., Kidwell, J. P., Matin, A. 1991. The putative σ factor KatF has a central role in development of starvation-mediated general resistance in *Escherichia coli. J. Bacteriol.* 173: 4188–94
57. Moreno, F., San Millán, J. L., del Castillo, I., Gómez, J. M., Rodríguez-Sáinz, M. C., et al. 1992. *Escherichia coli* genes regulating the production of microcins MCCB17 and MCCC7. In *Bacteriocins, Microcins, and Lantibiotics,* ed. R. James, C. Lazdunski, F. Pattus, pp. 3–14. Heidelberg: Springer-Verlag
58. Morita, R. Y. 1986. Autoecological studies and marine ecosystems. In *Microbial Autoecology: A Method for Environmental Studies,* ed. R. L. Tate III, pp. 147–81. New York: Wiley & Sons
59. Moyer, C. L., Morita, R. Y. 1989. Effect of growth rate and starvation-survival on the viability and stability of a psychrophilic marine bacterium. *Appl. Environ. Microbiol.* 55:1122–27
60. Mulvey, M. R., Loewen, P. C. 1989. Nucleotide sequence of *katF* of *Escherichia coli* suggests KatF protein is a novel σ transcription factor. *Nucleic Acids Res.* 17:9979–91
61. Mulvey, M. R., Switala, J., Borys, A., Loewen, P. C. 1990. Regulation of transcription of *katE* and *katF* in *Escherichia coli. J. Bacteriol.* 172: 6713–20
62. Nilsson, L., Verbeek, H., Vijgenboom, E., van Drunen, C., Vanet, A., Bosch, L. 1992. FIS-dependent *trans* activation of stable RNA operon of *Escherichia coli. J. Bacteriol.* 174:921–29
63. Ninneman, O., Koch, C., Kahmann, R. 1992. The *E. coli fis* promoter is subject to stringent control and autoregulation. *EMBO J.* 11:1075–83
64. Novick, A., Szilard, L. 1950. Experiments with the chemostat on spontaneous mutations of bacteria. *Proc. Natl. Acad. Sci. USA* 36:708–19
65. Nyström, T., Albertson, N., Kjelleberg, S. 1988. Synthesis of membrane and periplasmic proteins during starvation of a marine *Vibrio* sp. *J. Gen. Microbiol.* 134:1645–51
66. Nyström, T., Flärdh, K., Kjelleberg, S. 1990. Responses to multiple-nutrient starvation in marine *Vibrio* sp. strain CCUG15956. *J. Bacteriol.* 172:7085–97
67. Nyström, T., Kjelleberg, S. 1989. Role of protein synthesis in the cell division and starvation resistance to autolysis of marine *Vibrio* during the initial phases of starvation. *J. Gen. Microbiol.* 135:1599–1606
68. Olsén, A., Arnqvist, A., Hammer, M., Sukupolvi, S., Normark, S. 1993. The RpoS sigma factor relieves H-NS mediated transcriptional repression of *csgA,* the subunit gene of fibronectin binding curli in *Escherichia coli. Mol. Microbiol.* 7:523–36
69. Olsén, A., Jonsson, A., Normark, S. 1989. Fibronectin binding mediated by a novel class of surface organelles on *Escherichia coli. Nature* 338:652–55
70. Postgate, J. R. 1967. Viability measurements and the survival of microbes under minimal stress. *Adv. Microbiol. Physiol.* 1:2–23
71. Postgate, J. R., Hunter, J. R. 1962. The survival of starved bacteria. *J. Gen. Microbiol.* 29:233–63
72. Reeve, C. A., Amy, P. S., Matin, A. 1984. Role of protein synthesis in the survival of carbon-starved *Escherichia coli* K-12. *J. Bacteriol.* 160:1041–46
73. Romeo, T., Gong, M., Liu, M. Y., Brun, A. M. 1993. Characterization of *csrA,* a gene that affects glycogen biosynthesis, gluconeogenesis, cell size and surface properties: evidence for a new global regulon in *Escherichia coli* K-12. *J. Bacteriol.* Submitted
74. Ryan, F. J. 1959. Bacterial mutation in stationary phase and the question of cell turnover. *J. Gen. Microbiol.* 21:503–39
75. Sak, B. D., Eisenstark, A., Touati, D. 1989. Exonuclease III and the hydroperoxidase II in *Escherichia coli* are both regulated by the *katF* product. *Proc. Natl. Acad. Sci. USA* 86:3271–75
76. Schellhorn, H. E., Stones, V. L. 1992. Regulation of *katF* and *katE* in *Escherichia coli* K-12 by weak acids. *J. Bacteriol* 174:4769–76
77. Schultz, J. E., Latter, G. I., Matin, A. 1988. Differential regulation by cyclic AMP of starvation protein synthesis in *Escherichia coli. J. Bacteriol.* 170:3903–9
78. Schultz, J. E., Matin, A. 1991. Molecular and functional characterization of a carbon starvation gene of *Escherichia coli. J. Mol. Biol.* 218:129–40
79. Siegele, D. A., Almirón, M., Kolter, R. 1993. Approaches to the study of survival and death in stationary phase

Escherichia coli. In *Starvation in Bacteria,* ed. S. Kjelleberg. New York: Plenum. In press
80. Siegele, D. A., Kolter, R. 1992. Life after log. *J. Bacteriol.* 174:345–48
81. Deleted in press
82. Smith, I. 1989. Initiation of sporulation. See Ref. 82a, pp. 185–210
82a. Smith, I., Slepecky, R. A., Setlow, P., eds. 1989. *Regulation of Procaryotic Development*. Washington, DC: Am. Soc. Microbiol.
83. Sonenshein, A. L. 1989. Metabolic regulation of sporulation and other stationary phase phenomena. See Ref. 82a, pp. 109–30
84. Spassky, A., Rimsky, S., Garreau, H., Buc, H. 1984. H1a, an *E. coli* DNA-binding protein which accumulates in stationary phase, strongly compacts DNA in vitro. *Nucleic Acids Res.* 12:5321–40
85. Spector, M. P., Aliabadi, Z., Gonzalez, T., Foster, J. W. 1986. Global control in *Salmonella typhimurium:* two-dimensional electrophoretic analysis of starvation-, anaerobiosis-, and heat shock–inducible proteins. *J. Bacteriol.* 168:420–24
86. Spector, M. P., Cubitt, C. L. 1992. Starvation-inducible loci of *Salmonella typhimurium:* regulation and roles in starvation-survival. *Mol. Microbiol.* 6: 1467–76
87. Spector, M. P., Park, Y. K., Tirgari, S., Gonzalez, T., Foster, J. W. 1988. Identification and characterization of starvation-regulated genetic loci in *Salmonella typhimurium* by using Mud-directed *lacZ* operon fusions. *J. Bacteriol.* 170:345–51
88. Steinhaus, E. A., Birkeland, J. M. 1939. Studies on the life and death of bacteria. I. The senescent phase in aging cultures and the probable mechanisms involved. *J. Bacteriol.* 38:249–61
89. Tanaka, K., Takayanasi, Y., Fujita, N., Ishihama, A., Takahashi, H. 1993. Heterogeneity of principal sigma factor in *Escherichia coli:* the *rpoS* gene product, σ^{38}, is a principal sigma factor of RNA polymerase in stationary phase *Escherichia coli*. *Proc. Natl. Acad. Sci. USA* 90:3511–15
90. Tormo, A., Almirón, M., Kolter, R. 1990. *surA,* an *Escherichia coli* gene essential for survival in stationary phase. *J. Bacteriol.* 172:4339–47
91. Touati, E., Dassa, E., Boquet, P. L. 1986. Pleiotropic mutations in *appR* reduce pH 2.5 acid phosphatase expression and restore succinate utilization in CRP-deficient strains of *Escherichia coli*. *Mol. Gen. Genet.* 202:257–64
92. Tuomanen, E., Markiewicz, Z., Tomasz, A. 1988. Autolysis-resistant peptidoglycan of anomalous composition in amino-acid-starved *Escherichia coli*. *J. Bacteriol.* 170:1373–76
93. Tuveson, R. W. 1981. The interaction of a gene (*nur*) controlling near-UV sensitivity and the *polA1* gene in strains of *E. coli* K-12. *Photochem. Photobiol.* 33:919–23
94. Vicente, M., Kushner, S. R., Garrido, T., Aldea, M. 1991. The role of the "gearbox" in the transcription of essential genes. *Mol. Microbiol.* 5:2085–91
95. Yura, T., Nagai, H., Mori, H. 1993. Regulation of the heat-shock response in bacteria. *Annu. Rev. Microbiol.* 47: 321–50
96. Zambrano, M. M., Siegele, D. A., Almirón, M., Tormo, A., Kolter, R. 1993. Microbial competition: *Escherichia coli* mutants that take over stationary phase cultures. *Science* 259: 1757–60

Annu. Rev. Microbiol. 1993. 47:875–912

POLYKETIDE SYNTHESIS: Prospects for Hybrid Antibiotics

Leonard Katz and Stefano Donadio[1]
Abbott Laboratories, Abbott Park, Illinois 60064

KEY WORDS: biosynthesis, fatty acid synthesis, multienzyme complex, multifunctional enzyme, *Streptomyces* spp.

CONTENTS

ABSTRACT

Polyketides fall into two structural classes: aromatic and complex. The former are built mainly from acetate units through a reiterative process wherein the β-carbonyl groups formed after each condensation cycle are left largely unreduced. Complex polyketides are composed of acetates, propionates, or butyrates, and the extent of β-carbonyl reduction varies from one cycle to the next. Two themes for polyketide synthases are emerging. Aromatic PKSs are determined by four to six genes encoding mono- or bifunctional enzymes; one

[1]Present address: Lepetit Research Center, via R. Lepetit 34, 21040 Gerenzano (VA), Italy.

0066-4227/93/1001-0875$02.00

PKS complex is used for all synthesis steps. Complex PKSs are composed of several mulitfunctional polypeptides that contain enzymatic domains for the condensation and reduction steps; each domain is used at a unique step in the pathways, and the extent of β-carbonyl processing depends on the functional domains operating at that cycle. Mutations rendering certain domains nonfunctional have been introduced into genes for complex polyketides, resulting in the production of novel molecules.

INTRODUCTION

The term *polyketide* defines a class of molecules produced through the successive condensation of small carboxylic acids. This diverse group includes plant flavonoids, fungal aflatoxins, and hundreds of compounds of different structures that exhibit antibacterial, antifungal, antitumor, and anthelmintic properties. Some polyketides produced by fungi and bacteria are associated with sporulation or other developmental pathways; others do not yet have an ascribed function. Some polyketides have more than one pharmacological effect: the macrocyclic compound FK506, isolated as an immunosuppressive agent, has antifungal activity (67); the antifungal compound thysanone is also an inhibitor of rhinovirus protease (105); and the antibiotic erythromycin induces gastric contractions (58) and can act as a motilin agonist (92).

The diversity of polyketide structures reflects the wide variety of their biological properties. Among the smallest are 6-methylsalicylic acid (Figure 1), a six-carbon chain that cyclizes to form an aromatic ring, and the antibiotic furanomycin, a linear seven-carbon chain that undergoes epoxidation and transamination (62). The largest known polyketide is brevitoxin B with 50 carbon atoms in its chain. Many acetate-derived polyketides cyclize to form aromatic rings (e.g. anthracyclines) or lactone rings (e.g. macrolides, polyenes). Many cyclized polyketides undergo glycosidation at one or more sites (e.g. macrolides, anthracyclines), and virtually all are modified during their synthesis through hydroxylation, reduction, epoxidation, etc. The majority of polyketides are produced by the actinomycetes as secondary metabolites. The compounds are subdivided into structural subgroups: macrolides (e.g. spiramycin), polyethers (e.g. monensin), polyenes (e.g. amphothericin), and aromatic compounds (e.g. thysanone), as exemplified in Figure 1.

Because of the clinical importance of polyketides, numerous programs have been developed to obtain new structures with enhanced biological or pharmacological properties. The pioneering work of D. A. Hopwood and his colleagues in developing vectors and methods to introduce DNA into actinomycetes, along with their studies on the genetics of biosynthesis of actinorhodin, prompted interest in the rational design of novel or hybrid compounds through genetic manipulation. In 1990, Hopwood & Sherman

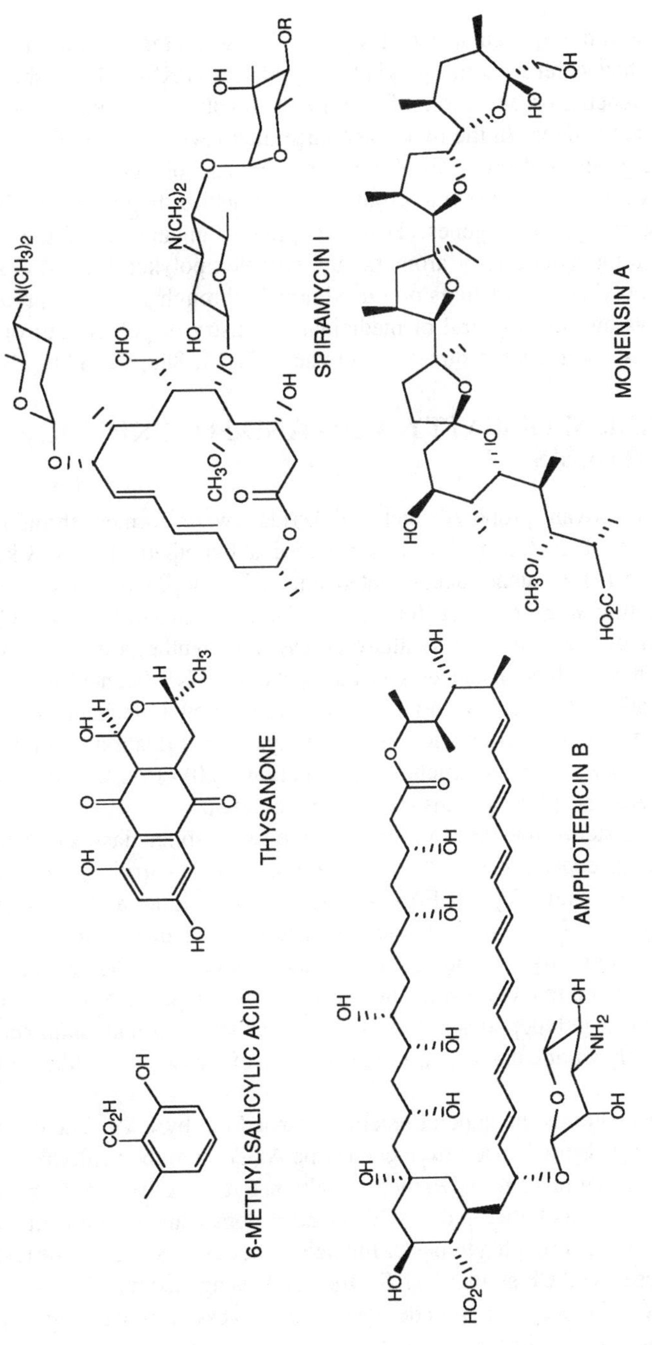

6-METHYLSALICYLIC ACID

THYSANONE

SPIRAMYCIN I

AMPHOTERICIN B

MONENSIN A

Figure 1 Structures of selected polyketides.

(53) reviewed the molecular genetics of polyketide synthesis with a focus on the genes that determine the polyketide synthases (PKSs). They showed a common genetic organization of the PKSs involved in the synthesis of aromatic polyketides. In the two years since their review appeared, information on the genetics of synthesis of the more complex polyketides has emerged and attempts have been made to produce new structures by mixing or altering polyketide biosynthesis genes. Here we provide an update of the recent findings on the synthesis of aromatic and complex polyketides and describe the novel polyketide structures produced largely through genetic approaches. For reviews on the structural or medicinal chemistry of polyketides or their biosynthesis, the reader is directed elsewhere (7, 18, 88a, 91, 111, 118).

MECHANISM OF FATTY ACID AND POLYKETIDE BIOSYNTHESIS

Birch & Donovan proposed that polyketides were formed through the condensation of acetate residues to produce a hypothetical poly-β-ketone (reviewed in 14). Evidence accumulated since this first proposal has indicated a substantial analogy between formation of long chain fatty acids (LCFA), carried out by the fatty acid synthase (FAS), and synthesis of polyketides, although in very few instances has the particular PKS been characterized biochemically (53). Here we briefly describe the known mechanism of FAS and apply a similar mechanistic model to polyketide formation and refer the reader to reviews on FAS structure and function (120) or on the comparison between FAS and PKS systems (53) for more detail.

All FAS systems thus far examined fall into two distinct classes (120). The type I systems consist of a multifunctional polypeptide carrying the required activities as domains. Type I FAS is usually a homodimer and is typical of animal systems. Type II systems, characteristic of plants and bacteria, comprise several discrete polypeptides, each carrying a distinct activity, loosely associated in a complex. The yeast FAS is composed of two different multifunctional polypeptides (84, 97). In *Brevibacterium ammoniagenes,* it is a single polypeptide that appears to be a fusion of the two yeast-like proteins (83).

Figure 2 diagrams the general reactions carried out by FAS. The extender unit, malonate, is transferred from coenzyme A (CoA) to the pantotheine arm on the acyl carrier protein (ACP) by the acyltransferase (AT). Decarboxylative condensation occurs between the ACP-bound malonate and the nascent chain, which is attached through a thioester linkage to the active-site cysteine residue of the β-ketoacyl ACP synthase (KS), the condensing enzyme. The resulting ACP-bound β-ketoacyl chain undergoes three successive processing steps, a β-ketoreduction, a dehydration, and an enoylreduction, by the action of the

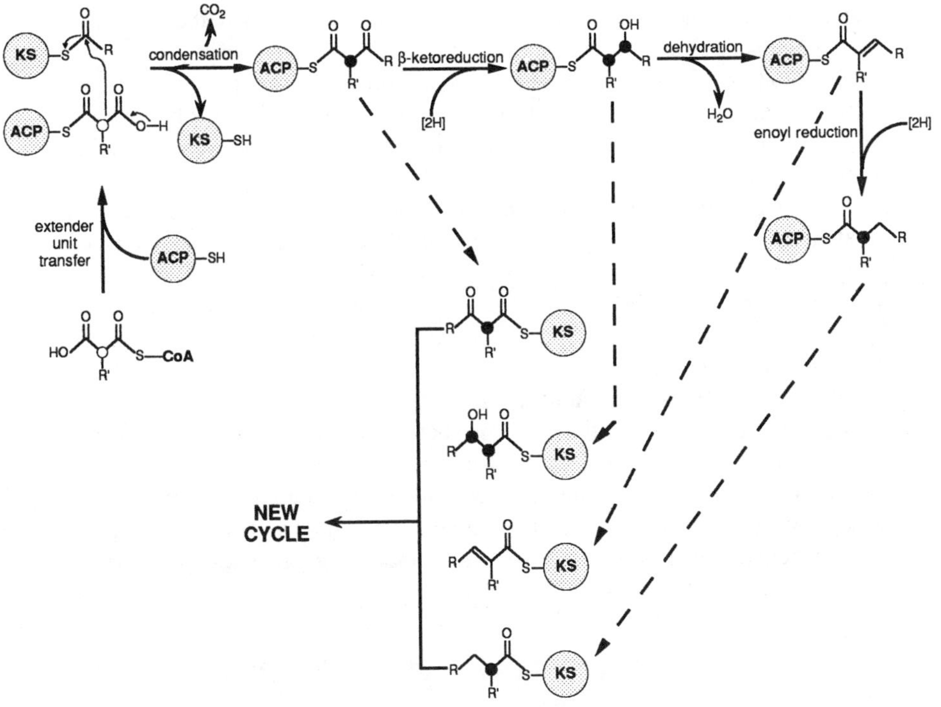

Figure 2 Reactions in fatty acid and polyketide formation. Chiral centers are denoted by a dot. Note change from open to filled dot at the α-position indicating inversion of configuration.

β-ketoreductase (KR), dehydratase (DH), and enoylreductase (ER), respectively. The elongated chain is then transacylated to the KS, and a new cycle can initiate. This process is repeated until the desired chain length is reached. In animal FAS, chain length is controlled by the narrow specificity of the thioesterase (TE), so that a limited spectrum of chains are released (74).

Polyketides are believed to be produced by an analogous mechanism, employing similar enzymatic activities. Polyketide synthesis, however, differs from LCFA formation in four aspects: (*a*) different starter units (linear or branched carboxylic acids, aromatic and aliphatic rings, etc) are used for polyketides, whereas acetate, or occasionally propionate or branched-chain carboxylates, are employed for LCFA (53); (*b*) the extent of processing may not be constant throughout synthesis of the polyketide, and a new cycle may initiate with an acyl chain containing a β-keto, β-hydroxy, α,β-ene- or fully reduced β-carbon (Figure 2); (*c*) chiral centers are introduced during the synthesis, from either the presence of a side chain on the extender unit or the maintenance of the β-hydroxyl group in the polyketide, and the absolute stereochemistry of identical substituents may vary along the chain (Figure 2); (*d*) termination of synthesis usually is accompanied by other processes, such as folding and cyclization, lactonization, or formation of an amide bond with an amino acid, all of which are believed to occur while the acyl chain is still enzyme bound.

These differences notwithstanding, all biochemical investigations on 6-methylsalicylic acid synthase (MSAS), the only PKS extensively characterized to date, have indicated an overall similarity to FAS (33, 60, 107). Despite these similarities, the MSAS PKS must also be programmed to perform β-ketoreduction and dehydration only during synthesis step 2 and to cyclize the resulting seven-carbon chain (Figure 1). For the longer, more complex polyketides, PKS programming is necessary to ensure that the correct molecular structure, out of the numerous permutations possible, is produced. We show below that two distinct types of enzyme systems operate for aromatic or complex polyketide synthesis, that different programming rules are emerging for these two classes of compounds, and how an understanding of PKS programming has allowed the use of rational approaches for altering polyketide pathways in a predicted manner.

BIOSYNTHESIS OF AROMATIC POLYKETIDES

Actinorhodin

The best-studied aromatic polyketide is the benzoisochromanequinone antibiotic actinorhodin (Figure 3) produced by *Streptomyces coelicolor* A3(2). Labeling experiments have established that the monomeric unit is made from eight acetate residues (43). Mutants blocked at each step of the pathway have been generated (96), and the chemical structures of most of the accumulated intermediates or shunt products have been determined (27, 131). The genes for the entire pathway have been cloned (79) and sequenced (20, 40, 41, 48), and when introduced into heterologous hosts, they directed the production of actinorhodin (78).

Figure 4 shows the biochemical pathway and corresponding gene map. The eight acetate-derived units condense to form a 16-carbon acyl chain, which is presumably still attached to the PKS via a thioester linkage. The earliest intermediate identified contains a reduced keto group at C-9, introduced by the *act*III gene product (48), either during chain assembly or after synthesis of the C_{16} intermediate. The *act*I, *act*III, and *act*VII loci (41, 101) are required for synthesis of the first intermediate and determine some of the components of the PKS, as discussed in detail below. *act*VII mutants do not accumulate the proposed pathway intermediate but undergo spontaneous C-6:C-15 ring

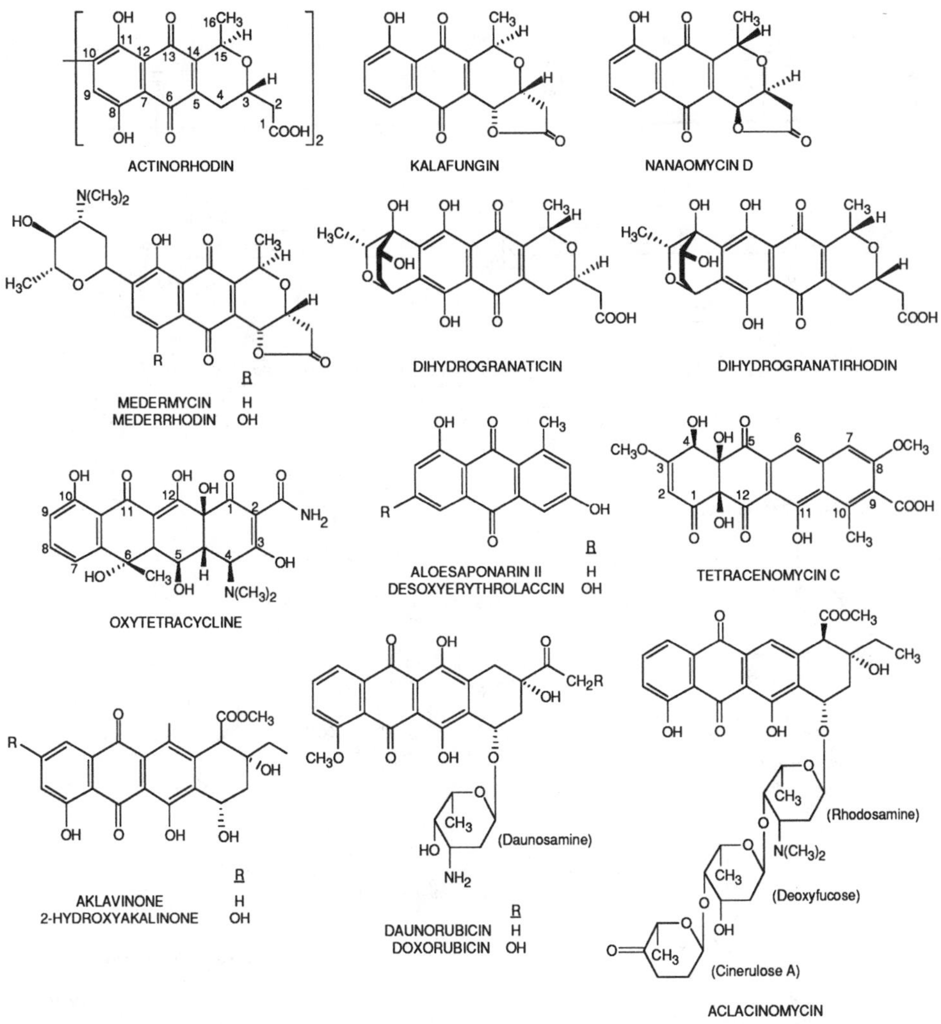

Figure 3 Structures of aromatic polyketides.

closure and fail to complete dehydration of C-8:C-9, yielding the shunt product mutactin (131).

The next step, C-5:C-14 dehydration, is determined by *act*IV (41, 101); this is followed by reduction at C-3, which is required for the formation of the pyran ring and governed by the as yet uncharacterized *act*VI locus. (After ring closure, C-3 and C-15 are chiral centers with the protons in the *trans* configuration.) In this case also, *act*VI mutants do not accumulate the pathway

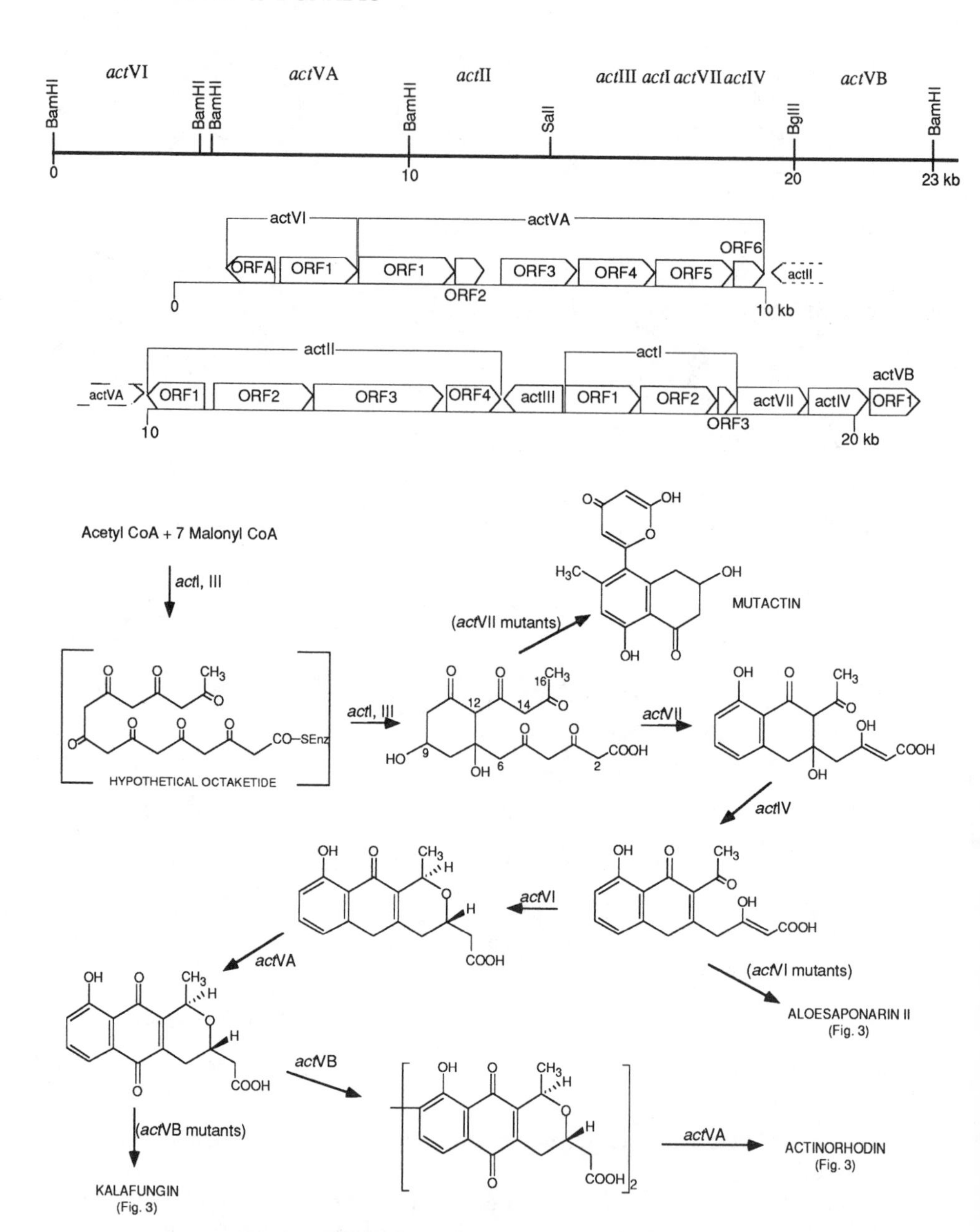

Figure 4 Organization of the *act* cluster and pathway for actinorhodin formation. Empty arrows in the two enlargements indicate ORFs.

intermediate but produce the shunt product aloesaponarin II (Figure 3). Prior to dimerization, the molecule is hydroxylated at C-6, and a final hydroxylation at C-8 yields actinorhodin. Hydroxylations at C-6 and C-8 are determined by *act*VA, which contains six ORFs (20). ORF1 appears to encode a transmembrane protein that may function to export the antibiotic and may play a role in self-resistance. ORFs 2 and 6 are likely involved in the two ring hydroxylations as the corresponding polypeptides resemble known or proposed hydroxylases. Dimerization is determined by *act*VB, where one ORF has been sequenced and proposed to encode a dimerase (41). Because mutants of *act*VB produce the shunt product kalafungin, which lacks the C-8 hydroxyl (Figure 3), the *act*VA-encoded function(s) responsible for C-8 hydroxylation must recognize only the dimer as substrate.

The production of kalafungin in *act*VB mutants differs from the synthesis of the other shunt products, mutactin and aloesaponarin II. Whereas in these two cases, the actinorhodin pathway intermediates are chemically unstable (and thus undergo spontaneous rearrangements), the pathway intermediate accumulated in an *act*VB mutant strain is not. The lactone ring formed in kalafungin results from oxidation at C-4, which is unlikely to occur spontaneously. Because dihydrogranaticin, nanaomycin D (Figure 3), and many other aromatic polyketides contain a similar lactone ring produced through oxidation at C-4, it will be of interest to determine whether *S. coelicolor* contains a C-4 oxidase gene in the *act* cluster or elsewhere. Genes for the synthesis of kalafungin in *Streptomyces tanashiensis* appear to be ordered differently from their *act* homologues in *S. coelicolor* (61).

Actinorhodin biosynthesis in *S. coelicolor* is under complex genetic regulation. The *act*II locus plays a role in regulation, export, and self-resistance (40). This pathway is also directly or indirectly regulated by a series of other genes that globally control secondary metabolic and/or developmental pathways (1, 2, 42, 50, 57, 71).

Other Isochromanequinones

Table 1 schematically shows the proposed pathways for the synthesis of polyketides related in structure to actinorhodin. In each instance, synthesis of the polyketide chain employs an acetate starter and seven malonate extenders, with ketoreduction taking place at a position equivalent to C-9 in actinorhodin. The postpolyketide reactions required to complete the synthesis are also indicated. Overall, many steps are common, but structures of the compounds differ enough to be unique. Dihydrogranaticin and nanaomycin D have the same C-3, C-15 configurations but are stereoisomers of kalafungin, suggesting that the specificity of the C-3 reductase or the C-3:C-15 cyclase determines the stereochemistry at these positions. Table 2 summarizes the available information about the genes involved in the biosynthesis of these compounds.

Table 1 Biosynthesis of selected aromatic polyketides[a]

Compound	Starter	No. malonate extenders	C-9 reduction and dehydration	Post-polyketide modification
Actinorhodin	Acetate	7	Yes	C-5 : C-14, C-7 : C12 cyclization and dehydration; C-6, C-8 hydroxylation; C-3 reduction; C-3 : C-15 dehydration; dimerization
Kalafungin	Acetate	7	Yes	C-5 : C-14, C-7 : C12 cyclization and dehydration; C-6 hydroxylation; C-3 reduction; C-3 : C-15 dehydration; C-4 oxidation; C-1 : C-4 lactonization
Nanomycin D	Acetate	7	Yes	C-5 : C-14, C-7 : C12 cyclization and dehydration; C-6 hydroxylation; C-3 reduction; C-3 : C-15 dehydration; C-4 oxidation; C-1 : C-4 lactonization
Dihydrogranaticin	Acetate	7	Yes	C-5 : C-14, C-7 : C12 cyclization and dehydration; C-6, C-8 hydroxylation; C-3 reduction; C-3 : C-15 dehydration; C-4 oxidation; C-1 : C-4 lactonization; C-9 : C-10 glycosidation
Oxytetracycline	Malonamide	8	Yes (C-8)	C-5a : C-11a, C-6a : C-10a cyclization and dehydration; C-2 : C-3, C-4a : C-12a cyclization; C-6 methylation; C-4 amination and *N*-dimethylation; C-5, C-6 hydroxylation
Tetracenomycin C	Acetate	9	No	C-4a : C-12a, C-5a : C-11a, C-6a : C-10a, C-9 : C-10 cyclization and dehydration; C-4, C-4a, C-5, C-12a hydroxylation; C-3, C-8 *O*-methylation
Daunorubicin	Propionate	9	Yes (C-2)	Synthesis of aklavinone [C-6a : C-10a, C-5a : C-11a, C4a : C-12a cyclization and dehydration; C-9 : C-10 cyclization; C-12 hydroxylation; C-10 (carboxy) *O*-methylation]; C-11, C-13 hydroxylation; C-10 decarbomethoxylation; C-7 glycosidation (daunosamine); C-4 *O*-methylation
Aclacinomycin A	Propionate	9	Yes (C-2)	Synthesis of aklavinone; C-7 (rhodosamine), C-4' (2-deoxyfucose), C-4'' (cinerulose A) glycosidation

[a] Carbon atoms numbered as shown in Figures 3 and 5. Post-polyketide modifications are not listed in sequence.

Table 2 Aromatic pKS genes from actinomycetes

Designation[a]	Host[b]	Description	References
act	*S. coelicolor* (A3(2)	Actinorhodin biosynthesis, export, and resistance cluster within 22-kb segment (see Figure 4).	20, 40, 41, 48
"*akl*"	*S. galileus* 3AR-33	Putative aklavinone genes, 3.4-kb *act*I-, *act*III-hybridizing fragment.	116
"*cin*"	*S. cinnamonensis*	*act*I-hybridizing 4.8-kb fragment; five ORFs; putative functions: KS-1[c], KS-2, ACP, cyclase/dehydratase, KR. No apparent phenotype from disruption of ORFs 1 and 2. Polyketide structure not known.	6
"*cur*"	*S. curacoi*	8.5-kb *act*I-hybridizing fragment; seven ORFs, organization and sequence similar to *whiE*. Brown pigment from expression in *S. lividans*. Polyketide structure not known.	10
dnr	*S. peucetius*	Daunorubicin biosynthesis and resistance cluster >40 kb; 14 ORFs sequenced. PKS genes identified by hybridization to *act*I probe. Putative functions: KS-1, KS-2, ACP. *dnr* genes identified for steps in sugar synthesis and post-polyketide modification. Four ORFs encode putative regulatory functions.	93, 112, 113; C. R. Hutchinson, personal communication
"*fren*"	*S. roseofulvus*	*act*I-hybridizing segment. Sequence shows putative KS-1, KS-2, ACP, KR, cyclase/dehydratase.	51
gra	*S. violaceoruber*	Dihydrogranaticin biosynthesis genes identified by *act*I hybridization and gene disruption. 6.8-kb PKS region has 6 ORFs; putative functions: KS-1, KS-2, ACP, cyclase/dehydratase, 2 KRs.	103
"*hir*"	*Saccharopolyspora hirsuta* 367	*act*I-hybridizing 11.7-kb fragment, 5.1-kb sequenced, Six ORFs; putative functions: KS-1, KS-2, ACP, biotin carboxyl carrier protein, KR. Polyketide structure not known.	72
"*kal*"	*S. tanashiensis*	*act*I-hybridizing 14-kb fragment contains seven kalafungin biosynthesis loci. Fragment shows hybridization to *act*I, -III, -VA, and -VI, not to *act*II, -IV, -VB and -VII. Gene order differs from *act* cluster.	61
otc	*S. rimosus*	Oxytetracycline biosynthesis and resistance gene cluster on 47-kb fragment. *act*I-hybridizing segment within cluster; six ORFs sequenced; putative functions: KS-1, KS-2, ACP, acyl CoA ligase, KR, cyclase/dehydratase.	13, 19, 51, 53, 115

Table 2 (*Continued*)

Designation[a]	Host[b]	Description	References
sch	*S. halstedii*	*act*I-hybridizing 5.2-kb fragment; seven ORFs, organization and sequence similar to *whiE*. Disruption causes loss of spore color. Polyketide structure not known.	15, 16
tcm	*S. glaucescens*	Tetracenomycin C biosynthesis and resistance gene cluster spans 24 kb and contains 12 ORFs. *tcmK, L, M,* and *N* encode PKS: KS-1, KS-2, ACP, cyclase/*O*-methyltransferase. Genes for post-polyketide modification steps characterized. Self-resistance through presumed export of tetracenomycin.	12, 30, 46, 47, 56, 86, 114
whiE	*S. coelicolor* A3(2)	5.7-kb fragment complements *whiE* (white spore) mutant. Hybridizes with *act*I probe. Seven ORFs; putative functions: KS-1, KS-2, ACP, cyclase. Polyketide structure not known. Brown pigment from expression in *S. lividans*.	29, 54

[a] Designations in quotations are used for simplification and do not appear in the literature.
[b] All species are *Streptomyces* unless otherwise indicated.
[c] KS-1 and KS-2 are homologs of *act*I Orf1 and Orf2.

Hybrid Compounds

DNA segments containing all or part of the *act* cluster have been introduced into several aromatic polyketide-producing actinomycetes. Results describing the use of the *act* PKS genes are discussed below. Here we describe new structures made through the introduction of genes involved in postpolyketide modifications.

When clones containing *act*VA were introduced into *Streptomyces* sp. AM-7161, the producer of the antibiotic medermycin (Figure 3), the host produced mederrhodin A, which differs from the parent compound by the presence of an additional hydroxyl group at C-8 (52). In all likelihood, since actinorhodin contains a C-8 hydroxyl on each of its units, the *act*VA-encoded enzyme that carries out the hydroxylation on 8,8′-dideoxoactinorhodin also used medermycin as a substrate. On the other hand, *act*VB mutants impaired in dimerization fail to add this hydroxyl to the shunt metabolite kalafungin, which is otherwise identical to medermycin except for the lack of the carbon-bound sugar at C-10. Perhaps C-8 hydroxylation requires the presence of a bulky group at C-10.

In a second instance, introduction of a plasmid carrying the entire *act* cluster into *Streptomyces violaceoruber* Tü22, the producer of dihydrogranaticin, resulted in the production of dihydrogranatirhodin (52) (Figure 3). This molecule has the stereochemistry of dihydrogranaticin at C-15 and that of actinorhodin at C-3. The production of this hybrid compound may reflect an interaction of the *act*- and *gra*-encoded C-3 reductases/dehydratases during the synthesis of dihydrogranatirhodin to yield a hybrid stereochemistry.

Anthracyclines

Anthracyclines (e.g. dauno- and doxorubicin, the aclacinomycins, and tetracenomycins; Figure 3) contain an anthraquinone-like structure generally attached to a sugar at C-7 or C-10. Labeling experiments (26, 70, 89, 90) have confirmed the polyketide origin of the aglycone (anthracyclinone). Pathways for the formation of anthracyclines are similar to those for actinorhodin. These include: synthesis of the unreduced (or singly reduced) decaketide chain, folding, aldol condensations and dehydrations to form the aromatic rings, hydroxylation to form the quinone, and further postpolyketide modifications sometimes including glycosidation (Table 1). All anthracyclines use malonate as extenders but differ in the use of acetate (tetracenomycin), propionate (daunorubicin), butyrate (feudomycin), or isobutyrate (13-methyl-aclacinomycin A) as starter. The best-characterized pathway is that for tetracenomycin C in *Streptomyces glaucescens,* from the work of C. R. Hutchinson and his colleagues (unpublished results). Mutants for each of the seven biosynthetic steps have been isolated (87), and the accumulated intermediates identified (130). The entire set of biosynthesis genes (*tcm*) has been cloned and sequenced (12, 30, 46, 47, 56, 86, 114) as summarized in Table 2. The biosynthesis of aclacinomycins (Figure 3) proceeds through the synthesis of the aglycone aklavinone (127), followed by the modifications described in Table 1. A DNA segment from an aclacinomycin-producing *Streptomyces galilaeus* strain has been cloned (116).

The identification of intermediates accumulated in blocked mutants and bioconversion experiments has suggested the daunorubicin (Figure 3) pathway in *Streptomyces peucetius*. This pathway consists of: formation of aklavinone, C-11 hydroxylation, C-7 glycosidation, decarbomethoxylation at C-10, C-13 hydroxylation and oxidation (yielding carminomycin), and finally C-4 *O*-methylation. C-14 oxidation yields doxorubicin (Figure 3). A cluster of genes for daunorubicin synthesis has been identified and partially sequenced (93, 112, 113), as summarized in Table 2. Self-resistance is determined by a locus that contains an ORF related to the *mdr1* gene of mammalian tumor cells (45). Daunorubicin analogues have been found in variant strains. 11-Deoxodaunorubicin is produced by a different *S. peucetius* strain (5) and 13-dihydrodaunorubicin by a strain of *Streptomyces coeruleorubidus* (17), suggesting that the C-11 and C-13 hydroxylation steps can be bypassed en route to the final product. Novel anthracyclines have been produced through bioconversion of modified aglycones fed to strains blocked in the synthesis of aklavinone. *S. galilaeus* KE303 could attach the three sugars, rhodosamine, deoxyfucose, and cinerulose A, not only to aklavinone (to produce aclacinomycin A), but also to a series of modified anthracyclinones to produce the corresponding anthracyclines (80, 89), indicating the loose aglycone specificity of the glycosidation enzymes. Similarly, a mutant of *S. coeruleorubidis* that normally produces daunorubicin could convert 1-hydroxyanthracyclinones to the corresponding 1-hydroxyanthracyclines (126, 127). In addition, the mutant could convert carminomycinone to daunorubicin, indicating that glycosidation could be postponed from an early to a later step in the pathway. Taken together, these findings illustrate the wide variety of new structures that can be made and underscore the potential for engineering novel anthracycline biosynthesis pathways by heterologous gene cloning.

AROMATIC POLYKETIDE SYNTHASES

Gene Organization

By strict analogy to FAS, the aromatic PKS is responsible for assembly of the polyketide chain. In the case of actinorhodin, assembly is carried out by the products of *act*I and *act*III, because mutants of *act*VII, which encodes the putative cyclase/dehydratase (41, 101), produce an octaketide. Translational coupling has been found between *act*I-ORF3 and *act*VII (41), but not in the equivalent ORFs in other systems. For the purposes of this discussion we consider the cyclases as components of aromatic PKSs, although this supposition has not been established.

The *act*I and *tcmKLM* loci strongly hybridize to genomic DNA from many actinomycete producers of polyketides (77). A DNA probe containing *act*I has been used for the cloning or identification of PKS genes from 13 different actinomycetes (summarized in Table 2). Though the sequence has been determined for 12 sets of PKS genes, for only six has the *act*I-hybridizing segment been implicated in the synthesis of a compound of known structure. For *cur* (10), *sch* (16), and *whiE* (29), the structures of the polyketides have not been elucidated but are known from genetic experiments to be pigments associated with *Streptomyces* spores. Introduction of the *hir* segment into *Streptomyces lividans* resulted in the production of a new antibiotic (72). For *cin,* neither the structure nor the role of the corresponding polyketide has been elucidated (6).

The PKS-encoding segments thus far characterized show remarkable similarity: they all encode homologues of *act*I ORFs 1, 2, and 3, and are arranged in a similar order (Figure 4). The deduced polypeptides from ORFs 1 and 2 show extensive similarity to type II KS functions, although the Orf2 polypeptides lack, in the highly conserved GPXXXXXXXCXSXL motif present at the active site of type I and II KSs, the cysteine residue required for thioester linkage to the acyl chain. The Orf3 polypeptides are all less than 100 amino acids in length, contain the highly conserved DLXGYDS motif characteristic of the 4′-phosphopantotheine binding site of ACPs, and show good end-to-end similarity (65).

The remaining ORFs that determine PKS functions are less conserved in both sequence and relative location. The KR-encoding gene lies upstream of ORF1 and is transcribed divergently in *act* (48), *gra* (103), and *hir* (72), but is downstream of ORF3 in *otc, fren* (D. H. Sherman, personal communication), and *cin* (6). The cyclase is believed to reside in a bifunctional polypeptide that also carries a DH domain in *act, cin, fren, gra,* and *otc* and an *O*-methyltransferase in *tcm.* In *act, cur, gra, sch, tcm,* and *whiE,* the cyclase-encoding gene lies immediately downstream of ORF3. In *cin, fren,* and *otc,* the cyclase-encoding gene lies downstream of the *act*III homologue. Other features are summarized in Table 2.

Programming Polyketide Synthesis

As described above, the components of the PKS are involved in acyl chain assembly and cyclization. To build the correct polyketide structure, the PKS must be programmed to choose the correct starter (malonates are always used as extenders), count the number of condensations, properly fold the completed acyl chain, cyclize it into the correct number of rings, and release the resulting structure. How this programming is accomplished is not yet understood, although it clearly must reside in the PKS components themselves.

The aromatic PKSs are comprised of five polypeptides (six in *gra*): a KS-like function (Orf1), Orf2 [actual biochemical role not known but required for actinorhodin synthesis (41, 102)], the ACP (Orf3), the KR (absent in the *tcm* PKS), and the bifunctional cyclase/dehydratase (in the case of *tcm*, cyclase/*O*-methyltransferase). These PKS systems, therefore, largely resemble the type II bacterial or plant FAS systems described above, but FAS systems also contain an AT required for transfer of malonyl CoA to the ACP and a TE for chain release. However, no ORF encoding an AT or TE has been uncovered in any of the 12 PKS sequences examined. Fernandez-Moreno et al (41) have recently pointed out that all Orf1 polypeptides contain the GHSXG motif, typical of AT functions, in their C-terminal halves, but whether an AT or a TE (which also contains this motif) is associated with Orf1 has not been established. The presence of the KS-like sequence Orf2 appears to be characteristic of type II PKSs. The current model for aromatic PKSs calls for formation of a heterodimer between Orf1 and Orf2 (53). Exactly how the correct starter unit is selected, how the ACP is charged with malonate, and which mechanisms determine folding and chain release are not currently known, however. Chain length is addressed below.

Other questions concerning PKS programming have arisen. Because all bacterial cells produce fatty acids, do the type II PKS and FAS systems share components? In some cases a host carries two sets of PKS genes, such as *act* and *whiE* in *S. coelicolor*. How are the corresponding PKS components kept separate? Can they substitute for one another under certain conditions?

Investigators have begun to explore these questions. PKS genes from one producer have been introduced into cells producing a different compound, either by plasmid-borne complementation of the corresponding host mutation or by replacement of the host counterpart. Table 3 compiles the results from these experiments. Replacement of *act*I-ORF3 with ACP genes from *gra, fren, otc,* or *tcm,* or with the FAS ACP gene from *Saccharopolyspora erythraea,* allowed production of a blue pigment, presumably actinorhodin, or of octaketides similar to actinorhodin or its intermediates or shunt products (64, 65). *S. coelicolor act*I-ORF1 mutants produced a blue pigment when the ORF1 counterpart from *gra* (102), *hir* (72), *otc,* or *whiE* (66) was introduced. The *Escherichia coli fabB* gene, encoding KS I, also complemented *act*I-ORF1 (66). Only the *cin* segment failed to complement an *act*I-ORF1 mutation, but it did complement an *act*III mutation (6). However, an *act*I-ORF2 mutation was complemented only by *gra*-ORF2 (102) and not by the counterparts from *otc* or *whiE* or by *fabB* (66).

The act PKS genes have also been introduced into various *S. galilaeus*

Table 3 Complementation and gene replacements by aromatic PKS genes

Incoming DNA fragment[a]	Host	Polyketide produced[b]	References
Complementation:			
gra ORF1	*S. coelicolor act*I ORF1−	Actinorhodin*	102
whiE ORFIII	*S. coelicolor act*I ORF1−	Actinorhodin*	66
otc ORF1	*S. coelicolor act*I ORF1−	Actinorhodin*	66
E. coli fabB	*S. coelicolor act*I ORF1−	Actinorhodin*	66
"*hir*"	*S. coelicolor act*I ORF1−	Actinorhodin*	72
gra ORF2	*S. coelicolor act*I ORF1−	None	102
"*cin*"	*S. coelicolor act*I ORF1−	None	6
gra ORF2	*S. coelicolor act*I ORF2−	Actinorhodin*	102
whiE ORFIII, IV, V	*S. coelicolor act*I ORF2−	None	66
otc ORF2	*S. coelicolor act*I ORF2−	None	66
E. coli fabB	*S. coelicolor act*I ORF2−	None	66
gra ORF5	*S. coelicolor act*III−	Actinorhodin*	102
"*cin*"	*S. coelicolor act*III−	Actinorhodin*	6
"*hir*"	*S. lividans*	Bioactivity	72
*act*I, III, IV, VII	*S. azureus* 14921	Aloesaponarin II	8, 111
—	*S. galilaeus* 31133	Aklavinone	8, 111
*act*I ORF1, 2	*S. galilaeus* 31133	Aloesaponarin II	8, 111
—	*S. galilaeus* 31671	2-Hydroxyaklavinone	8, 111
*act*I, III, IV	*S. galilaeus* 31671	Aklavinone,	8, 111
		Aloesaponarin II	8, 111
*act*I ORF1, 2	*S. galilaeus* 31671	Desoxyerythrolaccin	8, 111
*act*III	*S. galilaeus* 31671	Aklavinone	8, 111
—	*S. galilaeus* ANR58	2-Hydroxyaklavinone	8, 111
"*akl*"	*S. galilaeus* ANR58	Aklavinone	110
*act*III	*S. galilaeus* ANR58	Aklavinone	8, 111
Gene Replacement:			
gra ORF3	*S. coelicolor*, replace *act*I ORF3	Actinorhodin*	64, 65
"*fren*" ORF3	*S. coelicolor*, replace *act*I ORF3	Octaketide	64, 65
otc ORF3	*S. coelicolor*, replace *act*I ORF3	Octaketide	64, 65
tcm ORF3	*S. coelicolor*, replace *act*I ORF3	Octaketide	64, 65
Saccharopolyspora erythraea FAS ACP	*S. coelicolor*, replace *act*I ORF3	Actinorhodin*	64, 95

[a] Dashes indicate there is no incoming DNA.
[b] Asterisks indicate that the polyketide is presumed to be actinorhodin.

strains that produce anthracyclinones. ANR58 produces 2-hydroxyaklavinone rather than aklavinone, which is produced by the parent strain. Introduction of a plasmid carrying *act*III into ANR58 resulted in the restoration of aklavinone production, indicating that the 2-hydroxy form is the result of a failure of the acyl chain to undergo β-ketoreduction during synthesis (111). This finding also proves that the KR involved in the synthesis of the octaketide precursor of actinorhodin can associate with the PKS components that produce a decaketide.

Other components of two PKS systems also appear to be interchangeable. As mentioned above, *act*I, *act*III, *act*IV, and *act*VII are required for the synthesis of the octaketide aloesaponarin II. Introduction of a fragment containing *act*I-ORFs1 and 2 into wild-type *S. galilaeus* resulted in aloesaponarin II production (8, 111). The same fragment in ANR58 led to the production of the hydroxylated counterpart of aloesaponarin II, desoxyerythrolaccin (Figure 3), as expected (8). Strohl and coworkers (111) have pointed out that, if *S. galilaeus* does not produce another, yet to be identified octaketide, these findings indicate that the ACP, KR, cyclase, and DH(s) involved in synthesis of aklavinone can interact with the Orf1 and Orf2 proteins involved in the synthesis of the actinorhodin precursor and that the length of the polyketide chain is programmed by Orf1 and Orf2 exclusively. If so, the dendrogram (Figure 5) showing the relationship among the various Orfs 1 or 2 may be predictive of polyketide structure. Orfs for polyketides of the same length, Act and Gra, Dnr and Tcm, are most closely related and the relative order is more or less the same for both Orf1 and Orf2. The spore pigment PKS components form their own branch, suggesting that the corresponding polyketides are structurally related (15). In addition, this

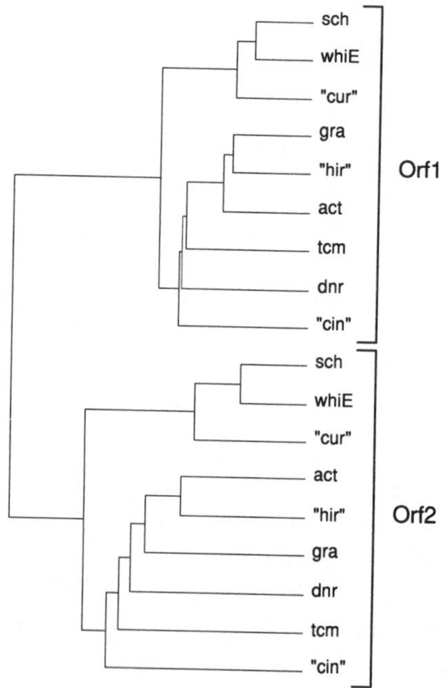

Figure 5 Dendrogram, using PILEUP program of GCG (31), showing relationships of type I PKS components Orf1 and Orf2.

analysis suggests that the *hir* polyketide is closely related to actinorhodin and dihydrogranaticin, whereas *cin*, which falls in the *tcm-dnr* group, may produce an anthracycline. Perhaps this is the basis for the failure of the *cin* segment to complement an *act*I-Orf1 mutation; the segment did complement a *tcmKL* mutant (C. R. Hutchinson, personal communication). As more sequences for aromatic PKSs become available, and the corresponding structures are determined, dendrograms similar to that shown in Figure 5 may become a valuable aid for predicting the structure of an unknown polyketide from its PKS sequence.

Purification of the individual components of the PKS and assembly and synthesis of polyketides in vitro is probably the best way to decipher PKS programming. The tetracenomycin Orf1 and Orf2 components have been identified and overproduced (44), and studies have shown that Orf3 is modified with pantotheine (100). An in vitro polyketide-synthesizing system capable of directing formation of tetracenomycin F2, an early intermediate, has been constructed using overproduced TcmKLMN polypeptides (56).

COMPLEX POLYKETIDES

Biosynthesis

The class known as complex polyketides includes some natural products consisting of a polyketide skeleton more greatly reduced than the poly-β-keto intermediate that produces aromatic polyketides. These molecules consist of long alkyl carboxylates carrying various reduced functions at odd-numbered carbon atoms (keto, hydroxyl, enoyl, or alkane) and different side chains at even-numbered carbons (hydrogen, methyl, ethyl, or propyl). In addition, the polyketide chain may incorporate an unusual moiety at the ω position (e.g. *m*-C_7N unit in ansamycins), or this moiety may be linked to the carboxylate (e.g. pipecolate in FK506). Finally, the polyketide-derived unit can appear in the finished product as a linear chain (e.g. polyethers), a lactone (e.g. macrolides), or as a lactam (e.g. macrolactams).

Addition of appropriately labeled small molecule precursors to growing cultures established the general rule that the polyketide backbone results from the successive condensation of small carboxylic acids (91, 104). A significant difference is generally found between the two major producers of complex polyketides, the actinomycetes and the fungi: in the former, the methyl and ethyl side chains usually result from incorporation of propionate and butyrate units, respectively (91), whereas in fungi, the methyl side chains are methionine-derived, the result of incorporation of acetate into the polyketide chain followed by C-methylation (e.g. 82). Evidence that synthesis of polyketides is processive came initially from the elegant work of Cane & Yang (25) and Yue et al (129), who showed that N-acylcysteamine thioesters five to seven carbons long exhibiting the appropriate processing level and stereochemistry were specifically incorporated into the corresponding positions of the erythromycin and tylosin aglycones when fed to the producing cultures *Saccharopolyspora erythraea* and *Streptomyces fradiae,* respectively (Figure 6). These studies have since been extended to a series of other polyketides produced by actinomycetes and fungi (Figure 6). Intact incorporation of intermediate(s) was observed with the five-carbon diketide for nargenicin (24), the five- and seven-carbon di- and triketides for methymycin (23), the eight-carbon tetratketide for nonactin (106), and the four-, six-, and eight-carbon intermediates for aspyrone (109). Incorporation of di- and tetraketide intermediates into dehydrocurvularin was achieved through the use of blocked mutants (128) or inhibitors of β-oxidation of fatty acids (73). Thus compounds mimicking the hypothetical acyl-ACP intermediates predicted to form after one, two, or three cycles could be incorporated intact if they had already undergone the adjustments for the correct stereochemistry and the degree of reduction encountered in the final product.

Analysis of the fermentation broths of mutants of the mycinamycin producer *Micromonospora griseorubida* led to the identification of a series of branched chain fatty acids and a decarboxylated derivative that matched, in the stereochemistry of the side chains and in the extent of reduction, the carboxyl-distal segment of the mycinamycin aglycone (68, 69): the mycinonic acids recovered corresponded to the intermediates expected from the third through sixth condensation cycle (Figure 6). Analogous work carried out on some *S. fradiae* mutants defective in tylosin synthesis (Figure 6) identified putative intermediates in tylactone formation halted after the fourth and fifth cycle and showed them to be intermediates in polyketide formation, rather than degradation products of the aglycone (55).

Taken together, these results point to a stepwise processive mechanism for complex polyketide synthesis in which the stereochemistry of the side chains and the appropriate degree of processing of the β-carbons are fixed into the acyl chain prior to subsequent elongation cycles. In this respect, synthesis of complex polyketides is similar to LCFA formation, in which processing precedes future elongation. However, whereas the latter consists of a series of identical iterative steps, in the formation of complex polyketides each cycle is usually different from the previous one, so that at each step the enzyme must determine the correct stereochemistry of the side chain and the degree and stereochemistry of processing of the β-carbonyl. Insight into how the PKS makes these determinations has come from analysis of the primary sequence of the PKS as deduced from the nucleotide sequence of the corresponding genes.

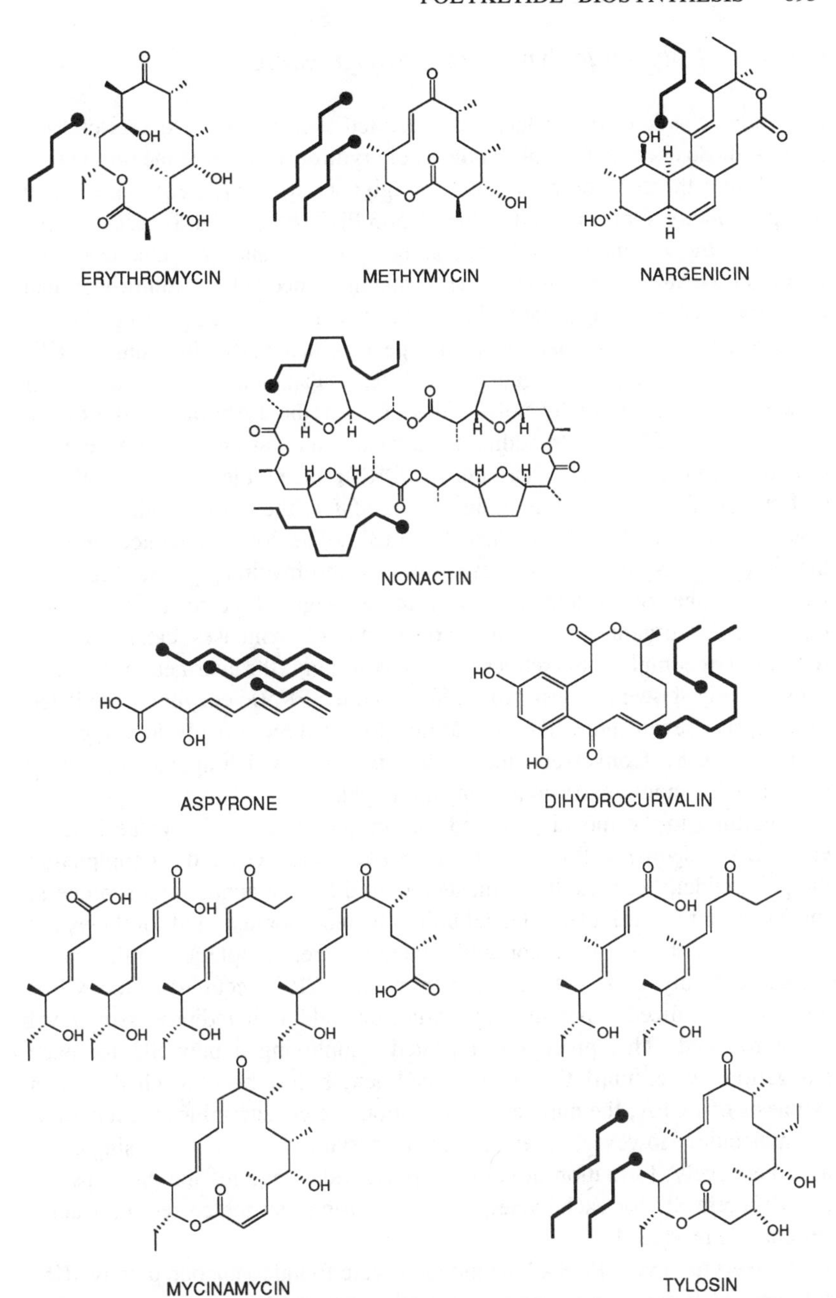

Figure 6 Proposed intermediates in complex polyketide formation. Incorporated compounds are indicated by the thick segmented lines next to each structure, with the dot denoting the thioester. Accumulated intermediates are shown above the corresponding structure.

Complex Polyketide Synthases: Programming

Molecular characterization has been reported thus far for two sets of PKS genes—those encoding the biosynthesis of erythromycin aglycone (*eryA*) (11, 28, 37) and those encoding avermectin aglycone (*avr*) (75, 76). The 3′ end of a gene likely to encode part of the FK506 PKS (85), and a *Bacillus subtilis* gene of unknown function (99) resembling *eryA*, have also been characterized. A common feature among these genes is that they encode large multifunctional polypeptides containing putative FAS-like activities, generally arranged in the same relative order as that found in type I animal FASs and fungal PKSs (Figure 7). This type of organization differs substantially from the type II PKS systems for aromatic polyketides and for the FAS likely to exist in actinomycetes (56, 87). In addition, the genes thus far characterized consist of two (three *eryA* genes; at least one *avr* ORF), three (at least two *avr* ORFs), or four (*pksX*) repeated units designated *modules;* the corresponding protein segments are called *synthase units* (SUs) (37). The correspondence between the six cycles required for synthesis of the erythromycin aglycone (Figure 8) and the number of modules present in the *eryA* genes led to the suggestion that modular organization is a characteristic of synthases catalyzing the formation of complex polyketides (37). Accordingly, the avermectin PKS was subsequently shown to consist of 12 SUs, and the FK506 enzyme is predicted to comprise seven more SUs in addition to the three already identified (cf Figure 7 and 8). Conversely, the *B. subtilis* PksX, which appears to consist of four SUs, should carry four elongation cycles.

According to the model proposed for complex polyketide synthesis (37), each SU is responsible for one of the FAS-like cycles required for completing the polyketide; it carries the elements required for the condensation process, for selecting the particular extender unit to be incorporated, and for the extent of processing that the β-carbon will undergo. After completion of the cycle, the nascent polyketide is transferred from the ACP it occupies to the KS of the next SU utilized, where the appropriate extender unit and processing level are introduced. This process is repeated, employing a new SU for each elongation cycle, until the programmed length has been reached. As in synthesis of LCFA, the number of elongation cycles determines the length of the molecule. However, whereas fatty acid synthesis involves a single SU used iteratively, formation of complex polyketides requires the participation of a different SU for each cycle, thereby ensuring that the correct molecular structure is produced.

In the erythomycin PKS all the modules were found to encode putative KS, AT, and ACP domains responsible for all of the condensation processes but a variable number of functions involved in the processing of the β-carbons. Modules 1, 2, 5, and 6 encode a KR domain only; module 3 lacks a functional

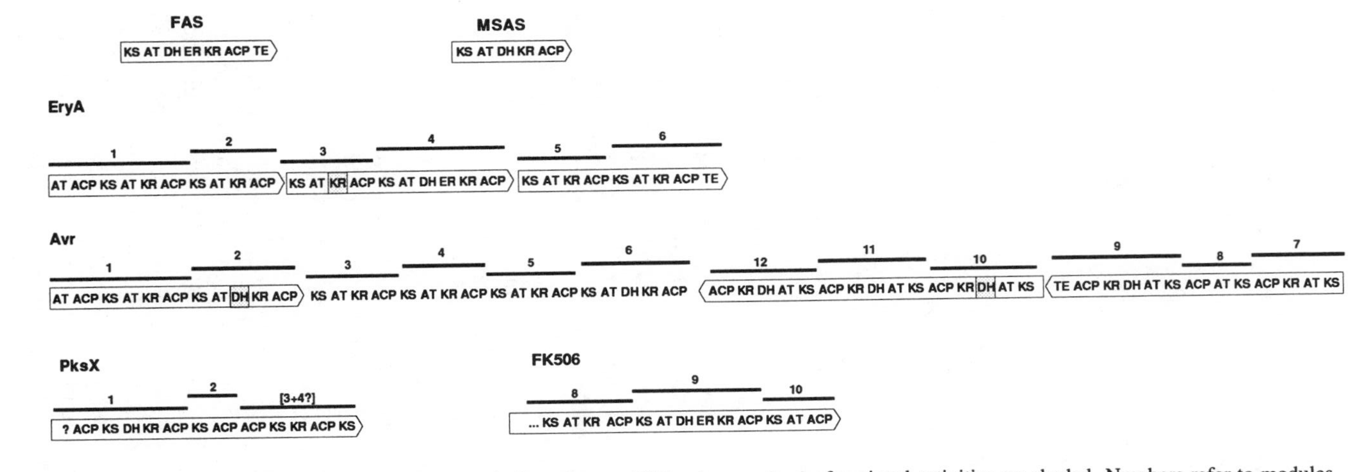

Figure 7 Organization of type I systems. Arrows indicate known ORFs. Apparently dysfunctional activities are shaded. Numbers refer to modules. FAS and MSAS from Ref. 3, 9, and 98.

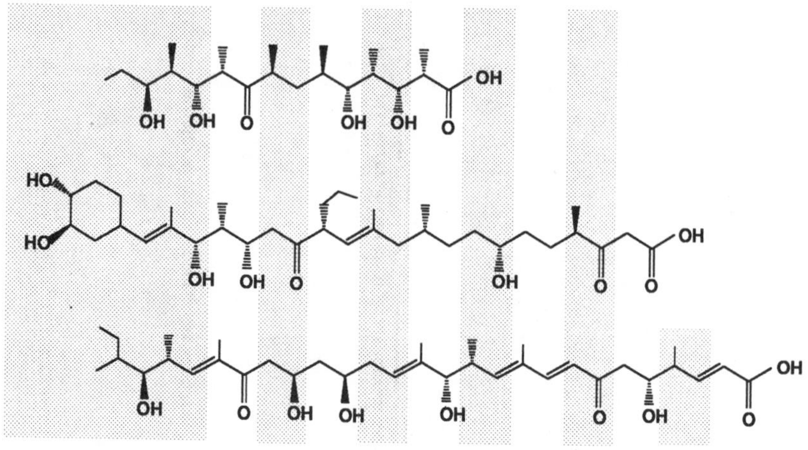

Starter + 1 3 5 7 9 11

Figure 8 Structures of hypothetical linear polyketide intermediates of erythromycin, FK506, and avermectin. Stippling denotes carbons added during odd-numbered steps.

KR, although much of the sequence is preserved; and module 4 carries the full complement of KR, DH, and ER domains. In addition, extra AT and ACP domains were detected at the N terminus of SU 1, and a TE domain was found and at the C terminus of SU 6 (11, 28, 37). The variable composition of the modules, which correlates with the asymmetry in the synthesis of the polyketide precursor, enabled a specific step to be assigned to each module (37). This observation suggested, for the erythromycin PKS, colinearity between the order of the modules and the synthesis steps that they govern (cf Figures 7 and 8), which has been confirmed in part by the experiments described below.

The *avr* PKS genes are organized into two converging blocks of ORFs (Figure 7). Assignment of each module to a specific step in the synthesis has not as yet been possible, but the modules seem to be ordered (76; D. J. MacNeil, personal communication). In the case of the FK506 PKS, the available information suggests that modules 8–10 govern the corresponding steps of this pathway (85; H. Motamedi, personal communication). These observations indicate that, although colinearity between genetic and biochemical order may not be a general rule, the modules are not scattered randomly in the genetic locus encoding the PKS but are arranged in an order closely correlated to the synthesis steps for the polyketide.

Examination of the composition of a variety of SUs has provided important information on the domain organization of type I systems. This information has in turn led to proposals for the extents and putative active-site residues of the DH (11, 34) and ER (11, 34, 124) domains, which had not been conclusively identified in FAS systems (120). An interesting feature emerging is the difference in the composition of the SUs for achieving the equivalent extent of β-carbon processing. Retention of the keto group in step 3 of the erythromycin polyketide synthesis is achieved by a SU that carries an entire KR domain believed to be dysfunctional (11, 37), whereas in SU 8 of the *avr* PKS, the KR domain is completely absent (76). Similarly, six DH-like domains are encoded by the *avr* PKS locus, although only four dehydration steps are predicted from the avermectin structure: two of these domains may be inactive because they differ in the putative DH signature sequence (D. J. MacNeil, personal communication).

Another feature observed is the variable extent of similarity among AT domains of the avermectin PKS, which points to less conservation among AT domains belonging to SUs involved in malonate addition than among those specific for methylmalonate (D. J. MacNeil, personal communication). A more dramatic example is encountered in PksX, in which AT domains are absent altogether from each SU (Figure 7); however, a PKS-encoding ORF found downstream from *pksX* seems to encode AT domains (A. M. Albertini, personal communication). Other unusual features of PksX are the existence of two consecutive ACP domains and a polypeptide ending with a KS domain. Because the structure of the PksX-produced polyketide is not currently known, it is not clear whether this unusual organization is a consequence of the existence of four SUs in one polypeptide or if it represents a new type of PKS organization.

An important question about the mechanism of PKS fidelity needs to be resolved. The choice of the incorporated extender unit may rest on the specificity of the AT domain (37), although only indirect evidence—formation of 2-norerythromycin by a mechanism as yet unknown in a strain carrying a dysfunctional AT from module 6 (81)—supports this view. Therefore, it will be particularly interesting to examine how unusual extender units, such as the octanoate moiety incorporated at the C-2 position of fungichromin (88), are handled by PKSs and how cells activate such unusual extender units.

Altering the Specificity of the PKS: Reprogramming

If each enzymatic activity is involved in a single biochemical step in the pathway, loss of any one activity should affect only a single step in the synthesis. Knowledge of the correlation between the structure of the polyketide and the organization of the PKS genes should, in principle, enable selected altered genes to produce a polyketide derivative with predicted structure. Because the degree of processing appears to depend on the presence of functional domains in a particular SU, inactivation of a KR, DH, or ER should result in a polyketide less processed at a single site, but only if the altered chain thus produced can be utilized as a substrate for the subsequent synthesis steps. Thus the inactivation result of one of these domains should be the formation of a polyketide retaining a ketone, hydroxyl, or unsaturation at the corresponding position. This rationale has led to the successful production of altered erythromycin derivatives from strains in which a KR or an ER domain had been inactivated.

Erythromycin synthesis involves several modifications of the polyketide-derived lactone ring: C-6 hydroxylation of the aglycone; attachment of the glucose-derived deoxysugars mycarose and desosamine at the C-3 and C-5 hydroxyls, respectively; hydroxylation of the lactone at C-12; and *O*-methylation of mycarose at C-3″ (Figure 9). The pathway for erythromycin formation, which has been the subject of extensive biochemical and genetic studies in *Saccharopolyspora erythraea,* represents the best-understood pathway for a complex polyketide. Work undertaken in several laboratories has led to the identification and sequencing of the ~56-kb segment comprising the *ery* cluster (Figure 9) (11, 28, 32, 35, 37, 38, 49, 94, 108, 117, 119, 121–123).

The erythromycin PKS was reprogrammed by introducing changes at two sites in the *eryA* genes corresponding to functions involved in processing β-carbonyl groups. In the first instance, a strain lacking the KR domain of SU 5 was made. This domain was proposed to reduce the β-keto group of the C_{13} intermediate (Figure 8), resulting in the presence of the β-hydroxyl group at the C-5 position of 6-deoxyerythronolide B (DEB) (Figure 9). A strain devoid of this activity was constructed by deleting the entire KR domain from the chromosomal copy of module 5 (37). According to the model, failure to reduce the β-keto group of the C_{13} intermediate would result in the production of the 5-oxo derivative of DEB, in which the keto group has been retained through the next synthesis step (Figure 10). The major metabolites accumulated by this mutant were the 5-oxo derivative of DEB and 3-α-mycarosyl-5-deoxy-5-oxo-DEB, consistent with the prediction that the KR of SU 5 is responsible for the ketoreduction at the fifth condensation step (37).

The second change involved the methylene group appearing at C-7 in DEB, which is presumably introduced through reduction of the α,β unsaturation formed during the fourth synthesis step (Figure 8). This reaction was predicted to be carried out by the ER domain encoded by module 4; thus, if this activity were abolished, the resulting PKS would produce the 6,7 anhydro derivative of DEB, where that double bond has been retained through synthesis steps 4, 5, and 6 (Figure 10). The segment encoding the putative NAD(P)H binding motif in the ER domain of EryAII was mutated by introducing the same two amino acid changes encountered in the equivalent sequence present in the dysfunctional KR domain from module 3. A strain carrying this mutation

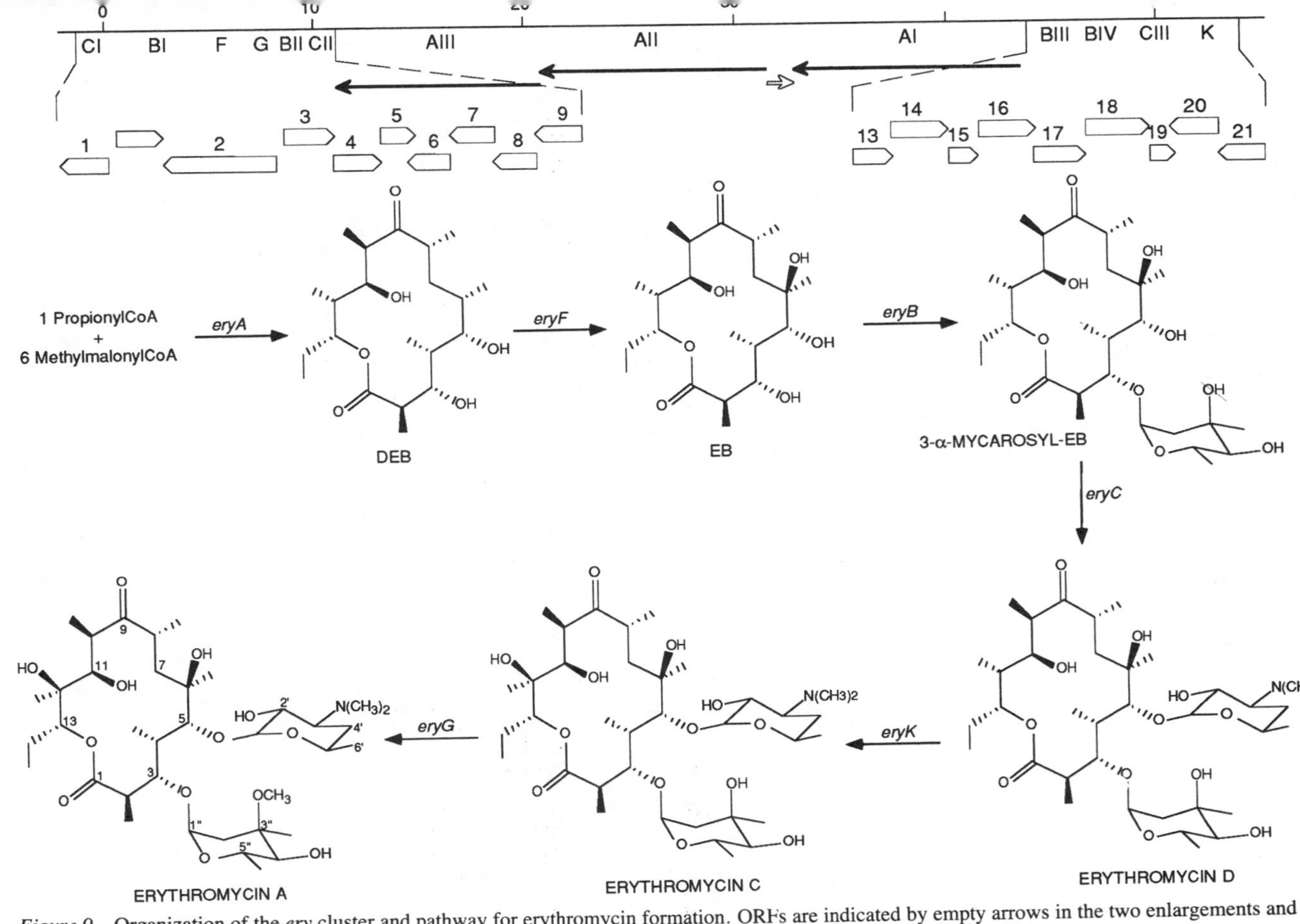

Figure 9 Organization of the *ery* cluster and pathway for erythromycin formation. ORFs are indicated by empty arrows in the two enlargements and by filled arrows for AI, AII, and AIII.

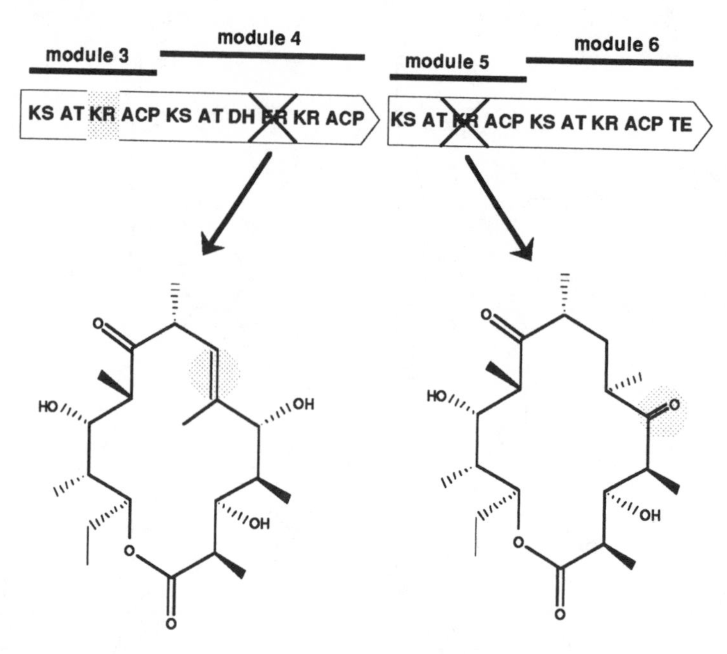

Figure 10 Altered polyketides produced though genetic intervention. The domain inactivated is crossed out, whereas the naturally inactive KR is stippled. The differences in the lactones produced are highlighted.

accumulated 6,7-anhydroerythromycin C as the final metabolite, consistent with the involvement of the ER domain in reducing the double bond during synthesis step 4 (35, 132).

These two examples illustrate how altered erythromycin lactones can be produced by genetic intervention of the PKS genes. This enzyme system seems to be able to tolerate, at least to some extent, alteration in its programming and can process altered polyketide chains for one or two further cycles. How flexible are PKSs in accepting alterations in their programming? Because all type I PKS examined thus far appear to be structurally related, this answer can be addressed by determining the minimum number of mutations that a PKS gene must undergo in nature to produce a derivative of the original polyketide. A first approach would be the analysis of sequences of PKSs involved in formation of closely related molecules. The macrolide sporeamicin, structurally equivalent to 11-oxo-erythromycin, was recently isolated from a *Saccharopolyspora* strain (125). This compound may have arisen by mutation in the KR domain of module 2, resulting in the failure of the erythromycin-like PKS to perform the β-ketoreduction during synthesis step 2. The PKS locus from the sporeamicin producer was found to be closely related to that from *Saccharopolyspora erythraea* (36), which would indicate that, if the KR from the sporeamicin module 2 is indeed dysfunctional, major alterations in the PKS are not required for processing an altered polyketide through four elongation cycles. Analogous work (76) has indicated that the PKSs for avermectin and nemadectin are closely related, although structural differences exist between these two molecules. In addition, genes for nemadectin biosynthesis can complement *avr* mutants (D. J. MacNeil, personal communication).

The ability to produce altered polyketides suggests that the transfer of the growing chain from one SU to the next may occur without regard to substrate structure, possibly by a precise steric juxtapositioning of ACP and KS domains from the two SUs (11, 63). On the other hand, if the structure of the nascent polyketide plays an important role in chain transfer, only a few changes to chain structure will be processed into the completed polyketide. The observation that unprocessed or partially processed diketide intermediates of tylosin (129) and aspyrone (59) could also be incorporated suggests that feeding *N*-acylcysteamine derivatives to selected blocked mutants may shed light on the mechanism and specificity of chain transfer from ACP to KS.

Biochemical studies, initiated thus far with the purified *eryA*-encoded polypeptides or portions thereof (21, 22, 71a), will therefore be essential for understanding structure and function in type I PKSs and will aid in designing in vivo pathways for the formation of novel polyketides with desired features. However, the large size of type I PKS polypeptides (in the 300- to 500-kDa range) and their possible association into higher-order structures render biochemical analysis a formidable task.

Postpolyketide Modifications

Because the polyketide skeleton made by the PKS seldom represents the final product of a pathway, information on the genes governing postpolyketide modifications is important as these reactions may be affected by altered chains generated by reprogramming of the PKS. In fact, the first novel structures produced by targeted genetic manipulations were obtained through alteration of postpolyketide pathways.

The *Streptomyces thermotolerans carE* gene, encoding the acylase responsible for adding an isovaleryl or butyryl group to the 4″ hydroxyl of carbomycin, was introduced into the spiramycin producer *Streptomyces ambofaciens*. The resulting strain produced the novel metabolite 4″-isovalerylspiramycin, confirming earlier evidence that different macrolide substrates could be acylated when fed to *S. thermotolerans* (39). In wild-type *Saccharopolyspora erythraea,* addition of the sugar mycarose to the lactone ring occurs after C-6 hydroxylation. However, this step was efficiently bypassed in a strain in which the C-6 hydroxylase gene, *eryF,* had been inactivated by targeted gene disruption (Figure 9), resulting in the formation of the acid-stable antibiotic 6-deoxyerythromycin (122).

Thus we can engineer polyketide pathways by genetic intervention of the PKS and by adding or eliminating modification steps. Many of the enzymes involved in postpolyketide modifications do not seem to have absolute specificity for a particular structure. In the erythromycin pathway, mycarose and desosamine can be attached to erythronolide B (EB), as in the wild-type strain; to 6-deoxy-EB, when *eryF* is inactivated (122); or to the altered lactones 2-nor-EB (81) and 6,7-anhydro-EB (35). In addition, 5,6-dideoxy-5-oxo-EB can also serve as a substrate for the mycarosyl transferase (37). Whereas C-12 hydroxylation has been thus far observed with all the altered macrolides produced, *O*-methylation of mycarose did not occur on a 6,7-anhydro-EB substrate (35). Similarly, 5,6-dideoxy-5-oxo-EB, a very poor substrate for the C-6 hydroxylase in vitro (4), was hydroxylated at very low frequency in vivo (J. B. McAlpine, personal communication). Thus, addition of the two sugars apparently can occur despite the presence of altered features at carbons very close to the site of linkage, whereas C-6 hydroxylation and perhaps *O*-methylation of mycarose may have narrower substrate specificities and thus are substantially affected by changes in the polyketide structure.

CONCLUSIONS AND PROSPECTS

Molecular genetic analysis of the PKS genes has revealed two distinct classes of enzymes operating for different polyketides: (*a*) The aromatics, which are made through an essentially iterative process have a type II system consisting of fewer polypeptides than those present in type II FAS systems. (*b*) The complex polyketides have a type I system comprising several repeats of the same activities arranged in few very large polypeptides. Genes for other type II aromatic PKSs have been successfully isolated with *act*I-based probes, but these have failed to detect genes for type I systems. On the other hand, *eryA*-based probes strongly hybridize with genomic DNAs from macrolide producers (117), although it remains to be seen whether they will be useful for obtaining genes for complex polyketides other than macrolides. The high similarity existing between the various PKS genes should enable isolation of genes for related polyketides through a judicious choice of hybridization probes within actinomycetes. As more PKS systems are analyzed, it will be interesting to determine the organization of the enzymes responsible for the synthesis of polyketides consisting of a symmetrical chain linked to an asymmetrical one (such as polyenes). Do they combine elements from type I and type II systems? Is a type I PKS also used in an iterative manner?

The information available on the programming of type II PKS systems indicates that aromatic polyketides offer a modest opportunity for novel structures. The small number of discrete polypeptides renders the isolation and genetic manipulation of type II PKS gene clusters easier to perform than that of their type I counterparts. Similarly, the likely small size of a type II PKS should facilitate in vitro studies with the purified enzymatic components. However, until the biochemical rules governing these steps are elucidated, a rational approach to novel structures will not be possible.

The situation is somehow reversed with type I PKSs. Here, the basic rules of programming are understood, although several aspects await clarification. These systems offer, in principle, enormous possibilities for reprogramming: the ketone, hydroxyl, olefin, and fully saturated groups introduced during synthesis could be interchanged; hydrogen, methyl, ethyl, or longer side chains could be conveniently placed; and ultimately, length and cyclization of the chain could also be varied. However, the large size of type I PKS genes and enzymes complicates genetic manipulation and in vitro studies.

The best chance of producing novel structures is through genetic alteration of the postpolyketide pathways, as one can envisage elimination of unwanted groups (as for 6-deoxyerythromycin), addition of specific moieties (as in the case of isovalerylspiramycin), or functional groups (as for mederrhodin and dihydrogranatirhodin). Expansion of these efforts requires a better understanding of the specificities of the enzymes involved in postpolyketide modifications. Still, prospects for selecting the desired components from a library of polyketide and postpolyketide biosynthesis genes and combining them to produce novel structures appear promising.

ACKNOWLEDGMENTS

We thank Alessandra Albertini, David Cane, Keith Chater, Claude Dery, David Hopwood, Dick Hutchinson, Chaitan Khosla, Doug MacNeil, Jim McAlpine, Haideh Motamedi, John Robinson, and David Sherman for sharing their results with us prior to publication. We are indebted to Dick Hutchinson and David Hopwood for critical review of the manuscript.

Literature Cited

1. Adamitis, T., Champness, W. 1992. Genetic analysis of *absB*, a *Streptomyces coelicolor* locus involved in global antibiotic regulation. *J. Bacteriol.* 174:4622–38
2. Adamitis, T., Riggle, P., Champness, W. 1992. Mutations in a new *Streptomyces coelicolor* locus which globally block antibiotic synthesis but not sporulation. *J. Bacteriol.* 172:2962–69
3. Amy, C. M., Witkowski, A., Naggert, J., Williams, B., Randhawa, Z., Smith, S. 1989. Molecular cloning and sequencing of cDNAs encoding the entire rat fatty acid synthase. *Proc. Natl. Acad. Sci. USA* 86:3114–18
4. Andersen, J. F., Tatsuta, K., Gunji, H., Ishiyama, T., Hutchinson, C. R. 1992. Substrate specificity of 6-deoxyerythronolide B hydroxylase, a soluble bacterial cytochrome P450. *Biochemistry* 32:1905–13
5. Arcamone, F., Cassinelli, G., Dimatteo, F., Forenza, S., Ripamonte, M., et al. 1980. Structures of novel anthracycline antitumor antibiotics from

Micromonospora peucetica. *J. Am. Chem. Soc*. 102:1462–63
6. Arrowsmith, T. J., Malpartida, F., Sherman, D. H., Birch, A. W., Hopwood, D. A., Robinson, J. A. 1992. Characterization of *actI*-homologous DNA encoding polyketide synthase genes from the monensin producer *Streptomyces cinnamonensis*. *Mol. Gen. Genet*. 234:254–64
6a. Baltz, R. H., Ingolia, T., Hegeman, G., eds. 1993 *Genetics and Molecular Biology of Industrial Microorganisms*. Washington DC: Am. Soc. Microbiol. In press
7. Baltz, R. H., Seno, E. T. 1988. Genetics of *Streptomyces fradiae* and tylosin biosynthesis. *Annu. Rev. Microbiol*. 42:547–74
8. Bartel, P. L., Zhu, C.-B., Lampel, J. S., Dosch, D. C., Conners, N. C., et al. 1990. Biosynthesis of anthraquinones by interspecies cloning of actinorhodin biosynthesis genes in streptomycetes: clarification of actinorhodin gene functions. *J. Bacteriol*. 172: 4816–26
9. Beck, J., Ripka, S., Siegner, A., Schiltz, E., Schweizer, E. 1990. The multifunctional 6-methylsalicylic acid synthase gene of *Penicillium patulum:* its gene structure relative to that of other polyketide synthases. *Eur. J. Biochem*. 192:487–98
10. Bergh, S., Uhlen, M. 1992. Analysis of a polyketide synthesis-encoding gene cluster of *Streptomyces curacoi*. *Gene* 117:131–36
11. Bevitt, D. J., Cortes, J., Haydock, S. F., Leadlay, P. F. 1992. 6-Deoxyerythronolide B synthase from *Saccharopolyspora erythraea*. Cloning of the structural gene, sequence analysis and inferred domain structure of the multifunctional enzyme. *Eur. J. Biochem*. 204:39–49
12. Bibb, M. J., Biro, S., Motamedi H., Collins, J. F., Hutchinson, C. R. 1989. Analysis of the nucleotide sequence of the *Streptomyces glaucescens tcmI* genes provides key information about the enzymology of polyketide tetracenomycin C antibiotic biosynthesis. *EMBO J*. 8:2727–36
13. Binnie, D., Warren, M., Butler, M. J. 1989. Cloning and heterologous expression in *Streptomyces lividans* of *Streptomyces rimosus* genes involved in oxytetracycline biosynthesis. *J. Bacteriol*. 171:887–95
14. Birch, A. J. 1967. Biosynthesis of polyketides and related compounds. *Science* 156:202–6
15. Blanco, G., Brian, P., Pereda, A., Mendez, C., Salas, J. A., Chater, K. F. 1993. Hybridization and DNA sequence analyses suggest an early divergence of related biosynthetic gene sets for polyketide antibiotics and spore pigments in *Streptomyces* spp. *Gene* In press
16. Blanco, G., Pereda, A., Mendez, C., Salas, J. A. 1992. Cloning and disruption of a fragment of *Streptomyces halstedii* DNA involved in the biosynthesis of a spore pigment. *Gene* 112:59–65
17. Blumauerova, M., Kralovcova, E., Mateju, J., Jizba, J., Vanek, Z. 1979. Biotransformations of anthracyclinones in *Streptomyces coeruleorubidus* and *Streptomyces galilaeus*. *Folia Microbiol*. 24:117–27
18. Brown, J. R., Imam, S. H. 1984. Recent studies on doxorubicin and its analogues. *Prog. Med. Chem*. 21:169–236
19. Butler, M. J., Friend, E. J., Hunter, I. S., Kaczmarek, F. S., Sugden, D. A. Warren, M. 1989. Molecular cloning of resistance genes and architecture of a linked gene cluster involved in biosynthesis of oxytetracycline by *Streptomyces rimosus*. *Mol. Gen. Genet*. 215:231–38
20. Caballero, J., Martinez, E., Malpartida, F., Hopwood, D. A. 1991. Organization and functions of the *act*VA region of the actinorhodin biosynthetic gene cluster of *Streptomyces coelicolor*. *Mol. Gen. Genet*. 230:401–12
21. Caffrey, P., Bevitt, D. J., Staunton, J., Leadlay, P. F. 1992. Identification of DEBS1, DEBS2 and DEBS3, the multienzyme polypeptides of the erythromycin-producing polyketide synthase from *Saccharopolyspora erythraea*. *FEBS Lett*. 304:225–28
22. Caffrey, P., Green, B., Packman, L., Rawlings, B. J., Staunton, J., Leadlay, P. F. 1991. An acyl-carrier-protein-thioesterase domain from the 6-deoxyerythronolide B synthase of *Saccharopolyspora erythraea:* high level production, purification and characterisation in *Escherichia coli*. *Eur. J. Biochem*. 195:823–30
23. Cane, D. E., Lambalot, R. H., Prabhakaren, P. C., Ott, W. R. 1993. Macrolide biosynthesis 7. Incorporation of polyketide chain elongation intermediates in methymycin. *J. Am. Chem. Soc*. 115:522–26
24. Cane, D. E., Ott, W. R. 1988. Macrolide biosynthesis. 5. Intact incorporation of a chain-elongation inter-

mediate into nargenicin. *J. Am. Chem. Soc.* 110: 1 4840–42
25. Cane, D. E., Yang, C. 1987. Macrolide biosynthesis. 4. Intact incorporation of a chain-elongation intermediate into erythromycin. *J. Am. Chem. Soc.* 109: 1255–57
26. Casey, M. L., Paulick, R. C., Whitlock, H. W. 1978. Carbon-13 nuclear magnetic resonance study of the biosynthesis of daunomycin and islandicin. *J. Org. Chem.* 43:1627–34
27. Cole, S. P., Rudd, B. A. M., Hopwood, D. A., Chang, C. J., Floss, H. G. 1987. Biosynthesis of the antibiotic actinorhodin. Analysis of blocked mutants of *Streptomyces coelicolor*. *J. Antibiot.* 40:340–47
28. Cortes, J., Haydock, S. F., Roberts, G. A., Bevitt, D. J., Leadlay, P. F. 1990. An unusually large multifunctional polypeptide in the erythromycin-polyketide synthase of *Saccharopolyspora erythraea*. *Nature* 346: 176–78
29. Davis, N. K., Chater, K. F. 1990. Spore colour in *Streptomyces coelicolor* A3(2) involves the developmentally regulated synthesis of a compound biosynthetically related to polyketide antibiotics. *Mol. Microbiol.* 4:1679–92
30. Decker, H., Motamedi, H., Hutchinson, C. R. 1993. Nucleotide sequences and heterologous expression of *tcmG* and *tcmP*, biosynthetic genes for tetracenomycin C synthesis in *Streptomyces glaucescens*. *J. Bacteriol.* 175: 3876–86
31. Devereux, J., Haeberli, P., Smithies, O. 1984. A comprehensive set of sequence analysis programs for the VAX. *Nucleic Acids Res.* 12:387–95
32. Dhillon, N., Hale, R. S., Cortes, J., Leadlay, P. F. 1989. Molecular characterization of a gene from *Saccharopolyspora erythraea* (*Streptomyces erythraeus*) which is involved in erythromycin biosynthesis. *Mol. Microbiol.* 3:1405–14
33. Dimroth, P., Walter, H., Lynen, F. 1978. 6-Methylsalicylic acid synthetase from *Penicillium patulum*—some catalytic properties of enzyme and its relation to fatty-acid synthetase. *Eur. J. Biochem.* 13:98–110.
34. Donadio, S., Katz, L. 1992. Organization of the enzymatic domains in the multifunctional polyketide synthase involved in erythromycin biosynthesis in *Saccharopolyspora erythraea*. *Gene* 111:51–60
35. Donadio, S., Stassi, D., McAlpine, J. B., Staver, M. J., Sheldon, P. J., et al. 1993. Recent developments in the genetics of erythromycin formation. See Ref. 6a, In press
36. Donadio, S., Staver, M. J. 1993. IS*1136*, an insertion element in the erythromycin gene cluster of *Saccharopolyspora erythraea*. *Gene* In press
37. Donadio, S., Staver, M. J., McAlpine, J. B., Swanson, S. J., Katz, L. 1991. Modular organization of genes required for complex polyketide biosynthesis. *Science* 252:675–79
38. Donadio, S., Staver, M. J., McAlpine, J. B, Swanson, S. J., Katz, L. 1992. Biosynthesis of the erythromycin macrolactone and a rational approach for producing hybrid macrolides. *Gene* 115:97–103
39. Epp, J. K., Huber, M. L. B., Turner, J. R., Goodson, T., Schoner,. B. E. 1989. Production of a hybrid macrolide antibiotic in *Streptomyces ambofaciens* and *Streptomyces lividans* by introduction of a cloned carbomycin biosynthetic gene from *Streptomyces thermotolerans*. *Gene* 85:293–301
40. Fernandez-Moreno, M. A., Caballero, J. L., Hopwood, D. A., Malpartida, F. 1991. The act cluster contains regulatory and antibiotic export genes, direct targets for translational control by the bldA transfer RNA gene of Streptomyces. *Cell* 66:769–80
41. Fernandez-Moreno, M. A., Martinez, E., Boto, L., Hopwood, D. A., Malpartida, F. 1992. Nucleotide sequence and deduced functions of a set of co-transcribed genes of *Streptomyces coelicolor* A3(2) including the polyketide synthase for the antibiotic actinorhodin. *J. Biol. Chem.* 267:19278–90
42. Fernandez-Moreno, M. A., Martin-Triana, A. J., Martinez, E., Niemi, J., Kieser, H. M., et al. 1992. *abaA,* a new pleiotropic regulatory locus for antibiotic production in *Streptomyces coelicolor*. *J. Bacteriol.* 174:2958–67
43. Gorst-Allman, C. P., Rudd, B. A. M., Chang, C.-J., Floss, H. G. 1981. Biosynthesis of actinorhodin. Point of dimerization. *J. Org. Chem.* 46:455–56
44. Gramajo, H. C., White, J., Hutchinson, C. R., Bibb, M. J. 1991. Overproduction and localization of components of the polyketide synthase of *Streptomyces glaucescens* involved in the production of the antibiotic tetracenomycin C. *J. Bacteriol.* 173:6475–83
45. Guilfoile, P. G., Hutchinson, C. R. 1991. A bacterial analog of the MDR gene of mammalian tumor-cells is pres-

ent in *Streptomyces peucetius,* the producer of daunorubicin and doxorubicin. *Proc. Natl. Acad. Sci. USA* 88:8553–57
46. Guilfoile, P. G., Hutchinson, C. R. 1992. Sequence and transcriptional analysis of the *Streptomyces glauscencens tcmAR* tetracenomycin C resistance and repressor gene loci. *J. Bacteriol.* 174:3651–58
47. Guilfoile, P. G., Hutchinson, C. R. 1992. The *Streptomyces glaucescens* TcmR protein represses transcription of the divergently oriented *tcmR* and *tcmA* genes by binding to an intergenic operator region. *J. Bacteriol.* 174:3659–66
48. Hallam, S. E., Malpartida F., Hopwood, D. A. 1988. Nucleotide sequence, transcription and deduced function of a gene involved in polyketide antibiotic synthesis in *Streptomyces coelicolor. Gene* 74:305–20
49. Haydock, S. F., Dowson, J. A., Dhillon, N., Roberts, G. A., Cortes, J., Leadlay, P. F. 1991. Cloning and sequence analysis of genes involved in erythromycin biosynthesis in *Saccharopolyspora erythraea;* sequence similarities between EryG and a family of S-adenosylmethionine-dependent methyltransferases. *Mol. Gen. Genet.* 230: 120–28
50. Hong, S.-K., Kito, M., Beppu, T., Horinouchi, S. 1991. Phosphorylation of the AfsR product, a global regulatory protein for secondary metabolite formation in *Streptomyces coelicolor* A3(2). *J. Bacteriol.* 173:2311–18
51. Hopwood, D. A., Khosla, C. 1992. Genes for polyketide secondary metabolic pathways in microorganisms and plants. In *Secondary Metabolites: Their Function and Evolution. Ciba Foundation Symposium 17.* pp. 88–112. Chichester: Wiley
52. Hopwood, D. A., Malpartida, F., Kieser, H. M., Ikeda, H., Duncan, J., et al. 1985. Production of 'hybrid' antibiotics by genetic engineering. *Nature* 314:642–44
53. Hopwood, D. A., Sherman, D. H. 1990. Molecular genetics of polyketides and its comparison to fatty acid biosynthesis. *Annu. Rev. Genet.* 24:37–66
54. Horinuchi, S., Beppu, T. 1985. Construction and application of a promoter-probe plasmid that allows chromogenic identification in *Streptomyces lividans. J. Bacteriol.* 162:406–12
55. Huber, M. L. B., Paschal, J. W., Leeds, J. P., Kirst, H. A., Wind, J. A., et al. 1990. Branched chain fatty acids produced by mutants of *Streptomyces fradiae,* putative precursors of the lactone ring of tylosin. *Antimicrob. Agents Chemother.* 34:1535–41
56. Hutchinson, C. R., Decker, H., Motamedi, H., Shen, B., Summers, R. G., et al. 1993. Molecular genetics and biochemistry of a bacterial type II polyketide synthase. See Ref. 6a, In press
57. Ishizuka, H., Horinouchi, S., Kieser, H. M., Hopwood, D. A., Beppu, T. 1992. A putative two-component regulatory system involved in secondary metabolism in *Streptomyces* spp. *J. Bacteriol.* 174:7585–94
58. Itoh, Z., Suzuki, T., Nakaya, M., Inoue, M., Mitsuhashi, S. 1984. Gastrointestinal motor-stimulating activity of macrolide antibiotics and analysis of their side effects on the canine gut. *Antimicrob. Agents Chemother.* 26: 863–69
59. Jacobs, A., Staunton, J., Sutkowski, A. C. 1991. Aspyrone biosynthesis in *Aspergillus melleus:* identification of the intermediates formed on the polyketide synthase (PKS) in the first chain extension cycle leading to crotonate. *J. Chem. Soc. Chem. Commun.* 1113–14
60. Jordan, P. M., Spencer, J. B. 1991. Mechanistic and stereochemical investigations of fatty acid and polyketide biosynthesis using chiral malonates. *Tetrahedron* 47:6015–28
61. Kakinuma, S., Takada, Y., Ikeda, H., Tanaka, H., Omura, S., Hopwood, D. 1991. Cloning of large DNA fragments, which hybridize with actinorhodin biosynthesis genes from kalafungin and nanaomycin A methyl ester producers and identification of genes for kalafungin biosynthesis of the kalafungin producer. *J. Antibiot.* 44:995–1005
62. Katagiri, K., Tori, K., Kimura, Y., Yoshida, T., Nagasaki, T., Minato, H. 1967. A new antibiotic. Furanomycin, an isoleucine antagonist. *J. Med. Chem.* 10:1149–54
63. Katz, L., Donadio, S. 1993. Macrolides. In *Biochemistry and Genetics of Antibiotic Biosynthesis,* ed. C. Stuttard, L. C. Vining, Butterworth-Heineman, pp. 71–104
64. Khosla, C., Ebert-Khosla, S., Hopwood, D. A. 1992. Targeted gene replacements in a *Streptomyces* polyketide synthase gene cluster: role for the acyl carrier protein. *Mol. Microbiol.* 6:3237–49
65. Khosla, C., McDaniel, R., Ebert-Khosla, S., Torres, R., Sherman, D. H., et. al. 1993. Genetic construction

and functional analysis of hybrid polyketide synthases containing heterologous acyl carrier proteins. *J. Bacteriol.* 175:2197–2204

66. Kim, E.-S., Sherman, D. H. 1992. Trans-complementation of actinorhodin non-producing mutants with oxytetracycline, *whiE* spore pigment, and fatty acid biosynthesis genes. *Am. Soc. Microbiol. Conf. Genet. of Ind. Microorg., 5th, Bloomington, Ind.* A-24 (Abstr.)
67. Kino, T., Hatanaka, H., Miyata, S., Inamura, N., Nishiyama, M., et al. 1987. FK-506, a novel immunosuppressant isolated from a *Streptomyces*. II. Immunosuppressive effect of FK-506 *in vitro*. *J. Antibiot.* 40:1256–65
68. Kinoshita, K., Takenaka, S., Hayashi, M. 1988. Isolation of proposed intermediates in the biosynthesis of mycinamicins. *J. Chem. Soc. Chem. Commun.* 943–49
69. Kinoshita, K., Takenaka, S., Hayashi, M. 1991. Mycinamycin biosynthesis: isolation and structural elucidation of mycinonic acids, proposed intermediates for formation of mycinamycins. X-ray molecular structure of *p*-bromophenacyl-5-hydroxy-4-methylhept-2-enoate. *J. Chem. Soc. Perkin Trans.* 1:2247–53
70. Kitamura, I., Tobe, H., Yoshimoto, A., Oki, T., Naganawa, H., et al. 1981. Biosynthesis of aklavinone and aclacinomycins. *J. Antibiot.* 34:1498–1500
71. Lawlor, E. J., Baylis, H. A., Chater, K. F. 1987. Pleiotropic morphological and antibiotic deficiencies result from mutations in a gene encoding a tRNA-like product in *Streptomyces coelicolor* A3(2). *Genes Dev.* 1:1305–10

71a. Leadlay, P. F., Stanton, J., Aparacio, J. F., Bevitt, D. J., Caffrey, P., et al. 1993. The erythromycin-producing polyketide synthase. *Biochem. Soc. Trans.* 21:218–22

72. Le Gouill, C., Desmarais, D., Dery, C. V. 1993. *Saccharopolyspora hirsuta* 367 encodes clustered polyketide synthases, polyketide reductase, acyl carrier protein, and biotin carboxyl carrier protein homologous genes. *Mol. Gen. Genet.* In press
73. Li, Z., Martin, F. M., Vederas, J. C. 1992. Biosynthetic incorporation of labeled tetraketide intermediates into dehydrocurvularin, a phytotoxin from *Alternaria cineraria,* with assistance of β-oxidation inhibitors. *J. Am. Chem. Soc.* 114:1531–33
74. Lin, C. Y., Smith, S. 1978. Properties of thioesterase component obtained by limited trypsinization of fatty-acid synthase multienzyme complex. *J. Biol. Chem.* 253:1954–63
75. MacNeil, D. J., Occi, J. L., Gewain, K. M., MacNeil, T., Gibbons, P. H., et al. 1992. Complex organization of the *Streptomyces avermitilis* genes encoding the avermectin polyketide synthase. *Gene* 115:119–25
76. MacNeil, D. J., Occi, J. L., Gewain, K. M., MacNeil, T., Gibbons, P. H., et al. 1993. A comparison of the genes encoding the polyketide synthases for avermectin, erythromycin, and nemadectin. See Ref. 6a, In press
77. Malpartida, F., Hallam, S. E., Kieser, H. M., Motamedi, H., Hopwood, D. A., et al. 1987. Homology between *Streptomyces* genes coding for synthesis of different polyketides used to clone antibiotic biosynthetic genes. *Nature* 325:818–21
78. Malpartida, F., Hopwood, D. A. 1984. Molecular cloning of the whole biosynthetic pathway of a *Streptomyces* antibiotic and its expression in a heterologous host. *Nature* 309:462–64
79. Malpartida, F., Hopwood, D. A. 1986. Physical and genetic characterisation of the gene cluster for the antibiotic actinorhodin in *Streptomyces coelicolor* A3(2). *Mol. Gen. Genet.* 205:66–73
80. Matzuzawa, Y., Yoshimoto, A., Shibamoto, N., Tobe., H., Oki, T., et al. 1981. New anthracycline metabolites from mutant strains of *Streptomyces galilaeus* MA144-M1. II. Structure of 2-hydroxyaklavinone and new aklavinone glycosides. *J. Antibiot.* 34: 959–64
81. McAlpine, J. B., Tuan, J. S., Brown, D. P., Grebner, K. D., Whittern, D. N., et al. 1987. New antibiotics from genetically engineered actinomycetes. I. 2-Norerythromycins, isolation and structural determinations. *J. Antibiot.* 40:1115–22
82. McInnes, A. G. 1974. Tenellin and bassianin, metabolites of *Beauveria* species—structure elucidation with N-15-enriched and doubly C-13-enriched compounds using C-13 nuclear magnetic resonance spectroscopy. *J. Chem. Soc. Chem. Commun.* 281–82
83. Meurer, G., Biermann, G., Schutz, A., Harth, S., Schweizer, E. 1992. Molecular structure of the multifunctional fatty acid synthetase gene of *Brevibacterium ammoniagenes:* its sequence of catalytic domains is formally consistent with a head to tail fusion

of two yeast genes. *Mol. Gen. Genet.* 232:106–16

84. Mohammed, A. H., Chirala, S. S., Mody, N., Huang, W.-K., Wakil, S. 1988. Primary structure of the multifunctional α subunit of yeast fatty acid synthase derived from *FAS2* gene sequence. *J. Biol. Chem.* 263:12315–25
85. Motamedi, H., Cai, S. J., Shafiee, A. 1992. Early genes involved in the biosynthesis of macrolactone ring of immunosuppressive drug FK506 from *Streptomyces* sp. MA6548. *Am. Soc. Microbiol. Conf. Genet. of Ind. Microorg., 5th, Bloomington, Ind.* A-19 (Abstr.)
86. Motamedi, H., Hutchinson, C. R. 1987. Cloning and heterologous expression of a gene cluster for the biosynthesis of tetracenomycin C, the anthracycline antitumor antibiotic of *Streptomyces glaucescens*. *Proc. Natl. Acad. Sci. USA* 84:4445–49
87. Motamedi, H., Wendt-Pienkowski, E., Hutchinson, C. R. 1986. Isolation of tetracenomycin C-nonproducing *Streptomyces glaucescens* mutants. *J. Bacteriol.* 167:575–80
88. Noguchi, H., Harrison, P. H., Arai, K., Nakashima, T. T., Trimble, L. A., Vederas, J. C. 1988. Biosynthesis and full NMR assignment of fungichromin, a polyene antibiotic from *Streptomyces cellulosae*. *J. Am. Chem. Soc.* 110:2938–45

88a. O'Hagan, D. 1992. Biosynthesis of polyketide metabolites. *Nat. Prod. Rep.* 9:447–79

89. Oki, T., Yoshimoto, A., Matsuzawa, Y., Takeuchi, T., Umezawa, H. 1980. Biosynthesis of anthracycline antibiotics by *Streptomyces galilaeus*. I. Glycosidation of various anthracyclinones by an aclacinomycin-negative mutant and biosynthesis of aclacinomycins from aklavinone. *J. Antibiot.* 33:1331–40
90. Ollis, W. D., Sutherland, I. O. 1960. The incorporation of propionate in the biosynthesis of ϵ-pyrromycinone (rutilantinone). *Proc. Chem. Soc. (London)* 347:34–38
91. Omura, S. 1984. *Macrolide Antibiotics: Chemistry, Biology and Practice*. Orlando: Academic
92. Omura, S. 1992. The expanded horizon for microbial metabolites—a review. *Gene* 115:141–49
93. Otten, S. L., Stutzman-Engwall, K. L., Hutchinson, C. R. 1990. Cloning and expression of daunorubicin biosynthesis genes from *Streptomyces peucetius* and *S. peucetius* subsp. *caesius*. *J. Bacteriol.* 172:3427–34
94. Paulus T. J., Tuan, J. S., Luebke, V. E., Maine, G. T., DeWitt, J. P., Katz, L. 1990. Mutation and cloning of *eryG*, the structural gene for erythromycin *O*-methyltransferase from *Saccharopolyspora erythraea* and expression of *eryG* in *Escherichia coli*. *J. Bacteriol.* 172:2541–46
95. Revill, W. P., Leadlay, P. F. 1991. Cloning, characterization and high level expression in *Escherichia coli* of the *Saccharopolyspora erythraea* gene encoding an acyl carrier protein potentially involved in fatty acid biosynthesis. *J. Bacteriol.* 173:4379–85
96. Rudd, B. A. M., Hopwood, D. A. 1979. Genetics of actinorhodin biosynthesis by *Streptomyces coelicolor* A3(2). *J. Gen Microbiol.* 114:35–43
97. Schweizer, M., Roberts, L. M., Holtke, H.-M., Takabayashi, K., Hollerer, E., et al. 1986. The pentafunctional FAS1 gene of yeast: its nucleotide sequence and order of the catalytic domains. *Mol. Gen. Genet.* 203:479–86
98. Schweizer, M., Takabayashi, K., Laux, T., Beck, K.-F., Schreglmann, R. 1989. Rat mammary gland fatty acid synthase: localization of the constituent domains and two functional polyadenylation/termination signals in the cDNA. *Nucleic Acids Res.* 17:567–86
99. Scotti, C., Piatti, M., Cuzzoni, A., Perani, P., Tognoni, A., et al. 1993. A *Bacillus subtilis* ORF coding for a polypeptide highly similar to polyketide synthases. *Gene* In press
100. Shen, B., Summers, R. G., Gramajo, H. C., Bibb, M. J., Hutchinson, C. R. 1992. Purification and characterization of the acyl carrier protein of the *Streptomyces glaucescens* tetracenomycin C polyketide synthase. *J. Bacteriol.* 174:3818–21
101. Sherman, D. H., Bibb, M. J., Simpson, T. J., Johnson, D., Malpartida, F., et al. 1991. Molecular genetic analysis reveals a putative bifunctional polyketide cyclase/dehydrase gene from *Streptomyces coelicolor* and *Streptomyces violaceoruber*, and a cyclase/O-methyltransferase from *Streptomyces glaucescens*. *Tetrahedron* 47:6029–43
102. Sherman, D. H., Kim, E.-S., Bibb, M. J., Hopwood, D. A. 1992. Functional replacement of genes for individual polyketide synthase components in *Streptomyces coelicolor* A3(2) by heterologous genes from a different polyketide pathway. *J. Bacteriol.* 174: 6184–90
103. Sherman, D. H., Malpartida, F., Bibb M. J., Kieser, H. M., Bibb, M. J.,

Hopwood, D. A. 1989. Structure and deduced function of the granaticin-producing polyketide synthase gene cluster of *Streptomyces violaceoruber* TU22. *EMBO J.* 8:2717–25

104. Simpson, T. J. 1991. The biosynthesis of polyketides. *Nat. Prod. Rep.* 8:573–602
105. Singh, S. B., Cordingly, M. G., Ball, R. G., Smith, J. L., Dombrowski, A. W., Goetz, M. G. 1991. Structure and stereochemistry of thysanone: a novel human rhinovirus 3C-protease inhibitor from *Thysanophora penicilloides*. *Tetrahedron Lett.* 32:5279–82
106. Spavold, Z. M., Robinson, J. A. 1988. Nonactin biosynthesis: on the role of (6*R*,8*R*)- and (6*S*,8*S*)-2-methyl-6,8-dihydroxynon-2*E*-enoic acids in the formation of nonactin. *J. Chem. Soc. Chem. Commun.* 4–6
107. Spencer, J. B., Jordan, P. M. 1992. Investigation of the mechanism and steric course of the reaction catalyzed by 6-methylsalicylic acid synthase from *Penicillium patulum* using *(R)*-[1-^{13}C;2-^{2}H]- and *(S)*-[1-^{13}C;2-^{2}H]malonates. *Biochemistry* 31:9107–16
108. Stassi, D., Donadio, S., Staver, M. J., Katz, L. 1993. Identification of a *Saccharopolyspora erythraea* gene required for the final hydroxylation step in erythromycin biosynthesis. *J. Bacteriol.* 175:182–89
109. Staunton, J., Sutkowski, A. C. 1991. The polyketide synthase (PKS) of aspyrone biosynthesis: evidence for the enzyme bound intermediates from incorporation studies with *N*-acetylcysteamine thioesters in intact cells of *Aspergillus melleus*. *J. Chem. Soc. Chem. Commun.* pp. 1110–12
110. Strohl, W. R., Bartel, P. L., Conners, N. C., Zhu, C.-B., Dosch, D. C., et al. 1989. Biosynthesis of natural and hybrid polyketides by anthracycline-producing streptomycetes. In *Genetics and Molecular Biology of Industrial Microorganisms*, ed. C. L. Hershberger, S. W. Queener, G. Hegeman, pp. 68–84. Washington DC: Am. Soc. Microbiol.
111. Strohl, W. R., Conners, N. C. 1992. Significance of anthraquinone formation resulting from the cloning of actinorhodin genes in heterologous streptomycetes. *Mol. Microbiol.* 6:147–52
112. Stutzman-Engwall, K. J., Hutchinson, C. R. 1989. Multigene families for anthracycline antibiotic production in *Streptomyces peucetius*. *Proc. Natl. Acad. Sci. USA* 86:3135–39
113. Stutzman-Engwall, K. J., Otten, S. L., Hutchinson, C. R. 1992. Regulation of secondary metabolism in *Streptomyces* spp. and overproduction of daunorubicin in *Streptomyces peucetius*. *J. Bacteriol.* 174:144–54
114. Summers, R. G., Wendt-Pienkowski, E., Motamedi, H., Hutchinson, C. R. 1992. Nucleotide sequence of the *tcmII-tcmIV* region of the tetracenomycin biosynthetic gene cluster of *Streptomyces glaucescens* and evidence that the *tcmN* gene encodes a multifunctional cyclase-dehydratase-O-methyltransferase. *J. Bacteriol.* 174:1810–20
115. Thamchaipenet, A., Linton, K. J., Hunter, I. S. 1992. Further characterization of the gene cluster for oxytetracycline production from *Streptomyces rimosus*. *Am. Soc. Microbiol. Conf. Genet. of Ind. Microorg., 5th, Bloomington, Ind* B-8 (Abstr.)
116. Tsukamoto, N., Fujii, I., Ebizuka, Y., Sankawa, U. 1992. Cloning of the aklavinone biosynthesis genes from *Streptomyces galilaeus*. *J. Antibiot.* 45: 1286–94
117. Tuan, J. S., Weber, J. M., Staver, M. J., Leung, J. O., Donadio, S., Katz, L. 1990. Cloning of genes involved in erythromycin biosynthesis from *Saccharopolyspora erythraea* using a novel actinomycete–*Escherichia coli* cosmid. *Gene* 90:21–29
118. Vanek, Z., Tax, J., Komersova, I., Sedmera, P., Vokoun, J. 1977. Anthracyclines. *Folia Microbiol.* 22:139–59
119. Vara, J. A., Lewandowska-Skarbek, M., Wang, Y.-G., Donadio, S., Hutchinson C. R. 1989. Cloning of genes governing the deoxysugar portion of the erythromycin biosynthesis pathway in *Saccharopolyspora erythraea* (*Streptomyces erythreus*). *J. Bacteriol.* 171: 5872–81
120. Wakil, S. J. 1989. Fatty acid synthase, a proficient multifunctional enzyme. *Biochemistry* 28:4523–30
121. Weber, J. M., Leung, J. O., Maine, G. T., Potenz, R. H. B., Paulus, T. J., DeWitt, J. P. 1990. Organization of a cluster of erythromycin genes in *Saccharopolyspora erythraea*. *J. Bacteriol.* 172:2372–82
122. Weber, J. M., Leung, J. O., Swanson, S. J., Idler, K. B., McAlpine, J. B. 1991. An erythromycin derivative produced by targeted gene disruption in *Saccharopolyspora erythraea*. *Science* 252:114–17
123. Weber, J. M., Schoner, B., Losick, R. 1989. Identification of a gene re-

quired for the terminal step in erythromycin biosynthesis in *Saccharopolyspora erythraea* (*Streptomyces erythreus*). *Gene* 75:235–41

124. Witkowski, A., Rangan, V. S., Randhawa, Z. I., Amy, C. M., Smith, S. 1991. Structural organization of the multifunctional animal fatty-acid synthase. *Eur. J. Biochem.* 198:571–79
125. Yaginuma, S., Morishita, A., Ishizawa, K., Murofushi, S., Hayashi., M., Mutoh, N. 1992. Sporeamicin A, a new macrolide antibiotic. I. Taxonomy, fermentation, isolation and characterization. *J. Antibiot.* 45:599–606
126. Yoshimoto, A., Johdo, O., Tone, H., Okamoto, R., Naganawa, H., et al. 1992. Production of new anthracycline antibiotics 1-hydroxyoxaunomycin and 6-deoxyoxaunomycin by limited biosynthetic conversion using a daunorubicin-negative mutant. *J. Antibiot.* 45:1609–17
127. Yoshimoto, A., Oki, T., Takeuchi, T., Umezawa, H. 1980. Microbial conversion of anthracyclinones to daunomycin by blocked mutants of *Streptomyces coeruleorubidus*. *J. Antibiot.* 33: 1158–66
128. Yoshizawa, Y., Li, Z., Reese, P. B., Vederas, J. C. 1990. Intact incorporation of acetate-derived di- and tetraketides during biosynthesis of dehydrocurvularin, a macrolide phytotoxin from *Alternaria cineraria*. *J. Am. Chem. Soc.* 112:3212–13
129. Yue, S., Duncan, J. S., Yamamoto, Y., Hutchinson, C. R. 1987. Macrolide biosynthesis. Tylactone formation involves the processive addition of three carbon units. *J. Am. Chem. Soc.* 109: 1253–55
130. Yue, S., Motamedi, H., Wendt-Pienkowski, E., Hutchinson, C. R. 1986. Anthracycline metabolites of tetracenomycin C-non-producing *Streptomyces glaucescens* mutants. *J. Bacteriol.* 167: 581–86
131. Zhang, H.-L., He, X.-G., Adefarati A., Gallucci, J., Cole, S. P., et al. 1990. Mutactin, a novel polyketide from *Streptomyces coelicolor*. Structure and biosynthetic relationship to actinorhodin. *J. Org. Chem.* 55:1682–44
132. Donadio, S., McAlpine, J. B., Sheldon, P. J., Jackson, M., Katz, L. 1993. An erythromycin analog produced by reprogramming of polyketide synthesis. *Proc. Natl. Acad. Sci. USA* 90:In press

Annu. Rev. Microbiol. 1993. 47:913–44

RELEASE OF RECOMBINANT MICROORGANISMS

M. Wilson and S. E. Lindow
Department of Plant Pathology, University of California, Berkeley, California 94720

KEY WORDS: field testing, fungi, bacteria, bioremediation, biological control

---

CONTENTS

ABSTRACT

This review addresses current environmental applications of naturally occurring, nonrecombinant microorganisms and potential future genetic modifications of such organisms, as well as releases of recombinant microorganisms that have occurred to date. Awareness of the current uses of nonrecombinant microorganisms provides insight into the diversity of habitats in which recombinant microorganisms may be released in the future, while an examination of potential and realized genetic modifications provides insight into the variety of applications for which recombinant microorganisms may be used.

0066-4227/93/1001-0913$02.00

Analysis of the behavior, persistence, and dispersal of nonrecombinant strains further provides valuable information required for the assessment of the risk involved in release of recombinant derivatives of those strains. Approximately 27 distinct releases of recombinant microorganisms have occurred to date. This review assesses what has been learned from such releases regarding persistence, dispersal, and potential deleterious environmental effects.

INTRODUCTION

Microorganisms have evolved adaptations permitting them to colonize most aquatic and terrestrial habitats. Thus, the adaptations of naturally occurring microorganisms are highly diverse and represent an invaluable genetic resource. While many naturally occurring microorganisms with desirable traits are already used for various purposes, molecular techniques allow the manipulation of strains for more effective or varied applications. Genetic techniques are most often used to make subtle changes in microorganisms that already have many of the attributes required for a particular use. Current molecular techniques allow directed modification of microorganisms by removal or addition of one or a few genes (109). Advances are being made in the manipulation of fungi and yeasts, but genetic manipulation of these organisms lags behind that of bacteria. This review emphasizes current environmental applications of nonrecombinant microorganisms to try to anticipate likely future applications of recombinant organisms, primarily bacteria, that will require their release into the open environment. Because very few recombinant microbes have yet been introduced into the open environment, this activity remains primarily a speculative venture. We therefore analyze the releases of both unaltered and recombinant microorganisms to better address the risks associated with the numerous introductions of recombinant microbes to natural habitats that are anticipated to occur in the future. We attempt to place the release of recombinant microbes in perspective with the many unaltered microbes that have been and will be field tested.

CURRENT ENVIRONMENTAL APPLICATIONS OF NATURALLY OCCURRING MICROORGANISMS AND ANTICIPATED USES OF RECOMBINANT STRAINS

Biological Control of Plant Diseases

CONTROL OF DISEASES ON THE AERIAL SURFACES OF PLANTS The control of diseases on the aerial surfaces of plants is limited by the availability of effective chemicals, particularly in the case of bacterial diseases. Few bactericides are available; growers rely on copper compounds and antibiotics such as streptomycin and oxytetracycline. Resistance to these bactericides is becoming more widespread and alternatives are now being sought. Biological control of aerial diseases, particularly those caused by phytopathogenic bacteria, should be an important future application of recombinant microorganisms.

Many naturally occurring microorganisms have been used to control diseases on the aerial surfaces of plants (7, 17, 179). The more common bacterial species that have been used for the control of diseases in the phyllosphere include *Pseudomonas syringae, Pseudomonas fluorescens, Pseudomonas cepacia, Erwinia herbicola,* and *Bacillus subtilis*. One strain of *P. fluorescens* has recently been registered for the commercial control of fire blight on pear. Fungal genera that have been used for the control of diseases in the phyllosphere include *Trichoderma, Ampelomyces,* and the yeasts *Tilletiopsis* and *Sporobolomyces*. The mechanisms of action proposed for these biological-control agents, which include competition for sites and/or nutrients, antibiosis, and hyperparasitism, represent targets for strain improvement through genetic manipulation.

Several phytopathogenic bacteria exhibit an epiphytic phase prior to invasion, during which time they are susceptible to competition from other microorganisms. While preemptive competitive exclusion of phytopathogenic bacteria in the phyllosphere can be achieved using naturally occurring strains, avirulent mutants of the pathogen, in which deleterious phenotypic traits have been removed, may be more effective because they occupy the same niche as the parental strain. Phytopathogenic bacteria possess several genes that encode phenotypes that allow them to parasitize plants and overcome defense responses elicited by the plant (132). In addition, phytopathogenic bacteria possess pathogenicity genes such as *hrp* (191). Isogenic, avirulent mutants can be produced by insertional inactivation of genes involved in pathogenicity. A nonpathogenic strain of *Pseudomonas syringae* pv. *tomato,* produced by Tn*5* insertional mutagenesis, prevented growth of pathogenic strains in the tomato phyllosphere, presumably by preemptive competitive exclusion (35). Nonpathogenic mutants of *Erwinia amylovora,* produced by transposon mutagenesis, have also been used in the biological control of fire blight (125).

Antibiosis has been proposed as the mechanism of control of several bacterial (183) and fungal (103) diseases in the phyllosphere. Molecular techniques could possibly be used to enhance the efficacy of biological-control agents whose primary mode of action is antibiosis. For example, the transcriptional regulation of genes conferring antibiotic production could be altered by substitution of a constitutive regulatory region or by replacing the regulatory region with promoters known to direct high levels of transcription. It may also be possible to transfer the genes required for antibiotic production from an organism that colonizes the phyllosphere only poorly to one that colonizes more aggressively. Although genes responsible for antibiotic production in *E. herbicola* have been cloned (90), they have not yet been expressed in other epiphytic species.

Biological-control agents must normally achieve a high population in the phyllosphere in order to control other strains, but colonization by the agent may be reduced by competition with the indigenous microflora. Application of a bactericide to which most members of the indigenous microflora are sensitive, but to which the biological-control agent is resistant, can maximize colonization by the biological-control agent. Integration of chemical pesticides and biological-control agents has been reported with *Trichoderma* spp. (71); and *P. syringae* pv. *tomato* (35). Biological-control agents tolerant to specific bactericides could be constructed using molecular techniques. Copper-tolerance determinants occur as a single operon in *P. syringae* (15), which could be transferred to a copper-sensitive organism (164), thereby producing a biological-control agent that can be applied simultaneously with a copper bactericide. Molecular techniques may eventually be used to transfer several beneficial traits, such as the production of one or more antibiotics and pesticide resistance, to an aggressive phyllosphere colonist, which also possesses a conditional suicide system restricting it to one specific host plant.

BIOLOGICAL CONTROL OF FROST INJURY Frost injury of sensitive crop plants is incited by ice nucleation–active (Ice^+) bacterial species, particularly *P. syringae,* which inhabit the aerial surfaces of plants (107). Such plants can avoid damage by supercooling if Ice^+ bacterial populations are low or absent. Biological control of frost injury can be achieved by the prophylactic application of naturally occurring Ice^- strains to the uncolonized blossom or leaf tissue, in a manner similar to that employed in the biological control of phytopathogenic bacteria. Preemptive competitive exclusion of Ice^+ strains from the aerial surfaces by the Ice^- strains reduces the probability of freezing injury. *P. fluorescens* strain A506 has recently been registered commercially for the control of frost injury of pear and is available as Frostban B).

While biological control of frost injury can be achieved using naturally occurring Ice^- strains, isogenic Ice^- *P. syringae* mutants were hypothesized to be most effective in the exclusion of a parental Ice^+ *P. syringae* strain. The gene conferring ice nucleation activity in *P. syringae* was cloned and deletions internal to the structural gene were produced in vitro (130). Reciprocal exchange of the deletion-containing *ice* gene for the native chromosomal gene was accomplished by a process of homologous recombination (108). The resulting Ice^- *P. syringae* mutants were phenotypically and behaviorally identical to the Ice^+ parental strains, except for the inability to nucleate ice (108). These Ice^- *P. syringae* mutants effectively reduced the populations of isogenic Ice$^+$ *P. syringae* strains both in the laboratory and the field (112–114, 117). Although molecular techniques have been successful in constructing a biological-control agent of frost injury, biological control of a heterogeneous mixture of indigenous Ice$^+$ strains can be accomplished with a naturally occurring strain such as *P. fluorescens* A506. The use of mixtures of complementary recombinant Ice$^-$ *P. syringae* strains for biological frost control may improve control.

BIOLOGICAL CONTROL OF SOIL-BORNE DISEASES Chemical control of soil-borne plant diseases is frequently ineffective because of the physical and chemical heterogeneity of the soil, which may prevent effective concentrations of the chemical from reaching the pathogen. Chemicals used in the soil are also generally resistant to degradation and therefore tend to persist as environmental pollutants. Biological-control agents in contrast specifically colonize the rhizosphere, the site requiring protection, and leave no toxic residues. The localization of the biological-control agent in the rhizosphere allows the application of a smaller quantity of the biological agent than the more broadly spread chemicals.

Microorganisms have been used extensively for the biological control of soil-borne plant diseases and also for plant-growth promotion (12, 169, 187). Fluorescent pseudomonads are the most frequently used bacteria for biological control and plant-growth promotion, but *Bacillus* and *Streptomyces* species have also been commonly used. *Trichoderma, Gliocladium,* and *Coniothyrium* species are the most frequently used fungal biological-control agents. Perhaps the most successful biological-control agent of a soil-borne pathogen is *Agrobacterium radiobacter* strain K84, used against crown gall disease caused by *Agrobacterium tumefaciens* (53). A great deal of interest has focused on the mechanisms of action of these biological-control agents, particularly the role played by antibiotics. Consequently, many workers in this area have reported modifications aimed at efficacy enhancement. Biological control with *A. radiobacter* strain K84 is mediated primarily by the bacteriocin agrocin 84 synthesis, which is directed by genes on a plasmid, pAgK84. This plasmid also bears genes for resistance to agrocin 84 and conjugal transfer capacity. Consequently, pAgK84 may be transferred to *A. tumefaciens,* which then becomes resistant to agrocin 84. To prevent this resistance, a transfer-deficient mutant of strain K84 was constructed. *A. radiobacter* strain K1026 is identical to the parental strain, except that the agrocin-producing plasmid, pAgK1026, has had the conjugal-transfer region deleted (88). This strain is now used commercially worldwide for the control of crown gall disease (152, 184).

Competition as a mechanism of biological control has been exploited with soil-borne plant pathogens as with pathogens on the aerial surfaces of plants.

Naturally occurring, nonpathogenic strains of *Fusarium oxysporum* have been used to control wilt diseases caused by pathogenic Fusarium spp. (4). Molecular techniques have been used to remove various deleterious traits of soil-borne phytopathogenic bacteria to construct a competitive antagonist of the pathogen. Random Tn*5* insertions into the genome of *Pseudomonas solanacearum* (174, 175) or insertion of an interposon (Ω-km) into the *hrp* cluster (60) produced avirulent mutants. The avirulent mutants exhibited various levels of invasiveness of tomato plants and provided protection against bacterial wilt disease caused by the virulent pathogen (60). The phytopathogenic bacterium *Erwinia carotovora* subsp. *carotovora* secretes various extracellular enzymes, including pectinases, cellulases, and proteases. The pectinases are known to be a major pathogenicity determinant in soft rot disease of potato. *E. carotovora* subsp. *carotovora* mutants defective in the production of pectate lyase (140) have been used in the biological control of this disease (161). Out^- mutants of *E. carotovora* subsp. *betavasculorum,* which produce but do not secrete pectinases, have been constructed (36) and may also have potential for the biological control of potato soft rot.

Molecular techniques have also facilitated the introduction of beneficial traits into rhizosphere-competent organisms to produce potential biological-control agents. Chitin is a major structural component of many plant pathogenic fungi. Biological control of some soil-borne fungal diseases has been correlated with chitinase production (23) and chitinolytic bacteria exhibit antagonism in vitro against fungi (66). The importance of chitinase activity was further demonstrated by the loss of biological-control efficacy in *Serratia marcescens* mutants in which the *chiA* gene had been inactivated (89). Chitinase genes have been cloned from several genera including *Streptomyces* (139), *Vibrio* (193), and *Cellvibrio* (194). A recombinant *Escherichia coli* containing the *chiA* gene from *S. marcescens* was effective in reducing disease incidence caused by *Sclerotium rolfsii* and *Rhizoctonia solani* (129, 148). In another study, chitinase genes from *S. marcescens* were expressed in *Pseudomonas* sp. conferring the ability to control the pathogens *F. oxysporum* f.sp. *redolens* and *Gauemannomyces graminis* var. *tritici* (162, 163).

Various extracellular antibiotics produced by *Pseudomonas* spp. have been shown to be involved in the biological-control ability of soil-borne plant pathogens (58), including phenazine-1-carboxylic acid (PCA) (168, 170), oomycin A (74, 87), pyoluteorin (PLT) (75, 91), and 2,4-diacetyl-phloroglucinol (PHL) (75, 91). In systems where antibiosis has been shown to play a primary role, molecular techniques can be used to enhance biological-control efficacy by increasing levels of antibiotic synthesis, either by increasing the copy number of the biosynthetic genes or by achieving constitutive synthesis. For example, increased production of PLT and PHL and superior control of *Pythium ultimum* damping-off of cucumber was achieved by increasing the number of antibiotic biosynthesis genes in *P. fluorescens* strain CHA0 (75, 119). Constitutive synthesis of oomycin A in *P. fluorescens* strain HV37a was achieved by insertion of a strong promoter from *E. coli* upstream of the *afuE* locus. The recombinant *P. fluorescens* strain containing the *tac-afuE* construct produced significantly higher levels of oomycin than the parental strain and provided greater control of *P. ultimum* infection (72, 73). Alternatively, biosynthetic genes can be introduced into a strain deficient in antibiotic production, or into one that produces a different antibiotic in order to increase the spectrum of activity. Cloned PCA biosynthetic genes were transferred from *P. fluorescens* 2-79, which exhibits poor rhizosphere competence, into *P. putida* and *P. fluorescens* strains that exhibit superior rhizosphere competence. The recombinant strains, which synthesized PCA in vitro, are potentially superior biological-control agents because of their ability to colonize the rhizosphere (22). In a similar study, a cloned genomic fragment from *Pseudomonas* F113 was transferred into various *Pseudomonas* strains, one of which was subsequently able to synthesize PHL and inhibit *P. ultimum* damping-off of sugar beet (56). The spectrum of activity through expression of an additional antibiotic increased when genes conferring PHL synthesis were mobilized from *P. aureofaciens* into *P. fluorescens* 2-79, which normally only produces PCA. This procedure increased activity against *G. graminis* var. *tritici, P. ultimum,* and *R. solani* (185).

Genetic modifications of biological-control fungi so far have been limited by the availability of appropriate methodologies. While transformation systems are being developed (153), genetic manipulation of *Trichoderma* and *Gliocladium* species is currently restricted to physical and chemical mutation and protoplast fusion (77, 79, 159). *Trichoderma harzianum* isolates tolerant to fungicides have been generated by spontaneous mutation (1) and UV exposure (134). In another study, improved root colonizing ability and biological-control efficacy was achieved by protoplast fusion with two *T. harzianum* strains (78).

Molecular techniques permit the construction of a superior biological-control agent by such approaches as the introduction of one or more beneficial traits into an organism exhibiting high levels of rhizosphere competence or inactivation of the pathogenicity genes in a pathogen. A biological-control agent could potentially be tailored to each host/pathogen system. However, the superior ability of such strains under field conditions remains to be demonstrated.

Biological Control of Insect Pests and Vectors

While many insect pests can be controlled through the use of chemical pesticides, bioaccumulation of pesticide residues and the contamination of soils and groundwater is a major environmental concern. In a rapidly developing field, investigators have focused on biological alternatives to pesticides, particularly the use of *Bacillus* toxins and of entomopathogenic fungi.

Several genera of entomopathogenic fungi have shown potential for the control of insect pests of agricultural crops, including *Aschersonia, Beauveria, Hirsutella, Metarhizium,* and *Verticillium* (26, 67, 120). *Beauveria bassiana,* one of the most extensively studied mycoinsecticides, has been used to control European corn borer (*Ostrinia nubilalis*), Colorado potato beetle (*Leptinotarsa decemlineata*), and codling moth (*Cydia pomonella*) (67). Although entomopathogenic fungi have been subjected to physical and chemical mutagenesis, the literature reports few examples of transformation (80). The first step toward enhancing the efficacy of entomopathogenic fungi lies in the identification of traits determining pathogenicity such as spore adhesion and germination, penetration and production of cuticle-degrading enzymes, production of toxins and enzymes, and avoidance of host recognition (80). Biological-control activity could possibly be enhanced through genetic manipulations affecting such features as sporulation, spore dispersal, and stress tolerance of spores (80). The development of fungal transformation vectors (153, 196) may allow the transfer of genes conferring, for example, lipase or protease activity to a strain that exhibits near optimal pathogenicity but is deficient in a single trait (80). Fungicide resistance will be an important trait for any mycoinsecticide planned for an agricultural application. The entomopathogen *Metarhizium anisopliae* has been transformed to benomyl resistance using the *benA3* gene from *Aspergillus nidulans* (68). To date, no recombinant entomopathogenic fungi have been released in the field.

Biological pest control agents based on *Bacillus thuringiensis* are produced commercially and used worldwide (40, 54). *B. thuringiensis* strains produce insecticidal crystal proteins (ICPs) that exhibit specific activity against different insect orders, including Diptera (*B. thuringiensis* subsp. *israelensis*), Lepidoptera (*B. thuringiensis* subsp. *kurstaki*), and Coleoptera (*B. thuringiensis* subsp. *tenebrionis*) (40). Another species, *Bacillus sphaericus,* exhibits toxic activity against mosquitoes of the genera *Aedes, Anopheles,* and *Culex* (14, 133). The crystal protein (*cry*) genes have been studied extensively and their potential for genetic manipulation reviewed (64, 85, 195).

B. thuringiensis strains exhibiting novel ICP gene combinations have been produced by plasmid curing and conjugal transfer (25, 64). This methodology has been used to produce the Condor® and Foil® bioinsecticides. In the case of the Foil® bioinsecticide, conjugal transfer was used to develop a *B. thuringiensis* strain with a broader spectrum of activity. Plasmids encoding activity against the European corn borer (*O. nubilalis*) were combined with a plasmid-encoding activity against the Colorado potato beetle (*L.*

decemlineata) (25, 64). Field trials with the Foil® bioinsecticide demonstrated the superior performance of this product compared with other *B. thuringiensis*–based products (64, 65). Recently, the development of *B. thuringiensis* cloning vectors (63, 64, 195) and an efficient transformation system for *B. thuringiensis* based on electroporation (19) has permitted the construction of recombinant *B. thuringiensis* strains. A broader spectrum of activity can be achieved in recombinant *B. thuringiensis* strains expressing both the native and cloned ICPs (37, 102).

The poor persistence of *B. thuringiensis* products in the environment has prompted the development of various carrier or cellular delivery systems, in which the ICP genes are expressed in a novel host chosen to enhance persistence of the toxin in the environment of the target pest (62). Effectiveness of mosquito control using *B. sphaericus* and *B. thuringiensis* subsp. *israelensis* is limited by sedimentation of the spores out of the larval feeding zone (14, 133). Expression of the ICPs in various cyanobacterial species may improve persistence of the bioinsecticide in the larval feeding zone (8, 30, 41a, 167). *B. thuringiensis cry* genes have been expressed in several hosts for use in agricultural ecosystems (55, 62). Rhizobacterial species transformed to express *B. thuringiensis cry* genes include *P. fluorescens* (69, 127, 186), *P. cepacia* (160), and *Rhizobium* spp. (154).

A novel method of delivery of ICPs using endophytic bacteria was recently developed. The endophytic bacterium *Clavibacter xyli* subsp. *cynodontis* (Cxc), originally isolated from Bermuda grass, serves as a systemic endophyte in corn (136). Different ICP gene constructions from *B. thuringiensis* subsp. *kurstaki* HD73 were introduced into the chromosome of Cxc (42, 52, 181). The recombinant Cxc/Bt expresses the ICP of *B. thuringiensis* subsp. *kurstaki*, which is toxic to the European corn borer. Systemic colonization of corn plants by the recombinant Cxc/Bt (98, 137) prevented or reduced corn borer damage to corn in the field (52, 99, 173). Current efforts are directed at increasing the production of the ICP in recombinant Cxc/Bt in order to enhance potency of ingested bacterial cells (52). Recombinant *B. thuringiensis* strains and other epiphytic and rhizobacterial species expressing ICPs will undoubtedly be used for pest control in the future. The use of strains expressing multiple or hybrid ICPs may help to delay the development of pest resistance.

Biological Control of Weeds

While weed control with chemical herbicides has been largely successful, herbicides suffer from a lack of selectivity, and their persistent nature often makes them serious environmental pollutants. Biological weed-control agents can potentially be highly specific and generally leave no toxic residues. Many species of plant pathogenic fungi have been tested for their ability to control a variety of weeds, and several of these are now commercially available (29,

165, 166). *Phytophthora palmivora* (DeVine®) has been used commercially since 1981 to control strangler vine (*Morrenia odorata*) in citrus groves (165). *Colletotrichum gloeosporoides* f.sp. *aeschynomene* (Collego®) has been used commercially since 1982 for control of northern joint vetch (*Aeschynomene virginica*) (165). Other mycoherbicides that are either commercially available or under testing include *C. gloeosporoides* f.sp. *cuscutae* (Luboa 2) for control of dodder, *Alternaria cassiae* (CASST) for control of sickelpod, and *Colletotrichum coccodes* (VELGO) for control of velvet leaf (165).

The optimal pathogen for a mycoherbicide application should exhibit a high level of virulence and a narrow host range; however, few naturally occurring pathogens exhibit both attributes (142). Genetic manipulation may be used to enhance the virulence of a narrow-host-range pathogen, or alternatively reduce the host range of a highly virulent pathogen. Chemical and physical mutagenesis of *Sclerotinia sclerotiorum,* a highly virulent pathogen of Canada thistle, produced mutants with reduced host ranges that have potential for control of this weed (142). Several authors have suggested possible approaches to enhance the virulence of fungal pathogens (28, 70, 92, 165), including introduction of genes for the production of specific phytotoxins, or enzymes, such as cutinases, pectinases, and cellulases. Such modifications, however, will not be possible until appropriate fungal transformation systems have been developed (196) and genes for the biosynthesis of these virulence factors have been cloned.

Symbiotic Nitrogen Fixation

Leguminous crops form symbiotic associations with *Rhizobium* and *Bradyrhizobium* that fix atmospheric nitrogen in a form that can be used by the plant. Rhizobial inoculants have been used for several years in attempts to increase legume productivity. Genetic modification of rhizobial strains has focused on three areas: host range modification, enhanced nitrogen fixation, and enhanced competition with indigenous strains for nodulation (84, 110, 131, 158). Several types of genes are involved in symbiotic nitrogen fixation, including nodulation (*nod*), nitrogen fixation (*fix*), and nitrogenase (*nif*) genes. In *Rhizobium* species these genes reside on *sym* plasmids, while in *Bradyrhizobium* species functional analogues of these genes reside on the chromosome (43).

Host-range modifications have been achieved by mutation of the common nodulation genes (*nodDABC*). The operon *nodABC* is activated by the gene product of *nodD* in combination with host-secreted phenolics (43). Chemical mutation of *nodD* in *Rhizobium leguminosarum* bv. *trifolii* produced mutants with inducer-independent ability to activate *nod* gene expression, thereby extending the host range compared with the parental strain (121). In a similar study, a flavonoid-independent hybrid NodD protein constitutively activated *nod* genes in *R. leguminosarum* and *R. meliloti,* increasing the host range of these strains (156). Host-range modifications have also been achieved by manipulation of the host-specific nodulation genes (*nodEFGH*) (41, 44). The host range of *R. leguminosarum* bv. *trifolii* was extended to include alfalfa plants following the integration of *nodEFG* and *nodH* genes from *R. meliloti* into its *sym* plasmid (57). Transfer of pSym genes into alternate hosts, such as *A. tumefaciens* and *E. coli,* may eventually permit nitrogen fixation in nonleguminous hosts (83, 180).

Two different approaches have been used to enhance nitrogen fixation—modified expression of the regulatory gene *nifA* and improved substrate transport through modified expression of the C4-dicarboxylate transport (*dct*) genes (24, 131, 141). In *Bradyrhizobium japonicum* and *R. meliloti,* the *nifA* product activates transcription of several genes involved in nitrogen fixation. Increased expression of *nifA* was achieved by chromosomal insertion of an enhancement cassette containing an additional copy of *nifA*. Recombinant *B. japonicum* and *R. meliloti* strains with additional *nifA* sequences improved legume yields in greenhouse and field trials (21, 24, 141). The rate of nitrogen fixation may be additionally limited by the rate of uptake of dicarboxylic acids, the primary source of energy. Transfer of *dct* genes from *R. meliloti* to *B. japonicum* resulted in higher nitrogenase activity (18). A recombinant *R. meliloti* with a chromosomally inserted enhancement cassette with both *dctABD* and *nifA* gave the largest yield increases in recent field trials (T. Wacek, personal communication).

Certain strains of rhizobia contain an uptake hydrogenase that can use H_2 to produce ATP. Hydrogen gas is a by-product of nitrogenase activity and strains that possess uptake hydrogenase (Hup^+) can recycle this hydrogen with the formation of ATP (158). The *hup* gene, encoding biosynthesis of uptake hydrogenase, has been cloned and used to transform naturally Hup^- strains. These Hup^+ recombinants have been shown to exhibit increased nitrogen fixation (131). However, most of the strains used commercially are Hup^+; hence such modifications may not be of value unless gene dosage effects are demonstrated (131).

Improvements in legume yield achieved with rhizobial inoculants are limited by competition between the inoculant and indigenous rhizobial strains that are frequently poor at fixing nitrogen (157, 178). Considerable effort has been directed to the determination of phenotypes and genotypes that contribute to superior nodulation competitiveness (178). These phenotypes include antibiosis, cell surface characteristics, and motility (178). *R. leguminosarum* bv. *trifolii* strain T24 produces a potent antibiotic, trifolitoxin, active against other rhizobia (177), but this strain produces nodules on clover that fix little nitrogen (176). Insertion of trifolitoxin biosynthesis (*tfx*) genes into the genome of a symbiotically effective strain of *R. leguminosarum* bv. *trifolii* produced a recombinant strain that was highly competitive with respect to a trifolitoxin-sensitive strain in greenhouse tests (176, 178). Modifications of symbiotically effective strains by addition of traits beneficial in nodulation competitiveness may ultimately produce inoculant strains that achieve acceptable frequencies of nodule occupancy, even when indigenous population levels are high.

Bioremediation

The bioremediation of environmental pollutants such as petroleum hydrocarbons or recalcitrant synthetics may in the future become an important application of recombinant microorganisms. Although some toxic wastes can be treated in contained bioreactors (49) or activated sludge systems, in situ bioremediation of petroleum spills or contaminated soils or aquifers necessitates the release of organisms into the environment. Nonrecombinants have been used in the bioremediation of petroleum hydrocarbons (9, 100) and xenobiotics (16). Although microcosm tests of recombinants have occurred (51, 126), and the EPA recently granted permission for a contained test of a recombinant *E. coli* for trichloroethylene (TCE) degradation (192; B. Ensley, personal communication), no field releases of recombinants have taken place.

Few naturally occurring organisms possess the necessary degradative pathways for complete mineralization of the more recalcitrant xenobiotics, such as pentachlorophenol (PCP), 2,4,5-trichlorophenoxyacetic acid (2,4,5-T) and polychlorinated biphenyls (PCB). Operons encoding steps in the mineralization of xenobiotics are frequently plasmid borne (20, 143). Metabolic pathways can be constructed by in vivo genetic manipulation through conjugational transfer in mixtures of strains, each possessing components of the required degradative pathway (20, 27). The degradative strains produced by these techniques have been extensively reviewed (76, 124, 138).

In vitro construction of degradative pathways through genetic engineering has great potential in systems for which sufficient genetic and biochemical information is available (2, 50, 61, 124). Molecular techniques are particularly useful for enzyme recruitment. Genes encoding enzymes with broad substrate specificity were added to extend the usefulness of the chlorocatechol pathway in several bacteria. *Pseudomonas* B13 has a pathway for the complete degradation of chlorocatechols, but the first enzyme in the chlorobenzoate pathway has narrow substrate specificity. Recruitment of a broad-specificity dioxygenase from a TOL plasmid permitted the degradation of a wider range of chlorobenzoates and chloroaromatics (101). In situ bioremediation may be unsuccessful if an inducer of a key degradative enzyme is absent. Molecular techniques have been used to alter the regulation of degradative enzymes requiring induction. Insertional mutagenesis in *P. cepacia* strain G4 produced mutants that constitutively metabolized TCE without aromatic induction (151).

Successful application of recombinant organisms for in situ bioremediation will depend not only on the construction and optimization of degradative pathways but also on the successful introduction, establishment, and containment of the organism, as well as the resolution of problems resulting from the heterogeneous distribution of the target compound, nutrients, and the degrading organism (16, 50, 115). Biological containment systems based on lethal genes, possibly coupled to the regulatory system of the substrate to be degraded, have been proposed (33, 50) and may be a regulatory requirement of field releases.

HISTORY AND ANALYSIS OF THE RELEASE OF NONRECOMBINANT MICROORGANISMS

Although much attention has focused recently on the release of recombinant microorganisms, the use of nonrecombinant microorganisms in the open environment has a long history and is pertinent to this issue. We have much to learn from studying the nature of previous uncontained releases of nonrecombinant microorganisms. A common assertion is that nonrecombinant strains have been released frequently without harm. We therefore try to assess not only the frequency with which nonrecombinant strains have been released but also address assessments of their fate and effects in an effort to apply this knowledge to the release of recombinant strains.

Frequency of Release of Nonrecombinant Microorganisms

The release of microorganisms into nature has long been an integral part of the research activities of several biological disciplines including plant pathology and entomology. For the purpose of this review, we have attempted to identify all releases of microorganisms within the past three years. This review of the literature was restricted to the introduction of bacteria and fungi into the open environment, specifically excluding viruses. Figure 1 summarizes research reports describing the introduction of microorganisms into the open environment. Plant pathogens are the most commonly released microorganisms, followed closely by nonpathogenic microorganisms used in the biological control of plant diseases. Mycorrhizal fungal species and many strains of *Rhizobium* have also been commonly released as part of research activities. Interestingly, very few bacteria have yet been released for the bioremediation of toxic compounds or agricultural pesticides. The actual number of yearly releases of nonrecombinant microorganisms may differ substantially from that reflected in Figure 1. We expect that the actual number of releases may in some cases be higher by a factor of 2 or more because many microorganisms may have been released incidentally to the main purpose of a study. Plant pathogens are usually released into field sites either to insure the presence of a pathogen so disease-control strategies can be evaluated or to measure the impact of a disease on crop productivity. Such studies are usually conducted where the previous occurrence of the disease insures the presence of sufficient pathogen inoculum, but such sites do not always exist. The uniform presence of a pathogen can be insured by its inoculation into a field site. Hundreds of releases of plant pathogens occur yearly. In 1992 alone, we identified at least 138 reported studies in which plant pathogens were released into field sites (Figure 1). Because only experiments that are successful are generally reported in the literature, substantially more than the reported number of releases probably occurred. Similarly, most reports of field studies include data from at least two years of field results to allow evaluation of the reproducibility of a result. Therefore, the number of actual

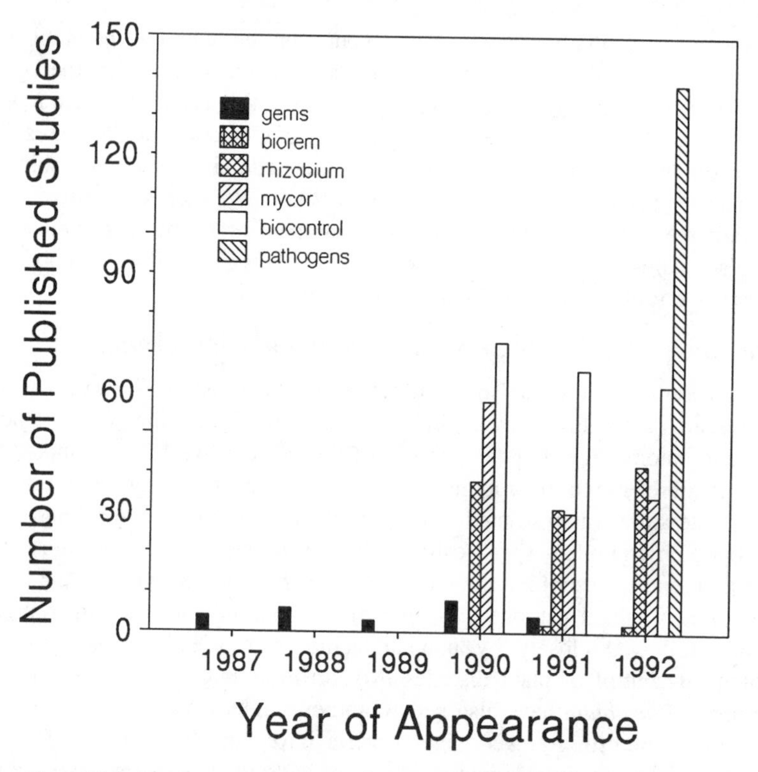

Figure 1 Published reports on the use of different microorganisms in field studies. The number of published studies involving (*left to right*) genetically engineered microorganisms (gems), microorganisms for bioremediation, *Rhizobium* or *Bradyrhizobium* species, mycorrhizal species, microorganisms used in biological control of plant pests, and plant pathogens is given for the first year of the experimental study reported. While all uses of gems are tabulated, releases of plant pathogens is reported only for 1992 and the use of the other organisms is reported only for the past three years for comparative purposes.

releases may be greater than the number of reports of a pathogen's introduction by a factor of two or more. Because field inoculations of plant pathogens have occurred since the turn of the century, thousands of releases of plant pathogens have been made.

The increased interest in biological control of plant diseases and insects as an alternative to chemical pesticides in the past 15 years (7, 17, 40, 67, 169, 179, 187) has led to the testing of many naturally occurring microorganisms. An average of about 55 studies have been reported annually from field tests of fungi and bacteria for the control of plant diseases; the numbers are fewer for the control of insects and weeds with pathogenic strains. The targets of biological-control efforts include both foliar and soil-borne plant pathogens, foliage- and root-feeding insects, and terrestrial and aquatic weed species. Therefore, indigenous microorganisms have been released into many different habitats. The long history of biological-control studies suggest that hundreds of field releases have occurred. For example, as of 1989, 109 pathogens that could infect a total of 69 different weed plant species were under consideration for biological weed control (29). Many of these pathogens had already been investigated under field settings to test their efficacy; many more probably will be tested for biological-control potential in the field in the future.

Mycorrhizal associations of roots continue to be intensively studied to determine their effects on plant nutrition and productivity as well as disease avoidance (189). Each of the past three years has seen an average of about 40 research reports on the inoculation of plants under field conditions with mycorrhizal fungi. A wide range of terrestrial plant species have been inoculated with one of several different mycorrhizal fungal species.

The infection of leguminous plants with *Rhizobium* species is an important feature of the nitrogen metabolism of such plants. Therefore, attention has focused on this interaction for over a century (157, 158). Earlier studies of *Rhizobium* species tended to focus on practical aspects of inoculation of plants with this species under field conditions, survival in soil, and productivity of inoculated plants. More recent studies of *Rhizobium* and *Bradyrhizobium* species have primarily addressed genetic and biochemical determinants of plant-microbe interactions. Nonetheless, an average of about 40 research reports involving field releases of *Rhizobium* species have been published in each of the past three years (Figure 1). Because inoculation of leguminous plants with *Rhizobium* and *Bradyrhizobium* species is now a common agricultural practice, many thousands of inoculations occur yearly around the world (157, 158).

Not only have more plant pathogens and microorganisms with the potential for biological control been released than other types of agents, the variety of these organisms and of the types of habitats into which they have been introduced is much wider than for other microorganisms. There are hundreds of different plant pathogenic fungal and bacterial species, and most have been tested under field situations—clearly considerable numbers of diverse microorganisms have been introduced to the open environment. Nearly every report of a released pathogen involves a genotypically unique strain. In contrast, many studies of mycorrhizae and *Rhizobium* species involve repeated inoculations of a few different strains into different habitats. Certainly, however, thousands of habitat-microbe genotype combinations have been tested as part of field testing of plant pathogens and biological-control agents in nature.

In addition to releases of microorganisms during research studies, several microorganisms are registered for commercial use as pesticides. The US Environmental Protection Agency (EPA) has registered 195 different strains of 11 different bacterial species for control of fungal and bacterial diseases of roots, control of bacterial-incited plant frost damage, and control of insect pests (6). The vast majority of these taxa are *B. thuringiensis* strains with different spectra of insecticidal activity. Eleven different strains of six fungal species are also registered by the EPA for biological control of weed plants, plant diseases, and insect larvae (6). Several of these registered pesticides are produced and used commercially on a substantial scale, thus resulting in innumerable applications to field settings. For example, the fungus *C. gloeosporoides* f.sp. *aeschynomene* is sold as the product Collego® and applied widely in the southern US to kill Northern joint vetch (*A. virginica*) in rice fields (29, 165, 166). Several commercial formulations of different *B. thuringiensis* subspecies are also commonly used worldwide on agricultural and forest crops to manage insect pests (29, 40, 54).

While most pathogens, biological-control agents, and mycorrhizal or *Rhizobium* strains that have been field tested represent a reintroduction into a habitat that is the same or similar to the one from which the organism was isolated, a few microorganisms have been introduced from nonindigenous sources. This distinction is clearly illustrated by the use of fungi as biological-control agents of weeds. Most fungal species inoculated onto weeds, such as *A. cassiae* and *C. gloeosporoides,* were isolated in North America and then reintroduced onto plants in large numbers to cause severe disease as a "mycoherbicide" (29, 165, 166). In contrast, the rust fungus, *Puccinia chondrillina,* was isolated in the Mediterranean region of southern Europe, the site of origin of skeleton weed (*Chrondrilla juncea*), and released in Australia and the western US to attack this weed because these areas lacked any of the plant's predators or parasites (3, 38, 39). Rust fungi are easily dispersed by wind, so these pathogens have spread far from the initial-introduction spot and have successfully reduced the aggressiveness of the weed plants that they attack; no unforeseen impact on the terrestrial habitats into which they were introduced have been observed (39). Three other fungi have been similarly introduced to various parts of the world in classical biological weed control programs. Nonindigenous bacteria have also been introduced for evaluation of the biological control of plant diseases (EPA, unpublished data). The EPA has authorized the field testing of 14 nonindigenous *B. thuringiensis* strains in the US. The introduction of nonindigenous microorganisms into the environment certainly must pose at least the same risk as the introduction of genetically engineered microorganisms because the nonindigenous strains have many novel traits not present in the indigenous microflora.

Assessment of the Fate of Released Nonrecombinant Microorganisms

Releases of nonrecombinant microorganisms have apparently not resulted in significant perturbations of the habitats into which they have been introduced, but the large majority of release studies have not sought to address this possibility. Most introductions of pathogens, for example, have been to limited areas, and no subsequent reports of disease of plants outside the area initially inoculated or of unexpected host plants have appeared. In most studies specifically reisolating the inoculated pathogen would be difficult because genetic markers such as antibiotic resistance traits are not present on the strains. Therefore, most studies apparently have not attempted to describe the fate of the introduced microorganisms. Several studies, however, have involved the release of pathogens for the purpose of determining their persistence or spread as part of other epidemiological studies (188). Most of these reveal that after inoculation in high numbers, the population size of bacterial plant pathogens drops to low, and frequently undetectable, levels within a few days to a few years.

In addition, the population dynamics of biological-control agents have often been measured. Because a biological-control agent of plant disease must colonize plants to achieve control, many studies have quantified antagonist populations both spatially and temporally (81, 116, 187). The dispersal of organisms, however, has not often been the focus of such studies. Obvious deleterious effects either on plant growth, such as stunting or pathogenicity, or on other components of a field site, such as arthropod mortality attributable to a biological-control agent, have not been reported in the literature. The widespread commercial use of plant pathogens as biological-control agents of weeds has not resulted in unexpected effects on nonsusceptible plant species (165). The difficulty in describing changes in the microbial communities into which microorganisms have been introduced explains why more detailed studies of perturbations resulting from released microbes have not been commonly made. The diversity of microorganisms present in most terrestrial soil or plant habitats, the difficulty in isolating all microbial components of the habitat, and their uncertain taxonomic position, make a quantitative measure of perturbation associated with released microorganisms difficult to realize (13, 59). Many studies addressing risks of recombinant microorganisms now focus on improvement of methods for such assessments (13, 45, 59, 144, 147, 182).

RELEASES OF RECOMBINANT MICROORGANISMS

Frequency of Release of Recombinant Microorganisms

Compared with the numerous releases of nonrecombinant microorganisms that occur yearly, the number of recombinant microorganisms that have been released remains small. For the purposes of this review, microorganisms altered by acquisition of a plasmid, by protoplast fusion, or by chemical or ultraviolet mutagenesis are not considered recombinant. Several microorganisms altered in such ways have been introduced into the open environment. For example, the EPA has authorized the field testing of 10 transconjugant *B. thuringiensis* strains containing a different complement of plasmids harboring ICP genes. The EPA has also authorized the release of two *Trichoderma* strains resulting from protoplast fusion and 10 strains of either *Verticillium lecanii, S. sclerotiorum,* or *T. harzianum* with chemical or ultraviolet-induced mutations. In contrast, since the first field release of a Tn*5*-containing strain of *P. fluorescens* in the Netherlands in 1986, in only 27 studies have recombinant microorganisms been tested in the open environment (Table 1). All field tests have involved recombinant bacteria. A large percentage of the tests involve simple inactivation of genes by transposon insertion (82); genetic tagging of bacteria with selectable or identifiable markers such as *lacZY* (47, 48, 135), the *lux* operon (150), or antibiotic-resistance markers (5, 10, 11); or deletion of genes by molecular genetic procedures in vitro (88, 105, 111, 117, 184). Moreover, many of the releases represent the introduction of the same altered strain to different sites.

Initial experiments with recombinant bacteria in the US, especially those of microorganisms meant to be tested as biological pesticides, were burdened with the necessity of obtaining an Experimental Use Permit (EUP) from the EPA (93, 95, 111). Acceptable documentation for an EUP, normally necessary only for large field tests of chemical pesticides, required a great deal of information and greatly delayed initial experiments (111). Also required for these experiments was the collection of detailed data on the dispersal and survival potential of the altered bacteria (93, 95, 111). The EPA has subsequently considerably relaxed such requirements, and though the procedures to obtain permits for the testing of recombinant microorganisms from the US Department of Agriculture, Animal, and Plant Health Inspection

Table 1 Releases of recombinant bacteria

Organism	Altered trait	Location released	Year released	Reference
Pseudomonas fluorescens	Tn*5* marker	The Netherlands	1986	10, 11
Pseudomonas syringae (2 strains)	Deletion of *ice* genes	California	1987	111, 117
Pseudomonas fluorescens	Deletion of *ice* gene	California	1987	105, 106
Pseudomonas syringae	Deletion of *ice* gene	California	1987	105, 106
Pseudomonas aureofaciens	Introduction of *lacZY* marker on Tn7	South Carolina	1987	93, 94, 96
Agrobacterium radiobacter	Deletion of *tra* region on pAgK84	Australia	1988	88, 184
Pseudomonas fluorescens	Introduction of *lacZY* marker on Tn7	Washington	1988	34
Rhizobium leguminosarum	Tn*5* marker on *sym* plasmid	England, France, Germany	1988	5
Rhizobium meliloti	Insertion of Ω in *ino* site	Wisconsin	1988	24
Rhizobium fredii	*lacZY*marker	Australia	?	123
Clavibacter xyli	*B. thuringiensis* endotoxin gene	Maryland	1988	97, 99
Rhizobium trifolii	Deletion of phenolic metabolism gene	Australia	?	123
Bradyrhizobium japonicum	Insertion of *nifA* and Ω in *ino* region	Louisiana	1989	21
Rhizobium meliloti	Insertion of *nifA* & Ω upstream of *nifD*	Wisconsin	1989	141
Clavibacter xyli	*B. thuriengensis* endotoxin gene	Nebraska	1989	97
Pseudomonas fluorescens	*lacZY* marker	Indiana	1990	46
Pseudomonas fluorescens	*lacZY* marker	Montana	1990	46
Pseudomonas corrugata	*lacZY* marker	Australia	1990	46
Bacillus thuringiensis	*B. thuringiensis wuhanensis* toxin gene and Tet^r added	Mississippi	1990	
Clavibacter xyli	*B. thuringiensis* toxin gene with enhanced activity	Maryland	1990	97
Rhizobium MU95	Tn*5* insertion	Australia	1990	
Xanthomonas campestris	*lux* operon marker added with transposon	Alabama	1990	150
Rhizobium leguminosarum	Trifolitoxin biosynthetic genes added	Wisconsin	1990	176
Pseudomonas syringae	Tn*5* insertion in *lemA* gene	Wisconsin	1991	82
Rhizobium meliloti	Tn*5* marker	Canada	1991	
Pseudomonas fluorescens	*lacZY* marker	Wisconsin	1991	135
Pseudomonas putida	*lux* operon marker added with transposon	Alabama	1991	149

Service (USDA-APHIS) still necessitate considerable effort, they have overall been greatly simplified (149).

Despite this apparent relaxation of requirements, the number of studies of genetically engineered microorganisms (GEMS) that have been initiated has not increased as rapidly as many projected (Figure 1). Although four field studies releasing genetically engineered bacteria began in 1987, still only four were initiated in 1991. The number of initiated studies peaked at eight in 1990, and we could find no evidence of new studies initiated in 1992 (Figure 1). The number of releases of recombinant microbes continues to be fewer than the number of conferences and review articles addressing the safety of such releases. We agree with the assessment of Shaw et al (149, p. 51) that:

> . . . restrictions on the introduction of GEMS in field experiments posed a major problem to the pursuit of such studies. Federal regulations and, to a lesser extent, state regulations often prohibit such releases unless extensive greenhouse and microcosm studies have been previously conducted. The burden to develop these data typically rests upon the researcher and is onerous to prohibitive. Even after generating the required data, there is no guarantee that a permit will be issued. For these reasons, academic scientists with limited resources have generally been discouraged from conducting such experimentation.

Assessment of the Fate of Released Recombinant Microorganisms

In contrast to most introductions of nonrecombinant bacteria, the fate of GEMS has been carefully monitored. A necessary and prominent feature of permits obtained from both the EPA and USDA-APHIS is an experimental design and procedures that enable careful description of the survival and dispersal of the introduced microorganism. For example, permits from EPA to conduct releases of ice nucleation–deficient (Ice^-) mutants of *P. syringae* and *P. fluorescens* in 1987 required extensive studies of the aerosol dispersal of spray-applied inoculum and frequent surveys for the presence of these antibiotic resistant–marked strains on plants, soil, insects, and water in and around the plot areas (111). Permits also required that all treated plant material be destroyed at the end of the experiment to test the feasibility of eradication of the introduced bacteria (111). The EPA also conducted extensive studies of the dispersal of bacteria during and after spray inoculation at field sites to document the dispersal potential of the Ice^- bacteria applied in such a manner and to test different detection methods (146). The results of monitoring released Ice^- *P. syringae* strains reveal that the rate of dispersal was closely predicted by studies of these strains conducted in the laboratory and greenhouse and by prior field experiments using nonrecombinant bacterial strains (106, 108, 111, 117).

Releases of several recombinant *Pseudomonas* strains have been designed specifically to evaluate their fate and dispersal in field sites. The use of *lacZY*-encoded β-galactosidase activity in combination with rifampicin resistance markers allowed rapid and unambiguous selection and identification of introduced strains (48, 94). Initial trials of *Pseudomonas aureofaciens* strains containing *lacZY* (which was introduced with a disarmed variant of Tn7), involved extensive studies of its survival and movement in the plot area (94, 95). Greenhouse and microcosm studies predicted that the *P. aureofaciens* strain would initially colonize inoculated wheat seeds and roots in large numbers but would slowly decrease in population size with time (93–96). These strains indeed decreased to undetectable levels within 31 weeks after inoculation into field sites in South Carolina (94). It was also anticipated that movement through soil by such strains or along the roots of wheat would be restricted, with the largest populations occurring near the inoculated seeds. The recombinant strains of *P. aureofaciens* behaved in a manner similar to that projected in each of several field studies (34). No transfer of Tn*7::lacZY* from the chromosomal insertion in *P. aureofaciens* into other soil microorganisms was detected (94).

The behavior of recombinant *C. xyli* subsp. *cynodontis* strains containing genes for an ICP from *B. thuringiensis* was also predicted from preliminary studies using nonrecombinant strains or recombinant strains studied in the laboratory. Greenhouse studies of *C. xyli* indicated that it could be moved from plant to plant by mechanical transmission but not by insects (97). Such findings indicated that the recombinant strain should be contained within the plants into which it was inoculated or in plants developing from infected seeds. To date, this has been the case in field studies of recombinant strains of this species (97). While large populations of recombinant *C. xyli* developed in inoculated corn plants, it was not detected on trap plants nearby (97).

Although the number of recombinant microorganisms that have been so far released into the open environment remains small, we agree with the statements by Kluepfel et al (94) regarding conclusions that can be drawn from the results of initial experiments (p. 351):

> The initial results of this planned release of a genetically engineered, soil-borne root-colonizing bacterium have shown that such a release can be conducted in a safe and responsible manner. Though we should guard against generalization we and others have not observed any adherent danger in the use of genetically engineered bacteria in the environment simply because the organism has been modified genetically. This, however, does not exclude the necessity of a case by case evaluation of the microorganism and its new genetic construct before its release into the environment. By continuing to build our database on the microbial ecology on both native and genetically engineered microorganisms we will move

toward the establishment of scientifically-based risk assessment procedures that will guide their future release.

CONCLUDING REMARKS

This review was not intended to address the voluminous literature on the risk assessment of releases of recombinant bacteria. Many excellent treatments of different aspects of this subject have appeared (31, 32, 59, 86, 155, 171, 172, 190). A deterrent to generalized methods of assessing risk has been the diversity of the microbes themselves, the large number of traits that potentially might be modified, and the diverse habitats into which they will be introduced. For this reason, different components of risk take on paramount importance for various recombinant microbes, while some risk components are common to nearly all GEMS. For example, the risk of horizontal gene transfer should be considered in all cases. Many studies have addressed the potential of genetic transfer by conjugation, transformation, transduction, and cell fusion (32, 122, 128, 145). Unfortunately, most studies have focused on horizontal transfer using a plasmid model, often extrapolating from studies of antibiotic resistance plasmids that can undergo strong selection in an antibiotic-containing environment. This model has been strongly criticized (32). Comeaux et al (32) provided a very lucid description of genetic transfer in relation to risk assessment of recombinant bacteria. We agree strongly with their conclusions that (p. 132):

> . . . for . . . GEMS, systems may be used that minimize genetic exchange. However, total elimination of genetic transfer from or to GEMS released into agriculture habitats is probably an unrealistic goal. Therefore, considerations for planned releases should document and evaluate gene movement from GEMS into indigenous bacterial populations and evaluate the ecological impact of that gene movement on community structure and function. Assessment of risks should be based on reasonable interpretations of the results of logical tests for potential hazards and perforce avoid tests for more esoteric potential hazards.

The potential effects of GEMS on a habitat into which they have been introduced are hard to assess. As discussed above, even describing a microbial community, except at a generalized functional level, is largely beyond our current capabilities. We encourage the development of new tools to enable sensitive assays of changes in community function as well as in describing community composition. Such tools will not only prove extremely useful in assessing potential risks associated with the release of genetically engineered microbes, but will be a great aid in all future microbial ecological studies.

A common concern in releases of GEMS is that they will out-compete indigenous microbes because of the novel trait that they carry. Because nonindigenous bacteria frequently do not become established when introduced into a new habitat, indigenous bacteria are typically genetically manipulated for environmental releases; hence, GEMS usually will be introduced into habitats in which genotypically and phenotypically similar organisms already exist. In many cases, a study of the behavior of the parental strain in a given habitat should be sufficient to evaluate the prospects of the GEM. Successful colonization of the parental strain in a field setting should indicate the basic capability of the genetically altered strain in a given habitat. Comparative microcosm studies can then be done to determine the contribution of the new trait(s) introduced into the GEM to predict any differences in testable behavior of the GEM in the field. Therefore, one will not have to predict the field behavior of a genetically engineered microbe from laboratory features of the GEM alone but, instead, can rely on extensive knowledge of the parental strain or knowledge that can be gained from the release of the parental strain into a given habitat to assist in prediction of the propensity of a GEM to occupy a given site and to have effects on indigenous populations because of such occupation.

We intended in this review to provide insight into the opportunities and needs for the release of GEMS to advance our knowledge of microbial ecology and to benefit society. An analysis of previous work demonstrates that GEMS will likely be used for the biological control of plant diseases and pests as well as nitrogen fixation in the future. We anticipate that the future will also see a great demand for releases of GEMS for bioremediation. The absence of careful quantitative examination of the fate of many previously introduced nonengineered organisms weakens the foundation for understanding the fate of released GEMS. However, carefully conducted studies with released GEMS suggests that their release will have little deleterious effect. By summarizing the history of releases of nonengineered microbes and the current status of the releases of GEMS, we hope to have shown that GEMS can be released without deleterious perturbation of a given habitat, and that their release, in combination with continued cautious and thorough study, can contribute to superior strategies for their effective application to solve field-based problems.

ACKNOWLEDGMENTS

We thank B. Ensley, T. Wacek, C. Gawron-Burke, S. Kostka, T. Yamamoto, and D. Drahos for providing useful information, including unpublished data for this review, as well as J. Payne from USDA-APHIS and S. Matten from the EPA for providing compilations of regulatory actions relating to the release of GEMS, which were helpful in assembling the list provided here. We also thank G. Beattie for helpful manuscript suggestions.

Literature Cited

1. Abd El-Moity, T. H., Papavizas, G. C., Shatla, M. N. 1982. Induction of new isolates of *Trichoderma harzianum* tolerant to fungicides and their experimental use for control of white rot of onion. *Phytopathology* 72:396–400
2. Abramowicz, D. A. 1990. Aerobic and anaerobic biodegradation of PCBs: a review. *Crit. Rev. Biotechnol.* 10:241–51
3. Adams, E. B., Line, R. F. 1984. Epidemiology and host morphology in the parasitism of rush skeleton weed by *Puccinia chondrillina. Phytopathology* 74:745–48
4. Alabouvette, C., Couteaudier, Y. 1992. Biological control of *Fusarium* wilts with non-pathogenic fusaria. See Ref. 171a, pp. 415–26
5. Amarger, N., Delgutte, D. 1990. Monitoring genetically manipulated *Rhizobium leguminosarum* biovar *viciae* released in the field. See Ref. 117a, pp. 221–28
6. Anderson, E. L., Betz, F. S. 1990. EPA perspective on risk assessment for environmental introductions: progression from small-scale testing to commercial use. See Ref. 117a, pp. 119–26
7. Andrews, J. H. 1992. Biological control in the phyllosphere. *Annu. Rev. Phytopathol.* 30:603–35
8. Angsuthanasombat, C., Panyim, S. 1989. Biosynthesis of 130-kilodalton mosquito larvicide in the cyanobacterium *Agmonellum quadriplicatum* PR-6. *Appl. Environ. Microbiol.* 55: 2428–30
9. Atlas, R. M., Atlas, M. C. 1991. Biodegradation of oil and bioremediation of oil spills. *Curr. Opin. Biotechnol.* 2:440–43

9a. Baker, R., Dunn, P. E., eds. 1990. *New Directions in Biological Control: Alternatives for Suppressing Agricultural Pests and Diseases.* New York: Liss

10. Bakker, P. A. H. M., Salentijn, E., Hoekstra, W. P. M., Schippers, B. 1990. Fate of transposon Tn*5* labelled *Pseudomonas fluorescens* in the field. In *Proc. Int. Workshop on Plant Growth Promoting Rhizobacteria, 2nd,* pp. 412–13. Interlaken, Switzerland: Int. Union of Biological Sciences
11. Bakker, P. A. H. M., Schippers, B., Hoekstra, W. P. M., Salentijn, E. 1990. Survival and stability of a Tn*5* transposon derivative of *Pseudomonas fluorescens* WCS374 in the field. See Ref. 117a, pp. 201–4
12. Bakker, P. A. H. M., van Peer, R., Schippers, B. 1991. Suppression of soil-borne plant pathogens by fluorescent pseudomonads: mechanisms and prospects. In *Biotic Interactions and Soil-Borne Diseases,* ed. A. B. R. Beemster, G. J. Bollen, M. Gerlagh, M. A. Ruissen, B. Schippers, A. Tempel, pp. 217–30. Amsterdam: Elsevier
13. Bej, A. K., Perlin, M., Atlas, R. M. 1992. Impact of introducing genetically modified microorganisms on soil microbial community diversity. See Ref. 159a, pp. 137–39
14. Berry, C., Hindley, J., Oei, C. 1991. The *Bacillus sphaericus* toxins and their potential for biotechnological development. See Ref. 117b, pp. 35–51
15. Bender, C. L., Cooksey, D. A. 1987. Molecular cloning of copper resistance genes from *Pseudomonas syringae* pv. *tomato. J. Bacteriol.* 169:470–74
16. Bewley, R. J. F. 1992. Bioremediation and waste management. See Ref. 159a, pp. 33–45
17. Blakeman, J. P., Fokkema, N. J. 1982. Potential for biological control of plant diseases on the phylloplane. *Annu. Rev. Phytopathol.* 20:167–92
18. Birkenhead, K., Manian, S. S., O'Gara, F. 1988. Dicarboxylic acid transport in *Bradyrhizobium japonicum:* use of *Rhizobium meliloti dct* gene(s) to enhance nitrogen fixation. *J. Bacteriol.* 170:184–89
19. Bone, E. J., Ellar, D. J. 1989. Transformation of *Bacillus thuringiensis* by electroporation. *FEMS Microbiol. Lett.* 58:171–78
20. Boyle, M. 1992. The importance of genetic exchange in degradation of xenobiotic chemicals. In *Environmental Microbiology,* ed. R. Mitchell, pp. 319–33. New York: Wiley-Liss
21. Breitenbeck, G. A., Hankinson, T. 1992. Dispersal and persistence of recombinant *Bradyrhizobium japonicum. Agron. J.* 84:259
22. Bull, C. T., Weller, D. M., Tomashow, L. S. 1991. Relation between root colonization and suppression of *Gauemannomyces graminis* var. *tritici* by *Pseudomonas fluorescens* strain 2–79. *Phytopathology* 81:954–59

22a. Burge, M. N., ed. 1988. *Fungi in Biological Control Systems.* Manchester: Manchester University Press

23. Buxton, E. W., Khalifa, O., Ward,

V. 1965. Effects of soil amendment with chitin on pea wilt caused by *Fusarium oxysporum* f.sp. *pisi*. *Ann. appl. Biol.* 55:83–88

24. Cannon, F. C., Beynon, J., Hankinson, T., Kwiatlowski, R., Legocki, R. P., et al. 1988. Increasing biological nitrogen fixation by genetic manipulation. In *Nitrogen Fixation: Hundred Years After*, ed. H. Bothe, F. J. de Bruijn, W. E. Newton, pp. 735–40. New York: Gustav Fischer
25. Carlton, B. C. 1988. Development of genetically improved strains of *Bacillus thuringiensis:* a biological insecticide. In *Biotechnology for Crop Protection. Symposium Series 379,* ed. P. A. Hedin, J. J. Menn, R. M. Hollingworth, pp. 260–79. Washington, DC: Am. Chem. Soc.
26. Carruthers, R. I., Hural, K. 1990. Fungi as naturally occurring entomopathogens. See Ref. 9a, pp. 115–38
27. Chakrabarty, A. M., Friello, D. A., Bopp, L. H. 1978. Transposition of plasmid DNA segments specifying hydrocarbon degradation and their expression in various microorganisms. *Proc. Natl. Acad. Sci. USA* 75:3109–12
28. Charudattan, R. 1985. The use of natural and genetically altered strains of pathogens for weed control. In *Biological Control In Agricultural IPM Systems,* ed. M. A. Hoy, D. C. Herzog, pp. 347–72. Orlando: Academic
29. Charudattan, R. 1991. The mycoherbicide approach with plant pathogens. See Ref. 164b, pp. 24–57
30. Chungjatupornchai, W. 1990. Expression of the mosquitocidal-protein genes of *Bacillus thuringiensis* subsp. *israelensis* and the herbicide-resistance gene *bar* in *Synechocystis* PCC6803. *Curr. Microbiol.* 21:283–88
31. Colwell, R. R. 1991. Risk assessment in environmental biotechnology. *Curr. Opin. Biotechnol.* 2:470–75
32. Comeaux, J. L., Pooranampillai, C. D., Lacy, G. H., Stromberg, V. K. 1990. Transfer of genes to other populations, and analysis of associated potential risks. See Ref. 118, pp. 132–45
33. Contreras, A., Molin, S., Ramos, J. L. 1991. Conditional-suicide containment system for bacteria which mineralize aromatics. *Appl. Environ. Microbiol.* 57:1504–8
34. Cook, R. J., Weller, D. M., Kovacevich, P., Drahos, D., Hemming, B., Barnes, G., Pierson, E. L. 1990. Establishment, monitoring, and termination of field tests with genetically altered bacteria applied to wheat for biological control of take-all. See Ref. 117a, pp. 177–87
35. Cooksey, D. A. 1988. Reduction of infection by *Pseudomonas syringae* pv. *tomato* using a non-pathogenic, copper-resistant strain combined with a copper bactericide. *Phytopathology* 78: 601–3
36. Costa, J. M., Loper, J. E. 1992. Out-minus mutants of *Erwinia carotovora* subsp. *betavasculorum* with potential for biological control of potato soft rot. *Phytopathology* 82:1121
37. Crickmore, N., Nicholls, C., Earp, D. J., Hodgman, T. C., Ellar, D. J. 1990. The construction of *Bacillus thuringiensis* strains expressing novel entomocidal δ-endotoxin combinations. *Biochem. J.* 270:133–36
38. Cullen, J. M. 1976. Evaluating the success of the programme for the biological control of *Chondrilla juncea* L. In *Proc. Int. Symp. on Biological Control of Weeds, 4th,* ed. T. E. Freeman, pp. 117–21. Gainesville: Univ. Fla.
39. Cullen, J. M., Kable, P. F., Katt, M. 1973. Epidemic spread of a rust imported for biological control. *Nature* 244:462–64
40. Currier, T. C., Gawron-Burke, C. 1990. Commercial development of *Bacillus thuringiensis* bioinsecticide products. See Ref. 124a, pp. 111–43
41. Debelle, F., Maillet, F., Vasse, J., Rosenberg, C., de Billy, F., et al. 1988. Interference between *Rhizobium meliloti* and *Rhizobium trifolii* nodulation genes: genetic basis of *Rhizobium meliloti* dominance. *J. Bacteriol.* 170: 5718–27

41a. de Marsac, N. T., de la Torre, F., Szulmajster, J. 1987. Expression of the larvicidal gene of Bacillus sphaericus 1593M in the cyanobacterium *Anacystis nidulans* R2. *Mol. Gen. Genet.* 209:396–98

42. Dimock, M. B., Beach, R. M., Carlson, P. S. 1989. Endophytic bacteria for the delivery of crop protection agents. In *Proceedings of Biotechnology, Biological Pesticides and Novel Plant-Pest Resistance for Insect Pest Management,* ed. D. W. Roberts, R. R. Granados, pp. 88–92. Ithaca, NY: Boyce Thompson Inst. for Plant Research
43. Djordevic, M. A., Gabriel, D. W., Rolfe, B. G. 1987. *Rhizobium*—the refined parasite of legumes. *Annu. Rev. Phytopathol.* 25:145–68

44. Djordevic, M. A., Schofield, P. R., Rolfe, B. G. 1985. Tn*5* mutagenesis of *Rhizobium trifolii* host-specific nodulation genes result in mutants with altered host-range ability. *Mol. Gen. Genet*. 200:463–71
45. Drahos, D. J. 1991. Current practices for monitoring genetically engineered microbes in the environment. *AgBiotech News Inf.* 3:39–48
46. Drahos, D. J. 1991. Field testing of genetically engineered microorganisms. *Biotechnol. Adv*. 9:157–71
47. Drahos, D., Barry, G., Hemming, B., Brandt, E. 1988. Field-testing of *lacZY*-marked recombinant soil bacteria. *Phytopathology* 78:1544
48. Drahos, D. J., Hemming, B. C., McPherson, S. 1986. Tracking recombinant organisms in the environment: β-galactosidase as a selectable non-antibiotic marker for fluorescent pseudomonads. *Bio/Technology* 4:439–44
49. Drahos, D. J., Mueller, J. G., Lantz, S. E., Head, C. S., Middaugh, D. P., Pritchard, P. H. 1992. Microbial degradation of high molecular weight polyaromatic hydrocarbons at the American Creosote Works superfund site, Pensacola, Florida. In *Proc. Int. Symp. on the Biosafety Results of Field Tests of Genetically Modified Plants and Microorganisms, 2nd,* ed. R. Casper, J. Landsmann, pp. 153–62. (Abstr.)
50. Duque, E., Ramos-Gonzalez, M., Delgado, A., Contreras, A., Molin, S., Ramos, J. L. 1992. Genetically engineered *Pseudomonas* strains for mineralization of aromatics: survival, performance, gene transfer, and biological containment. See Ref. 62a, pp. 429–37
51. Dwyer, D. F., Rojo, F., Timmis, K. N. 1988. Fate and behavior in an activated sludge microcosm of a genetically-engineered micro-organism designed to degrade substituted aromatic compounds. See Ref. 164a, pp. 77–88
52. Fahey, J. W., Dimock, M. B., Tomasino, S. F., Taylor, J. M., Carlson, P. 1992. Genetically engineered endophytes as biocontrol agents: a case study from industry. In *Microbial Ecology of Leaves,* ed. S. S. Hirano, J. H. Andrews, pp. 401–11. New York: Springer Verlag
53. Farrand, S. K. 1990. *Agrobacterium radiobacter* strain K84: a model biocontrol system. See Ref. 9a, pp. 679–91
54. Feitelson, J. S., Payne, J., Kim, L. 1992. *Bacillus thuringiensis:* insects and beyond. *Bio/Technology* 10:271–75
55. Feitelson, J. S., Quick, T. C., Gaertner, F. 1990. Alternative hosts for *Bacillus thuringiensis* delta-endotoxin genes. See Ref. 9a, pp. 561–71
56. Fenton, A. M., Stephens, P. M., Crowley, J., O'Callaghan, M., O'Gara, F. 1992. Exploitation of gene(s) involved in 2,4-diacetylphloroglucinol biosynthesis to confer a new biocontrol capability to a *Pseudomonas* strain. *Appl. Environ. Microbiol*. 58:3873–78
57. Folch, J. L. 1992. *Extension of* Rhizobium leguminosarum *biovar* trifolii *RS1051 host range to alfalfa plants by integration into its symbiotic plasmid of* nodEFG *and* nodH *genes of* Rhizobium meliloti. Presented at Int. Symp. Mol. Plant-Microbe Interact., Seattle, Wash.
58. Fravel, D. R. 1988. Role of antibiosis in the biocontrol of plant diseases. *Annu. Rev. Phytopathol*. 26:75–91
59. Fredrickson, J. K., Hagedorn, C. 1992. Overview: identifying the ecological effects from the release of genetically engineered microorganisms and microbial pest control agents. In *Microbial Ecology: Principles, Methods, and Applications,* ed. M. A. Levin, R. J. Seidler, M. Rogul, pp. 559–78. New York: McGraw-Hill
60. Frey, P., Prior, P., Trigalet-Demery, D., Marie, C., Kotoujansky, A., Trigalet, A. 1993. Advances in biological control of tomato bacterial wilt using genetically engineered avirulent mutants of *Pseudomonas solanacearum*. In *Proc. Int. Conf. Plant Pathogenic Bacteria, 8th, Versailles, France* In press
61. Furukawa, K., Suzuki, H. 1988. Gene manipulation of catabolic activities for production of intermediates of various biphenyl compounds. *Appl. Microbiol. Biotechnol*. 29:363–69
62. Gaertner, F. 1990. Cellular delivery systems for insecticidal proteins: living and non-living microorganisms. In *Controlled Delivery of Crop-Protection Agents,* ed. R. M. Wilkins, pp. 245–57. London: Taylor and Francis
62a. Galli, E., Silver, S., Witholt, B., eds. 1992. Pseudomonas: *Molecular Biology and Technology*. FEMS Symp. No. 60. Washington: Am. Soc. Microbiol.
63. Gamela, P. H., Piot, J.-C. 1992. Characterization and properties of a novel plasmid vector for *Bacillus thuringiensis* displaying compatibility with host plasmids. *Gene* 120:17–26
64. Gawron-Burke, C., Baum, J. A. 1991. Genetic manipulation of *Bacillus*

thuringiensis insecticidal crystal protein genes in bacteria. In *Genetic Engineering*, ed. J. K. Setlow, 13:237–63. New York: Plenum

65. Gawron-Burke, C., Johnson, T. B. 1993. Development of *Bacillus thuringiensis* based pesticides for the control of potato insect pests. *Advances in Potato Pest Biology and Management*. In press
66. Gay, P. A., Saikumar, K. V., Cleveland, T. E., Tuzun, S. 1992. Antagonistic effect of chitinolytic bacteria against toxin-producing fungi. *Phytopathology* 82:1074
67. Gillespie, A. T. 1988. Use of fungi to control pests of agricultural importance. See Ref. 22a, pp. 37–60
68. Goettel, M. S., St. Leger, R. J., Bhairi, S., Roberts, D. W., Staples, R. C. 1989. Transformation of the entomopathogenic fungus *Metarhizium anisopliae* using the *benA3* gene from *Aspergillus nidulans*. *J. Cellular Biochem. Suppl*. 13A:170
69. Graham, T. L., Watrud, L. S., Perlak, F. J., Tran, M. T., Lavrick, P. B., et al. 1986. A model genetically engineered pesticide: cloning and expression of the *Bacillus thuringiensis* subsp. *kurstaki* δ-endotoxin into *Pseudomonas fluorescens*. In *Recognition in Microbe-Plant Symbiotic and Pathogenic Interactions*, ed. B. Lugtenberg, pp. 385–93. Berlin: Springer-Verlag
70. Greaves, M. P., Bailey, J. A., Hargreaves, J. A. 1989. Mycoherbicides: opportunities for genetic manipulation. *Pestic. Sci*. 26:93–101

70a. Gresshoff, P. M., Roth, L. E., Stacey, G., Newton, W. E., eds. 1990. *Nitrogen Fixation: Achievements and Objectives*. New York: Chapman and Hall

71. Gullino, M. L., Garibaldi, A. 1988. Biological and integrated control of grey mould of grapevine: results in Italy. *EPPO Bull*. 18:9–12
72. Gutterson, N. 1990. Microbial fungicides: recent approaches to elucidating mechanisms. *CRC Rev. Biotechnol*. 10: 69–91
73. Gutterson, N., Howie, W., Suslow, T. 1990. Enhancing efficiencies of biocontrol agents by use of biotechnology. See Ref. 9a, pp. 749–65
74. Gutterson, N. I., Layton, T. J., Ziegle, J. S., Warren, G. J. 1986. Molecular cloning of genetic determinants for inhibition of fungal growth by a fluorescent pseudomonad. *J. Bacteriol*. 165:696–703
75. Haas, D., Keel, C., Laville, J., Maurhofer, M., Oberhansli, T., et al. 1991. Secondary metabolites of *Pseudomonas fluorescens* strain CHAO involved in suppression of root diseases. In *Advances in Molecular Genetics of Plant-Microbe Interactions*, ed. H. Hennecke, D. P. S. Verma, 1:450–56. Amsterdam: Kluwer Academic
76. Haggblom, M. 1990. Mechanisms of bacterial degradation and transformation of chlorinated monoaromatic compounds. *J. Basic Microbiol*. 30:115–41

76a. Halvorson, H. O., Pramer, D., Rogul, M., eds. 1985. *Engineered Organisms in the Environment: Scientific Issues*. Washington, DC: Am. Soc. Microbiol.

77. Harman, G. E., Stasz, T. E. 1991. Protoplast fusion for the production of superior biocontrol fungi. See Ref. 164b, pp. 171–86
78. Harman, G. E., Taylor, A. G., Stasz, T. E. 1989. Combining effective strains of *Trichoderma harzianum* and solid matrix priming to improve biological seed treatments. *Plant Dis*. 73:631–37
79. Hayes, C. K. 1992. Improvement of *Trichoderma* and *Gliocladium* by genetic manipulation. See Ref. 171a, pp. 277–86
80. Heale, J. B. 1988. The potential impacts of fungal genetics and molecular biology on biological control with particular reference to entomopathogens. See Ref. 22a, pp. 211–34
81. Hirano, S. S., Demars, S. J., Morris, C. E. 1981. Survival, establishment and dispersal of *Pseudomonas syringae* on snap beans (*Phaseolus vulgaris*). *Phytopathology* 71:881
82. Hirano, S. S., Willis, D. K., Upper, C. D. 1992. Population dynamics of a Tn*5*-induced non-lesion forming mutant of *Pseudomonas syringae* pv. *syringae* on bean plants in the field. *Phytopathology* 82:1067
83. Hirsch, A. M., Wilson, K. J., Jones, J. D. G., Bang, M., Walker, V. V., Ausubel, F. M. 1984. *Rhizobium meliloti* nodulation genes allow *Agrobacterium tumefaciens* and *Escherichia coli* to form pseudonodules on alfalfa. *J. Bacteriol*. 158:1133–43
84. Hodgson, A. L. M., Stacey, G. 1986. Potential for *Rhizobium* improvement. *Crit. Rev. Biotechnol*. 4:1–74
85. Hofte, H., Whiteley, H. R. 1989. Insecticidal crystal proteins of *Bacillus thuringiensis*. *Microbiol. Rev*. 53:242–55
86. Hollander, A. K. 1991. Environmental impacts of genetically engineered microbial and viral biocontrol agents. See Ref. 117b, pp. 251–66

87. Howie, W. J., Suslow, T. V. 1991. Role of antibiotic biosynthesis in the inhibition of *Pythium ultimum* in the cotton spermosphere and rhizosphere by *Pseudomonas fluorescens*. *Mol. Plant-Microbe Interact*. 4:393–99
88. Jones, D. A., Ryder, M. H., Clare, B. G., Farrand, S. K., Kerr, A. 1988. Construction of a Tra⁻ deletion mutant of pAgK84 to safeguard the biological control of crown gall. *Mol. Gen. Genet*. 212:207–14
89. Jones, J. D. G., Grady, K. L., Suslow, T. V., Bedbrook, J. R. 1986. Isolation and characterization of genes encoding two chitinase enzymes from *Serratia marcescens*. *EMBO J*. 5:467–73
90. Kearns, L. P., Mahanty, H. K. 1993. Identification and cloning of *Erwinia herbicola* DNA responsible for suppression of *Erwinia amylovora*. *Acta Hortic*. In press
91. Keel, C., Schnider, U., Maurhofer, M., Voisard, C., Laville, J., et al. 1992. Suppression of root diseases by *Pseudomonas fluorescens* CHA0: importance of the bacterial secondary metabolite 2,4-diacetylphloroglucinol. *Mol. Plant-Microbe Interact*. 5:4–13
92. Kistler, H. C. 1991. Genetic manipulation of plant pathogenic fungi. See Ref. 164b, pp. 152–69
92a. Klement, Z., ed. 1990. *Proceedings of the Seventh International Conference on Plant Pathogenic Bacteria*. Budapest: Akedimia Kiado
93. Kluepfel, D. A., Kline, E. L., Hughes, T., Skipper, H., Gooden, D., et al. 1990. Field testing of a genetically engineered rhizosphere inhabiting pseudomonad: development of a model system. See Ref. 117a, pp. 189–99
94. Kluepfel, D. A., Kline, E. J., Skipper, H. D., Hughes, T. A., Gooden, D. T., et al. 1991. The release and tracking of genetically engineered bacteria in the environment. *Phytopathology* 81: 348–52
95. Kluepfel, D. A., Tonkyn, D. W. 1990. Release of soil-borne genetically modified bacteria: biosafety implications from contained experiments. See Ref. 117a, pp. 55–65
96. Kluepfel, D. A., Tonkyn, D. W. 1992. The ecology of genetically altered bacteria in the rhizosphere. See Ref. 171a, pp. 407–13
97. Kostka, S. J. 1990. The design and execution of successive field releases of genetically engineered microorganisms. See Ref. 117a, pp. 167–76
98. Kostka, S. J., Reeser, P. W., Miller, D. P. 1988. Experimental host range of *Clavibacter xyli* subsp. *cynodontis* (CXC) and a CXC/*Bacillus thuringiensis* recombinant (CXC/BT). *Phytopathology* 78:1540
99. Kostka, S. J., Tomasino, S. F., Turner, J. T., Reeser, P. W. 1988. Field release of a transformed strain of *Clavibacter xyli* subsp. *cynodontis* containing a delta-endotoxin gene from *Bacillus thuringiensis* subsp. *kurstaki* (BT). *Phytopathology* 78:1540
100. Leahy, J. G., Colwell, R. R. 1990. Microbial degradation of hydrocarbons in the environment. *Microbiol. Rev*. 54:305–15
101. Lehrbach, P. R., Zeyer, J., Reineke, W., Knackmuss, H.-J., Timmis, K. N. 1984. Enzyme recruitment in vitro: use of cloned genes to extend the range of haloaromatics degraded by *Pseudomonas* sp. strain B13. *J. Bacteriol*. 158:1025–32
102. Lereclus, D., Vallade, M., Chafaux, J., Arantes, O., Rambaud, S. 1992. Expansion of insecticidal host range of *Bacillus thuringiensis* by in vivo genetic recombination. *Bio/Technology* 10:418–21
103. Levy, E., Gough, F. J., Berlin, K. D., Guiana, P. W., Smith, T. J. 1992. Inhibition of *Septoria tritici* and other phytopathogenic fungi and bacteria by *Pseudomonas fluorescens* and its antibiotics. *Plant Pathol*. 41:335–41
104. Deleted in proof
105. Lindemann, J., Suslow, T. V. 1987. Competition between ice nucleation–active wild type and ice nucleation–deficient deletion mutant strains of *Pseudomonas syringae* and *Pseudomonas fluorescens* biovar I and biological control of frost injury on strawberry blossoms. *Phytopathology* 77:882–86
106. Lindemann, J., Suslow, T. V. 1987. Characteristics relevant to the question of environmental fate of genetically engineered INA-deletion mutant strains of *Pseudomonas*. *Curr. Plant. Sci. Biotechnol. Agric*. 4:1005–12
107. Lindow, S. E. 1983. The role of bacterial ice nucleation in frost injury to plants. *Annu. Rev. Phytopathol*. 21: 363–84
108. Lindow, S. E. 1985. Ecology of *Pseudomonas syringae* relevant to the field use of Ice⁻ deletion mutants constructed *in vitro* for plant frost control. See Ref. 76a, pp. 23–35
109. Lindow, S. E. 1986. In vitro construction of biological control agents. In *BCPC Monograph No. 34. Biotechnology and Crop Improvement and Pro-*

tection, pp. 185–98. London: Br. Crop Prot. Council
110. Lindow, S. E. 1989. Genetic engineering of bacteria from managed and natural habitats. *Science* 244:1300–7
111. Lindow, S. E. 1990. Design and results of field tests of recombinant Ice$^-$ *Pseudomonas syringae* strains. See Ref. 118, pp. 61–69
112. Lindow, S. E. 1990. Use of genetically altered bacteria to achieve plant frost control. See Ref. 124a, pp. 85–110
113. Lindow, S. E. 1990. Environmental use of genetically engineered organisms. In *Advances in Biotechnology*, ed. E. Heseltine, pp. 101–14. Stockholm: AB Boktryck HBG
114. Lindow, S. E. 1992. Ice$^-$ strains of *Pseudomonas syringae* introduced to control ice nucleation active strains on potato. See Ref. 171a, pp. 169–74
115. Lindow, S. E. 1992. Environmental release of pseudomonads: potential benefits and risks. See Ref. 62a, pp. 399–407
116. Lindow, S. E., Knudsen, G. R., Seidler, R. J., Walter, M. V., Lambou, V. W., et al. 1988. Aerial dispersal and epiphytic survival of *Pseudomonas syringae* during a pretest for the release of genetically engineered strains into the environment. *Appl. Environ. Microbiol.* 54:1557–63
117. Lindow, S. E., Panopoulos, N. J. 1988. Field test of recombinant Ice$^-$ *Pseudomonas syringae* for biological frost control in potato. See Ref. 164a, pp. 121–38
117a. MacKenzie, D. R., Henry, S. C., eds. 1990. *Biological Monitoring of Genetically Engineered Plants and Microbes*. Bethesda: Agricultural Research Institute
117b. Maramorosch, K., ed. 1991. *Biotechnology for Biological Control of Pests and Vectors*. Boca Raton: CRC Press
118. Marois, J., Bruening, J. 1990. *Proceedings of the International Conference on Risk Assessment in Agricultural Biotechnology*. Oakland: Univ. Calif. Press
119. Maurhofer, M., Keel, C., Schnider, U., Voisard, C., Haas, D., Defago, G. 1992. Influence of enhanced antibiotic production in *Pseudomonas fluorescens* strain CHA0 on its disease suppressive capacity. *Phytopathology* 82:190–95
120. McCoy, C. W. 1990. Entomogenous fungi as microbial pesticides. See Ref. 9a, pp. 139–59
121. McIver, J., Djordevic, M. A., Weinman, J. J., Bender, G. L., Rolfe, B. G. 1989. Extension of host range of *Rhizobium leguminosarum* bv. *trifolii* caused by point mutations in *nodD* that result in alterations in regulatory function and recognition of inducer molecules. *Mol. Plant-Microbe Interact.* 2:97–106
122. Miller, R. V., Levy, S. B. 1989. Horizontal gene transfer in relation to environmental release of genetically engineered microorganisms. In *Gene Transfer in the Environment*, ed. S. B. Levy, R. V. Miller, pp. 405–20. New York: McGraw-Hill
123. Millis, N. F. 1992. Australian experience with the release of microorganisms. In *Proc. Int. Symp. on the Biosafety Results of Field Tests of Genetically Modified Plants and Microorganisms, 2nd, Goslar, Germany*. (Abstr.)
124. Mulbry, W., Kearney, P. C. 1991. Degradation of pesticides by microorganisms and the potential for genetic manipulation. *Crop Prot.* 10: 334–46
124a. Nakas, J. P., Hagedorn, C., eds. 1990. *Biotechnology of Plant-Microbe Interactions*. New York: McGraw-Hill
125. Norelli, J. L., Gilbert, M. T., Aldwinckle, H. S., Zumoff, C. H., Beer, S. V. 1990. Population dynamics of nonpathogenic mutants of *Erwinia amylovora* in apple host tissue. *Acta Hortic.* 273:239–40
126. Nusslein, K., Maris, D., Timmis, K., Dwyer, D. F. 1992. Expression and transfer of engineered catabolic pathways harbored by *Pseudomonas* spp. introduced into activated sludge microcosms. *Appl. Environ. Microbiol.* 58: 3380–86
127. Obukowicz, M. G., Perlak, F. J., Kusano-Kretzmer, K., Mayer, E. J., Bolten, S. L., Watrud, L. S. 1986. Tn*5*-mediated integration of the delta-endotoxin gene from *Bacillus thuringiensis* into the chromosome of root-colonizing pseudomonads. *J. Bacteriol.* 168:982–89
128. Olson, B. H., Ogunseitan, O. A., Rochelle, P. A., Tebbe, C. C., Tsai, Y. L. 1991. The implications of horizontal gene transfer for the environmental impact of genetically engineered microorganisms. In *Risk Assessment in Genetic Engineering*, ed. M. A. Levin, H. S. Strauss, pp. 163–88. New York: McGraw-Hill
129. Oppenheim, A. B., Chet, I. 1992. Cloned chitinases in fungal plant-pathogen control strategies. *Trends Biotechnol.* 10:392–94

130. Orser, C. S., Staskawicz, B. J., Panopoulos, N. J., Dahlbeck, D., Lindow, S. E. 1985. Cloning and expression of bacterial ice nucleation genes in *Escherichia coli*. *J. Bacteriol.* 164:359–66

131. Paau, A. S. 1991. Improvement of *Rhizobium* inoculants by mutation, genetic engineering and formulation. *Biotechnol. Adv.* 9:173–84

132. Panopoulos, N. J., Peet, R. C. 1985. The molecular genetics of plant pathogenic bacteria. *Annu. Rev. Phytopathol.* 23:381–419

133. Pantuwatana, S. 1991. Control of mosquito vectors by genetically engineered *Bacillus thuringiensis* and *B. sphaericus* in the tropics. See Ref. 117b, pp.149–61

134. Papavizas, G. C., Lewis, J. A., Abd-El Moity, T. H. 1982. Evaluation of new biotypes of *Trichoderma harzianum* tolerant to benomyl and enhanced biocontrol capabilities. *Phytopathology* 72:126–32

135. Parke, J. L., Zablotowicz, R. M., Rand, R. E. 1992. Tracking a *lacZY*-marked strain of *Pseudomonas fluorescens* in a Wisconsin field trial. *Phytopathology* 82:1177–78

136. Prunier, J. P., Kostka, S. J., Labonne, G. 1990. *Clavibacter xyli* subsp. *cynodontis:* an endophytic bacterium from corn. See Ref. 92a, pp. 305–10

137. Reeser, P. W., Kostka, S. J. 1988. Population dynamics of *Clavibacter xyli* subsp. *cynodontis* (Cxc) and a Cxc/*Bacillus thuringiensis* subsp. *kurstaki* (Bt) recombinant in corn (*Zea mays*). *Phytopathology* 78:1540

138. Reineke, W., Knackmuss, H.-J. 1988. Microbial degradation of haloaromatics. *Annu. Rev. Microbiol.* 42:263–87

139. Robbins, P. W., Albright, C., Benfield, B. 1988. Cloning and expression of a *Streptomyces plicatus* chitinase (chitinase-63) in *Escherichia coli*. *J. Biol. Chem.* 263:443–47

140. Roberts, D. P., Berman, P. M., Allen, C., Stromberg, V. K., Lacy, G. H., Mount, M. S. 1986. *Erwinia carotovora:* molecular cloning of a 3.4 kilobase DNA fragment mediating production of pectate lyases. *Can. J. Plant Pathol.* 8:17–27

141. Ronson, C. W., Bosworth, A., Genova, M., Gudbransen, S., Hankinson, T., et al. 1990. Field release of genetically-engineered *Rhizobium meliloti* and *Bradyrhizobium japonicum* strain. See Ref. 70a, pp. 397–403

142. Sands, D. C., Ford, E. J., Miller, R. V. 1990. Genetic manipulation of broad host-range fungi for biological control of weeds. *Weed Technology* 4:471–74

143. Sayler, G. S., Hooper, S. W., Layton, A. C., King, J. M. H. 1990. Catabolic plasmids of environmental and ecological significance. *Microb. Ecol.* 19:1–20

144. Sayre, P. G. 1990. Assessment of genetically engineered microorganisms under the Toxic Substances Control Act: considerations prior to small-scale release. See Ref. 70a, pp. 405–15

145. Sayre, P., Miller, R. V. 1991. Bacterial mobile genetic elements: importance in assessing the environmental fate of genetically engineered sequences. *Plasmid* 26:151–71

146. Seidler, R. 1988. *EPA Special Report: release of Ice-minus recombinant bacteria at California test sites*. Environmental Research Laboratory, Corvallis, Oregon

147. Seidler, R. J. 1992. Evaluation of methods for detecting ecological effects from genetically engineered microorganisms and microbial pest control agents in terrestrial ecosystems. *Biotechnol. Adv.* 10:149–78

148. Shapira, R., Ordentlich, A., Chet, I., Oppenheim, A. B. 1989. Control of plant diseases by chitinases expressed from cloned DNA in *Escherichia coli*. *Phytopathology* 79:1246–49

149. Shaw, J. J., Beauchamp, C., Dane, F., Kriel, R. J. 1992. Securing a permit from the United States Department of Agriculture for field work with genetically engineered microbes: a nonprohibitory process. *Microb. Releases* 1:51–53

150. Shaw, J. J., Dane, F., Geiger, D., Kloepper, J. W. 1992. Use of bioluminescence for detection of genetically engineered microorganisms released into the environment. *Appl. Environ. Microbiol.* 58:267–73

151. Shields, M. S., Reagin, M. J. 1992. Selection of a *Pseudomonas cepacia* strain constitutive for the degradation of trichloroethylene. *Appl. Environ. Microbiol.* 58:3977–83

152. Shim, J.-S., Farrand, S. K., Kerr, A. 1987. Biological control of crown gall: construction and testing of new biocontrol agents. *Phytopathology* 77:463–66

153. Sivan, A., Stasz, T. E., Hemmat, M., Hayes, C. K., Harman, G. E. 1992. Transformation of *Trichoderma* spp. with plasmids conferring hygromycin B resistance. *Mycologia* 84:687–94

154. Skot, L., Harrison, S. P., Nath, A., Mytton, L. R., Clifford, B. C. 1990.

Expression of insecticidal in *Rhizobium* containing the δ-endotoxin gene cloned from *Bacillus thuringiensis* subsp. *tenebrionis*. *Plant Soil* 27:285–95
155. Smit, E., van Elsas, J. D., van Veen, J. A. 1992. Risks associated with the application of genetically modified microorganisms in terrestrial ecosystems. *FEMS Microbiol. Rev.* 88:263–78
156. Spaink, H. P., Okker, R. J. H., Wijffelman, C. A., Tak, T., Goosen-de Roo, L., et al. 1989. Symbiotic properties of *Rhizobia* containing a flavonoid-independent hybrid *nodD* product. *J. Bacteriol.* 171:4045–53
157. Stacey, G. 1985. The *Rhizobium* experience. See Ref. 76a, pp. 109–21
158. Stacey, G., Upchurch, R. G. 1984. *Rhizobium* inoculation of legumes. *Trends Biotechnol.* 2:65–70
159. Stasz, T. E. 1990. Genetic improvement of fungi by protoplast fusion for biological control of plant pathogens. *Can. J. Plant Pathol.* 12:322–27
159a. Stewart-Tull, D. E. S., Sussman, M., eds. 1992. *The Release of Genetically Modified Microorganisms—REGEM 2*. New York: Plenum
160. Stock, C. A., McLoughlin, T. J., Klein, J. A., Adang, M. J. 1990. Expression of a *Bacillus thuringiensis* crystal protein gene in *Pseudomonas cepacia* 526. *Can. J. Microbiol.* 36: 879–84
161. Stromber, V. K., Orvos, D. R., Scanferlato, V. S. 1990. *In planta* competition among cell-degrading enzyme mutants and wild-type strains of *Erwinia carotovora*. See Ref. 92a, pp. 721–25
162. Sundheim, L. 1990. Biocontrol of *Fusarium oxysporum* with a chitinase encoding gene from *Serratia marcescens* on a stable plasmid in *Pseudomonas*. *J. Cell. Biochem. Suppl.* 13A:171
163. Sundheim, L. 1992. Effect of chitinase encoding genes in biocontrol *Pseudomonas* spp. See Ref. 171a, pp. 331–33
164. Sundin, G. W., Jones, A. L., Fulbright, D. W. 1989. Copper resistance in *Pseudomonas syringae* pv. *syringae* from cherry orchards and its associated transfer in vitro with a plasmid. *Phytopathology* 79:861–65
164a. Sussman, M., Collins, C. H., Skinner, F. A., Stewart-Tull, D. E., eds. 1988. *The Release of Genetically-Engineered Micro-Organisms*. London: Academic
164b. Te Beest, D. O., ed. 1991. *Microbial Control of Weeds*. New York: Chapman and Hall
165. Te Beest, D. O., Yang, X. B., Cisar, C. R. 1992. The status of biological control of weeds with fungal pathogens. *Annu. Rev. Phytopathol.* 30:637–57
166. Templeton, G. E., Heiny, D. K. 1990. Mycoherbicides. In *New Directions in Biological Control: Alternatives for Suppressing Agricultural Pests and Diseases*, pp. 279–86. New York: Liss
167. Thanabalu, T., Hindley, J., Brenner, S., Oei, C., Berry, C. 1992. Expression of the mosquitocidal toxins of *Bacillus sphaericus* and *Bacillus thuringiensis* subsp. *israelensis* by recombinant *Caulobacter crescentus*, a vehicle for biological control of aquatic insect larvae. *Appl. Environ. Microbiol.* 58:905–10
168. Thomashow, L. S., Weller, D. M. 1988. Role of a phenazine antibiotic from *Pseudomonas fluorescens* in biological control of *Gauemannomyces graminis* var. *tritici*. *J. Bacteriol.* 170:3499–3508
169. Thomashow, L. S., Weller, D. M. 1990. Application of fluorescent pseudomonads to control diseases of wheat and some mechanisms of disease suppression. In *Biological Control of Soil-Borne Plant Pathogens*, ed. D. Hornby, pp. 109–22. Wallingford, UK: CAB International
170. Thomashow, L. S., Weller, D. M., Bonsall, R. F., Pierson, L. S. III. 1990. Production of the antibiotic phenazine-1-carboxylic acid by fluorescent *Pseudomonas* species in the rhizosphere of wheat. *Appl. Environ. Microbiol.* 56:908–12
171. Tiedje, J. M., Colwell, R. K., Grossman, Y. L., Hodson, R. E., Lenski, R. E., et al. 1989. The planned introduction of genetically engineered organisms: ecological considerations and recommendations. *Ecology* 70:298–315
171a. Tjamos, E. C., Papavizas, G. C., Cook, R. J., eds. 1992. *Biological Control of Plant Diseases: Progress and Challenges for the Future*. New York: Plenum
172. Tolin, S. A., Vidaver, A. K. 1989. Guidelines and regulations for research with genetically modified organisms: a view from academe. *Annu. Rev. Phytopathol.* 27:551–81
173. Tomasino, S. F., Beach, R. M., Leister, R. J. 1992. Control of European corn borer in field corn using a genetically engineered endophyte. *Phytopathology* 82:1128
174. Trigalet, A., Demery, D. 1986. Invasiveness in tomato plants of Tn*5*-induced avirulent mutants of *Pseudomonas solanacearum*. *Physiol. Mol. Plant Pathol.* 28:423–30

175. Trigalet, A., Trigalet-Demery, D. 1990. Use of avirulent mutants of *Pseudomonas solanacearum* for the biological control of bacterial wilt of tomato plants. *Physiol. Mol. Plant Pathol.* 36:27–38
176. Triplett, E. W. 1990. Construction of a symbiotically effective strain of *Rhizobium leguminosarum* bv. *trifolii* with increased nodulation competitiveness. *Appl. Environ. Microbiol.* 56:98–103
177. Triplett, E. W., Barta, T. M. 1987. Trifolitoxin production and nodulation are necessary for the expression of superior nodulation competitiveness by *Rhizobium leguminosarum* bv. *trifolii* strain T24 on clover. *Plant Physiol.* 85:335–42
178. Triplett, E. W., Sadowsky, M. J. 1992. Genetics of competition for nodulation of legumes. *Annu. Rev. Microbiol.* 46:399–428
179. Tronsmo, A. 1992. Leaf and blossom epiphytes and endophytes as biological control agents. See Ref. 171a, pp. 43–54
180. Truchet, G., Rosenberg, G., Vasse, J., Julliot, J.-S., Camut, S., Denarie, J. 1984. Transfer of *Rhizobium meliloti* pSym genes into *Agrobacterium tumefaciens* host-specific nodulation by atypical infection. *J. Bacteriol.* 157: 134–42
181. Turner, J. T., Lampel, J. S., Stearman, R. S., Sundin, G. W., Gunyuzlu, P., Anderson, J. J. 1991. Stability of the δ-endotoxin gene from *Bacillus thuringiensis* subsp. *kurstaki* in a recombinant strain of *Clavibacter xyli* subsp. *cynodontis*. *Appl. Environ. Microbiol.* 57:3522–28
182. van Elsas, J. D., Heijnen, C. E., van Veen, J. A. 1990. The fate of introduced genetically engineered microorganisms (GEM's) in soil, in microcosm, and field: impact of soil textural aspects. See Ref. 117a, pp. 67–79
183. Vanneste, J. L., Yu, J., Beer, S. V. 1992. Role of antibiotic production by *Erwinia herbicola* Eh252 in biological control of *Erwinia amylovora*. *J. Bacteriol.* 174:2785–96
184. Vicedo, B., Penalver, R., Asins, M. J., Lopez, M. M. 1993. Biological control of *Agrobacterium tumefaciens*, colonization, and pAgK84 transfer with *Agrobacterium radiobacter* and the Tra^- mutant strain K1026. *Appl. Environ. Microbiol.* 59:309–15
185. Vincent, M. N., Harrison, L. A., Brackin, J. M., Kovacevich, P. A., Mukerji, P., et al. 1991. Genetic analysis of the antifungal activity of a soilborne *Pseudomonas aureofaciens* strain. *Appl. Environ. Microbiol.* 57:2928–34
186. Waalwijk, C., Dullemans, A., Maat, C. 1991. Construction of a bioinsecticidal rhizosphere isolate of *Pseudomonas fluorescens*. *FEMS Microbiol. Lett.* 77:257–64
187. Weller, D. M. 1988. Biological control of soilborne plant pathogens in the rhizosphere with bacteria. *Annu. Rev. Phytopathol.* 26:379–407
188. Weller, D. M., Saettler, A. W. 1978. Rifampin-resistant *Xanthomonas phaseoli* var. *fuscans* and *Xanthomonas phaseoli:* tools for study of bean blight bacteria. *Phytopathology* 68:778–81
189. Wilcox, H. E. 1990. Mycorrhizal associations. See Ref. 124a, pp. 227–55
190. Williamson, M. 1992. Environmental risks from the release of genetically modified organisms (GMOs)—the need for molecular ecology. *Mol. Ecol.* 1:3–8
191. Willis, D. K., Rich, J. R., Hrabak, E. M. 1991. *hrp* genes of phytopathogenic bacteria. *Mol. Plant-Microbe Interact.* 4:132–38
192. Winter, R. B., Yen, K., Ensley, B. D. 1989. Efficient degradation of trichlorethylene by a recombinant *Escherichia coli*. *Bio/Technology* 7:282–85
193. Wortman, A. T., Somerville, C. C., Colwell, R. R. 1986. Chitinase determinants of *Vibrio vulnificus:* gene cloning and applications of a chitinase probe. *Appl. Environ. Microbiol.* 52: 142–45
194. Wynne, E. C., Pemberton, J. M. 1986. Cloning of a gene cluster from *Cellvibrio mixtus* which encodes for cellulase, chitinase, amylase, and pectinase. *Appl. Environ. Microbiol.* 52:1362–67
195. Yamamoto, T., Powell, G. K. 1993. *Bacillus thuringiensis* crystal protein: recent advancements in understanding the insecticidal activity. In *Advanced Engineered Pesticides,* ed. L. Kim. New York: Marcel Dekker In press
196. Yoder, O. C., Weltring, K., Turgeon, B. G., Garber, R. C., Van Etten, H. D. 1986. Technology for molecular cloning of fungal virulence genes. In *Biology and Molecular Biology of Plant-Pathogen Interactions,* ed. J. Bailey, pp. 371–84. New York: Plenum

Annu. Rev. Microbiol. 1993. 47:945–63

THE TN5 TRANSPOSON

W. S. Reznikoff
Department of Biochemistry, College of Agricultural and Life Sciences, University of Wisconsin-Madison, Madison, Wisconsin 53706

KEY WORDS: transposase, end sequences, gene regulation, *cis*-activity

---

CONTENTS

ABSTRACT

The bacterial transposon Tn*5* encodes two proteins, the transposase and a related protein, the transposition inhibitor, whose relative abundance determines, in part, the frequency of Tn*5* transposition. The synthesis of these proteins is programmed by a complex set of genetic regulatory elements. The host DNA methylation function, *dam,* inhibits transposase promoter recognition and indirectly enhances the transposition inhibitor promoter. The inhibitor lacks the N-terminal 55 amino acids of the transposase, suggesting that this sequence plays a key role in the transposition process. An intact N-terminal sequence is required for the transposase's recognition of the 19-bp end DNA sequences. This is the first critical step in the transposition process. Transposase–end DNA interaction is itself regulated by an intricate series of reactions involving several host proteins: DnaA, Dam, and Fis. The transposase is a unique protein in that it acts primarily *in cis* and inhibits its own activity *in trans*. Models to explain these properties are described. Finally circumstantial evidence suggests that transposition occurs preferentially from newly replicated DNA that has yet to be partitioned to progeny cells. This

0066-4227/93/1001-0945$02.00

timing of transposition is likely to have a selective advantage for the host and the transposable element.

Introduction

Transposition is a recombination process in which DNA sequences termed transposable elements move from an original site on a DNA molecule to a new site on the same or on a different DNA molecule. In addition, transposable elements can cause, and are associated with, other types of genetic rearrangements such as deletions, inversions, and chromosome fusions. The genomes of prokaryotic and eukaryotic organisms contain these elements. One could consider them as an ancient genetic machinery for causing genomic rearrangements and, therefore, for facilitating genome evolution. In addition, their associated biochemical reactions are likely to be similar to other interesting events involving the interaction of proteins and DNA. For these reasons transposable elements are of considerable interest.

The transposable elements found in *Escherichia coli* fall into three general classes determined by their mechanisms of transposition. Transposons such as Tn*3* and γδ transpose through a two-step replicative mechanism in which a cointegrate (fused replicon) structure is an intermediate. Transposons Tn*10* and Tn*903* transpose through a conservative cut-and-paste mechanism. Bacteriophage Mu and related viruses represent the third class of transposable element. In these cases the transposition can occur through either of the above two mechanisms, depending upon the proteins involved and the precise nature of the DNA strand cutting after an intermediate is formed between the transposable element and the target DNA sequence. Tn*5* is generally assumed to transpose via a conservative mechanism (2); however, this presumption has not been critically tested. The reader is encouraged to examine the models and evidence for these transposition mechanisms in the recent monograph by Berg & Howe (3).

The conservative and replicative mechanisms of transposition share many basic characteristics. The transposable element encodes two critical functions required for the process—the end sequences and a protein termed the transposase. The element is defined by the specific sequences at its end. Transposition and related events remove the transposable element from its original sequence context precisely at the ends of these sequences. Changes in any base pair of these sequences typically reduces the frequency of or abolishes transposition. The transposable element also encodes a protein called the transposase. The transposase is a critical participant in many transposition functions including: specifically binding to the end sequences, bringing the two ends together through a protein oligomerization process, cutting or nicking the DNA adjacent to the end sequences, and inserting the transposable element DNA into a DNA target site.

Host proteins also play critical roles in the transposition process such as facilitating the end-sequence binding of the transposase, nucleating the higher-order structure in which the ends are brought together, performing the necessary repair or replication functions, and regulating several steps in the transposition process.

Transposition is, in general, a quite rare, highly regulated process. Such tight regulation might enable the host cell to strike a balance between insuring proliferation of the transposable element and insuring the cell's own genetic survival—the very process of transposition causes chromosome breakage and rearrangements. The mechanisms of the regulation vary remarkably among the many cases studied, although some aspects are common to all, and the strategies, perhaps not surprisingly, are similar to ones found for other bacterial genetic systems. In addition to the role of host functions, transposable element encoded–functions often play a critical role in this regulation.

Implicit in the above general description is the fact that several very interesting molecular events are involved in and regulate the transposition process. Studying these events will help us understand other genetic processes. For instance, elucidating how transposition is regulated in a given system will tell us how complex DNA metabolizing processes might be controlled and give yet more examples of how gene expression can be modulated. The transposase is a protein that performs multiple complex functions. Protein structure/function studies would seek answers to the questions about the organization of the peptide domains that perform these functions and how they interact. The terminal DNA sequences are at first glance simply an example of a target for protein binding. However, their functionality is considerably more complex because they are often the target for more than one protein, and the protein–end sequence interaction is associated with at least two events: protein binding and DNA strand scission. Work done on other systems suggests that short DNA sequences can dictate several extremely intricate reactions, and transposable element end sequences may be a perfect example of such compact complexity. Transposable elements use host functions like biochemical parasites. Understanding how they use these functions will tell us much about the transposition process and about the functionality of the host functions themselves. Because many of these host functions are ones involved in host DNA metabolism, understanding their role in the transposition process may give us insights as to how host DNA metabolism is organized and regulated.

Finally, there are always surprising connections in biology. By studying transposition we embark on a largely unknown path into the cell's metabolism. For instance, if the host functions that Tn*5* uses are organized in the cell in a unique spatial fashion, studying Tn*5* transposition may reveal that arrangement.

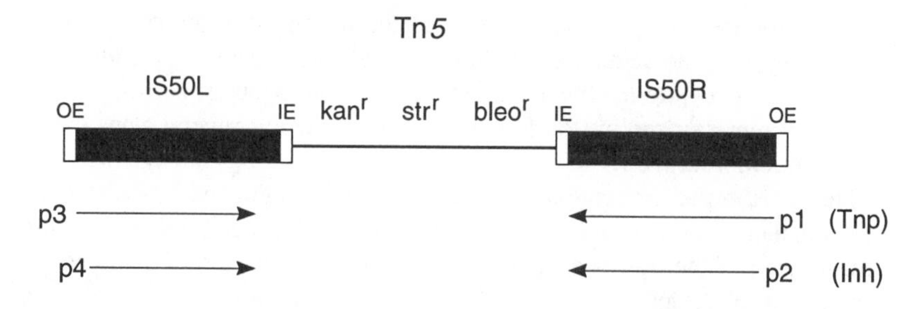

Figure 1 Transposon Tn*5*. Tn*5* is a composite transposon in which genes encoding three antibiotic resistance proteins are bracketed by two IS*50* elements, IS*50*L and IS*50*R. Both IS*50* elements are delineated by 19-bp sequences, the inside end (IE) and the outside end (OE). IS*50*R encodes the transposase (Tnp or p1) and a second protein that inhibits transposition (Inh or p2). Tnp and Inh are translated in the same reading frame, but the Inh AUG is 55 codons downstream from the Tnp AUG. IS*50*L contains an ochre codon that results in the synthesis of p3 and p4, nonfunctional analogues of Tnp and Inh.

My laboratory and several other investigators have been studying the bacterial transposable element Tn*5* as a model system. Tn*5* is an example of a composite transposon in which antibiotic resistance genes are flanked by two nearly identical insertion sequences, IS*50*R and IS*50*L (see Figure 1 for a schematic). IS*50*R is a fully functional transposable element, while IS*50*L contains an ochre codon that results in the synthesis of inactive proteins (33).

IS*50*R encodes the transposase (p1 or Tnp) (15, 18, 19, 33), and the Tn*5* Tnp has two opposing activities. *In cis* it acts as a transposase, catalyzing the transposition of the Tn*5* or IS*50* sequences from which it was encoded (15, 19). However, *in trans* it primarily acts as an inhibitor of transposition (42). The paradox of Tnp inhibiting the activity of other Tnp molecules is discussed in greater detail below. The Tnp inhibitory activity is one means by which Tn*5* transposition is down-regulated.

A second protein [the inhibitor (Inh or p2)] is also encoded by IS*50*R (15, 18, 19, 45). Inh is translated in the same reading frame as Tnp but lacks the N-terminal 55 amino acids. Inh's only known function is to inhibit Tn*5* transposition, which is thought to be the major means of down-regulating this process. The properties of Inh relative to the Tnp suggest that the N-terminal 55 amino acids of Tnp play an important role in its activity. The possible function of this sequence is an important topic of investigation, and this review discusses our current understanding of this function.

The relative abundance of Tnp and Inh plays a major role in determining the Tn*5* transposition frequency. Interestingly, separate, apparently competing, promoters program Tnp and Inh syntheses (21, 43). Thus the regulation of these promoter activities becomes a crucial question. As discussed below, it is the host that regulates the relative activity of these two promoters, and the regulatory mechanism that has evolved appears to link the occurrence of transposition to the DNA replication process.

Tn*5* has also evolved a mechanism for preventing the spurious synthesis of Tnp by virtue of accidental placement within an active transcription unit (21). The translation initiation signals were designed such that they will only function as part of a correctly initiated Tnp mRNA (21, 37). The strategy to accomplish this translation control may be widespread and is discussed below.

Each IS*50* is bounded by two unique 19-bp end sequences [termed outside (OE) and inside (IE) ends] that are critical for transposition (17, 34). The distinguishing feature between IS*50* and Tn*5* transposition is the choice of the transposable element ends. Tn*5* transposition utilizes two OEs, whereas IS*50* transposition uses an OE and an IE sequence. Tnp binds to both of these sequences during the transposition process, and they likely perform other functions vital for Tnp activity. In addition, they are the sequences recognized by host functions. As we shall see, a host function binds to the OE to enhance transposition while the host functions affecting the IE down-regulate transposition. Two of the relevant host functions (DnaA and Dam) also link the transposition process to DNA replication.

Finally, we are just now beginning to obtain clues as to the possible relationship of Tn*5* transposition with overall cell processes. Some of these clues, which point to DNA replication, have been suggested above. Others will become obvious during the discussion of the lethal effect of overproducing Tnp.

Douglas Berg (whose laboratory has contributed much of what we know about Tn*5*) recently published an excellent review of Tn*5* (2). Therefore, this review is not a comprehensive treatment of the subject. Rather, this chapter concentrates on areas of current and future research interest raised by the questions implied above. The specific topics covered are:

1. How are the syntheses of Tnp and Inh regulated?
2. What are the possible functions for the Tnp N-terminal sequence?
3. What sort of protein recognition reactions occur at the 19-bp terminal sequences?
4. What is the molecular basis for the *cis*-active nature of the transposase?
5. How might transposition be linked to chromosome replication and/or cell division?

The Regulation of Transposase and Inhibitor Synthesis

Studies by Biek & Roth (4) first indicated that the frequency of Tn*5* transposition was regulated by Tn*5*-encoded functions. Tn*5* sequences that were newly introduced into a cell transposed at dramatically lower frequencies in a cell already containing Tn*5* as opposed to a cell lacking Tn*5*. In addition, we (18) found that the Tn*5* transposition frequency was constant regardless of the copy number of Tn*5* (hence the transposition frequency of an individual Tn*5* decreased in the presence of additional Tn*5*s). We now know that this down-regulation is primarily (but not entirely; see below) a consequence of the Tn*5*-encoded Inh protein. Inh functions *in trans* to block transposition. Moreover, the frequency of transposition is in part set by the abundance of Tnp and the ratio of Tnp to Inh. Thus the incoming Tn*5* in the Biek & Roth experiment (4) encountered a preexisting pool of inhibiting Inh; as the copy number of Tn*5* increases, the concentration of *trans*-acting Inh increases, but the amount of *cis*-acting Tnp per Tn*5* remains constant.

What then determines the abundance of Tnp and Inh? The answers to this question are not only interesting for Tn*5*, they also reveal genetic regulatory motifs found in other systems.

Tnp and Inh are expressed from overlapping, probably competitive, promoters. This conclusion was deduced by a deletion analysis of in vivo promoter activities (21). As shown in Figure 2, the T2 transcript can only encode the Inh while the T1 transcript encodes both proteins [but the Inh is translated inefficiently from the T1 message (37)]. Such overlapping promoters are found in other systems (e.g. see 30) and can program contradictory functions. Thus by studying this Tn*5* arrangement we will elucidate a more general regulation strategy.

DNA (*dam*) methylation down-regulates the synthesis of the T1 (Tnp) transcript and appears to up-regulate the synthesis of the T2 transcript (43). The frequency of Tn*5* transposition is 10-fold higher in *dam* hosts, and this change in transposition seems to mirror (and presumably is the consequence of) a four- to fivefold increase in T1 mRNA and a twofold decrease in T2 mRNA. The opposite effect on promoter activity suggests that the promoters compete for RNA polymerase. An inspection of the DNA sequence indicates, and site-specific mutation studies prove, that the relevant GATC *dam*

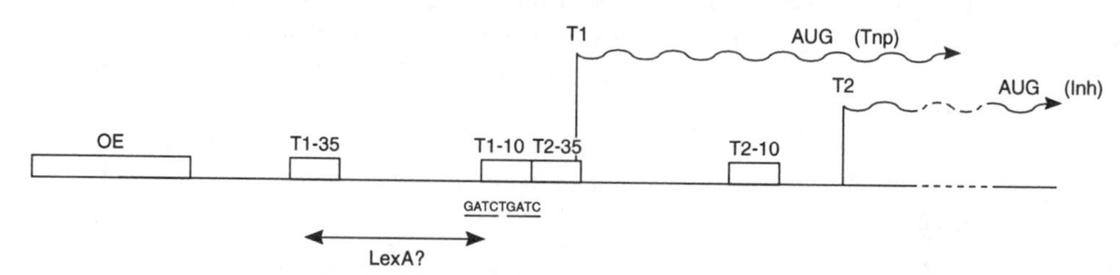

Figure 2 Controlling elements for Tn*5*/IS*50* Tnp and Inh synthesis. The 5′ end of IS*50* is defined by the 19-bp OE sequence (see Figure 4, below). The promoter for transposase synthesis (T1) initiates transcription 66 bp from the end of IS*50* (21). Between the −35 and −10 regions of T1 is a weak LexA binding site (23). Overlapping the T1 −10 region are two Dam methylation sites that when methylated down regulate Tnp mRNA synthesis (43). The T2 (Inh) promoter overlaps T1 (21). The AUGs for Tnp and Inh are indicated (21).

methylation sites overlap with the −10 region of the T1 promoter. The specific relevance of *dam* regulation of Tnp (and Inh) synthesis is that it should couple transposition to DNA replication (about which more is discussed below) because newly replicated DNA is hemimethylated. Also, this regulation should stimulate transposition off of DNA that is newly introduced into the cell, which is exactly the observed result (27, 32).

Other transposon systems [in particular Tn*10* (31)] also display Dam methylation control of Tnp synthesis and of transposition. Moreover, these studies demonstrate in a quite precise manner a type of gene regulation displayed in many systems—chemical modification of specific DNA sequences to modulate protein binding. This mechanism reappears in the discussion of protein recognition of the IS*50* IE sequence.

Tn*5* could transpose into highly expressed genes, thereby giving rise to read-through transcripts into the transposase gene. Production of these transcripts might result in spurious transposase synthesis. However, a set of experiments (21) has demonstrated that read-through expression of transposase does not occur because such messages do not program the translation of Tnp.

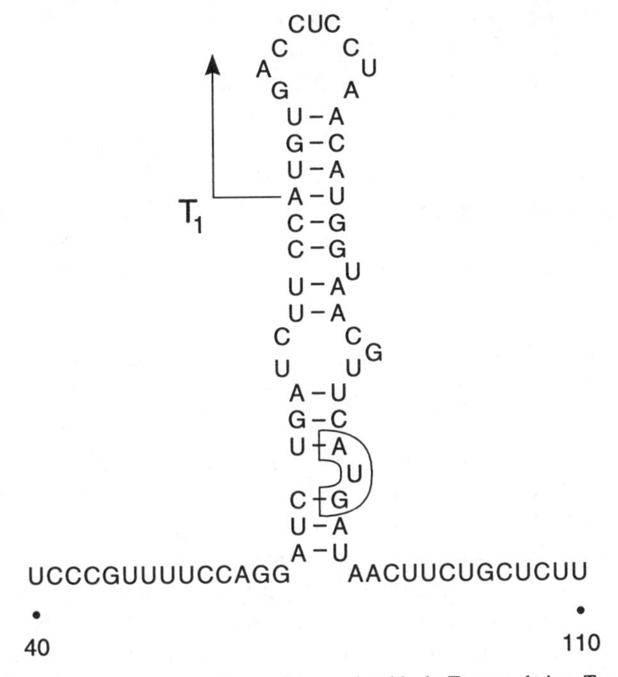

Figure 3 Secondary structure in read-through transcripts blocks Tnp translation. Transcripts that read through the end of IS*50* do not express Tnp because of a secondary structure that occludes the Shine-Dalgarno sequence end and the AUG (21, 37). The start of the T1 transcript (which will not form a secondary structure) and the Tnp AUG are indicated. This figure is similar to one shown in Ref. 21.

An RNA secondary structure in these read-through RNAs occludes the Tnp translation initiation signals (37). As shown in Figure 3, RNA sequences upstream of the T1 transcript start site are necessary for the formation of this secondary structure. This mechanism prevents read-through expression of Tn*5* Tnp. The same mechanism appears to be shared by several other transposable elements [IS*2* (13), IS*3* (39), IS*4* (20), IS*5* (22), IS*150* (38), and Tn*10* (5)] and may be a general motif for *Escherichia coli* gene systems [for instance, the same pattern is seen for *lac* (36, 46) and *gal* (26)].

Do other mechanisms exist that regulate transposase synthesis? Recent reports present conflicting evidence regarding the possible role of LexA in regulating transposase synthesis. Our laboratory has found that LexA has no significant control over transposase levels (40), while Kuan & Tessman postulate that LexA represses transposase synthesis approximately fivefold, presumably by binding to a weak LexA binding site upstream of the T1 promoter (23) (see Figure 2). These studies differ in two respects: (*a*) Weinreich et al performed direct tests of a LexA effect (40), but the Kuan & Tessman studies were indirect (23), and (*b*) the sequence contexts of the two Tn*5* systems were different. The former difference is a matter of technique, while the latter may suggest an interesting molecular phenomenon. Many genetic regulatory proteins are enhanced in their DNA-binding activities through cooperative binding to a secondary site (1). For instance, binding of the *lac* repressor to its operator is thought to be stabilized through a bond to a secondary binding site, thereby forming a looped DNA structure (10, 11). Future studies might examine whether such a nearby LexA binding site is present in the Kuan-Tessman system, but not in the one which we studied, and if so, whether such a fortuitous nearby LexA binding site could facilitate LexA binding to its weak Tn*5* site.

The Transposase N-Terminal Sequence Is Critical for Sequence-Specific DNA Binding and Other Transposase Activities

The transposase and its inhibitor are identical in sequence except that the Inh is missing 55 N-terminal amino acids (20). Because Inh does not promote transposition, it is missing one or more critical Tnp activities (15–17, 33, 45), suggesting that the N-terminal 55 amino acids encode a domain of importance. In vitro analyses of purified Tnp and purified Inh have shown that at least one property missing in the Inh preparations is a DNA sequence–specific binding activity. The Tnp, on the other hand, binds to OE and IE end sequences specifically and shows the expected sensitivity to OE sequence mutations and Dam methylation of the IE (N. de la Cruz, M. Weinreich, T. Wiegand, M. Krebs & W. Reznikoff, submitted; R. A. Jilk & W. Reznikoff, unpublished results). Recent studies have demonstrated that deletion of 11

N-terminal amino acids or introduction of 4 different N-terminal missense mutations destroys the DNA-binding activity of Tnp (M. D. Weinreich & W. Reznikoff, unpublished results). These experiments show that the N-terminal 55 amino acids are necessary for the sequence-specific DNA-binding activity and imply that the DNA-binding domain is in this region. Determining the precise DNA-binding domain will be instructive because the Tn*5* transposase does not contain a previously classified DNA-binding motif, and yet the activity is clearly there.

Another surprising property requires the N-terminal three amino acids of Tnp. Overproduction of the transposase (in the absence of transposition) kills host cells. Deletion of as little as three amino acids prevents this killing (M. D. Weinreich & W. Reznikoff, unpublished results). A possible biological

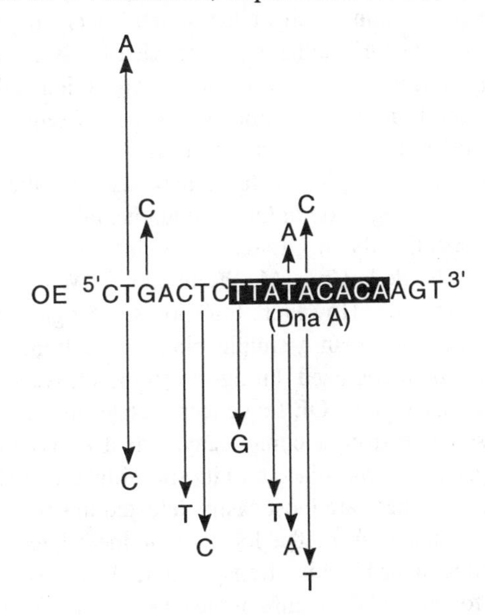

Figure 4 (*Top*) Outside and inside end sequences. The DnaA box is indicated for the outside end sequence (12, 17, 34, 44). Downward-pointing arrows denote mutations that act like a deletion in the adjacent deletion experiment. Upward-pointing arrows denote mutations that constitute the second class in the adjacent deletion experiment (16). The 11A change gave rise to one deletion in the second class. (*Bottom*) For the inside end sequence, the Dam sites and the overlapping Fis binding site are indicated (34, 41). Dam site methylation regulates both Tnp binding (43) and Fis binding (41). The lower-case letters indicate base pairs outside of the required 19-bp sequence. The Tnp/Inh termination codon is the complement to positions 12–14.

significance for the cell killing will be described in the section on chromosome replication below.

Recognition of Terminal 19–Base Pair Sequences

The Tn*5* and IS*50* transposable elements are defined by the terminal 19-bp sequences. Tn*5* is delineated by two inverted OE sequences and IS*50* by an OE and an IE sequence (see Figure 4) (17, 34). All transposase-mediated genetic events (save one mentioned below) occur at the precise boundary of the relevant 19-bp sequences, and almost all changes in these sequences greatly reduce or abolish the frequency of transposition (24, 29). Presumably the importance of these sequences lies in the fact that they are recognized by proteins during the transposition process.

Although we are not certain of all of the salient facts, the protein-DNA interactions occurring at the OE and IE sequences clearly display a remarkable degree of complexity; at least two proteins recognize the OE and three proteins recognize the IE. In addition, different portions of the OE sequence may play roles in different steps in the transposition process.

The OE end is recognized by both the transposase and the host protein DnaA (17, 29, 44). Tnp recognition of OE is fundamental to the transposition process. Tnp binds specifically to OE sequences in vitro as judged by gel retardation assays (42; N. de la Cruz, M. Weinreich, T. Wiegand, M. Krebs & W. Reznikoff, submitted). However, transposase recognition of the OE may be much more complex than a simple binding reaction. For instance, some base pairs could be recognized during the strand-cleavage reaction.

The first evidence that various OE sequences dictate different steps in the transposition process came from a complicated set of in vivo observations (16). Tnp can catalyze deletions adjacent to its site of insertion starting at the end of the OE sequence. These are most easily detected in situations in which the OE far removed (distal) from the location of the adjacent deletion is defective, greatly reducing or blocking transposition. Two classes of adjacent deletions have been found, and the nature of the class depends upon the precise change in the distal OE. If the distal OE is deleted, then the adjacent deletions start immediately adjacent to the functional OE. Some OE point mutations (which totally block transposition) also fall into this class. Mutant OE sequences of this class that have been tested for transposase binding in vitro bind transposase with a lower affinity than the wild-type OE (R. A. Jilk & W. Reznikoff, unpublished results). The second class of adjacent deletion is unusual in that this type starts one bp removed from the OE. These adjacent deletions are found in association with another subset of distal OE mutations, some of which are only partially defective in transposition and do not seem to impair transposase binding in vitro (16; R. A. Jilk & W. Reznikoff, unpublished results). Figure 4 displays the distribution of bp defects. Because different subsets of OE point mutations give rise to different classes of adjacent deletions, this implies a functional fine structure to the OE. This fine structure is also seen in the binding of Tnp to the OE. Some OE mutations disrupt Tnp binding while others do not. A possible explanation for some of the mutations that do not impair transposase binding is that they are recognized by a different protein, DnaA. However, this explanation does not fit some cases (2A, 3C) because the mutations are located outside of the DnaA box as defined by sequence homology (12, 44). On the other hand, some mutations that presumably alter transposase binding (8G, 10T, 11A) are within the DnaA box, suggesting that these base pairs may be recognized by both proteins.

A closer examination of the results summarized in Figure 4 suggests an additional level of complexity. At positions 2 and 12 different mutations are associated with different classes of adjacent deletion formation—perhaps in these two cases the same base pair is recognized in different ways at different steps in the transposition process. Possibly both DnaA and transposase recognize position 12, and different base pair changes may discriminate differentially between these two protein-recognition processes. For position 2, Tnp might make different molecular contacts with the same base pair at different steps in the overall reaction.

DnaA also recognizes the OE sequence. This observation was first made through footprinting studies of the OE (12). Later experiments tested the relevance of this binding, showing that Tn*5* and IS*50* transposition occurs at a reduced efficiency in *dna* hosts, and that a Tn*5* derivative with a mutation in the OE DnaA box did not appear to manifest sensitivity to the host *dna* genotype (29, 44). Current experiments are examining the effect of DnaA on transposase binding and are attempting to determine the in vitro properties of DnaA binding to various OE mutant sequences.

Thus, some end-sequence base pairs are recognized in very complex ways, at different steps in the transposition reaction and/or by different proteins. A similar complexity for end sequences has been suggested for other transposable elements (7, 14).

The IE sequence is even more complicated. It contains recognition sites for two proteins (transposase and Dam DNA methylase); it overlaps a binding site for a third protein, Fis, and also encodes the termination codon for the transposase (see Figure 4). The IE sequence and the OE sequence differ at 7 out of 19 positions, yet the transposase binds in vitro to an unmethylated IE with an affinity resembling its binding to the OE (R. A. Jilk & W. Reznikoff, unpublished results). This Tnp-IE binding result is consistent with the observed transposition preferences (9, 25). This result is quite surprising because it suggests considerable flexibility in the DNA recognition domain of Tnp. Alternatively, we could hypothesize that the positions that differ are not recognized by Tnp either because they do not play a critical protein-recognition role [position 4 may be an example of this (24)] or because they are recognized by some other protein (e.g. DnaA). However, the in vivo adjacent deletion studies suggest that mutations at some of these dissimilar positions do reduce Tnp binding directly (16). Thus we conclude that Tnp specifically recognizes two quite diverse sequences—another reflection of the complex substructures of IE and OE.

Dam DNA methylation inhibits transposase binding to the IE. Transposition of IS*50* is sensitive to Dam DNA methylation because of the inhibition of two separate sequence-specific protein-DNA interactions. As described above, transcription initiation programmed by the T1 (transposase) promoter is directly inhibited by Dam DNA methylation of sites overlapping the −10 region (43). However, Dam DNA methylation also inhibits IS*50* (but not Tn*5*) transposition even when the transposase synthesis is programmed by a Dam-insensitive promoter (*tac*), indicating a direct effect of Dam methylation on IE recognition during transposition (24, 29, 43). This effect is presumably mediated through the Dam sites within the IE (see Figure 4). We recently showed that Dam methylation of IE DNA fragments reduces transposase binding in vitro (R. A. Jilk & W. Reznikoff, unpublished experiments). This observation is consistent with direct Dam DNA methylation control over transposition reactions using the IE. Very similar effects of DNA methylation that regulate Tn*10*/IS*10* transposition have also been observed (31). As noted below, DNA methylation regulation of transposition may suggest a coupling between transposition and DNA replication.

The role of Dam DNA methylation regulation of IS*50* transposition (utilizing the IE) is complicated by the fact that a second protein (Fis) binds to a sequence overlapping the IE, and this binding (which is also blocked by Dam DNA methylation) is thought to compete with transposase-IE binding. Some in vivo experiments designed to study the frequency of IS*50* transposition actually used various recombinant DNA constructs that have the same general structure as IS*50* (OE–transposase gene–reporter gene–IE) but differ in the size of the IE sequence. These studies gave quite different results depending upon whether a 19- or a 24-bp sequence was used for IE. The 24-bp IE construct demonstrates much lower transposition frequencies in a *dam* host (9, 25). We now know that the 24-bp sequence (but not the 19-bp sequence) contains the overlapping Fis-binding site, and that Fis acts to inhibit transposition in a Dam-dependent manner (41). In vitro gel retardation and footprinting studies confirm that Fis binds to this sequence and that it binds efficiently only to the unmethylated site (41). The physiological significance of this Fis effect is unclear, but because Fis abundance varies with the stage of bacterial growth, it may act to dampen IS*50* transposition from newly replicated DNA during the exponential phase, when it is most abundant.

The translation termination codon for the transposase (and its inhibitor) is located within the IE at positions 14–12. We have no information as to whether this influences transposition utilizing an IE or whether Tnp bound to the IE can influence the synthesis of the transposase.

p1 Is a Cis-*Active Transposase (and a* Trans-*Active Transposition Inhibitor)*

The Tn*5* transposase functions primarily *in cis,* acting preferentially on OE-OE or OE-IE sequences located close to the site of transposase synthesis on the same relicon (15, 19). This observation has also been made for the transposase proteins encoded by other transposons such as Tn*10* and Tn*903* (8, 28). This preferential *cis* activity could result from either a chemical or functional instability in the transposase, and/or from a sequestration of the transposase during or shortly after synthesis. The Tn*903* transposase is known to be chemically unstable (8), but various studies have indicated that Tn*5* Tnp is, by and large, chemically stable in vivo (see 33).

My current hypothesis (pictured in Figure 5) is that the Tn*5* Tnp is *cis*-active owing to protein conformation changes that are regulated by two factors; by the release of the Tnp protein from the translation apparatus and by the oligomerization of Tnp with monomers of Inh or Tnp. We propose that the nearly complete, tethered Tnp has a high affinity end-sequence binding activity and can initiate the transposition process prior to completion of its translation. Such a property would obviously lead to *cis* but not *trans* activity. The released completed Tnp is proposed to have two conformations in equilibrium, a low-abundance active conformation similar to the tethered protein and a high-abundance inactive conformation. Furthermore, the formation of Tnp-Tnp or Tnp-Inh oligomers probably occurs with the inactive conformation (or oligomerization inactivates the active conformation). The

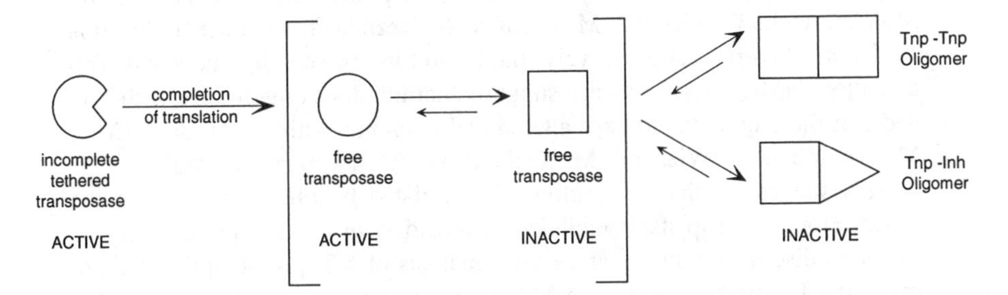

Figure 5 Transposase acts *in cis*. The Tn*5* transposase functions primarily *in cis* (14, 18). Presumably, Tnp tethered to the translation apparatus can initiate the transposition reaction. Once Tnp synthesis is complete, the molecule favors an inactive conformation, and Tnp-Tnp or Tnp-Inh oligomerization prior to end-sequence binding will also inactivate Tnp.

longer a Tnp monomer exists without end-sequence binding the higher is the probability of oligomerization. Thus this property would also lead to *cis* activity.

The proposed model includes aspects of sequestration (the incomplete tethered Tnp is held in close proximity to its gene) and functional instability (a conformational change facilitated by oligomerization). The evidence for this model is quite circumstantial, but several of its predictions are clear.

The primary evidence supporting the notion that the tethered nascent protein may bind end sequences with an enhanced affinity came from gel retardation experiments examining the binding of Tnp subfragments to OE DNA. An in vitro synthesized Tnp subfragment lacking 100 carboxyl terminal amino acids binds specifically to OE DNA and does so with an affinity ~10-fold higher than that of full-length Tnp (L. Mahnke & W. Reznikoff, unpublished results). This observation is consistent with the OE-binding domain residing at the N terminus (see section 2 of this review) and suggests that this OE binding domain is occluded in most of the intact Tnp.

However, the completed Tnp is not completely inactive as Tnp does have some *trans* activity. We have recently described Tnp mutants that increase the frequency of transposition both *in cis* and *in trans* (42). These mutant proteins may have altered the equilibrium between active and inactive conformations of Tnp, and the mutations are believed to have a conformational effect because the mutant Tnp proteins are more sensitive to proteolytic degradation (T. Wiegand & W. Reznikoff, unpublished results).

In vivo studies suggest that protein oligomerization regulates Tnp activity. The Inh protein is known to inhibit transposase activity *in trans* (15, 18, 19, 45). Two simple models explain this inhibitory activity. Inh might bind to the end sequences forming inactive Inh–end sequence complexes, thereby blocking Tnp access to these sequences. However, Inh and other N-terminal deletions of Tnp cannot bind end-sequence DNA efficiently (N. de la Cruz, M. Weinreich, T. Wiegand, M. Krebs & W. Reznikoff, submitted); thus this model is not correct. Alternatively, Inh could form mixed oligomers with Tnp and alter Tnp activity. Evidence suggests that Inh does oligomerize with Tnp and that the oligomers do have altered DNA-binding activity (N. de la Cruz, M. Weinreich, T. Wiegand, M. Krebs & W. Reznikoff, submitted).

Also suggesting that Tnp activity is regulated by oligomerization is the observation that Tnp itself inhibits transposition *in trans*. This property of Tnp was discovered during an in vivo analysis of a Tnp mutant that fails to make the Inh protein (42). The MA56 mutant destroys the Inh start codon and introduces an alanine in place of methionine at codon 56 of Tnp. This mutant is fully functional for transposition *in cis* but inhibits transposition *in trans*.

Although the model described above and in Figure 5 is consistent with all the observations, direct experiments are needed to test it. For instance, do the hypertransposing mutants alter the conformation of Tnp? Will tethered, incomplete Tnp bind to OE DNA with a high affinity? Does oligomerization favor the inactive Tnp conformation?

Transposition May Be Linked to Chromosome Replication

Some circumstantial evidence might link Tn5 transposition to the process of chromosome replication and partition. In particular the following host functions are known to influence the frequency of Tn5 transposition:

Dam: a protein catalyzing postreplicative DNA methylation; *dam* cells have elevated transposition frequencies (43).

DnaA: an *oriC*-binding protein required for the initiation of DNA replication (29, 44); *dnaA* cells have reduced transposition frequencies.

SulA: a protein that inhibits cell division; *sulA* cells have elevated transposition frequencies (35).

In addition, DeLong & Syvanen (6) reported that a small fraction of transposase is associated with the inner membrane fraction upon cell lysis. These correlations are intriguing but do not show a mechanistic link.

Observations that suggest a functional link between transposition and chromosome partition came from experiments analyzing the cellular consequences of transposase over expression. Cell death occurs in the absence of transposition as well as any other Tn5-encoded function (such as Inh or the end sequences) (M. D. Weinreich, unpublished results). The dying cells form long multinuclear filaments, indicating a failure in partition-dependent septation. *E. coli* mutants resistant to transposase overproduction killing continue to divide when transposase is overproduced. The mutations mapped using Hfr crosses reside at more than one locus. The one that has been most closely examined maps near the *fts* locus at 76 min (M. D. Weinreich & W. Reznikoff, unpublished results). *fts* genes encode functions required for septation.

These results point to a possible link between transposition and chromosome-partition function. This link is in addition to the preferential transposition of Tn5 or IS*50* sequences off of newly replicated DNA (which results from *dam* regulatory effects) and to the sharing of DnaA in both transposition and replication processes. However, at least the link to chromosome partition can be bypassed; transposition does occur in the mutants resistant to transposase overproduction (M. D. Weinreich & W. Reznikoff, unpublished results).

How is this coupling accomplished and what is the advantage of this arrangement? We hope an analysis of the mutants mentioned above will suggest an answer to the first question. As for the second, many of the host functions participating in Tn5 transposition are assumed to be components of a sequestered organelle involved in host-chromosome replication. One possible advantage to Tn*5* accruing from such an association is that a relationship between Tn*5* and the same organelle might have evolved in order to insure that the organelle's transposition apparatus would have efficient access to these functions.

Alternatively, the proposed coupling of Tn*5* transposition to chromosome replication and partition might decrease the risk of genomic "suicide" occurring as a result of the conservative cut-and-paste transposition process. Transposition is most likely to happen soon after DNA replication (when there are two copies of the donor DNA sequence) due to the influence of *dam* DNA methylation on transposase synthesis and inside-end usage. If, however, transposase inhibits chromosome partitioning, then this might delay cell division in the very cells in which transposition is most likely to have occurred, e.g. cells that contain transposase, assuring that the ongoing chromosome replication would be completed prior to cell division. Thus transposition would be biased towards cells in which a copy of the parental genome is maintained and can be inherited.

Conclusion

The Tn*5*/IS*50* system has been an amazingly productive object of experimental inquiry. The work to date has elucidated (and future work will continue to elucidate) the mechanisms by which transposition occurs and is regulated and will also continue to provide insights into important aspects of molecular biology.

One of the interesting aspects of Tn*5* molecular biology is the density of information encoded within a short sequence. The IS*50* sequence is 1533 bp in length. This element encodes all of the Tn*5* functions required for transposition and all of the Tn*5* regulatory mechanisms. Thus in the first 259 bp, starting at the OE, we find a complex Tnp-specific sequence, a DnaA-binding site, an imperfect symmetry element that blocks the expression of read-through mRNA, a promoter for Tnp mRNA synthesis, a possible LexA binding site, Dam methylation sites regulating the Tnp promoter activity, a promoter for Inh mRNA synthesis, translation initiation signals for Tnp and Inh, and sequences encoding the important N terminus of Tnp. Most of the rest of the sequence up to the Fis site, which overlaps the IE, is involved with encoding Tnp and Inh, although *cis*-active sites may also be in this region [e.g. a Fis binding site of unknown function (41)].

I have not discussed the bulk of the protein-coding sequences in this review because the required Tnp (and Inh) structure-function studies have not been performed. For instance, the Tnp presumably executes the following operations: binding of OE and IE, bringing the ends together (or inactivating the unbound Tnp) through oligomerization, cutting the DNA next to the end sequences, cleaving target DNA to give 9-bp 5′ overhangs, and inserting the transposable element into target DNA. Each of the functions requires particular Tnp domains of critical importance. Where are they? How are they organized in a three-dimensional structure? Understanding the molecular basis of Tnp *cis* activity should also reveal fascinating insights into this protein's structure-function relationships. These and other questions will be the object of future experiments, which will require detailed genetic, biochemical, and structural analyses.

Even among the transposition operations for which we have substantial information, much work is left to be done. For instance, how does DnaA binding to the OE influence the Tnp-OE interaction? How do the different base pairs of the OE interact with Tnp and at what steps of the transposition reaction does this occur? How do the different base pairs of the IE function with regard to Tnp activity? How does Fis perturb the Tnp-IE interaction? What is the active form of the Tnp, and how many molecules are involved in synaptic complex formation?

Finally, research efforts should be directed at determining if transposition is coupled to the cell-division cycle and how that linkage is accomplished.

ACKNOWLEDGMENTS

The research in the author's laboratory has been supported by grants from the NIH (GM19670), the NSF (DMB-9020517), and the American Cancer Society (MV-554). The author is indebted to the many members of his laboratory who have contributed their imagination and work, and the editor of this volume for very helpful suggestions.

Literature Cited

1. Bellomy, G. R., Record, M. T. Jr. 1990. DNA loop formation: role in gene regulation and implications for DNA structure. In *Progress in Nucleic Acid Research and Molecular Biology,* ed. W. Cohn, K. Moldave, 39:81–128. New York: Academic
2. Berg, D. E. 1989. Transposon Tn*5*. See Ref. 3, pp. 185–210
3. Berg, D. E., Howe, M. M. 1989. *Mobile DNA*. Washington, DC: Am. Soc. Microbiol.
4. Biek, D., Roth, J. R. 1990. Regulation of Tn*5* transposition in *Salmonella typhimurium*. *Proc. Natl. Acad. Sci. USA* 77:6047–51
5. Davis, M. A., Simons, R. W., Kleckner, N. 1985. Tn10 protects itself at two levels from fortuitous activation by external promoters. *Cell* 43:379–97
6. DeLong, A., Syvanen, M. 1990. Membrane association of the Tnp and Inh proteins of IS*50*R. *J. Bacteriol.* 172: 5516–19
7. Derbyshire, K. M., Hwang, L., Grindley, N. D. F. 1987. Genetic analysis of the interaction of the insertion sequence IS*903* transposase with its terminal inverted repeats. *Proc. Natl. Acad. Sci. USA* 84:8049–53
8. Derbyshire, K. M., Kramer, M., Grindley, N. D. F. 1990. Role of instability in the *cis* action of the insertion sequence IS*903* transposase. *Proc. Natl. Acad. Sci. USA* 87:4048–52
9. Dodson, K. W., Berg, D. E. 1989. Factors affecting transposition activity of IS*50* and Tn*5* ends. *Gene* 76:207–13
10. Eismann, E., von Wilcken-Bergmann, B., Muller-Hill, B. 1987. Specific destruction of the second *lac* operator decreases repression of the *lac* operon in *Escherichia coli* fivefold. *J. Mol. Biol.* 195:949–52

11. Flashner, Y., Gralla, J. D. 1988. Dual mechanism of repression at a distance in the *lac* operon. *Proc. Natl. Acad. Sci. USA* 85:8968–72
12. Fuller, R. S., Funnell, B. E., Kornberg, A. 1984. The dnaA protein complex with the E. coli chromosomal replicative origin (oriC) and other DNA sites. *Cell* 38:889–900
13. Ghosal, D., Sommer, H., Saedler, H. 1979. Nucleotide sequence of the transposable DNA-element IS2. *Nucleic Acids Res.* 6:1111–22
14. Huisman, O., Errada, P. R., Signon, L., Kleckner, N. 1989. Mutational analysis of IS10's outside end. *EMBO J.* 8:2101–9
15. Isberg, R. R., Lazaar, A. L., Syvanen, M. 1982. Regulation of Tn5 by the right repeat proteins: control at the level of the transposition reactions? *Cell* 30:883–92
16. Jilk, R. A., Makris, J. C., Borchardt, L., Reznikoff, W. S. 1993. Implications of Tn*5* associated adjacent deletions. *J. Bacteriol.* 175:1264–71
17. Johnson, R. C., Reznikoff, W. S. 1983. DNA sequences at the ends of transposon Tn*5* required for transposition. *Nature* 304:280–82
18. Johnson, R. C., Reznikoff, W. S. 1984. The role of the IS*50*R proteins in the promotion and control of Tn*5* transposition. *J. Mol. Biol.* 177:645–61
19. Johnson, R. C., Yin, J. C.-P., Reznikoff, W. S. 1982. The control of Tn5 transposition in Escherichia coli is mediated by protein from the right repeat. *Cell* 30:873–82
20. Klaer, R., Kuhn, S., Tillmann, E., Fritz, H.-J., Starlinger, P. 1981. The sequence of IS*4*. *Mol. Gen. Genet.* 181:169–75
21. Krebs, M. P., Reznikoff, W. S. 1986. Transcriptional and translational sites of IS*50*. Control of transposase and inhibitor expression. *J. Mol. Biol.* 192: 781–91
22. Kroger, M., Hobom, G. 1982. Structural analysis of insertion sequence IS*5*. *Nature* 297:159–62
23. Kuan, C.-T., Tessman, I. 1991. LexA protein of *Escherichia coli* represses expression of the Tn*5* transposase gene. *J. Bacteriol.* 173:6406–10
24. Makris, J. C., Nordmann, P. L., Reznikoff, W. S. 1988. Mutational analysis of IS*50* and Tn*5* ends. *Proc. Natl. Acad. Sci. USA* 85:2224–28
25. Makris, J. C., Reznikoff, W. S. 1989. Orientation of IS*50* transposase gene and transposition. *J. Bacteriol.* 171: 5212–14
26. Merril, C. R., Gottesman, M. E., Adhya, S. L. 1981. *Escherichia coli gal* operon proteins made after prophage lambda induction. *J. Bacteriol.* 147: 875–87
27. McCommas, S. A., Syvanen, M. 198. Temporal control of Tn*5* transposition. *J. Bacteriol.* 170:889–94
28. Morisato, D., Way, J. C., Kim, H.-J., Kleckner, N. 1983. Tn10 transposase acts preferentially on nearby transposon ends in vivo. *Cell* 32:799–807
29. Phadnis, S. H., Berg, D. E. 1987. Identification of base pairs in the outside end of insertion sequence IS*50* that are needed for IS*50* and Tn*5* transposition. *Proc. Natl. Acad. Sci. USA* 84:9118–22
30. Reznikoff, W. S., Bertrand, K., Donnelly, C., Krebs, M., Maquat, L. E., et al. 1987. Complex promoters. In *RNA Polymerase and the Regulation of Transcription,* ed. W. S. Reznikoff, R. R. Burgess, J. E. Dahlberg, C. A. Gross, M. T. Record, Jr., M. P. Wickens, pp. 105–13. New York: Elsevier
31. Roberts, D. E., Hoopes, B. C., McClure, W. R., Kleckner, N. 1985. IS10 transposition is regulated by DNA adenine methylation. *Cell* 43:117–30
32. Rosetti, O. L., Altman, R., Young, R. 1984. Kinetics of Tn*5* transposition. *Gene* 32:91–8
33. Rothstein, S. J., Reznikoff, W. S. 1981. The functional differences in the inverted repeats of the Tn5 are caused by a single base pair nonhomology. *Cell* 23:191–99
34. Sasakawa, C., Carle, G. F., Berg, D. E. 1983. Sequences essential for transposition at the termini of IS*50*. *Proc. Natl. Acad. Sci. USA* 80:7293–97
35. Sasakawa, C., Uno, Y., Yoshikawa, M. 1987. *lon-sulA* regulatory function affects the efficiency of transposition of Tn*5* from λ b221 cI857 Pam Oam to the chromosome. *Biochem. Biophys. Res. Commun.* 142:879–84
36. Schulz, V. P., Reznikoff, W. S. 1990. In vitro secondary structure analysis of mRNA from *lacZ* translation initiation mutants. *J. Mol. Biol.* 211:427–45
37. Schulz, V. P., Reznikoff, W. S. 1991. Translation initiation of IS*50*R readthrough transcripts. *J. Mol. Biol.* 221: 65–80
38. Schwartz, E., Kroger, M., Rak, B. 1988. IS*150:* distribution, nucleotide sequence and phylogenetic relationships of a new *E. coli* insertion element. *Nucleic Acids Res.* 16:6789–99
39. Timmerman, K. P., Tu, C.-P. D.

1985. Complete sequence of IS*3*. *Nucleic Acids Res*. 13:2127–39

40. Weinreich, M. D., Makris, J. C., Reznikoff, W. S. 1991. Induction of the SOS response in *Escherichia coli* inhibits Tn*5* and IS*50* transposition. *J. Bacteriol*. 173:6910–18
41. Weinreich, M. D., Reznikoff, W. S. 1992. Fis plays a role in Tn*5* and IS*50* transposition. *J. Bacteriol*. 174:4530–37
42. Wiegand, T. W., Reznikoff, W. S. 1992. Characterization of two hypertransposing Tn*5* mutants. *J. Bacteriol*. 174:1229–39
43. Yin, J. C.-P., Krebs, M. P., Reznikoff, W. S. 1988. The effect of *dam* methylation on Tn*5* transposition. *J. Mol. Biol*. 199:35–46
44. Yin, J. C.-P., Reznikoff, W. S. 1987. *dnaA*, an essential host gene, and Tn*5* mutants. *J. Bacteriol*. 169:4637–45
45. Yin, J. C.-P., Reznikoff, W. S. 1988. p2 and the inhibition of Tn*5* transposition. *J. Bacteriol*. 170:3008–15
46. Yu, X. M., Munson, L. M., Reznikoff, W. S. 1984. Molecular cloning and sequence analysis of *trp-lac* fusion deletions. *J. Mol. Biol*. 172:355–62

SUBJECT INDEX

R

U

V

W

X

Y

Z

CUMULATIVE INDEXES

CONTRIBUTING AUTHORS, VOLUMES 43–47

CHAPTER TITLES, VOLUMES 43–47

IMMUNOLOGY

MORPHOLOGY, ULTRASTRUCTURE, AND DIFFERENTIATION

PHYSIOLOGY, GROWTH, AND NUTRITION

ANNUAL REVIEWS SERIES *Volumes not listed are no longer in print*		Prices, postpaid, per volume. USA / other countries (incl. Canada)	Regular Order Please send Volume(s):	Standing Order Begin with Volume:
Annual Review of	**BIOPHYSICS AND BIOMOLECULAR STRUCTURE**			
Vols. 1-20	(1972-1991)............................	$55.00/$60.00		
Vol. 21	(1992).....................................	$59.00/$64.00		
Vol. 22	(avail. June 1993)..................	$59.00/$64.00	Vol(s). ______	Vol.______
Annual Review of	**CELL BIOLOGY**			
Vols. 1-7	(1985-1991)............................	$41.00/$46.00		
Vol. 8	(1992).....................................	$46.00/$51.00		
Vol. 9	(avail. Nov. 1993)..................	$46.00/$51.00	Vol(s). ______	Vol.______
Annual Review of	**COMPUTER SCIENCE**			
Vols. 1-2	(1986-1987)............................	$41.00/$46.00		
Vols. 3-4	(1998-1989/1990)..................	$47.00/$52.00	Vol(s). ______	Vol.______
Series suspended until further notice. Purchase the complete set for the special promotional price of $100.00 USA / $115.00 other countries, when all four volumes are ordered at the same time. Orders at the special price must be prepaid.				
Annual Review of	**EARTH AND PLANETARY SCIENCES**			
Vols. 1-19	(1973-1991)............................	$55.00/$60.00		
Vol. 20	(1992).....................................	$59.00/$64.00		
Vol. 21	(avail. May 1993)..................	$59.00/$64.00	Vol(s). ______	Vol.______
Annual Review of	**ECOLOGY AND SYSTEMATICS**			
Vols. 2-12, 14-22	(1971-1981, 1983-1991).........	$40.00/$45.00		
Vol. 23	(1992).....................................	$44.00/$49.00		
Vol. 24	(avail. Nov. 1993)..................	$44.00/$49.00	Vol(s). ______	Vol.______
Annual Review of	**ENERGY AND THE ENVIRONMENT**			
Vols. 1-16	(1976-1991)............................	$64.00/$69.00		
Vol. 17	(1992).....................................	$68.00/$73.00		
Vol. 18	(avail. Oct. 1993)..................	$68.00/$73.00	Vol(s). ______	Vol.______
Annual Review of	**ENTOMOLOGY**			
Vols. 10-16, 18	(1965-1971, 1973)			
20-36	(1975-1991)............................	$40.00/$45.00		
Vol. 37	(1992)	$44.00/$49.00		
Vol. 38	(avail. Jan. 1993)	$44.00/$49.00	Vol(s). ______	Vol.______
Annual Review of	**FLUID MECHANICS**			
Vols. 2-4, 7, 9-11	(1970-1972, 1975, 1977-1979)			
14-23	(1982-1991)	$40.00/$45.00		
Vol. 24	(1992)	$44.00/$49.00		
Vol. 25	(avail. Jan. 1993)	$44.00/$49.00	Vol(s). ______	Vol.______
Annual Review of	**GENETICS**			
Vols. 1-12, 14-25	(1967-1978, 1980-1991)	$40.00/$45.00		
Vol. 26	(1992).....................................	$44.00/$49.00		
Vol. 27	(avail. Dec. 1993)..................	$44.00/$49.00	Vol(s). ______	Vol.______
Annual Review of	**IMMUNOLOGY**			
Vols. 1-9	(1983-1991)	$41.00/$46.00		
Vol. 10	(1992)	$45.00/$50.00		
Vol. 11	(avail. April 1993)	$45.00/$50.00	Vol(s). ______	Vol.______
Annual Review of	**MATERIALS SCIENCE**			
Vols. 1, 3-19	(1971, 1973-1989)..................	$68.00/$73.00		
Vols. 20-22	(1990-1992)	$72.00/$77.00		
Vol. 23	(avail. Aug. 1993)	$72.00/$77.00	Vol(s). ______	Vol.______